www.physioex.com

Log on

Explore

Succeed

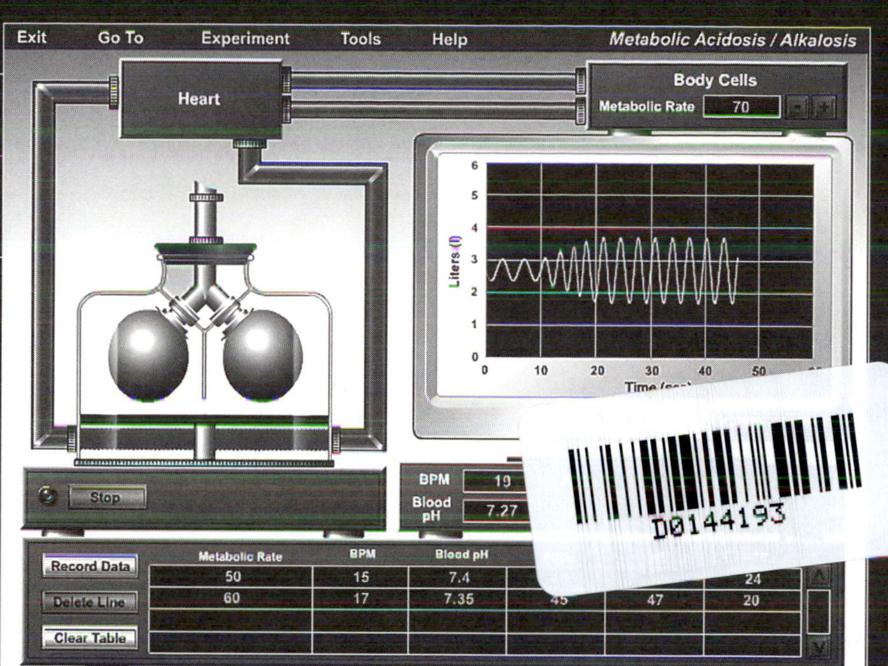

To help you succeed in your course, your professor has arranged for you to enjoy access to a great media resource, PhysioEx™ V4.0 Laboratory Simulations in Physiology (www.physioex.com). You'll find that PhysioEx™ V4.0 will enhance your course materials and help you make the grade! A CD-ROM version of PhysioEx™ V4.0 is also available with each new copy of Marieb's ***Human Anatomy & Physiology Laboratory Manual Updates***.

Got technical questions?

For technical support, please visit www.aw.com/techsupport, send an email to online.support@pearsoned.com (for website questions), or send an email to media.support@pearsoned.com (for CD-ROM questions) with a detailed description of your computer system and the technical problem. You can also call our tech support hotline at 1-800-677-6337 Monday-Friday, 8 a.m. to 5 p.m. CST.

What your system needs to use these media resources:

WINDOWS
Pentium I/ 266MHz processor or faster
Windows 95, 98, NT, 2000, ME, XP
64 MB RAM, (128 MB RAM recommended)
800 x 600 screen resolution; millions of colors
4x CD-ROM drive (if using CD version)
Browser: Netscape Navigator 4.6+ or Internet Explorer 5.0+
Plug-Ins: Macromedia Flash 6
Printer

MACINTOSH
604/300 MHz or G3/233 MHz processor
OS 8.6 or higher
64 MB RAM, (128 MB RAM recommended)
800 x 600 screen resolution; millions of colors
4x CD-ROM drive (if using CD version)
Browser: Netscape Navigator 4.6+ or Internet Explorer 5.0+
Plug-Ins: Macromedia Flash 6
Printer

Here's your personal ticket to success:

How to log on to www.physioex.com

1. Go to www.physioex.com.
2. Click "Register Here."
3. Enter your pre-assigned access code exactly as it appears below.
4. Complete online registration form to create your own personal Login Name and Password.
5. Once your Login Name and Password are confirmed by email, go back to www.physioex.com, type in your new Login Name and Password, and click "Enter."

Your Access Code is:

USPHX-GNAWS-EYING-COWED-AFOOT-MOOSE

Record your new user ID and password on the back of this card.

Cut out this card and keep it handy. It's your ticket to valuable information.

Important: Please read the License Agreement, located on the launch screen, before using **www.physioex.com** or the **PhysioEx™ V4.0 CD-ROM**. By using the website or CD-ROM, you indicate that you have read, understood, and accepted the terms of this agreement.

Human Anatomy & Physiology Laboratory Manual

Main Version

Sixth Edition Update

Elaine N. Marieb, R.N., Ph.D.,
Holyoke Community College

with contributions by
Linda S. Kollett, Ph.D.,
Massasoit Community College

PhysioEx™ Version 4.0 authored by
Peter Z. Zao,
North Idaho College, and
Timothy Stabler, Ph.D.,
Indiana University Northwest

with contributions by
Marcia C. Gibson
University of Wisconsin-Madison

San Francisco Boston New York
Cape Town Hong Kong London Madrid Mexico City
Montreal Munich Paris Singapore Sydney Tokyo Toronto

Publisher: Daryl Fox
Managing Editor, Editorial: Kay Ueno
Project Editor: Susan Teahan
PhysioEx Project Editor: Barbara Yien
Associate Project Editor: Mary Ann Murray
Publishing Assistant: David Stalder
Managing Editor, Production: Wendy Earl
Production Supervisors: Sharon Montooth / Michele Mangelli
Cover Designer: f. tani hasegawa
Marketing Manager: Lauren Harp

Library of Congress Cataloging-in-Publication Data
Marieb, Elaine Nicpon,
Human anatomy and physiology laboratory manual / Elaine N. Marieb, Linda S. Kollett; PhysioEx exercises authored by Peter Z. Zao, Timothy Stabler.—6th ed., main version.
p. cm.—(The Benjamin Cummings series in human anatomy and physiology)
ISBN 0-8053-5358-5
1. Human physiology—Laboratory manuals. 2. Human anatomy—Laboratory manuals I. Kollett, Linda S. II. Zao, Peter Z. III. Stabler, Timothy. IV. Title. V. Series.
QP44 .M34 2001b
612'.0078—dc21 2001028465

Benjamin Cummings gratefully acknowledges Carolina Biological Supply for the use of numerous histology images found on the PhysioEx CD-ROM.

By R.A. Chase
The Bassett Atlas of Human Anatomy (1989)

By Kapit/Elson
The Anatomy Coloring Book, third edition (2002)

By Kapit/Macey/Meisami
The Physiology Coloring Book, second edition (2000)

By E.N. Marieb
Human Anatomy and Physiology, fifth edition (2001)

Human Anatomy and Physiology, Study Guide, fifth edition (2001)

Human Anatomy and Physiology Laboratory Manual, Cat Version, seventh edition update (2003)

Human Anatomy and Physiology Laboratory Manual, Fetal Pig Version, seventh edition update (2003)

Human Anatomy Laboratory Manual with Cat Dissections, third edition (2001)

Essentials of Human Anatomy and Physiology, seventh edition (2003)

The A&P Coloring Workbook: A Complete Study Guide, seventh edition (2003)

By E.N. Marieb and J. Mallatt
Human Anatomy, third edition update (2003)

The Author and Publisher believe that the lab experiments described in this publication, when conducted in conformity with the safety precautions described herein and according to the school's laboratory safety procedures, are reasonably safe for the student to whom this manual is directed. Nonetheless, many of the described experiments are accompanied by some degree of risk, including human error, the failure or misuses of laboratory or electrical equipment, mismeasurement, chemical spills, and exposure to sharp objects, heat, bodily fluids, blood, or other biologics. The Author and Publisher disclaim any liability arising from such risks in connection with any of the experiments contained in this manual. If students have any questions or problems with materials, procedures, or instructions on any experiment, they should always ask their instructor for help before proceeding.

ISBN 0-8053-5358-5

1 2 3 4 5 6 7 8 9 10—VHP—06 05 04 03 02

www.aw.com/bc

Contents

Preface to the Instructor

The philosophy behind the sixth edition update of this manual mirrors that of all earlier editions. It reflects a still-developing sensibility for the way teachers teach and students learn engendered by years of teaching the subject, and by listening to the suggestions of other instructors as well as those of students enrolled in multifaceted health-care programs. *Human Anatomy and Physiology Laboratory Manual: Main Version* was originally developed to facilitate and enrich the laboratory experience for both teachers and students. This, its sixth edition, retains those same goals.

This manual, intended for students in introductory human anatomy and physiology courses, presents a wide range of laboratory experiences for students concentrating in nursing, physical therapy, dental hygiene, pharmacology, respiratory therapy, health and physical education, as well as biology and premedical programs. It differs from *Human Anatomy and Physiology Laboratory Manual, Cat and Fetal Pig* versions (seventh edition updates, 2003) in that it does not contain detailed guidelines for dissecting a laboratory animal. The manual's coverage is intentionally broad, allowing it to serve both one- and two-semester courses.

Basic Pedagogical Approach

The generous variety of experiments in this manual provides flexibility that enables instructors to gear their laboratory approach to specific academic programs, or to their own teaching preferences. The manual is still independent of any textbook, so it contains the background discussions and terminology necessary to perform all experiments. Such a self-contained learning aid eliminates the need for students to bring a textbook into the laboratory.

Each of the 46 exercises leads students toward a coherent understanding of the structure and function of the human body. The manual begins with anatomical terminology and an orientation to the body, which together provide the necessary tools for studying the various body systems. The exercises that follow reflect the dual focus of the manual—both anatomical and physiological aspects receive considerable attention. As the various organ systems of the body are introduced, the initial exercises focus on organization, from the cellular to the organ system level. As indicated by the table of contents, the anatomical exercises are usually followed by physiological experiments that familiarize students with various aspects of body functioning and promote the critical understanding that function follows structure. Homeostasis is continually emphasized as a requirement for optimal health. Pathological conditions are viewed as a loss of homeostasis; these discussions can be recognized by the homeostasis imbalance logo within the descriptive material of each exercise. This holistic approach encourages an integrated understanding of the human body.

Features and Changes

In this revision, I have continued to try to respond to reviewers' and users' feedback concerning trends that are having an impact on the anatomy and physiology laboratory experience, most importantly:

- the growing reluctance of students to perform experiments using living laboratory animals and the declining popularity of animal dissection exercises
- the increased use of computers in the laboratory, and hence the subsequent desire for more computer simulation exercises
- the replacement of older recording equipment with computerized data acquisition and compilation systems
- the continued importance of visual learning for today's student
- the need to reinforce writing, computation, and critical thinking skills across the curriculum

The specific changes implemented to address these trends fall neatly into two areas: pedagogical and multimedia. Changes made in each of these areas are described next.

Pedagogical Features

1. Design Enhancements

A brilliantly color-coded heading design enhances the lab manual pedagogy and distinguishes important features of the text. Opening pages of exercises are enriched with colored background screens that highlight descriptions of exercise objectives and lists of required materials. Blue *Activity* heads are used throughout the manual, alerting the student that "hands-on" learning is to follow. *Activity* heads for experiments involving PowerLab® and other apparatus are also set off in blue. A scissors icon and a green *Dissection* head herald sections that entail dissection of isolated organs. The conclusion of each *Activity* and *Dissection* is indicated by a block symbol, which is of the same hue as the section heading. Tables and charts are framed in vivid blue and placed near relevant text.

2. Art Program Revisions

A completely revitalized art program is offered with this new edition. Virtually all the illustrations have been revised for more detail, improved line quality, and stronger color. Muscle, bone, and joint illustrations are especially improved, and selected tables have been embellished with new full-color drawings. Several new photographs have been added to accompany diagrammatic figures, and many new photomicrographs are offered throughout the main exercises and in the Histology Atlas. Two new bone plates with photographs of the pectoral girdle and right upper limb and the pelvic girdle and right lower limb are now part of the Human Anatomy Atlas. All new sharp full-color photographs are offered with the rat dissection activities.

3. Updated Anatomical Terminology

The anatomical terminology in this sixth edition has been updated to match that in Human Anatomy & Physiology, fifth edition (main text authored by Elaine Marieb).

4. Organization Changes

As in the previous edition of the manual, the principal laboratory lessons, Exercises 1 through 46, appear first and are followed by corresponding Review Sheets and the excellent Histology and Human Anatomy Atlases. Within the main section are ten exercises designated with the letter "A." The "A" indicates that the exercise has a correlating "B" exercise—a PhysioEx™ computer simulation that can be used along with or in place of a wet lab activity. All 12 PhysioEx™ modules, as well as PhysioEx™ Review Sheets, are contained in a separate section located near the end of the manual.

Multimedia Features

1. PhysioEx™ Version 4.0 Computer Simulations

The PhysioEx™ CD-ROM, shrink-wrapped with every lab manual, has been expanded in the sixth edition update of the manual to include new labs on nerve impulse physiology, endocrine system physiology, and acid-base balance. Unlike the typical tutorial-based computer supplements that usually target anatomy, the 10 easy-to-use physiology experiments on PhysioEx™ Version 4.0 allow students to explore with different variables while being guided through the process of discovery within the structure and security of a written lab exercise. In addition, a new Histology Review Supplement is now included, with 40 new slides keyed to topics in the lab simulations and accompanying worksheets. Particularly advantageous is the fact that the students can conduct or review the experiments and slides at home on a personal computer. PhysioEx™ Version 4.0 also provides convenient "laboratory access" for students enrolled in Internet-based distance education courses and now is available online at www.physioex.com. (Use the access code found at the front of your lab manual to log onto the site.)

PhysioEx™ Version 4.0 topics include:

- Exercise 5B, *The Cell—Transport Mechanisms and Permeability: Computer Simulation.* Explores how substances cross the cell's membrane. Simple and facilitated diffusion, osmosis, filtration, and active transport are covered.
- Exercise 6B, *Histology Tutorial.* Includes over 200 histology images, viewable at various magnifications, with accompanying descriptions and labels.
- Exercise 16B, *Skeletal Muscle Physiology: Computer Simulation.* Provides insights into the complex physiology of skeletal muscle. Electrical stimulation, isometric contractions, and isotonic contractions are investigated.
- Exercise 18B, *Neurophysiology of Nerve Impulses: Computer Simulation.* Investigates stimuli that elicit action potentials, stimuli that inhibit action potentials, and factors affecting nerve conduction velocity.
- Exercise 28B, *Endocrine System Physiology: Computer Simulation.* Investigates the relationship between hormones and metabolism; the effect of estrogen replacement therapy; and the effect of insulin on diabetes.
- Exercise 33B, *Cardiovascular Dynamics: Computer Simulation.* Allows students to perform experiments that would be difficult if not impossible to do in a traditional laboratory. Topics of inquiry include vessel resistance and pump (heart) mechanics.
- Exercise 34B, *Frog Cardiovascular Physiology: Computer Simulation.* Variables influencing heart activity are examined. Topics include setting up and recording baseline heart activity, the refractory period of cardiac muscle, and an investigation of physical and chemical factors that affect enzyme activity.
- Exercise 37B, *Respiratory System Mechanics: Computer Simulation.* Investigates physical and chemical aspects of pulmonary function. Students collect data simulating normal lung volumes. Other activities examine factors such as airway resistance and the effect of surfactant on lung function.
- Exercise 39B, *Chemical and Physical Processes of Digestion: Computer Simulation.* Turns the student's computer into a virtual chemistry lab where enzymes, reagents, and incubation conditions can be manipulated (in compressed time) to examine factors that affect enzyme activity.
- Exercise 41B, *Renal Physiology—The Function of the Nephron: Computer Simulation.* Simulates the function of a single nephron. Topics include factors influencing glomerular filtration, the effect of hormones on urine function, and glucose transport maximum.
- Exercise 47, *Acid-Base Balance: Computer Simulation.* Topics include respiratory and metabolic acidosis/alkalosis, and renal and respiratory compensation.

2. PowerLab® Instructions

Instructions for the use of the PowerLab® data acquisition and compilation system are included in the lab manual: Exercises 16A, 22, 31, 33A, 34A, and 37A.

3. Biopac® Instructions

Instructions for use of the Biopac® Student Lab System for Exercises 16A, 18A, 22, 31, 33A, 34A, and 37A can be found in Appendix B of the Instructor's Guide.

4. Intelitool® Instructions Update

Four physiological experiments (Exercises 16i, 22i, 31i, and 37i) using Intelitool® equipment have been updated and are available in the *Instructor's Guide.* Instructors using Intelitool® equipment in their laboratory may copy these exercises for student handouts.

5. Videotapes

Human Anatomy and Physiology videotapes are available to qualified adopters. Produced by University Media Services and scripted by Rose Leigh Vines, Rosalee Carter, and Ann

Motekaitis of California State University, Sacramento, these excellent videotapes reinforce many of the concepts covered in this manual and will represent a valuable addition to any multimedia library.

Special Features Retained

Virtually all the special features appreciated by the adopters of the last edition are retained.

- The prologue, "Getting Started—What to Expect, The Scientific Method, and Metrics," explains the scientific method, the logical, practical, and reliable way of approaching and solving problems in the laboratory and reviews metric units and interconversions. A format for writing lab reports is also included.
- Each exercise begins with learning objectives.
- Key terms appear in boldface print, and each term is defined when introduced.
- Illustrations are large and of exceptional quality. Full-color photographs and drawings highlight, differentiate, and focus student attention on important structures.
- Body structures are studied from simple-to-complex levels, and physiological experiments allow ample opportunity for student observation and experimentation.
- The numerous physiological experiments for each organ system range from simple experiments that can be performed without specialized tools to more complex experiments using laboratory equipment computers and instrumentation techniques.
- Tear-out laboratory review sheets, located toward the end of the manual, are designed to accompany each lab exercise. The review sheets provide space for recording and interpreting experimental results and require students to label diagrams and answer multiple-choice and short-answer questions.
- In addition to the figures, isolated animal organs such as the sheep heart and pig kidney are employed because of their exceptional similarity to human organs.
- The Histology Atlas has 63 color photomicrographs, and the edges of its pages are colored purple for quick location. The photomicrographs selected are those deemed most helpful to students because they correspond closely with slides typically viewed in the lab. Most such tissues are stained with hematoxylin and eosin (H & E), but a few depicted in the Histology Atlas are stained with differential stains to allow selected cell populations to be identified in a given tissue. Line drawings, corresponding exactly to selected plates in the Histology Atlas, appear in appropriate places in the text and add to the utility and effectiveness of the atlas. These diagrams can be colored by the student to replicate the stains of the slides they are viewing; thus they provide a valuable learning aid.
- All exercises involving body fluids (blood, urine, saliva) incorporate current Centers for Disease Control (CDC) guidelines for handling human body fluids. Because it is important that nursing students, in particular, learn how to safely handle bloodstained articles, the human focus has been retained. However, the decision to allow testing of human (student) blood or to use animal blood in the laboratory is left to the discretion of the instructor in accordance with institutional guidelines. The CDC guidelines for handling body fluids are reinforced by the laboratory safety procedures described on the inside front cover of this text, in Exercise 29: Blood, and in the *Instructor's Guide*. The inside cover can be photocopied and posted in the lab to help students become well versed in laboratory safety.
- Appendix C correlates some of the required anatomical laboratory observations with the corresponding sections of A.D.A.M.® Interactive Anatomy. Using A.D.A.M.® to complement the printed manual descriptions of anatomical structures provides an extremely useful study method for visually oriented students.
- Four logos alert students to special features or instructions. These include:

The dissection scissors icon appears at the beginning of activities that entail the dissection of isolated animal organs.

The homeostasis imbalance icon directs the student's attention to conditions representing a loss of homeostasis.

A safety icon notifies students that specific safety precautions must be observed when using certain equipment or conducting particular lab procedures. (For example, when working with ether, a hood is to be used, or when handling body fluids such as blood, urine, or saliva, gloves are to be worn.)

A1A The A.D.A.M.® icon indicates where use of the A.D.A.M.® software would enhance the study and comprehension of laboratory topics.

Supplements

- The *Instructor's Guide* that accompanies all versions of the *Human Anatomy and Physiology Laboratory Manual* contains a wealth of information for those teaching this course. Instructors can find help in planning the experiments, ordering equipment and supplies, anticipating pitfalls and problem areas, and locating audiovisual material. The probable in-class time required for each lab is indicated by an hourglass icon. Other useful resources are the Trends in Instrumentation section that describes the latest laboratory equipment and technological teaching tools available and the directions for using Biopac® Student Lab System and Intelitool® instrumentation. Additional supplements include the following videos, which are available free of charge to qualified adopters:
- *Selected Actions of Hormones and Other Chemical Messengers* videotape by Rose Leigh Vines and Juanita Barrena (0-8053-4155-2)
- *Human Musculature* videotape by Rose Leigh Vines and Allan Hinderstein (0-8053-0106-2)

- The *Human Cardiovascular System: The Heart* videotape by Rose Leigh Vines and Rosalee Carter, University Media Services, California State University, Sacramento (0-8053-4289-3)
- *The Human Cardiovascular System: The Blood Vessels* videotape by Rose Leigh Vines, University Media Services, California State University, Sacramento (0-8053-4297-4)
- The *Human Nervous System: Human Brain and Cranial Nerves* videotape by Rose Leigh Vines and Rosalee Carter, University Media Services, California State University, Sacramento (0-8053-4012-2)
- The *Human Nervous System: The Spinal Cord and Nerves* videotape by Rose Leigh Vines and Rosalee Carter, University Media Services, California State University, Sacramento (0-8053-4013-0)
- *The Human Respiratory System* videotape by Rose Leigh Vines and Ann Motekaitis (0-8053-4822-0)
- *The Human Digestive System* videotape by Rose Leigh Vines and Ann Motekaitis (0-8053-4823-9)
- *The Human Urinary System* videotape by Rose Leigh Vines and Ann Motekaitis (0-8053-4915-4)
- *The Human Reproductive Systems* videotape by Rose Leigh Vines and Ann Motekaitis(0-8053-4914-6)
- Student Video Series Vol. I (0-8053-4110-2)
- Student Video Series Vol. II (0-8053-6115-4)

A.D.A.M.® Software

Available for purchase from Benjamin Cummings to enhance student learning are the following:

Student Workbook for A.D.A.M.® Standard
(ISBN 0-8053-2115-2)

A.D.A.M.® Interactive Anatomy Student Package, Second Edition
(Win: ISBN 0-8053-5043-8; Mac: ISBN 0-8053-5044-6)

A.D.A.M.® Interactive Anatomy Student Lab Guide
(ISBN 0-8053-5049-7)

A.D.A.M.® Anatomy Practice CD-ROM
(ISBN 0-8053-9680-2)

Contact your Benjamin Cummings sales representative for more information on these titles, or visit our web site at www.awl.com/bc.

Acknowledgments

I wish to thank the following reviewers for their contributions to this edition:

Theresa M. Auburn (Palo Alto College), Jeff Blodig (Johnson County Community College), Teresa Brandon (Dona Ana Branch Community College), Pamela J. Carlton (College of Staten Island), Cynthia Conaway (Mount Union College), Alfred M. Dufty, Jr. (Boise State University), Jenny L. McFarland (Edmonds Community College), Herbert W. House (Elon College), Shelley Jones (Florida Community College@Jacksonville), Earl F. Lindberg (Davidson Community College), Pablo Mendoza (El Paso Community College), Diane Pelletier (Green River Community College), Judy Penn (Shoreline Community College), Frances Ragsdale (Winona State University), Sandra Stewart (Vincennes University), James Swan (University of New Mexico and Albuquerque T-VI A Community College), Shawn Zeltwanger (Emory University)

My continued thanks to my colleagues and friends at Benjamin Cummings who worked with me in the production of this edition, especially Kay Ueno, managing editor who got the project perking, and Susan Teahan, project editor who efficiently shepherded the manuscript every millimeter of the way. Applause also to Barbara Yien, media project editor who managed the new version of PhysioEx™. Speaking of PhysioEx™, its excellence reflects the expertise of Peter Zao and that of Timothy Stabler, who came aboard with the revision. They generated the ideas behind the equipment graphics and envisioned the animations that would be needed. Credit also goes to the team at Cadre Design for the expert progamming and wonderful graphics produced in PhysioEx™.

Kudos also to the folks at, or working with, Wendy Earl Productions, who did their usual great job. Sharon Montooth, my production editor for this project, got the job done in jig time. Kelly Murphy, art editor, was in charge of overseeing the entire art program. A just-right cover image was designed by f. tani Hasegawa. As usual I want to thank Anita Wagner, a very competent copy editor who is now an old hand at my books. Last but not least, Linda Kollett worked with me on the edition. PowerLab® procedure instructions are her handy work alone. She extensively reviewed the last edition and contributed ideas for revising selected activities, illustrations, and photographs for this edition. Linda also revised the Instructor's Guide to accompany the lab manual. I expect that Linda will be an even stronger voice in the next edition.

Preface to the Student

Hopefully, your laboratory experiences will be exciting times for you. Like any unfamiliar experience, it really helps if you know in advance what to expect and what will be expected of you.

Laboratory Activities

The A&P laboratory exercises in this manual are designed to help you gain a broad understanding of both anatomy and physiology. You can anticipate examining models, dissecting an animal, and using a microscope to look at tissue slides (anatomical approaches). You will also investigate chemical conditions or observe changes in both living and nonliving systems, manipulate variables in computer simulations, and conduct experiments that examine responses of living organisms to various stimuli (physiological approaches).

Because some students question the use of animals in the laboratory setting, their concerns need to be addressed. Be assured that the preserved organ specimens used in the anatomy and physiology labs are *not* harvested from animals raised specifically for dissection purposes. Organs that are of no use to the meat packing industry (such as the brain, heart, or lungs) are sent from slaughterhouses to biological supply houses for preparation.

Every effort is being made to find alternative methods that do not use living animals to study physiological concepts. For example, included in this edition is the PhysioEx™ CD-ROM. The ten simulation exercises on this CD allow you to convert a computer into a virtual laboratory. You will be able to manipulate variables to investigate physiological phenomena. Such computer-based simulations provide you with alternatives to the use of real animals.

There is little doubt that computer simulations offer certain advantages: (1) they allow you to experiment at length without time constraints of traditional experiments, and (2) they make it possible to investigate certain concepts that would be difficult or impossible to explore in traditional exercises. Yet, the main disadvantage of computer simulations is that the real-life aspects of experimentation are sacrificed. An animated frog muscle or heart on a computer screen is not really a substitute for observing the responses of actual muscle tissue. Consequently, living animal experiments remain an important part of the approach of this manual to the study of human anatomy and physiology. However, wherever possible, the minimum number of animals needed to demonstrate a particular point is used. Furthermore, some instructor-delivered and videotaped demonstrations of live animal experiments are suggested.

If you use living animals for experiments, you will be expected to handle them humanely. Inconsiderate treatment of laboratory animals will not be tolerated in your anatomy and physiology laboratory.

A.D.A.M.® Interactive Anatomy

If the A.D.A.M.® CD-ROM software is available for your use, Appendix C of the manual will help you link the various laboratory topics with specific frames of the A.D.A.M.® software to help you in your studies.

Icons/Visual Mnemonics

I have tried to make this manual very easy for you to use, and to this end two colored section heads and four different icons (visual mnemonics) are used throughout:

The *Dissection* head is green and is accompanied by the **dissection scissors icon** at the beginning of activities that require you to dissect isolated animal organs.

The *Activity* head is blue. Because most exercises have some explanatory background provided before the experiment(s), this visual cue alerts you that your lab involvement is imminent.

The **homeostasis imbalance icon** appears where a clinical disorder is described to indicate what happens when there is a structural abnormality or physiological malfunction (e.g., a loss of homeostasis).

AIA The **A.D.A.M.® Interactive Anatomy icon** alerts you where the use of the A.D.A.M.® CD-ROM would enhance your laboratory experience.

The **safety icon** alerts you to special precautions that should be taken when handling lab equipment or conducting certain procedures. For example, it alerts you to use a ventilating hood when using volatile chemicals and signifies that you should take special measures to protect yourself when handling blood or other body fluids (e.g., saliva, urine).

Hints for Success in the Laboratory

With the possible exception of those who have photographic memories, most students can use helpful hints and guidelines to ensure that they have successful lab experiences.

1. Perhaps the best bit of advice is to attend all your scheduled labs and to participate in all the assigned exercises. Learning is an *active* process.

2. Scan the scheduled lab exercise and the questions in the review section in the back of the manual that pertain to it *before* going to lab.

3. Be on time. Most instructors explain what the lab is about, pitfalls to avoid, and the sequence or format to be followed at the beginning of the lab session. If you are late, not only will you miss this information, you will not endear yourself to the instructor.

4. Follow the instructions in the order in which they are given. If you do not understand a direction, ask for help.

5. Review your lab notes after completing the lab session to help you focus on and remember the important concepts.

6. Keep your work area clean and neat. Move books and coats out of the way. This reduces confusion and accidents.

7. Assume that all lab chemicals and equipment are sources of potential danger to you. Follow directions for equipment use and observe the laboratory safety guidelines provided inside the front cover of this manual.

8. Keep in mind the real value of the laboratory experience—a place for you to observe, manipulate, and experience hands-on activities that will dramatically enhance your understanding of the lecture presentations.

I really hope that you enjoy your A&P laboratories and that this lab manual makes learning about intricate structures and functions of the human body a fun and rewarding process. I'm always open to constructive criticism and suggestions for improvement in future editions. If you have any, please write to me.

Elaine N. Marieb
Anatomy and Physiology
Benjamin Cummings
1301 Sansome Street
San Francisco, CA 94111

Getting Started—What to Expect, The Scientific Method, and Metrics

Two hundred years ago science was largely a plaything of wealthy patrons, but today's world is dominated by science and its technology. Whether or not we believe that such domination is desirable, we all have a responsibility to try to understand the goals and methods of science that have seeded this knowledge and technological explosion.

The biosciences are very special and exciting because they open the doors to an understanding of all the wondrous workings of living things. A course in human anatomy and physiology (a minute subdivision of bioscience) provides such insights in relation to your own body. Although some experience in scientific studies is helpful when beginning a study of anatomy and physiology, perhaps the single most important prerequisite is curiosity.

Gaining an understanding of science is a little like becoming acquainted with another person. Even though a written description can provide a good deal of information about the person, you can never really know another unless there is personal contact. And so it is with science—if you are to know it well, you must deal with it intimately.

The laboratory is the setting for "intimate contact" with science. It is where scientists test their ideas (do research), the essential purpose of which is to provide a basis from which predictions about scientific phenomena can be made. Likewise, it will be the site of your "intimate contact" with the subject of human anatomy and physiology as you are introduced to the methods and instruments used in biological research.

For many students, human anatomy and physiology is taken as an introductory-level course; and their scientific background exists, at best, as a dim memory. If this is your predicament, this prologue may be just what you need to fill in a few gaps and to get you started on the right track before your actual laboratory experiences begin. So—let's get to it!

The Scientific Method

Science would quickly stagnate if new knowledge were not continually derived from and added to it. The approach commonly used by scientists when they investigate various aspects of their respective disciplines is called the **scientific method.** This method is *not* a single rigorous technique that must be followed in a lockstep manner. It is nothing more or less than a logical, practical, and reliable way of approaching and solving problems of every kind—scientific or otherwise—to gain knowledge. It comprises five major steps.

Step 1: Observation of Phenomena

The crucial first step involves observation of some phenomenon of interest. In other words, before a scientist can investigate anything, he or she must decide on a *problem* or focus for the investigation. In most college laboratory experiments, the problem or focus has been decided for you. However, to illustrate this important step, we will assume that you want to investigate the true nature of apples, particularly green apples. In such a case you would begin your studies by making a number of different observations concerning apples.

Step 2: Statement of the Hypothesis

Once you have decided on a focus of concern, the next step is to design a significant question to be answered. Such a question is usually posed in the form of a **hypothesis,** an unproven conclusion that attempts to explain some phenomenon. (At its crudest level, a hypothesis can be considered to be a "guess" or an intuitive hunch that tentatively explains some observation.) Generally, scientists do not restrict themselves to a single hypothesis; instead, they usually pose several and then test each one systematically.

We will assume that, to accomplish step 1, you go to the supermarket and randomly select apples from several bins. When you later eat the apples, you find that the green apples are sour, but the red and yellow apples are sweet. From this observation, you might conclude (*hypothesize*) that "green apples are sour." This statement would represent your current understanding of green apples. You might also reasonably predict that if you were to buy more apples, any green ones you buy will be sour. Thus, you would have gone beyond your initial observation that "these" green apples are sour to the prediction that "all" green apples are sour.

Any good hypothesis must meet several criteria. First, *it must be testable*. This characteristic is far more important than its being correct. The test data may or may not support the hypothesis or new information may require that the hypothesis be modified. Clearly the accuracy of a prediction in any scientific study depends on the accuracy of the initial information on which it is based.

In our example, no great harm will come from an inaccurate prediction—that is, were we to find that some green apples are sweet. However, in some cases human life may depend on the accuracy of the prediction: thus, (1) Repeated testing of scientific ideas is important, particularly because scientists working on the same problem do not always agree in their conclusions. (2) Conclusions drawn from scientific tests are only as accurate as the information on which they are based; therefore, careful observation is essential, even at the very outset of a study.

A second criterion is that, even though hypotheses are guesses of a sort, *they must be based on measurable, describable facts. No mysticism can be theorized*. We cannot conjure up, to support our hypothesis, forces that have not been shown to exist. For example, as scientists, we cannot say that the tooth fairy took Johnny's tooth unless we can prove that the tooth fairy exists!

Third, a hypothesis *must not be anthropomorphic*. Human beings tend to anthropomorphize—that is, to relate all experiences to human experience. Whereas we could state that bears instinctively protect their young, it would be anthropomorphic to say that bears love their young, because love is a human emotional response. Thus, the initial hypothesis must be stated without interpretation.

Step 3: Data Collection

Once the initial hypothesis has been stated, scientists plan experiments that will provide data (or evidence) to support or disprove their hypotheses—that is, they *test* their hypotheses. Data are accumulated by making qualitative or quantitative observations of some sort. The observations are often aided by the use of various types of equipment such as cameras, microscopes, stimulators, or various electronic devices that allow chemical and physiologic measurements to be made.

Observations referred to as **qualitative** are those we can make with our senses—that is, by using our vision, hearing, or sense of taste, smell, or touch. For some quick practice in qualitative observation, compare and contrast* an orange and an apple.

Whereas the differences between an apple and an orange are obvious, this is not always the case in biological observations. Quite often a scientist tries to detect very subtle differences that cannot be determined by qualitative observations; data must be derived from measurements. Such observations based on precise measurements of one type or another are **quantitative observations.** Examples of quantitative observations include careful measurements of body or organ dimensions such as mass, size, and volume; measurement of volumes of oxygen consumed during metabolic studies; determination of the concentration of glucose in urine; and determination of the differences in blood pressure and pulse under conditions of rest and exercise. An apple and an orange could be compared quantitatively by analyzing the relative amounts of sugar and water in a given volume of fruit flesh, the pigments and vitamins present in the apple skin and orange peel, and so on.

A valuable part of data gathering is the use of experiments to support or disprove a hypothesis. An **experiment** is a procedure designed to describe the factors in a given situation that affect one another (that is, to discover cause and effect) under certain conditions.

Two general rules govern experimentation. The first of these rules is that the experiment(s) should be conducted in such a manner that every **variable** (any factor that might affect the outcome of the experiment) is under the control of the experimenter. The **independent variables** are manipulated by the experimenter. For example, if the goal is to determine the effect of body temperature on breathing rate, the independent variable is body temperature. The effect observed or value measured (in this case breathing rate,) is called the **dependent** or **response variable.** Its value "depends" on the value chosen for the independent variable. The ideal way to perform such an experiment is to set up and run a series of tests that are all identical, except for one specific factor that is varied.

One specimen (or group of specimens) is used as the **control** against which all other experimental samples are compared. The importance of the control sample cannot be overemphasized. The control group provides the "normal standard" against which all other samples are compared relative to the dependent variable. Taking our example one step further, if we wanted to investigate the effects of body temperature (the independent variable) on breathing rate (the dependent variable), we could collect data on the breathing rate of individuals with "normal" body temperature (the implicit control group), and compare these data to breathing-rate measurements obtained from groups of individuals with higher and lower body temperatures.

The second rule governing experimentation is that valid results require that testing be done on large numbers of subjects. It is essential to understand that it is nearly impossible to control all possible variables in biological tests. Indeed, there is a bit of scientific wisdom that mirrors this truth—that is, that laboratory animals, even in the most rigidly controlled and carefully designed experiments, "will do as they damn well please." Thus, stating that the testing of a drug for its pain-killing effects was successful after having tested it on only one postoperative patient would be scientific suicide. Large numbers of patients would have to receive the drug and be monitored for a decrease in postoperative pain before such a statement could have any scientific validity. Then, other researchers would have to be able to uphold those conclusions by running similar experiments. *Repeatability* is an important part of the scientific method and is the primary basis for support or rejection of many hypotheses.

* *Compare* means to emphasize the similarities between two things, whereas *contrast* means that the differences are to be emphasized.

During experimentation and observation, data must be carefully recorded. Usually, such initial, or raw, data are recorded in table form. The table should be labeled to show the variables investigated and the results for each sample. At this point, *accurate recording* of observations is the primary concern. Later, these raw data will be reorganized and manipulated to show more explicitly the outcome of the experimentation.

Some of the observations that you will be asked to make in the anatomy and physiology laboratory will require that a drawing be made. Don't panic! The purpose of making drawings (in addition to providing a record) is to force you to observe things very closely. You need not be an artist (most biological drawings are simple outline drawings), but you do need to be neat and as accurate as possible. It is advisable to use a 4H pencil to do your drawings because it is easily erased and doesn't smudge. Before beginning to draw, you should examine your specimen closely, studying it as though you were going to have to draw it from memory. For example, when looking at cells you should ask yourself questions such as "What is their shape—the relationship of length and width? How are they joined together?" Then decide precisely what you are going to show and how large the drawing must be to show the necessary detail. After making the drawing, add labels in the margins and connect them by straight lines (leader lines) to the structures being named.

Step 4: Manipulation and Analysis of Data

The form of the final data varies, depending on the nature of the data collected. Usually, the final data represent information converted from the original measured values (raw data) to some other form. This may mean that averaging or some other statistical treatment must be applied, or it may require conversions from one kind of units to another. In other cases, graphs may be needed to display the data.

Elementary Treatment of Data

Only very elementary statistical treatment of data is required in this manual. For example, you will be expected to understand and/or compute an average (mean), percentages, and a range.

Two of these statistics, the mean and the range, are useful in describing the *typical* case among a large number of samples evaluated. Let us use a simple example. We will assume that the following heart rates (in beats/min) were recorded during an experiment: 64, 70, 82, 94, 85, 75, 72, 78. If you put these numbers in numerical order, the **range** is easily computed, because the range is the difference between the highest and lowest numbers obtained (highest number minus lowest number). The **mean** is obtained by summing the items and dividing the sum by the number of items. What is the range and the mean for the set of numbers just provided:

1. ____*

The word *percent* comes from the Latin meaning "for 100"; thus *percent,* indicated by the percent sign, %, means parts per 100 parts. Thus, if we say that 45% of Americans have type O blood, what we are really saying is that among each group of 100 Americans, 45 (45/100) can be expected to have type O blood. Any ratio can be converted to a percent by multiplying by 100 and adding the percent sign.

$$.25 \times 100 = 25\% \qquad 5 \times 100 = 500\%$$

It is very easy to convert any number (including decimals) to a percent. The rule is to move the decimal point two places to the right and add the percent sign. If no decimal point appears, it is *assumed* to be at the end of the number; and zeros are added to fill any empty spaces. Two examples follow:

$$0.25 = 0.25 = 25\%$$

$$5 = 5\_\_ = 500\%$$

Change the following to percents:

2. 38 = __________ **4.** 1.6 = __________

3. .75 = __________

Note that although you are being asked here to convert numbers to percents, percents by themselves are meaningless. We always speak in terms of a percentage *of* something.

To change a percent to decimal form, remove the percent sign, and divide by 100. Change the following percents to whole numbers or decimals:

5. 800% = __________ **6.** 0.05% = __________

Making and Reading Line Graphs

For some laboratory experiments you will be required to show your data (or part of them) graphically. Simple line graphs allow relationships within the data to be shown interestingly and allow trends (or patterns) in the data to be demonstrated. An advantage of properly drawn graphs is that they save the reader's time because the essential meaning of large numbers of statistical data can be visualized at a glance.

To aid in making accurate graphs, graph paper (or a printed grid in the manual) is used. Line graphs have both horizontal (X) and vertical (Y) axes with scales. Each scale should have uniform intervals—that is, each unit measured on the scale should require the same distance along the scale as any other. Variations from this rule may be misleading and result in false interpretations of the data. By convention, the condition that is manipulated (the independent variable) in the experimental series is plotted on the X-axis (the horizontal axis); and the value that we then measure (the dependent variable) is plotted on the Y-axis (the vertical axis). To plot the data, a dot or a small **x** is placed at the precise point where the two variables (measured for each sample) meet; and then a line (this is called the **curve**) is drawn to connect the plotted points.

* Answers are given on page xviii.

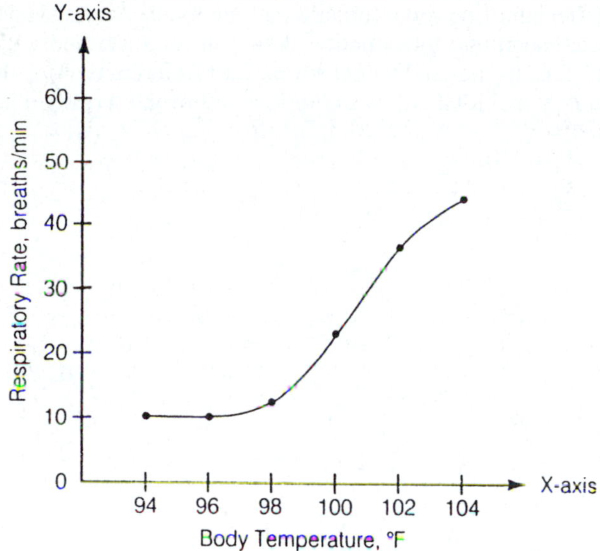

Figure G.1 Example of graphically presented data.

Sometimes, you will see the curve on a line graph extended beyond the last plotted point. This is (supposedly) done to predict "what comes next." When you see this done, be skeptical. The information provided by such a technique is only slightly more accurate than that provided by a crystal ball! When constructing a graph, be sure to label the X-axis and Y-axis and give the graph a legend (See Figure G.1).

To read a line graph, pick any point on the line, and match it with the information directly below on the X-axis and with that directly to the left of it on the Y-axis. Figure G.1 is a graph that illustrates the relationship between breaths per minute (respiratory rate) and body temperature. Answer the following questions about this graph:

7. What was the respiratory rate at a body temperature of 96°F? ______________

8. Between which two body temperature readings was the increase in breaths per minute greatest?

____

Step 5: Reporting Conclusions of the Study

Drawings, tables, and graphs alone do not suffice as the final presentation of scientific results. The final step requires that you provide a straightforward description of the conclusions drawn from your results. If possible, your findings should be compared to those of other investigators working on the same problem. (For laboratory investigations conducted by students, these comparative figures are provided by classmates.)

It is important to realize that scientific investigations do not always yield the anticipated results. If there are discrepancies between your results and those of others, or what you expected to find based on your class notes or textbook readings, this is the place to try to explain those discrepancies.

Results are often only as good as the observation techniques used. Depending on the type of experiment conducted, several questions may need to be answered. Did you weigh the specimen carefully enough? Did you balance the scale first? Was the subject's blood pressure actually as high as you recorded it, or did you record it inaccurately? If you did record it accurately, is it possible that the subject was emotionally upset about something, which might have given falsely high data for the variable being investigated? Attempting to explain an unexpected result will often teach you more than you would have learned from anticipated results.

When the experiment produces results that are consistent with the hypothesis, then the hypothesis can be said to have reached a higher level of certainty. The probability that the hypothesis is correct is greater.

A hypothesis that has been validated by many different investigators is called a **theory.** Theories are useful in two important ways. First, they link sets of data; and second, they make predictions that may lead to additional avenues of investigation. (Okay, we know this with a high degree of certainty; what's next?)

When a theory has been repeatedly verified and appears to have wide applicability in biology, it may assume the status of a **biological principle.** A principle is a statement that applies with a high degree of probability to a range of events. For example, "Living matter is made of cells or cell products" is a principle stated in many biology texts. It is a sound and useful principle, and will continue to be used as such—unless new findings prove it wrong.

We have been through quite a bit of background concerning the scientific method and what its use entails. Because it is important that you remember the phases of the scientific method, they are summarized here:

1. Observation of some phenomenon

2. Statement of a hypothesis (based on the observations)

3. Collection of data (testing the hypothesis with controlled experiments)

4. Manipulation and analysis of the data

5. Reporting of the conclusions of the study (routinely done by preparing a lab report—see p. xvi)

Writing a Lab Report Based on the Scientific Method

A laboratory report is not the same as a scientific paper, but it has some of the same elements and is a formal way to report the results of a scientific experiment. The report should have a cover page that includes the title of the experiment, the author's name, the name of the course, the instructor, and the date. The report should include five separate clearly marked sections: Introduction, Materials and Methods, Results, Discussion and Conclusions, and References. Use the template provided below to guide you through writing a lab report.

Lab Report

Cover Page

- Title of Experiment
- Author's Name
- Course
- Instructor
- Date

Introduction

- Provide background information.
- Describe any relevant observations.
- State hypotheses clearly.

Materials and Methods

- List equipment or supplies needed.
- Provide step-by-step directions for conducting the experiment.

Results

- Present data using a drawing (figure), table, or graph.
- Summarize findings briefly.

Discussion and Conclusions

- Analyze data.
- Conclude whether data gathered support or do not support hypotheses.
- Include relevant information from other sources.
- Explain any uncontrolled variables or unexpected difficulties.
- Make suggestions for further experimentation.

Reference List

- Cite the source of any material used to support this report.

Table G.1 Metric System

A. Commonly used units		B. Fractions and their multiples		
Measurement	**Unit**	**Fraction or multiple**	**Prefix**	**Symbol**
Length	Meter (m)	10^6 one million	mega	M
Volume	Liter (L; l with prefix)	10^3 one thousand	kilo	k
Mass	Gram (g)	10^{-1} one tenth	deci	d
Time*	Second (s)	10^{-2} one hundredth	centi	c
Temperature	Degree Celsius (°C)	10^{-3} one thousandth	milli	m
		10^{-6} one millionth	micro	μ
		10^{-9} one billionth	nano	n

* The accepted standard for time is the second; and thus hours and minutes are used in scientific, as well as everyday, measurement of time. The only prefixes generally used are those indicating *fractional portions* of seconds—for example, millisecond and microsecond.

Metrics

No matter how highly developed our ability to observe, observations have scientific value only if they can be communicated to others. This necessitates the use of the widely accepted system of metric measurements.

Without measurement, we would be limited to qualitative description. For precise and repeatable communication of information, the agreed-upon system of measurement used by scientists is the **metric system.**

A major advantage of the metric system is that it is based on units of 10. This allows rapid conversion to workable numbers so that neither very large nor very small figures need be used in calculations. Fractions or multiples of the standard units of length, volume, mass, time, and temperature have been assigned specific names. Table G.1 above shows the commonly used units of the metric system, along with the prefixes used to designate fractions and multiples thereof.

To change from smaller units to larger units, you must *divide* by the appropriate factor of 10 (because there are fewer of the larger units). For example, a milliunit (milli = one thousandth), such as a millimeter, is one step smaller than a centiunit (centi = one hundredth), such as a centimeter. Thus to change milliunits to centiunits, you must divide by 10. On the other hand, when converting from larger units to smaller ones, you must *multiply* by the appropriate factor of 10. A partial scheme for conversions between the metric units is shown below.

The objectives of the sections that follow are to provide a brief overview of the most-used measurements in science or health professions and to help you gain some measure of confidence in dealing with them. (A listing of the most frequently used conversion factors, for conversions between British and metric system units, is provided in Appendix A.)

Length Measurements

The metric unit of length is the **meter (m)**. Smaller objects are measured in centimeters or millimeters. Subcellular structures are measured in micrometers.

To help you picture these units of length, some equivalents follow:

One meter (m) is slightly longer than one yard (1 m = 39.37 in.).

One centimeter (cm) is approximately the width of a piece of chalk. (Note: there are 2.54 cm in 1 in.)

One millimeter (mm) is approximately the thickness of the wire of a paper clip or of a mark made by a No. 2 pencil lead.

One micrometer (μm) is extremely tiny and can be measured only microscopically.

Make the following conversions between metric units of length:

9. 12 cm = ______ mm

10. 2000 μm = ______ mm

Now, circle the answer that would make the most sense in each of the following statements:

11. A match (in a matchbook) is (0.3, 3, 30) cm long.

12. A standard-size American car is about 4 (mm, cm, m, km) long.

$$\text{microunit} \underset{\times 1000}{\overset{\div 1000}{\rightleftharpoons}} \text{milliunit} \underset{\times 10}{\overset{\div 10}{\rightleftharpoons}} \text{centiunit} \underset{\times 100}{\overset{\div 100}{\rightleftharpoons}} \text{unit} \underset{\times 1000}{\overset{\div 1000}{\rightleftharpoons}} \text{kilounit}$$

$$\text{smallest} \rightleftharpoons \text{largest}$$

Volume Measurements
The metric unit of volume is the liter. A **liter (l,** or sometimes **L,** especially without a prefix) is slightly more than a quart (1 L = 1.057 quarts). Liquid volumes measured out for lab experiments are usually measured in milliliters (ml). (The terms *ml* and *cc*, cubic centimeter, are used interchangeably in laboratory and medical settings.)

To help you visualize metric volumes, the equivalents of some common substances follow:

A 12-oz can of soda is just slightly more than 360 ml.

A fluid ounce is 30 ml (cc).

A teaspoon of vanilla is about 5 ml (cc)

Compute the following:

13. How many 5-ml injections can be prepared from 1 liter of a medicine?

____

14. A 450-ml volume of alcohol is _______ L.

Mass Measurements
Although many people use the terms *mass* and *weight* interchangeably, this usage is inaccurate. **Mass** is the amount of matter in an object; and an object has a constant mass, regardless of where it is—that is, on earth, or in outer space. However, weight varies with gravitational pull; the greater the gravitational pull, the greater the weight. Thus, our astronauts are said to be weightless* when in outer space, but they still have the same mass as they do on earth.

The metric unit of mass is the **gram (g).** Medical dosages are usually prescribed in milligrams (mg) or micrograms (μg); and in the clinical agency, body weight (particularly of infants) is typically specified in kilograms (kg) (1 kg = 2.2 lb).

The following examples are provided to help you become familiar with the masses of some common objects:

Two aspirin tablets have a mass of approximately 1 g.

A nickel has a mass of 5 g.

The mass of an average woman (132 lb) is 60 kg.

* Astronauts are not *really* weightless. It is just that they and their surroundings are being pulled toward the earth at the same speed; and so, in reference to their environment, they appear to float.

Make the following conversions:

15. 300 g = _______ mg = _______ μg

16. 4000 μg = _______ mg = _______ g

17. A nurse must administer to her patient, Mrs. Smith, 5 mg of a drug per kg of body mass. Mrs. Smith weighs 140 lb. How many grams of the drug should the nurse administer to her patient?

_______ g

Temperature Measurements
In the laboratory and in the clinical agency, temperature is measured both in metric units (degrees Celsius, °C) and in British units (degrees Fahrenheit, °F). Thus it helps to be familiar with both temperature scales.

The temperatures of boiling and freezing water can be used to compare the two scales:

The freezing point of water is 0°C and 32°F.

The boiling point of water is 100°C and 212°F.

As you can see, the range from the freezing point to the boiling point of water on the Celsius scale is 100 degrees, whereas the comparable range on the Fahrenheit scale is 180 degrees. Hence, one degree on the Celsius scale represents a greater change in temperature. Normal body temperature is approximately 98.6°F or 37°C.

To convert from the Celsius scale to the Fahrenheit scale, the following equation is used:

$$°C = \frac{5(°F - 32)}{9}$$

To convert from the Fahrenheit scale to the Celsius scale, the following equation is used:

$$°F = 9/5\ °C + 32$$

Perform the following temperature conversions:

18. Convert 38°C to °F: _______

19. Convert 158°F to °C: _______

Answers

1. range of 94–64 or 30 beats/min; mean 77.5

2. 3800%

3. 75%

4. 160%

5. 8

6. 0.0005

7. 10 breaths/min

8. interval between 100–102° (went from 22 to 36 breaths/min)

9. 12 cm = 120 mm

10. 2000 μm = 2 mm

11. 3 cm long

12. 4 m long

13. 200

14. 0.45 L

15. 300 g = 3×10^5 mg = 3×10^8 μg

16. 4000 μg = 4 mg = 4×10^{-3} g (0.004g)

17. 0.32 g

18. 100.4°F

19. 70°C

The Language of Anatomy

exercise

1

Objectives

1. To describe the anatomical position verbally or by demonstration.
2. To use proper anatomical terminology to describe body directions, planes, and surfaces.
3. To name the body cavities and indicate the important organs in each.

Materials

- ❑ Human torso model (dissectible)
- ❑ Human skeleton
- ❑ Demonstration: sectioned and labeled kidneys [three separate kidneys uncut or cut so that (a) entire, (b) transverse section, and (c) longitudinal sectional views are visible]
- ❑ Modeling clay
- ❑ Scalpel

A1A See Appendix C, Exercise 1 for links to A.D.A.M.® Interactive Anatomy.

Most of us are naturally curious about our bodies. This fact is amply demonstrated by infants, who are fascinated with their own waving hands or their mother's nose. The study of the gross anatomy of the human body elaborates on this fascination. Unlike the infant, however, the student of anatomy must learn to observe and identify the dissectible body structures formally. The purpose of any gross anatomy experience is to examine the three-dimensional relationships of body structures—a goal that can never completely be achieved by using illustrations and models, regardless of their excellence.

When beginning the study of any science, the student is often initially overcome by jargon unique to the subject. The study of anatomy is no exception. But without this specialized terminology, confusion is inevitable. For example, what do *over, on top of, superficial to, above,* and *behind* mean in reference to the human body? Anatomists have an accepted set of reference terms that are universally understood. These allow body structures to be located and identified with a minimum of words and a high degree of clarity.

This exercise presents some of the most important anatomical terminology used to describe the body and introduces you to basic concepts of **gross anatomy,** the study of body structures visible to the naked eye.

Anatomical Position

When anatomists or doctors refer to specific areas of the human body, they do so in accordance with a universally accepted standard position called the **anatomical position.** It is essential to understand this position because much of the body terminology employed in this book refers to this body positioning, regardless of the position the body happens to be in. In the anatomical position the human body is erect, with the feet only slightly apart, head and toes pointed forward, and arms hanging at the sides with palms facing forward (Figure 1.1).

- Assume the anatomical position, and notice that it is not particularly comfortable. The hands are held unnaturally forward rather than hanging partially cupped toward the thighs.

Surface Anatomy

Body surfaces provide a wealth of visible landmarks for study of the body.

Axial: relating to head, neck, and trunk, the axis of the body.

Appendicular: relating to limbs and their attachments to the axis.

Anterior Body Landmarks

Note the following regions in Figure 1.2a:

Abdominal: Pertaining to the anterior body trunk region inferior to the ribs

Antebrachial: Pertaining to the forearm

Antecubital: Pertaining to the anterior surface of the elbow

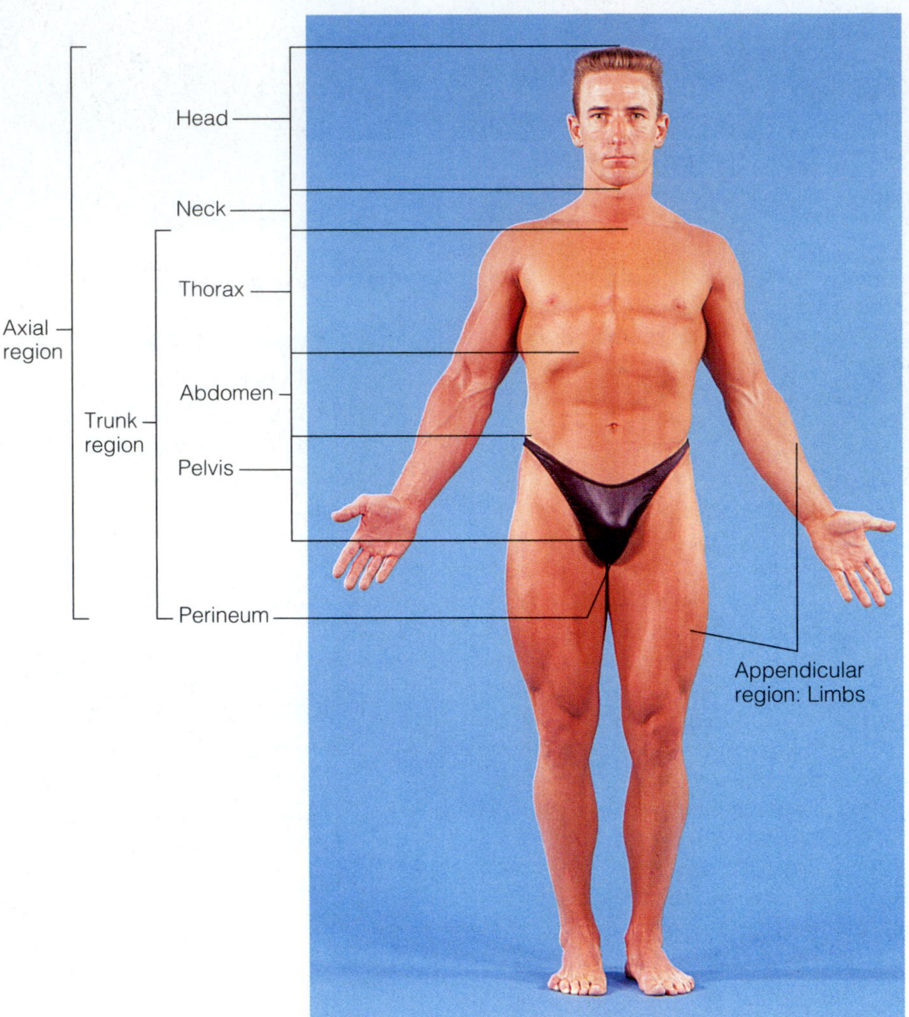

Figure 1.1 Anatomical position.

Axillary: Pertaining to the armpit
Brachial: Pertaining to the arm
Buccal: Pertaining to the cheek
Carpal: Pertaining to the wrist
Cervical: Pertaining to the neck region
Coxal: Pertaining to the hip
Crural: Pertaining to the leg
Digital: Pertaining to the fingers or toes
Femoral: Pertaining to the thigh
Frontal: Pertaining to the forehead
Hallux: Pertaining to the great toe
Inguinal: Pertaining to the groin
Mammary: Pertaining to the breast
Mental: Pertaining to the chin
Nasal: Pertaining to the nose
Oral: Pertaining to the mouth
Orbital: Pertaining to the bony eye socket (orbit)
Palmar: Pertaining to the palm of the hand
Patellar: Pertaining to the anterior knee (kneecap) region
Pedal: Pertaining to the foot
Pelvic: Pertaining to the pelvis region
Fibular (Peroneal): Pertaining to the side of the leg
Pollex: Pertaining to the thumb
Pubic: Pertaining to the genital region
Sternal: Pertaining to the region of the breastbone
Tarsal: Pertaining to the ankle
Thoracic: Pertaining to the chest
Umbilical: Pertaining to the navel

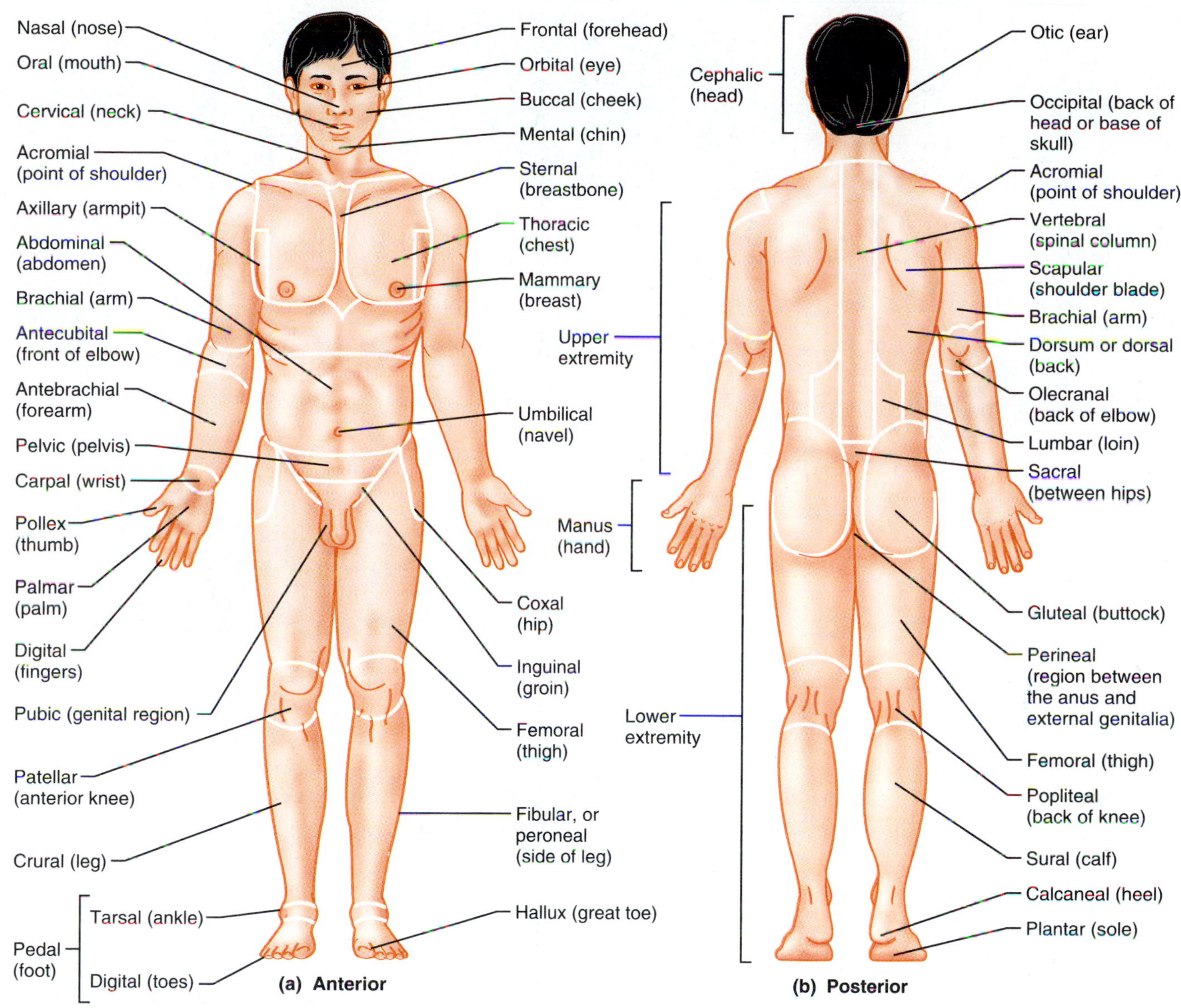

Figure 1.2 **Surface anatomy.** **(a)** Anterior body landmarks. **(b)** Posterior body landmarks.

Posterior Body Landmarks

Note the following body surface regions in Figure 1.2b:

Acromial: Pertaining to the point of the shoulder

Calcaneal: Pertaining to the heel of the foot

Cephalic: Pertaining to the head

Dorsum: Pertaining to the back

Gluteal: Pertaining to the buttocks or rump

Lumbar: Pertaining to the area of the back between the ribs and hips; the loin

Manus: Pertaining to the hand

Occipital: Pertaining to the posterior aspect of the head or base of the skull

Olecranal: Pertaining to the posterior aspect of the elbow

Otic: Pertaining to the ear

Perineal: Pertaining to the region between the anus and external genitalia

Plantar: Pertaining to the sole of the foot

Popliteal: Pertaining to the back of the knee

Sacral: Pertaining to the region between the hips (overlying the sacrum)

Scapular: Pertaining to the scapula or shoulder blade area

Sural: Pertaining to the calf or posterior surface of the leg

Vertebral: Pertaining to the area of the spinal column

Activity 1: Locating Body Regions

Locate the anterior and posterior body landmarks on yourself, your lab partner, and a torso model before continuing. ■

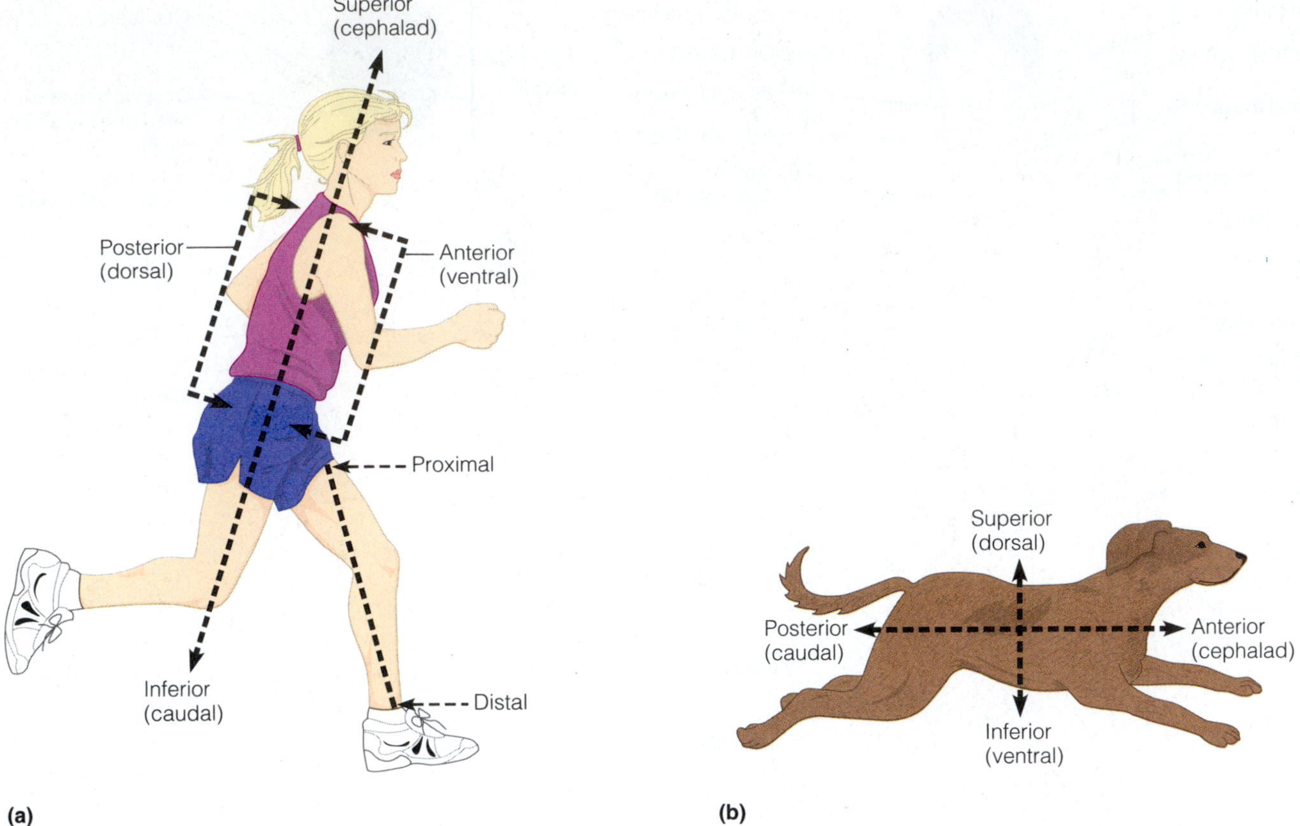

Figure 1.3 Anatomical terminology describing body orientation and direction. **(a)** With reference to a human. **(b)** With reference to a four-legged animal.

Body Orientation and Direction

Study the terms below, referring to Figure 1.3. Notice that certain terms have a different meaning for a four-legged animal than they do for a human.

Superior/inferior (*above/below*): These terms refer to placement of a structure along the long axis of the body. Superior structures always appear above other structures, and inferior structures are always below other structures. For example, the nose is superior to the mouth, and the abdomen is inferior to the chest.

Anterior/posterior (*front/back*): In humans the most anterior structures are those that are most forward—the face, chest, and abdomen. Posterior structures are those toward the backside of the body. For instance, the spine is posterior to the heart.

Medial/lateral (*toward the midline/away from the midline or median plane*): The sternum (breastbone) is medial to the ribs; the ear is lateral to the nose.

The terms of position described above depend on an assumption of anatomical position. The next four term pairs are more absolute. Their applicability is not relative to a particular body position, and they consistently have the same meaning in all vertebrate animals.

Cephalad (cranial)/caudal (*toward the head/toward the tail*): In humans these terms are used interchangeably with *superior* and *inferior.* But in four-legged animals they are synonymous with *anterior* and *posterior,* respectively.

Dorsal/ventral (*backside/belly side*): These terms are used chiefly in discussing the comparative anatomy of animals, assuming the animal is standing. *Dorsum* is a Latin word meaning "back." Thus, *dorsal* refers to the animal's back or the *back*side of any other structures; e.g., the posterior surface of the human leg is its dorsal surface. The term *ventral* derives from the Latin term *venter,* meaning "belly," and always refers to the belly side of animals. In humans the terms *ventral* and *dorsal* are used interchangeably with the terms *anterior* and *posterior,* but in four-legged animals *ventral* and *dorsal* are synonymous with *inferior* and *superior,* respectively.

Proximal/distal (*nearer the trunk or attached end/farther from the trunk or point of attachment*): These terms are used primarily to locate various areas of the body limbs. For example, the fingers are distal to the elbow; the knee is proximal to the toes.

Superficial (external)/deep (internal) (*toward or at the body surface/away from the body surface*): These terms locate body organs according to their relative closeness to the body surface. For example, the skin is superficial to the skeletal muscles, and the lungs are deep to the rib cage.

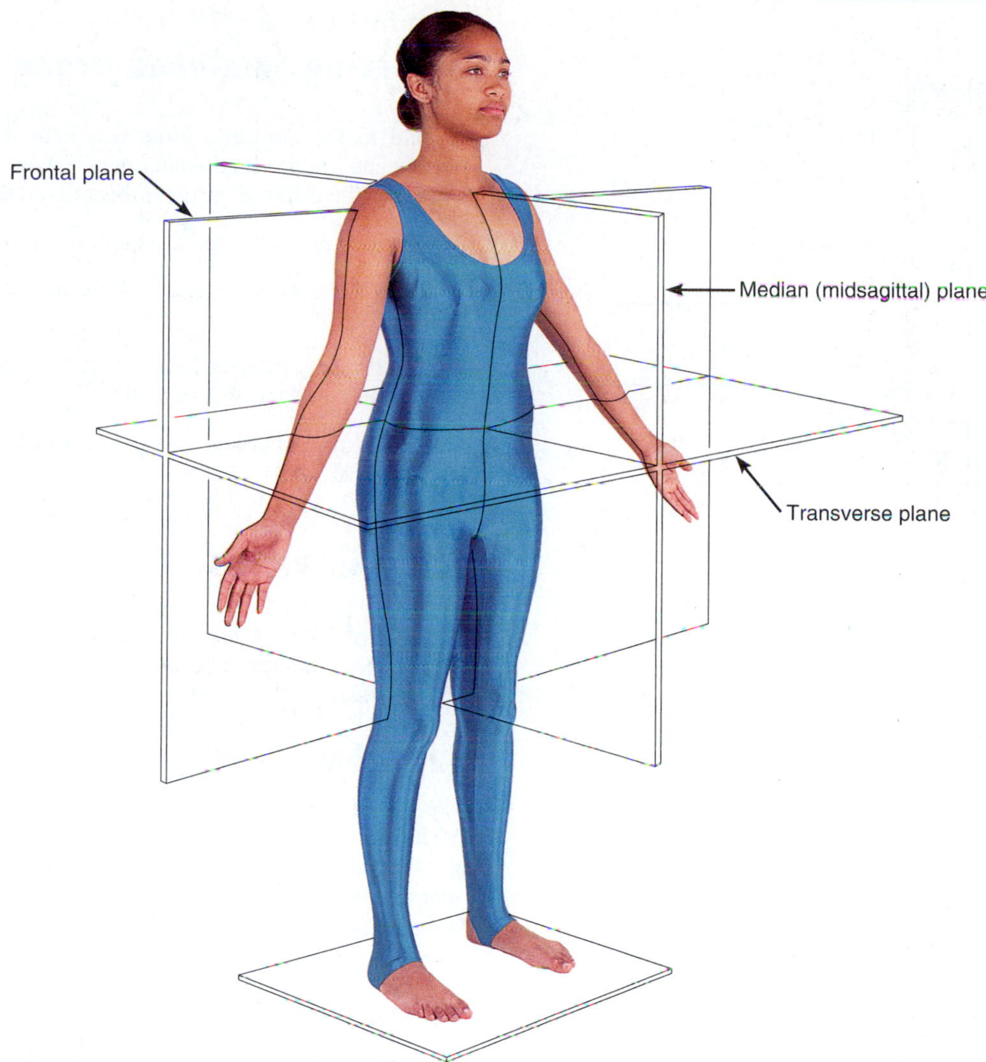

Figure 1.4 Planes of the body.

Activity 2: Practicing Using Correct Anatomical Terminology

Before continuing, use a human torso model, a skeleton, or your own body to specify the relationship between the following structures.

1. The wrist is proximal to the hand.
2. The trachea (windpipe) is anterior to the spine.
3. The brain is superior to the spinal cord.
4. The kidneys are posterior to the liver.
5. The nose is medial to the cheekbones. ■

Body Planes and Sections

The body is three-dimensional, and in order to observe its internal structures, it is often helpful and necessary to make use of a **section,** or cut. When the section is made through the body wall or through an organ, it is made along an imaginary surface or line called a **plane.** Anatomists commonly refer to three planes (Figure 1.4) or sections that lie at right angles to one another.

Sagittal plane: A plane that runs longitudinally and divides the body into right and left parts is referred to as a sagittal plane. If it divides the body into equal parts, right down the median plane of the body, it is called a **median,** or **midsagittal, plane.** All other sagittal planes are referred to as **parasagittal planes.**

Frontal plane: Sometimes called a **coronal plane,** the frontal plane is a longitudinal plane that divides the body (or an organ) into anterior and posterior parts.

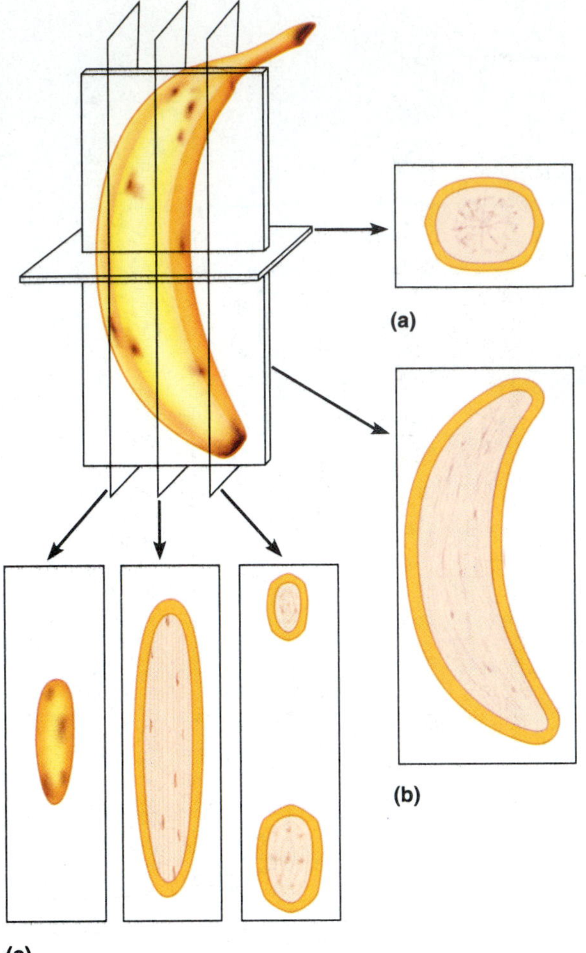

Figure 1.5 Objects can look odd when viewed in section. This banana has been sectioned in three different planes **(a–c),** and only in one of these planes is it easily recognized as a banana (plane b). In order to recognize human organs in section, one must anticipate how the organs will look when cut that way. If one cannot recognize a sectioned organ, it is possible to reconstruct its shape from a series of successive cuts, as from three serial sections in **(c).**

Transverse plane: A transverse plane runs horizontally, dividing the body into superior and inferior parts. When organs are sectioned along the transverse plane, the sections are commonly called **cross sections.**

On microscope slides, the abbreviation for a longitudinal section (sagittal or frontal) is l.s. Cross sections are abbreviated x.s. or c.s.

As shown in Figure 1.5, a sagittal or frontal plane section of any nonspherical object, be it a banana or a body organ, provides quite a different view than a transverse section. Parasagittal sections provide still different views.

Activity 3: Observing Sectioned Organ Specimens

1. Go to the demonstration area and observe the transversely and longitudinally cut organ specimens. Pay close attention to the different structural details in the samples.

2. Obtain some clay and a scalpel from the supply area.

3. Use the clay to produce three similar models of the kidney.

4. Section the models to demonstrate cuts along the sagittal, frontal, and transverse planes.

5. Verify the accuracy of your sectioned models with your instructor. ■

Body Cavities

The axial portion of the body has two large cavities that provide different degrees of protection to the organs within them (Figure 1.6).

Dorsal Body Cavity

The dorsal body cavity can be subdivided into the **cranial cavity,** in which the brain is enclosed within the rigid skull, and the **vertebral** or **spinal cavity,** within which the delicate spinal cord is protected by the bony vertebral column. Because the spinal cord is a continuation of the brain, these cavities are continuous with each other.

Ventral Body Cavity

Like the dorsal cavity, the ventral body cavity is subdivided. The superior **thoracic cavity** is separated from the rest of the ventral cavity by the dome-shaped diaphragm. The heart and lungs, located in the thoracic cavity, are afforded some measure of protection by the bony rib cage. The cavity inferior to the diaphragm is often referred to as the **abdominopelvic cavity.** Although there is no further physical separation of the ventral cavity, some prefer to describe the abdominopelvic cavity in terms of a superior **abdominal cavity,** the area that houses the stomach, intestines, liver, and other organs, and an inferior **pelvic cavity,** the region that is partially enclosed by the bony pelvis and contains the reproductive organs, bladder, and rectum. Notice in Figure 1.6 that the abdominal and pelvic cavities are not continuous with each other in a straight plane but that the pelvic cavity is tipped away from the perpendicular.

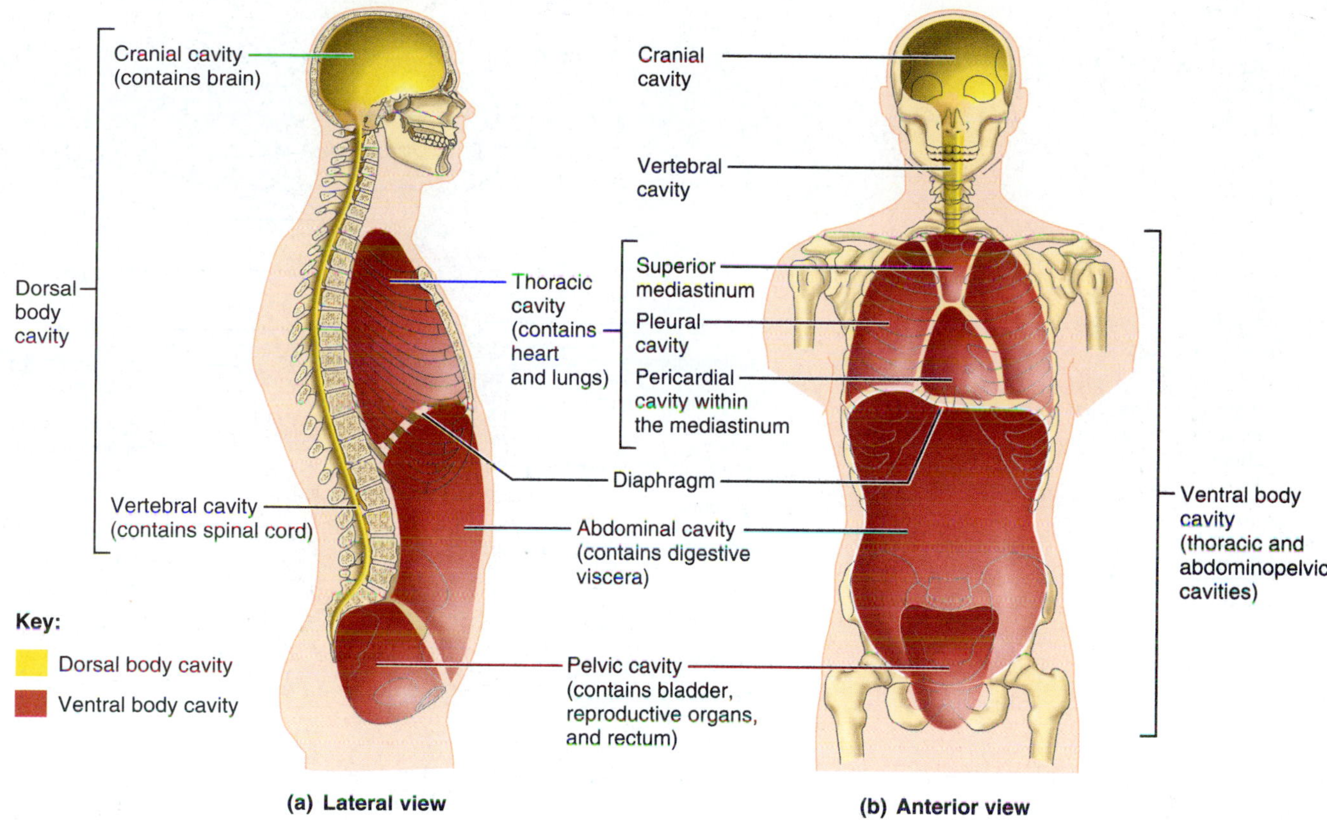

Figure 1.6 Body cavities and their subdivisions.

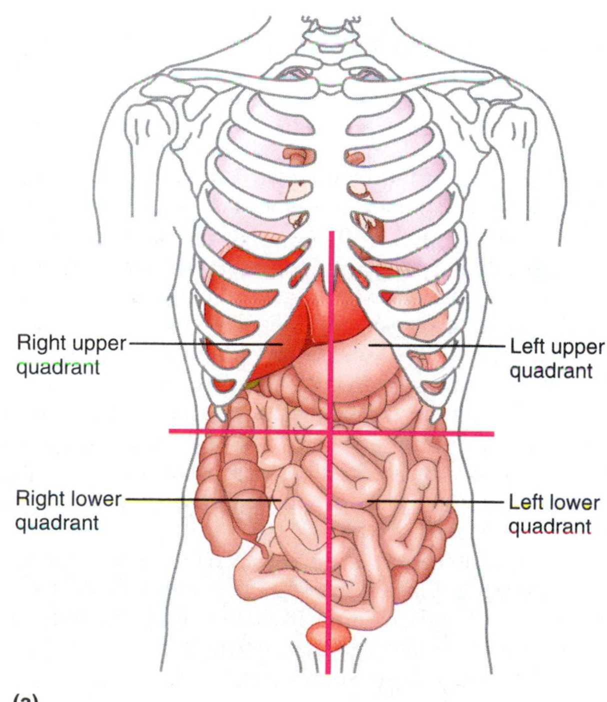

Figure 1.7 Abdominopelvic surface and cavity. (a) The four quadrants, showing superficial organs in each quadrant. *(continues on page 8)*

Abdominopelvic Quadrants and Regions Because the abdominopelvic cavity is quite large and contains many organs, it is helpful to divide it up into smaller areas for discussion or study.

A scheme, used by most physicians and nurses, divides the abdominal surface (and the abdominopelvic cavity deep to it) into four approximately equal regions called **quadrants.** These quadrants are named according to their relative position—that is, *right upper quadrant, right lower quadrant, left upper quadrant,* and *left lower quadrant* (see Figure 1.7a). Note that the terms left and right refer to the left and right of the figure, not your own. The left and right of the figure are referred to as **anatomical left and right.**

A different scheme commonly used by anatomists divides the abdominal surface and abdominopelvic cavity into nine separate regions by four planes, as shown in Figure 1.7b. Although the names of these nine regions are unfamiliar to you now, with a little patience and study they will become easier to remember. As you read through the descriptions of these nine regions and locate them in Figure 1.7b, also look at Figure 1.7c to note the organs the regions contain.

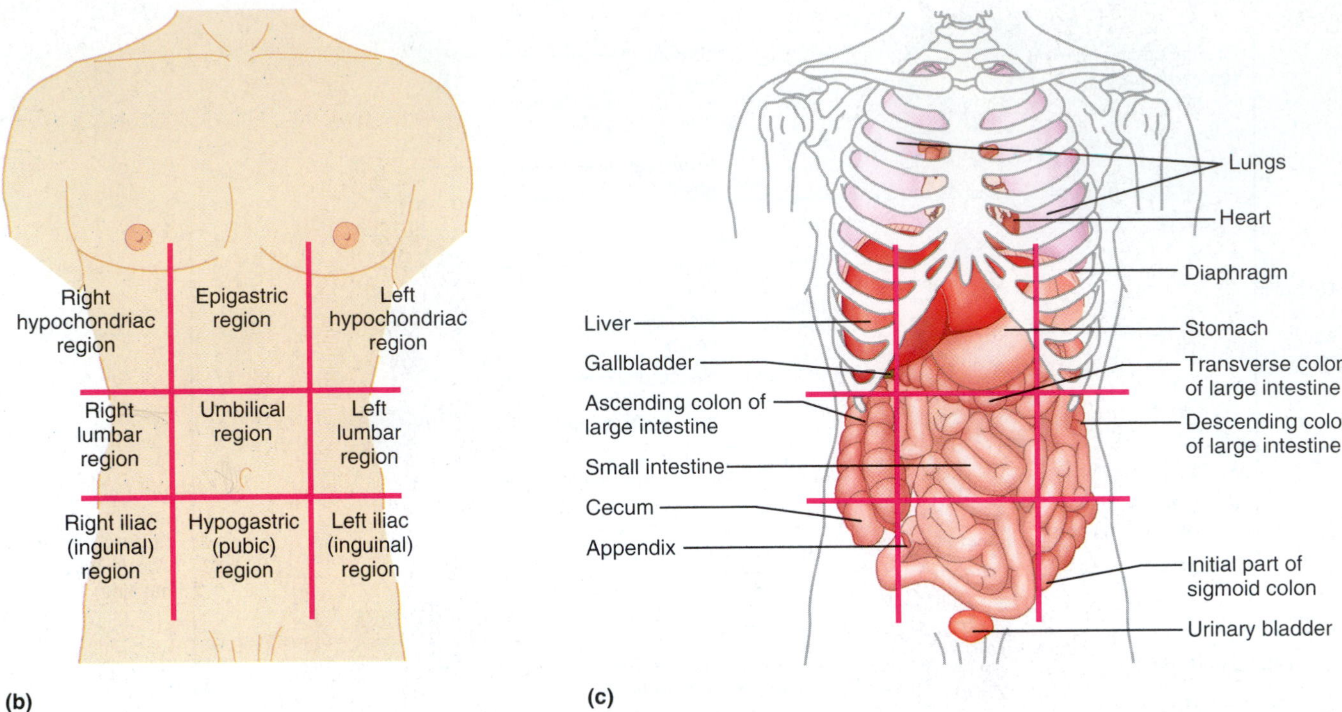

Figure 1.7 (*continued*) **Abdominopelvic surface and cavity. (b)** Nine regions delineated by four planes. The superior horizontal plane is just inferior to the ribs; the inferior horizontal plane is at the superior aspect of the hip bones. The vertical planes are just medial to the nipples. **(c)** Anterior view of the abdominopelvic cavity showing superficial organs.

Umbilical region: The centermost region, which includes the umbilicus

Epigastric region: Immediately superior to the umbilical region; overlies most of the stomach

Hypogastric (pubic) region: Immediately inferior to the umbilical region; encompasses the pubic area

Iliac regions: Lateral to the hypogastric region and overlying the superior parts of the hip bones

Lumbar regions: Between the ribs and the flaring portions of the hip bones; lateral to the umbilical region

Hypochondriac regions: Flanking the epigastric region laterally and overlying the lower ribs

Activity 4: Locating Abdominal Surface Regions

Locate the regions of the abdominal surface on a torso model and on yourself before continuing. ■

Serous Membranes of the Ventral Body Cavity The walls of the ventral body cavity and the outer surfaces of the organs it contains are covered with an exceedingly thin, double-layered membrane called the **serosa,** or **serous membrane.** The part of the membrane lining the cavity walls is referred to as the **parietal serosa,** and it is continuous with a similar membrane, the **visceral serosa,** covering the external surface of the organs within the cavity. These membranes produce a thin lubricating fluid that allows the visceral organs to slide over one another or to rub against the body wall without friction. Serous membranes also compartmentalize the various organs so that infection of one organ is prevented from spreading to others.

The specific names of the serous membranes depend on the structures they envelop. Thus the serosa lining the abdominal cavity and covering its organs is the **peritoneum,** that enclosing the lungs is the **pleura,** and that around the heart is the **pericardium** (Figure 1.8).

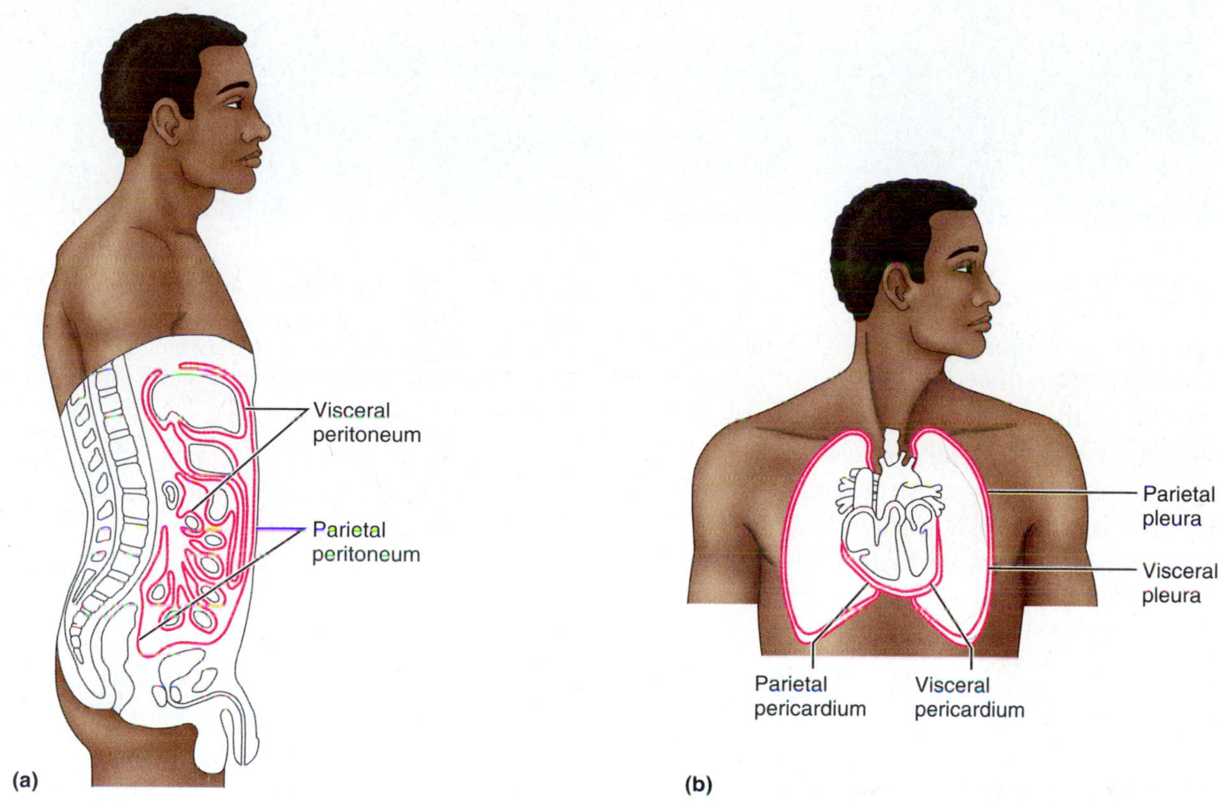

Figure 1.8 Serous membranes. (a) The peritoneums surround the digestive viscera in the abdominopelvic cavity. **(b)** Serosae in the thoracic cavity are the pleurae surrounding the lungs and the pericardia surrounding the heart.

exercise

2 Organ Systems Overview

Objectives

1. To name the human organ systems and indicate the major functions of each.
2. To list two or three organs of each system, and categorize the various organs by organ system.
3. To identify these organs in a dissected rat or human cadaver or on a dissectible human torso model.
4. To identify the correct organ system for each organ when presented with a list of organs (as studied in the laboratory).

Materials

- ❑ Freshly killed or preserved rat predissected by instructor as a demonstration or for student dissection (one rat for every two to four students) or predissected human cadaver
- ❑ Dissecting pans
- ❑ Twine or large dissecting pins
- ❑ Scissors
- ❑ Probes
- ❑ Forceps
- ❑ Disposable plastic gloves
- ❑ Human torso model (dissectible)

A1A See Appendix C, Exercise 2 for links to A.D.A.M.® Interactive Anatomy.

The basic unit or building block of all living things is the **cell.** Cells fall into four different categories according to their structures and functions. Each of these corresponds to one of the four tissue types: epithelial, muscular, nervous, and connective. A **tissue** is a group of cells that are similar in structure and function. An **organ** is a structure composed of two or more tissue types that performs a specific function for the body. For example, the small intestine, which digests and absorbs nutrients, is composed of all four tissue types.

An **organ system** is a group of organs that act together to perform a particular body function. For example, the organs of the digestive system work together to ensure that food moving through the digestive system is properly broken down and that the end products are absorbed into the bloodstream to provide nutrients and fuel for all the body's cells. In all, there are 11 organ systems, which are described in Table 2.1. The lymphatic system also encompasses a *functional system* called the immune system, which is composed of an army of mobile *cells* that act to protect the body from foreign substances.

Read through this summary of the body's organ systems before beginning your rat dissection or examination of the predissected human cadaver. If a human cadaver is not available, Figures 2.3 through 2.6 will serve as a partial replacement.

Dissection and Identification: The Organ Systems of the Rat

Now you will have a chance to observe the size, shape, location, and distribution of the organs and organ systems. Many of the external and internal structures of the rat are quite similar in structure and function to those of the human, so a study of the gross anatomy of the rat should help you understand our own physical structure.

The following instructions have been written to complement and direct your dissection and observation of a rat, but the descriptions for organ observations from the procedure entitled "Examining the Ventral Body Cavity," which begins on p. 13, apply as well to superficial observations of a previously dissected human cadaver. In addition, the general instructions for observing external structures can easily be extrapolated to serve human cadaver observations, and photographs provided in Figures 2.3 to 2.5 will provide visual aids.

Note that four of the organ systems listed in Table 2.1 will not be studied at this time (integumentary, skeletal, muscular, and nervous), as they require microscopic study or more detailed dissection.

Activity 1: Observing External Structures

1. If your instructor has provided a predissected rat, go to the demonstration area to make your observations. Alternatively, if you and/or members of your group will be dissecting the specimen, obtain a preserved or freshly killed rat (one for every two to four students), a dissecting pan, dissecting pins, scissors, probe, forceps, and disposable gloves and bring them to your laboratory bench.

If a predissected human cadaver is available, obtain a probe, forceps, and disposable gloves before going to the demonstration area.

Table 2.1 Overview of Organ Systems of the Body

Organ system	Major component organs	Function
Integumentary (Skin)	Epidermal and dermal regions; cutaneous sense organs and glands	• Protects deeper organs from mechanical, chemical, and bacterial injury, and desiccation (drying out) • Excretes salts and urea • Aids in regulation of body temperature • Produces vitamin D
Skeletal	Bones, cartilages, tendons, ligaments, and joints	• Body support and protection of internal organs • Provides levers for muscular action • Cavities provide a site for blood cell formation
Muscular	Muscles attached to the skeleton	• Primary function is to contract or shorten; in doing so, skeletal muscles allow locomotion (running, walking, etc.), grasping and manipulation of the environment, and facial expression • Generates heat
Nervous	Brain, spinal cord, nerves, and sensory receptors	• Allows body to detect changes in its internal and external environment and to respond to such information by activating appropriate muscles or glands • Helps maintain homeostasis of the body via rapid transmission of electrical signals
Endocrine	Pituitary, thymus, thyroid, parathyroid, adrenal, and pineal glands; ovaries, testes, and pancreas	• Helps maintain body homeostasis, promotes growth and development; produces chemical "messengers" (hormones) that travel in the blood to exert their effect(s) on various "target organs" of the body
Cardiovascular	Heart, blood vessels, and blood	• Primarily a transport system that carries blood containing oxygen, carbon dioxide, nutrients, wastes, ions, hormones, and other substances to and from the tissue cells where exchanges are made; blood is propelled through the blood vessels by the pumping action of the heart • Antibodies and other protein molecules in the blood act to protect the body
Lymphatic/ Immunity	Lymphatic vessels, lymph nodes, spleen, thymus, tonsils, and scattered collections of lymphoid tissue	• Picks up fluid leaked from the blood vessels and returns it to the blood • Cleanses blood of pathogens and other debris • Houses lymphocytes that act via the immune response to protect the body from foreign substances (antigens)
Respiratory	Nasal passages, pharynx, larynx, trachea, bronchi, and lungs	• Keeps the blood continuously supplied with oxygen while removing carbon dioxide
Digestive	Oral cavity, esophagus, stomach, small and large intestines, and accessory structures (teeth, salivary glands, liver, and pancreas)	• Breaks down ingested foods to minute particles, which can be absorbed into the blood for delivery to the body cells • Undigested residue removed from the body as feces
Urinary	Kidneys, ureters, bladder, and urethra	• Rids the body of nitrogen-containing wastes (urea, uric acid, and ammonia), which result from the breakdown of proteins and nucleic acids by body cells • Maintains water, electrolyte, and acid-base balance of blood
Reproductive	Male: testes, prostate gland, scrotum, penis, and duct system, which carries sperm to the body exterior	• Provides germ cells (sperm) for perpetuation of the species
	Female: ovaries, uterine tubes, uterus, mammary glands, and vagina	• Provides germ cells (eggs); the female uterus houses the developing fetus until birth; mammary glands provide nutrition for the infant

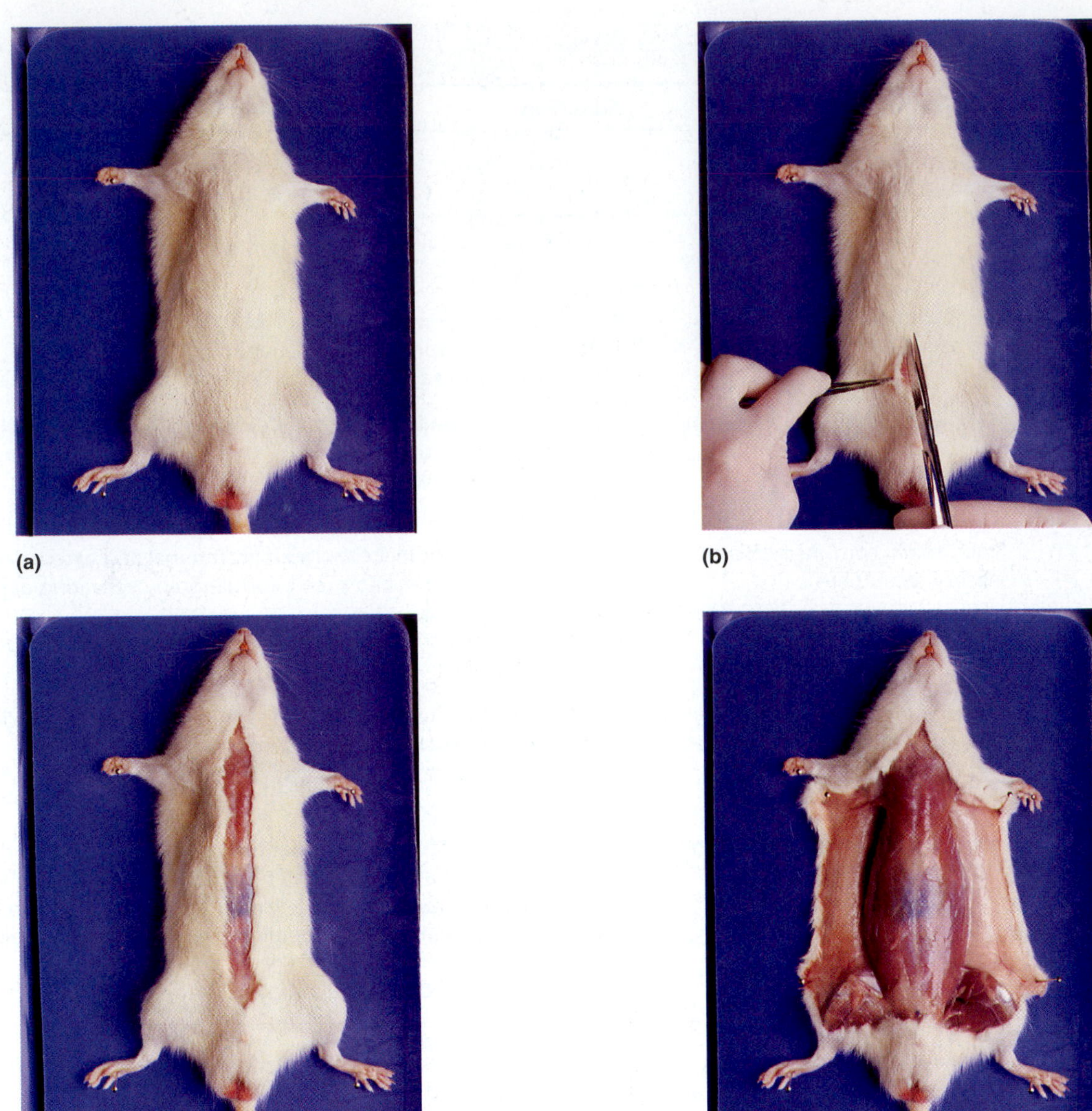

Figure 2.1 Rat dissection: Securing for dissection and the initial incision. (a) Securing the rat to the dissection tray with dissecting pins. **(b)** Using scissors to make the incision on the median line of the abdominal region. **(c)** Completed incision from the pelvic region to the lower jaw. **(d)** Reflection (folding back) of the skin to expose the underlying muscles.

2. Don the gloves before beginning your observations. This precaution is particularly important when handling freshly killed animals, which may harbor internal parasites.

3. Observe the major divisions of the body—head, trunk, and extremities. If you are examining a rat, compare these divisions to those of humans. ■

Activity 2: Examining the Oral Cavity

Examine the structures of the oral cavity. Identify the teeth and tongue. Observe the extent of the hard palate (the portion underlain by bone) and the soft palate (immediately posterior to the hard palate, with no bony support). Notice that the posterior end of the oral cavity leads into the throat, or pharynx. The pharynx is a passageway used by both the digestive and respiratory systems. ■

Activity 3: Opening the Ventral Body Cavity

1. Pin the animal to the wax of the dissecting pan by placing its dorsal side down and securing its extremities to the wax with large dissecting pins as shown in Figure 2.1a. If the

dissecting pan is not waxed, you will need to secure the animal with twine as follows. (Some may prefer this method in any case.) Obtain the roll of twine. Make a loop knot around one upper limb, pass the twine under the pan, and secure the opposing limb. Repeat for the lower extremities.

2. Lift the abdominal skin with a forceps, and cut through it with the scissors (Figure 2.1b). Close the scissor blades and insert them under the cut skin. Moving in a cephalad direction, open and close the blades to loosen the skin from the underlying connective tissue and muscle. Once this skin-freeing procedure has been completed, cut the skin along the body midline, from the pubic region to the lower jaw (Figure 2.1c). Make a lateral cut about halfway down the ventral surface of each limb. Complete the job of freeing the skin with the scissor tips, and pin the flaps to the tray (Figure 2.1d). The underlying tissue that is now exposed is the skeletal musculature of the body wall and limbs. It allows voluntary body movement. Notice that the muscles are packaged in sheets of pearly white connective tissue (fascia), which protect the muscles and bind them together.

3. Carefully cut through the muscles of the abdominal wall in the pubic region, avoiding the underlying organs. Remember, to *dissect* means "to separate"—not mutilate! Now, hold and lift the muscle layer with a forceps and cut through the muscle layer from the pubic region to the bottom of the rib cage. Make two lateral cuts through the rib cage (Figure 2.2). A thin membrane attached to the inferior boundary of the rib cage should be obvious; this is the **diaphragm,** which separates the thoracic and abdominal cavities. Cut the diaphragm away to loosen the rib cage. You can now lift the ribs to view the contents of the thoracic cavity. ■

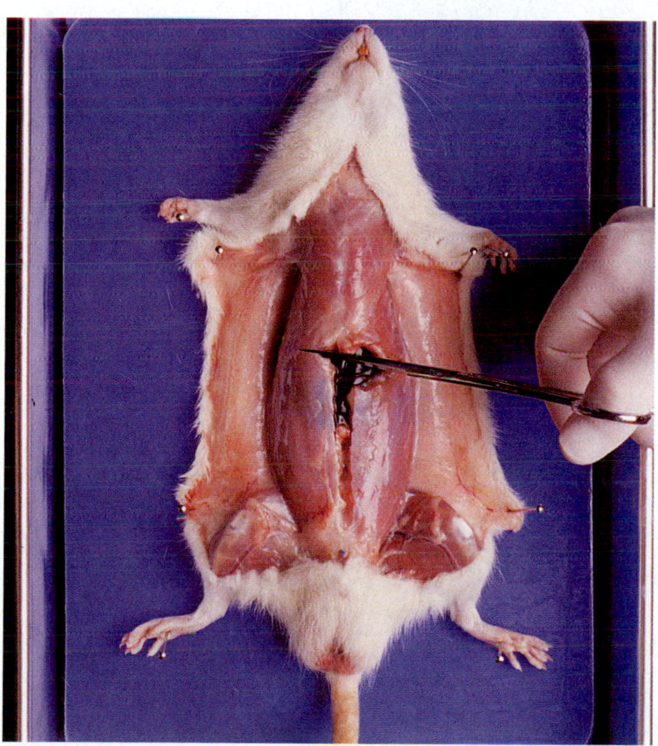

Figure 2.2 Rat dissection: Making lateral cuts at the base of the rib cage.

Activity 4:
Examining the Ventral Body Cavity

1. Examine the structures of the thoracic cavity, starting with the most superficial structures and working deeper. As you work, refer to Figure 2.3, which shows the superficial organs. Choose the appropriate view (a or b) depending on whether you are examining a rat or a human cadaver.

Thymus: An irregular mass of glandular tissue overlying the heart (not illustrated in the human cadaver photograph).

With the probe, push the thymus to the side to view the heart.

Heart: Medial oval structure enclosed within the pericardium (serous membrane sac).

Lungs: Flanking the heart on either side.

Now observe the throat region to identify the trachea.

Trachea: Tubelike "windpipe" running medially down the throat; part of the respiratory system.

Follow the trachea into the thoracic cavity; notice where it divides into two branches. These are the bronchi.

Bronchi: Two passageways that plunge laterally into the tissue of the two lungs.

To expose the esophagus, push the trachea to one side.

Esophagus: A food chute; the part of the digestive system that transports food from the pharynx (throat) to the stomach.

Diaphragm: A thin muscle attached to the inferior boundary of the rib cage; separates the thoracic and abdominal cavities.

Follow the esophagus through the diaphragm to its junction with the stomach.

Stomach: A curved organ important in food digestion and temporary food storage.

2. Examine the superficial structures of the abdominopelvic cavity. Lift the *greater omentum,* an extension of the peritoneum that covers the abdominal viscera. Continuing from the stomach, trace the rest of the digestive tract (Figure 2.4).

Small intestine: Connected to the stomach and ending just before the saclike cecum.

Large intestine: A large muscular tube connected to the small intestine and ending at the anus.

Cecum: The initial portion of the large intestine.

Follow the course of the large intestine to the rectum, which is partially covered by the urinary bladder.

Rectum: Terminal part of the large intestine; continuous with the anal canal.

Anus: The opening of the digestive tract (through the anal canal) to the exterior.

Now lift the small intestine with the forceps to view the mesentery.

Text continues on page 16

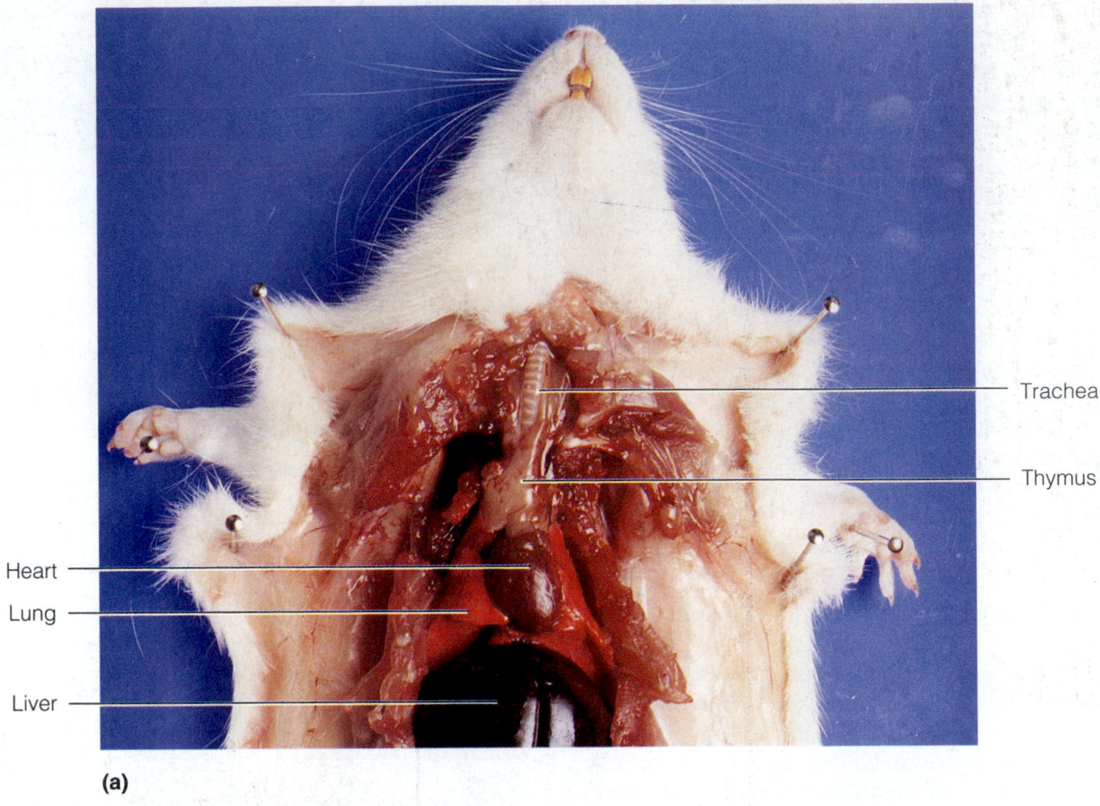

(a)

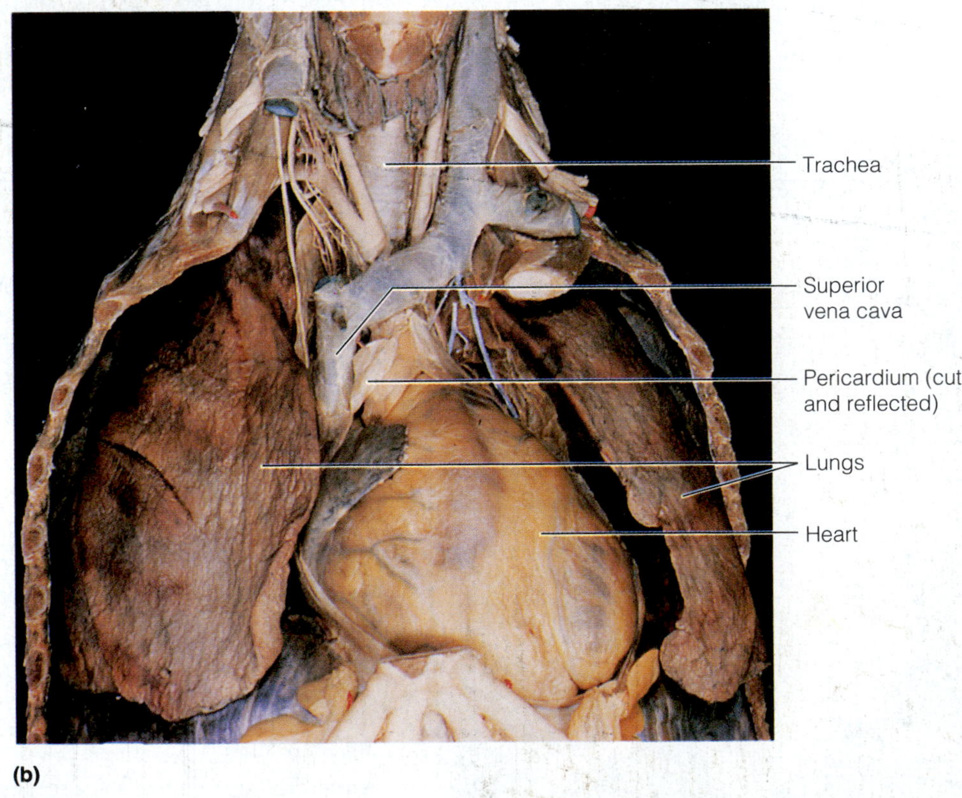

(b)

Figure 2.3 Superficial organs of the thoracic cavity. **(a)** Dissected rat. **(b)** Human cadaver.

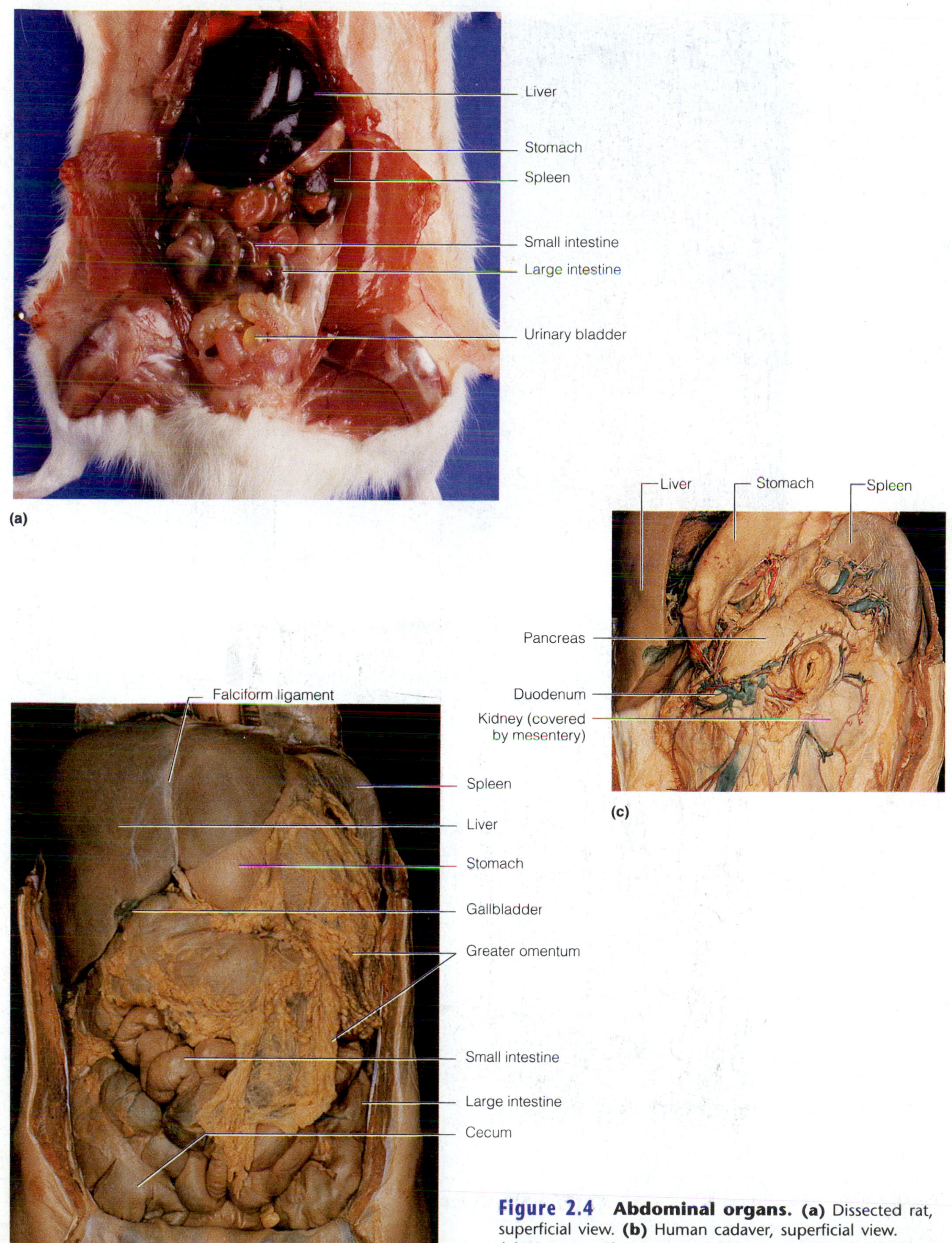

Figure 2.4 Abdominal organs. (a) Dissected rat, superficial view. **(b)** Human cadaver, superficial view. **(c)** Human cadaver, intermediate view. Stomach reflected superiorly to reveal the pancreas, spleen, and duodenum.

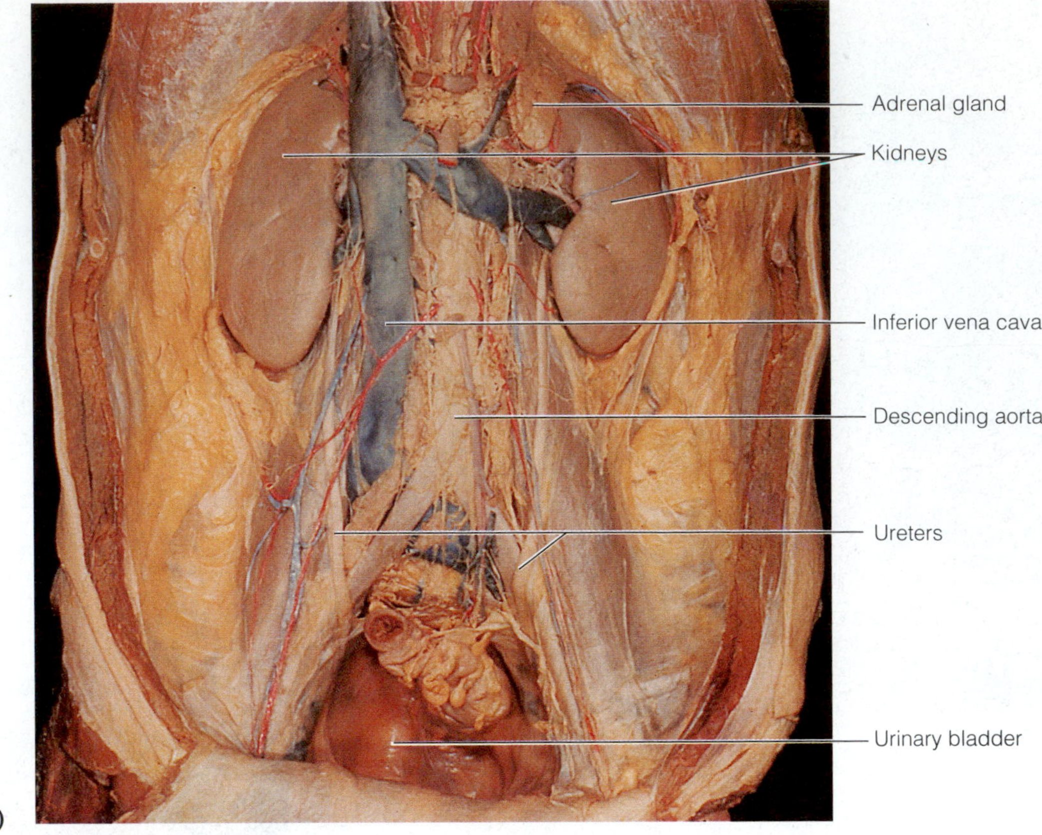

(a)

Figure 2.5 Deep structures of the abdominopelvic cavity. (a) Human cadaver.

Mesentery: An apronlike serous membrane; suspends many of the digestive organs in the abdominal cavity. Notice that it is heavily invested with blood vessels and, more likely than not, riddled with large fat deposits.

Locate the remaining abdominal structures.

Pancreas: A diffuse gland; rests dorsal to and in the mesentery between the first portion of the small intestine and the stomach. You will need to lift the stomach to view the pancreas.

Spleen: A dark red organ curving around the left lateral side of the stomach; considered part of the lymphatic system and often called the red blood cell graveyard.

Liver: Large and brownish red; the most superior organ in the abdominal cavity, directly beneath the diaphragm.

3. To locate the deeper structures of the abdominopelvic cavity, move the stomach and the intestines to one side with the probe.

Examine the posterior wall of the abdominal cavity to locate the two kidneys (Figure 2.5).

Kidneys: Bean-shaped organs; retroperitoneal (behind the peritoneum).

Adrenal glands: Large endocrine glands that sit astride the superior margin of each kidney; considered part of the endocrine system.

Carefully strip away part of the peritoneum with forceps and attempt to follow the course of one of the ureters to the bladder.

Ureter: Tube running from the indented region of a kidney to the urinary bladder.

Urinary bladder: The sac that serves as a reservoir for urine.

4. In the midline of the body cavity lying between the kidneys are the two principal abdominal blood vessels. Identify each.

Inferior vena cava: The large vein that returns blood to the heart from the lower regions of the body.

Descending aorta: Deep to the inferior vena cava; the largest artery of the body; carries blood away from the heart down the midline of the body.

5. Only a cursory examination of reproductive organs will be done. If you are working with a rat, first determine if the animal is a male or female. Observe the ventral body surface beneath the tail. If a saclike scrotum and an opening for the anus are visible, the animal is a male. If three body openings—urethral, vaginal, and anus—are present, it is a female.

Male Animal Make a shallow incision into the **scrotum.** Loosen and lift out the oval **testis.** Exert a gentle pull on the testis to identify the slender **vas deferens,** or sperm duct, which carries sperm from the testis superiorly into the abdominal cavity and joins with the urethra. The urethra runs through the penis of the male and carries both urine and sperm out of the body. Identify the **penis,** extending from the bladder to the ventral body wall. Figure 2.5b indicates other glands of the male rat's reproductive system, but they need not be identified at this time.

Text continues on page 18

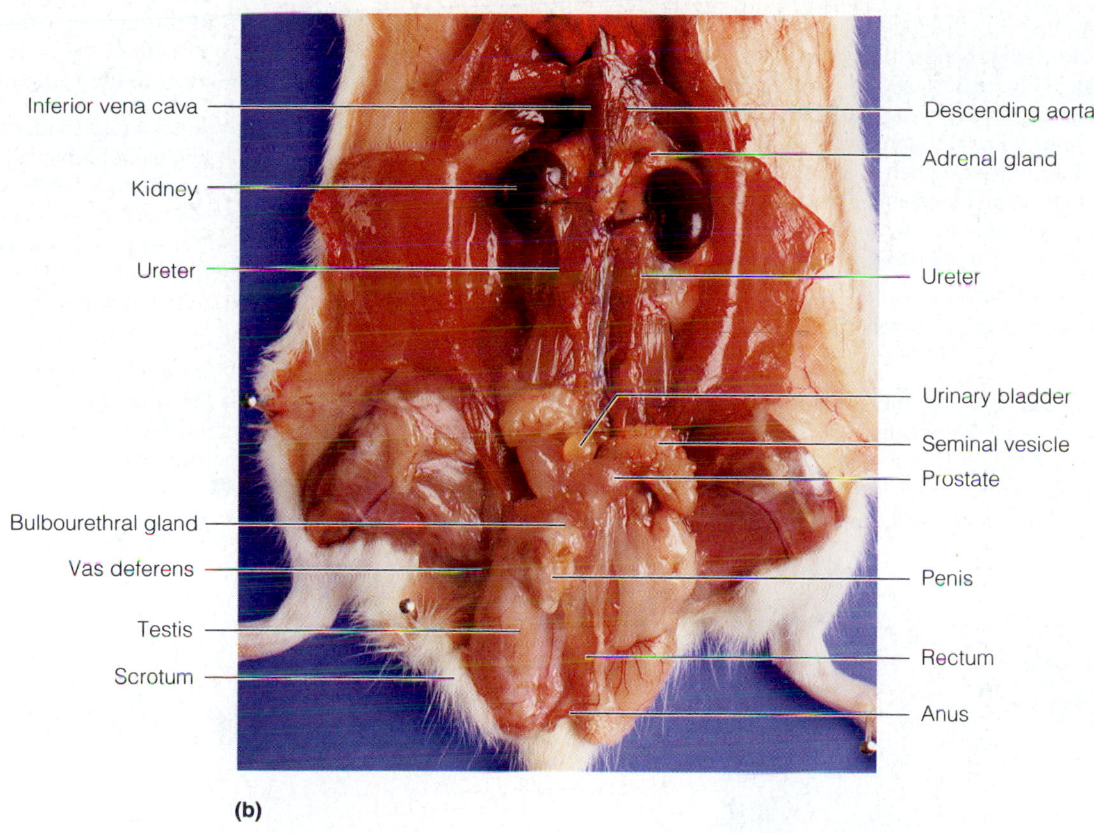

(b)

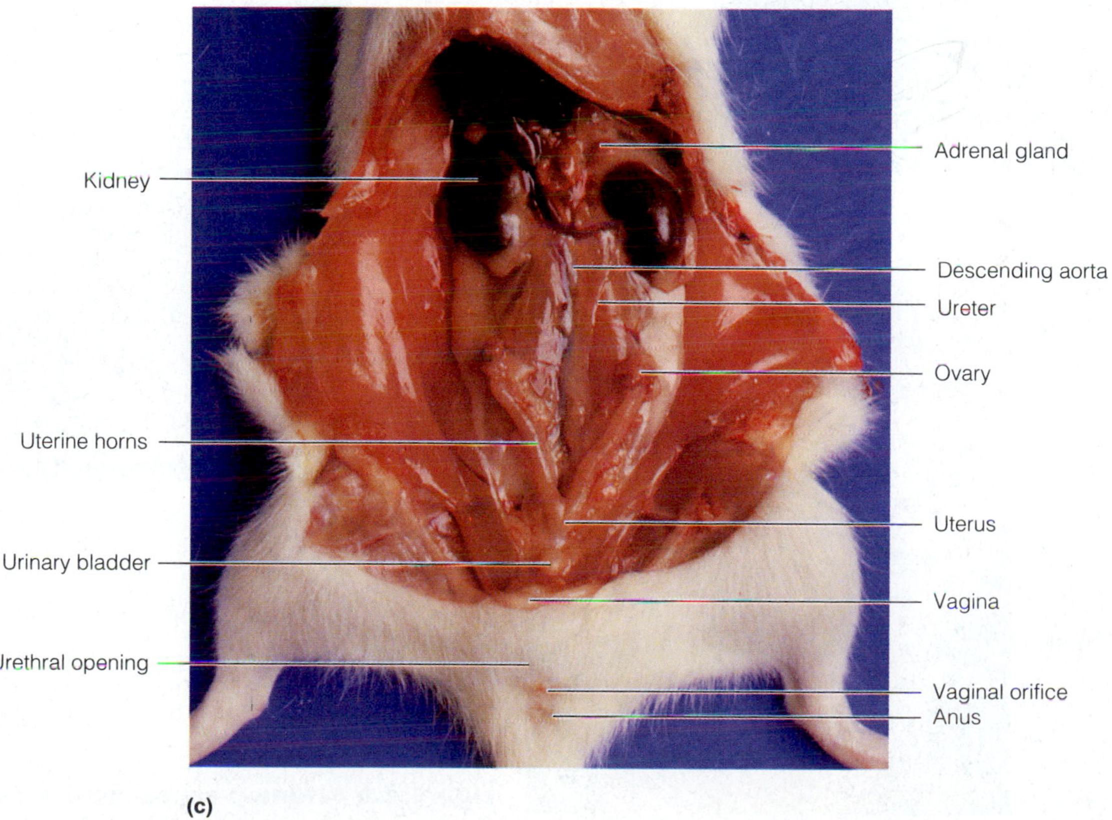

(c)

Figure 2.5 (*continued*) **(b)** Dissected male rat. (Some reproductive structures also shown.) **(c)** Dissected female rat. (Some reproductive structures also shown.)

Female Animal Inspect the pelvic cavity to identify the Y-shaped **uterus** lying against the dorsal body wall and beneath the bladder (Figure 2.5c). Follow one of the uterine horns superiorly to identify an **ovary,** a small oval structure at the end of the uterine horn. (The rat uterus is quite different from the uterus of a human female, which is a single-chambered organ about the size and shape of a pear.) The inferior undivided part of the rat uterus is continuous with the vagina, which leads to the body exterior. Identify the **vaginal orifice** (external vaginal opening).

If you are working with a human cadaver, proceed as indicated next.

Male Cadaver Make a shallow incision into the **scrotum** (Figure 2.6a). Loosen and lift out the oval **testis**. Exert a gentle pull on the testis to identify the slender **ductus (vas) deferens,** or sperm duct, which carries sperm from the testis superiorly into the abdominal cavity and joins with the urethra (Figure 2.6b). The urethra runs through the penis of the male and carries both urine and sperm out of the body. Identify the **penis,** extending from the bladder to the ventral body wall.

Female Cadaver Inspect the pelvic cavity to identify the pear-shaped **uterus** lying against the dorsal body wall and beneath the bladder. Follow one of the **uterine tubes** superiorly to identify an **ovary,** a small oval structure at the end of the uterine tube (Figure 2.6c). The inferior part of the uterus is continuous with the **vagina,** which leads to the body exterior. Identify the **vaginal orifice** (external vaginal opening).

6. When you have finished your observations, rewrap or store the dissection animal or cadaver according to your instructor's directions. Wash the dissecting tools and equipment with laboratory detergent. Dispose of the gloves. Then wash and dry your hands before continuing with the examination of the torso model. ■

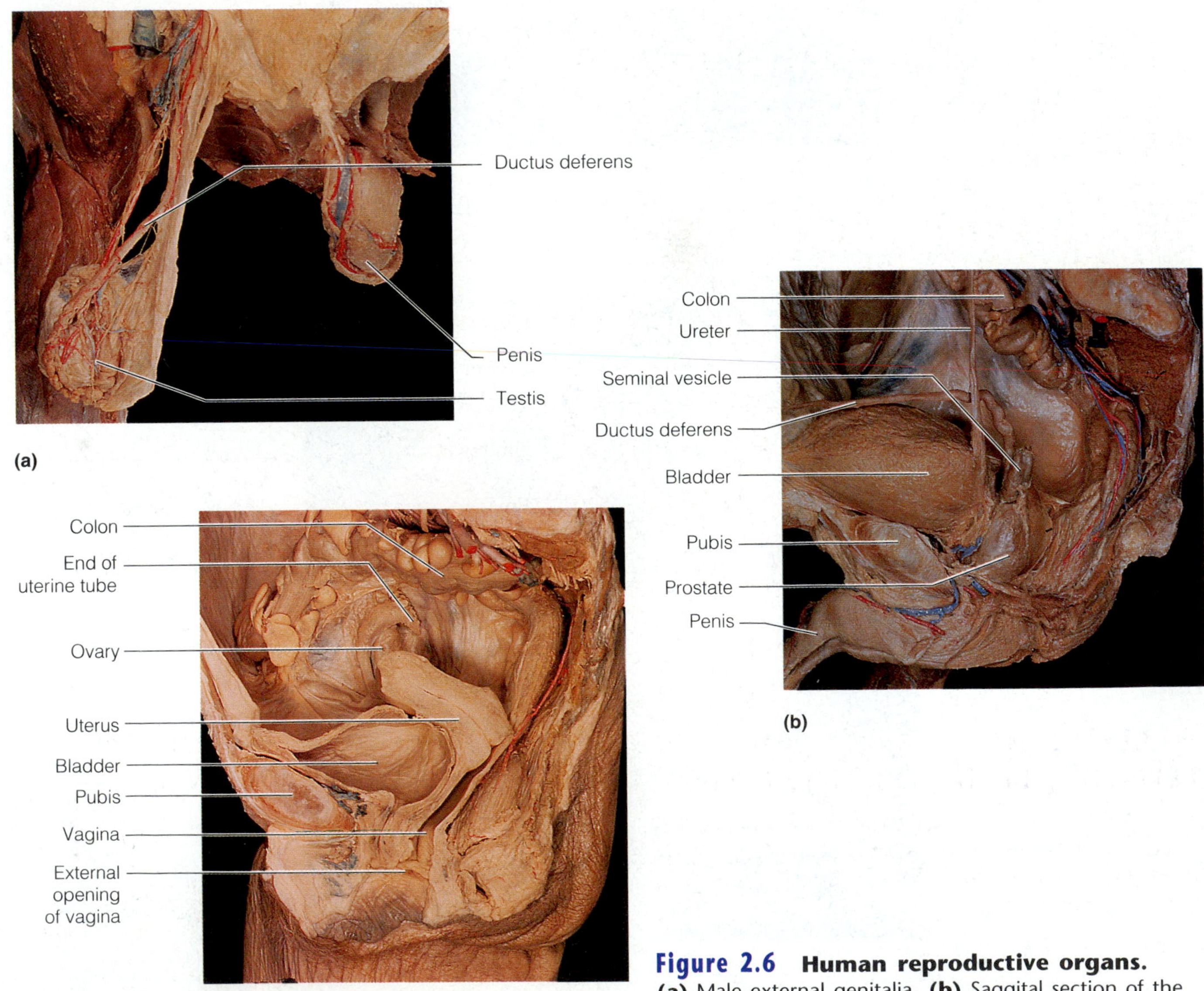

Figure 2.6 Human reproductive organs. **(a)** Male external genitalia. **(b)** Saggital section of the male pelvis. **(c)** Saggital section of the female pelvis.

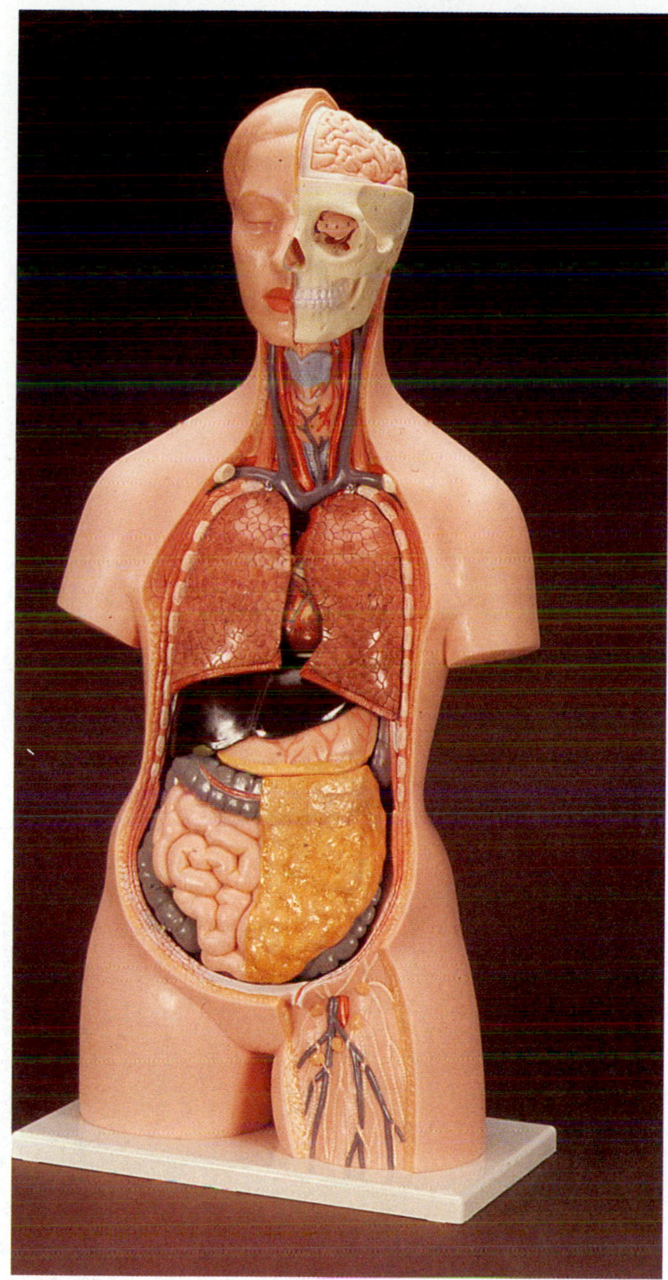

Figure 2.7 Human torso model.

Activity 5: Examining the Human Torso Model

1. Examine a human torso model to identify the organs listed. (Note: If a torso model is not available, Figure 2.7 may be used for this part of the exercise.)

Adrenal gland	Liver
Aortic arch	Lungs
Brain	Pancreas
Bronchi	Rectum
Descending aorta	Small intestine
Diaphragm	Spinal cord
Esophagus	Spleen
Heart	Stomach
Inferior vena cava	Trachea
Kidneys	Ureters
Large intestine	Urinary bladder

2. Place each of these organs in the correct body cavity or cavities. In the case of organs found in the abdominopelvic cavity, also indicate which region of the abdominal surfaces they underlie using the anatominal scheme.

Dorsal body cavity ____________________

Thoracic cavity ____________________

Abdominopelvic cavity ____________________

3. Determine which organs are found in each abdominopelvic region.

1. Umbilical region: ____________________
2. Epigastric region: ____________________
3. Hypogastric region: ____________________
4. Right iliac region: ____________________
5. Left iliac region: ____________________
6. Right lumbar region: ____________________
7. Left lumbar region: ____________________
8. Right hypochondriac region: ____________________
9. Left hypochondriac region: ____________________

Now, assign each of the organs just identified to one of the organ system categories listed below.

Digestive: ______________________________

____

Urinary: ______________________________

____

Cardiovascular: ______________________________

____

Endocrine: ______________________________

____

Reproductive: ______________________________

____

Respiratory: ______________________________

____

Lymphatic/Immunity: ______________________________

____

Nervous: ______________________________

____

10X is 100X objective

working distance from objective to specimen

red 4X objective (scanning)
yellow 10X low power objective
blue 40X high power objective

IRIS DIAPHRAM regulates LIGHT

MAG ↑ LIGHT ↓

FIELD SIZE DECREASE AS YOU INCREASE MAGNIFICATION

TEST KNOW PG 10

The Microscope

exercise

3

With the invention of the microscope, biologists gained a valuable tool to observe and study structures (like cells) that are too small to be seen by the unaided eye. The information gained helped in establishing many of the theories basic to the understanding of biological sciences. This exercise will familiarize you with the workhorse of microscopes—the compound microscope—and provide you with the necessary instructions for its proper use.

Objectives

1. To identify the parts of the microscope and list the function of each.
2. To describe and demonstrate the proper techniques for care of the microscope.
3. To define *total magnification* and *resolution.*
4. To demonstrate proper focusing technique.
5. To define *parfocal, field,* and *depth of field.*
6. To estimate the size of objects in a field.

Materials

- ❑ Compound microscope
- ❑ Stereomicroscope
- ❑ Millimeter ruler
- ❑ Prepared slides of the letter *e* or newsprint
- ❑ Immersion oil
- ❑ Lens paper
- ❑ Prepared slide of grid ruled in millimeters (grid slide)
- ❑ Prepared slide of 3 crossed colored threads
- ❑ Clean microscope slide and coverslip
- ❑ Toothpicks (flat-tipped)
- ❑ Physiologic saline in a dropper bottle
- ❑ Methylene blue stain (dilute) in a dropper bottle
- ❑ Filter paper or paper towels
- ❑ Prepared slide of cheek epithelial cells
- ❑ Coins
- ❑ Beaker containing fresh 10% household bleach solution for wet mount disposal
- ❑ Disposable autoclave bag

Note to the Instructor: The slides and coverslips used for viewing cheek cells are to be soaked for 2 hours (or longer) in 10% bleach solution and then drained. The slides and disposable autoclave bag (containing coverslips, lens paper, and used toothpicks) are to be autoclaved for 15 min at 121°C and 15 pounds pressure to ensure sterility. After autoclaving, the disposable autoclave bag may be discarded in any disposal facility and the slides and glassware washed with laboratory detergent and reprepared for use. These instructions apply as well to any bloodstained glassware or disposable items used in other experimental procedures.

Care and Structure of the Compound Microscope

The **compound microscope** is a precision instrument and should always be handled with care. At all times you must observe the following rules for its transport, cleaning, use, and storage:

- When transporting the microscope, hold it in an upright position with one hand on its arm and the other supporting its base. Avoid jarring the instrument when setting it down.
- Use only special grit-free lens paper to clean the lenses. Clean all lenses before and after use.
- Always begin the focusing process with the lowest-power objective lens in position, changing to the higher-power lenses as necessary.
- Use the coarse adjustment knob only with the lowest power lens.
- Always use a coverslip with temporary (wet mount) preparations.
- Before putting the microscope in the storage cabinet, remove the slide from the stage, rotate the lowest-power objective lens into position, and replace the dust cover or return the microscope to the appropriate storage area.
- Never remove any parts from the microscope; inform your instructor of any mechanical problems that arise.

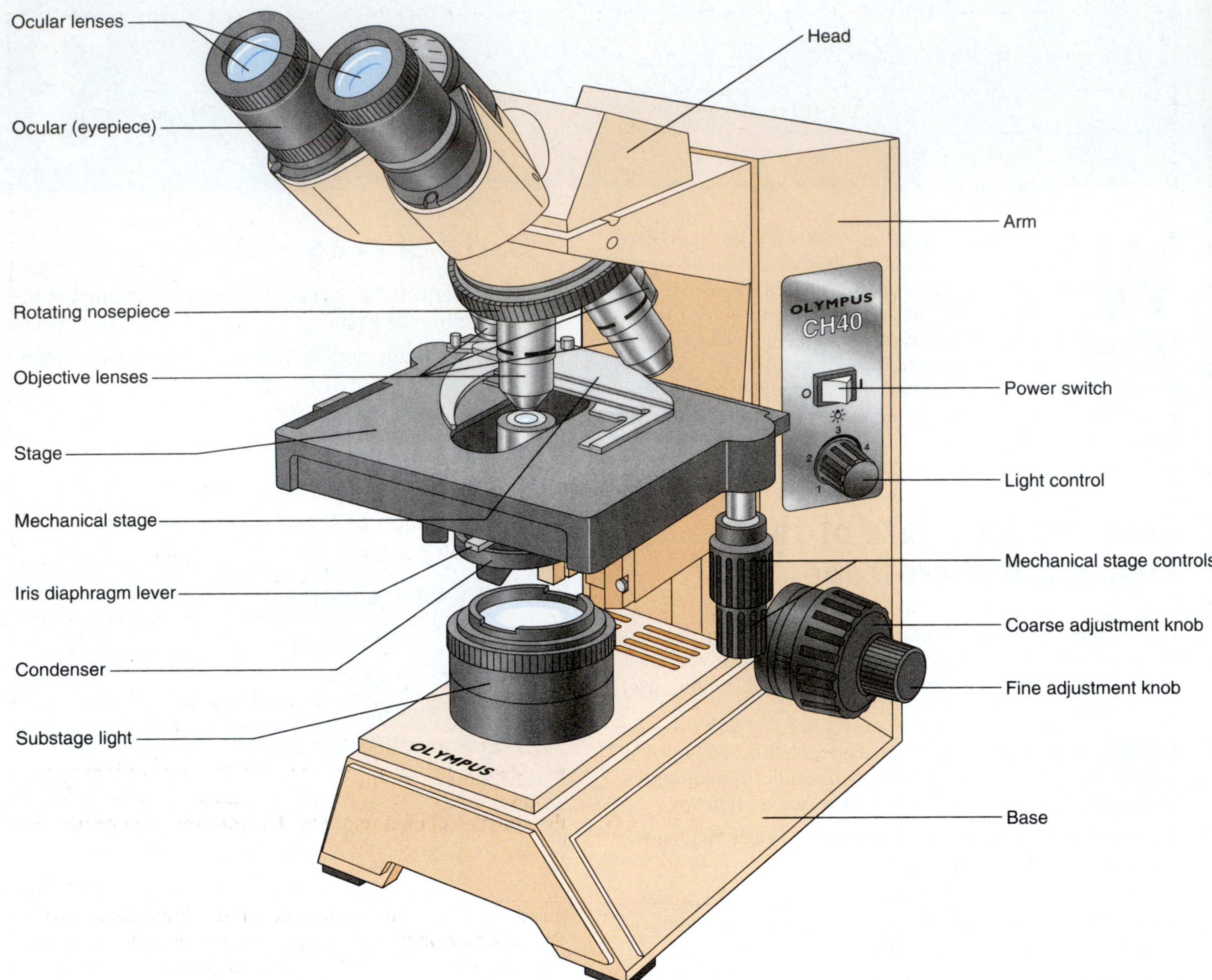

Figure 3.1 Compound microscope and its parts.

Activity 1:
Identifying the Parts of a Microscope

1. Obtain a microscope and bring it to the laboratory bench. (Use the proper transport technique!)

- Record the number of your microscope in the summary chart on p. 23.

Compare your microscope with the illustration in Figure 3.1 and identify the following microscope parts:

Base: Supports the microscope. (Note: Some microscopes are provided with an inclination joint, which allows the instrument to be tilted backward for viewing dry preparations.)

Substage light (or *mirror*): Located in the base. In microscopes with a substage light source, the light passes directly upward through the microscope. If a mirror is used, light must be reflected from a separate free-standing lamp.

Stage: The platform the slide rests on while being viewed. The stage has a hole in it to permit light to pass through both it and the specimen. Some microscopes have a stage equipped with *spring clips;* others have a clamp-type *mechanical stage* as shown in Figure 3.1. Both hold the slide in position for viewing; in addition, the mechanical stage permits precise movement of the specimen.

Condenser: Concentrates the light on the specimen. The condenser may have a height-adjustment knob that raises and lowers the condenser to vary light delivery. Generally, the best position for the condenser is close to the inferior surface of the stage.

Iris diaphragm lever: Arm attached to the condenser that regulates the amount of light passing through the condenser. The iris diaphragm permits the best possible contrast when viewing the specimen.

Coarse adjustment knob: Used to focus on the specimen.

Summary Chart for Microscope # ___________

	Scanning	Low power	High power	Oil immersion
Magnification of objective lens	×	×	×	×
Total magnification	×	×	×	×
Working distance	mm	mm	mm	mm
Detail observed Letter *e*				
Field size (diameter)	mm µm	mm µm	mm µm	mm µm

Fine adjustment knob: Used for precise focusing once coarse focusing has been completed.

Head or **body tube:** Supports the objective lens system (which is mounted on a movable nosepiece), and the ocular lens or lenses.

Arm: Vertical portion of the microscope connecting the base and head.

Ocular (or *eyepiece*): Depending on the microscope, there are one or two lenses at the superior end of the head or body tube. Observations are made through the ocular(s). An ocular lens has a magnification of 10×. (It increases the apparent size of the object by ten times or ten diameters). If your microscope has a **pointer** (used to indicate a specific area of the viewed specimen), it is attached to one ocular and can be positioned by rotating the ocular lens.

Nosepiece: Generally carries three or four objective lenses and permits sequential positioning of these lenses over the light beam passing through the hole in the stage. Use the nosepiece to change the objective lenses. Do not directly grab the lenses.

Objective lenses: Adjustable lens system that permits the use of a **scanning lens,** a **low-power lens,** a **high-power lens,** or an **oil immersion lens.** The objective lenses have different magnifying and resolving powers.

2. Examine the objective lenses carefully; note their relative lengths and the numbers inscribed on their sides. On many microscopes, the scanning lens, with a magnification between 4× and 5×, is the shortest lens. If there is no scanning lens, the low-power objective lens is the shortest and typically has a magnification of 10×. The high-power objective lens is of intermediate length and has a magnification range from 40× to 50×, depending on the microscope. The oil immersion objective lens is usually the longest of the objective lenses and has a magnifying power of 95× to 100×. Some microscopes lack the oil immersion lens.

- Record the magnification of each objective lens of your microscope in the first row of the chart above. Also, cross out the column relating to a lens that your microscope does not have.

3. Rotate the lowest power objective lens until it clicks into position, and turn the coarse adjustment knob about 180 degrees. Notice how far the stage (or objective lens) travels during this adjustment. Move the fine adjustment knob 180 degrees, noting again the distance that the stage (or the objective lens) moves. ■

Magnification and Resolution

The microscope is an instrument of magnification. In the compound microscope, magnification is achieved through the interplay of two lenses—the ocular lens and the objective lens. The objective lens magnifies the specimen to produce a **real image** that is projected to the ocular. This real image is magnified by the ocular lens to produce the **virtual image** seen by your eye (Figure 3.2).

The **total magnification** (TM) of any specimen being viewed is equal to the power of the ocular lens multiplied by the power of the objective lens used. For example, if the ocular lens magnifies 10× and the objective lens being used magnifies 45×, the total magnification is 450× (10 × 45).

- Determine the total magnification you may achieve with each of the objectives on your microscope and record the figures on the second row of the chart.

The compound light microscope has certain limitations. Although the level of magnification is almost limitless, the

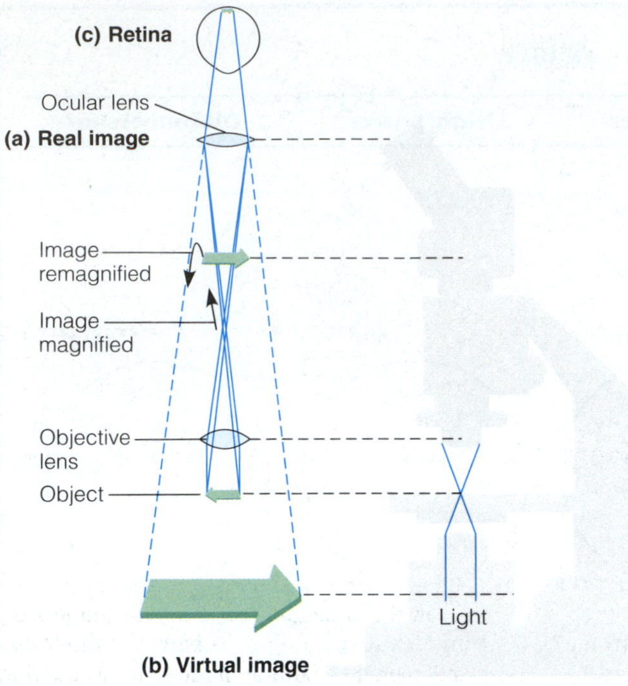

Figure 3.2 Image formation in light microscopy. (a) Light passing through the objective lens forms a real image. **(b)** The real image serves as the object for the ocular lens, which remagnifies the image and forms the virtual image. **(c)** The virtual image passes through the lens of the eye and is focused on the retina.

resolution (or resolving power), that is, the ability to discriminate two close objects as separate, is not. The human eye can resolve objects about 100 μm apart, but the compound microscope has a resolution of 0.2 μm under ideal conditions. Objects closer than 0.2 μm are seen as a single fused image.

Resolving power is determined by the amount and physical properties of the visible light that enters the microscope. In general, the more light delivered to the objective lens, the greater the resolution. The size of the objective lens aperture (opening) decreases with increasing magnification, allowing less light to enter the objective. Thus, you will probably find it necessary to increase the light intensity at the higher magnifications.

Activity 2: Viewing Objects Through the Microscope

1. Obtain a millimeter ruler, a prepared slide of the letter *e* or newsprint, a dropper bottle of immersion oil, and some lens paper. Adjust the condenser to its highest position and switch on the light source of your microscope. (If the light source is not built into the base, use the curved surface of the mirror to reflect the light up into the microscope.)

2. Secure the slide on the stage so that you can read the slide label and the letter *e* is centered over the light beam passing through the stage. If you are using a microscope with spring clips, make sure the slide is secured at both ends. If your microscope has a mechanical stage, open the jaws of its slide retainer (holder) by using the control lever (typically) located at the rear left corner of the mechanical stage. Insert the slide squarely within the confines of the slide retainer. Check to see that the slide is resting on the stage (and not on the mechanical stage frame) before releasing the control lever.

3. With your lowest power (scanning or low-power) objective lens in position over the stage, use the coarse adjustment knob to bring the objective lens and stage as close together as possible.

4. Look through the ocular lens and adjust the light for comfort using the iris diaphragm. Now use the coarse adjustment knob to focus slowly away from the *e* until it is as clearly focused as possible. Complete the focusing with the fine adjustment knob.

5. Sketch the letter *e* in the circle on the summary chart (p. 23) just as it appears in the **field** (the area you see through the microscope).

What is the total magnification? _______ ×

How far is the bottom of the objective lens from the specimen? In other words, what is the **working distance**? Use a millimeter ruler to make this measurement summary.

_______ mm

Record the TM detail observed and the working distance in the summary chart (p. 23).

How has the apparent orientation of the *e* changed top to bottom, right to left, and so on?

____

____

6. Move the slide slowly away from you on the stage as you view it through the ocular lens. In what direction does the image move?

____

Move the slide to the left. In what direction does the image move?

____

At first this change in orientation may confuse you, but with practice you will learn to move the slide in the desired direction with no problem.

7. Today most good laboratory microscopes are **parfocal**; that is, the slide should be in focus (or nearly so) at the higher magnifications once you have properly focused. *Without touching the focusing knobs,* increase the magnification by rotating the next higher magnification lens (low-power or

high-power) into position over the stage. Make sure it clicks into position. Using the fine adjustment only, sharpen the focus.* Note the decrease in working distance. As you can see, focusing with the coarse adjustment knob could drive the objective lens through the slide, breaking the slide and possibly damaging the lens. Sketch the letter *e* in the summary chart (p. 23). What new details become clear?

____

____

What is the total magnification now? ______________ ×

Record the TM, detail observed, and working distance in the summary chart (p. 23).

As best you can, measure the distance between the objective and the slide (the working distance) and record it on the chart (p. 23).

Is the image larger or smaller? ______________

Approximately how much of the letter *e* is visible now?

____

Is the field larger or smaller? ______________

Why is it necessary to center your object (or the portion of the slide you wish to view) before changing to a higher power?

____

Move the iris diaphragm lever while observing the field. What happens?

____

Is it more desirable to increase *or* decrease the light when changing to a higher magnification?

______________ Why? ______________

____

8. If you have just been using the low-power objective, repeat the steps given in direction 7 using the high-power objective lens.

Record the TM, detail observed, and working distance in the summary chart (p. 23).

9. Without touching the focusing knob, rotate the high-power lens out of position so that the area of the slide over the opening in the stage is unobstructed. Place a drop of immersion oil over the *e* on the slide and rotate the oil immersion lens into position. Set the condenser at its highest point (closest to the stage), and open the diaphragm fully. Adjust the fine focus and fine-tune the light for the best possible resolution.

*If you are unable to focus with a new lens, your microscope is not parfocal. Do not try to force the lens into position. Consult your instructor.

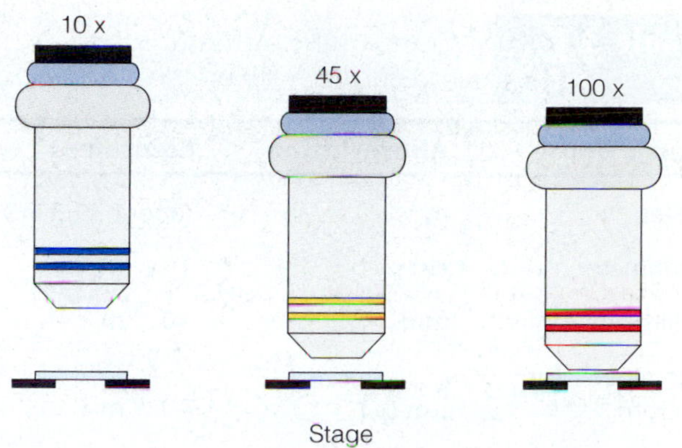

Figure 3.3 Relative working distances of the 10×, 45×, and 100× objectives.

Note: If for some reason the specimen does not come into view after adjusting the fine focus, do not go back to the 40× lens to recenter. You do not want oil from the oil immersion lens to cloud the 40× lens. Turn the revolving nosepiece in the other direction to the low-power lens and recenter and refocus the object. Then move the immersion lens back into position, again avoiding the 40× lens.

Is the field again decreased in size? ______________

What is the total magnification with the oil immersion lens?

______________ ×

Is the working distance less *or* greater than it was when the high-power lens was focused?

____

Compare your observations on the relative working distances of the objective lenses with the illustration in Figure 3.3. Explain why it is desirable to begin the focusing process in the lowest power.

____

____

____

10. Rotate the oil immersion lens slightly to the side and remove the slide. Clean the oil immersion lens carefully with lens paper and then clean the slide in the same manner with a fresh piece of lens paper. ■

Table 3.1 Comparison of Metric Units of Length*

Metric unit	Abbreviation	Equivalent
Meter	m	(about 39.3 in.)
Centimeter	cm	10^{-2} m
Millimeter	mm	10^{-3} m
Micrometer (or micron)	μm (μ)	10^{-6} m
Nanometer (or millimicrometer or millimicron)	nm (mμ)	10^{-9} m
Angstrom	Å	10^{-10} m

*Refer to the "Getting Started" exercise (p. xii) for tips on metric conversions.

Size of the Microscope Field

By this time you should know that the size of the microscope field decreases with increasing magnification. For future microscope work, it will be useful to determine the diameter of each of the microscope fields. This information will allow you to make a fairly accurate estimate of the size of the objects you view in any field. For example, if you have calculated the field diameter to be 4 mm and the object being observed extends across half this diameter, you can estimate the length of the object to be approximately 2 mm.

Microscopic specimens are usually measured in micrometers and millimeters, both units of the metric system. You can get an idea of the relationship and meaning of these units from Table 3.1. A more detailed treatment appears in Appendix A.

Activity 3: Determining the Size of the Microscope Field

1. Obtain a grid slide (a slide prepared with graph paper ruled in millimeters). Each of the squares in the grid is 1 mm on each side. Use your lowest power objective to bring the grid lines into focus.

2. Move the slide so that one grid line touches the edge of the field on one side, and then count the number of squares you can see across the diameter of the field. If you can see only part of a square, as in the accompanying diagram, estimate the part of a millimeter that the partial square represents.

~2.5 mm

Record this figure in the appropriate space marked "field size" on the summary chart (p. 23). (If you have been using the scanning lens, repeat the procedure with the low-power objective lens.)

Complete the chart by computing the approximate diameter of the high-power and oil immersion fields. The general formula for calculating the unknown field diameter is:

$$\text{Diameter of field } A \times \text{ total magnification of field } A = \text{diameter of field } B \times \text{ total magnification of field } B$$

where A represents the known or measured field and B represents the unknown field.

This can be simplified to

$$\text{Diameter of field } B = \frac{\text{diameter of field } A \times \text{ total magnification of field } A}{\text{total magnification of field } B}$$

For example, if the diameter of the low-power field (field A) is 2 mm and the total magnification is 50×, you would compute the diameter of the high-power field (field B) with a total magnification of 100× as follows:

$$\text{Field diameter B} = (2 \text{ mm} \times 50)/100$$
$$\text{Field diameter B} = 1 \text{ mm}$$

3. Estimate the length (longest dimension) of the following microscopic objects. *Base your calculations on the field sizes you have determined for your microscope.*

a. Object seen in low-power field:
approximate length:

_______ mm

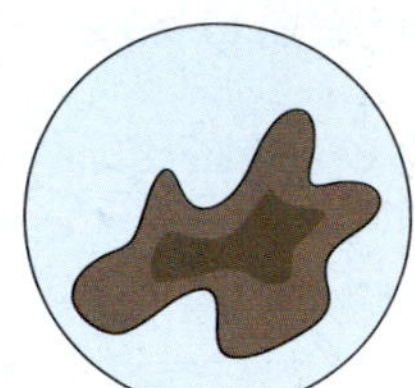

b. Object seen in high-power field:
approximate length:

_______ mm

or _______ μm

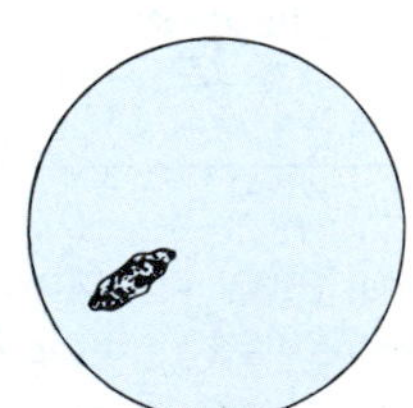

c. Object seen in oil immersion field:
approximate length:

_______ μm

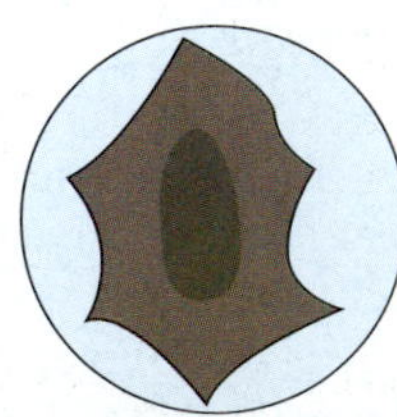

4. If an object viewed with the oil immersion lens looked as it does in the field depicted just below, could you determine its approximate size from this view?

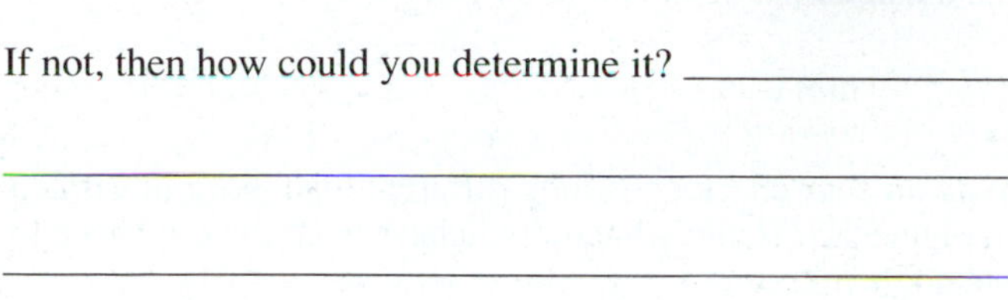

If not, then how could you determine it? ________________

____

____

____ ■

Perceiving Depth

Any microscopic specimen has depth as well as length and width; it is rare indeed to view a tissue slide with just one layer of cells. Normally you can see two or three cell thicknesses. Therefore, it is important to learn how to determine relative depth with your microscope. In microscope work the **depth of field** (the depth of the specimen clearly in focus) is greater at lower magnifications.

Activity 4:
Perceiving Depth

1. Obtain a slide with colored crossed threads. Focusing at low magnification, locate the point where the three threads cross each other.

2. Use the iris diaphragm lever to greatly reduce the light, thus increasing the contrast. Focus down with the coarse adjustment until the threads are out of focus, then slowly focus upward again, noting which thread comes into clear focus first. (You will see two or even all three threads, so you must be very careful in determining which one first comes into clear focus.) Observe: As you rotate the adjustment knob forward (away from you), does the stage rise or fall? If the stage rises, then the first clearly focused thread is the top one; the last clearly focused thread is the bottom one.

If the stage falls, how is the order affected? ______________

____

Record your observations, relative to which thread is uppermost, middle, or lowest:

Top thread ______________________________________

Middle thread ____________________________________

Bottom thread ___________________________________ ■

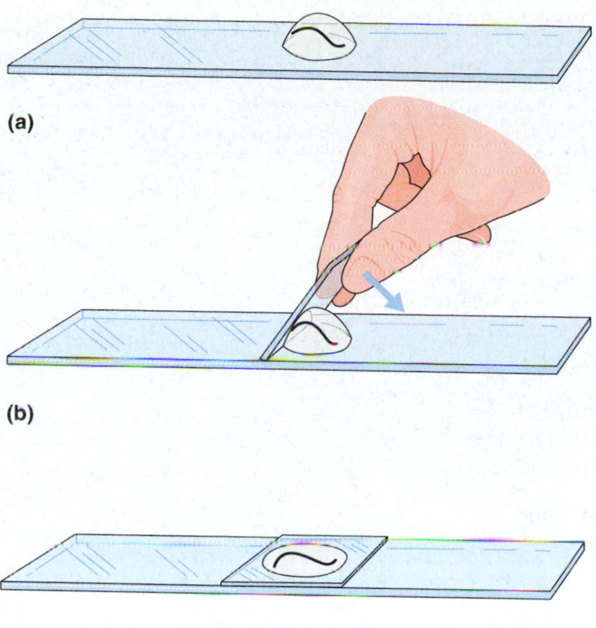

Figure 3.4 Procedure for preparation of a wet mount. **(a)** The object is placed in a drop of water (or saline) on a clean slide, **(b)** a coverslip is held at a 45° angle with the fingertips, and **(c)** it is lowered carefully over the water and the object.

Viewing Cells Under the Microscope

There are various ways to prepare cells for viewing under a microscope. Cells and tissues can look very different with different stains and preparation techniques. One method of preparation is to mix the cells in physiologic saline (called a wet mount) and stain them with methylene blue stain.

If you are not instructed to prepare your own wet mount, obtain a prepared slide of epithelial cells to make the observations in step 10 of Activity 5.

Activity 5:
Preparing and Observing a Wet Mount

1. Obtain the following: a clean microscope slide and coverslip, two flat-tipped toothpicks, a dropper bottle of physiologic saline, a dropper bottle of methylene blue stain, and filter paper (or paper towels). Handle only your own slides through-out the procedure.

2. Place a drop of physiologic saline in the center of the slide. Using the flat end of the toothpick, *gently* scrape the inner lining of your cheek. Agitate the end of the toothpick containing the cheek scrapings in the drop of saline (Figure 3.4a).

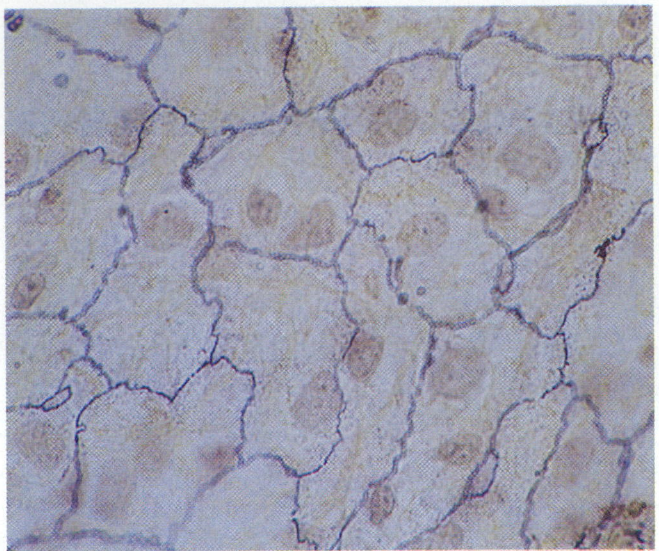

Figure 3.5 Epithelial cells of the cheek cavity (surface view, 488×).

 *Immediately* discard the used toothpick in the disposable autoclave bag provided at the supplies area.

3. Add a tiny drop of the methylene blue stain to the preparation. (These epithelial cells are nearly transparent and thus difficult to see without the stain, which colors the nuclei of the cells and makes them look much darker than the cytoplasm.) Stir again and then dispose of the toothpick as described above.

4. Hold the coverslip with your fingertips so that its bottom edge touches one side of the fluid drop (Figure 3.4b), then *carefully* lower the coverslip onto the preparation (Figure 3.4c). *Do not just drop the coverslip,* or you will trap large air bubbles under it, which will obscure the cells. *A coverslip should always be used with a wet mount* to prevent soiling the lens if you should misfocus.

5. Examine your preparation carefully. The coverslip should be closely apposed to the slide. If there is excess fluid around its edges, you will need to remove it. Obtain a piece of filter paper, fold it in half, and use the folded edge to absorb the excess fluid. (You may use a twist of paper towel as an alternative.)

 Before continuing, discard the filter paper in the disposable autoclave bag.

6. Place the slide on the stage and locate the cells in low power. You will probably want to dim the light with the iris diaphragm to provide more contrast for viewing the lightly stained cells. Furthermore, a wet mount will dry out quickly in bright light because a bright light source is hot.

7. Cheek epithelial cells are very thin, six-sided cells. In the cheek, they provide a smooth, tilelike lining, as shown in Figure 3.5.

8. Make a sketch of the epithelial cells that you observe.

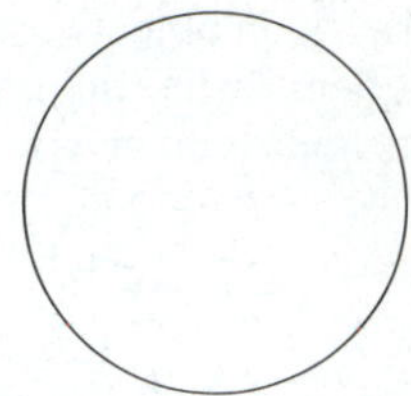

Use information on your summary chart (p. 23) to estimate the diameter of cheek epithelial cells.

_______ mm

Why do *your* cheek cells look different than those illustrated in Figure 3.5? (Hint: what did you have to *do* to your cheek to obtain them?)

____

____

⚠ 9. When you complete your observations of the wet mount, dispose of your wet mount preparation in the beaker of bleach solution, and put the coverslips in an autoclave bag.

10. Obtain a prepared slide of cheek epithelial cells and view them under the microscope.

Estimate the diameter of one of these cheek epithelial cells using information from the summary chart (p. 23).

_______ mm

Why are these cells more similar to those seen in Figure 3.5 and easier to measure than those of the wet mount?

____

____

11. Before leaving the laboratory, make sure all other materials are properly discarded or returned to the appropriate laboratory station. Clean the microscope lenses and put the dust cover on the microscope before you return it to the storage cabinet. ■

The Stereomicroscope (Dissecting Microscope)

Occasionally biologists look at specimens too large to observe with the compound microscope but too small to observe easily with the unaided eye. The stereomicroscope, sometimes called a dissecting microscope, can be helpful in these situations. It works basically like a large magnifying glass.

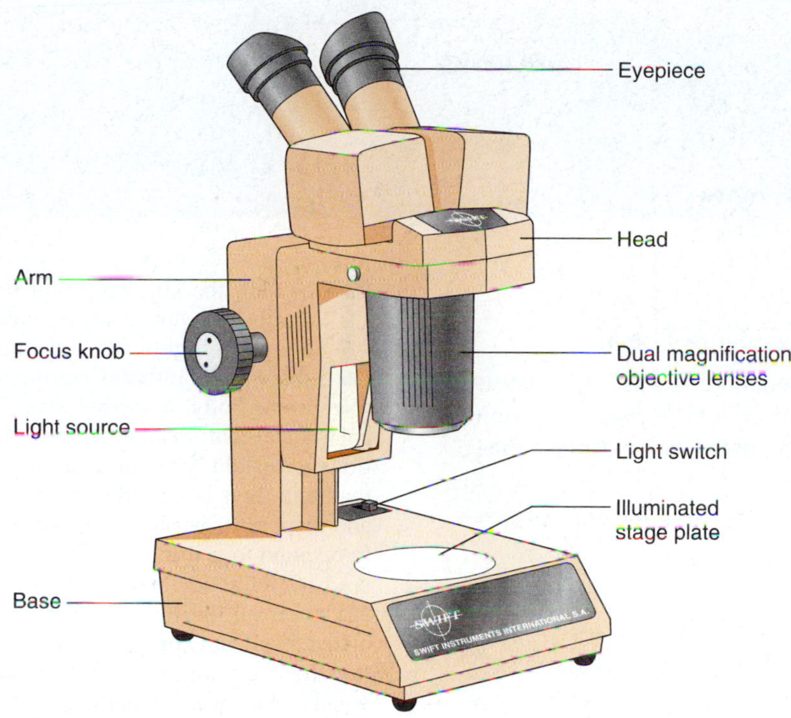

Figure 3.6 Stereomicroscope and its parts.

Activity 6: Identifying the Parts of a Stereomicroscope

1. Obtain the microscope and put it on the lab bench.

2. Using what you have learned about the compound microscope and Figure 3.6, identify the following parts:

Adjustment focus knob: Used to focus on the specimen

Arm: Connects the base to the head of the microscope

Base: Supports the microscope

Eyepieces or **oculars:** Magnify the images from the objective lenses

Head or **body tube:** Supports the ocular and objective lenses

Light control: Switch that allows you to choose transmitted light, reflected light, or both

Objective lenses: Adjustable lens system used to increase or decrease magnification

Stage: Platform that holds the specimen

Substage light: Located in the base; sends light up through the specimen; the source of transmitted light

Upper light source: Located above the stage; directs light onto the surface of the specimen; the source of reflected light ■

Activity 7: Using the Stereomicroscope

1. Place a coin on the stage of the microscope. Turn on the light source, and adjust the oculars until you can see a single image of the specimen.

2. Focus the microscope on the coin.

3. Experiment with transmitted and reflected light until you get the best image. Which works best in this situation, the transmitted or reflected light?

____

4. Increase and decrease the magnification to familiarize yourself with the controls.

5. Each coin has a small letter indicating where it was minted. See if you can determine the initial of the mint that produced your coin.

Where was your coin produced? ______________

What is the total magnification you used? ____________ ■

exercise

4 The Cell—Anatomy and Division

Objectives

1. To define *cell, organelle,* and *inclusion.*
2. To identify on a cell model or diagram the following cellular regions and to list the major function of each: nucleus, cytoplasm, and plasma membrane.
3. To identify and list the major functions of the various organelles studied.
4. To compare and contrast specialized cells with the concept of the "generalized cell."
5. To define *interphase, mitosis,* and *cytokinesis.*
6. To list the stages of mitosis and describe the events of each stage.
7. To identify the mitotic phases on slides or appropriate diagrams.
8. To explain the importance of mitotic cell division and its product.

Materials

- ❑ Three-dimensional model of the "composite" animal cell or laboratory chart of cell anatomy
- ❑ Modeling clay or Play-doh®
- ❑ Prepared slides of simple squamous epithelium ($AgNO_3$ stain), teased smooth muscle, human blood cell smear, and sperm
- ❑ Compound microscope
- ❑ Prepared slides of whitefish blastulae
- ❑ Three-dimensional models of mitotic stages
- ❑ Video of mitosis

Note to the Instructor: See directions for handling wet mount preparations and disposable supplies on p. 27, Exercise 3. For suggestions on video of mitosis, see Instructor's Guide.

The **cell,** the structural and functional unit of all living things, is a very complex entity. The cells of the human body are highly diverse, and their differences in size, shape, and internal composition reflect their specific roles in the body. Nonetheless, cells do have many common anatomical features, and all cells must perform certain functions to sustain life. For example, all cells have the ability to maintain their boundaries, to metabolize, to digest nutrients and dispose of wastes, to grow and reproduce, to move, and to respond to a stimulus. Most of these functions are considered in detail in later exercises. This exercise focuses on structural similarities that typify the "composite," or "generalized," cell and considers only the function of cell reproduction (cell division). Transport mechanisms (the means by which substances cross the plasma membrane) are dealt with separately in Exercise 5.

Anatomy of the Composite Cell

In general, all cells have three major regions, or parts, that can readily be identified with a light microscope: the **nucleus,** the **plasma membrane,** and the **cytoplasm.** The nucleus is usually seen as a round or oval structure near the center of the cell. It is surrounded by cytoplasm, which in turn is enclosed by the plasma membrane. Since the advent of the electron microscope, even smaller cell structures—organelles—have been identified. Figure 4.1a is a diagrammatic representation of the fine structure of the composite cell; Figure 4.1b depicts cellular structure (particularly that of the nucleus) as revealed by the electron microscope.

Nucleus

The nucleus is often described as the control center of the cell and is necessary for cell reproduction. A cell that has lost or ejected its nucleus (for whatever reason) is literally programmed to die because the nucleus is the site of the "genes," or genetic material—DNA.

When the cell is not dividing, the genetic material is loosely dispersed throughout the nucleus in a threadlike form called **chromatin.** When the cell is in the process of dividing

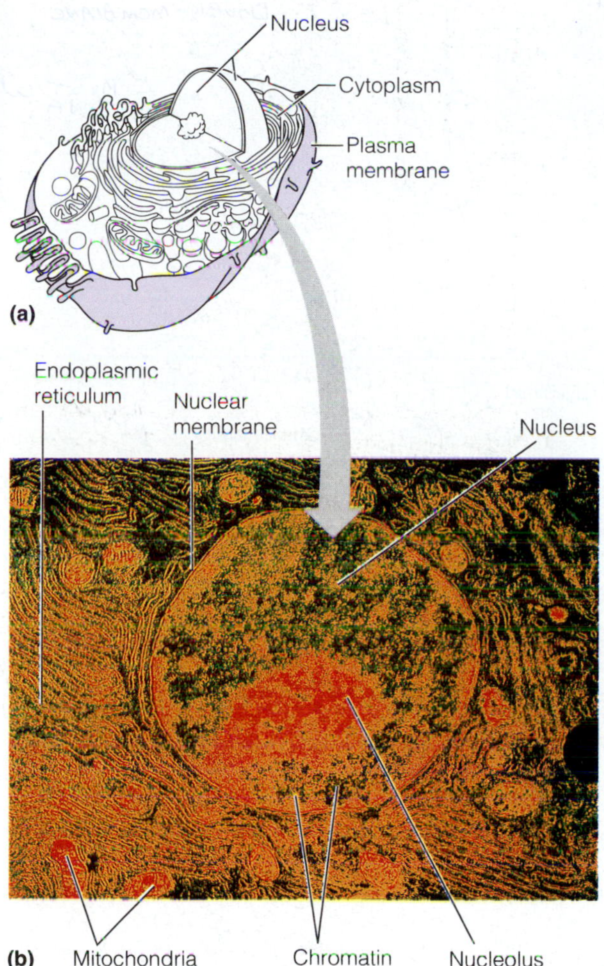

Figure 4.1 Anatomy of the composite animal cell. (a) Diagrammatic view. **(b)** Transmission electron micrograph (10,000×).

to form daughter cells, the chromatin coils and condenses to form dense, darkly staining rodlike bodies called **chromosomes**—much in the way a stretched spring becomes shorter and thicker when it is released. (Cell division is discussed later in this exercise.) Notice the appearance of the nucleus carefully—it is somewhat nondescript when a cell is healthy. When the nucleus appears dark and the chromatin becomes clumped, this is an indication that the cell is dying and undergoing degeneration.

The nucleus also contains one or more small round bodies, called **nucleoli,** composed primarily of proteins and ribonucleic acid (RNA). The nucleoli are assembly sites for ribosomal particles (particularly abundant in the cytoplasm), which are the actual protein-synthesizing "factories."

The nucleus is bound by a double-layered porous membrane, the **nuclear envelope**. The nuclear envelope is similar in composition to other cellular membranes, but it is distinguished by its large *nuclear pores*. Although they are spanned by diaphragms, these pores permit easy passage of protein and RNA molecules.

Activity 1: Identifying Parts of a Cell

Identify the nuclear membrane, chromatin, nucleoli, and the nuclear pores in Figure 4.1a and b and Figure 4.3. ■

Plasma Membrane

The **plasma membrane** separates cell contents from the surrounding environment. Its main structural building blocks are phospholipids (fats) and globular protein molecules, but some of the externally facing proteins and lipids have sugar (carbohydrate) side chains attached to them that are important in cellular interactions (Figure 4.2). Described by the fluid-mosaic model, the membrane appears to have a bilayer of phospholipid molecules that the protein molecules float in. Occasional cholesterol molecules dispersed in the fluid phospholipid bilayer help stabilize it.

Besides providing a protective barrier for the cell, the plasma membrane plays an active role in determining which substances may enter or leave the cell and in what quantity. Because of its molecular composition, the plasma membrane is selective about what passes through it. It allows nutrients to enter the cell but keeps out undesirable substances. By the same token, valuable cell proteins and other substances are kept within the cell, and excreta or wastes pass to the exterior. This property is known as **selective permeability.** Transport through the plasma membrane occurs in two basic ways. In *active transport,* the cell must provide energy (ATP) to power the transport process. In *passive transport,* the transport process is driven by concentration or pressure differences. Additionally, the plasma membrane maintains a resting potential that is essential to normal functioning of excitable cells, and plays a vital role in cell signaling and cell-to-cell interactions. In some cells the membrane is thrown into minute fingerlike projections or folds called **microvilli,** which greatly increase the surface area of the cell available for absorption or passage of materials, and binding of signaling molecules.

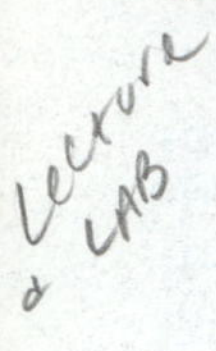

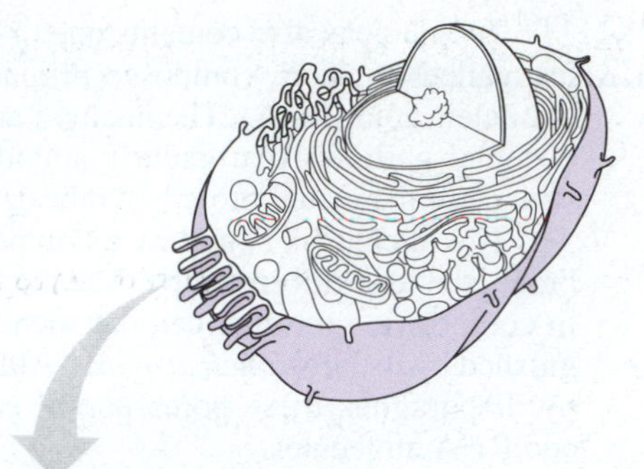

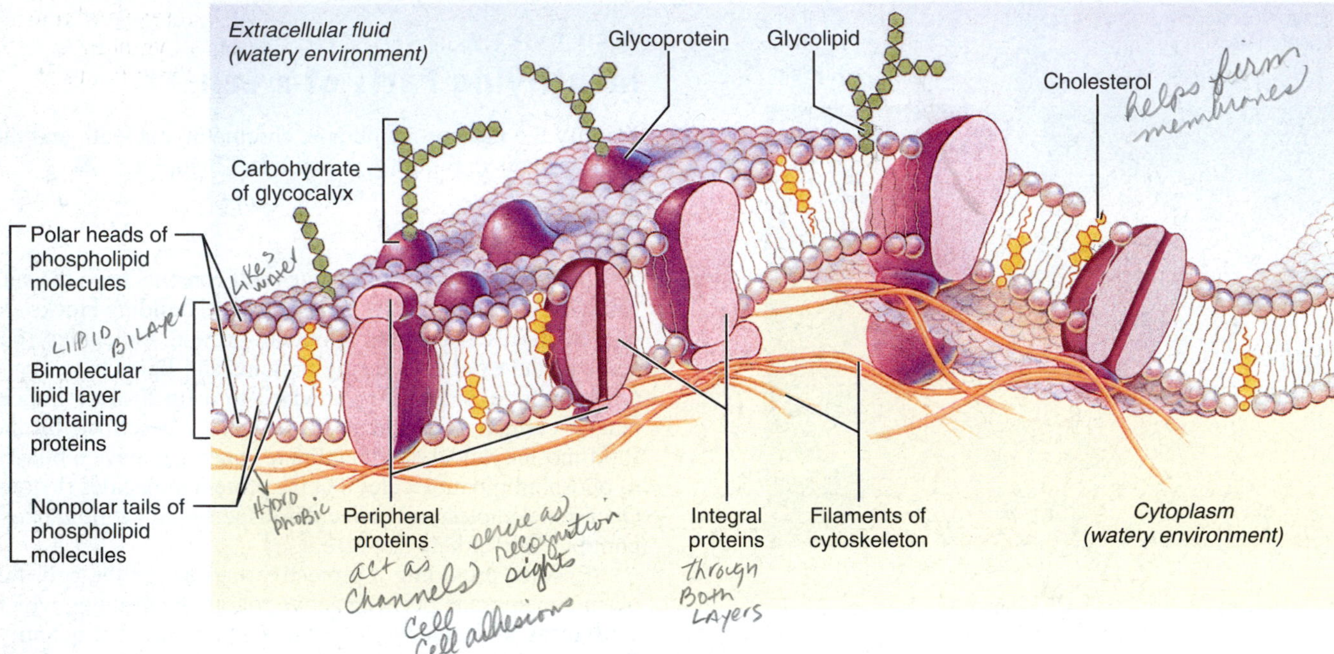

Figure 4.2 Structural details of the plasma membrane.

Activity 2:
Identifying Components of a Plasma Membrane

Identify the phospholipid and protein portions of the plasma membrane in Figure 4.2. Also locate the sugar side chains and cholesterol molecules. Identify the microvilli in the diagram at the top of Figure 4.2 and in Figure 4.3. ■

Cytoplasm and Organelles

The cytoplasm consists of the cell contents outside the nucleus. It is the major site of most activities carried out by the cell. Suspended in the **cytosol,** the fluid cytoplasmic material, are many small structures called **organelles** (literally, "small organs"). The organelles are the metabolic machinery of the cell, and they are highly organized to carry out specific functions for the cell as a whole. The organelles include the ribosomes, endoplasmic reticulum, Golgi apparatus, lysosomes, peroxisomes, mitochondria, cytoskeletal elements, and centrioles.

Activity 3:
Locating Organelles

Each organelle type is summarized in Table 4.1 and described briefly next. Read through this material and then, as best you can, locate the organelles in both Figures 4.1b and 4.3. ■

- The **ribosomes** are densely staining spherical bodies composed of RNA and protein. They are the actual sites of protein synthesis. They are seen floating free in the cytoplasm or attached to a membranous structure. When they are attached, the whole ribosome-membrane complex is called the *rough endoplasmic reticulum.*

- The **endoplasmic reticulum (ER)** is a highly folded system of membranous tubules and cisternae (sacs) that extends throughout the cytoplasm. The ER is continuous with the nuclear membrane. Thus it is assumed that the ER provides a system of channels for the transport of cellular substances (primarily proteins) from one part of the cell to another. The ER exists in two forms; a particular cell may have both or only one, depending on its specific functions. The

Table 4.1 Cytoplasmic Organelles

Organelle		Location and function
Ribosomes	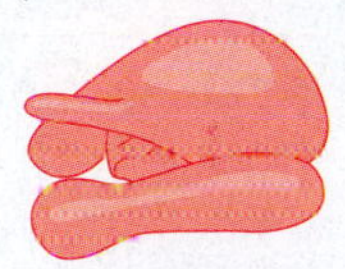	Tiny spherical bodies composed of RNA and protein; actual sites of protein synthesis; floating free or attached to a membranous structure (the rough ER) in the cytoplasm
Endoplasmic reticulum (ER)	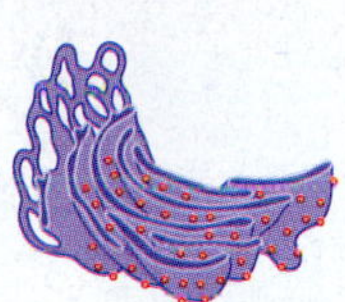	Membranous system of tubules that extends throughout the cytoplasm; two varieties: rough ER—studded with ribosomes (tubules of the rough ER provide an area for storage and transport of the proteins made on the ribosomes to other cell areas; external face synthesizes phospholipids and cholesterol); smooth ER—has no function in protein synthesis (a site of steroid and lipid synthesis, lipid metabolism, and drug detoxification)
Golgi apparatus	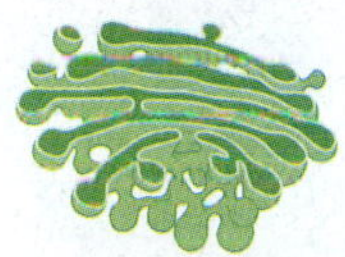	Stack of flattened sacs with bulbous ends and associated small vesicles; found close to the nucleus; role in packaging proteins or other substances for export from the cell or incorporation into the plasma membrane and in packaging lysosomal enzymes
Lysosomes	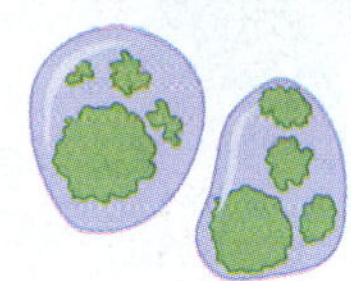	Various-sized membranous sacs containing digestive enzymes (acid hydrolases); function to digest worn-out cell organelles and foreign substances that enter the cell; have the capacity of total cell destruction if ruptured
Peroxisomes	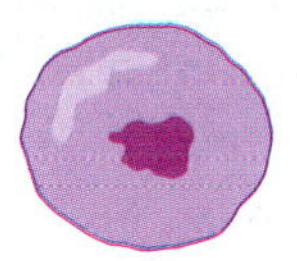	Small lysosome-like membranous sacs containing oxidase enzymes that detoxify alcohol, hydrogen peroxide, and other harmful chemicals
Mitochondria	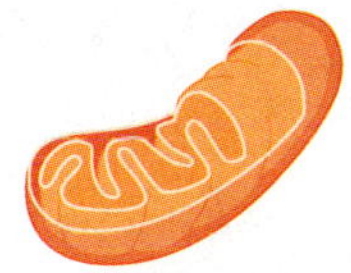	Generally rod-shaped bodies with a double-membrane wall; inner membrane is thrown into folds, or cristae; contain enzymes that oxidize foodstuffs to produce cellular energy (ATP); often referred to as "powerhouses of the cell"
Centrioles	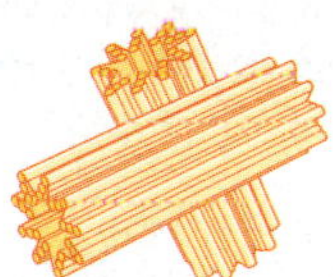	Paired, cylindrical bodies lie at right angles to each other, close to the nucleus; direct the formation of the mitotic spindle during cell division; form the bases of cilia and flagella
Cytoskeletal elements: microtubules, intermediate filaments, and microfilaments	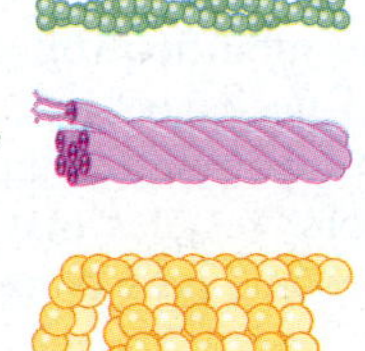	Provide cellular support; function in intracellular transport; microtubules form the internal structure of the centrioles and help determine cell shape; intermediate filaments, stable elements composed of a variety of proteins, resist mechanical forces acting on cells; microfilaments are formed largely of actin, a contractile protein, and thus are important in cell mobility (particularly in muscle cells)

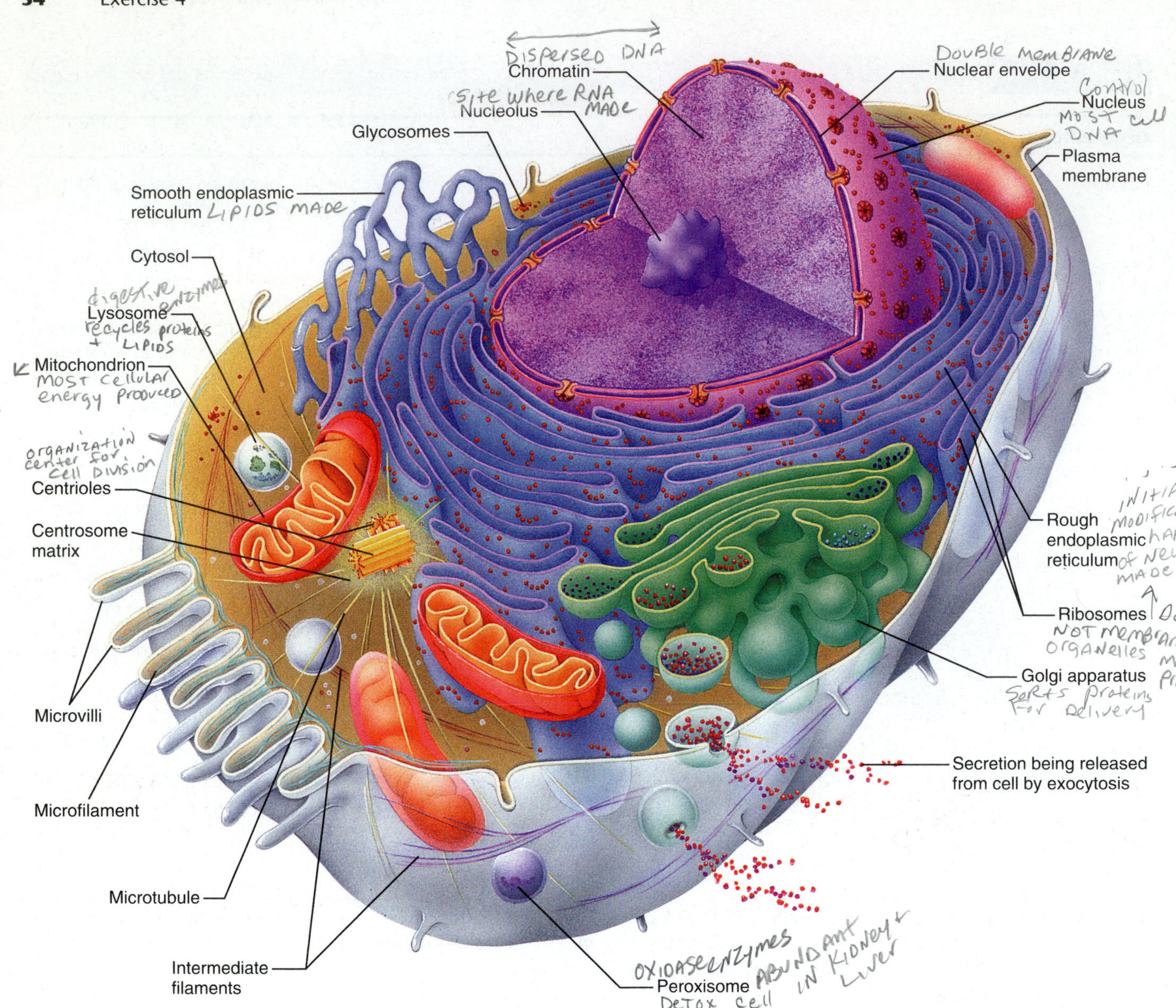

Figure 4.3 Structure of the generalized cell. No cell is exactly like this one, but this composite illustrates features common to many human cells. Note that not all organelles are drawn to the same scale in this illustration.

rough ER, as noted earlier, is studded with ribosomes. Its cisternae modify and store the newly formed proteins and dispatch them to other areas of the cell. The external face of the rough ER is involved in phospholipid and cholesterol synthesis. The amount of rough ER is closely correlated with the amount of protein a cell manufactures and is especially abundant in cells that make protein products for export—for example, the pancreas cells that produce digestive enzymes destined for the small intestine. The **smooth ER** has no protein synthesis–related function but is present in conspicuous amounts in cells that produce steroid-based hormones—for example, the interstitial cells of the testes, which produce testosterone. Smooth ER is also abundant in cells that are highly active in lipid metabolism and drug detoxification activities—liver cells, for instance.

- The **Golgi apparatus** is a stack of flattened sacs with bulbous ends that is generally found close to the nucleus. Within its cisternae, the proteins delivered to it by transport vesicles from the rough ER are modified (by attachment of sugar groups), segregated, and packaged into membranous vesicles that ultimately (1) are incorporated into the plasma membrane, (2) become secretory vesicles that release their contents from the cell, or (3) become lysosomes.

- The **lysosomes,** which appear in various sizes, are membrane-bound sacs containing an array of powerful digestive enzymes. A product of the packaging activities of the Golgi apparatus, the lysosomes contain *acid hydrolase* enzymes capable of digesting worn-out cell structures and foreign substances that enter the cell via vesicle formation through phagocytosis or bulk-phase endocytosis (see Exercise 5). Lysosomes are also involved in some of the changes that occur during menstruation, when the uterine lining is sloughed off. Because they have the capacity of total cell destruction, the lysosomes are often referred to as the "suicide sacs" of the cell.

- **Peroxisomes,** like lysosomes, are enzyme-containing sacs. However, their *oxidase* enzymes have a different task. Using oxygen, they detoxify a number of harmful substances, most importantly free radicals. Peroxisomes are particularly abundant in kidney and liver cells, cells that are actively involved in detoxification.

- The **mitochondria** are generally rod-shaped bodies with a double-membrane wall; the inner membrane is thrown into folds, or *cristae*. Oxidative enzymes on or within the mitochondria catalyze the reactions of the Krebs cycle and the electron transport chain (collectively called oxidative respiration), in which foods are broken down to produce energy. The released energy is captured in the bonds of ATP (adenosine triphosphate) molecules, which are then transported out of the mitochondria to provide a ready energy supply to power the cell. Every living cell requires a constant supply of ATP for its many activities. Because the mitochondria provide the bulk of this ATP, they are referred to as the powerhouses of the cell.

- The **cytoskeletal elements** ramify throughout the cytoplasm, forming an internal scaffolding called the *cytoskeleton* that supports and moves substances within the cell. The **microtubules** are basically slender tubules formed of proteins called *tubulins,* which have the ability to aggregate and then disaggregate spontaneously. Microtubules organize the cytoskeleton and direct formation of the spindle formed by the centrioles during cell division. They also act in the transport of substances down the length of elongated cells (such as neurons), suspend organelles, and help maintain cell shape by providing rigidity to the soft cellular substance. **Intermediate filaments** are *stable* proteinaceous cytoskeletal elements that act as internal guy wires to resist mechanical (pulling) forces acting on cells. **Microfilaments,** ribbon or cordlike elements, are formed of contractile proteins. Because of their ability to shorten and then relax to assume a more elongated form, these are important in cell mobility and are very conspicuous in cells that are specialized to contract (such as muscle cells). A cross-linked network of microfilaments braces and strengthens the internal face of the plasma membrane.

The cytoskeletal structures are changeable and minute. With the exception of the microtubules of the spindle, which are very obvious during cell division (see pp. 37–39), and the microfilaments of skeletal muscle cells (see p. 132), they are rarely seen, even in electron micrographs, and are not depicted in Figure 4.1b. However, special stains can reveal the plentiful supply of these very important organelles.

- The paired **centrioles** lie close to the nucleus in all animal cells capable of reproducing themselves. They are rod-shaped bodies that lie at right angles to each other. Internally each centriole is composed of nine triplets of microtubules. During cell division, the centrioles direct the formation of the mitotic spindle. Centrioles also form the basis for cell projections called cilia and flagella.

The cell cytoplasm contains various other substances and structures, including stored foods (glycogen granules and lipid droplets), pigment granules, crystals of various types, water vacuoles, and ingested foreign materials. But these are not part of the active metabolic machinery of the cell and are therefore called **inclusions.**

Activity 4:
Examining the Cell Model

Once you have located all of these structures in Figure 4.3, examine the cell model (or cell chart) to repeat and reinforce your identifications. ■

Activity 5:
Preparing Clay Models of Cell Structures

Use modeling clay or Play-doh® to make models of the plasma membrane, nucleus, and organelles. ■

Differences and Similarities in Cell Structure

Activity 6:
Observing Various Cell Structures

1. Obtain a compound microscope and prepared slides of simple squamous epithelium, sperm, smooth muscle cells (teased), and human blood.

2. Observe each slide under the microscope, carefully noting similarities and differences in the cells. (The oil immersion lens will be needed to observe blood and sperm.) Distinguish the limits of the individual cells, and notice the shape and position of the nucleus in each case. When you look at the human blood smear, direct your attention to the red blood cells, the pink-stained cells that are most numerous. The color photomicrographs illustrating a blood smear (Plate 58) and sperm (Plate 53) that appear in the Histology Atlas may be helpful in this cell structure study. Sketch your observations in the circles provided on p. 36.

Simple squamous epithelium

Diameter ________

Human red blood cells

Diameter ________

Sperm cells

Length ________

Diameter ________

Teased smooth muscle cells

Length ________

Diameter ________

3. Measure the length and/or diameter of each cell and record below the appropriate sketch.

4. How do these four cell types differ in shape and size?

____

____

____

____

How might cell shape affect cell function?

____

____

____

____

____

Which cells have visible projections?

____

How do these projections relate to the function of these cells?

____

____

____

Do any of these cells lack a plasma membrane? ________

A nucleus? ________________________________

In the cells with a nucleus, can you discern nucleoli?

____

Were you able to observe any of the organelles in these cells?

________ Why or why not? ________

____

____

____

____ ■

Cell Division: Mitosis and Cytokinesis

A cell's *life cycle* is the series of changes it goes through from the time it is formed until it reproduces itself. It encompasses two stages—**interphase,** the longer period during which the cell grows and carries out its usual activities, and **cell division,** when the cell reproduces itself by dividing. In an interphase cell about to divide, the genetic material (DNA) is replicated (duplicated exactly). Once this important event has occurred, cell division ensues.

Cell division in all cells other than bacteria consists of a series of events collectively called mitosis and cytokinesis. **Mitosis** is nuclear division; **cytokinesis** is the division of the cytoplasm, which begins after mitosis is nearly complete. Although mitosis is usually accompanied by cytokinesis, in some instances cytoplasmic division does not occur, leading to the formation of binucleate (or multinucleate) cells. This is relatively common in the human liver.

The process of **mitosis** results in the formation of two daughter nuclei that are genetically identical to the mother nucleus. This distinguishes mitosis from **meiosis,** a specialized type of nuclear division that occurs only in the reproductive organs (testes or ovaries). Meiosis, which yields four daughter nuclei that differ genetically in composition from the mother nucleus, is used only for the production of eggs and sperm (gametes) for sexual reproduction. The function of cell division, including mitosis and cytokinesis in the body, is to increase the number of cells for growth and repair while maintaining their genetic heritage.

The stages of mitosis illustrated in Figure 4.4 include the following events:

Prophase (Figure 4.4b and c): At the onset of cell division, the chromatin threads coil and shorten to form densely staining, short, barlike **chromosomes.** By the middle of prophase the chromosomes appear as double-stranded structures (each strand is a **chromatid**) connected by a small median body called a **centromere.** The centrioles separate from one another and act as focal points for the assembly of two systems of microtubules: the **mitotic spindle,** which forms between the centrioles, and the **asters,** which radiate outward from the ends of the spindle and anchor it to the plasma membrane. Some of the spindle microtubules, the *kinetochore microtubules,* attach to special protein complexes on each chromosome's centromere. Spindle fibers that do not attach to the chromosomes are called *polar microtubules.* The spindle acts as a scaffolding for the attachment and movement of the chromosomes during later mitotic stages. Meanwhile, the nuclear membrane and the nucleolus break down and disappear.

Metaphase (Figure 4.4d): A brief stage, during which the chromosomes migrate to the central plane or equator of the spindle and align along that plane in a straight line (the so-called *metaphase plate*) from the superior to the inferior region of the spindle (lateral view). Viewed from the poles of the cell (end view), the chromosomes appear to be arranged in a "rosette," or circle, around the widest dimension of the spindle.

Anaphase (Figure 4.4e): During anaphase, the centromeres split, and the chromatids (now called chromosomes again) separate from one another and then progress slowly toward opposite ends of the cell. The chromosomes are pulled by the kinetochore microtubules attached to their centromeres, their "arms" dangling behind them. Anaphase is complete when poleward movement ceases.

Telophase (Figure 4.4f): During telophase, the events of prophase are essentially reversed. The chromosomes clustered at the poles begin to uncoil and resume the chromatin form, the spindle breaks down and disappears, a nuclear membrane forms around each chromatin mass, and nucleoli appear in each of the daughter nuclei.

Mitosis is essentially the same in all animal cells, but depending on the type of tissue, it takes from 5 minutes to several hours to complete. In most cells, centriole replication is deferred until interphase of the next cell cycle.

Cytokinesis, or the division of the cytoplasmic mass, begins during telophase (Figure 4.4f), and provides a good guideline for where to look for the mitotic figures of telophase. In animal cells, a *cleavage furrow* begins to form approximately over the equator of the spindle, and eventually splits or pinches the original cytoplasmic mass into two portions. Thus at the end of cell division two daughter cells exist, each smaller in cytoplasmic mass than the mother cell but genetically identical to it. The daughter cells grow and carry out the normal spectrum of metabolic processes until it is their turn to divide.

Cell division is extremely important during the body's growth period. Most cells undergo mitosis until puberty, when normal body size is achieved and overall body growth ceases. After this time in life, only certain cells carry out cell division routinely—for example, cells subjected to abrasion (epithelium of the skin and lining of the gut). Other cell populations—such as liver cells—stop dividing but retain this ability should some of them be removed or damaged. Skeletal muscle, cardiac muscle, and mature neurons completely lose this ability to divide and thus are severely handicapped by injury. Throughout life, the body retains its ability to repair cuts and wounds and to replace some of its aged cells.

Activity 7:
Identifying the Mitotic Stages

1. Watch a video presentation of mitosis (if available).

2. Using the three-dimensional models of dividing cells provided, identify each of the mitotic states described above.

3. Obtain a prepared slide of whitefish blastulae to study the stages of mitosis. The cells of each *blastula* (a stage of embryonic development consisting of a hollow ball of cells) are at approximately the same mitotic stage, so it may be necessary to observe more than one blastula to view all the mitotic stages. The exceptionally high rate of mitosis observed in this tissue is typical of embryos, but if it occurs in specialized tissues it can indicate cancerous cells, which also have an extraordinarily high mitotic rate. Examine the slide carefully, identifying the four mitotic stages and the process of cytokinesis. Compare your observations with Figure 4.4, and verify your identifications with your instructor. ■

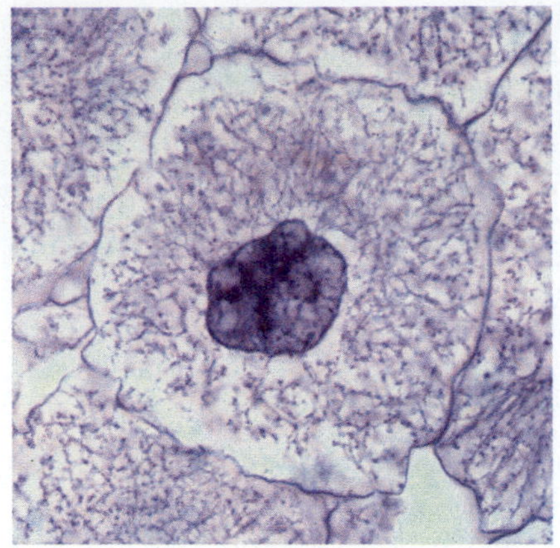

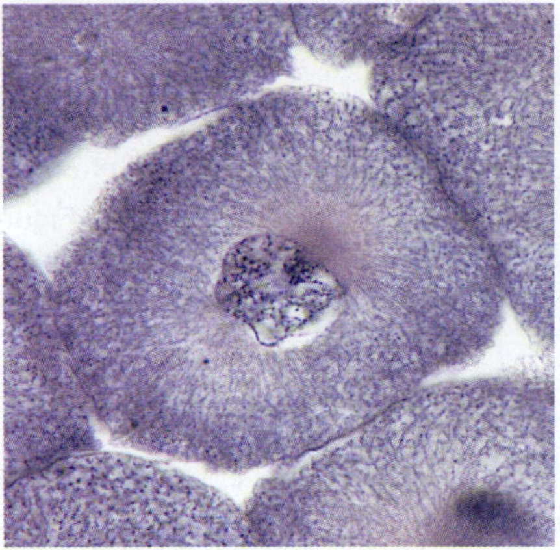

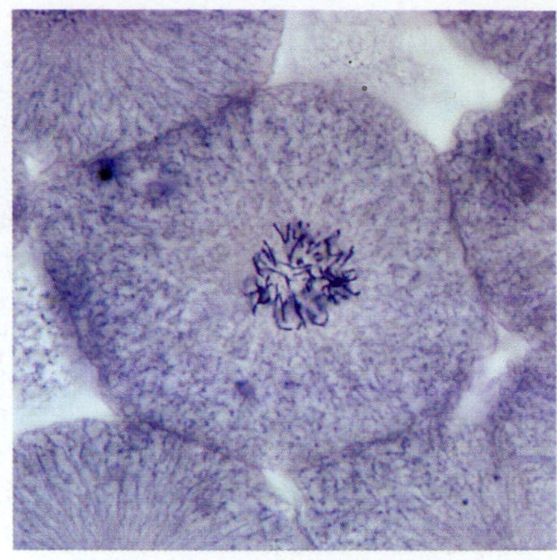

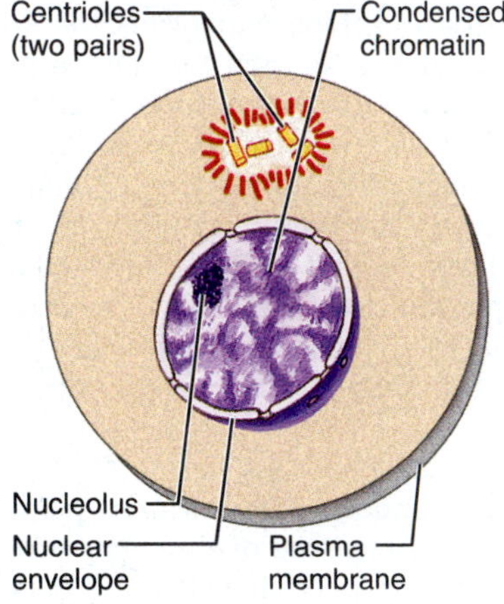

(a)

Interphase

Interphase is the period of a cell's life when it is carrying out its normal metabolic activities and growing. During interphase, the chromosomal material is seen in the form of extended and condensed chromatin, and the nuclear membrane and nucleolus are intact and visible. Microtubule arrays (asters) are seen extending from the centrosomes. During various periods of this phase, the centrioles begin replicating (G_1 through G_2), DNA is replicated (S), and the final preparations for mitosis are completed (G_2). The centriole pair finishes replicating into two pairs during G_2.

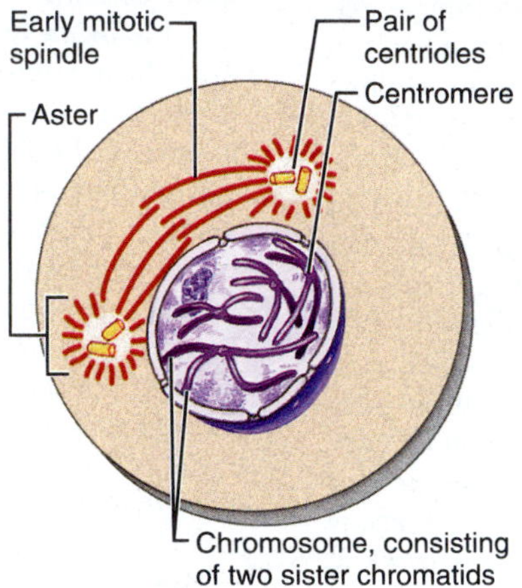

(b)

Early prophase

As mitosis begins, microtubule arrays called *asters* ("stars") are seen extending from the centrosome matrix around the centrioles. Early in *prophase,* the first and longest phase of mitosis, the chromatin threads coil and condense, forming barlike *chromosomes* that are visible with a light microscope. Since DNA replication has occurred during interphase, each chromosome is actually made up of two identical chromatin threads, now called *chromatids.* The chromatids of each chromosome are held together by a small, buttonlike body called a *centromere.* After the chromatids separate, each is considered a new chromosome.

As the chromosomes appear, the nucleoli disappear, and the cytoskeletal microtubules disassemble. The centriole pairs separate from one another. The centrioles act as focal points for growth of a new assembly of microtubules called the **mitotic spindle.** As these microtubules lengthen, they push the centrioles farther and farther apart, propelling them toward opposite ends (poles) of the cell.

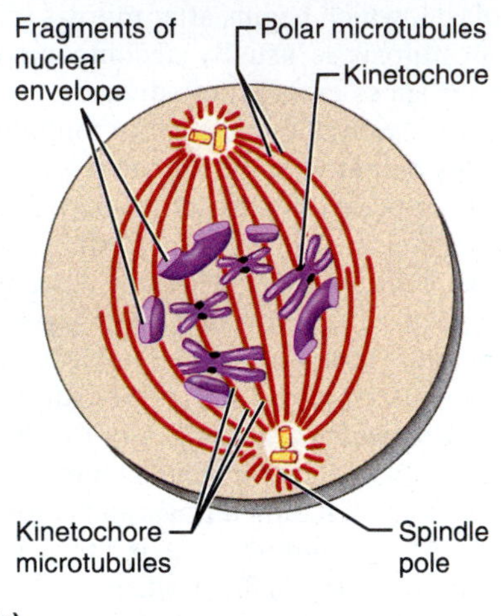

(c)

Late prophase

While the centrioles are still moving away from each other, the nuclear membrane fragments, allowing the spindle to occupy the center of the cell and to interact with the chromosomes. Meanwhile, some of the growing spindle microtubules attach to special protein–DNA complexes, called *kinetochores* (ki-ne′to-korz), on each chromosome's centromere. Such microtubules are called *kinetochore microtubules.* The remaining spindle microtubules, which do not attach to any chromosomes, are called *polar microtubules.* The tips of the polar microtubules are linked near the center; these push against each other forcing the poles apart. The kinetochore microtubules, on the other hand, pull on each chromosome from both poles, resulting in a tug-of-war that ultimately draws the chromosomes to the middle of the cell.

Figure 4.4 The interphase cell and the stages of mitosis. The cells shown are from an early embryo of a whitefish. Photomicrographs are above; corresponding diagrams are below. (Micrographs approximately 600×).

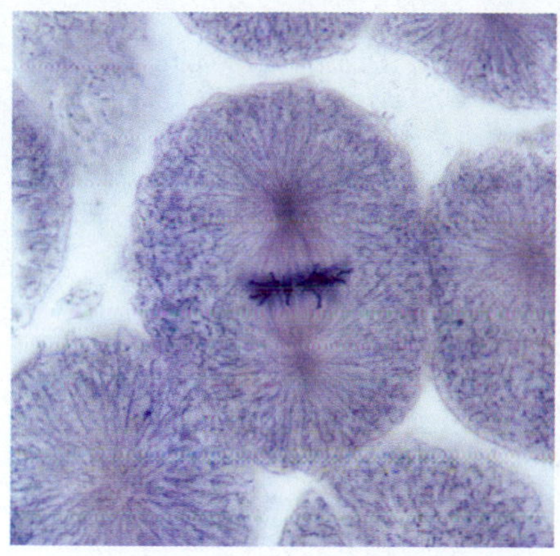

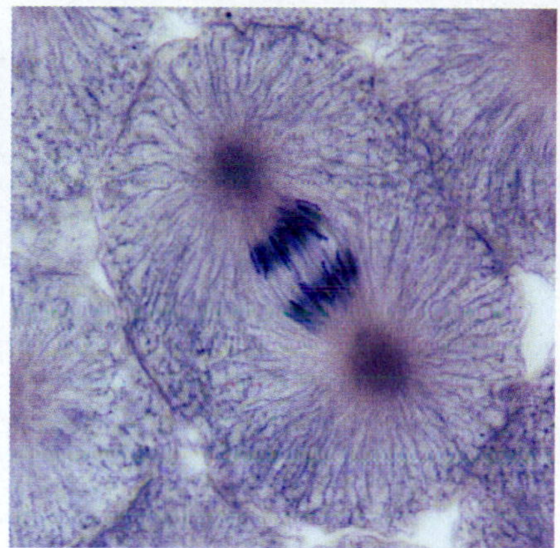

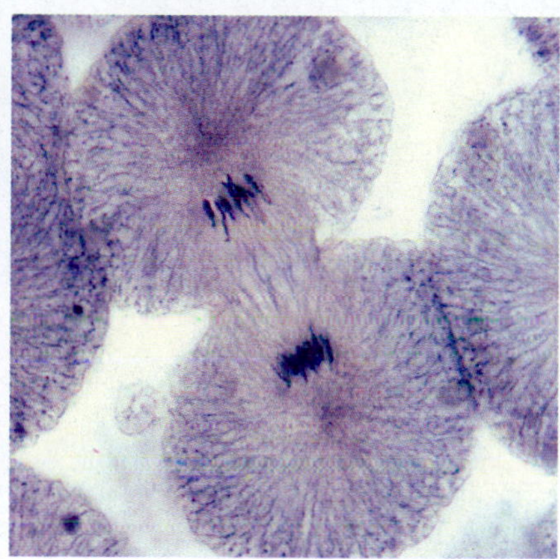

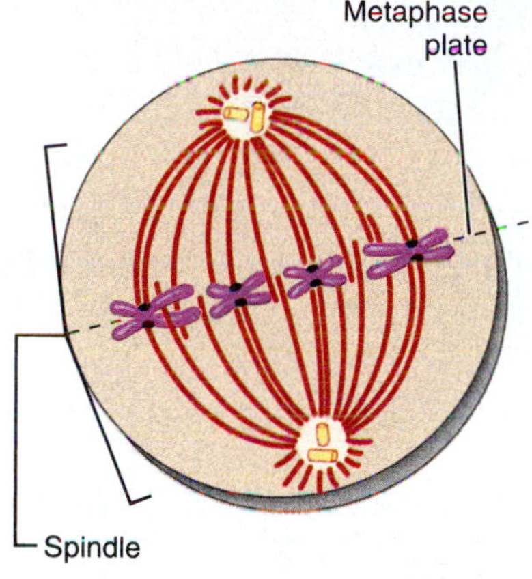

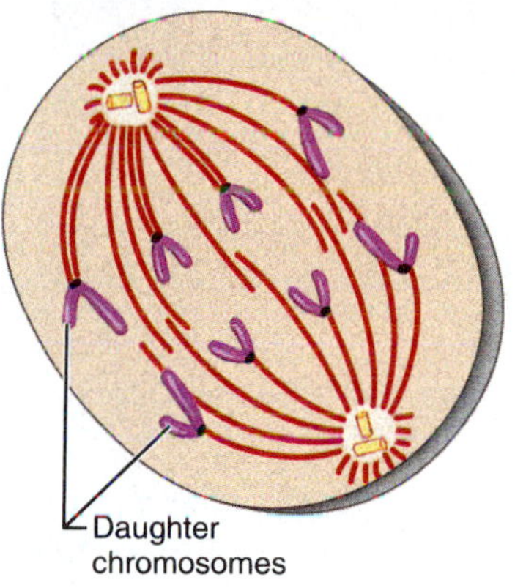

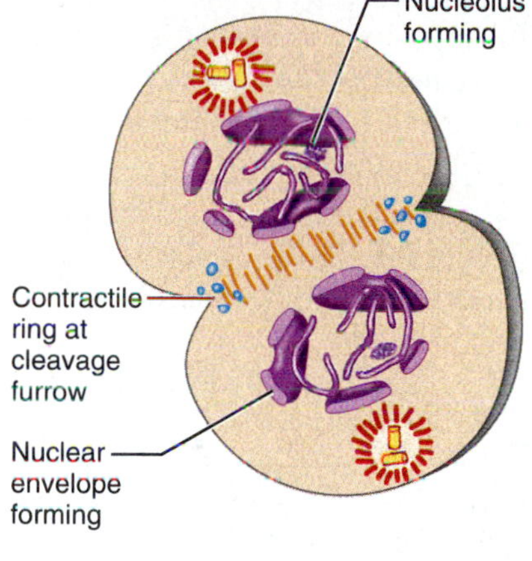

(d)

Metaphase

Metaphase is the second phase of mitosis. The chromosomes cluster at the middle of the cell, with their centromeres precisely aligned at the exact center, or *equator,* of the spindle. This arrangement of the chromosomes along a plane midway between the poles is called the *metaphase* plate.

(e)

Anaphase

Anaphase, the third phase of mitosis, begins abruptly as the centromeres of the chromosomes split, and each chromatid now becomes a chromosome in its own right. The kinetochore fibers, moved along by motor proteins in the kinetochores, rapidly disassemble at their kinetochore ends by removing tubulin subunits, and gradually pull each chromosome toward the pole it faces. By contrast the polar microtubules slide past each other and lengthen (a process presumed to be driven by kinesin motor molecules), and push the two poles of the cell apart, causing the cell to elongate. Anaphase is easy to recognize because the moving chromosomes look V-shaped. The centromeres, which are attached to the kinetochore microtubules, lead the way, and the chromosomal "arms" dangle behind them. Anaphase is the shortest stage of mitosis; it typically lasts only a few minutes.

This process of moving and separating the chromosomes is helped by the fact that the chromosomes are short, compact bodies. Diffuse threads of extended chromatin would tangle, trail, and break, which would damage the genetic material and result in its imprecise "parceling out" to the daughter cells.

(f)

Telophase and cytokinesis

Telophase begins as soon as chromosomal movement stops. This final phase is like prophase in reverse. The identical sets of chromosomes at the opposite poles of the cell uncoil and resume their threadlike extended-chromatin form. A new nuclear membrane, derived from the rough ER, reforms around each chromatin mass. Nucleoli reappear within the nuclei, and the spindle breaks down and disappears. Mitosis is now ended. The cell, for just a brief period, is binucleate (has two nuclei) and each new nucleus is identical to the original mother nucleus.

As a rule, as mitosis draws to a close, *cytokinesis* completes the division of the cell into two daughter cells. Cytokinesis occurs as a contractile ring of peripheral microfilaments forms at the *cleavage furrow* and squeezes the cells apart. Cytokinesis actually begins during late anaphase and continues through and beyond telophase.

Figure 4.4 (*continued*)

exercise

5A The Cell—Transport Mechanisms and Cell Permeability: Wet Lab

Objectives

1. To define *differential permeability; diffusion* (*simple diffusion* and *osmosis*); *isotonic, hypotonic,* and *hypertonic solutions; passive processes of active transport; bulk-phase endocytosis; phagocytosis;* and *solute pump.*
2. To describe the processes that account for the movement of substances across the plasma membrane and to indicate the driving force for each.
3. To determine which way substances will move passively through a differentially permeable membrane (given appropriate information on concentration differences).

Materials

For passive process experiments:

- ❑ Clean slides and coverslips
- ❑ Wax marking pencil
- ❑ Forceps
- ❑ Glass stirring rods
- ❑ 25-ml graduated cylinders
- ❑ Compound microscopes
- ❑ Hot plate and large beaker for hot water bath

Diffusion:

- ❑ Carmine dye crystals
- ❑ Petri plate containing 12 ml of 1.5% agar-agar
- ❑ 3.5% solution methylene blue (approximately 0.1 *M*)
- ❑ 1.6% solution potassium permanganate (approximately 0.1 *M*)
- ❑ Millimeter rulers

- ❑ Medicine dropper
- ❑ Four dialysis sacs or small Hefty "alligator" sandwich bags
- ❑ Beakers (250 ml)
- ❑ Distilled water
- ❑ 40% glucose solution
- ❑ Fine twine or dialysis tubing clamps
- ❑ 10% NaCl solution
- ❑ Boiled starch solution
- ❑ Laboratory balance
- ❑ Benedict's solution in dropper bottle
- ❑ Test tubes in racks, test tube holder
- ❑ Wax marker
- ❑ Small funnel
- ❑ Silver nitrate ($AgNO_3$) in dropper bottle
- ❑ Lugol's iodine solution in dropper bottle
- ❑ Paper towels

- ❑ Vials of animal (mammalian) blood obtained from a biological supply house or veterinarian—at option of instructor
- ❑ Physiologic (mammalian) saline solution in dropper bottle
- ❑ 1.5% sodium chloride (NaCl) solution in dropper bottle
- ❑ Distilled water in dropper bottle
- ❑ Medicine dropper
- ❑ Basin and wash bottles containing 10% household bleach solution
- ❑ Filter paper
- ❑ Test tubes in racks, test tube holder
- ❑ Disposable autoclave bag
- ❑ Disposable plastic gloves

Diffusion demonstrations:

1: Diffusion of a dye through water

Prepared the morning of the laboratory session with setup time noted. Potassium permanganate crystals are placed in a 1000-ml graduated cylinder, and distilled water is added slowly and with as little turbulence as possible to fill to the 1000-ml mark.

2: Osmometer

Just before the laboratory begins, the broad end of a thistle tube is closed with a differentially permeable dialysis membrane, and the tube is secured to a ring stand. Molasses is added to approximately 5 cm above the thistle tube bulb and the bulb is immersed in a beaker of distilled water. At the beginning of the lab session, the level of the molasses in the tube is marked with a wax pencil.

Filtration:

- ❑ Ring stand, ring, clamp
- ❑ Filter paper, funnel
- ❑ Solution containing a mixture of uncooked starch, powdered charcoal, and copper sulfate ($CuSO_4$)
- ❑ 10-ml graduated cylinder
- ❑ 100 ml beaker
- ❑ Lugol's iodine in a dropper bottle

Active transport:

- ❑ Culture of starved amoeba (*Amoeba proteus*)
- ❑ Depression slide
- ❑ Coverslip (glass)
- ❑ *Tetrahymena pyriformis* culture
- ❑ Compound microscope

PhysioEx™ 4.0 Computer Simulation on p. P-4

Note to the Instructor: See directions for handling wet mount preparations and disposable supplies on p. 27, Exercise 3.

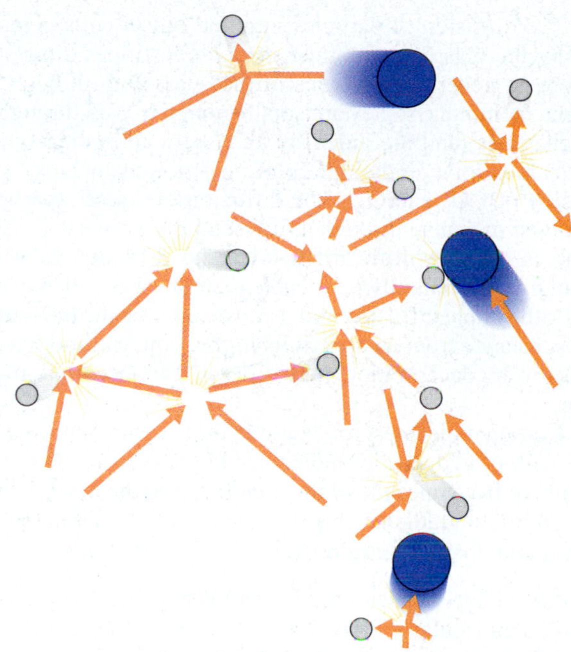

Figure 5A.1 Random movement and numerous collisions cause molecules to become evenly distributed. The small spheres represent water molecules; the large spheres represent glucose molecules.

Because of its molecular composition, the plasma membrane is selective about what passes through it. It allows nutrients to enter the cell but keeps out undesirable substances. By the same token, valuable cell proteins and other substances are kept within the cell, and excreta or wastes pass to the exterior. This property is known as **differential,** or **selective, permeability.** Transport through the plasma membrane occurs in two basic ways. In **active processes,** the cell provides energy (ATP) to power the transport process. In **passive processes,** concentration or pressure differences drive the movement.

Passive Processes

The two important passive processes of membrane transport are *diffusion* and *filtration*. Diffusion is an important transport process for every cell in the body. By contrast, filtration usually occurs only across capillary walls. Each of these will be considered in turn.

Recall that all molecules possess *kinetic energy* and are in constant motion. At a specific temperature, given molecules have about the same average kinetic energy. Since kinetic energy is directly related to both mass and velocity ($KE = \frac{1}{2}mv^2$), smaller molecules tend to move faster. As molecules move about randomly at high speeds, they collide and ricochet off one another, changing direction with each collision (Figure 5A.1).

Although individual molecules cannot be seen, the random motion of small particles suspended in water can be observed. This is called Brownian movement.

Activity 1:
Observing Brownian Movement

1. Using the blunt end of a toothpick, add a small amount of water-insoluble carmine dye to a drop of water on a microscope slide and stir to mix.
2. Add a coverslip and observe under high power (400×).

Does the movement appear to be directed or random?

___

What effect would a change in temperature have on the speed of the movement?

___

Diffusion When a **concentration gradient** (difference in concentration) exists, the net effect of this random molecular movement is that the molecules eventually become evenly distributed throughout the environment, that is, the process called diffusion occurs. Hence, **diffusion** is the movement of molecules from a region of their higher concentration to a region of their lower concentration. Its driving force is the kinetic energy of the molecules themselves.

There are many examples of diffusion in nonliving systems. For example, if a bottle of ether was uncorked at the front of the laboratory, very shortly thereafter you would be nodding as the ether molecules become distributed throughout the room. The ability to smell a friend's cologne shortly after he or she has entered the room is another example.

The diffusion of particles into and out of cells is modified by the plasma membrane, which constitutes a physical barrier. In general, molecules diffuse passively through the plasma membrane if they are small enough to pass through its ion channels (and are aided by an electrical gradient), or if they can dissolve in the lipid portion of the membrane (as in the case of CO_2 and O_2). The diffusion of solutes (particles dissolved in water) through a differentially permeable membrane is called **simple diffusion.** The diffusion of water through a differentially permeable membrane is called **osmosis.** Both simple diffusion and osmosis involve the movement of a substance from an area of its higher concentration to one of its lower concentration, that is, down its concentration gradient.

Certain molecules, for example glucose, are able to combine with protein carrier molecules in the plasma membrane and move from one side of the membrane to the other down a concentration gradient. This process is called **facilitated diffusion** and does not require ATP.

Diffusion of Dye Through Agar Gel and Water The relationship between molecular weight and the rate of diffusion can be examined easily by observing the diffusion of the molecules of two different types of dye through an agar gel. The dyes used in this experiment are methylene blue, which has a molecular weight of 320 and is deep blue in color, and potassium permanganate, a purple dye with a molecular weight of 158. Although the agar gel appears quite solid, it is primarily (98.5%) water and allows free movement of the diffusing dye molecules through it.

Activity 2:
Observing Diffusion of Dye Through Agar Gel

Avoid contact between your skin and the dye crystals by using the forceps to pick up the crystals.

1. Work with members of your group to formulate a hypothesis about the rates of diffusion of methylene blue and potassium permanganate through the agar gel. Justify your hypothesis.

2. Obtain a petri dish containing agar gel, a millimeter ruler, a wax marking pencil, dropper bottles of methylene blue and potassium permanganate, and a medicine dropper. (See Figure 5A.2.)

3. Using the wax marking pencil, draw a line on the bottom of the petri dish dividing it into two sections.

4. Create a well in the center of each section using the medicine dropper. To do this, squeeze the bulb of the medicine dropper, and push it down into the agar. Release the bulb as you slowly pull the dropper out of the agar. This should remove an agar plug, leaving a well in the agar.

5. Carefully fill one well with the methylene blue solution and the other well with the potassium permanganate solution. Record the time.

____

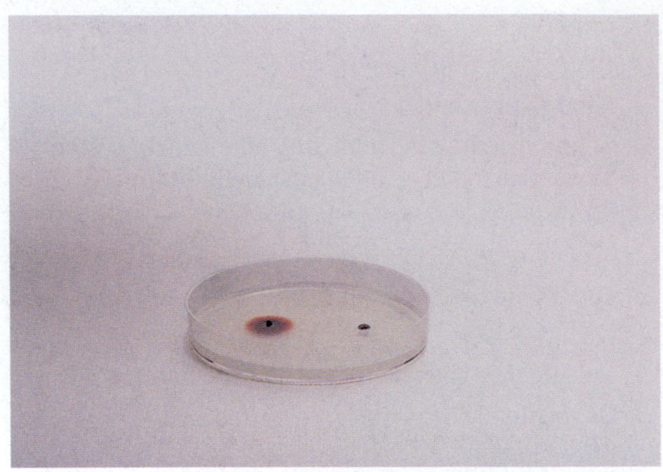

Figure 5A.2 Comparing diffusion rates. Agar-plated petri dish as it appears after the diffusion of 0.1*M* methylene blue placed in one well and 0.1*M* potassium permanganate placed in another.

Time (min)	Diffusion of methylene blue (mm)	Diffusion of potassium permanganate (mm)
15		
30		
45		
60		
75		
90		

6. At 15-minute intervals, use the millimeter ruler to measure the distance the dye has diffused from each well. These observations should be continued for 1½ hours, and the results recorded in the chart above.

Which dye diffused more rapidly? ______________

What is the relationship between molecular weight and rate of molecular movement (diffusion)?

____

Why did the dye molecules move? ______________

____

Compute the rate of diffusion of the potassium permanganate molecules in millimeters per minute (mm/min) and record.

_______ mm/min

Compute the rate of diffusion of the methylene blue molecules in mm/min and record.

_______ mm/min

7. Suggest an additional variable to test. Form a hypothesis. Set up and conduct the experiment. Record your results.

8. Prepare a lab report for these experiments. (See Gettting Started: Writing a Lab Report, p. xii.) ■

Make a mental note to yourself to go to demonstration area 1 at the end of the laboratory session to observe the extent of diffusion of the potassium permanganate dye through water. At that time, follow the directions given next.

Activity 3: Observing Diffusion of Dye Through Water

1. Measure the number of millimeters the dye has diffused from the bottom of the graduated cylinder and record.

_______ mm

2. Record the time the demonstration was set up and the time of your observation. Then compute the rate of the dye's diffusion through water and record below.

Time of setup _______

Time of observation _______

Rate of diffusion _______ mm/min

3. Does the potassium permanganate dye move (diffuse) more rapidly through water or the agar gel? (Explain your answer.)

____

____ ■

Activity 4: Observing Diffusion Through Nonliving Membranes

The following experiment provides information on the diffusion of water and solutes through differentially permeable membranes, which may be applied to the study of transport mechanisms in living membrane-bound cells.

1. Read through the experiments in this activity, and develop a hypothesis for each part.

2. Obtain four dialysis sacs,* a small funnel, a 25-ml graduated cylinder, a wax marker, fine twine or dialysis tubing clamps, and four beakers (250 ml). Number the beakers 1 to 4 with the wax marker, and half fill all of them with distilled water except beaker 2, to which you should add 40% glucose solution.

3. Prepare the dialysis sacs one at a time. Using the funnel, half fill each with 20 ml of the specified liquid (see below). Press out the air, fold over the open end of the sac, and tie it securely with fine twine or clamp it. Before proceeding to the next sac, quickly and carefully blot the sac dry by rolling it on a paper towel, and weigh it with a laboratory balance. Record the weight in the data chart, and then drop the sac into the corresponding beaker. Be sure the sac is completely covered by the beaker solution, adding more solution if necessary.

- Sac 1: 40% glucose solution. Weight: _______________ g
- Sac 2: 40% glucose solution. Weight: _______________ g
- Sac 3: 10% NaCl solution. Weight: _______________ g
- Sac 4: boiled starch solution. Weight: _______________ g

Allow sacs to remain undisturbed in the beakers for 1 hour. (Use this time to continue with other experiments.)

4. After an hour, get a beaker of water boiling on the hot plate. Obtain the supplies you will need to determine your experimental results: dropper bottles of Benedict's solution, silver nitrate solution, and Lugol's iodine, a test tube rack, four test tubes, and a test tube holder.

5. Quickly and gently blot sac 1 dry and weigh it. (Note: Do not squeeze the sac during the blotting process.) Record in the data chart.

Weight of sac 1: _______ g

Has there been any change in weight? _______________

Conclusions? _______________________________________

____

____

Place 5 ml of Benedict's solution in each of two test tubes. Put 4 ml of the beaker fluid into one test tube and 4 ml of the sac fluid into the other. Mark the tubes for identification and then place them in a beaker containing boiling water. Boil 2 minutes. Cool slowly. If a green, yellow, or rusty red precipitate forms, the test is positive, meaning that glucose is pres-

*Dialysis sacs are differentially permeable membranes with pores of a particular size. The selectivity of living membranes depends on more than just pore size, but using the dialysis sacs will allow you to examine selectivity due to this factor.

Data from Experiments on Diffusion Through Nonliving Membranes						
Beaker	**Contents of sac**	**Initial weight**	**Final weight**	**Weight change**	**Tests—beaker fluid**	**Tests—sac fluid**
Beaker 1 ½ filled with distilled water	20 ml 40% glucose solution				Benedict's test:	Benedict's test:
Beaker 2 ½ filled with 40% glucose solution	20 ml 40% glucose solution					
Beaker 3 ½ filled with distilled water	20 ml 10% NaCl solution				$AgNO_3$ test:	
Beaker 4 ½ filled with distilled water	20 ml boiled starch solution				Lugol's test:	

ent. If the solution remains the original blue color, the test is negative. Record results in the data chart.

Was glucose still present in the sac? ________________

Was glucose present in the beaker? ________________

Conclusions? ________________

____

____

6. Blot gently and weigh sac 2: ________________ g
Record weight in the data chart.

Was there an *increase* or *decrease* in weight? ________

With 40% glucose in the sac and 40% glucose in the beaker, would you expect to see any net movements of water (osmosis) or of glucose molecules (simple diffusion)?

________ Why or why not? ________________

____

7. Blot gently and weigh sac 3: ________________ g
Record weight in the data chart.

Was there any change in weight? ________________

Conclusions? ________________

____

Take a 5-ml sample of beaker 3 solution and put it in a clean test tube. Add a drop of silver nitrate. The appearance of a white precipitate or cloudiness indicates the presence of AgCl, which is formed by the reaction of $AgNO_3$ with NaCl (sodium chloride). Record results in the data chart.

Results? ________________

Conclusions? ________________

8. Blot gently and weigh sac 4: ________________ g
Record weight in the data chart.

Was there any change in weight? ________________

Conclusions? ________________

____

Take a 5-ml sample of beaker 4 solution and add a couple of drops of Lugol's iodine solution. The appearance of a black color is a positive test for the presence of starch. Record results in the data chart. Did any starch diffuse from the sac into the beaker?

________ Explain: ________________

____

9. In which of the test situations did net osmosis occur?

____

In which of the test situations did net simple diffusion occur?

____

What conclusions can you make about the relative size of glucose, starch, NaCl, and water molecules?

___

___

___

___

With what cell structure can the dialysis sac be compared?

___

10. Prepare a lab report for the experiment. (See Getting Started: Writing a Lab Report, p. xii.) Be sure to include in your discussion the answers to the questions proposed in this activity.

11. Before leaving the laboratory, observe demonstration 2, the *osmometer demonstration* set up before the laboratory session to follow the movement of water through a membrane (osmosis). Measure the distance the water column has moved during the laboratory period and record below. (The position of the meniscus in the thistle tube at the beginning of the laboratory period is marked with wax pencil.)

Distance the meniscus has moved: _______ mm ■

Activity 5: Investigating Diffusion Through Living Membranes

To examine permeability properties of plasma membranes, conduct the following two experiments. Begin by developing hypotheses.

Experiment 1:

1. The following supplies should be available at your laboratory bench to conduct this experimental series: a clean slide and coverslip, vial of animal blood, medicine dropper, physiologic saline, 1.5% sodium chloride solution, three test tubes, test tube rack, glass stirring rod, 15-ml graduated cylinder, filter paper, and plastic gloves.

2. Label three test tubes A, B, and C, and prepare them as follows:

- A: add 2 ml physiologic saline
- B: add 2 ml 1.5% sodium chloride solution
- C: add 2 ml distilled water

⚠ 3. Don the gloves, and use a medicine dropper to add 5 drops of animal blood to each test tube. Stir each test tube with the glass rod, rinsing between each sample. Do not remove the gloves until the slides have been appropriately disposed of.

4. Hold each test tube in front of this printed page. *Record* the clarity of print seen through the fluid in each tube.

Test tube A ___

Test tube B ___

Test tube C ___

Experiment 2: Now you will conduct a microscopic study of red blood cells suspended in the same three solutions. The objective is to determine if these solutions have any effect on cell shape by promoting net osmosis.

1. Place a very small drop of physiologic saline on a slide. Using the medicine dropper, add a small drop of animal blood to the saline on the slide. Tilt the slide to mix, cover with a coverslip, and immediately examine the preparation under the high-power lens. Notice that the red blood cells retain their normal smooth disclike shape (see Figure 5A.3a). This is because the physiologic saline is **isotonic** to the cells. That is, it contains a concentration of nonpenetrating solutes (e.g., proteins and some ions) equal to that in the cells (same solute-solvent ratio). Consequently, the cells neither gain nor lose water by osmosis.

2. Prepare another wet mount of animal blood, but this time use 1.5% saline solution as the suspending medium. After 5 minutes, carefully observe the red blood cells under high power. What is happening to the normally smooth disc shape of the red blood cells?

___

___

___

This crinkling-up process, called **crenation,** is due to the fact that the 1.5% sodium chloride solution is slightly hypertonic to the cytosol of the red blood cell. A **hypertonic** solution contains more nonpenetrating solutes (thus less water) than are present in the cell. Under these circumstances, water tends to leave the cells by osmosis. Compare your observations to Figure 5A.3b.

3. Add a drop of distilled water to the edge of the coverslip. Fold a piece of filter paper in half and place its folded edge at the opposite edge of the coverslip; it will absorb the saline solution and draw the distilled water across the cells. Watch the red blood cells as they float across the field. After about 5 minutes have passed, describe the change in their appearance.

___

___

Distilled water contains *no* solutes (it is 100% water). Distilled water and *very* dilute solutions (that is, those containing less than 0.9% nonpenetrating solutes) are **hypotonic** to the cell. In a hypotonic solution, the red blood cells first "plump up" (Figure 5A.3c) but then they suddenly start to disappear. The red blood cells burst as the water floods into them, leaving "ghosts" in their wake—a phenomenon called **hemolysis.**

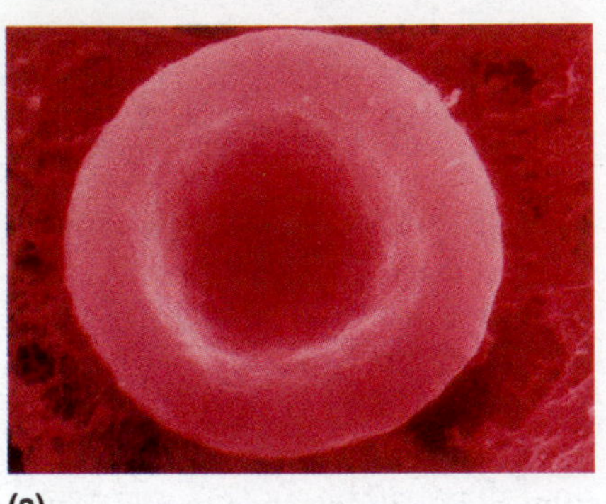

(a)

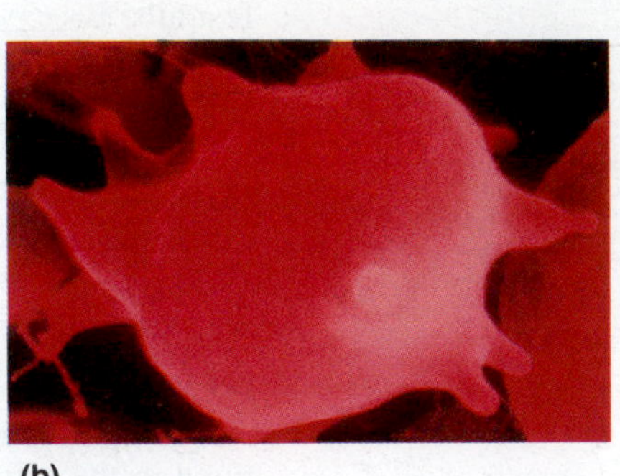

(b)

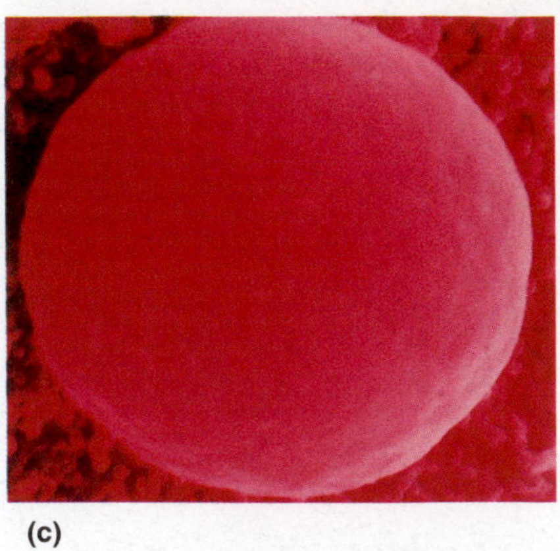

(c)

Figure 5A.3 Influence of isotonic, hypertonic, and hypotonic solutions on red blood cells. **(a)** Red blood cells suspended in an isotonic solution, where the cells retain their normal size and shape. **(b)** Red blood cells suspended in hypertonic solution. As the cells lose water to the external environment, they shrink and become prickly, a phenomenon called crenation. **(c)** Red blood cells suspended in a hypotonic solution. Notice their spherical bloated shape, a result of excessive water intake.

How do your observations of test tube C in Experiment 1 correlate with what you just observed under the microscope?

____

____

____

4. Place the blood-soiled slides and test tube in the bleach-containing basin. Put the coverslips you used into the disposable autoclave bag. Obtain a wash (squirt) bottle containing 10% bleach solution and squirt the bleach liberally over the bench area where blood was handled. Wipe the bench down with a paper towel wet with the bleach solution and allow it to dry before continuing. Remove gloves, and discard in the autoclave bag.

5. Prepare a lab report for experiments 1 and 2. (See Getting Started: Writing a Lab Report, p. xii.) Be sure to include in the discussion answers to the questions proposed in this activity. ■

Filtration

Filtration is the process by which water and solutes are forced through a membrane from an area of higher hydrostatic (fluid) pressure into an area of lower hydrostatic pressure. Like diffusion, it is a passive process. For example, fluids and solutes filter out of the capillaries in the kidneys into the kidney tubules because the blood pressure in the capillaries is greater than the fluid pressure in the tubules. Filtration is not a selective process. The amount of filtrate (fluids and solutes) formed depends almost entirely on the pressure gradient (difference in pressure on the two sides of the membrane) and on the size of the membrane pores.

Activity 6: Observing the Process of Filtration

1. Obtain the following equipment: a ring stand, ring, and ring clamp; a funnel; a piece of filter paper; a beaker; a 10-ml graduated cylinder; a solution containing uncooked starch, powdered charcoal, and copper sulfate; and a dropper bottle of Lugol's iodine. Attach the ring to the ring stand with the clamp.

2. Fold the filter paper in half twice, open it into a cone, and place it in a funnel. Place the funnel in the ring of the ring stand and place a beaker under the funnel. Shake the starch solution, and fill the funnel with it to just below the top of the filter paper. When the steady stream of filtrate changes to countable filtrate drops, count the number of drops formed in 10 seconds and record.

______________ drops

When the funnel is half empty, again count the number of drops formed in 10 seconds and record the count.

______________ drops

3. After all the fluid has passed through the filter, check the filtrate and paper to see which materials were retained by the paper. (Note: If the filtrate is blue, the copper sulfate passed. Check both the paper and filtrate for black particles to see if the charcoal passed. Finally, using a 10-ml graduated cylinder, put a 2-ml filtrate sample into a test tube. Add several drops of Lugol's iodine. If the sample turns blue/black when iodine is added, starch is present in the filtrate.)

Passed: ______________________________

Retained: ______________________________.

What does the filter paper represent? ______________

During which counting interval was the filtration rate

greatest? ______________________________

Explain: ______________________________

____

____

What characteristic of the three solutes determined whether or not they passed through the filter paper?

______________________________ ■

Active Processes

Whenever a cell uses the bond energy of ATP to move substances across its boundaries, the process is referred to as an *active process*. Substances moved by active means are generally unable to pass by diffusion. They may be too large to pass through the membrane channels; they may not be lipid soluble; or they may have to move against rather than with a concentration gradient. There are two types of active processes: **active transport** and **vesicular transport.**

Active Transport

Active transport requires carrier proteins that combine specifically with the transported substance, which is similar to enzyme-substrate interactions described in your text. Active transport may be primary, driven directly by hydrolysis of ATP, or secondary, acting with a primary transport system as a coupled system. In many cases the substances move against concentration or electrochemical gradients or both. Some of the substances that are moved into the cells by such carriers, commonly called **solute pumps,** are amino acids and some sugars. Both solutes are lipid insoluble and too large to pass through the membrane channels but necessary for cell life. On the other hand, sodium ions (Na^+) are ejected from cells by active transport. There is more Na^+ outside the cell than inside, so the Na^+ tends to remain in the cell unless actively transported out.

Vesicular Transport

Large particles and molecules are transported across the membrane by vesicular transport. Movement may be into the cell **(endocytosis)** or out of the cell **(exocytosis). In bulk-phase endocytosis,** the cell membrane sinks beneath the material to form a small vesicle, which then pinches off into the cell interior (see Figure 5A.4a). Bulk-phase endocytosis is most common for taking in liquids containing protein or fat.

In **phagocytosis** (cell eating), parts of the plasma membrane and cytoplasm expand and flow around a relatively large or solid material (for example, bacteria or cell debris) and engulf it (Figure 5A.4b). The membranous sac thus

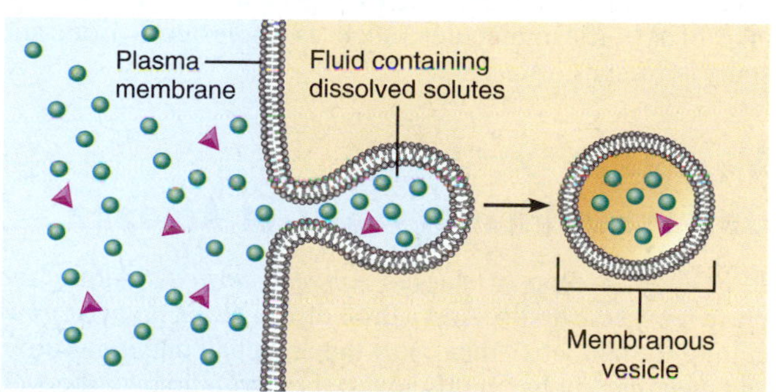

(a) Bulk-phase endocytosis

Figure 5A.4 Three types of endocytosis.
(a) In bulk-phase endocytosis, dissolved proteins gather on the external surface of the plasma membrane, causing the membrane to invaginate and to incorporate a droplet of the fluid.

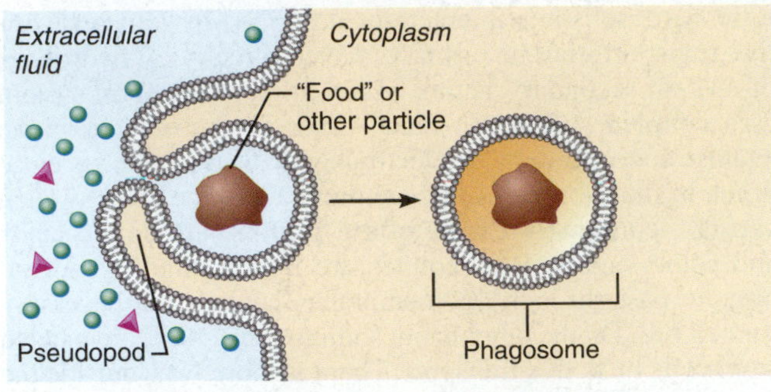

(b) Phagocytosis

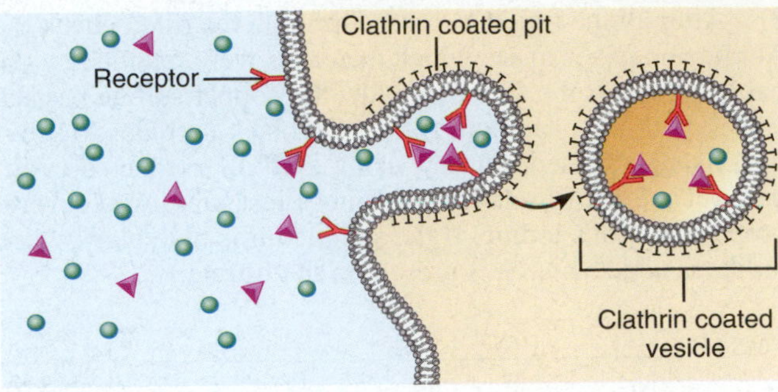

(c) Receptor-mediated endocytosis

Figure 5A.4 *(continued)* Three types of endocytosis. (b) In phagocytosis, cellular extensions (pseudopodia) flow around the external particle and enclose it within a vacuole. **(c)** In receptor-mediated endocytosis, plasma membrane proteins bind only with certain substances in a clathrin protein coated pit.

formed, called a *phagosome*, is then fused with a lysosome and its contents are digested. In the human body, phagocytic cells are mainly found among the white blood cells and macrophages that act as scavengers and help protect the body from disease-causing microorganisms and cancer cells.

A more selective type of endocytosis uses plasma membrane receptors and is called **receptor-mediated endocytosis** (Figure 5A.4c). As opposed to the phagocytosis used by the body's scavenger cells (see below), this type of endocytosis is exquisitely selective and is used primarily for cellular uptake of specific molecules, such as cholesterol, iron, and some hormones.

Activity 7: Observing Phagocytosis in Amoeba

1. Obtain a drop of starved *Amoeba proteus* culture and place it on a coverslip. Add a drop of *Tetrahymena pyriformis* culture (an amoeba "meal") to the amoeba-containing drop, and then quickly but gently invert the coverslip over the well of a depression slide.

2. Locate an amoeba under low power. Keep the light as dim as possible; otherwise the amoeba will "ball up" and begin to disintegrate.

3. Watch as the amoeba phagocytizes the *Tetrahymena* by forming pseudopods that engulf it. As mentioned, in unicellular organisms like the amoeba, phagocytosis is an important food-getting mechanism, but in higher organisms, it is more important as a protective device.

4. Return all equipment to the appropriate supply areas and rinse glassware used. ■

Note: If you have not already done so, complete Activity 3: Observing Diffusion of Dye Through Water, p. 43.

Classification of Tissues

exercise **6A**

Exercise 4 describes cells as the building blocks of life and the all-inclusive functional units of unicellular organisms. But in higher organisms cells do not usually operate as isolated, independent entities. In humans and other multicellular organisms, cells depend on one another and cooperate to maintain homeostasis in the body.

With a few exceptions (parthenogenetic organisms), even the most complex animal starts out as a single cell, the fertilized egg, which divides almost endlessly. The trillions of cells that result become specialized for a particular function; some become supportive bone, others the transparent lens of the eye, still others skin cells, and so on. Thus a division of labor exists, with certain groups of cells highly specialized to perform functions that benefit the organism as a whole. Cell specialization carries with it certain hazards, because when a small specific group of cells is indispensable, any inability to function on its part can paralyze or destroy the entire body.

Groups of cells that are similar in structure and function are called **tissues.** The four primary tissue types—epithelium, connective tissue, nervous tissue, and muscle—have distinctive structures, patterns, and functions. The four primary tissues are further divided into subcategories, as described shortly.

To perform specific body functions, the tissues are organized into **organs** such as the heart, kidneys, and lungs. Most organs contain several representatives of the primary tissues, and the arrangement of these tissues determines the organ's structure and function. Thus **histology,** the study of tissues, complements a study of gross anatomy and provides the structural basis for a study of organ physiology.

The main objective of this exercise is to familiarize you with the major similarities and dissimilarities of the primary tissues, so that when the tissue makeup of an organ is described, you will be able to more easily understand (and perhaps even predict) the organ's major function. Because epithelium and some types of connective tissue will not be considered again, they are emphasized more than muscle, nervous tissue, and bone (a connective tissue), which are covered in more depth in later exercises. In Activity 5, p. 67, you will be constructing a concept map to solidify your knowledge of tissue types.

Objectives

1. To name the four major types of tissues in the human body and the major subcategories of each.
2. To identify the tissue subcategories through microscopic inspection or inspection of an appropriate diagram or projected slide.
3. To state the location of the various tissue types in the body.
4. To list the general functions and structural characteristics of each of the four major tissue types.

Materials

- ❑ Compound microscope
- ❑ Immersion oil
- ❑ Prepared slides of simple squamous, simple cuboidal, simple columnar, stratified squamous (nonkeratinized), stratified cuboidal, stratified columnar, pseudostratified ciliated columnar, and transitional epithelium
- ❑ Prepared slides of mesenchyme; of adipose, areolar, reticular, and dense (both regular [tendon] and irregular [dermis]) connective tissues); of hyaline and elastic cartilage; of fibrocartilage; of bone (cross section); and of blood
- ❑ Prepared slides of skeletal, cardiac, and smooth muscle (longitudinal sections)
- ❑ Prepared slide of nervous tissue (spinal cord smear)

PhysioEx™ 4.0 Computer Simulation on p. P-15

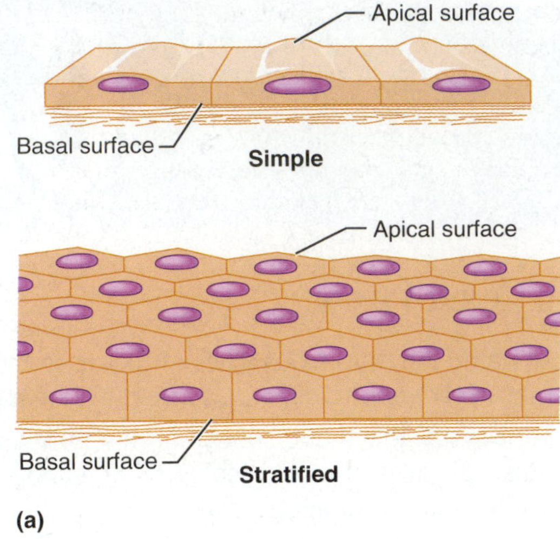

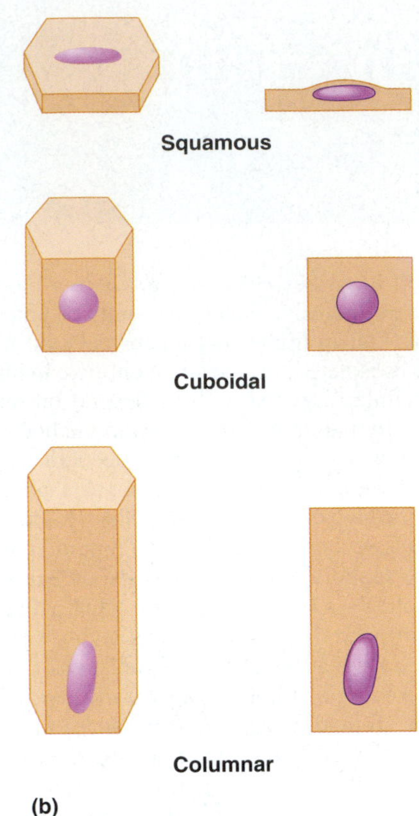

Figure 6.1 Classification of epithelia. (a) Classification on the basis of arrangement (relative number of layers). **(b)** Classification on the basis of cell shape. For each category, a whole cell is shown on the left and a longitudinal section is shown on the right.

Epithelial Tissue

Epithelial tissue, or **epithelium,** covers surfaces. For example, epithelium covers the external body surface (as the epidermis), lines its cavities and tubules, and generally marks off our "insides" from our outsides. Since the various endocrine (hormone-producing) and exocrine glands of the body almost invariably develop from epithelial membranes, glands too are logically classed as epithelium.

Epithelial functions include protection, absorption, filtration, excretion, secretion, and sensory reception. For example, the epithelium covering the body protects against bacterial invasion and chemical damage; that lining the respiratory tract is ciliated to sweep dust and other foreign particles away from the lungs. Epithelium specialized to absorb substances lines the stomach and small intestine. In the kidney tubules, the epithelium absorbs, secretes, and filters. Secretion is a specialty of the glands.

The following characteristics distinguish epithelial tissues from other types:

- Cellularity and specialized contacts. Cells fit closely together to form membranes, or sheets of cells, and are bound together by specialized junctions.
- Polarity. The membranes always have one free surface, called the *apical surface.*
- Supported by connective tissue. The cells are attached to and supported by an adhesive **basement membrane,** which is an amorphous material secreted partly by the epithelial cells (*basal lamina*) and connective tissue cells (*reticular lamina*) that lie adjacent to each other.
- Avascularity. Epithelial tissues have no blood supply of their own (are avascular), but depend on diffusion of nutrients from the underlying connective tissue. (Glandular epithelia, however, are very vascular.)
- Regeneration. If well nourished, epithelial cells can easily regenerate themselves. This is an important characteristic because many epithelia are subjected to a good deal of friction.

The covering and lining epithelia are classified according to two criteria—arrangement or relative number of layers and cell shape (Figure 6.1). On the basis of arrangement, there are **simple** epithelia, consisting of one layer of cells attached to the basement membrane, and **stratified** epithelia, consisting of two or more layers of cells. The general types based on shape are **squamous** (scalelike), **cuboidal** (cubelike), and **columnar** (column-shaped) epithelial cells. The terms denoting shape and arrangement of the epithelial cells are combined to describe the epithelium fully. *Stratified epithelia are named according to the cells at the apical surface of the epithelial membrane,* not those resting on the basement membrane.

There are, in addition, two less easily categorized types of epithelia. **Pseudostratified epithelium** is actually a simple columnar epithelium (one layer of cells), but because its cells vary in height, and the nuclei lie at different levels above the basement membrane, it gives the false appearance

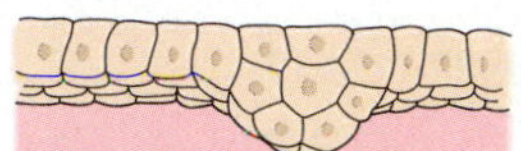

(a)

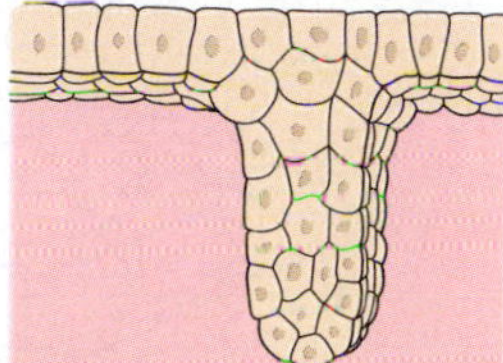

(b)

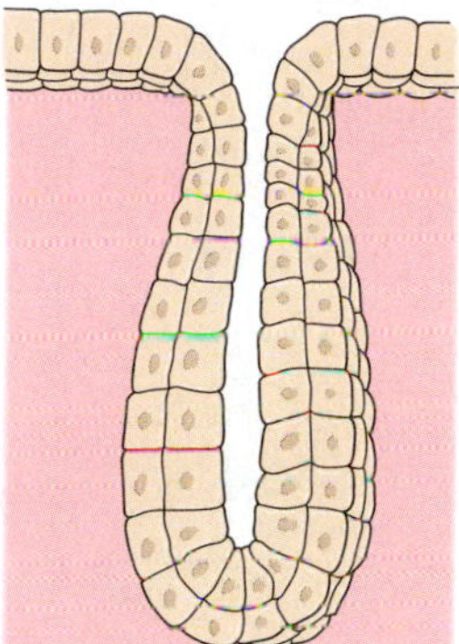

(c) Exocrine gland

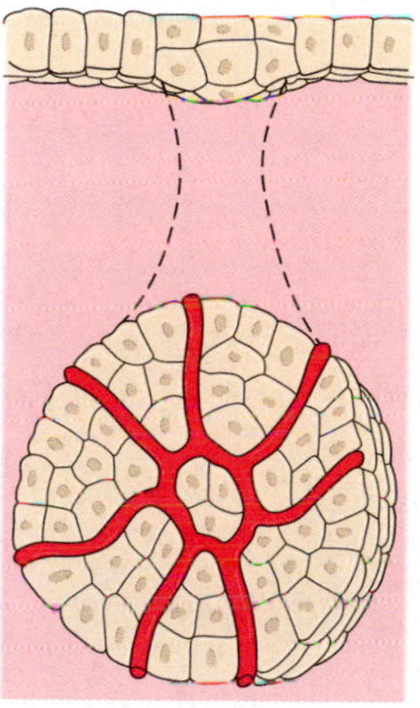

(d) Endocrine gland

Figure 6.2 Formation of endocrine and exocrine glands from epithelial sheets. **(a)** Epithelial cells grow and push into the underlying tissue. **(b)** A cord of epithelial cells forms. **(c)** In an exocrine gland, a lumen (cavity) forms. The inner cells form the duct, the outer cells produce the secretion. **(d)** In a forming endocrine gland, the connecting duct cells atrophy, leaving the secretory cells with no connection to the epithelial surface. However, they do become heavily invested with blood and lymphatic vessels that receive the secretions.

of being stratified. This epithelium is often ciliated. **Transitional epithelium** is a rather peculiar stratified squamous epithelium formed of rounded, or "plump," cells with the ability to slide over one another to allow the organ to be stretched. Transitional epithelium is found only in urinary system organs subjected to periodic distension, such as the bladder. The superficial cells are flattened (like true squamous cells) when the organ is distended and rounded when the organ is empty.

Epithelial cells forming glands are highly specialized to remove materials from the blood and to manufacture them into new materials, which they then secrete. There are two types of glands, as shown in Figure 6.2. **Endocrine glands** lose their surface connection (duct) as they develop; thus they are referred to as ductless glands. Their secretions (all hormones) are extruded directly into the blood or the lymphatic vessels that weave through the glands. **Exocrine glands** retain their ducts, and their secretions empty through these ducts to an epithelial surface. The exocrine glands—including the sweat and oil glands, liver, and pancreas—are both external and internal; they will be discussed in conjunction with the organ systems to which their products are functionally related.

The most common types of epithelia, their most common locations in the body, and their functions are described in Figure 6.3.

Activity 1: Examining Epithelial Tissue Under the Microscope

Obtain slides of simple squamous, simple cuboidal, simple columnar, stratified squamous (nonkeratinized), pseudostratified ciliated columnar, stratified cuboidal, stratified columnar, and transitional epithelia. Examine each carefully, and notice how the epithelial cells fit closely together to form intact sheets of cells, a necessity for a tissue that forms linings or covering membranes. Scan each epithelial type for modifications for specific functions, such as cilia (motile cell projections that help to move substances along the cell surface), and microvilli, which increase the surface area for absorption. Also be alert for goblet cells, which secrete lubricating mucus (see Plate 1 of the Histology Atlas). Compare your observations with the photomicrographs in Figure 6.3.

While working, check the questions in the laboratory review section for this exercise. A number of the questions there refer to some of the observations you are asked to make during your microscopic study. ■

Text continues on page 56

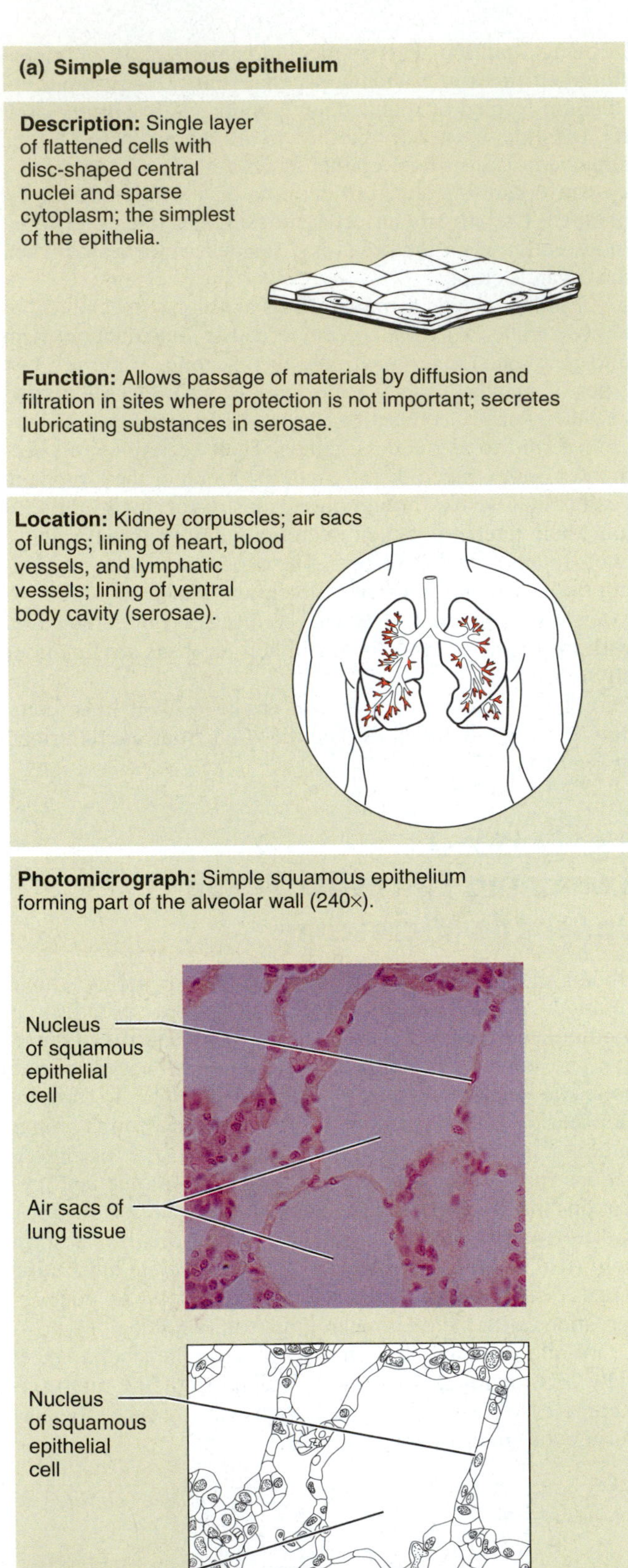

(a) Simple squamous epithelium
Description: Single layer of flattened cells with disc-shaped central nuclei and sparse cytoplasm; the simplest of the epithelia.
Function: Allows passage of materials by diffusion and filtration in sites where protection is not important; secretes lubricating substances in serosae.
Location: Kidney corpuscles; air sacs of lungs; lining of heart, blood vessels, and lymphatic vessels; lining of ventral body cavity (serosae).
Photomicrograph: Simple squamous epithelium forming part of the alveolar wall (240×).
Nucleus of squamous epithelial cell
Air sacs of lung tissue
Nucleus of squamous epithelial cell
Air sacs of lung tissue

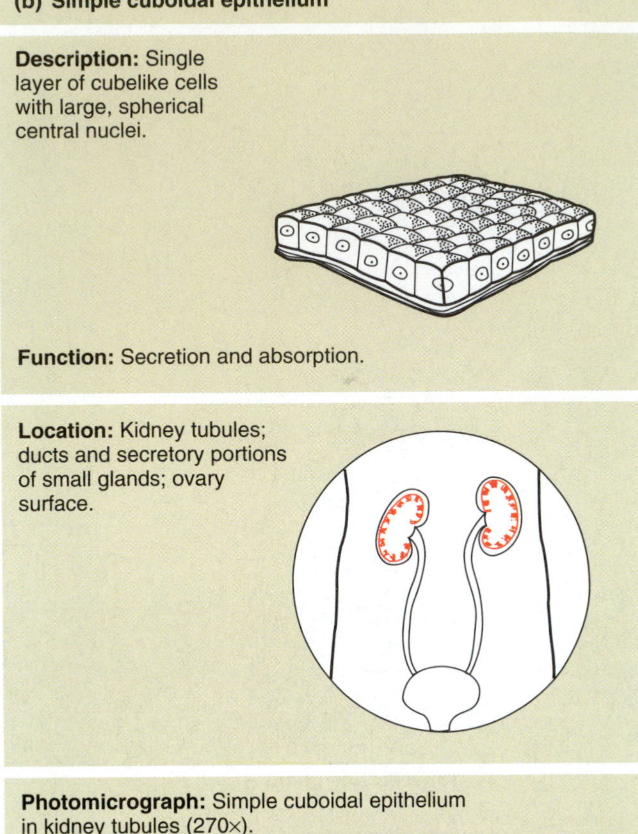

(b) Simple cuboidal epithelium
Description: Single layer of cubelike cells with large, spherical central nuclei.
Function: Secretion and absorption.
Location: Kidney tubules; ducts and secretory portions of small glands; ovary surface.

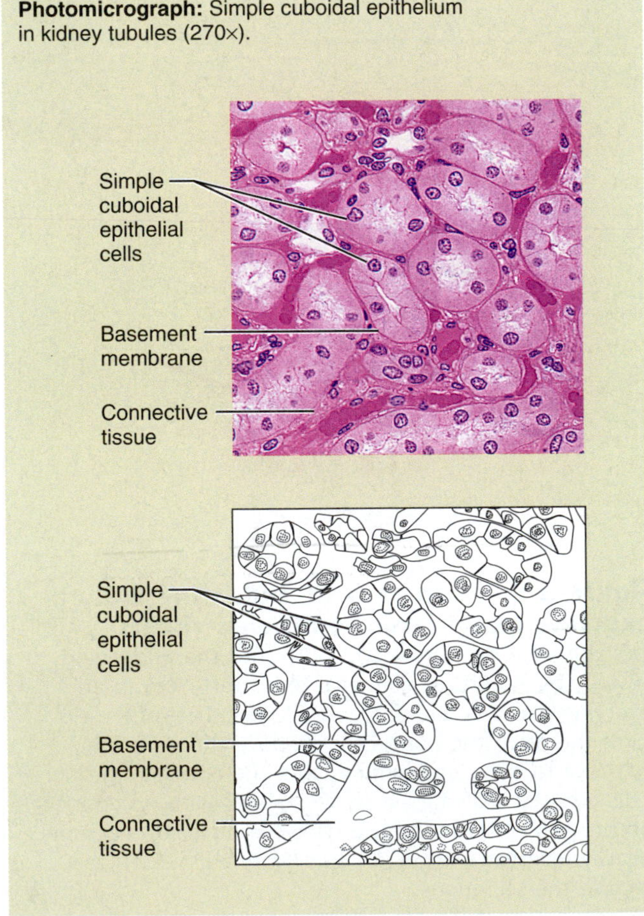

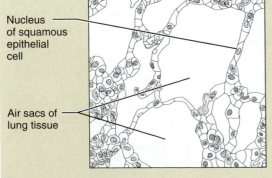

Figure 6.3 Epithelial tissues. Simple epithelia (**a** and **b**).

(c) Simple columnar epithelium

Description: Single layer of tall cells with *oval* nuclei; some cells bear cilia; layer may contain mucus-secreting unicellular glands (goblet cells).

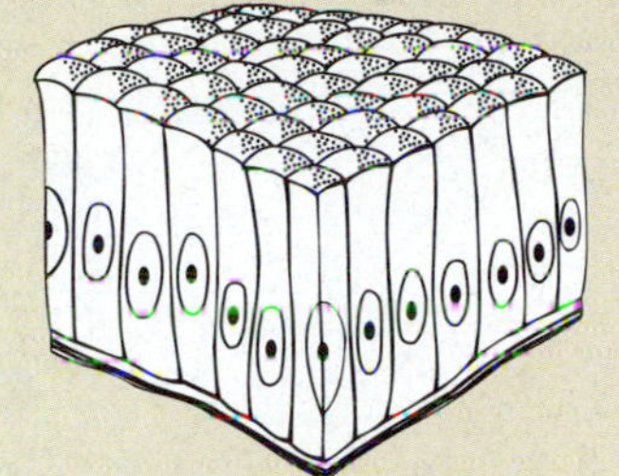

Function: Absorption; secretion of mucus, enzymes, and other substances; ciliated type propels mucus (or reproductive cells) by ciliary action.

Location: Nonciliated type lines most of the digestive tract (stomach to anal canal), gallbladder and excretory ducts of some glands; ciliated variety lines small bronchi, uterine tubes, and the uterus.

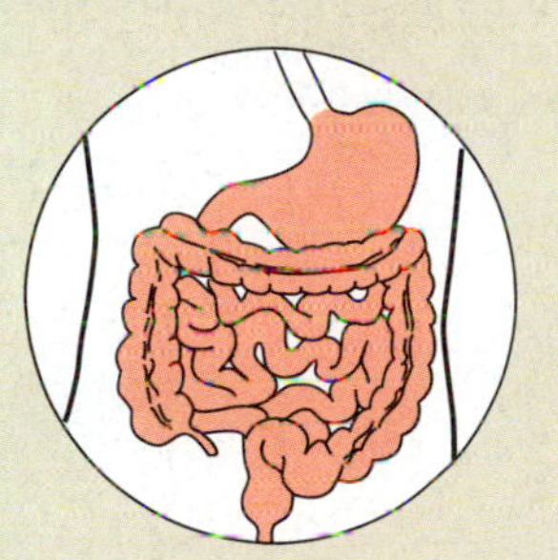

Photomicrograph: Simple columnar epithelium of the stomach mucosa (500×).

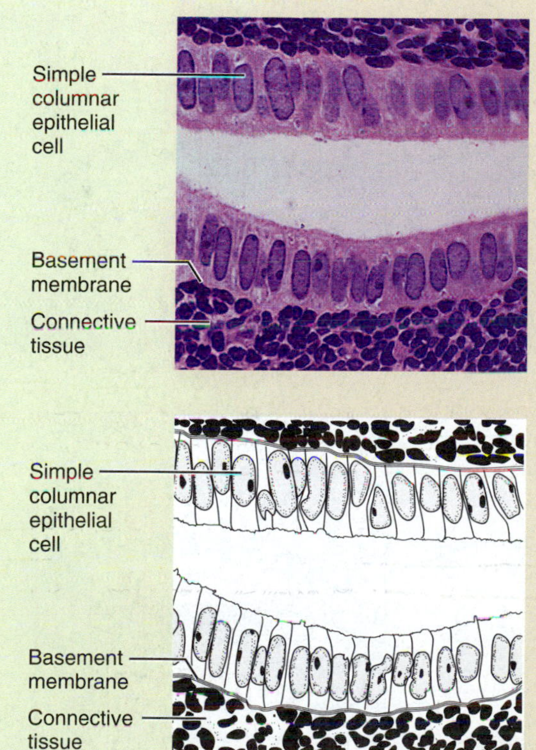

(d) Pseudostratified columnar epithelium

Description: Single layer of cells of differing heights, some not reaching the free surface; nuclei seen at different levels; may contain goblet cells and bear cilia.

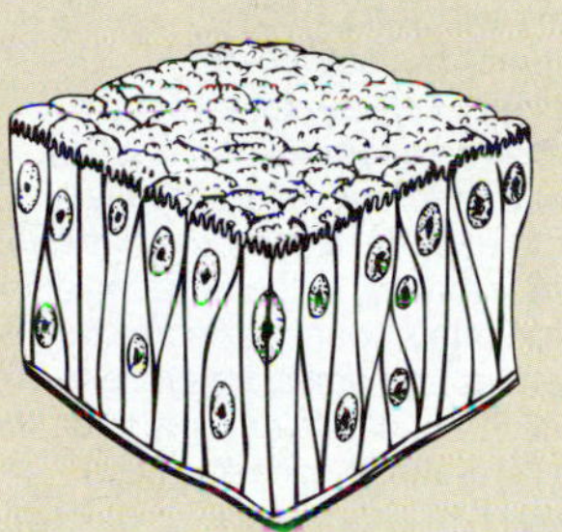

Function: Secretion, particularly of mucus; propulsion of mucus by ciliary action.

Location: Nonciliated type in male's sperm–carrying ducts and ducts of large glands; ciliated variety lines the trachea, most of the upper respiratory tract.

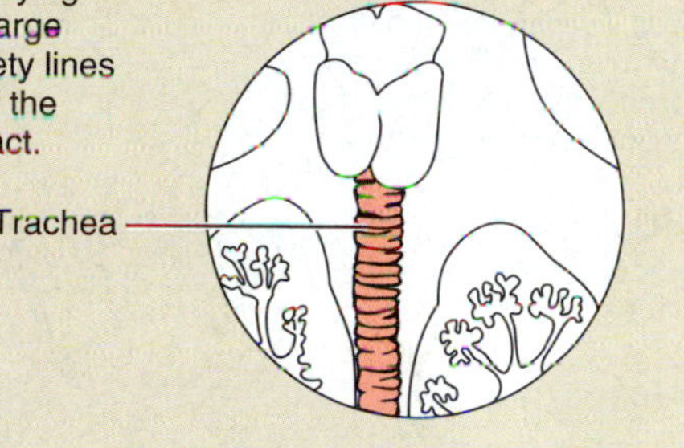

Photomicrograph: Pseudostratified ciliated columnar epithelium lining the human trachea (800×).

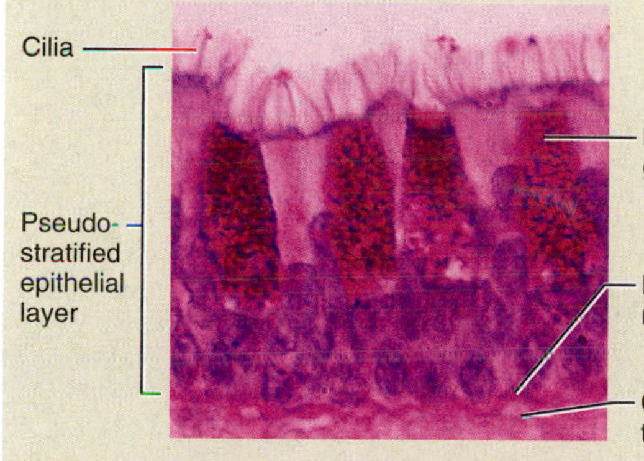

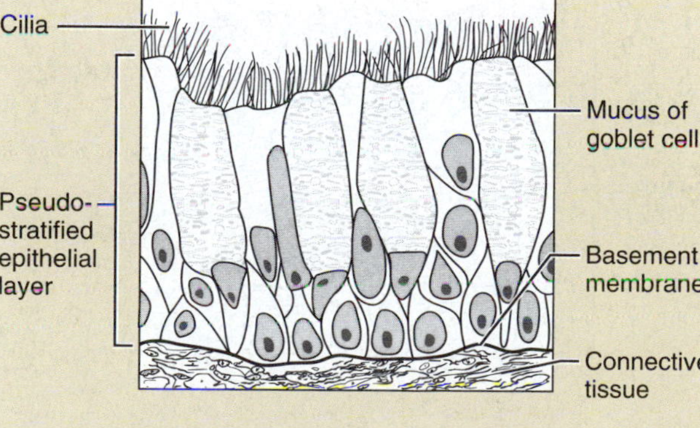

Figure 6.3 *(continued)* Simple epithelia **(c** and **d).**

(e) Stratified squamous epithelium

Description: Thick membrane composed of several cell layers; basal cells are cuboidal or columnar and metabolically active; surface cells are flattened (squamous); in the keratinized type, the surface cells are full of keratin and dead; basal cells are active in mitosis and produce the cells of the more superficial layers.

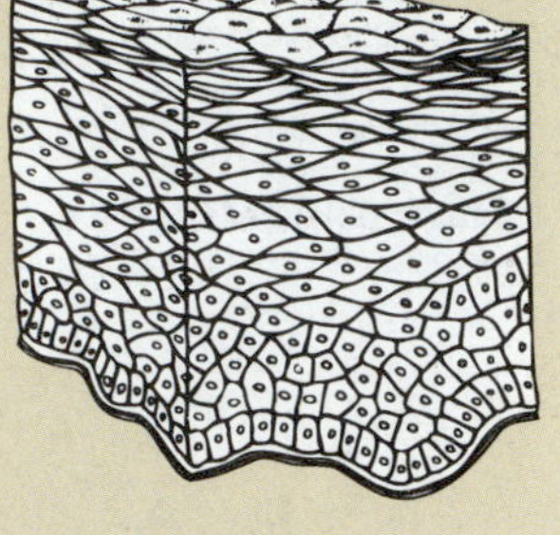

Function: Protects underlying tissues in areas subjected to abrasion.

Location: Nonkeratinized type forms the moist linings of the esophagus, mouth, and vagina; keratinized variety forms the epidermis of the skin, a dry membrane.

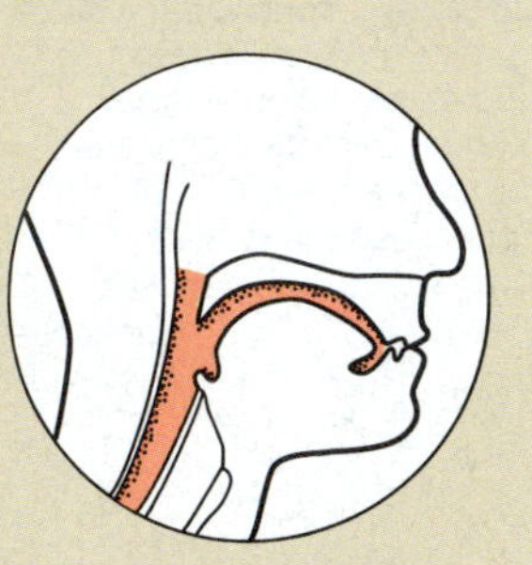

Photomicrograph: Stratified squamous epithelium lining of the esophagus (158×).

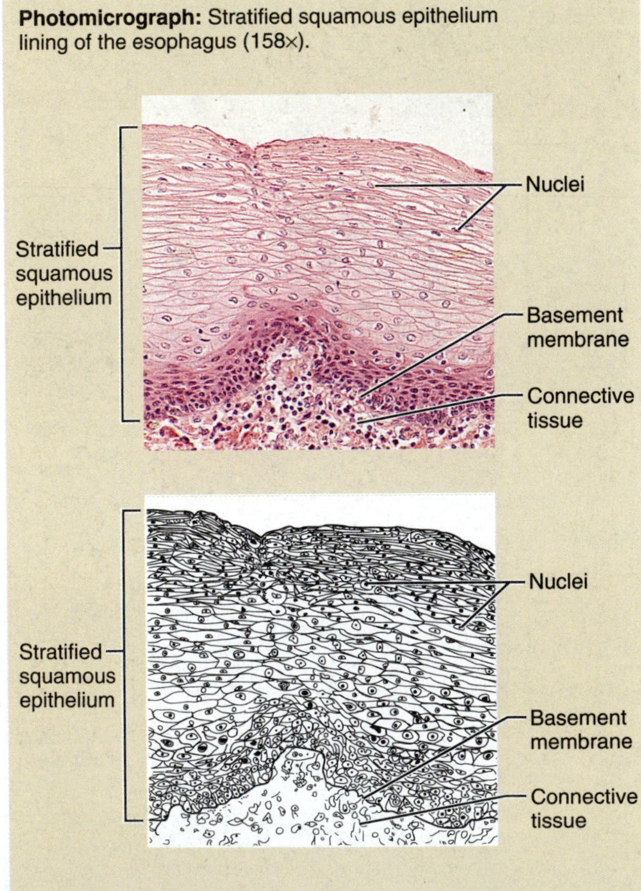

(f) Stratified cuboidal epithelium

Description: Generally two layers of cube-like cells.

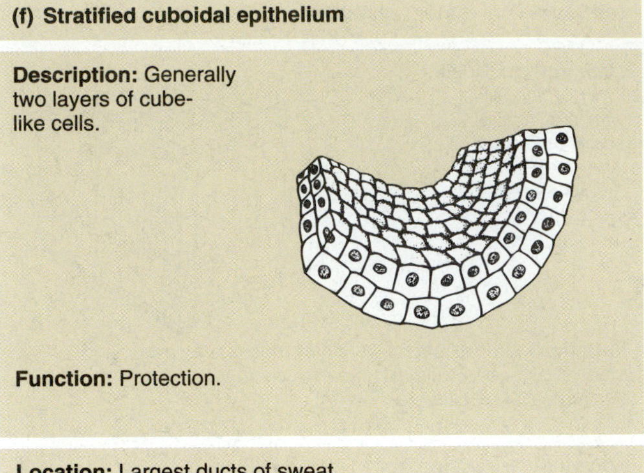

Function: Protection.

Location: Largest ducts of sweat glands, mammary glands, and salivary glands.

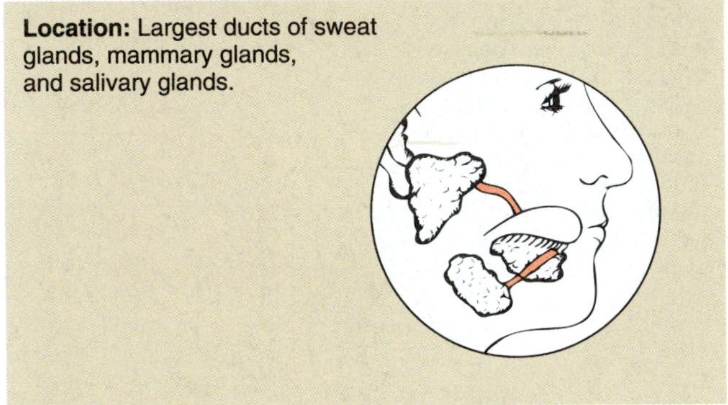

Photomicrograph: Stratified cuboidal epithelium forming a salivary gland duct (150×).

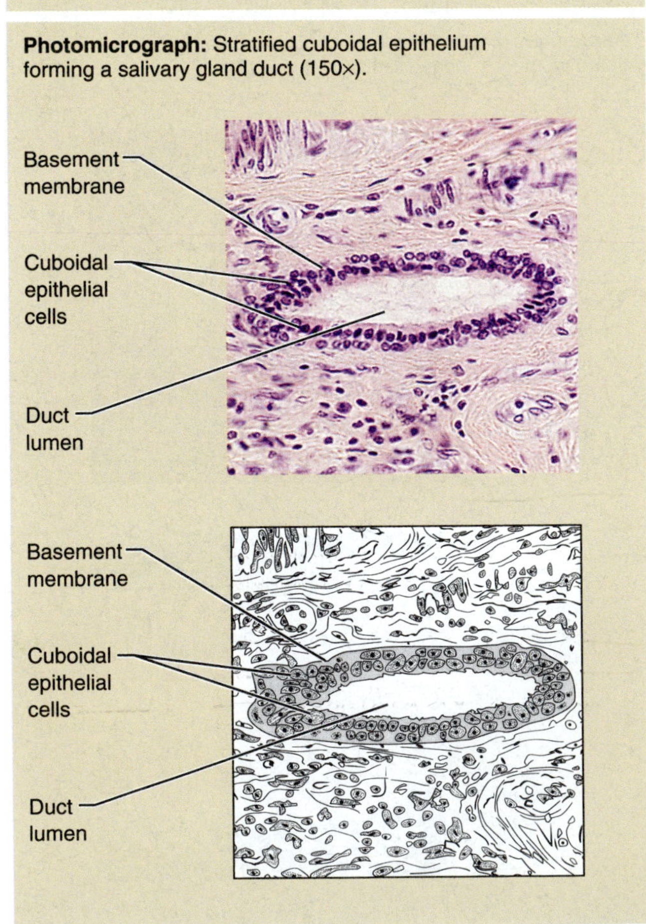

Figure 6.3 *(continued)* Epithelial tissues. Stratified epithelia **(e** and **f)**.

(g) Stratified columnar epithelium

Description: Several cell layers; basal cells usually cuboidal; superficial cells elongated and columnar.

Function: Protection; secretion.

Location: Rare in the body; small amounts in male urethra and in large ducts of some glands.

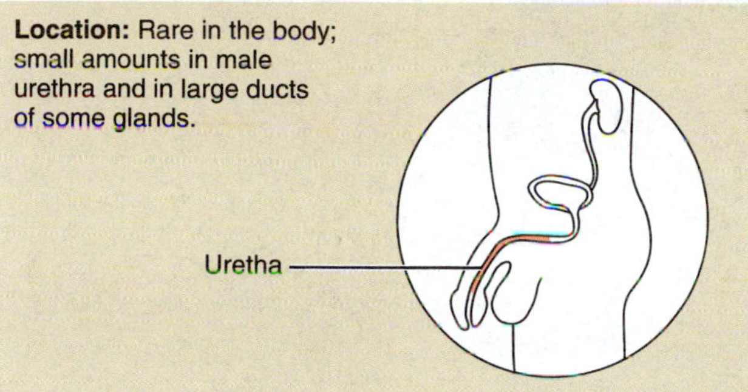

Photomicrograph: Stratified columnar epithelium lining of the male urethra (700×).

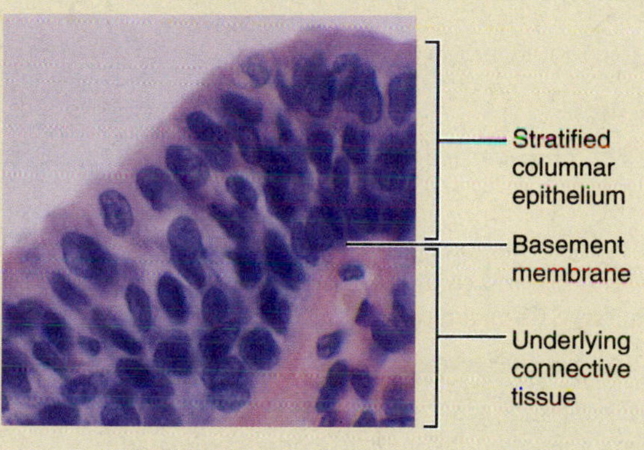

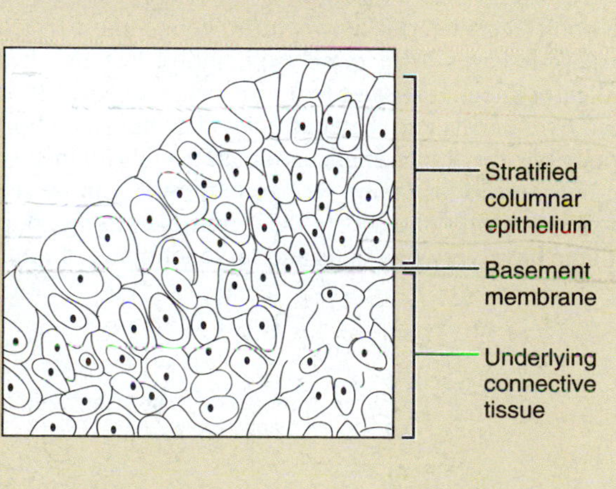

(h) Transitional epithelium

Description: Resembles both stratified squamous and stratified cuboidal; basal cells cuboidal or columnar; surface cells dome-shaped or squamous-like, depending on degree of organ stretch.

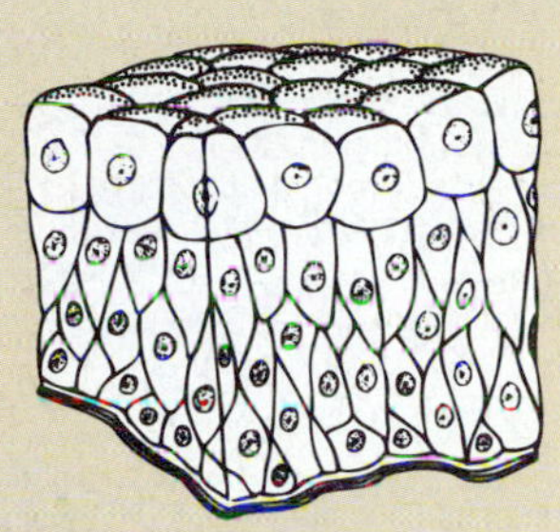

Function: Stretches readily and permits distension of urinary organ by contained urine.

Location: Lines the ureters, bladder, and part of the urethra.

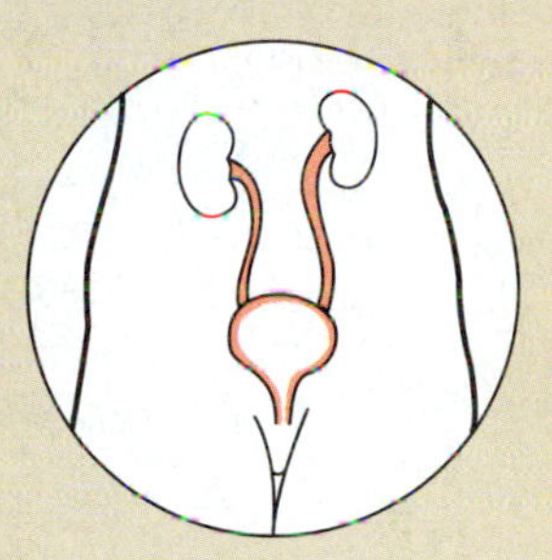

Photomicrograph: Transitional epithelium lining of the bladder, relaxed state (277×); note the bulbous, or rounded, appearance of the cells at the surface; these cells flatten and become elongated when the bladder is filled with urine.

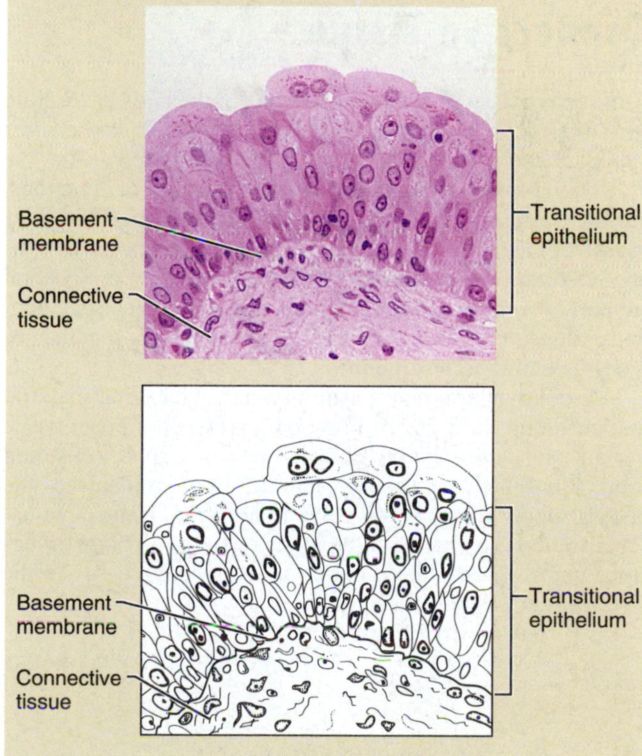

Figure 6.3 (*continued*) Stratified epithelia **(g** and **h)**.

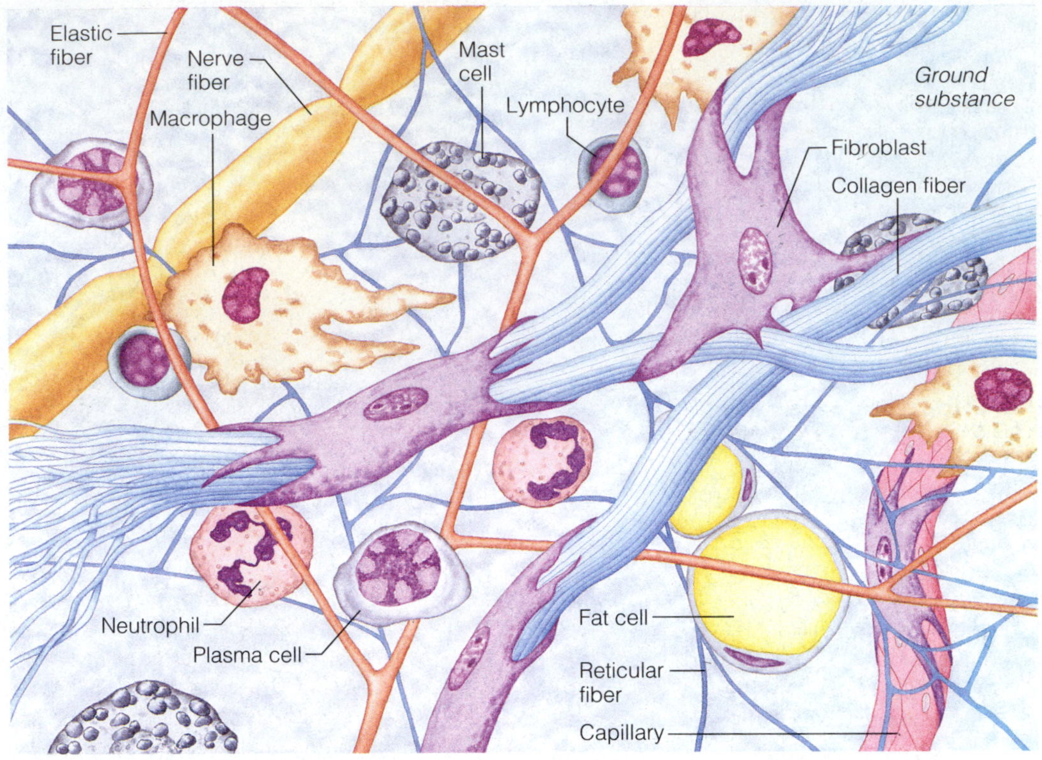

Figure 6.4 Areolar connective tissue: A prototype (model) connective tissue. Note the various cell types and the three classes of fibers (collagen, reticular, elastic) embedded in the ground substance.

Connective Tissue

Connective tissue is found in all parts of the body as discrete structures or as part of various body organs. It is the most abundant and widely distributed of the tissue types.

The **connective tissues** perform a variety of functions, but they primarily protect, support, and bind together other tissues of the body. For example, bones are composed of connective tissue **(bone** or **osseous tissue),** and they protect and support other body tissues and organs. The ligaments and tendons **(dense connective tissue)** bind the bones together or bind skeletal muscles to bones.

Areolar connective tissue is a soft packaging material that cushions and protects body organs. Fat **(adipose)** tissue provides insulation for the body tissues and a source of stored food. Blood-forming **(hematopoietic)** tissue replenishes the body's supply of red blood cells. In addition, connective tissue also serves a vital function in the repair of all body tissues since many wounds are repaired by connective tissue in the form of scar tissue.

The characteristics of connective tissue include the following:

- With a few exceptions (cartilages, which are avascular, and tendons and ligaments, which are poorly vascularized), connective tissues have a rich supply of blood vessels.
- Connective tissues are composed of many types of cells.
- There is a great deal of noncellular, nonliving material (matrix) between the cells of connective tissue.

The nonliving material between the cells—the extracellular matrix—deserves a bit more explanation because it distinguishes connective tissue from all other tissues. It is produced by the cells and then extruded. The matrix is primarily responsible for the strength associated with connective tissue, but there is variation. At one extreme, adipose tissue is composed mostly of cells. At the opposite extreme, bone and cartilage have very few cells and large amounts of matrix.

The matrix has two components—ground substance and fibers. The **ground substance** is composed chiefly of interstitial fluid, cell adhesion proteins, and proteoglycans. Depending on its specific composition, the ground substance may be liquid, semisolid, gel-like, or very hard. When the matrix is firm, as in cartilage and bone, the connective tissue cells reside in cavities in the matrix called *lacunae*. The fibers, which provide support, include **collagen** (white) **fibers, elastic** (yellow) **fibers,** and **reticular** (fine collagen) **fibers.** Of these, the collagen fibers are most abundant.

Generally speaking, the ground substance functions as a molecular sieve, or medium, through which nutrients and other dissolved substances can diffuse between the blood capillaries and the cells. The fibers in the matrix hinder diffusion somewhat and make the ground substance less pliable. The properties of the connective tissue cells and the makeup and arrangement of their matrix elements vary tremendously, accounting for the amazing diversity of this tissue type. Nonetheless, the connective tissues have a common structural plan seen best in *areolar connective tissue* (Figure 6.4), a soft packing tissue that occurs throughout the body. Since all other connective tissues are variations of areolar, it is considered the model or prototype of the connective tissues. Notice in Figure 6.4 that areolar tissue has all three varieties of fibers, but they are sparsely arranged in its transparent gel-like ground substance. The cell type that secretes its matrix is the *fibroblast,* but a wide variety of other cells including phagocytic cells like macrophages and certain white blood cells and mast cells that act in the inflammatory response are present as well. The more durable connective tissues, such as bone, cartilage, and the dense fibrous varieties, characteristically have a firm ground substance and many more fibers.

There are four main types of adult connective tissue, all of which typically have large amounts of matrix. These are **connective tissue proper** (which includes areolar, adipose, reticular, and dense [fibrous] connective tissues), **cartilage, bone,** and **blood.** All of these derive from an embryonic tissue called *mesenchyme*. Figure 6.5 lists the general characteristics, location, and function of some of the connective tissues found in the body.

Activity 2:
Examining Connective Tissue Under the Microscope

Obtain prepared slides of mesenchyme; of adipose, areolar, reticular, dense regular and irregular connective tissue; of hyaline and elastic cartilage and fibrocartilage; of osseous connective tissue (bone); and of blood. Compare your observations with the views illustrated in Figure 6.5.

Distinguish between the living cells and the matrix and pay particular attention to the denseness and arrangement of the matrix. For example, notice how the matrix of the dense fibrous connective tissues, making up tendons and the dermis of the skin, is packed with collagen fibers, and that in the *regular* variety (tendon), the fibers are all running in the same direction, whereas in the dermis (a dense *irregular* connective tissue) they appear to be running in many directions.

While examining the areolar connective tissue, a soft "packing tissue," notice how much empty space there appears to be, and distinguish between the collagen fibers and the coiled elastic fibers. Also, try to locate a **mast cell,** which has large, darkly staining granules in its cytoplasm (*mast* = stuffed full of granules). This cell type releases histamine that makes capillaries quite permeable during inflammatory reactions and allergies and thus is partially responsible for that "runny nose" of allergies.

In adipose tissue, locate a signet ring cell in which the nucleus can be seen pushed to one side by the large, fat-filled vacuole that appears to be a large empty space. Also notice how little matrix there is in fat or adipose tissue. Distinguish between the living cells and the matrix in the dense fibrous, bone, and hyaline cartilage preparations.

Scan the blood slide at low and then high power to examine the general shape of the red blood cells. Then, switch to the oil immersion lens for a closer look at the various types of white blood cells. How does blood differ from all other connective tissues?

____

____ ■

Text continues on page 63

Embryonic connective tissue

(a) Mesenchyme

Description: Embryonic connective tissue; gel-like ground substance containing fibers; star-shaped mesenchymal cells.

Function: Gives rise to all other connective tissue types.

Location: Primarily in embryo.

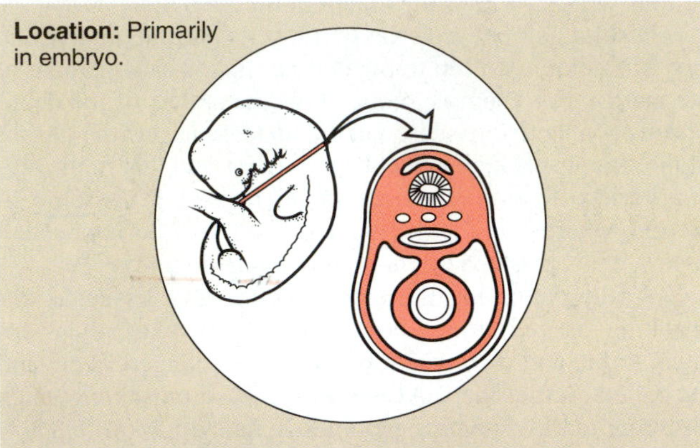

Photomicrograph: Mesenchymal tissue, an embryonic connective tissue (523×); the clear-appearing background is the fluid ground substance of the matrix; notice the fine, sparse fibers.

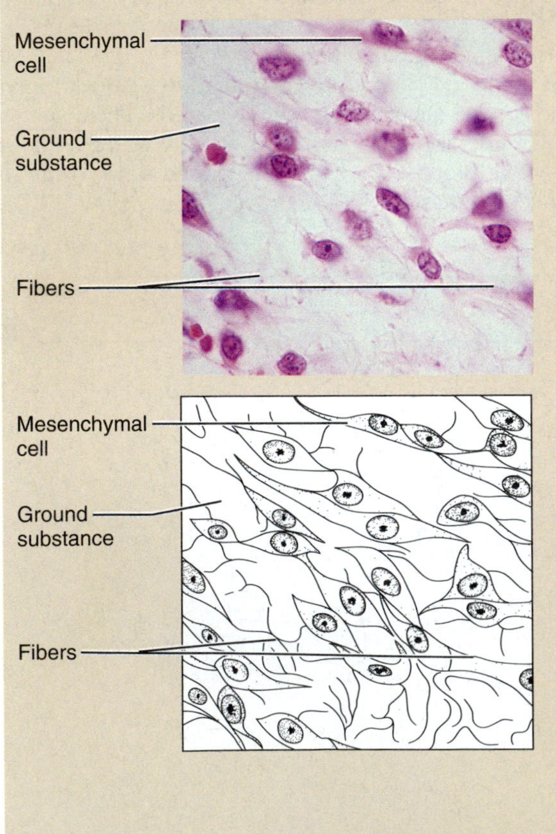

Connective tissue proper: Loose connective tissue (b to d)

(b) Areolar connective tissue

Description: Gel-like matrix with all three fiber types; cells: fibroblasts, macrophages, mast cells, and some white blood cells.

Function: Wraps and cushions organs; its macrophages phagocytize bacteria; plays important role in inflammation; holds and conveys tissue fluid.

Location: Widely distributed under epithelia of body, e.g, forms lamina propria of mucous membranes; packages organs; surrounds capillaries.

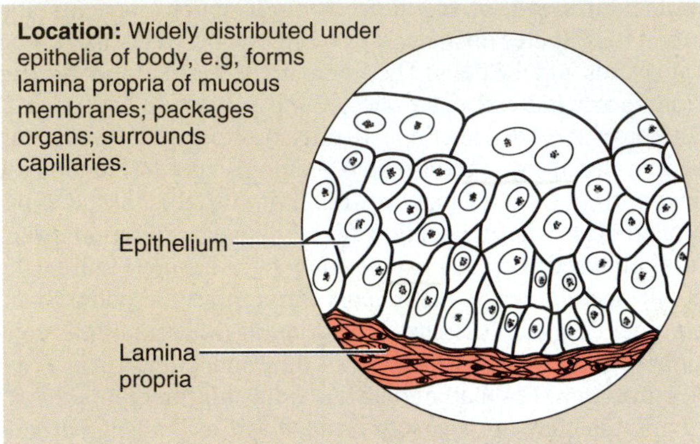

Photomicrograph: Areolar connective tissue, a soft packaging tissue of the body (200×).

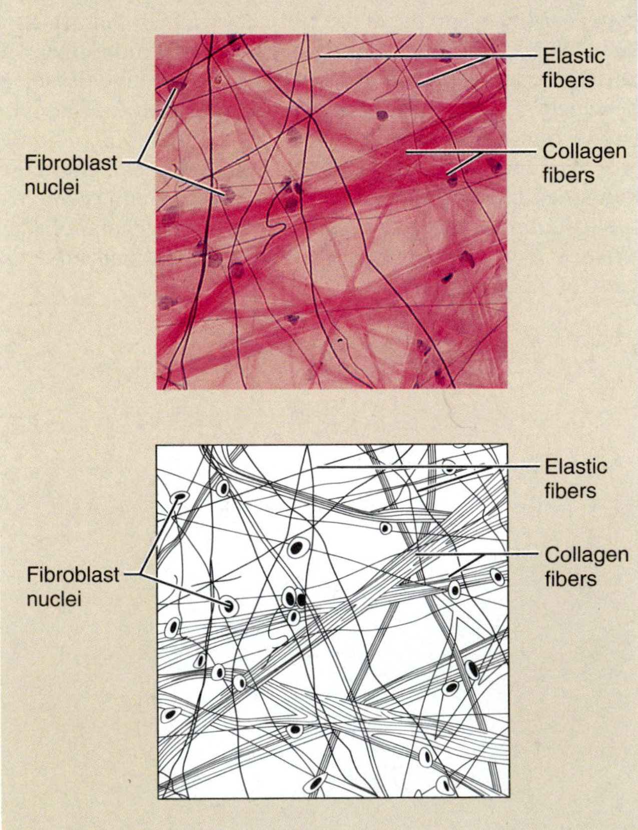

Figure 6.5 Connective tissues. Embryonic connective tissue **(a)** and connective tissue proper **(b)**.

Loose connective tissue (*continued*)

(c) Adipose tissue

Description: Matrix as in areolar, but very sparse; closely packed adipocytes, or fat cells, have nucleus pushed to the side by large fat droplet.

Function: Provides reserve food fuel; insulates against heat loss; supports and protects organs.

Location: Under skin; around kidneys and eyeballs; within abdomen; in breasts.

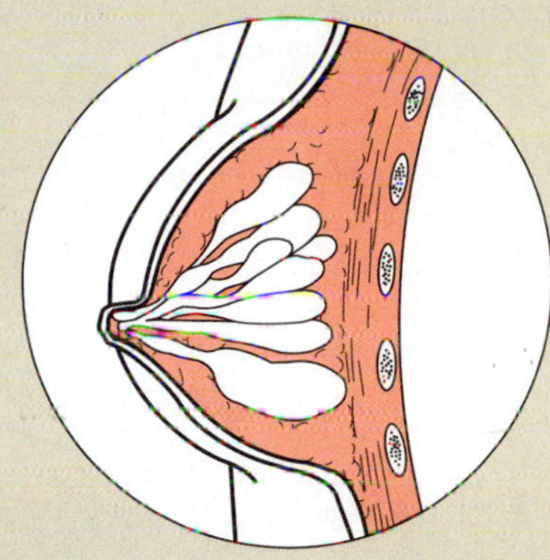

Photomicrograph: Adipose tissue from the subcutaneous layer under the skin (170×).

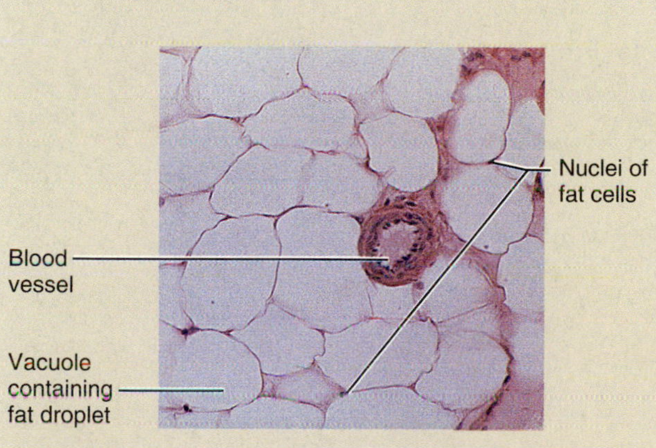

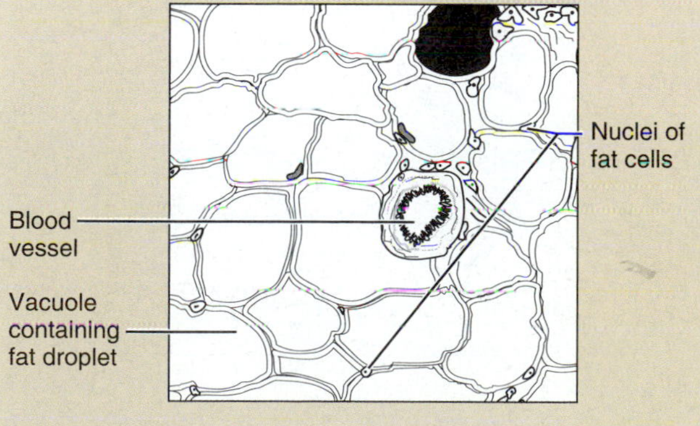

Loose connective tissue (*continued*)

(d) Reticular connective tissue

Description: Network of reticular fibers in a typical loose ground substance; reticular cells lie on the network.

Function: Fibers form a soft internal skeleton (stroma) that supports other cell types.

Location: Lymphoid organs (lymph nodes, bone marrow, and spleen).

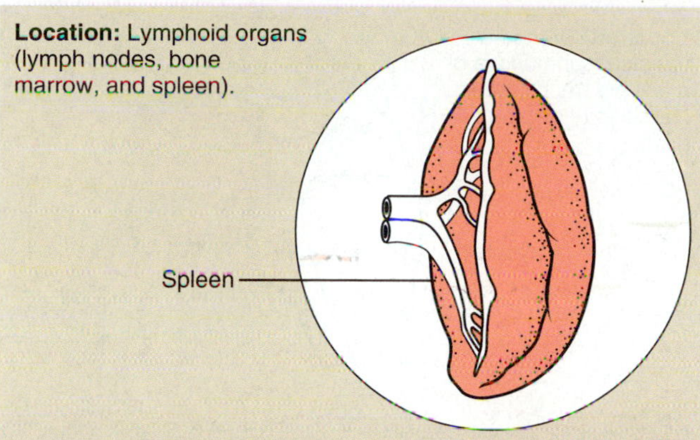

Photomicrograph: Dark-staining network of reticular connective tissue fibers forming the internal skeleton of the spleen (650×).

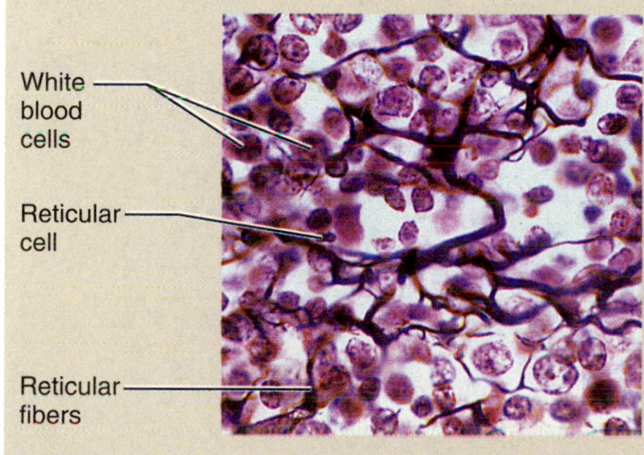

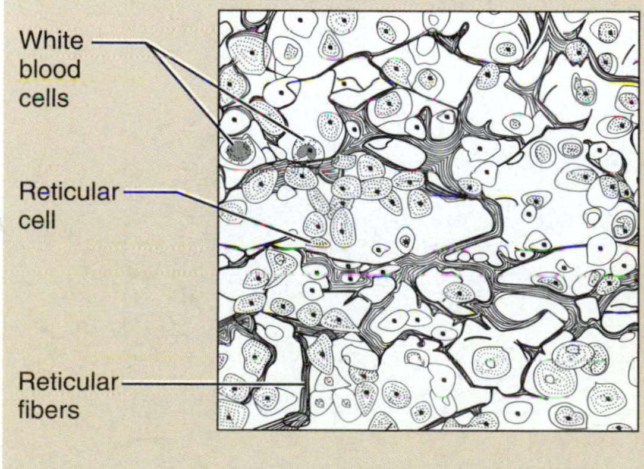

Figure 6.5 (*continued*) Connective tissue proper **(c** and **d).**

Connective tissue proper: Dense connective tissue (e and f)

(e) Dense irregular connective tissue

Description: Primarily irregularly arranged collagen fibers; some elastic fibers; major cell type is the fibroblast.

Function: Able to withstand tension exerted in many directions; provides structural strength.

Location: Dermis of the skin; submucosa of digestive tract; fibrous capsules of organs and of joints.

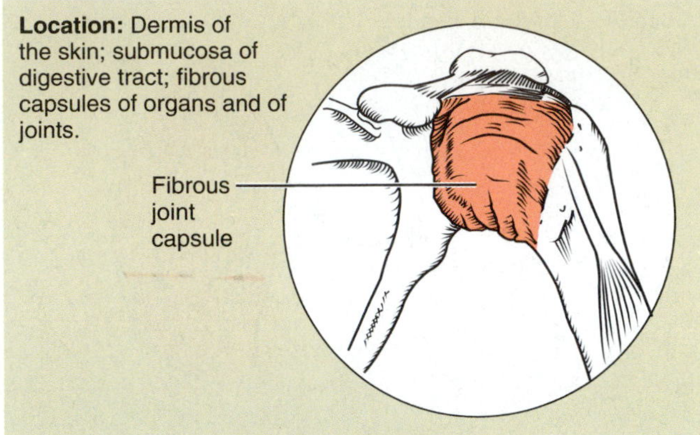

Photomicrograph: Dense irregular connective tissue from the dermis of the skin (550×).

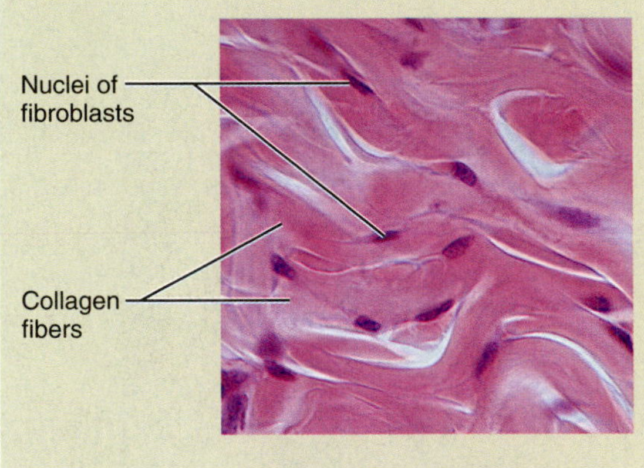

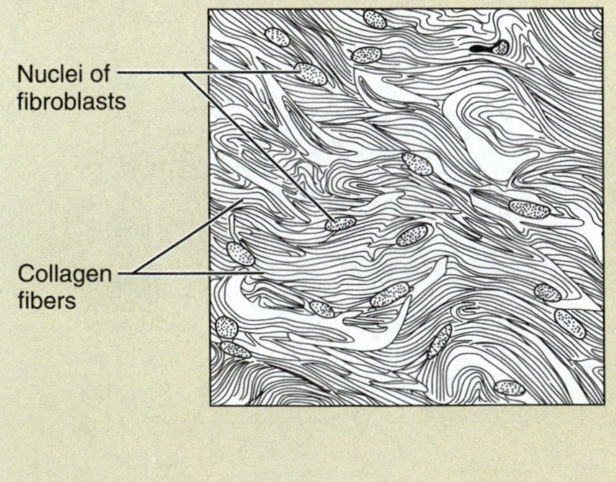

Dense connective tissue *(continued)*

(f) Dense regular connective tissue

Description: Primarily parallel collagen fibers; a few elastin fibers; major cell type is the fibroblast.

Function: Attaches muscles to bones or to muscles; attaches bones to bones; withstands great tensile stress when pulling force is applied in one direction.

Location: Tendons, most ligaments, aponeuroses.

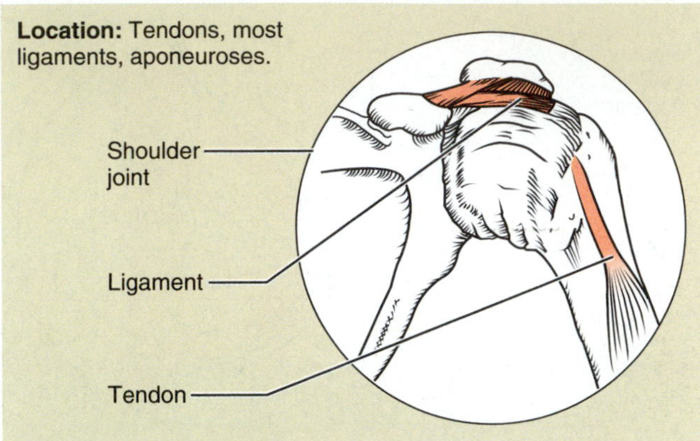

Photomicrograph: Dense regular connective tissue from a tendon (200×).

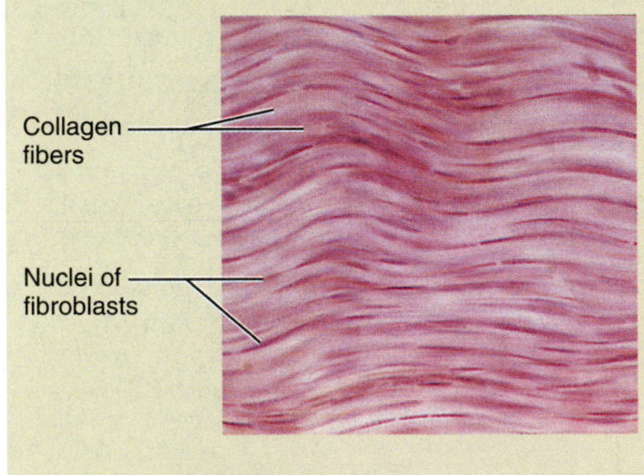

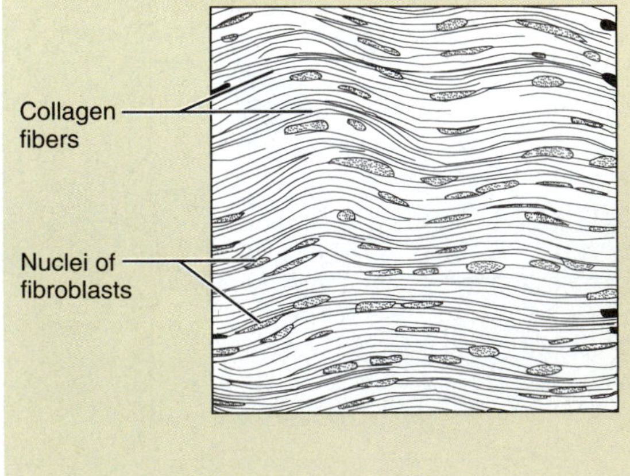

Figure 6.5 *(continued)* **Connective tissues.** Connective tissue proper **(e** and **f).**

Cartilage: (g to i)

(g) Hyaline cartilage

Description: Amorphous but firm matrix; collagen fibers form an imperceptible network; chondroblasts produce the matrix and when mature (chondrocytes) lie in lacunae.

Function: Supports and reinforces; has resilient cushioning properties; resists compressive stress.

Location: Forms most of the embryonic skeleton; covers the ends of long bones in joint cavities; forms costal cartilages of the ribs; cartilages of the nose, trachea, and larynx.

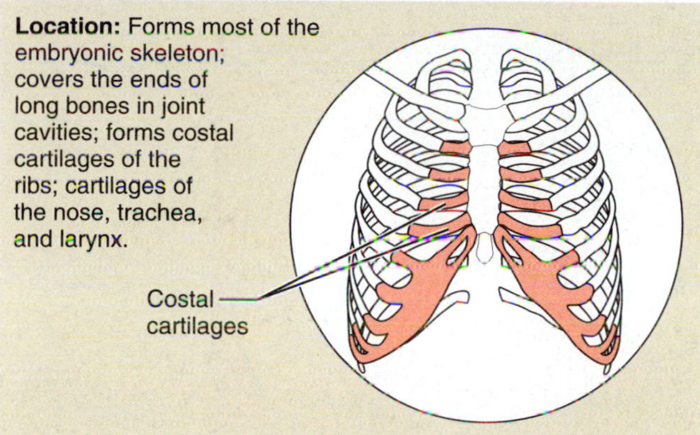

Photomicrograph: Hyaline cartilage from the trachea (375×).

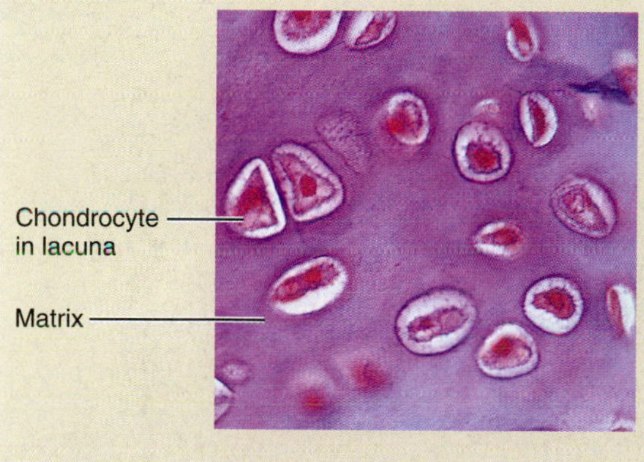

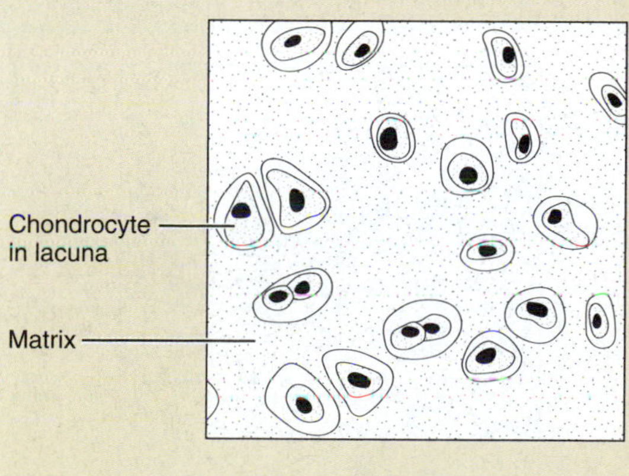

Cartilage *(continued)*

(h) Elastic cartilage

Description: Similar to hyaline cartilage, but more elastic fibers in matrix.

Function: Maintains the shape of a structure while allowing great flexibility.

Location: Supports the external ear (pinna); epiglottis.

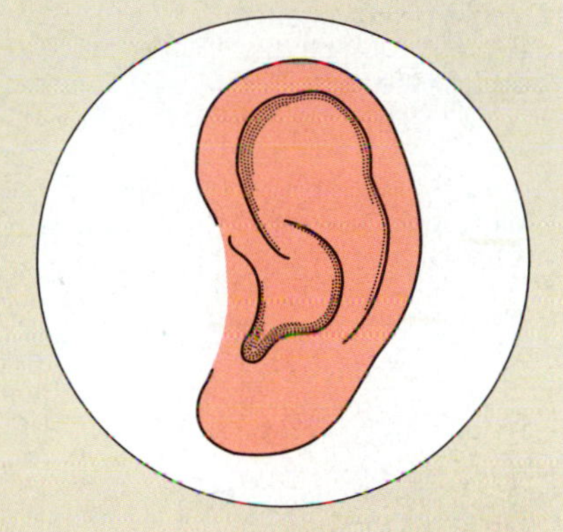

Photomicrograph: Elastic cartilage from the human ear pinna; forms the flexible skeleton of the ear (158×).

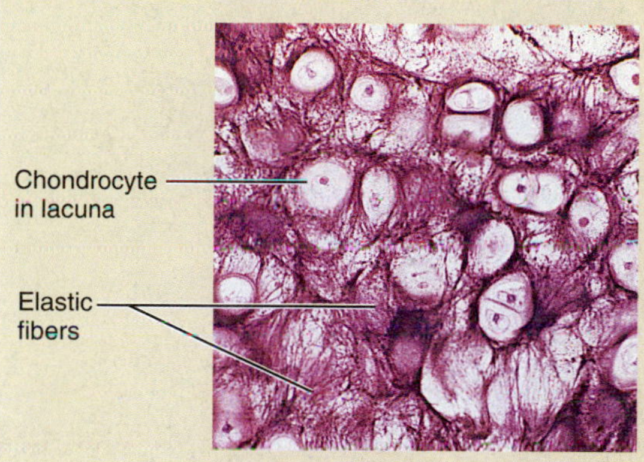

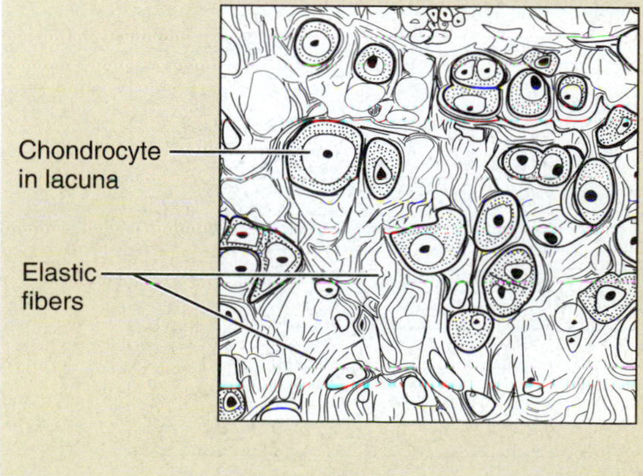

Figure 6.5 *(continued)* Cartilage **(g** and **h)**.

Cartilage (*continued*)

(i) Fibrocartilage

Description: Matrix similar but less firm than in hyaline cartilage; thick collagen fibers predominate.

Function: Tensile strength with the ability to absorb compressive shock.

Location: Intervertebral discs; pubic symphysis; discs of knee joint.

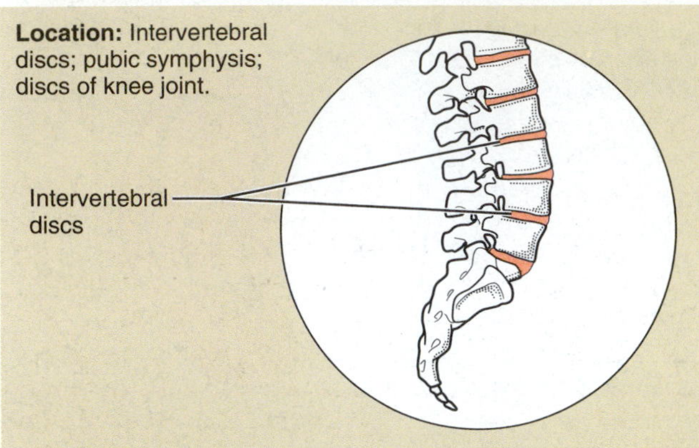

Photomicrograph: Fibrocartilage of an intervertebral disc (200×).

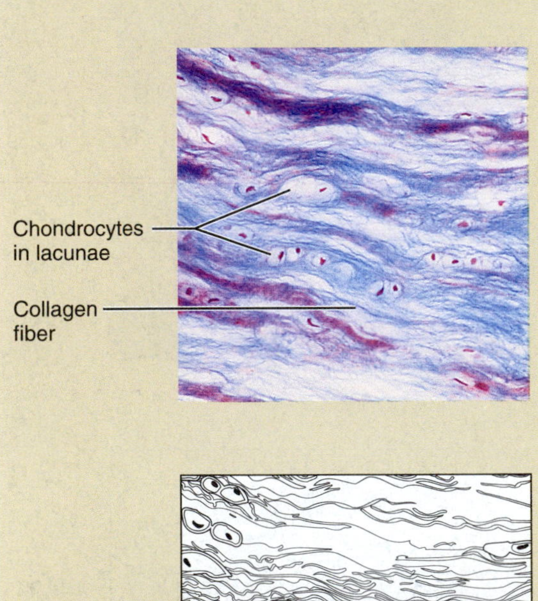

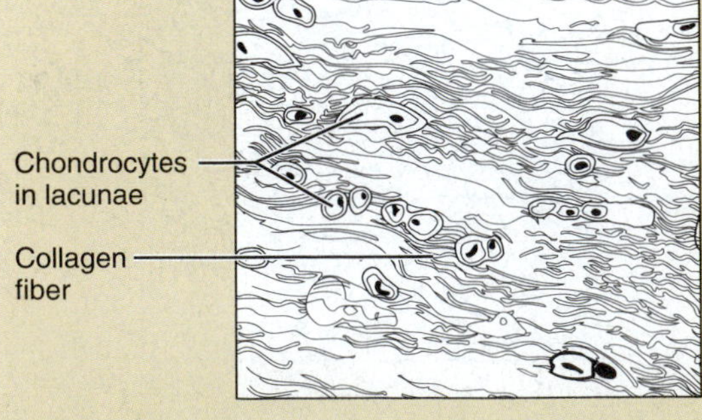

Others: (j and k)

(j) Bone (osseous tissue)

Description: Hard, calcified matrix containing many collagen fibers; osteocytes lie in lacunae. Very well vascularized.

Function: Bone supports and protects (by enclosing); provides levers for the muscles to act on; stores calcium and other minerals and fat; marrow inside bones is the site for blood cell formation (hematopoiesis).

Location: Bones

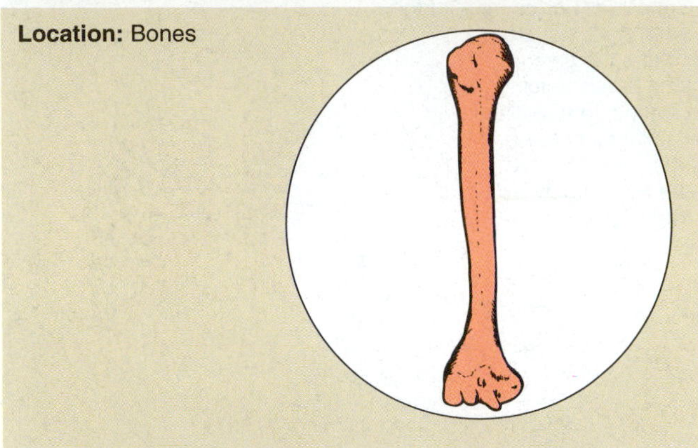

Photomicrograph: Cross-sectional view of bone (110×).

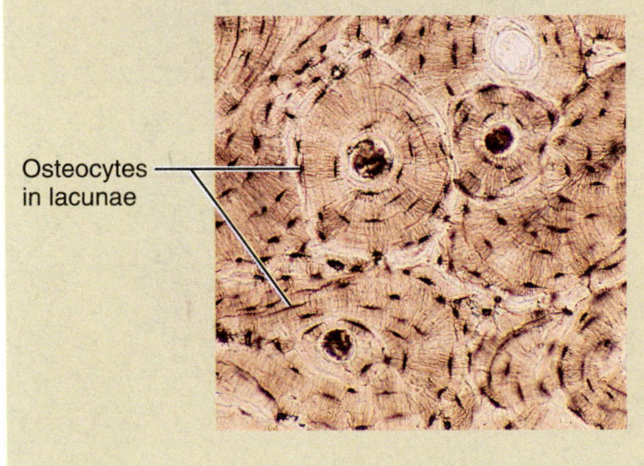

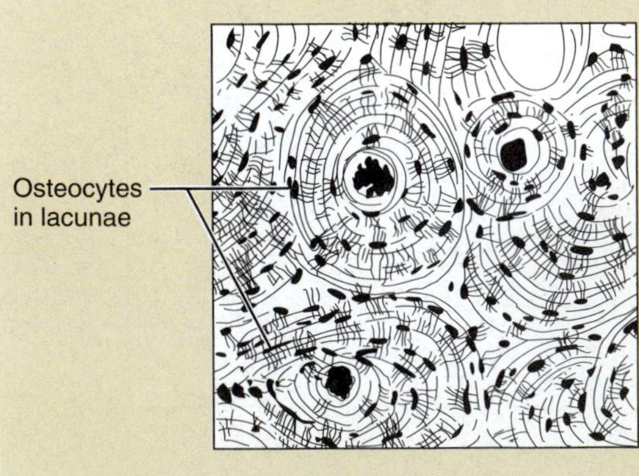

Figure 6.5 (*continued*) Connective tissues. Cartilage **(i)** and bone **(j)**.

Others *(continued)*

(k) Blood

Description: Red and white blood cells in a fluid matrix (plasma).

Function: Transport of respiratory gases, nutrients, wastes and other substances.

Location: Contained within blood vessels.

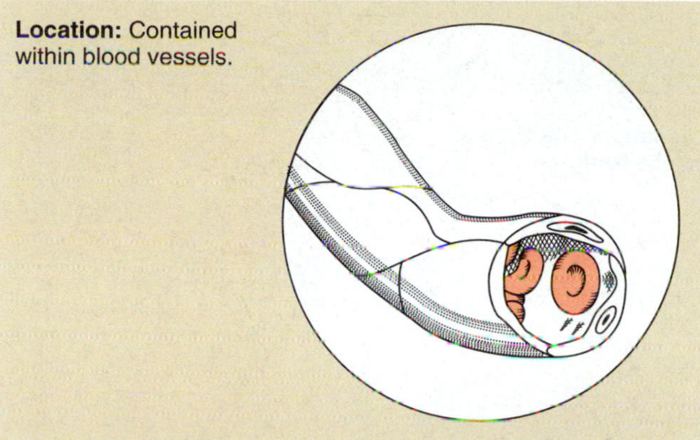

Photomicrograph: Smear of human blood (1100×); two white blood cells (neutrophil in upper left and lymphocyte in lower right) are seen surrounded by red blood cells.

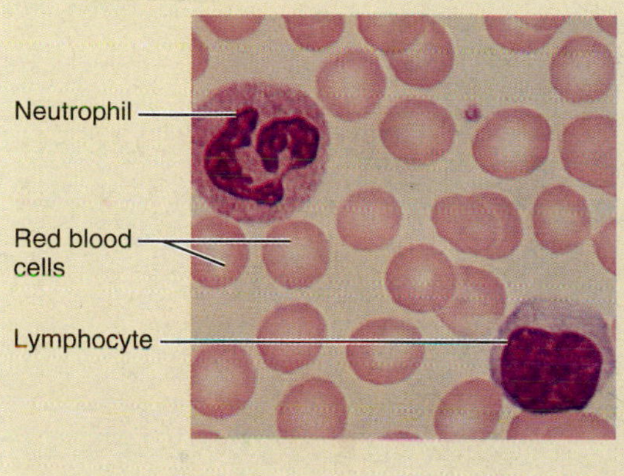

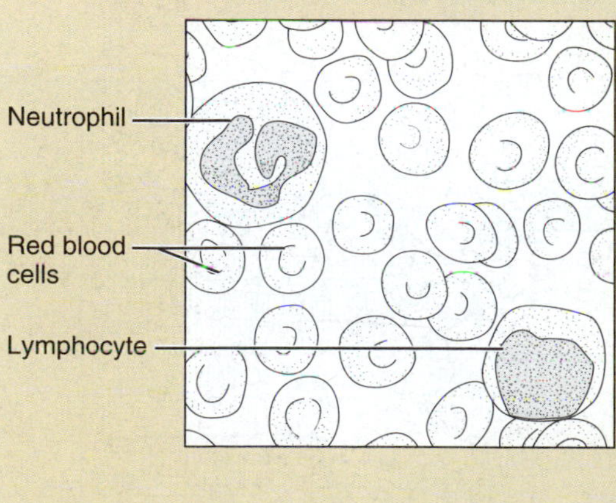

Figure 6.5 *(continued)* Blood **(k).**

Muscle Tissue

Muscle tissue is highly specialized to contract and produces most types of body movement. As you might expect, muscle cells tend to be quite elongated, providing a long axis for contraction. The three basic types of muscle tissue are described briefly here. Cardiac and skeletal muscles are treated more completely in later exercises.

Skeletal muscle, the "meat," or flesh, of the body, is attached to the skeleton. It is under voluntary control (consciously controlled), and its contraction moves the limbs and other external body parts. The cells of skeletal muscles are long, cylindrical, and multinucleate (several nuclei per cell); they have obvious *striations* (stripes).

Cardiac muscle is found only in the heart. As it contracts, the heart acts as a pump, propelling the blood into the blood vessels. Cardiac muscle, like skeletal muscle, has striations. But cardiac cells are branching uninucleate cells that interdigitate (fit together) at junctions called **intercalated discs.** These structural modifications allow the cardiac muscle to act as a unit. Cardiac muscle is under involuntary control, which means that we cannot voluntarily or consciously control the operation of the heart.

Text continues on page 65

(a) Skeletal muscle

Description: Long, cylindrical, multinucleate cells; obvious striations.

Function: Voluntary movement; locomotion; manipulation of the environment; facial expression; voluntary control.

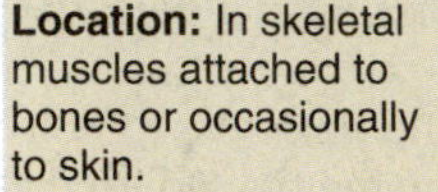

Location: In skeletal muscles attached to bones or occasionally to skin.

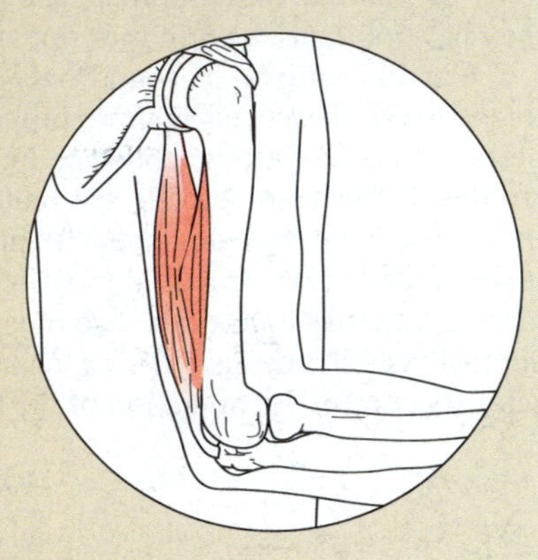

Photomicrograph: Skeletal muscle (approx. 130×). Notice the obvious banding pattern and the fact that these large cells are multinucleate.

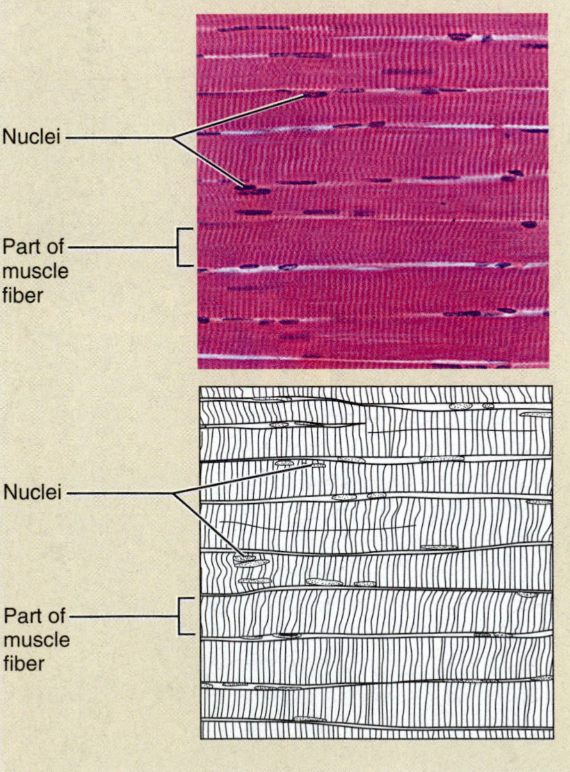

(b) Cardiac muscle

Description: Branching, striated, generally uninucleate cells that interdigitate at specialized junctions (intercalated discs).

Function: As it contracts, it propels blood into the circulation; involuntary control.

Location: The walls of the heart.

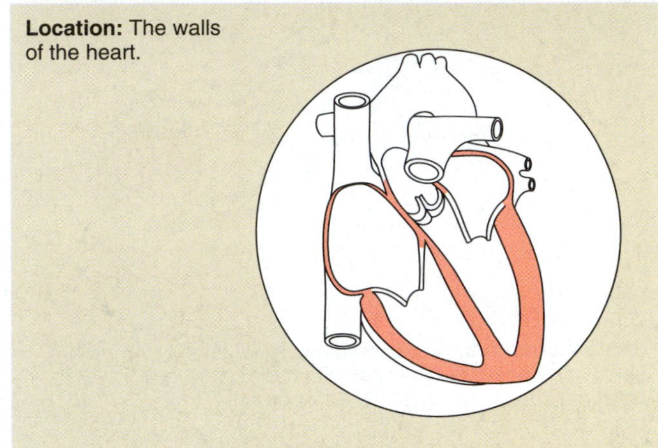

Photomicrograph: Cardiac muscle (275×); notice the striations, branching of cells, and the intercalated discs.

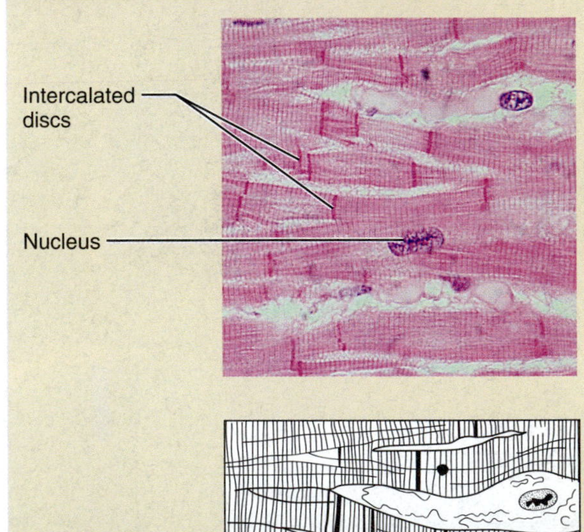

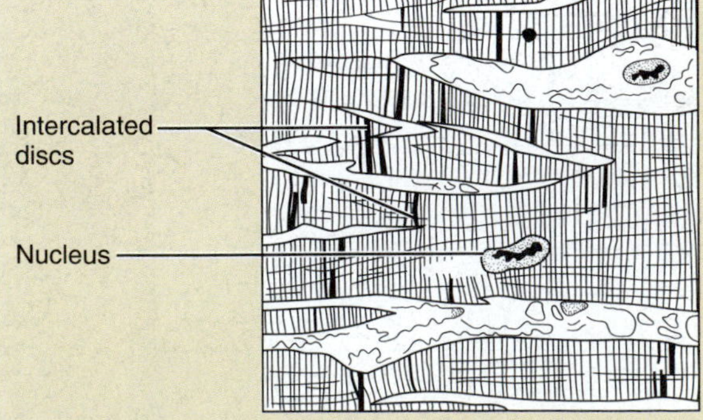

Figure 6.6 Muscle tissues (a and b).

(c) Smooth muscle

Description: Spindle-shaped cells with central nuclei; cells arranged closely to form sheets; no striations.

Function: Propels substances or objects (foodstuffs, urine, a baby) along internal passageways; involuntary control.

Location: Mostly in the walls of hollow organs.

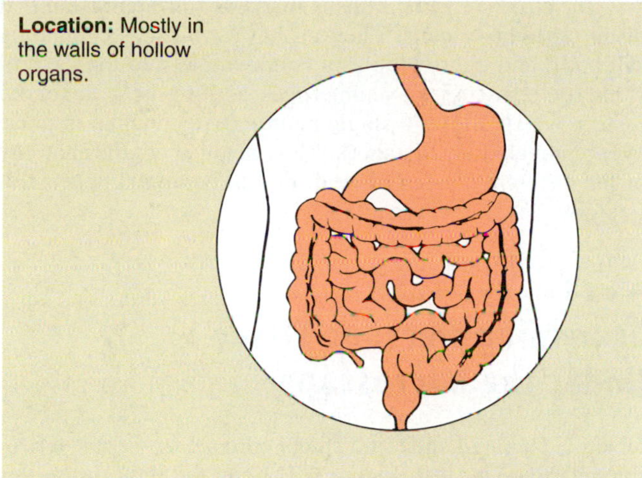

Photomicrograph: Sheet of smooth muscle (approx. 300×).

Smooth muscle cell

Nuclei

Smooth muscle cell

Nuclei

Figure 6.6 *(continued)* **Muscle tissues (c).**

Smooth muscle, or *visceral muscle,* is found mainly in the walls of hollow organs (digestive and urinary tract organs, uterus, blood vessels). Typically it has two layers that run at right angles to each other; consequently its contraction can constrict or dilate the lumen (cavity) of an organ and propel substances along predetermined pathways. Smooth muscle cells are quite different in appearance from those of skeletal or cardiac muscle. No striations are visible, and the uninucleate smooth muscle cells are spindle-shaped.

Activity 3: Examining Muscle Tissue Under the Microscope

Obtain and examine prepared slides of skeletal, cardiac, and smooth muscle. Notice their similarities and dissimilarities in your observations and in the illustrations in Figure 6.6. Teased smooth muscle shows individual cell shape clearly (see Histology Atlas, Plate 3). ■

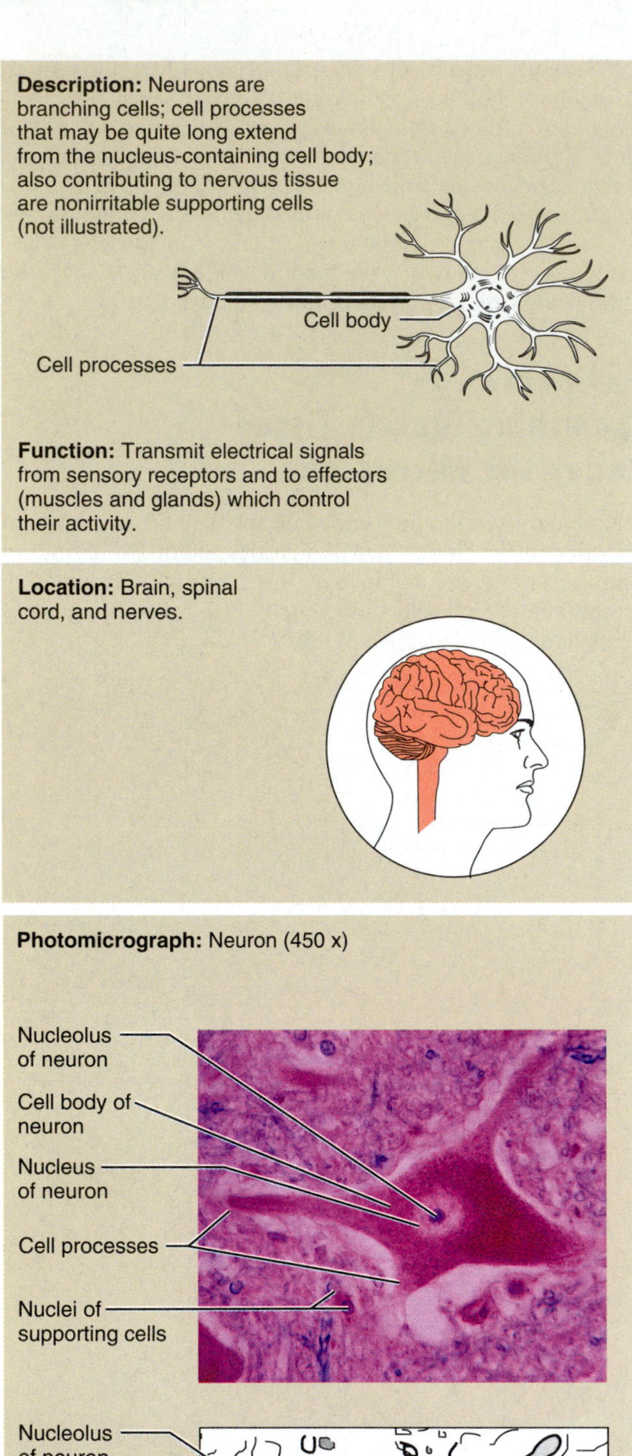

Description: Neurons are branching cells; cell processes that may be quite long extend from the nucleus-containing cell body; also contributing to nervous tissue are nonirritable supporting cells (not illustrated).
Cell body
Cell processes
Function: Transmit electrical signals from sensory receptors and to effectors (muscles and glands) which control their activity.
Location: Brain, spinal cord, and nerves.
Photomicrograph: Neuron (450 x)
Nucleolus of neuron
Cell body of neuron
Nucleus of neuron
Cell processes
Nuclei of supporting cells
Nucleolus of neuron
Cell body of neuron
Nucleus of neuron
Cell processes
Nuclei of supporting cells

Figure 6.7 Nervous tissue.

Nervous Tissue

Nervous tissue is composed of two major cell populations. The **neuroglia** are special supporting cells that protect, support, and insulate the more delicate neurons. The **neurons** are highly specialized to receive stimuli (irritability) and to conduct waves of excitation, or impulses, to all parts of the body (conductivity). They are the cells that are most often associated with nervous system functioning.

The structure of neurons is markedly different from that of all other body cells. They all have a nucleus-containing cell body, and their cytoplasm is drawn out into long extensions (cell processes)—sometimes as long as 1 m (about 3 feet), which allows a single neuron to conduct an impulse over relatively long distances. More detail about the anatomy of the different classes of neurons and neuroglia appears in Exercise 17.

Activity 4:
Examining Nervous Tissue Under the Microscope

Obtain a prepared slide of a spinal cord smear. Locate a neuron and compare it to Figure 6.7. Keep the light dim—this will help you see the cellular extensions of the neurons. Also see Plates 5 and 6 in the Histology Atlas.

Constructing a **concept map** of the tissues will help you to organize the tissues logically and will be a useful tool for looking at slides throughout the course. A concept map aids in organism identification by a process of elimination based on observable traits. Each step of the map is a question with a yes or no answer. For example, tissues, our topic here, are separated based on observations made through the microscope. ■

Activity 5: Constructing a Concept Map of the Tissues

Using the following steps, prepare a concept map that separates the tissues based on what is observed in the photomicrographs in Figures 6.3, 6.5, 6.6, and 6.7. Your instructor will give you a list of the tissue types to be included.

1. Read the sections on epithelial, connective, muscle, and nervous tissues. Carefully review the characteristics of the assigned tissues.
2. Prepare a series of questions based on features observed through the microscope that
   a. will have only two possible answers, yes or no.
   b. will separate the tissues in a logical manner. Figure 6.8 provides an example of a concept map separating out simple squamous epithelium.
3. A helpful first question is "Is there a free edge?" This question separates epithelial tissue from connective, muscle, and nervous tissue.
4. A branch of the concept map is complete when only a single tissue type is alone at the end of a branch.
5. When your concept map is complete, use it to help identify tissue types on prepared slides.

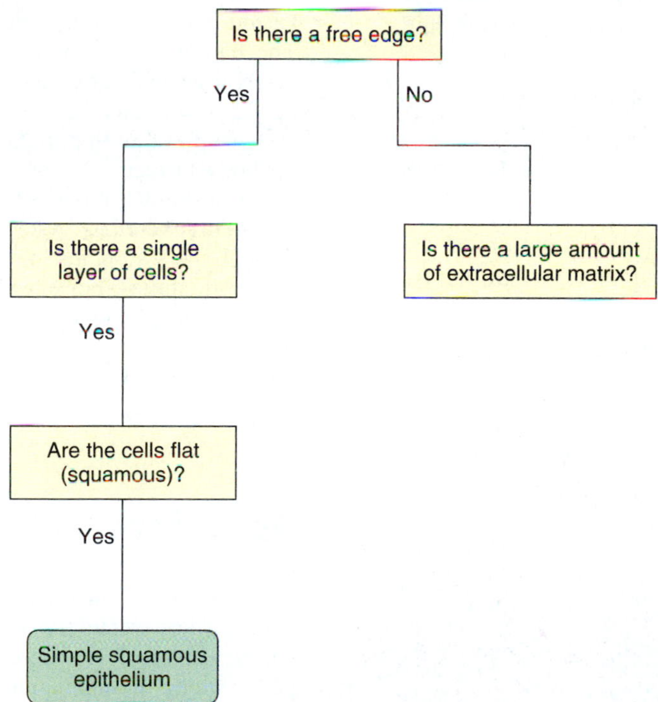

Figure 6.8 A concept map separating tissues based on observable characteristics. A map of simple squamous epithelium has been completed as an example. The map should continue until each tissue type is alone at the end of a branch.

exercise

7 The Integumentary System

Objectives

1. To recount several important functions of the skin, or integumentary system.
2. To recognize and name during observation of an appropriate model, diagram, projected slide, or microscopic specimen the following skin structures: epidermis (and indicate relative positioning of its strata), dermis (papillary and reticular layers), hair follicles and hair, sebaceous glands, and sweat glands.
3. To name the layers of the epidermis and describe the characteristics of each.
4. To compare the properties of the epidermis to those of the dermis.
5. To describe the distribution and function of the skin derivatives—sebaceous glands, sweat glands, and hairs.
6. To differentiate between eccrine and apocrine sweat glands.
7. To enumerate the factors determining skin color.
8. To describe the function of melanin.
9. To identify the major regions of nails.

Materials

- ❑ Skin model (three-dimensional, if available)
- ❑ Compound microscope
- ❑ Prepared slide of human scalp
- ❑ Prepared slide of skin of palm or sole
- ❑ Sheet of #20 bond paper ruled to mark off cm^2 areas
- ❑ Scissors
- ❑ Betadine swabs, or Lugol's iodine and cotton swabs
- ❑ Adhesive tape
- ❑ Data collection sheet for plotting distribution of sweat glands
- ❑ Dissecting microscope
- ❑ Slides and coverslips

The **skin,** or **integument,** is considered an organ system because of its extent and complexity. It is much more than an external body covering; architecturally the skin is a marvel. It is tough yet pliable, a characteristic that enables it to withstand constant insult from outside agents.

The skin has many functions, most (but not all) concerned with protection. It insulates and cushions the underlying body tissues and protects the entire body from mechanical damage (bumps and cuts), chemical damage (acids, alkalis, and the like), thermal damage (heat), and bacterial invasion (by virtue of its acid mantle and continuous surface). The hardened uppermost layer of the skin (the cornified layer) prevents water loss from the body surface. The skin's abundant capillary network (under the control of the nervous system) plays an important role in regulating heat loss from the body surface.

The skin has other functions as well. For example, it acts as a mini-excretory system; urea, salts, and water are lost through the skin pores in sweat. The skin also has important metabolic duties. For example, like liver cells, it carries out some chemical conversions that activate or inactivate certain drugs and hormones, and it is the site of vitamin D synthesis for the body. Finally, the cutaneous sense organs are located in the dermis.

Basic Structure of the Skin

The skin has two distinct regions—the superficial *epidermis* composed of epithelium and an underlying connective tissue *dermis* (Figure 7.1). These layers are firmly "cemented" together along an undulating border. But friction, such as the rubbing of a poorly fitting shoe, may cause them to separate, resulting in a blister. Immediately deep to the dermis is the **hypodermis** or **superficial fascia** (primarily adipose tissue), which is not considered part of the skin. The main skin areas and structures are described below.

Activity 1: Locating Structures on a Skin Model

As you read, locate the following structures in Figure 7.1 and on a skin model. ■

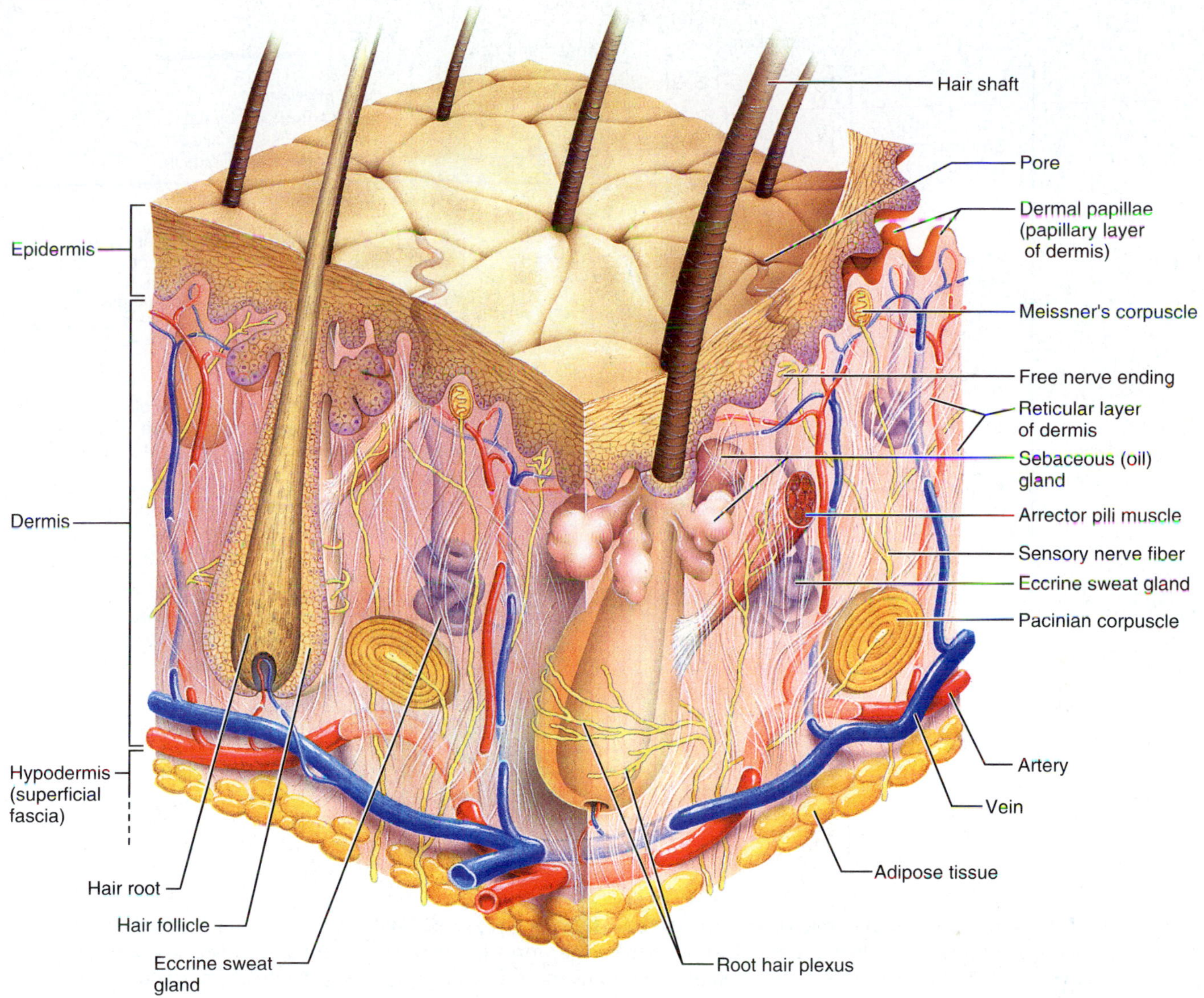

Figure 7.1 Skin structure. Three-dimensional view of the skin and the underlying hypodermis. The epidermis and dermis have been pulled apart at the right corner to reveal the dermal papillae.

Epidermis

Structurally, the avascular epidermis is a keratinized stratified squamous epithelium consisting of four distinct cell types and four or five distinct layers.

Cells of the Epidermis

- **Keratinocytes** (literally, keratin cells): The most abundant epidermal cells, they function mainly to produce keratin fibrils. **Keratin** is a fibrous protein that gives the epidermis its durability and protective capabilities. Keratinocytes are tightly connected to each other by desmosomes.

Far less numerous are the following types of epidermal cells (Figure 7.2):

- **Melanocytes:** Spidery black cells that produce the brown-to-black pigment called **melanin.** The skin tans because melanin production increases when the skin is exposed to sunlight. The melanin provides a protective pigment umbrella over the nuclei of the cells in the deeper epidermal layers, thus shielding their genetic material (DNA) from the damaging effects of ultraviolet radiation. A concentration of melanin in one spot is called a *freckle.*
- **Langerhans' cells:** Also called *epidermal dendritic cells,* these phagocytic cells (macrophages) play a role in immunity.
- **Merkel cells:** Occasional spiky hemispheres that, in conjunction with sensory nerve endings, form sensitive touch receptors called *Merkel discs* located at the epidermal-dermal junction.

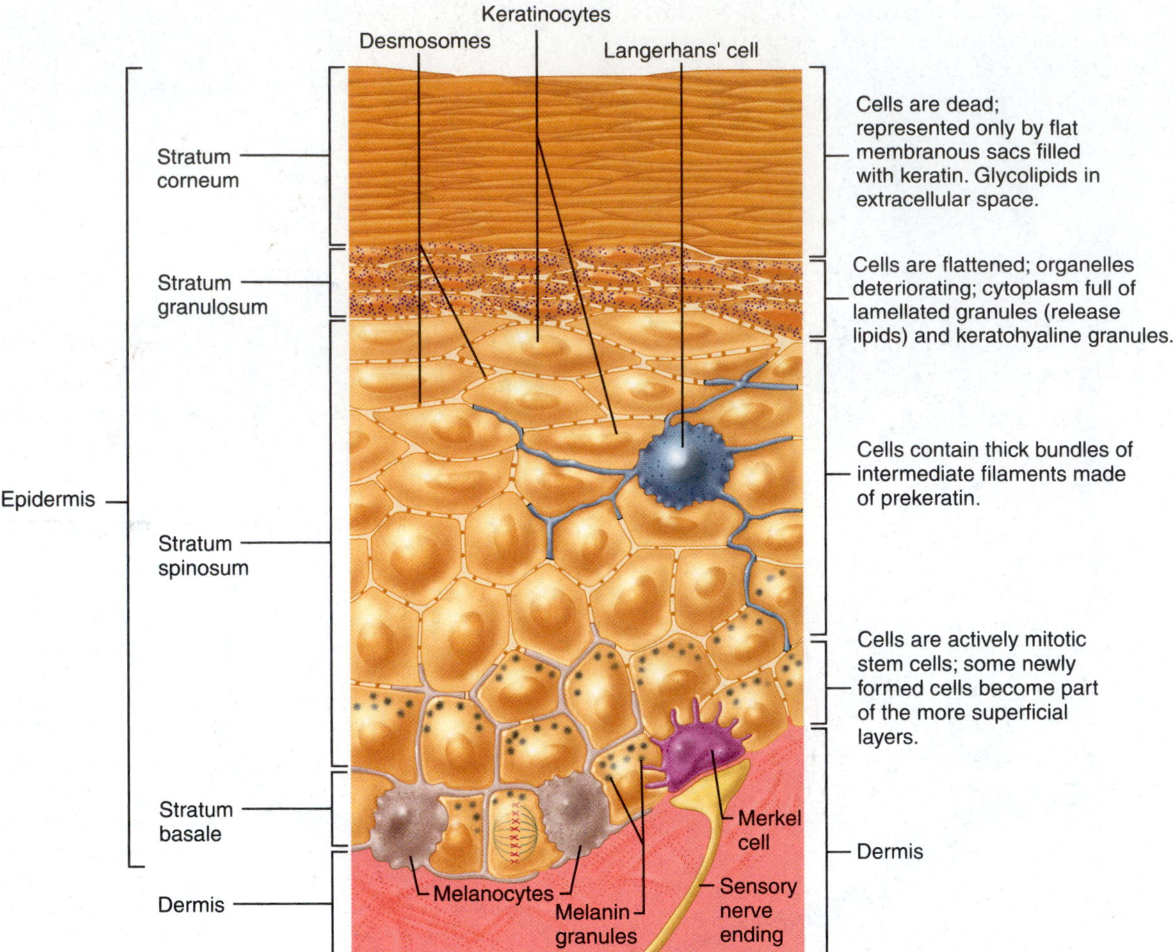
Keratinocytes
Desmosomes
Langerhans' cell
Stratum corneum
Cells are dead; represented only by flat membranous sacs filled with keratin. Glycolipids in extracellular space.
Stratum granulosum
Cells are flattened; organelles deteriorating; cytoplasm full of lamellated granules (release lipids) and keratohyaline granules.
Epidermis
Stratum spinosum
Cells contain thick bundles of intermediate filaments made of prekeratin.
Cells are actively mitotic stem cells; some newly formed cells become part of the more superficial layers.
Stratum basale
Merkel cell
Dermis
Dermis
Melanocytes
Melanin granules
Sensory nerve ending

Figure 7.2 Diagram showing the main features—layers and relative numbers of the different cell types—in epidermis of thin skin. Keratinocytes (beige) form the bulk of the epidermis. Melanocytes (gray) produce the pigment melanin. Langerhans' cells (blue) function as macrophages. Merkel cell (purple) associates with a sensory nerve ending (yellow) that extends from the dermis. The pair forms a Merkel disc (touch receptor). Notice that the keratinocytes, but not the other cell types, are joined by numerous desmosomes. The stratum lucidum, present in thick skin, is not shown here.

Layers of the Epidermis From deep to superficial, the layers of the epidermis are the stratum basale, stratum spinosum, stratum granulosum, stratum lucidum, and stratum corneum (Figure 7.2).

- **Stratum basale** (basal layer): A single row of cells immediately adjacent to the dermis. Its cells are constantly undergoing mitotic cell division to produce millions of new cells daily, hence its alternate name *stratum germinativum.* From 10 to 25% of the cells in this stratum are melanocytes, which thread their processes through this and the adjacent layers of keratinocytes (see Figure 7.2).

- **Stratum spinosum** (spiny layer): A stratum consisting of several cell layers immediately superficial to the basal layer. Its cells contain thick weblike bundles of intermediate filaments made of a prekeratin protein. The stratum spinosum cells appear spiky (hence their name) because as the skin tissue is prepared for histological examination, they shrink but their desmosomes hold tight. Cells divide fairly rapidly in this stratum, but less so than in the stratum basale. Cells in the basal and spiny layers are the only ones to receive adequate nourishment (via diffusion of nutrients from the dermis). So as their daughter cells are pushed upward and away from the source of nutrition, they gradually die.

- **Stratum granulosum** (granular layer): A thin layer named for the abundant granules its cells contain. These granules are of two types: (1) *lamellated granules,* which contain a waterproofing glycolipid that is secreted into the extracellular space; and (2) *keratohyaline granules,* which combine with the intermediate filaments in the more superficial layers to form the keratin fibrils. At the upper border of this layer, the cells are beginning to die.

- **Stratum lucidum** (clear layer): A very thin translucent band of flattened dead keratinocytes with indistinct boundaries. It is not present in regions of thin skin.

- **Stratum corneum** (horny layer): This outermost epidermal layer consists of some 20 to 30 cell layers, and accounts for the bulk of the epidermal thickness. Cells in this layer, like those in the stratum lucidum (where it exists), are dead and their flattened scalelike remnants are fully keratinized. They are constantly rubbing off and being replaced by division of the deeper cells.

Dermis

The dense irregular connective tissue making up the dermis consists of two principal regions—the papillary and reticular areas. Like the epidermis, the dermis varies in thickness. For example, the skin is particularly thick on the palms of the hands and soles of the feet and is quite thin on the eyelids.

- **Papillary layer:** The more superficial dermal region composed of areolar connective tissue. It is very uneven and has fingerlike projections from its superior surface, the **dermal papillae,** which attach it to the epidermis above. These projections lie on top of the larger dermal ridges. In the palms of the hands and soles of the feet, they produce the fingerprints, unique patterns of *epidermal ridges* that remain unchanged throughout life. Abundant capillary networks in the papillary layer furnish nutrients for the epidermal layers and allow heat to radiate to the skin surface. The pain and touch receptors (Meissner's corpuscles) are also found here.
- **Reticular layer:** The deepest skin layer. It is composed of dense irregular connective tissue, and contains many arteries and veins, sweat and sebaceous glands, and pressure receptors.

Both the papillary and reticular layers are heavily invested with collagenic and elastic fibers. The elastic fibers give skin its exceptional elasticity in youth. In old age, the number of elastic fibers decreases and the subcutaneous layer loses fat, which leads to wrinkling and inelasticity of the skin. Fibroblasts, adipose cells, various types of macrophages (which are important in the body's defense), and other cell types are found throughout the dermis.

The abundant dermal blood supply allows the skin to play a role in the regulation of body temperature. When body temperature is high, the arterioles serving the skin dilate, and the capillary network of the dermis becomes engorged with the heated blood. Thus body heat is allowed to radiate from the skin surface. If the environment is cool and body heat must be conserved, the arterioles constrict so that blood bypasses the dermal capillary networks.

Any restriction of the normal blood supply to the skin results in cell death and, if severe enough, skin ulcers (Figure 7.3). **Bedsores (decubitus ulcers)** occur in bedridden patients who are not turned regularly enough. The weight of the body exerts pressure on the skin, especially over bony projections (hips, heels, etc.), which leads to restriction of the blood supply and death of tissue. ■

The dermis is also richly provided with lymphatic vessels and a nerve supply. Many of the nerve endings bear highly specialized receptor organs that, when stimulated by environmental changes, transmit messages to the central nervous system for interpretation. Some of these receptors—bare nerve endings (pain receptors), a Meissner's corpuscle, a Pacinian corpuscle, and a root hair plexus—are shown in Figure 7.1. (These receptors are discussed in depth in Exercise 23.)

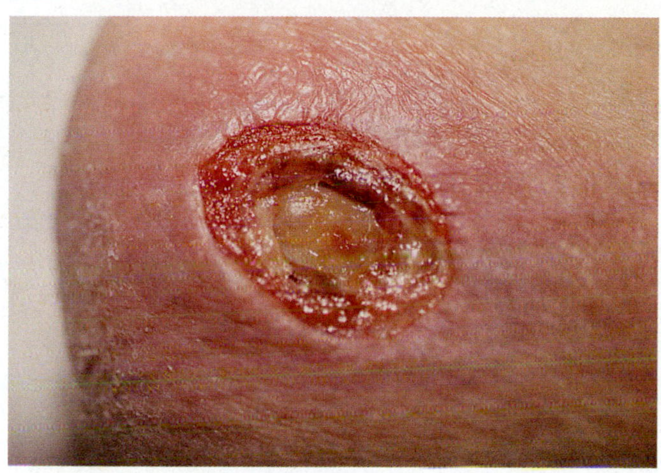

Figure 7.3 Photograph of a deep (stage III) decubitus ulcer.

Skin Color

Skin color is a result of three factors—the relative amount of two pigments (melanin and carotene) in skin and the degree of oxygenation of the blood. People who produce large amounts of melanin have brown-toned skin. In light-skinned people, who have less melanin, the dermal blood supply flushes through the rather transparent cell layers above, giving the skin a rosy glow. *Carotene* is a yellow-orange pigment present primarily in the stratum corneum and in the adipose tissue of the hypodermis. Its presence is most noticeable when large amounts of carotene-rich foods (carrots, for instance) are eaten.

Skin color may be an important diagnostic tool. For example, flushed skin may indicate hypertension, fever, or embarrassment, whereas pale skin is typically seen in anemic individuals. When the blood is inadequately oxygenated, as during asphyxiation and serious lung disease, both the blood and the skin take on a bluish or cyanotic cast. **Jaundice,** in which the tissues become yellowed, is almost always diagnostic for liver disease, whereas a bronzing of the skin hints that a person's adrenal cortex is hypoactive **(Addison's disease).** ■

Accessory Organs of the Skin

The accessory organs of the skin—cutaneous glands, hair, and nails—are all derivatives of the epidermis, but they reside in the dermis. They originate from the stratum basale and grow downward into the deeper skin regions.

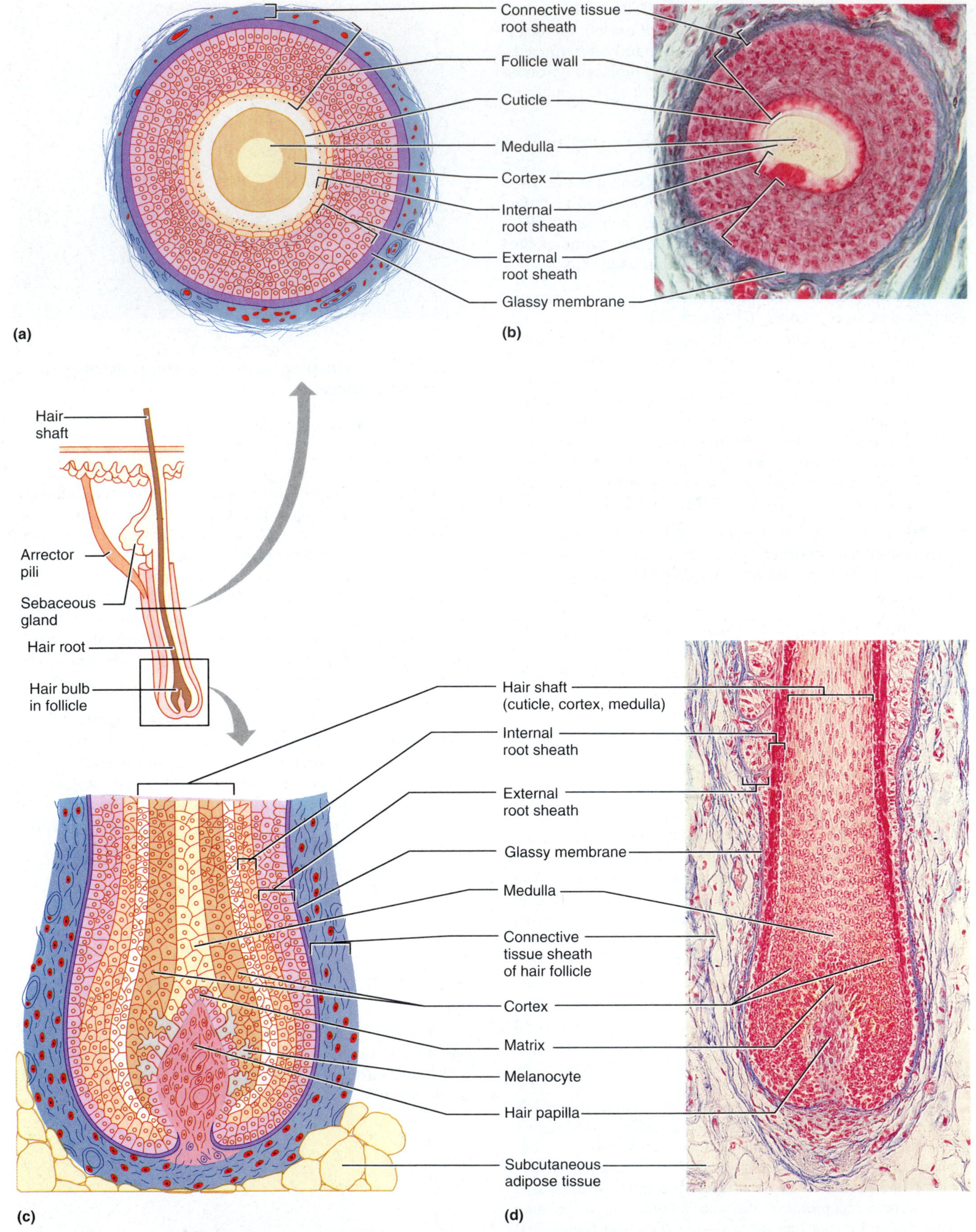

Figure 7.4 Structure of a hair and hair follicle. (a) Diagrammatic view. **(b)** Photomicrograph of a cross section of a hair and hair follicle (147×). **(c)** Diagrammatic view of longitudinal section of a hair in its follicle, and expanded view of the hair bulb. **(d)** Photomicrograph of longitudinal view of hair bulb (164×).

Hairs and Associated Structures

Hairs, enclosed in hair follicles, are found all over the entire body surface, except for thick-skinned areas (the palms of the hands and the soles of the feet), parts of the external genitalia, the nipples, and the lips.

- **Hair:** Structure consisting of a medulla, a central region surrounded first by the *cortex* and then by a protective *cuticle* (Figure 7.4). Abrasion of the cuticle results in split ends. Hair color is a manifestation of the amount and kind of melanin pigment within the hair cortex. The portion of the hair enclosed within the follicle is called the **root;** that portion projecting from the scalp surface is called the **shaft.** The **hair bulb** is a collection of well-nourished germinal epithelial cells at the basal end of the follicle. As the daughter cells are pushed farther away from the growing region, they die and become keratinized; thus the bulk of the hair shaft, like the bulk of the epidermis, is dead material.

- **Follicle:** A structure formed from both epidermal and dermal cells (see Figure 7.4). Its inner epithelial root sheath, with two parts (internal and external), is enclosed by the connective tissue root sheath, which is essentially dermal tissue. A small nipple of dermal tissue that protrudes into the hair bulb from the connective tissue sheath and provides nutrition to the growing hair is called the **papilla.**

- **Arrector pili muscle:** Small bands of smooth muscle cells connect each hair follicle to the papillary layer of the dermis (Figure 7.1). When these muscles contract (during cold or fright), the slanted hair follicle is pulled upright, dimpling the skin surface with goose bumps. This phenomenon is especially dramatic in a scared cat, whose fur actually stands on end to increase its apparent size. The activity of the arrector pili muscles also exerts pressure on the sebaceous glands surrounding the follicle, causing a small amount of sebum to be released.

Activity 2: Thick and Thin Skin Microscopically

1. Obtain a prepared slide of the human scalp, and study it carefully under the microscope. Compare your tissue slide to the view shown in Figure 7.5a, and identify as many of the structures diagrammed in Figure 7.1 as possible.

How is this stratified squamous epithelium different from that observed in Exercise 6?

____

____

How do these differences relate to the functions of these two similar epithelia?

____

____

____

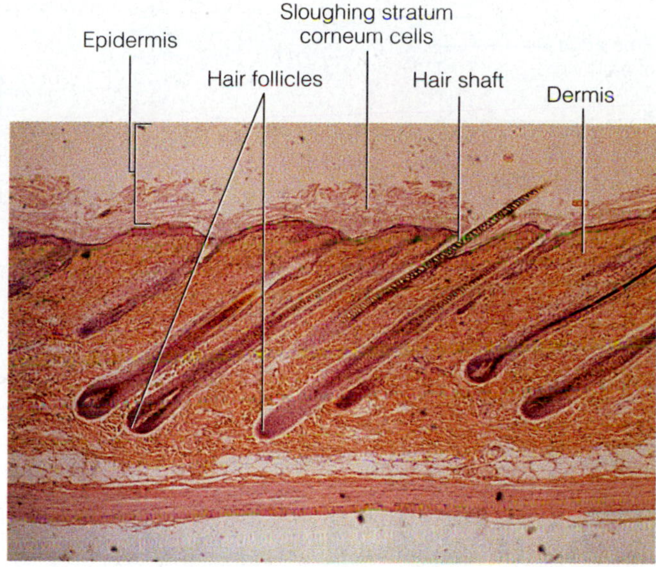

(a)

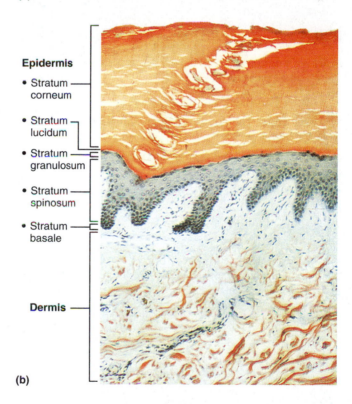

Epidermis
- Stratum corneum
- Stratum lucidum
- Stratum granulosum
- Stratum spinosum
- Stratum basale

Dermis

(b)

Figure 7.5 Photomicrographs of skin. **(a)** Thin skin with hairs (34×). **(b)** Thick hairless skin (150×).

2. Obtain a prepared slide of hairless skin of the palm or sole (Figure 7.5b). Compare the slide to Figure 7.5a. In what ways does the thick skin of the palm or sole differ from the thin skin of the scalp?

____

____ ■

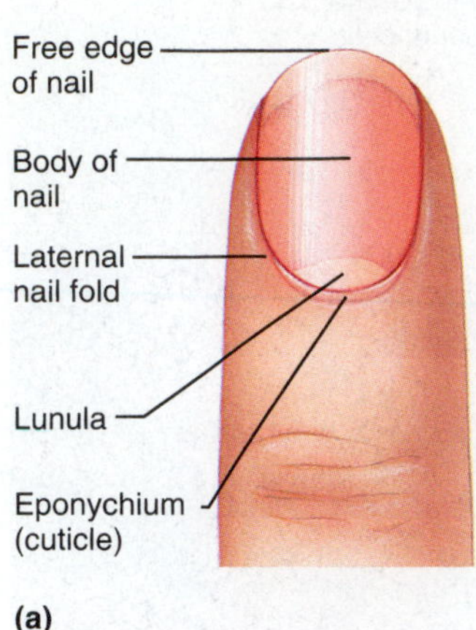

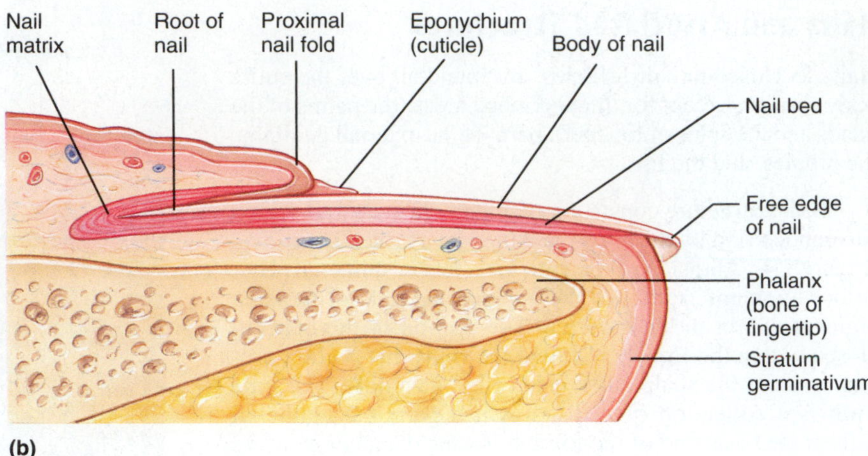

Figure 7.6 Structure of a nail. (a) Surface view of the distal part of a finger showing nail parts. The nail matrix that forms the nail lies beneath the lunula; the epidermis of the nail bed underlies the nail. **(b)** Sagittal section of the fingertip.

Nails

Nails are hornlike derivatives of the epidermis (Figure 7.6). Their named parts are:

- **Body:** The visible attached portion.
- **Free edge:** The portion of the nail that grows out away from the body.
- **Root:** The part that is embedded in the skin and adheres to an epithelial nail bed.
- **Nail folds:** Skin folds that overlap the borders of the nail.
- **Eponychium:** The thick proximal nail fold commonly called the cuticle.
- **Nail matrix:** The thickened proximal part of the nail bed containing germinal cells responsible for nail growth. As the matrix produces the nail cells, they become heavily keratinized and die. Thus nails, like hairs, are mostly nonliving material.
- **Lunula:** The proximal region of the thickened nail matrix, which appears as a white crescent. Everywhere else, nails are transparent and nearly colorless, but they appear pink because of the blood supply in the underlying dermis. When someone is cyanotic due to a lack of oxygen in the blood, the nail beds take on a blue cast.

Activity 3:
Identifying Nail Structures

Identify the nail structures shown in Figure 7.6 on yourself or your lab partner. ■

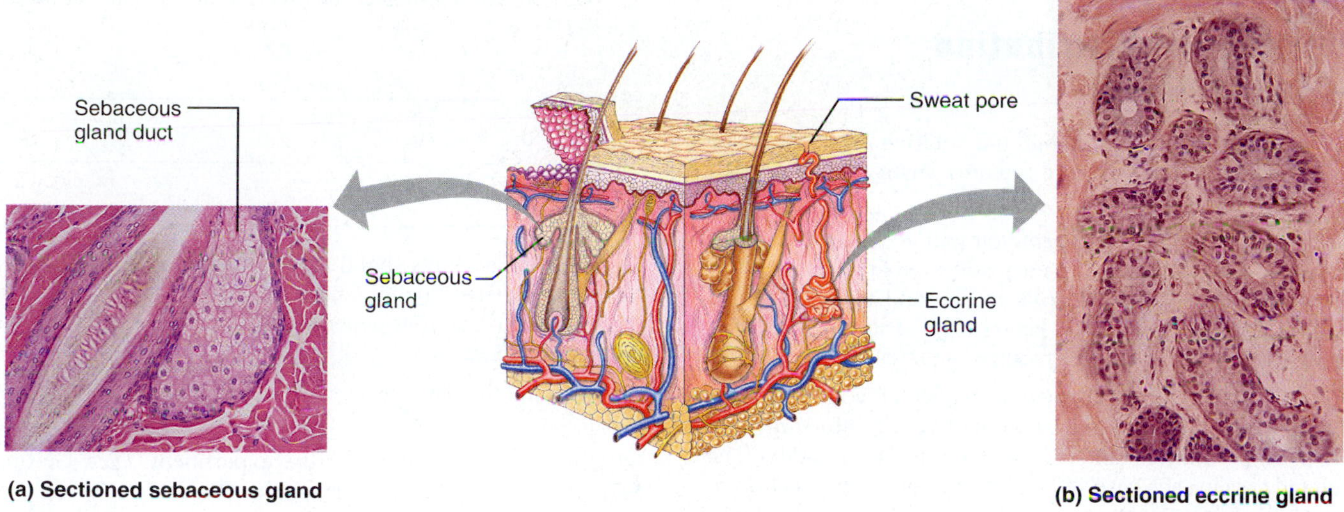

Figure 7.7 Cutaneous glands. (a) Photomicrograph of a sebaceous gland (104×). **(b)** Photomicrograph of eccrine sweat glands (148×).

Cutaneous Glands

The cutaneous glands fall primarily into two categories: the sebaceous glands and the sweat glands (Figure 7.1 and Figure 7.7).

Sebaceous glands The sebaceous glands are found nearly all over the skin, except for the palms of the hands and the soles of the feet. Their ducts usually empty into a hair follicle, but some open directly on the skin surface.

Sebum is the product of sebaceous glands. It is a mixture of oily substances and fragmented cells that acts as a lubricant to keep the skin soft and moist (a natural skin cream) and keeps the hair from becoming brittle. The sebaceous glands become particularly active during puberty when more male hormones (androgens) begin to be produced; thus the skin tends to become oilier during this period of life.

Blackheads are accumulations of dried sebum, bacteria, and melanin from epithelial cells in the oil duct. **Acne** is an active infection of the sebaceous glands. ■

Sweat (sudoriferous) glands These exocrine glands are widely distributed all over the skin. Outlets for the glands are epithelial openings called *pores*. Sweat glands are categorized by the composition of their secretions.

- **Eccrine glands:** Glands distributed all over the body that produce clear perspiration consisting primarily of water, salts (mostly NaCl), and urea. Eccrine sweat glands, under the control of the nervous system, are an important part of the body's heat-regulating apparatus. They secrete perspiration when the external temperature or body temperature is high. When this water-based substance evaporates, it carries excess body heat with it. Thus evaporation of greater amounts of perspiration provides an efficient means of dissipating body heat when the capillary cooling system is not sufficient or is unable to maintain body temperature homeostasis.

- **Apocrine glands:** Found predominantly in the axillary and genital areas, these glands secrete a milky protein- and fat-rich substance (also containing water, salts, and urea) that is an excellent nutrient medium for the microorganisms typically found on the skin.

Activity 4:
Differentiating Sebaceous and Sweat Glands Microscopically

Using the slide *thin skin with hairs* used in Activity 2, and Figure 7.7 as a guide, identify sebaceous and eccrine sweat glands. What characteristics relating to location or gland structure allow you to differentiate these glands?

____

____

____ ■

Activity 5: Plotting the Distribution of Sweat Glands

1. Form a hypothesis about the relative distribution of sweat glands on the palm and forearm. Justify your hypothesis.

2. For this simple experiment you will need two squares of bond paper (each 1 cm × 1 cm), adhesive tape, and a Betadine (iodine) swab *or* Lugol's iodine and a cotton-tipped swab. (The bond paper has been preruled in cm^2—just cut along the lines to obtain the required squares.)

3. Paint an area of the medial aspect of your left palm (avoid the crease lines) and a region of your left forearm with the iodine solution, and allow it to dry thoroughly. The painted area in each case should be slightly larger than the paper squares to be used.

4. Have your lab partner *securely* tape a square of bond paper over each iodine-painted area, and leave them in place for 20 minutes. (If it is very warm in the laboratory while this test is being conducted, good results may be obtained within 10 to 15 minutes.)

5. After 20 minutes, remove the paper squares, and count the number of blue-black dots on each square. The presence of a blue-black dot on the paper indicates an active sweat gland. (The iodine in the pore is dissolved in the sweat and reacts chemically with the starch in the bond paper to produce the blue-black color.) Thus "sweat maps" have been produced for the two skin areas.

6. Which skin area tested has the greater density of sweat glands?

____

____

____

7. Tape your results (bond paper squares) to a data collection sheet labeled "palm" and "forearm" at the front of the lab. Be sure to put your paper squares in the correct columns on the data sheet.

8. Once all the data has been collected, review the class results.

9. Prepare a lab report for the experiment. (See Getting Started: Writing a Lab Report, p. xii) ■

Classification of Body Membranes

exercise

8

Objectives

1. To compare the structure and function of the major membrane types.
2. To list the general functions of each membrane type and indicate its location in the body.
3. To recognize by microscopic examination cutaneous, mucous, and serous membranes.

Materials

- ❑ Compound microscope
- ❑ Prepared slides of trachea (x.s.), esophagus (x.s.), and small intestine (x.s.)
- ❑ Prepared slide of serous membrane (e.g., mesentery artery and vein (x.5.) or small intestine (x.s.))
- ❑ Longitudinally cut fresh beef joint (if available)

A1A See Appendix C, Exercise 8 for links to A.D.A.M.® Interactive Anatomy.

The body membranes, which cover surfaces, line body cavities, and form protective (and often lubricating) sheets around organs, fall into two major categories. These are the so-called *epithelial membranes* and the *synovial membranes*.

Epithelial Membranes

The term "epithelial membrane" is used in various ways. Here we will define an **epithelial membrane** as a simple organ consisting of an epithelial sheet bound to an underlying layer of connective tissue proper. Most of the covering and lining epithelia take part in forming one of the three common varieties of epithelial membranes: cutaneous, mucous, or serous.

The **cutaneous membrane** (Figure 8.1a) is the skin, a dry membrane with a keratinizing epithelium (the epidermis). Since the skin is discussed in some detail in Exercise 7, the mucous and serous membranes will receive our attention here.

Mucous Membranes

The **mucous membranes (mucosae)*** are composed of epithelial cells resting on a layer of loose connective tissue called the **lamina propria.** They line all body cavities that open to the body exterior—the respiratory, digestive (Figure 8.1b), and urogenital tracts. In most cases mucosae are "wet" membranes, which are continuously bathed by secretions (or, in the case of urinary mucosa, urine). Although mucous membranes often secrete mucus, this is not a requirement. The mucous membranes of both the digestive and respiratory tracts secrete mucus; that of the urinary tract does not.

Activity 1: Examining the Microscopic Structure of Mucous Membranes

Using Plates 33, 36, and 39 of the Histology Atlas as guides, examine slides made from cross sections of the trachea, esophagus, and small intestine. Draw the mucosa of each in the appropriate circle, and fully identify each epithelial type. Remember to look for the epithelial cells at the free surface. Also search the epithelial sheets for **goblet cells**—columnar epithelial cells with a large mucus-containing vacuole (goblet) in their apical cytoplasm. Which mucosae contain goblet cells?

____

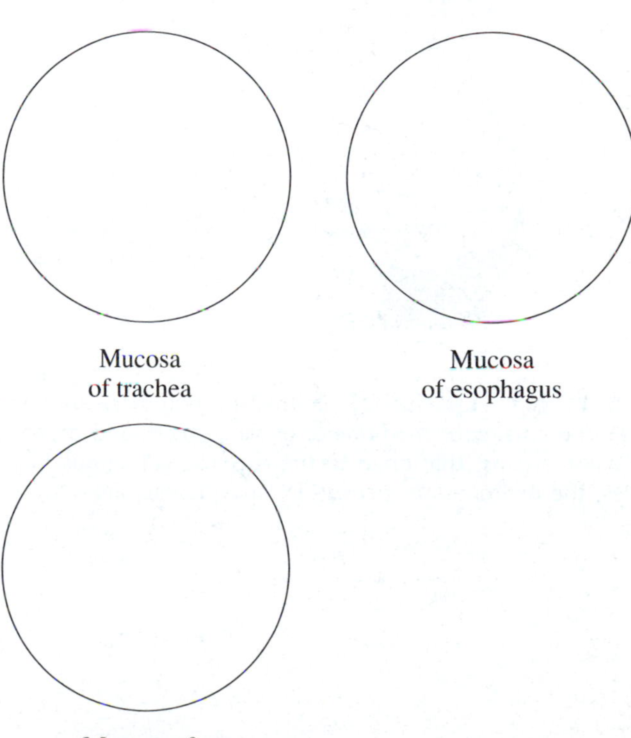

*Notice the spelling difference between *mucous,* the membrane type, and *mucus,* the product of glands.

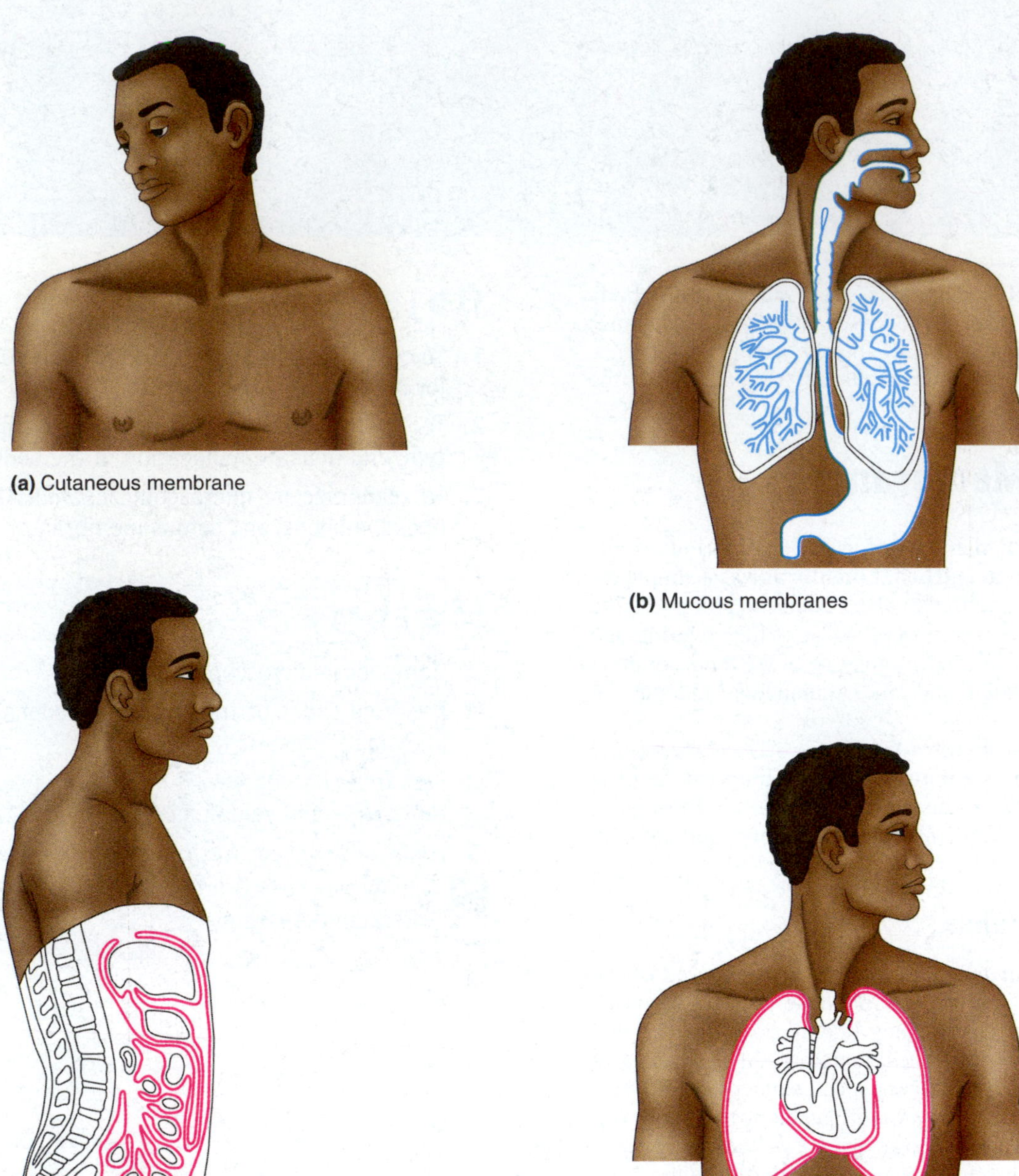

Figure 8.1 Epithelial. Epithelial membranes are composite membranes with epithelial and connective tissue elements. **(a)** The cutaneous membrane, or skin, covers and protects the body surface. **(b)** Mucous membranes line body cavities (hollow organs) that open to the exterior. **(c)** Serous membranes line the closed ventral cavity of the body. Three examples, the peritoneums, pericardia, and pleurae, are illustrated here.

Compare and contrast the roles of these three mucous membranes.

___

___

___

___ ■

Serous Membranes

The **serous membranes (serosae)** are also epithelial membranes (Figure 8.1c). They are composed of a layer of simple squamous epithelium on a scant amount of areolar connective tissue. The serous membranes generally occur in twos and are actually continuous. The parietal layer lines a body cavity, and the visceral layer covers the outside of the organs in that cavity (see also Figure 1.8, p. 9). In contrast to the mucous membranes, which line open body cavities, the serous membranes line body cavities that are closed to the exterior (with the exception of the female peritoneal cavity and the dorsal body cavity). These double-layered serosae secrete a thin fluid (serous fluid) that lubricates the organs and body walls and thus reduces friction as the organs slide across one another and against the body cavity walls.

A serous membrane also lines the interior of blood vessels (endothelium) and the heart (endocardium). In capillaries, the entire wall is composed of serosa that serves as a selectively permeable membrane between the blood and the tissue fluid of the body.

Activity 2:
Examining the Microscopic Structure of a Serous Membrane

Using Plates 27 and 39 of the Histology Atlas as guides, examine a prepared slide of a serous membrane and diagram it in the circle provided here.

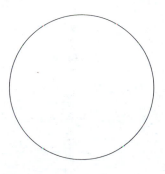

What are the specific names of the serous membranes covering the heart and lining the cavity in which it resides (respectively)?

___

The abdominal viscera and visceral cavity (respectively)?

___ ■

Synovial Membranes

Synovial membranes, unlike mucous and serous membranes, are composed entirely of connective tissue; they contain no epithelial cells. These membranes line the cavities surrounding the joints, providing a smooth surface and secreting a lubricating fluid. They also line smaller sacs of connective tissue (bursae and tendon sheaths), which cushion structures moving against each other, as during muscle activity. Figure 8.2 illustrates the positioning of a synovial membrane in the joint cavity.

Activity 3:
Examining the Gross Structure of a Synovial Membrane

If a freshly sawed beef joint is available, visually examine the interior surface of the joint capsule to observe the smooth texture of the synovial membrane. ■

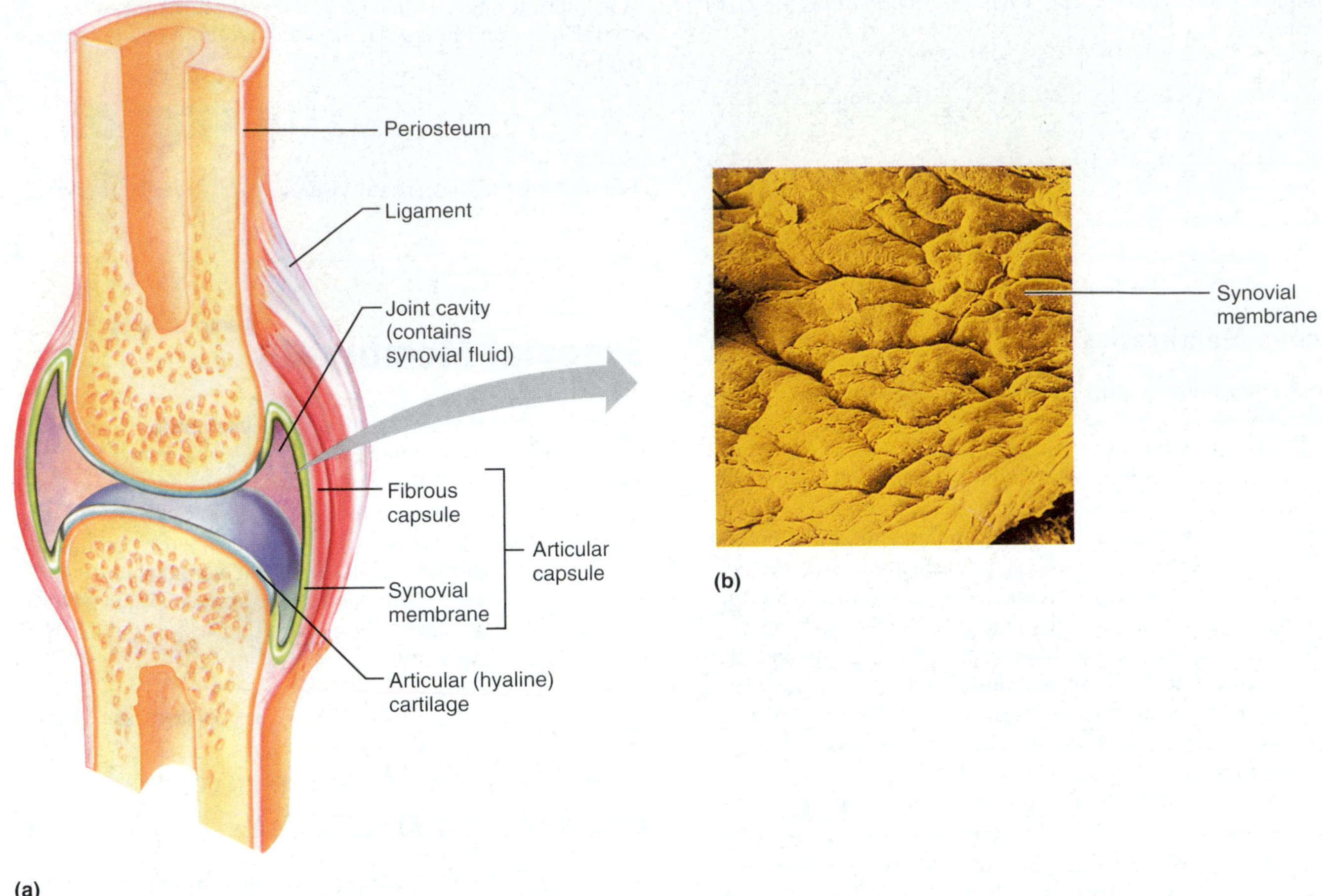

Figure 8.2 General structure of a synovial joint. (a) The joint cavity is lined with a synovial membrane, derived solely from connective tissue. **(b)** Scanning electron micrograph of synovial membrane (13×).

Overview of the Skeleton: Classification and Structure of Bones and Cartilages

exercise 9

Objectives

1. To list five functions of the skeletal system.
2. To identify the four main kinds of bones.
3. To identify surface bone markings and functions.
4. To identify the major anatomical areas on a longitudinally cut long bone (or diagram of one).
5. To identify the major regions and structures of an osteon in a histologic specimen of compact bone (or diagram of one).
6. To explain the role of the inorganic salts and organic matrix in providing flexibility and hardness to bone.
7. To locate and identify the three major types of skeletal cartilages.

Materials

- ❑ Disarticulated bones (identified by name or number) that demonstrate classic examples of the four bone classifications (long, short, flat, and irregular)
- ❑ Long bone sawed longitudinally (beef bone from a slaughterhouse, if possible, or prepared laboratory specimen)
- ❑ Disposable plastic gloves
- ❑ Long bone soaked in 10% hydrochloric acid (HCl) (or vinegar) until flexible
- ❑ Long bone baked at 250°F for more than 2 hours
- ❑ Compound microscope
- ❑ Prepared slides of ground bone (x.s.), hyaline cartilage, elastic cartilage, and fibrocartilage
- ❑ 3-D model of microscopic structure of compact bone
- ❑ Articulated skeleton

A.I.A. See Appendix C, Exercise 9 for links to A.D.A.M.® Interactive Anatomy.

The **skeleton,** the body's framework, is constructed of two of the most supportive tissues found in the human body—cartilage and bone. In embryos, the skeleton is predominantly composed of hyaline cartilage, but in the adult, most of the cartilage is replaced by more rigid bone. Cartilage persists only in such isolated areas as the bridge of the nose, the larynx, the trachea, joints, and parts of the rib cage.

Besides supporting and protecting the body as an internal framework, the skeleton provides a system of levers with which the skeletal muscles work to move the body. In addition, the bones store lipids and many minerals (most importantly calcium). Finally, the red marrow cavities of bones provide a site for hematopoiesis (blood cell formation).

The skeleton is made up of bones that are connected at joints, or articulations. The skeleton is subdivided into two divisions: the **axial skeleton** (those bones that lie around the body's center of gravity) and the **appendicular skeleton** (bones of the limbs, or appendages) (Figure 9.1).

Before beginning your study of the skeleton, imagine for a moment that your bones have turned to putty. What if you were running when this metamorphosis took place? Now imagine your bones forming a continuous metal framework within your body, somewhat like a network of plumbing pipes. What problems could you envision with this arrangement? These images should help you understand how well the skeletal system provides support and protection, as well as facilitating movement.

Bone Markings

Even a casual observation of the bones will reveal that bone surfaces are not featureless smooth areas but are scarred with an array of bumps, holes, and ridges. These **bone markings** reveal where bones form joints with other bones, where muscles, tendons, and ligaments were attached, and where blood vessels and nerves passed. Bone markings fall into two categories: projections, or processes that grow out from the bone and serve as sites of muscle attachment or help form joints; and depressions or cavities, indentations or openings in the bone that often serve as conduits for nerves and blood vessels. The bone markings are summarized in Table 9.1.

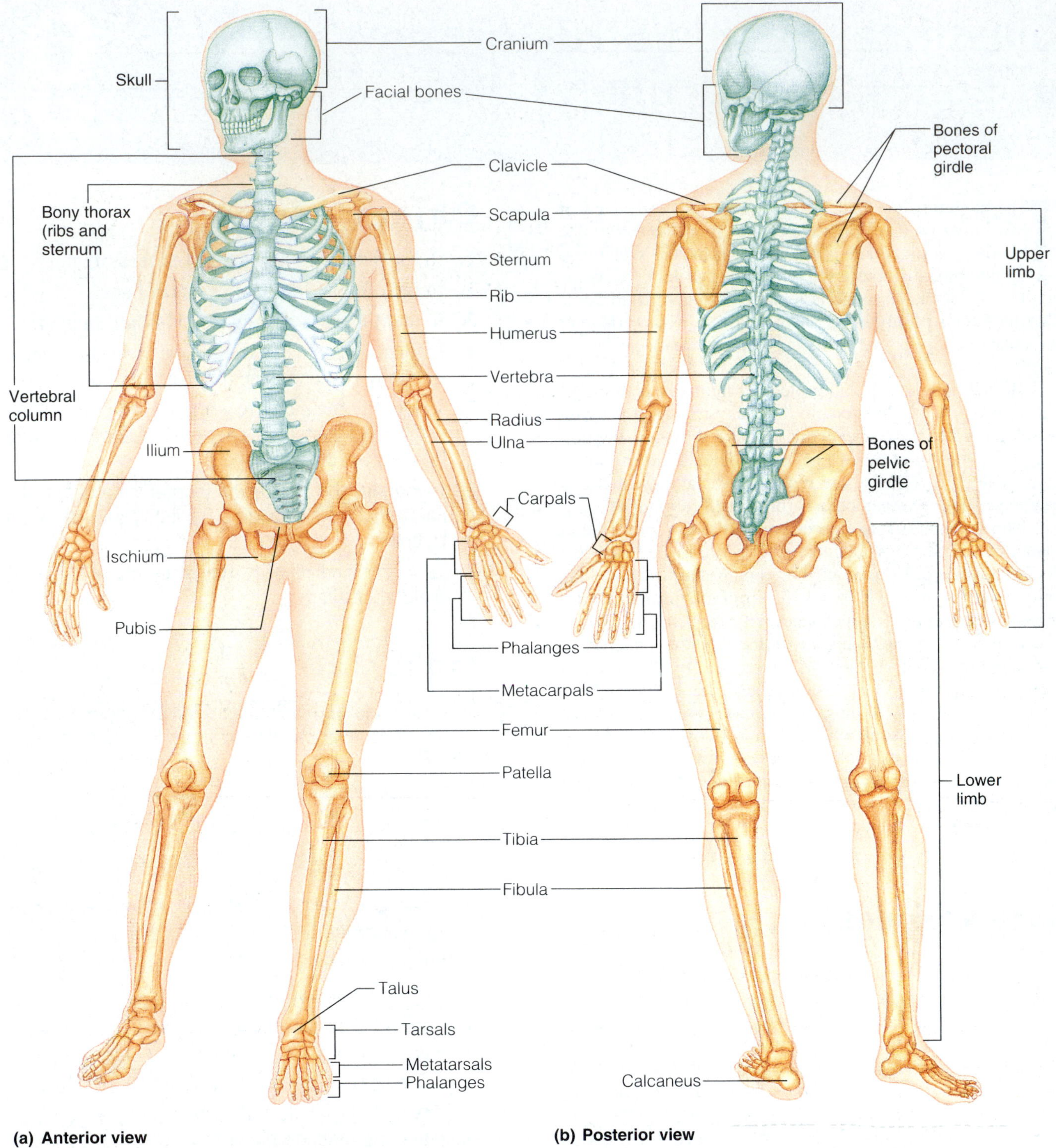

Figure 9.1 The human skeleton. The bones of the axial skeleton are colored green to distinguish them from the bones of the appendicular skeleton.

Table 9.1 Bone Markings

Name of bone marking	Description
Projections that are sites of muscle and ligament attachment	
Tuberosity (too″bĕ-ros′ĭ-te)	Large rounded projection; may be roughened
Crest	Narrow ridge of bone; usually prominent
Trochanter (tro-kan′ter)	Very large, blunt, irregularly shaped process. (The only examples are on the femur.)
Line	Narrow ridge of bone; less prominent than a crest
Tubercle (too′ber-kl)	Small rounded projection or process
Epicondyle (ep″ĭ-kon′dīl)	Raised area on or above a condyle
Spine	Sharp, slender, often pointed projection
Process	Prominence or projection
Projections that help to form joints	
Head	Bony expansion carried on a narrow neck
Facet	Smooth, nearly flat articular surface
Condyle (kon′dīl)	Rounded articular projection
Ramus (ra′mus)	Armlike bar of bone
Cavities	
Sinus	Space within a bone, filled with air and lined with mucous membrane
Depressions and openings allowing blood vessels and nerves to pass	
Meatus (me-a′tus)	Canal-like passageway
Fossa (fos′ah)	Shallow, basinlike depression in a bone, often serving as an articular surface
Groove	Furrow
Fissure	Narrow, slitlike opening
Foramen (fo-ra′men) (plural: foramina)	Round or oval opening through a bone

Illustration

Classification of Bones

The 206 bones of the adult skeleton are composed of two basic kinds of osseous tissue that differ in their texture. **Compact** bone looks smooth and homogeneous; **spongy** (or *cancellous*) bone is composed of small trabeculae (bars) of bone and lots of open space.

Bones may be classified further on the basis of their relative gross anatomy into four groups: long, short, flat, and irregular bones.

Long bones, such as the femur (Figure 9.1), are much longer than they are wide, generally consisting of a shaft with heads at either end. Long bones are composed predominantly of compact bone. **Short bones** are typically cube shaped, and they contain more spongy bone than compact bone. See the tarsals and carpals in Figure 9.1.

Flat bones are generally thin, with two waferlike layers of compact bone sandwiching a layer of spongy bone between them. Although the name "flat bone" implies a structure that is level or horizontal, many flat bones are curved (for example, the bones of the skull). Bones that do not fall into one of the preceding categories are classified as **irregular bones.** The vertebrae are irregular bones (see Figure 9.1).

Some anatomists also recognize two other subcategories of bones. **Sesamoid bones** are special types of short bones formed in tendons. The patellas (kneecaps) are sesamoid bones. **Wormian** or **sutural bones** are tiny bones between cranial bones. Except for the patellas, the sesamoid and Wormian bones are not included in the bone count of 206 because they vary in number and location in different individuals.

Activity 1:
Examining and Classifying Bones

Examine the isolated (disarticulated) bones on display. See if you can find specific examples of the bone markings described in Table 9.1. Then classify each of the bones into one of the four anatomical groups by recording its name or number in the accompanying chart. Verify your identifications with your instructor before leaving the laboratory. ■

Long	Short	Flat	Irregular

Gross Anatomy of the Typical Long Bone

Activity 2:
Examining a Long Bone

1. Obtain a long bone that has been sawed along its longitudinal axis. If a cleaned dry bone is provided, no special preparations need be made.

Note: If the bone supplied is a fresh beef bone, don plastic gloves before beginning your observations.

With the help of Figure 9.2, identify the shaft, or diaphysis. Observe its smooth surface, which is composed of compact bone. If you are using a fresh specimen, carefully pull away the periosteum, or fibrous membrane covering, to view the bone surface. Notice that many fibers of the periosteum penetrate into the bone. These fibers are called Sharpey's fibers. Blood vessels and nerves travel through the periosteum and invade the bone. Osteoblasts (bone-forming cells) and osteoclasts (bone-destroying cells) are found on the inner or osteogenic layer of the periosteum.

2. Now inspect the **epiphysis,** the end of the long bone. Notice that it is composed of a thin layer of compact bone that encloses spongy bone.

3. Identify the **articular cartilage,** which covers the epiphyseal surface in place of the periosteum. Because it is composed of glassy hyaline cartilage, it provides a smooth surface to prevent friction at joint surfaces.

4. If the animal was still young and growing, you will be able to see the **epiphyseal plate,** a thin area of hyaline cartilage that provides for longitudinal growth of the bone during youth. Once the long bone has stopped growing, these areas are replaced with bone and appear as thin, barely discernible remnants—the **epiphyseal lines.**

5. In an adult animal, the central cavity of the shaft (*medullary cavity*) is essentially a storage region for adipose tissue, or **yellow marrow.** In the infant, this area is involved in forming blood cells, and so **red marrow** is found in the marrow cavities. In adult bones, the red marrow is confined to the interior of the epiphyses, where it occupies the spaces between the trabeculae of spongy bone.

6. If you are examining a fresh bone, look carefully to see if you can distinguish the delicate **endosteum** lining the shaft. The endosteum also covers the trabeculae of spongy bone and lines the canals of compact bone. Like the periosteum, the endosteum contains both osteoblasts and osteoclasts. As the bone grows in diameter on its external surface, it is constantly being broken down on its inner surface. Thus the thickness of the compact bone layer composing the shaft remains relatively constant.

7. If you have been working with a fresh bone specimen, return it to the appropriate area and properly dispose of your gloves, as designated by your instructor. Wash your hands before continuing on to the microscope study. ■

Longitudinal bone growth at epiphyseal plates (growth plates) follows a predictable sequence and provides a reliable indicator of the age of children exhibiting normal growth. In cases in which problems of long-bone growth are suspected (for example, pituitary dwarfism), X rays are taken to view the width of the growth plates. An abnormally thin epiphyseal plate indicates growth retardation. ■

Chemical Composition of Bone

Bone is one of the hardest materials in the body. Although relatively light, bone has a remarkable ability to resist tension and shear forces that continually act on it. An engineer would tell you that a cylinder (like a long bone) is one of the strongest structures for its mass. Thus nature has given us an extremely strong, exceptionally simple (almost crude), and flexible supporting system without sacrificing mobility.

The hardness of bone is due to the inorganic calcium salts deposited in its ground substance. Its flexibility comes from the organic elements of the matrix, particularly the collagen fibers.

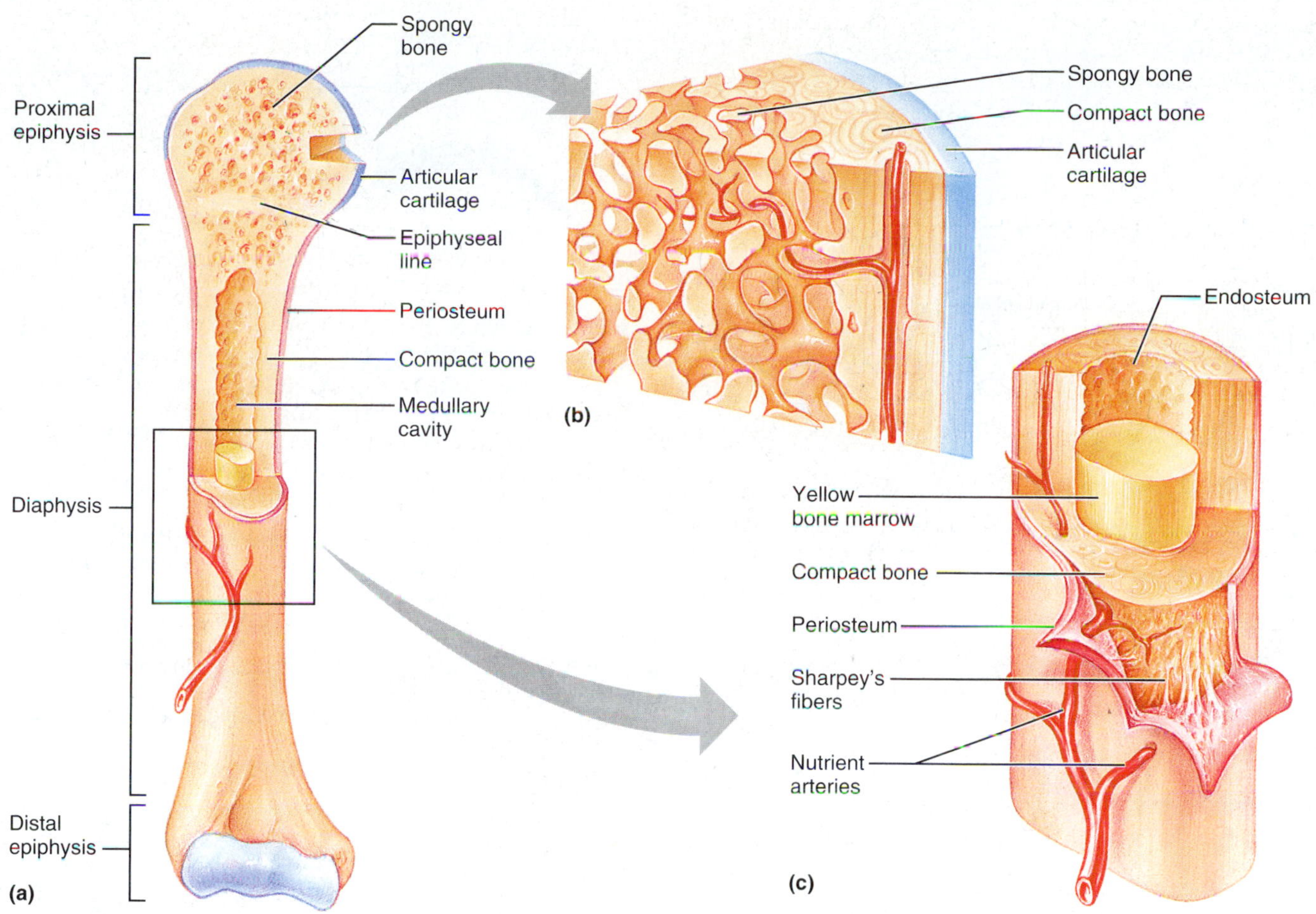

Figure 9.2 The structure of a long bone (humerus of the arm). **(a)** Anterior view with longitudinal section cut away at the proximal end. **(b)** Pie-shaped, three-dimensional view of spongy bone and compact bone of the epiphysis. **(c)** Cross section of shaft (diaphysis). Note that the external surface of the diaphysis is covered by a periosteum, but the articular surface of the epiphysis is covered with hyaline cartilage.

Activity 3: Examining the Effects of Heat and Hydrochloric Acid on Bones

Obtain a bone sample that has been soaked in hydrochloric acid (HCl) (or vinegar) and one that has been baked. Heating removes the organic part of bone, while acid dissolves out the minerals. Do the treated bones retain the structure of untreated specimens?

____

Gently apply pressure to each bone sample. What happens to the heated bone?

____

____

What happens to the bone treated with acid?

____

____

What does the acid appear to remove from the bone?

____

What does baking appear to do to the bone?

____

____

In rickets, the bones are not properly calcified. Which of the demonstration specimens would more closely resemble the bones of a child with rickets?

____ ■

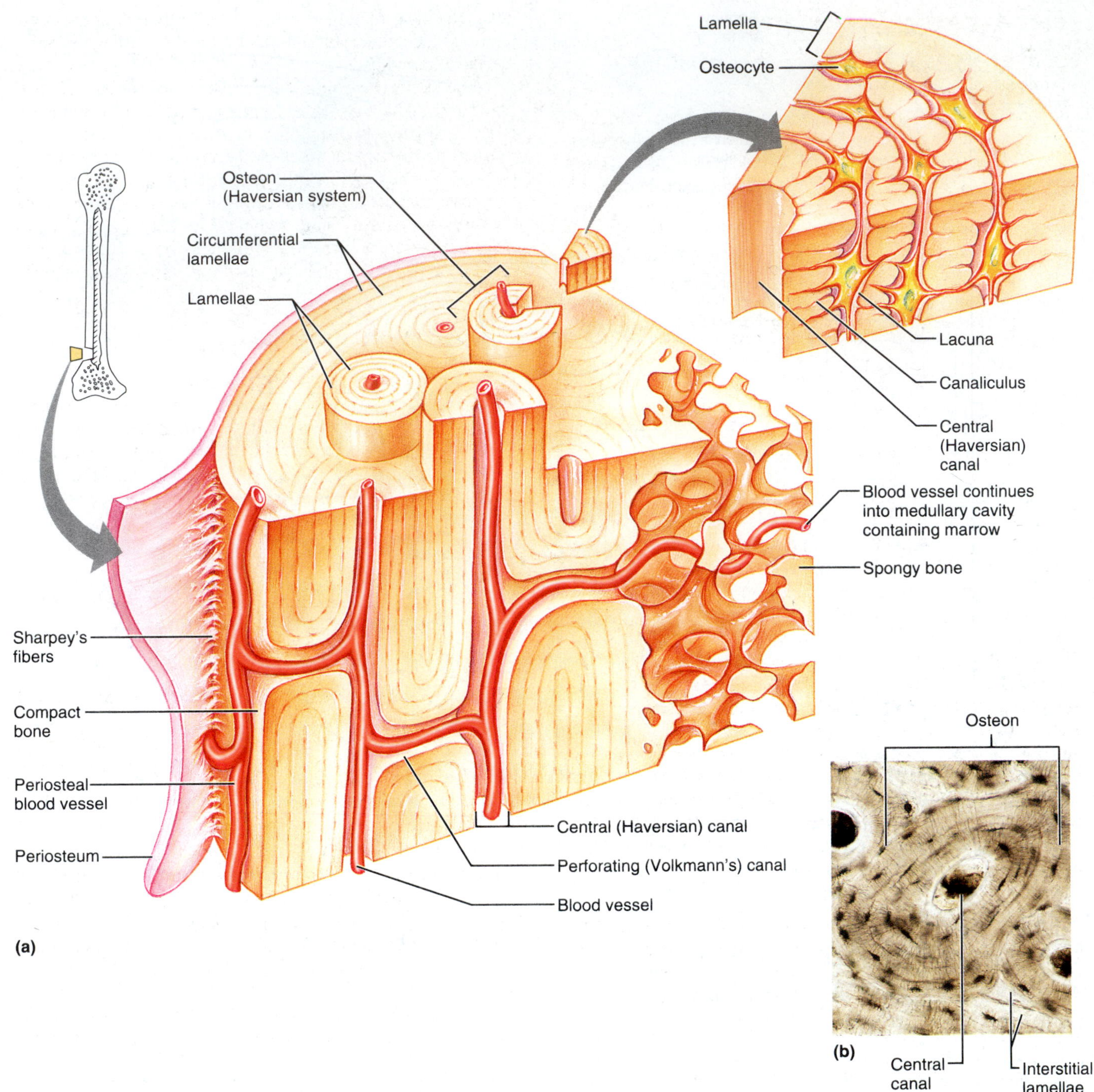

Figure 9.3 Microscopic structure of compact bone. (a) Diagrammatic view of a pie-shaped segment of compact bone, illustrating its structural units (osteons). The inset shows a more highly magnified view of a portion of one osteon. Notice the position of osteocytes in lacunae (cavities of the matrix). **(b)** Photomicrograph of a cross-sectional view of one osteon (90×).

Microscopic Structure of Compact Bone

As you have seen, spongy bone has a spiky, open-work appearance, resulting from the arrangement of the **trabeculae** that compose it, whereas compact bone appears to be dense and homogeneous. However, microscopic examination of compact bone reveals that it is riddled with passageways carrying blood vessels, nerves, and lymphatic vessels that provide the living bone cells with needed substances and a way to eliminate wastes. Indeed, bone histology is much easier to understand when you recognize that bone tissue is organized around its blood supply.

Activity 4: Examining the Microscopic Structure of Compact Bone

1. Obtain a prepared slide of ground bone and examine it under low power. Using Figure 9.3 as a guide, focus on a central canal. The **central (Haversian) canal** runs parallel to the long axis of the bone and carries blood vessels, nerves, and lymph vessels through the bony matrix. Identify the **osteocytes** (mature bone cells) in **lacunae** (chambers), which are arranged in concentric circles (concentric **lamellae**) around the central canal. A central canal and all the concentric lamellae surrounding it are referred to as an **osteon** or **Haversian system.** Also identify **canaliculi,** tiny canals radiating outward from a central canal to the lacunae of the first lamella and then from lamella to lamella. The canaliculi form a dense transportation network through the hard bone matrix, connecting all the living cells of the osteon to the nutrient supply. The canaliculi allow each cell to take what it needs for nourishment and to pass along the excess to the next osteocyte. You may need a higher-power magnification to see the fine canaliculi.

2. Also note the **perforating (Volkmann's) canals** in Figure 9.3. These canals run into the compact bone and marrow cavity from the periosteum, at right angles to the shaft. With the central canals, the perforating canals complete the communication pathway between the bone interior and its external surface.

3. If a model of bone histology is available, identify the same structures on the model. ■

Cartilages of the Skeleton

Location and Basic Structure

As mentioned earlier, cartilaginous regions of the skeleton have a fairly limited distribution in adults (Figure 9.4). The most important of these skeletal cartilages are (1) **articular cartilages,** which cover the bone ends at movable joints; (2) **costal cartilages,** found connecting the ribs to the sternum (breastbone); (3) **laryngeal cartilages,** which largely construct the larynx (voice box); (4) **tracheal** and **bronchial cartilages,** which reinforce other passageways of the respiratory system; (5) **nasal cartilages,** which support the external nose; (6) **intervertebral discs,** which separate and cushion bones of the spine (vertebrae); and (7) the cartilage supporting the external ear.

The skeletal cartilages consist of some variety of *cartilage tissue,* which typically consists primarily of water and is fairly resilient. Additionally, cartilage tissues are distinguished by the fact that they contain no nerves or blood vessels. Like bones, each cartilage is surrounded by a covering of dense connective tissue, called a *perichondrium* (rather than a periosteum). The perichondrium acts like a girdle to resist distortion of the cartilage when the cartilage is subjected to pressure. It also plays a role in cartilage growth and repair.

Classification of Cartilage

The skeletal cartilages have representatives from each of the three cartilage tissue types—hyaline, elastic, and fibrocartilage. Although you have already studied cartilage tissues (Exercise 6), some of that information will be recapped briefly here and you will have a chance to review the microscopic structure unique to each cartilage type.

Activity 5: Observing the Microscopic Structure of Different Types of Cartilage

Obtain prepared slides of hyaline cartilage, elastic cartilage, and fibrocartilage and bring them to your laboratory bench for viewing.

As you read through the descriptions of these cartilage types, keep in mind that the bulk of cartilage tissue consists of a nonliving *matrix* (containing a jellylike ground substance and fibers) secreted by chondrocytes.

Hyaline Cartilage **Hyaline cartilage** looks like frosted glass when viewed by the unaided eye. As easily seen in Figure 9.4, most skeletal cartilages are composed of hyaline cartilage. Its chondrocytes, snugly housed in lacunae, appear spherical, and collagen fibers are the only fiber type in its matrix. Hyaline cartilage provides sturdy support with some resilience or "give." Draw a small section of hyaline cartilage in the circle below. Label the chrondrocytes, lacunae, and the cartilage matrix. Compare your drawing to Figure 6.5g, p. 61.

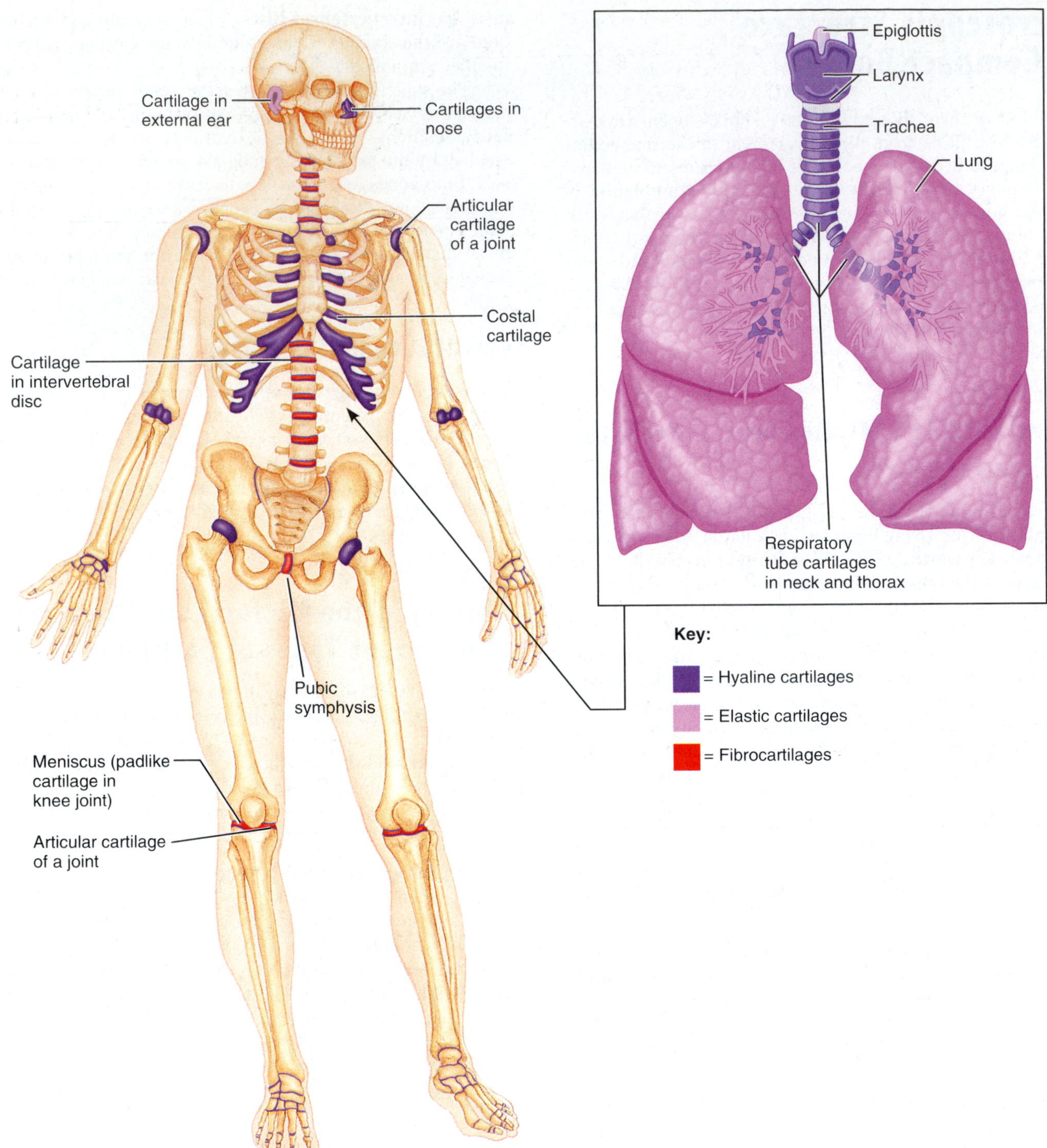

Figure 9.4 Cartilages in the adult skeleton and body. The cartilages that support the respiratory tubes and larynx are shown separately above.

Elastic Cartilage **Elastic cartilage** can be envisioned as "hyaline cartilage with more elastic fibers." Consequently, it is much more flexible than hyaline cartilage and it tolerates repeated bending better. Essentially, only the cartilages of the external ear and the epiglottis (which flops over and covers the larynx when we swallow) are made of elastic cartilage. Focus on how this cartilage differs from hyaline cartilage as you diagram it in the circle below. Compare your illustration to Figure 6.5h, p. 61.

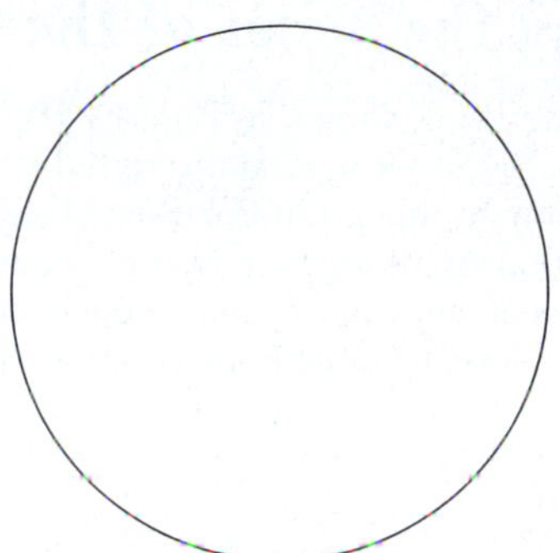

Fibrocartilage **Fibrocartilage** consists of rows of chondrocytes alternating with rows of thick collagen fibers. This tissue looks like a cartilage-dense regular connective tissue hybrid, and it is always found where hyaline cartilage joins a tendon or ligament. Fibrocartilage has great tensile strength and can withstand heavy compression. Hence, its use to construct the intervertebral discs and the cartilages within the knee joint makes a lot of sense (see Figure 9.4). Sketch a section of this tissue in the circle provided and then compare your sketch to the view seen in Figure 6.5i, p. 62. ■

exercise

10 The Axial Skeleton

Objectives

1. To identify the three bone groups composing the axial skeleton.
2. To identify the bones composing the axial skeleton, either by examining isolated bones or by pointing them out on an articulated skeleton or a skull, and to name the important bone markings on each.
3. To distinguish the different types of vertebrae.
4. To discuss the importance of intervertebral discs and spinal curvatures.
5. To distinguish three abnormal spinal curvatures.

Materials

- ❑ Intact skull and Beauchene skull
- ❑ X rays of individuals with scoliosis, lordosis, and kyphosis (if available)
- ❑ Articulated skeleton, articulated vertebral column
- ❑ Isolated cervical, thoracic, and lumbar vertebrae, sacrum, and coccyx
- A1A See Appendix C, Exercise 10 for links to A.D.A.M.® Interactive Anatomy.

The **axial skeleton** (the green portion of Figure 9.1 on p. 82) can be divided into three parts: the skull, the vertebral column, and the bony thorax.

The Skull

The **skull** is composed of two sets of bones. Those of the **cranium** enclose and protect the fragile brain tissue. The **facial bones** present the eyes in an anterior position and form the base for the facial muscles, which make it possible for us to present our feelings to the world. All but one of the bones of the skull are joined by interlocking joints called *sutures*. The mandible, or lower jawbone, is attached to the rest of the skull by a freely movable joint.

Activity 1: Identifying the Bones of the Skull

The bones of the skull, shown in Figures 10.1 through 10.7, are described below. As you read through this material, identify each bone on an intact (and/or Beauchene*) skull. Note that important bone markings are listed beneath the bones on which they appear and that a color-coded dot before each bone name corresponds to the bone color in the figures. ■

The Cranium

The cranium may be divided into two major areas for study—the **cranial vault** or **calvaria,** forming the superior, lateral, and posterior walls of the skull, and the **cranial floor** or **base,** forming the skull bottom. Internally, the cranial floor has three distinct concavities, the **anterior, middle,** and **posterior cranial fossae** (see Figure 10.3). The brain sits in these fossae, completely enclosed by the cranial vault.

Eight large flat bones construct the cranium. *With the exception of two paired bones (the parietals and the temporals), all are single bones.* Sometimes the six ossicles of the middle ear are also considered part of the cranium. Because the ossicles are functionally part of the hearing apparatus, their consideration is deferred to Exercise 25, Special Senses: Hearing and Equilibrium.

● **Frontal** See Figures 10.1, 10.3, and 10.6. Anterior portion of cranium; forms the forehead, superior part of the orbit, and floor of anterior cranial fossa.

Supraorbital foramen (notch): Opening above each orbit allowing blood vessels and nerves to pass.

Glabella: Smooth area between the eyes.

● **Parietal** See Figures 10.1 and 10.6. Posterolateral to the frontal bone, forming sides of cranium.

Sagittal suture: Midline articulation point of the two parietal bones.

Coronal suture: Point of articulation of parietals with frontal bone.

● **Temporal** See Figures 10.1 through 10.3 and 10.6. Inferior to parietal bone on lateral skull. The temporals can be divided into four major parts: the **squamous region** abuts the parietals; the **tympanic region** surrounds the external ear opening; the **mastoid region** is the area posterior to the ear; and the **petrous region** forms the lateral region of the skull base.

*Two views (Plates A and B) of a Beauchene skull are provided in the Human Anatomy Atlas.

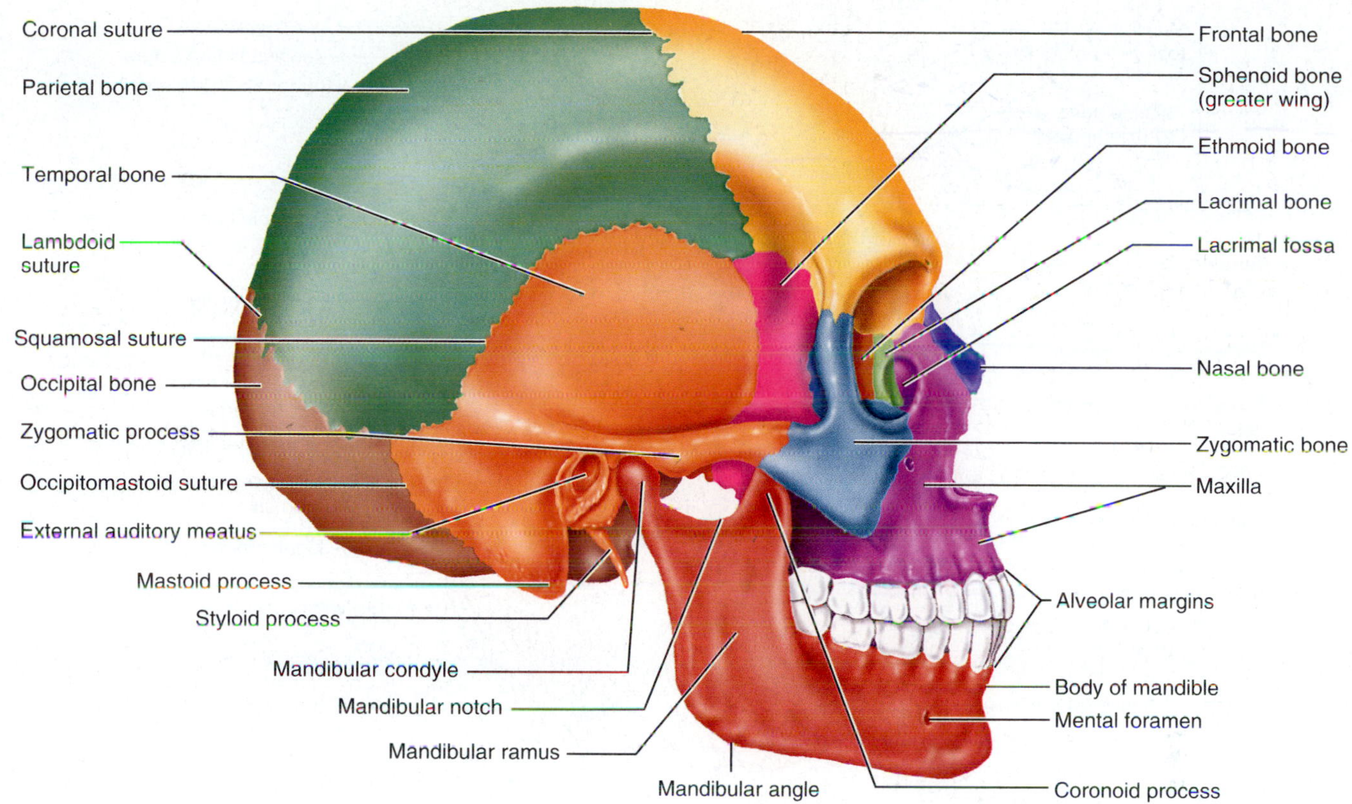

Figure 10.1 External anatomy of the right lateral aspect of the skull.

Important markings associated with the flaring squamous region (Figures 10.1 and 10.2) include:

Squamosal suture: Point of articulation of the temporal bone with the parietal bone.

Zygomatic process: A bridgelike projection joining the zygomatic bone (cheekbone) anteriorly. Together these two bones form the *zygomatic arch*.

Mandibular fossa: Rounded depression on the inferior surface of the zygomatic process (anterior to the ear); forms the socket for the mandibular condyle, the point where the mandible (lower jaw) joins the cranium.

Tympanic region markings (Figures 10.1 and 10.2) include:

External auditory meatus: Canal leading to eardrum and middle ear.

Styloid (*stylo*=stake, pointed object) **process:** Needlelike projection inferior to external auditory meatus; attachment point for muscles and ligaments of the neck. This process is often broken off demonstration skulls.

Prominent structures in the mastoid region (Figures 10.1 and 10.2) are:

Mastoid process: Rough projection inferior and posterior to external auditory meatus; attachment site for muscles.

The mastoid process, full of air cavities and so close to the middle ear—a trouble spot for infections—often becomes infected too, a condition referred to as **mastoiditis.** Because the mastoid area is separated from the brain by only a thin layer of bone, an ear infection that has spread to the mastoid process can inflame the brain coverings or the meninges. The latter condition is known as **meningitis.** ■

Stylomastoid foramen: Tiny opening between the mastoid and styloid processes through which cranial nerve VII leaves the cranium.

The petrous region (Figures 10.2 and 10.3), which helps form the middle and posterior cranial fossae, exhibits several obvious foramina with important functions:

Jugular foramen: Opening medial to the styloid process through which the internal jugular vein and cranial nerves IX, X, and XI pass.

Carotid canal: Opening medial to the styloid process through which the internal carotid artery passes into the cranial cavity.

Internal acoustic meatus: Opening on posterior aspect (petrous portion) of temporal bone allowing passage of cranial nerves VII and VIII (Figure 10.3).

Foramen lacerum: A jagged opening between the petrous temporal bone and the sphenoid providing passage for a number of small nerves, and for the internal carotid artery to enter the middle cranial fossa (after it passes through part of the temporal bone).

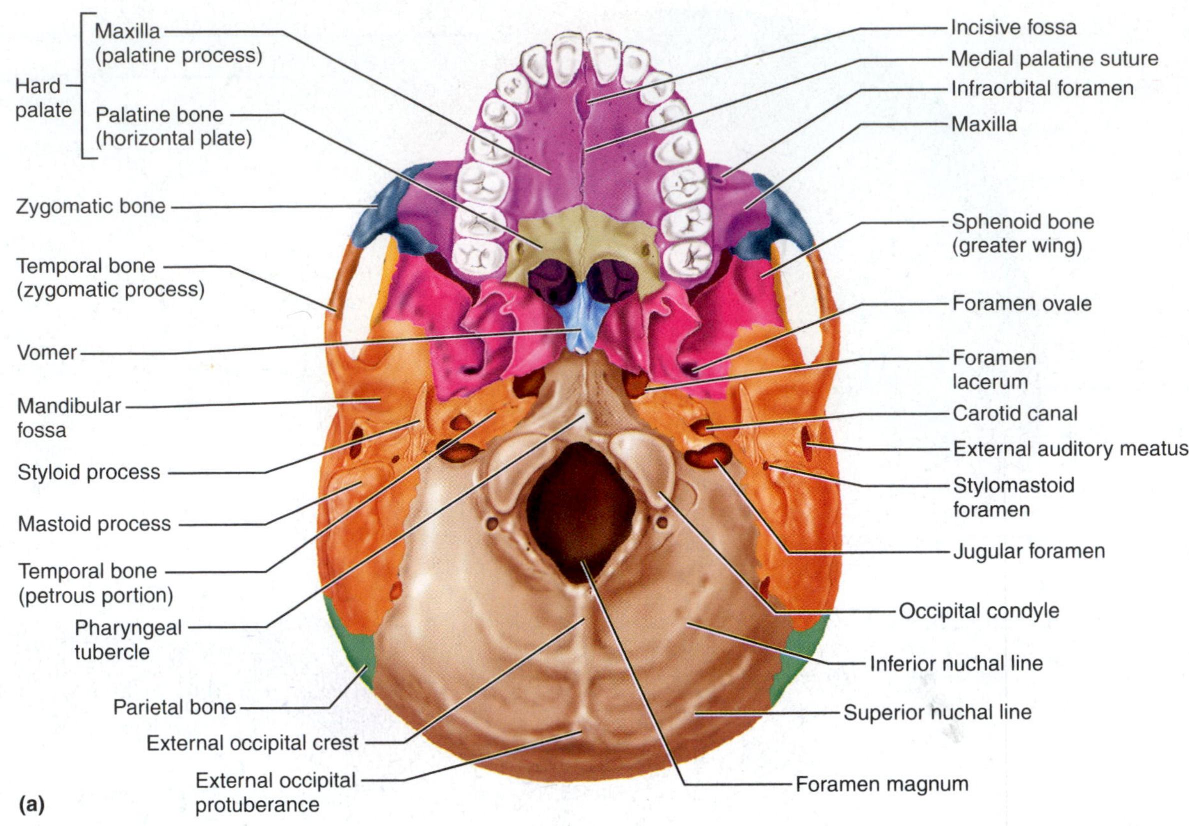

Figure 10.2 Inferior superficial view of the skull, mandible removed.

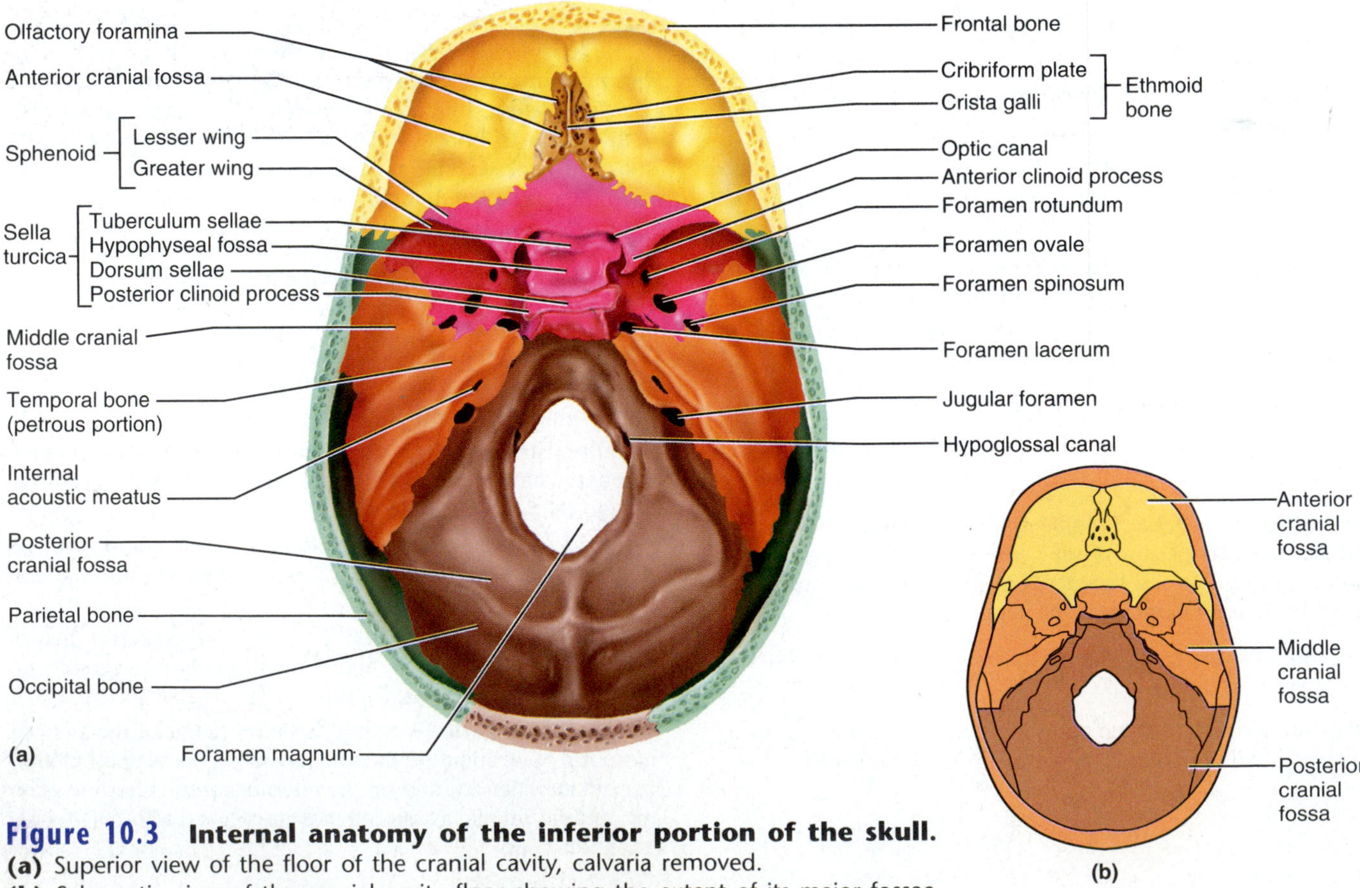

Olfactory foramina
Anterior cranial fossa
Sphenoid
Lesser wing
Greater wing
Sella turcica
Tuberculum sellae
Hypophyseal fossa
Dorsum sellae
Posterior clinoid process
Middle cranial fossa
Temporal bone (petrous portion)
Internal acoustic meatus
Posterior cranial fossa
Parietal bone
Occipital bone
(a)
Foramen magnum
Frontal bone
Cribriform plate
Crista galli
Ethmoid bone
Optic canal
Anterior clinoid process
Foramen rotundum
Foramen ovale
Foramen spinosum
Foramen lacerum
Jugular foramen
Hypoglossal canal
Anterior cranial fossa
Middle cranial fossa
Posterior cranial fossa
(b)

Figure 10.3 Internal anatomy of the inferior portion of the skull.
(a) Superior view of the floor of the cranial cavity, calvaria removed.
(b) Schematic view of the cranial cavity floor showing the extent of its major fossae.

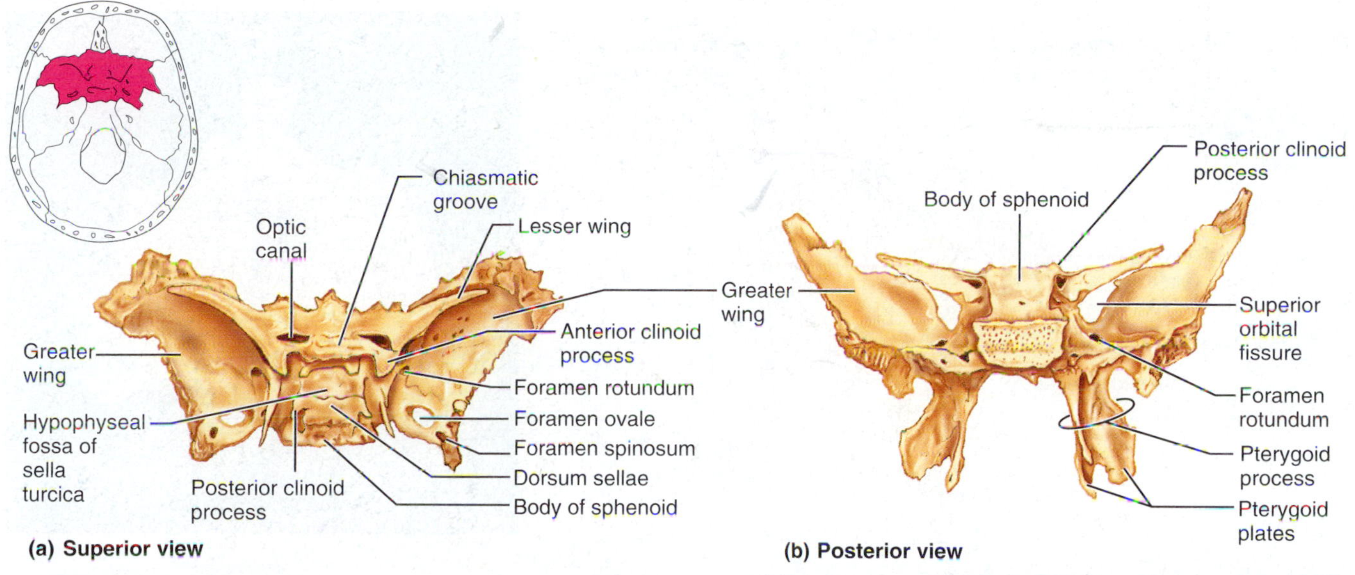

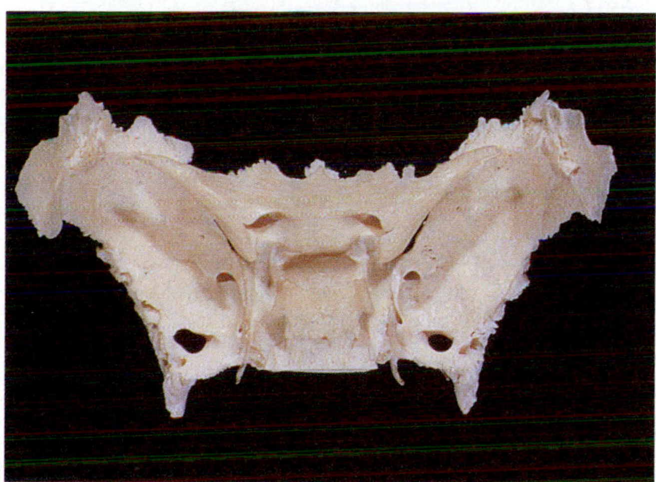

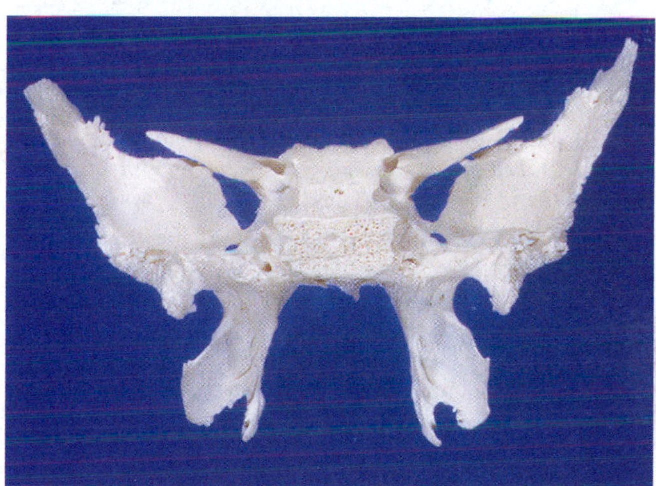

Figure 10.4 **The sphenoid bone.**

Occipital See Figures 10.1, 10.2, 10.3, and 10.6. Most posterior bone of cranium—forms floor and back wall. Joins sphenoid bone anteriorly via its narrow basioccipital region.

Lambdoid suture: Site of articulation of occipital bone and parietal bones.

Foramen magnum: Large opening in base of occipital, which allows the spinal cord to join with the brain.

Occipital condyles: Rounded projections lateral to the foramen magnum that articulate with the first cervical vertebra (atlas).

Hypoglossal canal: Opening medial and superior to the occipital condyle through which the hypoglossal nerve (cranial nerve XII) passes.

External occipital crest and protuberance: Midline prominences posterior to the foramen magnum.

Sphenoid See Figures 10.1 through 10.4 and 10.6. Bat-shaped bone forming the anterior plateau of the middle cranial fossa across the width of the skull.

Greater wings: Portions of the sphenoid seen exteriorly anterior to the temporal and forming a portion of the orbits of the eyes.

Superior orbital fissures: Jagged openings in orbits providing passage for cranial nerves III, IV, V, and VI to enter the orbit where they serve the eye.

The sphenoid bone can be seen in its entire width if the top of the cranium (calvaria) is removed (Figure 10.3).

Sella turcica (Turk's saddle): A saddle-shaped region in the sphenoid midline which nearly encloses the pituitary gland in a living person. The pituitary gland sits in the **hypophyseal fossa** portion of the sella turcica. This fossa is abutted fore and aft respectively by the **tuberculum sellae** and the **dorsum sellae.** The dorsum sellae terminates laterally in the **posterior clinoid processes.**

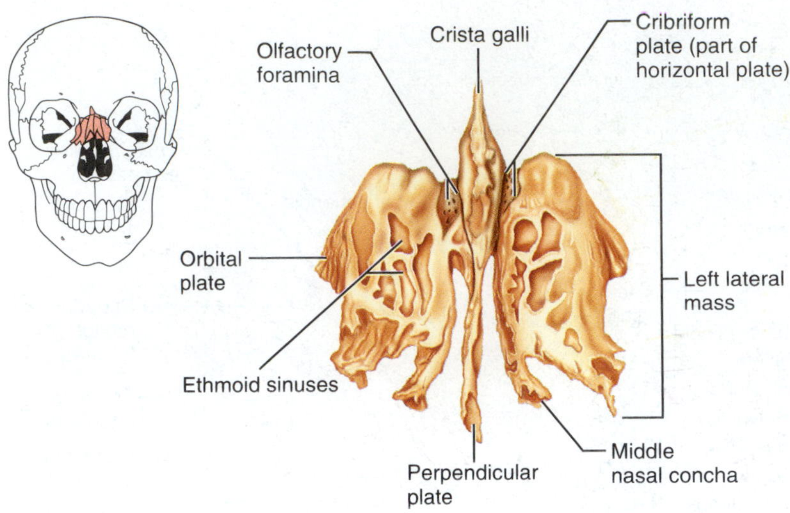

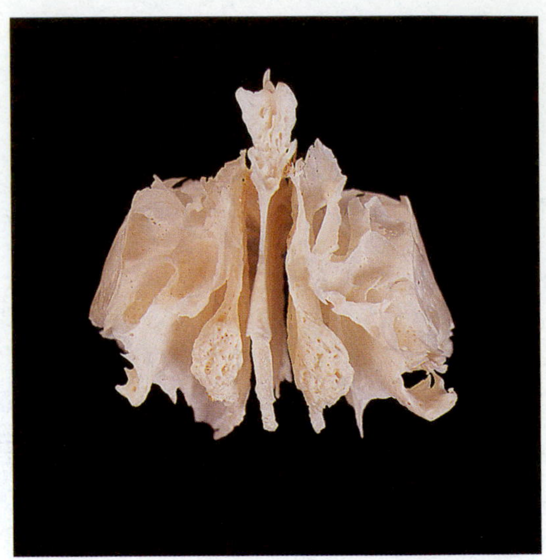

Figure 10.5 The ethmoid bone. Anterior view.

Lesser wings: Bat-shaped portions of the sphenoid anterior to the sella turcica. Posteromedially these terminate in the pointed **anterior clinoid processes,** which provide an anchoring site for the dura mater (outermost membrane covering of the brain).

Optic canals: Openings in the bases of the lesser wings through which the optic nerves enter the orbits to serve the eyes; these foramina are connected by the *chiasmatic groove*.

Foramen rotundum: Opening lateral to the sella turcica providing passage for a branch of the fifth cranial nerve. (This foramen is not visible on an inferior view of the skull.)

Foramen ovale: Opening posterior to the sella turcica that allows passage of a branch of the fifth cranial nerve.

● **Ethmoid** See Figures 10.1, 10.3, 10.5, and 10.6. Irregularly shaped bone anterior to the sphenoid. Forms the roof of the nasal cavity, upper nasal septum, and part of the medial orbit walls.

Crista galli (cock's comb): Vertical projection providing a point of attachment for the dura mater, helping to secure the brain within the skull.

Cribriform plates: Bony plates lateral to the crista galli through which olfactory fibers pass to the brain from the nasal mucosa. Together the cribriform plates and the midline crista galli form the *horizontal plate* of the ethmoid bone.

Perpendicular plate: Inferior projection of the ethmoid that forms the superior part of the nasal septum.

Lateral masses: Irregularly shaped thin-walled bony regions flanking the perpendicular plate laterally. Their lateral surfaces (*orbital plates*) shape part of the medial orbit wall.

Superior and middle nasal conchae (turbinates): Thin, delicately coiled plates of bone extending medially from the lateral masses of the ethmoid into the nasal cavity. The conchae make air flow through the nasal cavity more efficient and greatly increase the surface area of the mucosa that covers them, thus increasing the mucosa's ability to warm and humidify incoming air.

Facial Bones

Of the 14 bones composing the face, 12 are paired. *Only the mandible and vomer are single bones.* An additional bone, the hyoid bone, although not a facial bone, is considered here because of its location.

● **Mandible** See Figures 10.1, 10.6, and 10.7. The lower jawbone, which articulates with the temporal bones in the only freely movable joints of the skull.

Body: Horizontal portion; forms the chin.

Ramus: Vertical extension of the body on either side.

Mandibular condyle: Articulation point of the mandible with the mandibular fossa of the temporal bone.

Coronoid process: Jutting anterior portion of the ramus; site of muscle attachment.

Angle: Posterior point at which ramus meets the body.

Mental foramen: Prominent opening on the body (lateral to the midline) that transmits the mental blood vessels and nerve to the lower jaw.

Mandibular foramen: Open the lower jaw of the skull to identify this prominent foramen on the medial aspect of the mandibular ramus. This foramen permits passage of the nerve involved with tooth sensation (mandibular branch of cranial nerve V) and is the site where the dentist injects Novocain to prevent pain while working on the lower teeth.

Alveolar margin: Superior margin of mandible; contains sockets in which the teeth lie.

Mandibular symphysis: Anterior median depression indicating point of mandibular fusion.

● **Maxillae** See Figures 10.1, 10.2, 10.6, and 10.7. Two bones fused in a median suture; form the upper jawbone and part of the orbits. All facial bones, except the mandible, join the maxillae. Thus they are the main, or keystone, bones of the face.

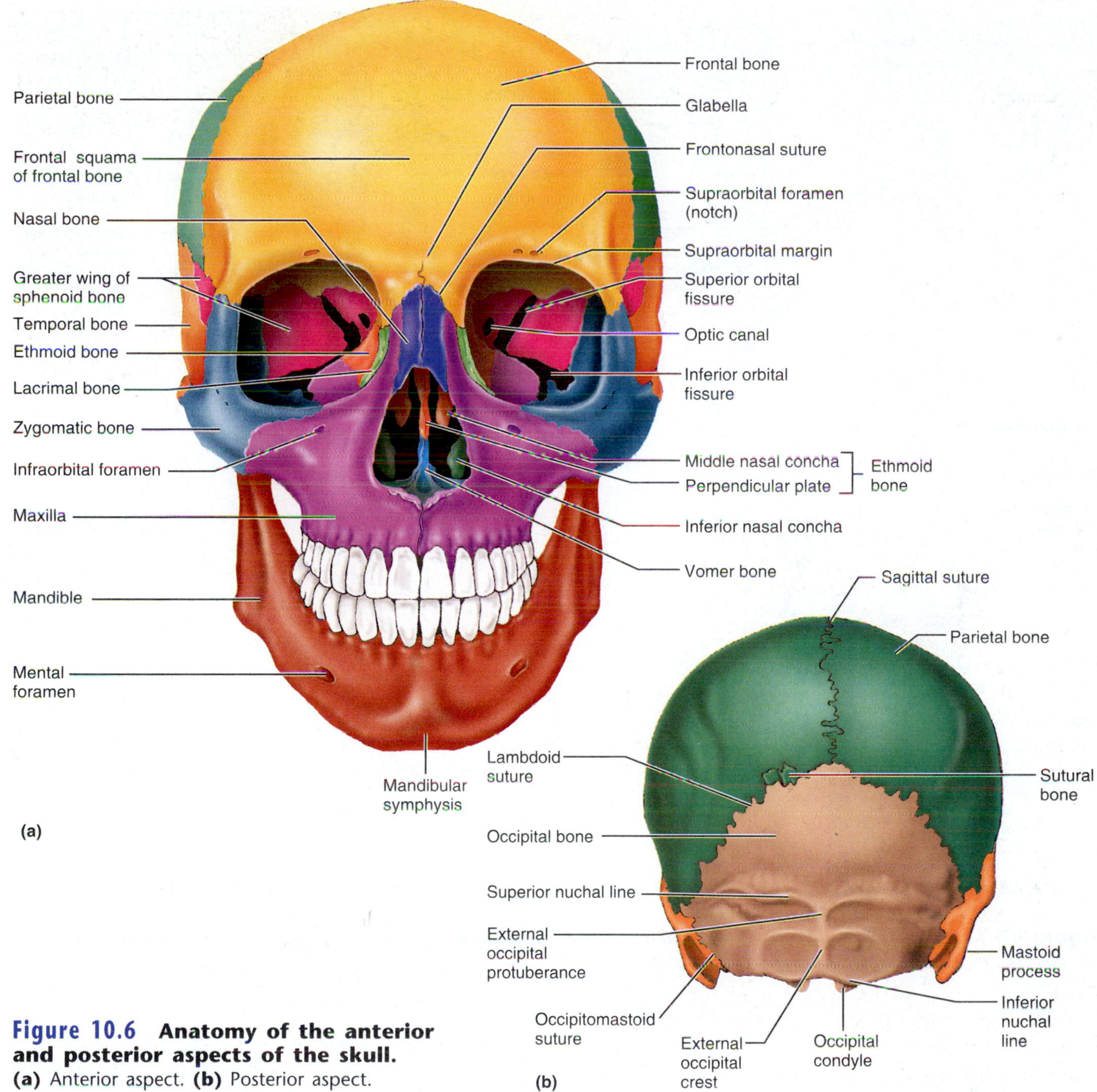

Figure 10.6 Anatomy of the anterior and posterior aspects of the skull. **(a)** Anterior aspect. **(b)** Posterior aspect.

Alveolar margin: Inferior margin containing sockets (alveoli) in which teeth lie.

Palatine processes: Form the anterior hard palate.

Infraorbital foramen: Opening under the orbit carrying the infraorbital nerves and blood vessels to the nasal region.

Incisive fossa: Large bilateral opening located posterior to the central incisor tooth of the maxilla and piercing the hard palate; transmits the nasopalatine arteries and blood vessels.

- **Palatine** See Figures 10.2 and 10.7. Paired bones posterior to the palatine processes; form posterior hard palate and part of the orbit.

- **Zygomatic** See Figures 10.1, 10.2, and 10.6. Lateral to the maxilla; forms the portion of the face commonly called the cheekbone, and forms part of the lateral orbit. Its three processes are named for the bones with which they articulate.

- **Lacrimal** See Figures 10.1 and 10.6. Fingernail-sized bones forming a part of the medial orbit walls between the maxilla and the ethmoid. Each lacrimal bone is pierced by an opening, the **lacrimal fossa,** which serves as a passageway for tears (*lacrima* means "tear").

- **Nasal** See Figures 10.1 and 10.6. Small rectangular bones forming the bridge of the nose.

Articulates with frontal bone
Frontal process
Orbital surface
Zygomatic process
Maxillary bone
Infraorbital foramen
Anterior nasal spine
Alveolar margin

(b) Maxilla

Temporomandibular joint
Mandibular fossa of temporal bone
Mandibular notch
Mandibular condyle
Coronoid process
Mandibular foramen
Ramus of mandible
Alveolar margin
Mental foramen
Mandibular angle
Body of mandible

Orbital process
Perpendicular plate
Sphenoidal process
Horizontal plate
Pyramidal process

(a) Mandible

(c) Palatine bones (posterior aspect)

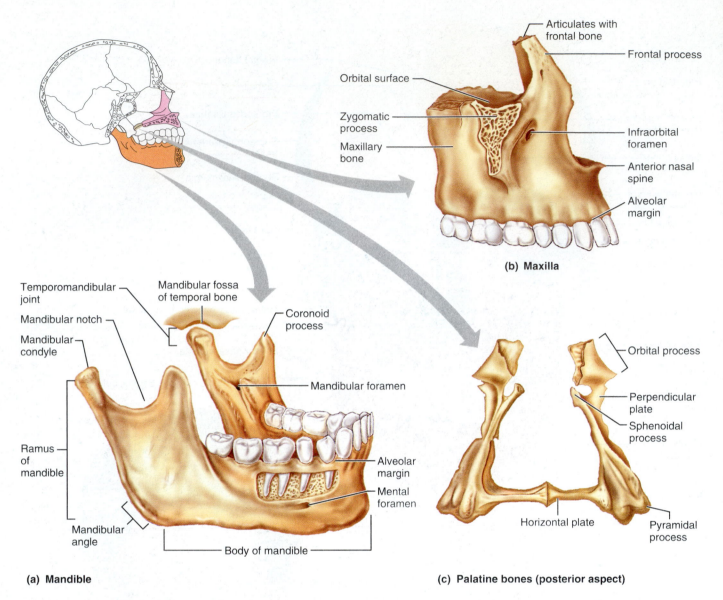

Figure 10.7 Detailed anatomy of some isolated facial bones. (Note that the mandible, maxilla, and palatine bones are not drawn in proportion to each other.)

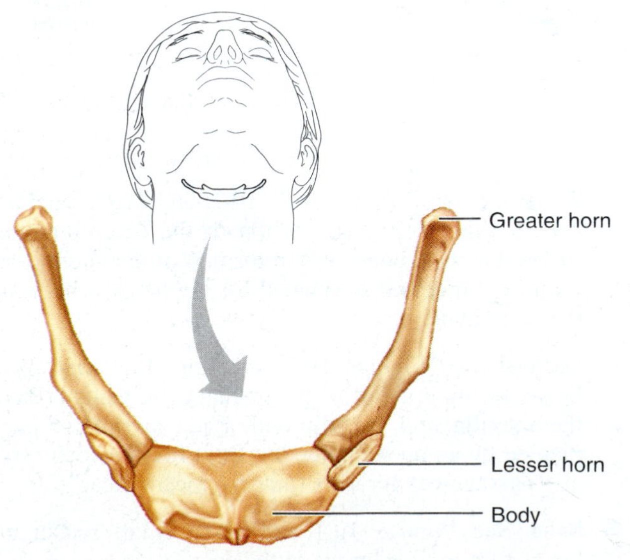

Figure 10.8 Hyoid bone.

- **Vomer (*vomer* = plow)** See Figures 10.2 and 10.6. Blade-shaped bone in median plane of nasal cavity that forms the posterior and inferior nasal septum.

- **Inferior Nasal Conchae (turbinates)** See Figure 10.6. Thin curved bones protruding medially from the lateral walls of the nasal cavity; serve the same purpose as the turbinate portions of the ethmoid bone (described earlier).

Hyoid Bone

Not really considered or counted as a skull bone. Located in the throat above the larynx (Figure 10.8); serves as a point of attachment for many tongue and neck muscles. Does not articulate with any other bone, and is thus unique. Horseshoe shaped with a body and two pairs of **horns,** or **cornua.**

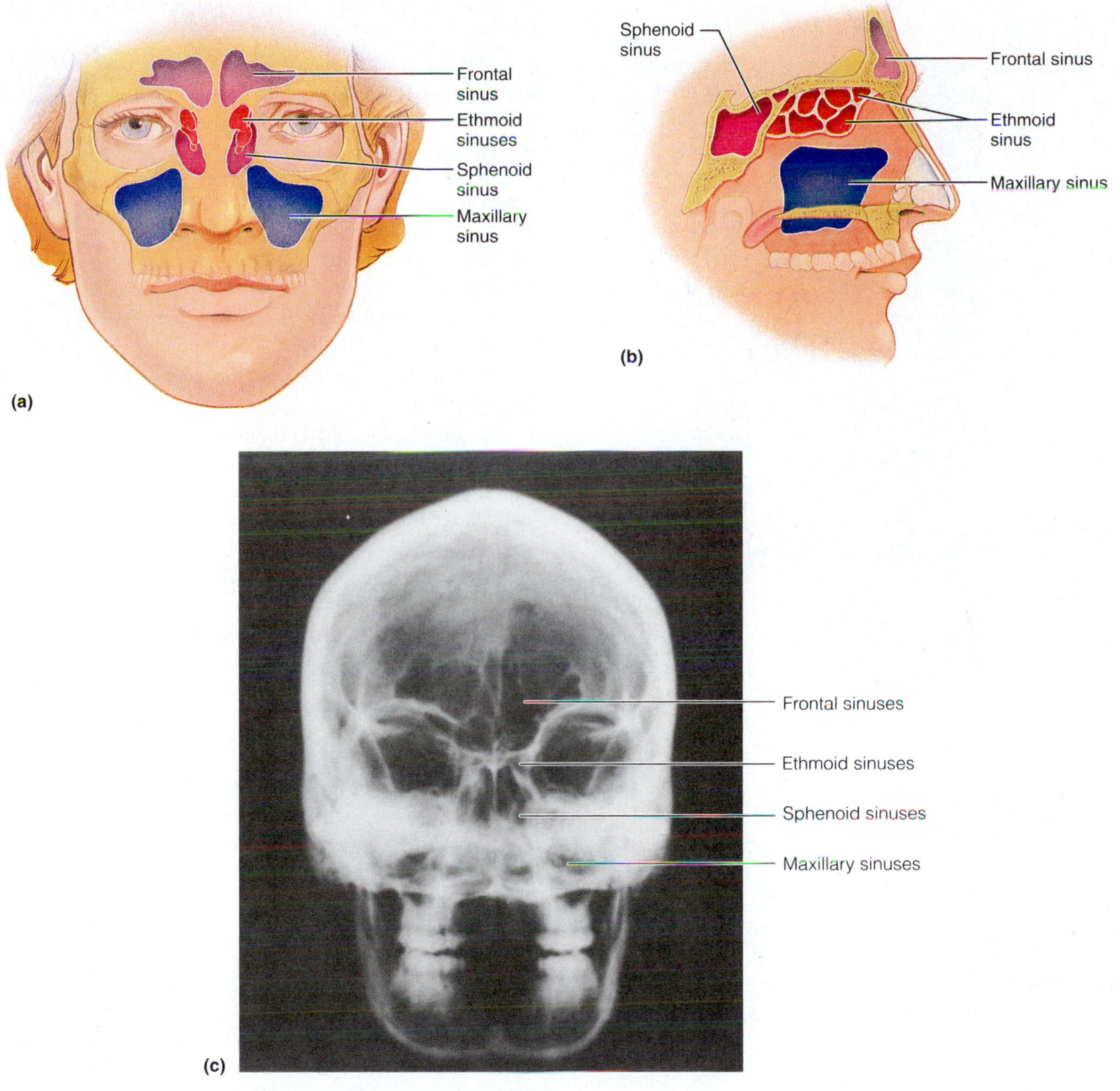

Figure 10.9 Paranasal sinuses. (a) Anterior "see-through" view. **(b)** As seen in a sagittal section of the head. **(c)** Skull X ray showing three of the paranasal sinuses, anterior view.

Paranasal Sinuses

Four skull bones—maxillary, sphenoid, ethmoid, and frontal—contain sinuses (mucosa-lined air cavities), which lead into the nasal passages (see Figure 10.9). These paranasal sinuses lighten the facial bones and may act as resonance chambers for speech. The maxillary sinus is the largest of the sinuses found in the skull.

Sinusitis, or inflammation of the sinuses, sometimes occurs as a result of an allergy or bacterial invasion of the sinus cavities. In such cases, some of the connecting passageways between the sinuses and nasal passages may become blocked with thick mucus or infectious material. Then, as the air in the sinus cavities is absorbed, a partial vacuum forms. The result is a sinus headache localized over the inflamed sinus area. Severe sinus infections may require surgical drainage to relieve this painful condition. ■

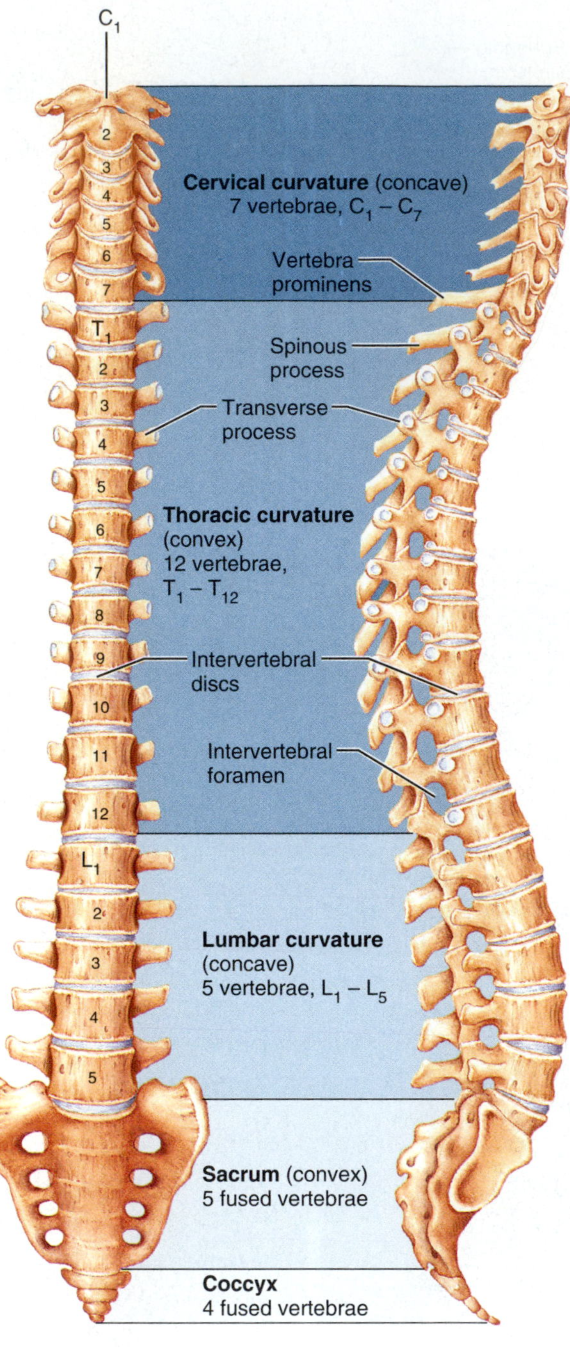

Figure 10.10 The vertebral column. Notice the curvatures in the lateral view. (The terms *convex* and *concave* refer to the curvature of the posterior aspect of the vertebral column.)

Activity 2: Palpating Skull Markings

Palpate the following areas on yourself:

- Zygomatic bone and arch. (The most prominent part of your cheek is your zygomatic bone. Follow the posterior course of the zygomatic arch to its junction with your temporal bone.)
- Mastoid process (the rough area behind your ear).
- Temporomandibular joints. (Open and close your jaws to locate these.)
- Greater wing of sphenoid. (Find the indentation posterior to the orbit and superior to the zygomatic arch on your lateral skull.)
- Superior orbital foramen. (Apply firm pressure along the superior orbital margin to find the indentation resulting from this foramen.)
- Inferior orbital foramen. (Apply firm pressure along the inferomedial border of the orbit to locate this large foramen.)
- Mandibular angle (most inferior and posterior aspect of the mandible).
- Mandibular symphysis (midline of chin).
- Nasal bones. (Run your index finger and thumb along opposite sides of the bridge of your nose until they "slip" medially at the inferior end of the nasal bones.)
- External occipital protuberance. (This midline projection is easily felt by running your fingers up the furrow at the back of your neck to the skull.)
- Hyoid bone. (Place a thumb and index finger beneath the chin just anterior to the mandibular angles, and squeeze gently. Exert pressure with the thumb, and feel the horn of the hyoid with the index finger.) ■

The Vertebral Column

The **vertebral column,** extending from the skull to the pelvis, forms the body's major axial support. Additionally, it surrounds and protects the delicate spinal cord while allowing the spinal nerves to issue from the cord via openings between adjacent vertebrae. The term *vertebral column* might suggest a rather rigid supporting rod, but this is far from the truth. The vertebral column consists of 24 single bones called **vertebrae** and two composite, or fused, bones (the sacrum and coccyx) that are connected in such a way as to provide a flexible curved structure (Figure 10.10). Of the 24 single vertebrae, the seven bones of the neck are called *cervical vertebrae;* the next 12 are *thoracic vertebrae;* and the 5 supporting the lower back are *lumbar vertebrae.* Remembering common mealtimes for breakfast, lunch, and dinner (7 A.M., 12 noon, and 5 P.M.) may help you to remember the number of bones in each region.

The vertebrae are separated by pads of fibrocartilage, **intervertebral discs,** that cushion the vertebrae and absorb shocks. Each disc is composed of two major regions, a central gelatinous *nucleus pulposus* that behaves like a fluid, and an outer ring of encircling collagen fibers called the *annulus fibrosus* that stabilizes the disc and contains the pulposus.

As a person ages, the water content of the discs decreases (as it does in other tissues throughout the body), and the discs become thinner and less compressible. This situation, along with other degenerative changes such as

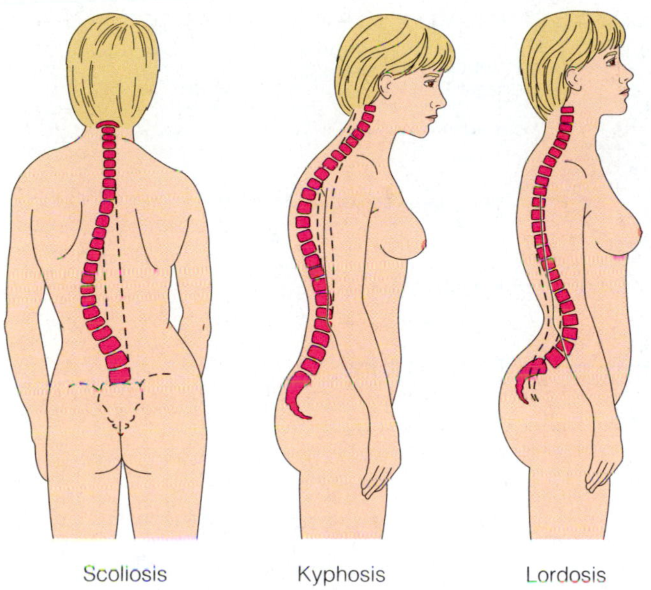

Figure 10.11 Abnormal spinal curvatures.

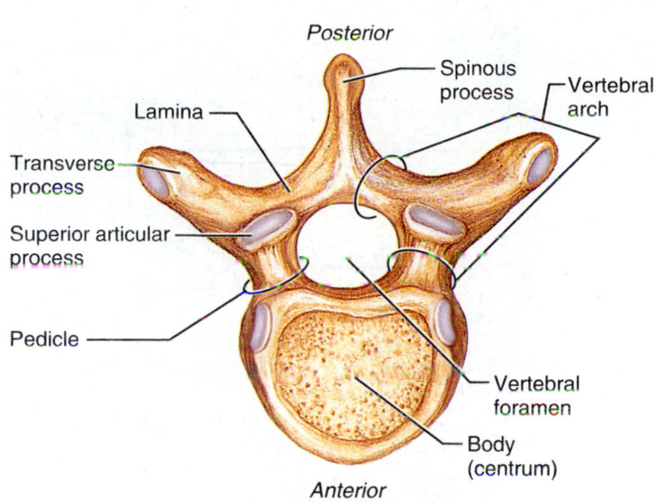

Figure 10.12 A typical vertebra, superior view. Inferior articulating surfaces not shown.

weakening of the ligaments and tendons of the vertebral column, predisposes older people to **ruptured discs.** In a **ruptured disc,** the nucleus pulposus herniates through the annulus portion and typically compresses adjacent nerves. ■

The presence of the discs and the S-shaped or springlike construction of the vertebral column prevent shock to the head in walking and running and provide flexibility to the body trunk. The thoracic and sacral curvatures of the spine are referred to as *primary curvatures* because they are present and well developed at birth. Later the *secondary curvatures* are formed. The cervical curvature becomes prominent when the baby begins to hold its head up independently, and the lumbar curvature develops when the baby begins to walk.

Activity 3: Examining Spinal Curvatures

1. Observe the normal curvature of the vertebral column in your laboratory specimen, and compare it to Figure 10.10. Then examine Figure 10.11, which depicts three abnormal spinal curvatures—*scoliosis, kyphosis,* and *lordosis.* These abnormalities may result from disease or poor posture. Also examine X rays, if they are available, showing these same conditions in a living patient.

2. Then, using an articulated vertebral column (or an articulated skeleton), examine the freedom of movement between two lumbar vertebrae separated by an intervertebral disc.

When the fibrous disc is properly positioned, are the spinal cord or peripheral nerves impaired in any way?

____

Remove the disc and put the two vertebrae back together. What happens to the nerve?

____

What would happen to the spinal nerves in areas of malpositioned or "slipped" discs?

____

____ ■

Structure of a Typical Vertebra

Although they differ in size and specific features, all vertebrae have some features in common (Figure 10.12).

Body (or centrum): Rounded central portion of the vertebra, which faces anteriorly in the human vertebral column.

Vertebral arch: Composed of pedicles, laminae, and a spinous process, it represents the junction of all posterior extensions from the vertebral body.

Vertebral foramen: Opening enclosed by the body and vertebral arch; a conduit for the spinal cord.

Transverse processes: Two lateral projections from the vertebral arch.

Spinous process: Single medial and posterior projection from the vertebral arch.

Superior and inferior articular processes: Paired projections lateral to the vertebral foramen that enable articulation with adjacent vertebrae. The superior articular processes typically face toward the spinous process, whereas the inferior articular processes face away from the spinous process.

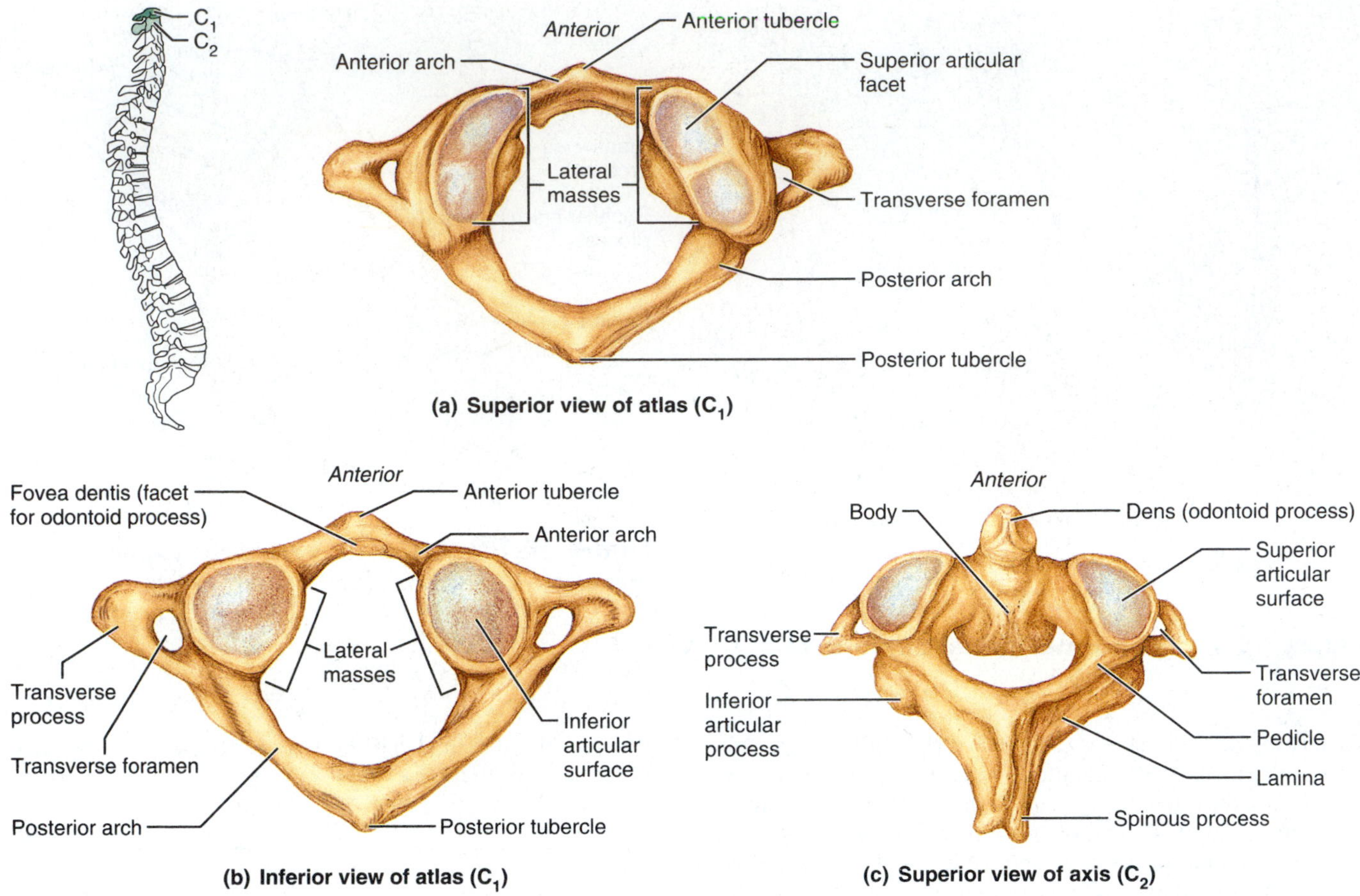

Figure 10.13 Cervical vertebrae C_1 and C_2.

Intervertebral foramina: The right and left pedicles have notches on their inferior and superior surfaces that create openings, the intervertebral foramina, for spinal nerves to leave the spinal cord between adjacent vertebrae.

Figures 10.13 and 10.14 and Table 10.1 show how specific vertebrae differ; refer to them as you read the following sections.

Cervical Vertebrae

The seven cervical vertebrae (referred to as C_1 through C_7) form the neck portion of the vertebral column. The first two cervical vertebrae (atlas and axis) are highly modified to perform special functions (see Figure 10.13). The **atlas** (C_1) lacks a body, and its lateral processes contain large concave depressions on their superior surfaces that receive the occipital condyles of the skull. This joint enables you to nod "yes." The **axis** (C_2) acts as a pivot for the rotation of the atlas (and skull) above. It bears a large vertical process, the **odontoid process,** or **dens,** that serves as the pivot point. The articulation between C_1 and C_2 allows you to rotate your head from side to side to indicate "no."

The more typical cervical vertebrae (C_3 through C_7) are distinguished from the thoracic and lumbar vertebrae by several features (see Table 10.1). They are the smallest, lightest vertebrae and the vertebral foramen is triangular. The spinous process is short and often bifurcated, or divided into two branches. The spinous process of C_7 is not branched, however, and is substantially longer than that of the other cervical vertebrae. Because the spinous process of C_7 is visible through the skin, it is called the *vertebra prominens* (Figure 10.10) and is used as a landmark for counting the vertebrae. Transverse processes of the cervical vertebrae are wide, and they contain foramina through which the vertebral arteries pass superiorly on their way to the brain. Any time you see these foramina in a vertebra, you can be sure that it is a cervical vertebra.

- Palpate your vertebra prominens.

Thoracic Vertebrae

The 12 thoracic vertebrae (referred to as T_1 through T_{12}) may be recognized by the following structural characteristics. As shown in Table 10.1, they have a larger body than the cervical vertebrae. The body is somewhat heart shaped, with two small articulating surfaces, or *costal demifacets,* on each side (one superior, the other inferior) close to the origin of the vertebral arch. These demifacets articulate with the heads of the corresponding ribs. The vertebral foramen is oval or round, and the spinous process is long, with a sharp downward hook. The closer the thoracic vertebra is to the lumbar region, the less sharp and shorter the spinous process. Articular facets on the transverse processes articulate with the tubercles of the

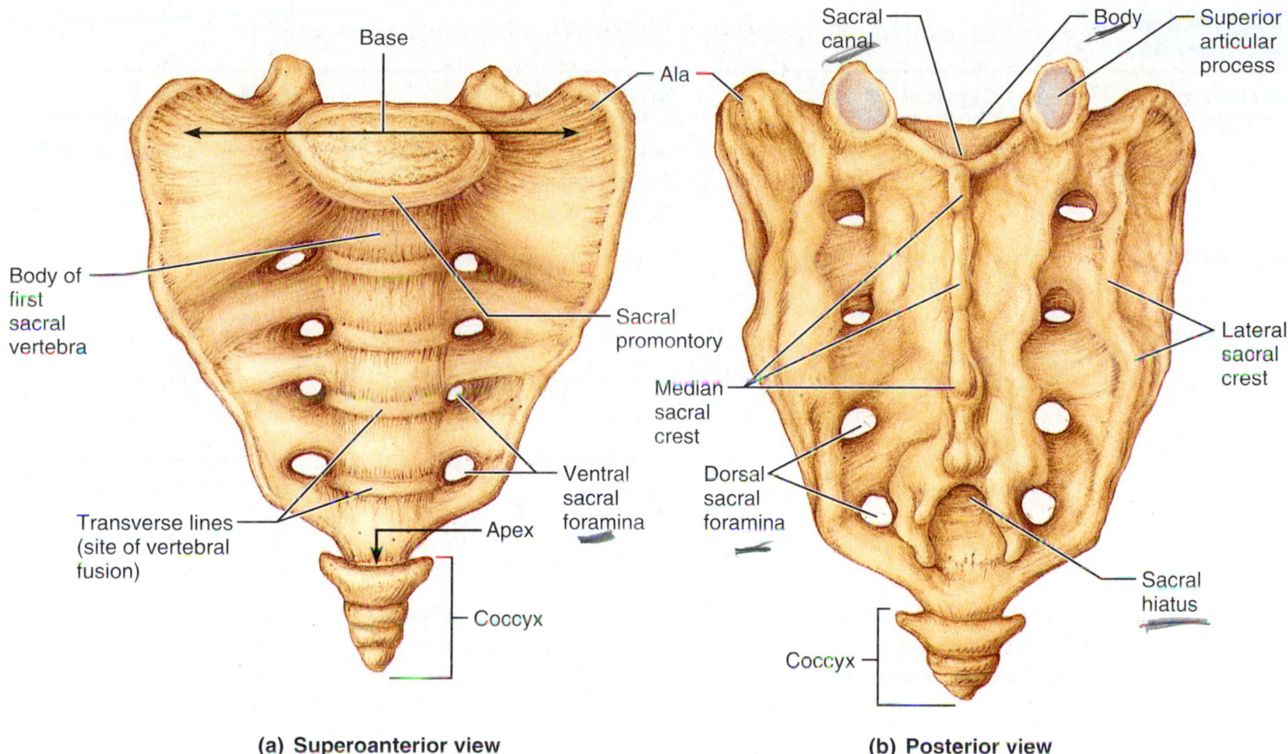

Figure 10.14 Sacrum and coccyx.

ribs. Besides forming the thoracic part of the spine, these vertebrae form the posterior aspect of the bony thoracic cage (rib cage). Indeed, they are the only vertebrae that articulate with the ribs.

Lumbar Vertebrae

The five lumbar vertebrae (L_1 through L_5) have massive blocklike bodies and short, thick, hatchet-shaped spinous processes extending directly backward (see Table 10.1). The superior articular facets face posteromedially; the inferior ones are directed anterolaterally. These structural features reduce the mobility of the lumbar region of the spine. Since most stress on the vertebral column occurs in the lumbar region, these are also the sturdiest of the vertebrae.

The spinal cord ends at the superior edge of L_2, but the outer covering of the cord, filled with cerebrospinal fluid, extends an appreciable distance beyond. Thus a *lumbar puncture* (for examination of the cerebrospinal fluid) or the administration of "saddle block" anesthesia for childbirth is normally done between L_3 and L_4 or L_4 and L_5, where there is little or no chance of injuring the delicate spinal cord.

The Sacrum

The **sacrum** (Figure 10.14) is a composite bone formed from the fusion of five vertebrae. Superiorly it articulates with L_5, and inferiorly it connects with the coccyx. The **median sacral crest** is a remnant of the spinous processes of the fused vertebrae. The winglike **alae,** formed by fusion of the transverse processes, articulate laterally with the hip bones. The sacrum is concave anteriorly and forms the posterior border of the pelvis. Four ridges (lines of fusion) cross the anterior part of the sacrum, and **sacral foramina** are located at either end of these ridges. These foramina allow blood vessels and nerves to pass. The vertebral canal continues inside the sacrum as the **sacral canal** and terminates near the coccyx via an enlarged opening called the **sacral hiatus.** The **sacral promontory** (anterior border of the body of S_1) is an important anatomical landmark for obstetricians.

• Attempt to palpate the median sacral crest of your sacrum. (This is more easily done by thin people and obviously in privacy.)

The Coccyx

The **coccyx** (see Figure 10.14) is formed from the fusion of three to five small irregularly shaped vertebrae. It is literally the human tailbone, a vestige of the tail that other vertebrates have. The coccyx is attached to the sacrum by ligaments.

Activity 4:
Examining Vertebral Structure

Obtain examples of each type of vertebra and examine them carefully, comparing them to Figures 10.13, 10.14, and Table 10.1, and to each other. ■

Table 10.1 Regional Characteristics of Cervical, Thoracic, and Lumbar Vertebrae

Characteristic	(a) Cervical (3–7)	(b) Thoracic	(c) Lumbar
Body	Small, wide side to side	Larger than cervical; heart shaped; bears two costal demifacets	Massive; kidney shaped
Spinous process	Short; bifid; projects directly posteriorly	Long; sharp; projects inferiorly	Short; blunt; projects directly posteriorly
Vertebral foramen	Triangular	Circular	Triangular
Transverse processes	Contain foramina	Bear facets for ribs (except T_{11} and T_{12})	Thin and tapered
Superior and inferior articulating processes	Superior facets directed superoposteriorly	Superior facets directed posteriorly	Superior facets directed posteromedially (or medially)
	Inferior facets directed inferoanteriorly	Inferior facets directed anteriorly	Inferior facets directed anterolaterally (or laterally)
Movements allowed	Flexion and extension; lateral flexion; rotation; the spine region with the greatest range of movement	Rotation; lateral flexion possible but limited by ribs; flexion and extension prevented	Flexion and extension; some lateral flexion; rotation prevented

Superior view

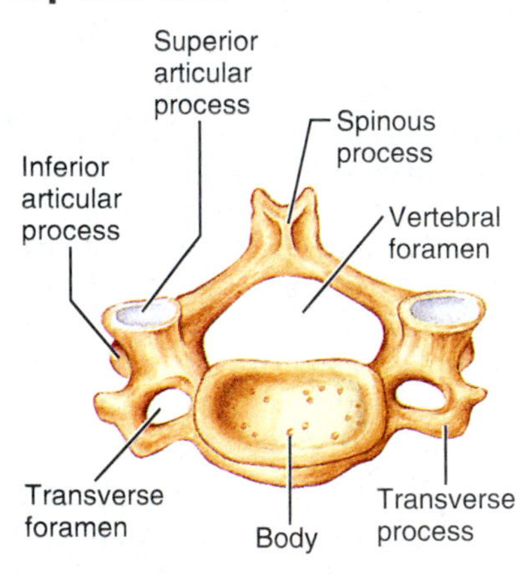

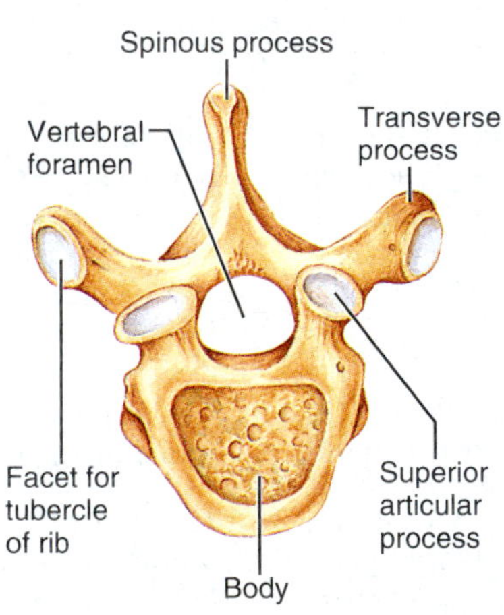

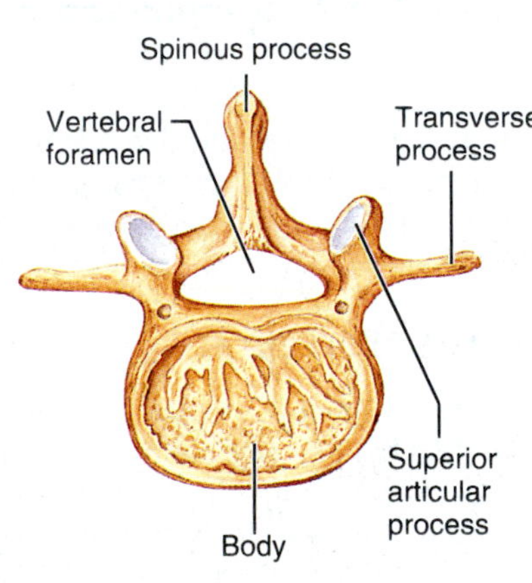

Right lateral view

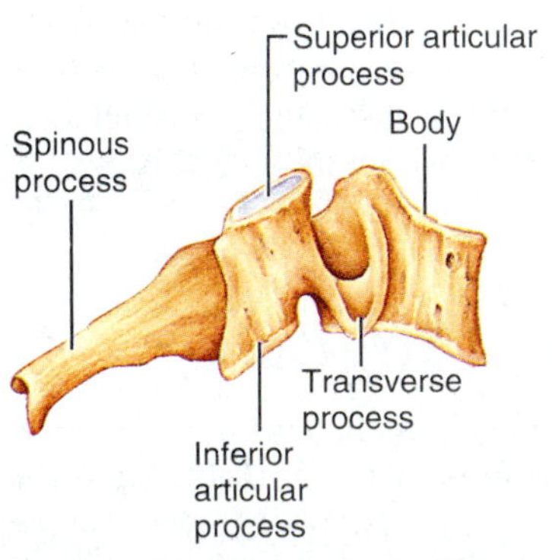

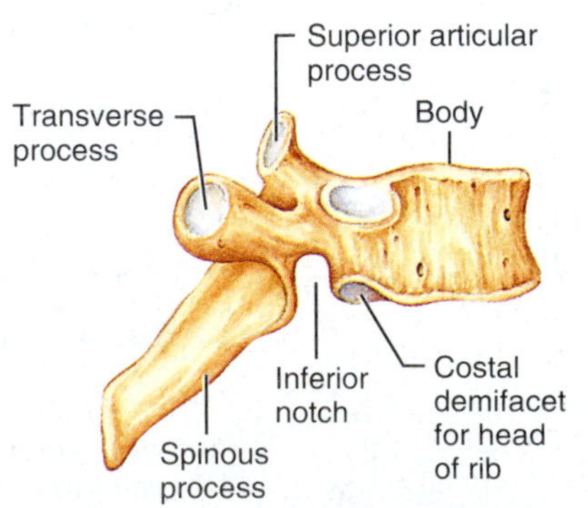

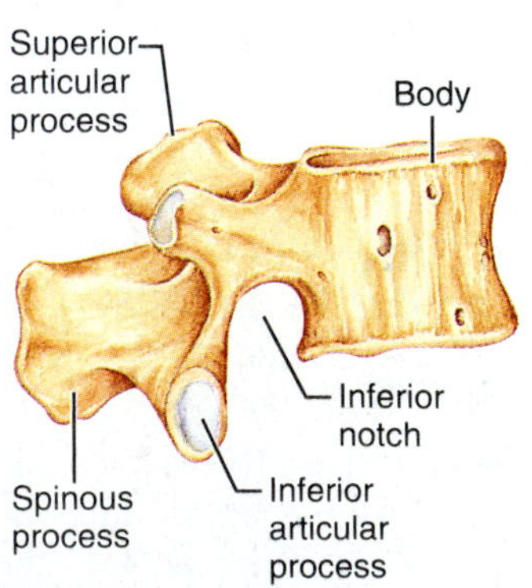

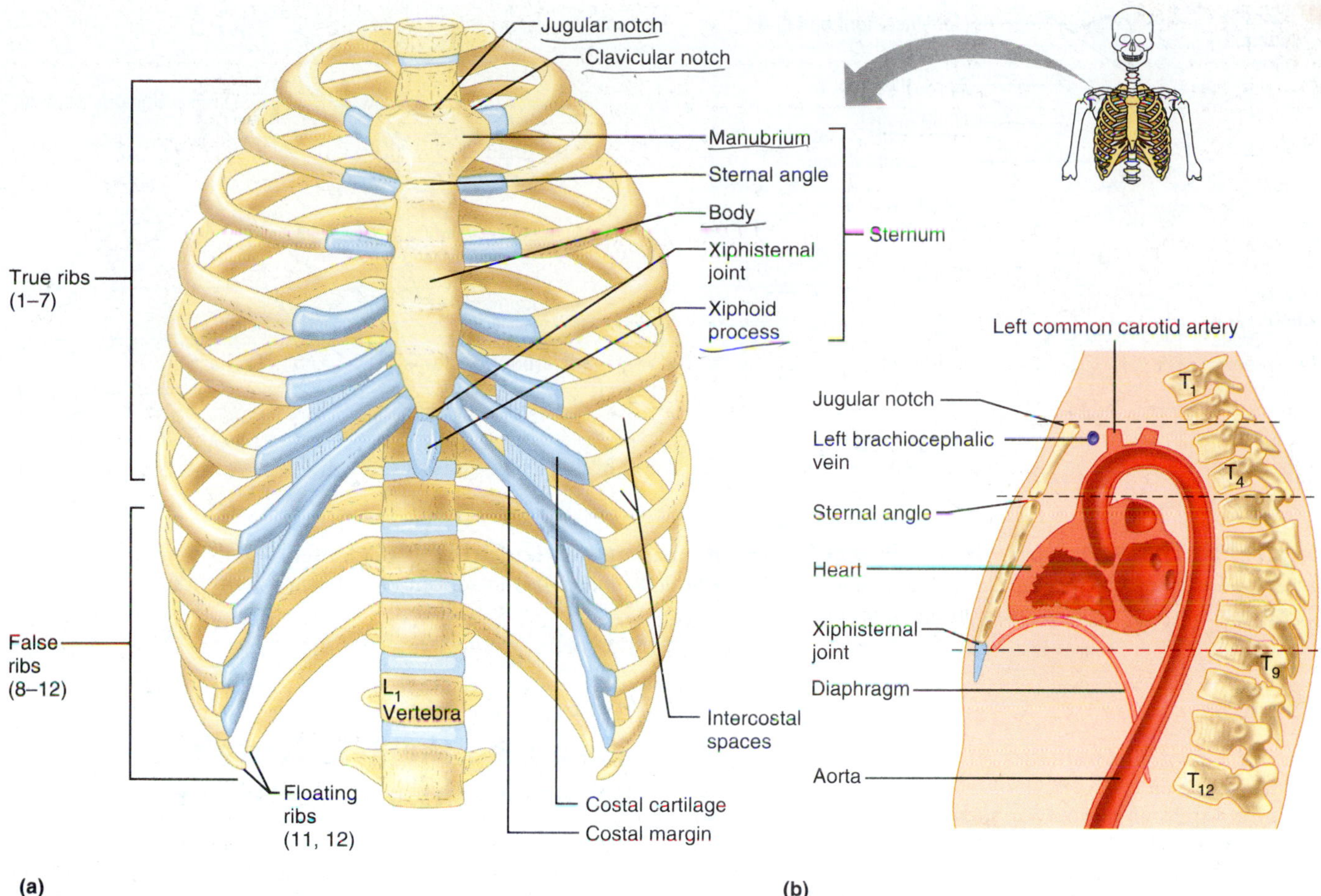

Figure 10.15 The bony thorax. (a) Skeleton of the bony thorax, anterior view (costal cartilages are shown in blue). **(b)** Left lateral view of the thorax, illustrating the relationship of the surface anatomical landmarks of the thorax to the vertebral column (thoracic portion).

The Bony Thorax

The **bony thorax** is composed of the sternum, ribs, and thoracic vertebrae (Figure 10.15). It is also referred to as the **thoracic cage** because of its appearance and because it forms a protective cone-shaped enclosure around the organs of the thoracic cavity (heart and lungs, for example).

The Sternum

The **sternum** (breastbone), a typical flat bone, is a result of the fusion of three bones—the manubrium, body, and xiphoid process. It is attached to the first seven pairs of ribs. The superiormost **manubrium** looks like the knot of a tie; it articulates with the clavicle (collarbone) laterally. The **body (gladiolus)** forms the bulk of the sternum. The **xiphoid process** constructs the inferior end of the sternum and lies at the level of the fifth intercostal space. Although it is made of hyaline cartilage in children, it is usually ossified in adults.

In some people, the xiphoid process projects dorsally. This may present a problem because physical trauma to the chest can push such a xiphoid into the heart or liver (both immediately deep to the process), causing massive hemorrhage. ■

The sternum has three important bony landmarks—the jugular notch, the sternal angle, and the xiphisternal joint. The **jugular notch** (concave upper border of the manubrium) can be palpated easily; generally it is at the level of the third thoracic vertebra. The **sternal angle** is a result of the manubrium and body meeting at a slight angle to each other, so that a transverse ridge is formed at the level of the second ribs. It provides a handy reference point for counting ribs to locate the second intercostal space for listening to certain heart valves, and is an important anatomical landmark for thoracic surgery. The **xiphisternal joint,** the point where the sternal body and xiphoid process fuse, lies at the level of the ninth thoracic vertebra.

- Palpate your sternal angle and jugular notch.

Because of its accessibility, the sternum is a favored site for obtaining samples of blood-forming (hematopoietic) tissue for the diagnosis of suspected blood diseases. A needle is inserted into the marrow of the sternum and the sample withdrawn (sternal puncture).

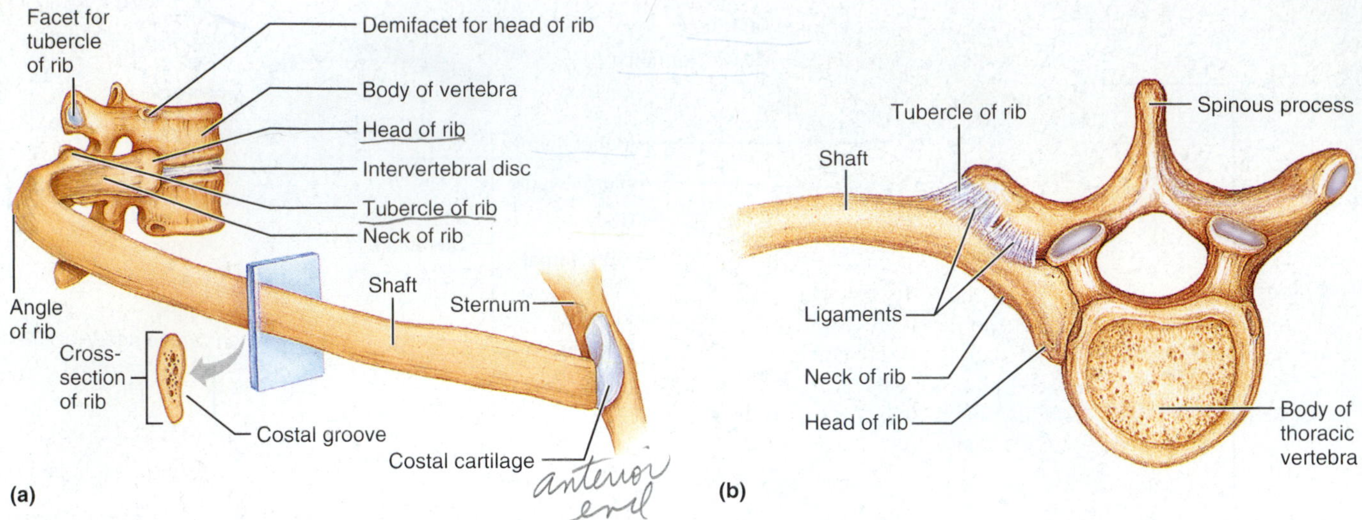

Figure 10.16 Structure of a "typical" true rib and its articulations. **(a)** Vertebral and sternal articulations of a typical true rib. **(b)** Superior view of the articulation between a rib and a thoracic vertebra, with costovertebral ligaments shown on left side only.

The Ribs

The 12 pairs of **ribs** form the walls of the thoracic cage (see Figures 10.15 and 10.16). All of the ribs articulate posteriorly with the vertebral column via their heads and tubercles and then curve downward and toward the anterior body surface. The first seven pairs, called the *true*, or *vertebrosternal, ribs*, attach directly to the sternum by their "own" costal cartilages. The next five pairs are called *false ribs;* they attach indirectly to the sternum or entirely lack a sternal attachment. Of these, rib pairs 8–10, which are also called *vertebrochondral ribs*, have indirect cartilage attachments to the sternum via the costal cartilage of rib 7. The last two pairs, called *floating*, or *vertebral, ribs*, have no sternal attachment.

Activity 5:

Examining the Relationship Between Ribs and Vertebrae

First take a deep breath to expand your chest. Notice how your ribs seem to move outward and how your sternum rises. Then examine an articulated skeleton to observe the relationship between the ribs and the vertebrae.

Refer to Activity 3: Palpating Landmarks of the Trunk and Activity 4: Palpating Landmarks of the Abdomen in Exercise 46, Surface Anatomy Roundup (pp. 464–467). ■

The Appendicular Skeleton

exercise **11**

Objectives

1. To identify on an articulated skeleton the bones of the pectoral and pelvic girdles and their attached limbs.
2. To arrange unmarked, disarticulated bones in proper relative position to form the entire skeleton.
3. To differentiate between a male and a female pelvis.
4. To discuss the common features of the human appendicular girdles (pectoral and pelvic), and to note how their structure relates to their specialized functions.
5. To identify specific bone markings in the appendicular skeleton.

Materials

- ❑ Articulated skeletons
- ❑ Disarticulated skeletons (complete)
- ❑ Articulated pelves (male and female for comparative study)
- ❑ X rays of bones of the appendicular skeleton

A1A See Appendix C, Exercise 11 for links to A.D.A.M.® Interactive Anatomy.

The **appendicular skeleton** (the gold-colored portion of Figure 9.1) is composed of the 126 bones of the appendages and the pectoral and pelvic girdles, which attach the limbs to the axial skeleton. Although the upper and lower limbs differ in their functions and mobility, they have the same fundamental plan, with each limb composed of three major segments connected together by freely movable joints.

Activity 1:

Examining and Identifying Bones of the Appendicular Skeleton

Carefully examine each of the bones described throughout this exercise and identify the characteristic bone markings of each. The markings aid in determining whether a bone is the right or left member of its pair. *This is a very important instruction because you will be constructing your own skeleton to finish this laboratory exercise.* Additionally, when corresponding X rays are available, compare the actual bone specimen to its X ray image. ■

Bones of the Pectoral Girdle and Upper Extremity

The Pectoral (Shoulder) Girdle

The paired **pectoral,** or **shoulder, girdles** (Figure 11.1) each consist of two bones—the anterior clavicle and the posterior scapula. The shoulder girdles function to attach the upper limbs to the axial skeleton. In addition, the shoulder girdles serve as attachment points for many trunk and neck muscles.

The **clavicle,** or collarbone, is a slender doubly curved bone—convex forward on its medial two-thirds and concave laterally. Its *sternal* (medial) *end,* which attaches to the sternal manubrium, is rounded or triangular in cross section. The sternal end projects above the manubrium and can be felt and (usually) seen forming the lateral walls of the *jugular notch* (see Figure 10.15, p. 103). The *acromial* (lateral) *end* of the clavicle is flattened where it articulates with the scapula to form part of the shoulder joint. On its posteroinferior surface is the prominent **conoid tubercle** (Figure 11.2a). This projection anchors a ligament and provides a handy landmark for determining whether a given clavicle is from the right or left side of the body. The clavicle serves as an anterior brace, or strut, to hold the arm away from the top of the thorax.

The **scapulae** (Figure 11.2), or shoulder blades, are generally triangular and are commonly called the "wings" of humans. Each scapula has a flattened body and two important processes—the **acromion** (the enlarged end of the spine of the scapula) and the beaklike **coracoid process** (*corac* = crow, raven). The acromion connects with the clavicle; the coracoid process points anteriorly over the tip of the shoulder joint and serves as an attachment point for some of the upper limb muscles. The **suprascapular notch** at the base of the coracoid process allows nerves to pass. The scapula has no direct attachment to the axial skeleton but is loosely held in place by trunk muscles.

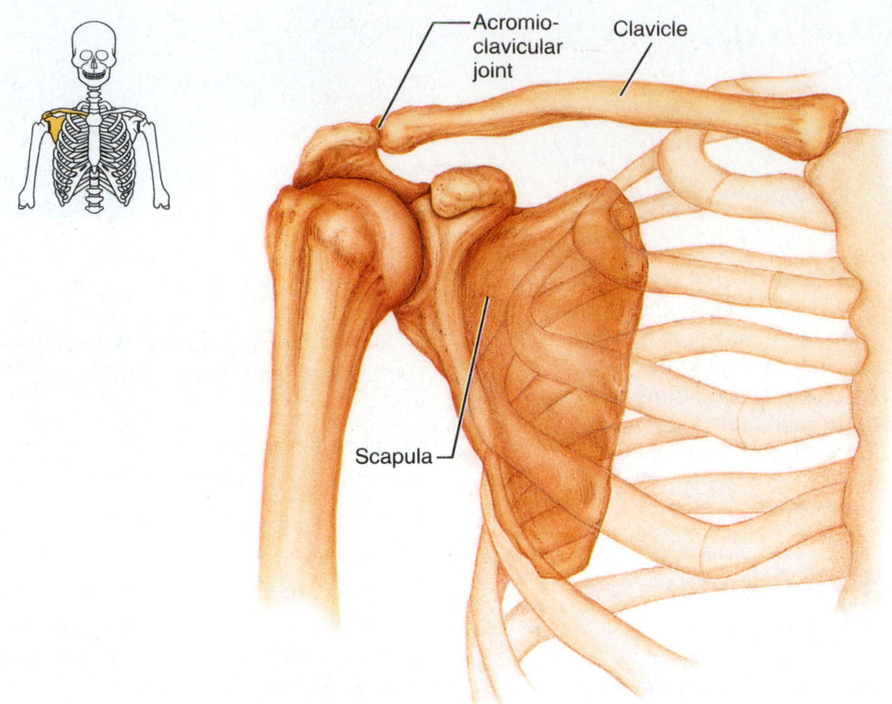

Figure 11.1 Articulated bones of the pectoral (shoulder) girdle. The right pectoral girdle is articulated to show the relationship of the girdle to the bones of the thorax and arm.

The scapula has three angles: superior, inferior, and lateral. The inferior angle provides a landmark for auscultating (listening to) lung sounds. The scapula also has three named borders: superior, medial (vertebral), and lateral (axillary). Several shallow depressions (fossae) appear on both sides of the scapula and are named according to location; there are the anterior *subscapular fossa* and the posterior *infraspinous* and *supraspinous fossae*. The **glenoid cavity,** a shallow socket that receives the head of the arm bone (humerus), is located in the lateral angle.

The shoulder girdle is exceptionally light and allows the upper limb a degree of mobility not seen anywhere else in the body. This is due to the following factors:

- The sternoclavicular joints are the *only* site of attachment of the shoulder girdles to the axial skeleton.
- The relative looseness of the scapular attachment allows it to slide back and forth against the thorax with muscular activity.
- The glenoid cavity is shallow, and does little to stabilize the shoulder joint.

However, this exceptional flexibility exacts a price: the arm bone (humerus) is very susceptible to dislocation, and fracture of the clavicle disables the entire upper limb.

The Arm

The arm (Figure 11.3) consists of a single bone—the **humerus,** a typical long bone. Proximally its rounded *head* fits into the shallow glenoid cavity of the scapula. The head is separated from the shaft by the *anatomical neck* and the more constricted *surgical neck,* which is a common site of fracture. Opposite the head are two prominences, the **greater** and **lesser tubercles** (from lateral to medial aspect), separated by a groove (the **intertubercular** or **bicipital groove**) that guides the tendon of the biceps muscle to its point of attachment (the superior rim of the glenoid cavity). In the midpoint of the shaft is a roughened area, the **deltoid tuberosity,** where the large fleshy shoulder muscle, the deltoid, attaches. Nearby, the **radial groove** runs obliquely, indicating the pathway of the radial nerve.

At the distal end of the humerus are two condyles—the medial **trochlea** (looking rather like a spool), which articulates with the ulna, and the lateral **capitulum,** which articulates with the radius of the forearm. This condyle pair is flanked medially by the **medial epicondyle** and laterally by the **lateral epicondyle.**

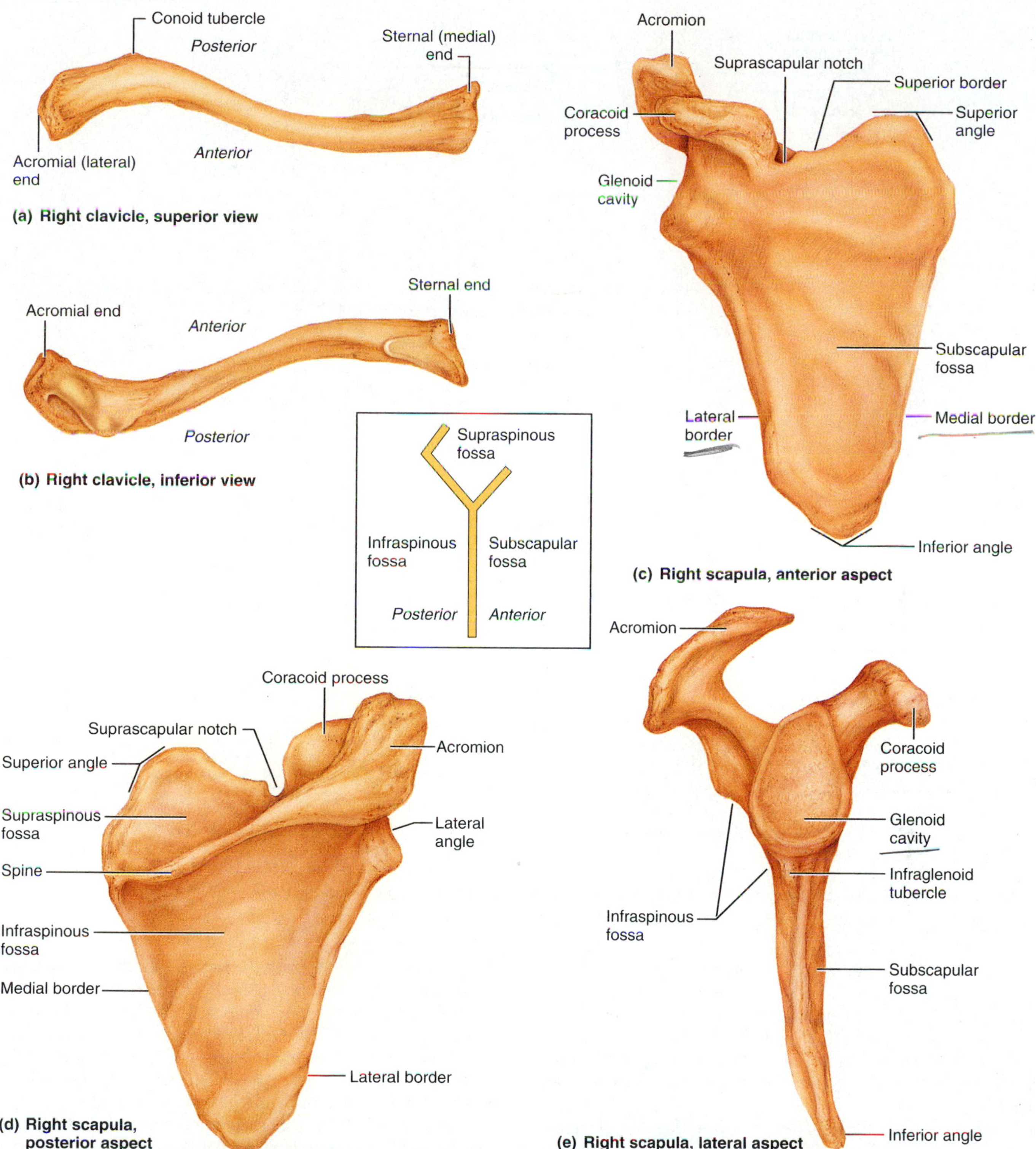

Figure 11.2 Individual bones of the pectoral (shoulder) girdle. View **(e)** is accompanied by a schematic representation of its orientation.

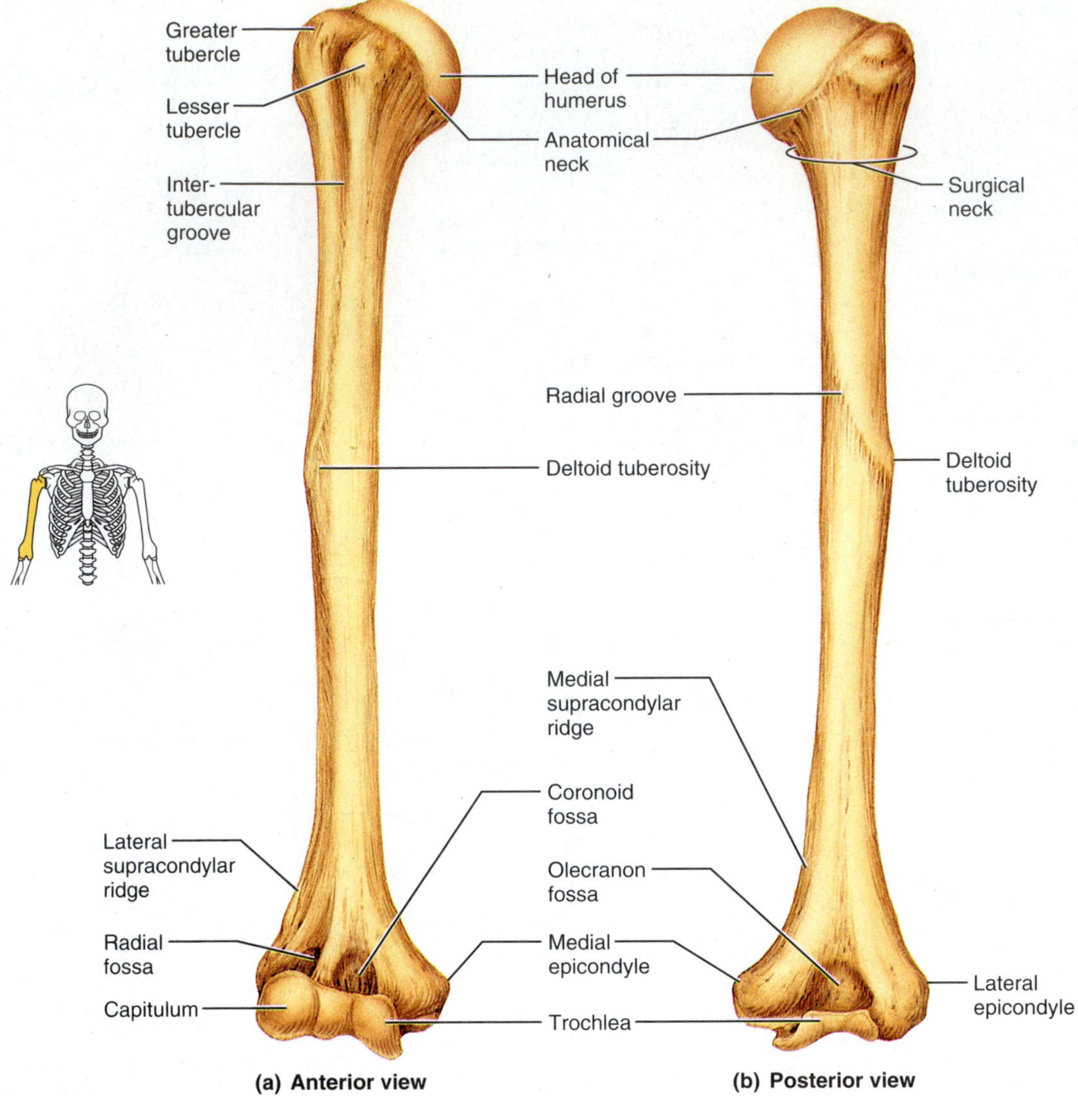

Figure 11.3 Bone of the right arm. Humerus, **(a)** anterior view, **(b)** posterior view.

The medial epicondyle is commonly called the "funny bone." The large ulnar nerve runs in a groove beneath the medial epicondyle, and when this region is sharply bumped, we are likely to experience a temporary, but excruciatingly painful, tingling sensation. This event is called "hitting the funny bone," a strange expression, because it is certainly *not* funny!

Above the trochlea on the anterior surface is a depression, the **coronoid fossa;** on the posterior surface is the **olecranon fossa.** These two depressions allow the corresponding processes of the ulna to move freely when the elbow is flexed and extended. The small **radial fossa,** lateral to the coronoid fossa, receives the head of the radius when the elbow is flexed.

The Forearm

Two bones, the radius and the ulna, compose the skeleton of the forearm, or antebrachium (see Figure 11.4). When the body is in the anatomical position, the **radius** is in the lateral position in the forearm and the radius and ulna are parallel. Proximally, the disc-shaped head of the radius articulates with the capitulum of the humerus. Just below the head, on the medial aspect of the shaft, is a prominence called the **radial tuberosity,** the point of attachment for the tendon of the biceps muscle of the arm. Distally, the small **ulnar notch** reveals where it articulates with the end of the ulna.

The **ulna** is the medial bone of the forearm. Its proximal end bears the anterior **coronoid process** and the posterior **olecranon process,** which are separated by the **trochlear notch.** Together these processes grip the trochlea of the humerus in a plierslike joint. The small **radial notch** on the lateral side of the coronoid process articulates with the head of the radius. The slimmer distal end, the ulnar **head,** bears a small medial **styloid process,** which serves as a point of attachment for the ligaments of the wrist.

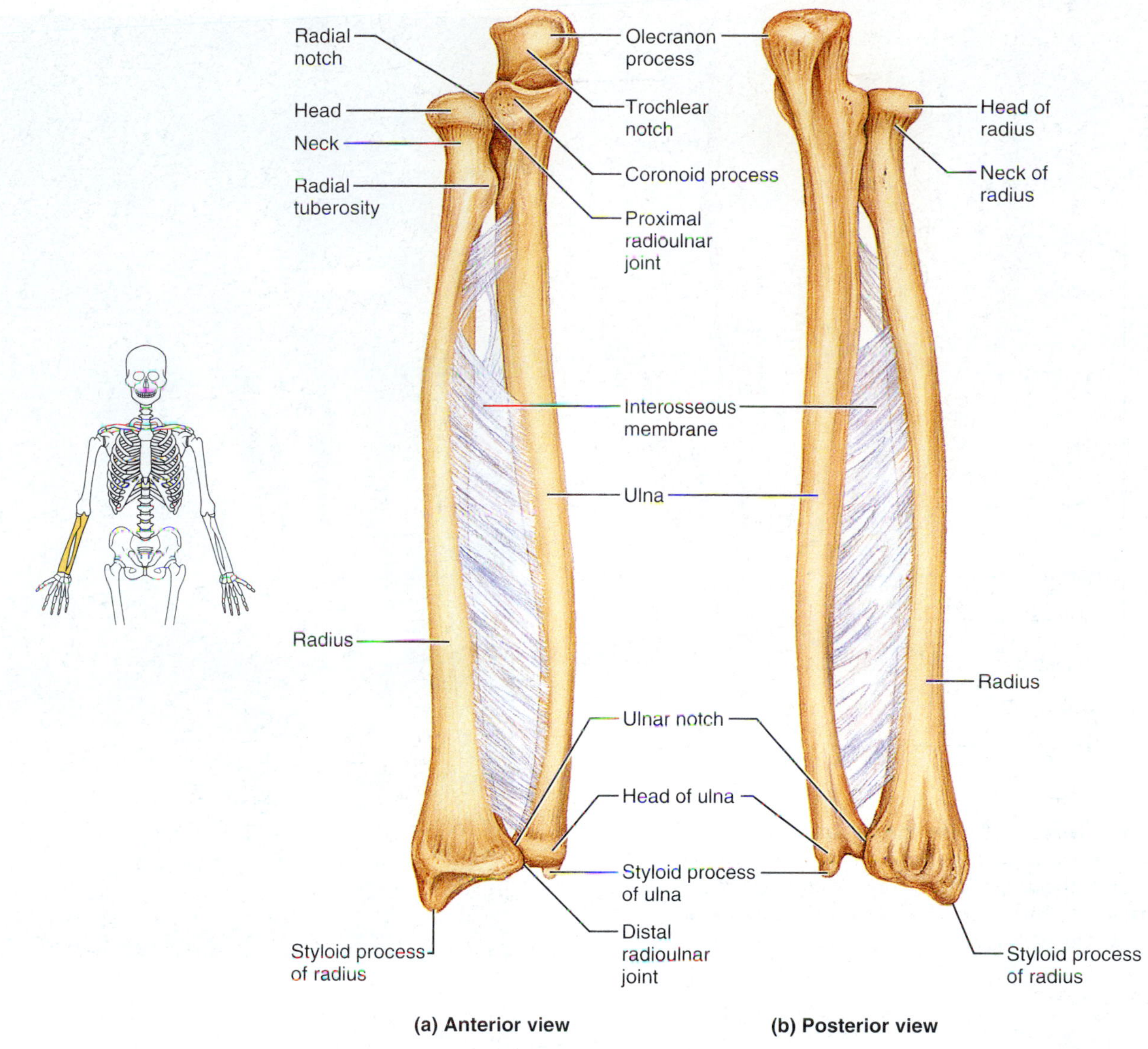

Figure 11.4 Bones of the right forearm. Radius and ulna, **(a)** anterior view, **(b)** posterior view.

The Hand

The skeleton of the hand, or manus (Figure 11.5), includes three groups of bones, those of the carpus (wrist), the metacarpals (bones of the palm), and the phalanges (bones of the fingers).

The wrist is the proximal portion of the hand. It is referred to anatomically as the **carpus;** the eight bones composing it are the **carpals.** (So you actually wear your wristwatch over the distal part of your forearm.) The carpals are arranged in two irregular rows of four bones each, which are illustrated in Figure 11.5. In the proximal row (lateral to medial) are the scaphoid, lunate, triquetral, and pisiform bones; the scaphoid and lunate articulate with the distal end of the radius. In the distal row are the trapezium, trapezoid, capitate, and hamate. The carpals are bound closely together by ligaments, which restrict movements between them.

The **metacarpals,** numbered 1 to 5 from the thumb side of the hand toward the little finger, radiate out from the wrist like spokes to form the palm of the hand. The *bases* of the metacarpals articulate with the carpals of the wrist; their more bulbous *heads* articulate with the phalanges of the fingers distally. When the fist is clenched, the heads of the metacarpals become prominent as the knuckles.

Like the bones of the palm, the fingers are numbered from 1 to 5, beginning from the thumb (*pollex*) side of the hand. The 14 bones of the fingers, or digits, are miniature long bones, called **phalanges** (s. phalanx) as noted above. Each finger contains three phalanges (proximal, middle, and distal) except the thumb, which has only two (proximal and distal).

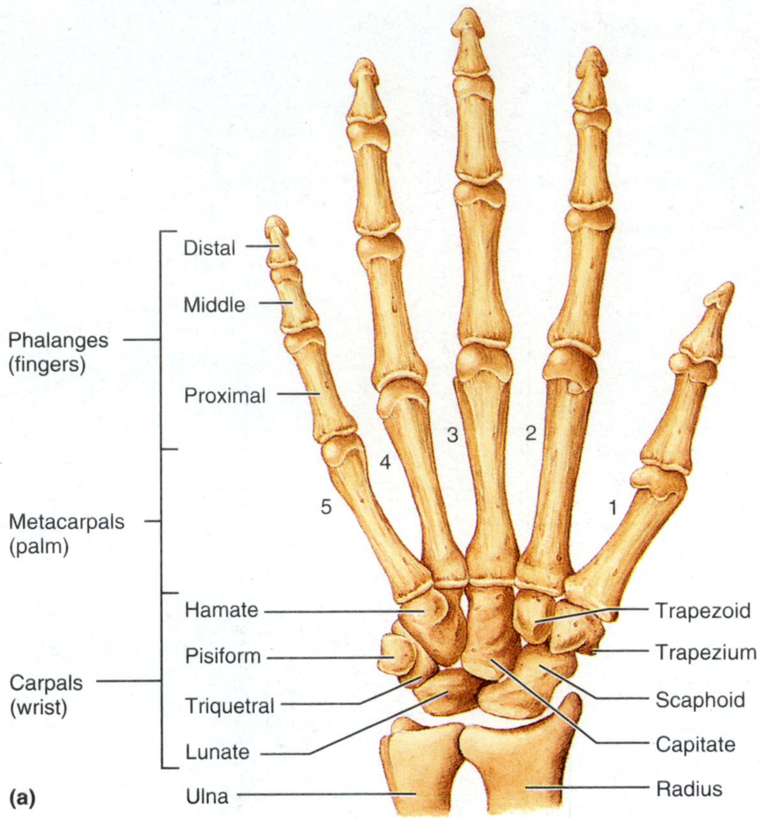

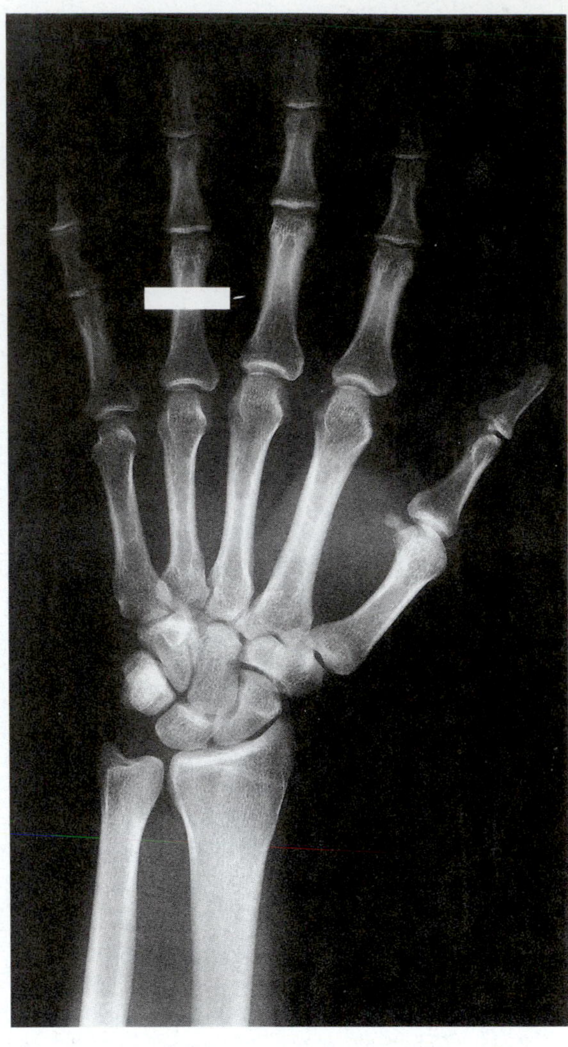

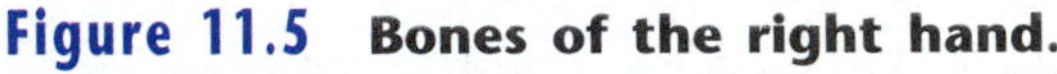

Figure 11.5 Bones of the right hand.
(a) Anterior view showing the relationships of the carpals, metacarpals, and phalanges. **(b)** X ray. White bar on the proximal phalanx of the ring finger shows the position at which a ring would be worn.

Activity 2:

Palpating the Surface Anatomy of the Pectoral Girdle and the Upper Limb

Before continuing on to study the bones of the pelvic girdle, take the time to identify the following bone markings on the skin surface of the upper limb. It is usually preferable to observe and palpate the bone markings on your lab partner, particularly since many of these markings can only be seen from the dorsal aspect.

- Clavicle: Palpate the clavicle along its entire length from sternum to shoulder.
- Acromioclavicular joint: The high point of the shoulder, which represents the junction point between the clavicle and the acromion of the scapular spine.
- Spine of the scapula: Extend your arm at the shoulder so that your scapula moves posteriorly. As you do this, your scapular spine will be seen as a winglike protrusion on your dorsal thorax and can be easily palpated by your lab partner.
- Lateral epicondyle of the humerus: The inferiormost projection at the lateral aspect of the distal humerus. After you have located the epicondyle, run your finger posteriorly into the hollow immediately dorsal to the epicondyle. This is the site where the extensor muscles of the hand are attached and is a common site of the excruciating pain of tennis elbow, a condition in which those muscles and their tendons are abused physically.
- Medial epicondyle of the humerus: Feel this medial projection at the distal end of the humerus.
- Olecranon process of the ulna: Work your elbow—flexing and extending—as you palpate its dorsal aspect to feel the olecranon process of the ulna moving into and out of the olecranon fossa on the dorsal aspect of the humerus.
- Styloid process of the ulna: With the hand in the anatomical position, feel out this small inferior projection on the medial aspect of the distal end of the ulna.
- Styloid process of the radius: Find this projection at the distal end of the radius (lateral aspect). It is most easily located by moving the hand medially at the wrist. Once you have palpated the styloid process, move your fingers just medially onto the anterior wrist. Press firmly and then let up slightly on the pressure. You should be able to feel your pulse at this pressure point, which lies over the radial artery (radial pulse).
- Pisiform: Just distal to the styloid process of the ulna, feel the rounded pealike pisiform bone.

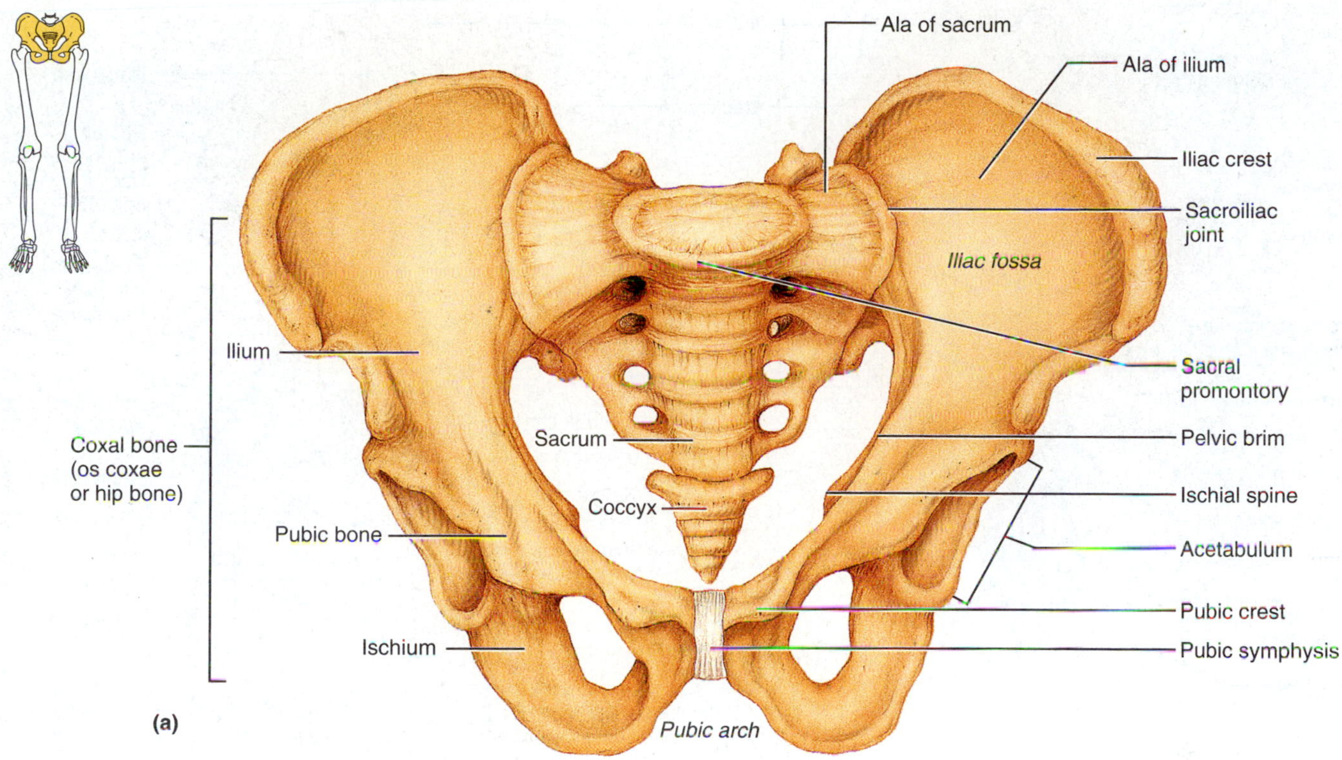

Figure 11.6 Bones of the pelvic girdle. (a) Articulated bony pelvis, showing the two coxal bones, which together compose the pelvic girdle, the sacrum, and the coccyx. *(Figure continues on page 112)*

- Metacarpophalangeal joints (knuckles): Clench your fist and find the first set of flexed-joint protrusions beyond the wrist—these are your metacarpophalangeal joints. ■

Bones of the Pelvic Girdle and Lower Limb

The Pelvic (Hip) Girdle

As with the bones of the pectoral girdle and upper limb, pay particular attention to bone markings needed to identify right and left bones.

The **pelvic girdle,** or **hip girdle** (Figure 11.6), is formed by the two **coxal** (*coxa* = hip) **bones** (also called the **ossa coxae** or hip bones). The two coxal bones together with the sacrum and coccyx form the **bony pelvis.** In contrast to the bones of the shoulder girdle, those of the pelvic girdle are heavy and massive, and they attach securely to the axial skeleton. The sockets for the heads of the femurs (thigh bones) are deep and heavily reinforced by ligaments to ensure a stable, strong limb attachment. The ability to bear weight is more important here than mobility and flexibility. The combined weight of the upper body rests on the pelvis (specifically, where the hip bones meet the sacrum).

Each coxal bone is a result of the fusion of three bones—the ilium, ischium, and pubis—which are distinguishable in the young child. The **ilium,** a large flaring bone, forms the major portion of the coxal bone. It connects posteriorly, via its **auricular surface,** with the sacrum at the **sacroiliac joint.** The superior margin of the iliac bone, the **iliac crest,** is rough; when you rest your hands on your hips, you are palpating your iliac crests. The iliac crest terminates anteriorly in the **anterior superior spine** and posteriorly in the **posterior superior spine.** Two inferior spines are located below these. The shallow **iliac fossa** marks its internal surface, and a prominent ridge, the **arcuate line,** outlines the pelvic inlet, or pelvic brim.

The **ischium** is the "sit-down" bone, forming the most inferior and posterior portion of the coxal bone. The most outstanding marking on the ischium is the rough **ischial tuberosity,** which receives the weight of the body when sitting. The **ischial spine,** superior to the ischial tuberosity, is an important anatomical landmark of the pelvic cavity. (See Comparison of the Male and Female Pelves, Table 11.1 on page 113.) The obvious **lesser** and **greater sciatic notches** allow nerves and blood vessels to pass to and from the thigh. The sciatic nerve passes through the latter.

The **pubis** is the most anterior portion of the coxal bone. Fusion of the **rami** of the pubic bone anteriorly and the ischium posteriorly forms a bar of bone enclosing the **obturator foramen,** through which blood vessels and nerves run from the pelvic cavity into the thigh. The pubic bones of each hip bone meet anteriorly at the **pubic crest** to form a cartilaginous joint called the **pubic symphysis.** At the lateral end of the pubic crest is the *pubic tubercle* (see Figure 11.6c) to which the important *inguinal ligament* attaches.

The ilium, ischium, and pubis fuse at the deep hemispherical socket called the **acetabulum** (literally, "vinegar cup"), which receives the head of the thigh bone.

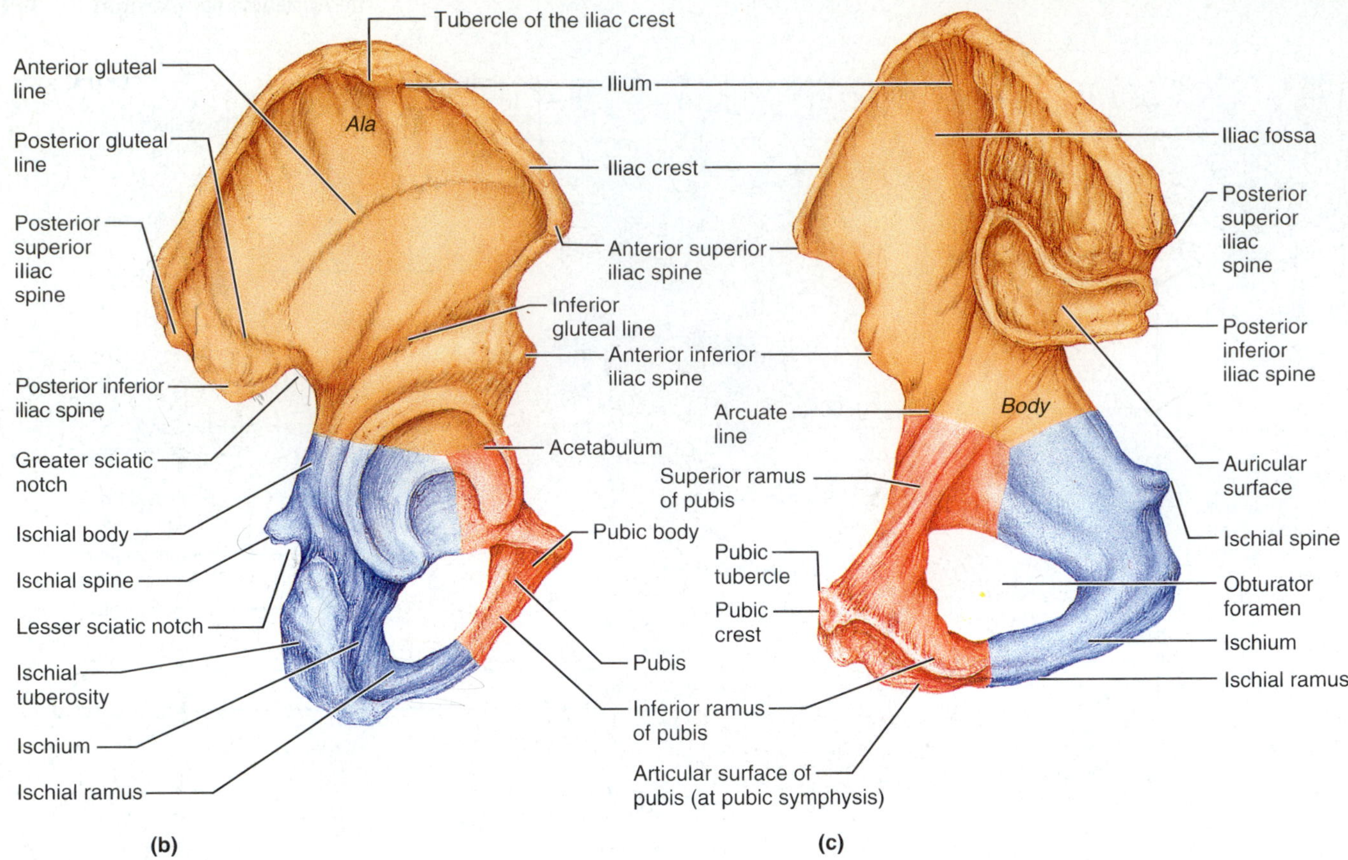

Figure 11.6 *(continued)* **Bones of the pelvic girdle.** **(b)** Right coxal bone, lateral view, showing the point of fusion of the ilium, ischium, and pubic bones. **(c)** Right coxal bone, medial view.

Activity 3: Observing Pelvic Articulations

Before continuing with the bones of the lower limbs, take the time to examine an articulated pelvis. Notice how each coxal bone articulates with the sacrum posteriorly and how the two coxal bones join at the pubic symphysis. The sacroiliac joint is a common site of lower back problems because of the pressure it must bear. ■

Comparison of the Male and Female Pelves Although bones of males are usually larger, heavier, and have more prominent bone markings, the male and female skeletons are very similar. The exception to this generalization is pelvic structure.

The female pelvis reflects modifications for childbearing. Generally speaking, the female pelvis is wider, shallower, lighter, and rounder than that of the male. Not only must her pelvis support the increasing size of a fetus, but it must also be large enough to allow the infant's head (its largest dimension) to descend through the birth canal at birth.

To describe pelvic sex differences, we need to introduce a few more terms. Anatomically, the pelvis is described in terms of a false pelvis and a true pelvis. The **false pelvis** is that portion superior to the arcuate line; it is bounded by the alae of the ilia laterally and the sacral promontory and lumbar vertebrae posteriorly. Although the false pelvis supports the abdominal viscera, it does not restrict childbirth in any way. The **true pelvis** is the region inferior to the arcuate line that is almost entirely surrounded by bone. Its posterior boundary is formed by the sacrum. The ilia, ischia, and pubic bones define its limits laterally and anteriorly.

The dimensions of the true pelvis, particularly its inlet and outlet, are critical if delivery of a baby is to be uncomplicated; and they are carefully measured by the obstetrician. The **pelvic inlet,** or **pelvic brim,** is the opening delineated by the sacral promontory posteriorly and the arcuate lines of the ilia anterolaterally. It is the superiormost margin of the true pelvis. Its widest dimension is from left to right, that is, along the frontal plane. The **pelvic outlet** is the inferior margin of the true pelvis. It is bounded anteriorly by the pubic arch, laterally by the ischia, and posteriorly by the sacrum and coccyx. Since both the coccyx and the ischial spines protrude into the outlet opening, a sharply angled coccyx or large, sharp ischial spines can dramatically narrow the outlet. The largest dimension of the outlet is the anterior-posterior diameter.

The major differences between the male and female pelves are summarized in Table 11.1.

Activity 4: Comparing Male and Female Pelves

Examine male and female pelves for the following differences:

- The female inlet is larger and more circular.
- The female pelvis as a whole is shallower, and the bones are lighter and thinner.
- The female sacrum is broader and less curved, and the pubic arch is more rounded.

Table 11.1 Comparison of the Male and Female Pelves

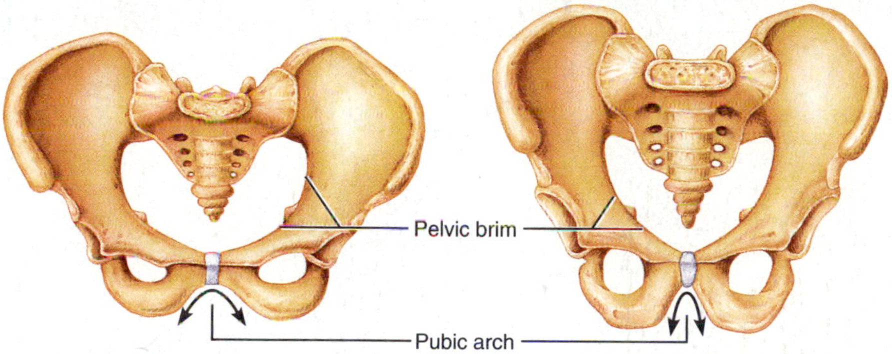

Characteristic	Female	Male
General structure and functional modifications	Tilted forward; adapted for childbearing; true pelvis defines the birth canal; cavity of the true pelvis is broad, shallow, and has a greater capacity	Tilted less far forward; adapted for support of a male's heavier build and stronger muscles; cavity of the true pelvis is narrow and deep
Bone thickness	Less; bones lighter, thinner, and smoother	Greater; bones heavier and thicker, and markings are more prominent
Acetabula	Smaller; farther apart	Larger; closer
Pubic angle/arch	Broader (80°–90°); more rounded	More acute (50°–60°)
Anterior view		
Sacrum	Wider; shorter; sacral curvature is accentuated	Narrow; longer; sacral promontory more ventral
Coccyx	More movable; straighter	Less movable; curves ventrally
Left lateral view	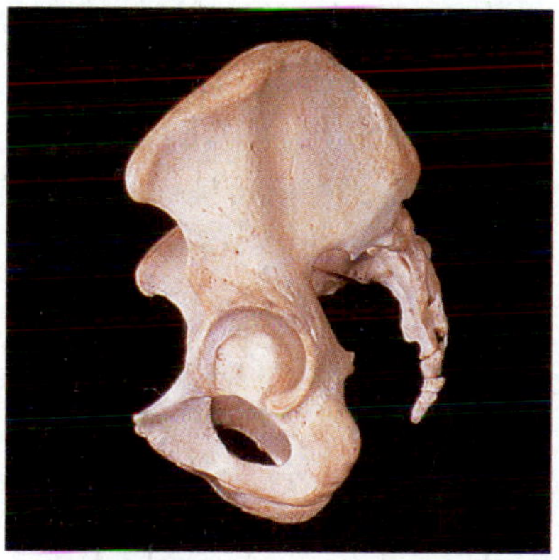	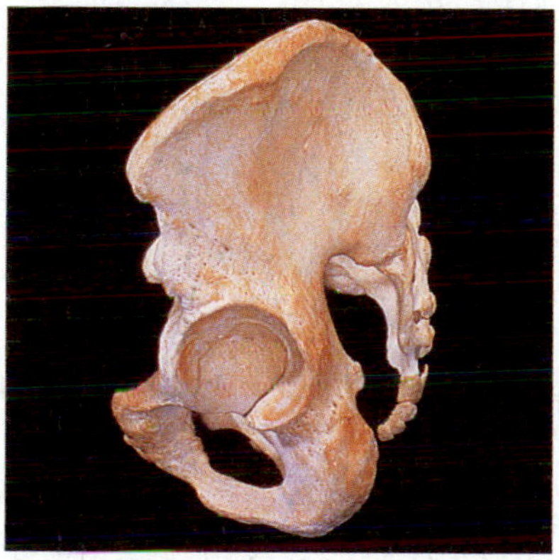
Pelvic inlet (brim)	Wider; oval from side to side	Narrow; basically heart shaped
Pelvic outlet	Wider; ischial tuberosities shorter, farther apart and everted	Narrower; ischial tuberosities longer, sharper, and point more medially
Posteroinferior view	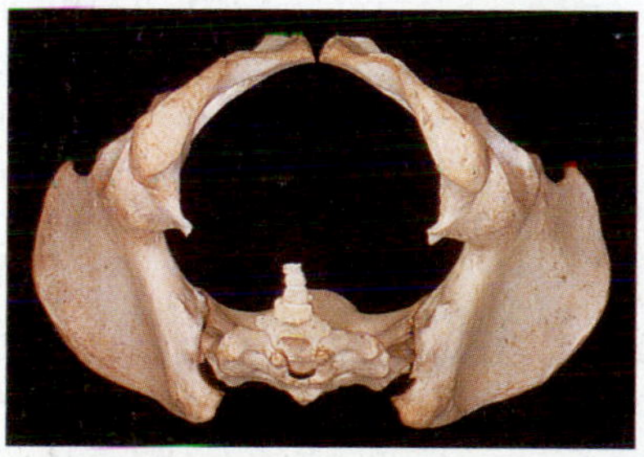	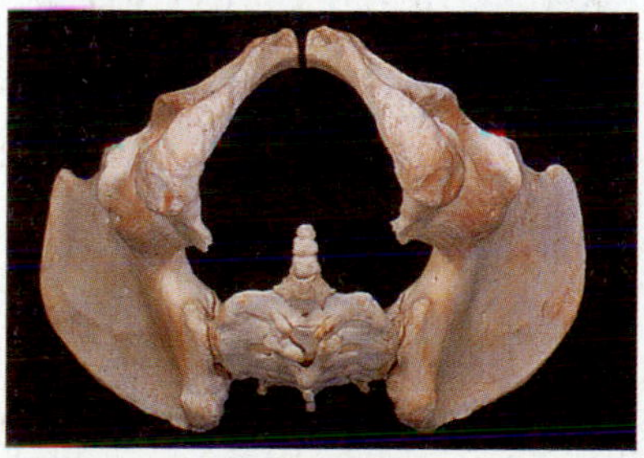

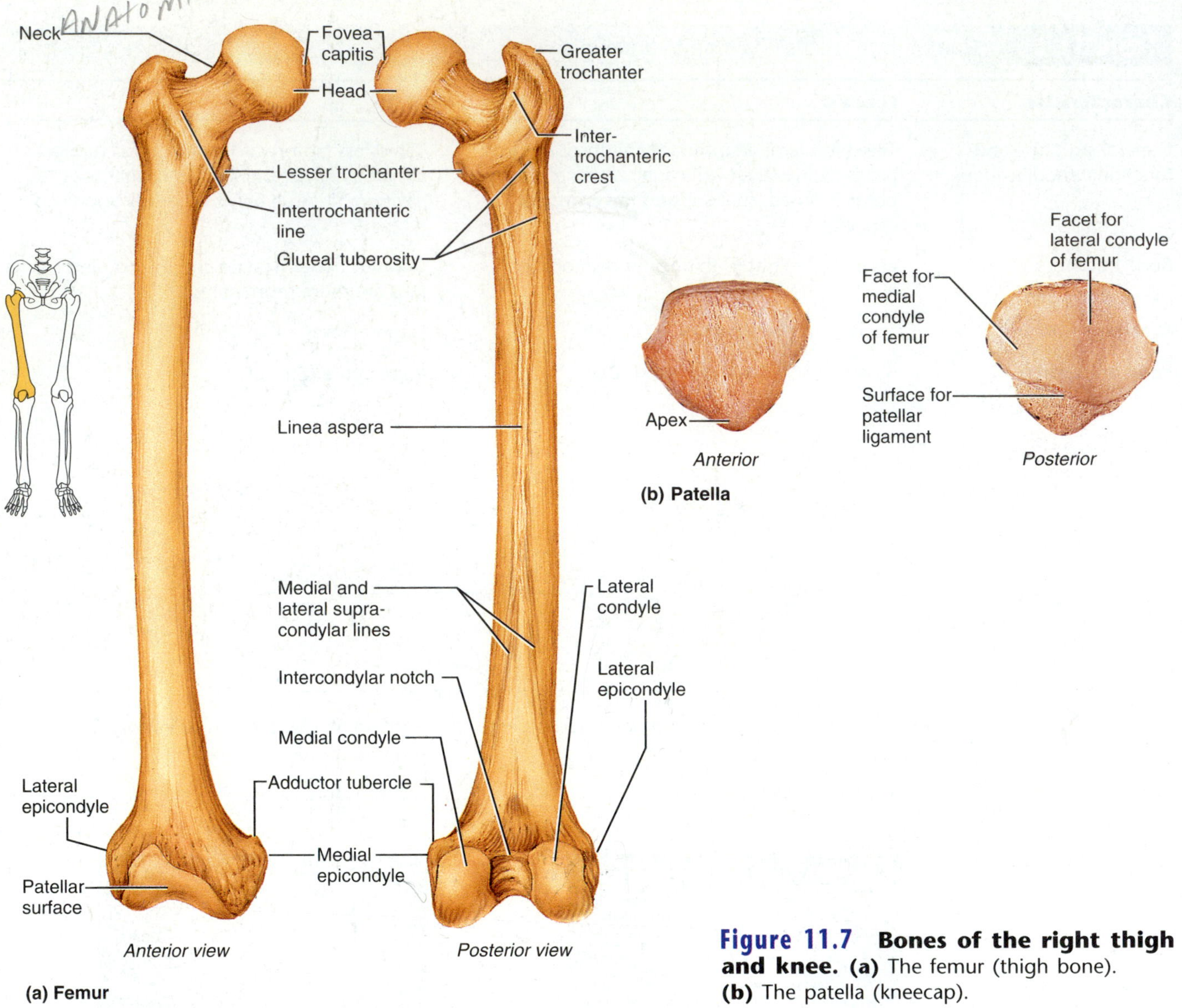

Figure 11.7 Bones of the right thigh and knee. (a) The femur (thigh bone). **(b)** The patella (kneecap).

- The female acetabula are smaller and farther apart, and the ilia flare more laterally.
- The female ischial spines are shorter, farther apart, and everted, thus enlarging the pelvic outlet. ■

The Thigh

The **femur,** or thigh bone (Figure 11.7a), is the sole bone of the thigh. It is the heaviest, strongest bone in the body. The ball-like head of the femur articulates with the hip bone via the deep, secure socket of the acetabulum. Obvious in the femur's head is a small central pit called the **fovea capitis** ("pit of the head") from which a small ligament runs to the acetabulum. The head of the femur is carried on a short, constricted *neck,* which angles laterally to join the shaft. The neck is the weakest part of the femur and is a common fracture site (an injury called a broken hip), particularly in the elderly. At the junction of the shaft and neck are the **greater** and **lesser trochanters** (separated posteriorly by the **intertrochanteric crest** and anteriorly by the **intertrochanteric line**). The trochanters and trochanteric crest, as well as the **gluteal tuberosity** and the **linea aspera** located on the shaft, are sites of muscle attachment.

The femur inclines medially as it runs downward to the leg bones; this brings the knees in line with the body's center of gravity, or maximum weight. The medial course of the femur is more noticeable in females because of the wider female pelvis.

Distally, the femur terminates in the **lateral** and **medial condyles,** which articulate with the tibia below, and the **patellar surface,** which forms a joint with the patella (kneecap) anteriorly. The **lateral** and **medial epicondyles,** just superior to the condyles, are separated by the **intercondylar notch.** On the superior part of the medial epicondyle is a bump, the **adductor tubercle,** to which the large adductor magnus muscle attaches.

The **patella** (Figure 11.7b) is a triangular sesamoid bone enclosed in the (quadriceps) tendon that secures the anterior thigh muscles to the tibia. It guards the knee joint anteriorly and improves the leverage of the thigh muscles acting across the knee joint.

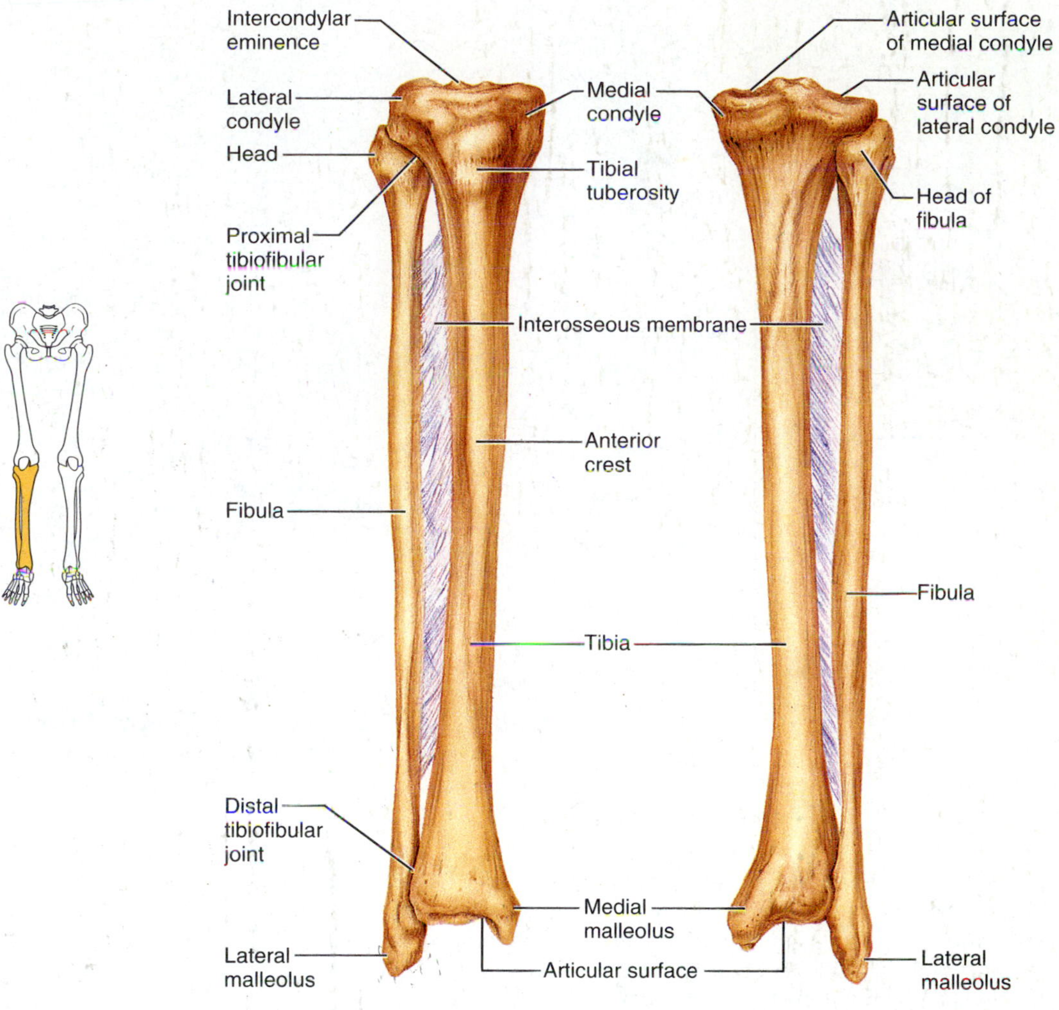

Figure 11.8 Bones of the right leg. Tibia and fibula, anterior view on left; posterior view on right.

The Leg

Two bones, the tibia and the fibula, form the skeleton of the leg (see Figure 11.8). The **tibia,** or *shinbone,* is the larger and more medial of the two leg bones. At the proximal end, the **medial** and **lateral condyles** (separated by the **intercondylar eminence**) receive the distal end of the femur to form the knee joint. The **tibial tuberosity,** a roughened protrusion on the anterior tibial surface (just below the condyles), is the site of attachment of the patellar (kneecap) ligament. Small facets on its superior and inferior lateral surface articulate with the fibula. Distally, a process called the **medial malleolus** forms the inner (medial) bulge of the ankle, and the smaller distal end articulates with the talus bone of the foot. The anterior surface of the tibia bears a sharpened ridge that is relatively unprotected by muscles. This so-called **anterior crest** is easily felt beneath the skin.

The **fibula,** which lies parallel to the tibia, takes no part in forming the knee joint. Its proximal head articulates with the lateral condyle of the tibia. The fibula is thin and sticklike with a sharp anterior crest. It terminates distally in the **lateral malleolus,** which forms the outer part, or lateral bulge, of the ankle.

The Foot

The bones of the foot include the 7 **tarsal** bones, 5 **metatarsals,** which form the instep, and 14 **phalanges,** which form the toes (see Figure 11.9). Body weight is concentrated on the two largest tarsals, which form the posterior aspect of the foot, the *calcaneus* (heel bone) and the *talus,* which lies between the tibia and the calcaneus. The other tarsals are named and identified in Figure 11.9. Like the fingers of the hand, each toe has 3 phalanges except the great toe, which has 2.

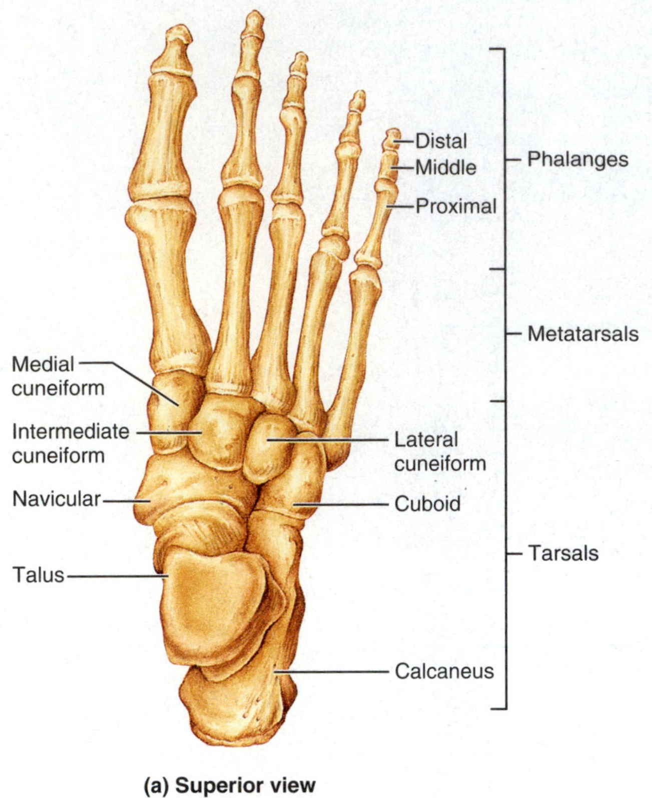

(a) Superior view

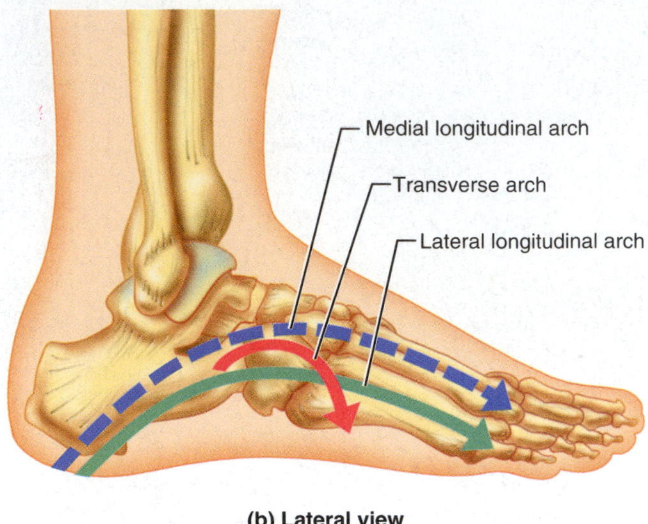

(b) Lateral view

Figure 11.9 Bones of the right foot. **(a)** Superior view. **(b)** Lateral view showing arches of the foot.

The bones in the foot are arranged to produce three strong arches—two longitudinal arches (medial and lateral) and one transverse arch (Figure 11.9b). Ligaments, binding the foot bones together, and tendons of the foot muscles hold the bones firmly in the arched position but still allow a certain degree of give. Weakened arches are referred to as fallen arches or flat feet.

Activity 5: Palpating the Surface Anatomy of the Pelvic Girdle and Lower Limb

Locate and palpate the following bone markings on yourself and/or your lab partner.

- Iliac crest and anterior superior iliac spine: Rest your hands on your hips—they will be overlying the iliac crests. Trace the crest as far posteriorly as you can and then follow it anteriorly to the anterior superior iliac spine. This latter bone marking is easily felt in almost everyone, and is clearly visible through the skin (and perhaps the clothing) of very slim people. (The posterior superior iliac spine is much less obvious and is usually indicated only by a dimple in the overlying skin. Check it out in the mirror tonight.)
- Greater trochanter of the femur: This is easier to locate in females than in males because of the wider female pelvis; also it is more likely to be clothed by bulky muscles in males. Try to locate it on yourself as the most lateral point of the proximal femur. It typically lies about 6–8 inches below the iliac crest.
- Patella and tibial tuberosity: Feel your kneecap and palpate the ligaments attached to its borders. Follow the inferior patellar ligament to the tibial tuberosity.
- Medial and lateral condyles of the femur and tibia: As you move from the patella inferiorly on the medial (and then the lateral) knee surface, you will feel first the femoral and then the tibial condyle.
- Medial malleolus: Feel the medial protrusion of your ankle, the medial malleolus of the distal tibia.
- Lateral malleolus: Feel the bulge of the lateral aspect of your ankle, the lateral malleolus of the fibula.
- Calcaneus: Attempt to follow the extent of your calcaneus or heel bone. ■

Activity 6: Constructing a Skeleton

1. When you finish examining yourself and the disarticulated bones of the appendicular skeleton, work with your lab partner to arrange the disarticulated bones on the laboratory bench in their proper relative positions to form an entire skeleton. Careful observations of the bone markings should help you distinguish between right and left members of bone pairs.
2. When you believe that you have accomplished this task correctly, ask the instructor to check your arrangement to ensure that it is correct. If it is not, go to the articulated skeleton and check your bone arrangements. Also review the descriptions of the bone markings as necessary to correct your bone arrangement. ■

The Fetal Skeleton

exercise 12

Objectives

1. To define *fontanel* and discuss the function and fate of fontanels in the fetus.
2. To demonstrate important differences between the fetal and adult skeletons.

Materials

- ❑ Isolated fetal skull
- ❑ Fetal skeleton
- ❑ Adult skeleton

A human fetus about to be born has 275 bones, many more than the 206 bones found in the adult skeleton. This is because many of the bones described as single bones in the adult skeleton (for example, the coxal bone, sternum, and sacrum) have not yet fully ossified and fused in the fetus.

Activity:
Examining a Fetal Skull and Skeleton

1. Obtain a fetal skull and study it carefully. Make observations as needed to answer the following questions.

- Does it have the same bones as the adult skull?
- How does the size of the fetal face relate to the cranium?
- How does this compare to what is seen in the adult?

2. Indentations between the bones of the fetal skull, called **fontanels,** are fibrous membranes. These areas will become bony (ossify) as the fetus ages, completing the process by the age of 20 to 22 months. The fontanels allow the fetal skull to be compressed slightly during birth and also allow for brain growth during late fetal life. Locate the following fontanels on the fetal skull with the aid of Figure 12.1: anterior (or frontal) fontanel, mastoid fontanel, sphenoidal fontanel, and posterior (or occipital) fontanel.

3. Notice that some of the cranial bones have conical protrusions. These are growth (ossification) centers. Notice also that the frontal bone is still bipartite, and the temporal bone is incompletely ossified, little more than a ring of bone in the fetus.

4. Obtain a fetal skeleton or use Figure 12.2, and examine it carefully, noting differences between it and an adult skeleton. Pay particular attention to the vertebrae, sternum, frontal bone of the cranium, patellae (kneecaps), coxal bones, carpals and tarsals, and rib cage.

5. Check the questions in the review section before completing this study to ensure that you have made all of the necessary observations. ■

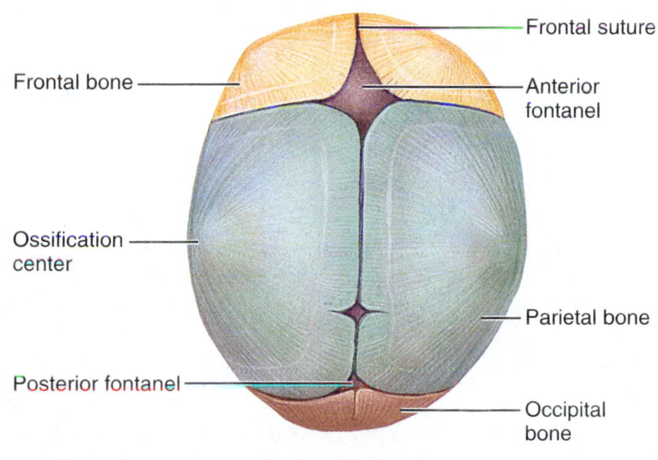

(a) Superior view

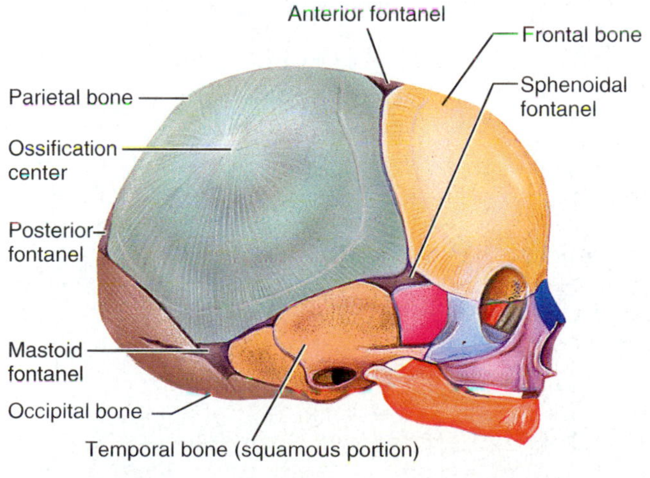

(b) Lateral view

Figure 12.1 The fetal skull.

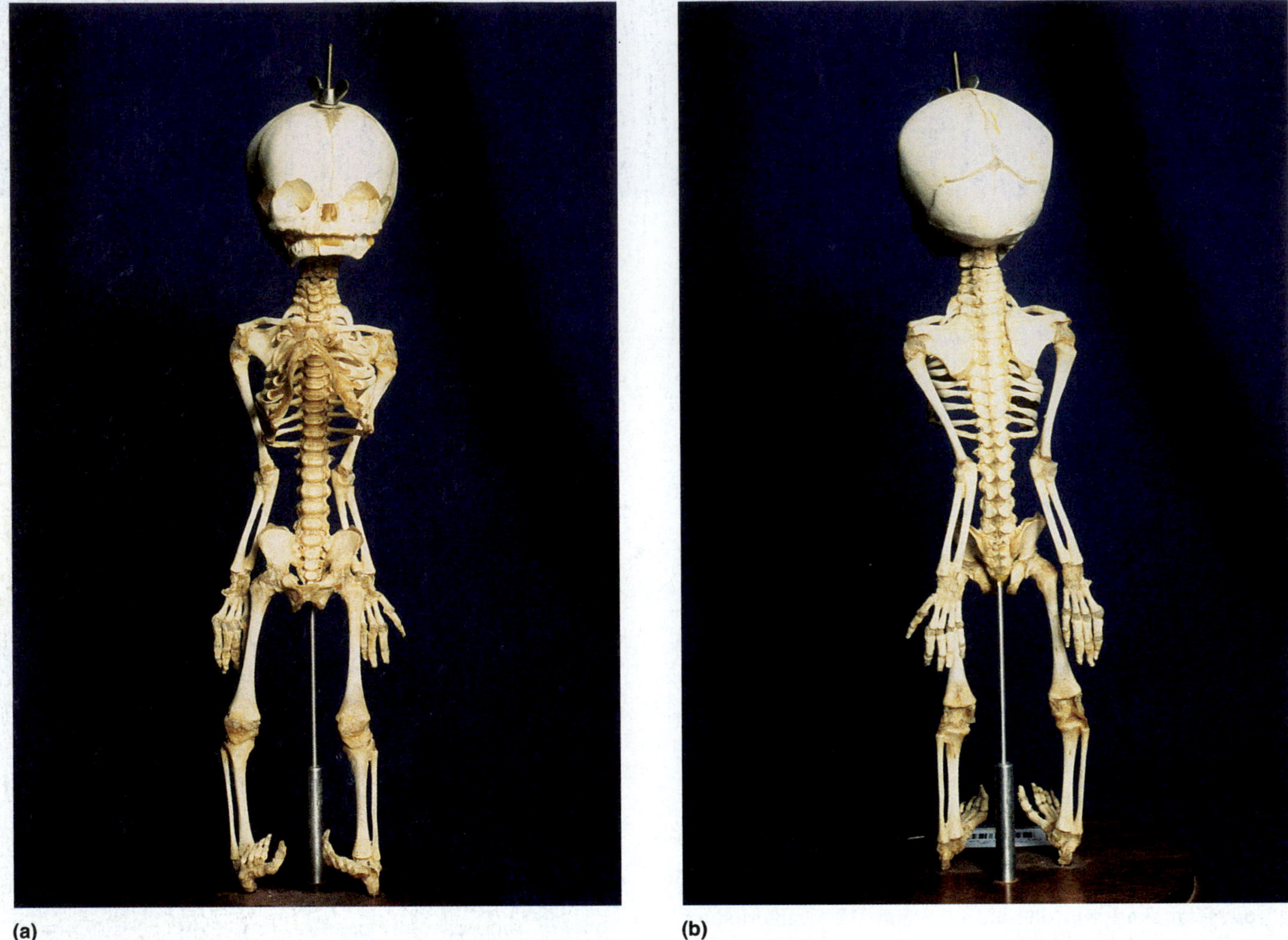

(a) (b)

Figure 12.2 **Fetal skeleton.** **(a)** Anterior view. **(b)** Posterior view.

Articulations and Body Movements

exercise 13

Objectives

1. To name and describe the three functional categories of joints.
2. To name and describe the three structural categories of joints, and to compare their structure and mobility.
3. To identify the types of synovial joints.
4. To define *origin* and *insertion* of muscles.
5. To demonstrate or identify the various body movements.

Materials

- ❑ Articulated skeleton
- ❑ Skull
- ❑ Diarthrotic beef joint (fresh or preserved), preferably a knee joint
- ❑ Disposable gloves
- ❑ Anatomical chart of joint types (if available)
- ❑ X rays of normal and arthritic joints (if available)

AIA See Appendix C, Exercise 13 for links to A.D.A.M.® Interactive Anatomy.

With rare exceptions, every bone in the body is connected to, or forms a joint with, at least one other bone. **Articulations,** or joints, perform two functions for the body. They (1) hold the bones together and (2) allow the rigid skeletal system some flexibility so that gross body movements can occur.

Joints may be classified structurally or functionally. The structural classification is based on the presence of connective tissue fiber, cartilage, or a joint cavity between the articulating bones. Structurally, there are *fibrous, cartilaginous,* and *synovial joints.*

The functional classification focuses on the amount of movement allowed at the joint. On this basis, there are **synarthroses,** or immovable joints; **amphiarthroses,** or slightly movable joints; and **diarthroses,** or freely movable joints. Freely movable joints predominate in the limbs, whereas immovable and slightly movable joints are largely restricted to the axial skeleton, where firm bony attachments and protection of enclosed organs are a priority.

As a general rule, fibrous joints are immovable, and synovial joints are freely movable. Cartilaginous joints offer both rigid and slightly movable examples. Since the structural categories are more clear-cut, we will use the structural classification here and indicate functional properties as appropriate.

Fibrous Joints

In **fibrous joints,** the bones are joined by fibrous tissue. No joint cavity is present. The amount of movement allowed depends on the length of the fibers uniting the bones. Although some fibrous joints are slightly movable, most are synarthrotic and permit virtually no movement.

The two major types of fibrous joints are sutures and syndesmoses. In **sutures** (Figure 13.1d) the irregular edges of the bones interlock and are united by very short connective tissue fibers, as in most joints of the skull. In **syndesmoses** the articulating bones are connected by short ligaments of dense fibrous tissue; the bones do not interlock. The joint at the distal end of the tibia and fibula is an example of a syndesmosis (Figure 13.1e). Although this syndesmosis allows some give, it is classed functionally as a synarthrosis.

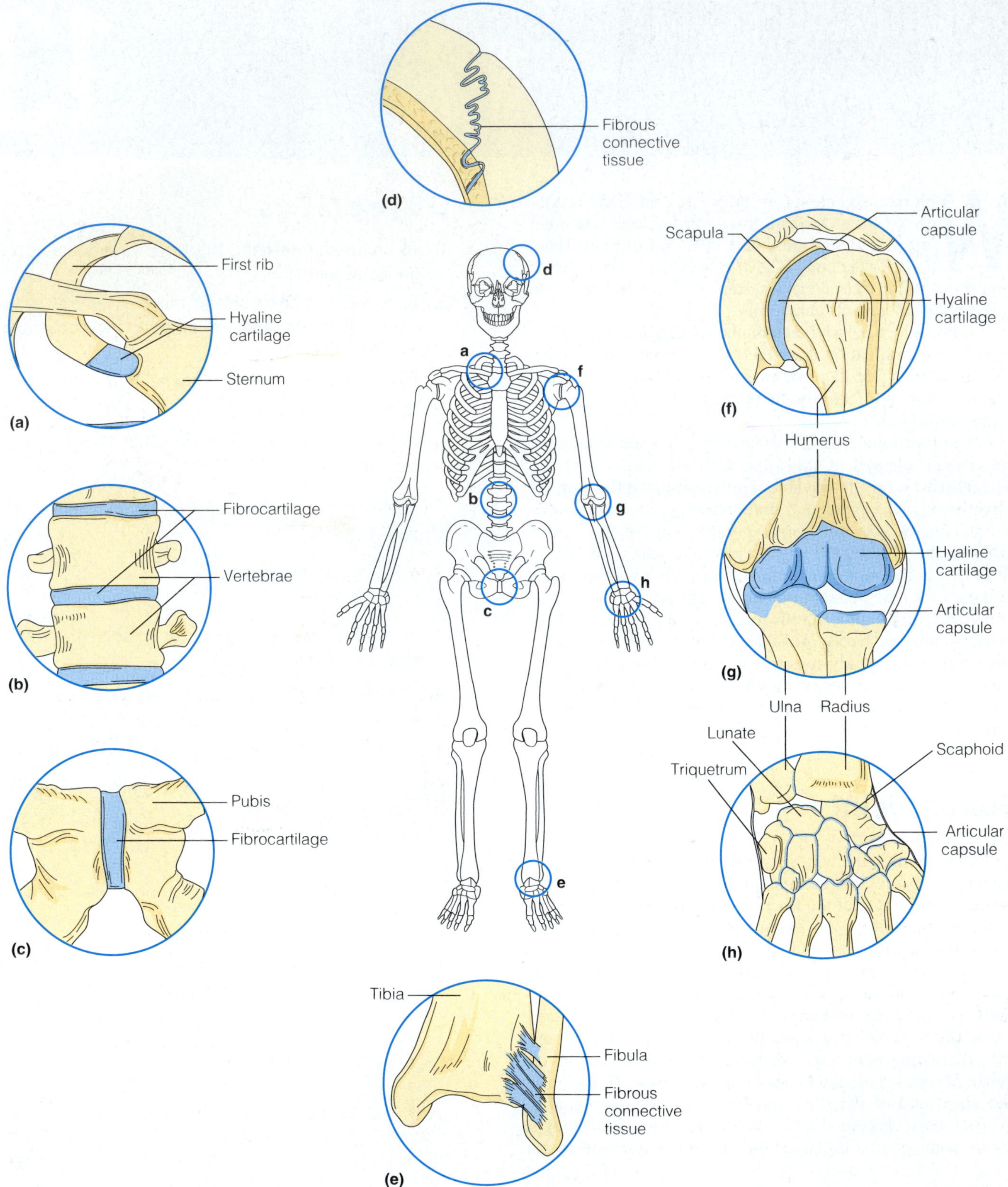

Figure 13.1 Types of joints. Joints to the left of the skeleton are cartilaginous joints; joints above and below the skeleton are fibrous joints; joints to the right of the skeleton are synovial joints. **(a)** Synchondrosis (joint between costal cartilage of rib 1 and the sternum). **(b)** Symphyses (intervertebral discs of fibrocartilage connecting adjacent vertebrae). **(c)** Symphysis (fibrocartilaginous pubic symphysis connecting the pubic bones anteriorly). **(d)** Suture (fibrous connective tissue connecting interlocking skull bones). **(e)** Syndesmosis (fibrous connective tissue connecting the distal ends of the tibia and fibula). **(f)** Synovial joint (multiaxial shoulder joint). **(g)** Synovial joint (uniaxial elbow joint). **(h)** Synovial joints (biaxial intercarpal joints of the hand).

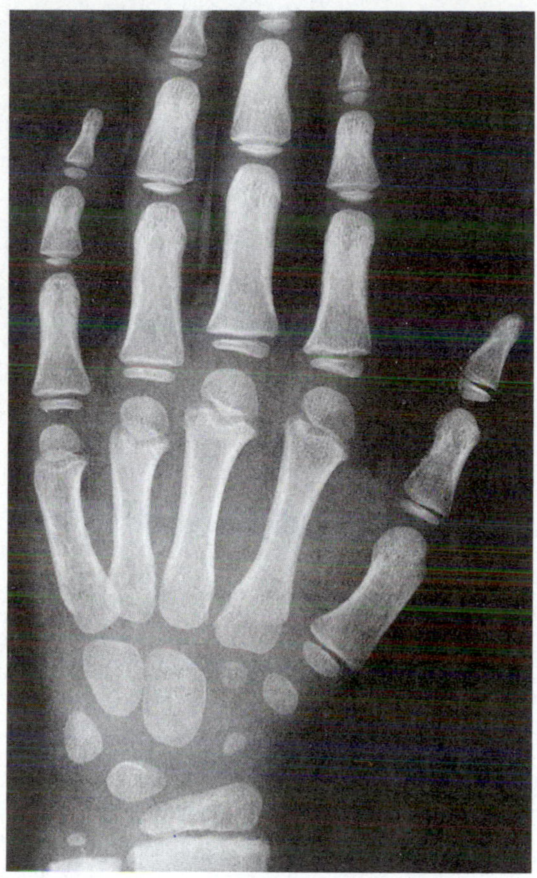

Figure 13.2 X ray of the hand of a child. Notice the cartilaginous epiphyseal plates, examples of temporary synchondroses.

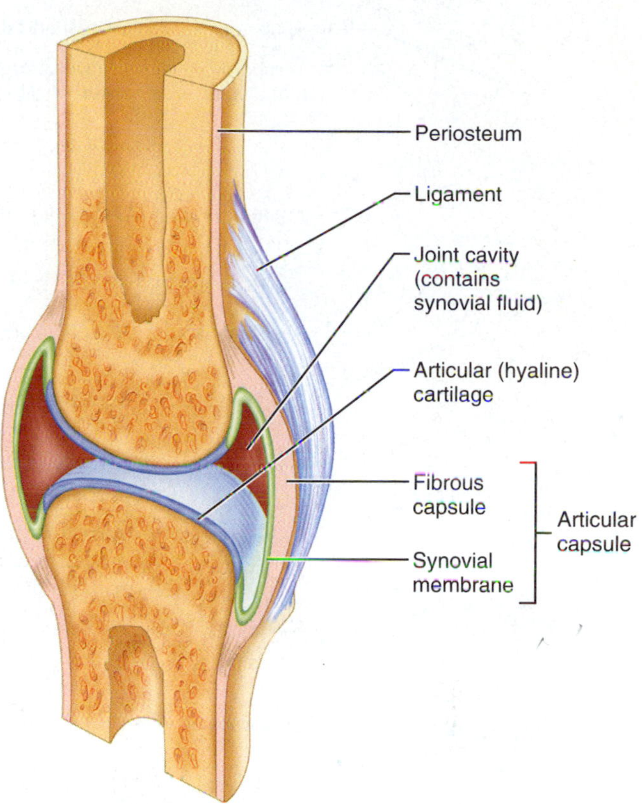

Figure 13.3 Major structural features of a synovial joint.

Activity 1:
Identifying Fibrous Joints

Examine a human skull again. Notice that adjacent bone surfaces do not actually touch but are separated by fibrous connective tissue. Also examine a skeleton and anatomical chart of joint types and Table 13.1 for examples of fibrous joints. ■

Cartilaginous Joints

In **cartilaginous joints,** the articulating bone ends are connected by a plate or pad of cartilage. No joint cavity is present. The two major types of cartilaginous joints are synchondroses and symphyses. Although there is variation, most cartilaginous joints are *slightly movable* (amphiarthroses) functionally. In **symphyses** (*symphysis* means "a growth together") the bones are connected by a broad, flat disc of fibrocartilage. The intervertebral joints and the pubic symphysis of the pelvis are symphyses (see Figure 13.1b and c). In **synchondroses** the bony portions are united by hyaline cartilage. The articulation of the costal cartilage of the first rib with the sternum (Figure 13.1a) is a synchondrosis, but perhaps the best examples of synchondroses are the epiphyseal plates seen in the long bones of growing children (Figure 13.2). The epiphyseal plates are flexible during childhood but eventually they are totally ossified. See Figure 11.5b on p. 110 for an X ray of an adult hand.

Activity 2:
Identifying Cartilaginous Joints

Identify the cartilaginous joints on a human skeleton, Table 13.1, and on an anatomical chart of joint types. ■

Synovial Joints

Synovial joints are those in which the articulating bone ends are separated by a joint cavity containing synovial fluid (see Figure 13.1f–h). All synovial joints are diarthroses, or freely movable joints. Their mobility varies, however; some synovial joints can move in only one plane, and others can move in several directions (multiaxial movement). Most joints in the body are synovial joints.

All synovial joints have the following structural characteristics (Figure 13.3):

- The joint surfaces are enclosed by a two-layered *articular capsule* (a sleeve of connective tissue), creating a joint cavity.

TABLE 13.1 Structural and Functional Characteristics of Body Joints

Illustration	Joint	Articulating bones	Structural type*	Functional type; movements allowed
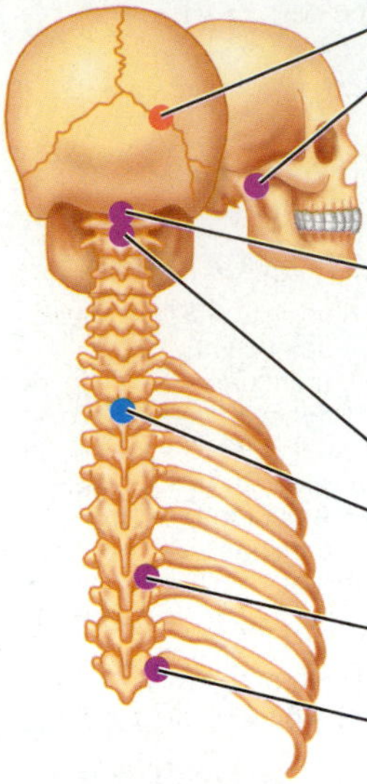	Skull	Cranial and facial bones	Fibrous; suture	Synarthrotic; no movement
	Temporo-mandibular	Temporal bone of skull and mandible	Synovial; modified hinge† (contains articular disc)	Diarthrotic; gliding and uniaxial rotation; slight lateral movement, elevation, depression, protraction, and retraction of mandible
	Atlanto-occipital	Occipital bone of skull and atlas	Synovial; condyloid	Diarthrotic; biaxial; flexion, extension, abduction, adduction, circumduction of head on neck
	Atlantoaxial	Atlas (C_1) and axis (C_2)	Synovial; pivot	Diarthrotic; uniaxial; rotation of the head
	Intervertebral	Between adjacent vertebral bodies	Cartilaginous; symphysis	Amphiarthrotic; slight movement
	Intervertebral	Between articular processes	Synovial; plane	Diarthrotic; gliding
	Vertebrocostal	Vertebrae (transverse processes or bodies) and ribs	Synovial; plane	Diarthrotic; gliding of ribs
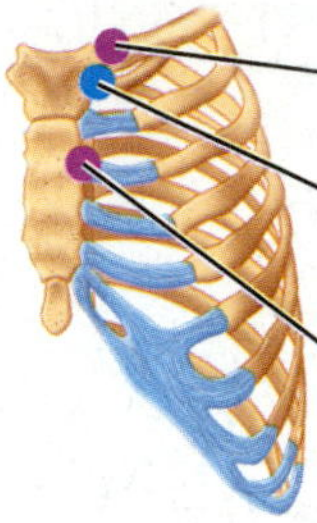	Sternoclavicular	Sternum and clavicle	Synovial; shallow saddle (contains articular disc)	Diarthrotic; multiaxial (allows clavicle to move in all axes)
	Sternocostal	Sternum and rib 1	Cartilaginous; synchondrosis	Synarthrotic; no movement
	Sternocostal	Sternum and ribs 2–7	Synovial; double plane	Diarthrotic; gliding
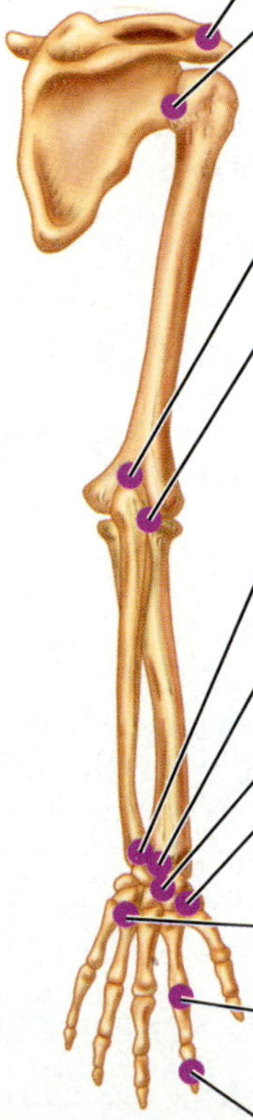	Acromioclavicular	Acromion of scapula and clavicle	Synovial; plane	Diarthrotic; gliding and rotation of scapula on clavicle
	Shoulder (glenohumeral)	Scapula and humerus	Synovial; ball and socket	Diarthrotic; multiaxial; flexion, extension, abduction, adduction, circumduction, rotation of humerus
	Elbow	Ulna (and radius) with humerus	Synovial; hinge	Diarthrotic; uniaxial; flexion, extension of forearm
	Radioulnar (proximal)	Radius and ulna	Synovial; pivot	Diarthrotic; uniaxial; rotation of radius around long axis of forearm to allow pronation and supination
	Radioulnar (distal)	Radius and ulna	Synovial; pivot (contains articular disc)	Diarthrotic; uniaxial; rotation (convex head of ulna rotates in ulnar notch of radius)
	Wrist (radiocarpal)	Radius and proximal carpals	Synovial; condyloid	Diarthrotic; biaxial; flexion, extension, abduction, adduction, circumduction of hand
	Intercarpal	Adjacent carpals	Synovial; plane	Diarthrotic; gliding
	Carpometacarpal of digit 1 (thumb)	Carpal (trapezium) and metacarpal 1	Synovial; saddle	Diarthrotic; biaxial; flexion, extension, abduction, adduction, circumduction, opposition of metacarpal 1
	Carpometacarpal of digits 2–5	Carpal(s) and metacarpal(s)	Synovial; plane	Diarthrotic; gliding of metacarpals
	Knuckle (metacarpophalangeal)	Metacarpal and proximal phalanx	Synovial; condyloid	Diarthrotic; biaxial; flexion, extension, abduction, adduction, circumduction of fingers
	Finger (interphalangeal)	Adjacent phalanges	Synovial; hinge	Diarthrotic; uniaxial; flexion, extension of fingers

TABLE 13.1 Structural and Functional Characteristics of Body Joints *(continued)*

Illustration	Joint	Articulating bones	Structural type*	Functional type; movements allowed
	Sacroiliac	Sacrum and coxal bone	Synovial; plane	Diarthrotic; little movement, slight gliding possible (more during pregnancy)
	Pubic symphysis	Pubic bones	Cartilaginous; symphysis	Amphiarthrotic; slight movement (enhanced during pregnancy)
	Hip (coxal)	Hip bone and femur	Synovial; ball and socket	Diarthrotic; multiaxial; flexion, extension, abduction, adduction, rotation, circumduction of thigh
	Knee (tibiofemoral)	Femur and tibia	Synovial; modified hinge† (contains articular discs)	Diarthrotic; biaxial; flexion, extension of leg, some rotation allowed
	Knee (femoropatellar)	Femur and patella	Synovial; plane	Diarthrotic; gliding of patella
	Tibiofibular	Tibia and fibula (proximally)	Synovial; plane	Diarthrotic; gliding of fibula
	Tibiofibular	Tibia and fibula (distally)	Fibrous; syndesmosis	Synarthrotic; slight "give" during dorsiflexion
	Ankle	Tibia and fibula with talus	Synovial; hinge	Diarthrotic; uniaxial; dorsiflexion, and plantar flexion of foot
	Intertarsal	Adjacent tarsals	Synovial; plane	Diarthrotic; gliding; inversion and eversion of foot
	Tarsometatarsal	Tarsal(s) and metatarsal(s)	Synovial; plane	Diarthrotic; gliding of metatarsals
	Metatarso-phalangeal	Metatarsal and proximal phalanx	Synovial; condyloid	Diarthrotic; biaxial; flexion, extension, abduction, adduction, circumduction of great toe
	Toe (interphalangeal)	Adjacent phalanges	Synovial; hinge	Diarthrotic; uniaxial; flexion, extension of toes

*__Fibrous joints__ indicated by orange circles; **cartilaginous joints** by blue circles; **synovial joints** by purple circles.

†These modified hinge joints are structurally bicondylar.

- The inner layer is a smooth connective tissue membrane, called *synovial membrane*, which produces a lubricating fluid (synovial fluid) that reduces friction. The outer layer or *fibrous capsule* is dense irregular connective tissue.
- *Articular* (hyaline) cartilage covers the surfaces of the bones forming the joint.
- The articular capsule is typically reinforced with ligaments and may contain bursae (fluid-filled sacs that reduce friction where tendons cross bone).
- Fibrocartilage pads *(articular discs)* may be present within the capsule.

Activity 3:
Examining Synovial Joint Structure

Examine a beef joint to identify the general structural features of diarthrotic joints.

⚠ If the joint is freshly obtained from the slaughterhouse and you will be handling it, don plastic gloves before beginning your observations. ■

Types of Synovial Joints

Because there are so many types of synovial joints, they have been divided into the following subcategories on the basis of movements allowed (Figure 13.4):

- Plane (Gliding): Articulating surfaces are flat or slightly curved, allowing sliding movements in one or two planes. Examples are the intercarpal and intertarsal joints and the vertebrocostal joints of ribs 2 through 7.
- Hinge: The rounded process of one bone fits into the concave surface of another to allow movement in one plane (uniaxial), usually flexion and extension. Examples are the elbow and interphalangeal joints.

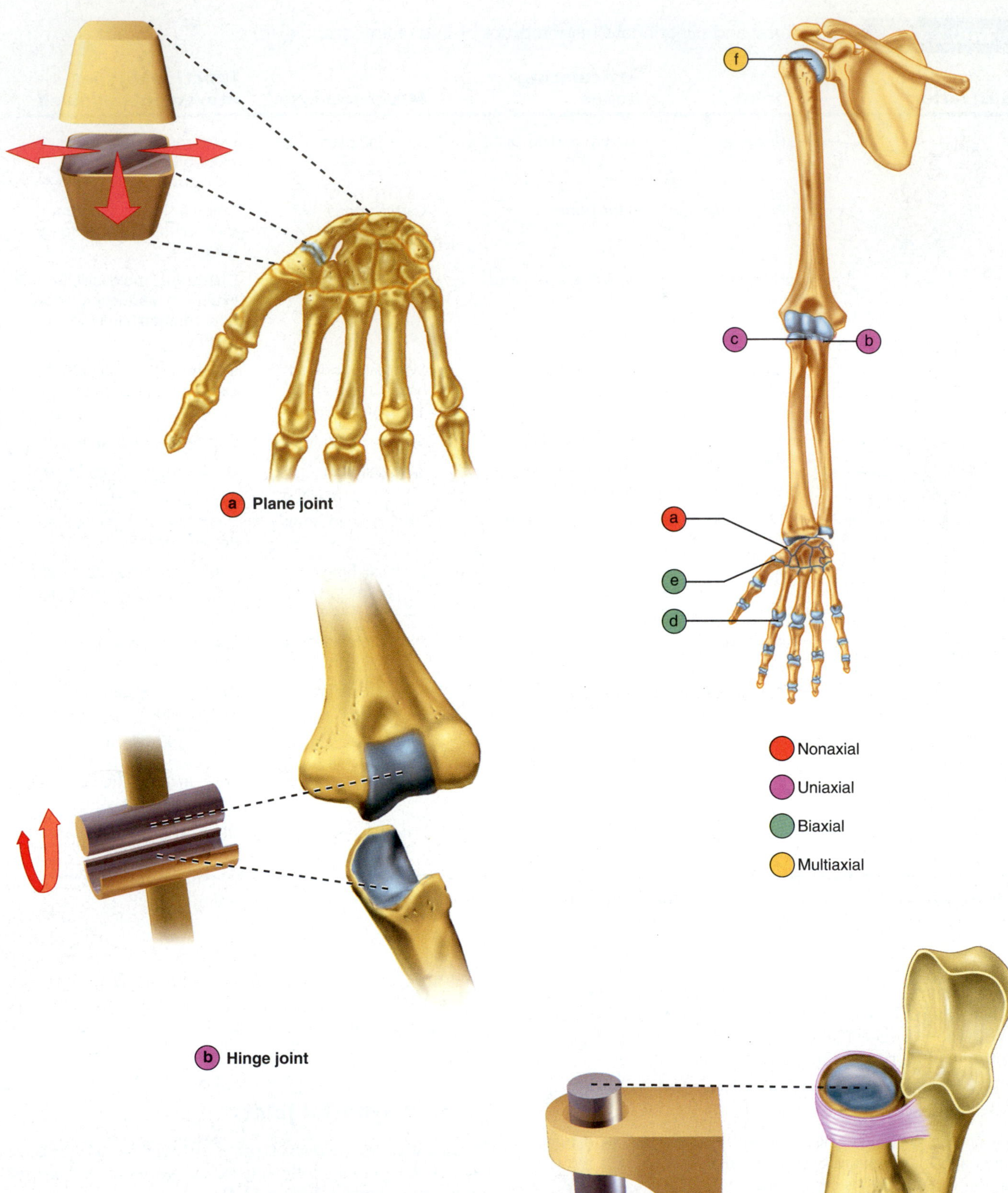

Figure 13.4 Types of synovial joints. Dashed lines indicate the articulating bones. **(a)** Plane joint (e.g., intercarpal and intertarsal joints). **(b)** Hinge joint (e.g., elbow joints and interphalangeal joints). **(c)** Pivot joint (e.g., proximal radioulnar joint).

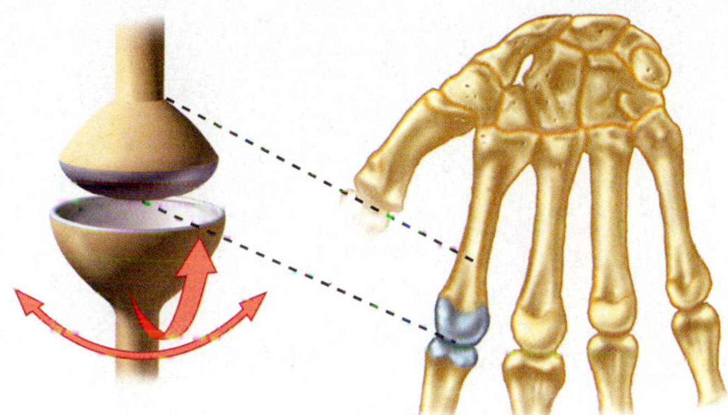

(d) Condyloid joint

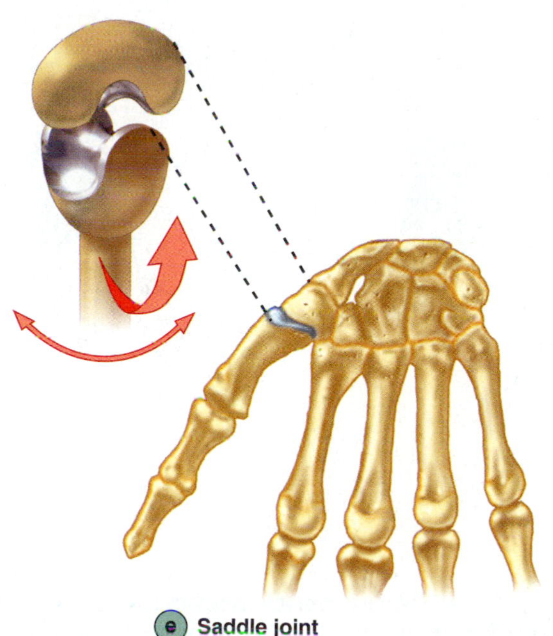

(e) Saddle joint

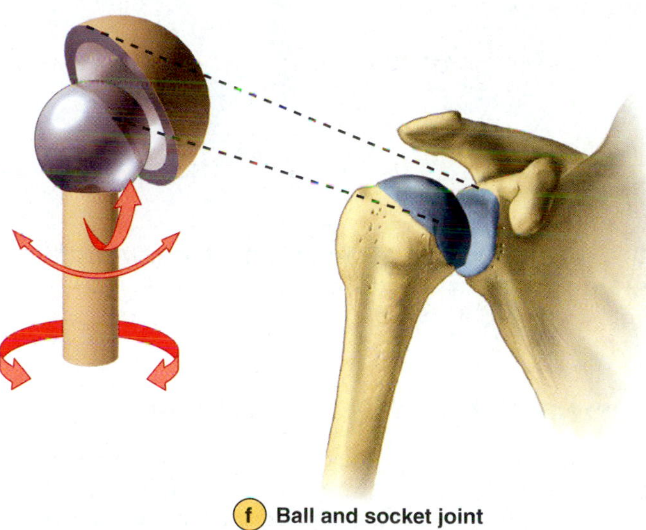

(f) Ball and socket joint

Figure 13.4 (*continued*) **(d)** Condyloid joint (e.g., metacarpophalangeal joints). **(e)** Saddle joint (e.g., carpometacarpal joint of the thumb). **(f)** Ball-and-socket joint (e.g., shoulder joint).

- Pivot: The rounded or conical surface of one bone articulates with a shallow depression or foramen in another bone. Pivot joints allow uniaxial rotation, as in the joint between the atlas and axis (C_1 and C_2).
- Condyloid (Ellipsoidal): The oval condyle of one bone fits into an ellipsoidal depression in another bone, allowing biaxial (two-way) movement. The radiocarpal (wrist) joint and the metacarpalphalangeal joints (knuckles) are examples.
- Saddle: Articulating surfaces are saddle shaped; the articulating surface of one bone is convex, and the reciprocal surface is concave. Saddle joints, which are biaxial, include the joint between the thumb metacarpal and the trapezium of the wrist.
- Ball and socket: The ball-shaped head of one bone fits into a cuplike depression of another. These are multiaxial joints, allowing movement in all directions and pivotal rotation. Examples are the shoulder and hip joints.

Activity 4: Identifying Types of Synovial Joints

Examine the articulated skeleton, anatomical charts, Table 13.1, and yourself to identify the subcategories of synovial joints. Make sure you understand the terms *uniaxial, biaxial,* and *multiaxial*. ■

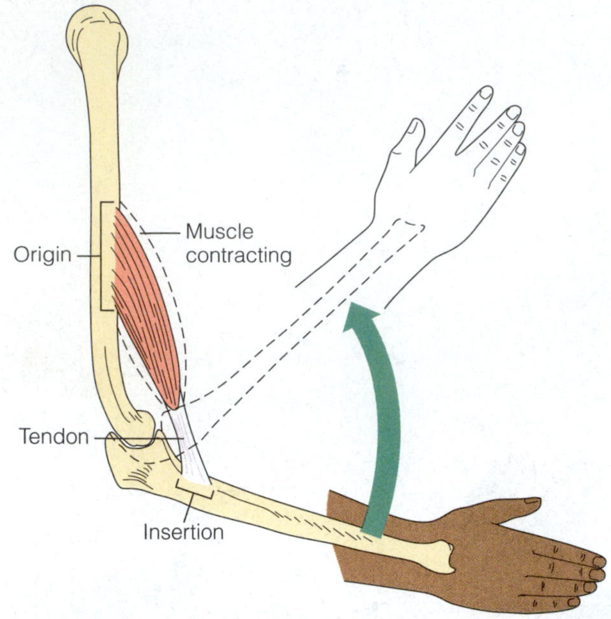

Figure 13.5 Muscle attachments (origin and insertion). When a skeletal muscle contracts, its insertion moves toward its origin.

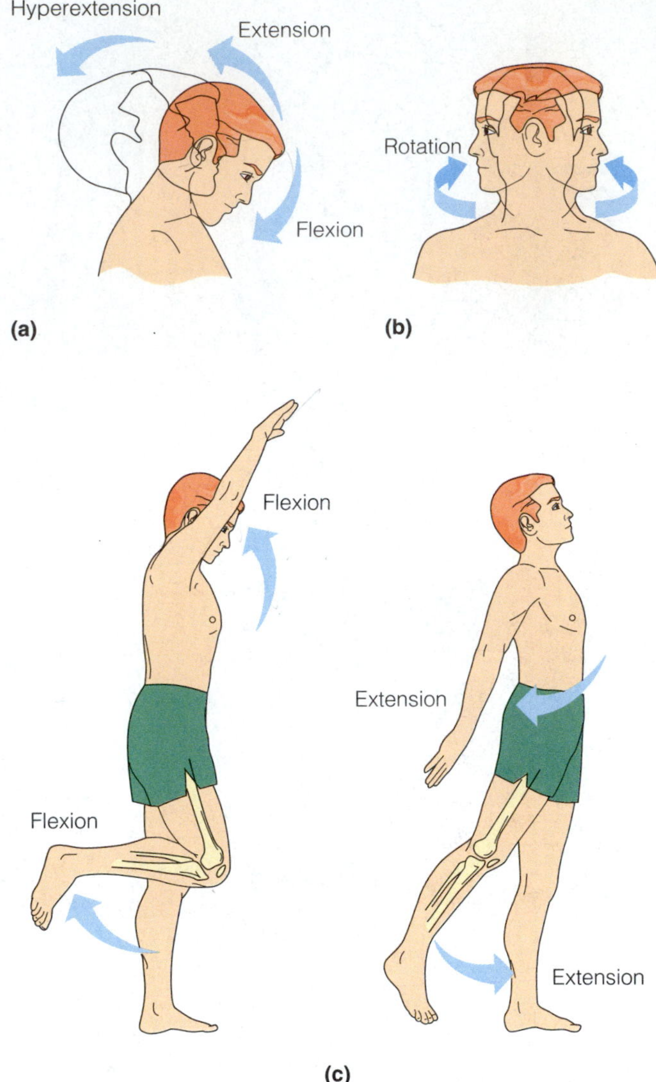

Figure 13.6 Movements occurring at synovial joints of the body. (a) Flexion and extension of the head. **(b)** Rotation of the head. **(c)** Flexion and extension of the knee and shoulder.

Movements Allowed by Synovial Joints

Every muscle of the body is attached to bone (or other connective tissue structures) at two points–the **origin** (the stationary, immovable, or less movable attachment) and the **insertion** (the movable attachment). Body movement occurs when muscles contract across diarthrotic synovial joints (Figure 13.5). When the muscle contracts and its fibers shorten, the insertion moves toward the origin. The type of movement depends on the construction of the joint (uniaxial, biaxial, or multiaxial) and on the placement of the muscle relative to the joint. The most common types of body movements are described below and illustrated in Figure 13.6.

Activity 5: Demonstrating Movements of Synovial Joints

Attempt to demonstrate each movement as you read through the following material:

Flexion (Figure 13.6a and c): A movement, generally in the sagittal plane, that decreases the angle of the joint and reduces the distance between the two bones. Flexion is typical of hinge joints (bending the knee or elbow), but is also common at ball-and-socket joints (bending forward at the hip).

Extension (Figure 13.6a and c): A movement that increases the angle of a joint and the distance between two bones or parts of the body (straightening the knee or elbow). Extension is the opposite of flexion. If extension is greater than 180 degrees (bending the trunk backward), it is termed *hyperextension*.

Abduction (Figure 13.6d): Movement of a limb away from the midline or median plane of the body, generally on the frontal plane, or the fanning movement of fingers or toes when they are spread apart.

Adduction (Figure 13.6d): Movement of a limb toward the midline of the body. Adduction is the opposite of abduction.

Rotation (Figure 13.6b and e): Movement of a bone around its longitudinal axis without lateral or medial displacement. Rotation, a common movement of ball-and-socket joints,

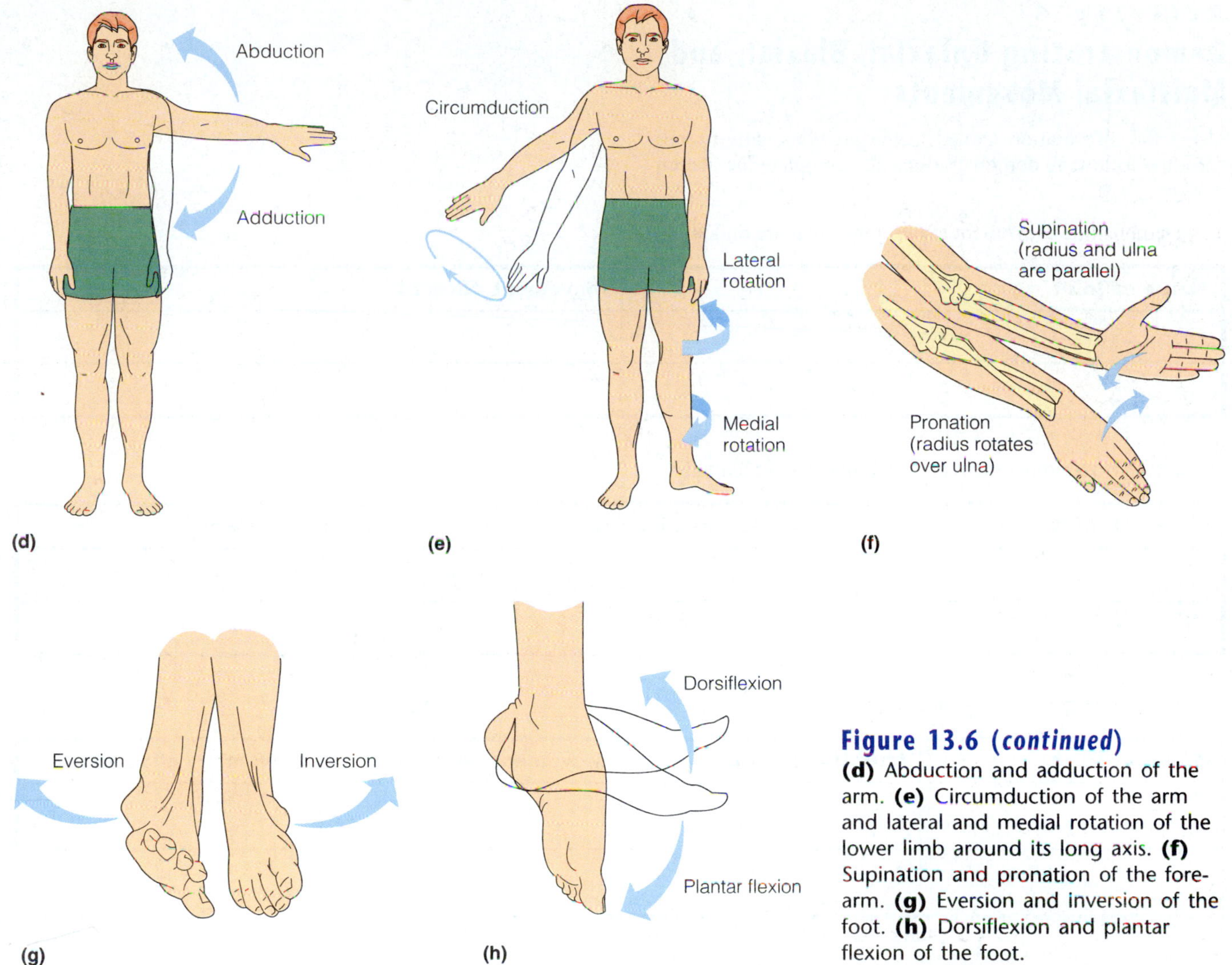

Figure 13.6 (*continued*) **(d)** Abduction and adduction of the arm. **(e)** Circumduction of the arm and lateral and medial rotation of the lower limb around its long axis. **(f)** Supination and pronation of the forearm. **(g)** Eversion and inversion of the foot. **(h)** Dorsiflexion and plantar flexion of the foot.

also describes the movement of the atlas around the odontoid process of the axis.

Circumduction (Figure 13.6e): A combination of flexion, extension, abduction, and adduction commonly observed in ball-and-socket joints like the shoulder. The proximal end of the limb remains stationary, and the distal end moves in a circle. The limb as a whole outlines a cone. Condyloid and saddle joints also allow circumduction.

Pronation (Figure 13.6f): Movement of the palm of the hand from an anterior or upward-facing position to a posterior or downward-facing position. This action moves the distal end of the radius across the ulna.

Supination (Figure 13.6f): Movement of the palm from a posterior position to an anterior position (the anatomical position). Supination is the opposite of pronation. During supination, the radius and ulna are parallel.

The last four terms refer to movements of the foot:

Inversion (Figure 13.6g): A movement that results in the medial turning of the sole of the foot.

Eversion (Figure 13.6g): A movement that results in the lateral turning of the sole of the foot; the opposite of inversion.

Dorsiflexion (Figure 13.6h): A movement of the ankle joint in a dorsal direction (standing on one's heels).

Plantar flexion (Figure 13.6h): A movement of the ankle joint in which the foot is flexed downward (standing on one's toes or pointing the toes). ■

Activity 6:
Demonstrating Uniaxial, Biaxial, and Multiaxial Movements

Using the information gained in the previous activity, perform the following demonstrations and complete the accompanying charts:

1. Demonstrate movement at two joints that are uniaxial.

Name of joint	Movement allowed

2. Demonstrate movement at two joints that are biaxial.

Name of joint	Movement allowed	Movement allowed

3. Demonstrate movement at two joints that are multiaxial.

Name of joint	Movement allowed	Movement allowed	Movement allowed

Activity 7:
Examining Selected Synovial Joints

Now you will have the opportunity to compare and contrast the structure of the hip and knee joints using Figures 13.7 and 13.8 as your guides. Both of these joints are large weight-bearing joints of the lower limb, but they differ substantially in their security. Read through the brief descriptive material below, and look at the questions in the review section that pertain to this exercise before beginning your comparison.

The Knee Joint The knee is the largest and most complex joint in the body. Three joints in one (Figure 13.7), it allows extension, flexion, and a little rotation. The tibiofemoral joint, actually a duplex joint between the femoral condyles above and the menisci (semilunar cartilages) of the tibia below, is functionally a hinge joint, a very unstable one made slightly more secure by the menisci. Some rotation occurs when the knee is partly flexed, but during extension, rotation and side-to-side movements are counteracted by the menisci and ligaments. The other joint is the femoropatellar joint, the intermediate joint anteriorly. The knee is unique in that it is only partly enclosed by an articular capsule. The capsule is reinforced by three broad ligaments, the *patellar ligament* and the *medial* and *lateral patellar retinacula,* which merge with the capsule. Extracapsular ligaments including the *fibular* and *tibial collateral ligaments* (which prevent rotation during extension) and the *oblique popliteal* and *arcuate popliteal ligaments* are crucial in reinforcing the knee. The *cruciate ligaments,* which are intracapsular ligaments, prevent anterior-posterior displacement of the joint and overflexion and hyperextension of the joint.

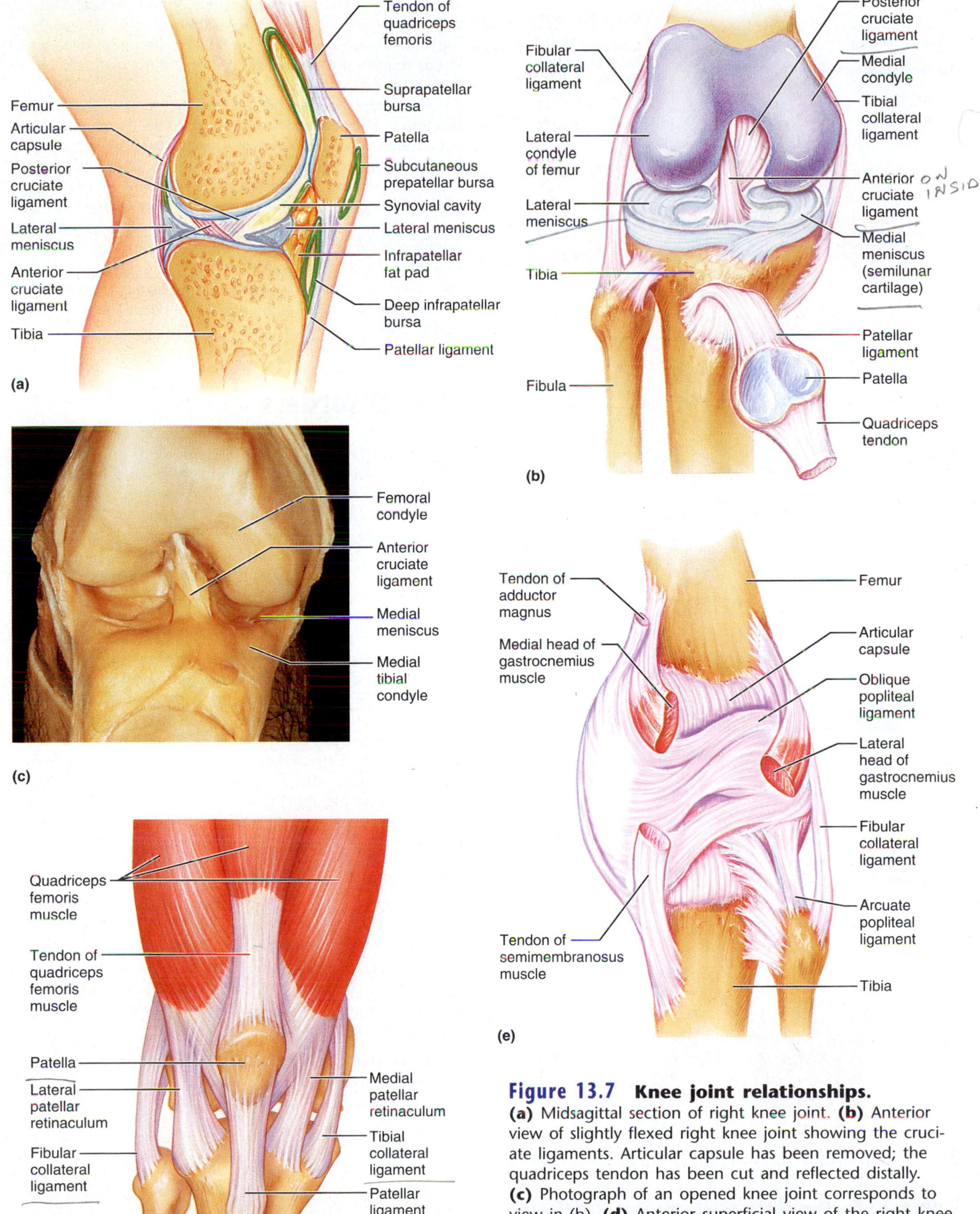

Figure 13.7 Knee joint relationships. **(a)** Midsagittal section of right knee joint. **(b)** Anterior view of slightly flexed right knee joint showing the cruciate ligaments. Articular capsule has been removed; the quadriceps tendon has been cut and reflected distally. **(c)** Photograph of an opened knee joint corresponds to view in (b). **(d)** Anterior superficial view of the right knee. **(e)** Posterior superficial view of the ligaments clothing the knee joint.

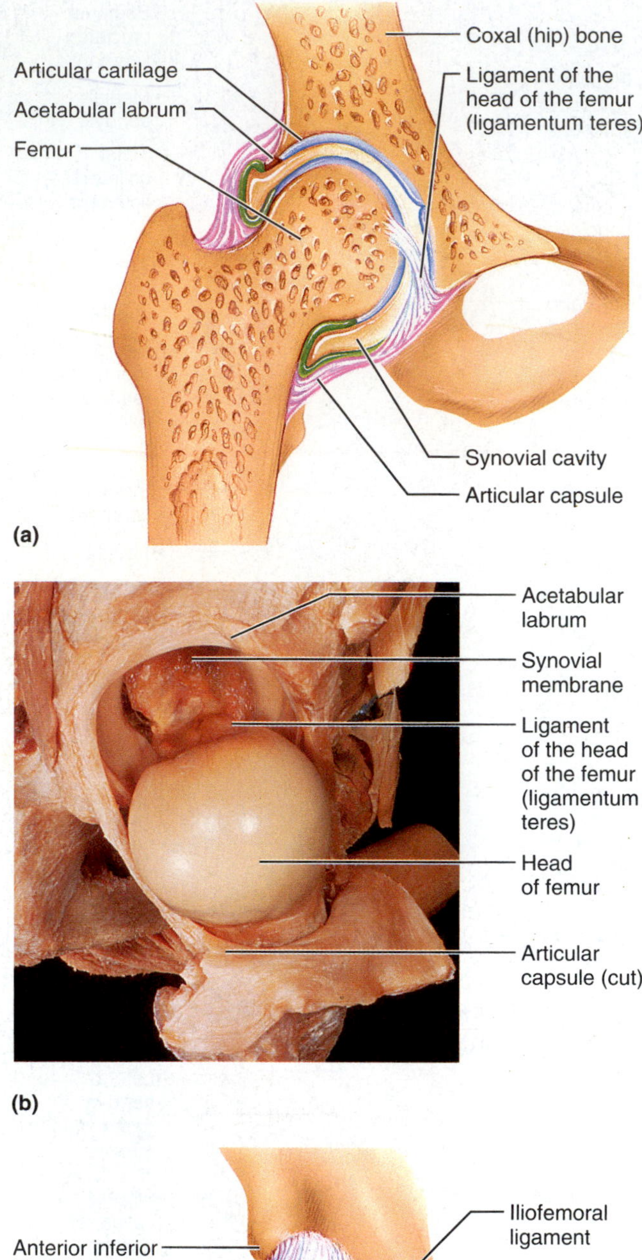

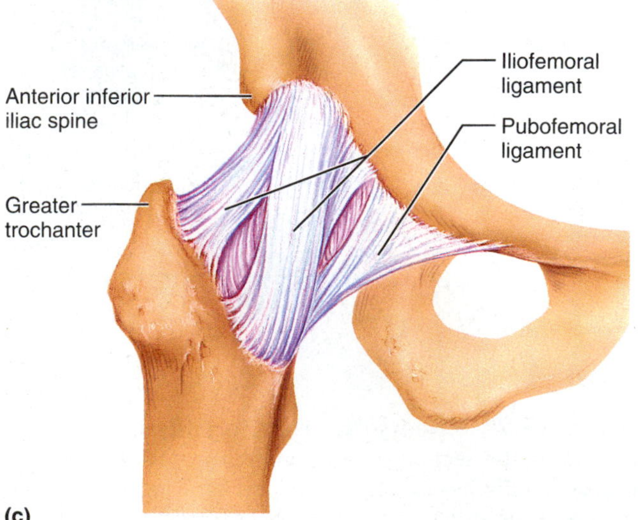

Figure 13.8 Hip joint relationships. (a) Frontal section through the right hip joint. **(b)** Photograph of the interior of the hip joint, lateral view. **(c)** Anterior superficial view of the right hip joint.

The Hip Joint The hip joint is a ball-and-socket joint, so movements can occur in all possible planes. However, its movements are definitely limited by its deep socket and strong reinforcing ligaments, the two factors that account for its exceptional stability (Figure 13.8).

The deeply cupped acetabulum that receives the head of the femur is enhanced by a circular rim of fibrocartilage called the *acetabular labrum.* Because the diameter of the labrum is smaller than that of the femur's head, dislocations of the hip are rare. A short ligament, *the ligament of the head of the femur* or *ligamentum teres,* runs from the pitlike *fovea capitis* on the femur head to the acetabulum where it helps to secure the femur. Several strong ligaments, including the *iliofemoral* and *pubofemoral* anteriorly and the *ischiofemoral* that spirals posteriorly (not shown), are arranged so that they "screw" the femur head into the socket when a person stands upright. ■

Joint Disorders

Most of us don't think about our joints until something goes wrong with them. Joint pains and malfunctions are caused by a variety of things. For example, a hard blow to the knee can cause a painful bursitis, known as "water on the knee," due to damage to, or inflammation of, the patellar bursa. Slippage of a fibrocartilage pad or the tearing of a ligament may result in a painful condition that persists over a long period, since these poorly vascularized structures heal so slowly.

Sprains and dislocations are other types of joint problems. In a **sprain,** the ligaments reinforcing a joint are damaged by excessive stretching or are torn away from the bony attachment. Since both ligaments and tendons are cords of dense connective tissue with a poor blood supply, sprains heal slowly and are quite painful. **Dislocations** occur when bones are forced out of their normal position in the joint cavity. They are normally accompanied by torn or stressed ligaments and considerable inflammation. The process of returning the bone to its proper position, called reduction, should be done only by a physician. Attempts by the untrained person to "snap the bone back into its socket" are often more harmful than helpful.

Advancing years also take their toll on joints. Weight-bearing joints in particular eventually begin to degenerate. *Adhesions* (fibrous bands) may form between the surfaces where bones join, and extraneous bone tissue (*spurs*) may grow along the joint edges. Such degenerative changes lead to the complaint so often heard from the elderly: "My joints are getting so stiff. . . ."

- If possible, compare an X ray of an arthritic joint to one of a normal joint. ■

Microscopic Anatomy, Organization, and Classification of Skeletal Muscle

exercise **14**

Objectives

1. To describe the structure of skeletal muscle from gross to microscopic levels.
2. To define and explain the role of the following:

actin	myofilament	tendon
myosin	perimysium	endomysium
fiber	aponeurosis	epimysium
myofibril		

3. To describe the structure of a neuromuscular junction and to explain its role in muscle function.
4. To define *agonist* (prime mover), *antagonist, synergist, fixator, origin,* and *insertion.*
5. To cite criteria used in naming skeletal muscles.

Materials

- ❑ Three-dimensional model of skeletal muscle cells (if available)
- ❑ Forceps
- ❑ Dissecting needles
- ❑ Microscope slides and coverslips
- ❑ 0.9% saline solution in dropper bottles
- ❑ Chicken breast or thigh muscle (freshly obtained from the meat market)
- ❑ Compound microscope
- ❑ Histologic slides of skeletal muscle (longitudinal and cross-sectional) and skeletal muscle showing neuromuscular junctions
- ❑ Three-dimensional model of skeletal muscle showing neuromuscular junction (if available)

The bulk of the body's muscle is called **skeletal muscle** because it is attached to the skeleton (or associated connective tissue structures). Skeletal muscle influences body contours and shape, allows you to grin and frown, provides a means of locomotion, and enables you to manipulate the environment. The balance of the body's muscle—smooth and cardiac muscle—as the major component of the walls of hollow organs and the heart, respectively, is involved with the transport of materials within the body.

Each of the three muscle types has a structure and function uniquely suited to its task in the body. However, because the term *muscular system* applies specifically to skeletal muscle, the primary objective of this unit is to investigate the structure and function of skeletal muscle.

Skeletal muscle is also known as *voluntary muscle* (because it can be consciously controlled) and as *striated muscle* (because it appears to be striped). As you might guess from both of these alternative names, skeletal muscle has some very special characteristics. Thus an investigation of skeletal muscle should begin at the cellular level.

The Cells of Skeletal Muscle

Skeletal muscle is composed of relatively large, long cylindrical cells, sometimes called **fibers,** ranging from 10 to 100 μm in diameter and up to 6 cm in length. However, the cells of large, hard-working muscles like the antigravity muscles of the hip are extremely coarse, ranging up to 25 cm in length, and can be seen with the naked eye.

Text continues on page 133

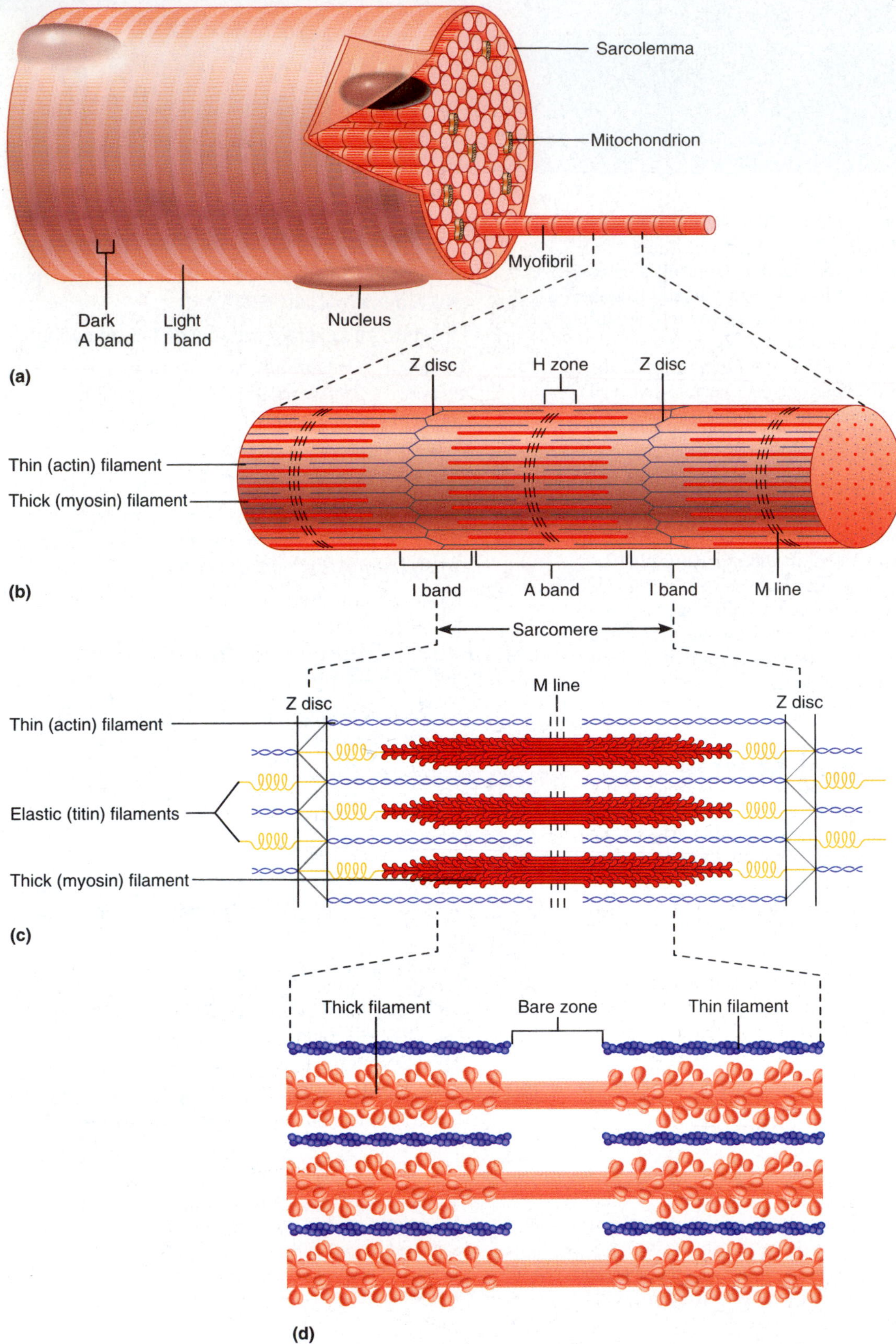

Figure 14.1 Anatomy of a skeletal muscle cell (fiber). (a) A muscle fiber. One myofibril has been extended. **(b)** Enlarged view of a myofibril showing its banding pattern. **(c)** Enlarged view of one sarcomere (contractile unit) of a myofibril showing its banding pattern. **(d)** Structure of the thick and thin myofilaments found in the sarcomeres.

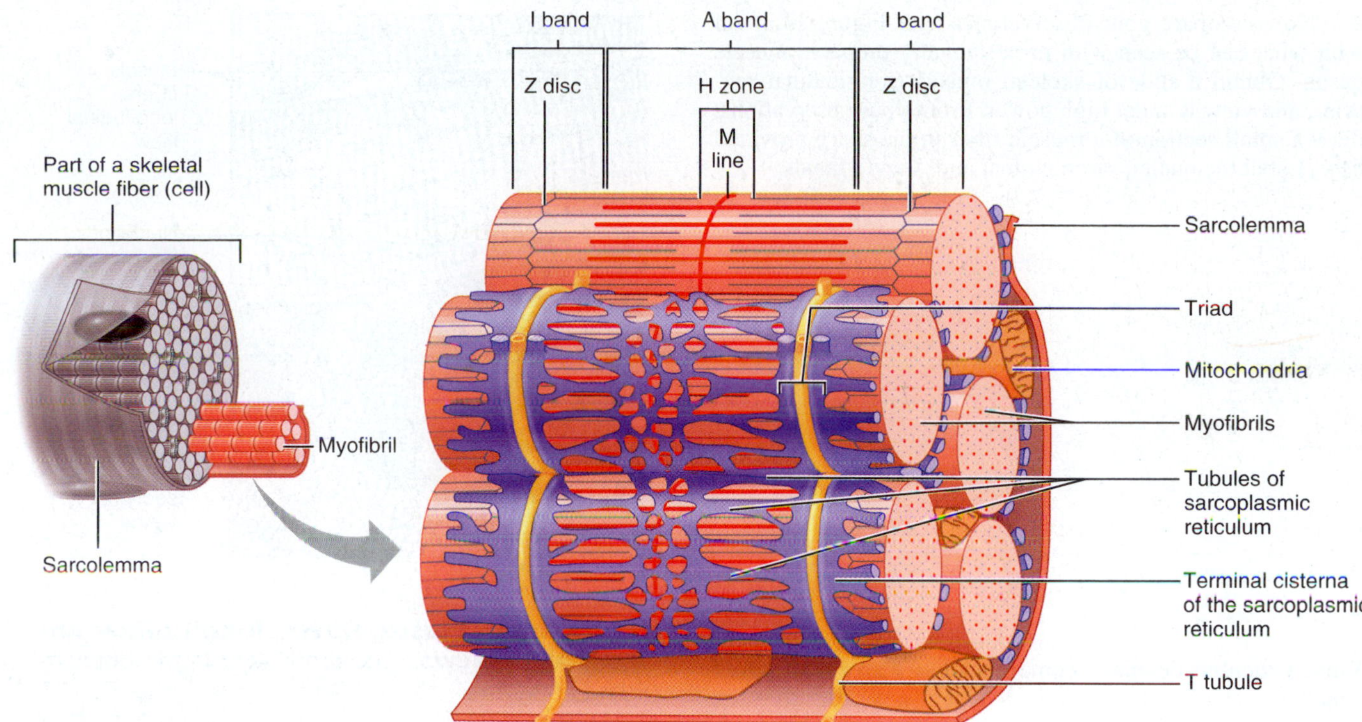

Figure 14.2 Relationship of the sarcoplasmic reticulum and T tubules to the myofibrils of skeletal muscle.

Skeletal muscle cells (Figure 14.1a) are multinucleate; multiple oval nuclei can be seen just beneath the plasma membrane (called the *sarcolemma* in these cells). The nuclei are pushed peripherally by the longitudinally arranged **myofibrils,** which nearly fill the sarcoplasm (Figure 14.1b). Alternating light (I) and dark (A) bands along the length of the perfectly aligned myofibrils give the muscle fiber as a whole its striped appearance.

Electron microscope studies have revealed that the myofibrils are made up of even smaller threadlike structures called **myofilaments** (Figure 14.1c and d). The myofilaments are composed largely of two varieties of contractile proteins—**actin** and **myosin**—which slide past each other during muscle activity to bring about shortening or contraction of the muscle cells. It is the highly specific arrangement of the myofilaments within the myofibrils that is responsible for the banding pattern in skeletal muscle. The actual contractile units of muscle, called **sarcomeres,** extend from the middle of one I band (its Z disc) to the middle of the next along the length of the myofibrils. (See Figure 14.1b and c.) At each junction of the A and I bands, the sarcolemma indents into the muscle cell, forming a **transverse tubule (T tubule).** These tubules run deep into the muscle cell between cross channels or **terminal cisternae** of the elaborate smooth endoplasmic reticulum called the **sarcoplasmic reticulum** (Figure 14.2).

Activity 1: Examining Skeletal Muscle Cell Anatomy

1. Look at the three-dimensional model of skeletal muscle cells, noting the relative shape and size of the cells. Identify the nuclei, myofibrils, and light and dark bands.

2. Obtain forceps, two dissecting needles, slide and coverslip, and a dropper bottle of saline solution. With forceps, remove a very small piece of muscle from the chicken breast (or thigh). Place the tissue on a clean microscope slide, and add a drop of the saline solution.

3. Pull the muscle fibers apart with the dissecting needles (tease them) until you have a fluffy-looking mass of tissue. Cover the teased tissue with a coverslip, and observe under the high-power lens of a microscope. Look for the banding pattern. Regulate the light carefully to obtain the highest possible contrast.

4. Now compare your observations with Figure 14.3 and with what can be seen with professionally prepared muscle tissue. Obtain a slide of skeletal muscle (longitudinal section), and view it under high power. From your observations, draw a small section of a muscle fiber in the space provided here. Label the nuclei, sarcolemma, and A and I bands.

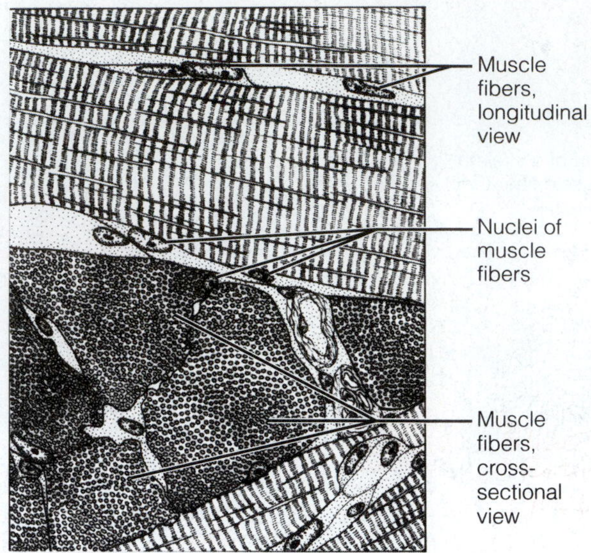

Figure 14.3 Muscle fibers, longitudinal and transverse views. (See also Plate 2 in the Histology Atlas.)

What structural details become apparent with the prepared slide?

____

____

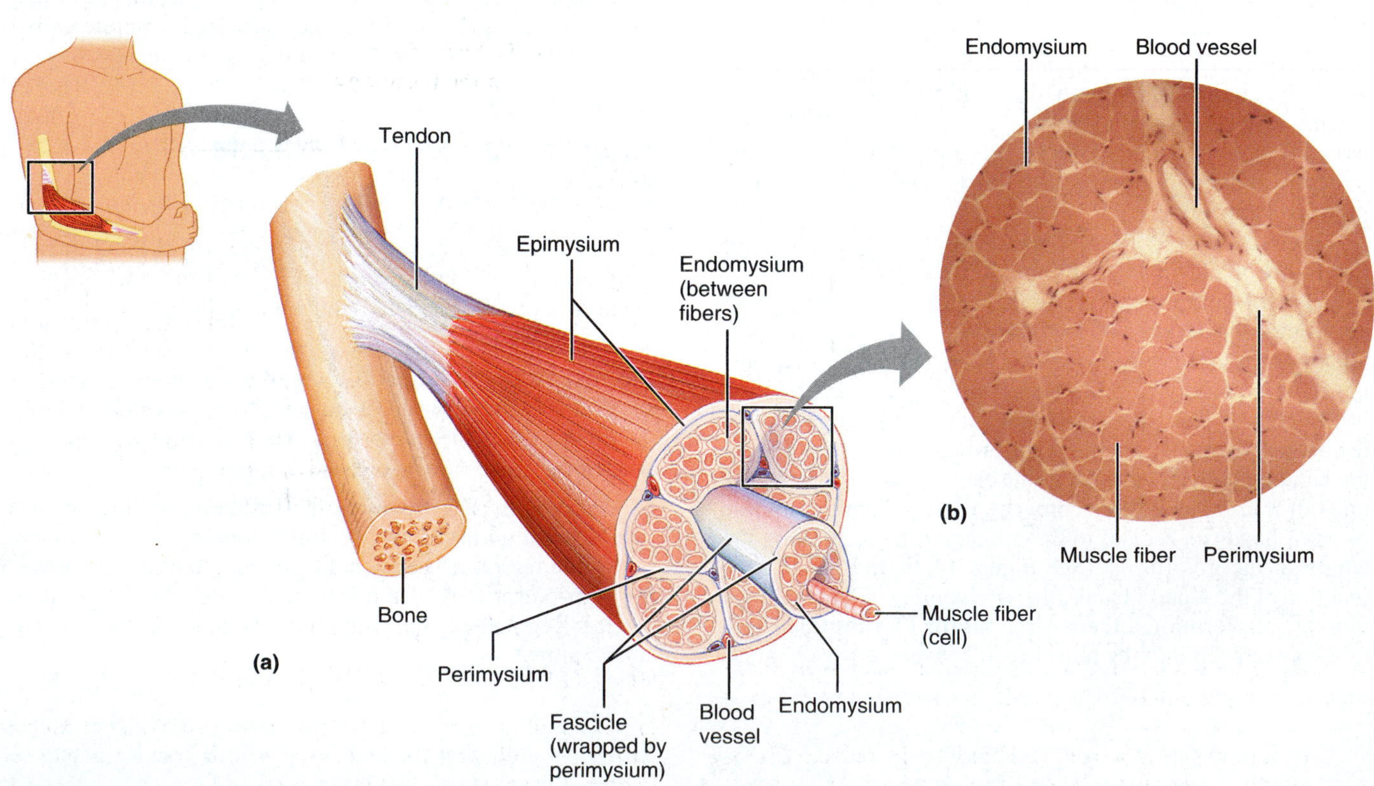

Figure 14.4 Connective tissue coverings of skeletal muscle. **(a)** Diagrammatic view. **(b)** Photomicrograph of a cross section of skeletal muscle (73×).

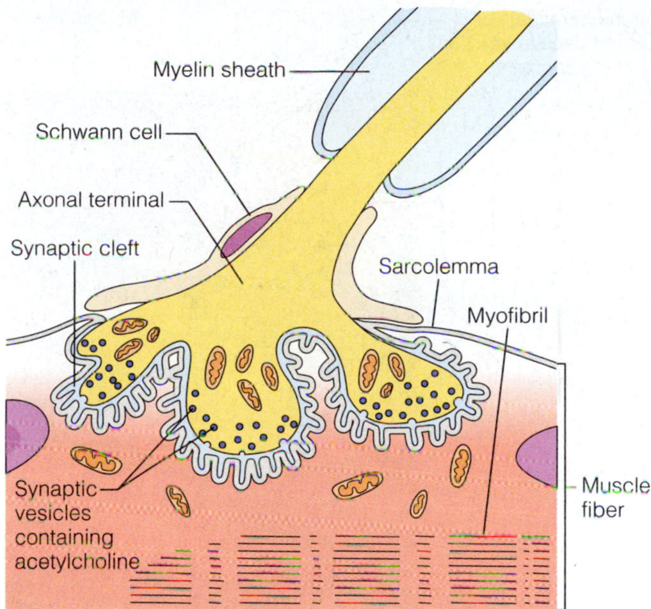

Figure 14.5 The neuromuscular junction.

Organization of Skeletal Muscle Cells into Muscles

Muscle fibers are soft and surprisingly fragile. Thus thousands of muscle fibers are bundled together with connective tissue to form the organs we refer to as skeletal muscles (Figure 14.4). Each muscle fiber is enclosed in a delicate, areolar connective tissue sheath called **endomysium.** Several sheathed muscle fibers are wrapped by a collagenic membrane called **perimysium,** forming a bundle of fibers called a **fascicle,** or **fasciculus.** A large number of fascicles are bound together by a substantially coarser "overcoat" of dense connective tissue called an **epimysium,** which sheathes the entire muscle. These epimysia blend into the **deep fascia,** still coarser sheets of dense connective tissue that bind muscles into functional groups, and into strong cordlike **tendons** or sheetlike **aponeuroses,** which attach muscles to each other or indirectly to bones. As noted in Exercise 13, a muscle's more movable attachment is called its *insertion* whereas its fixed (or immovable) attachment is the *origin.*

Tendons perform several functions, two of the most important being to provide durability and to conserve space. Because tendons are tough collagenic connective tissue, they can span rough bony prominences that would destroy the more delicate muscle tissues. Because of their relatively small size, more tendons than fleshy muscles can pass over a joint.

In addition to supporting and binding the muscle fibers, and providing strength to the muscle as a whole, the connective tissue wrappings provide a route for the entry and exit of nerves and blood vessels that serve the muscle fibers. The larger, more powerful muscles have relatively more connective tissue than muscles involved in fine or delicate movements.

As we age, the mass of the muscle fibers decreases, and the amount of connective tissue increases; thus the skeletal muscles gradually become more sinewy, or "stringier." ■

Activity 2:
Observing the Structure of Skeletal Muscle Tissue

Obtain a slide showing a cross section of skeletal muscle tissue. Using Figure 14.4 as a reference, identify the muscle fibers, endomysium, perimysium, and epimysium (if visible). ■

The Neuromuscular Junction

The voluntary skeletal muscle cells are always stimulated by motor neurons via nerve impulses. The junction between a nerve fiber (axon) and a muscle cell is called a **neuromuscular,** or **myoneural, junction** (Figure 14.5).

Each motor axon breaks up into many branches called *axonal terminals* as it approaches the muscle, and each of these branches participates in forming a neuromuscular junction with a single muscle cell. Thus a single neuron may stimulate many muscle fibers. Together, a neuron and all the muscle cells it stimulates make up the functional structure called the **motor unit.** Part of a motor unit is shown in Figure 14.6 and in Plate 4 of the Histology Atlas.

The neuron and muscle fiber membranes, close as they are, do not actually touch. They are separated by a small fluid-filled gap called the **synaptic cleft** (see Figure 14.5).

Within the axonal terminals are many mitochondria and vesicles containing a neurotransmitter chemical called acetylcholine (ACh). When a nerve impulse reaches the axonal endings, some of these vesicles release their contents into the synaptic cleft. The acetylcholine rapidly diffuses across the junction and combines with the receptors on the sarcolemma. If sufficient acetylcholine has been released, a transient change in the permeability of the sarcolemma briefly allows more sodium ions to diffuse into the muscle fiber. The result is depolarization of the sarcolemma and subsequent contraction of the muscle fiber.

Activity 3:
Studying the Structure of a Neuromuscular Junction

1. If possible, examine a three-dimensional model of skeletal muscle cells that illustrates the neuromuscular junction. Identify the structures just described.

2. Obtain a slide of skeletal muscle stained to show a portion of a motor unit. Examine the slide under high power to identify the axonal fibers extending leashlike to the muscle cells. Follow one of the axonal fibers to its terminus to identify the oval-shaped axonal terminal. Compare your observations to Figure 14.6. Sketch a small section in the space provided, labeling the motor axon, its terminal branches, and muscle fibers. ■

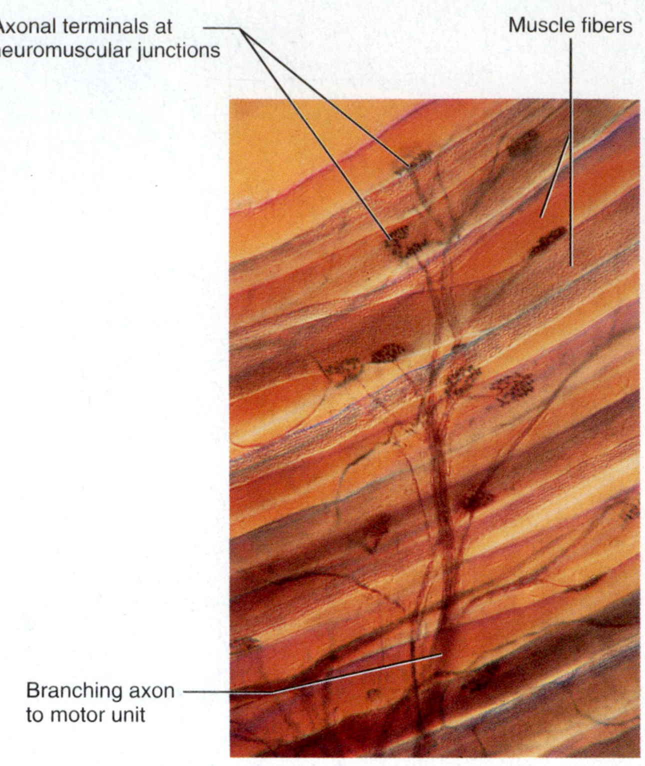

Figure 14.6 A portion of a motor unit. Photomicrograph of a portion of a motor unit (126×). (See also Plate 4 in the Histology Atlas.)

Classification of Skeletal Muscles

Naming Skeletal Muscles

Remembering the names of the skeletal muscles is a monumental task, but certain clues help. Muscles are named on the basis of the following criteria:

- **Direction of muscle fibers:** Some muscles are named in reference to some imaginary line, usually the midline of the body or the longitudinal axis of a limb bone. A muscle with fibers (and fascicles) running parallel to that imaginary line will have the term *rectus* (straight) in its name. For example, the rectus abdominis is the straight muscle of the abdomen. Likewise, the terms *transverse* and *oblique* indicate that the muscle fibers run at right angles and obliquely (respectively) to the imaginary line.
- **Relative size of the muscle:** Terms such as *maximus* (largest), *minimus* (smallest), *longus* (long), and *brevis* (short) are often used in naming muscles—as in gluteus maximus and gluteus minimus.
- **Location of the muscle:** Some muscles are named for the bone with which they are associated. For example, the frontalis muscle overlies the frontal bone.
- **Number of origins:** When the term *biceps, triceps,* or *quadriceps* forms part of a muscle name, you can generally assume that the muscle has two, three, or four origins (respectively). For example, the biceps muscle of the arm has two heads, or origins.
- **Location of the muscle's origin and insertion:** For example, the sternocleidomastoid muscle has its origin on the sternum (*sterno*) and clavicle (*cleido*), and inserts on the mastoid process of the temporal bone.
- **Shape of the muscle:** For example, the deltoid muscle is roughly triangular (*deltoid* = triangle), and the trapezius muscle resembles a trapezoid.
- **Action of the muscle:** For example, all the adductor muscles of the anterior thigh bring about its adduction, and all the extensor muscles of the wrist extend the wrist.

Types of Muscles

Most often, body movements are not a result of the contraction of a single muscle but instead reflect the coordinated action of several muscles acting together. Muscles that are primarily responsible for producing a particular movement are called **prime movers,** or **agonists.**

Muscles that oppose or reverse a movement are called **antagonists.** When a prime mover is active, the fibers of the antagonist are stretched and in the relaxed state. The antagonist can also regulate the prime mover by providing some resistance, to prevent overshoot or to stop its action.

It should be noted that antagonists can be prime movers in their own right. For example, the biceps muscle of the arm (a prime mover of elbow flexion) is antagonized by the triceps (a prime mover of elbow extension).

Synergists aid the action of agonists by reducing undesirable or unnecessary movement. Contraction of a muscle crossing two or more joints would cause movement at all joints spanned if the synergists were not there to stabilize them. For example, you can make a fist without bending your wrist only because synergist muscles stabilize the wrist joint and allow the prime mover to exert its force at the finger joints.

Fixators, or fixation muscles, are specialized synergists. They immobilize the origin of a prime mover so that all the tension is exerted at the insertion. Muscles that help maintain posture are fixators; so too are muscles of the back that stabilize or "fix" the scapula during arm movements.

Gross Anatomy of the Muscular System

exercise **15**

Objectives

1. To name and locate the major muscles of the human body (on a torso model, a human cadaver, lab chart, or diagram) and state the action of each.
2. To explain how muscle actions are related to their location.
3. To name muscle origins and insertions as required by the instructor.
4. To identify antagonists of the major prime movers.

Materials

- ❑ Human torso model or large anatomical chart showing human musculature
- ❑ Human cadaver for demonstration (if available)
- ❑ Disposable gloves
- ❑ *Human Musculature* videotape*

A.I.A. See Appendix C, Exercise 15 for links to A.D.A.M.® Interactive Anatomy.

For instructions on animal dissections, see the dissection exercises starting on p. 709 in the cat and fetal pig editions of this manual.

*Available to qualified adopters from Benjamin Cummings.

Identification of Human Muscles

Muscles of the Head and Neck

The muscles of the head serve many specific functions. For instance, the muscles of facial expression differ from most skeletal muscles because they insert into the skin (or other muscles) rather than into bone. As a result, they move the facial skin, allowing a wide range of emotions to be shown on the face. Other muscles of the head are the muscles of mastication, which manipulate the mandible during chewing, and the six extrinsic eye muscles located within the orbit, which aim the eye. (Orbital muscles are studied in Exercise 24.)

Activity 1: Identifying Head and Neck Muscles

Neck muscles are primarily concerned with the movement of the head and shoulder girdle. Figures 15.1 and 15.2 are summary figures illustrating the superficial musculature of the body as a whole. Head and neck muscles are discussed in Tables 15.1 and 15.2 and shown in Figures 15.3 and 15.4.

While reading the tables and identifying the head and neck muscles in the figures, try to visualize what happens when the muscle contracts. Then, use a torso model or an anatomical chart to again identify as many of these muscles as possible. (If a human cadaver is available for observation, specific instructions for muscle examination will be provided by your instructor.) Then carry out the following palpations on yourself:

- To demonstrate the temporalis, clench your teeth. The masseter can also be palpated now at the angle of the jaw. ■

Muscles of the Trunk

The trunk musculature includes muscles that move the vertebral column; anterior thorax muscles that act to move ribs, head, and arms; and muscles of the abdominal wall that play a role in the movement of the vertebral column but more importantly form the "natural girdle," or the major portion of the abdominal body wall.

Activity 2: Identifying Muscles of the Trunk

The trunk muscles are described in Tables 15.3 and 15.4 and shown in Figures 15.5 through 15.8. As before, identify the muscles in the figure as you read the tabular descriptions and then identify them on the torso or laboratory chart. ■

Activity 3: Demonstrating Operation of Trunk Muscles

Now, work with a partner to demonstrate the operation of the following muscles. One of you can demonstrate the movement (the following steps are addressed to this partner). The other can supply resistance and palpate the muscle being tested.

1. Fully abduct the arm and extend the elbow. Now adduct the arm against resistance. You are using the *latissimus dorsi*.
2. To observe the *deltoid*, attempt to abduct your arm against resistance. Now attempt to elevate your shoulder against resistance; you are contracting the upper portion of the *trapezius*.
3. The *pectoralis major* is used when you press your hands together at chest level with your elbows widely abducted. ■

Text continues on page 151

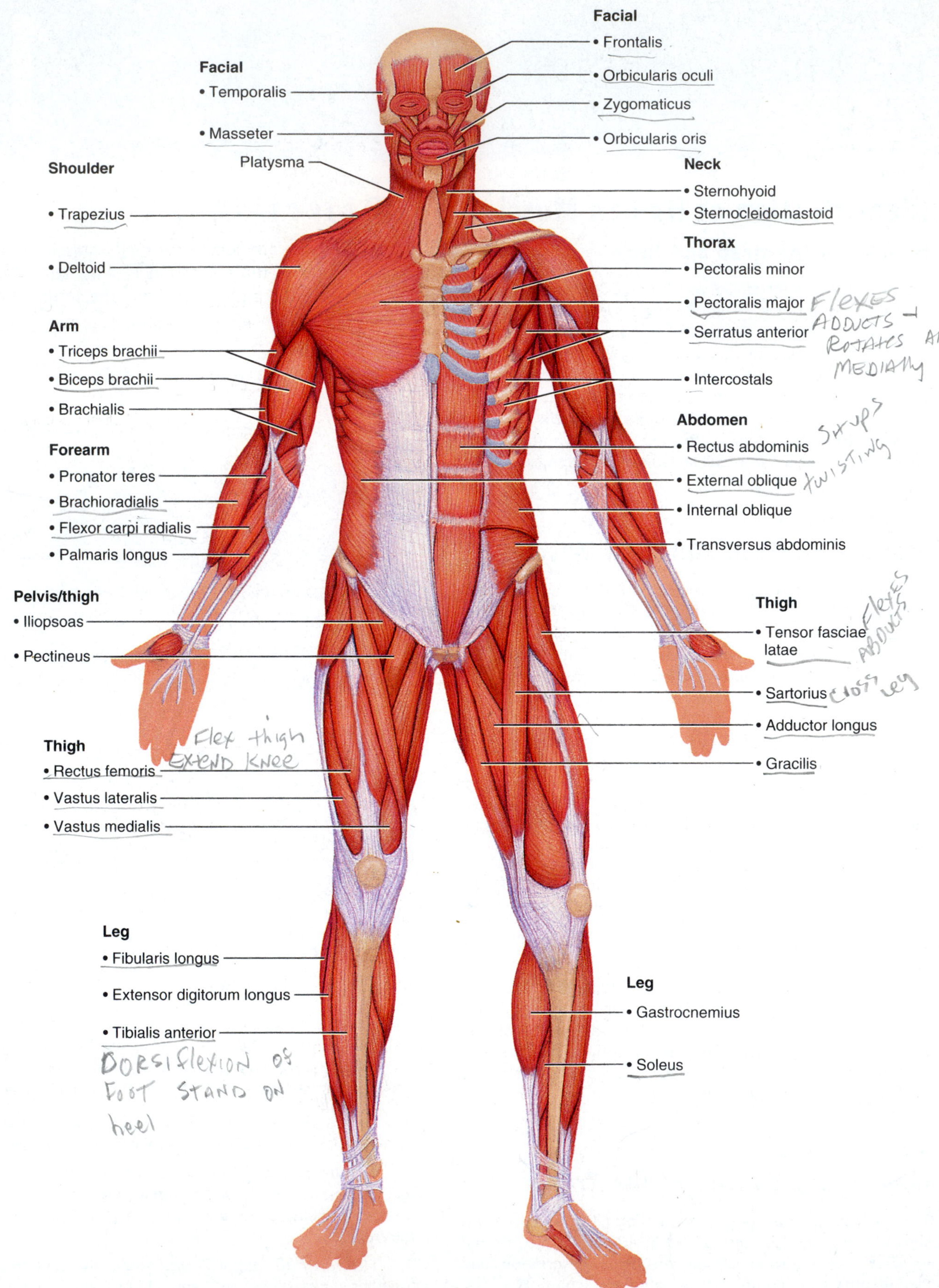

Figure 15.1 Anterior view of superficial muscles of the body. The abdominal surface has been partially dissected on the left side of the body to show somewhat deeper muscles.

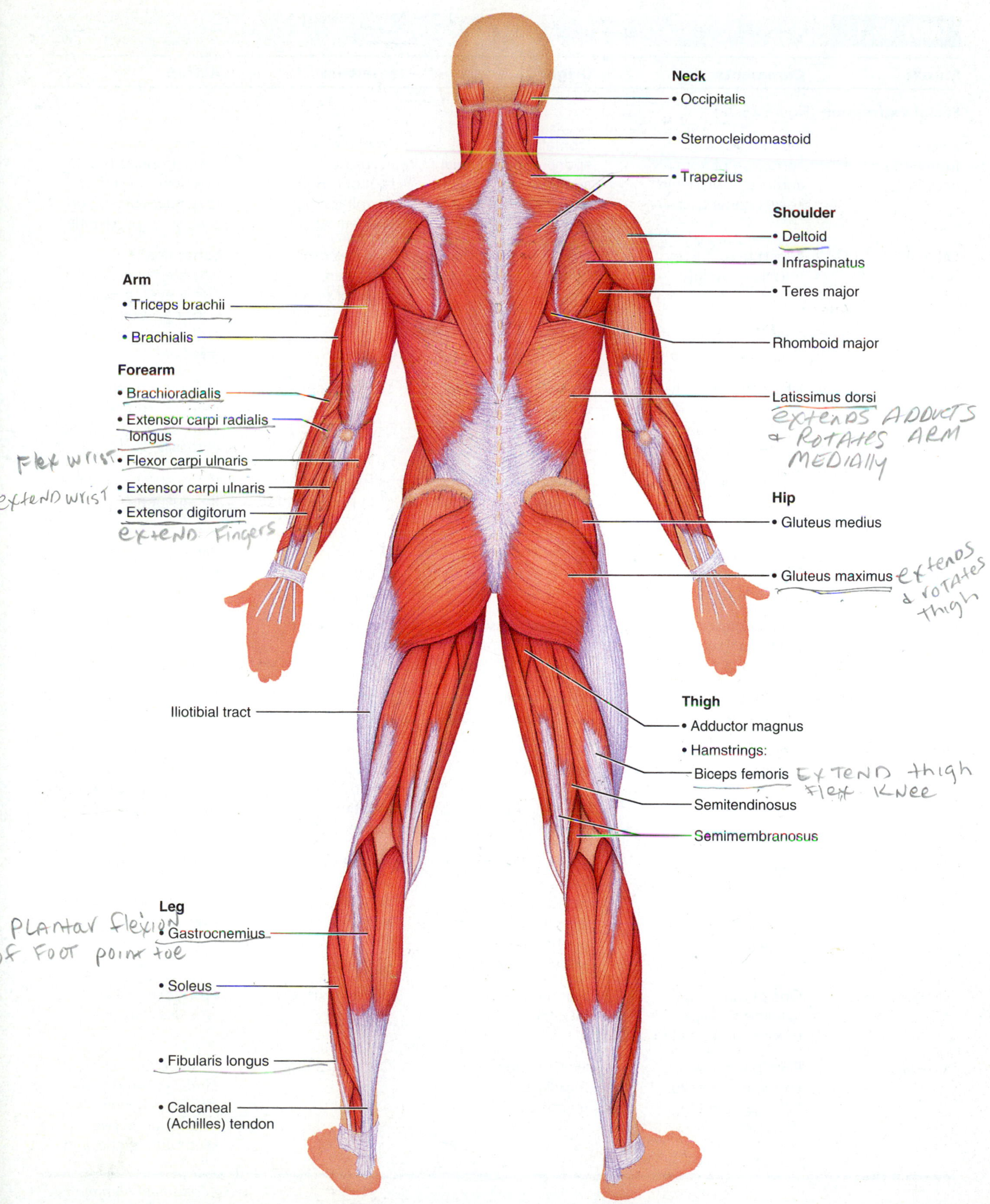

Figure 15.2 **Posterior view of superficial muscles of the body.**

Table 15.1 Major Muscles of Human Head (see Figure 15.3)

Muscle	Comments	Origin	Insertion	Action
Facial Expression (Figure 15.3a)				
Epicranius—frontalis and occipitalis	Bipartite muscle consisting of frontalis and occipitalis, which covers dome of skull	Frontalis: galea aponeurotica (cranial aponeurosis); occipitalis: occipital and temporal bones	Frontalis: skin of eyebrows and root of nose; occipitalis: galea aponeurotica	With aponeurosis fixed, frontalis raises eyebrows; occipitalis fixes aponeurosis and pulls scalp posteriorly
Orbicularis oculi	Tripartite sphincter muscle of eyelids	Frontal and maxillary bones and ligaments around orbit	Encircles orbit and inserts in tissue of eyelid	Various parts can be activated individually; closes eyes, produces blinking, squinting, and draws eyebrows inferiorly
Corrugator supercilii	Small muscle; activity associated with that of orbicularis oculi	Arch of frontal bone above nasal bone	Skin of eyebrow	Draws eyebrows medially and inferiorly; wrinkles skin of forehead vertically
Levator labii superioris	Thin muscle between orbicularis oris and inferior eye margin	Zygomatic bone and infraorbital margin of maxilla	Skin and muscle of upper lip and border of nostril	Raises and furrows upper lip; opens lips
Zygomaticus—major and minor	Extends diagonally from corner of mouth to cheekbone	Zygomatic bone	Skin and muscle at corner of mouth	Raises lateral corners of mouth upward (smiling muscle)
Risorius	Slender muscle; runs inferior and lateral to zygomaticus	Fascia of masseter muscle	Skin at angle of mouth	Draws corner of lip laterally; tenses lip; zygomaticus synergist
Depressor labii inferioris	Small muscle from lower lip to mandible	Body of mandible lateral to its midline	Skin and muscle of lower lip	Draws lower lip inferiorly
Depressor anguli oris	Small muscle lateral to depressor labii inferioris	Body of mandible below incisors	Skin and muscle at angle of mouth below insertion of zygomaticus	Zygomaticus antagonist; draws corners of mouth downward and laterally
Orbicularis oris	Multilayered muscle of lips with fibers that run in many different directions; most run circularly	Arises indirectly from maxilla and mandible; fibers blended with fibers of other muscles associated with lips	Encircles mouth; inserts into muscle and skin at angles of mouth	Closes mouth; purses and protrudes lips (kissing and whistling muscle)
Mentalis	One of muscle pair forming V-shaped muscle mass on chin	Mandible below incisors	Skin of chin	Protrudes lower lip; wrinkles chin
Buccinator	Principal muscle of cheek; runs horizontally, deep to the masseter	Molar region of maxilla and mandible	Orbicularis oris	Draws corner of mouth laterally; compresses cheek (as in whistling); holds food between teeth during chewing

(continued on page 142)

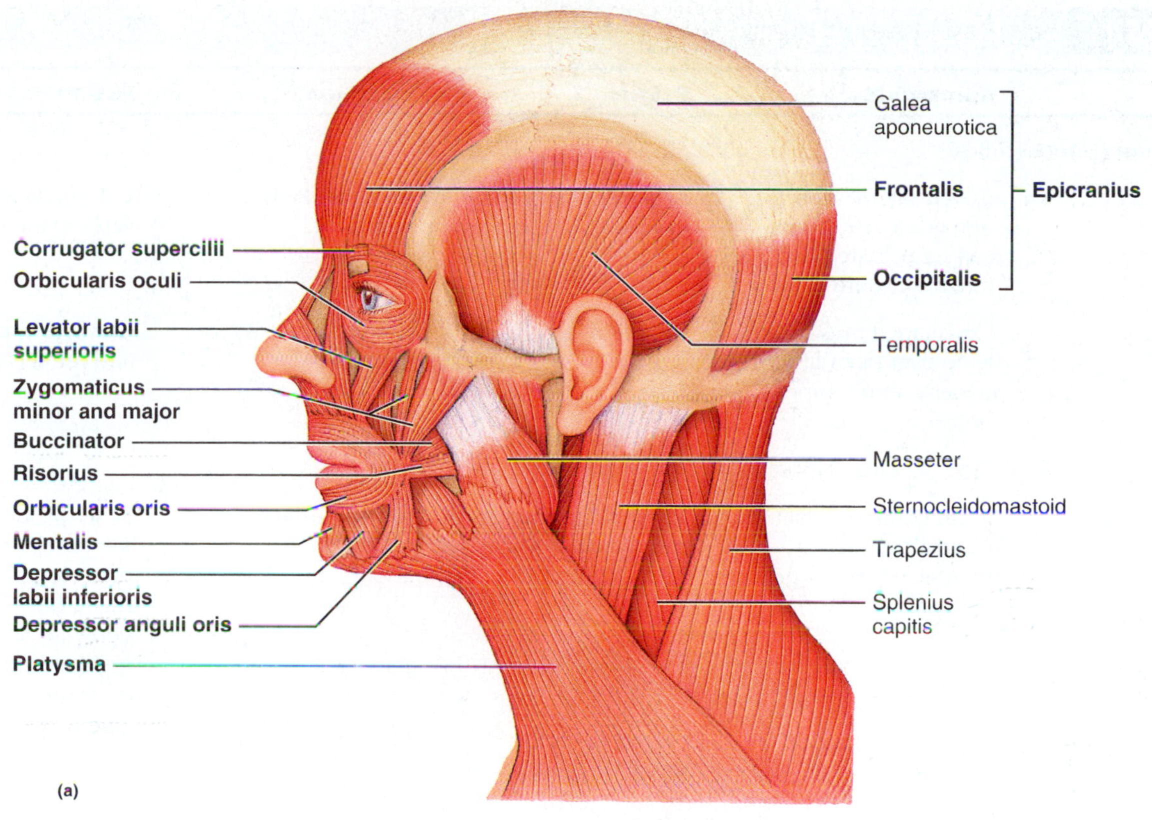

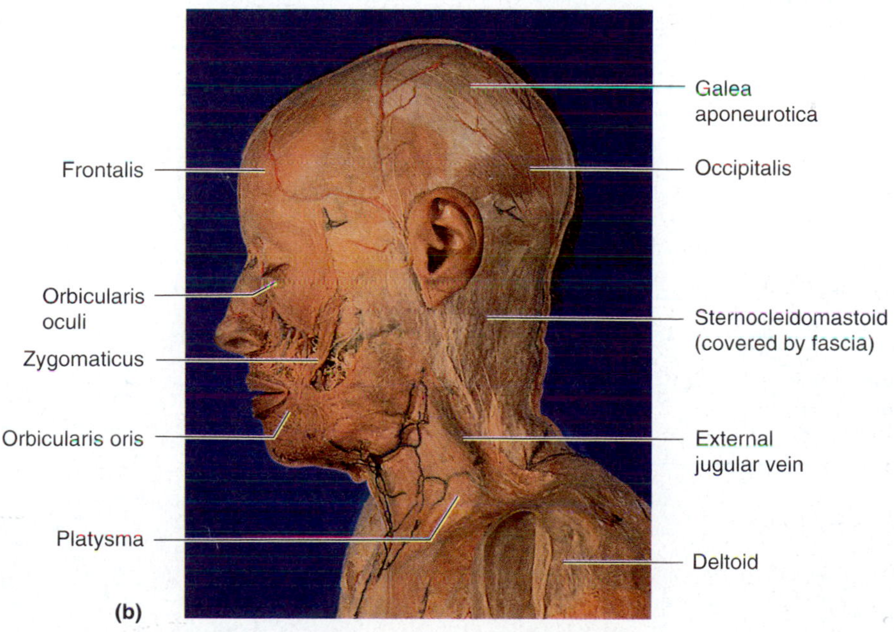

Figure 15.3 Muscles of the scalp, face, and neck; left lateral view. **(a)** Superficial muscles. **(b)** Photo of superficial structures of head and neck.

Table 15.1 Major Muscles of Human Head *(continued)*

Muscle	Comments	Origin	Insertion	Action
Mastication (Figure 15.3c,d)				
Masseter	Covers lateral aspect of mandibular ramus; can be palpated on forcible closure of jaws	Zygomatic arch and maxilla	Angle and ramus of mandible	Closes jaw and elevates mandible
Temporalis	Fan-shaped muscle lying over parts of frontal, parietal, and temporal bones	Temporal fossa	Coronoid process of mandible	Closes jaw; elevates and retracts mandible
Buccinator	(See muscles of facial expression.)			
Medial pterygoid	Runs along internal (medial) surface of mandible (thus largely concealed by that bone)	Sphenoid, palatine, and maxillary bones	Medial surface of mandible, near its angle	Synergist of temporalis and masseter; elevates mandible; in conjunction with lateral pterygoid, aids in grinding movements
Lateral pterygoid	Superior to medial pterygoid	Greater wing of sphenoid bone	Mandibular condyle	Protracts jaw (moves it anteriorly); in conjunction with medial pterygoid, aids in grinding movements of teeth

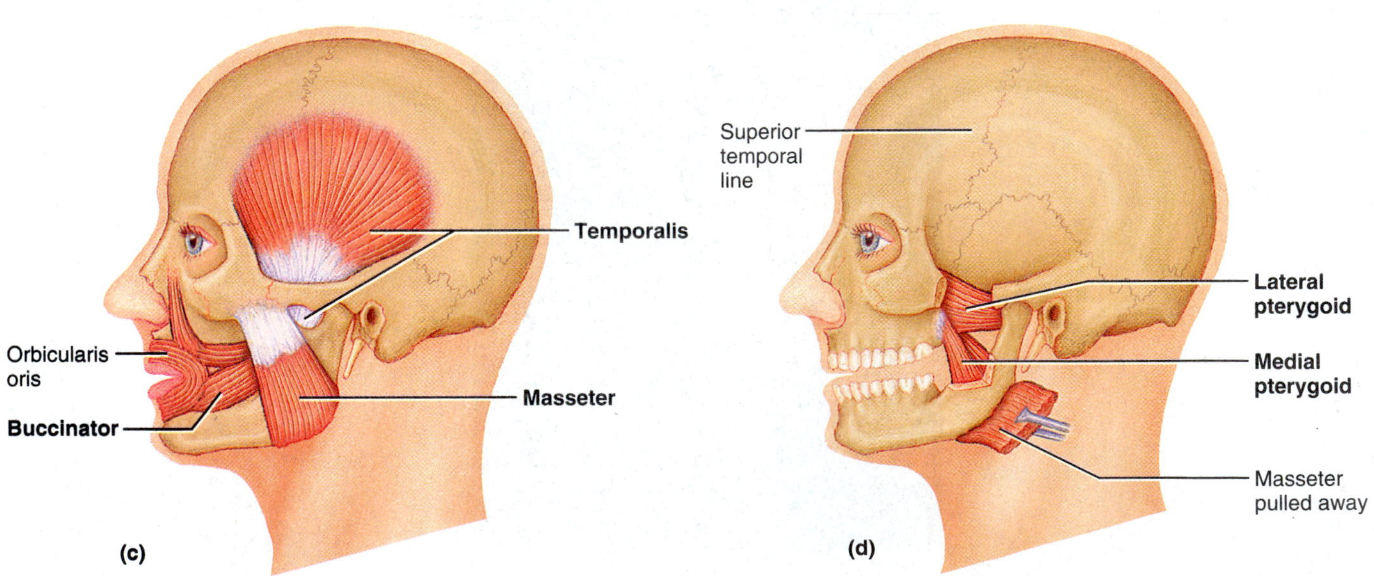

Figure 15.3 *(continued)* **Muscles promoting mastication. (c)** Lateral view of the temporalis, masseter, and buccinator muscles. **(d)** Lateral view of the deep chewing muscles, the medial and lateral pterygoid muscles.

Table 15.2 Anterolateral Muscles of Human Neck (see Figure 15.4)

Muscle	Comments	Origin	Insertion	Action
Superficial				
Platysma	Unpaired muscle: thin, sheetlike superficial neck muscle, not strictly a head muscle but plays role in facial expression (see also Fig. 15.3a)	Fascia of chest (over pectoral muscles) and deltoid	Lower margin of mandible, skin, and muscle at corner of mouth	Depresses mandible; pulls lower lip back and down; i.e., produces downward sag of the mouth
Sternocleidomastoid	Two-headed muscle located deep to platysma on anterolateral surface of neck; fleshy parts on either side indicate limits of anterior and posterior triangles of neck	Manubrium of sternum and medial portion of clavicle	Mastoid process of temporal bone and superior nuchal line of occipital bone	Simultaneous contraction of both muscles of pair causes flexion of neck forward, generally against resistance (as when lying on the back); acting independently, rotate head toward shoulder on opposite side
Scalenes—anterior, middle, and posterior	Located more on lateral than anterior neck; deep to platysma and sterno-cleidomastoid (see Fig. 15.4c)	Transverse processes of cervical vertebrae	Anterolaterally on first two ribs	Flex and slightly rotate neck; elevate first two ribs (aid in inspiration)

(continued)

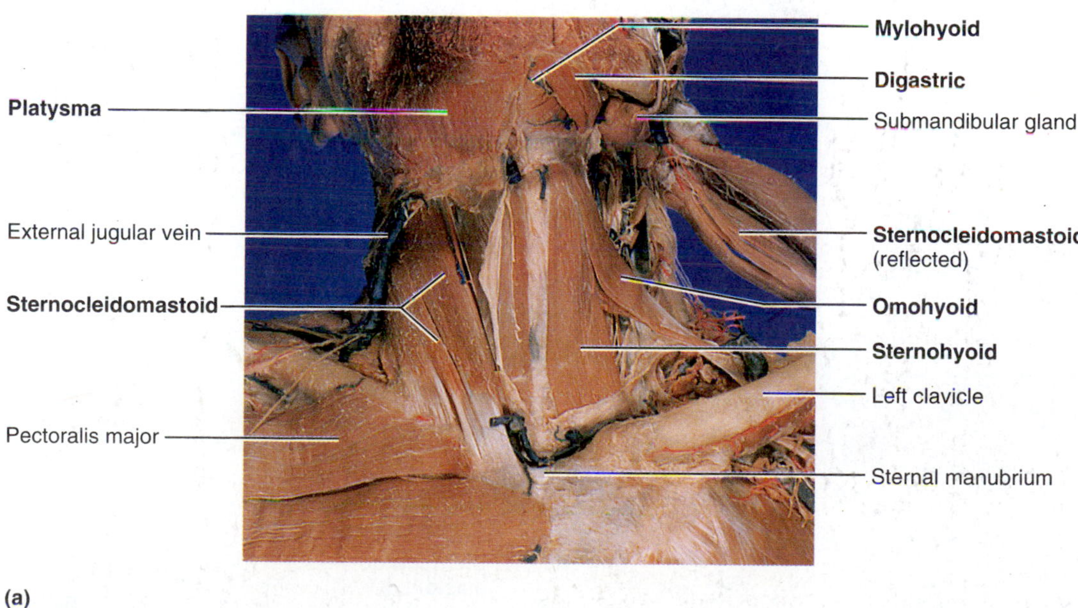

Figure 15.4 Muscles of the anterolateral neck and throat. (a) Photo of the anterior and lateral regions of the neck. The fascia has been partially removed (left side of the photo) to expose the sternocleidomastoid muscle. On the right side of the photo, the sternocleidomastoid muscle is reflected to expose the sternohyoid and omohyoid muscles.

Table 15.2 Anterolateral Muscles of Human Neck *(continued)*

Muscle	Comments	Origin	Insertion	Action
Deep (Figure 15.4a,b)				
Digastric	Consists of two bellies united by an intermediate tendon; assumes a V-shaped configuration under chin	Lower margin of mandible (anterior belly) and mastoid process (posterior belly)	By a connective tissue loop to hyoid bone	Acting in concert, elevate hyoid bone; open mouth and depress mandible
Stylohyoid	Slender muscle parallels posterior border of digastric; below angle of jaw	Styloid process of temporal	Hyoid bone	Elevates and retracts hyoid bone
Mylohyoid	Just deep to digastric; forms floor of mouth	Medial surface of mandible	Hyoid bone and median raphe	Elevates hyoid bone and base of tongue during swallowing
Sternohyoid	Runs most medially along neck; straplike	Manubrium and medial end of clavicle	Lower margin of body of hyoid bone	Acting with sternothyroid and omohyoid (all inferior to hyoid bone), depresses larynx and hyoid bone if mandible is fixed; may also flex skull
Sternothyroid	Lateral and deep to sternohyoid	Posterior surface of manubrium	Thyroid cartilage of larynx	(See Sternohyoid above)
Omohyoid	Straplike with two bellies; lateral to sternohyoid	Superior surface of scapula	Hyoid bone; inferior border	(See Sternohyoid above)
Thyrohyoid	Appears as a superior continuation of sternothyroid muscle	Thyroid cartilage	Hyoid bone	Depresses hyoid bone; elevates larynx if hyoid is fixed

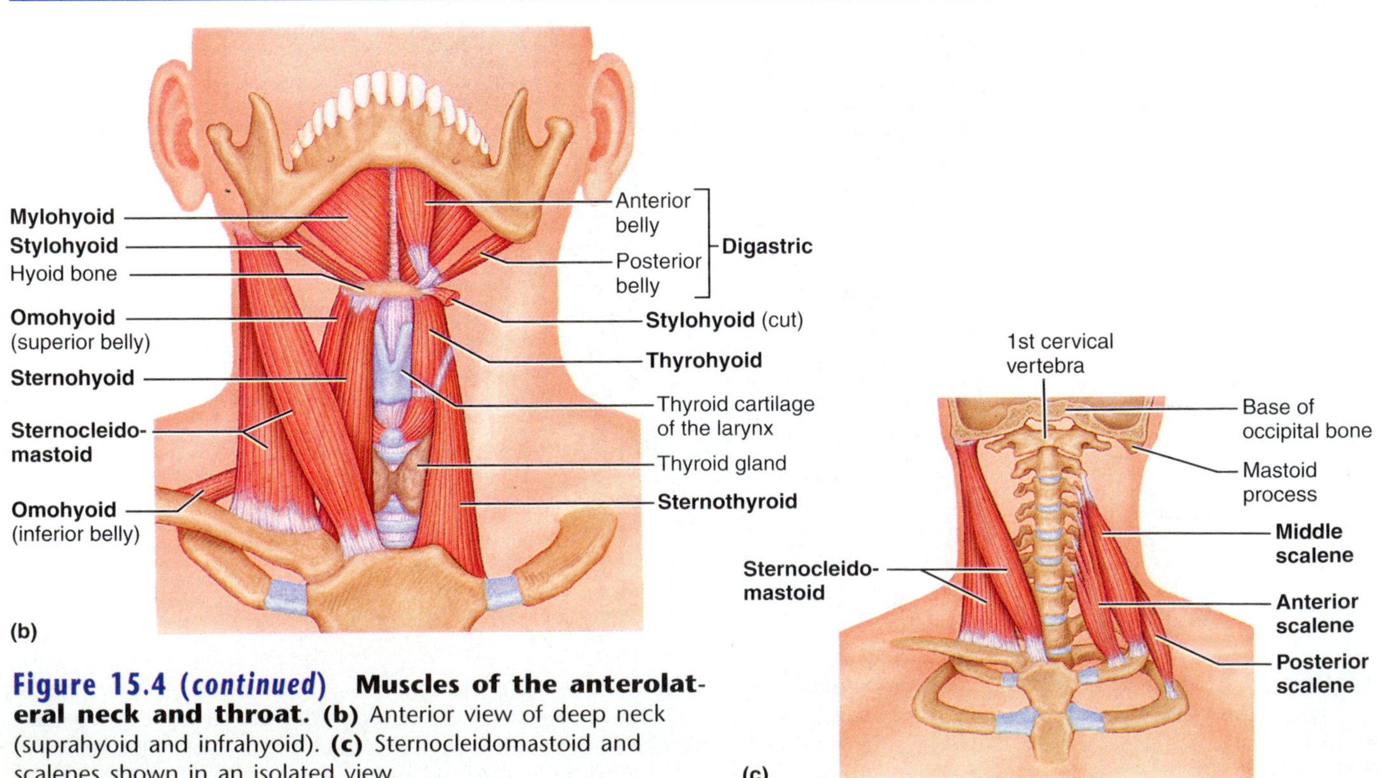

Figure 15.4 *(continued)* **Muscles of the anterolateral neck and throat. (b)** Anterior view of deep neck (suprahyoid and infrahyoid). **(c)** Sternocleidomastoid and scalenes shown in an isolated view.

Table 15.3 Anterior Muscles of Human Thorax, Shoulder, and Abdominal Wall (see Figures 15.5, 15.6, and 15.7)

Muscle	Comments	Origin	Insertion	Action
Thorax and Shoulder, Superficial (Figure 15.5)				
Pectoralis major	Large fan-shaped muscle covering upper portion of chest	Clavicle, sternum, cartilage of first six (or seven) ribs, and aponeurosis of external oblique muscle	Fibers converge to insert by short tendon into intertubercular groove of humerus	Prime mover of arm flexion; adducts, medially rotates arm; with arm fixed, pulls chest upward (thus also acts in forced inspiration)
Serratus anterior	Deep to scapula; beneath and inferior to pectoral muscles on lateral rib cage	Lateral aspect of first to eighth (or ninth) ribs	Vertebral border of anterior surface of scapula	Moves scapula forward toward chest wall; rotates scapula, causing inferior angle to move laterally and upward; abduction and raising of arm
Deltoid	Fleshy triangular muscle forming shoulder muscle mass; intramuscular injection site	Lateral third of clavicle; acromion and spine of scapula	Deltoid tuberosity of humerus	Acting as a whole, prime mover of arm abduction; when only specific fibers are active, can aid in flexion, extension, and rotation of humerus

(continued)

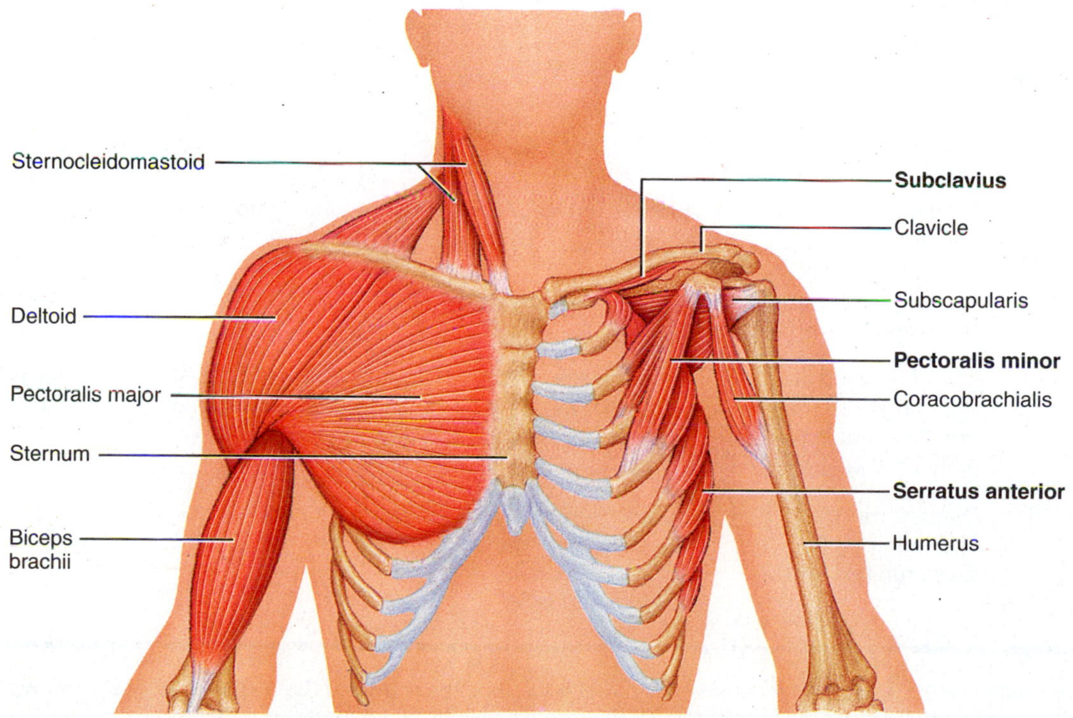

Figure 15.5 Superficial muscles of the thorax and shoulder acting on the scapula and arm, anterior view. The superficial muscles, which effect arm movements, are shown on the left. These muscles have been removed on the right side of the figure to show the muscles that stabilize or move the pectoral girdle.

Table 15.3 Anterior Muscles of Human Thorax, Shoulder, and Abdominal Wall *(continued)*

Muscle	Comments	Origin	Insertion	Action
Pectoralis minor	Flat, thin muscle directly beneath and obscured by pectoralis major	Anterior surface of third, fourth, and fifth ribs, near their costal cartilages	Coracoid process of scapula	With ribs fixed, draws scapula forward and inferiorly; with scapula fixed, draws rib cage superiorly
Thorax, Deep: Muscles of Respiration (Figure 15.6)				
External intercostals	11 pairs lie between ribs; fibers run obliquely downward and forward toward sternum	Inferior border of rib above (not shown in figure)	Superior border of rib below	Pulls ribs toward one another to elevate rib cage; aids in inspiration
Internal intercostals	11 pairs lie between ribs; fibers run deep and at right angles to those of external intercostals	Superior border of rib below	Inferior border of rib above (not shown in figure)	Draws ribs together to depress rib cage; aids in forced expiration; antagonistic to external intercostals
Diaphragm	Broad muscle; forms floor of thoracic cavity; dome shaped in relaxed state; fibers converge from margins of thoracic cage toward a central tendon	Inferior border of rib and sternum, costal cartilages of last six ribs and lumbar vertebrae	Central tendon	Prime mover of inspiration flattens on contraction, increasing vertical dimensions of thorax; increases intra-abdominal pressure
Abdominal Wall (Figure 15.7a and b)				
Rectus abdominis	Medial superficial muscle, extends from pubis to rib cage; ensheathed by aponeuroses of oblique muscles; segmented	Pubic crest and symphysis	Xiphoid process and costal cartilages of fifth through seventh ribs	Flexes and rotates vertebral column; increases abdominal pressure; fixes and depresses ribs; stabilizes pelvis during walking
External oblique	Most superficial lateral muscle; fibers run downward and medially; ensheathed by an aponeurosis	Anterior surface of last eight ribs	Linea alba,* pubic crest and tubercles, and iliac crest	See Rectus abdominis, above; also aids muscles of back in trunk rotation and lateral flexion
Internal oblique	Most fibers run at right angles to those of external oblique, which it underlies	Lumbar fascia, iliac crest, and inguinal ligament	Linea alba, pubic crest, and costal cartilages of last three ribs	As for External oblique
Transversus abdominis	Deepest muscle of abdominal wall; fibers run horizontally	Inguinal ligament, iliac crest, cartilages of last five or six ribs, and lumbar fascia	Linea alba and pubic crest	Compresses abdominal contents

*The linea alba ("white line") is a narrow, tendinous sheath that runs along the middle of the abdomen from the sternum to the pubic symphysis. It is formed by the fusion of the aponeurosis of the external oblique and transversus muscles.

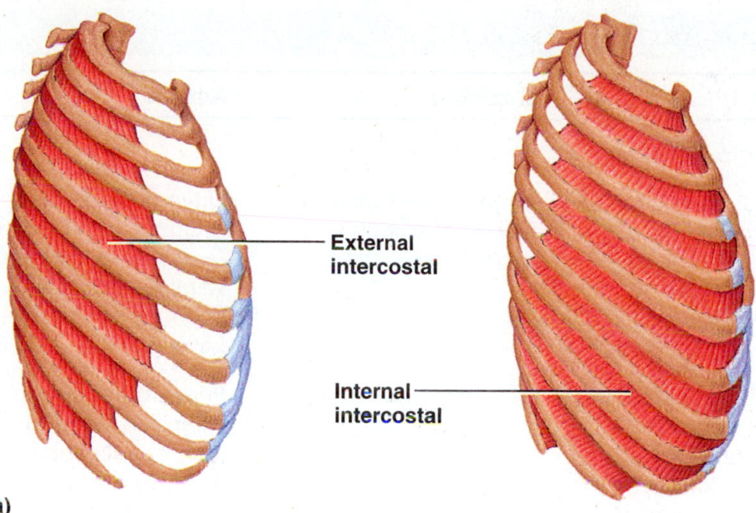

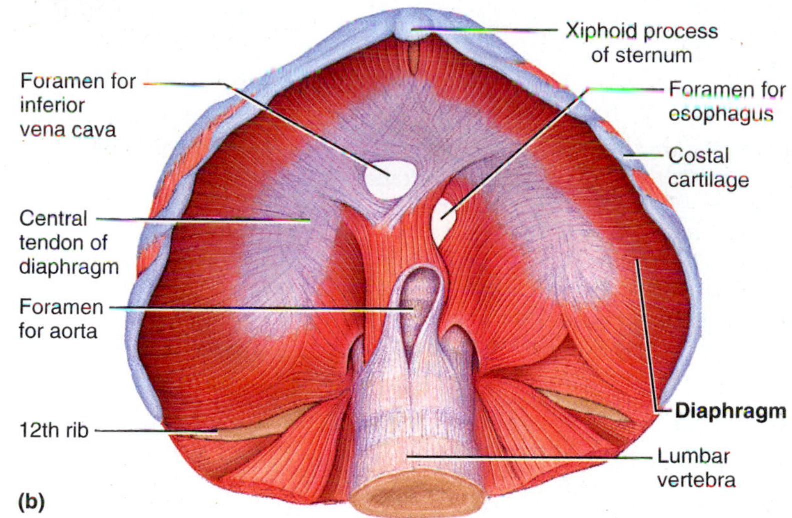

Figure 15.6 Deep muscles of the thorax: muscles of respiration. **(a)** The external intercostals (inspiratory muscles) are shown on the left and the internal intercostals (expiratory muscles) are shown on the right. These two muscle layers run obliquely and at right angles to each other. **(b)** Inferior view of the diaphragm, the prime mover of inspiration. Notice that its muscle fibers converge toward a central tendon, an arrangement that causes the diaphragm to flatten and move inferiorly as it contracts. The diaphragm and its tendon are pierced by the great vessels (aorta and inferior vena cava) and the esophagus.

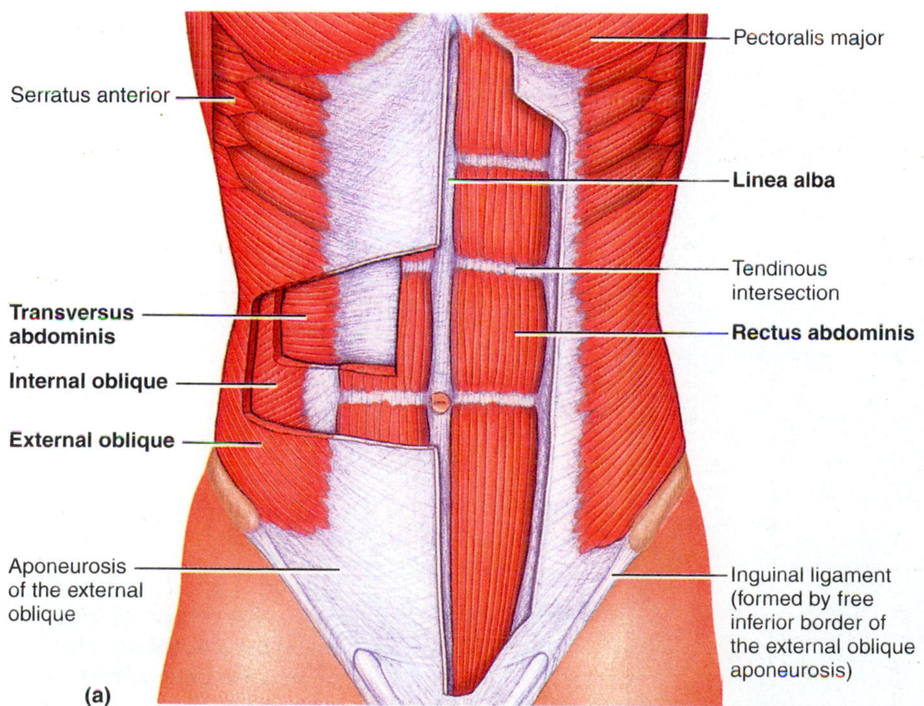

Figure 15.7 Anterior view of the muscles forming the anterolateral abdominal wall. **(a)** The superficial muscles have been partially cut away on the left side of the diagram to reveal the deeper internal oblique and transversus abdominis muscles.

Table 15.4 Posterior Muscles of Human Trunk (see Figure 15.8)

Muscle	Comments	Origin	Insertion	Action
Muscles of the Neck, Shoulder, and Thorax (Figure 15.8a)				
Trapezius	Most superficial muscle of posterior thorax; very broad origin and insertion	Occipital bone; ligamentum nuchae; spines of C_7 and all thoracic vertebrae	Acromion and spinous process of scapula; lateral third of clavicle	Extends head; raises, rotates, and retracts (adducts) scapula and stabilizes it; upper fibers elevate scapula; lower fibers depress it
Latissimus dorsi	Broad flat muscle of lower back (lumbar region); extensive superficial origins	Indirect attachment to spinous processes of lower six thoracic vertebrae, lumbar vertebrae, lower 3 to 4 ribs, and iliac crest	Floor of intertubercular groove of humerus	Prime mover of arm extension; adducts and medially rotates arm; depresses scapula; brings arm down in power stroke, as in striking a blow
Infraspinatus	Partially covered by deltoid and trapezius; a rotator cuff muscle	Infraspinous fossa of scapula	Greater tubercle of humerus	Lateral rotation of humerus; helps hold head of humerus in glenoid cavity; stabilizes shoulder
Teres minor	Small muscle inferior to infraspinatus; a rotator cuff muscle	Lateral margin of scapula	Greater tubercle of humerus	As for infraspinatus
Teres major	Located inferiorly to teres minor	Posterior surface at inferior angle of scapula	Intertubercular groove of humerus	Extends, medially rotates, and adducts humerus; synergist of latissimus dorsi

(continued)

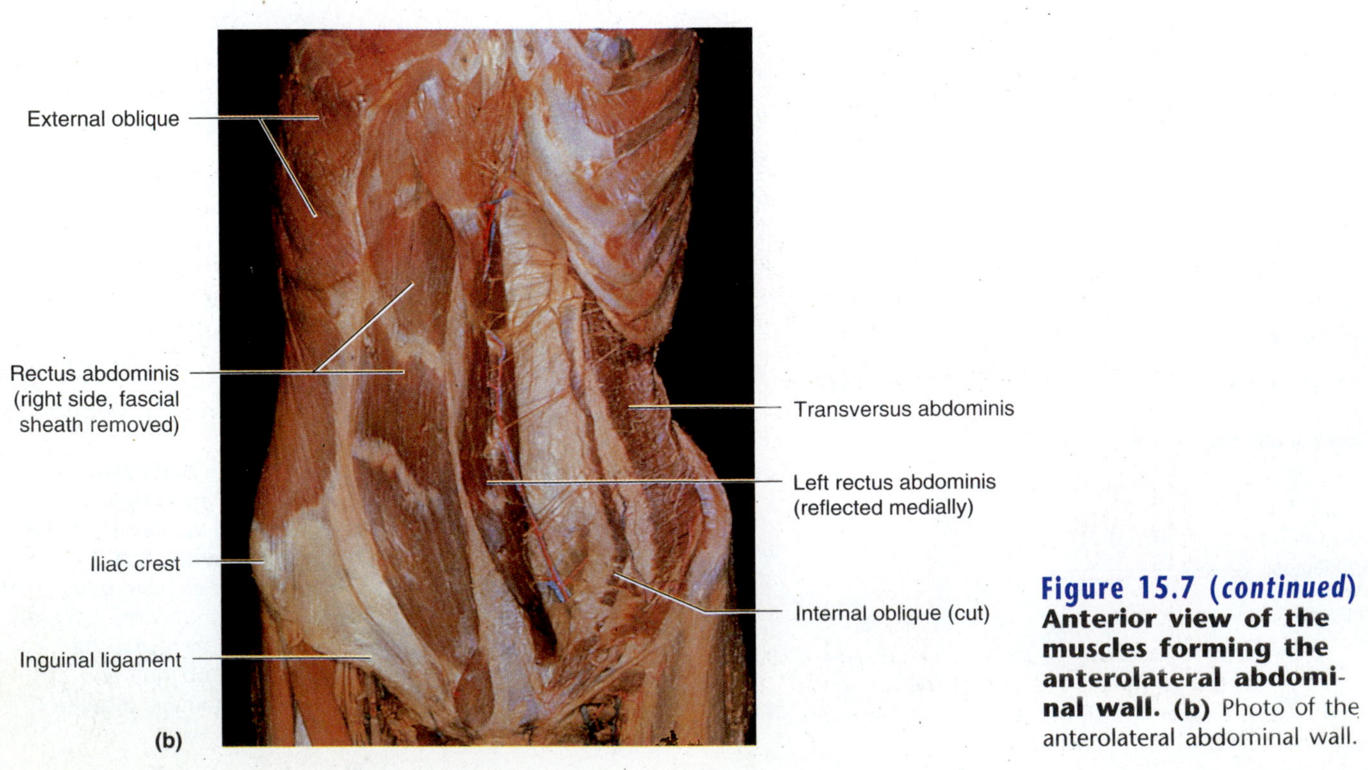

Figure 15.7 *(continued)* Anterior view of the muscles forming the anterolateral abdominal wall. (b) Photo of the anterolateral abdominal wall.

Table 15.4 (continued)

Muscle	Comments	Origin	Insertion	Action
Supraspinatus	Obscured by trapezius; a rotator cuff muscle	Supraspinous fossa of scapula	Greater tubercle of humerus	Assists abduction of humerus; stabilizes shoulder joint
Levator scapulae	Located at back and side of neck, deep to trapezius	Transverse processes of C_1 through C_4	Medial border of scapula superior to spine	Elevates and adducts scapula; with fixed scapula, flexes neck to the same side
Rhomboids—major and minor	Beneath trapezius and inferior to levator scapulae; rhomboid minor is the more superior muscle	Spinous processes of C_7 and T_1 through T_5	Medial border of scapula	Pulls scapula medially (retraction); stabilizes scapula; rotates glenoid cavity downward
Muscles Associated with the Vertebral Column (Figure 15.8b)				
Semispinalis	Deep composite muscle of the back—thoracis, cervicis, and capitis portions	Transverse processes of C_7–T_{12}	Occipital bone and spinous processes of cervical vertebrae and T_1–T_4	Acting together, extend head and vertebral column; acting independently (right vs. left) causes rotation toward the opposite side

(continued)

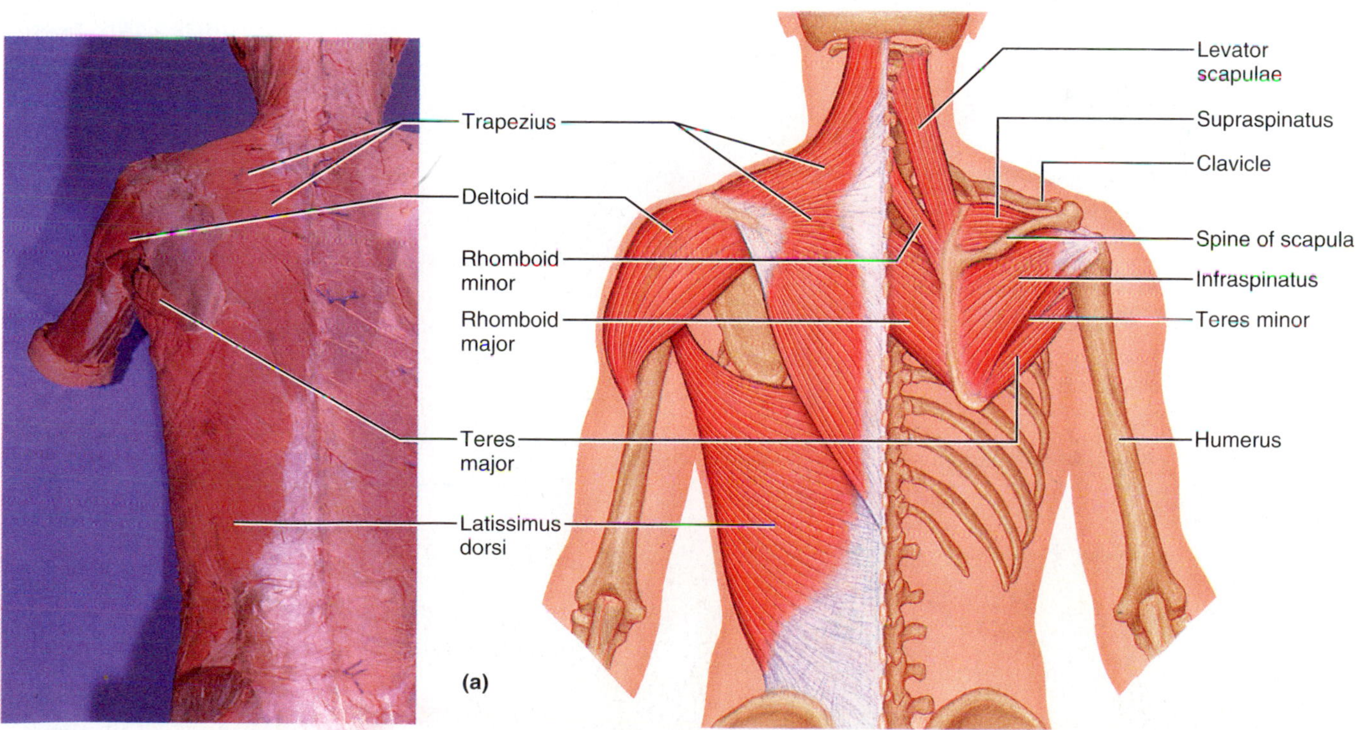

Figure 15.8 Muscles of the neck, shoulder, and thorax, posterior view. (a) The superficial muscles of the back are shown for the left side of the body, with a corresponding photograph. The superficial muscles are removed on the right side of the illustration to reveal the deeper muscles acting on the scapula and the rotator cuff muscles that help to stabilize the shoulder joint.

Table 15.4 Posterior Muscles of Human Trunk *(continued)*

Muscle	Comments	Origin	Insertion	Action
Erector spinae	A long tripartite muscle composed of iliocostalis (lateral), longissimus, and spinalis (medial) muscle columns; superficial to semispinalis muscles; extends from pelvis to head	Iliac crest, transverse processes of lumbar, thoracic, and cervical vertebrae, and/or ribs 3–6 depending on specific part	Ribs and transverse processes of vertebrae about six segments above origin. Longissimus also inserts into mastoid process	Extend and bend the vertebral column laterally; fibers of the longissimus also extend head

(continued)

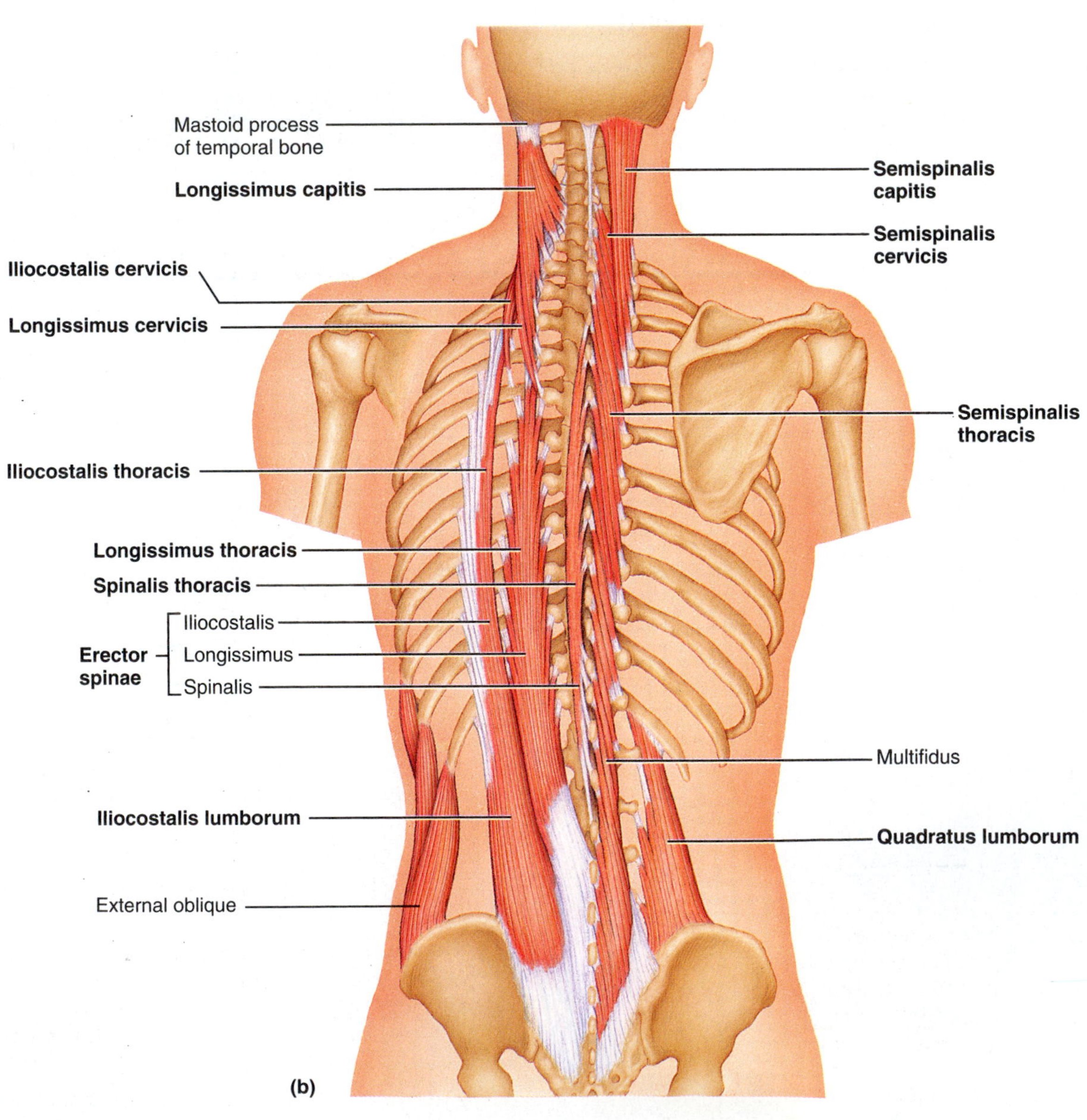

Figure 15.8 *(continued)* **Muscles of the neck, shoulder, and thorax, posterior view. (b)** The erector spinae and semispinalis muscles, which respectively form the intermediate and deep muscle layers of the back associated with the vertebral column.

Table 15.4 *(continued)*

Muscle	Comments	Origin	Insertion	Action
Splenius (see Figure 15.8c)	Superficial muscle (capitis and cervicis parts) extending from upper thoracic region to skull	Ligamentum nuchae and spinous processes of C_7–T_6	Mastoid process, occipital bone, and transverse processes of C_2–C_4	As a group, extend or hyperextend head; when only one side is active, head is rotated and bent toward the same side
Quadratus lumborum	Forms greater portion of posterior abdominal wall	Iliac crest and lumbar fascia	Inferior border twelfth rib; transverse processes of lumbar vertebrae	Each flexes vertebral column laterally; together extend the lumbar spine and fix the twelfth rib; maintains upright posture

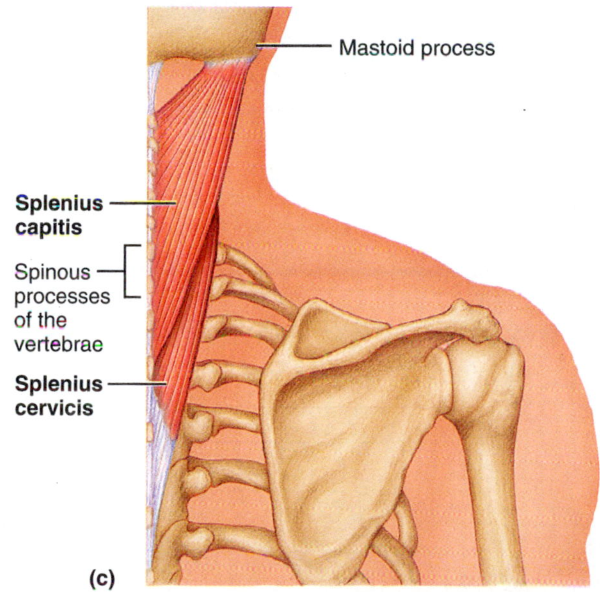

Figure 15.8 (*continued*) Muscles of the neck, shoulder, and thorax, posterior view. (c) Deep (splenius) muscles of the posterior neck. Superficial muscles have been removed.

Muscles of the Upper Limb

The muscles that act on the upper limb fall into three groups: those that move the arm, those causing movement at the elbow, and those effecting movements of the wrist and hand.

The muscles that cross the shoulder joint to insert on the humerus and move the arm (subscapularis, supraspinatus and infraspinatus, deltoid, and so on) are primarily trunk muscles that originate on the axial skeleton or shoulder girdle. These muscles are included with the trunk muscles.

The second group of muscles, which cross the elbow joint and move the forearm, consists of muscles forming the musculature of the humerus. These muscles arise primarily from the humerus and insert in forearm bones. They are responsible for flexion, extension, pronation, and supination. The origins, insertions, and actions of these muscles are summarized in Table 15.5 and the muscles are shown in Figure 15.9.

The third group composes the musculature of the forearm. For the most part, these muscles insert on the digits and produce movements at the wrist and fingers. In general, muscles acting on the wrist and hand are more easily identified if their insertion tendons are located first. These muscles are described in Table 15.6 and illustrated in Figure 15.10.

Activity 4: Identifying Muscles of the Upper Limb

First study the tables and figures, then see if you can identify these muscles on a torso model, anatomical chart, or cadaver. Complete this portion of the exercise with palpation demonstrations as outlined next.

- To observe the *biceps brachii,* attempt to flex your forearm (hand supinated) against resistance. The insertion tendon of this biceps muscle can also be felt in the lateral aspect of the antecubital fossa (where it runs toward the radius to attach).
- If you acutely flex your elbow and then try to extend it against resistance, you can demonstrate the action of your *triceps brachii.*
- Strongly flex your wrist and make a fist. Palpate your contracting wrist flexor muscles (which originate from the medial epicondyle of the humerus) and their insertion tendons, which can be easily felt at the anterior aspect of the wrist.
- Flare your fingers to identify the tendons of the *extensor digitorum* muscle on the dorsum of your hand. ■

Text continues on page 156

Table 15.5 Muscles of Human Humerus That Act on the Forearm (see Figure 15.9)

Muscle	Comments	Origin	Insertion	Action
Triceps brachii	Sole, large fleshy muscle of posterior humerus; three-headed origin	Long head: inferior margin of glenoid cavity; lateral head: posterior humerus; medial head: distal radial groove on posterior humerus	Olecranon process of ulna	Powerful forearm extensor; antagonist of forearm flexors (brachialis and biceps brachii)
Anconeus	Short triangular muscle blended with triceps	Lateral epicondyle of humerus	Lateral aspect of olecranon process of ulna	Abducts ulna during forearm pronation; extends elbow
Biceps brachii	Most familiar muscle of anterior humerus because this two-headed muscle bulges when forearm is flexed	Short head: coracoid process; tendon of long head runs in intertubercular groove and within capsule of shoulder joint	Radial tuberosity	Flexion (powerful) of elbow and supination of forearm; "it turns the corkscrew and pulls the cork"; weak arm flexor
Brachioradialis	Superficial muscle of lateral forearm; forms lateral boundary of antecubital fossa	Lateral ridge at distal end of humerus	Base of styloid process of radius	Synergist in forearm flexion
Brachialis	Immediately deep to biceps brachii	Distal portion of anterior humerus	Coronoid process of ulna	A major flexor of forearm

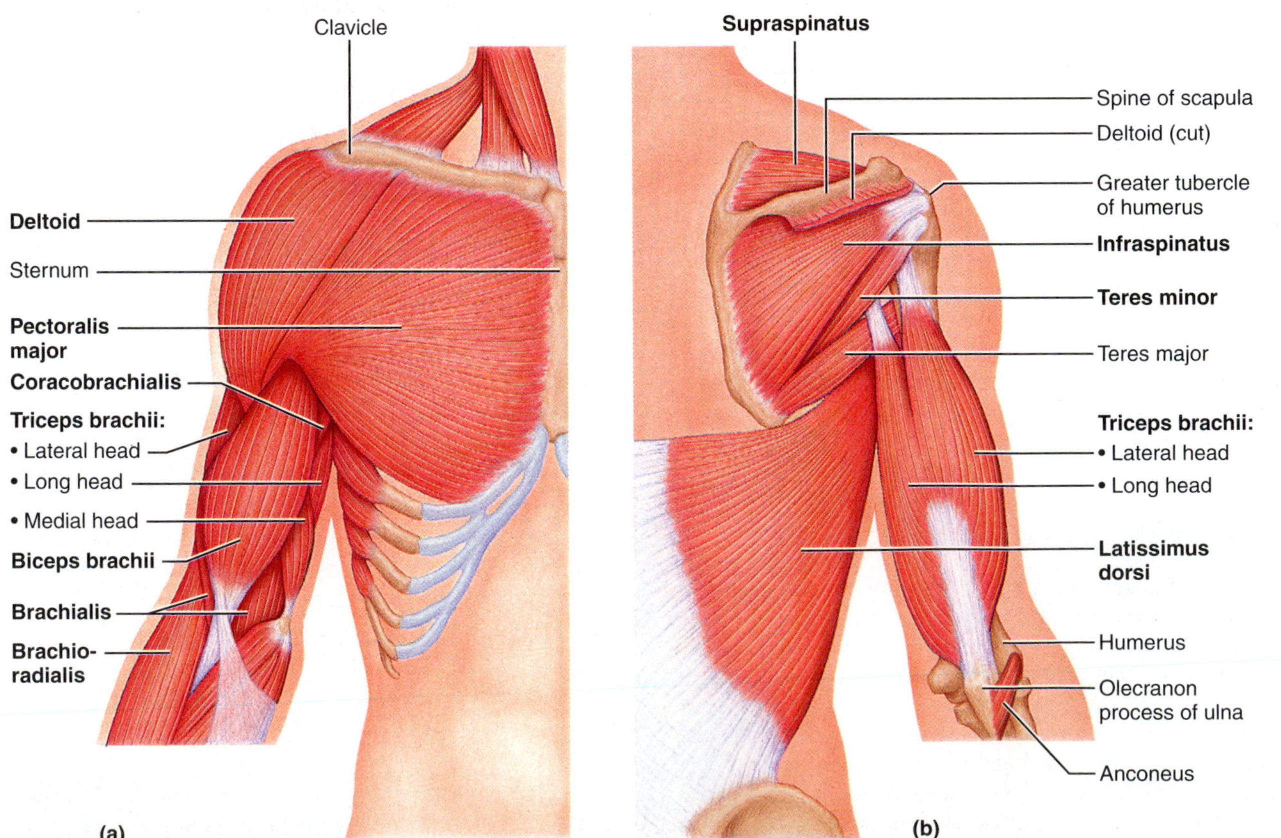

Figure 15.9 **Muscles causing movements of the forearm.** **(a)** Superficial muscles of the anterior thorax, shoulder, and arm, anterior view. **(b)** Posterior aspect of the arm showing the lateral and long heads of the triceps brachii muscle.

Table 15.6 Muscles of Human Forearm That Act on Hand and Fingers (see Figure 15.10)

Muscle	Comments	Origin	Insertion	Action
Anterior Compartment (Figure 15.10a, b, c)				
Superficial				
Pronator teres	Seen in a superficial view between proximal margins of brachioradialis and flexor carpi radialis	Medial epicondyle of humerus and coronoid process of ulna	Midshaft of radius	Acts synergistically with pronator quadratus to pronate forearm; weak elbow flexor
Flexor carpi radialis	Superficial; runs diagonally across forearm	Medial epicondyle of humerus	Base of second and third metacarpals	Powerful flexor of wrist; abducts hand
Palmaris longus	Small fleshy muscle with a long tendon; medial to flexor carpi radialis	Medial epicondyle of humerus	Palmar aponeurosis; skin and fascia of palm	Flexes wrist (weak); tenses skin and fascia of palm

(continued)

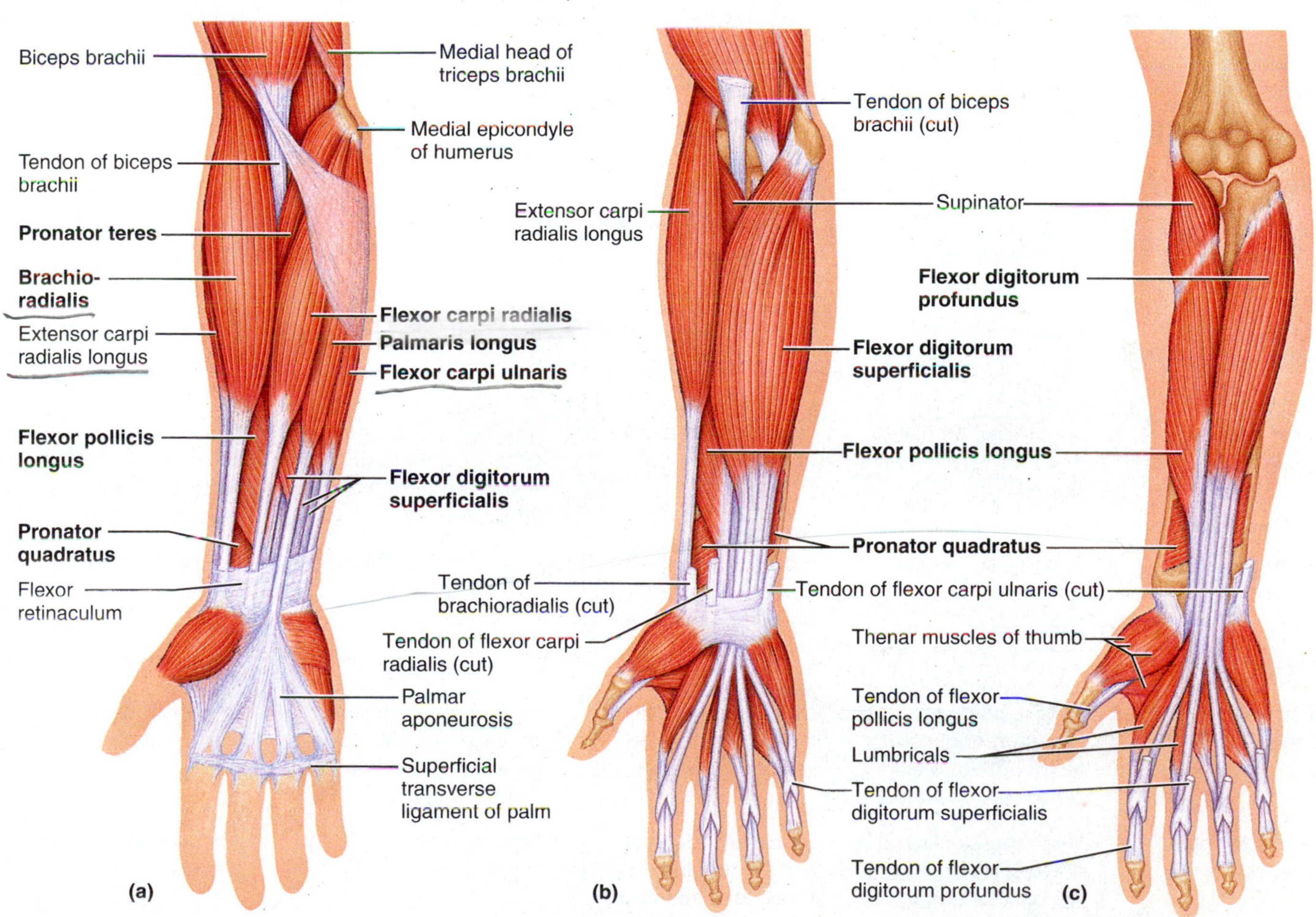

Figure 15.10 Muscles of the forearm and wrist. (a) Superficial anterior view of right forearm and hand. **(b)** The brachioradialis, flexors carpi radialis and ulnaris, and palmaris longus muscles have been removed to reveal the position of the somewhat deeper flexor digitorum superficialis. **(c)** Deep muscles of the anterior compartment. Superficial muscles have been removed. Note: The thenar muscles of the thumb and the lumbricals that help move the fingers are illustrated here but are not described in Table 15.6.

Table 15.6 Muscles of Human Forearm That Act on Hand and Fingers *(continued)*

Muscle	Comments	Origin	Insertion	Action
Flexor carpi ulnaris	Superficial; medial to palmaris longus	Medial epicondyle of humerus and olecranon process and posterior surface of ulna	Base of fifth metacarpal; pisiform and hamate bones	Powerful flexor of wrist; adducts hand
Flexor digitorum superficialis	Deeper muscle; deep to muscles named above; visible at distal end of forearm	Medial epicondyle of humerus, coronoid process of ulna, and shaft of radius	Middle phalanges of second through fifth fingers	Flexes wrist and middle phalanges of second through fifth fingers
Deep				
Flexor pollicis longus	Deep muscle of anterior forearm; distal to and paralleling lower margin of flexor digitorum superficialis	Anterior surface of radius, and interosseous membrane	Distal phalanx of thumb	Flexes thumb (*pollix* is Latin for "thumb")
Flexor digitorum profundus	Deep muscle; overlain entirely by flexor digitorum superficialis	Anteromedial surface of ulna, interosseous membrane, and coronoid process	Distal phalanges of second through fifth fingers	Sole muscle that flexes distal phalanges; assists in wrist flexion
Pronator quadratus	Deepest muscle of distal forearm	Distal portion of anterior ulnar surface	Anterior surface of radius, distal end	Pronates forearm
Posterior Compartment (Figure 15.10d, e, f)				
Superficial				
Extensor carpi radialis longus	Superficial; parallels brachioradialis on lateral forearm	Lateral supracondylar ridge of humerus	Base of second metacarpal	Extends and abducts wrist
Extensor carpi radialis brevis	Deep to extensor carpi radialis longus	Lateral epicondyle of humerus	Base of third metacarpal	Extends and abducts wrist; steadies wrist during finger flexion
Extensor digitorum	Superficial; medial to extensor carpi radialis brevis	Lateral epicondyle of humerus	By four tendons into distal phalanges of second through fifth fingers	Prime mover of finger extension; extends wrist; can flare (abduct) fingers
Extensor carpi ulnaris	Superficial; medial posterior forearm	Lateral epicondyle of humerus; posterior border of ulna	Base of fifth metacarpal	Extends and adducts wrist
Deep				
Extensor pollicis longus and brevis	Muscle pair with a common origin and action; deep to extensor carpi ulnaris	Dorsal shaft of ulna and radius, interosseous membrane	Base of distal phalanx of thumb (longus) and proximal phalanx of thumb (brevis)	Extends thumb
Abductor pollicis longus	Deep muscle; lateral and parallel to extensor pollicis longus	Posterior surface of radius and ulna; interosseous membrane	First metacarpal and trapezium	Abducts and extends thumb
Supinator	Deep muscle at posterior aspect of elbow	Lateral epicondyle of humerus; proximal ulna	Proximal end of radius	Acts with biceps brachii to supinate forearm; antagonist of pronator muscles

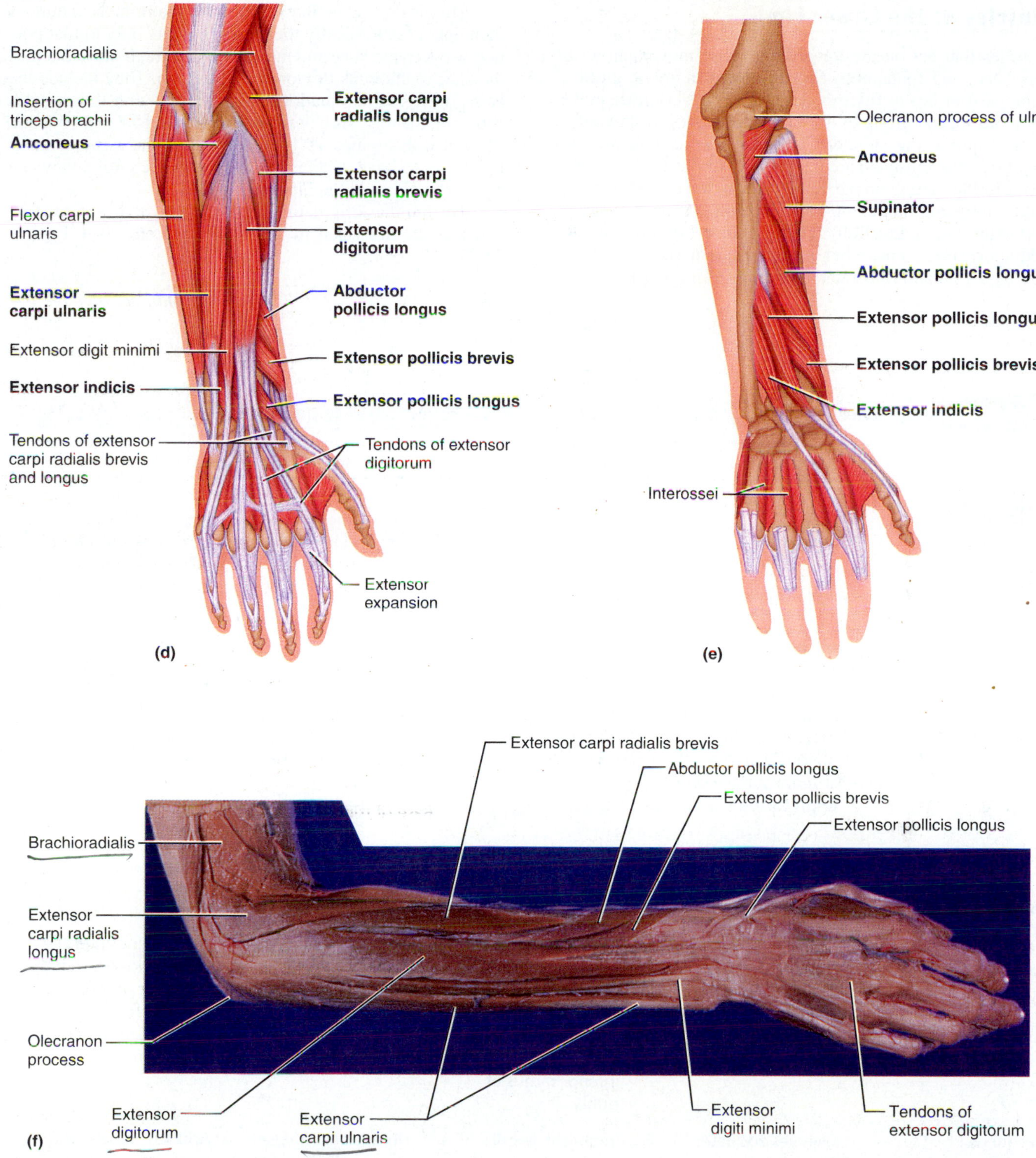

Brachioradialis
Insertion of triceps brachii
Anconeus
Flexor carpi ulnaris
Extensor carpi ulnaris
Extensor digit minimi
Extensor indicis
Tendons of extensor carpi radialis brevis and longus
Extensor carpi radialis longus
Extensor carpi radialis brevis
Extensor digitorum
Abductor pollicis longus
Extensor pollicis brevis
Extensor pollicis longus
Tendons of extensor digitorum
Extensor expansion
(d)

Olecranon process of ulna
Anconeus
Supinator
Abductor pollicis longus
Extensor pollicis longus
Extensor pollicis brevis
Extensor indicis
Interossei
(e)

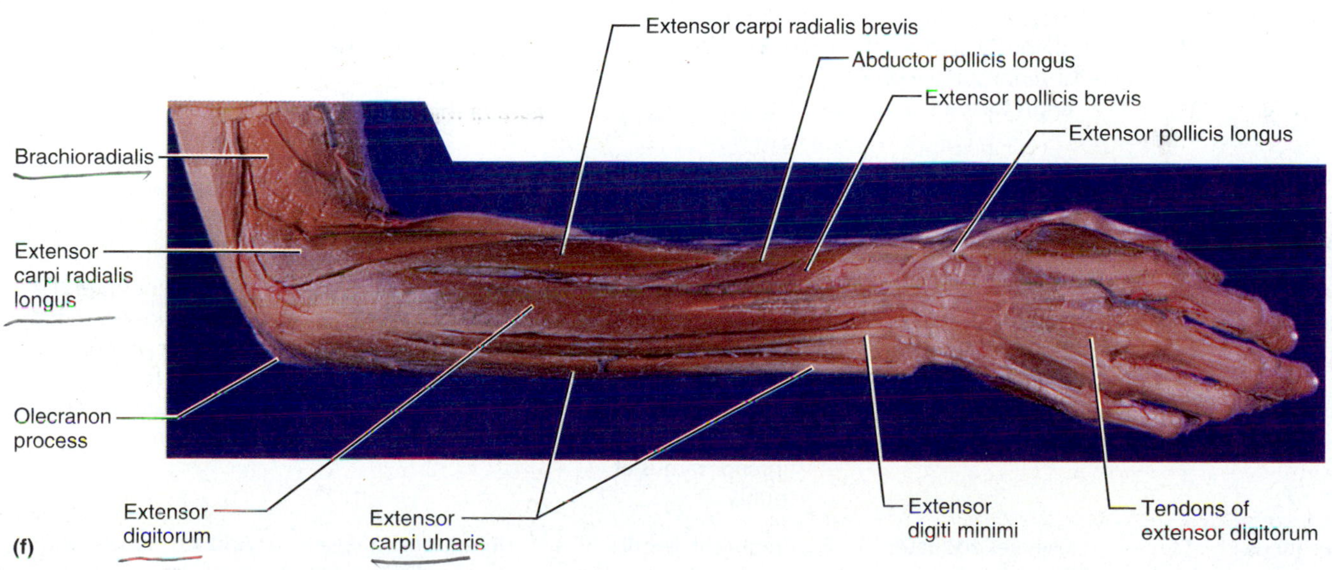

Figure 15.10 (*continued*) Muscles of the forearm and wrist. (d) Superficial muscles, posterior view. **(e)** Deep posterior muscles; superficial muscles have been removed. The interossei, the deepest layer of instrinsic hand muscles, are also illustrated. **(f)** Photo of deep posterior muscles of the right forearm. The superficial muscles have been removed.

Muscles of the Lower Limb

Muscles that act on the lower limb cause movement at the hip, knee, and foot joints. Since the human pelvic girdle is composed of heavy fused bones that allow very little movement, no special group of muscles is necessary to stabilize it. This is unlike the shoulder girdle, where several muscles (mainly trunk muscles) are needed to stabilize the scapulae.

Muscles acting on the thigh (femur) cause various movements at the multiaxial hip joint (flexion, extension, rotation, abduction, and adduction). These include the iliopsoas, the adductor group, and other muscles summarized in Tables 15.7 and 15.8 and illustrated in Figures 15.11 and 15.12.

Muscles acting on the leg form the major musculature of the thigh. (Anatomically the term *leg* refers only to that portion between the knee and the ankle.) The thigh muscles cross the knee to allow its flexion and extension. They include the hamstrings and the quadriceps and, along with the muscles acting on the thigh, are described in Tables 15.7 and 15.8 and illustrated in Figures 15.11 and 15.12. Since some of these muscles also have attachments on the pelvic girdle, they can cause movement at the hip joint.

The muscles originating on the leg and acting on the foot and toes are described in Table 15.9 and shown in Figures 15.13 and 15.14.

Text continues on page 164

Table 15.7 Muscles Acting on Human Thigh and Leg, Anterior and Medial Aspects (see Figure 15.11)

Muscle	Comments	Origin	Insertion	Action
Origin on the Pelvis				
Iliopsoas—iliacus and psoas major	Two closely related muscles; fibers pass under inguinal ligament to insert into femur via a common tendon; iliacus is more lateral	Iliacus: iliac fossa and crest, lateral sacrum; psoas major: transverse processes, bodies, and discs of T_{12} and lumbar vertebrae	On and just below lesser trochanter of femur	Flex trunk on thigh; flex thigh; lateral flexion of vertebral column (psoas)
Sartorius	Straplike superficial muscle running obliquely across anterior surface of thigh to knee	Anterior superior iliac spine	By an aponeurosis into medial aspect of proximal tibia	Flexes, abducts, and laterally rotates thigh; flexes knee; known as "tailor's muscle" because it helps effect cross-legged position in which tailors are often depicted
Medial Compartment				
Adductors—magnus, longus, and brevis	Large muscle mass forming medial aspect of thigh; arise from front of pelvis and insert at various levels on femur	Magnus: ischial and pubic rami and ischial tuberosity; longus: pubis near pubic symphysis; brevis: body and inferior ramus of pubis	Magnus: linea aspera and adductor tubercle of femur; longus and brevis: linea aspera	Adduct and medially rotate and flex thigh; posterior part of magnus is also a synergist in thigh extension
Pectineus	Overlies adductor brevis on proximal thigh	Pectineal line of pubis (and superior ramus)	Inferior from lesser trochanter to linea aspera of femur	Adducts, flexes, and medially rotates thigh
Gracilis	Straplike superficial muscle of medial thigh	Inferior ramus and body of pubis	Medial surface of tibia just inferior to medial condyle	Adducts thigh; flexes and medially rotates leg, especially during walking

(continued on page 158)

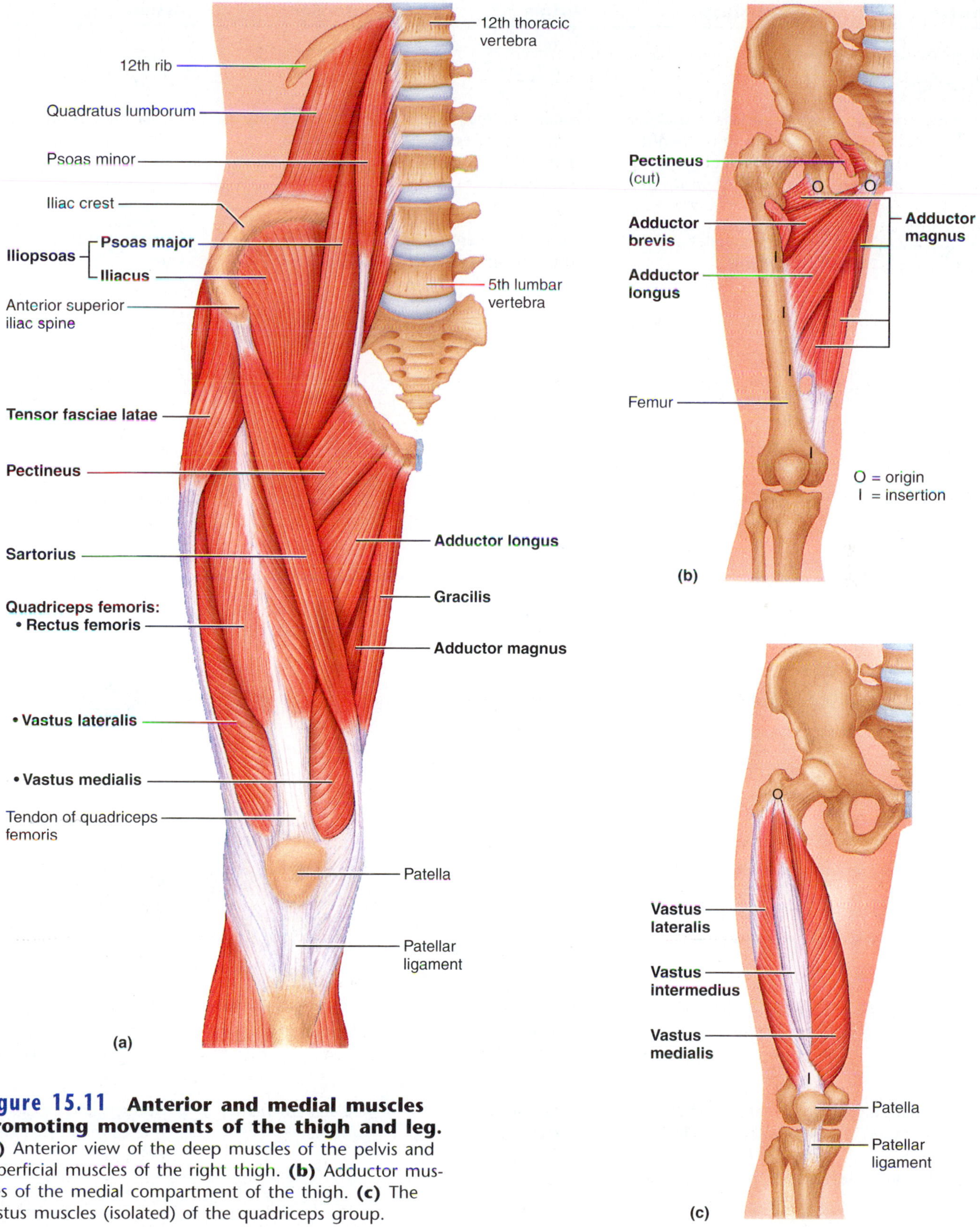

12th rib
12th thoracic vertebra
Quadratus lumborum
Psoas minor
Iliac crest
Iliopsoas
Psoas major
Iliacus
Anterior superior iliac spine
5th lumbar vertebra
Tensor fasciae latae
Pectineus
Sartorius
Adductor longus
Gracilis
Quadriceps femoris:
• Rectus femoris
Adductor magnus
• Vastus lateralis
• Vastus medialis
Tendon of quadriceps femoris
Patella
Patellar ligament
(a)

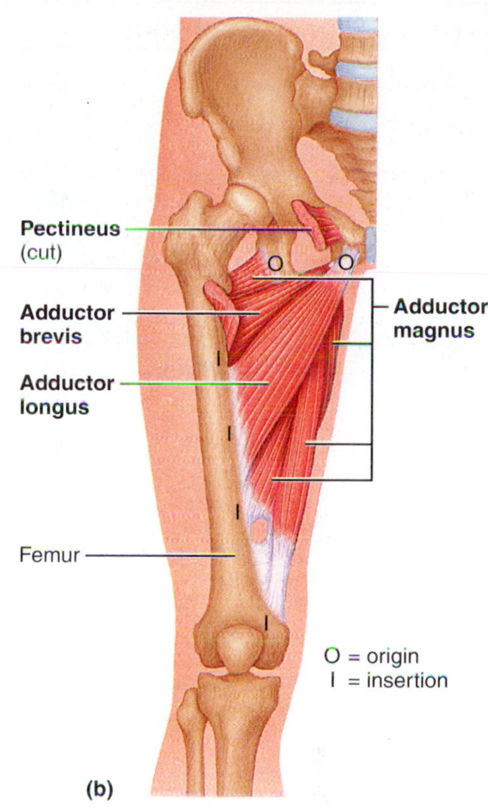

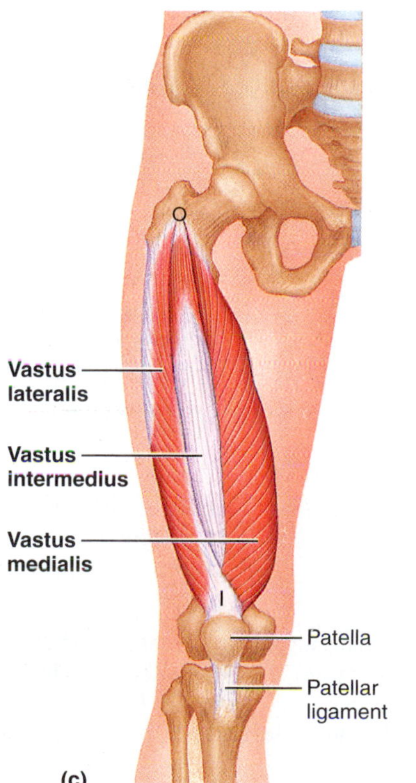

Figure 15.11 Anterior and medial muscles promoting movements of the thigh and leg. **(a)** Anterior view of the deep muscles of the pelvis and superficial muscles of the right thigh. **(b)** Adductor muscles of the medial compartment of the thigh. **(c)** The vastus muscles (isolated) of the quadriceps group.

Table 15.7 Muscles Acting on Human Thigh and Leg, Anterior and Medial Aspects *(continued)*

Muscle	Comments	Origin	Insertion	Action
Anterior Compartment				
Quadriceps*				
Rectus femoris	Superficial muscle of thigh; runs straight down thigh; only muscle of group to cross hip joint	Anterior inferior iliac spine and superior margin of acetabulum	Tibial tuberosity and patella	Extends knee and flexes thigh at hip
Vastus lateralis	Forms lateral aspect of thigh	Greater trochanter, intertrochanteric line, and linea aspera	Tibial tuberosity and patella	Extends and stabilizes knee
Vastus medialis	Forms inferomedial aspect of thigh	Linea aspera and intertrochanteric line	Tibial tuberosity and patella	Extends knee; stabilizes patella
Vastus intermedius	Obscured by rectus femoris; lies between vastus lateralis and vastus medialis on anterior thigh	Anterior and lateral surface of femur (not shown in figure)	Tibial tuberosity and patella	Extends knee
Tensor fasciae latae	Enclosed between fascia layers of thigh	Anterior aspect of iliac crest and anterior superior iliac spine	Iliotibial tract (lateral portion of fascia lata)	Flexes, abducts, and medially rotates thigh; steadies trunk

*The quadriceps form the flesh of the anterior thigh and have a common insertion in the tibial tuberosity via the patellar tendon. They are powerful leg extensors, enabling humans to kick a football, for example.

Table 15.8 Muscles Acting on Human Thigh and Leg, Posterior Aspect (see Figure 15.12)

Muscle	Comments	Origin	Insertion	Action
Origin on the Pelvis				
Gluteus maximus	Largest and most superficial of gluteal muscles (which form buttock mass); important injection site	Dorsal ilium, sacrum, and coccyx	Gluteal tuberosity of femur and iliotibial tract*	Complex, powerful thigh extensor (most effective when thigh is flexed, as in climbing stairs—but not as in walking); antagonist of iliopsoas; laterally rotates and abducts thigh
Gluteus medius	Partially covered by gluteus maximus; important injection site	Upper lateral surface of ilium	Greater trochanter of femur	Abducts and medially rotates thigh; steadies pelvis during walking
Gluteus minimus (not shown in figure)	Smallest and deepest gluteal muscle	External inferior surface of ilium	Greater trochanter of femur	Abducts and medially rotates thigh; steadies pelvis
Posterior Compartment				
Hamstrings†				
Biceps femoris	Most lateral muscle of group; arises from two heads	Ischial tuberosity (long head); linea aspera and distal femur (short head)	Tendon passes laterally to insert into head of fibula and lateral condyle of tibia	Extends thigh; laterally rotates leg; flexes knee

(continued)

Table 15.8 *(continued)*

Muscle	Comments	Origin	Insertion	Action
Semitendinosus	Medial to biceps femoris	Ischial tuberosity	Medial aspect of upper tibial shaft	Extends thigh; flexes knee; medially rotates leg
Semimembranosus	Deep to semitendinosus	Ischial tuberosity	Medial condyle of tibia; lateral condyle of femur	Extends thigh; flexes knee; medially rotates leg

*The iliotibial tract, a thickened lateral portion of the fascia lata, ensheathes all the muscles of the thigh. It extends as a tendinous band from the iliac crest to the knee.

†The hamstrings are the fleshy muscles of the posterior thigh. The name comes from the butchers' practice of using the tendons of these muscles to hang hams for smoking. As a group, they are strong extensors of the hip; they counteract the powerful quadriceps by stabilizing the knee joint when standing.

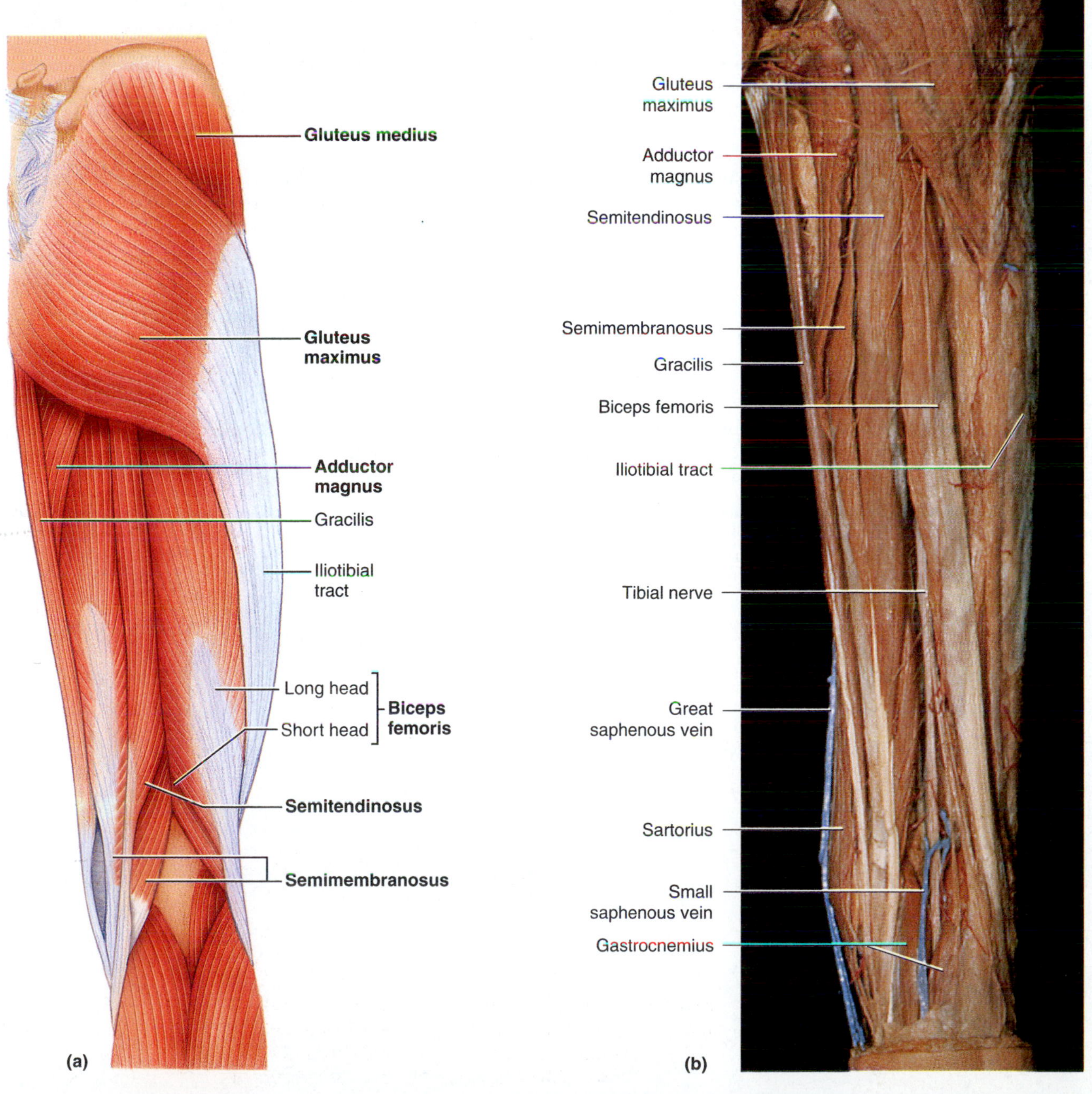

Figure 15.12 Muscles of the posterior aspect of the right hip and thigh. (a) Superficial view showing the gluteus muscles of the buttock and hamstring muscles of the thigh. **(b)** Photo of muscles of the posterior thigh.

Table 15.9 Muscles Acting on Human Foot and Ankle (see Figures 15.13 and 15.14)

Muscle	Comments	Origin	Insertion	Action
Lateral Compartment (Figure 15.13a, b and Figure 15.14c)				
Fibularis (peroneus) longus	Superficial lateral muscle; overlies fibula	Head and upper portion of fibula	By long tendon under foot to first metatarsal and medial cuneiform	Plantar flexes and everts foot; helps keep foot flat on ground
Fibularis (peroneus) brevis	Smaller muscle; deep to fibularis longus	Distal portion of fibula shaft	By tendon running behind lateral malleolus to insert on proximal end of fifth metatarsal	Plantar flexes and everts foot, as part of peronei group
Anterior Compartment (Figure 15.13a, b)				
Tibialis anterior	Superficial muscle of anterior leg; parallels sharp anterior margin of tibia	Lateral condyle and upper 2/3 of tibia; interosseous membrane	By tendon into inferior surface of first cuneiform and metatarsal 1	Prime mover of dorsiflexion; inverts foot; supports longitudinal arch of foot
Extensor digitorum longus	Anterolateral surface of leg; lateral to tibialis anterior	Lateral condyle of tibia; proximal 3/4 of fibula; interosseous membrane	Tendon divides into four parts; insert into middle and distal phalanges of toes 2–5	Prime mover of toe extension; dorsiflexes foot
Fibularis (peroneus) tertius	Small muscle; often fused to distal part of extensor digitorum longus	Distal anterior surface of fibula and interosseous membrane	Tendon inserts on dorsum of fifth metatarsal	Dorsiflexes and everts foot
Extensor hallucis longus	Deep to extensor digitorum longus and tibialis anterior	Anteromedial shaft of fibula and interosseous membrane	Tendon inserts on distal phalanx of great toe	Extends great toe; dorsiflexes foot

(continued on page 162)

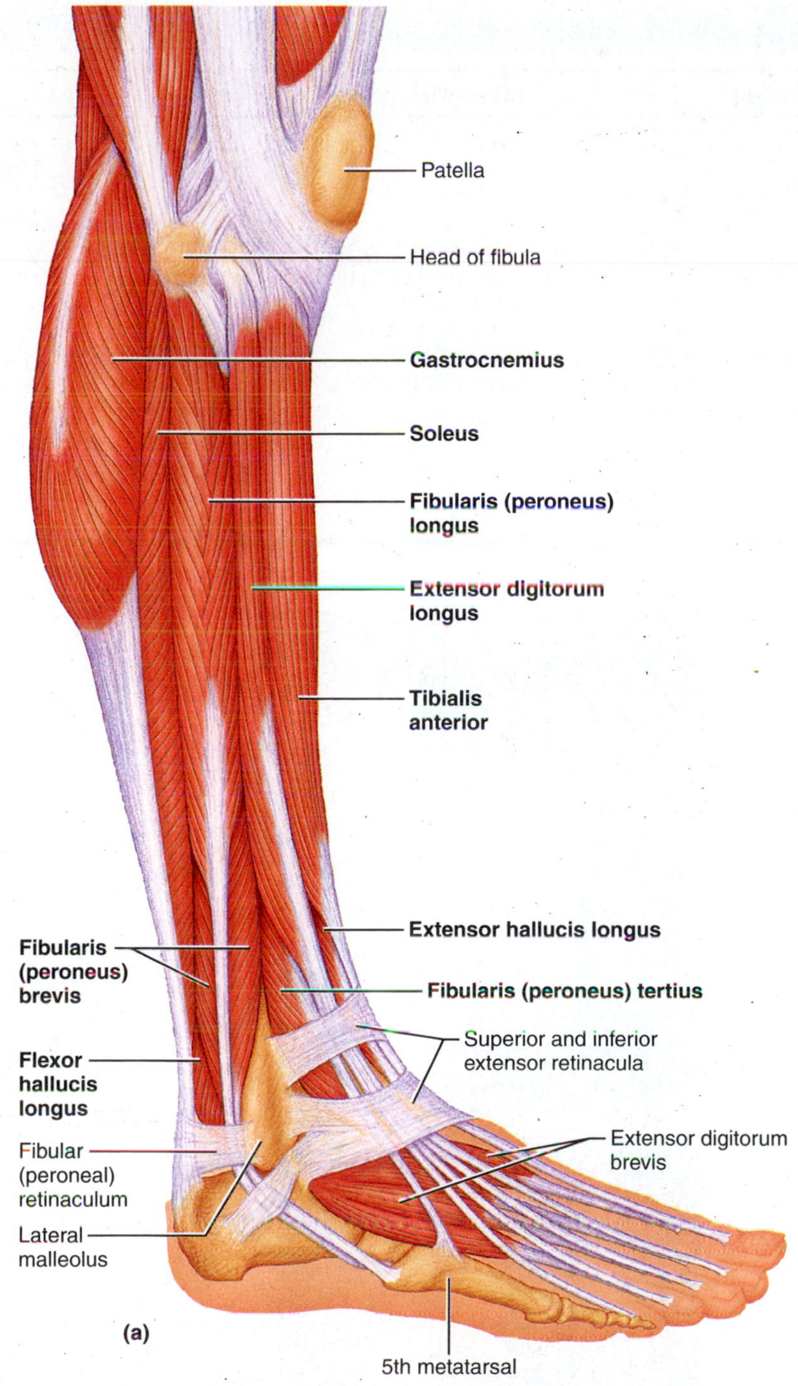

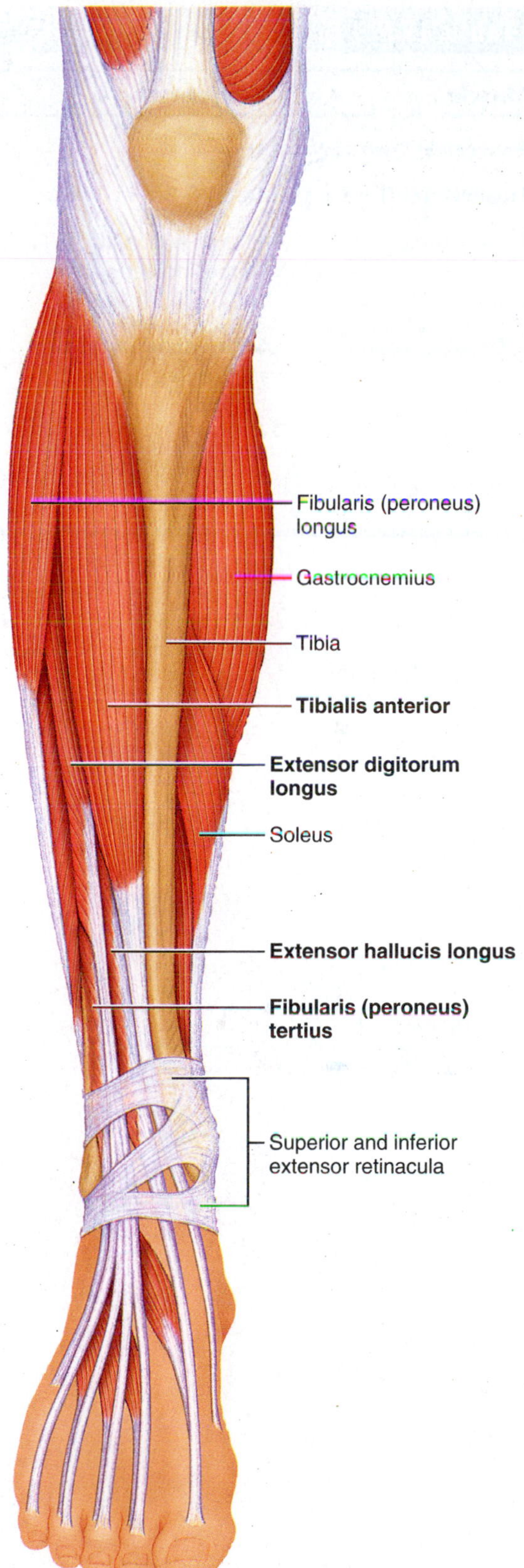

Figure 15.13 Muscles of the anterolateral aspect of the right leg. (a) Superficial view of lateral aspect of the leg, illustrating the positioning of the lateral compartment muscles (fibularis longus and brevis) relative to anterior and posterior leg muscles. **(b)** Superficial view of anterior leg muscles.

Table 15.9 Muscles Acting on Human Foot and Ankle *(continued)*

Muscle	Comments	Origin	Insertion	Action
Posterior Compartment				
Superficial (Figure 15.14a, b)				
Triceps surae	Muscle pair that shapes posterior calf		Via common tendon (calcaneal or Achilles) into heel	Plantar flex foot
Gastrocnemius	Superficial muscle of pair; two prominent bellies	By two heads from medial and lateral condyles of femur	Calcaneus via calcaneal tendon	Plantar flexes foot when knee is extended; crosses knee joint; thus can flex knee (when foot is dorsiflexed)

(continued)

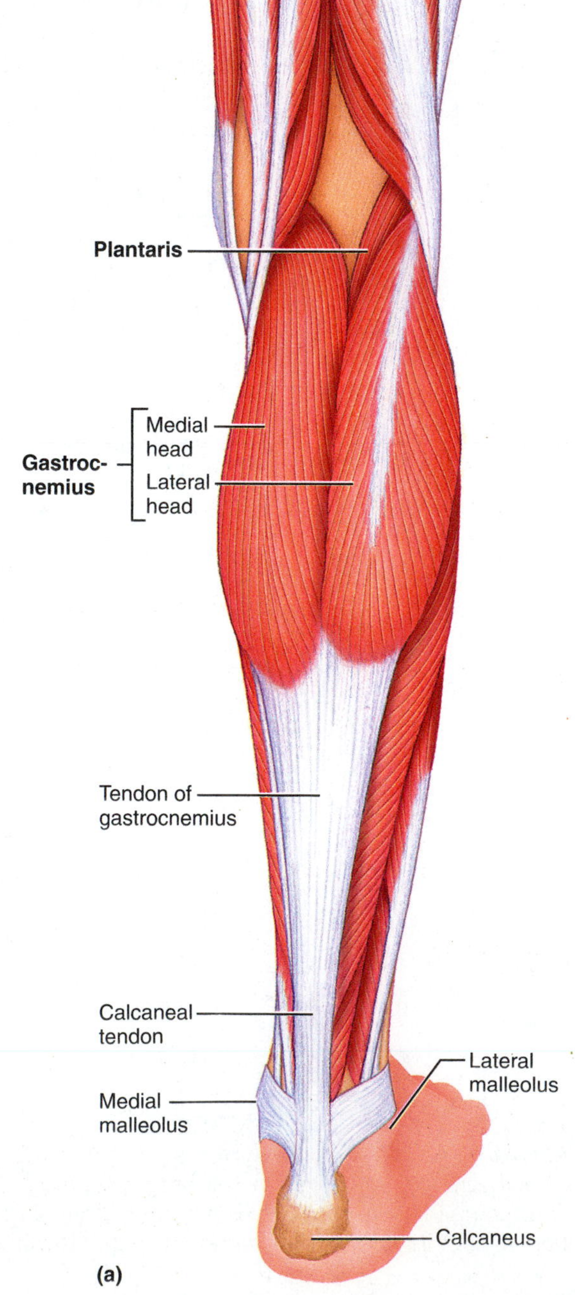

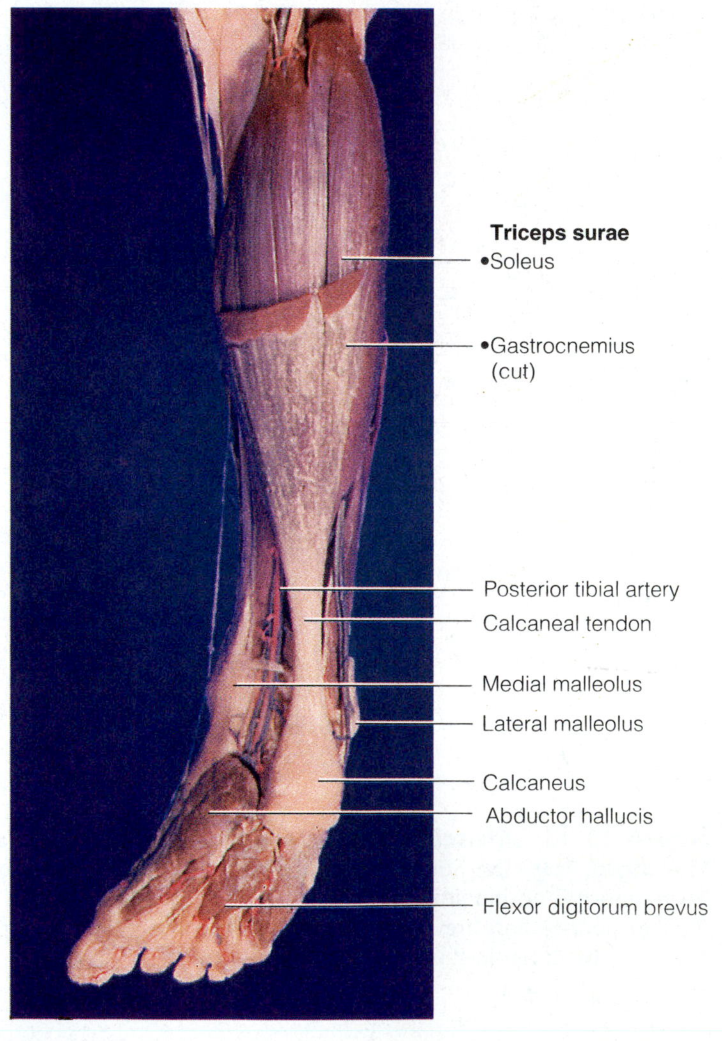

Figure 15.14 Muscles of the posterior aspect of the right leg. (a) Superficial view of the posterior leg. **(b)** Photo of posterior aspect of right leg. The gastrocnemius has been transected and its superior part removed.

Table 15.9 *(continued)*

Muscle	Comments	Origin	Insertion	Action
Soleus	Deep to gastrocnemius	Proximal portion of tibia and fibula; interosseous membrane	Calcaneus via calcaneal tendon	Plantar flexion; is an important muscle for locomotion
Deep (Figure 15.14c, d)				
Popliteus	Thin muscle at posterior aspect of knee	Lateral condyle of femur and lateral meniscus	Proximal tibia	Flexes and rotates leg medially to "unlock" extended knee when knee flexion begins

(continued on page 164)

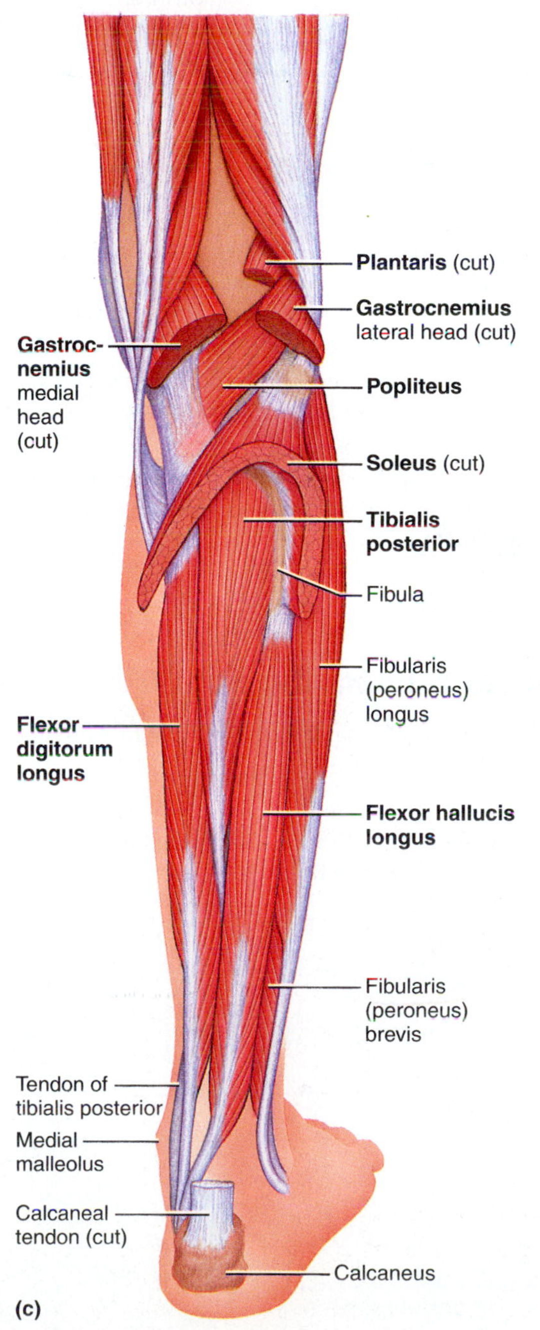

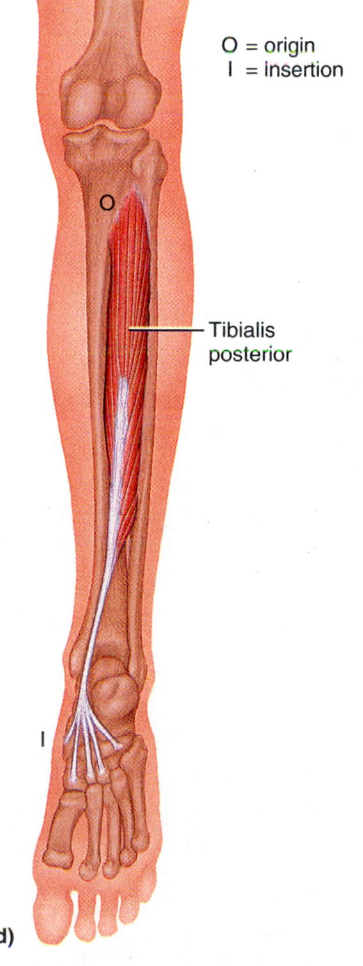

Figure 15.14 (*continued*) Muscles of the posterior aspect of the right leg. (c) The triceps surae has been removed to show the deep muscles of the posterior compartment. **(d)** Tibialis posterior shown in isolation so that its origin and insertion may be visualized.

Table 15.9 Muscles Acting on Human Foot and Ankle *(continued)*

Muscle	Comments	Origin	Insertion	Action
Tibialis posterior	Thick muscle deep to soleus	Superior portion of tibia and fibula and interosseous membrane	Tendon passes obliquely behind medial malleolus and under arch of foot; inserts into several tarsals and metatarsals 2–4	Prime mover of foot inversion; plantar flexes foot; stabilizes longitudinal arch of foot
Flexor digitorum longus	Runs medial to and partially overlies tibialis posterior	Posterior surface of tibia	Distal phalanges of second through fifth toes	Flexes toes; plantar flexes and inverts foot
Flexor hallucis longus (see also Figure 15.14)	Lies lateral to inferior aspect of tibialis posterior	Middle portion of fibula shaft; interosseous membrane	Tendon runs under foot to distal phalanx of great toe	Flexes great toe; plantar flexes and inverts foot; the "push-off muscle" during walking

Activity 5: Identifying Muscles of the Lower Limb

Identify the muscles acting on the thigh, leg, foot, and toes as instructed previously for other muscle groups. ■

Activity 6: Palpating Muscles of the Hip and Lower Limb

Complete this exercise by performing the following palpation demonstrations with your lab partner.

- Go into a deep knee bend and palpate your own *gluteus maximus* muscle as you extend your hip to resume the upright posture.
- Demonstrate the contraction of the anterior *quadriceps femoris* by trying to extend your knee against resistance. Do this while seated and note how the patellar tendon reacts. The *biceps femoris* of the posterior thigh comes into play when you flex your knee against resistance.
- Now stand on your toes. Have your partner palpate the lateral and medial heads of the *gastrocnemius* and follow it to its insertion in the calcaneal tendon.
- Dorsiflex and invert your foot while palpating your *tibialis anterior* muscle (which parallels the sharp anterior crest of the tibia laterally). ■

Activity 7: Review of Human Musculature

Review the muscles by watching the Human Musculature videotape.

For instructions on animal dissections, see the dissection exercises starting on p. 709 in the cat and fetal pig editions of this manual.

Skeletal Muscle Physiology—Frog Experimentation: Wet Lab

exercise

16A

Objectives

1. To observe muscle contraction on the microscopic level and describe the role of ATP and various ions in muscle contraction.
2. To define and explain the physiological basis of the following:
   depolarization
   repolarization
   action potential
   absolute and relative refractory periods
   subthreshold stimulus
   threshold stimulus
   maximal stimulus
   treppe
   wave summation
   tetanus
   muscle fatigue
3. To trace the events that result from the electrical stimulation of a muscle.
4. To recognize that a graded response of skeletal muscle is a function of the number of muscle fibers stimulated and the frequency of the stimulus.
5. To name and describe the phases of a muscle twitch.
6. To distinguish between a muscle twitch and a sustained (tetanic) contraction and to describe their importance in normal muscle activity.
7. To demonstrate how a computer or physiograph can be used to obtain pertinent and representative recordings of various physiological events of skeletal muscle activity.
8. To explain the significance of muscle tracings obtained during experimentation.

Materials

- ❑ ATP muscle kits (glycerinated rabbit psoas muscle;* ATP and salt solutions obtainable from Carolina Biological Supply)
- ❑ Petri dishes
- ❑ Microscope slides
- ❑ Cover glasses
- ❑ Millimeter ruler
- ❑ Compound microscope
- ❑ Dissecting microscope
- ❑ Pointed glass probes (teasing needles)
- ❑ Small beaker (50 ml)
- ❑ Distilled water
- ❑ Glass-marking pencil
- ❑ Watch or timer
- ❑ Frog Ringer's solution
- ❑ Scissors
- ❑ Metal needle probes
- ❑ Medicine dropper
- ❑ Cotton thread
- ❑ Forceps
- ❑ Disposable gloves
- ❑ Glass or porcelain plate
- ❑ Pithed bullfrog†
- ❑ Apparatus A or B‡

  A: physiograph, physiograph paper and ink, myograph, pin and clip electrodes, stimulator output extension cable, transducer cable, straight pins, frog board, laboratory stand, clamp

Notes to the Instructor:

*At the beginning of the lab, the muscle bundle should be removed from the test tube and cut into ~ 2-cm lengths. Both the cut muscle segments and the entubed glycerol should be put into a petri dish. One muscle *segment* is sufficient for each two to four students making observations.

†Bullfrogs to be pithed by lab instructor as needed for student experimentation. (If instructor prefers that students pith their own specimens, an instructional sheet on that procedure suitable for copying for student handouts is provided in the Instructor's Guide.)

‡Instructions for the use of a kymograph can be found in the Instructor's Guide.

B: PowerLab® unit and interface cable, Windows PC or Macintosh computer with Chart software installed, femur clamp, displacement transducer and cable, transducer stand, stimulator and stimulator output cable, aluminum foil, long pin bent into an S shape, thread, event marker, PowerLab® User's Guide, Chart User's Guide, weight basket, 10-g weights, frog Ringer's solution, and isolated frog gastrocnemius muscle

PhysioEx™ 4.0 Computer Simulation on p. P-17

Muscle Activity

The contraction of skeletal and cardiac muscle fibers can be considered in terms of three events—electrical excitation of the muscle cell, excitation-contraction coupling, and shortening of the muscle cell due to sliding of the myofilaments within it.

At rest, all cells maintain a potential difference, or voltage, across their plasma membrane; the inner face of the membrane is approximately –60 to –90 millivolts (mV) compared with the cell exterior. This potential difference is a result of differences in membrane permeability to cations, most importantly sodium (Na^+) and potassium (K^+) ions. Intracellular potassium concentration is much greater than its extracellular concentration, and intracellular sodium concentration is considerably less than its extracellular concentration. Hence, steep concentration gradients across the membrane exist for both cations. However, because the plasma membrane is slightly more permeable to K^+ than to Na^+, Na^+ influx into the cell is inadequate to balance K^+ outflow. The result of this unequal Na^+-K^+ diffusion across the membrane establishes the cell's **resting membrane potential.** The resting membrane potential is of particular interest in excitable cells, like muscle cells and neurons, because changes in that voltage underlie their ability to do work (to contract or to signal in muscle cells and neurons, respectively).

Action Potential

When a muscle cell is stimulated, the sarcolemma becomes temporarily permeable to sodium, which rushes into the cell. This sudden influx of sodium ions alters the membrane potential. That is, the cell interior becomes less negatively charged at that point, an event called **depolarization.** When depolarization reaches a certain level and the sarcolemma momentarily changes its polarity, a depolarization wave travels along the sarcolemma. Even as the influx of sodium ions occurs, the sarcolemma becomes impermeable to sodium and permeable to potassium ions. Consequently, potassium ions leak out of the cell, restoring the resting membrane potential (but not the original ionic conditions), an event called **repolarization.** The repolarization wave follows the depolarization wave across the sarcolemma. This rapid depolarization and repolarization of the membrane that is propagated along the entire membrane from the point of stimulation is called the **action potential.**

While the sodium gates are still open, there is no possibility of another response and the muscle cell is said to be in the **absolute refractory period.** The **relative refractory period** is the period after the sodium gates have closed when potassium gates are open and repolarization is ongoing. Especially strong stimuli may provoke a contraction during this part of the refractory period. Repolarization restores the muscle cell's excitability. Temporarily, the sodium-potassium pump, which actively transports K^+ into the cell and Na^+ out of the cell, need not be "revved up." But if the cell is stimulated to contract again and again in rapid-fire order, the loss of potassium and gain of sodium occurring during action potential generation begins to reduce its ability to respond. And so, eventually the sodium-potassium pump must be activated to reestablish the ionic concentrations of the resting state.

Contraction

Propagation of the action potential along the sarcolemma causes the release of calcium ions (Ca^{2+}) from storage in the tubules of the sarcoplasmic reticulum within the muscle cell. When the calcium ions bind to regulatory proteins on the actin myofilaments, they act as an ionic trigger that initiates contraction, and the actin and myosin filaments slide past each other. Once the action potential ends, the calcium ions are almost immediately transported back into the tubules of the sarcoplasmic reticulum. Instantly the muscle cell relaxes.

The events of the contraction process can most simply be summarized as follows: muscle cell contraction is initiated by generation and transmission of an action potential along the sarcolemma. This electrical event is coupled to the sliding of the myofilaments—contraction—by the release of Ca^{2+}. Keep in mind this sequence of events as you conduct the experiments.

Activity 1:
Observing Muscle Fiber Contraction

In this simple observational experiment, you will have the opportunity to review your understanding of muscle cell anatomy and to watch fibers respond to the presence of ATP and/or a solution of potassium and magnesium ions.

This experiment uses preparations of glycerinated muscle. The glycerination process denatures troponin and tropomyosin. Consequently, calcium, so critical for contraction in vivo, is not necessary here. The role of magnesium and potassium salts as cofactors in the contraction process is not well understood, but magnesium and potassium salts seem to be required for ATP-ase activity in this system.

1. Talk with other members of your lab group to develop a hypothesis about muscle fiber contraction for this experiment. The hypothesis should have three parts: (a) salts only, (b) ATP only, and (c) salts and ATP.

2. Obtain the following materials from the supply area: 2 glass teasing needles, 6 glass microscope slides and 3 cover glasses, millimeter ruler, dropper vials containing the following solutions: (a) 0.25% ATP in triply distilled water; (b) 0.25% ATP plus 0.05 *M* KCl plus 0.001 *M* $MgCl_2$ in distilled water; and (c) 0.05 *M* KCl plus 0.001 *M* $MgCl_2$ in distilled water; a petri dish, a beaker of distilled water, a glass-marking pencil, and a small portion of a previously cut muscle

Salts only	**Muscle fiber 1**	**Muscle fiber 2**	**Muscle fiber 3**	**Average**
Initial length (mm)				
Contracted length (mm)				
% Contraction				
ATP only				
Initial length (mm)				
Contracted length (mm)				
% Contraction				
Salts and ATP				
Initial length (mm)				
Contracted length (mm)				
% Contraction				

bundle segment. While you are at the supply area, place the muscle fibers in the petri dish and pour a small amount of glycerol (the fluid in the supply petri dish) over your muscle cells. Also obtain both a compound and a dissecting microscope and bring them to your laboratory bench.

3. Using clean fine glass needles, tease the muscle segment to separate its fibers. The objective is to isolate *single* muscle cells or fibers for observation. Be patient and work carefully so that the fibers do not get torn during this isolation procedure.

4. Transfer one or more of the fibers (or the thinnest strands you have obtained) onto a clean microscope slide with a glass needle, and cover it with a cover glass. Examine the fiber under low- and then high-power magnifications to observe the striations and the smoothness of the fibers when they are in the relaxed state.

5. Clean three microscope slides well and rinse in distilled water. Label the slides A, B, and C.

6. Transfer three or four fibers to microscope slide A with a glass needle. Using the needle as a prod, carefully position the fibers so that they are parallel to one another and as straight as possible. Place this slide under a dissecting microscope and measure the length of each fiber by holding a millimeter ruler adjacent to it. Alternatively, you can rest the microscope slide *on* the millimeter ruler to make your length determinations. Record the data on the chart at the top of the page.

7. Flood the fibers (situated under the dissecting microscope) with several drops of the solution containing ATP, potassium ions, and magnesium ions. Watch the reaction of the fibers after adding the solution. After 30 seconds (or slightly longer), remeasure each fiber and record the observed lengths on the chart. Also, observe the fibers to see if any width changes have occurred. Calculate the degree (or percentage) of contraction by using the following simple formula, and record this data on the chart above.

$$\text{Initial length (mm)} - \text{contracted length (mm)} = \text{degree of contraction (mm)}$$

then:

$$\frac{\text{Degree of contraction (mm)}}{\text{initial length (mm)}} \times 100 = \underline{\qquad} \text{ \% contraction}$$

8. Carefully transfer one of the contracted fibers to a clean microscope slide, cover with a cover glass, and observe with the compound microscope. Mentally compare your initial observations with the view you are observing now. What differences do you see? (Be specific.)

____

____

____

What zones (or bands) have disappeared?

____

9. Repeat steps 4 to 8 twice more, using clean slides and fresh muscle cells. On slide B use the solution of ATP in distilled water (no salts). Then, on slide C use the solution containing only salts (no ATP) for the third series. Record data on the chart above.

10. Collect the data from all the groups in your laboratory and use this data to prepare a lab report. (See Getting Started: Writing a Lab Report, p. xii.) Include in your discussion the following questions:

What degree of contraction was observed when ATP was applied in the absence of potassium and magnesium ions?

____

What degree of contraction was observed when the muscle fibers were flooded with a K^+- and Mg^{2+}-containing solution that lacked ATP?

____

What conclusions can you draw about the importance of ATP and potassium and magnesium ions to the contractile process?

____

____

Can you draw exactly the same conclusions from the data provided by each group? List some variables that might have been introduced into the procedure and that might account for any differences.

____

____ ■

Activity 2: Demonstrating Muscle Fatigue in Humans

1. Work in small groups. In each group select a subject, a timer, and a recorder.

2. Obtain a copy of the laboratory manual and a copy of the textbook. Weigh each book separately and then record the weight of each in the chart above.

3. The subject is to extend an arm and forearm straight out, holding the position until the arm shakes or the muscle begins to ache. Record the time to fatigue on the chart.

4. Allow the subject to rest for several minutes. Now ask the subject to hold the laboratory manual while keeping the arm and forearm in the same position as in step 3 above. Record the time to fatigue on the chart.

5. Allow the subject to rest again for several minutes. Now ask the subject to hold the textbook while keeping arm and forearm in the same position as in steps 3 and 4 above. Record the time to fatigue on the chart.

6. Each person in the group should take a turn as subject and all data should be recorded.

Load	Weight of object	Time elapsed until fatigue Trial 1	Trial 2	Trial 3
Appendage	N. A.			
Lab Manual				
Textbook				

7. What can you conclude about the effect of load on muscle fatigue? Explain.

____

____

____

Activity 3: Inducing Contraction in the Frog Gastrocnemius Muscle

Physiologists have learned a great deal about the way muscles function by isolating muscles from laboratory animals and then stimulating these muscles to observe their responses. Various stimuli—electrical shock, temperature changes, extremes of pH, certain chemicals—elicit muscle activity, but laboratory experiments of this type typically use electrical shock. This is because it is easier to control the onset and cessation of electrical shock, as well as the strength of the stimulus.

Preparing a Muscle for Experimentation

The preparatory work that precedes the recording of muscle activity tends to be quite time-consuming. If you work in teams of two or three, the work can be divided. While one of you is setting up the recording apparatus, one or two students can dissect the frog leg (Figure 16A.1). Experimentation should begin as soon as the dissection is completed.

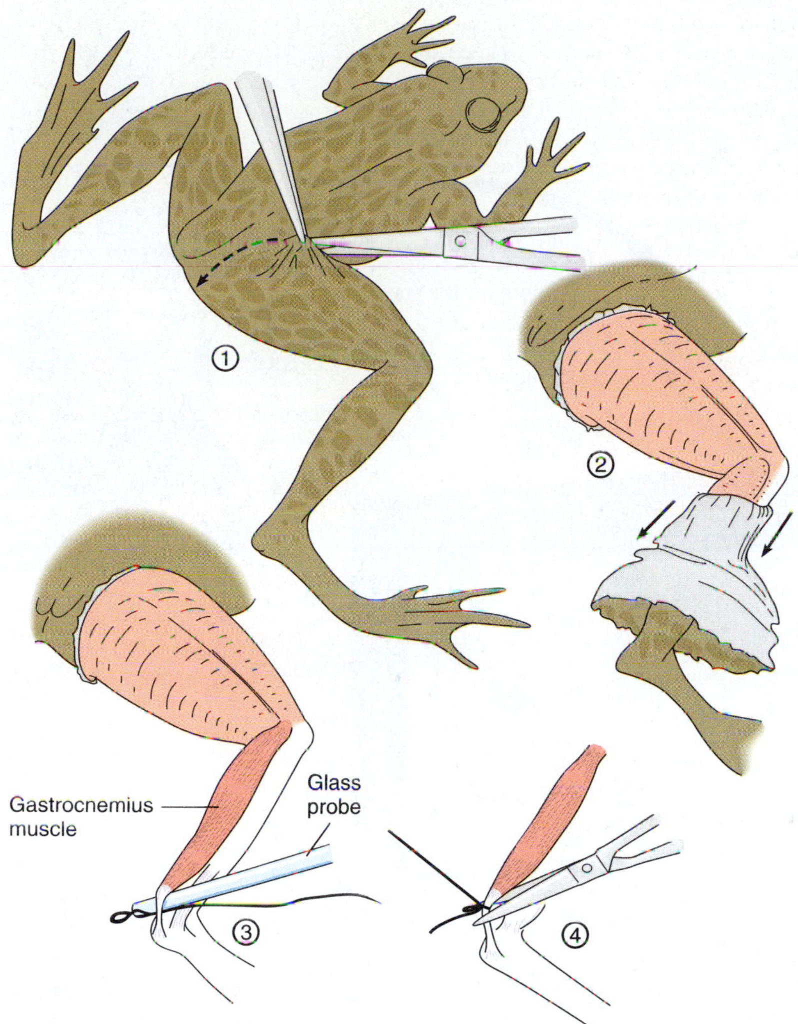

Figure 16A.1 Preparation of the frog gastrocnemius muscle. Numbers indicate the sequence of manipulation.

Various types of apparatus are used to record muscle contraction. All include a way to mark time intervals, a way to indicate exactly when the stimulus was applied, and a way to measure the magnitude of the contractile response. Instructions are provided here for setting up two different procedures: Procedure A for physiograph (Figure 16A.2) and Procedure B for PowerLab® (Figure 16A.3). Specific instructions for use of recording apparatus during recording will be provided by your instructor.

Dissection: Frog Hind Limb

1. Before beginning the frog dissection, have the following supplies ready at your laboratory bench: a small beaker containing 20 to 30 ml of frog Ringer's solution, scissors, a metal needle probe, a glass probe with a pointed tip, a medicine dropper, cotton thread, forceps, a glass or porcelain plate, and disposable gloves. While these supplies are being accumulated, one member of your team should notify the instructor that you are ready to begin experimentation, so that the frog can be prepared (pithed). Preparation of the frog in this manner renders it unable to feel pain and prevents reflex movements (like hopping) that would interfere with the experiments.

2. All students that will be handling the frog should obtain and don disposable plastic gloves. Obtain a pithed frog and place it ventral surface down on the glass plate. Make an incision into the skin approximately midthigh (Figure 16A.1), and then continue the cut completely around the thigh. Grasp the skin with the forceps and strip it from the leg and hindfoot. The skin adheres more at the joints, but a careful, persistent pulling motion—somewhat like pulling off a nylon stocking—will enable you to remove it in one piece. From this point on, the exposed muscle tissue should be kept moistened with the Ringer's solution to prevent spontaneous twitches.

3. Identify the gastrocnemius muscle (the fleshy muscle of the posterior calf) and the calcaneal (Achilles) tendon that secures it to the heel.

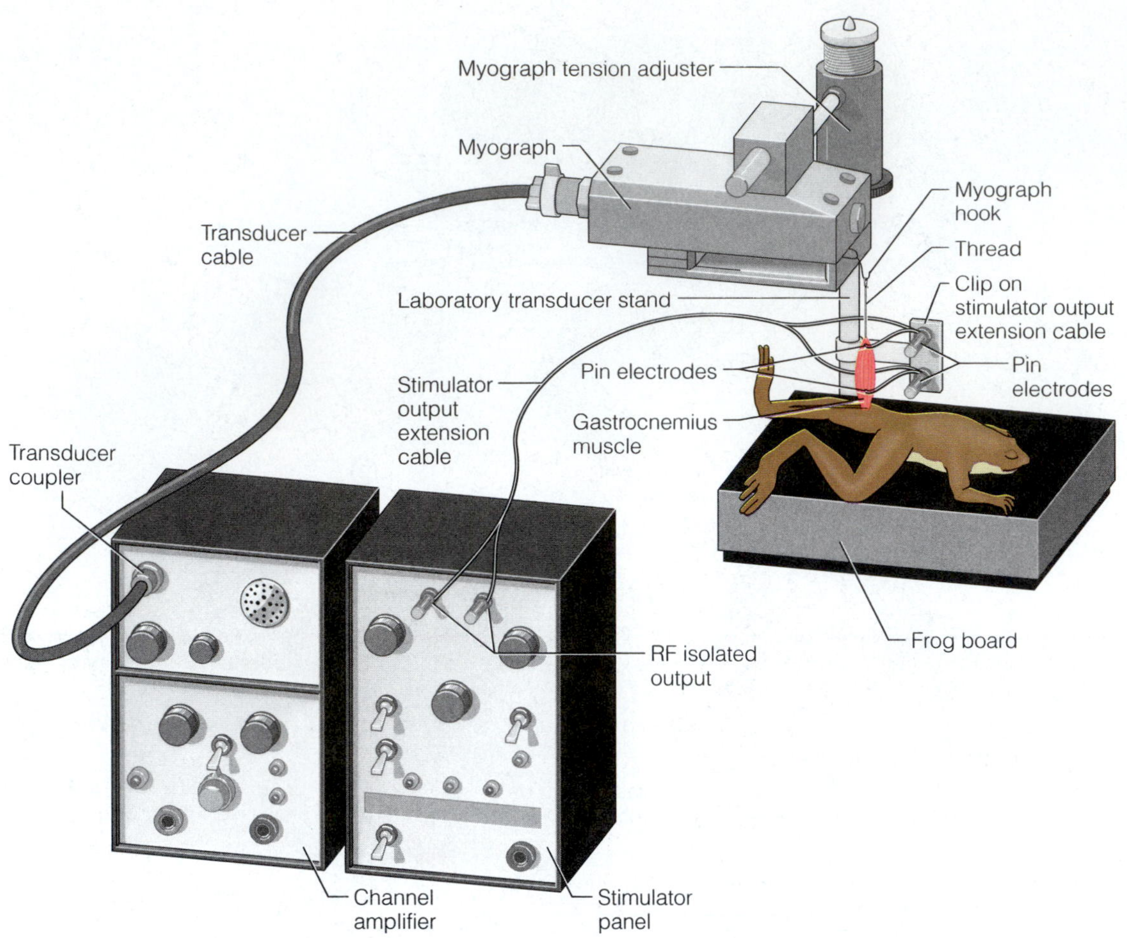

Materials:
Channel amplifier and stimulator transducer cable
Stimulator panel and stimulator output extension cable
Myograph
Myograph tension adjuster
Transducer stand
Two pin electrodes
Frog board and straight pins
Prepared frog (gastrocnemius muscle freed and calcaneal tendon ligated with thread)
Frog Ringer's solution

1. Connect myograph to transducer stand and attach frog board to stand.

2. Attach transducer cable to myograph and to input connection on channel amplifier.

3. Attach stimulator output extension cable to output on stimulator panel (red to red, black to black).

4. Using clip at opposite end of extension cable, attach cable to bottom of transducer stand adjacent to frog board.

5. Attach two pin electrodes securely to electrodes on clip.

6. Place knee of prepared frog in clip on frog board and secure by inserting a straight pin through tissues of frog. Keep frog muscle moistened with Ringer's solution.

7. Attach thread ligating the calcaneal tendon of frog to myograph leaf-spring hook.

8. Adjust position of myograph on stand to produce a constant tension on thread attached to muscle (taut but not tight). Gastrocnemius muscle should hang vertically directly below myograph hook.

9. Insert free ends of pin electrodes into the muscle, one at proximal end and other at distal end.

Figure 16A.2 Physiograph setup for frog gastrocnemius experiments.

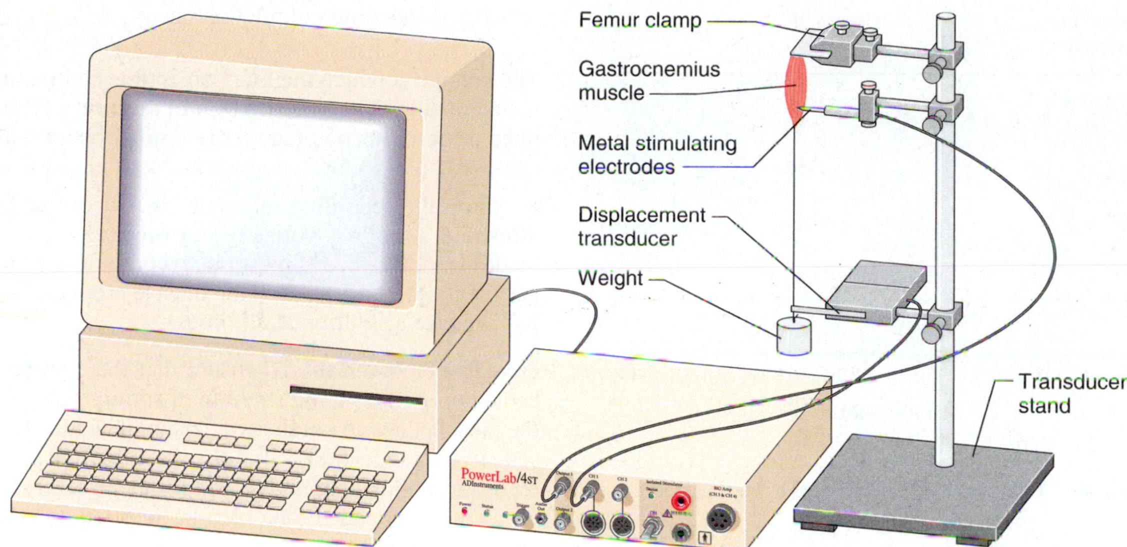

1. Attach a femur clamp and a displacement transducer to the transducer stand, and secure the femur in the femur clamp.
2. Insert an S-shaped pin into the tendon of the muscle.
3. Connect the muscle to the transducer by tying a thread to the pin and to the transducer. Be sure to keep the muscle moist with frog Ringer's solution throughout the experiment.
4. Readjust the transducer height to a position that leaves no slack in the cotton thread attached to the muscle.
5. Connect the transducer cable to CHANNEL 1 of the PowerLab® unit. Do not attempt to record while adjusting the transducer.
6. Connect wire leads to the positive (Out 1) and negative (Out 2) output terminals on the stimulator.
7. Attach one wire to the screw on the femur clamp.
8. Wrap the other wire in a thin strip of aluminum foil, and wrap the foil around the tendon to complete the circuit.
9. Make sure PowerLab® and computers are turned on.
10. Make sure that everything is connected properly.
11. Open the PowerLab® folder, and select the file named *Chart* or open the *Settings* file for this exercise.

    a. Turn off all the channels except CHANNEL 1. Expand CHANNEL 1 to fill the screen. Select the channel settings from the setup menu.

    b. Select a speed of 10s/*Division* and a *Range* of 20mV to begin the experiments.

    c. Select *Input Amplifier* from the CHANNEL 1 pop-up menu, and make sure that *Positive* and *AC* are both activated.

    d. Verify that the base line is at zero and at the bottom of the screen. If it is not, click and drag the vertical axis until the base line is at the bottom.
12. Set up the stimulator as follows:
    a. Power: On
    b. Output: Mono
    c. Mode: Off
    d. Polarity: Normal
    e. Stimulus: Regular
    f. Delay (if present): Minimal (0.02 msec) (*ms* and *msec* both mean *millisecond*)
    g. Volts: 0.1 to 2.0 V
    h. Duration: 10 msec

Figure 16A.3 PowerLab® setup for frog gastrocnemius experiments.

4. Slip a glass probe under the gastrocnemius muscle and run it along the entire length and under the calcaneal tendon to free them from the underlying tissues.

5. Cut a piece of thread about 10 in. long and use the glass probe to slide the thread under the calcaneal tendon. Knot the thread firmly around the tendon and then sever the tendon distal to the thread. Alternatively, you can bend a common pin into a Z-shape and insert the pin securely into the tendon. The thread is then attached to the opposite end of the pin. If you are using a physiograph, once the tendon has been tied or pinned, the frog is ready for experimentation (see Figure 16A.2). If you are using PowerLab®, the gastrocnemius muscle must be completely isolated, as described in step 6.

6. To prepare for a PowerLab® setup, cut away the fibulotibial bone just distal to the knee. Expose the femur of the thigh and cut it completely through at midthigh. Remove as much of the thigh muscle tissue as possible by carefully cutting it away with the scissors. The isolated gastrocnemius muscle can now be mounted in the femur clamp attached to the transducer stand, and the stimulating electrodes can be attached (see Figure 16A.3). About halfway through the laboratory period, dissect the second leg for use. ■

Recording Muscle Activity

Skeletal muscles consist of thousands of muscle cells and react to stimuli with graded responses. Thus muscle contractions can be weak or vigorous, depending on the require-

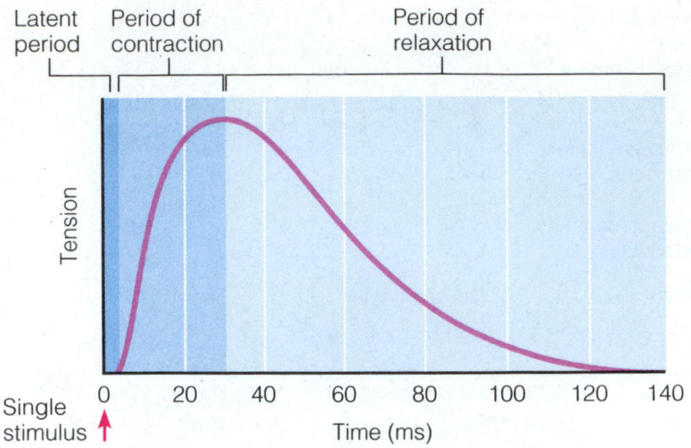

Figure 16A.4 Tracing of a muscle twitch.

ments of the task. Graded responses (different degrees of shortening) of a skeletal muscle depend on the number of muscle cells being stimulated. In the intact organism, the number of motor units firing at any one time determines how many muscle cells will be stimulated. In this laboratory, the frequency and strength of an electrical current determines the response.

A single contraction of skeletal muscle is called a **muscle twitch.** A tracing of a muscle twitch (Figure 16A.4) shows three distinct phases: latent, contraction, and relaxation. The **latent phase** is the interval from stimulus application until the muscle begins to shorten. Although no activity is indicated on the tracing during this phase, important electrical and chemical changes are occurring within the muscle. During the **contraction phase,** the muscle fibers shorten; the tracing shows an increasingly higher needle deflection and the tracing peaks. During the **relaxation phase,** represented by a downward curve of the tracing, the muscle fibers relax and lengthen. On a slowly moving recording surface, the single muscle twitch appears as a spike (rather than a bell-shaped curve, as in Figure 16A.4), but on a rapidly moving recording surface, the three distinct phases just described become recognizable.

Determining the Threshold Stimulus

Follow the steps in Procedure A if using the physiograph apparatus. Use the steps in Procedure B if using PowerLab®.

Procedure A: Physiograph

1. Assuming that you have already set up the recording apparatus, set the time marker to deliver one pulse per second and set the paper speed at a slow rate, approximately 0.1 cm per second.

2. Set the duration control on the stimulator between 7 and 15 msec, multiplier ×1 and the voltage control at 0 V, multiplier ×1. Turn the sensitivity control knob of the stimulator fully clockwise (lowest value, greatest sensitivity).

3. Administer single stimuli to the muscle at 1-minute intervals, beginning with 0.1 V and increasing each successive stimulus by 0.1 V until a contraction is obtained (shown by a spike on the paper).

At what voltage did contraction occur? ______________ V

The voltage at which the first perceptible contractile response is obtained is called the **threshold stimulus.** All stimuli applied prior to this point are termed **subthreshold stimuli,** because at those voltages no response was elicited.

4. Stop the recording and mark the record to indicate the threshold stimulus, voltage, and time. Do not remove the record from the recording surface; continue with the next experiment. Remember: keep the muscle preparation moistened with Ringer's solution at all times.

Procedure B: PowerLab® To ensure that the equipment is connected properly, set the strength of stimulus at 1 V. Stimulate the muscle once. A single twitch should be visible. Adjust the sensitivity range to a value that results in a muscle twitch recording about 2 cm high. Readjust the baseline if necessary.

Note: Unneeded data may be selected and deleted. You can scroll back and forth on the screen to see how much data has been collected.

To carry out the experiment, make sure the time interval between each stimulus is about the same (1 minute). This gives the muscle time to rest between responses. Keep the muscle moist with Ringer's solution at all times. Record the voltages as the experiment proceeds. Do not depend upon your memory.

1. To stimulate the muscle, click on the *Start* button on the lower right corner of the screen.

2. Set the voltage to 0.1 V, and click on the *Single Stimulation* switch once. Press *Stop* and *Save*. Remember to wait 1 minute between each stimulus, and to use *Stop* and *Save* after each stimulus.

The voltage at which the first perceptible contractile response is obtained is called the **threshold stimulus.** All stimuli applied prior to this point are termed **subthreshold stimuli** because at those voltages no response was elicited.

3. Mark the record *threshold stimulus* as a comment and indicate the voltage.

Observing Graded Muscle Response to Increased Stimulus Intensity

Follow the steps in Procedure A if using the physiograph apparatus. Use the steps in Procedure B if using PowerLab®.

Procedure A: Physiograph

1. Follow the previous setup instructions, but set the voltage control at the threshold voltage (as determined in the first experiment).

2. Deliver single stimuli at 1-minute intervals. Initially increase the voltage between shocks by 0.5 V; then increase the voltage by 1 to 2 V between shocks as the experiment continues, until contraction height increases no further. Stop the recording apparatus.

What voltage produced the highest spike (and thus the maximal strength of contraction)?

______________ V

This voltage, called the **maximal stimulus** (for *your* muscle specimen), is the weakest stimulus at which all muscle cells are being stimulated.

3. Mark the record *maximal stimulus*. Record the maximal stimulus voltage and the time you completed the experiment.

4. What is happening to the muscle as the voltage is increased?

___

___

What is another name for this phenomenon?

___

5. Explain why the strength of contraction does not increase once the maximal stimulus is reached.

___

___

Procedure B: PowerLab®

1. Follow the previous setup instructions, but set the voltage control at the threshold voltage (as determined in the first experiment).

2. For each trial, press *Start,* stimulate, and then press *Stop* and *Save*. Remember to allow the muscle to rest for 1 minute between trials.

3. Deliver single stimuli at 1-minute intervals. Initially increase the voltage between shocks by 0.5 V; then increase the voltage by 1 to 2 V between shocks as the experiment continues, until contraction height increases no further. Stop the recording apparatus.

What voltage produced the highest spike (thus the maximal strength of contraction)?

___ V

This voltage, called the **maximal stimulus** (for *your* muscle specimen), is the weakest stimulus at which all muscle cells are being stimulated.

4. Mark the recording *maximal stimulus* as a comment and record the voltage.

5. What is happening to the muscle as the voltage is increased? What is another name for this phenomenon?

___

___

___

6. Explain why the strength of the response does not increase once the maximal stimulus is reached.

___

___

___

Timing the Muscle Twitch

Follow the steps in Procedure A if using the physiograph apparatus. Use the steps in Procedure B if using PowerLab®.

Procedure A: Physiograph

1. Follow the previous setup directions, but set the voltage for the maximal stimulus (as determined in the preceding experiment) and set the paper advance or recording speed at maximum. Record the paper speed setting:

___ mm/sec

2. Determine the time required for the paper to advance 1 mm by using the formula

$$\frac{1 \text{ mm}}{\text{mm/sec (paper speed)}}$$

(Thus, if your paper speed is 25 mm/sec, each mm on the chart equals 0.04 sec.) Record the computed value:

1 mm = ___ sec

3. Deliver single stimuli at 1-minute intervals to obtain several "twitch" curves. Stop the recording.

4. Determine the duration of the latent, contraction, and relaxation phases of the twitches and record the data below.

Duration of latent period: ___ sec

Duration of contraction period: ___ sec

Duration of relaxation period: ___ sec

5. Label the record to indicate the point of stimulus, the beginning of contraction, the end of contraction, and the end of relaxation.

6. Allow the muscle to rest (but keep it moistened) before continuing with the next experiment.

Procedure B: PowerLab® If you have an external event marker and want to determine latent period, set up the event marker on one of the available PowerLab® inputs and add CHANNEL 2 to the screen.

1. Set the voltage for maximal stimulus as determined in the previous experiment, and set the speed to 100 ms/Division.

2. Deliver a single stimulus to obtain a "twitch" curve, and use the event marker if attached. *Stop* the recording and *Save*. Repeat two times, waiting 1 minute between stimuli.

3. Select one twitch and *Zoom* in on it.

4. Using the marker, measure the time from the start of the twitch until maximum tension is reached (period of contraction) and the time from the maximum tension to complete recovery (period of relaxation). If you have an event marker, measure the time from the initiation of the stimulus until the first observable response (latent period). Enter the data below.

Duration of latent period: ______________________ sec

Duration of contraction period: ______________________ sec

Duration of relaxation period: ______________________ sec

5. Label the record to indicate the point of stimulus, the beginning of contraction, the end of contraction, and the end of relaxation.

6. Allow the muscle to rest (but keep it moistened) before continuing with the next experiment.

Inducing Treppe, or the Staircase Phenomenon

As a muscle is stimulated to contract, a curious phenomenon is observed in the tracing pattern of the first few twitches. Even though the stimulus intensity is unchanged, the height of the individual spikes increases in a stepwise manner, producing a sort of staircase pattern called **treppe** (Figure 16A.5). This phenomenon is not well understood, but the following explanation has been offered: in the muscle cell's resting state, there is much less Ca^{2+} in the sarcoplasm and the enzyme systems in the muscle cell are less efficient than after it has contracted a few times. As the muscle cell begins to contract, the intracellular concentration of Ca^{2+} rises dramatically, and the heat generated by muscle activity increases the efficiency of the enzyme systems. As a result, the muscle becomes more efficient and contracts more vigorously. This is the physiological basis of the warm-up period prior to competition in sports events.

1. Set up the apparatus as in the previous experiment, again setting the voltage to the maximal stimulus determined earlier.

2. Deliver single stimuli at 1-sec intervals until the strength of contraction does not increase further.

3. Stop the recording apparatus and mark the record *treppe*. Note also the number of contractions (and seconds) required to reach the constant contraction magnitude. Record the voltage used and the time when you completed this experiment. Continue on to the next experiment.

Observing Graded Muscle Response to Increased Stimulus Frequency

Muscles subjected to frequent stimulation, without a chance to relax, exhibit two kinds of responses—wave summation and tetanus—depending on the level of stimulus frequency (Figure 16A.5).

Wave Summation: If a muscle is stimulated with a rapid series of stimuli of the same intensity before it has had a chance to relax completely, the response to the second and subsequent stimuli will be greater than to the first stimulus. This phenomenon, called **wave,** or **temporal, summation,** occurs because the muscle is already in a partially contracted state when subsequent stimuli are delivered.

Follow the steps in Procedure A if using the physiograph apparatus. Use the steps in Procedure B if using PowerLab®.

Procedure A: Physiograph

1. Set up the apparatus as in the previous experiment, setting the voltage to the maximal stimulus as determined earlier and the chart speed to maximum.

2. With the stimulator in single mode, deliver two successive stimuli as rapidly as possible.

3. Shut off the recorder and label the record as *wave summation*. Note also the time, the voltage, and the frequency. What did you observe?

____

____

Procedure B: Powerlab® Set the voltage to 0.3 to 0.5 V above maximal stimulus as determined previously, and set the chart speed at 100 ms/Division.

1. With the stimulator in single mode, deliver two successive stimuli as rapidly as you can. *Stop,* and *Save*.

2. Label the record *wave summation*. Note the time, the voltage, and the frequency.

What did you observe?

____

____

____

____

Tetanus: Stimulation of a muscle at an even higher frequency will produce a "fusion" (tetanization) of the summated twitches. In effect, a single sustained contraction is achieved in which no evidence of relaxation can be seen (Figure 16A.5). **Tetanus** is a feature of normal skeletal muscle functioning; the single muscle twitch is primarily a laboratory phenomenon.

Follow the steps in Procedure A if using the physiograph apparatus. Use the steps in Procedure B if using PowerLab®.

Procedure A: Physiograph

1. To demonstrate tetanus, maintain the conditions used for wave summation except for the frequency of stimulation. Set the stimulator to deliver 60 stimuli per second.

2. As soon as you obtain a single smooth, sustained contraction (with no evidence of relaxation), discontinue stimulation and shut off the recorder.

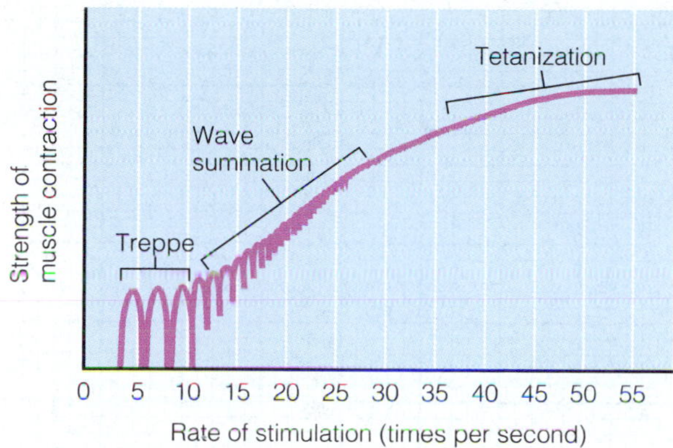

Figure 16A.5 Treppe, wave summation, and tetanization. Progressive summation of successive contractions occurs as the rate of stimulation is increased. Tetanization occurs when the rate of stimulation reaches approximately 35 per second, and maximum contraction force occurs at a stimulation rate of approximately 50 per second.

3. Label the tracing with the conditions of experimentation, the time, and the area of *tetanus*.

Procedure B: PowerLab® It is important to check the setup to be sure that there is room to record the results of this part of the experiment. Set the voltage to 0.3 to 0.5 V above the maximal voltage determined earlier. Set the scrolling speed to 200 ms/Division. Stimulate the muscle once and set the range so that the contraction is about 2 cm high on the screen. Press *Stop* and *Save*. If the contraction is too small, increase the sensitivity by decreasing the range from 20 mV to 10 mV and repeat. Adjust the baseline so that it is close to the bottom of the screen.

1. Set the stimulator mode to *Repeat* and the stimulator rate to 60 pulses per second (PPS).
2. Press *Start*, and record contractions for a few seconds, turn the stimulator off, *Stop*, and *Save*. Remember to keep the muscle moist at all times.
3. Mark the recording *tetanus*.

Inducing Muscle Fatigue

Muscle fatigue, the loss of the ability to contract, is believed to be a result of the oxygen debt that occurs in the tissue after prolonged activity (through the accumulation of such waste products as lactic acid as well as the depletion of ATP). True muscle fatigue rarely occurs in the body because it is most often preceded by a subjective feeling of fatigue. Furthermore, fatigue of the neuromuscular junctions typically precedes fatigue of the muscle.

1. To demonstrate muscle fatigue, set up an experiment like the tetanus experiment, but continue stimulation until the muscle completely relaxes and the contraction curve returns to the base line.
2. Measure the time interval between the beginning of complete tetanus and the beginning of fatigue (when the tracing begins its downward curve). Mark the record appropriately.
3. Determine the time required for complete fatigue to occur (the time interval from the beginning of fatigue until the return of the curve to the base line). Mark the record appropriately.
4. Allow the muscle to rest (keeping it moistened with Ringer's solution) for 10 min, and then repeat the experiment.

What was the effect of the rest period on the fatigued muscle?

___

___

What is the physiological basis for this reaction?

___

___

Determining the Effect of Load on Skeletal Muscle

When the fibers of a skeletal muscle are slightly stretched by a weight or tension, the muscle responds by contracting more forcibly and thus is capable of doing more work. When the actin and myosin barely overlap, sliding can occur along nearly the entire length of the actin filaments. If the load is increased beyond the optimum, the latent period becomes longer, contractile force decreases, and relaxation (fatigue) occurs more quickly. With excessive stretching, the muscle is unable to develop any tension and no contraction occurs. Since the filaments no longer overlap at all with this degree of stretching, the sliding force cannot be generated.

If your equipment allows you to add more weights to the muscle specimen or to increase the tension on the muscle, perform the following experiment to determine the effect of loading on skeletal muscle, and to develop a work curve for the frog's gastrocnemius muscle.

1. Set the stimulator to deliver the maximal voltage as previously determined.
2. Stimulate the unweighted muscle with single shocks at 1- to 2-sec intervals to achieve three or four muscle twitches.
3. Stop the recording apparatus and add 10 g of weight or tension to the muscle. Restart and advance the recording about 1 cm, and then stimulate again to obtain three or four spikes.
4. Repeat the previous step seven more times, increasing the weight by 10 g each time until the total load on the muscle is 80 g or the muscle fails to respond. If the calcaneal tendon tears, the weight will drop, thus ending the trial. In such cases, you will need to prepare another frog's leg to continue the experiments and the maximal stimulus will have to be determined for the new muscle preparation.
5. When these "loading" experiments are completed, discontinue recording. Mark the curves on the record to indicate the load (in grams).

Load (g)	Distance load lifted (mm)		Work done	
	Trial 1	Trial 2	Trial 1	Trial 2
0				
10				
20				
30				
40				
50				
60				
70				
80				

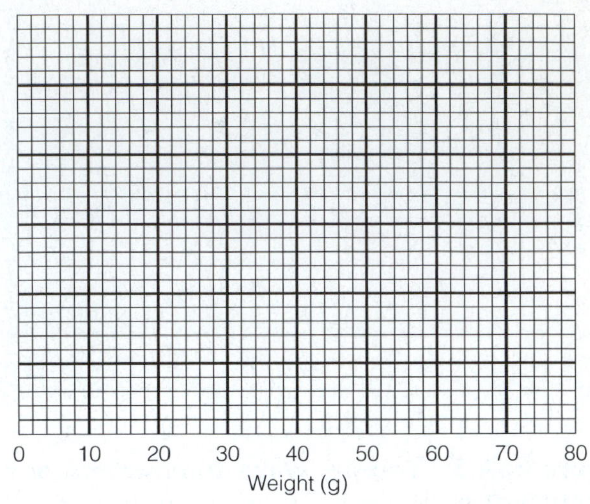

6. Measure the height of contraction (in millimeters) for each sequence of twitches obtained with each load, and insert this information on the chart above.

7. Compute the work done by the muscle for each twitch (load) sequence.

Weight of load (g) × distance load lifted (mm) = work done

Enter these calculations into the chart in the column labeled Work done, Trial 1.

8. Allow the muscle to rest for 5 minutes. Then conduct a second trial in the same manner (i.e., repeat steps 2 through 7). Record this second set of measurements and calculations in the columns labeled Trial 2. Be sure to keep the muscle well moistened with Ringer's solution during the resting interval.

9. Using two different colors, plot a line graph of work done against the weight on the accompanying grid for each trial. Label each plot appropriately.

10. Dismantle all apparatus and prepare the equipment for storage. Dispose of the frog remains in the appropriate container. Discard the gloves as instructed and wash and dry your hands.

11. Inspect your records of the experiments and make sure each is fully labeled with the experimental conditions, the date, and the names of those who conducted the experiments. For future reference, attach a tracing (or a copy of the tracing) for each experiment to this page. ■

Histology of Nervous Tissue

exercise

17

The nervous system is the master integrating and coordinating system, continuously monitoring and processing sensory information both from the external environment and from within the body. Every thought, action, and sensation is a reflection of its activity. Like a computer, it processes and integrates new "inputs" with information previously fed into it ("programmed") to produce an appropriate response ("readout"). However, no computer can possibly compare in complexity and scope to the human nervous system.

Despite its complexity, nervous tissue is made up of just two principal cell populations: neurons and their supporting cells. The supporting cells in the CNS (central nervous system: brain and spinal cord) are usually referred to as **neuroglia** or **glial cells**. The *neuroglia,* literally "nerve glue," include *astrocytes, oligodendrocytes, microglia,* and *ependymal cells* (Figure 17.1). The most important supporting cells in the PNS (peripheral nervous system), that is, in the neural structures outside the CNS, are *Schwann cells* and *satellite cells.* **Supporting cells** serve the needs of the neurons by acting as phagocytes and by bracing, protecting, and myelinating the delicate neurons. In addition, they play a role in capillary-neuron exchanges and control the chemical environment around neurons. Although neuroglia resemble neurons in some ways (they have fibrous cellular extensions), they are not capable of generating and transmitting nerve impulses, a capability that is highly developed in neurons. Our focus in this exercise is the highly excitable neurons.

Objectives

1. To differentiate between the functions of neurons and neuroglia.
2. To list four types of neuroglia cells.
3. To identify the important anatomical characteristics of a neuron on an appropriate diagram or projected slide.
4. To state the functions of axons, dendrites, axonal terminals, neurofibrils, and myelin sheaths.
5. To explain how a nerve impulse is transmitted from one neuron to another.
6. To explain the role of Schwann cells in the formation of the myelin sheath.
7. To classify neurons according to structure and function.
8. To distinguish between a nerve and a tract and between a ganglion and a nucleus.
9. To describe the structure of a nerve, identifying the connective tissue coverings (endoneurium, perineurium, and epineurium) and citing their functions.

Materials

- ❑ Model of a "typical" neuron (if available)
- ❑ Compound microscope
- ❑ Immersion oil
- ❑ Histologic slides of an ox spinal cord smear and teased myelinated nerve fibers
- ❑ Prepared slides of Purkinje cells (cerebellum), pyramidal cells (cerebrum), and a dorsal root ganglion
- ❑ Prepared slide of a nerve (cross section)

Neuron Anatomy

Neurons are the structural units of nervous tissue. They are highly specialized to transmit messages (nerve impulses) from one part of the body to another. Although neurons differ structurally, they have many identifiable features in common (Figure 17.2a and c). All have a **cell body** from which slender processes or fibers extend. Although neuron cell bodies are typically found in the CNS in clusters called **nuclei,** occasionally they reside in **ganglia** (collections of neuron cell bodies outside the CNS). They make up the gray matter of the nervous system. Neuron processes running through the CNS form **tracts** of white matter; in the PNS they form the peripheral **nerves.**

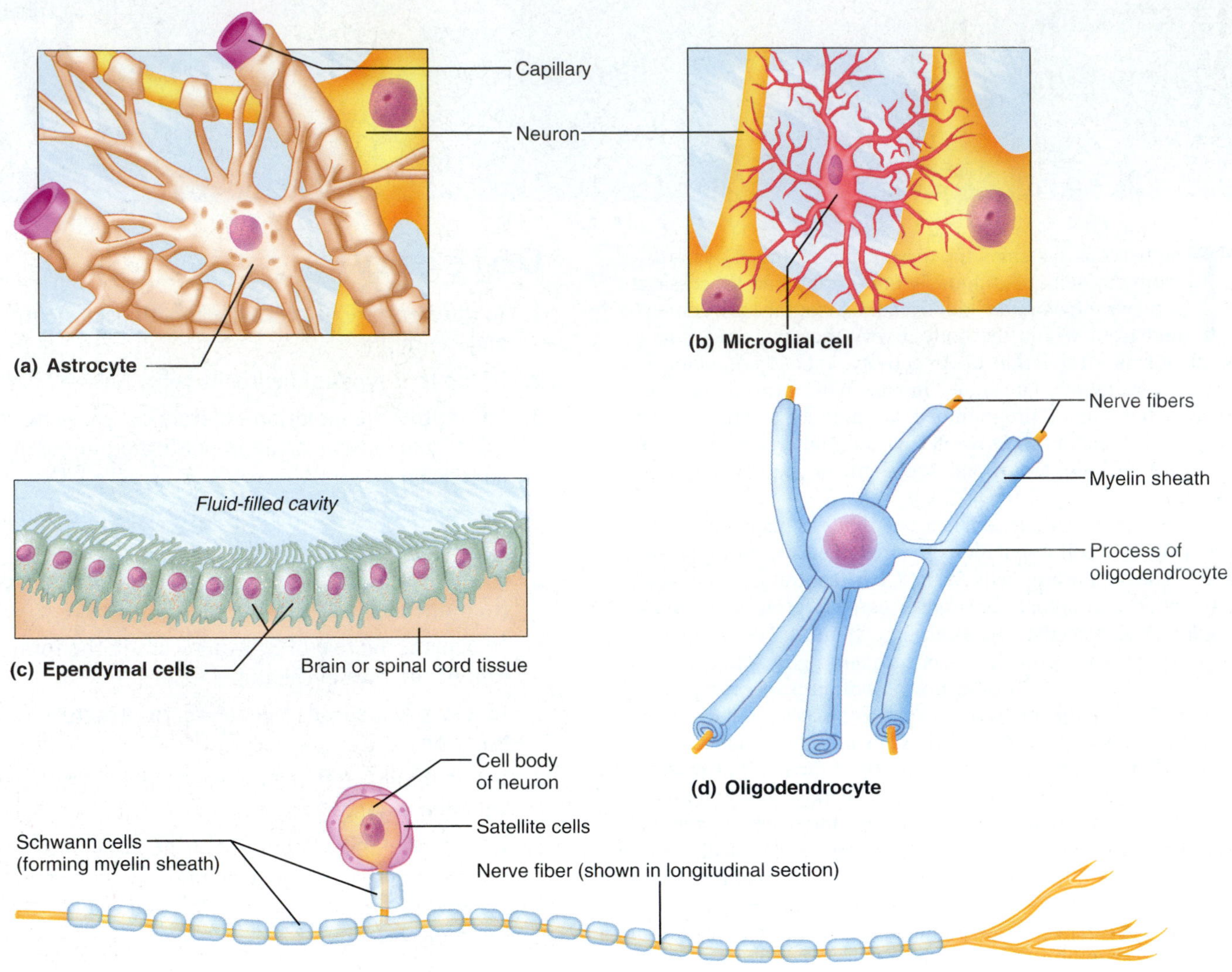

Figure 17.1 Supporting cells of nervous tissue. (a) Astrocyte. **(b)** Microglial cell. **(c)** Ependymal cells. **(d)** Oligodendrocyte. **(e)** Neuron with Schwann cells and satellite cells.

The neuron cell body contains a large round nucleus surrounded by cytoplasm (*neuroplasm*). The cytoplasm is riddled with neurofibrils and with darkly staining structures called Nissl bodies. **Neurofibrils,** the cytoskeletal elements of the neuron, have a support and intracellular transport function. **Nissl** (chromatophilic) **bodies,** an elaborate type of rough endoplasmic reticulum, are involved in the metabolic activities of the cell.

According to the older, traditional scheme, neuron processes that conduct electrical currents *toward* the cell body are called **dendrites;** and those that carry impulses *away from* the nerve cell body are called **axons.** When it was discovered that this functional scheme had pitfalls (some axons carry impulses *both* toward and away from the cell body), a newer functional definition of neuron processes was adopted. According to this scheme, dendrites are *receptive regions* (they bear receptors for neurotransmitters released by other neurons), whereas axons are *nerve impulse generators* and *transmitters.* Neurons have only one axon (which may branch into **collaterals**) but may have many dendrites, depending on the neuron type. Notice that the term *nerve fiber* is a synonym for axon and is thus quite specific.

In general, a neuron is excited by other neurons when their axons release neurotransmitters close to its dendrites or cell body. The electrical current produced travels across the cell body and (given a threshold stimulus) down the axon. As Figure 17.2a shows, the axon (in motor neurons) begins at a slightly enlarged cell body structure called the **axon hillock** and ends in many small structures called **axonal terminals,** or synaptic knobs. These terminals store the neurotransmitter chemical in tiny vesicles. Each axonal terminal is separated from the cell body or dendrites of the next (postsynaptic) neuron by a tiny gap called the **synaptic cleft** (Figure 17.2b). Thus, although they are close, there is no actual physical contact between neurons. When an impulse reaches the axonal terminals, some of the synaptic vesicles rupture and release

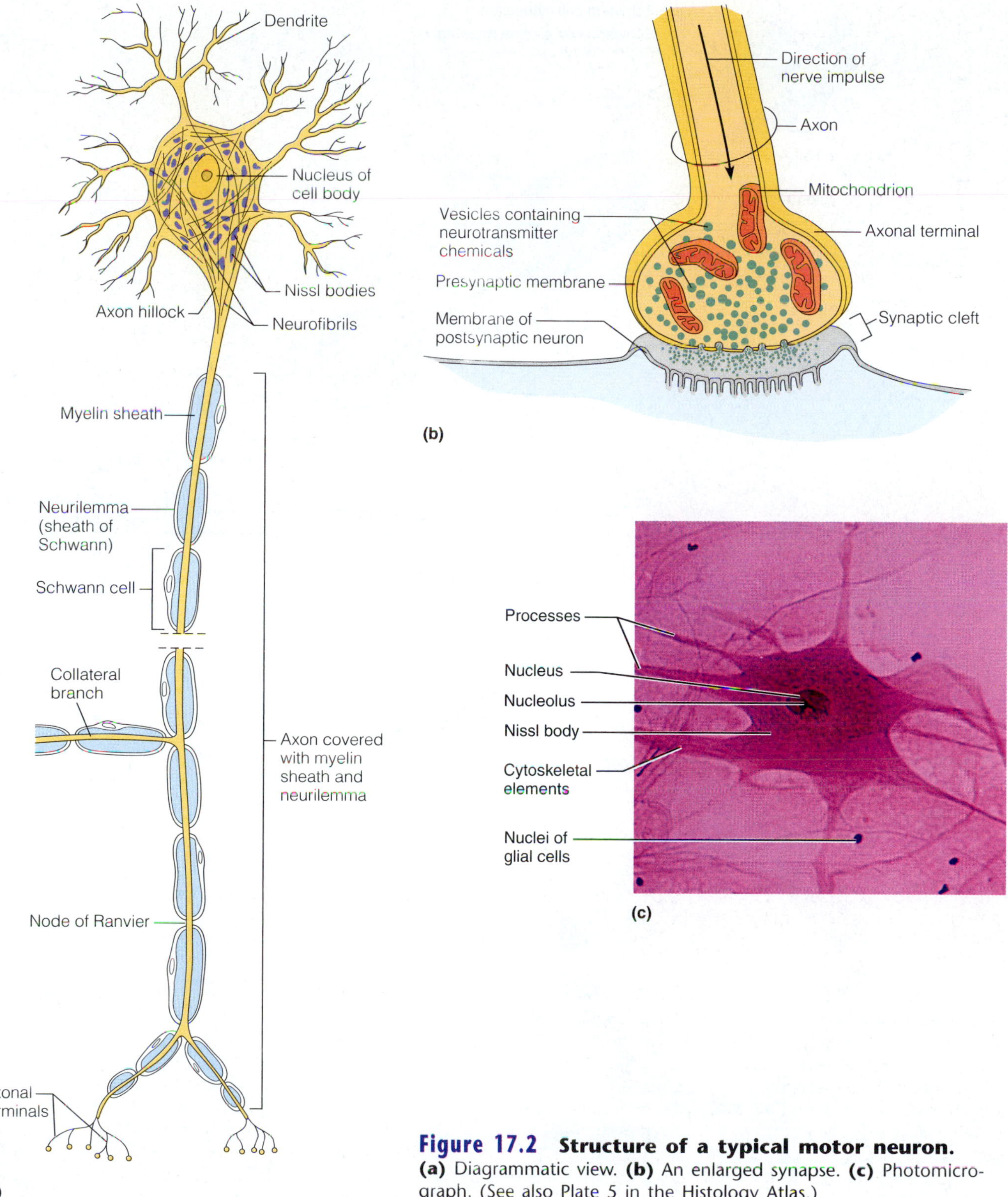

Figure 17.2 Structure of a typical motor neuron. **(a)** Diagrammatic view. **(b)** An enlarged synapse. **(c)** Photomicrograph. (See also Plate 5 in the Histology Atlas.)

neurotransmitter into the synaptic cleft. The neurotransmitter then diffuses across the synaptic cleft to bind to membrane receptors on the next neuron, initiating the action potential.*

* Specialized synapses in skeletal muscle are called neuromuscular junctions. They are discussed in Exercise 14.

Most long nerve fibers are covered with a fatty material called *myelin*, and such fibers are referred to as **myelinated fibers.** Axons in the peripheral nervous system are typically heavily myelinated by special supporting cells called **Schwann cells,** which wrap themselves tightly around the

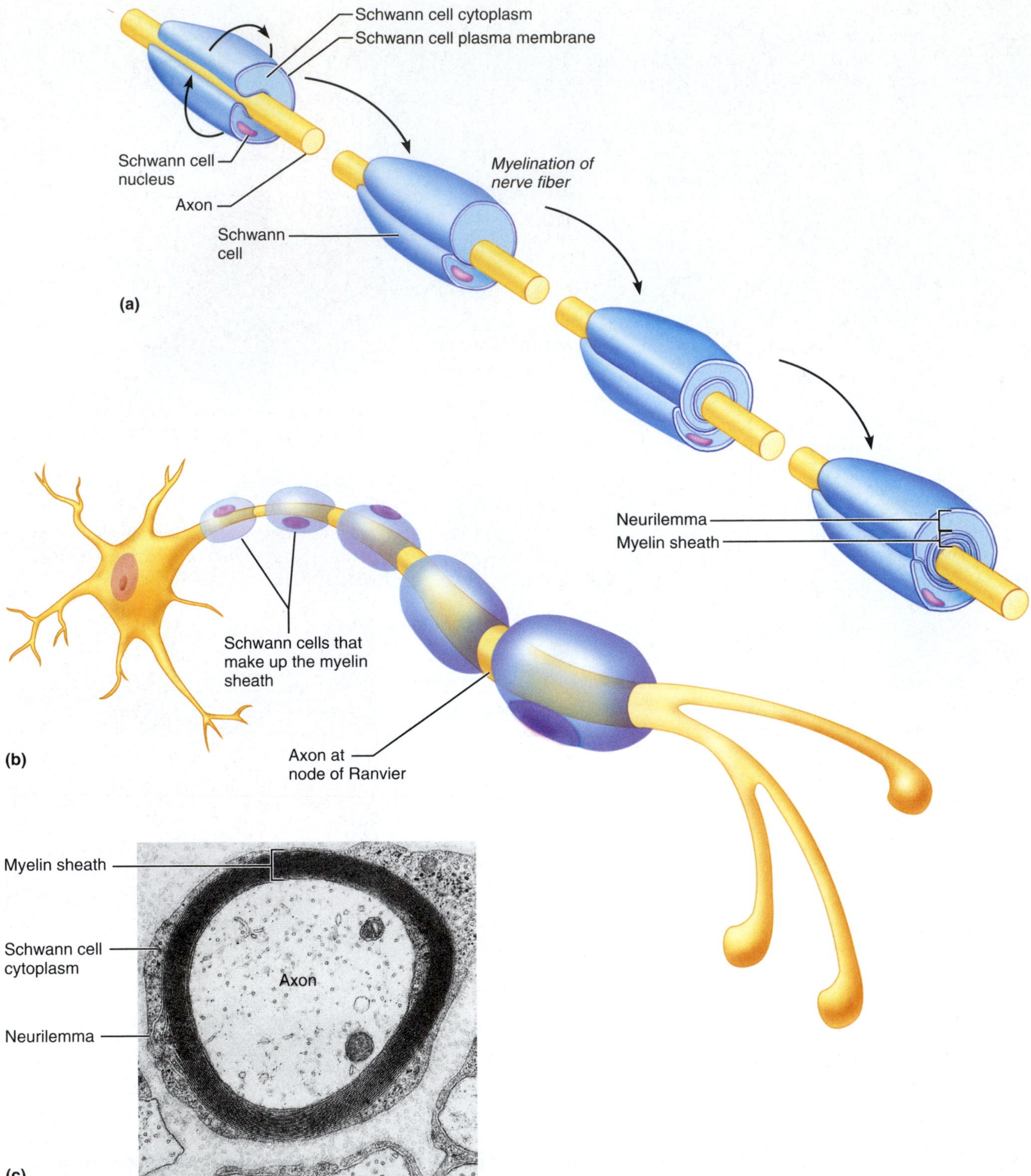

Figure 17.3 Myelination of neuron processes by individual Schwann cells. (a) A Schwann cell becomes apposed to an axon and envelops it in a trough. It then begins to rotate around the axon, wrapping it loosely in successive layers of its plasma membrane. Eventually, the Schwann cell cytoplasm is forced from between the membranes and comes to lie peripherally just beneath the exposed portion of the Schwann cell membrane. The tight membrane wrappings surrounding the axon form the myelin sheath. The area of Schwann cell cytoplasm and its exposed membrane are referred to as the neurilemma or sheath of Schwann. **(b)** Longitudinal view of myelinated axon showing portions of adjacent Schwann cells and the node of Ranvier between them. **(c)** Electron micrograph of cross section through a myelinated axon (20,000×).

axon jelly-roll fashion (Figure 17.3). During the wrapping process, the cytoplasm is squeezed from between adjacent layers of the Schwann cell membranes, so that when the process is completed a tight core of plasma membrane material (protein-lipoid material) encompasses the axon. This wrapping is the **myelin sheath.** The Schwann cell nucleus and the bulk of its cytoplasm ends up just beneath the outermost portion of its plasma membrane. This peripheral part of the Schwann cell and its exposed plasma membrane is referred to as the **neurilemma** or *sheath of Schwann*. Since the myelin sheath is formed by many individual Schwann cells, it is a discontinuous sheath. The gaps or indentations in the sheath are called **nodes of Ranvier** or **neurofibril nodes** (see Figure 17.2 and 17.3.).

Within the CNS, myelination is accomplished by glial cells called **oligodendrocytes** (see Figure 17.1d). These CNS sheaths do not exhibit the neurilemma seen in fibers myelinated by Schwann cells. Because of its chemical composition, myelin insulates the fibers and greatly increases the speed of neurotransmission by neuron fibers.

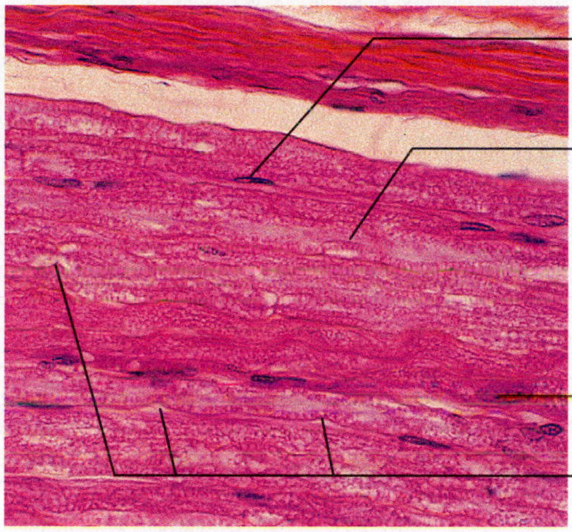

Figure 17.4 Photomicrograph of a small portion of a peripheral nerve in longitudinal section.

Activity 1:

Identifying Parts of a Neuron

1. Study the typical motor neuron shown in Figure 17.2, noting the structural details described above, and then identify these structures on a neuron model.

2. Obtain a prepared slide of the ox spinal cord smear, which has large, easily identifiable neurons. Study one representative neuron under oil immersion and identify the cell body; the nucleus; the large, prominent "owl's eye" nucleolus; and the granular Nissl bodies. If possible, distinguish the axon from the many dendrites.

Sketch the cell in the space provided here, and label the important anatomical details you have observed. Compare your sketch to Plate 5 of the Histology Atlas. Also reexamine Figure 17.2a, which differentiates the neuronal processes more clearly.

3. Obtain a prepared slide of teased myelinated nerve fibers. Using Figure 17.4 as a guide, identify the following: nodes of Ranvier, neurilemma, axis cylinder (the axon itself), Schwann cell nuclei, and myelin sheath.

Sketch a portion of a myelinated nerve fiber in the space provided here, illustrating two or three nodes of Ranvier. Label the axon, myelin sheath, nodes, and neurilemma.

Do the nodes seem to occur at consistent intervals, or are they irregularly distributed?

____

Explain the significance of this finding: ________________

____

____

____

____ ■

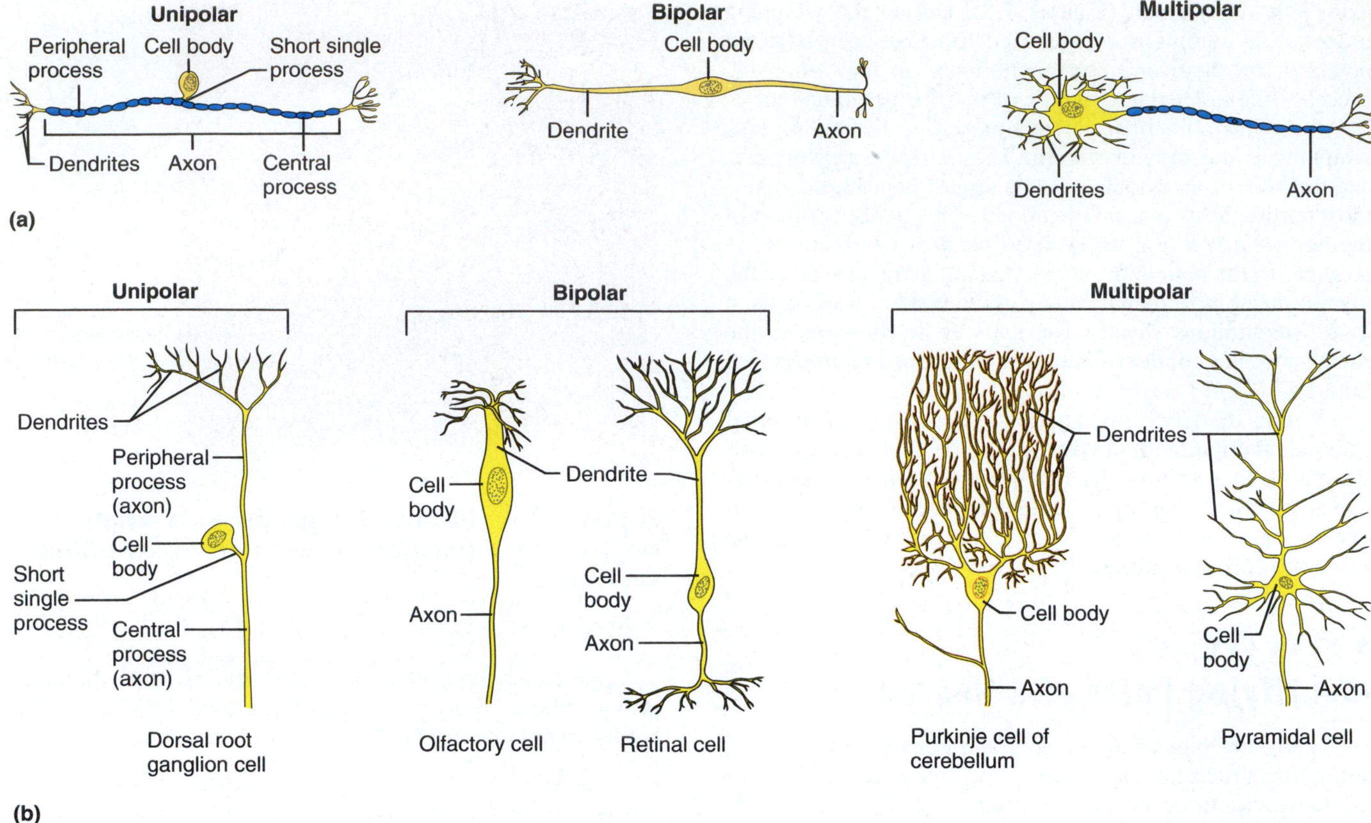

Figure 17.5 Classification of neurons according to structure. (a) Classification of neurons based on structure (number of processes extending from the cell body). **(b)** Structural variations within the classes.

Neuron Classification

Neurons may be classified on the basis of structure or of function.

Classification by Structure

Structurally, neurons may be differentiated according to the number of processes attached to the cell body (Figure 17.5a). In **unipolar neurons,** one very short process, which divides into *peripheral* and *central processes,* extends from the cell body. Functionally, only the most distal portions of the peripheral process act as dendrites; the rest acts as an axon along with the central process. Nearly all neurons that conduct impulses toward the CNS are unipolar.

Bipolar neurons have two processes—one axon and one dendrite—attached to the cell body. This neuron type is quite rare, typically found only as part of the receptor apparatus of the eye, ear, and olfactory mucosa.

Many processes issue from the cell body of **multipolar neurons,** all classified as dendrites except for a single axon. Most neurons in the brain and spinal cord (CNS neurons) and those whose axons carry impulses away from the CNS fall into this last category.

Activity 2: Studying the Microscopic Structure of Selected Neurons

Obtain prepared slides of Purkinje cells of the cerebellar cortex, pyramidal cells of the cerebral cortex, and a dorsal root ganglion. As you observe them under the microscope, try to pick out the anatomical details depicted in Figure 17.5b, and in Plates 6 and 7 of the Histology Atlas. Notice that the neurons of the cerebral and cerebellar tissues (both brain tissues) are extensively branched; in contrast, the neurons of the dorsal root ganglion are more rounded. You may also be able to identify astrocytes (a type of neuroglia) in the brain tissue slides if you examine them closely, and satellite cells can be seen surrounding the neurons in the dorsal root ganglion.

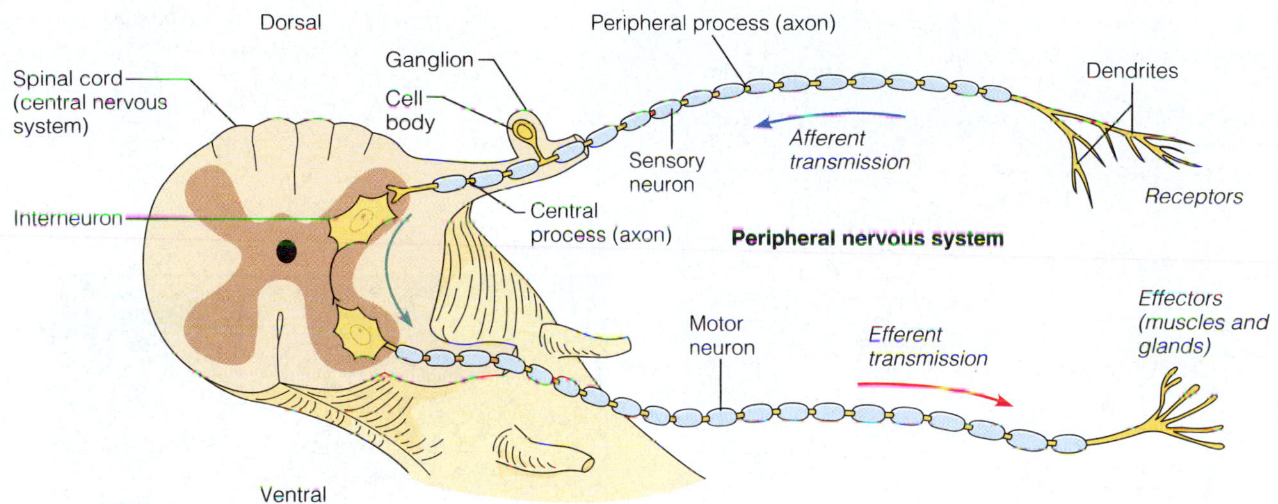

Figure 17.6 Classification of neurons on the basis of function. Sensory (afferent) neurons conduct impulses from the body's sensory receptors to the central nervous system; most are unipolar neurons with their nerve cell bodies in ganglia in the peripheral nervous system (PNS). Motor (efferent) neurons transmit impulses from the CNS to effectors such as muscles and glands. Association neurons (interneurons) complete the communication line between sensory and motor neurons. They are typically multipolar and their cell bodies reside in the CNS.

Which of these neuron types would be classified as multipolar neurons?

____

Which as unipolar?

____ ■

Classification by Function

In general, neurons carrying impulses from sensory receptors in the internal organs (viscera), the skin, skeletal muscles, joints, or special sensory organs are termed **sensory,** or **afferent, neurons** (see Figure 17.6). The dendritic endings of sensory neurons are often equipped with specialized receptors that are stimulated by specific changes in their immediate environment. The structure and function of these receptors is considered separately in Exercise 23 (General Sensation). The cell bodies of sensory neurons are always found in a ganglion outside the CNS, and these neurons are typically unipolar.

Neurons carrying activating impulses from the CNS to the viscera and/or body muscles and glands are termed **motor,** or **efferent, neurons.** Motor neurons are most often multipolar and their cell bodies are almost always located in the CNS.

The third functional category of neurons is the **association neurons,** or **interneurons,** which are situated between and contribute to pathways that connect sensory and motor neurons. Their cell bodies are always located within the CNS and they are multipolar neurons structurally.

Structure of a Nerve

A nerve is a bundle of neuron fibers or processes wrapped in connective tissue coverings that extends to and/or from the CNS and visceral organs or structures of the body periphery (such as skeletal muscles, glands, and skin).

Like neurons, nerves are classified according to the direction in which they transmit impulses. Nerves carrying both sensory (afferent) and motor (efferent) fibers are called **mixed nerves;** all spinal nerves are mixed nerves. Nerves that carry only sensory processes and conduct impulses only toward the CNS are referred to as **sensory,** or **afferent, nerves.** A few of the cranial nerves are pure sensory nerves, but the majority are mixed nerves. The ventral roots of the spinal cord, which carry only motor fibers, can be considered **motor,** or **efferent, nerves.**

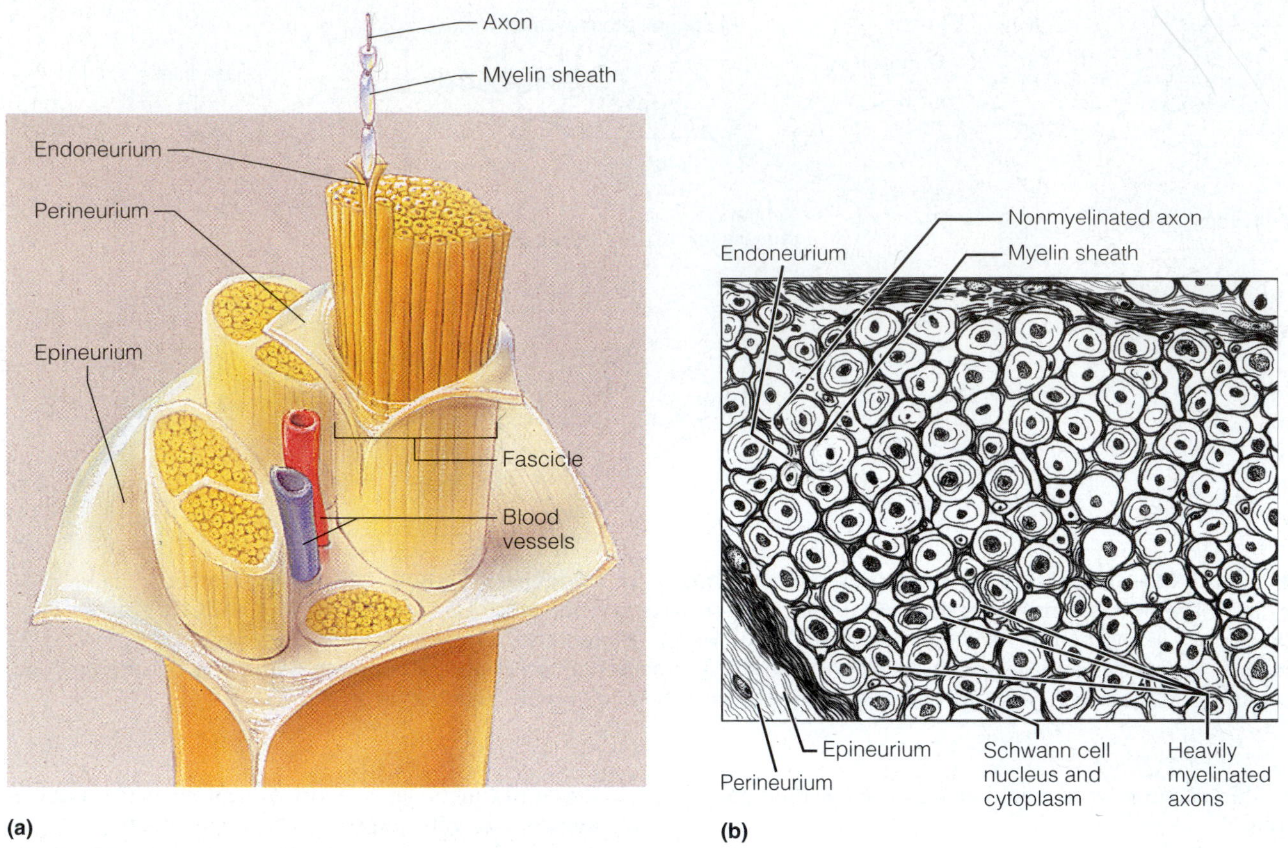

Figure 17.7 Structure of a nerve showing connective tissue wrappings. (a) Three-dimensional view of a portion of a nerve. **(b)** Cross-sectional view. (See also Plate 10 in the Histology Atlas.)

Within a nerve, each fiber is surrounded by a delicate connective tissue sheath called an **endoneurium,** which insulates it from the other neuron processes adjacent to it. The endoneurium is often mistaken for the myelin sheath; it is instead an additional sheath that surrounds the myelin sheath. Groups of fibers are bound by a coarser connective tissue, called the **perineurium,** to form bundles of fibers called **fascicles.** Finally, all the fascicles are bound together by a white, fibrous connective tissue sheath called the **epineurium,** forming the cordlike nerve (Figure 17.7). In addition to the connective tissue wrappings, blood vessels and lymphatic vessels serving the fibers also travel within a nerve.

Activity 3: Examining the Microscopic Structure of a Nerve

Examine under the compound microscope a prepared cross section of a peripheral nerve. Identify nerve fibers, myelin sheaths, fascicles, and endoneurium, perineurium, and epineurium sheaths. If desired, sketch the nerve in the space below. ■

Neurophysiology of Nerve Impulses: Wet Lab

exercise 18A

Objectives

1. To list the two major physiological properties of neurons.
2. To describe the polarized and depolarized states of the nerve cell membrane and to describe the events that lead to generation and conduction of a nerve impulse.
3. To explain how a nerve impulse is transmitted from one neuron to another.
4. To define *action potential, depolarization, repolarization, relative refractory period,* and *absolute refractory period.*
5. To list various substances and factors that can stimulate neurons.
6. To recognize that neurotransmitters may be either stimulatory or inhibitory in nature.
7. To state the site of action of the blocking agents ether and curare.

Materials

- ❑ *Rana pipiens**
- ❑ Dissecting instruments and tray
- ❑ Ringer's solution (frog) in dropper bottles, some at room temperature and some in an ice bath
- ❑ Thread
- ❑ Glass rods or probes
- ❑ Glass plates or slides
- ❑ Ring stand and clamp
- ❑ Stimulator; platinum electrodes
- ❑ Oscilloscope
- ❑ Nerve chamber
- ❑ Filter paper
- ❑ 0.01% hydrochloric acid (HCl) solution
- ❑ Sodium chloride (NaCl) crystals
- ❑ Heat-resistant mitts
- ❑ Bunsen burner
- ❑ Absorbent cotton

*Instructor to provide freshly pithed frogs (*Rana pipiens*) for student experimentation.

- ❑ Ether
- ❑ Pipettes
- ❑ 1-cc syringe with small-gauge needle
- ❑ 0.5% tubocurarine solution
- ❑ Frog board
- ❑ Disposable plastic gloves
- ❑ Safety goggles

PhysioEx™ 4.0 Computer Simulation on p. P-28

The Nerve Impulse

Neurons have two major physiological properties: **excitability,** or the ability to respond to stimuli and convert them into nerve impulses, and **conductivity,** the ability to transmit the impulse to other neurons, muscles, or glands. In a resting neuron (as in resting muscle cells), the exterior surface of the membrane is slightly more positively charged than the inner surface, as shown in Figure 18.1a. This difference in electrical charge on the two sides of the membrane results in a voltage across the plasma membrane referred to as the **resting membrane potential,** and a neuron in this state is said to be **polarized.** In the resting state, the predominant intracellular ion is potassium (K^+), and sodium ions (Na^+) are found in greater concentration in the extracellular fluids. The resting potential is maintained by a very active sodium-potassium pump, which transports Na^+ out of the cell and K^+ into the cell.

When the neuron is activated by a stimulus of adequate intensity—a **threshold stimulus**—the membrane at its *trigger zone,* typically the axon hillock (or the most peripheral part of a sensory neuron's axon), briefly becomes more permeable to sodium (sodium gates are opened). Sodium ions rush into the cell, increasing the number of positive ions inside the cell and reversing the polarity (Figure 18.1b). Thus the interior of the membrane becomes less negative at that point and the exterior surface becomes less positive—a phenomenon called **depolarization.** When depolarization reaches a certain point such that the local membrane polarity changes (momentarily the external face becomes negative and the internal face becomes positive), it initiates an **action potential**[†] (Figure 18.1c).

[†] If the stimulus is of less than threshold intensity, depolarization is limited to a small area of the membrane, and no action potential is generated.

Within a millisecond after the inward influx of sodium, the membrane permeability is again altered. As a result, Na^+ permeability decreases, K^+ permeability increases, and K^+ rushes out of the cell. Since K^+ ions are positively charged, their movement out of the cell reverses the membrane potential again, so that the external membrane surface is again positive relative to the internal membrane face (Figure 18.1d). This event, called **repolarization,** reestablishes the resting membrane potential. When the sodium gates are open, the neuron is totally insensitive to additional stimuli and is said to be in an **absolute refractory period.** During the time of repolarization, the neuron is nearly insensitive to further stimulation. A very strong stimulus may reactivate it, however; thus this period is referred to as the **relative refractory period.**

Once generated, the action potential is a self-propagating phenomenon that spreads rapidly along the entire length of the neuron. It is never partially transmitted; that is, it is an all-or-none response. This propagation of the action potential in neurons is also called the **nerve impulse.** When the nerve impulse reaches the axonal terminals, they release a neurotransmitter that acts either to stimulate or to inhibit the next neuron in the transmission chain. (Note that only stimulatory transmitters are considered here.)

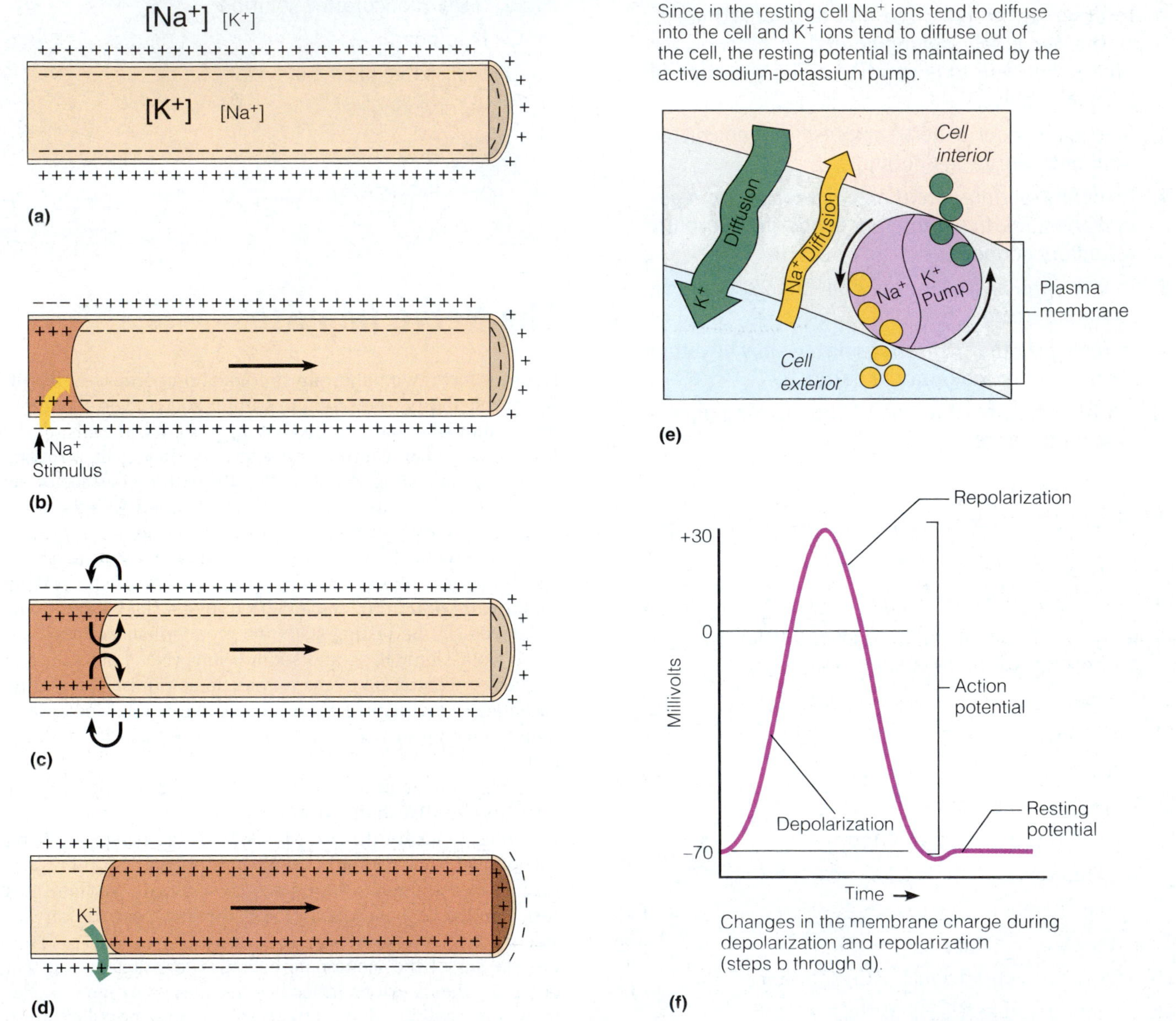

Figure 18.1 The nerve impulse. (a) Resting membrane potential (−85 mV). There is an excess of positive ions outside the cell, with Na^+ the predominant extracellular fluid ion and K^+ the predominant intracellular ion. The plasma membrane has a low permeability to Na^+. **(b)** Depolarization—reversal of the resting potential. Application of a stimulus changes the membrane permeability, and Na^+ ions are allowed to diffuse rapidly into the cell. **(c)** Generation of the action potential or nerve impulse. If the stimulus is of adequate intensity, the depolarization wave spreads rapidly along the entire length of the membrane. **(d)** Repolarization—reestablishment of the resting potential. The negative charge on the internal plasma membrane surface and the positive charge on its external surface are reestablished by diffusion of K^+ ions out of the cell, proceeding in the same direction as in depolarization. **(e)** The original ionic concentrations of the resting state are restored by the sodium-potassium pump. **(f)** A tracing of an action potential.

Because only minute amounts of sodium and potassium ions change places, once repolarization has been completed, the neuron can quickly respond again to a stimulus. In fact, thousands of impulses can be generated before ionic imbalances prevent the neuron from transmitting impulses. Eventually, however, it is necessary to restore the original ionic concentrations on the two sides of the membrane; this is accomplished by enhanced activity of the Na^+-K^+ pumps (Figure 18.1e).

Physiology of Nerve Fibers

In this laboratory session, you will investigate the functioning of nerve fibers by subjecting the sciatic nerve of a frog to various types of stimuli and blocking agents. Work in groups of two to four to lighten the work load.

Dissection: Isolating the Gastrocnemius Muscle and Sciatic Nerve

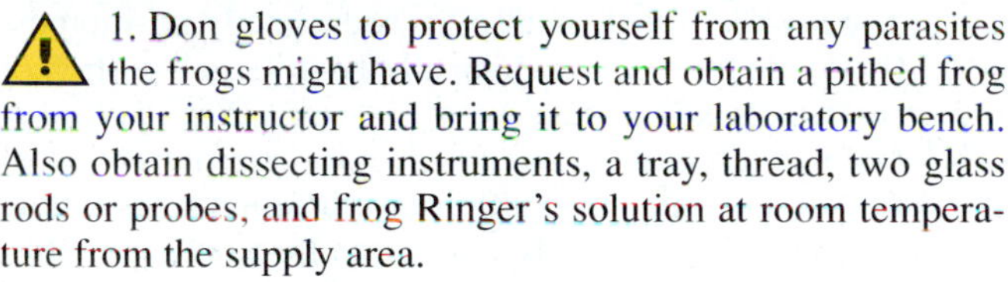

1. Don gloves to protect yourself from any parasites the frogs might have. Request and obtain a pithed frog from your instructor and bring it to your laboratory bench. Also obtain dissecting instruments, a tray, thread, two glass rods or probes, and frog Ringer's solution at room temperature from the supply area.

2. Prepare the sciatic nerve as illustrated in Figure 18.2. Place the pithed frog on the dissecting tray, dorsal side down. Make a cut through the skin around the circumference of the frog approximately halfway down the trunk, and then pull the skin down over the muscles of the legs. Open the abdominal cavity and push the abdominal organs to one side to expose the origin of the glistening white sciatic nerve, which arises from the last three spinal nerves. *Once the sciatic nerve has been exposed, it should be kept continually moist with room temperature Ringer's solution.*

3. Using a glass probe, slip a piece of thread moistened with Ringer's solution under the sciatic nerve close to its origin at the vertebral column. Make a single ligature (tie it firmly with the thread), and then cut through the nerve roots to free the proximal end of the sciatic nerve from its attachments. Using a glass rod or probe, carefully separate the posterior thigh muscles to locate and then free the sciatic nerve, which runs down the posterior aspect of the thigh.

4. Tie a piece of thread around the Achilles tendon of the gastrocnemius muscle, and then cut through the tendon distal to the ligature to free the gastrocnemius muscle from the heel. Using a scalpel, carefully release the gastrocnemius muscle from the connective tissue in the knee region. At this point you should have completely freed both the gastrocnemius muscle and the sciatic nerve, which innervates it. ■

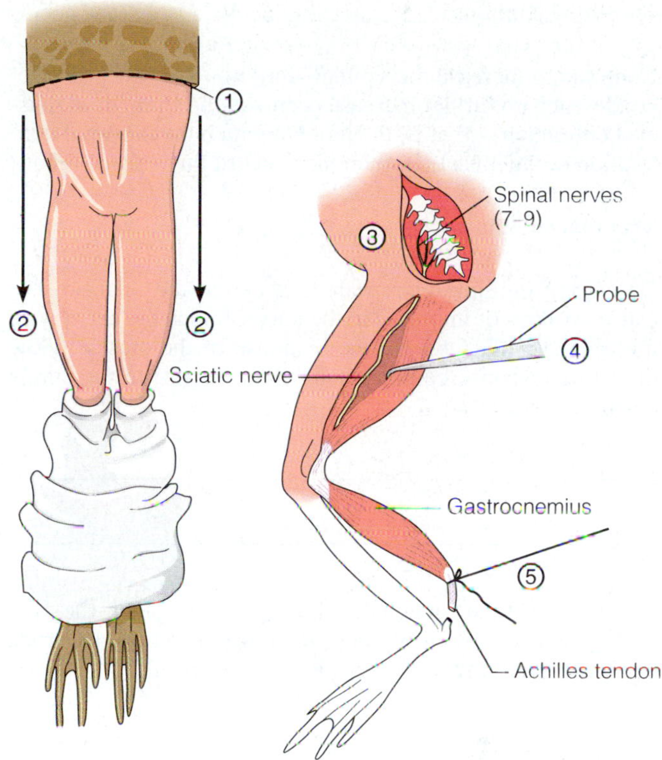

Figure 18.2 Removal of the sciatic nerve and gastrocnemius muscle. (1) Cut through the frog's skin around the circumference of the trunk. **(2)** Pull the skin down over the trunk and legs. **(3)** Make a longitudinal cut through the abdominal musculature and expose the roots of the sciatic nerve (arising from spinal nerves 7–9). Ligate the nerve and cut the roots proximal to the ligature. **(4)** Use a glass probe to expose the sciatic nerve beneath the posterior thigh muscles. **(5)** Ligate the Achilles tendon and cut it free distal to the ligature. Release the gastrocnemius muscle from the connective tissue of the knee region.

Activity 1: Eliciting a Nerve Impulse

In this first set of experiments, stimulation of the nerve and generation of the action potential will be indicated by contraction of the gastrocnemius muscle. Because you will make no mechanical recording (unless your instructor asks you to), you must keep complete and accurate records of all experimental procedures and results.

1. Obtain a glass slide or plate, ring stand and clamp, stimulator, electrodes, salt, Bunsen burner, and heat-resistant mitts. With glass rods, transfer the isolated muscle-nerve preparation to a glass plate or slide, and then attach the slide to a ring stand with a clamp. Allow the end of the sciatic nerve to hang over the free edge of the glass slide, so that it is easily accessible for stimulation. *Remember to keep the nerve moist at all times.*

2. You are now ready to investigate the response of the sciatic nerve to various stimuli, beginning with electrical stimulation. Using the stimulator and platinum electrodes, stimulate the sciatic nerve with single shocks, gradually increasing the intensity of the stimulus until the threshold stimulus is determined.

(The muscle as a whole will just barely contract at the threshold stimulus.) Record the voltage of this stimulus:

Threshold stimulus: _____________ V

Continue to increase the voltage until you find the point beyond which no further increase occurs in the strength of muscle contraction—that is, the point at which the maximal contraction of the muscle is obtained. Record this voltage below.

Maximal stimulus: _____________ V

Delivering multiple or repeated shocks to the sciatic nerve causes volleys of impulses in the nerve. Shock the nerve with multiple stimuli. Observe the response of the muscle. How does this response compare with the response to the single electrical shocks?

____

____

3. To investigate mechanical stimulation, pinch the free end of the nerve by firmly pressing it between two glass rods or by pinching it with forceps. What is the result?

____

4. Chemical stimulation can be tested by applying a small piece of filter paper saturated with 0.01% hydrochloric acid (HCl) solution (from the supply area) to the free end of the nerve. What is the result?

____

Drop a few grains of salt (NaCl) on the free end of the nerve. What is the result?

____

5. Now test thermal stimulation. Wearing the heat-resistant mitts, heat a glass rod for a few moments over a Bunsen burner. Then touch the rod to the free end of the nerve. What is the result?

____

What do these muscle reactions say about the irritability and conductivity of neurons?

____

____ ■

Although most neurons within the body are stimulated to the greatest degree by a particular stimulus (in many cases, a chemical neurotransmitter), a variety of other stimuli may trigger nerve impulses, as illustrated by the experimental series just conducted. Generally, no matter what type of stimulus is present, if the affected part responds by becoming activated, it will always react in the same way. Familiar examples are the well-known phenomenon of "seeing stars" when you receive a blow to the head or press on your eyeball (try it), both of which trigger impulses in your optic nerves.

Activity 2: Inhibiting the Nerve Impulse

Numerous physical factors and chemical agents can impair the ability of nerve fibers to function. For example, deep pressure and cold temperature both block nerve impulse transmission by preventing the local blood supply from reaching the nerve fibers. Local anesthetics, alcohol, and numerous other chemicals are also very effective at blocking nerve transmission. Ether, one such chemical blocking agent, will be investigated first.

⚠ Since ether is extremely volatile and explosive, perform this experiment in a vented hood. *Don safety glasses before beginning this procedure.*

1. Obtain another glass slide or plate, absorbent cotton, ether, and a pipette. Clamp the new glass slide to the ring stand slightly below the first slide of the apparatus setup for the previous experiment. With glass rods, gently position the sciatic nerve on this second slide, allowing a small portion of the nerve's distal end to extend over the edge. Place a piece of absorbent cotton soaked with ether under the midsection of the nerve on the slide, prodding it into position with a glass rod. Using a voltage slightly above the threshold stimulus, stimulate the distal end of the nerve at 2-min intervals until the muscle fails to respond. (If the cotton dries before this, rewet it with ether using a pipette.) How long did it take for anesthesia to occur?

_____________ sec

2. Once anesthesia has occurred, stimulate the nerve beyond the anesthetized area, between the ether-soaked pad and the muscle. What is the result?

____

3. Remove the ether-soaked pad and flush the nerve fibers with saline. Again stimulate the nerve at its distal end at 2-min intervals. How long does it take for recovery?

____

Does ether exert its blocking effect on the nerve fibers *or* on the muscle cells?

_________________________ Explain your reasoning.

____

____

If sufficient frogs are available and time allows, you may do the following classic experiment. In the 1800s Claude

Bernard described an investigation into the effect of curare on nerve-muscle interaction. *Curare* was used by some South American Indian tribes to tip their arrows. Victims struck with these arrows were paralyzed, but the paralysis was not accompanied by loss of sensation.

1. Prepare another frog as described in steps 1 through 3 of the dissection instructions. However, in this case position the frog ventral side down on a frog board. In exposing the sciatic nerve, take care not to damage the blood vessels in the thigh region, as the success of the experiment depends on maintaining the blood supply to the muscles of the leg.

2. Expose and gently tie the left sciatic nerve so that it can be lifted away from the muscles of the leg for stimulation. Slip another length of thread under the nerve, and then tie the thread tightly around the thigh muscles to cut off circulation to the leg. The sciatic nerve should be above the thread and *not in* the ligated tissue. Expose and ligate the sciatic nerve of the *right* leg in the same manner, but this time do *not* ligate the thigh muscles.

3. Obtain a syringe and needle, and a vial of 0.5% tubocurarine. Slowly and carefully inject 1 cc of the tubocurarine into the dorsal lymph sac of the frog.* The dorsal lymph sacs are located dorsally at the level of the scapulae, so introduce the needle of the syringe just beneath the skin between the scapulae and toward one side of the spinal column. *Handle the tubocurarine very carefully, because it is extremely poisonous.* Do not get any on your skin.

4. Wait 15 min after injection of the tubocurarine to allow it to be distributed throughout the body in the blood and lymphatic stream. Then stimulate electrically the left sciatic nerve. Be careful not to touch any of the other tissues with the electrode. Gradually increasing the voltage, deliver single shocks until the threshold stimulus is determined for this specimen.

Threshold stimulus: ______________ V

Now stimulate the right sciatic nerve with the same voltage intensity. Is there any difference in the reaction of the two muscles?

_______ If so, explain. ____________________________

____

____

If you did not find any difference, wait an additional 10 to 15 min and restimulate both sciatic nerves.

What is the result? ______________________________

____

* To obtain 1 cc of the tubocurarine, inject 1 cc of air into the vial through the rubber membrane, and then draw up 1 cc of the chemical into the syringe.

5. To determine the site at which tubocurarine acts, directly stimulate each gastrocnemius muscle. What is the result?

____

____

____

Explain the difference between the responses of the right and left sciatic nerves.

____

____

____

Explain the results when the muscles were stimulated directly.

____

____

____

At what site does tubocurarine (or curare) act?

____ ■

The Oscilloscope: An Experimental Tool

The *oscilloscope* is an instrument that visually displays the rapid but extremely minute changes in voltage that occur during an action potential. The oscilloscope is similar to a TV set in that the screen display is produced by a stream of electrons generated by an electron gun (cathode) at the rear of a tube. The electrons pass through the tube and between two sets of plates that lie alongside the beam pathway. When the electrons reach the fluorescent screen, they create a tiny glowing spot. The plates determine the placement of the glowing spot by controlling the vertical or horizontal sweep of the beam. Vertical movement represents the voltage of the input signal, and horizontal movement indicates the time base. When there is no electrical output signal to the oscilloscope, the electron beam sweeps horizontally (left to right) across the screen, but when the plates are electrically stimulated, the path of electrons is deflected vertically.

In this exercise, a frog's sciatic nerve will be electrically stimulated, and the action potentials generated will be observed on the oscilloscope. The dissected nerve will be placed in contact with two pairs of electrodes—*stimulating* and *recording*. The stimulating electrodes will be used to deliver a pulse of electricity to a point on the sciatic nerve. At another point on the nerve, a pair of recording electrodes connected to the oscilloscope will deliver the current to the plates

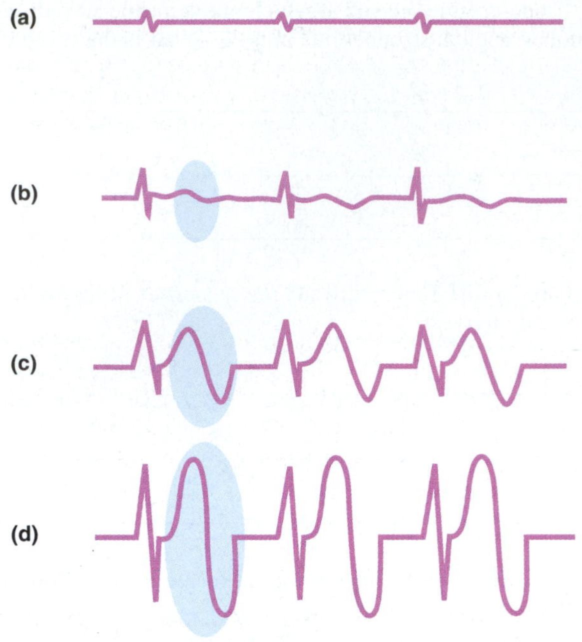

Figure 18.3 Oscilloscope scans of nerve stimulation using stimuli with increasing intensities. The first action potential in each scan is circled. **(a)** Stimulus artifacts only; no action potential produced. Subthreshold stimulation. **(b)** Threshold stimulation. **(c)** Submaximal stimulation. **(d)** Maximal stimulus.

inside the tube, and the electrical pulse will be recorded on the screen as a vertical deflection, or a *stimulus artifact* (Figure 18.3a). As the nerve is stimulated with increasingly higher voltage, the stimulus artifact increases in amplitude as well. When the stimulus voltage reaches a high enough level (threshold), an action potential will be generated by the nerve, and a *second* vertical deflection will appear on the screen, approximately 2 milliseconds after the stimulus artifact (Figure 18.3b–d). This second deflection reports the potential difference between the two recording electrodes—that is, between the first recording electrode which has already depolarized (and is in the process of repolarizing as the action potential travels along the nerve) and the second recording electrode.

Activity 3: Visualizing the Action Potential with an Oscilloscope

1. Obtain a nerve chamber, an oscilloscope, a stimulator, frog Ringer's solution (room temperature), a dissecting needle, and glass probes. Set up the experimental apparatus as illustrated in Figure 18.4. Connect the two stimulating electrodes to the output terminals of the stimulator and the two recording electrodes to the preamplifier of the oscilloscope.

2. Obtain another pithed frog, and prepare one of its sciatic nerves for experimentation as indicated on p. 187 in steps 1 through 3 in the dissection instructions. While working, be careful not to touch the nerve with your fingers, and do not allow the nerve to touch the frog's skin.

3. When you have freed the sciatic nerve to the knee region with the glass probe, slip another thread length beneath that end of the nerve and make a ligature. Cut the nerve distal to this tied thread and then carefully lift the cut nerve away from the thigh of the frog by holding the threads at the nerve's proximal and distal ends. Place the nerve in the nerve chamber so that it rests across all four electrodes (the two stimulating and two recording electrodes) as shown in Figure 18.4. Flush the nerve with room temperature frog Ringer's solution.

4. Adjust the horizontal sweep according to the instructions given in the manual or by your instructor, and set the stimulator duration, frequency, and amplitude to their lowest settings.

5. Begin to stimulate the nerve with single stimuli, slowly increasing the voltage until a threshold stimulus is achieved. The action potential will appear as a small rounded "hump"

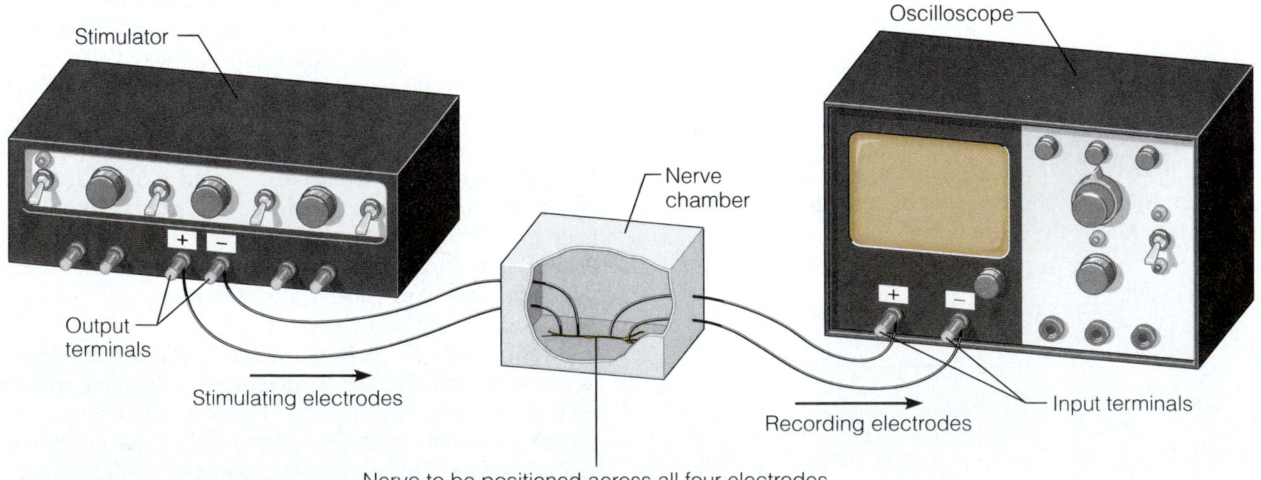

Figure 18.4 Setup for oscilloscope visualization of action potentials in a nerve.

immediately following the stimulus artifact. Record the voltage of the threshold stimulus:

Threshold stimulus: ______________ V

6. Flush the nerve with the Ringer's solution and continue to increase the voltage, watching as the vertical deflections produced by the action potentials become diphasic (show both upward and downward vertical deflections). Record the voltage at which the action potential reaches its maximal amplitude; this is the maximal stimulus:

Maximal stimulus: ______________ V

7. Set the stimulus voltage at a level just slightly lower than the maximal stimulus and gradually increase the frequency of stimulation. What is the effect on the size (amplitude) of the action potential?

____

8. Flush the nerve with saline once again, and allow it to sit for a few minutes while you obtain a bottle of Ringer's solution from the ice bath. Repeat steps 5 and 6 while your partner continues to flush the nerve preparation with the cold saline. Record the threshold and maximal stimuli and watch the oscilloscope pattern carefully to detect any differences in the velocity or speed of conduction from what was seen previously.

Threshold stimulus: ______________ V

Maximal stimulus: ______________ V

9. Flush the nerve preparation with room temperature Ringer's solution again and then gently lift the nerve by its attached threads and turn it around so that the end formerly resting on the stimulating electrodes now rests on the recording electrodes and vice versa. Stimulate the nerve. Is the impulse conducted in the opposite direction?

____

10. Dispose of the frog remains and gloves in the appropriate containers, and return your equipment to the proper supply area. ■

exercise 19

Gross Anatomy of the Brain and Cranial Nerves

Objectives

1. To identify the following brain structures on a dissected specimen, human brain model (or slices), or appropriate diagram, and to state their functions:
   - *Cerebral hemisphere structures:* lobes, important fissures, lateral ventricles, basal nuclei, corpus callosum, fornix, septum pellucidum
   - *Diencephalon structures:* thalamus, intermediate mass, hypothalamus, optic chiasma, pituitary gland, mammillary bodies, pineal body, choroid plexus of the third ventricle, interventricular foramen
   - *Brain stem structures:* corpora quadrigemina, cerebral aqueduct, cerebral peduncles of the midbrain, pons, medulla, fourth ventricle
   - *Cerebellum structures:* cerebellar hemispheres, vermis, arbor vitae
2. To describe the composition of gray and white matter.
3. To locate the well-recognized functional areas of the human cerebral hemispheres.
4. To define *gyri* and *fissures* (*sulci*).
5. To identify the three meningeal layers and state their function, and to locate the falx cerebri, falx cerebelli, and tentorium cerebelli.
6. To state the function of the arachnoid villi and dural sinuses.
7. To discuss the formation, circulation, and drainage of cerebrospinal fluid.
8. To identify at least four pertinent anatomical differences between the human brain and that of the sheep (or other mammal).
9. To identify the cranial nerves by number and name on an appropriate model or diagram, stating the origin and function of each.

Materials

- ❑ Human brain model (dissectible)
- ❑ 3-D model of ventricles
- ❑ Preserved human brain (if available)
- ❑ Coronally sectioned human brain slice (if available)
- ❑ Preserved sheep brain (meninges and cranial nerves intact)
- ❑ Dissecting tray and instruments
- ❑ Disposable gloves
- ❑ Materials as needed for cranial nerve testing (see Table 19.1): aromatic oils (e. g., vanilla and cloves), eye chart, ophthalmoscope, pen light, safety pin, mall probe (hot and cold), cotton, solutions of sugar, salt, vinegar, and quinine, ammonia, tuning fork, and tongue depressor
- ❑ *The Human Nervous System: The Brain and Cranial Nerves* videotape*

A1A See Appendix C, Exercise 19 for links to A.D.A.M.® Interactive Anatomy.

*Available to qualified adopters from Benjamin Cummings.

When viewed alongside all nature's animals, humans are indeed unique, and the key to their uniqueness is found in the brain. Only in humans has the brain region called the cerebrum become so elaborated and grown out of proportion that it overshadows other brain areas. Other animals are primarily concerned with informational input and response for the sake of survival and preservation of the species, but human beings devote considerable time to nonsurvival ends. They are the only animals who manipulate abstract ideas and search for knowledge for its own sake, who are capable of emotional response and artistic creativity, or who can anticipate the future and guide their lives according to ethical and moral values. For all this, humans can thank their overgrown cerebrum (cerebral hemispheres).

Each of us is a composite reflection of our brain's experience. If all past sensory input could mysteriously and suddenly be "erased," we would be unable to walk, talk, or communicate in any manner. Spontaneous movement would occur, as in a fetus, but no voluntary integrated function of any type would be possible. Clearly we would cease to be the same individuals.

Because of the complexity of the nervous system, its anatomical structures are usually considered in terms of two principal divisions: the central nervous system and the peripheral nervous system. The **central nervous system (CNS)** consists of the brain and spinal cord, which primarily interpret incoming sensory information and issue instructions based on past experience. The **peripheral nervous system (PNS)** consists of the cranial and spinal nerves, ganglia, and sensory receptors. These structures serve as communication lines as they carry impulses—from the sensory receptors to the CNS and from the CNS to the appropriate glands or muscles.

The PNS consists of two major subdivisions: the **sensory portion,** which consists of nerve fibers that conduct impulses

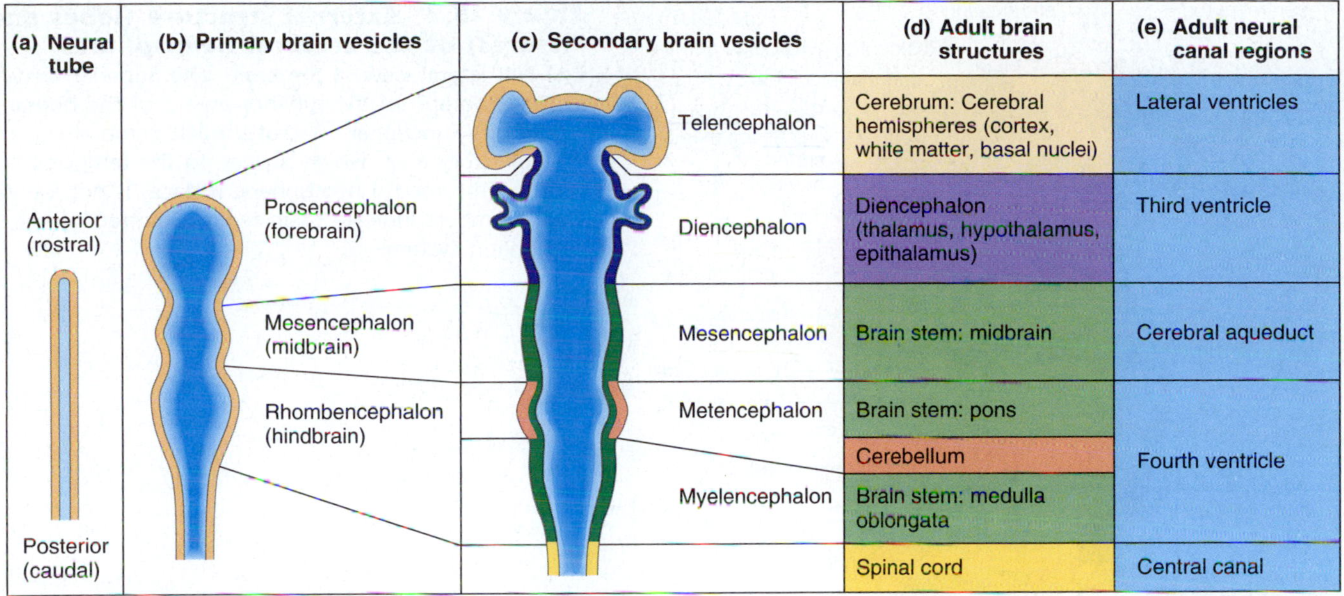

Figure 19.1 Embryonic development of the human brain. **(a)** The neural tube becomes subdivided into **(b)** the primary brain vesicles, which subsequently form **(c)** the secondary brain vesicles, which differentiate into **(d)** the adult brain structures. **(e)** The adult structures derived from the neural canal.

toward the CNS and the **motor arm,** which contains nerve fibers that conduct impulses away from the CNS. The motor arm, in turn, consists of the **somatic division,** sometimes called the **voluntary system,** which controls the skeletal muscles and the other subdivision, the **autonomic nervous system (ANS),** which controls smooth and cardiac muscles and glands. The ANS is often referred to as the involuntary nervous system. Its sympathetic and parasympathetic branches innervate smooth muscle, cardiac muscle, and glands, and play a major role in maintaining homeostasis.

In this exercise both CNS (brain) and PNS (cranial nerves) structures will be studied because of their close anatomical relationship.

The Human Brain

During embryonic development of all vertebrates, the CNS first makes its appearance as a simple tubelike structure, the **neural tube,** that extends down the dorsal median plane. By the fourth week, the human brain begins to form as an expansion of the anterior or rostral end of the neural tube (the end toward the head). Shortly thereafter, constrictions appear, dividing the developing brain into three major regions—**forebrain, midbrain,** and **hindbrain** (Figure 19.1). The remainder of the neural tube becomes the spinal cord.

During fetal development, two anterior outpocketings extend from the forebrain and grow rapidly to form the cerebral hemispheres. Because of space restrictions imposed by the skull, the cerebral hemispheres are forced to grow posteriorly and inferiorly, and finally end up enveloping and obscuring the rest of the forebrain and most midbrain structures. Somewhat later in development, the dorsal part of the hindbrain also enlarges to produce the cerebellum. The central canal of the neural tube, which remains continuous throughout the brain and cord, becomes enlarged in four regions of the brain, forming chambers called **ventricles** (see Figure 19.8a and b, p. 201).

Activity 1:
Identifying External Brain Structures

Identify external brain structures using the figures cited. Also use a model of the human brain and other learning aids as they are mentioned.

Generally, the brain is considered in terms of four major regions: the cerebral hemispheres, diencephalon, brain stem, and cerebellum. The relationship between these four anatomical regions and the structures of the forebrain, midbrain, and hindbrain is also outlined in Figure 19.1.

Cerebral Hemispheres

The **cerebral hemispheres** are the most superior portion of the brain (Figure 19.2). Their entire surface is thrown into elevated ridges of tissue called **gyri** that are separated by shallow grooves called **sulci** or deeper grooves called **fissures.** Many of the fissures and gyri are important anatomical landmarks.

The cerebral hemispheres are divided by a single deep fissure, the **longitudinal fissure.** The **central sulcus** divides the **frontal lobe** from the **parietal lobe,** and the **lateral sulcus** separates the **temporal lobe** from the parietal lobe. The **parieto-occipital sulcus,** which divides the **occipital lobe** from the parietal lobe, is not visible externally. Notice that the cerebral hemisphere lobes are named for the cranial bones that lie over them.

Some important functional areas of the cerebral hemispheres have also been located (Figure 19.2d). The **primary somatosensory area** is located in the **postcentral gyrus** of the parietal lobe. Impulses traveling from the body's sensory receptors (such as those for pressure, pain, and temperature) are

Precentral gyrus
Central sulcus
Postcentral gyrus
Frontal lobe
Parietal lobe
Parieto-occipital sulcus (on medial surface of hemisphere)
Lateral sulcus
Occipital lobe
Transverse fissure
Temporal lobe
Cerebellum
Pons
Medulla oblongata
Spinal cord
(a)

Cortex (gray matter)
Fissure (a deep sulcus)
Gyrus
Sulcus
White matter

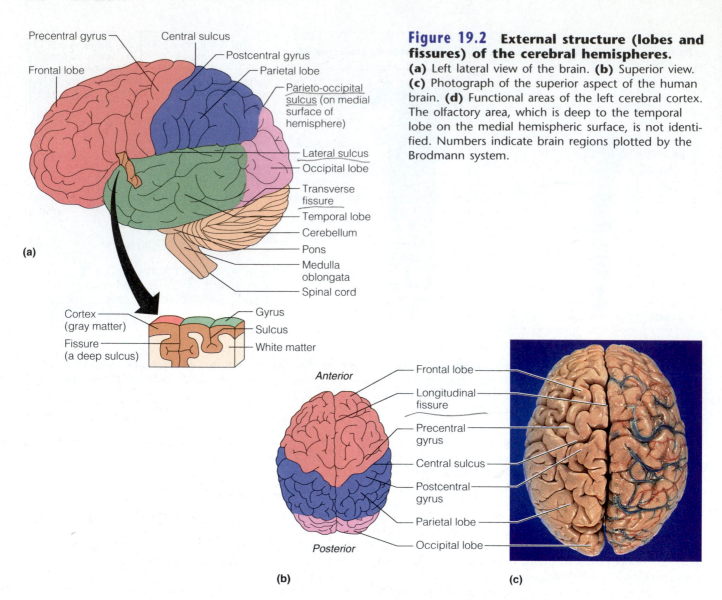

Figure 19.2 External structure (lobes and fissures) of the cerebral hemispheres. **(a)** Left lateral view of the brain. **(b)** Superior view. **(c)** Photograph of the superior aspect of the human brain. **(d)** Functional areas of the left cerebral cortex. The olfactory area, which is deep to the temporal lobe on the medial hemispheric surface, is not identified. Numbers indicate brain regions plotted by the Brodmann system.

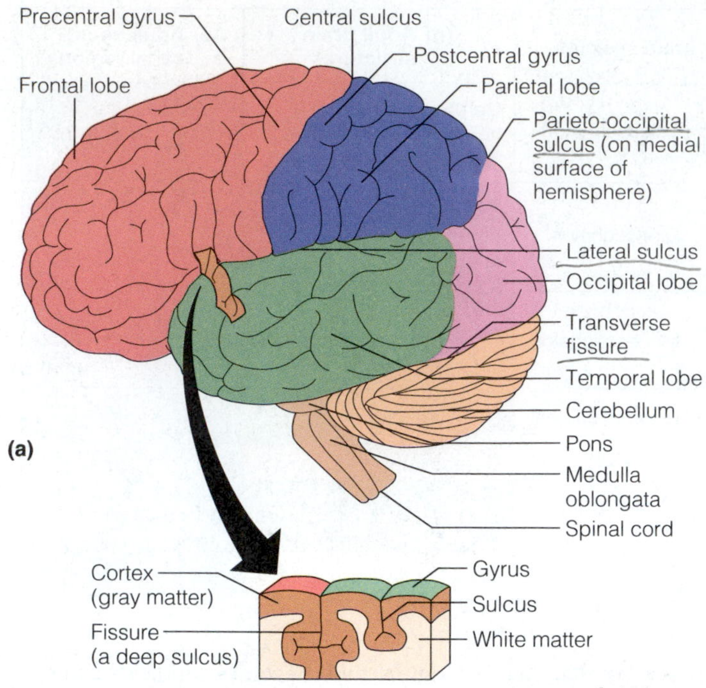

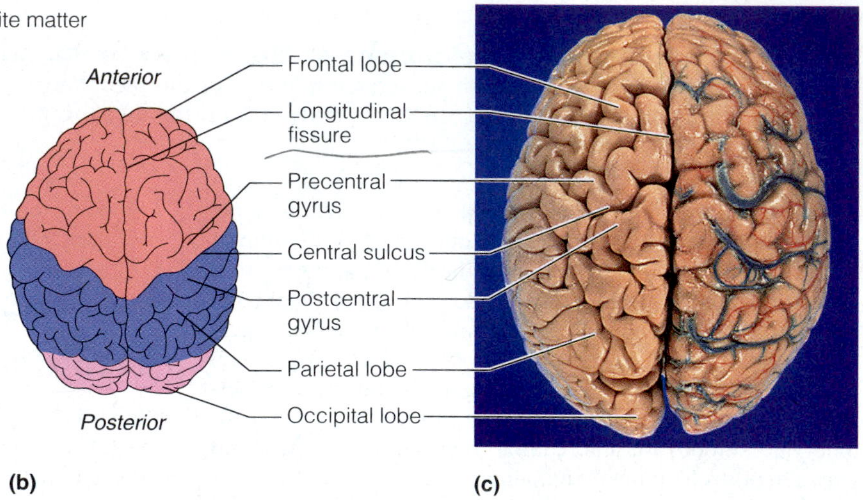

Anterior
Frontal lobe
Longitudinal fissure
Precentral gyrus
Central sulcus
Postcentral gyrus
Parietal lobe
Occipital lobe
Posterior

(b)

(c)

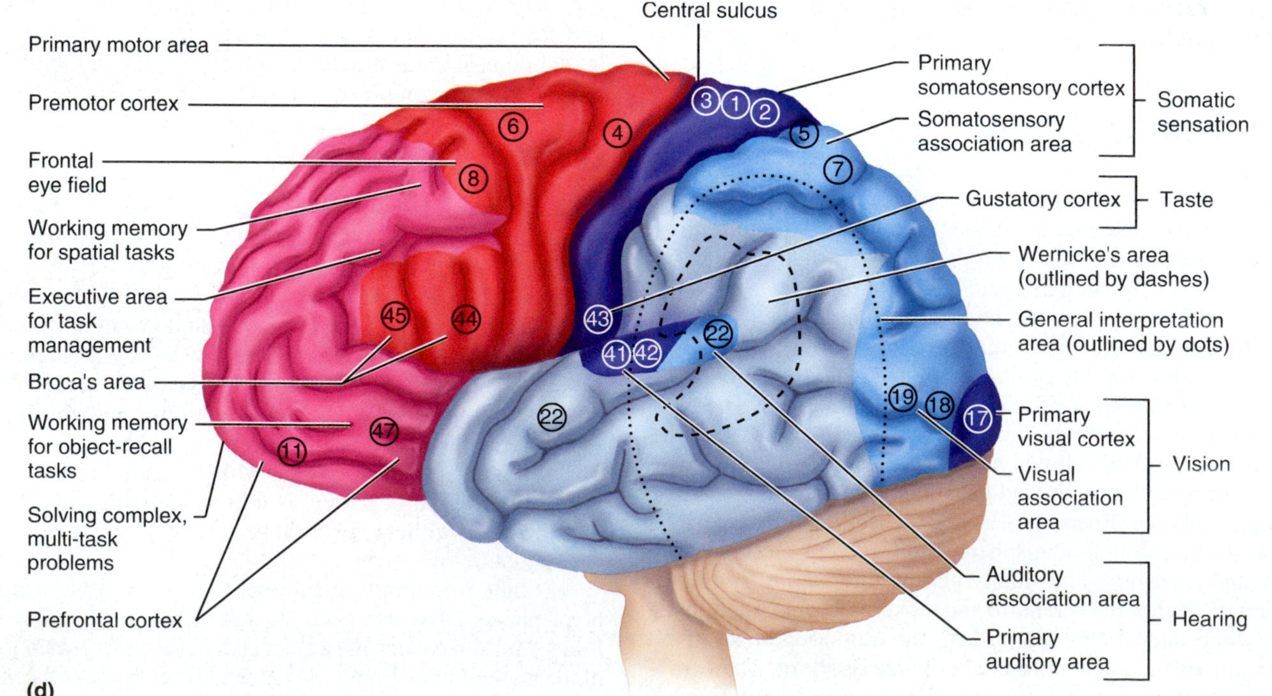

localized in this area of the brain. ("This information is from my big toe.") Immediately posterior to the primary somatosensory area is the **somatosensory association area,** in which the meaning of incoming stimuli is analyzed. ("Ouch! I have a *pain* there.") Thus, the somatosensory association area allows you to become aware of pain, coldness, a light touch, and the like.

Impulses from the special sense organs are interpreted in other specific areas also noted in Figure 19.2d. For example, the visual areas are in the posterior portion of the occipital lobe and the auditory area is located in the temporal lobe in the gyrus bordering the lateral sulcus. The olfactory area is deep within the temporal lobe along its medial surface, in a region called the **uncus** (see Figure 19.4a, p. 196).

The **primary motor area,** which is responsible for conscious or voluntary movement of the skeletal muscles, is located in the **precentral gyrus** of the frontal lobe. A specialized motor speech area called **Broca's area** is found at the base of the precentral gyrus just above the lateral sulcus. Damage to this area (which is located only in one cerebral hemisphere, usually the left) reduces or eliminates the ability to articulate words. Areas involved in intellect, complex reasoning, and personality lie in the anterior portions of the frontal lobes, in a region called the **prefrontal cortex.**

A rather poorly defined region at the junction of the parietal and temporal lobes is **Wernicke's area,** an area in which unfamiliar words are sounded out. Like Broca's area, Wernicke's area is located in one cerebral hemisphere only, typically the left.

Although there are many similar functional areas in both cerebral hemispheres, such as motor and sensory areas, each hemisphere is also a "specialist" in certain ways. For example, the left hemisphere is the "language brain" in most of us, because it houses centers associated with language skills and speech. The right hemisphere is more specifically concerned with abstract, conceptual, or spatial processes—skills associated with artistic or creative pursuits.

The cell bodies of cerebral neurons involved in these functions are found only in the outermost gray matter of the cerebrum, the area called the **cerebral cortex.** Most of the balance of cerebral tissue—the deeper **cerebral white matter**—is composed of fiber tracts carrying impulses to or from the cortex.

Using a model of the human brain (and a preserved human brain, if available), identify the areas and structures of the cerebral hemispheres described above.

Then continue using the model and preserved brain along with the figures as you read about other structures.

Diencephalon

The **diencephalon,** sometimes considered the most superior portion of the brain stem, is embryologically part of the forebrain, along with the cerebral hemispheres.

Turn the brain model so the ventral surface of the brain can be viewed. Using Figure 19.3 as a guide, start superiorly

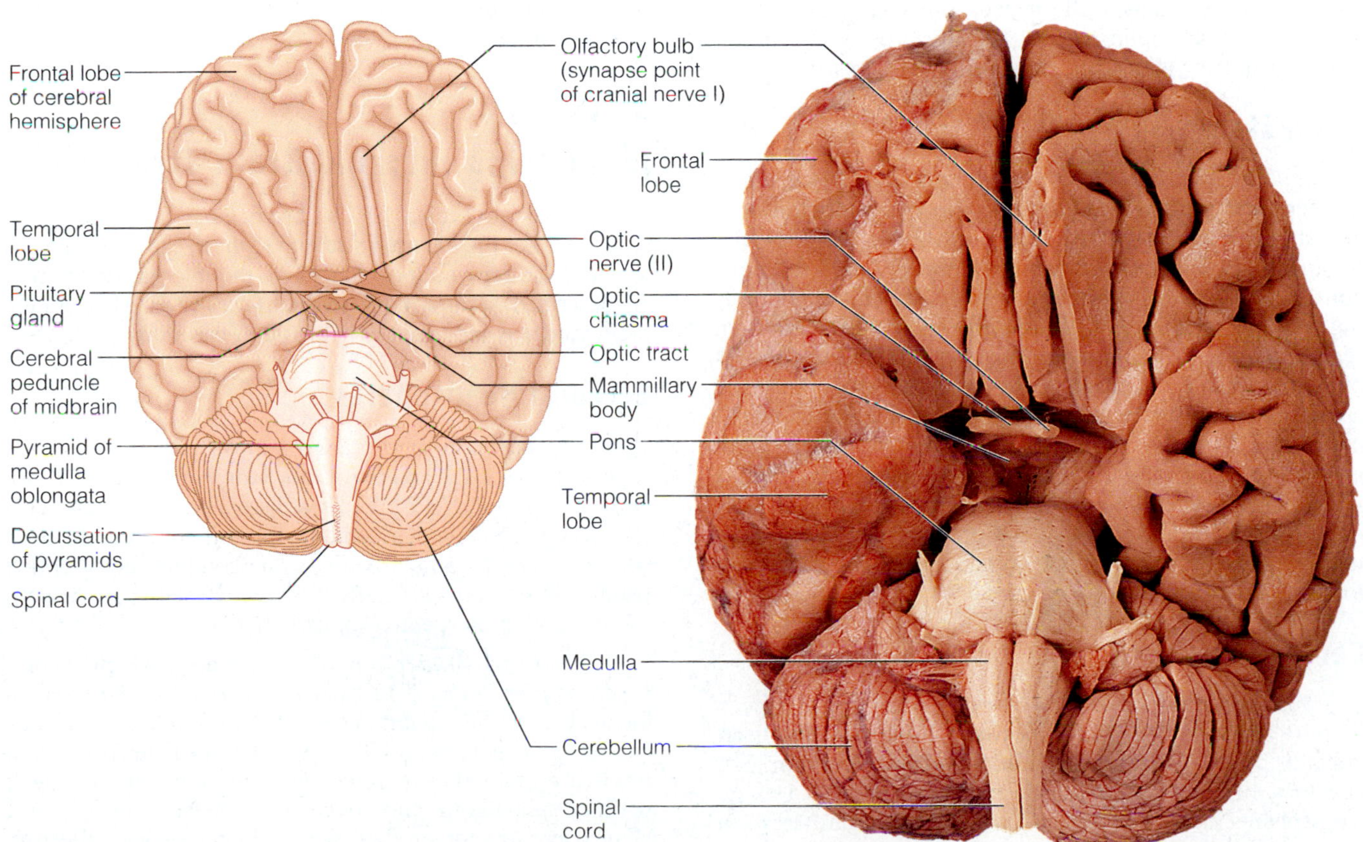

Figure 19.3 Ventral aspect of the human brain, showing the three regions of the brain stem. Only a small portion of the midbrain can be seen; the rest is surrounded by other brain regions. (See also Plate E in the Human Anatomy Atlas.)

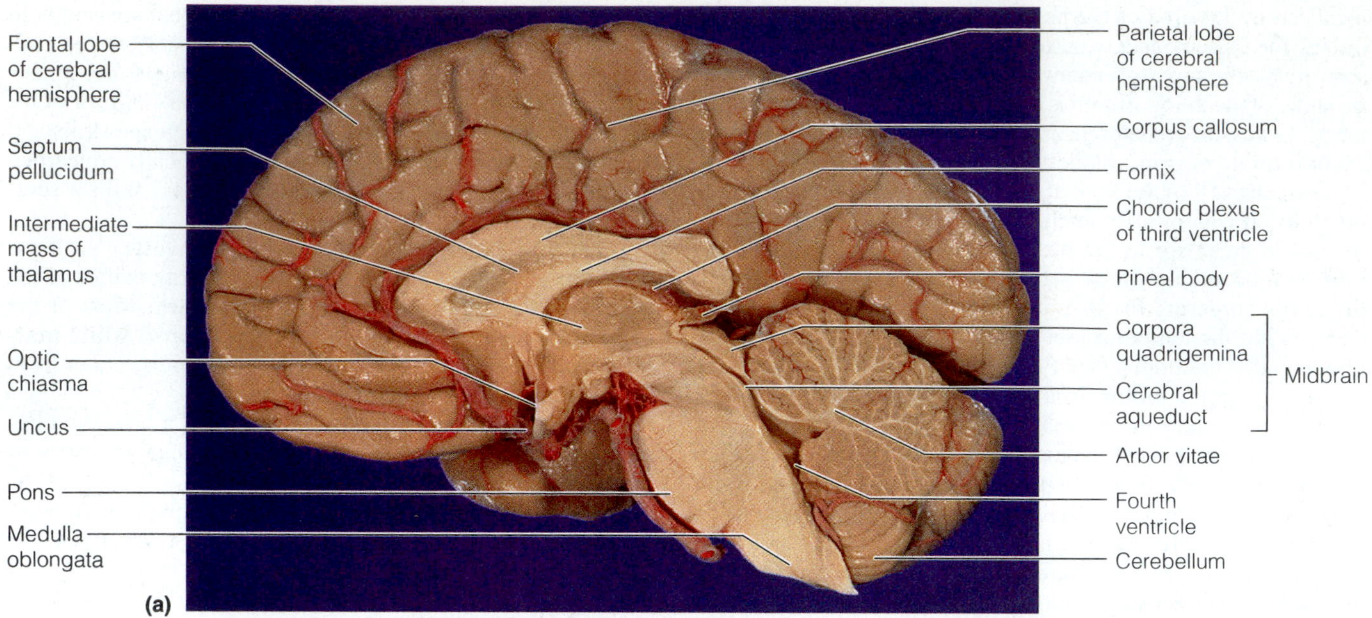

Figure 19.4 Diencephalon and brain stem structures as seen in a midsagittal section of the brain. (a) Photograph.

and identify the externally visible structures that mark the position of the floor of the diencephalon. These are the **olfactory bulbs** and **tracts, optic nerves, optic chiasma** (where the fibers of the optic nerves partially cross over), **optic tracts, pituitary gland,** and **mammillary bodies.**

Brain Stem

Continue inferiorly to identify the **brain stem** structures—the **cerebral peduncles** (fiber tracts in the **midbrain** connecting the pons below with cerebrum above), the pons, and the medulla oblongata. *Pons* means "bridge," and the **pons** consists primarily of motor and sensory fiber tracts connecting the brain with lower CNS centers. The lowest brain stem region, the **medulla oblongata,** is also composed primarily of fiber tracts. You can see the **decussation of pyramids,** a crossover point for the major motor tracts (pyramidal tracts) descending from the motor areas of the cerebrum to the cord, on the medulla's anterior surface. The medulla also houses many vital autonomic centers involved in the control of heart rate, respiratory rhythm, and blood pressure as well as involuntary centers involved in the initiation of vomiting, swallowing, and so on.

Cerebellum

1. Turn the brain model so you can see the dorsal aspect. Identify the large cauliflowerlike **cerebellum,** which projects dorsally from under the occipital lobe of the cerebrum. Notice that, like the cerebrum, the cerebellum has two major hemispheres and a convoluted surface (see Figure 19.6). It also has an outer cortex made up of gray matter with an inner region of white matter.

2. Remove the cerebellum to view the **corpora quadrigemina** (see Figure 19.4), located on the posterior aspect of the midbrain, a brain stem structure. The two superior prominences are the **superior colliculi** (visual reflex centers); the two smaller inferior prominences are the **inferior colliculi** (auditory reflex centers). ■

Activity 2:
Identifying Internal Brain Structures

The deeper structures of the brain have also been well mapped. Like the external structures, these can be studied in terms of the four major regions. As the internal brain areas are described, identify them on the figures cited. Also, use the brain model as indicated to help you in this study.

Cerebral Hemispheres

1. Take the brain model apart so you can see a median sagittal view of the internal brain structures (Figure 19.4). Observe the model closely to see the extent of the outer cortex (gray matter), which contains the cell bodies of cerebral neurons. (The pyramidal cells of the cerebral motor cortex [studied in Exercise 17, p. 182] are representative of the neurons seen in the precentral gyrus.)

2. Observe the deeper area of white matter, which is composed of fiber tracts. The fiber tracts found in the cerebral hemisphere white matter are called *association tracts* if they connect two portions of the same hemisphere, *projection tracts* if they run between the cerebral cortex and the lower brain or spinal cord, and *commissures* if they run from one hemisphere to another. Observe the large **corpus callosum,** the major commissure connecting the cerebral hemispheres. The corpus callosum arches above the structures of the diencephalon and roofs over the lateral ventricles. Note also the **fornix,** a bandlike fiber tract concerned with olfaction as

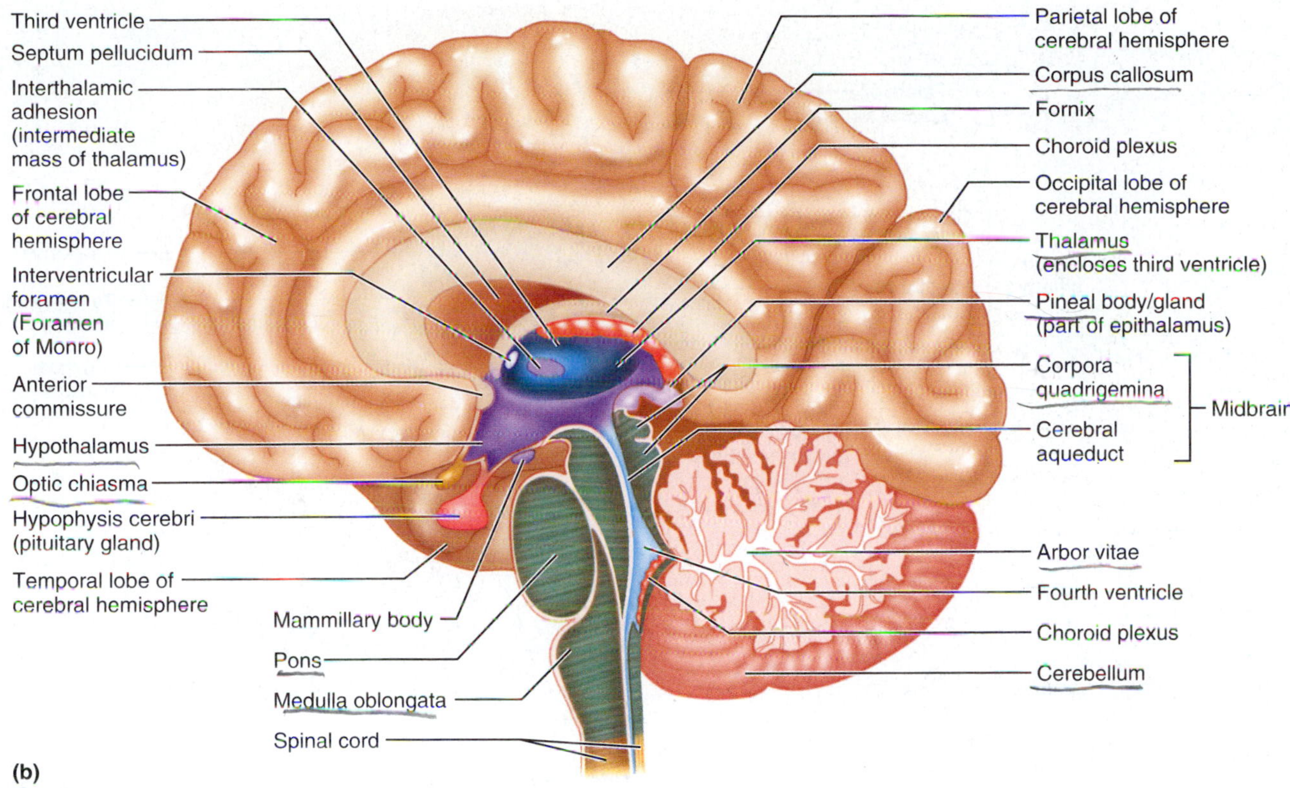

Figure 19.4 (*continued*) **(b)** Diagrammatic view.

well as limbic system functions, and the membranous **septum pellucidum,** which separates the lateral ventricles of the cerebral hemispheres.

3. In addition to the gray matter of the cerebral cortex, there are several "islands" of gray matter (clusters of neuron cell bodies) called **nuclei** buried deep within the white matter of the cerebral hemispheres. One important group of cerebral nuclei, called the **basal nuclei,*** flank the lateral and third ventricles. You can see the basal nuclei if you have an appropriate dissectible model or a coronally or cross-sectioned human brain slice. Otherwise, Figure 19.5 will suffice.

The basal nuclei, which are important subcortical motor nuclei (and part of the so-called *extrapyramidal system*), are involved in regulating voluntary motor activities. The most important of them are the arching, comma-shaped **caudate nucleus,** the **amygdaloid nucleus** (located at the tip of the caudate nucleus), and the **lentiform nucleus,** which is composed of the **putamen** and **globus pallidus nuclei.** The **corona radiata,** a spray of projection fibers coursing down from the precentral (motor) gyrus, combines with sensory fibers traveling to the sensory cortex to form a broad band of fibrous material called the **internal capsule**. The internal capsule passes between the diencephalon and the basal nuclei, and gives these basal nuclei a striped appearance. This is why the caudate nucleus and the lentiform nucleus are sometimes referred to collectively as the **corpus striatum,** or "striped body" (Figure 19.5a).

*The historical term for these nuclei is *basal ganglia*, a misleading term because ganglia are PNS structures.

4. Examine the relationship of the lateral ventricles and corpus callosum to the diencephalon structures; that is, thalamus, and third ventricle—from the cross-sectional viewpoint (see Figure 19.5b).

Diencephalon

1. The major internal structures of the diencephalon are the thalamus, hypothalamus, and epithalamus (Figure 19.4). The **thalamus** consists of two large lobes of gray matter that laterally enclose the shallow third ventricle of the brain. A slender stalk of thalamic tissue, the **intermediate mass,** or **interthalamic adhesion,** connects the two thalamic lobes and bridges the ventricle. The thalamus is a major integrating and relay station for sensory impulses passing upward to the cortical sensory areas for localization and interpretation. Locate also the **interventricular foramen** *(foramen of Monro)*, a tiny orifice connecting the third ventricle with the lateral ventricle on the same side.

2. The **hypothalamus** makes up the floor and the inferolateral walls of the third ventricle. It is an important autonomic center involved in regulation of body temperature, water balance, and fat and carbohydrate metabolism as well as in many other activities and drives (sex, hunger, thirst). Locate again the pituitary gland, or **hypophysis,** which hangs from the anterior floor of the hypothalamus by a slender stalk, the **infundibulum.** (The pituitary gland is usually not present in preserved brain specimens.) In life, the pituitary rests in the hypophyseal fossa of the sella turcica of the sphenoid bone. Its function is discussed in Exercise 27.

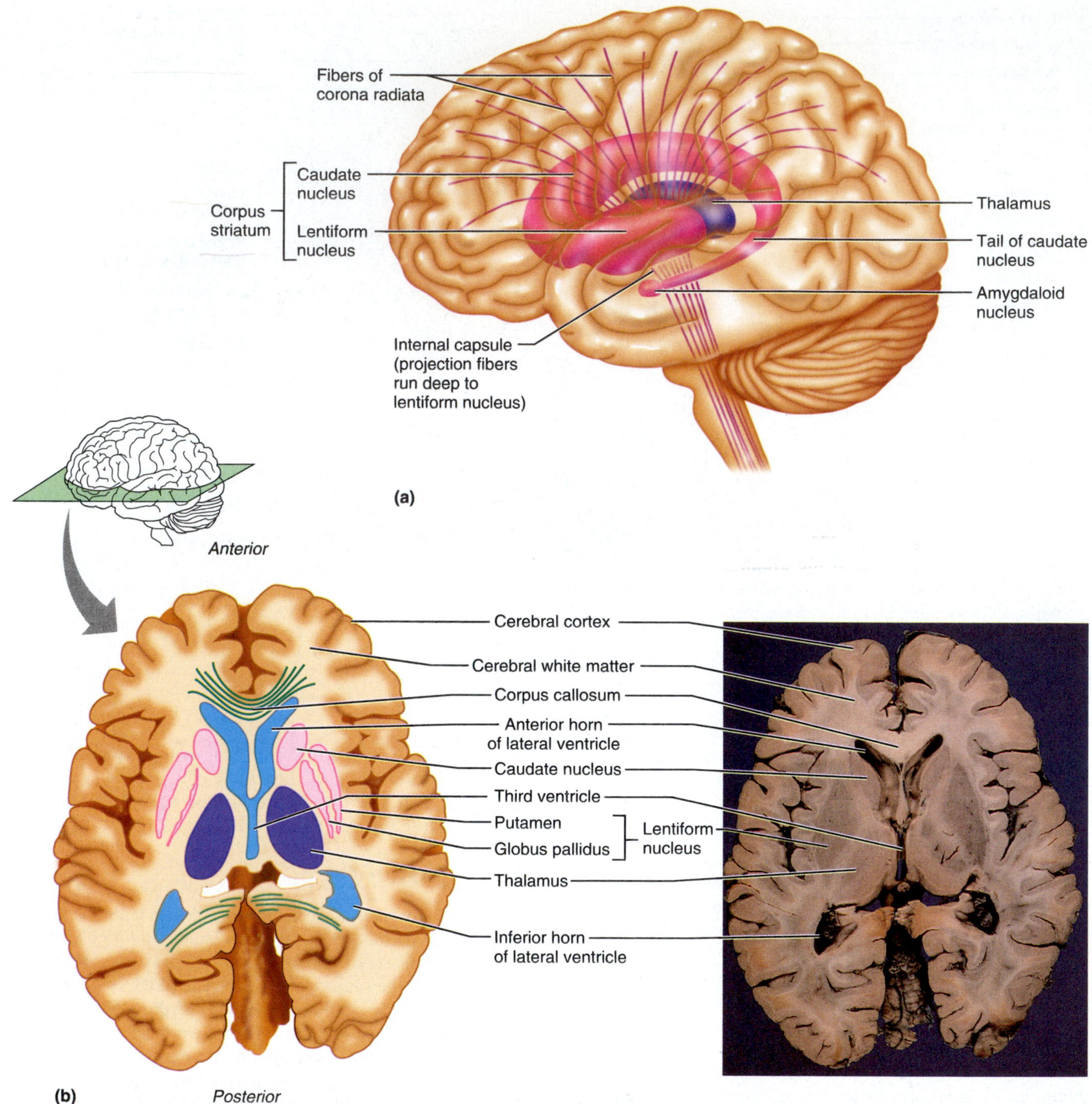

Figure 19.5 Basal nuclei. **(a)** Three-dimensional view of the basal nuclei showing their positions within the cerebrum. **(b)** A transverse section of the cerebrum and diencephalon showing the relationship of the basal nuclei to the thalamus and the lateral and third ventricles.

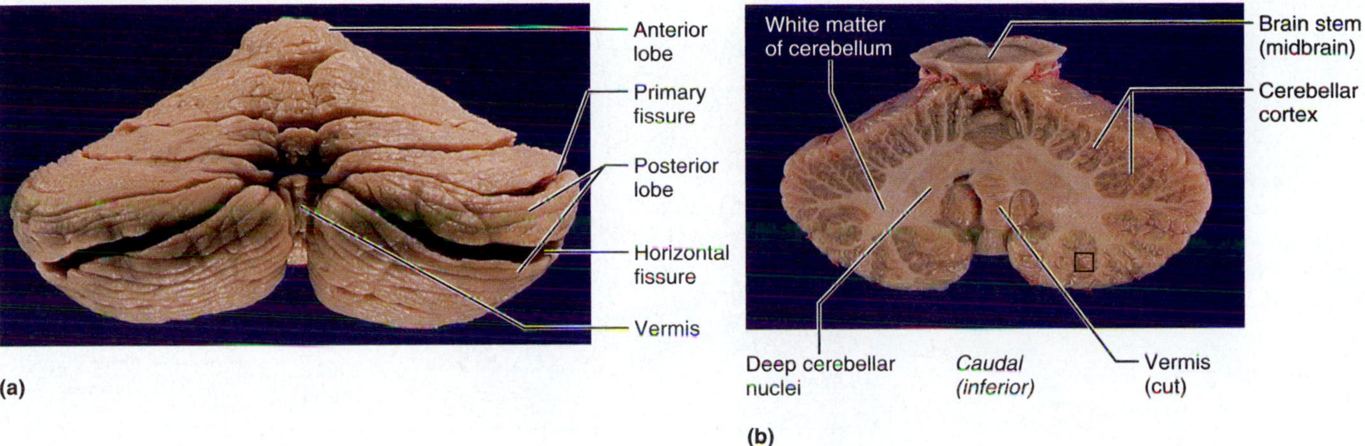

Figure 19.6 Cerebellum. (a) Posterior (dorsal) view. **(b)** The cerebellum, sectioned to reveal its cortex and medullary regions. (Note that the cerebellum is sectioned frontally and the brain stem is sectioned horizontally in this posterior view.)

Anterior to the pituitary, identify the optic chiasma portion of the optic pathway to the brain. The **mammillary bodies,** relay stations for olfaction, bulge exteriorly from the floor of the hypothalamus just posterior to the pituitary gland.

3. The **epithalamus** forms the roof of the third ventricle and is the most dorsal portion of the diencephalon. Important structures in the epithalamus are the **pineal body,** or **gland** (a neuroendocrine structure), and the **choroid plexus** of the third ventricle. The choroid plexuses, knotlike collections of capillaries within each ventricle, form the cerebrospinal fluid.

Brain Stem

1. Now trace the short midbrain from the mammillary bodies to the rounded pons below. Continue to refer to Figure 19.4. The **cerebral aqueduct** is a slender canal traveling through the midbrain; it connects the third ventricle to the fourth ventricle in the hindbrain below. The cerebral peduncles and the rounded corpora quadrigemina make up the midbrain tissue anterior and posterior (respectively) to the cerebral aqueduct.

2. Locate the hindbrain structures. Trace the rounded pons to the medulla oblongata below, and identify the fourth ventricle posterior to these structures. Attempt to identify the single median aperture and the two lateral apertures, three orifices found in the walls of the fourth ventricle. These apertures serve as conduits for cerebrospinal fluid to circulate into the subarachnoid space from the fourth ventricle.

Cerebellum

Examine the cerebellum. Notice that it is composed of two lateral hemispheres each with three lobes (*anterior, posterior,* and a deep *flocculonodular*) connected by a midline lobe called the **vermis** (Figure 19.6). As in the cerebral hemispheres, the cerebellum has an outer cortical area of gray matter and an inner area of white matter. The treelike branching of the cerebellar white matter is referred to as the **arbor vitae,** or "tree of life." The cerebellum is concerned with unconscious coordination of skeletal muscle activity and control of balance and equilibrium. Fibers converge on the cerebellum from the equilibrium apparatus of the inner ear, visual pathways, proprioceptors of the tendons and skeletal muscles, and from many other areas. Thus the cerebellum remains constantly aware of the position and state of tension of the various body parts. ■

Meninges of the Brain

The brain (and spinal cord) are covered and protected by three connective tissue membranes called **meninges** (Figure 19.7). The outermost meninx is the leathery **dura mater,** a

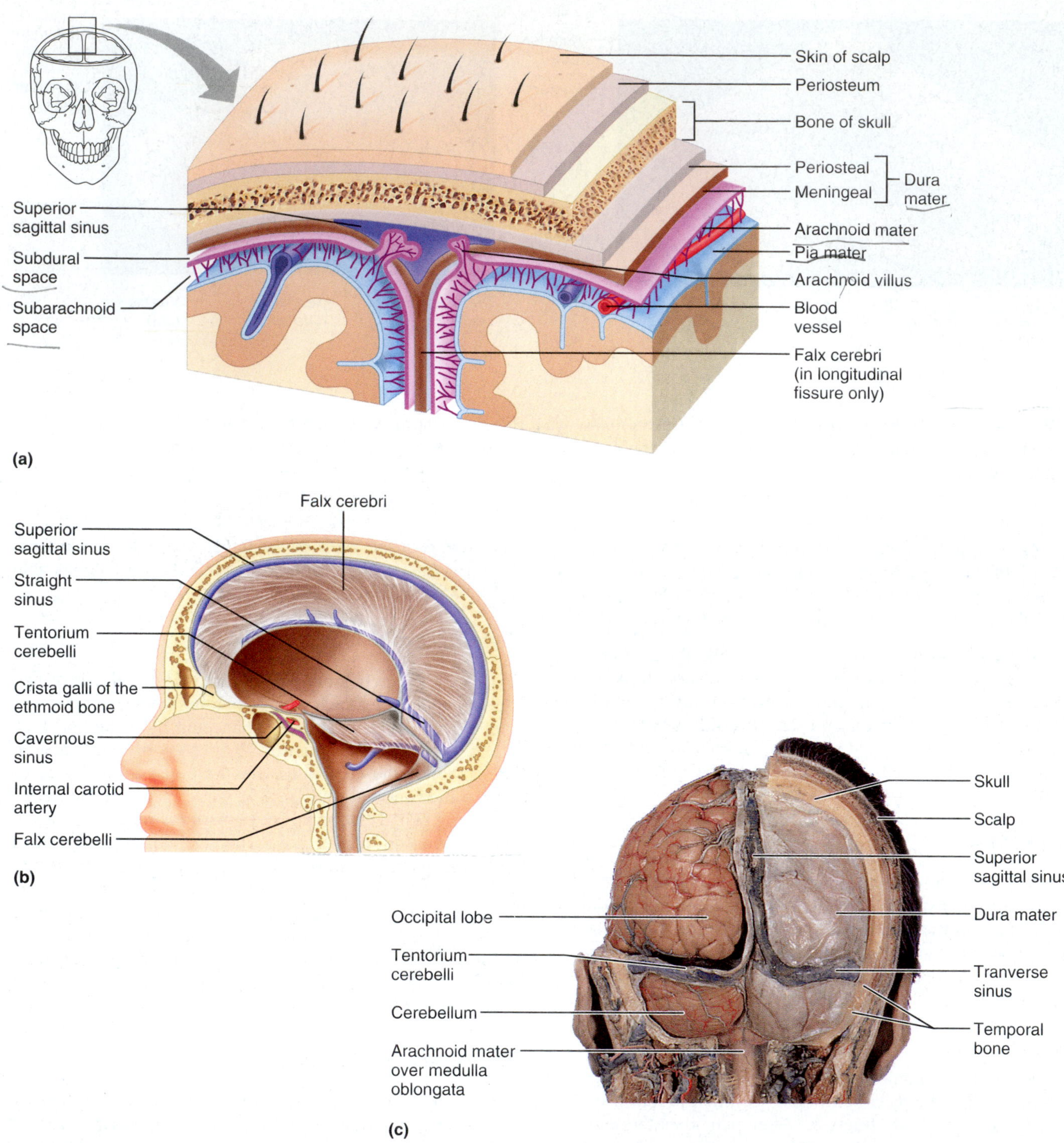

Figure 19.7 Meninges of the brain. (a) Three-dimensional frontal section showing the relationship of the dura mater, arachnoid mater, and pia mater. The meningeal dura forms the falx cerebri fold, which extends into the longitudinal fissure and attaches the brain to the ethmoid bone of the skull. A dural sinus, the superior sagittal sinus, is enclosed by the dural membranes superiorly. Arachnoid villi, which return cerebrospinal fluid to the dural sinus, are also shown. **(b)** Position of the dural folds, the falx cerebri, tentorium cerebelli, and falx cerebelli. **(c)** Posterior view of the brain in place, surrounded by the dura mater.

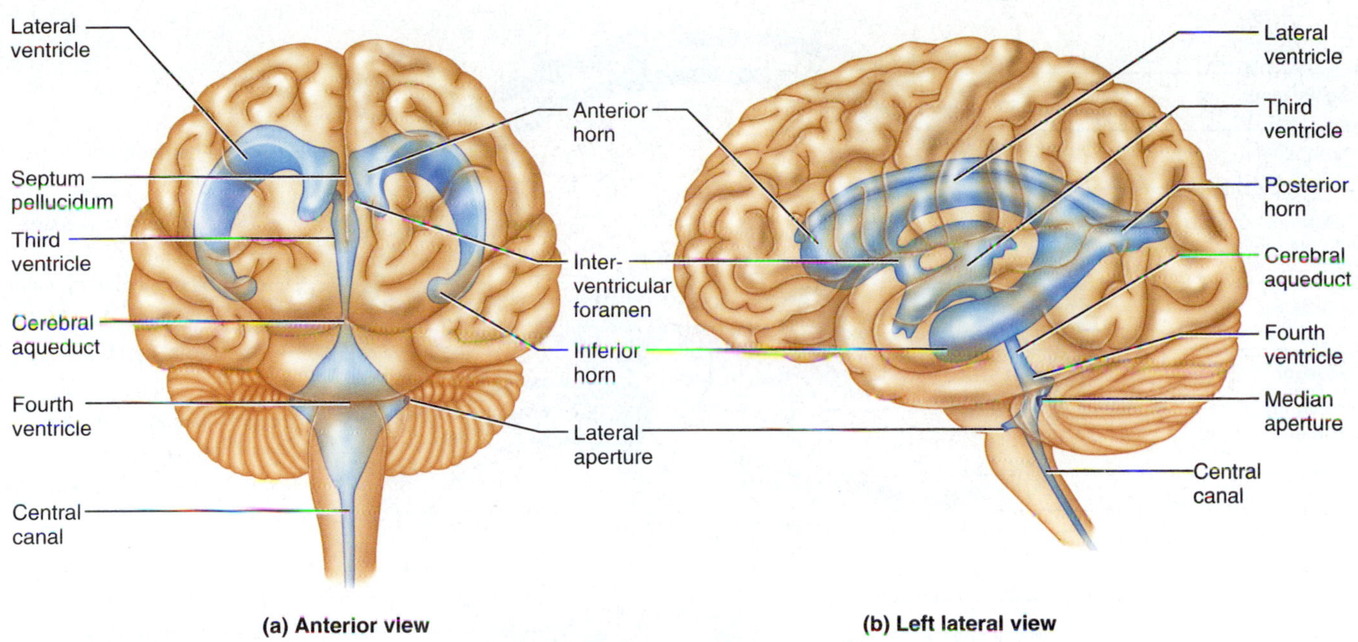

Figure 19.8 Location and circulatory pattern of cerebrospinal fluid. (a) Anterior view. **(b)** Lateral view. Note that different regions of the large lateral ventricles are indicated by the terms *anterior horn, posterior horn,* and *inferior horn.*

double-layered membrane. One of its layers (the *periosteal layer*) is attached to the inner surface of the skull, forming the periosteum. The other (the *meningeal layer*) forms the outermost brain covering and is continuous with the dura mater of the spinal cord.

The dural layers are fused together except in three places where the inner membrane extends inward to form a septum that secures the brain to structures inside the cranial cavity. One such extension, the **falx cerebri,** dips into the longitudinal fissure between the cerebral hemispheres to attach to the crista galli of the ethmoid bone of the skull. The cavity created at this point is the large **superior sagittal sinus,** which collects blood draining from the brain tissue. The **falx cerebelli,** separating the two cerebellar hemispheres, and the **tentorium cerebelli,** separating the cerebrum from the cerebellum below, are two other important inward folds of the inner dural membrane.

The middle meninx, the weblike **arachnoid mater,** underlies the dura mater and is partially separated from it by the **subdural space.** Threadlike projections bridge the **subarachnoid space** to attach the arachnoid to the innermost meninx, the **pia mater.** The delicate pia mater is highly vascular and clings tenaciously to the surface of the brain, following its convolutions.

In life, the subarachnoid space is filled with cerebrospinal fluid. Specialized projections of the arachnoid tissue called **arachnoid villi** protrude through the dura mater to allow the cerebrospinal fluid to drain back into the venous circulation via the superior sagittal sinus and other dural sinuses.

Meningitis, inflammation of the meninges, is a serious threat to the brain because of the intimate association between the brain and meninges. Should infection spread to the neural tissue of the brain itself, life-threatening **encephalitis** may occur. Meningitis is often diagnosed by taking a sample of cerebrospinal fluid from the subarachnoid space. ■

Cerebrospinal Fluid

The cerebrospinal fluid, much like plasma in composition, is continually formed by the **choroid plexuses,** small capillary knots hanging from the roof of the ventricles of the brain. The cerebrospinal fluid in and around the brain forms a watery cushion that protects the delicate brain tissue against blows to the head.

Within the brain, the cerebrospinal fluid circulates from the two lateral ventricles (in the cerebral hemispheres) into the third ventricle via the **interventricular foramina,** and then through the cerebral aqueduct of the midbrain into the fourth ventricle in the hindbrain (Figure 19.8). Some of the fluid reaching the fourth ventricle continues down the central canal of the spinal cord, but the bulk of it circulates into the subarachnoid space, exiting through the three foramina in the walls of the fourth ventricle (the two lateral and the single median apertures). The fluid returns to the blood in the dural sinuses via the arachnoid villi.

Ordinarily, cerebrospinal fluid forms and drains at a constant rate. However, under certain conditions—for example, obstructed drainage or circulation resulting from tumors or anatomical deviations—the cerebrospinal fluid accumulates and exerts increasing pressure on the brain which, uncorrected, causes neurological damage in adults. In infants, **hydrocephalus** (literally, "water on the brain") is indicated by a gradually enlarging head. Since the infant's skull is still flexible and contains fontanels, it can expand to accommodate the increasing size of the brain. ■

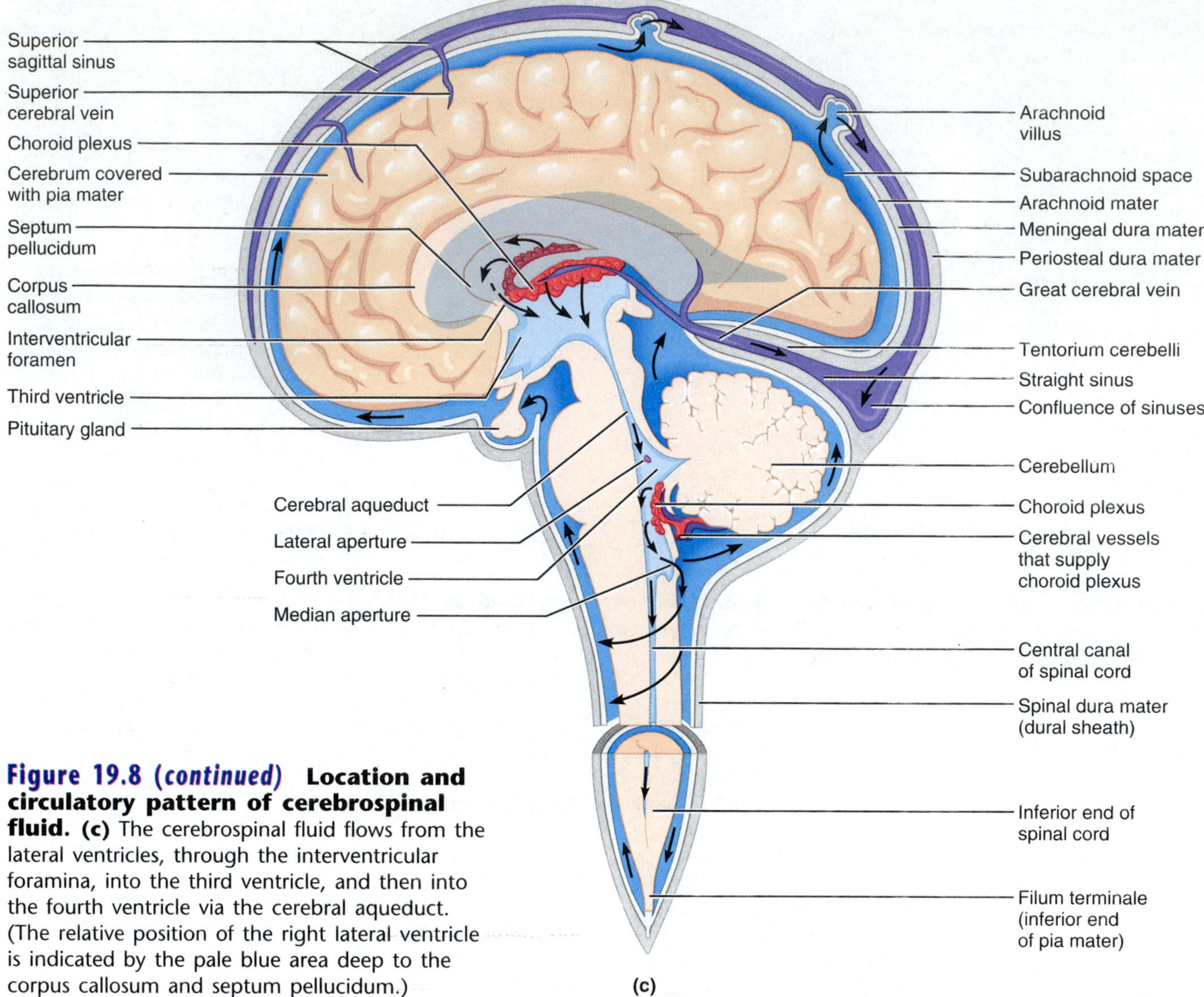

Figure 19.8 *(continued)* Location and circulatory pattern of cerebrospinal fluid. (c) The cerebrospinal fluid flows from the lateral ventricles, through the interventricular foramina, into the third ventricle, and then into the fourth ventricle via the cerebral aqueduct. (The relative position of the right lateral ventricle is indicated by the pale blue area deep to the corpus callosum and septum pellucidum.)

Cranial Nerves

The **cranial nerves** are part of the peripheral nervous system and not part of the brain proper, but they are most appropriately identified in conjunction with the study of brain anatomy. The 12 pairs of cranial nerves primarily serve the head and neck. Only one pair, the vagus nerves, extends into the thoracic and abdominal cavities. All but the first two pairs (olfactory and optic nerves) arise from the brain stem and pass through foramina in the base of the skull to reach their destination.

The cranial nerves are numbered consecutively, and in most cases their names reflect the major structures they control. The cranial nerves are described by name, number (Roman numeral), origin, course, and function in Table 19.1. This information should be committed to memory. A mnemonic device that might be helpful for remembering the cranial nerves in order is "*O*n *o*ccasion, *o*ur *t*rusty *t*ruck *a*cts *f*unny—*v*ery *g*ood *v*ehicle *a*ny*h*ow." The first letter of each word and the "a" and "h" of the final word "anyhow" will remind you of the first letter of the cranial nerve name.

Most cranial nerves are mixed nerves (containing both motor and sensory fibers). But close scrutiny of Table 19.1 will reveal that three pairs of cranial nerves (optic, olfactory, and vestibulocochlear) are purely sensory in function.

You may recall that the cell bodies of neurons are always located within the central nervous system (cortex or nuclei) or in specialized collections of cell bodies (ganglia) outside the CNS. Neuron cell bodies of the sensory cranial nerves are located in ganglia; those of the mixed cranial nerves are found both within the brain and in peripheral ganglia.

Table 19.1 The Cranial Nerves (see Figure 19.9)

Number and name	Origin and course	Function*	Testing
I. Olfactory	Fibers arise from olfactory epithelium and run through cribriform plate of ethmoid bone to synapse with olfactory bulbs.	Purely sensory—carries impulses associated with sense of smell.	Person is asked to sniff aromatic substances, such as oil of cloves and vanilla, and to identify each.
II. Optic	Fibers arise from retina of eye to form the optic nerve and pass through optic foramen in sphenoid bone. Fibers partially cross over at the optic chiasma and continue on to the thalamus as the optic tracts. Final fibers of this pathway travel from the thalamus to the optic cortex as the optic radiation.	Purely sensory—carries impulses associated with vision.	Vision and visual field are determined with eye chart and by testing the point at which the person first sees an object (finger) moving into the visual field. Fundus of eye viewed with ophthalmoscope to detect papilledema (swelling of optic disc, or point at which optic nerve leaves the eye) and to observe blood vessels.
III. Oculomotor	Fibers emerge from midbrain and exit from skull via superior orbital fissure to run to eye.	Primarily motor—somatic motor fibers to inferior oblique and superior, inferior, and medial rectus muscles, which direct eyeball, and to levator palpebrae muscles of eyelid; parasympathetic fibers to iris and smooth muscle controlling lens shape (reflex responses to varying light intensity and focusing of eye for near vision).	Pupils are examined for size, shape, and equality. Pupillary reflex is tested with penlight (pupils should constrict when illuminated). Convergence for near vision is tested, as is subject's ability to follow objects with the eyes.
IV. Trochlear	Fibers emerge from midbrain and exit from skull via superior orbital fissure to run to eye.	Primarily motor—provides somatic motor fibers to superior oblique muscle (an extrinsic eye muscle).	Tested in common with cranial nerve III.
V. Trigeminal	Fibers emerge from pons and form three divisions, which exit separately from skull: mandibular division through foramen ovale in sphenoid bone, maxillary division via foramen rotundum in sphenoid bone, and ophthalmic division through superior orbital fissure of eye socket.	Mixed—major sensory nerve of face; conducts sensory impulses from skin of face and anterior scalp, from mucosae of mouth and nose, and from surface of eyes; mandibular division also contains motor fibers that innervate muscles of mastication and muscles of floor of mouth.	Sensations of pain, touch, and temperature are tested with safety pin and hot and cold objects. Corneal reflex tested with wisp of cotton. Motor branch assessed by asking person to clench his teeth, open mouth against resistance, and move jaw side to side.
VI. Abducens	Fibers leave inferior region of pons and exit from skull via superior orbital fissure to run to eye.	Carries motor fibers to lateral rectus muscle of eye.	Tested in common with cranial nerve III.

*Does not include sensory impulses from proprioceptors.

Continues on p. 204

Table 19.1 The Cranial Nerves (see Figure 19.9) (*continued*)

Number and name	Origin and course	Function*	Testing
VII. Facial	Fibers leave pons and travel through temporal bone via internal acoustic meatus, exiting via stylomastoid foramen to reach the face.	Mixed—supplies somatic motor fibers to muscles of facial expression and parasympathetic motor fibers to lacrimal and salivary glands; carries sensory fibers from taste receptors of anterior portion of tongue.	Anterior two-thirds of tongue is tested for ability to taste sweet (sugar), salty, sour (vinegar), and bitter (quinine) substances. Symmetry of face is checked. Subject is asked to close eyes, smile, whistle, and so on. Tearing is assessed with ammonia fumes.
VIII. Vestibulocochlear	Fibers run from inner-ear equilibrium and hearing apparatus, housed in temporal bone, through internal acoustic meatus to enter pons.	Purely sensory—vestibular branch transmits impulses associated with sense of equilibrium from vestibular apparatus and semicircular canals; cochlear branch transmits impulses associated with hearing from cochlea.	Hearing is checked by air and bone conduction using tuning fork.
IX. Glossopharyngeal	Fibers emerge from medulla and leave skull via jugular foramen to run to throat.	Mixed—somatic motor fibers serve pharyngeal muscles, and parasympathetic motor fibers serve salivary glands; sensory fibers carry impulses from pharynx, tonsils, posterior tongue (taste buds), and pressure receptors of carotid artery.	Position of the uvula is checked. Gag and swallowing reflexes are checked. Subject is asked to speak and cough. Posterior third of tongue may be tested for taste.
X. Vagus	Fibers emerge from medulla and pass through jugular foramen and descend through neck region into thorax and abdomen.	Mixed—fibers carry somatic motor impulses to pharynx and larynx and sensory fibers from same structures; very large portion is composed of parasympathetic motor fibers, which supply heart and smooth muscles of abdominal visceral organs; transmits sensory impulses from viscera.	As for cranial nerve IX (IX and X are tested in common, since they both innervate muscles of throat and mouth).
XI. Accessory	Fibers arise from medulla and superior aspect of spinal cord and travel through jugular foramen to reach muscles of neck and back.	Mixed—provides somatic motor fibers to sternocleidomastoid and trapezius muscles and to muscles of soft palate, pharynx, and larynx (spinal and medullary fibers respectively).	Sternocleidomastoid and trapezius muscles are checked for strength by asking person to rotate head and shoulders against resistance.
XII. Hypoglossal	Fibers arise from medulla and exit from skull via hypoglossal canal to travel to tongue.	Mixed—carries somatic motor fibers to muscles of tongue.	Person is asked to protrude and retract tongue. Any deviations in position are noted.

*Does not include sensory impulses from proprioceptors.

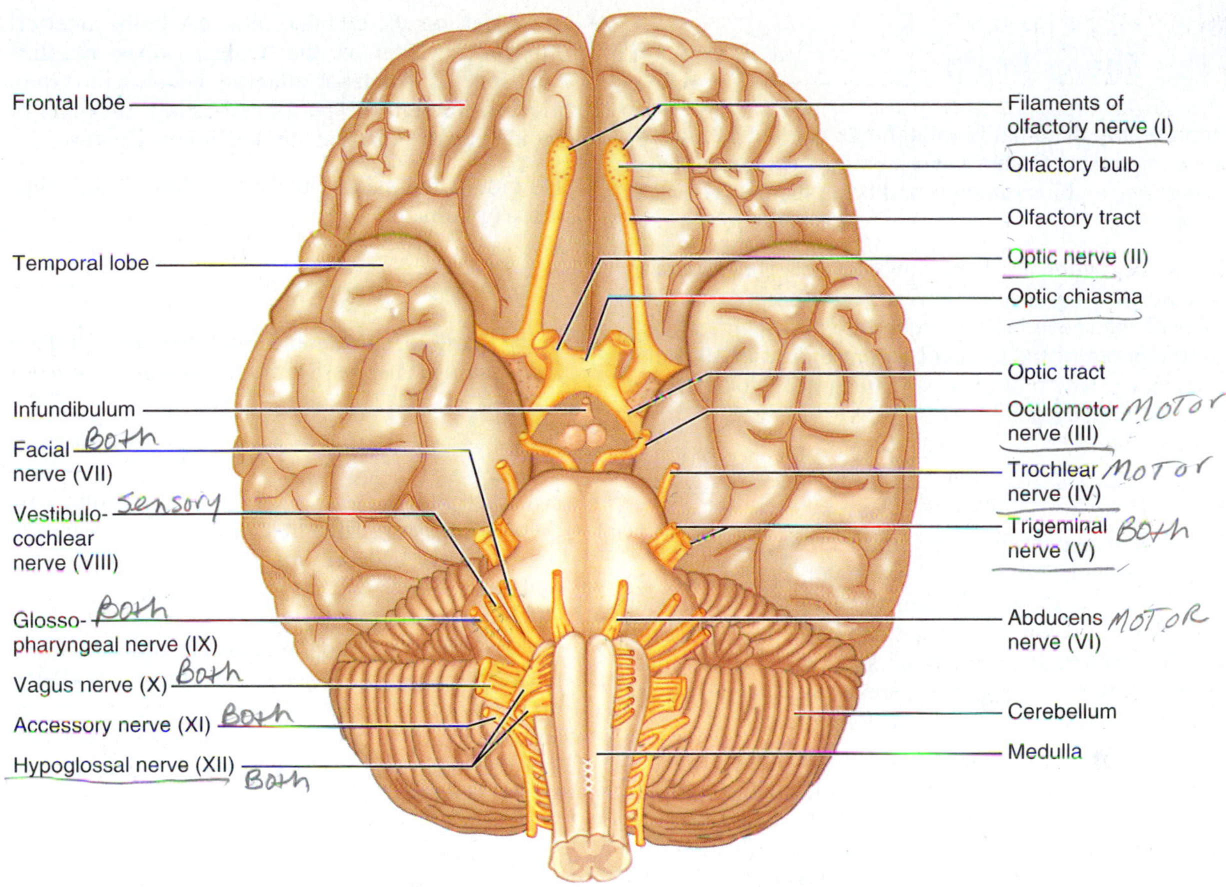

Figure 19.9 Ventral aspect of the human brain, showing the cranial nerves. (See also Plate E in the Human Anatomy Atlas.)

Activity 3: Identifying and Testing the Cranial Nerves

1. Observe the anterior surface of the brain model to identify the cranial nerves. Figure 19.9 may also aid you in this study. Notice that the first (olfactory) cranial nerves are not visible on the model because they consist only of short axons that run from the nasal mucosa through the cribriform plate of the ethmoid bone. (However, the synapse points of the first cranial nerves, the *olfactory bulbs,* are visible on the model.)

2. The last column of Table 19.1 describes techniques for testing cranial nerves, which is an important part of any neurological examination. This information may help you understand cranial nerve function, especially as it pertains to some aspects of brain function. Conduct tests of cranial nerve function following directions given in the "testing" column of the table.

3. Several cranial nerve ganglia are named here. *Using your textbook or an appropriate reference,* fill in the chart by naming the cranial nerve the ganglion is associated with and stating its location. ■

Cranial nerve ganglion	Cranial nerve	Site of ganglion
Trigeminal		
Geniculate		
Inferior		
Superior		
Spiral		
Vestibular		

Dissection: The Sheep Brain

The brain of any mammal is enough like the human brain to warrant comparison. Obtain a sheep brain, disposable gloves, dissecting tray, and instruments, and bring them to your laboratory bench.

1. Before beginning the dissection, turn your sheep brain so that you are viewing its left lateral aspect. Compare the various areas of the sheep brain (cerebrum, brain stem, cerebellum) to the photo of the human brain in Figure 19.10. Relatively speaking, which of these structures is obviously much larger in the human brain?

____

2. Place the intact sheep brain ventral surface down on the dissecting pan and observe the dura mater. Feel its consistency and note its toughness. Cut through the dura mater along the line of the longitudinal fissure (which separates the cerebral hemispheres) to enter the superior sagittal sinus. Gently force the cerebral hemispheres apart laterally to expose the corpus callosum deep to the longitudinal fissure.

3. Carefully remove the dura mater and examine the superior surface of the brain. Notice that, like the human brain, its surface is thrown into convolutions (fissures and gyri). Locate the arachnoid mater, which appears on the brain surface as a delicate "cottony" material spanning the fissures. In contrast, the innermost meninx, the pia mater, closely follows the cerebral contours.

Ventral Structures

Figure 19.11a and b shows the important features of the ventral surface of the brain. Turn the brain so that its ventral surface is uppermost.

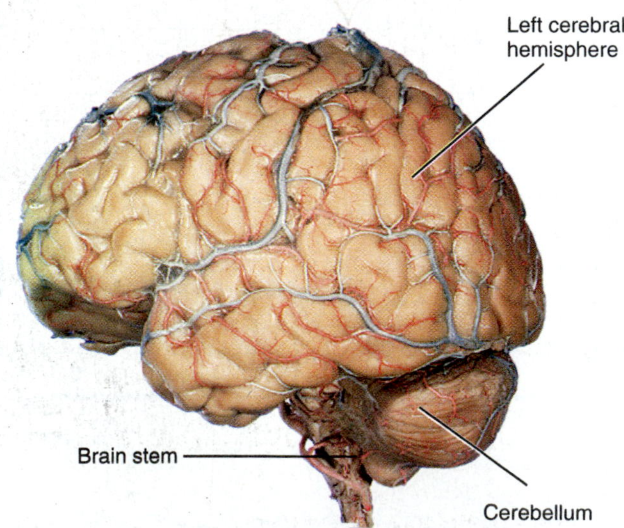

Figure 19.10 Photo of lateral aspect of the human brain.

1. Look for the clublike olfactory bulbs anteriorly, on the inferior surface of the frontal lobes of the cerebral hemispheres. Axons of olfactory neurons run from the nasal mucosa through the perforated cribriform plate of the ethmoid bone to synapse with the olfactory bulbs.

How does the size of these olfactory bulbs compare with those of humans?

____

Is the sense of smell more important as a protective and a food-getting sense in sheep or in humans?

____

2. The optic nerve (II) carries sensory impulses from the retina of the eye. Thus this cranial nerve is involved in the sense of vision. Identify the optic nerves, optic chiasma, and optic tracts.

3. Posterior to the optic chiasma, two structures protrude from the ventral aspect of the hypothalamus—the infundibulum (stalk of the pituitary gland) immediately posterior to the optic chiasma and the mammillary body. Notice that the sheep's mammillary body is a single rounded eminence. In humans it is a double structure.

4. Identify the cerebral peduncles on the ventral aspect of the midbrain, just posterior to the mammillary body of the hypothalamus. The cerebral peduncles are fiber tracts connecting the cerebrum and medulla. Identify the large oculomotor nerves (III), which arise from the ventral midbrain surface, and the tiny trochlear nerves (IV), which can be seen at the junction of the midbrain and pons. Both of these cranial nerves provide motor fibers to extrinsic muscles of the eyeball.

5. Move posteriorly from the midbrain to identify first the pons and then the medulla oblongata, both hindbrain structures composed primarily of ascending and descending fiber tracts.

6. Return to the junction of the pons and midbrain and proceed posteriorly to identify the following cranial nerves, all arising from the pons:

- Trigeminal nerves (V), which are involved in chewing and sensations of the head and face
- Abducens nerves (VI), which abduct the eye (and thus work in conjunction with cranial nerves III and IV)
- Facial nerves (VII), large nerves involved in taste sensation, gland function (salivary and lacrimal glands), and facial expression

7. Continue posteriorly to identify:

- Vestibulocochlear nerves (VIII), purely sensory nerves that are involved with hearing and equilibrium
- Glossopharyngeal nerves (IX), which contain motor fibers innervating throat structures and sensory fibers transmitting taste stimuli (in conjunction with cranial nerve VII)

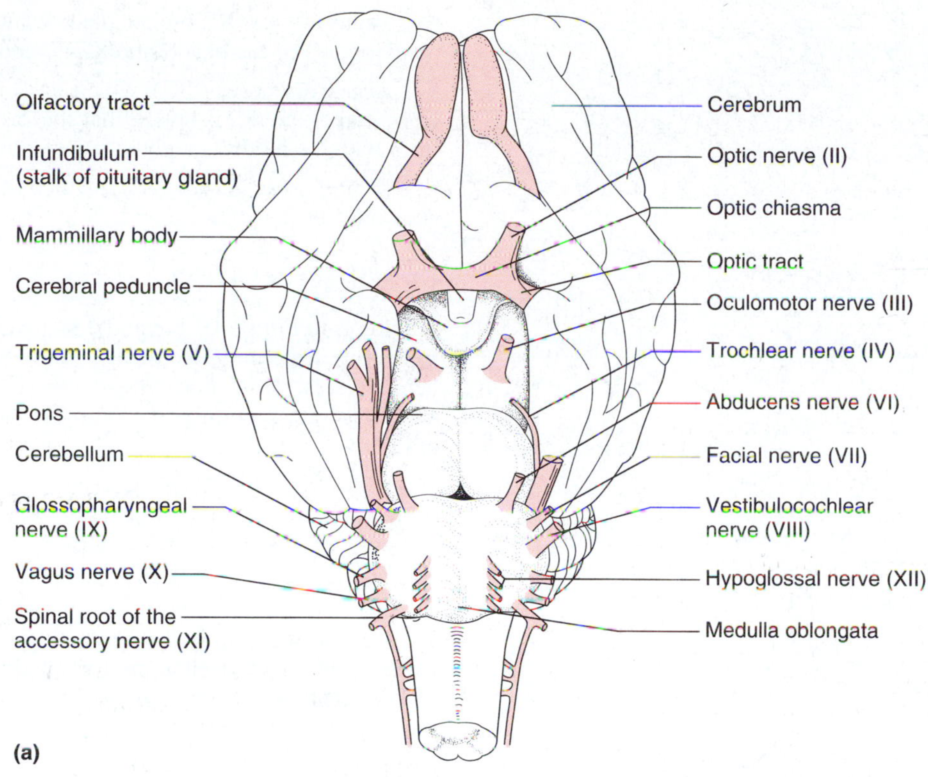

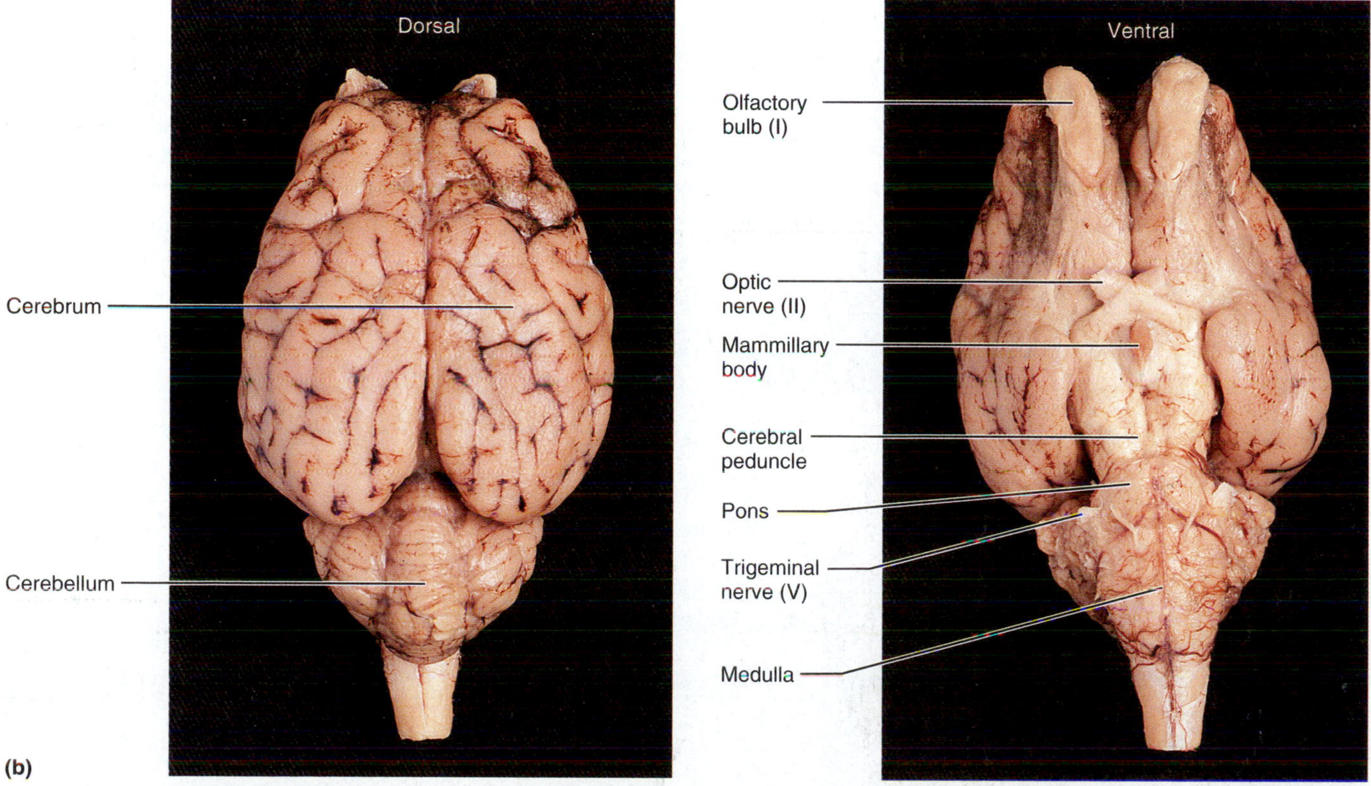

Figure 19.11 **Intact sheep brain.** **(a)** Diagrammatic ventral view. **(b)** Photographs showing ventral and dorsal views.

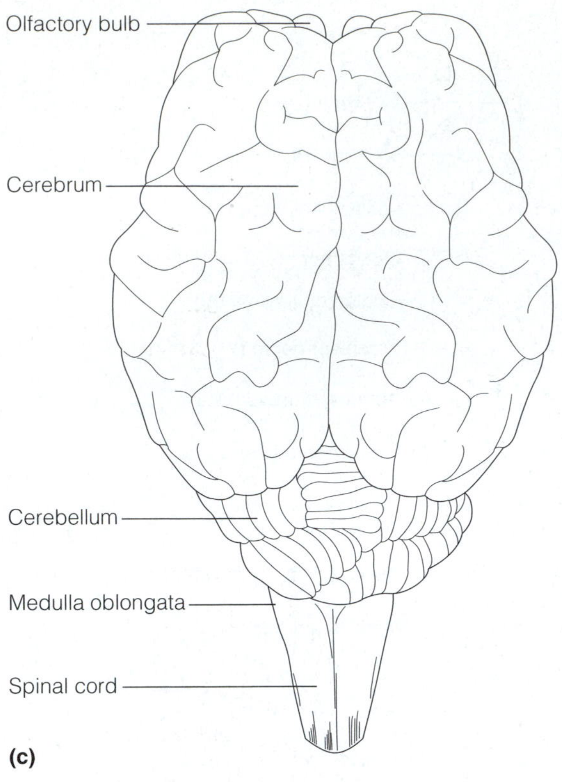

Figure 19.11 *(continued)* Intact sheep brain. **(c)** Diagrammatic dorsal view.

- Vagus nerves (X), often called "wanderers," which serve many organs of the head, thorax, and abdominal cavity
- Accessory nerves (XI), which serve muscles of the neck, larynx, and shoulder; notice that the accessory nerves arise from both the medulla and the spinal cord
- Hypoglossal nerves (XII), which stimulate tongue and neck muscles

Dorsal Structures

1. Refer to Figure 19.11c and d as a guide in identifying the following structures. Reidentify the now exposed cerebral hemispheres. How does the depth of the fissures in the sheep's cerebral hemispheres compare to that in the human brain?

____

____

2. Carefully examine the cerebellum. Notice that, in contrast to the human cerebellum, it is not divided longitudinally, and that its fissures are oriented differently. What dural falx is missing that is present in humans?

____

3. Locate the three pairs of cerebellar peduncles, fiber tracts that connect the cerebellum to other brain structures, by lifting the cerebellum dorsally away from the brain stem. The most posterior pair, the inferior cerebellar peduncles, connect the cerebellum to the medulla. The middle cerebellar peduncles attach the cerebellum to the pons, and the superior cerebellar peduncles run from the cerebellum to the midbrain.

4. To expose the dorsal surface of the midbrain, gently spread the cerebrum and cerebellum apart, as shown in Figure 19.12. Identify the corpora quadrigemina, which appear

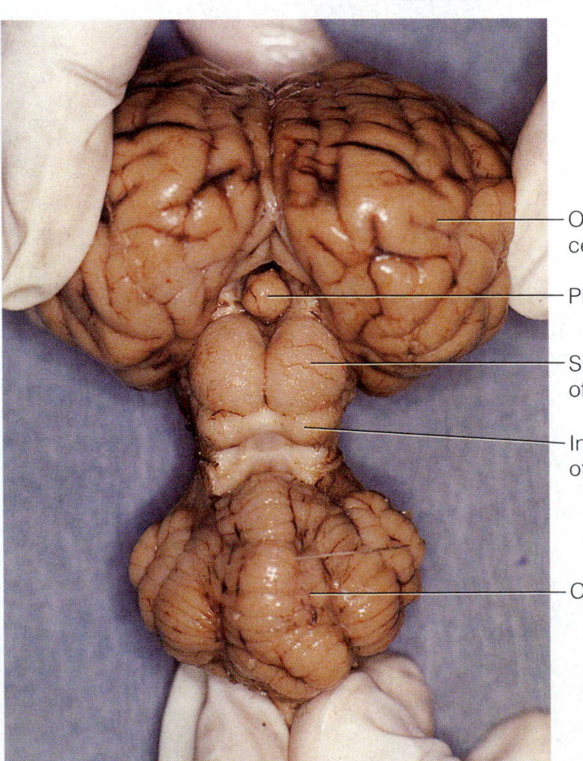

Figure 19.12 Means of exposing the dorsal midbrain structures of the sheep brain.

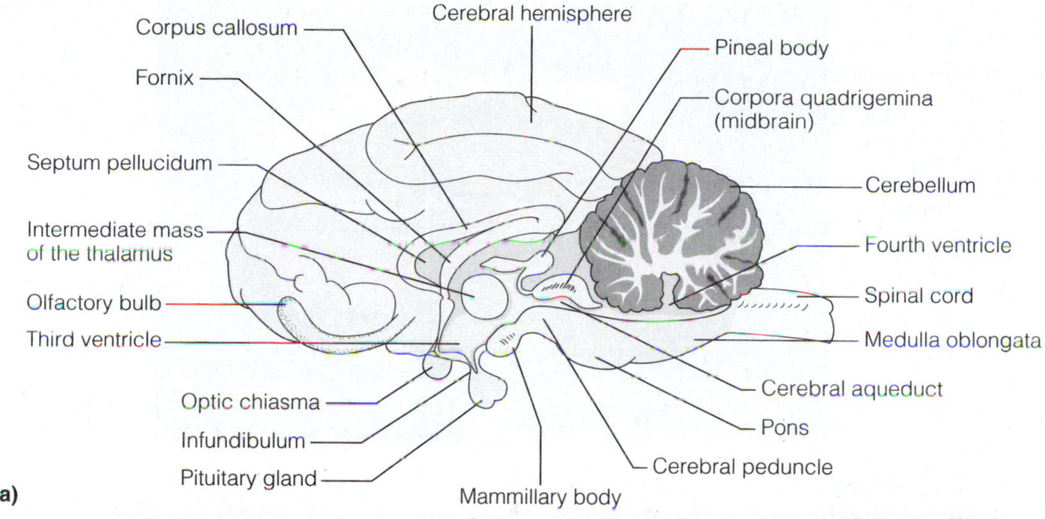

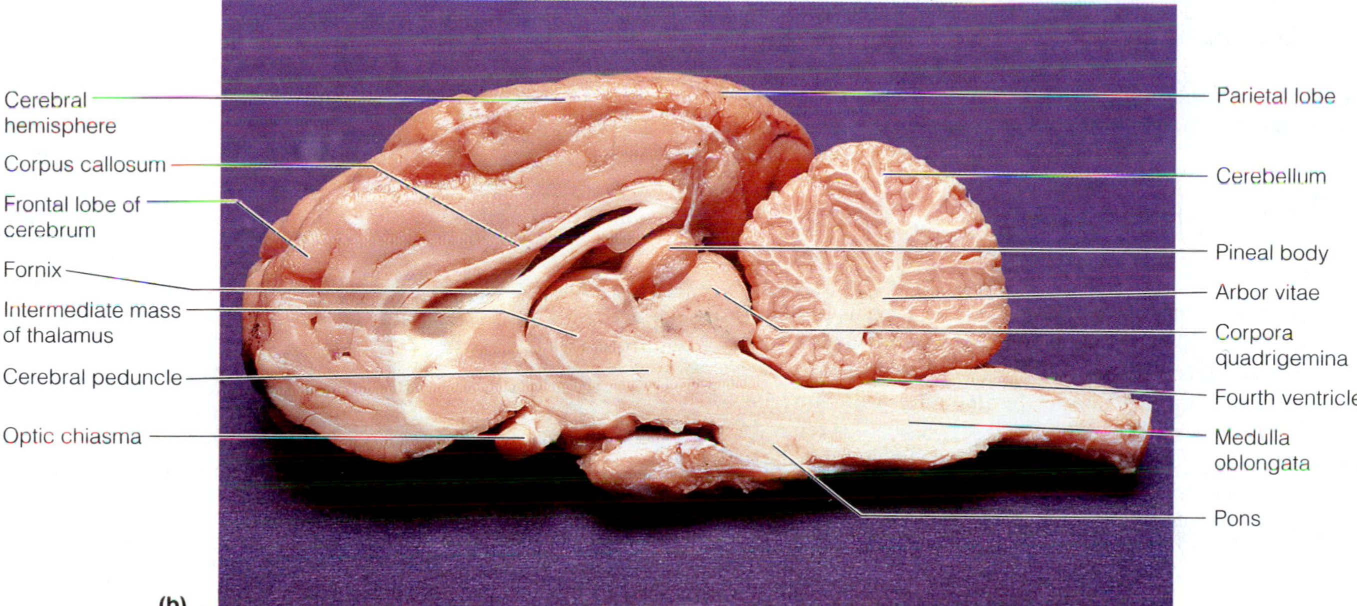

Figure 19.13 Sagittal section of the sheep brain showing internal structures. **(a)** Diagrammatic view. **(b)** Photograph.

as four rounded prominences on the dorsal midbrain surface. What is the function of the corpora quadrigemina?

____

____

Also locate the pineal body, which appears as a small oval protrusion in the midline just anterior to the corpora quadrigemina.

Internal Structures

1. The internal structure of the brain can only be examined after further dissection. Place the brain ventral side down on the dissecting tray and make a cut completely through it in a superior to inferior direction. Cut through the longitudinal fissure, corpus callosum, and midline of the cerebellum. Refer to Figure 19.13 as you work.

2. The thin nervous tissue membrane immediately ventral to the corpus callosum that separates the lateral ventricles is the septum pellucidum. Pierce this membrane and probe the lateral ventricle cavity. The fiber tract ventral to the septum pellucidum and anterior to the third ventricle is the fornix.

How does the relative size of the fornix in this brain compare with the human fornix?

____

____

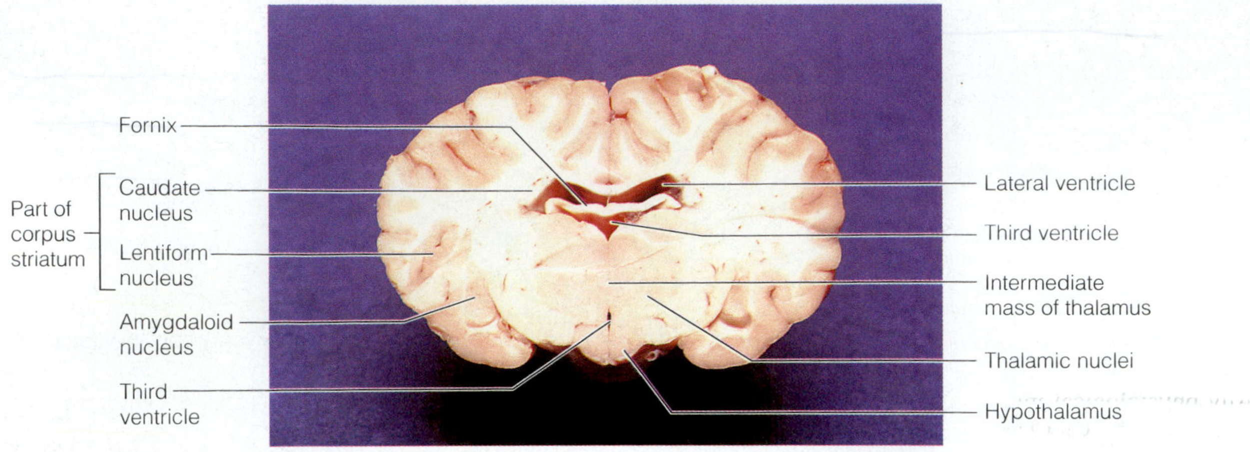

Figure 19.14 Frontal section of a sheep brain. Major structures revealed are the location of major basal nuclei in the interior, the thalamus, hypothalamus, and lateral and third ventricles.

Why do you suppose this is so? (Hint: What is the function of this band of fibers?)

____

____

3. Identify the thalamus, which forms the walls of the third ventricle and is located posterior and ventral to the fornix. The intermediate mass spanning the ventricular cavity appears as an oval protrusion of the thalamic wall. Anterior to the intermediate mass, locate the interventricular foramen, a canal connecting the lateral ventricle on the same side with the third ventricle.

4. The hypothalamus forms the floor of the third ventricle. Identify the optic chiasma, infundibulum, and mammillary body on its exterior surface. You can see the pineal body at the superoposterior end of the third ventricle, just beneath the junction of the corpus callosum and fornix.

5. Locate the midbrain by identifying the corpora quadrigemina that form its dorsal roof. Follow the cerebral aqueduct (the narrow canal connecting the third and fourth ventricles) through the midbrain tissue to the fourth ventricle. Identify the cerebral peduncles, which form its anterior walls.

6. Identify the pons and medulla, which lie anterior to the fourth ventricle. The medulla continues into the spinal cord without any obvious anatomical change, but the point at which the fourth ventricle narrows to a small canal is generally accepted as the beginning of the spinal cord.

7. Identify the cerebellum posterior to the fourth ventricle. Notice its internal treelike arrangement of white matter, the arbor vitae.

8. If time allows, obtain another sheep brain and section it along the frontal plane so that the cut passes through the infundibulum. Compare your specimen to the photograph in Figure 19.14, and attempt to identify all the structures shown in the figure.

9. Check with your instructor to determine if cow spinal cord sections (preserved) are available for the spinal cord studies in Exercise 21. If not, save the small portion of the spinal cord from your brain specimen. Otherwise, dispose of all the organic debris in the appropriate laboratory containers and clean the dissecting instruments and tray before leaving the laboratory. ■

Electroencephalography

exercise

20

Objectives

1. To define *electroencephalogram* and to discuss its clinical significance.
2. To describe or recognize typical tracings of the most common brain wave patterns (alpha, beta, theta, and delta waves) and to indicate the conditions under which each is most likely to be predominant.
3. To state the source of brain waves.
4. To define *alpha block.*
5. To monitor electroencephalography and recognize alpha rhythm.
6. To describe the effect of a sudden sound, mental concentration, and alkalosis on brain wave patterns.

Materials

- ❑ Oscilloscope and EEG lead-selector box or physiograph and high-gain preamplifier
- ❑ Cot
- ❑ Electrode gel
- ❑ EEG electrodes and leads
- ❑ Collodion gel or long elastic EEG straps

Brain Wave Patterns and the Electroencephalogram

Any physiological investigation of the brain can emphasize and expose only a very minute portion of its activity. Higher brain functions, such as consciousness and logical reasoning, are extremely difficult to investigate. It is obviously much easier to do experiments on the brain's input-output functions, some of which can be detected with appropriate recording equipment. Still, the ability to record brain activity does not necessarily guarantee an understanding of the brain.

The **electroencephalogram (EEG),** a record of the electrical activity of the brain, can be obtained through electrodes placed at various points on the skin or scalp of the head. This electrical activity, which is recorded as waves (Figure 20.1), is not completely understood at present but may be regarded as action potentials generated by brain neurons.

Certain characteristics of brain waves are known. They have a frequency of 1 to 30 hertz (Hz) or cycles per second, a dominant rhythm of 10 Hz, and an average amplitude (voltage) of 20 to 100 microvolts (μV). They vary in frequency in different brain areas, occipital waves having a lower frequency than those associated with the frontal and parietal lobes.

The first of the brain waves to be described by scientists were the alpha waves (or alpha rhythm). **Alpha waves** have an average frequency range of 8 to 13 Hz and are produced when the individual is in a relaxed state with the eyes closed. **Alpha block,** suppression of the alpha rhythm, occurs if the eyes are opened or if the individual begins to concentrate on some mental problem or visual stimulus. Under these conditions, the waves decrease in amplitude but increase in frequency. Under conditions of fright or excitement, the frequency increases still more.

Beta waves, closely related to alpha waves, are faster (14 to 25 Hz) and have a lower amplitude. They are typical of the attentive or alert state.

Very large (high-amplitude) waves with a frequency of 4 Hz or less that are seen in deep sleep are **delta waves. Theta waves** are large, abnormally contoured waves with a frequency of 4 to 7 Hz. Although theta waves are normal in children, they represent emotional problems or some sort of neural imbalance in adults.

Brain waves change with age, sensory stimuli, brain pathology or disease, and the chemical state of the body. (Glucose deprivation, oxygen poisoning, and sedatives all interfere with the rhythmic activity of brain output by disturbing the metabolism of the neurons.) Sleeping individuals and patients in a coma have EEGs that are slower (or lower frequency) than the alpha rhythm of normal adults. Fright, epileptic seizures, and various types of drug intoxication are associated with comparatively faster cortical activity. Thus impairment of cortical function is indicated by neuronal activity that is either too fast or too slow; unconsciousness occurs at both extremes of the frequency range.

Since spontaneous brain waves are always present, even during unconsciousness and coma, the absence of brain waves (a "flat" EEG) is taken as clinical evidence of death. The EEG is used clinically to diagnose and localize many types of brain lesions, including epileptic lesions, infections, abscesses, and tumors. ■

Activity:
Observing Brain Wave Patterns

If one electrode (the *active electrode*) is placed over a particular cortical area and another (the *indifferent electrode*) is placed over an inactive part of the head, such as the earlobe, all of the activity of the cortex underlying the active electrode will, theoretically, be recorded. The inactive area provides a

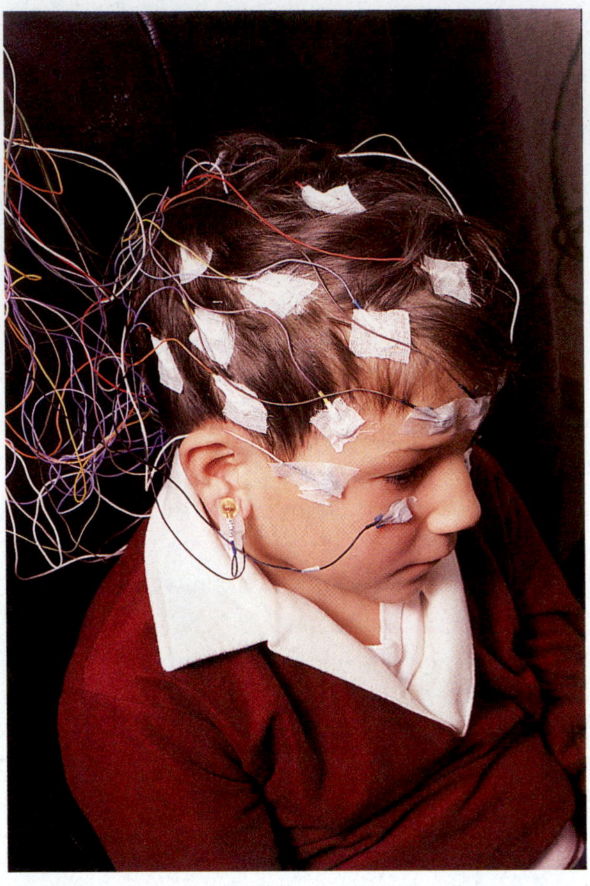

(a)

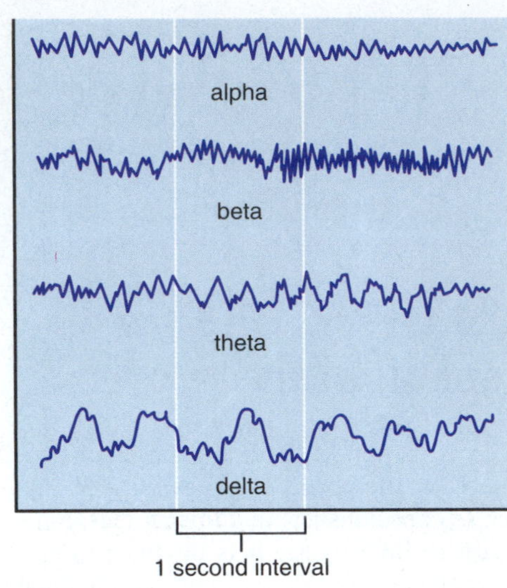

(b)

Figure 20.1 Electroencephalography and brain waves. (a) To obtain a recording of brain wave activity (an EEG), electrodes are positioned on the patient's scalp and attached to a recording device called an electroencephalograph. **(b)** Typical EEGs. Alpha waves are typical of the awake but relaxed state; beta waves occur in the awake, alert state; theta waves are common in children but not in normal awake adults; delta waves occur during deep sleep.

zero reference point or a baseline, and the EEG represents the difference between "activities" occurring under the two electrodes.

1. Connect the EEG selector box to the oscilloscope preamplifier, or connect the high-gain preamplifier to the physiograph channel amplifier. Adjust the horizontal sweep and sensitivity according to the directions given in the instrument manual or by your instructor.

2. Prepare the subject. The subject should lie undisturbed with eyes closed in a quiet, dimly lit area. (Someone who is able to relax easily makes a good subject.) Apply a small amount of electrode gel to the subject's forehead above the left eye and on the left earlobe. Press an electrode to each prepared area and secure each by (1) applying a film of collodion gel to the electrode surface and the adjacent skin or (2) using a long elastic EEG strap (knot tied at the back of the head). If collodion gel is used, allow it to dry before you continue.

3. Connect the active frontal lead (forehead) to the EEG selector box outlet marked "L Frontal." Connect the lead from the indifferent electrode (earlobe) to the ground outlet (or to the appropriate input terminal on the high-gain preamplifier).

4. Turn the oscilloscope or physiograph on, and observe the EEG pattern of the relaxed subject for a period of 5 min. If the subject is truly relaxed, you should see a typical alpha-wave pattern. (If the subject is unable to relax and the alpha-wave pattern does not appear in this time interval, test another subject.) Discourage all muscle movement during the monitoring period.*

5. Abruptly and loudly clap your hands. The subject's eyes should open and alpha block should occur. Observe the immediate brain wave pattern. How do the frequency and amplitude of the brain waves change?

____

____

Would you characterize this as beta rhythm? ____________

Why? ______________________________________

____

* Note that 60-cycle "noise" (appearing as fast, regular, low amplitude waves superimposed on the more irregular brain waves) may interfere with the tracings being made, particularly if the laboratory has a lot of electrical equipment.

6. Allow the subject about 5 min to achieve complete relaxation once again, then ask him or her to compute a number problem that requires concentration (for example, add 3 and 36, subtract 7, multiply by 2, add 50, etc.). Observe the brain wave pattern during the period of mental computation.

Observations: ______________________________

____

____

7. Once again allow the subject to relax until alpha rhythm resumes. Then, instruct him or her to hyperventilate for 3 min. *Be sure to tell the subject when to stop hyperventilating.* Hyperventilation rapidly flushes carbon dioxide out of the lungs, decreasing carbon dioxide levels in the blood and producing respiratory alkalosis.

Observe the changes in the rhythm and amplitude of the brain waves occurring during the period of hyperventilation.

Observations: ______________________________

____

____

8. Think of other stimuli that might affect brain wave patterns. Test your hypotheses. Describe what stimuli you tested and what responses you observed.

____

____

____

______________________________ ■

exercise

21 Spinal Cord and Spinal Nerves

Objectives

1. To identify important anatomical areas on a spinal cord model or appropriate diagram of the spinal cord, and to indicate the neuron type found in these areas (where applicable).
2. To indicate two major areas where the spinal cord is enlarged, and to explain the reasons for this anatomical characteristic.
3. To define *conus medullaris, cauda equina,* and *filum terminale.*
4. To locate on a diagram the fiber tracts in the spinal cord and to state their functional importance.
5. To list two major functions of the spinal cord.
6. To name the meningeal coverings of the spinal cord and state their function.
7. To describe the origin, fiber composition, and distribution of the spinal nerves, differentiating between roots, the spinal nerve proper, and rami, and to discuss the result of transecting these structures.
8. To discuss the distribution of the dorsal rami and ventral rami of the spinal nerves.
9. To identify the four major nerve plexuses, the major nerves of each, and their distribution.

Materials

- ❑ Spinal cord model (cross section)
- ❑ Laboratory charts of the spinal cord and spinal nerves and sympathetic chain
- ❑ Red and blue pencils
- ❑ Preserved cow spinal cord sections with meninges and nerve roots intact (or spinal cord segment saved from the brain dissection in Exercise 19)
- ❑ Dissecting tray and instruments
- ❑ Dissecting microscope
- ❑ Histologic slide of spinal cord (cross section)
- ❑ Disposable gloves
- ❑ Compound microscope
- ❑ *The Human Nervous System: The Spinal Cord and Spinal Nerves* videotape*

A1A See Appendix C, Exercise 21 for links to A.D.A.M.® Interactive Anatomy.

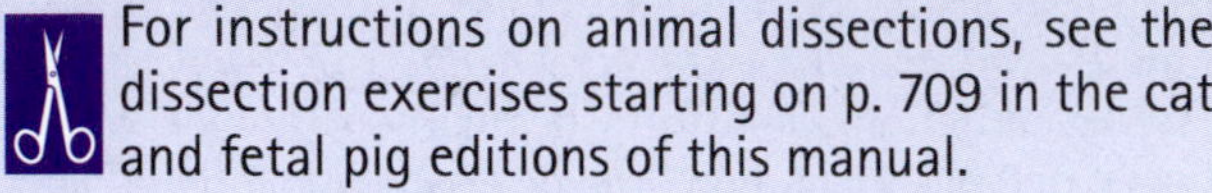
For instructions on animal dissections, see the dissection exercises starting on p. 709 in the cat and fetal pig editions of this manual.

*Available to qualified adopters from Benjamin Cummings

Anatomy of the Spinal Cord

The cylindrical **spinal cord,** a continuation of the brain stem, is an association and communication center. It plays a major role in spinal reflex activity and provides neural pathways to and from higher nervous centers. Enclosed within the vertebral canal of the spinal column, the spinal cord extends from the foramen magnum of the skull to the first or second lumbar vertebra, where it terminates in the cone-shaped **conus medullaris** (Figure 21.1). Like the brain, it is cushioned and protected by meninges. The dura mater and arachnoid meningeal coverings extend beyond the conus medullaris, approximately to the level of S_2, and the **filum terminale,** a fibrous extension of the pia mater, extends even farther (into the coccygeal canal) to attach to the posterior coccyx. *Denticulate ligaments,* saw-toothed shelves of pia mater, secure the spinal cord to the bony wall of the vertebral column all along its length (see Figure 21.1c).

The fact that the meninges, filled with cerebrospinal fluid, extend well beyond the end of the spinal cord provides an excellent site for removing cerebrospinal fluid for analysis (as when bacterial or viral infections of the spinal cord or its meningeal coverings are suspected) without endangering the delicate spinal cord. This procedure, called a *lumbar tap,* is usually performed below L_3. Additionally, "saddle block" or caudal anesthesia for childbirth is normally administered (injected) between L_3 and L_5.

In humans, 31 pairs of spinal nerves arise from the spinal cord and pass through intervertebral foramina to serve the body area at their approximate level of emergence. The cord is about the size of a finger in circumference for most of its length, but there are obvious enlargements in the cervical and lumbar areas where the nerves serving the upper and lower limbs issue from the cord.

Because the spinal cord does not extend to the end of the vertebral column, the spinal nerves emerging from the inferior end of the cord must travel through the vertebral canal for some distance before exiting at the appropriate intervertebral

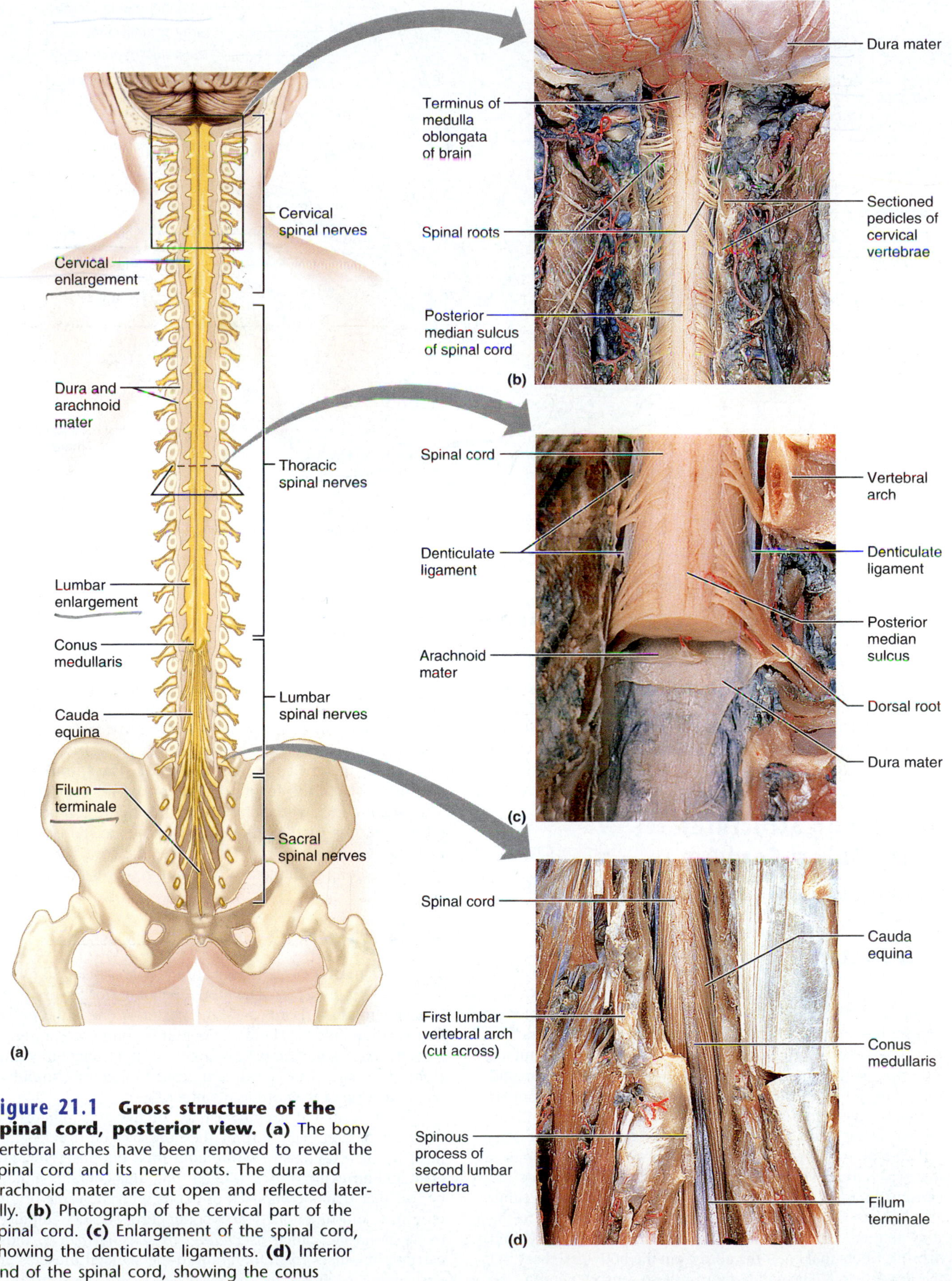

Figure 21.1 Gross structure of the spinal cord, posterior view. **(a)** The bony vertebral arches have been removed to reveal the spinal cord and its nerve roots. The dura and arachnoid mater are cut open and reflected laterally. **(b)** Photograph of the cervical part of the spinal cord. **(c)** Enlargement of the spinal cord, showing the denticulate ligaments. **(d)** Inferior end of the spinal cord, showing the conus medullaris, cauda equina, and filum terminale.

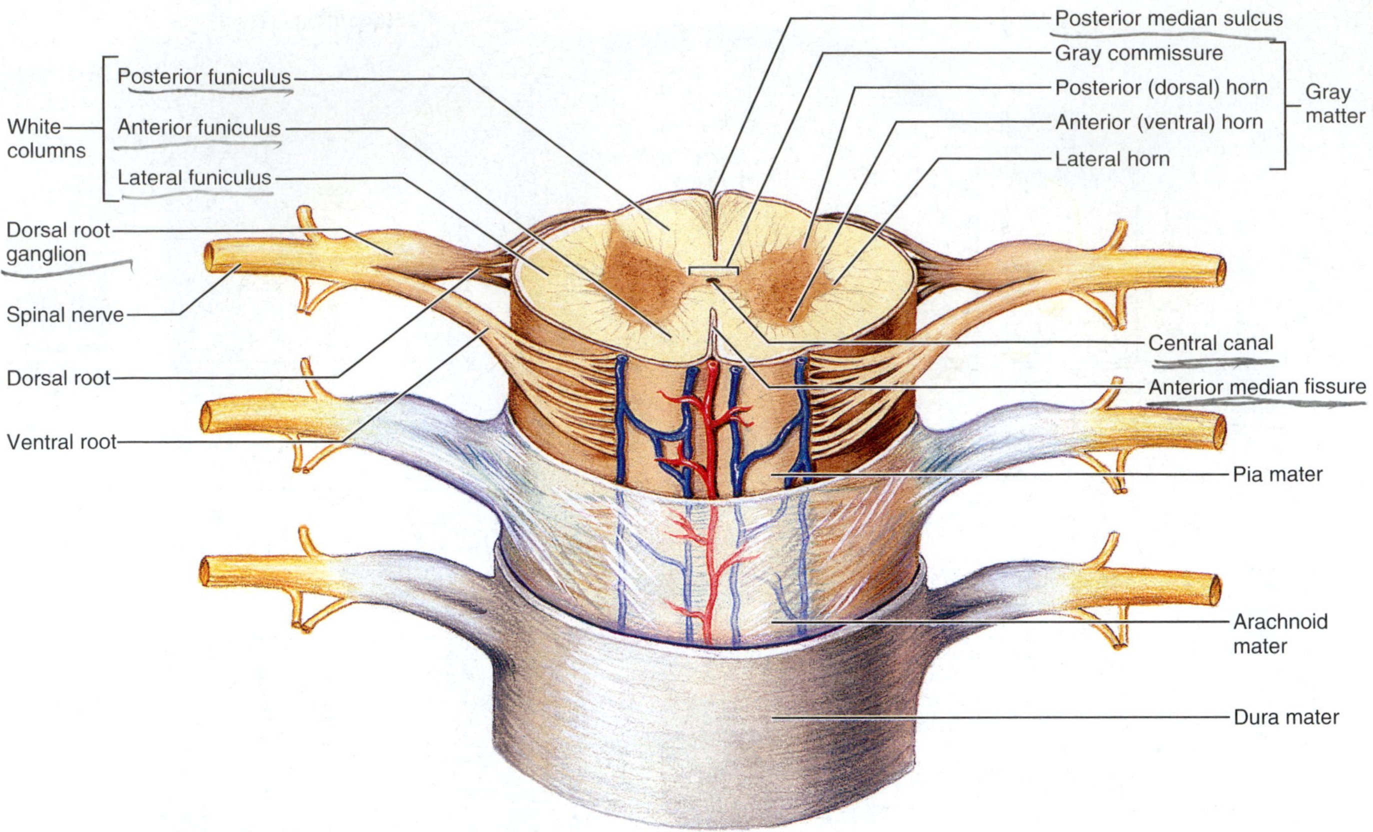

Figure 21.2 Anatomy of the human spinal cord (three-dimensional view).

foramina. This collection of spinal nerves traversing the inferior end of the vertebral canal is called the **cauda equina** (Figure 21.1d) because of its similarity to a horse's tail (the literal translation of *cauda equina*).

Activity 1: Identifying Structures of the Spinal Cord

Obtain a model of a cross section of a spinal cord and identify its structures as they are described next. ■

Gray Matter

In cross section, the **gray matter** of the spinal cord looks like a butterfly or the letter H (Figure 21.2). The two posterior projections are called the **posterior,** or **dorsal, horns.** The two anterior projections are the **anterior,** or **ventral, horns.** The tips of the anterior horns are broader and less tapered than those of the posterior horns. In the thoracic and lumbar regions of the cord, there is also a lateral outpocketing of gray matter on each side referred to as the **lateral horn.** The central area of gray matter connecting the two vertical regions is the **gray commissure.** The gray commissure surrounds the **central canal** of the cord, which contains cerebrospinal fluid.

Neurons with specific functions can be localized in the gray matter. The posterior horns contain interneurons and sensory fibers that enter the cord from the body periphery via the **dorsal root.** The cell bodies of these sensory neurons are found in an enlarged area of the dorsal root called the **dorsal root ganglion.** The anterior horns mainly contain cell bodies of motor neurons of the somatic nervous system (voluntary system), which send their axons out via the **ventral root** of the cord to enter the adjacent spinal nerve. The **spinal nerves** are formed from the fusion of the dorsal and ventral roots. The lateral horns, where present, contain nerve cell bodies of motor neurons of the autonomic nervous system (sympathetic division). Their axons also leave the cord via the ventral roots, along with those of the motor neurons of the anterior horns.

White Matter

The **white matter** of the spinal cord is nearly bisected by fissures (see Figure 21.2). The more open anterior fissure is the **anterior median fissure,** and the posterior one is the **posterior median sulcus.** The white matter is composed of myelinated fibers—some running to higher centers, some traveling from the brain to the cord, and some conducting impulses from one side of the cord to the other.

Because of the irregular shape of the gray matter, the white matter on each side of the cord can be divided into three primary regions or **white columns:** the **posterior, lateral,** and **anterior funiculi.** Each funiculus contains a number of fiber **tracts** composed of axons with the same origin, terminus, and function. Tracts conducting sensory impulses to the brain are called *ascending,* or *sensory, tracts;* those carrying impulses from the brain to the skeletal muscles are *descending,* or *motor, tracts.*

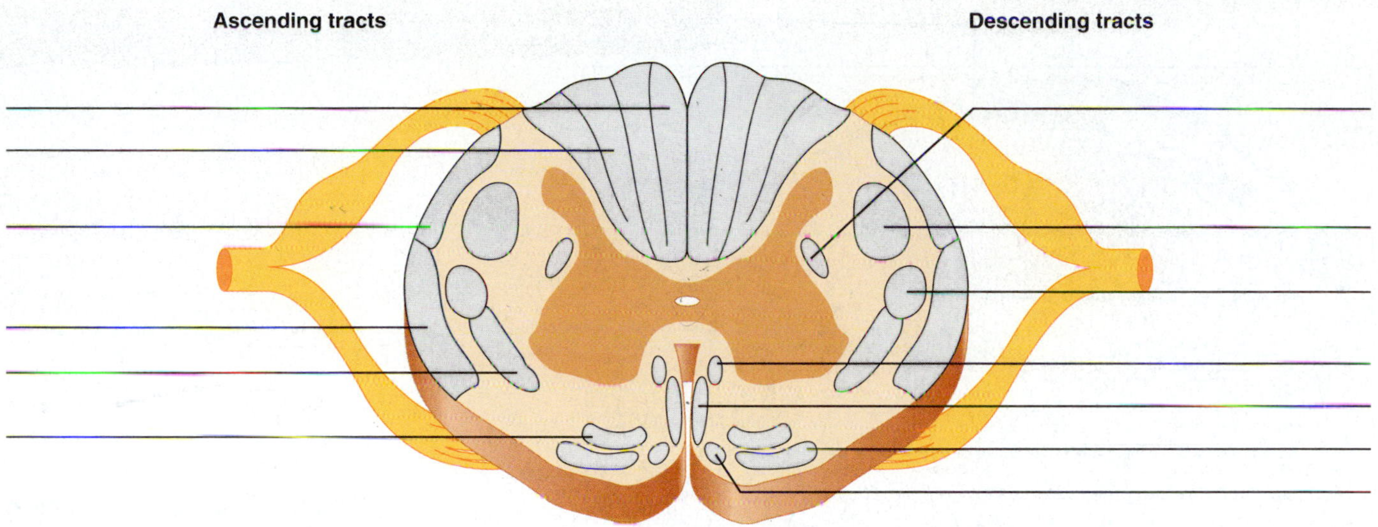

Figure 21.3 Cross section of the spinal cord showing the relative positioning of its major tracts.

Because it serves as the transmission pathway between the brain and the body periphery, the spinal cord is an extremely important functional area. Even though it is protected by meninges and cerebrospinal fluid in the vertebral canal, it is highly vulnerable to traumatic injuries, such as might occur in an automobile accident.

When the cord is transected (or severely traumatized), both motor and sensory functions are lost in body areas normally served by that (and lower) regions of the spinal cord. Injury to certain spinal cord areas may even result in a permanent flaccid paralysis of both legs (paraplegia) or of all four limbs (quadriplegia). ■

Activity 2:
Identifying Spinal Cord Tracts

With the help of your textbook, label Figure 21.3 with the tract names that follow. Each tract is represented on both sides of the cord, but for clarity, label the motor tracts on the right side of the diagram and the sensory tracts on the left side of the diagram. *Color ascending tracts blue and descending tracts red.* Then fill in the functional importance of each tract beside its name below. As you work, try to be aware of how the naming of the tracts is related to their anatomical distribution.

Fasciculus gracilis ______________________

Fasciculus cuneatus ______________________

Posterior spinocerebellar ______________________

Anterior spinocerebellar ______________________

Lateral spinothalamic ______________________

Anterior spinothalamic ______________________

Lateral corticospinal ______________________

Anterior corticospinal ______________________

Rubrospinal ______________________

Tectospinal ______________________

Vestibulospinal ______________________

Medial reticulospinal ______________________

Lateral reticulospinal ______________________ ■

Dissection:
Spinal Cord

1. Obtain a dissecting tray and instruments, disposable gloves, and a segment of preserved spinal cord (from a cow or saved from the brain specimen used in Exercise 19). Identify the tough outer meninx (dura mater) and the weblike arachnoid mater.

What name is given to the third meninx, and where is it found?

____

____

Peel back the dura mater and observe the fibers making up the dorsal and ventral roots. If possible, identify a dorsal root ganglion.

2. Cut a thin cross section of the cord and identify the anterior and posterior horns of the gray matter with the naked eye or with the aid of a dissecting microscope.

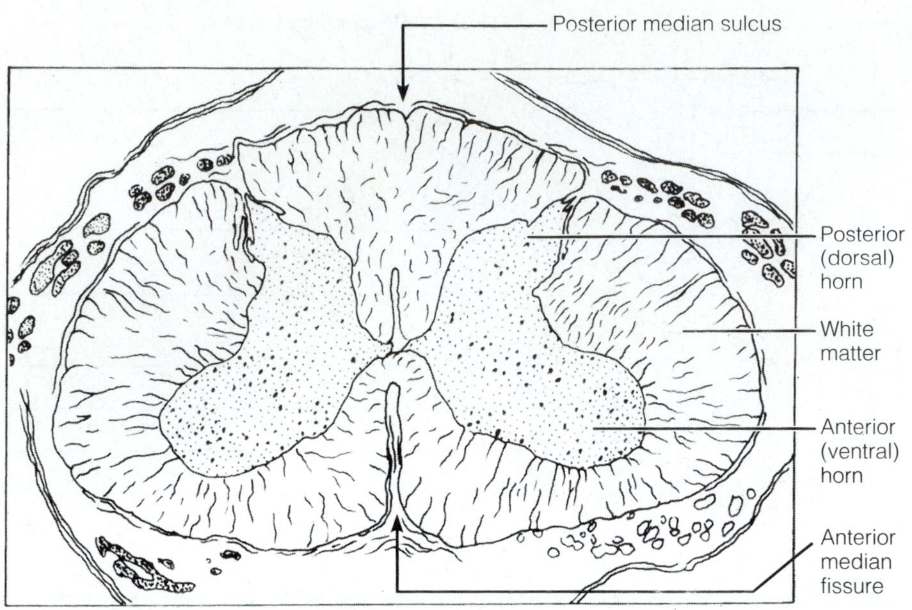

Figure 21.4 Cross section of the spinal cord. (See also Plate 9 in the Histology Atlas.)

How can you be certain that you are correctly identifying the anterior and posterior horns?

____

____

Also identify the central canal, white matter, anterior median fissure, posterior median sulcus, and posterior, anterior, and lateral funiculi.

3. Obtain a prepared slide of the spinal cord (cross section) and a compound microscope. Refer to Figure 21.4 as you examine the slide carefully under low power. Observe the shape of the central canal.

Is it basically circular or oval? ______________

Name the glial cell type that lines this canal. ______________

What would you expect to find in this canal in the living animal?

____

Can any neuron cell bodies be seen? ______________

Where? ______________________________

____

What type of neurons would these most likely be—motor, association, or sensory?

______________________________ ■

Spinal Nerves and Nerve Plexuses

The 31 pairs of human spinal nerves arise from the fusions of the ventral and dorsal roots of the spinal cord. Figure 21.5 shows how the nerves are named according to their point of issue. Because the ventral roots contain myelinated axons of motor neurons located in the cord and the dorsal roots carry sensory fibers entering the cord, all spinal nerves are **mixed nerves.** The first pair of spinal nerves leaves the vertebral canal between the base of the occiput and the atlas, but all the rest exit via the intervertebral foramina. The first through seventh pairs of cervical nerves emerge *above* the vertebra for which they are named. C_8 emerges between C_7 and T_1. (Notice that there are 7 cervical vertebrae, but 8 pairs of cervical nerves.) The remaining spinal nerve pairs emerge from the spinal cord *below* the same-numbered vertebra.

Text continues on page 220

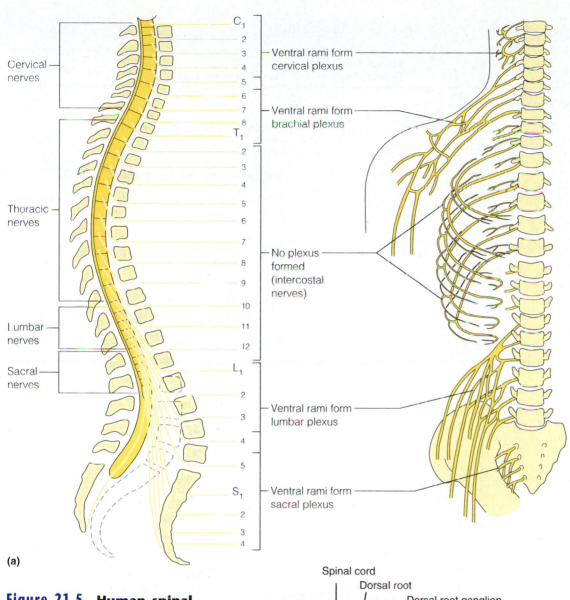

Cervical nerves
Thoracic nerves
Lumbar nerves
Sacral nerves
C_1 2 3 4 5 6 7 8
T_1 2 3 4 5 6 7 8 9 10 11 12
L_1 2 3 4 5
S_1 2 3 4
Ventral rami form cervical plexus
Ventral rami form brachial plexus
No plexus formed (intercostal nerves)
Ventral rami form lumbar plexus
Ventral rami form sacral plexus

(a)

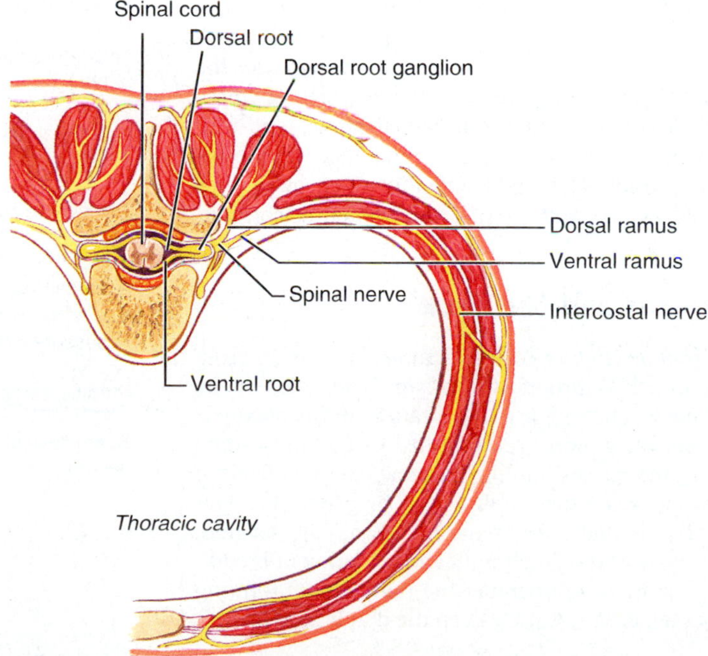

(b)

Figure 21.5 Human spinal nerves. (a) Relationship of spinal nerves to vertebrae (areas of plexuses formed by the ventral rami are indicated). **(b)** Relative distribution of the ventral and dorsal rami of a spinal nerve (cross section of left trunk).

Table 21.1 Branches of the Cervical Plexus (See Figure 21.6)

Nerves	Spinal roots (ventral rami)	Structures served
Cutaneous Branches (Superficial)		
Lesser occipital	C_2 (C_3)	Skin on posterolateral aspect of neck
Greater auricular	C_2, C_3	Skin of ear, skin over parotid gland
Transverse cutaneous	C_2, C_3	Skin on anterior and lateral aspect of neck
Supraclavicular (anterior, middle, and posterior)	C_3, C_4	Skin of shoulder and anterior aspect of chest
Motor Branches (Deep)		
Ansa cervicalis (superior and inferior roots)	C_1–C_3	Infrahyoid muscles of neck (omohyoid, sternohyoid, and sternothyroid)
Segmental and other muscular branches	C_1–C_5	Deep muscles of neck (geniohyoid and thyrohyoid) and portions of scalenes, levator scapulae, trapezius, and sternocleidomastoid muscles
Phrenic	C_3–C_5	Diaphragm (sole motor nerve supply)

Almost immediately after emerging, each nerve divides into **dorsal** and **ventral rami.** (Thus each spinal nerve is only about 1 or 2 cm long.) The rami, like the spinal nerves, contain both motor and sensory fibers. The smaller dorsal rami serve the skin and musculature of the posterior body trunk at their approximate level of emergence. The ventral rami of spinal nerves T_2–T_{12} pass anteriorly as the **intercostal nerves** to supply the muscles of intercostal spaces, and the skin and muscles of the anterior and lateral trunk. The ventral rami of all other spinal nerves form complex networks of nerves called **plexuses.** These plexuses serve the motor and sensory needs of the muscles and skin of the limbs. The fibers of the ventral rami unite in the plexuses (with a few rami supplying fibers to more than one plexus). From the plexuses the fibers diverge again to form peripheral nerves, each of which contains fibers from more than one spinal nerve. The four major nerve plexuses and their chief peripheral nerves are described and in Tables 21.1–21.4 and illustrated in Figures 21.6–21.9. Their names and site of origin should be committed to memory.

Cervical Plexus and the Neck

The **cervical plexus** (Figure 21.6 and Table 21.1) arises from the ventral rami of C_1 through C_5 to supply muscles of the shoulder and neck. The major motor branch of this plexus is the **phrenic nerve,** which arises from C_3–C_4 (plus some fibers from C_5) and passes into the thoracic cavity in front of the first rib to innervate the diaphragm. The primary danger of a broken neck is that the phrenic nerve may be severed, leading to paralysis of the diaphragm and cessation of breathing. A jingle to help you remember the rami (roots) forming the phrenic nerves is "C_3, C_4, C_5 keep the diaphragm alive."

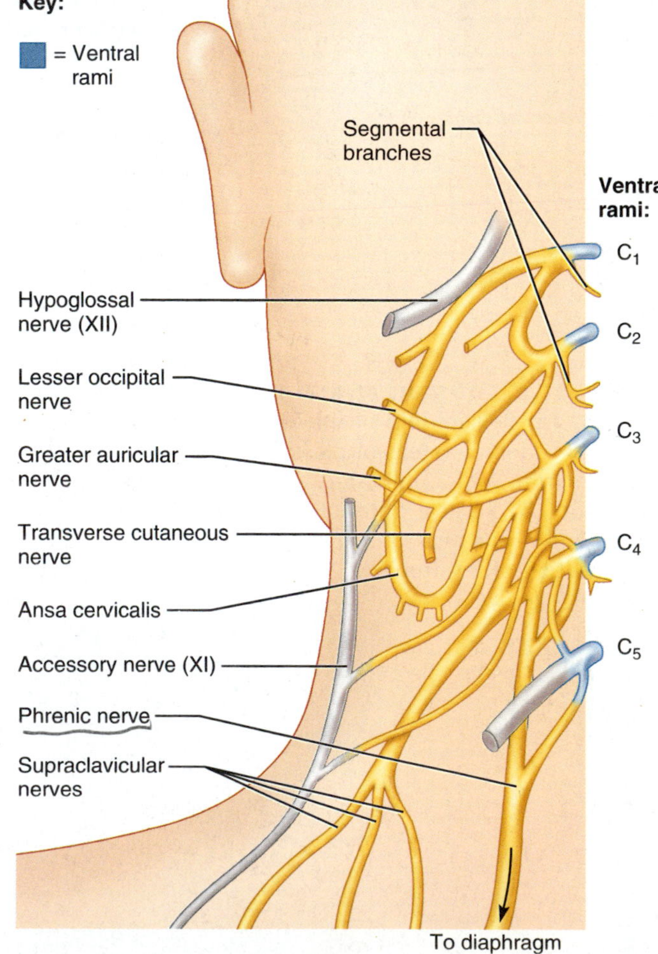

Figure 21.6 The cervical plexus. The nerves colored gray connect to the plexus but do not belong to it. (See Table 21.1.)

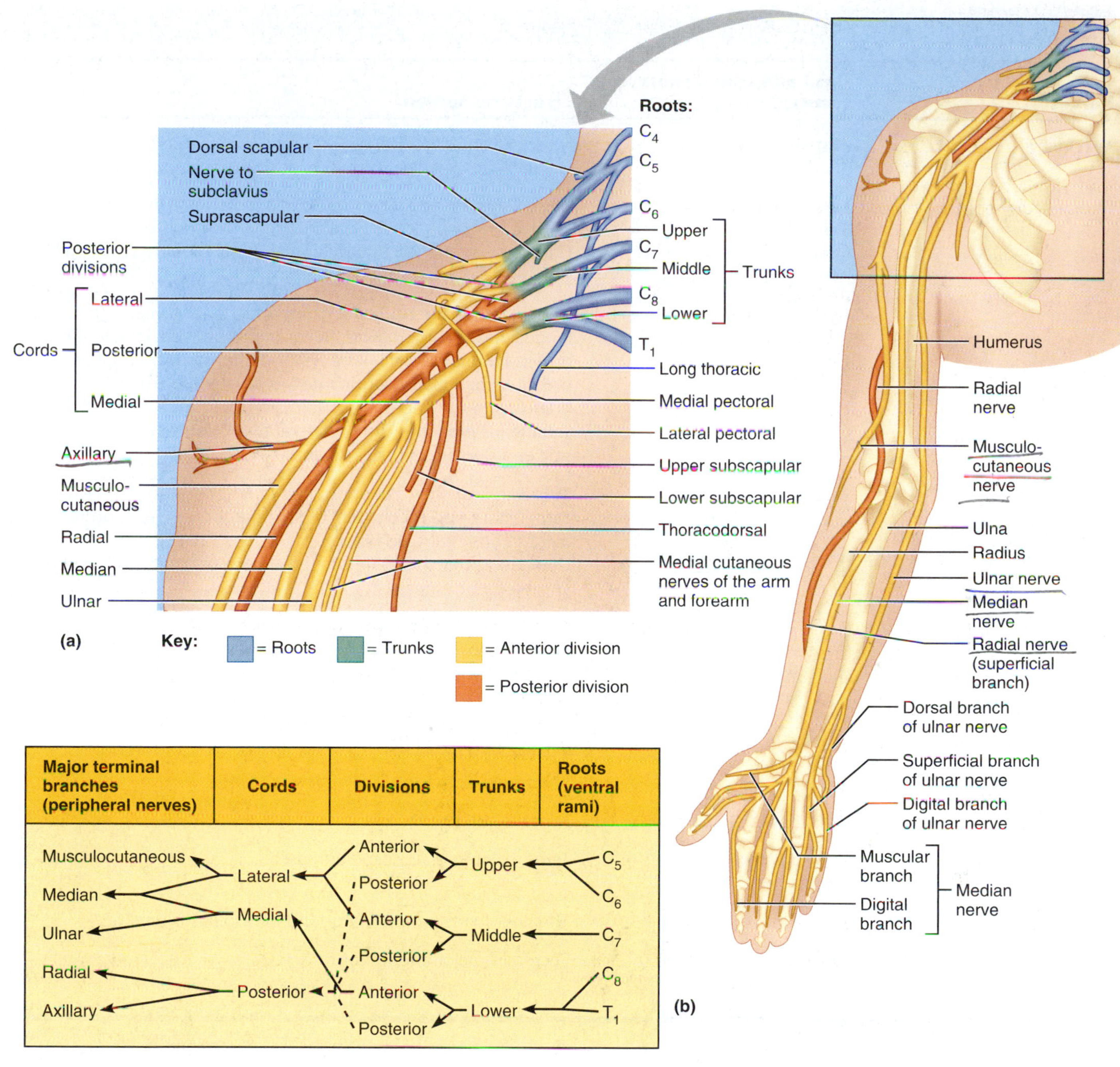

Figure 21.7 The brachial plexus. (a and b) Distribution of the major peripheral nerves of the upper limb. **(c)** Flowchart of consecutive branches in the brachial plexus from the major nerves formed from the cords to the spinal roots (ventral rami). (See Table 21.2. See also Plate G in the Human Anatomy Atlas.)

Brachial Plexus and the Upper Limb

The **brachial plexus** is large and complex, arising from the ventral rami of C_5 through C_8 and T_1 (Table 21.2). The plexus, after being rearranged consecutively into *trunks, divisions,* and *cords,* finally becomes subdivided into five major *peripheral nerves* (Figure 21.7 and Plate G of the Human Anatomy Atlas).

The **axillary nerve,** which serves the muscles and skin of the shoulder, has the most limited distribution. The large **radial nerve** passes down the posterolateral surface of the arm and forearm, supplying all the extensor muscles of the arm, forearm, and hand and the skin along its course. The radial nerve is often injured in the axillary region by the pressure of a crutch or by hanging one's arm over the back of a

Table 21.2 Branches of the Brachial Plexus (See Figure 21.7)

Nerves	Cord and spinal roots (ventral rami)	Structures served
Axillary	Posterior cord (C_5, C_6)	Muscular branches: deltoid and teres minor muscles Cutaneous branches: some skin of shoulder region
Musculocutaneous	Lateral cord (C_5–C_7)	Muscular branches: flexor muscles in anterior arm (biceps brachii, brachialis, coracobrachialis) Cutaneous branches: skin on anterolateral forearm (extremely variable)
Median	By two branches, one from medial cord (C_8, T_1) and one from the lateral cord (C_5–C_7)	Muscular branches to flexor group of anterior forearm (palmaris longus, flexor carpi radialis, flexor digitorum superficialis, flexor pollicis longus, lateral half of flexor digitorum profundus, and pronator muscles); intrinsic muscles of lateral palm and digital branches to the fingers Cutaneous branches: skin of lateral two-thirds of hand, palm side and dorsum of fingers 2 and 3
Ulnar	Medial cord (C_8, T_1)	Muscular branches: flexor muscles in anterior forearm (flexor carpi ulnaris and medial half of flexor digitorum profundus); most intrinsic muscles of hand Cutaneous branches: skin of medial third of hand, both anterior and posterior aspects
Radial	Posterior cord (C_5–C_8, T_1)	Muscular branches: posterior muscles of arm, forearm, and hand (triceps brachii, anconeus, supinator, brachioradialis, extensors carpi radialis longus and brevis, extensor carpi ulnaris, and several muscles that extend the fingers) Cutaneous branches: skin of posterolateral surface of entire limb (except dorsum of fingers 2 and 3)
Dorsal scapular	Branches of C_5 rami	Rhomboid muscles and levator scapulae
Long thoracic	Branches of C_5–C_7 rami	Serratus anterior muscle
Subscapular	Posterior cord; branches of C_5 and C_6 rami	Teres major and subscapular muscles
Suprascapular	Upper trunk (C_5, C_6)	Shoulder joint; supraspinatus and infraspinatus muscles
Pectoral (lateral and medial)	Branches of lateral and medial cords (C_5–T_1)	Pectoralis major and minor muscles

chair. The **median nerve** passes down the anteromedial surface of the arm to supply most of the flexor muscles in the forearm and several muscles in the hand (plus the skin of the lateral surface of the palm of the hand).

• Hyperextend your wrist to identify the long, obvious tendon of your palmaris longus muscle, which crosses the exact midline of the anterior wrist. Your median nerve lies immediately deep to that tendon, and the radial nerve lies just *lateral* to it.

The **musculocutaneous nerve** supplies the arm muscles that flex the forearm and the skin of the lateral surface of the forearm. The **ulnar nerve** travels down the posteromedial surface of the arm. It courses around the medial epicondyle of the humerus to supply the flexor carpi ulnaris, the ulnar head of the flexor digitorum profundus of the forearm, and all intrinsic muscles of the hand not served by the median nerve. It supplies the skin of the medial third of the hand, both the anterior and posterior surfaces. Trauma to the ulnar nerve, which often occurs when the elbow is hit, produces a smarting sensation commonly referred to as "hitting the funny bone."

Severe injuries to the brachial plexus cause weakness or paralysis of the entire upper limb. Such injuries may occur when the upper limb is pulled hard and the plexus is stretched (as when a football tackler yanks the arm of the halfback), and by blows to the shoulder that force the humerus inferiorly (as when a cyclist is pitched headfirst off his motorcycle and grinds his shoulder into the pavement). ■

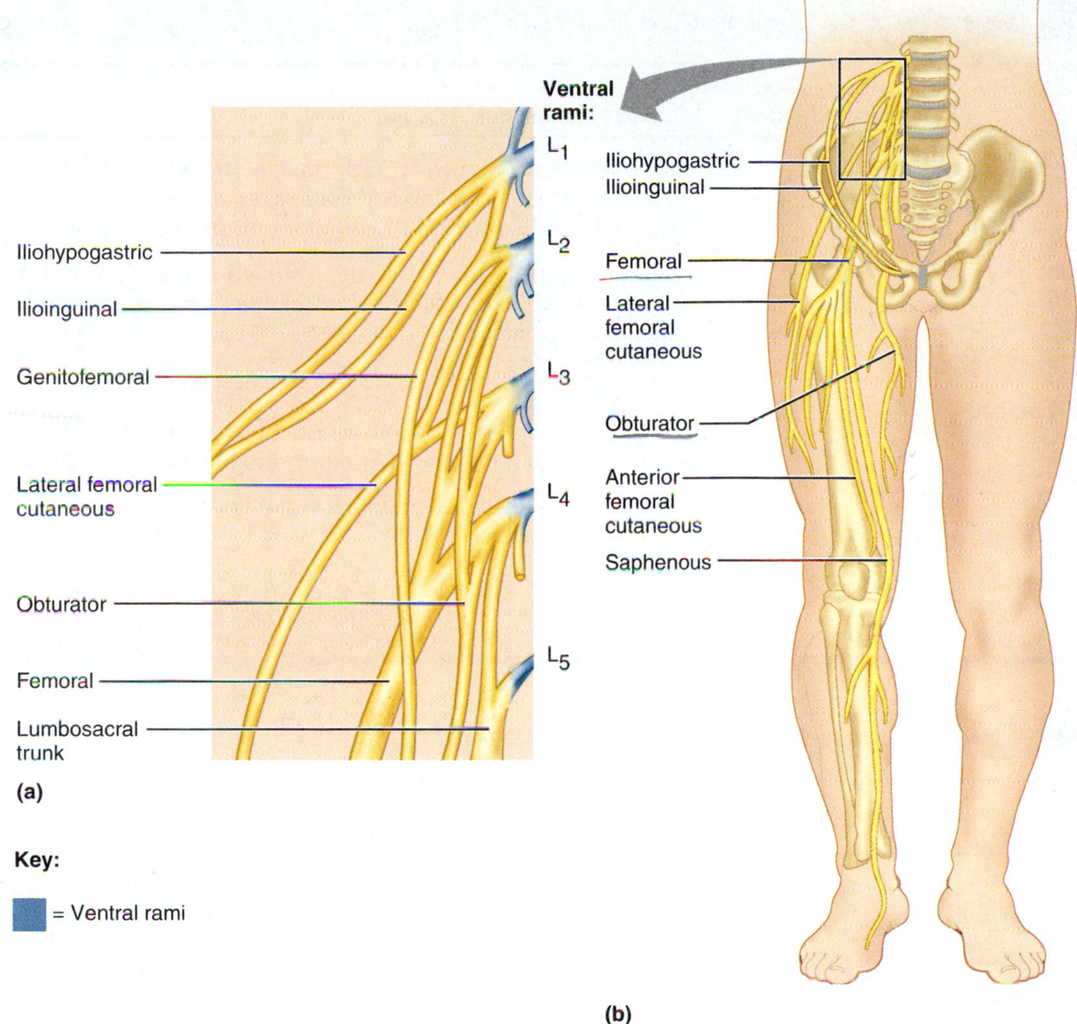

Figure 21.8 The lumbar plexus. (a) Spinal roots (anterior rami) and major branches of the lumbar plexus. **(b)** Distribution of the major peripheral nerves of the lumbar plexus in the lower limb. (See Table 21.3. See also Plate H in the Human Anatomy Atlas.)

Lumbosacral Plexus and the Lower Limb

The **lumbosacral plexus,** which serves the pelvic region of the trunk and the lower limbs, is actually a complex of two plexuses, the lumbar plexus and the sacral plexus (see Figures 21.8 and 21.9). These plexuses interweave considerably and many fibers of the lumbar plexus contribute to the sacral plexus.

The Lumbar Plexus The **lumbar plexus** arises from ventral rami of L_1 through L_4 (and sometimes T_{12}). Its nerves serve the lower abdominopelvic region and the anterior thigh (Table 21.3 and Figure 21.8). The largest nerve of this plexus is the **femoral nerve,** which passes beneath the inguinal ligament to innervate the anterior thigh muscles. The cutaneous branches of the femoral nerve (median and anterior femoral cutaneous and the saphenous nerves) supply the skin of the anteromedial surface of the entire lower limb. See Plate H in the Human Anatomy Atlas, p. 705.

Table 21.3 Branches of the Lumbar Plexus (See Figure 21.8)

Nerves	Spinal roots (ventral rami)	Structures served
Femoral	L_2–L_4	Skin of anterior and medial thigh via *anterior femoral cutaneous* branch; skin of medial leg and foot, hip and knee joints via *saphenous* branch; motor to anterior muscles (quadriceps and sartorius) of thigh; pectineus, iliacus
Obturator	L_2–L_4	Motor to adductor magnus (part), longus and brevis muscles, gracilis muscle of medial thigh, obturator externus; sensory for skin of medial thigh and for hip and knee joints
Lateral femoral cutaneous	L_2, L_3	Skin of lateral thigh; some sensory branches to peritoneum
Iliohypogastric	L_1	Skin of lower abdomen, lower back, and hip; muscles of anterolateral abdominal wall (obliques and transversus) and pubic region
Ilioinguinal	L_1	Skin of external genitalia and proximal medial aspect of the thigh; inferior abdominal muscles
Genitofemoral	L_1, L_2	Skin of scrotum in males, of labia majora in females, and of anterior thigh inferior to middle portion of inguinal region; cremaster muscle in males

Table 21.4 Branches of the Sacral Plexus (See Figure 21.9)

Nerves	Spinal roots (ventral rami)	Structures served
Sciatic nerve	L_4, L_5, S_1–S_3	Composed of two nerves (tibial and common fibular) in a common sheath that diverge just proximal to the knee
• Tibial (including sural branch and medial and lateral plantar branches)	L_4–S_3	Cutaneous branches: to skin of posterior surface of leg and sole of foot Motor branches: to muscles of back of thigh, leg, and foot (hamstrings [except short head of biceps femoris], posterior part of adductor magnus, triceps surae, tibialis posterior, popliteus, flexor digitorum longus, flexor hallucis longus, and intrinsic muscles of foot)
• Common fibular (superficial and deep branches)	L_4–S_2	Cutaneous branches: to skin of anterior surface of leg and dorsum of foot Motor branches: to short head of biceps femoris of thigh, fibularis muscles of lateral compartment of leg, tibialis anterior, and extensor muscles of toes (extensor hallucis longus, extensors digitorum longus and brevis)
Superior gluteal	L_4, L_5, S_1	Motor branches: to gluteus medius and minimus and tensor fasciae latae
Inferior gluteal	L_5–S_2	Motor branches: to gluteus maximus
Posterior femoral cutaneous	S_1–S_3	Skin of buttock, posterior thigh, and popliteal region; length variable; may also innervate part of skin of calf and heel
Pudendal	S_2–S_4	Supplies most of skin and muscles of perineum (region encompassing external genitalia and anus and including clitoris, labia, and vaginal mucosa in females, and scrotum and penis in males); external anal sphincter

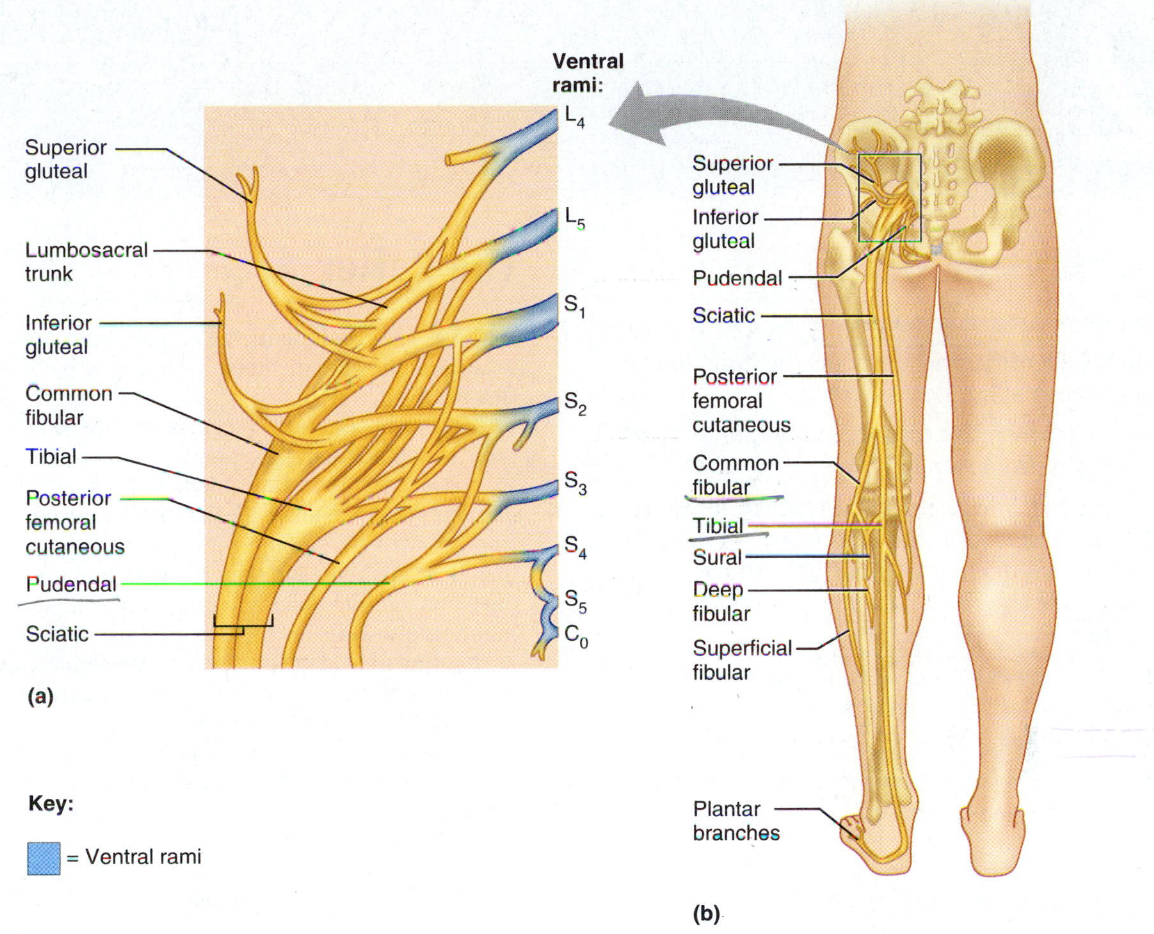

Figure 21.9 The sacral plexus. (a) The spinal roots (anterior rami) and major branches of the sacral plexus. **(b)** Distribution of the major peripheral nerves of the sacral plexus in the lower limb (posterior view). (See Table 21.4. See also Plates I and J in the Human Anatomy Atlas.)

The Sacral Plexus Arising from L_4 through S_4, the nerves of the **sacral plexus** supply the buttock, the posterior surface of the thigh, and virtually all sensory and motor fibers of the leg and foot (Table 21.4 and Figure 21.9). The major peripheral nerve of this plexus is the **sciatic nerve,** the largest nerve in the body. The sciatic nerve leaves the pelvis through the greater sciatic notch and travels down the posterior thigh, serving its flexor muscles and skin. In the popliteal region, the sciatic nerve divides into the **common fibular nerve** and the **tibial nerve,** which together supply the balance of the leg muscles and skin, both directly and via several branches. See Plates I and J in the Human Anatomy Atlas, p. 000.

Injury to the proximal part of the sciatic nerve, as might follow a fall or disc herniation, results in a number of lower limb impairments. *Sciatica* (si-at'ĭ-kah), characterized by stabbing pain radiating over the course of the sciatic nerve, is common. When the sciatic nerve is completely severed, the leg is nearly useless. The leg cannot be flexed and the foot drops into plantar flexion (dangles), a condition called *footdrop*. ■

Activity 3:
Identifying the Major Nerve Plexuses and Peripheral Nerves

Identify each of the four major nerve plexuses (and its major nerves) shown in Figures 21.6 to 21.9 on a large laboratory chart or model. Trace the courses of the nerves and relate those observations to the information provided in Tables 21.1 to 21.4. ■

exercise

22 Human Reflex Physiology

Objectives

1. To define *reflex* and *reflex arc.*
2. To name, identify, and describe the function of each element of a reflex arc.
3. To indicate why reflex testing is an important part of every physical examination.
4. To describe and discuss several types of reflex activities as observed in the laboratory; to indicate the functional or clinical importance of each; and to categorize each as a somatic or autonomic reflex action.
5. To explain why cord-mediated reflexes are generally much faster than those involving input from the higher brain centers.
6. To investigate differences in reaction time of reflexes and learned responses.

Materials

- ❑ Reflex hammer
- ❑ Sharp pencils
- ❑ Cot (if available)
- ❑ Absorbent cotton (sterile)
- ❑ Tongue depressor
- ❑ Metric ruler
- ❑ Reaction time ruler (if available)
- ❑ Flashlight
- ❑ 100- or 250-ml beaker
- ❑ 10- or 25-ml graduated cylinder
- ❑ Lemon juice in dropper bottle
- ❑ Wide-range pH paper
- ❑ Large laboratory bucket containing freshly prepared 10% household bleach solution (for saliva-soiled glassware)
- ❑ Disposable autoclave bag
- ❑ Wash bottle containing 10% bleach solution
- ❑ PowerLab® unit and computer
- ❑ Chart software
- ❑ Event marker and cable

The Reflex Arc

Reflexes are rapid, predictable, involuntary motor responses to stimuli; they are mediated over neural pathways called **reflex arcs.**

Reflexes can be categorized into one of two large groups: autonomic reflexes and somatic reflexes. **Autonomic** (or visceral) **reflexes** are mediated through the autonomic nervous system, and we are not usually aware of them. These reflexes activate smooth muscles, cardiac muscle, and the glands of the body and they regulate body functions such as digestion, elimination, blood pressure, salivation, and sweating. **Somatic reflexes** include all those reflexes that involve stimulation of skeletal muscles by the somatic division of the nervous system. An example of such a reflex is the rapid withdrawal of a hand from a hot object.

Reflex testing is an important diagnostic tool for assessing the condition of the nervous system. Distorted, exaggerated, or absent reflex responses may indicate degeneration or pathology of portions of the nervous system, often before other signs are apparent.

If the spinal cord is damaged, the easily performed reflex tests can help pinpoint the area (level) of spinal cord injury. Motor nerves above the injured area may be unaffected, whereas those at or below the lesion site may be unable to participate in normal reflex activity. ■

Components of a Reflex Arc

All reflex arcs have five essential components (Figure 22.1):

1. The *receptor* is the site of stimulus action.

2. The *sensory neuron* transmits the afferent impulses to the CNS.

3. The *integration center* consists of one or more synapses in the CNS.

4. The *motor neuron* conducts the efferent impulses from the integration center to an effector organ.

5. The *effector,* muscle fibers or glands, responds to the efferent impulses characteristically (by contracting or secreting, respectively).

The simple patellar or knee-jerk reflex shown in Figure 22.2a is an example of a simple, two-neuron, *monosynaptic* (literally, "one synapse") reflex arc. It will be demonstrated in the laboratory. However, most reflexes are more complex and *polysynaptic,* involving the participation of one or more association neurons in the reflex arc pathway. A three-neuron reflex arc (flexor reflex) is diagrammed in Figure 22.2b.

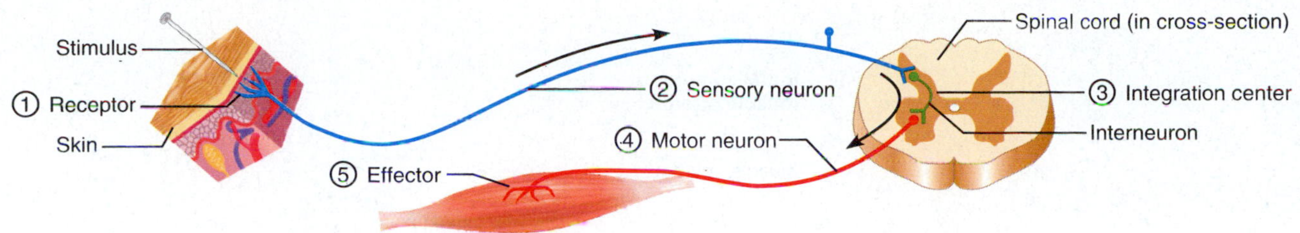

Figure 22.1 Simple reflex arcs. Components of all human reflex arcs: receptor, sensory neuron, integration center (one or more synapses in the CNS), motor neuron, and effector.

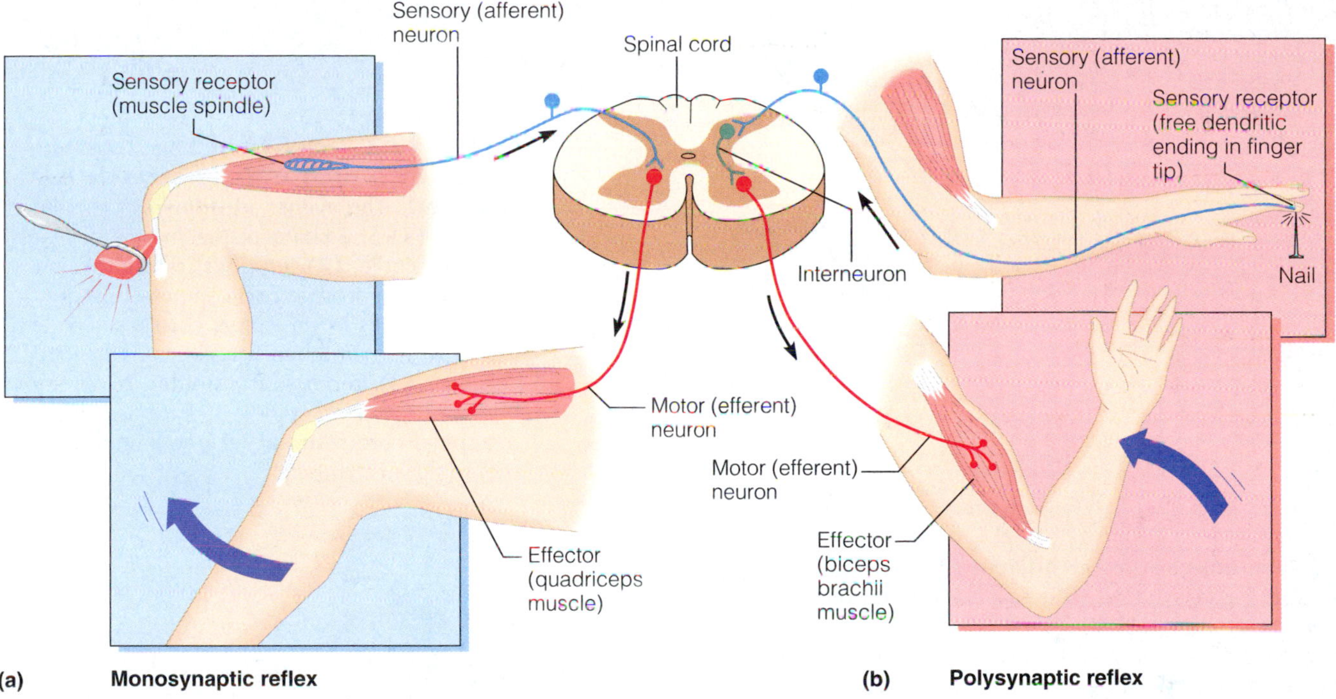

Figure 22.2 Monosynaptic and polysynaptic reflex arcs. The integration center is in the spinal cord, and in each example the receptor and effector are in the same limb. **(a)** The patellar reflex, a two-neuron monosynaptic reflex. **(b)** A flexor reflex, an example of a polysynaptic reflex.

Since delay or inhibition of the reflex may occur at the synapses, the more synapses encountered in a reflex pathway, the more time is required to effect the reflex.

Reflexes of many types may be considered programmed into the neural anatomy. Many *spinal reflexes* (reflexes that are initiated and completed at the spinal cord level, such as the flexor reflex) occur without the involvement of higher brain centers. These reflexes work equally well in decerebrate animals (those in which the brain has been destroyed), as long as the spinal cord is functional. Conversely, other reflexes require the involvement of functional brain tissue, since many different inputs must be evaluated before the appropriate reflex is determined. Superficial cord reflexes and pupillary responses to light are in this category. In addition, although many spinal reflexes do not require the involvement of higher centers, the brain is "advised" of spinal cord reflex activity and may alter it by facilitating or inhibiting the reflexes.

Somatic Reflexes

There are several types of somatic reflexes, including several that you will be eliciting during this laboratory session—the stretch, crossed extensor, superficial cord, corneal, and gag reflexes. Some require only spinal cord activity; others require brain involvement as well.

Spinal Reflexes

Stretch Reflexes **Stretch reflexes** are important postural reflexes, normally acting to maintain posture, balance, and locomotion. Stretch reflexes are initiated by tapping a tendon, which stretches the muscle the tendon is attached to. This stimulates the muscle spindles and causes reflex contraction of the stretched muscle or muscles, which resists further stretching. Even as the primary stretch reflex is occurring, impulses are being sent to other destinations as well. For

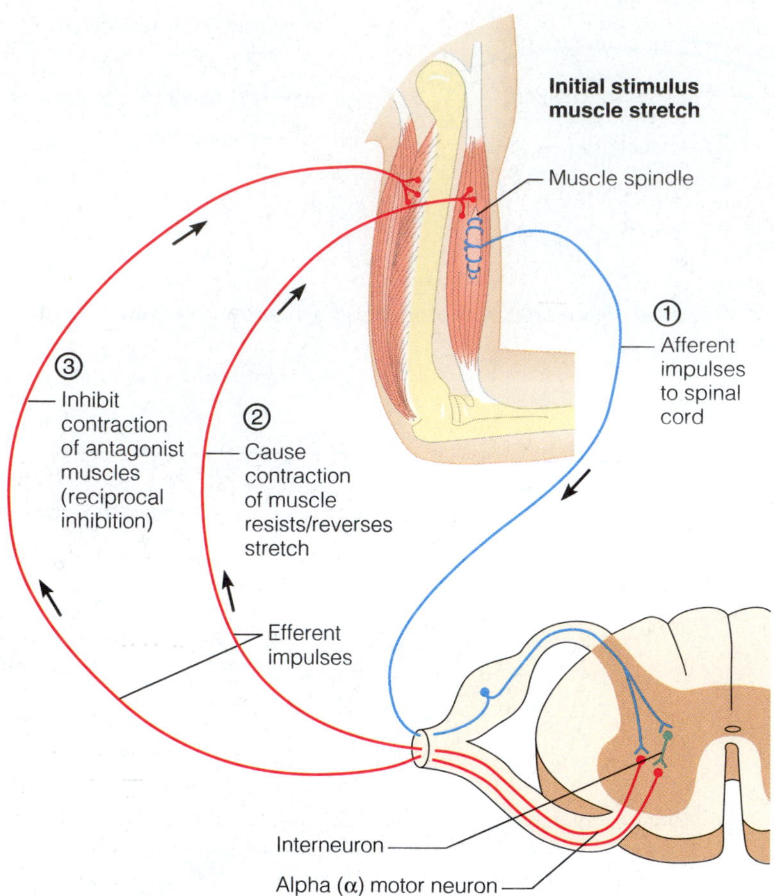

Figure 22.3 Events of the stretch reflex by which muscle stretch is damped. The events are shown in circular fashion. **(1)** Stretching of the muscle activates a muscle spindle. **(2)** Impulses transmitted by afferent fibers from muscle spindle to alpha motor neurons in the spinal cord result in activation of the stretched muscle, causing it to contract. **(3)** Impulses transmitted by afferent fibers from muscle spindle to interneurons in the spinal cord result in reciprocal inhibition of the antagonist muscle.

example, branches of the afferent fibers (from the muscle spindles) also synapse with association neurons (interneurons) controlling the antagonist muscles (Figure 22.3). The inhibition of the antagonist muscles that follows, called *reciprocal inhibition,* causes them to relax and prevents them from resisting (or reversing) the contraction of the stretched muscle caused by the main reflex arc. Additionally, impulses are relayed to higher brain centers (largely via the dorsal white columns) to advise of muscle length, speed of shortening, and the like—information needed to maintain muscle tone and posture. Stretch reflexes tend to be hypoactive or absent in cases of peripheral nerve damage or ventral horn disease, and hyperactive in corticospinal tract lesions. They are absent in deep sedation and coma.

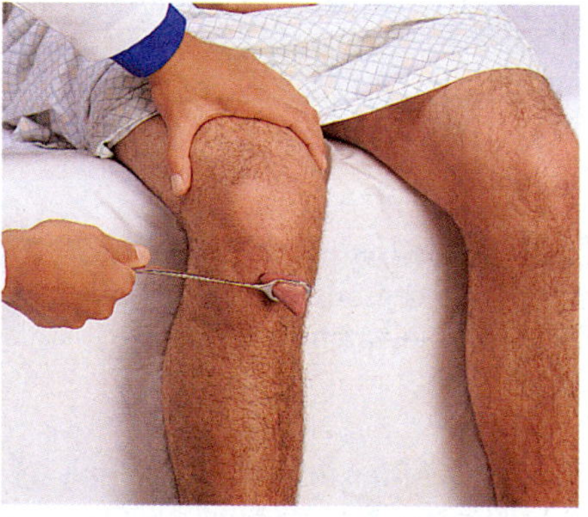

Figure 22.4 Testing the patellar reflex. The examiner supports the subject's knee so that the subject's muscles are relaxed, and then strikes the patellar ligament with the reflex hammer. The proper location may be ascertained by palpation of the patella.

Activity 1:
Initiating Stretch Reflexes

1. Test the **patellar,** or knee-jerk, **reflex** by seating a subject on the laboratory bench with legs hanging free (or with knees crossed). Tap the patellar ligament sharply with the reflex hammer just below the knee to elicit the knee-jerk response, which assesses the L_2–L_4 level of the spinal cord (Figure 22.4). Test both knees and record your observations. (Sometimes a reflex can be altered by your actions. If you encounter difficulty, consult your instructor for helpful hints.)

___

___

Which muscles contracted? ___

___

What nerve is carrying the afferent and efferent impulses?

___

2. Test the effect of mental distraction on the patellar reflex by having the subject add a column of three-digit numbers while you test the reflex again. Is the response greater than *or* less than the first response?

___

What are your conclusions about the effect of mental distraction on reflex activity?

___

___

3. Now test the effect of muscular activity occurring simultaneously in other areas of the body. Have the subject clasp the edge of the laboratory bench and vigorously attempt to pull it upward with both hands. At the same time, test the patellar reflex again. Is the response more or less vigorous than the first response?

___

4. Fatigue also influences the reflex response. The subject should jog in position until she or he is very fatigued (really fatigued—no slackers). Test the patellar reflex again and record whether it is more or less vigorous than the first response.

___

Would you say that nervous system activity *or* muscle function is responsible for the changes you have just observed?

___

Explain your reasoning. ___

___

___

___

5. The **Achilles,** or ankle-jerk, **reflex** assesses the first two sacral segments of the spinal cord. With your shoe removed and your foot dorsiflexed slightly to increase the tension of the gastrocnemius muscle, have your partner sharply tap your calcaneal (Achilles) tendon with the reflex hammer (Figure 22.5).

What is the result? ___

___

Does the contraction of the gastrocnemius normally result in the activity you have observed?

___ ■

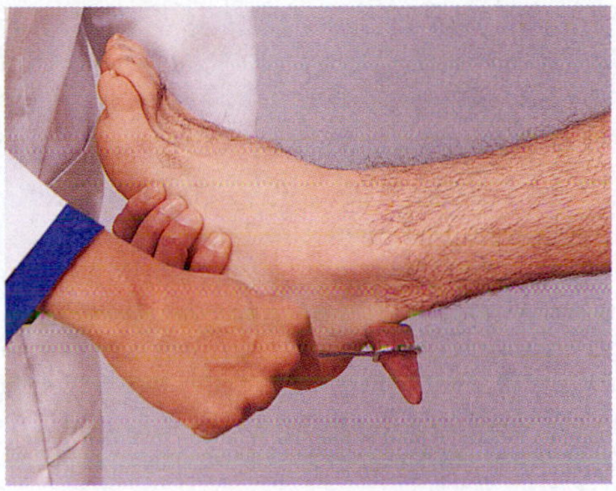

Figure 22.5 Testing the Achilles reflex. The examiner slightly dorsiflexes the subject's ankle by supporting the foot lightly in the hand, and then taps the Achilles tendon just above the ankle.

Crossed Extensor Reflex The **crossed extensor reflex** is more complex than the stretch reflex. It consists of a flexor, or withdrawal, reflex followed by extension of the opposite limb.

This reflex is quite obvious when, for example, a stranger suddenly and strongly grips one's arm. The immediate response is to withdraw the clutched arm and push the intruder away with the other arm. The reflex is more difficult to demonstrate in a laboratory because it is anticipated, and under these conditions the extensor part of the reflex may be inhibited.

Activity 2: Initiating the Crossed Extensor Reflex

The subject should sit with eyes closed and with the dorsum of one hand resting on the laboratory bench. Obtain a sharp pencil and suddenly prick the subject's index finger. What are the results?

___

___

___

Did the extensor part of this reflex seem to be slow compared to the other reflexes you have observed?

___

What are the reasons for this? ___

___

___

___ ■

The reflexes that have been demonstrated so far—the stretch and crossed extensor reflexes—are examples of reflexes in which the reflex pathway is initiated and completed at the spinal cord level.

Superficial Cord Reflexes The **superficial cord reflexes** (abdominal, cremaster, and plantar reflexes) result from pain and temperature changes. They are initiated by stimulation of receptors in the skin and mucosae. The superficial cord reflexes depend *both* on functional upper-motor pathways and on the cord-level reflex arc. Since only the plantar reflex can be tested conveniently in a laboratory setting, we will use this as our example.

The **plantar reflex,** an important neurological test, is elicited by stimulating the cutaneous receptors in the sole of the foot. In adults, stimulation of these receptors causes the toes to flex and move closer together. Damage to the pyramidal (or corticospinal) tract, however, produces *Babinski's sign,* an abnormal response in which the toes flare and the great toe moves in an upward direction. (In newborn infants, Babinski's sign is seen due to incomplete myelination of the nervous system.)

Activity 3: Initiating the Plantar Reflex

Have the subject remove a shoe and lie on the cot or laboratory bench with knees slightly bent and thighs rotated so that the lateral side of the foot rests on the cot. Alternatively, the subject may sit up and rest the lateral surface of the foot on a chair. Draw the handle of the reflex hammer firmly down the lateral side of the exposed sole from the heel to the base of the great toe (Figure 22.6).

What is the response? ______________________________

____

Is this a normal plantar reflex or Babinski's sign?

____ ■

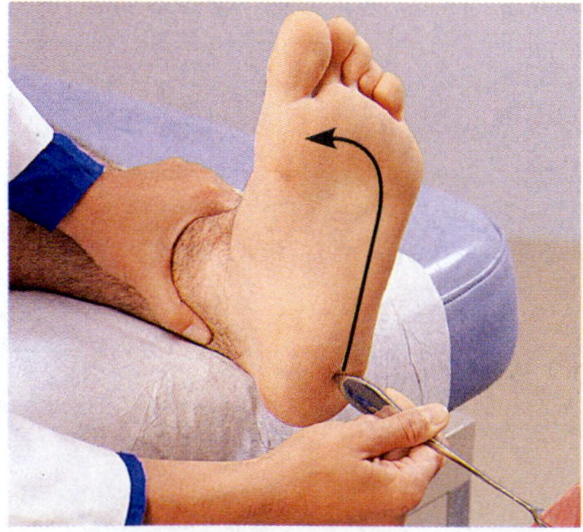

Figure 22.6 Testing the plantar reflex. Using a moderately sharp object, the examiner strokes the lateral border of the subject's sole, starting at the heel and continuing toward the big toe across the ball of the foot.

Cranial Nerve Reflex Tests

In these experiments, you will be working with your lab partner to illustrate two somatic reflexes mediated by cranial nerves.

Corneal Reflex The **corneal reflex** is mediated through the trigeminal nerve (cranial nerve V). The absence of this reflex is an ominous sign because it often indicates damage to the brain stem resulting from compression of the brain or other trauma.

Activity 4: Initiating the Corneal Reflex

Stand to one side of the subject; the subject should look away from you toward the opposite wall. Wait a few seconds and then quickly, *but gently,* touch the subject's cornea (on the side toward you) with a wisp of absorbent cotton. What reflexive reaction occurs when something touches the cornea?

____

What is the function of this reflex?

____

____

____ ■

Gag Reflex The **gag reflex** tests the somatic motor responses of cranial nerves IX and X. When the oral mucosa on the side of the uvula is stroked, each side of the mucosa should rise, and the amount of elevation should be equal.*

Activity 5: Initiating the Gag Reflex

For this experiment, select a subject who does not have a queasy stomach, because regurgitation is a possibility. Stroke the oral mucosa on each side of the subject's uvula with a tongue depressor. What happens?

____

____

⚠ Discard the used tongue depressor in the disposable autoclave bag before continuing. Do *not* lay it on the laboratory bench at any time. ■

* The uvula is the fleshy tab hanging from the roof of the mouth just above the root of the tongue.

Autonomic Reflexes

The autonomic reflexes include the pupillary, ciliospinal, and salivary reflexes, as well as a multitude of other reflexes. Work with your partner to demonstrate the four autonomic reflexes described next.

Pupillary Reflexes

There are several types of pupillary reflexes. The **pupillary light reflex** and the **consensual reflex** will be examined here. In both of these pupillary reflexes, the retina of the eye is the receptor, the optic nerve (cranial nerve II) contains the afferent fibers, the oculomotor nerve (cranial nerve III) is responsible for conducting efferent impulses to the eye, and the smooth muscle of the iris is the effector. Many central nervous system centers are involved in the integration of these responses. Absence of the normal pupillary reflexes is generally a late indication of severe trauma or deterioration of the vital brain stem tissue due to metabolic imbalance.

Activity 6: Initiating Pupillary Reflexes

1. Conduct the reflex testing in an area where the lighting is relatively dim. Before beginning, obtain a metric ruler to measure and record the size of the subject's pupils as best you can.

Right pupil: _______ mm Left pupil: _______ mm

2. Stand to the left of the subject to conduct the testing. The subject should shield his or her right eye by holding a hand vertically between the eye and the right side of the nose.

3. Shine a flashlight into the subject's left eye. What is the pupillary response?

___

Measure the size of the left pupil: _______________ mm

4. Observe the right pupil. Has the same type of change (called a *consensual response*) occurred in the right eye?

___

Measure the size of the right pupil: _______________ mm

The consensual response, or any reflex observed on one side of the body when the other side has been stimulated, is called a **contralateral response.** The pupillary light response, or any reflex occurring on the same side stimulated, is referred to as an **ipsilateral response.**

When a contralateral response occurs, what does this indicate about the pathways involved?

___

___

Was the sympathetic *or* the parasympathetic division of the autonomic nervous system active during the testing of these reflexes?

___

What is the function of these pupillary responses?

___

___

___ ■

Ciliospinal Reflex

The **ciliospinal reflex** is another example of reflex activity in which pupillary responses can be observed. This response may initially seem a little bizarre, especially in view of the consensual reflex just demonstrated.

Activity 7: Initiating the Ciliospinal Reflex

1. While observing the subject's eyes, gently stroke the skin (or just the hairs) on the left side of the back of the subject's neck, close to the hairline.

What is the reaction of the left pupil? _______________

The reaction of the right pupil? _______________

2. If you see no reaction, repeat the test using a gentle pinch in the same area.

The response you should have noted—pupillary dilation—is consistent with the pupillary changes occurring when the sympathetic nervous system is stimulated. Such a response may also be elicited in a single pupil when more impulses from the sympathetic nervous system reach it for any reason. For example, when the left side of the subject's neck was stimulated, sympathetic impulses to the left iris increased, resulting in the ipsilateral reaction of the left pupil.

On the basis of your observations, would you say that the sympathetic innervation of the two irises is closely integrated?

_________ Why or why not? _______________

___

___ ■

Salivary Reflex

Unlike the other reflexes, in which the effectors were smooth or skeletal muscles, the effectors of the **salivary reflex** are glands. The salivary glands secrete varying amounts of saliva in response to reflex activation.

Activity 8: Initiating the Salivary Reflex

1. Obtain a small beaker, a graduated cylinder, lemon juice, and wide-range pH paper. After refraining from swallowing for 2 minutes, the subject is to expectorate (spit) the accumulated saliva into a small beaker. Using the graduated cylinder, measure the volume of the expectorated saliva and determine its pH.

Volume: ______________ cc pH: ____________________

2. Now place 2 or 3 drops of lemon juice on the subject's tongue. Allow the lemon juice to mix with the saliva for 5 to 10 seconds, and then determine the pH of the subject's saliva by touching a piece of pH paper to the tip of the tongue.

pH: ____________________

As before, the subject is to refrain from swallowing for 2 minutes. After the 2 minutes is up, again collect and measure the volume of the saliva and determine its pH.

Volume: ______________ cc pH: ____________________

3. How does the volume of saliva collected after the application of the lemon juice compare with the volume of the first saliva sample?

____

How does the final saliva pH reading compare to the initial reading?

____

To that obtained 10 seconds after the application of lemon juice?

____

What division of the autonomic nervous system mediates the reflex release of saliva?

____

⚠ Dispose of the saliva-containing beakers and the graduated cylinders in the laboratory bucket that contains bleach and put the used pH paper into the disposable autoclave bag. Wash the bench down with 10% bleach solution before continuing. ■

Reaction Time of Basic and Learned or Acquired Reflexes

The time required for reaction to a stimulus depends on many factors—sensitivity of the receptors, velocity of nerve conduction, the number of neurons and synapses involved, and the speed of effector activation, to name just a few. Some reflexes are *basic* or inborn; others are *learned* or *acquired* reflexes, resulting from practice or repetition. There is no clear-cut distinction between basic and learned reflexes, as most reflex actions are subject to modification by learning or conscious effort. In general, however, if the response involves a specific reflex arc, the synapses are facilitated and the response time will be short. Learned reflexes involve a far larger number of neural pathways and many types of higher intellectual activities, including choice and decision making, which lengthens the response time.

There are various ways of testing reaction time of reflexes. The tests range from simple to ultrasophisticated. The following activities provide an opportunity to demonstrate the major time difference between simple and learned reflexes and to measure response time under various conditions.

Activity 9: Testing Reaction Time for Basic and Acquired Reflexes

1. Using a reflex hammer, elicit the patellar reflex in your partner. Note the relative reaction time needed for this basic reflex to occur.

2. Now test the reaction time for learned reflexes. The subject should hold a hand out, with the thumb and index finger extended. Hold a metric ruler so that its end is exactly 3 cm above the subject's outstretched hand. The ruler should be in the vertical position with the numbers reading from the bottom up. When the ruler is dropped, the subject should be able to grasp it between thumb and index finger as it passes, without having to change position. Have the subject catch the ruler five times, varying the time between trials. The relative speed of reaction can be determined by reading the number on the ruler at the point of the subject's fingertips.* (Thus if the number at the fingertips is 15 cm, the subject was unable to catch the ruler until 18 cm of length had passed through his or her fingers; 15 cm of ruler length plus 3 cm. to account for the distance of the ruler above the hand.)† Record the number

* Distance (d) can be converted to time (t) using the simple formula

$$d \text{ (in cm)} = (1/2)(980 \text{ cm/sec}^2)t^2$$

$$t^2 = d/(490 \text{ cm/sec}^2)$$

$$t = \sqrt{d/(490 \text{ cm/sec}^2)}$$

† An alternative would be to use a reaction time ruler, which converts distance to time (seconds).

of cm that pass through the subject's fingertips (or the number of seconds required for reaction) for each trial:

Trial 1:________ cm or sec Trial 4:________ cm or sec

Trial 2:________ cm or sec Trial 5:________ cm or sec

Trial 3:________ cm or sec

3. Perform the test again, but this time say a simple word each time you release the ruler. Designate a specific word as a signal for the subject to catch the ruler. On all other words, the subject is to allow the ruler to pass through his fingers. Trials in which the subject erroneously catches the ruler are to be disregarded. Record the distance the ruler travels (or the number of seconds required for reaction) in five *successful* trials:

Trial 1:________ cm or sec Trial 4:________ cm or sec

Trial 2:________ cm or sec Trial 5:________ cm or sec

Trial 3:________ cm or sec

Did the addition of a specific word to the stimulus increase or decrease the reaction time?

____

4. Perform the testing once again to investigate the subject's reaction to word association. As you drop the ruler, say a word—for example, *hot*. The subject is to respond with a word he or she associates with the stimulus word—for example, *cold*—catching the ruler while responding. If unable to make a word association, the subject must allow the ruler to pass through his or her fingers. Record the distance the ruler travels (or the number of seconds required for reaction) in five successful trials, as well as the number of times the ruler is not caught by the subject.

Trial 1:________ cm or sec Trial 4:________ cm or sec

Trial 2:________ cm or sec Trial 5:________ cm or sec

Trial 3:________ cm or sec

The number of times the subject was unable to catch the ruler:

____

You should have noticed quite a large variation in reaction time in this series of trials. Why is this so?

____

____

____

____ ■

Activity 10:
Using PowerLab® to Measure Reaction Time

1. To set up the equipment connect the PowerLab® unit to the computer. Plug the event marker into the socket on CHANNEL 1 (Figure 22.7a). Start the software by double-clicking on the PowerLab® folder.

Select the file named *Muscle #3*. When the Chart appears on the screen, CHANNEL 1 will be labeled *Event*.

2. Seat the subject so that he or she cannot see the computer screen and keyboard, and give the subject the event marker to hold.

The tester should hit the *Enter* key on the keyboard a few times to make sure that the subject can hear the tapping sound as the key is hit. Before beginning the test, instruct the subject to click the event marker as soon as he or she hears the sound.

3. To conduct the experiment click the *Start* button to start the Chart. The tester should tap the keyboard a total of ten times in a random fashion so that the subject cannot predict when the sound will occur.

Each time the keyboard is tapped, the subject should respond with the event marker as quickly as possible. Note that when the Enter key is tapped, a vertical dotted line appears on the screen. This is the comment marker (Figure 22.7b).

4. To analyze the data click and drag across the first response, and select the *Zoom* window. The comment marker line on the screen represents the time of the stimulus, and the event marker represents the time of the response. Move the cursor to the marker box, and place the marker on the beginning of the event trace. Using the cursor and the marker, measure the time between the stimulus and the response.

Repeat for each event.

Drop the longest and shortest values, and average the remaining results. Record in the chart on p. 234.

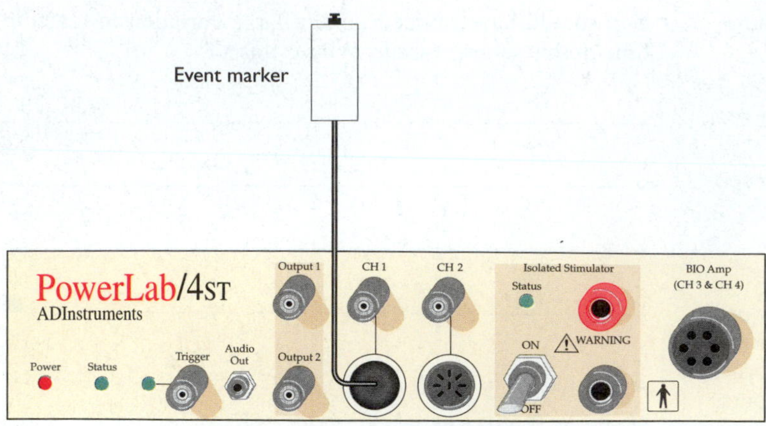

(a)

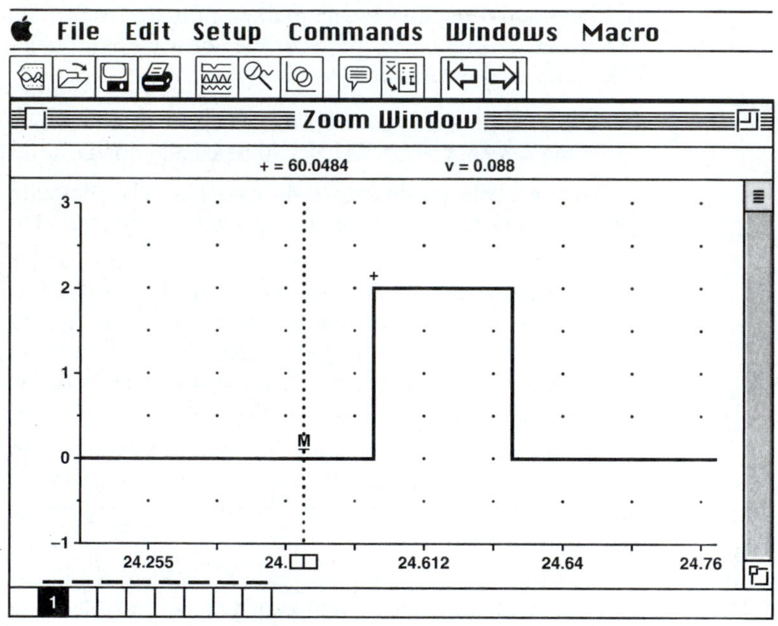

(b)

Figure 22.7 The PowerLab® setup. **(a)** The equipment used to measure reaction time. **(b)** Tapping the Enter key on the keyboard enters a comment (vertical dotted line). The response produces a trace. The marker and cursor are positioned to measure reaction time—the time between the two marks.

5. To expand the experiment choose another variable to test. Response to visual cues may be tested by having the subject block his or her ears and watch for the appearance of the comment marker on the screen when the Enter key is tapped. Many other variables can be tested for their effect on response time. Such variables include distraction, having the subject change hands, and conducting the hearing test blindfolded.

   Devise an experiment, conduct the test, and analyze the data as described in the steps above. Be sure to include a control. Record the average times on the chart.

6. Prepare a lab report for the experiment. (See Getting Started: Writing a Lab Report, p. xii.) Include in your discussion the relevance of this type of information to everyday life. ■

Reaction Times

Description of test	**Average response time**
1. Response to random sound production:	
2. Experimental control:	
3. Experimental variable:	

exercise 23

General Sensation

People are irritable creatures. Hold a sizzling steak before them and their mouths water. Flash your high beams in their eyes on the highway and they cuss. Tickle them and they giggle. These "irritants" (the steak, the light, and the tickle) and many others are stimuli that continually assault us.

The body's **sensory receptors** react to stimuli or changes within the body and in the external environment. The tiny sensory receptors of the *general senses* react to touch, pressure, pain, heat, cold, stretch, vibration, and changes in position and are distributed throughout the body. In contrast to these widely distributed *general sensory receptors*, the receptors of the special senses are large, complex *sense organs* or small, localized groups of receptors. The *special senses* include sight, hearing, equilibrium, smell, and taste.

Sensory receptors may be classified according to the source of their stimulus. **Exteroceptors** react to stimuli in the external environment, and typically they are found close to the body surface. Exteroceptors include the simple cutaneous receptors in the skin and the highly specialized receptor structures of the special senses (the vision apparatus of the eye, and the hearing and equilibrium receptors of the ear, for example). **Interoceptors** or **visceroceptors** respond to stimuli arising within the body. Interoceptors are found in the internal visceral organs, and include stretch receptors (in walls of hollow organs), chemoreceptors, and others. **Proprioceptors,** like interoceptors, respond to internal stimuli but are restricted to skeletal muscles, tendons, joints, ligaments, and connective tissue coverings of bones and muscles. They provide information on the position and degree of stretch of those structures.

The receptors of the special sense organs are complex and deserve considerable study. Thus the special senses (vision, hearing, equilibrium, taste, and smell) are covered separately in Exercises 24 through 26. Only the anatomically simpler **general sensory receptors**—cutaneous receptors and proprioceptors—will be studied in this exercise.

Objectives

1. To recognize various types of general sensory receptors as studied in the laboratory, and to describe the function and location of each type.
2. To define *exteroceptor, interoceptor,* and *proprioceptor.*
3. To demonstrate and relate differences in relative density and distribution of tactile and thermoreceptors in the skin.
4. To define *tactile localization* and describe how this ability varies in different areas of the body.
5. To explain the tactile two-point discrimination test, and to state its anatomical basis.
6. To define *referred pain.*
7. To define *adaptation, negative afterimage,* and *projection.*

Materials

- ❑ Compound microscope
- ❑ Immersion oil
- ❑ Histologic slides of Pacinian corpuscles (longitudinal section), Meissner's corpuscles, Golgi tendon organs, and muscle spindles
- ❑ Mall probes (obtainable from Carolina Biological Supply), or pointed metal rods of approximately 1-in. diameter and 2- to 3-in. length, with insulated blunt end
- ❑ Large beaker of ice water; chipped ice
- ❑ Hot water bath set at 45°C
- ❑ Laboratory thermometer
- ❑ Fine-point, felt-tipped markers (black, red, and blue)
- ❑ Small metric ruler
- ❑ Von Frey's hairs (or 1-in. lengths of horsehair glued to a matchstick) or sharp pencil
- ❑ Towel
- ❑ Caliper or esthesiometer
- ❑ Four coins (nickels or quarters)
- ❑ Three large finger bowls or 1000-ml beakers

Structure of General Sensory Receptors

You cannot become aware of changes in the environment unless your sensory neurons and their receptors are operating properly. Sensory receptors are modified dendritic endings (or specialized cells associated with the dendrites) that are sensitive to certain environmental stimuli. They react to such stimuli by initiating a nerve impulse. Several histologically distinct types of general sensory receptors have been identified in the skin. Their structures are depicted in Figure 23.1.

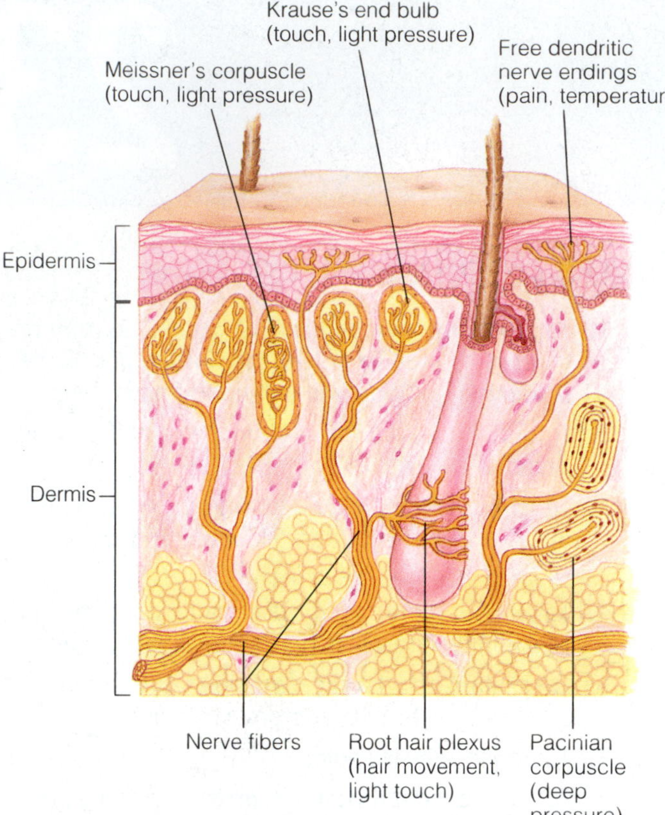

Figure 23.1 Cutaneous receptors. Free dendritic nerve endings, root hair plexus, Meissner's corpuscle, Pacinian corpuscle, and Krause's end bulb. (See also Plates 11, 12, and 13 in the Histology Atlas.)

Many references link each type of receptor to specific stimuli, but there is still considerable controversy about the precise qualitative function of each receptor. It may be that the responses of all these receptors overlap considerably. Certainly, intense stimulation of any of them is always interpreted as pain.

The least specialized of the cutaneous receptors are the **free** or **naked dendritic endings** of sensory neurons (Figure 23.1 and Plate 12 of the Histology Atlas), which respond chiefly to pain and temperature. (The pain receptors are widespread in the skin and make up a sizable portion of the visceral interoceptors.) Certain free dendritic endings associate with specific epidermal cells to form **Merkel discs,** or entwine in hair follicles to form **root hair plexuses.** Both Merkel discs and root hair plexuses function as light touch receptors.

The other cutaneous receptors are a bit more complex, with the dendritic endings *encapsulated* by connective tissue cells. **Meissner's corpuscles,** commonly referred to as *tactile receptors* because they respond to light touch, are located in the dermal papillae of the skin. **Krause's end bulbs** are a variation of Meissner's corpuscles. They are abundant in mucous membranes and may also be called **mucocutaneous corpuscles.** Ruffini's corpuscles appear to respond to deep pressure and stretch stimuli. As you inspect Figure 23.1, notice that all of the encapsulated receptors are quite similar, with the possible exception of the **Pacinian corpuscles,** which are anatomically more distinctive and lie deepest in the dermis. Pacinian corpuscles respond only when deep pressure is first applied. They are best suited to monitor vibrations.

Activity 1: Studying the Structure of Selected Sensory Receptors

1. Obtain histologic slides of Pacinian and Meissner's corpuscles. Locate, under low power, a Meissner's corpuscle in the dermal layer of the skin. As mentioned above, these are usually found in the dermal papillae. Then switch to the oil immersion lens for a detailed study. Notice that the naked dendritic fibers within the capsule are aligned parallel to the skin surface. Compare your observations to Figure 23.1 and Plate 11 of the Histology Atlas.

2. Next observe a Pacinian corpuscle located much deeper in the dermis. Try to identify the slender naked dendrite ending in the center of the receptor and the heavy capsule of connective tissue surrounding it (which looks rather like an onion cut lengthwise). Also, notice how much larger the Pacinian corpuscles are than the Meissner's corpuscles. Compare your observations to the views shown in Figure 23.1 and Plate 13 of the Histology Atlas.

3. Obtain slides of muscle spindles and Golgi tendon organs, the two major types of proprioceptors (see Figure 23.2). In the slide of **muscle spindles,** note that minute extensions of the dendrites of the sensory neurons coil around specialized slender skeletal muscle cells called **intrafusal cells,** or **fibers.** The **Golgi tendon organs** are composed of dendrites that ramify through the tendon tissue close to the muscle tendon attachment. Stretching of muscles or tendons excites these receptors, which then transmit impulses that ultimately reach the cerebellum for interpretation. Compare your observations to Figure 23.2. ■

Receptor Physiology

Sensory receptors act as *transducers,* changing environmental stimuli into afferent nerve impulses. Since the action potential generated in all nerve fibers is essentially identical, the stimulus is identified entirely by the area of the brain's sensory cortex that is stimulated (which, of course, differs for the various afferent nerves).

Four qualities of cutaneous sensations have traditionally been recognized: tactile (touch), heat, cold, and pain. Mapping these sensations on the skin has revealed that the sensory receptors for these qualities are not distributed uniformly. Instead, they have discrete locations and are characterized by clustering at certain points—**punctate distribution.**

The simple pain receptors, extremely important in protecting the body, are the most numerous. Touch receptors cluster where greater sensitivity is desirable, as on the hands and face. It may be surprising to learn that rather large areas of the skin are quite insensitive to touch because of a relative lack of touch receptors.

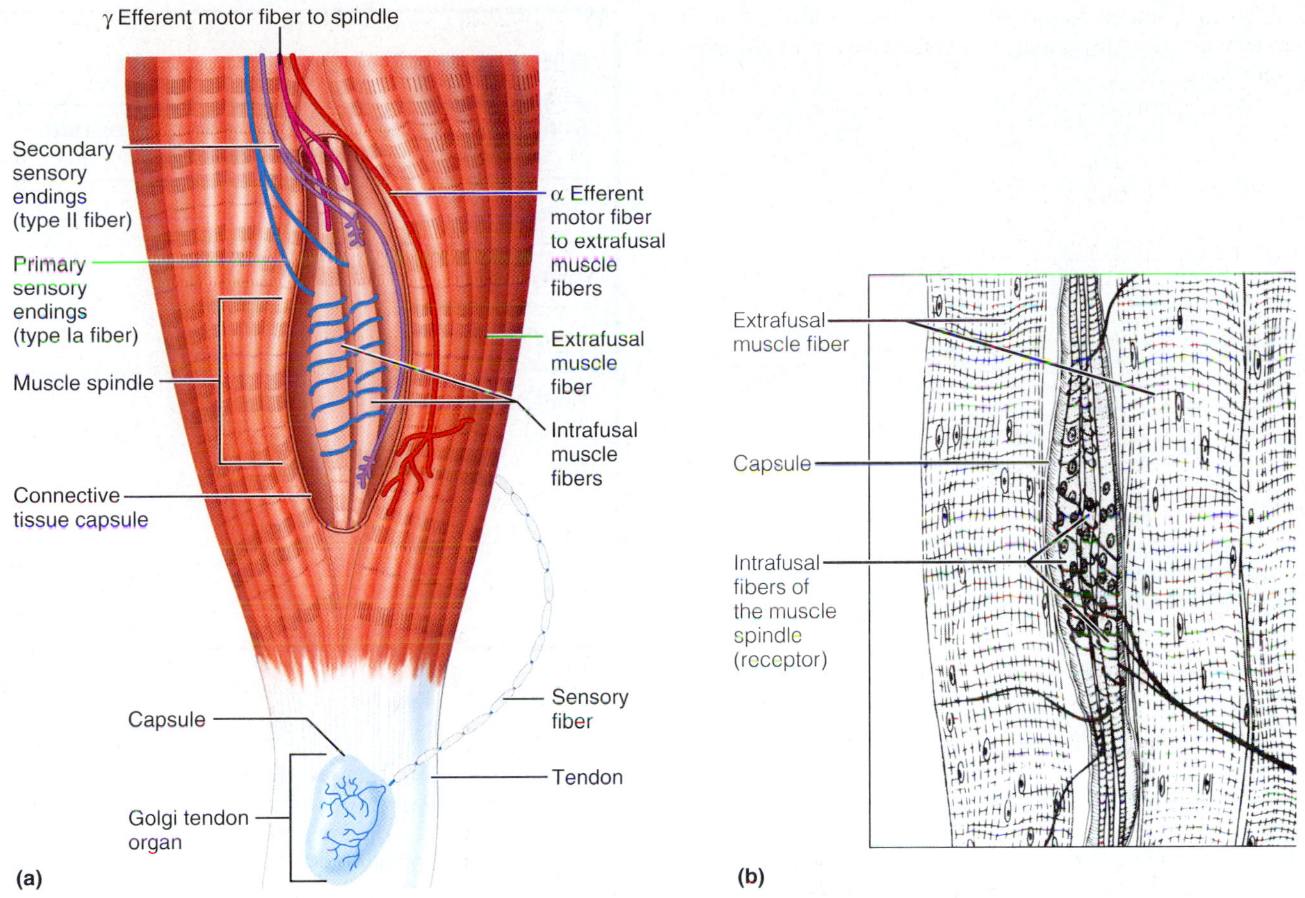

Figure 23.2 Proprioceptors. (a) Diagrammatic view of a muscle spindle and Golgi tendon organ. **(b)** Drawing of a photomicrograph of a muscle spindle. (See also Plate 14 in the Histology Atlas.)

There are several simple experiments you can conduct to investigate the location and physiology of cutaneous receptors. In each of the following activities, work in pairs with one person as the subject and the other as the experimenter. After you have completed an experiment, switch roles and go through the procedures again so that all class members obtain individual results. Keep an accurate account of each test that you perform.

Density and Location of Touch and Temperature Receptors

Activity 2: Plotting the Relative Density and Location of Touch and Temperature Receptors

1. Obtain 2 Mall probes (or temperature rods), metric ruler, Von Frey's hair (or a sharp pencil), and black, red, and blue felt-tipped markers and bring them to your laboratory bench.

2. Place one Mall probe (or temperature rod) in a beaker of ice water and the other in a water bath controlled at 45°C. With a felt marker, draw a square (2 cm on each side) on the ventral surface of the subject's forearm. During the following tests, the subject's eyes are to remain closed. The subject should tell the examiner when a stimulus is detected.

3. Working in a systematic manner from one side of the marked square to the other, gently touch the Von Frey's hair or a sharp pencil tip to different points within the square. The *hairs* should be applied with a pressure that just causes them to bend; use the same pressure for each contact. Do not apply deep pressure. The goal is to stimulate only the more superficially located Meissner's corpuscles (as opposed to the Pacinian corpuscles located in the subcutaneous tissue). Mark with a *black dot* all points at which the touch is perceived.

4. Remove the Mall probe (or temperature rod) from the ice water and quickly wipe it dry. Repeat the procedure outlined above, noting all points of cold perception with a *blue dot*.* Perception should be for temperature, not simply touch.

5. Remove the second Mall probe from the 45°C water bath and repeat the procedure once again, marking all points of heat perception with a *red dot*.*

*The Mall probes will have to be returned to the water baths approximately every 2 minutes to maintain the desired testing temperatures.

6. After each student has acted as the subject, draw a "map" of your own and your partner's receptor areas in the squares provided here.

2 cm

Student 1 ______________

2 cm

Student 2 ______________

How does the density of the heat receptors correspond to that of the touch receptors?

____

To that of the cold receptors? ______________

____

On the basis of your observations, which of these three types of receptors appears to be most abundant (at least in the area tested)?

____

Two-Point Discrimination Test

The density of the touch receptors varies significantly in different areas of the body. In general, areas that have the greatest density of tactile receptors have a heightened ability to "feel." These areas correspond to areas that receive the greatest motor innervation; thus they are also typically areas of fine motor control.

On the basis of this information, which areas of the body do you *predict* will have the greatest density of touch receptors?

____

____

Activity 3: Determining the Two-Point Threshold

1. Using a caliper or esthesiometer and a metric ruler, test the ability of the subject to differentiate two distinct sensations when the skin is touched simultaneously at two points. Beginning with the face, start with the caliper arms completely together. Gradually increase the distance between the arms, testing the subject's skin after each adjustment.

Determining Two-Point Threshold

Body area tested	Two-point threshold (millimeters)
Face	
Back of hand	
Palm of hand	
Fingertips	
Lips	
Back of neck	
Ventral forearm	

Continue with this testing procedure until the subject reports that *two points* of contact can be felt. This measurement, the smallest distance at which two points of contact can be felt, is the **two-point threshold.**

2. Repeat this procedure on the back and palm of the hand, fingertips, lips, back of the neck, and ventral forearm. Record your results in the chart above.

3. Which area has the smallest two-point threshold?

____

Tactile Localization

Tactile localization is the ability to determine which portion of the skin has been touched. The tactile receptor field of the body periphery has a corresponding "touch" field in the brain's somatosensory association area. Some body areas are well represented with touch receptors, allowing tactile stimuli to be localized with great accuracy, but density of touch receptors in other body areas allows only crude discrimination.

Activity 4: Testing Tactile Localization

1. The subject's eyes should be closed during the testing. The experimenter touches the palm of the subject's hand with a pointed black felt-tipped marker. The subject should then try to touch the exact point with his or her own marker, which should be of a different color. Measure the error of localization in millimeters.

2. Repeat the test in the same spot twice more, recording the error of localization for each test. Average the results of the three determinations and record it in the chart on p. 239.

Testing Tactile Location

Body area tested	Average error of localization (millimeters)
Palm of hand	
Fingertip	
Ventral forearm	
Back of hand	
Back of neck	

Does the ability to localize the stimulus improve the second time?

__________The third time? __________ Explain. __________

____

____

3. Repeat the above procedure on a fingertip, the ventral forearm, the back of a hand, and the back of the neck. Record the averaged results in the chart above.

4. Which area has the smallest error of localization?

____ ■

Adaptation of Sensory Receptors

The number of impulses transmitted by sensory receptors often changes both with the intensity of the stimulus and with the length of time the stimulus is applied. In many cases, when a stimulus is applied for a prolonged period, the rate of receptor discharge slows and conscious awareness of the stimulus declines or is lost until some type of stimulus change occurs. This phenomenon is referred to as **adaptation.** The touch receptors adapt particularly rapidly, which is highly desirable. Who, for instance, would want to be continually aware of the pressure of clothing on their skin? The simple experiments to be conducted next allow you to investigate the phenomenon of adaptation.

Activity 5: Demonstrating Adaptation of Touch Receptors

1. The subject's eyes should be closed. Place a coin on the anterior surface of the subject's forearm, and determine how long the sensation persists for the subject. Duration of the sensation:

______________ sec

2. Repeat the test, placing the coin at a different forearm location. How long does the sensation persist at the second location?

______________ sec

3. After awareness of the sensation has been lost at the second site, stack three more coins atop the first one.

Does the pressure sensation return? ______________

If so, for how long is the subject aware of the pressure in this instance?

______________ sec

Are the same receptors being stimulated when the four coins, rather than the one coin, are used?

__________ Explain. ______________________

____

4. To further illustrate the adaptation of touch receptors—in this case, the root hair plexuses of the hair follicles—gently and slowly bend one hair shaft with a pen or pencil until it springs back (away from the pencil) to its original position. Is the tactile sensation greater when the hair is being slowly bent or when it springs back?

____

Why is the adaptation of the touch receptors in the hair follicles particularly important to a woman who wears her hair in a ponytail? If the answer is not immediately apparent, consider the opposite phenomenon: what would happen, in terms of sensory input from her hair follicles, if these receptors did not exhibit adaptation?

____

____ ■

Activity 6: Demonstrating Adaptation of Temperature Receptors

Adaptation of the temperature receptors can be tested using some very unsophisticated methods.

1. Obtain three large finger bowls or 1000-ml beakers and fill the first with 45°C water. Have the subject immerse her or

his left hand in the water and report the sensation. Keep the left hand immersed for 1 min and then also immerse the right hand in the same bowl.

What is the sensation of the left hand when it is first immersed?

___

What is the sensation of the left hand after 1 min as compared to the sensation in the right hand just immersed?

___

Had adaptation occurred in the left hand? ___

2. Rinse both hands in tap water, dry them, and wait 5 min before conducting the next test. Just before beginning the test, refill the finger bowl with fresh 45°C water, fill a second with ice water, and fill a third with water at room temperature.

3. Place the *left* hand in the ice water and the *right* hand in the 45°C water. What is the sensation in each hand after 2 min as compared to the sensation perceived when the hands were first immersed?

___

___

Which hand seemed to adapt more quickly?

___

4. After reporting these observations, the subject should then place both hands simultaneously into the finger bowl containing the water at room temperature. Record the sensation in the left hand:

___

The right hand: ___

The sensations that the subject experiences when both hands were put into room-temperature water are called **negative afterimages.** They are explained by the fact that sensations of heat and cold depend on the rapidity of heat loss or gain by the skin and differences in the temperature gradient. ■

Referred Pain

Experiments on pain receptor localization and adaptation are commonly conducted in the laboratory. However, there are certain problems with such experiments. Pain receptors are densely distributed in the skin, and they adapt very little, if at all. (This lack of adaptability is due to the protective function of the receptors. The sensation of pain often indicates tissue damage or trauma to body structures.) Thus no attempt will be made in this exercise to localize the pain receptors or to prove their nonadaptability, since both would cause needless discomfort to those of you acting as subjects and would not add any additional insight.

However, the phenomenon of referred pain is easily demonstrated in the laboratory, and such experiments provide information that may be useful in explaining common examples of this phenomenon. **Referred pain** is a sensory experience in which pain is perceived as arising in one area of the body when in fact another, often quite remote area is receiving the painful stimulus. Thus the pain is said to be "referred" to a different area. The phenomenon of **projection,** the process by which the brain refers sensations to their *usual* point of stimulation, provides the most simple explanation of such experiences. Many of us have experienced referred pain as a radiating pain in the forehead after quickly swallowing an ice-cold drink. Referred pain is important in many types of clinical diagnosis because damage to many visceral organs results in this phenomenon. For example, inadequate oxygenation of the heart muscle often results in pain being referred to the chest wall and left shoulder (*angina pectoris*), and the reflux of gastric juice into the esophagus causes a sensation of intense discomfort in the thorax referred to as *heartburn.* In addition, amputees often report *phantom limb* pain—feelings of pain that appear to be coming from a part of the body that is no longer there.

Activity 7:
Demonstrating the Phenomenon of Referred Pain

Immerse the subject's elbow in a finger bowl containing ice water. In the chart on p. 241, record the quality (such as discomfort, tingling, or pain) and the quality progression of the sensations he or she reports for 2 min. Also record the location of the perceived sensations. The ulnar nerve, which serves the medial third of the hand, is involved in the phenomenon of referred pain experienced during this test. How does the localization of this referred pain correspond to the areas served by the ulnar nerve?

___

___

___

___ ■

Demonstrating Referred Pain

Time of observation	Quality of sensation	Localization of sensation
On immersion		
After 1 min		
After 2 min		

exercise

24 Special Senses: Vision

Objectives

1. To describe the structure and function of the accessory visual structures.
2. To identify the structural components of the eye when provided with a model, an appropriate diagram, or a preserved sheep or cow eye, and list the function(s) of each.
3. To describe the cellular makeup of the retina.
4. To discuss the mechanism of image formation on the retina.
5. To trace the visual pathway to the visual cortex and indicate the effects of damage to various parts of this pathway.
6. To define the following terms:

refraction	myopia
accommodation	hyperopia
convergence	cataract
astigmatism	glaucoma
emmetropia	conjunctivitis

7. To discuss the importance of the pupillary and convergence reflexes.
8. To explain the difference between rods and cones with respect to visual perception and retinal localization.
9. To state the importance of an ophthalmoscopic examination.

Materials

- ❑ Dissectible eye model
- ❑ Chart of eye anatomy
- ❑ Histologic section of an eye showing retinal layers
- ❑ Compound microscope
- ❑ Preserved cow or sheep eye
- ❑ Dissecting tray and instruments
- ❑ Disposable gloves
- ❑ Common straight pins
- ❑ Snellen eye chart (floor marked with chalk to indicate 20-ft distance from posted Snellen chart)
- ❑ Ishihara's color-blindness plates
- ❑ 1-inch-diameter discs of colored paper (white, red, blue, green)
- ❑ White, red, blue, and green chalk
- ❑ Metric ruler; meter stick
- ❑ Test tubes
- ❑ Laboratory lamp or penlight
- ❑ Ophthalmoscope

A.I.A. See Appendix C, Exercise 24 for links to A.D.A.M.® Interactive Anatomy.

Anatomy of the Eye

External Anatomy and Accessory Structures

The adult human eye is a sphere measuring about 2.5 cm (1 inch) in diameter. Only about one-sixth of the eye's anterior surface is observable (Figure 24.1); the remainder is enclosed and protected by a cushion of fat and the walls of the bony orbit.

The **lacrimal apparatus** consists of the **lacrimal gland, lacrimal canals, lacrimal sac,** and the **nasolacrimal duct.** The lacrimal glands are situated superior to the lateral aspect of each eye. They continually liberate a dilute salt solution (tears) that flows onto the anterior surface of the eyeball through several small ducts. The tears flush across the eyeball into the lacrimal canals medially, then into the lacrimal sac, and finally into the nasolacrimal duct, which empties into the nasal cavity. The lacrimal secretion also contains **lysozyme,** an antibacterial enzyme. Because it constantly flushes the eyeball, the lacrimal fluid cleanses and protects the eye surface as it moistens and lubricates it. As we age, our eyes tend to become dry due to decreased lacrimation, and thus are more vulnerable to bacterial invasion and irritation.

The anterior surface of each eye is protected by the **eyelids,** or **palpebrae.** (See Figure 24.1.) The medial and lateral junctions of the upper and lower eyelids are referred to as the **medial** and **lateral canthus** (respectively). The **caruncle,** a fleshy elevation at the medial canthus, produces a whitish oily secretion. A mucous membrane, the **conjunctiva,** lines the internal surface of the eyelids (as the *palpebral conjunctiva*) and continues over the anterior surface of the eyeball to its junction with the corneal epithelium (as the *ocular,* or *bulbar, conjunctiva*). The conjunctiva secretes mucus, which aids in lubricating the eyeball. Inflammation of the conjunctiva, often accompanied by redness of the eye, is called **conjunctivitis.**

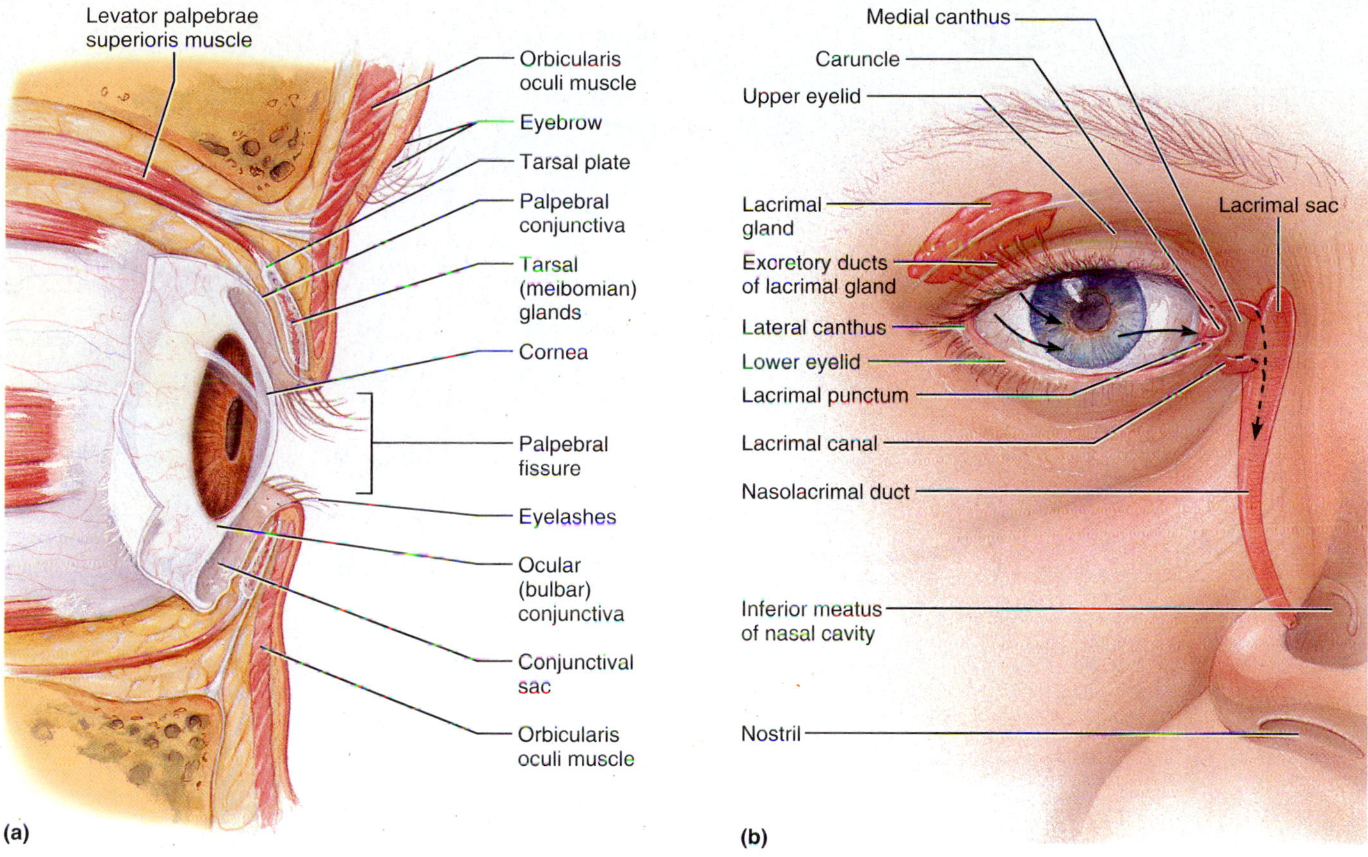

Figure 24.1 External anatomy of the eye and accessory structures. **(a)** Sagittal section. **(b)** Anterior view.

Projecting from the border of each eyelid is a row of short hairs, the **eyelashes.** The **ciliary glands,** modified sweat glands, lie between the eyelash hair follicles and help lubricate the eyeball. Small sebaceous glands associated with the hair follicles and the larger **meibomian (tarsal) glands,** located posterior to the eyelashes, secrete an oily substance. An inflammation of one of the ciliary glands or a small oil gland is called a **sty.**

Six **extrinsic eye muscles** attached to the exterior surface of each eyeball control eye movement and make it possible for the eye to follow a moving object. The names and positioning of these extrinsic muscles are noted in Figure 24.2. Their actions are given in the chart accompanying that figure.

Activity 1: Identifying Accessory Eye Structures

Observe the eyes of another student and identify as many of the accessory structures as possible. Ask the student to look to the left. What extrinsic eye muscles are responsible for this action?

Right eye ______________________________

Left eye ______________________________ ■

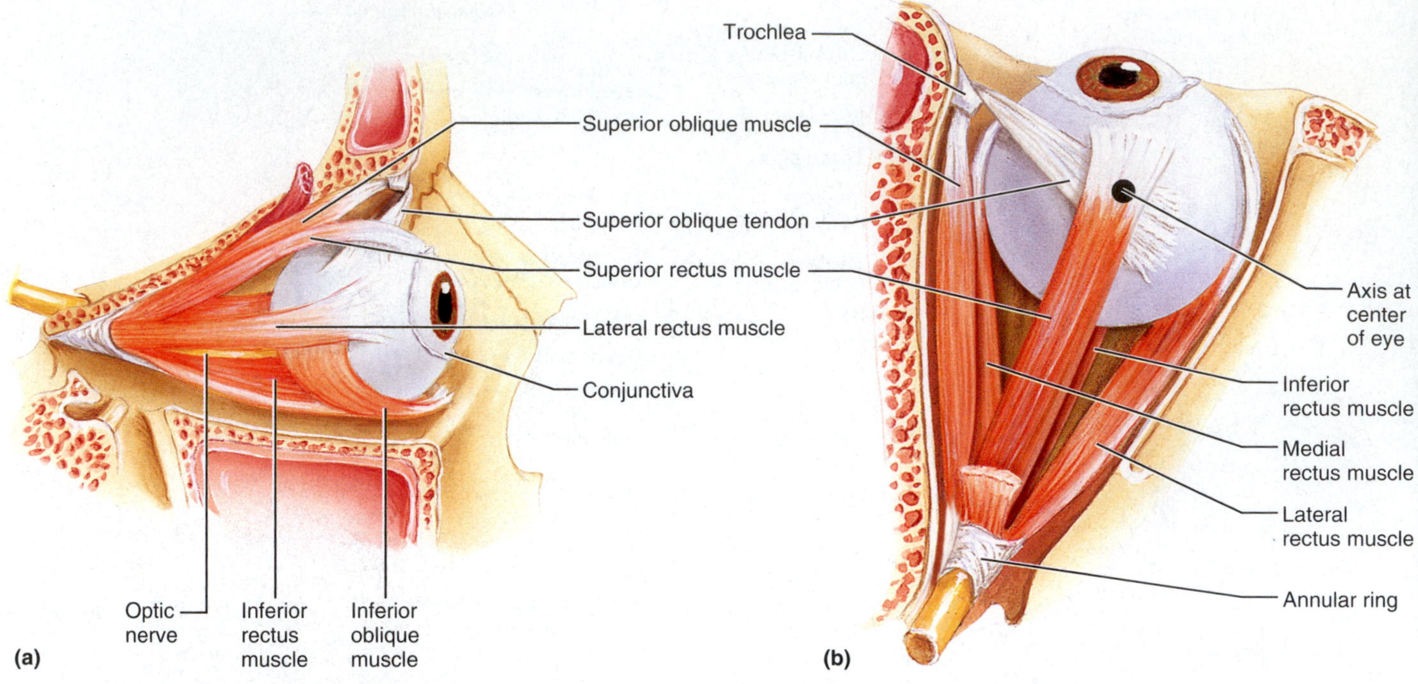

Name	Action	Controlling cranial nerve
Lateral rectus	Moves eye laterally	VI (abducens)
Medial rectus	Moves eye medially	III (oculomotor)
Superior rectus	Elevates eye	III (oculomotor)
Inferior rectus	Depresses eye	III (oculomotor)
Inferior oblique	Elevates eye and turns it laterally	III (oculomotor)
Superior oblique	Depresses eye and turns it laterally	IV (trochlear)

(c)

Figure 24.2 Extrinsic muscles of the eye. (a) Lateral view of the right eye. **(b)** Superior view of the right eye. **(c)** Summary of actions of the extrinsic eye muscles, and cranial nerves that control them.

Internal Anatomy of the Eye

Anatomically, the wall of the eye is constructed of three tunics, or coats. The outermost **fibrous tunic** is a protective layer composed of dense avascular connective tissue. It has two obviously different regions: The opaque white **sclera** forms the bulk of the fibrous tunic and is observable anteriorly as the "white of the eye." Its anteriormost portion is modified structurally to form the transparent **cornea,** through which light enters the eye. (See Figure 24.3.)

The middle tunic, called the **uvea,** is the **vascular tunic.** Its posteriormost part, the **choroid,** is a richly vascular nutritive layer containing a dark pigment that prevents light scattering within the eye. Anteriorly, the choroid is modified to form the **ciliary body,** which is chiefly composed of **ciliary muscles,** important in controlling lens shape, and **ciliary processes.** The ciliary processes secrete aqueous humor. The most anterior part of the uvea is the pigmented **iris.** The iris is incomplete, resulting in a rounded opening, the **pupil,** through which light passes.

The iris is composed of circularly and radially arranged smooth muscle fibers and acts as a reflexively activated diaphragm to regulate the amount of light entering the eye. In close vision and bright light, the circular muscles of the iris contract, and the pupil constricts. In distant vision and in dim light, the radial fibers contract, enlarging (dilating) the pupil and allowing more light to enter the eye.

The innermost **sensory tunic** of the eye is the delicate, two-layered **retina.** The outer **pigmented epithelial layer** abuts the choroid and extends anteriorly to cover the ciliary body and the posterior side of the iris. The transparent inner **neural (nervous) layer** extends anteriorly only to the ciliary body. It contains the photoreceptors, **rods** and **cones,** which begin the chain of electrical events that ultimately result in the transduction of light energy into nerve impulses that are transmitted to the optic cortex of the brain. Vision is the

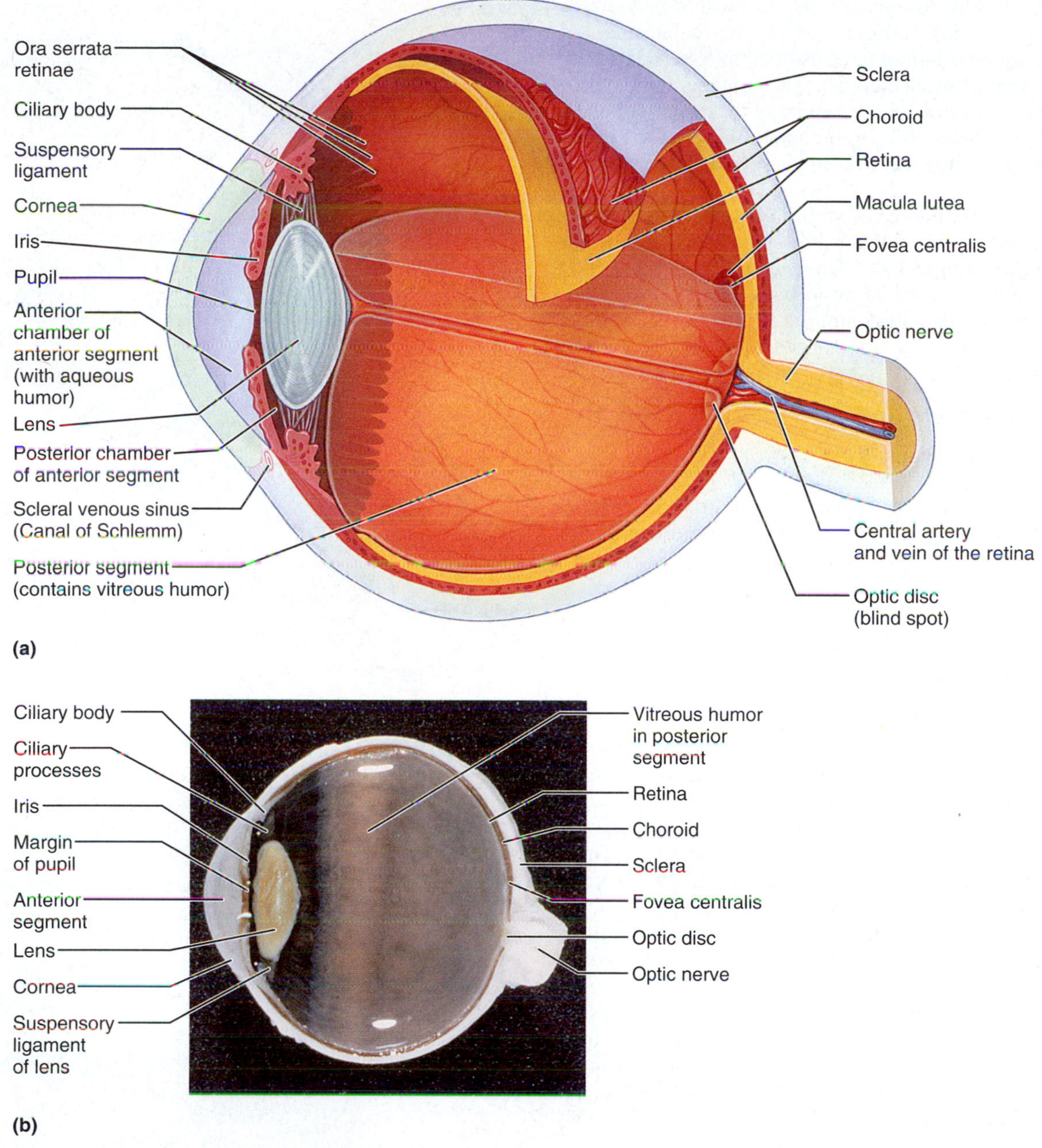

Figure 24.3 Sagittal section of the internal anatomy of the eye. (a) Diagrammatic view. The vitreous humor is illustrated only in the bottom half of the eyeball. **(b)** Photograph of the human eye.

result. The photoreceptor cells are distributed over the entire neural retina, except where the optic nerve leaves the eyeball. This site is called the **optic disc,** or *blind spot.* Lateral to each blind spot, and directly posterior to the lens, is an area called the **macula lutea** (yellow spot), an area of high cone density. In its center is the **fovea centralis,** a minute pit about 0.4 mm in diameter, which contains mostly cones and is the area of greatest visual acuity. Focusing for discriminative vision occurs in the fovea centralis.

Light entering the eye is focused on the retina by the **lens,** a flexible crystalline structure held vertically in the eye's interior by the **suspensory ligament** attached to the ciliary body. Activity of the ciliary muscle, which accounts for the bulk of ciliary body tissue, changes lens thickness to allow light to be properly focused on the retina.

In the elderly the lens becomes increasingly hard and opaque. **Cataracts,** which often result from this process, cause vision to become hazy or entirely obstructed. ■

The lens divides the eye into two segments: the **anterior segment** anterior to the lens, which contains a clear watery fluid called the **aqueous humor,** and the **posterior segment** behind the lens, filled with a gel-like substance, the **vitreous humor,** or **vitreous body.** The anterior segment is further divided into **anterior** and **posterior chambers,** located before and after the iris, respectively. The aqueous humor is continually formed by the capillaries of the **ciliary processes** of the ciliary body. It helps to maintain the intraocular pressure of the eye and provides nutrients for the avascular lens and cornea. The aqueous humor is reabsorbed into the **scleral venous sinus (canal of Schlemm).** The vitreous humor provides the major internal reinforcement of the posterior part of the eyeball, and helps to keep the neural layer of the retina pressed firmly against the wall of the eyeball. It is formed *only* before birth.

Anything that interferes with drainage of the aqueous fluid increases intraocular pressure. When intraocular pressure reaches dangerously high levels, the retina and optic nerve are compressed, resulting in pain and possible blindness, a condition called **glaucoma.** ■

Activity 2:
Identifying Internal Structures of the Eye

Obtain a dissectible eye model and identify its internal structures described above. As you work, also refer to Figure 24.3. ■

Microscopic Anatomy of the Retina

As described above, the retina consists of two main types of cells: a pigmented *epithelial* layer, which abuts the choroid, and an inner cell layer composed of *neurons,* which is in contact with the vitreous humor (Figure 24.4). The inner nervous layer is composed of three major neuronal populations. These are, from outer to inner aspect, the **photoreceptors** (rods and cones), the **bipolar cells,** and the **ganglion cells.**

The **rods** are the specialized receptors for dim light. Visual interpretation of their activity is in gray tones. The **cones** are color receptors that permit high levels of visual acuity, but they function only under conditions of high light intensity; thus, for example, no color vision is possible in moonlight. Mostly cones are found in the fovea centralis, and their number decreases as the retinal periphery is approached. By contrast, rods are most numerous in the periphery, and their density decreases as the macula is approached.

Light must pass through the ganglion cell layer and the bipolar neuron layer to reach and excite the rods and cones. As a result of a light stimulus, the photoreceptors undergo changes in their membrane potential that ultimately influence the bipolar neurons. These in turn stimulate the ganglion cells, whose axons leave the retina in the tight bundle of fibers known as the **optic nerve.** The retinal layer is thickest where the optic nerve attaches to the eyeball because an increasing number of ganglion cell axons converge at this point. It thins as it approaches the ciliary body.

Activity 3:
Studying the Microscopic Anatomy of the Retina

Obtain a histologic slide of a longitudinal section of the eye. Identify the retinal layers by comparing your view to Figure 24.4. ■

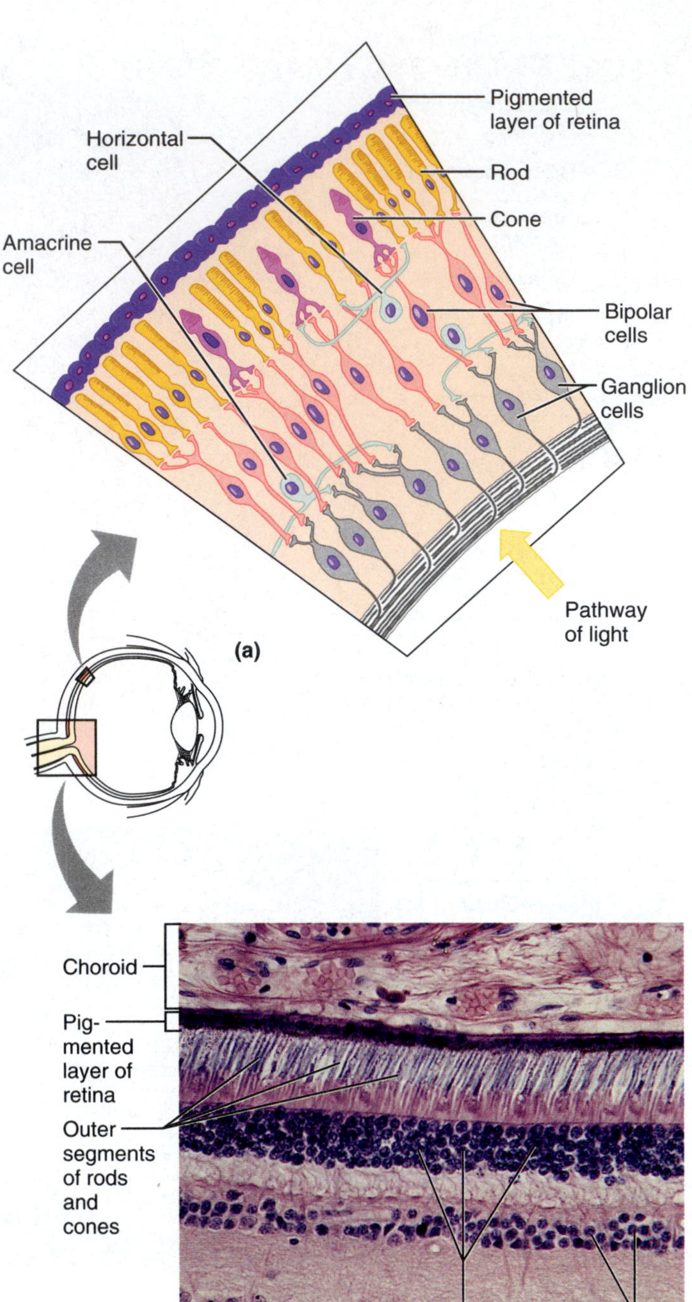

Figure 24.4 Microscopic anatomy of the cellular layers of the retina. (a) Diagrammatic view. **(b)** Photomicrograph of the retina (280×). (See also Plate 15 of the Histology Atlas.)

Visual Pathways to the Brain

The axons of the ganglion cells of the retina converge at the posterior aspect of the eyeball and exit from the eye as the optic nerve. At the **optic chiasma,** the fibers from the medial side of each eye cross over to the opposite side (Figure 24.5). The fiber tracts thus formed are called the **optic tracts.** Each optic tract contains fibers from the lateral side of the eye on the same side and from the medial side of the opposite eye.

The optic tract fibers synapse with neurons in the **lateral geniculate nucleus** of the thalamus, whose axons form the **optic radiation,** terminating in the **optic,** or **visual, cortex** in the occipital lobe of the brain. Here they synapse with the cortical cells, and visual interpretation occurs.

Activity 4: Predicting the Effects of Visual Pathway Lesions

After examining Figure 24.5, determine what effects lesions in the following areas would have on vision:

In the right optic nerve ______________________________

____

Through the optic chiasma ______________________________

____

In the left optic tract ______________________________

____

In the right cerebral cortex (visual area) ______________________________

______________________________ ■

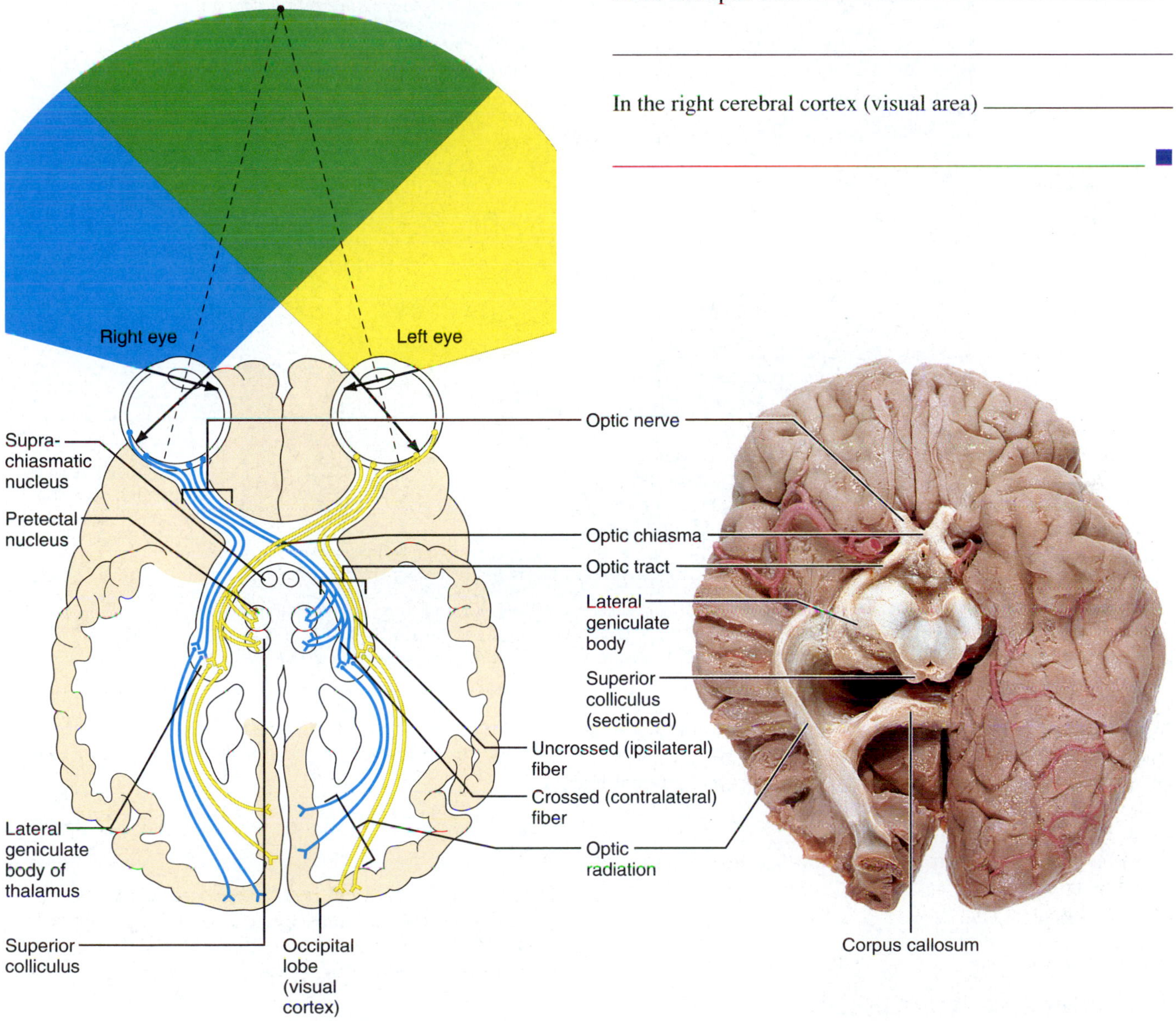

Right eye

Left eye

Supra-chiasmatic nucleus

Pretectal nucleus

Lateral geniculate body of thalamus

Superior colliculus

Occipital lobe (visual cortex)

Optic nerve

Optic chiasma

Optic tract

Lateral geniculate body

Superior colliculus (sectioned)

Uncrossed (ipsilateral) fiber

Crossed (contralateral) fiber

Optic radiation

Corpus callosum

(a)

(b)

Figure 24.5 Visual pathway to the brain. (a) Diagram. (Note that fibers from the lateral portion of each retinal field do not cross at the optic chiasma.) **(b)** Photograph.

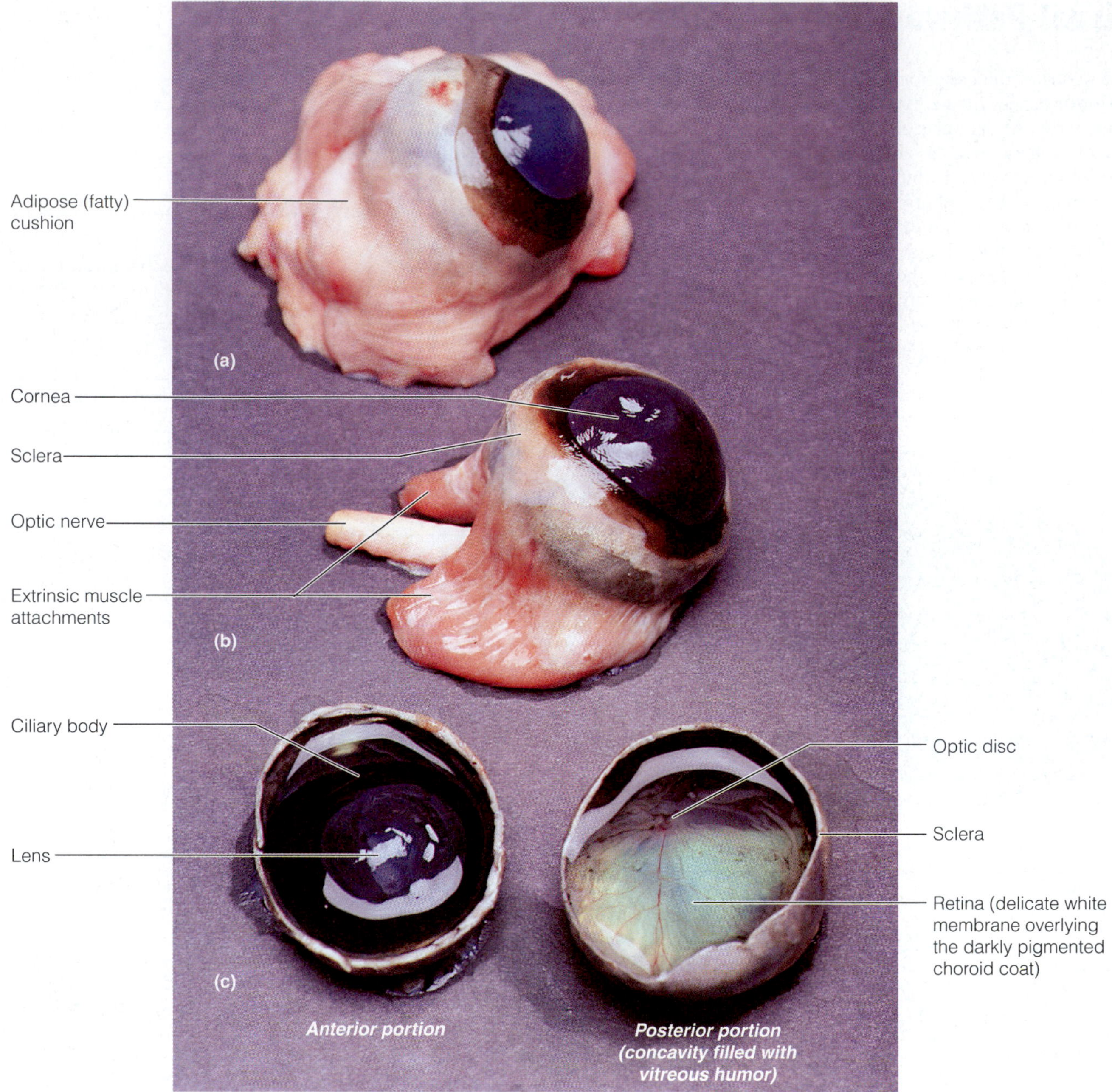

Figure 24.6 Anatomy of the cow eye. (a) Cow eye (entire) removed from orbit (notice the large amount of fat cushioning the eyeball). **(b)** Cow eye (entire) with fat removed to show the extrinsic muscle attachments and optic nerve. **(c)** Cow eye cut along the coronal plane to reveal internal structures.

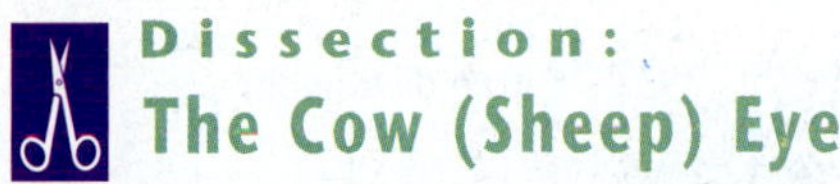

Dissection: The Cow (Sheep) Eye

1. Obtain a preserved cow or sheep eye, dissecting instruments, and a dissecting tray. Don disposable gloves.

2. Examine the external surface of the eye, noting the thick cushion of adipose tissue. Identify the optic nerve (cranial nerve II) as it leaves the eyeball, the remnants of the extrinsic eye muscles, the conjunctiva, the sclera, and the cornea. The normally transparent cornea is opalescent or opaque if the eye has been preserved. Refer to Figure 24.6 as you work.

3. Trim away most of the fat and connective tissue, but leave the optic nerve intact. Holding the eye with the cornea facing downward, carefully make an incision with a sharp scalpel into the sclera about 6 mm (1/4 inch) above the cornea. (The sclera of the preserved eyeball is *very* tough so you will have to apply substantial pressure to penetrate it.) Using scissors, complete the incision around the circumference of the eyeball paralleling the corneal edge.

4. Carefully lift the anterior part of the eyeball away from the posterior portion. Conditions being proper, the vitreous body should remain with the posterior part of the eyeball.

5. Examine the anterior part of the eye and identify the following structures:

Ciliary body: Black pigmented body that appears to be a halo encircling the lens.

Lens: Biconvex structure that is opaque in preserved specimens.

Suspensory ligament: A halo of delicate fibers attaching the lens to the ciliary body.

Carefully remove the lens and identify the adjacent structures:

Iris: Anterior continuation of the ciliary body penetrated by the pupil.

Cornea: More convex anteriormost portion of the sclera; normally transparent but cloudy in preserved specimens.

6. Examine the posterior portion of the eyeball. Carefully remove the vitreous humor, and identify the following structures:

Retina: The neural layer of the retina appears as a delicate yellowish-white, probably crumpled membrane that separates easily from the pigmented choroid.

Note its point of attachment. What is this point called?

____

Pigmented choroid coat: Appears iridescent in the cow or sheep eye owing to a special reflecting surface called the **tapetum lucidum.** This specialized surface reflects the light within the eye and is found in the eyes of animals that live under conditions of low-intensity light. It is not found in humans. ■

Visual Tests and Experiments

The Blind Spot

Activity 5: Demonstrating the Blind Spot

1. Hold Figure 24.7 about 46 cm (18 inches) from your eyes. Close your left eye, and focus your right eye on the X, which should be positioned so that it is directly in line with your right eye. Move the figure slowly toward your face, keeping your right eye focused on the X. When the dot focuses on the blind spot, which lacks photoreceptors, it will disappear.

2. Have your laboratory partner record in metric units the distance at which this occurs. The dot will reappear as the figure is moved closer. Distance at which the dot disappears:

Right eye ______________________________

Repeat the test for the left eye, this time closing the right eye and focusing the left eye on the dot. Record the distance at which the X disappears:

Left eye ______________________________ ■

Afterimages

When light from an object strikes **rhodopsin,** the purple pigment contained in the rods of the retina, it triggers a photochemical reaction that splits rhodopsin into its colorless precursor molecules (vitamin A and a protein called opsin). This event, called *bleaching of the pigment,* initiates a chain of events leading to impulse transmission along fibers of the optic nerve. Once bleaching has occurred in a rod, the photoreceptor pigment must be resynthesized before the rod can be restimulated. This takes a certain period of time. A similar but less well understood reaction occurs in the cones. Both phenomena—that is, the stimulation of the photoreceptor cells

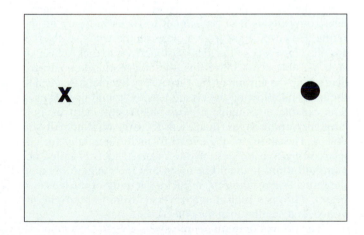

Figure 24.7 Blind spot test figure.

and their subsequent inactive period—can be demonstrated indirectly in terms of positive and negative afterimages.

Activity 6:
Demonstrating Afterimages

1. Stare at the American flag for a few seconds, and then gently close your eyes for approximately 1 minute.
2. Record, in sequence of occurrence, what you "saw" while your eyes were closed.

____

____

____

The bright image of the flag initially seen was a **positive afterimage** caused by the continued firing of the rods. The altered image that subsequently appeared against a lighter background was the **negative afterimage,** an indication that the visual pigment in the affected photoreceptor cells had been bleached. ■

Refraction, Visual Acuity, and Astigmatism

When light rays pass from one medium to another, their velocity, or speed of transmission, changes, and the rays are bent or **refracted.** Thus the light rays in the visual field are refracted as they encounter the cornea, lens, and vitreous humor of the eye.

The refractive index (bending power) of the cornea and vitreous humor are constant. But the lens's refractive index, or strength, can be varied by changing the lens's shape—that is, by making it more or less convex so that the light is properly converged and focused on the retina. The greater the lens convexity, or bulge, the more the light will be bent and the stronger the lens. Conversely, the less the lens convexity (the flatter it is), the less it bends the light.

In general, light from a distant source (over 6 m or 20 feet) approaches the eye as parallel rays, and no change in lens convexity is necessary for it to focus properly on the retina. However, light from a close source tends to diverge, and the convexity of the lens must increase to make close vision possible. To achieve this, the ciliary muscle contracts, decreasing the tension on the suspensory ligament attached to the lens and allowing the elastic lens to "round up." Thus, a lens capable of bringing a *close* object into sharp focus is stronger (more convex) than a lens focusing on a more distant object. The ability of the eye to focus differentially for objects of near vision (less than 6 m or 20 feet) is called **accommodation.** It should be noted that the image formed on the retina as a result of the refractory activity of the lens (see Figure 24.8) is a **real image** (reversed from left to right, inverted, and smaller than the object).

The normal or **emmetropic eye** is able to accommodate properly (Figure 24.9a). However, visual problems may result from (1) lenses that are too strong or too "lazy" (overconverging and underconverging, respectively), (2) from structural problems such as an eyeball that is too long or too short to provide for proper focusing by the lens, or (3) a cornea or lens with improper curvatures.

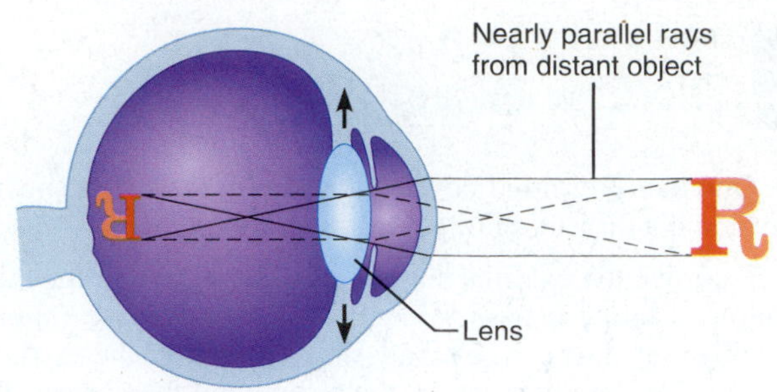

Figure 24.8 Refraction of light in the eye, resulting in the production of a real image on the retina.

Individuals in whom the image normally focuses in front of the retina are said to have **myopia,** or "nearsightedness" (Figure 24.9b); they can see close objects without difficulty, but distant objects are blurred or seen indistinctly. Correction requires a concave lens, which causes the light reaching the eye to diverge.

If the image focuses behind the retina, the individual is said to have **hyperopia** or farsightedness. Such persons have no problems with distant vision but need glasses with convex lenses to augment the converging power of the lens for close vision (Figure 24.9c).

Irregularities in the curvatures of the lens and/or the cornea lead to a blurred vision problem called **astigmatism.** Cylindrically ground lenses, which compensate for inequalities in the curvatures of the refracting surfaces, are prescribed to correct the condition. ■

Near-Point Accommodation The elasticity of the lens decreases dramatically with age, resulting in difficulty in focusing for near or close vision. This condition is called **presbyopia**—literally, old vision. Lens elasticity can be tested by measuring the **near point of accommodation.** The near point of vision is about 10 cm from the eye in young adults. It is closer in children and farther in old age.

Activity 7:
Determining Near Point of Accommodation

To determine your near point of accommodation, hold a common straight pin at arm's length in front of one eye. (If desired, the text in the lab manual can be used rather than a pin.) Slowly move the pin toward that eye until the pin image becomes distorted. Have your lab partner measure the distance from your eye to the pin at this point, and record the distance below. Repeat the procedure for the other eye.

Near point for right eye ________________________

Near point for left eye ________________________ ■

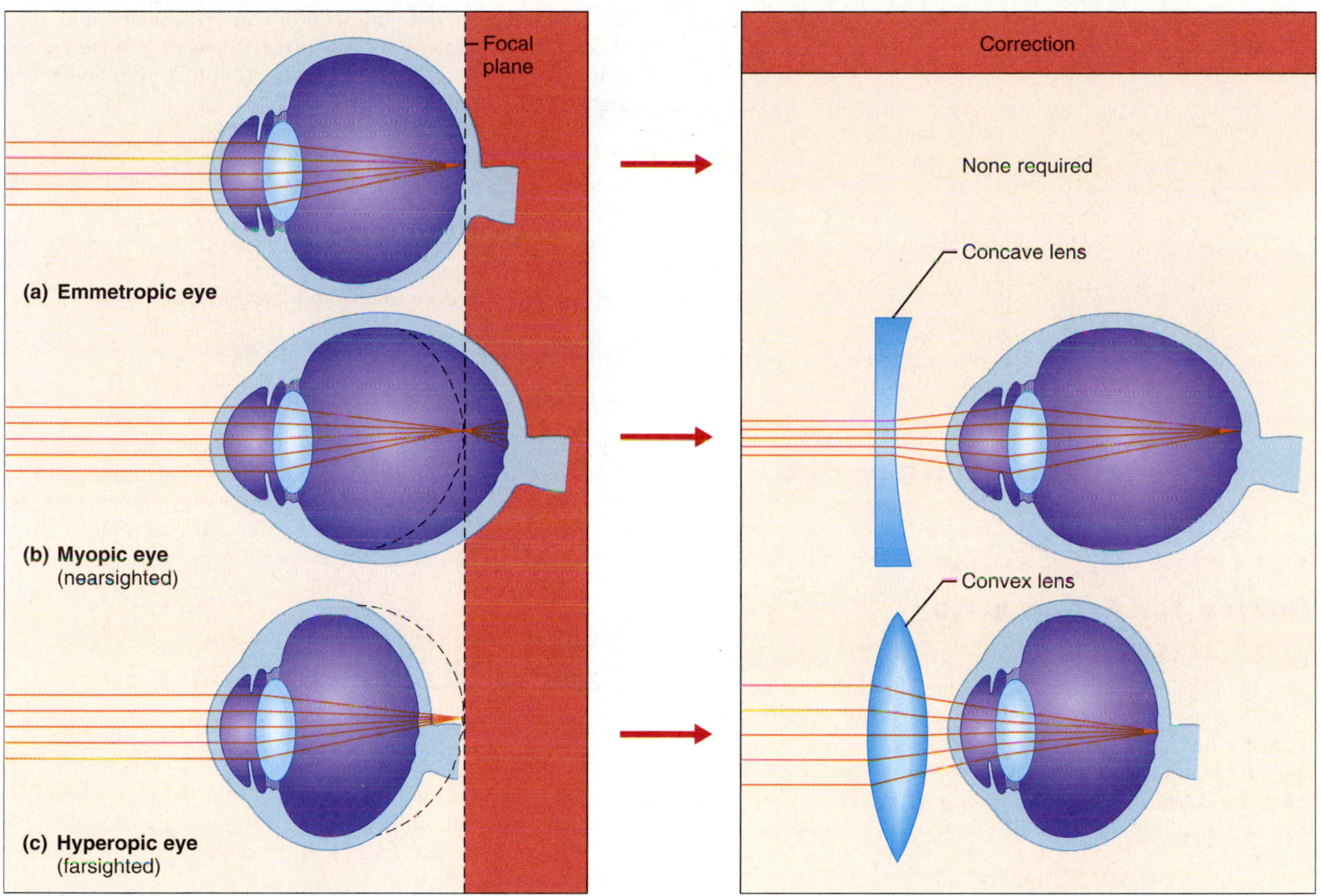

Figure 24.9 Problems of refraction. (a) In the emmetropic (normal) eye, light from both near and far objects is focused properly on the retina. **(b)** In a myopic eye, light from distant objects is brought to a focal point before reaching the retina. It then diverges. **(c)** In the hyperopic eye, light from a near object is brought to a focal point behind (past) the retina. (Refractory effect of cornea is ignored.)

Visual Acuity **Visual acuity,** or sharpness of vision, is generally tested with a Snellen eye chart, which consists of letters of various sizes printed on a white card. This test is based on the fact that letters of a certain size can be seen clearly by eyes with normal vision at a specific distance. The distance at which the normal, or emmetropic, eye can read a line of letters is printed at the end of that line.

Activity 8: Testing Visual Acuity

1. Have your partner stand 6 m or 20 feet from the posted Snellen eye chart and cover one eye with a card or hand. As your partner reads each consecutive line aloud, check for accuracy. If this individual wears glasses, give the test twice—first with glasses off and then with glasses on. *Do not remove contact lenses, but note that they were in place during the test.*

2. Record the number of the line with the smallest-sized letters read. If it is 20/20, the person's vision for that eye is normal. If it is 20/40, or any ratio with a value less than one, he or she has less than the normal visual acuity. (Such an individual is myopic.) If the visual acuity is 20/15, vision is better than normal, because this person can stand at 6 m or 20 feet from the chart and read letters that are only discernible by the normal eye at 4.5 m or 15 feet. Give your partner the number of the line corresponding to the smallest letters read, to record in step 4.

3. Repeat the process for the other eye.

4. Have your partner test and record your visual acuity. If you wear glasses, the test results *without* glasses should be recorded first.

Visual acuity, right eye without glasses ____________

Visual acuity, right eye with glasses ____________

Visual acuity, left eye without glasses ____________

Visual acuity, left eye with glasses ____________ ■

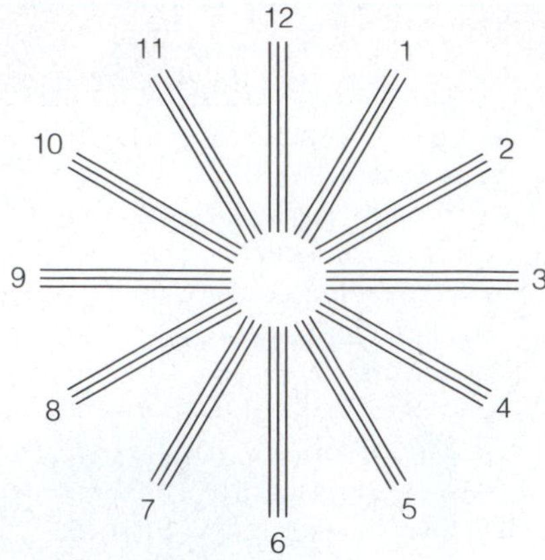

Figure 24.10 Astigmatism testing chart.

Activity 9: Testing for Astigmatism

The astigmatism chart (Figure 24.10) is designed to test for defects in the refracting surface of the lens and/or cornea.

View the chart first with one eye and then with the other, focusing on the center of the chart. If all the radiating lines appear equally dark and distinct, there is no distortion of your refracting surfaces. If some of the lines are blurred or appear less dark than others, at least some degree of astigmatism is present.

Is astigmatism present in your left eye? ____________

Right eye? ____________ ■

Color Blindness

Ishihara's color plates are designed to test for deficiencies in the cones or color photoreceptor cells. There are three cone types, each containing a different light-absorbing pigment. One type primarily absorbs the red wavelengths of the visible light spectrum, another the blue wavelengths, and a third the green wavelengths. Nerve impulses reaching the brain from these different photoreceptor types are then interpreted (seen) as red, blue, and green, respectively. Interpretation of the intermediate colors of the visible light spectrum is a result of overlapping input from more than one cone type.

Activity 10: Testing for Color Blindness

1. Locate the interpretation table that accompanies the Ishihara color plates, and prepare a sheet to record data for the test. Note which plates are patterns rather than numbers.

2. View the color plates in bright light or sunlight while holding them about .8 m or 30 inches away and at right angles to your line of vision. Report to your laboratory partner what you see in each plate. Take no more than 3 seconds for each decision.

3. Your partner should record your responses and then check their accuracy with the correct answers provided in the color plate book. Is there any indication that you have some degree of color blindness?

____________ If so, what type? ____________

____

Repeat the procedure to test your partner's color vision. ■

Relative Positioning of Rods and Cones on the Retina

The test subject for this demonstration should have shown no color vision problems during the test for color blindness. Students may work in pairs, or two students may perform the test for the entire class. White, red, blue, and green paper discs and chalk will be needed for this demonstration.

Activity 11: Mapping the Rods and Cones

1. Position the subject about .3 m or 1 foot away from the blackboard.

2. Make a small white chalk circle on the board immediately in front of the subject's right eye. Have the subject close the left eye and stare fixedly at the circle with the right eye throughout the test.

3. To map the extent of the rod field, begin to move a white paper disc into the field of vision from various sides of the visual field (beginning at least .6 m (2 feet) away from the white chalk circle) and plot with white chalk dots the points at which the disc first becomes visible to the test subject.

4. Repeat the procedure, using red, green, and blue paper discs and like-colored chalk to map the cone fields. Color (not object) identification is required.

5. After the test has been completed for all four discs, connect all dots of the same color. It should become apparent that each of the color fields has a different radius and that the rod and cone distribution on the retina is not uniform. (Normal fields have white outermost, followed by blue, red, and green as the fovea is approached.)

6. Record in the laboratory review section the rod and cone color field distribution you observed. Use appropriately colored pencils. ■

Binocular Vision

Humans, cats, predatory birds, and most primates are endowed with **binocular** (two-eyed) **vision.** Although both eyes look in approximately the same direction, they see slightly different views. Their visual fields, each about 170 degrees, overlap to a considerable extent; thus there is two-eyed vision at the overlap area (Figure 24.11).

In contrast, the eyes of many animals (rabbits, pigeons, and others) are more on the sides of their head. Such animals see in two different directions and thus have a panoramic field of view and **panoramic vision.**

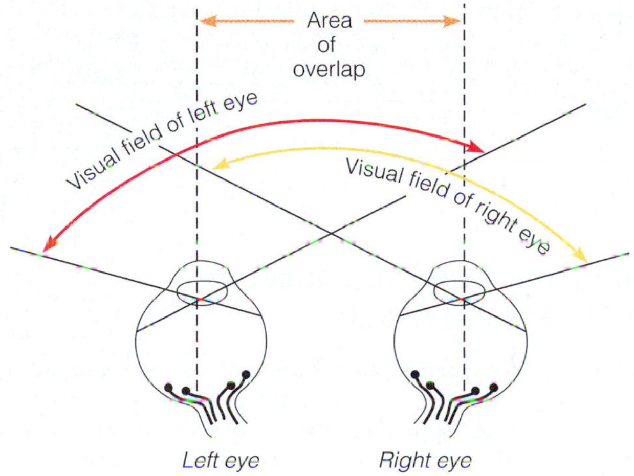

Figure 24.11 Overlapping of the visual fields.

Although both types of vision have their good points, binocular vision provides three-dimensional vision and an accurate means of locating objects in space. The slight differences between the views seen by the two eyes are fused by the higher centers of the visual cortex to give us *depth perception*. Because of the manner in which the visual cortex resolves these two different views into a single image, it is sometimes referred to as the "cyclopean eye of the binocular animal."

Activity 12: Tests for Binocular Vision

1. To demonstrate that a slightly different view is seen by each eye, perform the following simple experiment.

Close your left eye. Hold a pencil at arm's length directly in front of your right eye. Position another pencil directly beneath it and then move the lower pencil about half the distance toward you. As you move the lower pencil, make sure it remains in the *same plane* as the stationary pencil, so that the two pencils continually form a straight line. Then, without moving the pencils, close your right eye and open your left eye. Notice that with only the right eye open, the moving pencil stays in the same plane as the fixed pencil, but that when viewed with the left eye, the moving pencil is displaced laterally away from the plane of the fixed pencil.

2. To demonstrate the importance of two-eyed binocular vision for depth perception, perform this second simple experiment.

Have your laboratory partner hold a test tube erect about arm's length in front of you. With both eyes open, quickly insert a pencil into the test tube. Remove the pencil, bring it back close to your body, close one eye, and quickly and without hesitation insert the pencil into the test tube. (Do not feel for the test tube with the pencil!) Repeat with the other eye closed.

Was it as easy to dunk the pencil with one eye closed as with both eyes open?

______________ ■

Eye Reflexes

Both intrinsic (internal) and extrinsic (external) muscles are necessary for proper eye functioning. The *intrinsic muscles,* controlled by the autonomic nervous system, are those of the ciliary body (which alters the lens curvature in focusing) and the radial and circular muscles of the iris (which control pupillary size and thus regulate the amount of light entering the eye). The *extrinsic muscles* are the rectus and oblique muscles, which are attached to the eyeball exterior (see Figure 24.2). These muscles control eye movement and make it possible to keep moving objects focused on the fovea centralis. They are also responsible for **convergence,** or medial eye movements, which is essential for near vision. When convergence occurs, both eyes are directed toward the near object viewed. The extrinsic eye muscles are controlled by the somatic nervous system.

Activity 13: Demonstrating Reflex Activity of Intrinsic and Extrinsic Eye Muscles

Involuntary activity of both the intrinsic and extrinsic muscle types is brought about by reflex actions that can be observed in the following experiments.

Photopupillary Reflex

Sudden illumination of the retina by a bright light causes the pupil to constrict reflexively in direct proportion to the light intensity. This protective response prevents damage to the delicate photoreceptor cells.

Obtain a laboratory lamp or penlight. Have your laboratory partner sit with eyes closed and hands over his or her eyes. Turn on the light and position it so that it shines on the subject's right hand. After 1 minute, ask your partner to uncover and open the right eye. Quickly observe the pupil of that eye. What happens to the pupil?

____

Shut off the light and ask your partner to uncover and open the opposite eye. What are your observations?

____

Accommodation Pupillary Reflex

Have your partner gaze for approximately 1 minute at a distant object in the lab—*not* toward the windows or another light source. Observe your partner's pupils. Then hold some printed material 15 to 25 cm or 6 to 10 inches from his or her face, and direct him or her to focus on it.

How does pupil size change as your partner focuses on the printed material?

____

Explain the value of this reflex. ______________________

____

____

____

Convergence Reflex

Repeat the previous experiment, this time using a pen or pencil as the close object to be focused on. Note the position of your partner's eyeballs both while he or she is gazing at the distant and at the close object. Do they change position as the object of focus is changed?

__________ In what way? ______________________

____

____

____

Explain the importance of the convergence reflex.

____

____

____ ■

Ophthalmoscopic Examination of the Eye (Optional)

The ophthalmoscope is an instrument used to examine the *fundus,* or eyeball interior, to determine visually the condition of the retina, optic disc, and internal blood vessels. Certain pathologic conditions such as diabetes mellitus, arteriosclerosis, and degenerative changes of the optic nerve and retina can be detected by such an examination. The ophthalmoscope consists of a set of lenses mounted on a rotating disc (the **lens selection disc**), a light source regulated by a **rheostat control,** and a mirror that reflects the light so that the eye interior can be illuminated (Figure 24.12a).

The lens selection disc is positioned in a small slit in the mirror, and the examiner views the eye interior through this

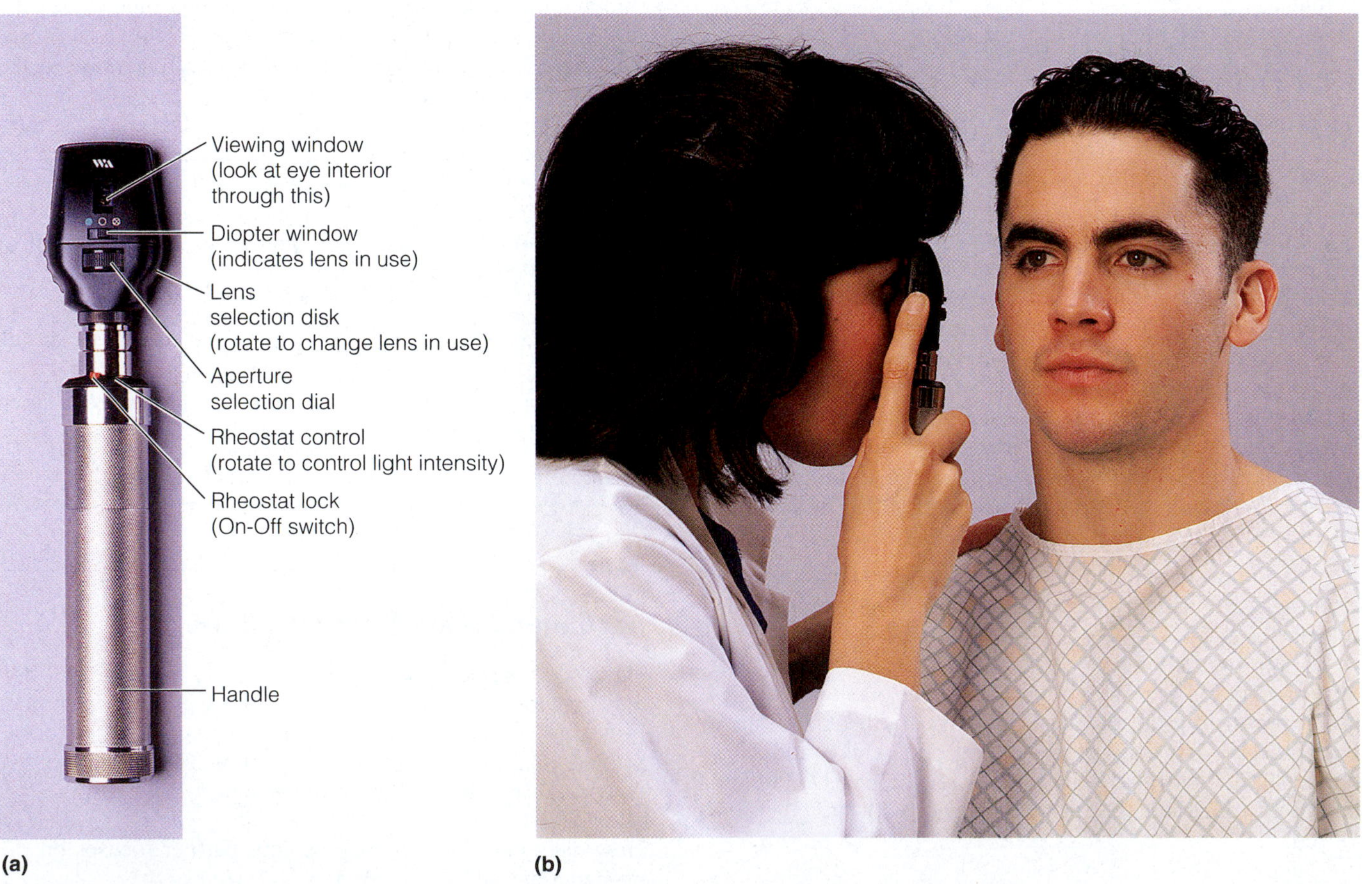

Figure 24.12 Structure and use of an ophthalmoscope. (a) Structure of an ophthalmoscope. **(b)** Proper position for beginning to examine the right eye with an ophthalmoscope.

slit, appropriately called the **viewing window.** The focal length of each lens is indicated in diopters preceded by a + sign if the lens is convex and by a – sign if the lens is concave. When the zero (0) is seen in the **diopter window,** there is no lens positioned in the slit. The depth of focus for viewing the eye interior is changed by changing the lens.

The light is turned on by depressing the red **rheostat lock button** and then rotating the rheostat control in the clockwise direction. The aperture selection disc on the front of the instrument allows the nature of the light beam to be altered. Generally, green light allows for clearest viewing of the blood vessels in the eye interior and is most comfortable for the subject.

Once you have examined the ophthalmoscope and have become familiar with it, you are ready to conduct an eye examination.

Activity 14: Conducting an Ophthalmoscopic Examination

1. Conduct the examination in a dimly lit or darkened room with the subject comfortably seated and gazing straight ahead. To examine the right eye, sit face-to-face with the subject, hold the instrument in your right hand, and use your right eye to view the eye interior (Figure 24.12b). You may want to steady yourself by resting your left hand on the subject's shoulder. To view the left eye, use your left eye, hold the instrument in your left hand, and steady yourself with your right hand.

2. Begin the examination with the 0 (no lens) in position. Grasp the instrument so that the lens disc may be rotated with the index finger. Holding the ophthalmoscope about 15 cm or 6 inches from the subject's eye, direct the light into the pupil at a slight angle—through the pupil edge rather than directly through its center. You will see a red circular area that is the illuminated eye interior.

3. Move in as close as possible to the subject's cornea (to within 5 cm or 2 in.) as you continue to observe the area. Steady your instrument-holding hand on the subject's cheek if necessary. If both your eye and that of the subject are normal, the fundus can be viewed clearly without further adjustment of the ophthalmoscope. If the fundus cannot be focused, slowly rotate the lens disc counterclockwise until the fundus can be clearly seen. When the ophthalmoscope is correctly set, the fundus of the right eye should appear as shown in Figure 24.13. (Note: If a positive [convex] lens is required and your eyes are normal, the subject has hyperopia. If a negative [concave] lens is necessary to view the fundus and your eyes are normal, the subject is myopic.)

When the examination is proceeding correctly, the subject can often see images of retinal vessels in his own eye that appear rather like cracked glass. If you are unable to achieve a sharp focus or to see the optic disc, move medially or laterally and begin again.

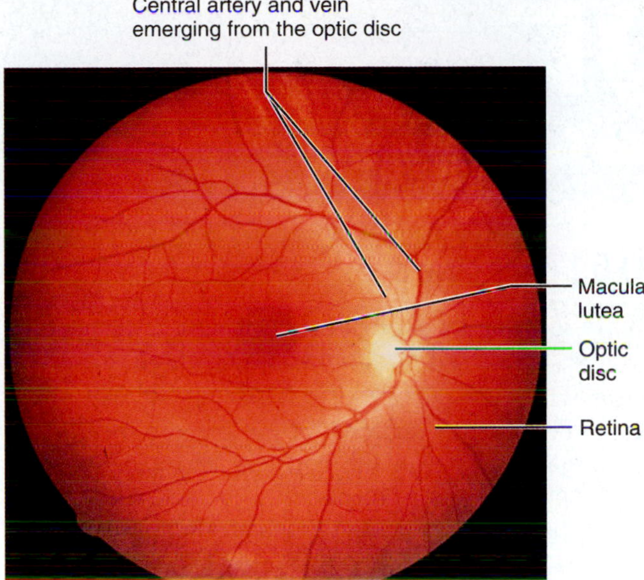

Figure 24.13 Posterior portion of right retina.

4. Examine the optic disc for color, elevation, and sharpness of outline, and observe the blood vessels radiating from near its center. Locate the macula, lateral to the optic disc. It is a darker area in which blood vessels are absent, and the fovea appears to be a slightly lighter area in its center. The macula is most easily seen when the subject looks directly into the light of the ophthalmoscope.

 Do not examine the macula for longer than 1 second at a time.

5. When you have finished examining your partner's retina, shut off the ophthalmoscope. Change places with your partner (become the subject) and repeat steps 1–4. ■

exercise

25 Special Senses: Hearing and Equilibrium

Objectives

1. To identify, by appropriately labeling a diagram, the anatomical structures of the outer, middle, and inner ear, and to explain their functions.
2. To describe the anatomy of the organ of hearing (organ of Corti in the cochlea) and explain its function in sound reception.
3. To describe the anatomy of the equilibrium organs of the inner ear (cristae ampullares and maculae), and to explain their relative function in maintaining equilibrium.
4. To define or explain *central deafness, conduction deafness,* and *nystagmus.*
5. To state the purpose of the Weber, Rinne, balance, Barany, and Romberg tests.
6. To explain how one is able to localize the source of sounds.
7. To describe the effects of acceleration on the semicircular canals.
8. To explain the role of vision in maintaining equilibrium.

Materials

- ❑ Three-dimensional dissectible ear model and/or chart of ear anatomy
- ❑ Histologic slides of the cochlea of the ear
- ❑ Compound microscope
- ❑ *Demonstration:* Microscope focused on a slide of a crista ampullaris receptor of a semicircular canal
- ❑ Otoscope (if available)
- ❑ Disposable otoscope tips (if available) and autoclave bag
- ❑ Alcohol swabs
- ❑ Absorbent cotton
- ❑ Pocket watch or clock that ticks
- ❑ Tuning forks (range of frequencies)
- ❑ Rubber mallet
- ❑ 12-inch ruler
- ❑ Metric ruler
- ❑ Audiometer
- ❑ Red and blue pencils

Anatomy of the Ear

Gross Anatomy

The ear is a complex structure containing sensory receptors for hearing and equilibrium. The ear is divided into three major areas: the *outer ear,* the *middle ear,* and the *inner ear* (Figure 25.1). The outer and middle ear structures serve the needs of the sense of hearing *only,* while inner ear structures function both in equilibrium and hearing reception.

Activity 1: Identifying Structures of the Ear

Obtain a dissectible ear model and identify the structures described below. Refer to Figure 25.1 as you work. ■

The **outer,** or **external, ear** is composed primarily of the pinna and the external auditory canal. The **pinna, or auricle,** is the skin-covered cartilaginous structure encircling the auditory canal opening. In many animals, it collects and directs sound waves into the external auditory canal. In humans this function of the pinna is largely lost. The portion of the pinna lying inferior to the external auditory canal is the **lobule.**

The **external auditory canal** is a short, narrow (about 2.5 cm long by 0.6 cm wide) chamber carved into the temporal bone. In its skin-lined walls are wax-secreting glands called **ceruminous glands.** The sound waves that enter the external auditory canal eventually encounter the **tympanic membrane,** or **eardrum,** which vibrates at exactly the same frequency as the sound wave(s) hitting it. The membranous eardrum separates the outer from the middle ear.

The **middle ear** is essentially a small chamber—the **tympanic cavity**—found within the temporal bone. The cavity is spanned by three small bones, collectively called the **ossicles** (**malleus, incus,** and **stapes**),* which articulate to form a lever system that amplifies and transmits the vibratory motion of the eardrum to the fluids of the inner ear via the **oval window.**

Connecting the middle ear chamber with the nasopharynx is the **pharyngotympanic,** or **auditory, tube.** Normally this tube is flattened and closed, but swallowing or yawning can cause it to open temporarily to equalize the pressure of the middle ear cavity with external air pressure. This is an important function. The eardrum does not vibrate properly unless the pressure on both of its surfaces is the same.

Because the mucosal membranes of the middle ear cavity and nasopharynx are continuous through the pharyngotympanic or auditory tube, **otitis media,** or inflam-

* The ossicles are often referred to by their common names, that is, hammer, anvil, and stirrup, respectively.

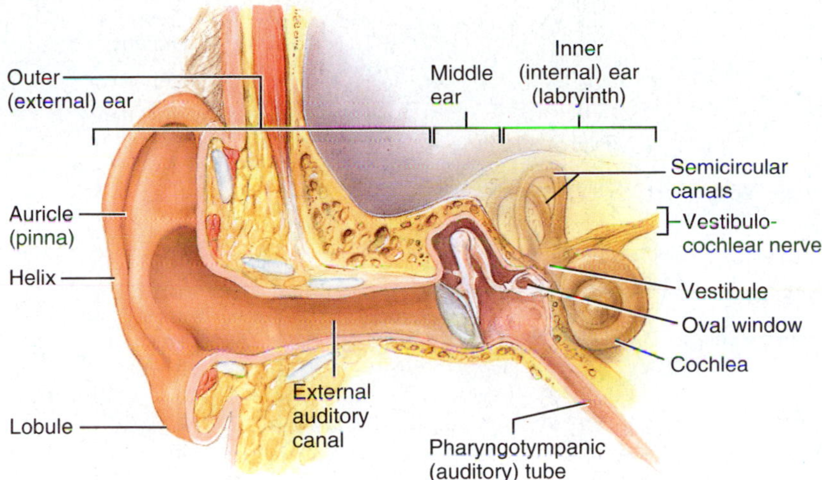

Figure 25.1 Anatomy of the ear.

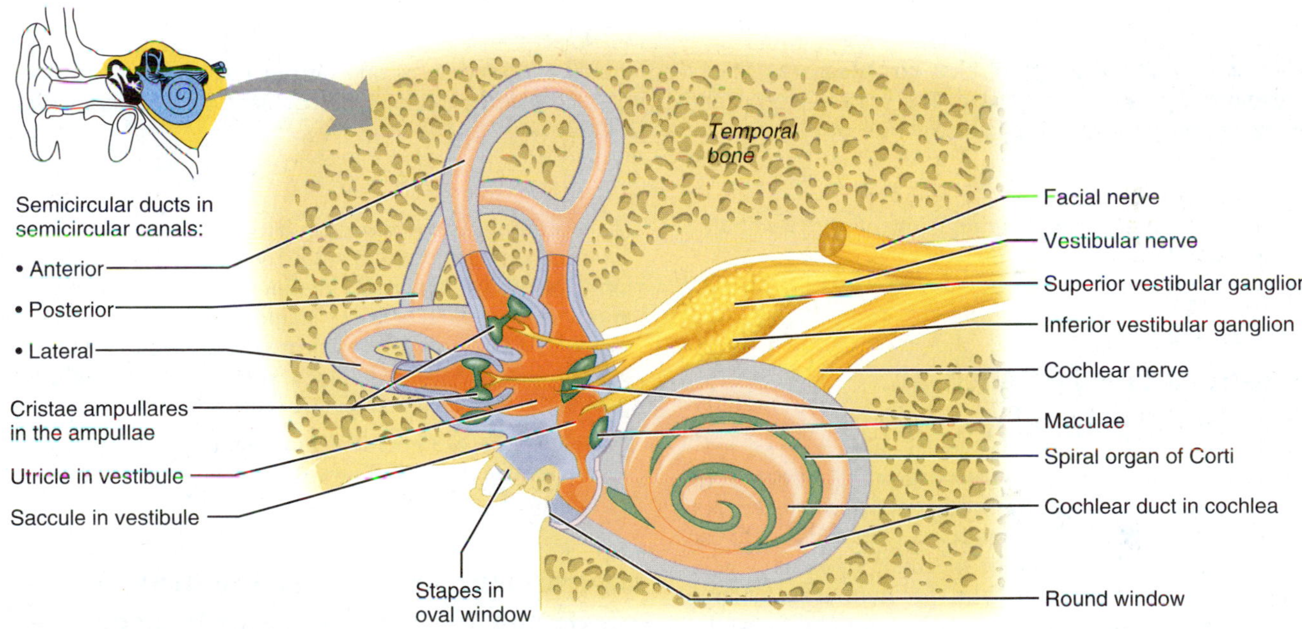

Figure 25.2 Inner ear. Right membranous labyrinth shown within the bony labyrinth.

mation of the middle ear, is a fairly common condition, especially among youngsters prone to sore throats. In cases where large amounts of fluid or pus accumulate in the middle ear cavity, an emergency myringotomy (lancing of the eardrum) may be necessary to relieve the pressure. Frequently, tiny ventilating tubes are put in during the procedure. ■

The **inner,** or **internal, ear** consists of a system of bony and rather tortuous chambers called the **osseous,** or **bony, labyrinth,** which is filled with an aqueous fluid called **perilymph** (Figure 25.2). Suspended in the perilymph is the **membranous labyrinth,** a system that mostly follows the contours of the osseous labyrinth. The membranous labyrinth is filled with a more viscous fluid called **endolymph.** The three subdivisions of the bony labyrinth are the cochlea, the vestibule, and the semicircular canals, with the vestibule situated between the cochlea and semicircular canals. The **vestibule** and the **semicircular canals** are involved with equilibrium.

The snail-like **cochlea** (see Figures 25.2 and 25.3) contains the sensory receptors for hearing. The cochlear membranous labyrinth, the **cochlear duct,** is a soft wormlike tube about 3.8 cm long. It winds through the full two and three-quarter turns of the cochlea and separates the perilymph-containing cochlear cavity into upper and lower chambers, the **scala vestibuli** and **scala tympani,** respectively. The scala vestibuli terminates at the oval window, which "seats" the foot plate of the stirrup located laterally in the tympanic cavity. The scala tympani is bounded by a membranous area called the **round window.** The cochlear duct, itself filled with endolymph, supports the **organ of Corti,** which contains the receptors for hearing—the sensory hair cells and

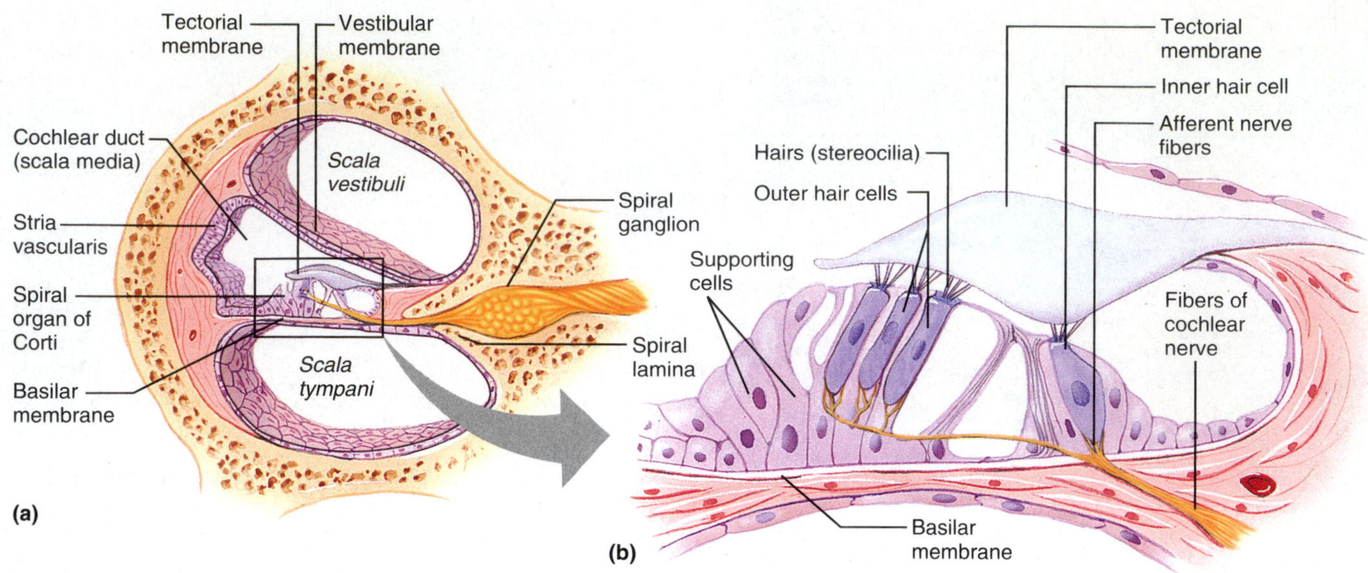

Figure 25.3 Anatomy of the cochlea. (a) Magnified cross-sectional view of one turn of the cochlea, showing the relationship of the three scalae. The scalae vestibuli and tympani contain perilymph; the cochlear duct (scala media) contains endolymph. **(b)** Detailed structure of the spiral organ of Corti.

nerve endings of the cochlear division of the vestibulocochlear nerve (VIII).

Activity 2: Examining the Ear with an Otoscope (Optional)

1. Obtain an otoscope and two alcohol swabs. Inspect your partner's ear canal and then select the largest-*diameter* (not length!) speculum that will fit comfortably into his or her ear to permit full visibility. Clean the speculum thoroughly with an alcohol swab, and then attach the speculum to the battery-containing otoscope handle. Before beginning, check that the otoscope light beam is strong. (If not, obtain another otoscope or new batteries.) Some otoscopes come with disposable tips. Be sure to use a new tip for each ear examined. Dispose of these tips in an autoclave bag after use.

2. When you are ready to begin the examination, hold the lighted otoscope securely between your thumb and forefinger (like a pencil), and rest the little finger of the otoscope-holding hand against your partner's head. This maneuver forms a brace that allows the speculum to move as your partner moves and prevents the speculum from penetrating too deeply into the ear canal during unexpected movements.

3. Grasp the ear pinna firmly and pull it up, back, and slightly laterally. If your partner experiences pain or discomfort when the pinna is manipulated, an inflammation or infection of the external ear may be present. If this occurs, do not attempt to examine the ear canal.

4. Carefully insert the speculum of the otoscope into the external auditory canal in a downward and forward direction only far enough to permit examination of the tympanic membrane or eardrum. Note its shape, color, and vascular network. The healthy tympanic membrane is pearly white. During the examination, notice if there is any discharge or redness in the canal and identify earwax.

5. After the examination, thoroughly clean the speculum with the second alcohol swab before returning the otoscope to the supply area. ■

Microscopic Anatomy of the Organ of Corti and the Mechanism of Hearing

The anatomical details of the organ of Corti are shown in Figure 25.3. The hair (auditory receptor) cells rest on the **basilar membrane**, which forms the floor of the cochlear duct, and their "hairs" (stereocilia) project into a gelatinous membrane, the **tectorial membrane,** that overlies them. The roof of the cochlear duct is called the **vestibular membrane.** The endolymph-filled chamber of the cochlear duct is the **scala media.**

Activity 3: Examining the Microscopic Structure of the Cochlea

Obtain a compound microscope and a prepared microscope slide of the cochlea and identify the areas shown in Figure 25.4. (See Plate 16 in the Histology Atlas). ■

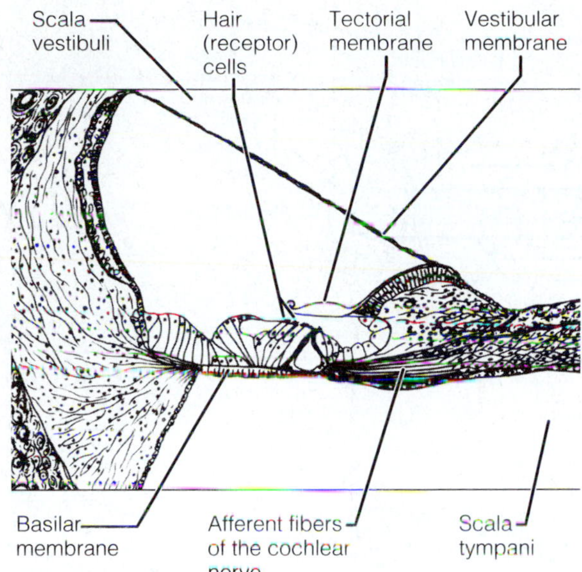

Figure 25.4 Microscopic view of the organ of Corti. (See also Plate 16 in the Histology Atlas.)

The mechanism of hearing begins as sound waves pass through the external auditory canal and through the middle ear into the inner ear, where the vibration eventually reaches the organ of Corti, which contains the receptors for hearing. Many theories have attempted to explain how the organ of Corti actually responds to sound.

The popular "traveling wave" hypothesis of Von Békésy suggests that vibration of the stirrup at the oval window initiates traveling waves that cause maximal displacements of the basilar membrane where they peak and stimulate the hair cells of the organ of Corti in that region. Since the area at which the traveling waves peak is a high-pressure area, the vestibular membrane is compressed at this point and, in turn, compresses the endolymph and the basilar membrane of the cochlear duct. The resulting pressure on the perilymph in the scala tympani causes the membrane of the round window to bulge outward into the middle ear chamber, thus acting as a relief valve for the compressional wave. Von Békésy found that high-frequency waves (high-pitched sounds) peaked close to the oval window and that low-frequency waves (low-pitched sounds) peaked farther up the basilar membrane near the apex of the cochlea. Although the mechanism of sound reception by the organ of Corti is not completely understood, we do know that hair cells on the basilar membrane are uniquely stimulated by sounds of various frequencies and amplitude and that once stimulated they depolarize and begin the chain of nervous impulses to the auditory centers of the temporal lobe cortex. This series of events results in the phenomenon we call hearing (Figure 25.5).

By the time most people are in their 60s, a gradual deterioration and atrophy of the organ of Corti begins, and leads to a loss in the ability to hear high tones and speech sounds. This condition, **presbycusis,** is a type of sensorineural deafness. Because many elderly people refuse to accept their hearing loss and resist using hearing aids, they begin to rely more and more on their vision for clues as to what is going on around them, and may be accused of ignoring people.

Although presbycusis is considered to be a disability of old age, it is becoming much more common in younger people as our world grows noisier. The damage (breakage of the "hairs" of the hair cells) caused by excessively loud sounds is progressive and cumulative. Each assault causes a bit more damage. Music played and listened to at deafening levels is definitely a contributing factor to the deterioration of hearing receptors. ■

Activity 4:
Conducting Laboratory Tests of Hearing

Perform the following hearing tests in a quiet area.

Acuity Test

Have your lab partner pack one ear with cotton and sit quietly with eyes closed. Obtain a ticking clock or pocket watch and hold it very close to his or her *unpacked* ear. Then slowly move it away from the ear until your partner signals that the ticking is no longer audible. Record the distance in centimeters at which ticking is inaudible.

Right ear ________________ Left ear ________________

Is the threshold of audibility sharp or indefinite?

____

Sound Localization

Ask your partner to close both eyes. Hold the pocket watch at an audible distance (about 15 cm) from his or her ear, and move it to various locations (front, back, sides, and above his or her head). Have your partner locate the position by pointing in each instance. Can the sound be localized equally well at all positions?

______________ If not, at what position(s) was the sound less easily located?

____

____

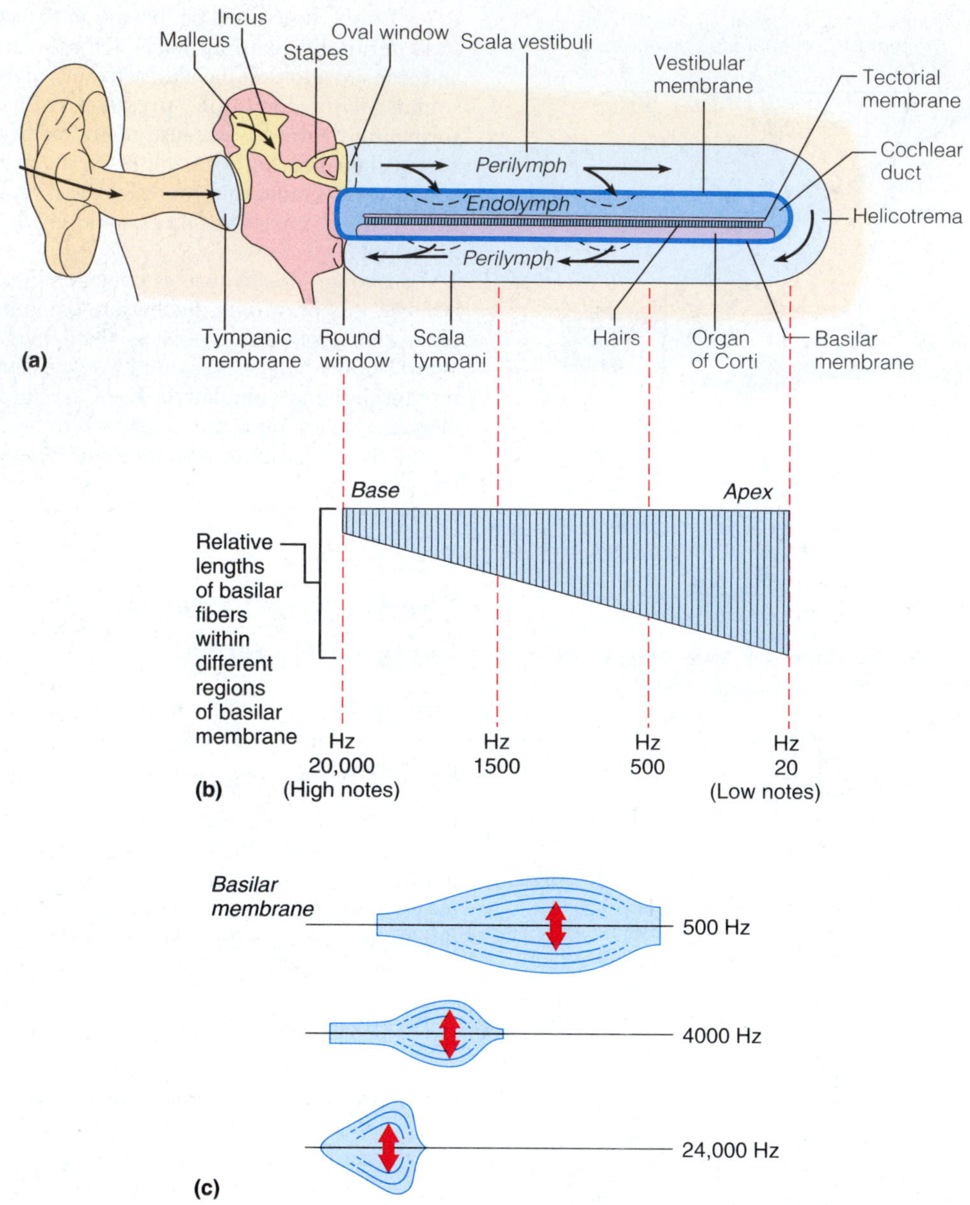

Figure 25.5 Resonance of the basilar membrane. (a) Fluid movement in the cochlea following the stirrup thrust at the oval window. Compressional wave created causes the round window to bulge into the middle ear. Pressure waves set up vibrations in the basilar membrane. **(b)** Fibers span the basilar membrane. The length of the fibers "tunes" specific regions to vibrate at specific frequencies. **(c)** Different frequencies of pressure waves in the cochlea stimulate particular hair cells and neurons.

The ability to localize the source of a sound depends on two factors—the difference in the loudness of the sound reaching each ear and the time of arrival of the sound at each ear. How does this information help to explain your findings?

___

___

Frequency Range of Hearing

Obtain three tuning forks: one with a low frequency (75 to 100 Hz [cps]), one with a frequency of approximately 1000 Hz, and one with a frequency of 4000 to 5000 Hz. Strike the lowest-frequency fork on the heel of your hand or with a rubber mallet, and hold it close to your partner's ear. Repeat with the other two forks.

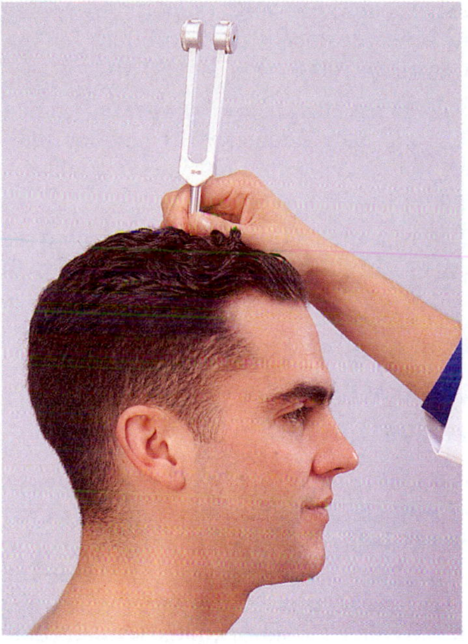

(a)

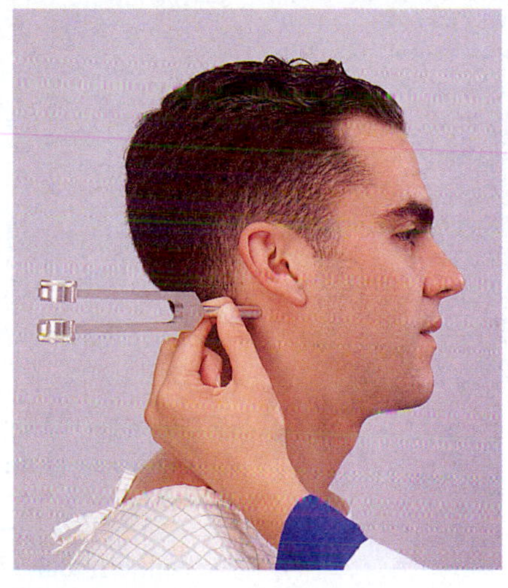

(b)

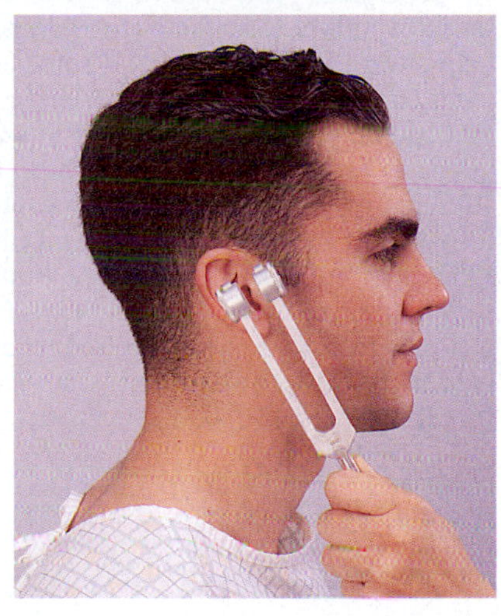

(c)

Figure 25.6 The Weber and Rinne tuning fork tests. (a) The Weber test to evaluate whether the sound remains centralized (normal) or lateralizes to one side or the other (indicative of some degree of conductive or sensorineural deafness). **(b** and **c)** The Rinne test to compare bone conduction and air conduction.

Which fork was heard most clearly and comfortably?

______________ Hz

Which was heard least well?___________ Hz

Weber Test to Determine Conductive and Sensorineural Deafness

Strike a tuning fork and place the handle of the tuning fork medially on your partner's head (see Figure 25.6a). Is the tone equally loud in both ears, or is it louder in one ear?

____

If it is equally loud in both ears, you have equal hearing or equal loss of hearing in both ears. If sensorineural deafness is present in one ear, the tone will be heard in the unaffected ear, but not in the ear with sensorineural deafness. If conduction deafness is present, the sound will be heard more strongly in the ear in which there is a hearing loss. Conduction deafness can be simulated by plugging one ear with cotton to interfere with the conduction of sound to the inner ear.

Rinne Test for Comparing Bone- and Air-Conduction Hearing

1. Strike the tuning fork, and place its handle on your partner's mastoid process (Figure 25.6b).

2. When your partner indicates that the sound is no longer audible, hold the still-vibrating prongs close to his auditory canal (Figure 25.6c). If your partner hears the fork again (by air conduction) when it is moved to that position, hearing is not impaired and the test result is to be recorded as positive (+). (Record below step 5.)

3. Repeat the test, but this time test air-conduction hearing first.

4. After the tone is no longer heard by air conduction, hold the handle of the tuning fork on the bony mastoid process. If the subject hears the tone again by bone conduction after hearing by air conduction is lost, there is some conductive deafness and the result is recorded as negative (−).

5. Repeat the sequence for the opposite ear.

Right ear ________________ Left ear ________________

Does the subject hear better by bone or by air conduction?

____ ■

Audiometry

When the simple tuning fork tests reveal a problem in hearing, audiometer testing is usually prescribed to determine the precise nature of the hearing deficit. An *audiometer* is an instrument (specifically, an electronic oscillator with earphones) used to determine hearing acuity by exposing each ear to sound stimuli of differing *frequencies* and *intensities.* The hearing range of human beings during youth is from 30 to 22,000 Hz, but hearing acuity declines with age, with reception for the high-frequency sounds lost first. Though this loss represents a major problem for some people, such as musicians, most of us tend to be fairly unconcerned until we begin to have problems hearing sounds in the range of 125 to 8000 Hz, the normal frequency range of speech.

The basic procedure of audiometry is to initially deliver tones of different frequencies to one ear of the subject at an intensity of 0 decibels (dB). (Zero decibels is not the complete absence of sound, but rather the softest sound intensity that can be heard by a person of normal hearing at each frequency.) If the subject cannot hear a particular frequency stimulus of 0 dB, the hearing threshold level control is adjusted until the subject reports that he or she can hear the tone. The number of decibels of intensity required above 0 dB is recorded as the hearing loss. For example, if the subject cannot hear a particular frequency tone until it is delivered at 30 dB intensity, then he or she has a hearing loss of 30 dB for that frequency.

Activity 5: Audiometry Testing

1. Before beginning the tests, examine the audiometer to identify the two tone controls: one to regulate frequency and a second to regulate the intensity (loudness) of the sound stimulus. Also identify the two output control switches that regulate the delivery of sound to one ear or the other (*red* to the right ear, *blue* to the left ear). Also find the *hearing threshold level control,* which is calibrated to deliver a basal tone of 0 dB to the subject's ears.

2. Place the earphones on the subject's head so that the red cord or ear-cushion is over the right ear and the blue cord or ear-cushion is over the left ear. Instruct the subject to raise one hand when he or she hears a tone.

3. Set the frequency control at 125 Hz and the intensity control at 0 dB. Press the red output switch to deliver a tone to the subject's right ear. If the subject does not respond, raise the sound intensity slowly by rotating the hearing level control counterclockwise until the subject reports (by raising a hand) that a tone is heard. Repeat this procedure for frequencies of 250, 500, 1000, 2000, 4000, and 8000.

4. Record the results in the chart below by marking a small red circle on the chart at each frequency-dB junction indicated. Then connect the circles with a red line to produce a hearing acuity graph for the right ear.

5. Repeat steps 3 and 4 for the left (blue) ear and record the results with blue circles and connecting lines on the chart. ■

Microscopic Anatomy of the Equilibrium Apparatus and Mechanisms of Equilibrium

The equilibrium apparatus of the inner ear, the **vestibular apparatus,** is in the vestibule and semicircular canals of the bony labyrinth. Their chambers are filled with perilymph, in which membranous labyrinth structures are suspended. The vestibule contains the saclike **utricle** and **saccule,** and the semicircular chambers contain **membranous semicircular ducts.** Like the cochlear duct, these membranes are filled with endolymph and contain receptor cells that are activated by the bending of their cilia.

The semicircular canals are centrally involved in the **mechanism of dynamic equilibrium.** They are 1.2 cm in circumference and are oriented in three planes—horizontal, frontal, and sagittal. At the base of each semicircular duct is an enlarged region, the **ampulla,** which communicates with the utricle of the vestibule. Within each ampulla is a receptor region called a **crista ampullaris,** which consists of a tuft of hair cells covered with a gelatinous cap, or **cupula** (Figure 25.7). When your head position changes in an angular direction, as when twirling on the dance floor or when taking a rough boat ride, the endolymph in the canal lags behind, pushing the cupula—like a swinging door—in a direction opposite to that of the angular motion (Figure 25.7c). This movement depolarizes the hair cells, resulting in enhanced

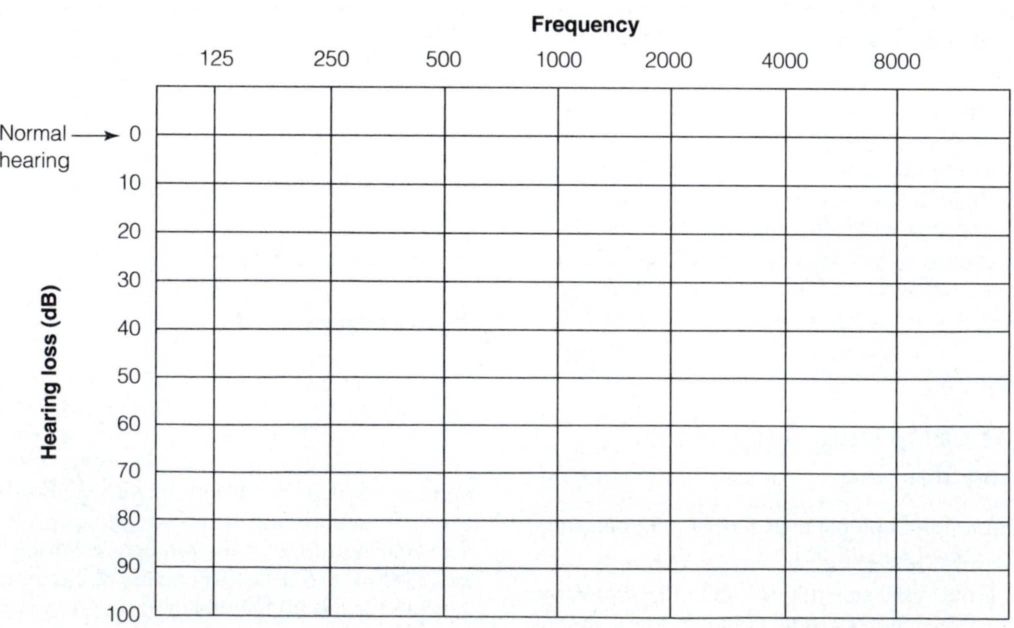

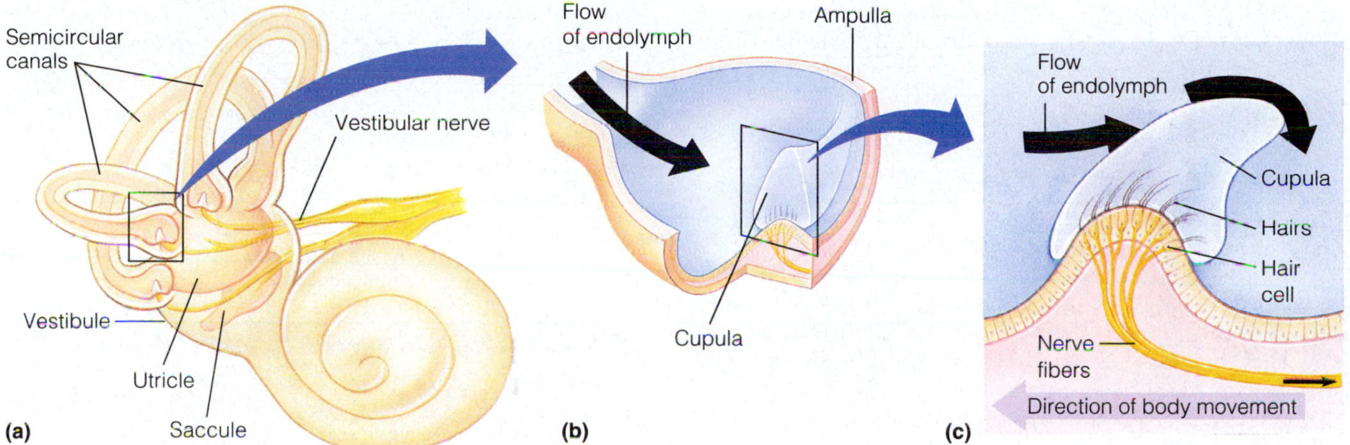

Figure 25.7 Structure and function of the crista ampullaris. **(a)** Arranged in the three spatial planes, the semicircular ducts in the semicircular canals each have a swelling called an ampulla at their base. **(b)** Each ampulla contains a crista ampullaris, a receptor that is essentially a cluster of hair cells with hairs projecting into a gelatinous cap called the cupula. **(c)** When the rate of rotation changes, inertia prevents the endolymph in the semicircular canals from moving with the head, so the fluid presses against the cupula, bending the hair cells in the opposite direction. The bending increases the frequency of action potentials in the sensory neurons in direct proportion to the amount of rotational acceleration. The mechanism adjusts quickly if rotation continues at a constant speed.

impulse transmission up the vestibular division of the eighth cranial nerve to the brain. Likewise, when the angular motion stops suddenly, the inertia of the endolymph causes it to continue to move, pushing the cupula in the same direction as the previous body motion. This movement again initiates electrical changes in the hair cells (in this case, it hyperpolarizes them). (This phenomenon accounts for the reversed motion sensation you feel when you stop suddenly after twirling.) If you begin to move at a constant rate of motion, the endolymph eventually comes to rest and the cupula gradually returns to its original position. The hair cells, no longer bent, send no new signals, and you lose the sensation of spinning. Thus the response of these dynamic equilibrium receptors is a reaction to *changes* in angular motion rather than to motion itself.

Activity 6: Examining the Microscopic Structure of the Crista Ampullaris

Go to the demonstration area and examine the slide of a crista ampullaris. Identify the areas depicted in Figure 25.7b and c. ■

Maculae in the vestibule contain the **hair cells,** receptors that are essential to the **mechanism of static equilibrium.** The maculae respond to gravitational pull, thus providing information on which way is up or down, and to linear or straightforward changes in speed. They are located on the walls of the saccule and utricle. The hair cells in each macula are embedded in the **otolithic membrane,** a gelatinous material containing small grains of calcium carbonate **(otoliths).** When the head moves, the otoliths move in response to variations in gravitational pull. As they deflect different hair cells, they trigger hyperpolarization or depolarization of the hair cells and modify the rate of impulse transmission along the vestibular nerve (Figure 25.8).

Although the receptors of the semicircular canals and the vestibule are responsible for dynamic and static equilibrium respectively, they rarely act independently. Complex interaction of many of the receptors is the rule. Processing is also complex and involves the brain stem and cerebellum as well as input from proprioceptors and the eyes.

Activity 7: Conducting Laboratory Tests on Equilibrium

The function of the semicircular canals and vestibule are not routinely tested in the laboratory, but the following simple tests illustrate normal equilibrium apparatus function as well as some of the complex processing interactions.

Balance Tests

1. Have your partner walk a straight line, placing one foot directly in front of the other.

Is he or she able to walk without undue wobbling from side to side ?

____

Did he or she experience any dizziness? ______________

The ability to walk with balance and without dizziness, unless subject to rotational forces, indicates normal function of the equilibrium apparatus.

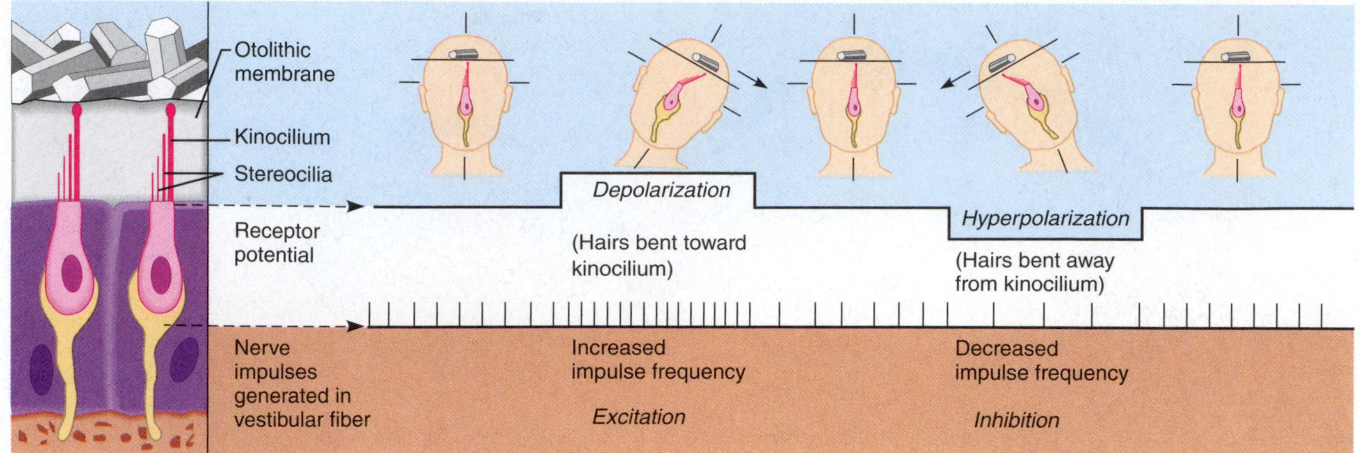

Figure 25.8 The effect of gravitational pull on a macula receptor in the utricle. When movement of the otolithic membrane (direction indicated by the arrow) bends the hair cells in the direction of the kinocilium, the vestibular fibers depolarize and generate action potentials more rapidly. When the hairs are bent in the direction away from the kinocilium, the hair cells become hyperpolarized, and the nerve fibers send impulses at a reduced rate (i.e., below the resting rate of discharge).

Was nystagmus* present? ______________________

2. Place three coins of different sizes on the floor. Ask your lab partner to pick up the coins, and carefully observe his or her muscle activity and coordination.

Did your lab partner have any difficulty locating and picking up the coins?

____

Describe your observations and your lab partner's observations during the test.

____

____

What kinds of interactions involving balance and coordination must occur for a person to move fluidly during this test?

____

____

3. If a person has a depressed nervous system, mental concentration may result in a loss of balance. Ask your lab partner to stand up and count backward from ten as rapidly as possible.

Did your lab partner lose balance? ______________

Barany Test (Induction of Nystagmus and Vertigo†)

This experiment evaluates the semicircular canals and should be conducted as a group effort to protect test subject(s) from possible injury.

⚠ The following precautionary notes should be read before beginning:

- The subject(s) chosen should not be easily inclined to dizziness during rotational or turning movements.
- Rotation should be stopped immediately if the subject feels nauseated.
- Because the subject(s) will experience vertigo and loss of balance as a result of the rotation, several classmates should be prepared to catch, hold, or support the subject(s) as necessary until the symptoms pass.

1. Instruct the subject to sit on a rotating chair or stool, and to hold on to the arms or seat of the chair, feet on stool rungs. The subject's head should be tilted forward approximately 30 degrees (almost touching the chest). The horizontal (lateral)

* **Nystagmus** is the involuntary rolling of the eyes in any direction or the trailing of the eyes slowly in one direction, followed by their rapid movement in the opposite direction. It is normal after rotation; abnormal otherwise. The direction of nystagmus is that of its quick phase on acceleration.

† **Vertigo** is a sensation of dizziness and rotational movement when such movement is not occurring or has ceased.

semicircular canal will be stimulated when the head is in this position. The subject's eyes are to remain *open* during the test.

2. Four classmates should position themselves so that the subject is surrounded on all sides. The classmate posterior to the subject will rotate the chair.

3. Rotate the chair to the subject's right approximately 10 revolutions in 10 seconds, then suddenly stop the rotation.

4. Immediately note the direction of the subject's resultant nystagmus; and ask him or her to describe the feelings of movement, indicating speed and direction sensation. Record below.

If the semicircular canals are operating normally, the subject will experience a sensation that the stool is still rotating immediately after it has stopped and *will* demonstrate nystagmus.

When the subject is rotated to the right, the cupula will be bent to the left, causing nystagmus during rotation in which the eyes initially move slowly to the left and then quickly to the right. Nystagmus will continue until the cupula has returned to its initial position. Then, when rotation is stopped abruptly, the cupula will be bent to the right, producing nystagmus with its slow phase to the right and its rapid phase to the left. In many subjects, this will be accompanied by a feeling of vertigo and a tendency to fall to the right.

Romberg Test

The Romberg test determines the integrity of the dorsal white column of the spinal cord, which transmits impulses to the brain from the proprioceptors involved with posture.

1. Have your partner stand with his or her back to the blackboard.

2. Draw one line parallel to each side of your partner's body. He or she should stand erect, with eyes open and staring straight ahead for 2 minutes while you observe any movements. Did you see any gross swaying movements?

---

3. Repeat the test. This time the subject's eyes should be closed. Note and record the degree of side-to-side movement.

---

4. Repeat the test with the subject's eyes first open and then closed. This time, however, the subject should be positioned with his or her left shoulder toward, but not touching, the board so that you may observe and record the degree of front-to-back swaying.

Do you think the equilibrium apparatus of the inner ear was operating equally well in all these tests?

The proprioceptors? ___

Why was the observed degree of swaying greater when the eyes were closed?

What conclusions can you draw regarding the factors necessary for maintaining body equilibrium and balance?

Role of Vision in Maintaining Equilibrium

To further demonstrate the role of vision in maintaining equilibrium, perform the following experiment. (Ask your lab partner to record observations and act as a "spotter.") Stand erect, with your eyes open. Raise your left foot approximately 30 cm off the floor, and hold it there for 1 minute.

Record the observations: ___

Rest for 1 or 2 minutes; and then repeat the experiment with the same foot raised, but with your eyes closed. Record the observations:

---

exercise

26 Special Senses: Olfaction and Taste

Objectives

1. To describe the location and cellular composition of the olfactory epithelium.
2. To describe the structure and function of the taste receptors.
3. To name the four basic qualities of taste sensation and list the chemical substances that elicit them.
4. To point out on a diagram of the tongue the predominant location of the basic types of taste receptors (salty, sweet, sour, bitter).
5. To explain the interdependence between the senses of smell and taste.
6. To name two factors other than olfaction that influence taste appreciation of foods.
7. To define *olfactory adaptation.*

Materials

- ❑ Prepared histologic slides: the tongue showing taste buds; nasal olfactory epithelium (longitudinal section)
- ❑ Compound microscope
- ❑ Small mirror
- ❑ Paper towels
- ❑ Granulated sugar
- ❑ Cotton-tipped swabs
- ❑ Disposable autoclave bag
- ❑ Paper cups; paper plates
- ❑ Prepared vials of 10% NaCl, 0.1% quinine or Epsom salt solution, 5% sucrose solution, and 1% acetic acid
- ❑ Beaker containing 10% bleach solution
- ❑ Prepared dropper bottles of oil of cloves, oil of peppermint, and oil of wintergreen or corresponding flavors found in the condiment section of a supermarket
- ❑ Equal-size food cubes of cheese, apple, raw potato, dried prunes, banana, raw carrot, and hard-cooked egg white (These prepared foods should be in an opaque container; a foil-lined egg carton would work well.)
- ❑ Toothpicks
- ❑ Plastic gloves
- ❑ Five numbered vials containing common household substances with strong odors (herbs, spices, etc.)
- ❑ Nose clips
- ❑ Chipped ice
- ❑ Absorbent cotton

The receptors for olfaction and taste are classified as **chemoreceptors** because they respond to chemicals in solution. Although four relatively specific types of taste receptors have been identified, the olfactory receptors are considered sensitive to a much wider range of chemical sensations. The sense of smell is the least understood of the special senses.

Localization and Anatomy of the Olfactory Receptors

The **olfactory epithelium** (organ of smell) occupies an area of about 5 cm^2 in the roof of the nasal cavity (Figure 26.1a). Since the air entering the human nasal cavity must make a hairpin turn to enter the respiratory passages below, the nasal epithelium is in a rather poor position for performing its function. This is why sniffing, which brings more air into contact with the receptors, intensifies the sense of smell.

The specialized receptor cells in the olfactory epithelium are surrounded by **supporting cells,** non-sensory epithelial cells. The **olfactory receptor cells** are bipolar neurons whose **olfactory hairs** (actually cilia) extend outward from the epithelium. Axonal nerve filaments emerging from their basal ends penetrate the cribriform plate of the ethmoid bone and proceed as the *olfactory nerves* to synapse in the olfactory bulbs lying on either side of the crista galli of the ethmoid bone. Impulses from neurons of the olfactory bulbs are then conveyed to the olfactory portion of the cortex (uncus) without synapsing in the thalamus.

Activity 1: Microscopic Examination of the Olfactory Epithelium

Obtain a longitudinal section of olfactory epithelium. Examine it closely, comparing it to Figure 26.1b and Plate 18 in the Histology Atlas. ■

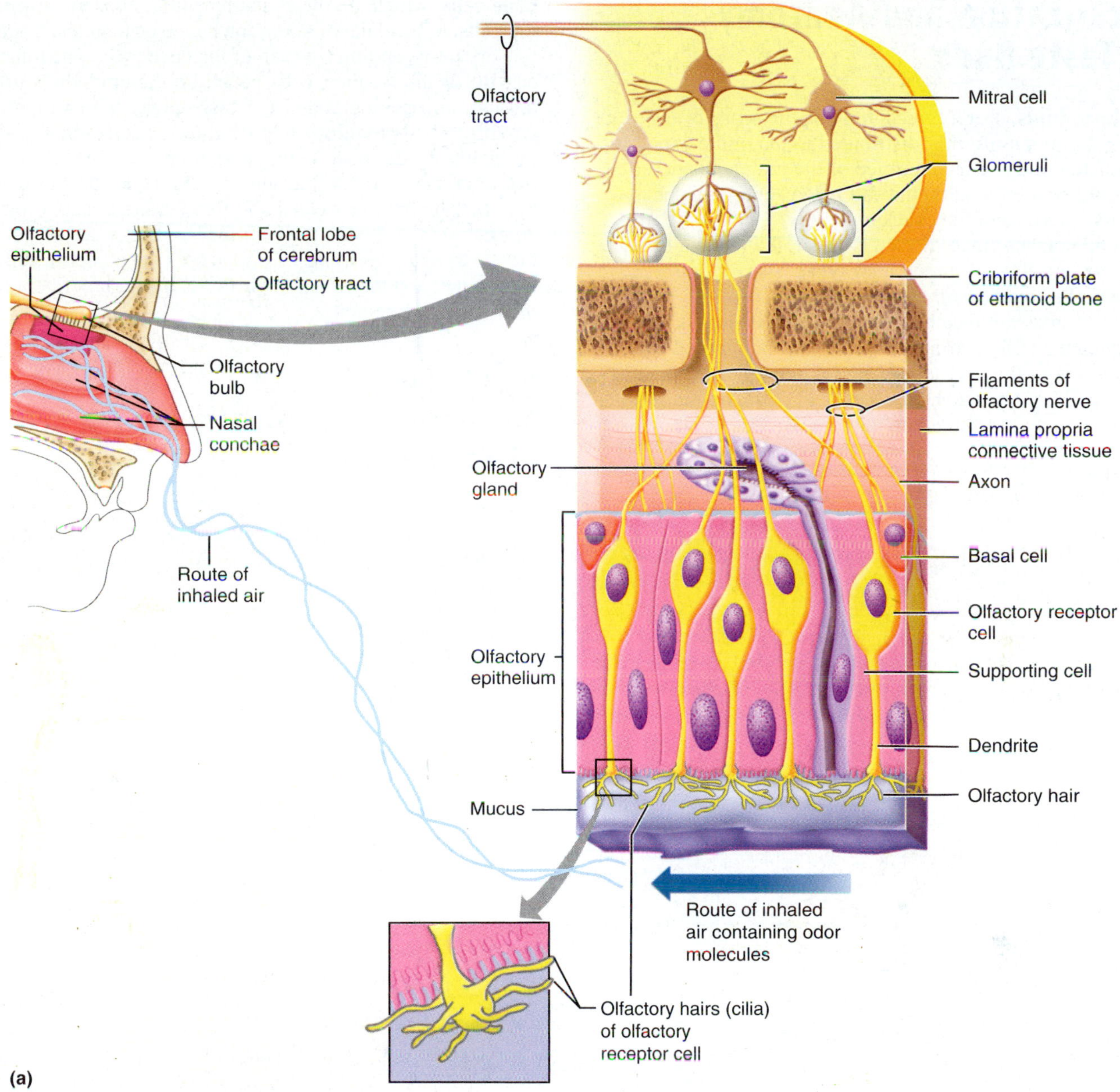

Olfactory tract
Mitral cell
Glomeruli
Olfactory epithelium
Frontal lobe of cerebrum
Olfactory tract
Cribriform plate of ethmoid bone
Olfactory bulb
Nasal conchae
Filaments of olfactory nerve
Lamina propria connective tissue
Olfactory gland
Axon
Basal cell
Route of inhaled air
Olfactory receptor cell
Olfactory epithelium
Supporting cell
Dendrite
Olfactory hair
Mucus
Route of inhaled air containing odor molecules
Olfactory hairs (cilia) of olfactory receptor cell
(a)

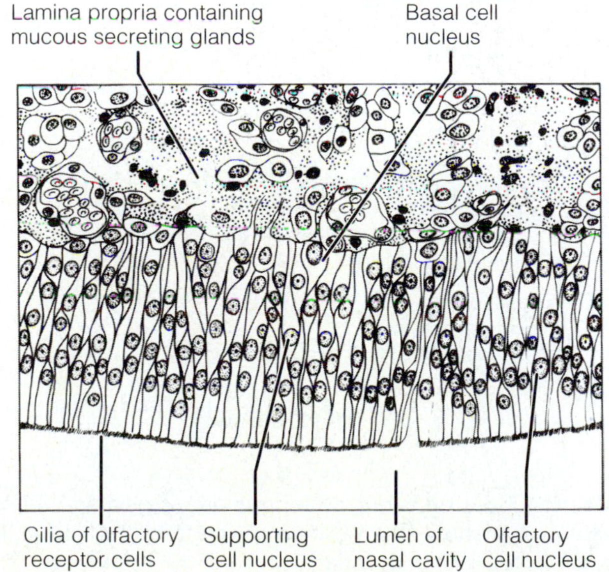

Figure 26.1 Location and cellular composition of olfactory epithelium. **(a)** Diagrammatic representation. **(b)** Microscopic view. (See also Plate 18 in the Histology Atlas.)

Localization and Anatomy of Taste Buds

The **taste buds,** specific receptors for the sense of taste, are widely but not uniformly distributed in the oral cavity. Most are located on the dorsal surface of the tongue (as described next). A few are found on the soft palate, epiglottis, pharynx, and inner surface of the cheeks.

The dorsal tongue surface is covered with small projections, or **papillae,** of three major types: sharp *filiform papillae* and the rounded *fungiform* and *circumvallate papillae.* The taste buds are located primarily on the sides of the circumvallate papillae (arranged in a V-formation on the posterior surface of the tongue) and on the more numerous fungiform papillae. The latter look rather like minute mushrooms and are widely distributed on the tongue. (See Figure 26.2.)

- Use a mirror to examine your tongue. Can you pick out the various papillae types?

______ If so, which? ____________________________

____

Each taste bud consists largely of a globular arrangement of two types of modified epithelial cells: the **gustatory,** or **taste cells,** which are the actual receptor cells, and **supporting cells.** Several nerve fibers enter each taste bud and supply sensory nerve endings to each of the taste cells. The long microvilli of the receptor cells penetrate the epithelial surface through an opening called the **taste pore.** When these microvilli, called **gustatory hairs,** contact specific chemicals in the solution, the taste cells depolarize. The afferent fibers from the taste buds to the sensory cortex in the postcentral gyrus of the brain are carried in three cranial nerves: the *facial nerve (VII)* serves the anterior two-thirds of the tongue; the *glossopharyngeal nerve (IX)* serves the posterior third of the tongue; and the *vagus nerve (X)* carries a few fibers from the pharyngeal region.

Activity 2: Microscopic Examination of Taste Buds

Obtain a microscope and a prepared slide of a tongue cross section. Use Figure 26.2b as a guide to aid you in locating the taste buds on the tongue papillae. Make a detailed study of one taste bud. Identify the taste pore and gustatory hairs if observed. Compare your observations to Plate 17 in the Histology Atlas. ■

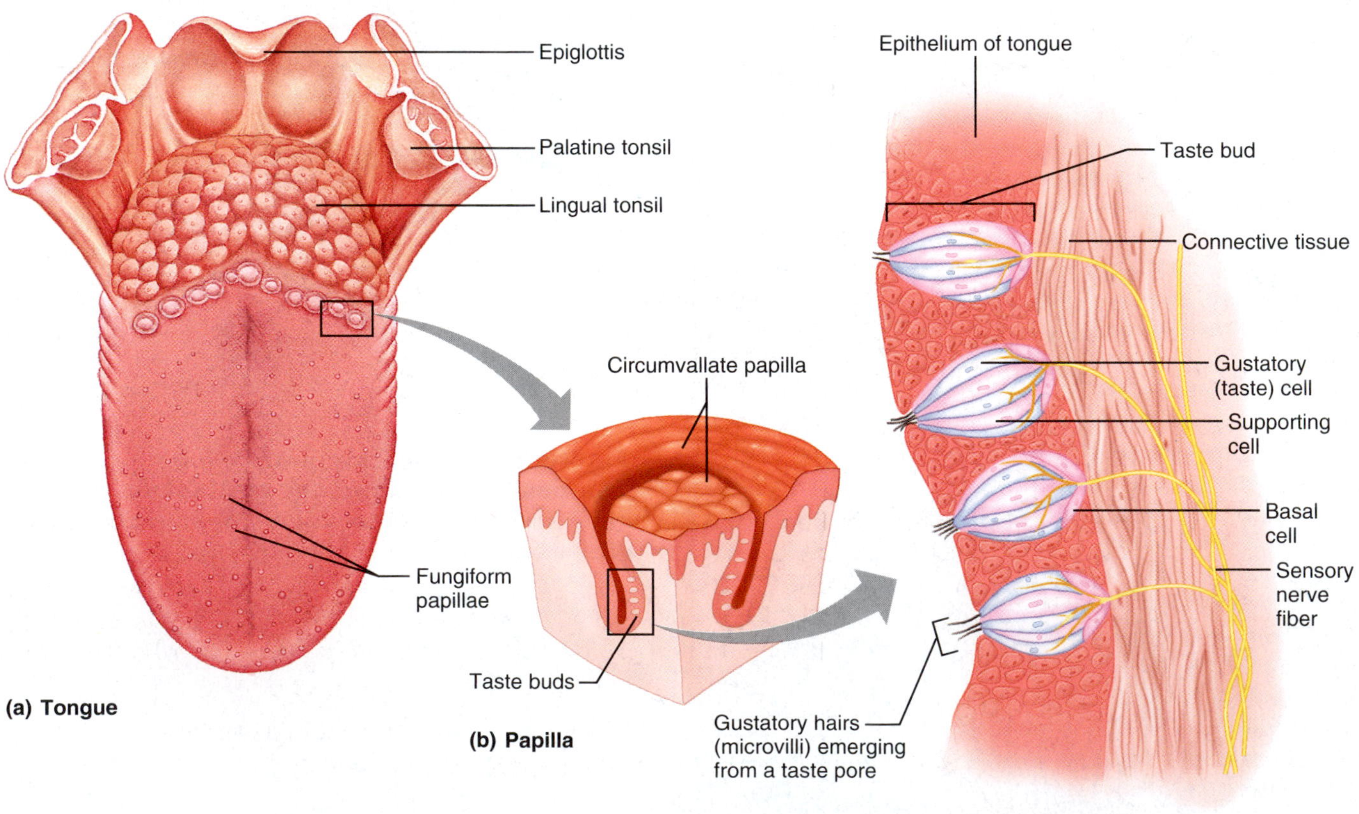

Figure 26.2 Location and structure of taste buds. (a) Taste buds on the tongue are associated with papillae, projections of the tongue mucosa. **(b)** A sectioned circumvallate papilla shows the position of the taste buds in its lateral walls. **(c)** An enlarged view of three taste buds. See also Plate 17 in the Histology Atlas.

Laboratory Experiments

Notify instructor of any food or scent allergies before beginning experiments.

Activity 3: Stimulating Taste Buds

1. Obtain several paper towels, a small amount of sugar, and a disposable autoclave bag and bring them to your bench.

2. With a paper towel, dry the dorsal surface of your tongue.

Immediately dispose of the paper towel in the autoclave bag.

3. Place a few sugar crystals on your dried tongue. Do *not* close your mouth. Time how long it takes to taste the sugar.

______________ sec

Why couldn't you taste the sugar immediately?

____ ■

Activity 4: Plotting Taste Bud Distribution

When taste is tested with pure chemical compounds, most taste sensations can be grouped into one of four basic qualities—sweet, sour, bitter, or salty. Although all taste buds are believed to respond in some degree to all four classes of chemical stimuli, each type responds optimally to only one.

The *sweet receptors* respond to a number of seemingly unrelated compounds such as sugars (fructose, sucrose, glucose), saccharine, and some amino acids. Both sweet and bitter responses are mediated by a G protein, gustducin. The sweet response causes K^+ channels to close, whereas the bitter response results in increased intracellular levels of Ca^{2+}. Sour receptors are mediated by hydrogen ions (H^+) and blockade of K^+ (or Na^+) channels. Salty taste seems to be due to influx of Na^+ through Na^+ channels.

1. Prepare to investigate the various taste receptors of your lab partner's tongue by obtaining the following: cotton-tipped swabs; one vial each of NaCl, quinine or Epsom salt solution, sucrose solution, and acetic acid; paper cups; and a flask of distilled or tap water.

2. Before each test, the subject should rinse his or her mouth thoroughly with water and lightly dry his or her tongue with a paper towel.

Dispose of used paper towels in the autoclave bag.

3. Generously moisten a swab with 5% sucrose solution and touch it to the center, back, tip, and sides of the dorsal surface of the subject's tongue.

4. Indicate, with O's on the tongue outline below, some locations of the sweet receptors.

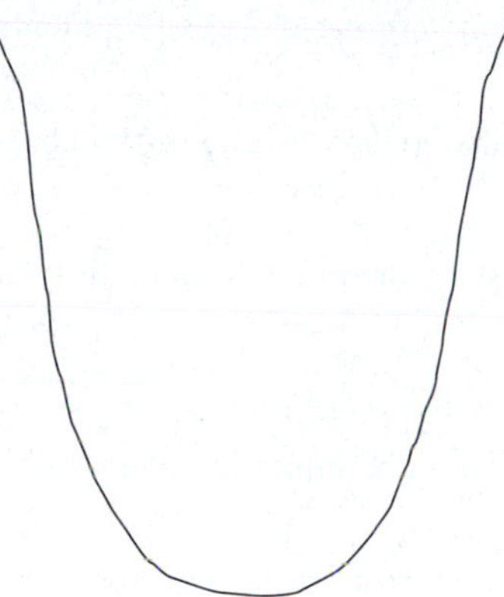

Put the used swab in the autoclave bag. Do not redip the swab into the sucrose solution.

5. Repeat the procedure with quinine (or Epsom salt solution) to indicate the location of the bitter receptors (use the symbol B), with NaCl to show the salt receptors (symbol +), and with acetic acid to mark the sour receptors (symbol −).

Use a fresh swab for each test, and properly dispose of the swabs immediately after use.

What area of the tongue dorsum seems to lack taste receptors?

____ ■

Activity 5: Examining the Combined Effects of Smell, Texture, and Temperature on Taste

Effects of Smell and Texture

1. Ask the subject to sit with eyes closed and to pinch his or her nostrils shut.

2. Using a paper plate, obtain samples of the food items listed in the chart below. At no time should the subject be allowed to see the foods being tested. Wear plastic gloves and use toothpicks to handle food.

3. Use an out-of-sequence order of food testing. For each test, place a cube of food in the subject's mouth and ask him or her to identify the food by using the following sequence of activities:

- First, manipulate the food with the tongue.
- Second, chew the food.

• Third, if a positive identification is not made with the first two techniques and the taste sense, ask the subject to release the pinched nostrils and to continue chewing with the nostrils open to determine if a positive identification can be made.

Record the results on the chart below by checking the appropriate column.

Was the sense of smell equally important in all cases?

___

Where did it seem to be important and why?

___

___

Discard gloves in autoclave bag. ■

Effect of Olfactory Stimulation

There is no question that what is commonly referred to as taste depends heavily on stimulation of the olfactory receptors, particularly in the case of strongly odoriferous substances. The following experiment should illustrate this fact.

1. Obtain vials of oil of wintergreen, peppermint, and cloves and some fresh cotton-tipped swabs. Ask the subject to sit so that he or she cannot see which vial is being used, and to dry the tongue and close the nostrils.

2. Apply a drop of one of the oils to the subject's tongue. Can he or she distinguish the flavor?

___

3. Have the subject open the nostrils, and record the change in sensation he or she reports.

___

4. Have the subject rinse the mouth well and dry the tongue.

5. Prepare two swabs, each with one of the two remaining oils.

6. Hold one swab under the subject's open nostrils, while touching the second swab to the tongue.

Record the reported sensations. ___

___

___

___

7. ⚠ Dispose of the used swabs and paper towels in the autoclave bag before continuing.

Which sense, taste or smell, appears to be more important in the proper identification of a strongly flavored volatile substance?

___

Effect of Temperature

In addition to the effect that olfaction and food texture play in determining our taste sensations, the temperature of foods also helps determine if the food is appreciated or even tasted. To illustrate this, have your partner hold some chipped ice on

Identification by Texture and Smell

Food	Texture only	Chewing with nostrils pinched	Chewing with nostrils open	Identification not made
Cheese	___	___	___	___
Apple	___	___	___	___
Raw potato	___	___	___	___
Banana	___	___	___	___
Dried prunes	___	___	___	___
Raw carrot	___	___	___	___
Hard-cooked egg white	___	___	___	___

the tongue for approximately a minute and then close his or her eyes. Immediately place any of the foods previously identified in his or her mouth and ask for an identification.

Results? ____

____

____ ■

Activity 6:

Assessing the Importance of Taste and Olfaction in Odor Identification

1. Go to the designated testing area. Close your nostrils with a nose clip, and breathe through your mouth. Breathing through your mouth only, attempt to identify the odors of common substances in the numbered vials at the testing area. Do not look at the substance in the container. Record your responses on the chart below.

Identification by Mouth and Nasal Inhalation

Identification with nose clips	Identification without nose clips	Other observations
1.		
2.		
3.		
4.		
5.		

2. Remove the nose clips, and repeat the tests using your nose to sniff the odors. Record your responses on the chart below.

3. Record any other observations you make as you conduct the tests.

4. Which method gave the best identification results?

____

What can you conclude about the effectiveness of the senses of taste and olfaction in identifying odors?

____ ■

Activity 7:

Demonstrating Olfactory Adaptation

Obtain some absorbent cotton and two of the following oils (oil of wintergreen, peppermint, or cloves). Press one nostril shut. Hold the bottle of oil under the open nostril and exhale through the mouth. Record the time required for the odor to disappear (for olfactory adaptation to occur).

______________ sec

Repeat the procedure with the other nostril.

______________ sec

Immediately test another oil with the nostril that has just experienced olfactory adaptation. What are the results?

____

____

What conclusions can you draw? ____

____

____ ■

exercise

27 Anatomy and Basic Function of the Endocrine Glands

Objectives

1. To identify and name the major endocrine glands and tissues of the body when provided with an appropriate diagram.
2. To list the hormones produced by the endocrine glands and discuss the general function of each.
3. To indicate the means by which hormones contribute to body homeostasis by giving appropriate examples of hormonal actions.
4. To cite the mechanism by which the endocrine glands are stimulated to release their hormones.
5. To describe the structural and functional relationship between the hypothalamus and the pituitary.
6. To describe a major pathological consequence of hypersecretion and hyposecretion of several of the hormones considered.
7. To correctly identify the histologic structure of the anterior and posterior pituitary, thyroid, parathyroid, adrenal cortex and medulla, pancreas, testis, and ovary by microscopic inspection or when presented with an appropriate photomicrograph or diagram. (Optional)
8. To name and point out the specialized hormone-secreting cells in the above tissues as studied in the laboratory. (Optional)

Materials

- ❑ Human torso model
- ❑ Anatomical chart of the human endocrine system
- ❑ Compound microscopes
- ❑ Colored pencils
- ❑ Histologic slides of the anterior pituitary,* posterior pituitary, thyroid gland, parathyroid glands, adrenal gland, and pancreas,* ovary, and testis tissue

AIA See Appendix C, Exercise 27 for links to A.D.A.M.® Interactive Anatomy.

For instructions on animal dissections, see the dissection exercises starting on p. 709 in the cat and fetal pig editions of this manual.

*Differential-staining if possible

The **endocrine system** is the second major controlling system of the body. Acting with the nervous system, it helps coordinate and integrate the activity of the body's cells. However, the nervous system employs electrochemical impulses to bring about rapid control, while the more slowly acting endocrine system employs chemical "messengers," or **hormones,** which are released into the blood to be transported throughout the body.

The term *hormone* comes from a Greek word meaning "to arouse." The body's hormones, which are steroids or amino acid–based molecules, arouse the body's tissues and cells by stimulating changes in their metabolic activity. These changes lead to growth and development and to the physiological homeostasis of many body systems. Although all hormones are bloodborne, a given hormone affects only the biochemical activity of a specific organ or organs. Organs that respond to a particular hormone are referred to as the **target organs** of that hormone. The ability of the target tissue to respond seems to depend on the ability of the hormone to bind with specific receptors (proteins) occurring on the cells' plasma membrane or within the cells.

Although the function of some hormone-producing glands (the anterior pituitary, thyroid, adrenals, parathyroids) is purely endocrine, the function of others (the pancreas and gonads) is mixed—both endocrine and exocrine. Both types of glands are derived from epithelium, but the endocrine, or ductless, glands release their product (always hormonal) directly into the blood or lymph. The exocrine glands release their products at the body's surface or outside an epithelial membrane via ducts. In addition, there are varied numbers of hormone-producing cells within the intestine, stomach, kidney, and placenta, organs whose functions are primarily nonendocrine. Only the major endocrine organs are considered here.

Gross Anatomy and Basic Function of the Endocrine Glands

Pituitary Gland (Hypophysis)

The *pituitary gland,* or *hypophysis,* is located in the concavity of the sella turcica of the sphenoid bone. It consists largely of two functional areas, the **adenohypophysis,** or **anterior pituitary,** and the **neurohypophysis,** consisting mainly of the **posterior pituitary.** The pituitary gland is attached to the hypothalamus by a stalk called the **infundibulum.**

Anterior Pituitary Hormones The anterior pituitary, or adenohypophysis, secretes a number of hormones. Four of these are **tropic** hormones. A tropic hormone stimulates its target organ, which is also an endocrine gland, to secrete its hormones. Target organ hormones then exert their effects on other body organs and tissues. The anterior pituitary tropic hormones include:

- The **gonadotropins—follicle-stimulating hormone (FSH)** and **luteinizing hormone (LH)**—regulate gamete production and hormonal activity of the gonads (ovaries and testes). The precise roles of the gonadotropins are described in Exercise 43 along with other considerations of reproductive system physiology.
- **Adrenocorticotropic hormone (ACTH)** regulates the endocrine activity of the cortex portion of the adrenal gland.
- **Thyroid stimulating hormone (TSH)** or **thyrotropin** influences the growth and activity of the thyroid gland.

The two other hormones produced by the anterior pituitary are not directly involved in the regulation of other endocrine glands of the body.

- **Growth hormone (GH)** is a general metabolic hormone that plays an important role in determining body size. It affects many tissues of the body; however, its major effects are exerted on the growth of muscle and the long bones of the body.

Hyposecretion results in pituitary dwarfism in children. Hypersecretion causes gigantism in children and **acromegaly** (overgrowth of bones in hands, feet, and face) in adults. ■

- **Prolactin (PRL)** stimulates breast development and promotes and maintains lactation by the mammary glands after childbirth. It may stimulate testosterone production in males.

The anterior pituitary controls the activity of so many other endocrine glands that it has often been called the *master endocrine gland.* However, the anterior pituitary is not autonomous in its control because release of the anterior pituitary hormones is controlled by neurosecretions, *releasing* or *inhibiting hormones,* produced by the hypothalamus. These hypothalamic hormones are liberated into the **hypophyseal portal system,** which serves the circulatory needs of the anterior pituitary (Figure 27.1).

Posterior Pituitary Hormones The posterior pituitary is not an endocrine gland in a strict sense because it does not synthesize the hormones it releases. (This relationship is also indicated in Figure 27.1.) Instead, it acts as a storage area for two hormones transported to it via the axons of neurons in the paraventricular and supraoptic nuclei of the hypothalamus.

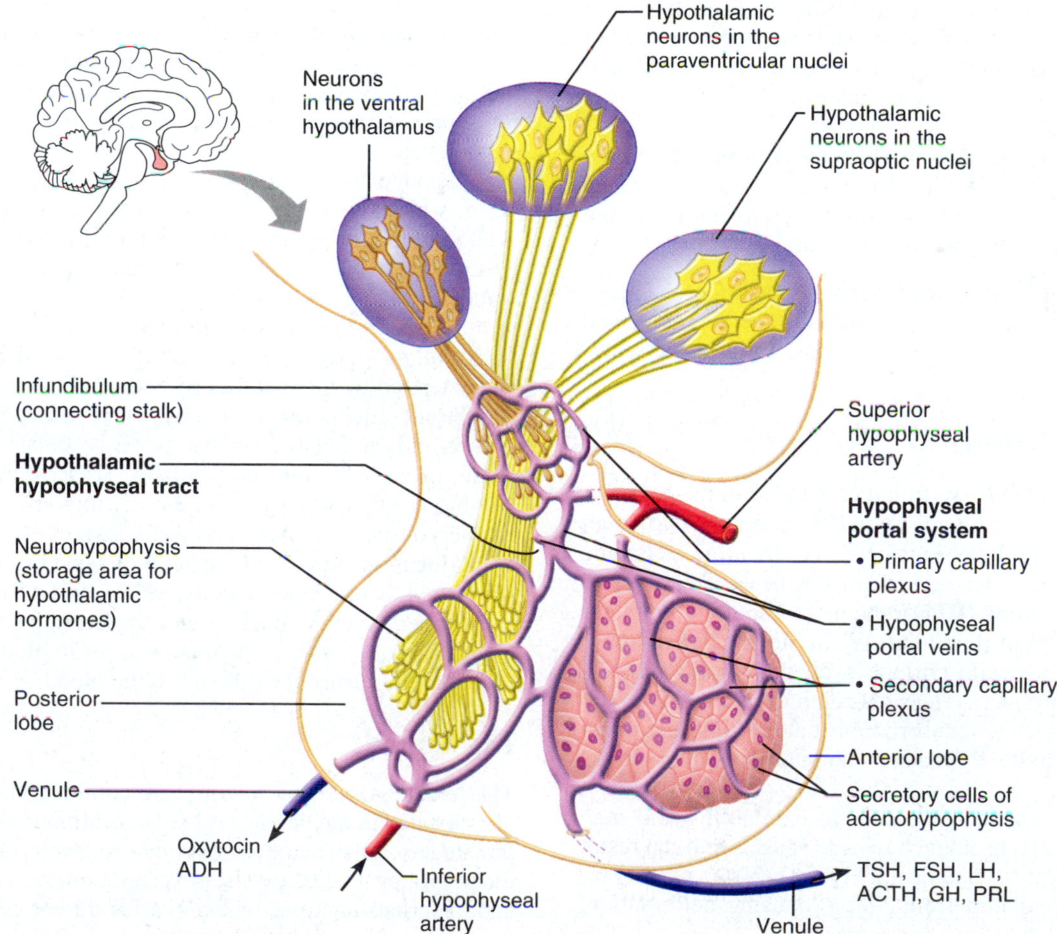

Figure 27.1 Neural and vascular relationships between the hypothalamus and the anterior and posterior lobes of the pituitary.

The hormones are released in response to nerve impulses from these neurons. The first of these hormones is **oxytocin,** which stimulates powerful uterine contractions during birth and coitus and also causes milk ejection in the lactating mother. The second, **antidiuretic hormone (ADH),** causes the distal and collecting tubules of the kidneys to reabsorb more water from the urinary filtrate, thereby reducing urine output and conserving body water. It also plays a minor role in increasing blood pressure because of its vasoconstrictor effect on the arterioles.

Hyposecretion of ADH results in dehydration from excessive urine output, a condition called **diabetes insipidus.** Individuals with this condition experience an insatiable thirst. Hypersecretion results in edema, headache, and disorientation. ■

Thyroid Gland

The *thyroid gland* is composed of two lobes joined by a central mass, or isthmus. It is located in the throat, just inferior to the larynx. It produces two major hormones, thyroid hormone and calcitonin.

Thyroid hormone (TH) is actually two physiologically active hormones known as **T_4 (thyroxine)** and **T_3 (triiodothyronine).** Because its primary function is to control the rate of body metabolism and cellular oxidation, TH affects virtually every cell in the body.

Hyposecretion of thyroxine leads to a condition of mental and physical sluggishness, which is called **myxedema** in the adult. Hypersecretion causes elevated metabolic rate, nervousness, weight loss, sweating, and irregular heartbeat. ■

Calcitonin (also called **thyrocalcitonin**) decreases blood calcium levels by stimulating calcium deposit in the bones. It acts antagonistically to parathyroid hormone, the hormonal product of the parathyroid glands.

• Try to palpate your thyroid gland by placing your fingers against your windpipe. As you swallow, the thyroid gland will move up and down on the sides and front of the windpipe.

Parathyroid Glands

The *parathyroid glands* are found embedded in the posterior surface of the thyroid gland. Typically, there are two small oval glands on each lobe, but there may be more and some may be located in other regions of the neck. They secrete **parathyroid hormone (PTH),** the most important regulator of calcium balance of the blood. When blood calcium levels decrease below a certain critical level, the parathyroids release PTH, which causes release of calcium from bone matrix and prods the kidney to reabsorb more calcium and less phosphate from the filtrate. PTH also stimulates the kidneys to activate vitamin D.

Hyposecretion increases neural excitability and may lead to **tetany,** prolonged muscle spasms that can result in respiratory paralysis and death. Hypersecretion of PTH results in loss of calcium from bones, causing deformation, softening, and spontaneous fractures. ■

Adrenal Glands

The two bean-shaped *adrenal,* or *suprarenal, glands* are located atop or close to the kidneys. Anatomically, the **adrenal medulla** develops from neural crest tissue and it is directly controlled by sympathetic nervous system neurons. The medullary cells respond to this stimulation by releasing **epinephrine** (80%) or **norepinephrine** (20%), which act in conjunction with the sympathetic nervous system to elicit the fight-or-flight response to stressors.

The **adrenal cortex** produces three major groups of steroid hormones, collectively called the **corticosteroids.** The **mineralocorticoids,** chiefly **aldosterone,** regulate water and electrolyte balance in the extracellular fluids, mainly by regulating sodium ion reabsorption by kidney tubules. The **glucocorticoids** (cortisone, hydrocortisone, and corticosterone) enable the body to resist long-term stressors, primarily by increasing blood glucose levels. The **gonadocorticoids,** or **sex hormones,** produced by the adrenal cortex are chiefly androgens (male sex hormones), but some estrogens (female sex hormones) are formed.

The gonadocorticoids are produced throughout life in relatively insignificant amounts; however, hypersecretion of these hormones produces abnormal hairiness (**hirsutism**), and masculinization occurs. ■

Pancreas

The *pancreas,* located partially behind the stomach in the abdomen, functions as both an endocrine and exocrine gland. It produces digestive enzymes as well as insulin and glucagon, important hormones concerned with the regulation of blood sugar levels.

Elevated blood glucose levels stimulate release of **insulin,** which decreases blood sugar levels, primarily by accelerating the transport of glucose into the body cells, where it is oxidized for energy or converted to glycogen or fat for storage.

Hyposecretion of insulin or some deficiency in the insulin receptors leads to **diabetes mellitus,** which is characterized by the inability of body cells to utilize glucose and the subsequent loss of glucose in the urine. Alterations of protein and fat metabolism also occur secondary to derangements in carbohydrate metabolism. Hypersecretion causes low blood sugar or **hypoglycemia**. Symptoms include anxiety, nervousness, tremors, and weakness. ■

Glucagon acts antagonistically to insulin. Its release is stimulated by low blood glucose levels, and its action is basically hyperglycemic. It stimulates the liver, its primary target organ, to break down its glycogen stores to glucose and subsequently to release the glucose to the blood.

The Gonads

The female *gonads,* or *ovaries,* are paired, almond-sized organs located in the pelvic cavity. In addition to producing the female sex cells (ova), the ovaries produce two steroid hormone groups, the estrogens and progesterone. The endocrine and exocrine functions of the ovaries do not begin until the onset of puberty, when the anterior pituitary gonadotropic hormones prod the ovary into action that produces rhythmic ovarian cycles in which ova develop and hormonal levels rise and fall. The **estrogens** are responsible for the development of the secondary sex characteristics of the female at puberty (primarily maturation of the reproductive organs and development of the breasts) and act with progesterone to bring about cyclic changes of the uterine lining that occur during the menstrual cycle. The estrogens also help prepare the mammary glands for lactation.

Progesterone, as already noted, acts with estrogen to bring about the menstrual cycle. During pregnancy it maintains the uterine musculature in a quiescent state and helps to prepare the breast tissue for lactation.

The paired oval *testes* of the male are suspended in a pouchlike sac, the scrotum, outside the pelvic cavity. In addition to the male sex cells, sperm, the testes produce the male sex hormone, **testosterone.** Testosterone promotes the maturation of the reproductive system accessory structures, brings about the development of the secondary sex characteristics, and is responsible for the male sexual drive, or libido. Both the endocrine and exocrine functions of the testes begin at puberty under the influence of the anterior pituitary gonadotropins. See Exercises 42 and 43.

Two glands not mentioned earlier as major endocrine glands should also be briefly considered here, the thymus and the pineal gland.

Thymus

The *thymus* is a bilobed gland situated in the superior thorax, posterior to the sternum and anterior to the heart and lungs. Conspicuous in the infant, it begins to atrophy at puberty, and by old age it is relatively inconspicuous. The thymus produces hormones called **thymosin** and **thymopoietin,** which help direct the maturation and specialization of a unique population of white blood cells called T lymphocytes or T cells. T lymphocytes are responsible for the cellular immunity aspect of body defense; that is, rejection of foreign grafts, tumors, or virus-infected cells.

Pineal Gland

The *pineal gland,* or *epiphysis cerebri,* is a small cone-shaped gland located in the roof of the third ventricle of the brain. Its major endocrine product is **melatonin.**

The endocrine role of the pineal body in humans is still controversial, but it is known to play a role in the biological rhythms (particularly mating and migratory behavior) of other animals. In humans, melatonin appears to exert some inhibitory effect on the reproductive system that prevents precocious sexual maturation.

Activity 1:
Identifying the Endocrine Organs

Locate the endocrine organs on Figure 27.2. Also locate these organs on the anatomical charts or torso. ■

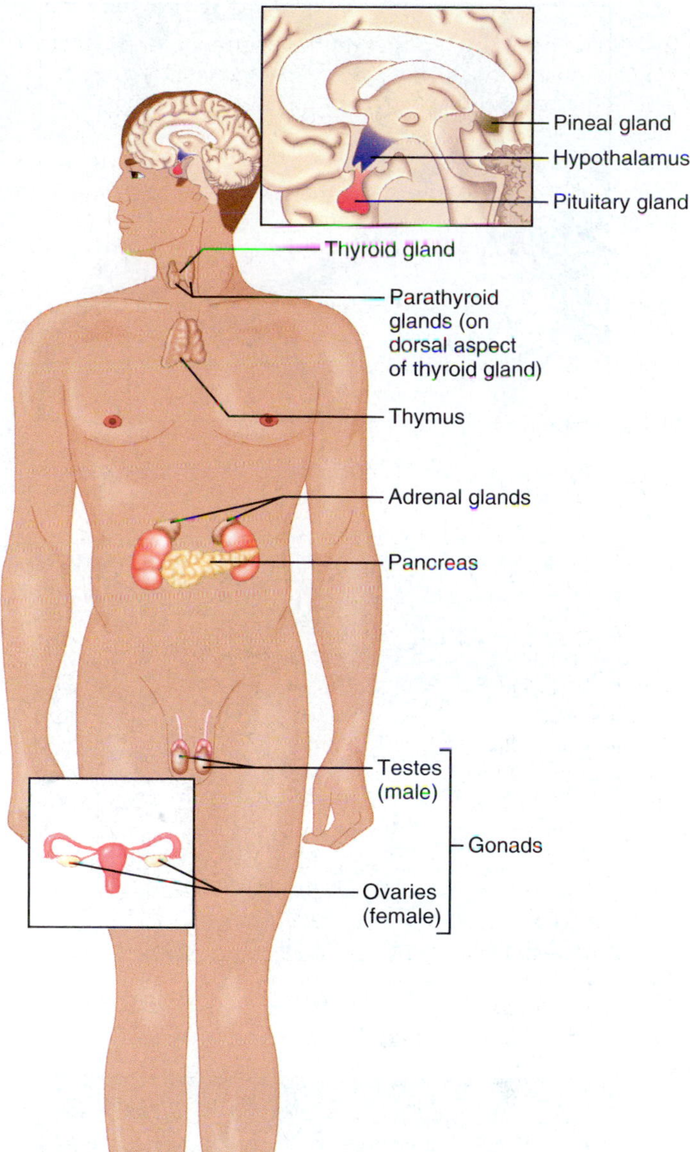

Figure 27.2 Human endocrine organs.

Microscopic Anatomy of Selected Endocrine Glands (Optional)

Activity 2:
Examining the Microscopic Structure of Endocrine Glands

To prepare for the histologic study of the endocrine glands, obtain a microscope, one of each slide on the materials list, and colored pencils. We will study only organs in which it is possible to identify the endocrine-producing cells. Compare your observations with the line drawings in Figure 27.3a–f of the endocrine tissue photomicrographs.

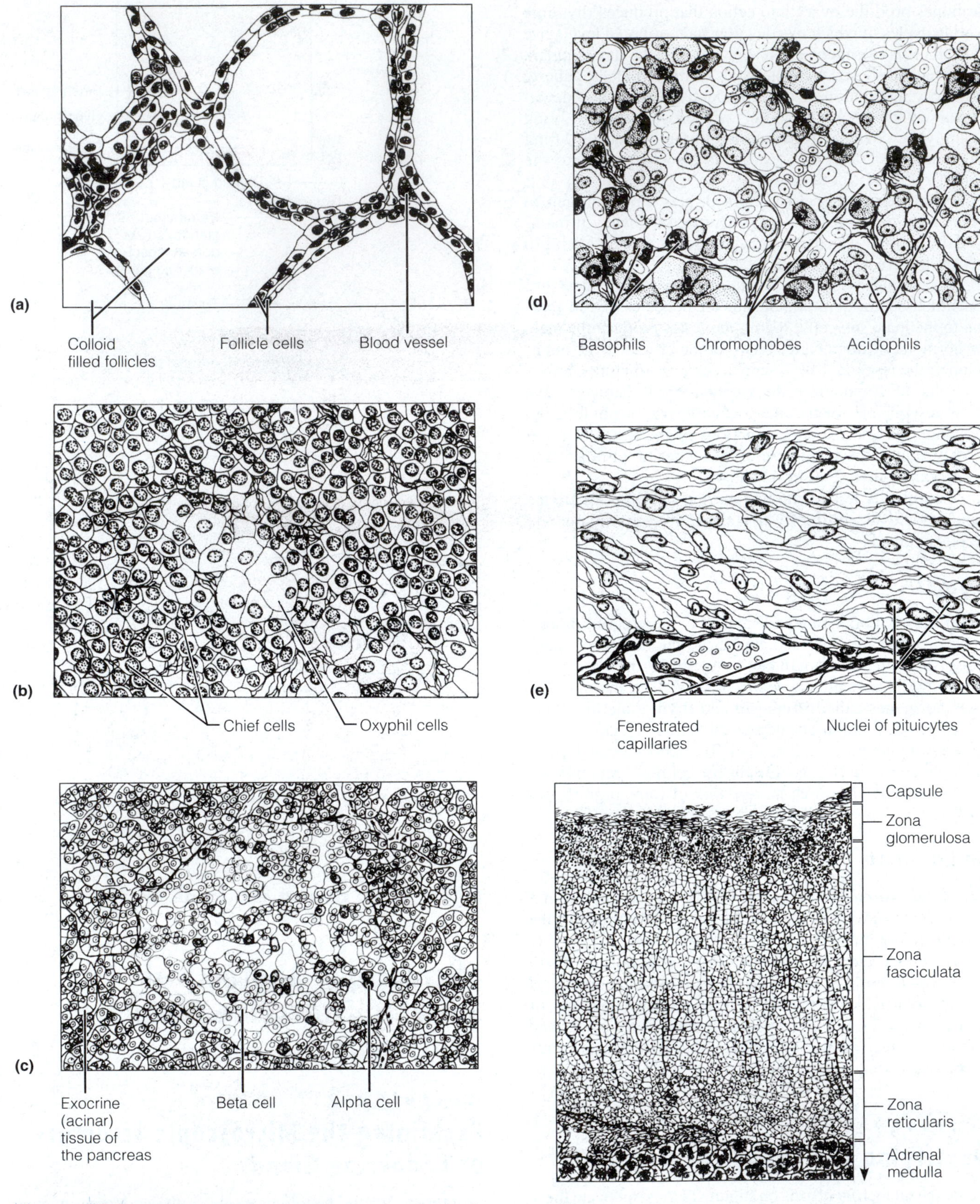

Figure 27.3 Microscopic anatomy of selected endocrine organs. (See also Plates 19–24 in the Histology Atlas.) **(a)** Thyroid; **(b)** parathyroid; **(c)** pancreas showing a pancreatic islet; **(d)** anterior pituitary; **(e)** posterior pituitary; **(f)** adrenal gland.

Thyroid Gland

1. Scan the thyroid under low power, noting the **follicles,** spherical sacs containing a pink-stained material (*colloid*). Stored T_3 and T_4 are attached to the protein colloidal material stored in the follicles as **thyroglobulin** and are released gradually to the blood. Compare the tissue viewed to Figure 27.3a and Plate 19 in the Histology Atlas.

2. Observe the tissue under high power. Note that the walls of the follicles are formed by simple cuboidal or squamous epithelial cells that synthesize the follicular products. The **parafollicular,** or **C, cells** you see between the follicles are responsible for calcitonin production.

3. Color appropriately two or three follicles in Figure 27.3a. Label the colloid and parafollicular cells.

When the thyroid gland is actively secreting, the follicles appear small, and the colloidal material has a ruffled border. When the thyroid is hypoactive or inactive, the follicles are large and plump and the follicular epithelium appears to be squamouslike. What is the physiological state of the tissue you have been viewing?

____

____

Parathyroid Glands

1. Observe the parathyroid tissue under low power to view its two major cell types, the chief cells and the oxyphil cells. Compare your observations to the view in Plate 20 of the Histology Atlas. The **chief cells,** which synthesize parathyroid hormone (PTH), are small and abundant, and arranged in thick branching cords. The function of the scattered, much larger **oxyphil cells** is unknown.

2. Color a small portion of the parathyroid tissue in Figure 27.3b. Label the chief cells and the oxyphil cells.

Pancreas

1. Observe pancreas tissue under low power to identify the roughly circular **pancreatic islets (islets of Langerhans),** the endocrine portions of the pancreas. The islets are scattered amid the more numerous acinar cells and stain differently (usually lighter), which makes their identification possible. (See Figure 38.14 [p. 404] and Plate 44 in the Histology Atlas.)

2. Focus on an islet and examine its cells under high power. Notice that the islet cells are densely packed and have no definite arrangement. In contrast, the cuboidal acinar cells are arranged around secretory ducts. Unless special stains are used, it will not be possible to distinguish the **alpha cells,** which tend to cluster at the periphery of the islets and produce glucagon, from the **beta cells,** which synthesize insulin. With these specific stains, the beta cells are larger and stain gray-blue, and the alpha cells are smaller and appear bright pink, as hinted at in Figure 27.3c and shown in Plate 21 in the Histology Atlas. What is the product of the acinar cells?

____

3. Draw a section of the pancreas in the space below. Label the islets and the acinar cells. If possible, differentiate the alpha and beta cells of the islets by color.

Pituitary Gland

1. Observe the general structure of the pituitary gland under low power to differentiate between the glandular anterior pituitary and the neural posterior pituitary. Figure 27.3d and e should help you get started.

2. Using the high-power lens, focus on the nests of cells of the anterior pituitary. It is possible to identify the specialized cell types that secrete the specific hormones when differential stains are used. Using Plate 22 in the Histology Atlas as a guide, locate the reddish-brown-stained **acidophil cells,** which produce growth hormone and prolactin, and the **basophil cells,** whose deep-blue granules are responsible for the production of the tropic hormones (TSH, ACTH, FSH, and LH). **Chromophobes,** the third cellular population, do not take up the stain and appear rather dull and colorless. The role of the chromophobes is controversial, but they apparently are not directly involved in hormone production.

3. Use appropriately colored pencils to identify acidophils, basophils, and chromophobes in Figure 27.3d.

4. Switch your focus to the posterior pituitary. Observe the nerve fibers (axons of hypophyseal neurons) that compose most of this portion of the pituitary. Also note the **pituicytes,** glial cells which are randomly distributed among the nerve fibers. Refer to Plate 23 in the Histology Atlas as you scan the slide.

What two hormones are stored here?

____

What is their source?

____

Adrenal Gland

1. Hold the slide of the adrenal gland up to the light to distinguish the outer cortex and inner medulla areas. Then scan the cortex under low power to distinguish the differences in cell appearance and arrangement in the three cortical areas. Refer to Figure 27.3f, Plate 24 in the Histology Atlas, and the following descriptions to identify the cortical areas:

- Connective tissue capsule of the adrenal gland.

- The outermost **zona glomerulosa,** where most mineralocorticoid production occurs and where the tightly packed cells are arranged in spherical clusters.
- The deeper intermediate **zona fasciculata,** which produces glucocorticoids. This is the thickest part of the cortex. Its cells are arranged in parallel cords.
- The innermost cortical zone, the **zona reticularis** abutting the medulla, which produces sex hormones and some glucocorticoids. The cells here stain intensely and form a branching network.

2. Switch focus to view the large, lightly stained cells of the adrenal medulla under high power. Notice their clumped arrangement.

What hormones are produced by the medulla?

______________________ and ______________________

3. Draw a representative area of each of the adrenal regions, indicating in your sketch the differences in relative cell size and arrangement.

Zona glomerulosa Zona fasciculata

Zona reticularis Adrenal medulla

Ovary

Because you will consider the *ovary* in greater histologic detail when you study the reproductive system, the objective in this laboratory exercise is just to identify the endocrine-producing parts of the ovary.

1. Scan an ovary slide under low power, and look for a **vesicular (Graafian) follicle,** a circular arrangement of cells enclosing a central cavity. See Figure 43.5 (p. 442) and Plate 25 in the Histology Atlas. This structure synthesizes estrogens.

2. Examine the vesicular follicle under high power, identifying the follicular cells that produce estrogens, the antrum (fluid-filled cavity), and developing ovum (if present). The ovum will be the largest cell in the follicle.

3. Draw and color a vesicular follicle below, labeling the antrum, follicle cells, and developing ovum.

4. Switch to low power, and scan the slide to find a **corpus luteum,** a large amorphous-looking area that produces progesterone (and some estrogens). A corpus luteum is shown in Plate 26 in the Histology Atlas.

Testis

1. Examine a section of a *testis* under low power. Identify the seminiferous tubules, which produce sperm, and the **interstitial cells,** which produce testosterone. The interstitial cells are scattered between the seminiferous tubules in the connective tissue matrix. The photomicrograph of seminiferous tubules, Plate 52 in the Histology Atlas, will be helpful here.

2. Draw a representative area of the testis in the space below. Label the seminiferous tubules and area of the interstitial cells. ■

Experiments on Hormonal Action: Wet Lab

exercise **28A**

Objectives

1. To describe and explain the effect of pituitary hormones on the ovary.
2. To describe and explain the effects of hyperinsulinism.
3. To describe and explain the effect of epinephrine on the heart.
4. To understand the physiological (and clinical) importance of metabolic rate measurement.
5. To investigate the effect of hypo-, hyper-, and euthyroid conditions on oxygen consumption and metabolic rate.
6. To assemble the necessary apparatus and properly use a manometer to obtain experimental results.
7. To calculate metabolic rate in terms of ml O_2/kg/hr.

Materials

Activity 1:

- ❑ Glass desiccator; manometer, 20-ml glass syringe, two-hole rubber stopper, and T-valve (1 for every 3 to 4 students)
- ❑ Soda lime (desiccant)
- ❑ Hardware cloth squares
- ❑ Petrolatum
- ❑ Rubber tubing; tubing clamps; scissors
- ❑ 7.6 cm (3-in.) pieces of glass tubing
- ❑ Animal balances
- ❑ Heavy animal-handling gloves
- ❑ Chart set up on chalkboard so that each student group can record its computed metabolic rate figures under the appropriate headings
- ❑ Young rats of the same sex, obtained 2 weeks prior to the laboratory session and treated as follows for 14 days:

  Group 1: control group—fed normal rat chow and water

  Group 2: experimental group A—fed normal rat chow, and drinking water containing 0.02% 6-n-propylthiouracil*

  Group 3: experimental group B—fed rat chow containing desiccated thyroid (2% by weight), and normal drinking water

Activity 2:†

- ❑ Female frogs (*Rana pipiens*)
- ❑ Disposable gloves
- ❑ Syringe (2-ml capacity)
- ❑ 20- to 25-gauge needle
- ❑ Frog pituitary extract
- ❑ Physiological saline
- ❑ Battery jars
- ❑ Spring or pond water
- ❑ Wax marking pencils

Activity 3:†

- ❑ 500- or 600-ml beakers
- ❑ 20% glucose solution
- ❑ Commercial insulin solution (400 IU per 100 ml H_2O)
- ❑ Finger bowls
- ❑ Small (4–5 cm or 1½–2 in.) freshwater fish (guppy, bluegill, or sunfish—listed in order of preference)
- ❑ Wax marking pencils

Activity 4:†

- ❑ Frog (*Rana pipiens*)
- ❑ 1:1000 epinephrine (Adrenalin) solution in dropper bottles
- ❑ Dissecting tray, instruments, and pins
- ❑ Frog Ringer's solution in dropper bottle
- ❑ Disposable gloves

PhysioEx™ 4.0 Computer Simulation on p. P-38

*Note to the Instructor: 6-n-propylthiouracil (PTU) is degraded by light and should be stored in light-resistant containers or in the dark.

†*The Selected Actions of Hormones and Other Chemical Messengers* videotape (available to qualified adopters from Benjamin Cummings) may be used in lieu of student participation in Experiments 2–4 of Exercise 28A.

The endocrine system exerts many complex and interrelated effects on the body as a whole, as well as on specific organs and tissues. Most scientific knowledge about this system is contemporary, and new information is constantly being presented. Many experiments on the endocrine system require relatively large laboratory animals; are time-consuming (requiring days to weeks of observation); and often involve technically difficult surgical procedures to remove the glands or parts of them, all of which makes it difficult to conduct more general types of laboratory experiments. Nevertheless, the four technically unsophisticated experiments presented here should illustrate how dramatically hormones affect body functioning.

To conserve laboratory specimens, experiments can be conducted by groups of four students. The use of larger working groups should not detract from benefits gained, since the major value of these experiments lies in observation.

Activity 1: Determining the Effect of Thyroid Hormone on Metabolic Rate

Metabolism is a broad term referring to all chemical reactions that are necessary to maintain life. It involves both *catabolism,* enzymatically controlled processes in which substances are broken down to simpler substances, and *anabolism,* processes in which larger molecules or structures are built from smaller ones. During catabolic reactions, energy is released as chemical bonds are broken. Some of the liberated energy is captured to make ATP, the energy-rich molecule used by body cells to energize all their activities; the balance is lost in the form of thermal energy or heat. Maintaining body temperature is critically related to the heat-liberating aspects of metabolism.

Various foodstuffs make different contributions to the process of metabolism. For example, carbohydrates, particularly glucose, are generally broken down or oxidized to make ATP, whereas fats are utilized to form cell membranes, myelin sheaths, and to insulate the body with a fatty cushion. (Fats are used secondarily for producing ATP, particularly when there are inadequate carbohydrates in the diet.) Proteins and amino acids tend to be conserved by body cells, and understandably so, since most structural elements of the body are proteinaceous in nature.

Thyroid hormone (collectively T_3 and T_4), produced by the thyroid gland, is the single most important hormone influencing the rate of cellular metabolism and body heat production.

Under conditions of excess thyroid hormone production (hyperthyroidism), an individual's basal metabolic rate (BMR), heat production, and oxygen consumption increase, and the individual tends to lose weight and become heat-intolerant and irritable. Conversely, hypothyroid individuals become mentally and physically sluggish, obese, and are cold-intolerant because of their low BMR. ■

Many factors other than thyroid hormone levels contribute to metabolic rate (for example, body size and weight, age, and activity level), but the focus of the following experiment is to investigate how differences in thyroid hormone concentration affect metabolism.

Three groups of laboratory rats will be used. The *control group* animals are assumed to be euthyroid and to have normal metabolic rates for their relative body weights. *Experimental group A* animals have received water containing the chemical 6-n-propylthiouracil, which counteracts or antagonizes the effects of thyroid hormone in the body. *Experimental group B* animals have been fed rat chow containing dried thyroid tissue, which contains thyroid hormone. The rates of oxygen consumption (an indirect means of determining metabolic rate) in the animals of the three groups will be measured and compared to investigate the effects of hyperthyroid, hypothyroid, and euthyroid conditions.

Oxygen consumption will be measured with a simple respirometer-manometer apparatus. Each animal will be placed in a closed chamber containing soda lime. As carbon dioxide is evolved and expired it will be absorbed by the soda lime; therefore, the pressure changes observed will indicate the volume of oxygen consumed by the animal during the testing interval. Students will work in groups of 3 to 4 to assemble the apparatus, make preliminary weight measurements on the animals, and record the data.

Preparing the Respirometer-Manometer Apparatus

1. Obtain a desiccator, a two-hole rubber stopper, a 20-ml glass syringe, a hardware cloth square, a T-valve, scissors, rubber tubing, two short pieces of glass tubing, soda lime, a manometer, a clamp, and petrolatum, and bring them to your laboratory bench. The apparatus will be assembled as illustrated in Figure 28A.1.

2. Shake soda lime into the bottom of the desiccator to thoroughly cover the glass bottom. Then place the hardware cloth on the ledge of the desiccator over the soda lime. The hardware cloth should be well above the soda lime, so that the animal will not be able to touch it. Soda lime is quite caustic and can cause chemical burns.

3. Lubricate the ends of the two pieces of glass tubing with petrolatum, and twist them into the holes in the rubber stopper until their distal ends protrude from the opposite side. *Do not plug the tubing with petrolatum.* Place the stopper into the desiccator cover, and set the cover on the desiccator temporarily.

4. Cut off a short (7.6-cm or 3-in.) piece of rubber tubing, and attach it to the top of one piece of glass tubing extending from the stopper. Cut and attach a 30-to 35-cm (12- to 14-in.) piece of rubber tubing to the other glass tubing. Insert the T-valve stem into the distal end of the longer-length tubing.

5. Cut another short piece of rubber tubing; attach one end to the T-valve and the other to the nib of the 20-ml syringe. Remove the plunger of the syringe and grease its sides generously with petrolatum. Insert the plunger back into the syringe barrel and work it up and down to evenly disperse the petrolatum on the inner walls of the syringe, then pull the plunger out to the 20-ml marking.

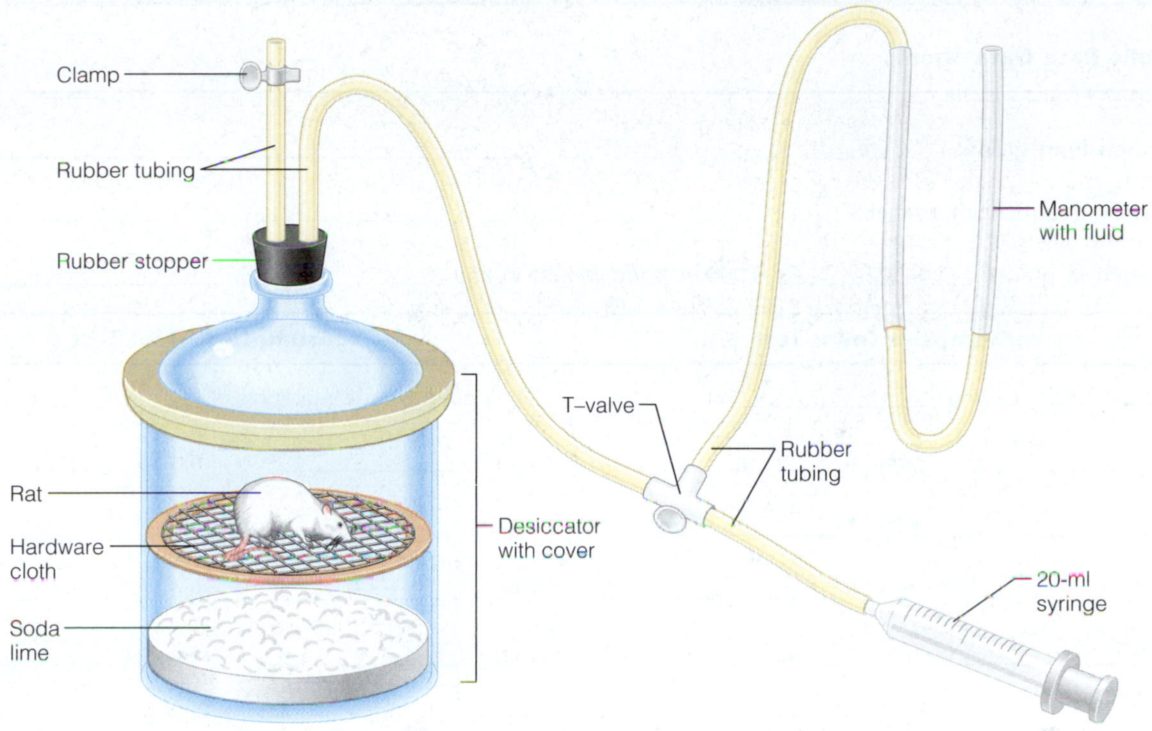

Figure 28A.1 Respirometer-manometer apparatus.

6. Cut a piece of rubber tubing long enough to reach from the third arm of the T-valve to one arm of the manometer. (The manometer should be partially filled with water so that a U-shaped water column is seen.) Attach the tubing to the T-valve and the manometer arm.

7. Remove the desiccator cover and generously grease the cover's bottom edge with petrolatum. Place the cover back on the desiccator and firmly move it from side to side to spread the lubricant evenly.

8. Test the system for leaks as follows: Firmly clamp the *short* length of rubber tubing extending from the stopper. Now gently push in on the plunger of the syringe. If the system is properly sealed, the fluid in the manometer will move away from the rubber tubing attached to its arm. If there is an air leak, the manometer fluid level will not change, or it will change and then quickly return to its original level. If either of these events occurs, check all glass-to-glass or glass-to-rubber tubing connections. Smear additional petrolatum on suspect areas and test again. The apparatus must be airtight before experimentation can begin.

9. After ensuring that there are no leaks in the system, unclamp the short rubber tubing, and remove the desiccator cover.

Preparing the Animal

1. Put on the heavy animal-handling gloves, and obtain one of the animals as directed by your instructor. Handling it gently, weigh the animal to the nearest 0.1 g on the animal balance.

2. Carefully place the animal on the hardware cloth in the desiccator. The objective is to measure oxygen usage at basal levels, so you do not want to prod the rat into high levels of activity, which would produce errors in your measurements.

3. Record the animal's group (control or experimental group A or B) and its weight in kilograms (that is, weight in grams/1000) on the data sheet on p. 282.

Equilibrating the Chamber

1. Place the lid on the desiccator, and move it slightly from side to side to seal it firmly.

2. Leave the short tubing unclamped for 7 to 10 minutes to allow for temperature equilibration in the chamber. (Since the animal's body heat will warm the air in the container, the air will expand initially. This must be allowed to occur before any measurements of oxygen consumption are taken. Otherwise, it would appear that the animal is evolving oxygen rather than consuming it.)

Determining Oxygen Consumption of the Animal

1. Once again clamp the short rubber tubing extending from the desiccator lid stopper.

2. Check the manometer to make sure that the fluid level is the same in both arms. (If not, manipulate the syringe plunger to make the fluid levels even.) Record the time and the position of the bottom of the plunger (use ml marking) in the syringe.

Metabolic Rate Data Sheet

Animal used from group: ______________________________

14-day prior treatment of animal: ______________________________

Body weight in grams: ______________ /1000 = body weight in kg: ______________

O_2 consumption/min: Test 1		**O_2 consumption/min: Test 2**	
Beginning syringe reading ______________ ml		Beginning syringe reading ______________ ml	
min 1 ______________	min 6 ______________	min 1 ______________	min 6 ______________
min 2 ______________	min 7 ______________	min 2 ______________	min 7 ______________
min 3 ______________	min 8 ______________	min 3 ______________	min 8 ______________
min 4 ______________	min 9 ______________	min 4 ______________	min 9 ______________
min 5 ______________	min 10 ______________	min 5 ______________	min 10 ______________
______________ Total O_2/10 min		______________ Total O_2/10 min	

Average ml O_2 consumed/10 min: ______________

Milliliters O_2 consumed/hr: ______________

Metabolic rate: ______________ ml O_2/kg/hr

Averaged class results:

Metabolic rate of control animals: ______________ ml O_2/kg/hr

Metabolic rate of experimental group A animals (PTU-treated): ______________ ml O_2/kg/hr

Metabolic rate of experimental group B animals (desiccated thyroid–treated): ______________ ml O_2/kg/hr

3. Observe the manometer fluid levels at 1-minute intervals. Each time make the necessary adjustment to bring the fluid levels even in the manometer by carefully pushing the syringe plunger further into the barrel. Determine the amount of oxygen used per minute in each interval by computing the difference in air volumes within the syringe. For example, if after the first minute, you push the plunger from the 20- to the 17-ml marking, the oxygen consumption is 3 ml/min. Then, if the plunger is pushed from 17 ml to 15 ml at the second minute reading, the oxygen usage during minute 2 would be 2 ml, and so on.

4. Continue taking readings (and recording oxygen consumption per minute interval on the data sheet) for 10 consecutive minutes or until the syringe plunger has been pushed nearly to the 0-ml mark. Then unclamp the short rubber tubing, remove the desiccator cover, and allow the apparatus to stand open for 2 to 3 minutes to flush out the stale air.

5. Repeat the recording procedures for another 10-minute interval. *Make sure that you equilibrate the temperature within the chamber before beginning this second recording series.*

6. After you have recorded the animal's oxygen consumption for two 10-minute intervals, unclamp the short rubber tubing, remove the desiccator lid, and carefully return the rat to its cage.

Computing Metabolic Rate

Metabolic rate calculations are generally reported in terms of $kcal/m^2/hr$ and require that corrections be made to present the data in terms of standardized pressure and temperature conditions. These more complex calculations will not be used here, since the object is simply to arrive at some generalized conclusions concerning the effect of thyroid hormone on metabolic rate.

1. Obtain the average figure for milliliters of oxygen consumed per 10-minute interval by adding up the minute-interval consumption figures for each 10-minute testing series and dividing the total by 2.

$$\frac{\text{Total ml } O_2 \text{ test series 1} + \text{total ml } O_2 \text{ test series 2}}{2}$$

2. Determine oxygen consumption per hour using the following formula, and record the figure on the data sheet:

$$\frac{\text{Average ml } O_2 \text{ consumed}}{10 \text{ min}} \times \frac{60 \text{ min}}{\text{hr}} = \text{ml } O_2/\text{hr}$$

3. To determine the metabolic rate in milliliters of oxygen consumed per kilogram of body weight per hour so that the results of all experiments can be compared, divide the figure just obtained in step 2 by the animal's weight in kilograms (kg = lb ÷ 2.2).

$$\text{Metabolic rate} = \frac{\text{ml } O_2/\text{hr}}{\text{weight (kg)}} = \underline{\qquad\qquad} \text{ ml } O_2/\text{kg/hr}$$

Record the metabolic rate on the data sheet and also in the appropriate space on the chart on the chalkboard.

4. Once all groups have recorded their final metabolic rate figures on the chalkboard, average the results of each animal grouping to obtain the mean for each experimental group. Also record this information on the data sheet. ■

Activity 2: Determining the Effect of Pituitary Hormones on the Ovary

As indicated in Exercise 27, anterior pituitary hormones called *gonadotropins,* specifically follicle-stimulating hormone (FSH) and luteinizing hormone (LH), regulate the ovarian cycles of the female. Although amphibians normally ovulate seasonally, many can be stimulated to ovulate "on demand" by injecting an extract of pituitary hormones.

⚠ 1. Don plastic gloves, and obtain two frogs. Place them in separate battery jars to bring them to your laboratory bench. Also bring back a syringe and needle, a wax marking pencil, pond or spring water, and containers of pituitary extract and physiological saline.

2. Before beginning, examine each frog for the presence of eggs. Hold the frog firmly with one hand and exert pressure on its abdomen toward the cloaca (in the direction of the legs). If ovulation has occurred, any eggs present in the oviduct will be forced out and will appear at the cloacal opening. If no eggs are present, continue with step 3. If eggs are expressed, return the animal to your instructor and obtain another frog for experimentation. Repeat the procedure for determining if eggs are present until two frogs that lack eggs have been obtained.

3. Aspirate 1 to 2 ml of the pituitary extract into a syringe. Inject the extract subcutaneously into the anterior abdominal (peritoneal) cavity of the frog you have selected to be the experimental animal. To inject into the peritoneal cavity, hold the frog with its ventral surface superiorly. Insert the needle through the skin and muscles of the abdominal wall in the lower quarter of the abdomen. Do not insert the needle far enough to damage any of the vital organs. With a wax marker, label its large battery jar "experimental," and place the frog in it. Add a small amount of pond water to the battery jar before continuing.

4. Aspirate 1 to 2 ml of normal saline into a syringe and inject it into the peritoneal cavity of the second frog—this will be the control animal. (Make sure you inject the same volume of fluid into both frogs.) Place this frog into the second battery jar, marked "control." Allow the animals to remain undisturbed for 24 hours.

5. After 24 hours,* again check each frog for the presence of eggs in the cloaca. (See step 2.) If no eggs are present, make arrangements with your laboratory instructor to return to the lab on the next day (at 48 hours after injection) to check your frogs for the presence of eggs.

6. Return the frogs to the terrarium before leaving or continuing with the lab.

In which of the prepared frogs was ovulation induced?

____

Specifically, what hormone in the pituitary extract causes ovulation to occur?

______________________________ ■

Activity 3: Observing the Effects of Hyperinsulinism

Many people with diabetes mellitus need injections of insulin to maintain blood sugar (glucose) homeostasis. Adequate levels of blood glucose are essential for proper functioning of the nervous system; thus, the administration of insulin must

* The student will need to inject the frog the day before the lab session or return to check results the day after the scheduled lab session.

be carefully controlled. If blood glucose levels fall precipitously, the patient will go into insulin shock.

A small fish will be used to demonstrate the effects of hyperinsulinism. Since the action of insulin on the fish parallels that in the human, this experiment should provide valid information concerning its administration to humans.

1. Prepare two finger bowls. Using a wax marker, mark one A and the other B. To finger bowl A, add 100 ml of the commercial insulin solution. To finger bowl B, add 200 ml of 20% glucose solution.

2. Place a small fish in finger bowl A and observe its actions carefully as the insulin diffuses into its bloodstream through the capillary circulation of its gills.

Approximately how long did it take for the fish to become comatose?

____

What types of activity did you observe in the fish before it became comatose?

____

____

____

____

3. When the fish is comatose, carefully transfer it to finger bowl B and observe its actions. What happens to the fish after it is transferred?

____

____

____

____

Approximately how long did it take for this recovery?

____

4. After all observations have been made and recorded, carefully return the fish to the aquarium. ■

Activity 4: Testing the Effect of Epinephrine on the Heart

As noted in Exercise 27, the adrenal medulla and the sympathetic nervous system are closely interrelated, specifically because the cells of the adrenal medulla and the postganglionic axons of the sympathetic nervous system both release catecholamines. This experiment demonstrates the effects of epinephrine on the frog heart.

⚠ 1. Obtain a frog, dissecting instruments and tray, disposable gloves, frog Ringer's solution, dropper bottle of 1:1000 epinephrine solution, and bring them to your laboratory bench. Don the gloves before beginning step 2.

2. Destroy the nervous system of the frog. (A frog used in the experiment on pituitary hormone effects may be used if your test results have already been obtained and are positive. Otherwise, obtain another frog.) Insert one blade of a scissors into its mouth as far as possible and quickly cut off the top of its head, posterior to the eyes. Then identify the spinal cavity and insert a dissecting needle into it to destroy the spinal cord.

3. Place the frog dorsal side down on a dissecting tray, and carefully open its ventral body cavity by making a vertical incision with the scissors.

4. Identify the beating heart, and carefully cut through the saclike pericardium to expose the heart tissue.

5. Visually count the heart rate for 1 minute, and record below. Keep the heart moistened with frog Ringer's solution during this interval.

Beats per minute: ______________________________

6. Flush the heart with epinephrine solution. Record the heartbeat rate per minute for 5 consecutive minutes.

minute 1 ______________ minute 4 ______________

minute 2 ______________ minute 5 ______________

minute 3 ______________

What was the effect of epinephrine on the heart rate?

____

Was the effect long-lived? ______________________________

7. Dispose of the frog in an appropriate container, and clean the dissecting tray and instruments before returning them to the supply area. ■

exercise

29

Blood

Objectives

1. To name the two major components of blood and state their average percentages in whole blood.
2. To describe the composition and functional importance of plasma.
3. To define *formed elements* and list the cell types composing them, cite their relative percentages, and describe their major functions.
4. To identify red blood cells, basophils, eosinophils, monocytes, lymphocytes, and neutrophils when provided with a microscopic preparation or appropriate diagram.
5. To provide the normal values for a total white blood cell count and a total red blood cell count and to state the importance of these tests.
6. To conduct the following blood test determinations in the laboratory, and to state their norms and the importance of each.

   hematocrit
   hemoglobin determination
   clotting time
   sedimentation rate
   differential white blood cell count
   ABO and Rh blood typing
   plasma cholesterol concentration
7. To discuss the reason for transfusion reactions resulting from the administration of mismatched blood.
8. To define *anemia, polycythemia, leukopenia, leukocytosis,* and *leukemia* and to cite a possible reason for each condition.

Materials

- ❑ Compound microscope
- ❑ Immersion oil
- ❑ Models and charts of blood cells
- ❑ Safety glasses (student provided)

Demonstration station:

- ❑ Microscopes set up with prepared slides demonstrating the following blood (or bone marrow) conditions:

  Macrocytic hypochromic anemia
  Microcytic hypochromic anemia
  Sickle-cell anemia
  Lymphocytic leukemia (chronic)
  Eosinophilia

General supply area:*

- ❑ Plasma (obtained from an animal hospital or prepared by centrifuging animal (e.g., cattle or sheep) blood obtained from a biological supply house
- ❑ Wide-range pH paper
- ❑ *Stained smears of human blood*:
  If desired by the instructor, heparinized animal blood obtained from a biological supply house or an animal hospital (e.g., dog blood), or EDTA-treated red cells (Reference cells[†]) with blood type labels obscured (available from Immunocor, Inc.)
- ❑ Assorted slides of white blood count pathologies labeled "Unknown Sample __"
- ❑ Clean microscope slides
- ❑ Sterile lancets
- ❑ Glass stirring rods
- ❑ Alcohol swabs (wipes)
- ❑ Absorbent cotton balls
- ❑ Wright stain in dropper bottle
- ❑ Distilled water in dropper bottle
- ❑ Test tubes
- ❑ Test tube racks
- ❑ Disposable gloves
- ❑ Pipette cleaning solutions—(1) 10% household bleach solution, (2) distilled water, (3) 70% ethyl alcohol, (4) acetone
- ❑ Bucket or large beaker containing 10% household bleach solution for slide and glassware disposal

*Note to the Instructor: See directions for handling of soiled glassware and disposable items on p. 21.

[†]The blood in these kits (each containing 4 blood cell types—A1, A2, B, and O—individually supplied in 10-ml vials) is used to calibrate cell counters and other automated clinical laboratory equipment. This blood has been carefully screened and can be safely used by students for blood typing and determining hematocrits. It is not usable for hemoglobin determinations or coagulation studies.

- ❑ Spray bottles containing 10% bleach solution
- ❑ Autoclave bag
- ❑ Designated lancet (sharps) disposal container
- ❑ Timers

Because many blood tests are to be conducted in this exercise, it is advisable to set up a number of appropriately labeled supply areas for the various tests, as designated below.

Note: Artificial blood prepared by Ward's Natural Science can be used for differential counts, hematocrit, and blood typing.

Hematocrit supply area:

- ❑ Heparinized capillary tubes
- ❑ Microhematocrit centrifuge and reading gauge (if the reading gauge is not available, a millimeter ruler may be used)
- ❑ Capillary tube sealer or modeling clay

Hemoglobin determination supply area:

- ❑ Hemoglobinometer, hemolysis applicator, and lens paper

Sedimentation rate supply area:

- ❑ Landau Sed-rate pipettes with tubing and rack*
- ❑ Wide-mouthed bottle of 5% sodium citrate
- ❑ Mechanical suction device
- ❑ Millimeter ruler

Coagulation time supply area:

- ❑ Capillary tubes (nonheparinized)
- ❑ Fine triangular file

Blood typing supply area:

- ❑ Blood typing sera (anti-A, anti-B, and anti-Rh [D])
- ❑ Rh typing box
- ❑ Wax marker
- ❑ Toothpicks
- ❑ Blood test cards or microscope slides
- ❑ Medicine dropper

Cholesterol-measurement supply area:

- ❑ Cholesterol test cards and color scale

*An alternative is the Westergren ESR method. See Instructor's Guide for ordering information.

In this exercise you will study plasma and formed elements of blood and conduct various hematological tests. These tests are extremely useful diagnostic tools for the physician because blood composition (number and types of blood cells, and chemical composition) reflects the status of many body functions and malfunctions.

⚠ **ALERT: Special precautions when handling blood.** This exercise provides information on blood from several sources: human, animal, human treated, and artificial blood. The decision to use animal blood for testing or to have students test their own blood will be made by the instructor in accordance with the educational goals of the student group. For example, for students in the nursing or laboratory technician curricula, learning how to safely handle human blood or other human wastes is essential. Whenever blood is being handled, special attention must be paid to safety precautions. These precautions should be used regardless of the source of the blood. This will both teach good technique and ensure the safety of the students.

Follow exactly the safety precautions listed below.

1. Wear safety gloves at all times. Discard appropriately.
2. Wear safety glasses throughout the exercise.
3. Handle only your own, freshly let (human) blood.
4. Be sure you understand the instructions and have all supplies on hand before you begin any part of the exercise.
5. Do not reuse supplies and equipment once they have been exposed to blood.
6. Keep the lab area clean. Do not let anything that has come in contact with blood touch surfaces or other individuals in the lab. Pay attention to the location of any supplies and equipment that come into contact with blood.
7. Dispose of lancets immediately after use in a designated disposal container. Do not put them down on the lab bench, even temporarily.
8. Dispose of all used cotton balls, alcohol swabs, blotting paper, and so forth in autoclave bags and place all soiled glassware in containers of 10% bleach solution.
9. Wipe down the lab bench with 10% bleach solution when you are finished.

Composition of Blood

The blood circulating to and from the body cells within the blood vessels is a rather viscous substance that varies from bright scarlet to a dull brick red, depending on the amount of oxygen it is carrying. The average volume of blood in the body is about 5–6 L in adult males and 4–5 L in adult females.

Blood is classified as a type of connective tissue because it consists of a nonliving fluid matrix (the **plasma**) in which living cells (**formed elements**) are suspended. The fibers typical of a connective tissue matrix become visible in blood only when clotting occurs. They then appear as fibrin threads, which form the structural basis for clot formation.

More than 100 different substances are dissolved or suspended in plasma (Figure 29.1), which is over 90% water. These include nutrients, gases, hormones, various wastes and

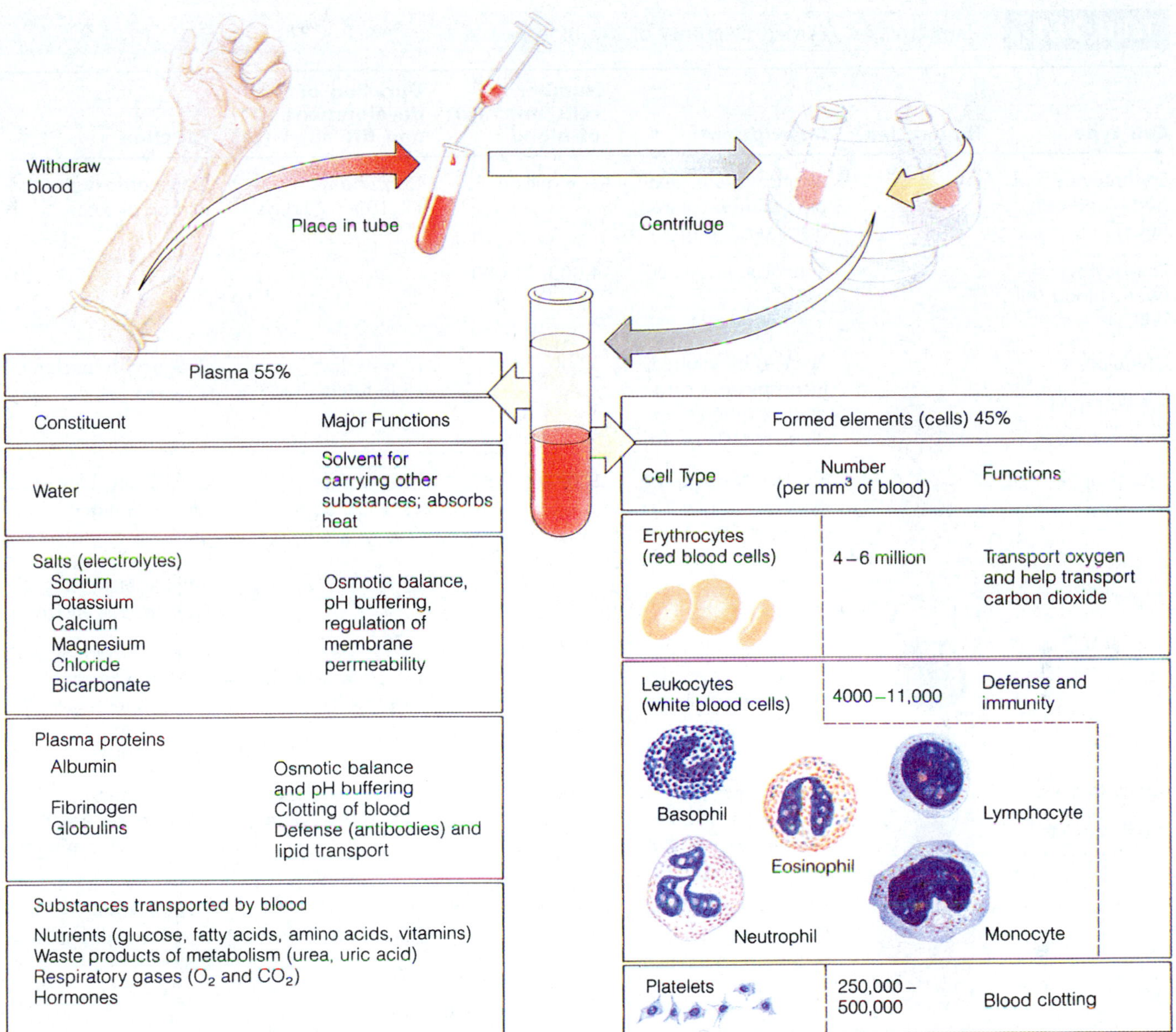

Figure 29.1 The composition of blood.

metabolites, many types of proteins, and electrolytes. The composition of plasma varies continuously as cells remove or add substances to the blood.

Three types of formed elements are present in blood (Table 29.1). The most numerous are the **erythrocytes,** or **red blood cells (RBCs),** which are literally sacs of hemoglobin molecules that transport the bulk of the oxygen carried in the blood (and a small percentage of the carbon dioxide). **Leukocytes,** or **white blood cells (WBCs),** are part of the body's nonspecific defenses and the immune system, and **platelets** function in hemostasis (blood clot formation). Formed elements normally constitute 45% of whole blood; plasma accounts for the remaining 55%.

Activity 1: Determining the Physical Characteristics of Plasma

Go to the general supply area and carefully pour a few milliliters of plasma into a test tube. Also obtain some wide-range pH paper and then return to your laboratory bench to make the following simple observations.

Table 29.1 Summary of Formed Elements of the Blood

Cell type	Illustration	Description*	Number of cells/mm^3 (μl) of blood	Duration of development (D) and life span (LS)	Function
Erythrocytes (red blood cells, RBCs)		Biconcave, anucleate disc; salmon-colored; diameter 7–8 μm	4–6 million	D: 5–7 days LS: 100–120 days	Transport oxygen and carbon dioxide
Leukocytes (white blood cells, WBCs)		Spherical, nucleated cells	4,000–11,000		
Granulocytes					
Neutrophil		Nucleus multilobed; inconspicuous cytoplasmic granules; diameter 10–14 μm	3,000–7,000	D: 6–9 days LS: 6 hours to a few days	Phagocytize bacteria
Eosinophil		Nucleus bilobed; red cytoplasmic granules; diameter 10–14 μm	100–400	D: 6–9 days LS: 8–12 days	Kill parasitic worms; destroy antigen-antibody complexes; inactivate some inflammatory chemicals of allergy
Basophil		Nucleus lobed; large blue-purple cytoplasmic granules; diameter 10–12 μm	20–50	D: 3–7 days LS: ? (a few hours to a few days)	Release histamine and other mediators of inflammation; contain heparin, an anticoagulant
Agranulocytes					
Lymphocyte		Nucleus spherical or indented; pale blue cytoplasm; diameter 5–17 μm	1,500–3,000	D: days to weeks LS: hours to years	Mount immune response by direct cell attack or via antibodies
Monocyte		Nucleus U-or kidney-shaped; gray-blue cytoplasm; diameter 14–24 μm	100–700	D: 2–3 days LS: months	Phagocytosis; develop into macrophages in tissues
Platelets		Discoid cytoplasmic fragments containing granules; stain deep purple; diameter 2–4 μm	250,000–500,000	D: 4–5 days LS: 5–10 days	Seal small tears in blood vessels; instrumental in blood clotting

*Appearance when stained with Wright's stain.

pH of Plasma

Test the pH of the plasma with wide-range pH paper. Record the pH observed.

____

Color and Clarity of Plasma

Hold the test tube up to a source of natural light. Note and record its color and degree of transparency. Is it clear, translucent, or opaque?

Color______________________________

Degree of transparency______________________________

Consistency

Dip your finger and thumb into the plasma and then press them firmly together for a few seconds. Gently pull them apart. How would you describe the consistency of plasma? Slippery, watery, sticky, or granular? Record your observations.

Activity 2:

Examining the Formed Elements of Blood Microscopically

In this section, you will observe blood cells on an already prepared (purchased) blood slide, or on a slide prepared from your own blood or blood provided by your instructor. Those using the purchased blood slide are to obtain a slide and begin their observations at step 6. Those testing blood provided by a biological supply source or an animal hospital are to obtain a tube of the supplied blood, disposable gloves, and the supplies listed in step 1, except for the lancets and alcohol swabs. After donning gloves, those students will jump to step 3b to begin their observations. If you are examining your own blood, you will perform all the steps described below *except* step 3b.

1. Obtain two glass slides, a glass stirring rod, dropper bottles of Wright stain and distilled water, two or three lancets, cotton balls, and alcohol swabs. Bring this equipment to the laboratory bench. Clean the slides thoroughly and dry them.

2. Open the alcohol swab packet and scrub your third or fourth finger with the swab. (Because the pricked finger may be a little sore later, it is better to prepare a finger on the hand used less often.) Circumduct your hand (swing it in a cone-shaped path) for 10 to 15 seconds. This will dry the alcohol and cause your fingers to become engorged with blood. Then, open the lancet packet and grasp the lancet by its blunt end. Quickly jab the pointed end into the prepared finger to produce a free flow of blood. It is *not* a good idea to squeeze or "milk" the finger, as this forces out tissue fluid as well as blood. If the blood is not flowing freely, another puncture should be made.

⚠ *Under no circumstances is a lancet to be used for more than one puncture*. Dispose of the lancets in the designated disposal container immediately after use.

3a. With a cotton ball, wipe away the first drop of blood; then allow another large drop of blood to form. Touch the blood to one of the cleaned slides approximately 1.3 cm or ½ inch from the end. Then quickly (to prevent clotting) use the second slide to form a blood smear as shown in Figure 29.2. When properly prepared, the blood smear is uniformly thin. If the blood smear appears streaked, the blood probably began to clot or coagulate before the smear was made, and another slide should be prepared. Continue at step 4.

3b. Dip a glass rod in the blood provided, and transfer a generous drop of blood to the end of a cleaned microscope slide. For the time being, lay the glass rod on a paper towel on the bench. Then, as described in step 3a, use the second slide to make your blood smear.

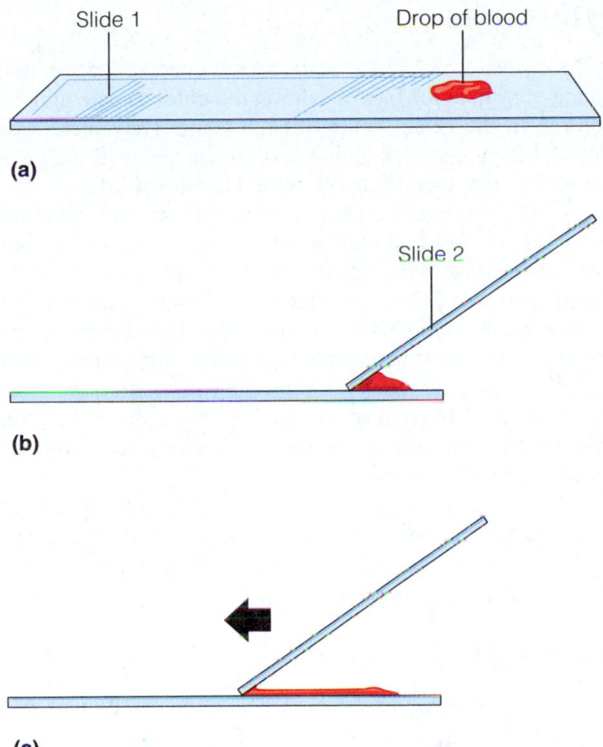

Figure 29.2 Procedure for making a blood smear. **(a)** Place a drop of blood on slide 1 approximately ½ inch from one end. **(b)** Hold slide 2 at a 30° to 40° angle to slide 1 (it should touch the drop of blood) and allow blood to spread along entire bottom edge of angled slide. **(c)** Smoothly advance slide 2 to end of slide 1 (blood should run out before reaching the end of slide 1).

4. Dry the slide by waving it in the air. When it is completely dry, it will look dull. Place it on a paper towel, and flood it with Wright stain. Count the number of drops of stain used. Allow the stain to remain on the slide for 3 to 4 minutes and then flood the slide with an equal number of drops of distilled water. Allow the water and Wright's stain mixture to remain on the slide for 4 or 5 minutes or until a metallic green film or scum is apparent on the fluid surface. Blow on the slide gently every minute or so to keep the water and stain mixed during this interval.

5. Rinse the slide with a stream of distilled water. Then flood it with distilled water, and allow it to lie flat until the slide becomes translucent and takes on a pink cast. Then stand the slide on its long edge on the paper towel, and allow it to dry completely. Once the slide is dry, you can begin your observations.

6. Obtain a microscope and scan the slide under low power to find the area where the blood smear is the thinnest. After scanning the slide in low power to find the areas with the largest numbers of nucleated WBCs, read the following descriptions of cell types, and find each one on Figure 29.1 and Table 29.1. (The formed elements are also shown in Plates 58 through 63 in the Histology Atlas.) Then, switch to the oil immersion lens and observe the slide carefully to identify each cell type.

Erythrocytes

Erythrocytes, or red blood cells, which average 7.5 μm in diameter, vary in color from a salmon red color to pale pink, depending on the effectiveness of the stain. They have a distinctive biconcave disk shape and appear paler in the center than at the edge (see Plate 59 in the Histology Atlas).

As you observe the slide, notice that the red blood cells are by far the most numerous blood cells seen in the field. Their number averages 4.5 million to 5.0 million cells per cubic millimeter of blood (for women and men, respectively).

Red blood cells differ from the other blood cells because they are anucleate when mature and circulating in the blood. As a result, they are unable to reproduce or repair damage and have a limited life span of 100 to 120 days, after which they begin to fragment and are destroyed in the spleen and other reticuloendothelial tissues of the body.

In various anemias, the red blood cells may appear pale (an indication of decreased hemoglobin content) or may be nucleated (an indication that the bone marrow is turning out cells prematurely). ■

Leukocytes

Leukocytes, or white blood cells, are nucleated cells that are formed in the bone marrow from the same stem cell (*hemocytoblast*) as red blood cells. They are much less numerous than the red blood cells, averaging from 4000 to 11,000 cells per cubic millimeter. Basically, white blood cells are protective, pathogen-destroying cells that are transported to all parts of the body in the blood or lymph. Important to their protective function is their ability to move in and out of blood vessels, a process called **diapedesis,** and to wander through body tissues by **amoeboid motion** to reach sites of inflammation or tissue destruction. They are classified into two major groups, depending on whether or not they contain conspicuous granules in their cytoplasm.

Granulocytes make up the first group. The granules in their cytoplasm stain differentially with Wright stain, and they have peculiarly lobed nuclei, which often consist of expanded nuclear regions connected by thin strands of nucleoplasm. There are three types of granulocytes:

Neutrophil: The most abundant of the white blood cells (40% to 70% of the leukocyte population); nucleus consists of 3 to 7 lobes and the pale lilac cytoplasm contains fine cytoplasmic granules, which are generally indistinguishable and take up both the acidic (red) and basic (blue) dyes (*neutrophil* = neutral loving); functions as an active phagocyte. The number of neutrophils increases exponentially during acute infections. (See Plates 58 and 59 in the Histology Atlas.)

Eosinophil: Represents 1% to 4% of the leukocyte population; nucleus is generally figure 8 or bilobed in shape; contains large cytoplasmic granules (elaborate lysosomes) that stain red-orange with the acid dyes in Wright stain (see Plate 62 in the Histology Atlas). Eosinophils are about the size of neutrophils and play a role in counterattacking parasitic worms. They also lessen allergy attacks by phagocytizing antigen-antibody complexes and inactivating some inflammatory chemicals.

Basophil: Least abundant leukocyte type representing less than 1% of the population; large U- or S-shaped nucleus with two or more indentations. Cytoplasm contains coarse, sparse granules that are stained deep purple by the basic dyes in Wright stain (see Plate 63 in the Histology Atlas). The granules contain several chemicals including histamine, a vasodilator which is discharged on exposure to antigens and helps mediate the inflammatory response. Basophils are about the size of neutrophils.

The second group, **agranulocytes,** or **agranular leukocytes,** contains no *visible* cytoplasmic granules. Although found in the bloodstream, they are much more abundant in lymphoid tissues. Their nuclei tend to be closer to the norm, that is, spherical, oval, or kidney shaped. Specific characteristics of the two types of agranulocytes are listed below.

Lymphocyte: The smallest of the leukocytes, approximately the size of a red blood cell (see Plates 58 and 60 in the Histology Atlas). The nucleus stains dark blue to purple, is generally spherical or slightly indented, and accounts for most of the cell mass. Sparse cytoplasm appears as a thin blue rim around the nucleus. Concerned with immunologic responses in body; one population, the B lymphocytes, oversees the production of antibodies that are released to blood. The second population, T lymphocytes, plays a regulatory role and destroys grafts, tumors, and virus-infected cells. Represents 20% to 45% of the WBC population.

Monocyte: The largest of the leukocytes; approximately twice the size of red blood cells (see Plate 61 in the Histology Atlas). Represents 4% to 8% of the leukocyte population. Dark blue nucleus is generally kidney shaped; abundant cytoplasm stains gray-blue. Once in the tissues, monocytes convert to macrophages, active phagocytes (the "long-term cleanup team"), increasing dramatically in number during chronic infections such as tuberculosis.

Students are often asked to list the leukocytes in order from the most abundant to the least abundant. The following silly phrase may help you with this task: *N*ever *l*et *m*onkeys *e*at *b*ananas (neutrophils, lymphocytes, monocytes, eosinophils, basophils).

Platelets

Platelets are cell fragments of large multinucleate cells (**megakaryocytes**) formed in the bone marrow. They appear as darkly staining, irregularly shaped bodies interspersed among the blood cells (see Plate 58 in the Histology Atlas). The normal platelet count in blood ranges from 250,000 to 500,000 per cubic millimeter. Platelets are instrumental in the clotting process that occurs in plasma when blood vessels are ruptured.

After you have identified these cell types on your slide, observe three-dimensional models of blood cells if these are available. Do not dispose of your slide, as it will be used later for the differential white blood cell count. ■

Hematologic Tests

When someone enters a hospital as a patient, several hematologic tests are routinely done to determine general level of health as well as the presence of pathologic conditions. You will be conducting the most common of these tests in this exercise.

Materials such as cotton balls, lancets, and alcohol swabs are used in nearly all of the following diagnostic tests. These supplies are at the general supply area and should be properly disposed of (glassware to the "bleach bucket", lancets in a designated disposal container, and disposable items to the autoclave bag) immediately after use.

Other necessary supplies and equipment are at specific supply areas marked according to the test with which they are used. Since nearly all of the tests require a finger stab, if you will be using your own blood it might be wise to quickly read through the tests to determine in which instances more than one preparation can be done from the same finger stab. For example, the hematocrit capillary tubes and sedimentation rate samples might be prepared at the same time. A little planning will save you the discomfort of a multiple punctured finger.

An alternative to using blood obtained from the finger stab technique is using heparinized blood samples supplied by your instructor. The purpose of using heparinized tubes is to prevent the blood from clotting. Thus blood collected and stored in such tubes will be suitable for all tests except coagulation time testing.

Total White and Red Blood Cell Counts

A **total WBC count** or **total RBC count** determines the total number of that cell type per unit volume of blood. Total WBC and RBC counts are a routine part of any physical exam. Most clinical agencies use computers to conduct these counts. Since the hand counting technique typically done in college labs is rather outdated, total RBC and WBC counts will not be done here, but the importance of such counts (both normal and abnormal values) is briefly described below.

Total White Blood Cell Count Since white blood cells are an important part of the body's defense system, it is essential to note any abnormalities in them.

Leukocytosis, an abnormally high WBC count, may indicate bacterial or viral infection, metabolic disease, hemorrhage, or poisoning by drugs or chemicals. A decrease in the white cell number below 4000/mm^3 (**leukopenia**) may indicate typhoid fever, measles, infectious hepatitis or cirrhosis, tuberculosis, or excessive antibiotic or X-ray therapy. A person with leukopenia lacks the usual protective mechanisms. **Leukemia**, a malignant disorder of the lymphoid tissues characterized by uncontrolled proliferation of abnormal WBCs accompanied by a reduction in the number of RBCs and platelets, is detectable not only by a total WBC count but also by a differential WBC count.

Total Red Blood Cell Count Since RBCs are absolutely necessary for oxygen transport, a doctor typically investigates any excessive change in their number immediately.

An increase in the number of RBCs (**polycythemia**) may result from bone marrow cancer or from living at high altitudes where less oxygen is available. A decrease in the number of RBCs results in anemia. (The term **anemia** simply indicates a decreased oxygen-carrying capacity of blood that may result from a decrease in RBC number or size or a decreased hemoglobin content of the RBCs.) A decrease in RBCs may result suddenly from hemorrhage or more gradually from conditions that destroy RBCs or hinder RBC production. ■

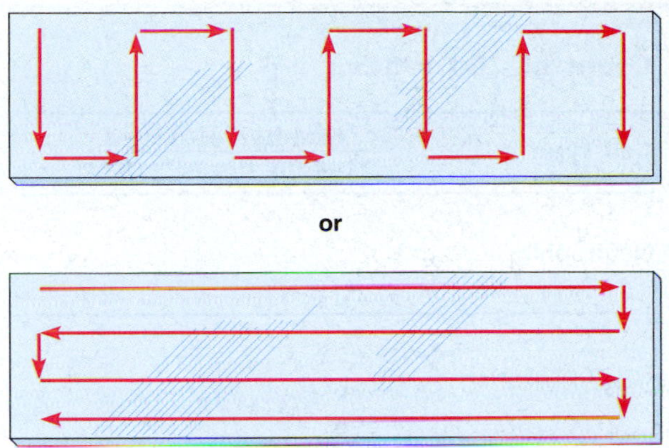

Figure 29.3 Alternative methods of moving the slide for a differential WBC count.

Differential White Blood Cell Count

To make a **differential white blood cell count,** 100 WBCs are counted and classified according to type. Such a count is routine in a physical examination and in diagnosing illness, since any abnormality or significant elevation in percentages of WBC types may indicate a problem or the source of pathology.

Activity 3:
Conducting a Differential WBC Count

1. Use the slide prepared for the identification of the blood cells in Activity 2. Begin at the edge of the smear and move the slide in a systematic manner on the microscope stage—either up and down or from side to side as indicated in Figure 29.3.

2. Record each type of white blood cell you observe by making a count on the chart at the top of p. 292 (for example, 𝍸𝍷𝍷 = 7 cells) until you have observed and recorded a total of 100 WBCs. Using the following equation, compute the percentage of each WBC type counted, and record the percentages on the data sheet on p. 292.

$$\text{Percent (\%)} = \frac{\text{\# observed}}{\text{Total \# counted (100)}} \times 100$$

3. Select a slide marked "Unknown sample," record the slide number, and use the count chart on p. 292 to conduct a differential count. Record the percentages on the data sheet at the bottom of p. 292.

How does the differential count from the unknown sample slide compare to a normal count?

____

____

Count of 100 WBCs

Cell type	Number observed Student blood smear
Neutrophils	
Eosinophils	
Basophils	
Lymphocytes	
Monocytes	

Using the text and other references, try to determine the blood pathology on the unknown slide. Defend your answer.

____

____

4. How does your differential white blood cell count correlate with the percentages given for each type on p. 290?

____

______________________________ ■

Hematocrit

The **hematocrit,** or **packed cell volume (PCV),** is routinely determined when anemia is suspected. Centrifuging whole blood spins the formed elements to the bottom of the tube, with plasma forming the top layer (see Figure 29.1). Since the blood cell population is primarily RBCs, the PCV is generally considered equivalent to the RBC volume, and this is the only value reported. However, the relative percentage of WBCs can be differentiated, and both WBC and plasma volume will be reported here. Normal hematocrit values for the male and female, respectively, are 47.0 ± 7 and 42.0 ± 5.

Activity 4: Determining the Hematocrit

The hematocrit is determined by the micromethod, so only a drop of blood is needed. If possible (and the centrifuge allows), all members of the class should prepare their capillary tubes at the same time so the centrifuge can be properly balanced and run only once.

1. Obtain two heparinized capillary tubes, capillary tube sealer or modeling clay, a lancet, alcohol swabs, and some cotton balls.

2. If you are using your own blood, cleanse a finger, and allow the blood to flow freely. Wipe away the first few drops

Hematologic Test Data Sheet

Differential WBC count:

WBC	Student blood smear	Unknown sample #____
% neutrophils	______	______
% eosinophils	______	______
% basophils	______	______
% monocytes	______	______
% lymphocytes	______	______

Hematocrit (PCV):

RBC ______ % of blood volume

WBC ______ % of blood volume } not generally reported

Plasma ______ % of blood } not generally reported

Hemoglobin content:

Hemoglobinometer (type: ______

______ g/100 ml blood; ______ % Hb

Ratio (PCV to grams Hb per 100 ml blood): ______

Sedimentation rate ______ mm/hr

Coagulation time ______

Blood typing:

ABO group ______ Rh factor ______

Cholesterol concentration ______ mg/dl blood

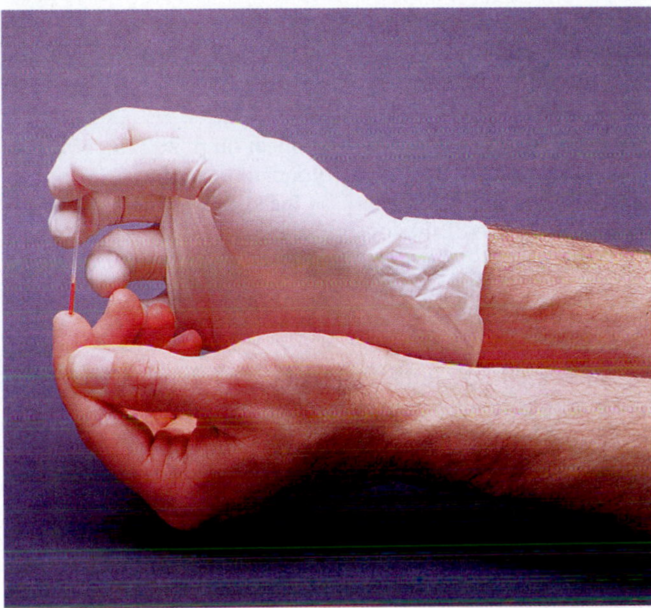

(a)

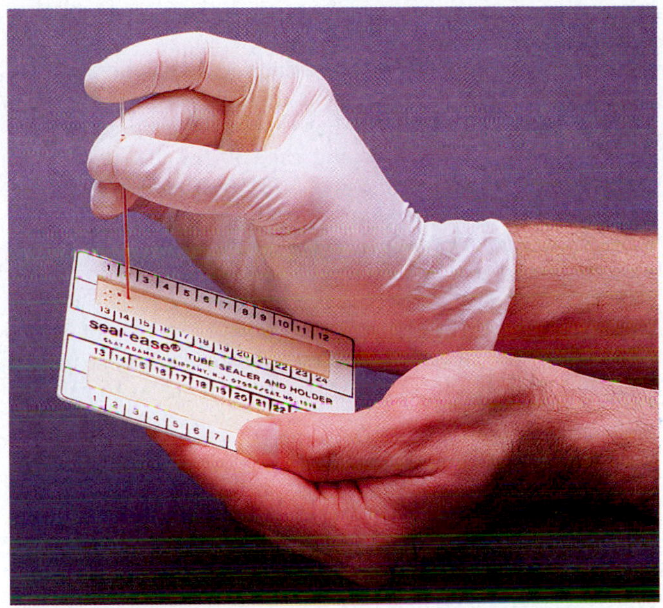

(b)

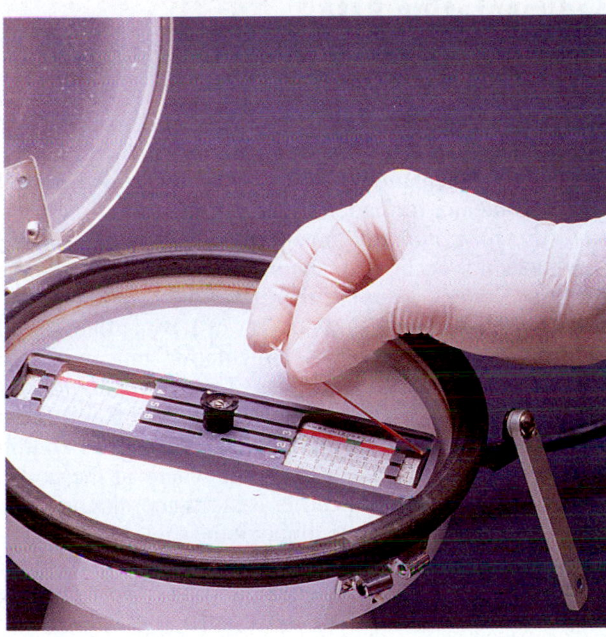

(c)

Figure 29.4 Steps in a hematocrit determination. (a) Load a heparinized capillary tube with blood. **(b)** Plug the blood-containing end of the tube with clay. **(c)** Place the tube in a microhematocrit centrifuge. (Centrifuge must be balanced.)

and, holding the red-line-marked end of the capillary tube to the blood drop, allow the tube to fill at least three-fourths full by capillary action (Figure 29.4a). If the blood is not flowing freely, the end of the capillary tube will not be completely submerged in the blood during filling, air will enter, and you will have to prepare another sample.

If you are using instructor-provided blood, simply immerse the red-marked end of the capillary tube in the blood sample and fill it three-quarters full as just described.

3. Plug the blood-containing end by pressing it into the capillary tube sealer or clay (Figure 29.4b). Prepare a second tube in the same manner.

4. Place the prepared tubes opposite one another in the radial grooves of the microhematocrit centrifuge with the sealed ends abutting the rubber gasket at the centrifuge periphery (Figure 29.4c). This loading procedure balances the centrifuge and prevents blood from spraying everywhere by centrifugal force. *Make a note of the numbers of the grooves your tubes are in.* When all the tubes have been loaded, make sure the centrifuge is properly balanced, and secure the centrifuge cover. Turn the centrifuge on, and set the timer for 4 or 5 minutes.

5. Determine the percentage of RBCs, WBCs, and plasma by using the microhematocrit reader. The RBCs are the bottom layer, the plasma is the top layer, and the WBCs are the buff-colored layer between the two. If the reader is not available, use a millimeter ruler to measure the length of the filled capillary tube occupied by each element, and compute its percentage by using the following formula:

$$\frac{\text{Height of the column composed of the element (mm)}}{\text{Height of the original column of whole blood (mm)}} \times 100$$

Record your calculations below and on the data sheet on p. 292.

% RBC ________ % WBC ________ % plasma ________

Usually WBCs constitute 1% of the total blood volume. How do your blood values compare to this figure and to the normal percentages for RBCs and plasma? (See p. 287.)

____

As a rule, a hematocrit is considered a more accurate test for determining the RBC composition of the blood than the total RBC count. A hematocrit within the normal range generally indicates a normal RBC number, whereas an abnormally high or low hematocrit is cause for concern. ■

Hemoglobin Concentration

As noted earlier, a person can be anemic even with a normal RBC count. Since hemoglobin is the RBC protein responsible for oxygen transport, perhaps the most accurate way of measuring the oxygen-carrying capacity of the blood is to determine its hemoglobin content. Oxygen, which combines reversibly with the heme (iron-containing portion) of the hemoglobin molecule, is picked up by the blood cells in the lungs and unloaded in the tissues. Thus, the more hemoglobin molecules the RBCs contain, the more oxygen they will be able to transport. Normal blood contains 12 to 18 g of hemoglobin per 100 ml of blood. Hemoglobin content in men is slightly higher (13 to 18 g) than in women (12 to 16 g).

Activity 5:
Determining Hemoglobin Concentration Using a Hemoglobinometer

Several techniques have been developed to estimate the hemoglobin content of blood, ranging from the old, rather inaccurate Tallquist method to expensive colorimeters, which are precisely calibrated and yield highly accurate results. Directions for use of a hemoglobinometer are provided here.

1. Obtain a hemoglobinometer, hemolysis applicator stick, alcohol swab, and lens paper and bring them to your bench. Test the hemoglobinometer light source to make sure it is working; if not, request new batteries before proceeding and test it again.

2. Remove the blood chamber from the slot in the side of the hemoglobinometer and disassemble the blood chamber by separating the glass plates from the metal clip. Notice as you do this that the larger glass plate has an H-shaped depression cut into it that acts as a moat to hold the blood, whereas the smaller glass piece is flat and serves as a coverslip.

3. Clean the glass plates with an alcohol swab and then wipe dry with lens paper. Hold the plates by their sides to prevent smearing during the wiping process.

4. Reassemble the blood chamber (remember: larger glass piece on the bottom with the moat up), but leave the moat plate about halfway out to provide adequate exposed surface to charge it with blood.

5. Obtain a drop of blood (from the provided sample or from your fingertip as before), and place it on the depressed area of the moat plate that is closest to you (Figure 29.5a).

6. Using the wood hemolysis applicator, stir or agitate the blood to rupture (lyse) the RBCs (Figure 29.5b). This usually takes 35 to 45 seconds. Hemolysis is complete when the blood appears transparent rather than cloudy.

7. Push the blood-containing glass plate all the way into the metal clip and then firmly insert the charged blood chamber back into the slot on the side of the instrument (Figure 29.5c).

8. Hold the hemoglobinometer in your left hand with your left thumb resting on the light switch located on the underside of the instrument. Look into the eyepiece and notice that there is a green area divided into two halves (a split field).

9. With the index finger of your right hand, slowly move the slide on the right side of the hemoglobinometer back and forth until the two halves of the green field match (Figure 29.5d).

10. Note and record on the data sheet on p. 292 the grams of Hb (hemoglobin)/100 ml of blood indicated on the uppermost scale by the index mark on the slide. Also record % Hb, indicated by one of the lower scales.

11. Disassemble the blood chamber once again, and carefully place its parts (glass plates and clip) into a bleach-containing beaker.

Generally speaking, the relationship between the PCV and grams of hemoglobin per 100 ml of blood is 3:1—for example, a PVC of 35 with 12 g of Hb per 100 ml of blood is a ratio of 3:1. How do your values compare?

____

Record on the data sheet (p. 292) the value obtained from your data. ■

Sedimentation Rate

The speed at which red blood cells settle to the bottom of a vertical tube when allowed to stand is called the **sedimentation rate.** The normal rate for adults is 0 to 6 mm/hr (averaging 3 mm/hr) and for children, 0 to 8 mm/hr (averaging 4 mm/hr). Sedimentation of RBCs apparently proceeds in three stages: rouleaux formation, rapid settling, and final packing. *Rouleaux formation* (alignment of RBCs like a stack of pennies) does not occur with abnormally shaped red blood cells (as in sickle-cell anemia); therefore, the sedimentation rate is decreased. The size and number of RBCs affect the packing phase. In anemia the sedimentation rate increases; in polycythemia the rate decreases. The sedimentation rate is greater than normal during menses and pregnancy, and very high sedimentation rates may indicate infectious conditions or tissue destruction occurring somewhere in the body. Although this test is nonspecific, it alerts the diagnostician to the need for further tests to pinpoint the site of pathology. The Landau micromethod, which uses just one drop of blood, is used here. An alternative method is the Westergren ESR method.

Activity 6:
Determining Sedimentation Rate

1. Obtain lancets, cotton balls, alcohol swabs, and Landau Sed-rate pipette and tubing, the Landau rack, a wide-mouthed bottle of 5% sodium citrate, a mechanical suction device, and a millimeter ruler.

2. Use the mechanical suction device to draw up the sodium citrate to the first (most distal) marking encircling the pipette. Prepare the finger for puncture, and produce a free flow of blood. Put the lancet in the disposal container.

3. Wipe off the first drop, and then draw the blood into the pipette until the mixture reaches the second encircling line. Keep the pipette tip immersed in the blood to avoid air bubbles.

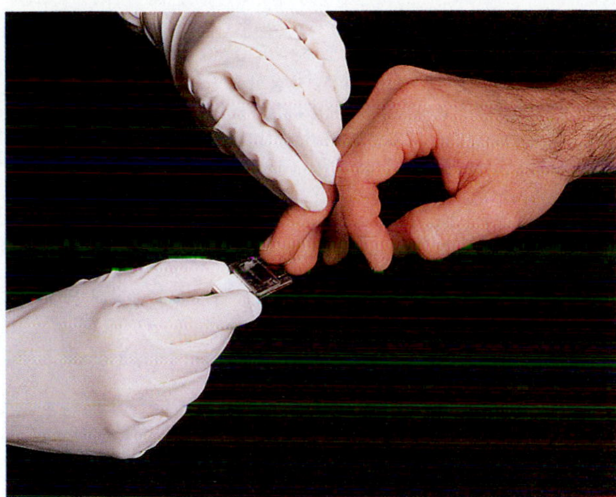

(a) A drop of blood is added to the moat plate of the blood chamber. The blood must flow freely.

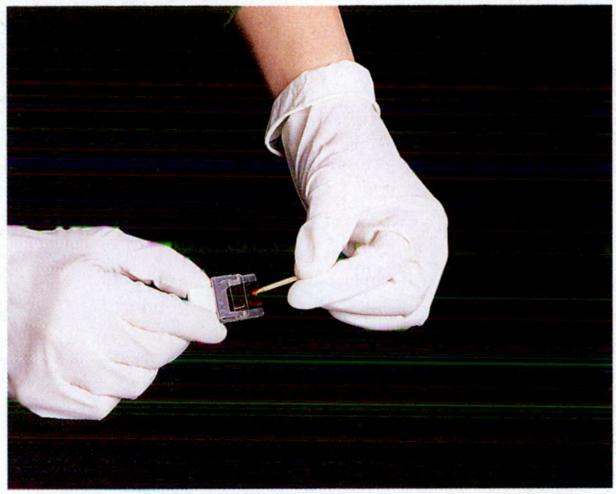

(b) The blood sample is hemolyzed with a wooden hemolysis applicator. Complete hemolysis requires 35 to 45 seconds.

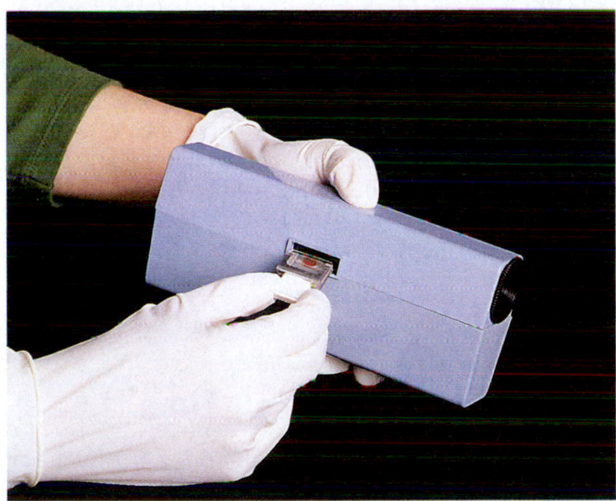

(c) The charged blood chamber is inserted into the slot on the side of the hemoglobinometer.

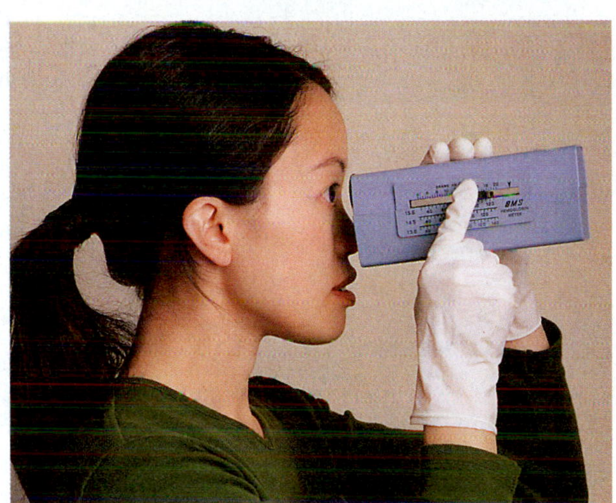

(d) The colors of the green split screen are found by moving the slide with the right index finger. When the two colors match in density, the grams/100 ml and %Hb are read on the scale.

Figure 29.5 Hemoglobin determination using a hemoglobinometer.

4. Thoroughly mix the blood with the citrate (an anticoagulant) by drawing the mixture into the bulb and then forcing it back down into the lumen. Repeat this mixing procedure six times, and then adjust the top level of the mixture as close to the zero marking as possible. *If any air bubbles are introduced during the mixing process, discard the sample and begin again.*

5. Seal the tip of the pipette by holding it tightly against the tip of your index finger, and then carefully remove the suction device from the upper end of the pipette.

6. Stand the pipette in an exactly vertical position on the Sed-rack with its lower end resting on the base of the rack. Record the time, and allow it to stand for exactly1 hour.

Time ______________________________

7. After 1 hour, measure the number of millimeters of visible clear plasma (which indicates the amount of settling of the RBCs), and record this figure here and on the data sheet.

Sedimentation rate ______________ mm/hr

8. Clean the pipette by using the mechanical suction device to draw up each of the cleaning solutions (available at the general supply area) in succession—bleach, distilled water, alcohol, and acetone. Discharge the contents each time before drawing up the next solution. Place the cleaned pipette in the bleach-containing beaker at the general supply area, after first drawing bleach up into the pipette. Put disposable items in the autoclave bag. ■

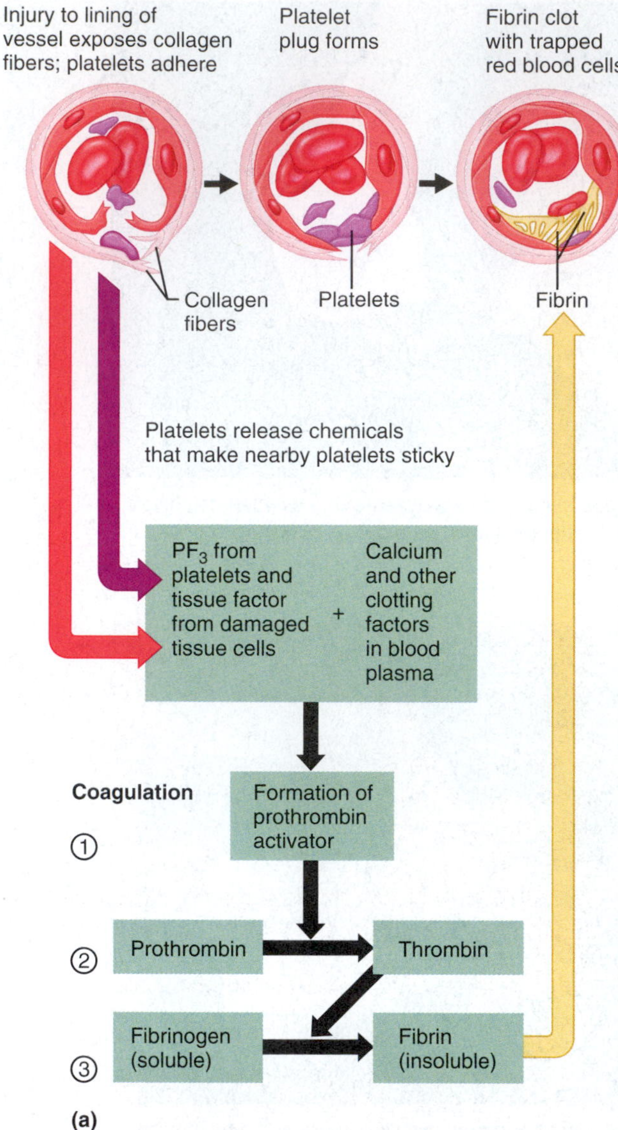

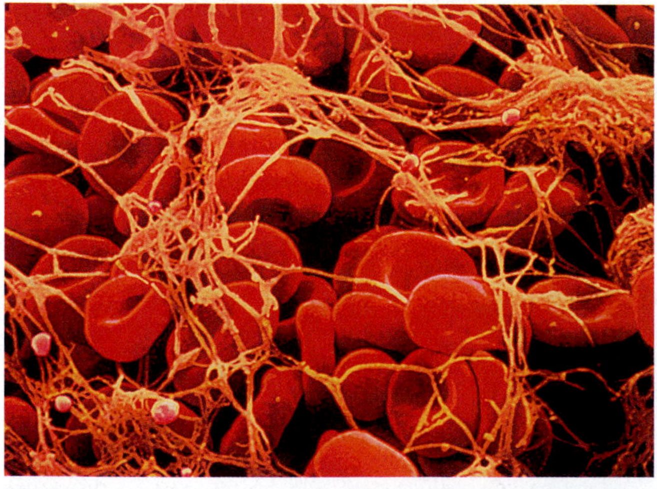

(b)

Figure 29.6 Events of hemostasis and blood clotting. (a) Simple schematic of events. Steps numbered 1–3 represent the major events of coagulation. **(b)** Photomicrograph of RBCs trapped in a fibrin mesh (3025×).

Bleeding Time

Normally a sharp prick of the finger or earlobe results in bleeding that lasts from 2 to 7 minutes (Ivy method) or 0 to 5 minutes (Duke method), although other factors such as altitude affect the time. How long the bleeding lasts is referred to as **bleeding time,** and tests the ability of platelets to stop bleeding in capillaries and small vessels. Absence of some clotting factors may affect bleeding time, but prolonged bleeding time is most often associated with deficient or abnormal platelets.

Coagulation Time

Blood clotting, or **coagulation,** is a protective mechanism that minimizes blood loss when blood vessels are ruptured. This process requires the interaction of many substances normally present in the plasma (clotting factors, or procoagulants) as well as some released by platelets and injured tissues. Basically hemostasis proceeds as follows (Figure 29.6a): The injured tissues and platelets release **tissue factor (TF)** and **PF_3** respectively, which trigger the clotting mechanism, or cascade. Tissue factor and PF_3 interact with other blood protein clotting factors and calcium ions to form **prothrombin activator,** which in turn converts **prothrombin** (present in plasma) to **thrombin.** Thrombin then acts enzymatically to polymerize the soluble **fibrinogen** proteins (present in plasma) into insoluble **fibrin,** which forms a meshwork of strands that traps the RBCs and forms the basis of the clot (Figure 29.6b). Normally, blood removed from the body clots within 2 to 6 minutes.

Activity 7:
Determining Coagulation Time

1. Obtain a *nonheparinized* capillary tube, a timer (or watch), a lancet, cotton balls, a triangular file, and alcohol swabs.

2. Clean and prick the finger to produce a free flow of blood. Discard the lancet in the disposal container.

3. Place one end of the capillary tube in the blood drop, and hold the opposite end at a lower level to collect the sample.

4. Lay the capillary tube on a paper towel.

Record the time. ____________________

5. At 30-sec intervals, make a small nick on the tube close to one end with the triangular file, and then carefully break the tube. Slowly separate the ends to see if a gel-like thread of fibrin spans the gap. When this occurs, record below and on the data sheet the time for coagulation to occur. Are your results within the normal time range?

____

6. Put used supplies in the autoclave bag. ■

Table 29.2 ABO Blood Typing

ABO blood type	Antigens present on RBC membranes	Antibodies present in plasma	% of U.S. population White	Black	Asian
A	A	Anti-B	40	27	28
B	B	Anti-A	11	20	27
AB	A and B	None	4	4	5
O	Neither	Anti-A and anti-B	45	49	40

Blood Typing

Blood typing is a system of blood classification based on the presence of specific glycoproteins on the outer surface of the RBC plasma membrane. Such proteins are called **antigens,** or **agglutinogens,** and are genetically determined. In many cases, these antigens are accompanied by plasma proteins, **antibodies** or **agglutinins,** that react with RBCs bearing different antigens, causing them to be clumped, agglutinated, and eventually hemolyzed. It is because of this phenomenon that a person's blood must be carefully typed before a whole blood or packed cell transfusion.

Several blood typing systems exist, based on the various possible antigens, but the factors routinely typed for are antigens of the ABO and Rh blood groups which are most commonly involved in transfusion reactions. Other blood factors, such as Kell, Lewis, M, and N, are not routinely typed for unless the individual will require multiple transfusions. The basis of the ABO typing is shown in Table 29.2.

Individuals whose red blood cells carry the Rh antigen are Rh positive (approximately 85% of the U.S. population); those lacking the antigen are Rh negative. Unlike ABO blood groups neither the blood of the Rh-positive (Rh^+) nor Rh-negative (Rh^-) individuals carries preformed anti-Rh antibodies. This is understandable in the case of the Rh-positive individual. However, Rh-negative persons who receive transfusions of Rh-positive blood become sensitized by the Rh antigens of the donor RBCs and their systems begin to produce anti-Rh antibodies. On subsequent exposures to Rh-positive blood, typical transfusion reactions occur, resulting in the clumping and hemolysis of the donor blood cells.

Although the blood of dogs and other mammals does react with some of the human agglutinins (present in the antisera), the reaction is not as pronounced and varies with the animal blood used. Hence, the most accurate and predictable blood typing results are obtained with human blood. The artificial blood kit does not use any body fluids and produces results that simulate but are not identical to results for human blood.

Activity 8:

Typing for ABO and Rh Blood Groups

Blood may be typed on glass slides or using blood test cards. Both methods are described next.

Typing Blood Using Glass Slides

1. Obtain two clean microscope slides, a wax marking pencil, anti-A, anti-B, and anti-Rh typing sera, toothpicks, lancets, alcohol swabs, medicine dropper, and the Rh typing box.

2. Divide slide 1 into halves with the wax marking pencil. Label the lower left-hand corner "anti-A" and the lower right-hand corner "anti-B." Mark the bottom of slide 2 "anti-Rh."

3. Place one drop of anti-A serum on the *left* side of slide 1. Place one drop of anti-B serum on the *right* side of slide 1. Place one drop of anti-Rh serum in the center of slide 2.

4. If you are using your own blood, cleanse your finger with an alcohol swab, pierce the finger with a lancet, and wipe away the first drop of blood. Obtain 3 drops of freely flowing blood, placing one drop on each side of slide 1 and a drop on slide 2. Immediately dispose of the lancet in a designated disposal container.

   If using instructor-provided animal blood or EDTA-treated red cells, use a medicine dropper to place one drop of blood on each side of slide 1 and a drop of blood on slide 2.

5. Quickly mix each blood-antiserum sample with a *fresh* toothpick. Then dispose of the toothpicks and used alcohol swab in the autoclave bag.

6. Place slide 2 on the Rh typing box and rock gently back and forth. (A slightly higher temperature is required for precise Rh typing than for ABO typing.)

7. After 2 minutes, observe all three blood samples for evidence of clumping. The agglutination that occurs in the positive test for the Rh factor is very fine and difficult to perceive; thus if there is any question, observe the slide under the microscope. Record your observations in the chart on p. 298.

8. Interpret your ABO results in light of the information in Figure 29.7. If clumping was observed on slide 2, you are Rh positive. If not, you are Rh negative.

9. Record your blood type on the chart on the bottom of p. 292.

10. Put the used slides in the bleach-containing bucket at the general supply area; put disposable supplies in the autoclave bag.

Blood Typing

Result	Observed (+)	Not observed (−)
Presence of clumping with anti-A		
Presence of clumping with anti-B		
Presence of clumping with anti-Rh		

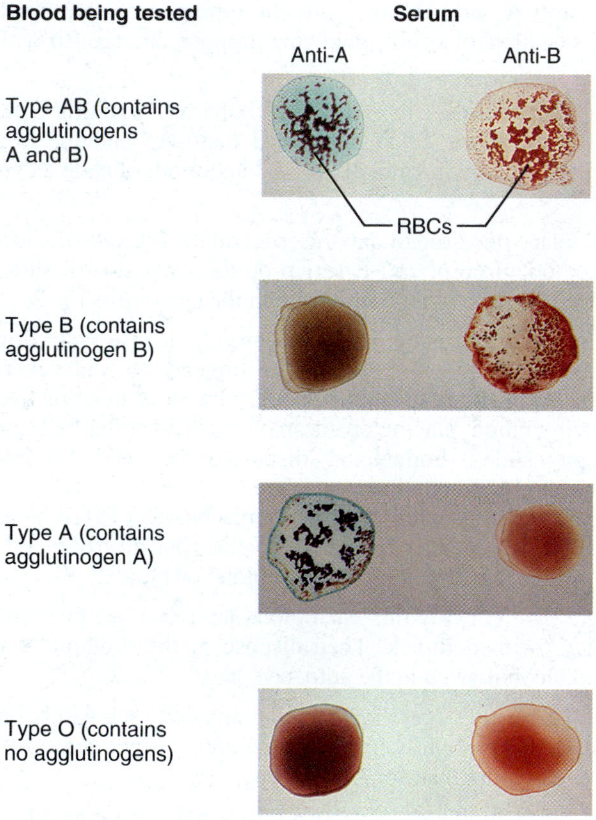

Figure 29.7 Blood typing of ABO blood types. When serum containing anti-A or anti-B agglutinins is added to a blood sample, agglutination will occur between the agglutinin and the corresponding agglutinogen (A or B). As illustrated, agglutination occurs with both sera in blood group AB, with anti-B serum in blood group B, with anti-A serum in blood group A, and with neither serum in blood group O.

Using Blood Typing Cards

1. Obtain a blood typing card marked A, B, and Rh, dropper bottles of anti-A serum, anti-B serum, and anti-Rh serum, toothpicks, lancets, and alcohol swabs.

2. Place a drop of anti-A serum in the spot marked anti-A, place a drop of anti-B serum on the spot marked anti-B, and place a drop of anti-Rh serum on the spot marked anti-Rh (or anti-D).

3. Carefully add a drop of blood to each of the spots marked "Blood" on the card. If you are using your own blood, refer to direction number 4 in Activity 8. Immediately discard the lancet in the designated disposal container.

4. Using a new toothpick for each test, mix the blood sample with the antibody. Dispose of the toothpicks appropriately.

5. Gently rock the card to allow the blood and antibodies to mix.

6. After two minutes, observe the card for evidence of clumping. The Rh clumping is very fine and may be difficult to observe. Record your observations in the chart on the left. Use Figure 29.7 to interpret your results.

7. Record your blood type on the chart at the bottom of p. 292, and discard the card in an autoclave bag. ■

Activity 9:
Observing Demonstration Slides

Before continuing on to the cholesterol determination, take the time to look at the slides of *macrocytic hypochromic anemia, microcytic hypochromic anemia, sickle-cell anemia, lymphocytic leukemia* (chronic), and *eosinophilia* that have been put on demonstration by your instructor. Record your observations in the appropriate section of the Review Sheets for Exercise 29. You can refer to your notes, the text, and other references later to respond to questions about the blood pathologies represented on the slides. ■

Cholesterol Concentration in Plasma

Atherosclerosis is the disease process in which the body's blood vessels become increasingly occluded by plaques. Because the plaques narrow the arteries, they can contribute to hypertensive heart disease. They also serve as focal points for the formation of blood clots (thrombi) which may break away and block smaller vessels farther downstream in the circulatory pathway, causing heart attacks or strokes.

Ever since medical clinicians discovered that cholesterol is a major component of the smooth muscle plaques formed during atherosclerosis, it has had a bad press. Today, virtually no physical examination of an adult is considered complete until cholesterol levels are assessed along with other lifestyle risk factors. A normal value for plasma cholesterol in adults ranges from 130 to 200 mg per 100 ml plasma; you will use blood to make such a determination.

Although the total plasma cholesterol concentration is valuable information, it may be misleading, particularly if a person's high-density lipoprotein (HDL) level is high and low-density lipoprotein (LDL) level is relatively low. Cholesterol, being water insoluble, is transported in the blood complexed to lipoproteins. In general, cholesterol bound into HDLs is destined to be degraded by the liver and then eliminated from the body, whereas that forming part of the LDLs is "traveling" to the body's tissue cells. When LDL levels are excessive, cholesterol is deposited in the blood vessel walls; hence, LDLs are considered to carry the "bad" cholesterol.

Activity 10:
Measuring Plasma Cholesterol Concentration

1. Go to the appropriate supply area, and obtain a cholesterol test card, lancet, and alcohol swab.

2. Prick finger with lancet, and place a drop of blood on the test area of the card. Put the lancet in the disposal container.

3. After 3 minutes, remove the blood sample strip from the card and discard in the autoclave bag.

4. Analyze the underlying test spot, using the included color scale. Record on the chart at the bottom of p. 292.

Cholesterol level ________ mg/dl

5. Before leaving the laboratory, use a paper towel saturated with bleach solution to wash down your laboratory bench. ■

exercise

30 Anatomy of the Heart

Objectives

1. To describe the location of the heart.
2. To name and locate the major anatomical areas and structures of the heart when provided with an appropriate model, diagram, or dissected sheep heart, and to explain the function of each.
3. To trace the pathway of blood through the heart.
4. To explain why the heart is called a double pump, and to compare the pulmonary and systemic circuits.
5. To explain the operation of the atrioventricular and semilunar valves.
6. To name and follow the functional blood supply of the heart.
7. To describe the histologic structure of cardiac muscle, and to note the importance of its intercalated discs and the spiral arrangement of its fibers in the heart.

Materials

- ❑ Torso model or laboratory chart showing heart anatomy
- ❑ Red and blue pencils
- ❑ Three-dimensional models of cardiac and skeletal muscle
- ❑ Heart model (three-dimensional)
- ❑ X ray of the human thorax for observation of the position of the heart *in situ;* X-ray viewing box
- ❑ Preserved sheep heart, pericardial sacs intact (if possible)
- ❑ Dissecting pan and instruments
- ❑ Pointed glass rods for probes
- ❑ Disposable gloves
- ❑ Compound microscope
- ❑ Histologic slides of cardiac muscle (longitudinal section)

Human Cardiovascular System: The Heart videotape*

A1A See Appendix C, Exercise 30 for links to A.D.A.M.® Interactive Anatomy.

*Available to qualified adopters from Benjamin Cummings.

The major function of the **cardiovascular system** is transportation. Using blood as the transport vehicle, the system carries oxygen, digested foods, cell wastes, electrolytes, and many other substances vital to the body's homeostasis to and from the body cells. The system's propulsive force is the contracting heart, which can be compared to a muscular pump equipped with one-way valves. As the heart contracts, it forces blood into a closed system of large and small plumbing tubes (blood vessels) within which the blood is confined and circulated. This exercise deals with the structure of the heart or circulatory pump. The anatomy of the blood vessels is considered separately in Exercise 32.

Gross Anatomy of the Human Heart

The **heart,** a cone-shaped organ approximately the size of a fist, is located within the mediastinum, or medial cavity, of the thorax. It is flanked laterally by the lungs, posteriorly by the vertebral column, and anteriorly by the sternum (Figure 30.1). Its more pointed **apex** extends slightly to the left and rests on the diaphragm, approximately at the level of the fifth intercostal space. Its broader **base,** from which the great vessels emerge, lies beneath the second rib and points toward the right shoulder. *In situ,* the right ventricle of the heart forms most of its anterior surface.

• If an X ray of a human thorax is available, verify the relationships described above; otherwise, Figure 30.1 should suffice.

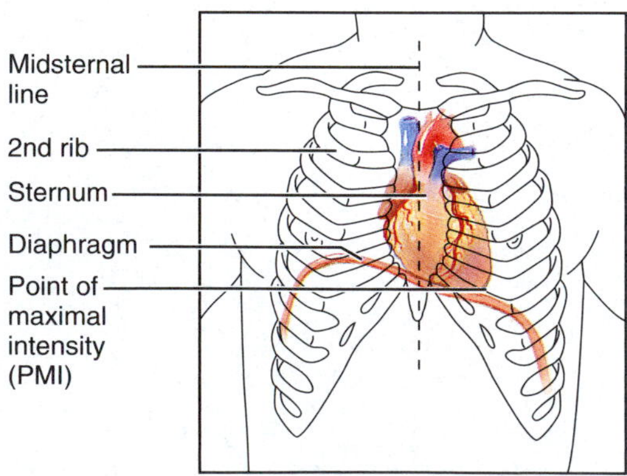

Figure 30.1 Location of the heart in the thorax.

The heart is enclosed within a double-walled fibroserous sac called the pericardium. The thin **visceral pericardium,** or **epicardium,** is closely applied to the heart muscle. It reflects downward at the base of the heart to form its companion serous membrane, the outer, loosely applied **parietal pericardium,** which is attached at the heart apex to the diaphragm. Serous fluid produced by these membranes allows the heart to beat in a relatively frictionless environment. The serous parietal pericardium, in turn, lines the loosely fitting superficial **fibrous pericardium** composed of dense connective tissue.

Inflammation of the pericardium, **pericarditis,** causes painful adhesions between the serous pericardial layers. These adhesions interfere with heart movements. ■

The walls of the heart are composed primarily of cardiac muscle—the **myocardium**—which is reinforced internally by a dense fibrous connective tissue network. This network—the *fibrous skeleton of the heart*—is more elaborate and thicker in certain areas, for example, around the valves and at the base of the great vessels leaving the heart.

Figure 30.2 shows two views of the heart—an external anterior view and a frontal section. As its anatomical areas are described in the text, consult the figure.

Heart Chambers

The heart is divided into four chambers: two superior **atria** and two inferior **ventricles,** each lined by thin serous endothelium called the **endocardium.** The septum that divides the heart longitudinally is referred to as the **interatrial** or **interventricular septum,** depending on which chambers it partitions. Functionally, the atria are receiving chambers and are relatively ineffective as pumps. Blood flows into the atria under low pressure from the veins of the body. The right atrium receives relatively oxygen-poor blood from the body via the **superior** and **inferior venae cavae** and the coronary sinus. Four **pulmonary veins** deliver oxygen-rich blood from the lungs to the left atrium.

The inferior thick-walled ventricles, which form the bulk of the heart, are the discharging chambers. They force blood out of the heart into the large arteries that emerge from its base. The right ventricle pumps blood into the **pulmonary trunk,** which routes blood to the lungs to be oxygenated. The left ventricle discharges blood into the **aorta,** from which all systemic arteries of the body diverge to supply the body tissues. Discussions of the heart's pumping action usually refer to ventricular activity.

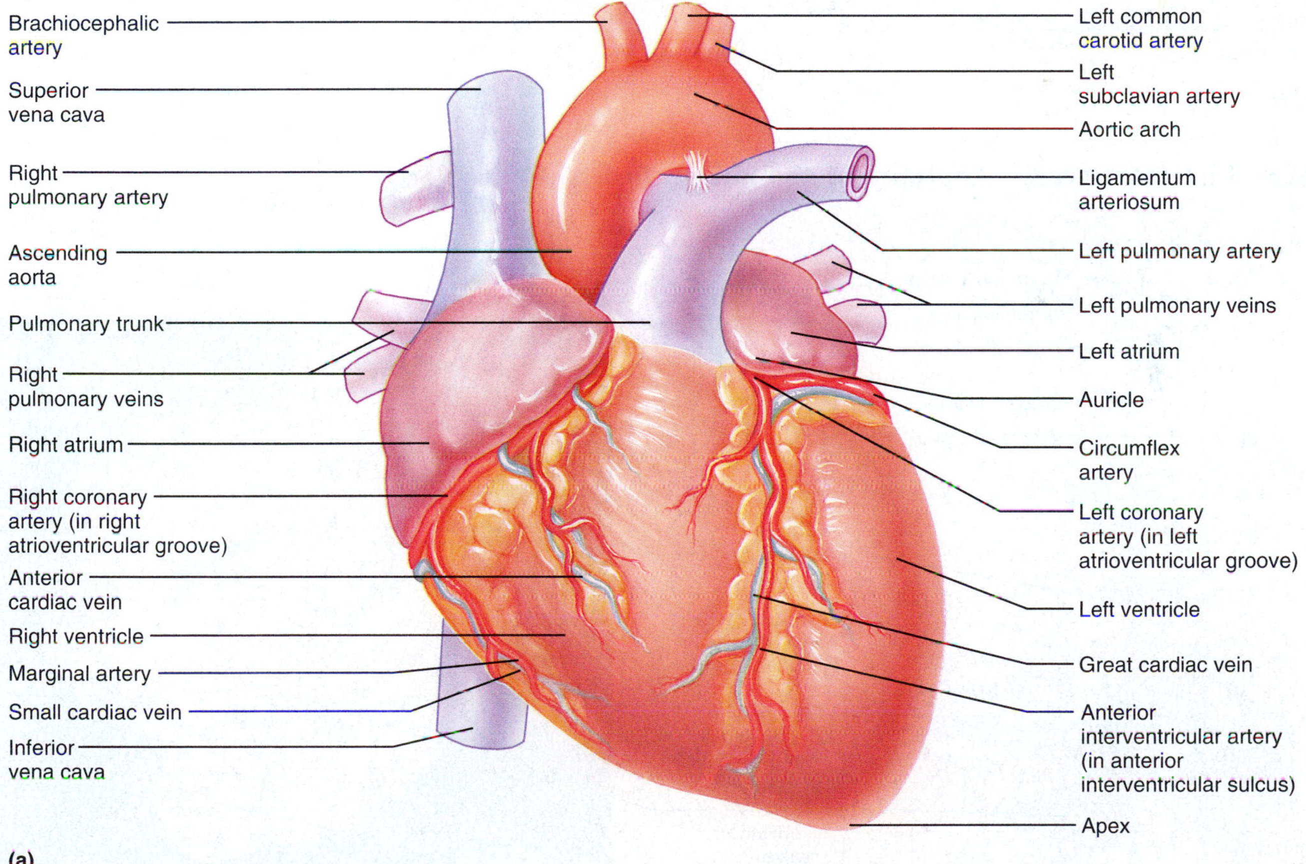

Figure 30.2 Anatomy of the human heart. (a) External anterior view.

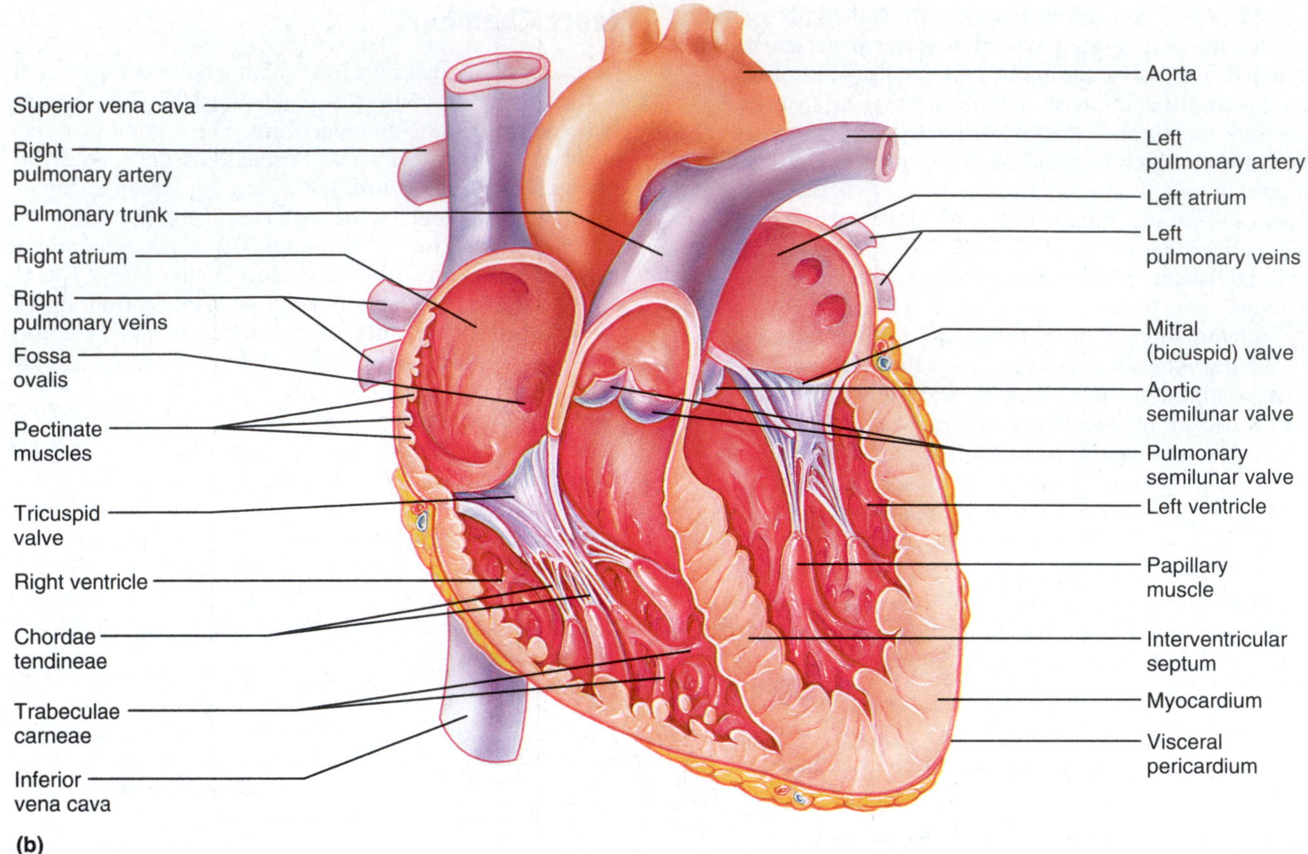

Figure 30.2 (*continued*) Anatomy of the human heart. (b) Frontal section.

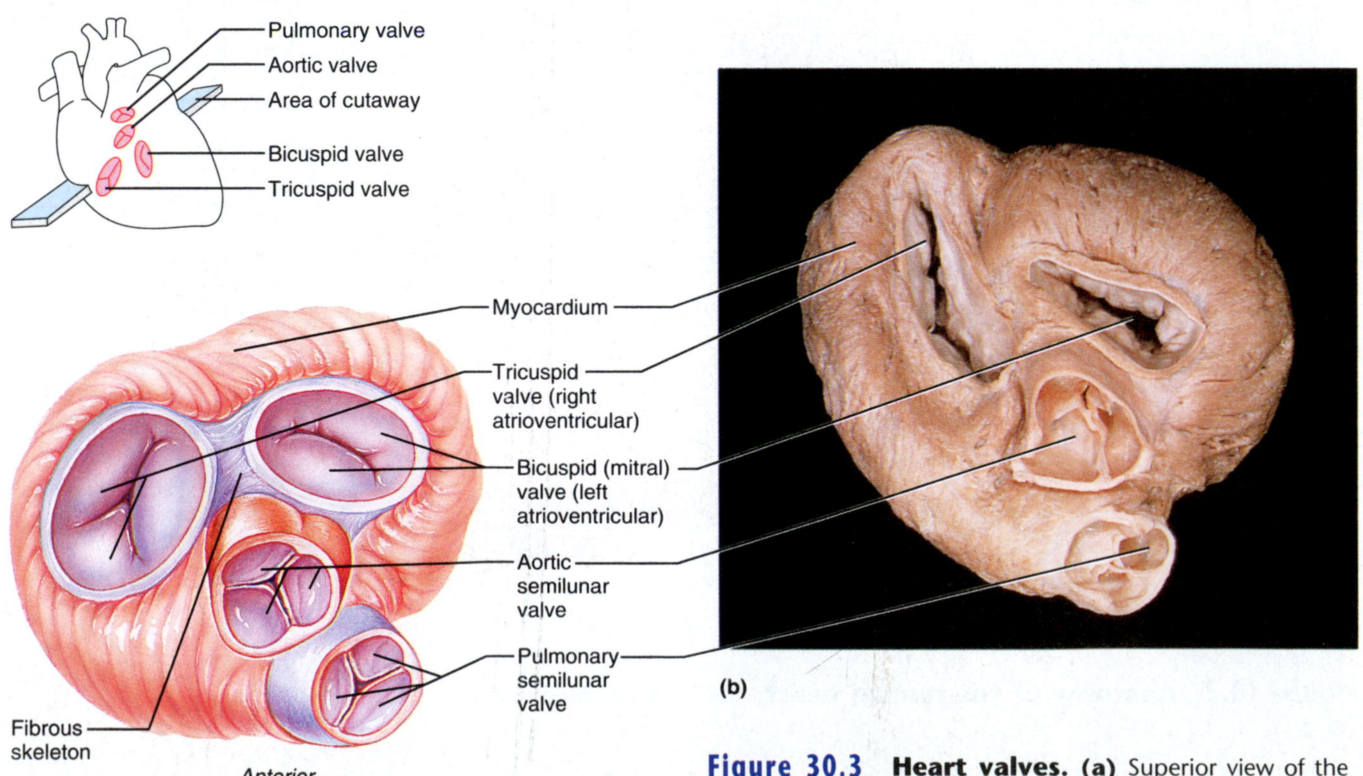

Figure 30.3 Heart valves. (a) Superior view of the two sets of heart valves (atria removed). **(b)** Photograph of the heart valves, superior view.

Heart Valves

Four valves enforce a one-way blood flow through the heart chambers. The **atrioventricular (AV) valves,** located between the atrial and ventricular chambers on each side, prevent backflow into the atria when the ventricles are contracting. The left atrioventricular valve, also called the **mitral** or **bicuspid valve,** consists of two cusps, or flaps, of endocardium. The right atrioventricular valve, the **tricuspid valve,** has three cusps (Figure 30.3). Tiny white collagenic cords called the **chordae tendineae** (literally, "heart strings") anchor the cusps to the ventricular walls. The chordae tendineae originate from small bundles of cardiac muscle, called **papillary muscles,** that project from the myocardial wall (see Figure 30.2b).

When blood is flowing passively into the atria and then into the ventricles during **diastole** (the period of ventricular filling), the AV valve flaps hang limply into the ventricular chambers and then are carried passively toward the atria by the accumulating blood. When the ventricles contract (**systole**) and compress the blood in their chambers, the intraventricular blood pressure rises, causing the valve flaps to be reflected superiorly, which closes the AV valves. The chordae tendineae, pulled taut by the contracting papillary muscles, anchor the flaps in a closed position that prevents backflow into the atria during ventricular contraction. If unanchored, the flaps would blow upward into the atria rather like an umbrella being turned inside out by a strong wind.

The second set of valves, the **pulmonary** and **aortic semilunar valves,** each composed of three pocketlike cusps, guards the bases of the two large arteries leaving the ventricular chambers. The valve cusps are forced open and flatten against the walls of the artery as the ventricles discharge their blood into the large arteries during systole. However, when the ventricles relax, blood flows backward toward the heart and the cusps fill with blood, closing the semilunar valves and preventing arterial blood from reentering the heart.

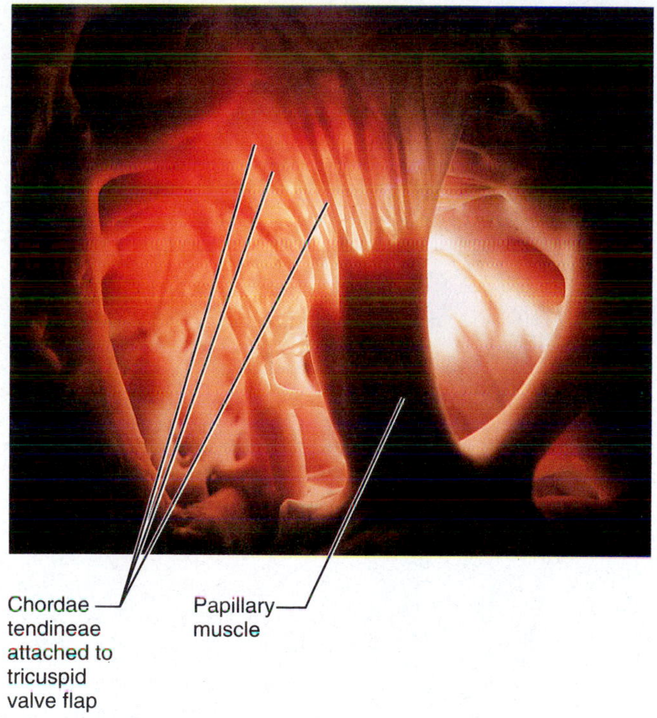

Activity 1:
Using the Heart Model to Study Heart Anatomy

When you have pinpointed in Figure 30.2 all the structures described above, observe the human heart model and reidentify the same structures without reference to the figure. ■

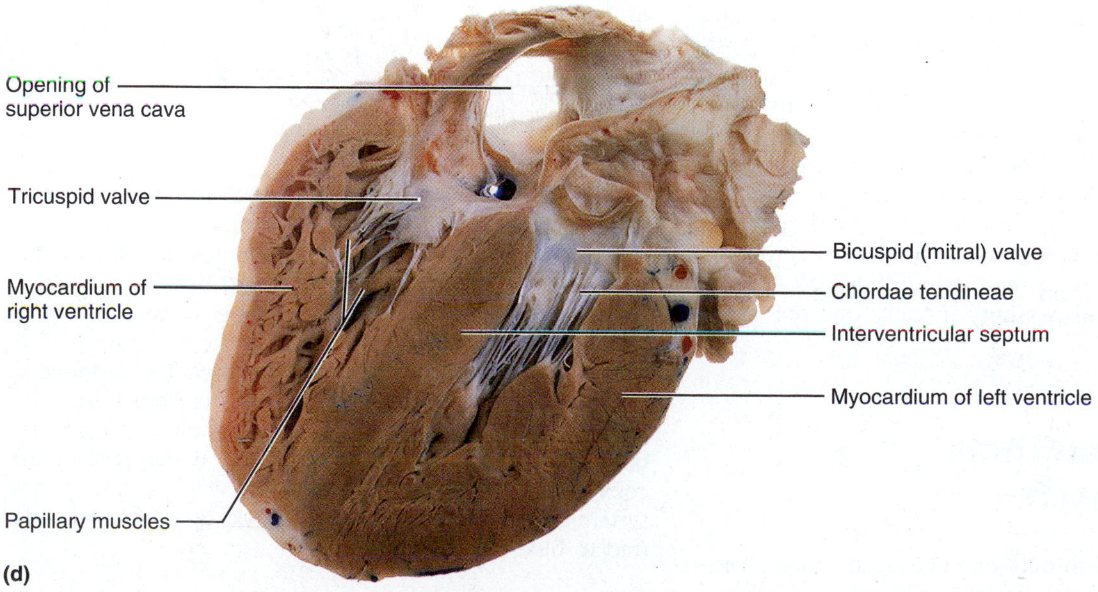

Figure 30.3 (*continued*) **(c)** Photograph of the right AV valve. View begins in the right ventricle, toward the right atrium. **(d)** Coronal section of the heart.

Pulmonary, Systemic, and Cardiac Circulations

Pulmonary and Systemic Circulations

The heart functions as a double pump. The right side serves as the **pulmonary circulation** pump, shunting the carbon dioxide–rich blood entering its chambers to the lungs to unload carbon dioxide and pick up oxygen, and then back to the left side of the heart (Figure 30.4). The function of this circuit is strictly to provide for gas exchange. The second circuit, which carries oxygen-rich blood from the left heart through the body tissues and back to the right heart, is called the **systemic circulation.** It supplies the functional blood supply to all body tissues.

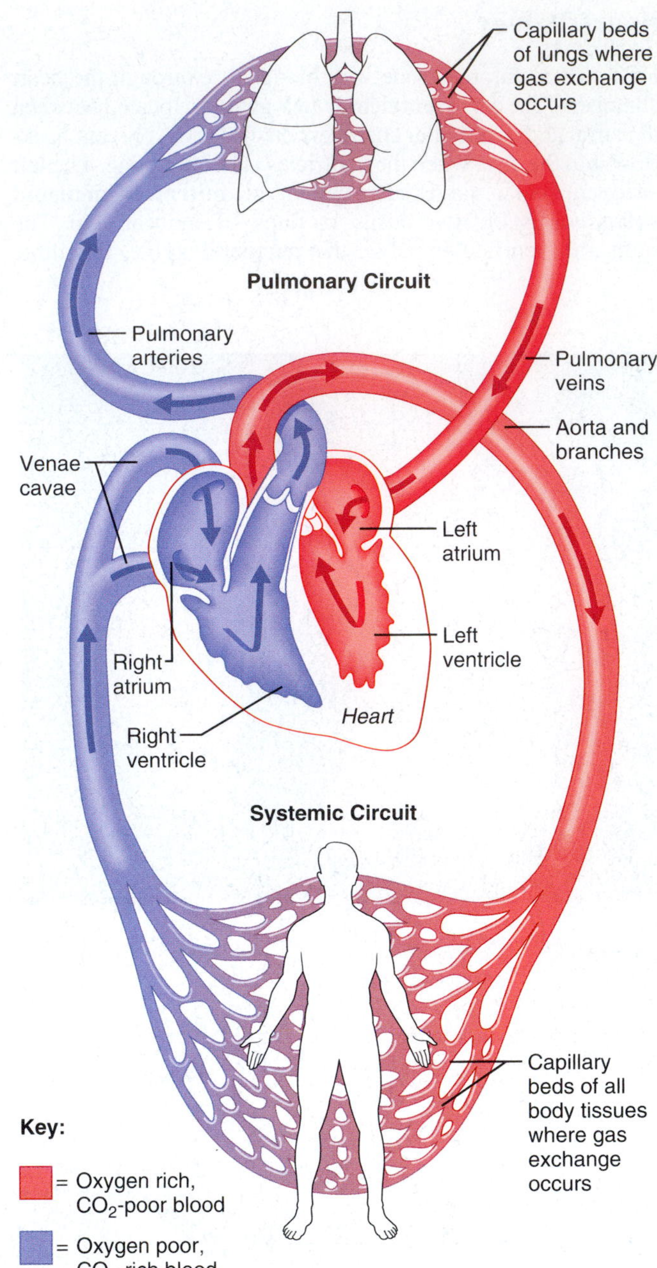

Figure 30.4 The systemic and pulmonary circuits. The heart is a double pump that serves two circulations. The right side of the heart pumps blood through the pulmonary circuit to the lungs and back to the left heart. (For simplicity, the actual number of two pulmonary arteries and four pulmonary veins has been reduced to one each.) The left heart pumps blood via the systemic circuit to all body tissues and back to the right heart. Notice that blood flowing through the pulmonary circuit gains oxygen and loses carbon dioxide as depicted by the color change from blue to red. Blood flowing through the systemic circuit loses oxygen and picks up carbon dioxide (red to blue color change).

Activity 2:
Tracing the Path of Blood Through the Heart

Trace the pathway of a red blood cell through the heart by adding arrows to the frontal section diagram (Figure 30.2b). Use red arrows for the oxygen-rich blood and blue arrows for the less oxygen-rich blood. ■

Cardiac Circulation

Even though the heart chambers are almost continually bathed with blood, this contained blood does not nourish the myocardium. The functional blood supply of the heart is provided by the right and left coronary arteries (see Figures 30.2 and 30.5). The **coronary arteries** issue from the base of the aorta just above the aortic semilunar valve and encircle the heart in the **atrioventricular groove** at the junction of the atria and ventricles. They then ramify over the heart's surface, the right coronary artery supplying the posterior surface of the ventricles and the lateral aspect of the right side of the heart, largely through its **posterior interventricular** and **marginal artery** branches. The left coronary artery supplies the anterior ventricular walls and the laterodorsal part of the left side of the heart via its two major branches, the **anterior interventricular artery** and the **circumflex artery.** The coronary arteries and their branches are compressed during systole and fill when the heart is relaxed. The myocardium is largely drained by the **great, middle,** and **small cardiac veins,** which empty into the **coronary sinus.** The coronary sinus, in turn, empties into the right atrium. In addition, several **anterior cardiac veins** empty directly into the right atrium (Figure 30.5).

Microscopic Anatomy of Cardiac Muscle

Cardiac muscle is found in only one place—the heart. The heart acts as a vascular pump, propelling blood to all tissues of the body; cardiac muscle is thus very important to life. Cardiac muscle is involuntary, ensuring a constant blood supply.

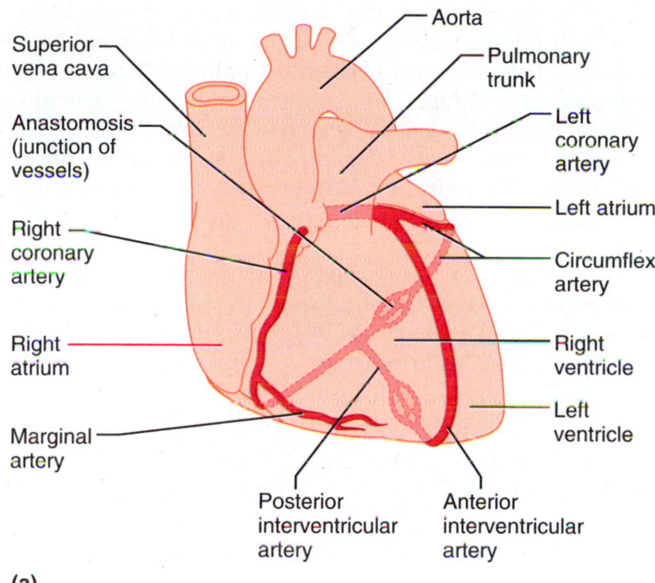

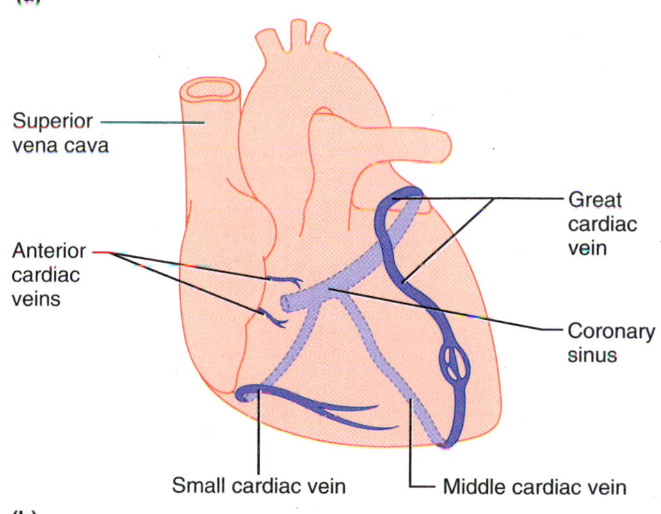

Figure 30.5 Cardiac circulation.

The cardiac cells, only sparingly invested in connective tissue, are arranged in spiral or figure-8-shaped bundles (Figure 30.6). When the heart contracts, its internal chambers become smaller (or are temporarily obliterated), forcing the blood into the large arteries leaving the heart.

Activity 3:
Examining Heart Anatomy

1. Observe the three-dimensional model of cardiac muscle, examining its branching cells and the areas where the cells interdigitate, the **intercalated discs.** These two structural features provide a continuity to cardiac muscle not seen in other muscle tissues and allow close coordination of heart activity.

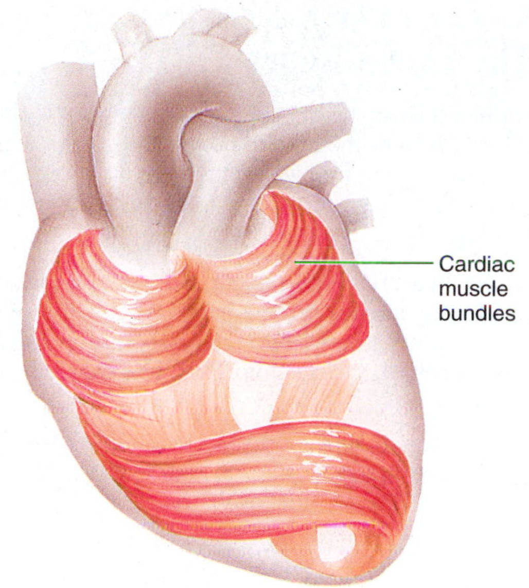

Figure 30.6 Longitudinal view of the heart chambers showing the spiral arrangement of the cardiac muscle fibers.

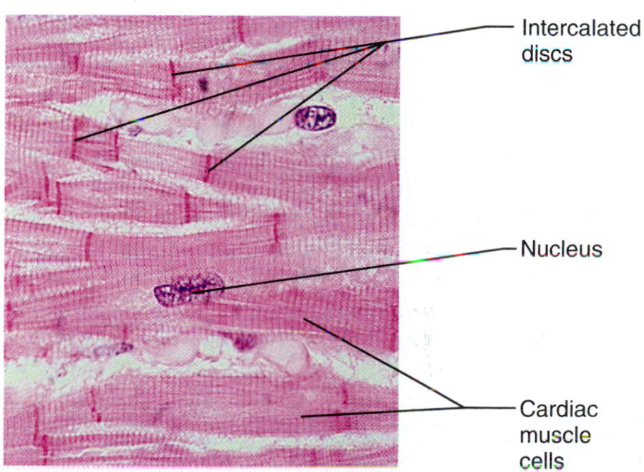

Figure 30.7 Photomicrograph of cardiac muscle (465×).

2. Compare the model of cardiac muscle to the model of skeletal muscle. Note the similarities and differences between the two kinds of muscle tissue.

3. Obtain and observe a longitudinal section of cardiac muscle under high power. Identify the nucleus, striations, intercalated discs, and sarcolemma of the individual cells and then compare your observations to the view seen in Figure 30.7. ■

Dissection: The Sheep Heart

Dissection of a sheep heart is valuable because it is similar in size and structure to the human heart. Also, a dissection experience allows you to view structures in a way not possible with models and diagrams. Refer to Figure 30.8 as you proceed with the dissection.

1. Obtain a preserved sheep heart, a dissection tray, dissecting instruments, a glass probe, and gloves. Rinse the sheep heart in cold water to remove excessive preservatives and to flush out any trapped blood clots. Now you are ready to make your observations.

2. Observe the texture of the pericardium. Also, note its point of attachment to the heart. Where is it attached?

___

3. If the serous pericardial sac is still intact, slit open the parietal pericardium and cut it from its attachments. Observe the visceral pericardium (epicardium). Using a sharp scalpel, carefully pull a little of this serous membrane away from the myocardium. How do its position, thickness, and apposition to the heart differ from those of the parietal pericardium?

___

___

___

4. Examine the external surface of the heart. Notice the accumulation of adipose tissue, which in many cases marks the separation of the chambers and the location of the coronary arteries that nourish the myocardium. Carefully scrape away some of the fat with a scalpel to expose the coronary blood vessels.

5. Identify the base and apex of the heart, and then identify the two wrinkled **auricles,** earlike flaps of tissue projecting from the atrial chambers. The balance of the heart muscle is ventricular tissue. To identify the left ventricle, compress the ventricular chambers on each side of the longitudinal fissures carrying the coronary blood vessels. The side that feels thicker and more solid is the left ventricle. The right ventricle feels much thinner and somewhat flabby when compressed. This difference reflects the greater demand placed on the left ventricle, which must pump blood through the much longer systemic circulation, a pathway with much higher resistance than the pulmonary circulation served by the right ventricle. Hold the heart in its anatomical position (Figure 30.8a), with the anterior surface uppermost. In this position the left ventricle composes the entire apex and the left side of the heart.

6. Identify the pulmonary trunk and the aorta extending from the superior aspect of the heart. The pulmonary trunk is more anterior, and you may see its division into the right and left pulmonary arteries if it has not been cut too closely to the heart. The thicker-walled aorta, which branches almost immediately, is located just beneath the pulmonary trunk. The first observable branch of the sheep aorta, the **brachiocephalic artery,** is identifiable unless the aorta has been cut immediately as it leaves the heart. The brachiocephalic artery splits to form the right carotid and subclavian arteries, which supply the right side of the head and right forelimb, respectively.

Carefully clear away some of the fat between the pulmonary trunk and the aorta to expose the **ligamentum arteriosum,** a cordlike remnant of the **ductus arteriosus.** (In the fetus, the ductus arteriosus allows blood to pass directly from the pulmonary trunk to the aorta, thus bypassing the nonfunctional fetal lungs.)

7. Cut through the wall of the aorta until you see the aortic semilunar valve. Identify the two openings to the coronary arteries just above the valve. Insert a probe into one of these holes to see if you can follow the course of a coronary artery across the heart.

8. Turn the heart to view its posterior surface. The heart will appear as shown in Figure 30.8b. Notice that the right and left ventricles appear equal-sized in this view. Try to identify the four thin-walled pulmonary veins entering the left atrium. Identify the superior and inferior venae cavae entering the right atrium. Due to the way the heart is trimmed, the pulmonary veins and superior vena cava may be very short or missing. If possible, compare the approximate diameter of the superior vena cava with the diameter of the aorta.

Which is larger? ___

Which has thicker walls? ___

Why do you suppose these differences exist?

___

___

___

9. Insert a probe into the superior vena cava, through the right atrium, and out the inferior vena cava. Use scissors to cut along the probe so that you can view the interior of the right atrium. Observe the right atrioventricular valve.

How many flaps does it have? ___

Pour some water into the right atrium and allow it to flow into the ventricle. *Slowly and gently* squeeze the right ventricle to watch the closing action of this valve. (If you squeeze too vigorously, you'll get a face full of water!) Drain the water from the heart before continuing.

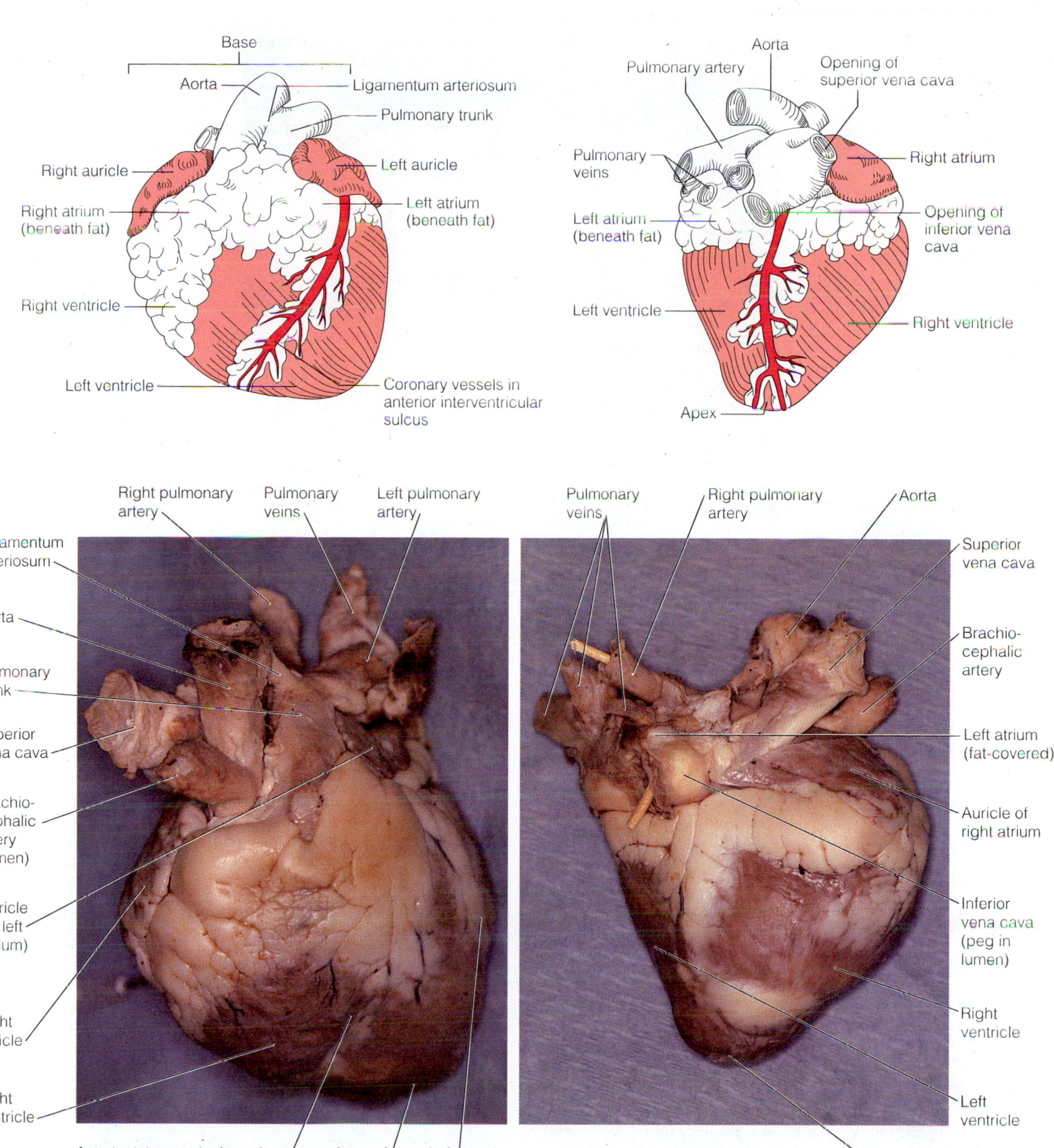

Figure 30.8 Anatomy of the sheep heart. **(a)** Anterior view. **(b)** Posterior view. Diagrammatic views at top; photographs at bottom.

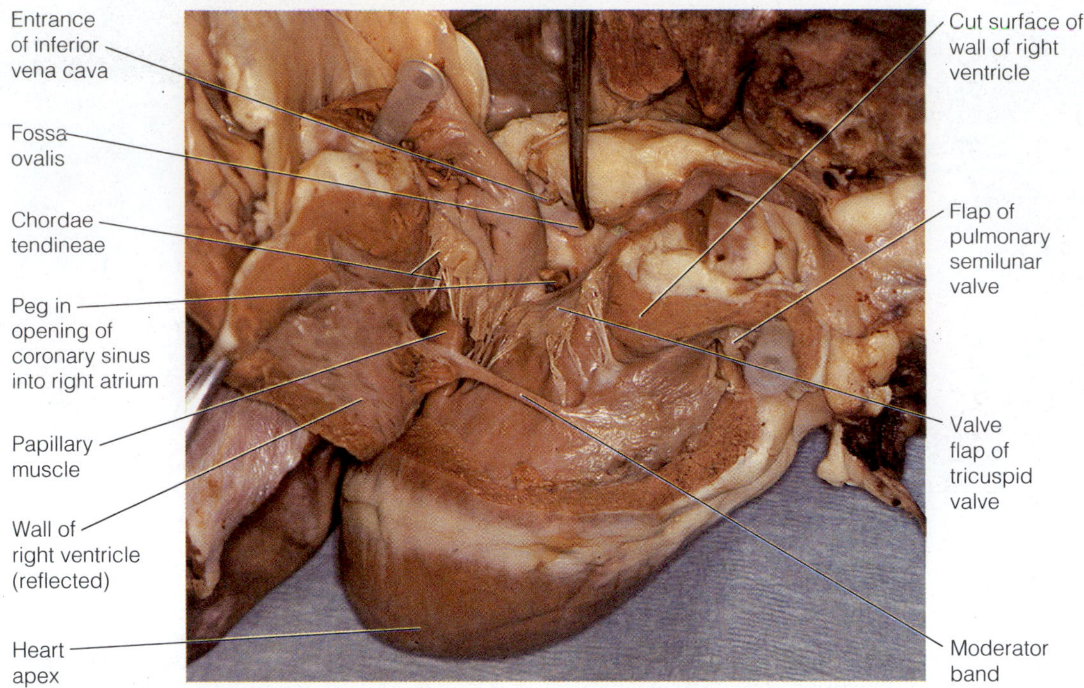

Figure 30.9 Right side of the sheep heart opened and reflected to reveal internal structures.

10. Return to the pulmonary trunk and cut through its anterior wall until you can see the pulmonary semilunar valve (Figure 30.9). Pour some water into the base of the pulmonary trunk to observe the closing action of this valve. How does its action differ from that of the atrioventricular valve?

___

___

___

After observing semilunar valve action, drain the heart once again. Extend the cut through the pulmonary trunk into the right ventricle. Cut down, around, and up through the atrioventricular valve to make the cut continuous with the cut across the right atrium (Figure 30.9).

11. Reflect the cut edges of the superior vena cava, right atrium, and right ventricle to obtain the view seen in Figure 30.9. Observe the comblike ridges of muscle throughout most of the right atrium. This is called **pectinate muscle** (*pectin* means "comb"). Identify, on the ventral atrial wall, the large opening of the inferior vena cava and follow it to its external opening with a probe. Notice that the atrial walls in the vicinity of the venae cavae are smooth and lack the roughened appearance (pectinate musculature) of the other regions of the atrial walls. Just below the inferior vena caval opening, identify the opening of the **coronary sinus,** which returns venous blood of the coronary circulation to the right atrium. Nearby, locate an oval depression, the **fossa ovalis,** in the interatrial septum. This depression marks the site of an opening in the fetal heart, the **foramen ovale,** which allows blood to pass from the right to the left atrium, thus bypassing the fetal lungs.

12. Identify the papillary muscles in the right ventricle, and follow their attached chordae tendineae to the flaps of the tricuspid valve. Notice the pitted and ridged appearance (**trabeculae carneae**) of the inner ventricular muscle.

13. Identify the **moderator band** (septomarginal band), a bundle of cardiac muscle fibers connecting the interventricular septum to anterior papillary muscles. It contains a branch of the atrioventricular bundle and helps coordinate contraction of the ventricle.

14. Make a longitudinal incision through the left atrium and continue it into the left ventricle. Notice how much thicker the myocardium of the left ventricle is than that of the right ventricle. Compare the *shape* of the left ventricular cavity to the shape of the right ventricular cavity. (See Figure 30.10.)

___

___

___

Are the papillary muscles and chordae tendineae observed in the right ventricle also present in the left ventricle?

___

Count the number of cusps in the left atrioventricular valve. How does this compare with the number seen in the right atrioventricular valve?

___

___

How do the sheep valves compare with their human counterparts?

___

15. Reflect the cut edges of the atrial wall, and attempt to locate the entry points of the pulmonary veins into the left atrium. Follow the pulmonary veins, if present, to the heart exterior with a probe. Notice how thin-walled these vessels are.

16. Properly dispose of the organic debris, clean the dissecting tray and instruments, and wash the lab bench before leaving the laboratory. ■

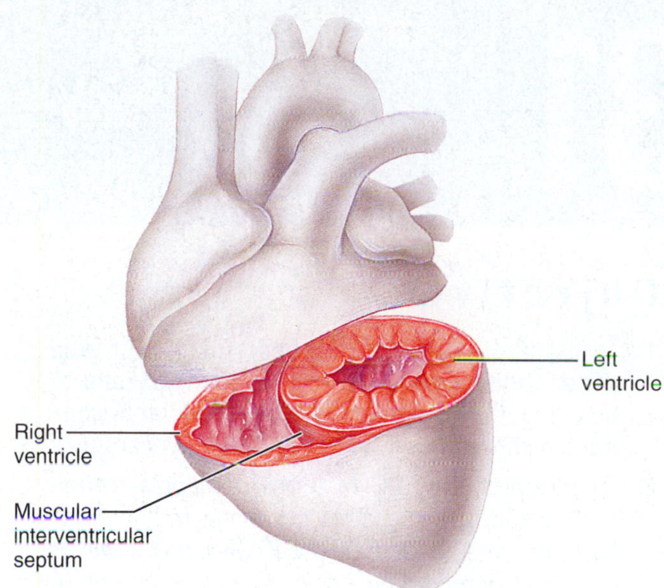

Figure 30.10 Anatomical differences in the right and left ventricles. The left ventricle has thicker walls, and its cavity is basically circular; by contrast, the right ventricle cavity is crescent-shaped and wraps around the left ventricle.

exercise

31 Conduction System of the Heart and Electrocardiography

Objectives

1. To list and localize the elements of the intrinsic conduction, or nodal, system of the heart; and to describe how impulses are initiated and conducted through this system and the myocardium.
2. To interpret the ECG in terms of depolarization and repolarization events occurring in the myocardium; and to identify the P, QRS, and T waves on an ECG recording using an ECG recorder or PowerLab®.
3. To calculate the heart rate, QRS interval, P-R interval, and Q-T interval from an ECG obtained during the laboratory period.
4. To define *tachycardia, bradycardia,* and *fibrillation.*

Materials

- ❑ Apparatus A or B
  A: ECG recording apparatus, electrode paste; alcohol swabs, cot
  B: PowerLab® unit and cable, computer with Chart software installed, ECG leads, electrodes, event marker

The Intrinsic Conduction System

Heart contraction results from a series of electrical potential changes (depolarization waves) that travel through the heart preliminary to each beat. Because cardiac muscle cells are electrically connected by gap junctions, the entire myocardium behaves like a single unit (a **functional syncytium**).

The ability of cardiac muscle to beat is intrinsic—it does not depend on impulses from the nervous system to initiate its contraction and will continue to contract rhythmically even if all nerve connections are severed. However, two types of controlling systems exert their effects on heart activity. One of these involves nerves of the autonomic nervous system, which accelerate or decrease the heartbeat rate depending on which division is activated. The second system is the **intrinsic conduction system,** or **nodal system,** of the heart, consisting of specialized noncontractile myocardial tissue. The intrinsic conduction system ensures that heart muscle depolarizes in an orderly and sequential manner (from atria to ventricles) and that the heart beats as a coordinated unit.

The components of the intrinsic conduction system include the **SA (sinoatrial) node,** located in the right atrium just inferior to the entrance to the superior vena cava; the **AV (atrioventricular) node** in the lower atrial septum at the junction of the atria and ventricles; the **AV bundle (bundle of His)** and right and left **bundle branches,** located in the interventricular septum; and the **Purkinje fibers,** which ramify within the muscle bundles of the ventricular walls. The Purkinje fiber network is much denser and more elaborate in the left ventricle because of the larger size of this chamber (Figure 31.1).

The SA node, which has the highest rate of discharge, provides the stimulus for contraction. Because it sets the rate of depolarization for the heart as a whole, the SA node is often referred to as the *pacemaker.* From the SA node, the impulse spreads throughout the atria and to the AV node. This electrical wave is immediately followed by atrial contraction. At the AV node, the impulse is momentarily delayed (approximately 0.1 sec), allowing the atria to complete their contraction. It then passes through the AV bundle, the right and left bundle branches, and the Purkinje fibers, finally resulting in ventricular contraction. Note that the atria and ventricles are separated from one another by a region of electrically inert connective tissue, so the depolarization wave can be transmitted to the ventricles only via the tract between the AV node and AV bundle. Thus, any damage to the AV node-bundle pathway partially or totally insulates the ventricles from the influence of the SA node. Although autorhythmic cells are found throughout the heart, their rates of spontaneous depolarization differ. The nodal system increases the rate of heart depolarization and synchronizes heart activity.

Electrocardiography

The conduction of impulses through the heart generates electrical currents that eventually spread throughout the body. These impulses can be detected on the body's surface and recorded with an instrument called an *electrocardiograph.* The graphic recording of the electrical changes (depolarization followed by repolarization) occurring during the cardiac cycle is called an **electrocardiogram (ECG)** (Figure 31.2). The typical ECG consists of a series of three recognizable waves called *deflection waves.* The first wave, the **P wave,** is a small wave that indicates depolarization of the atria immediately before atrial contraction. The large **QRS complex,** resulting from ventricular depolarization, has a complicated shape (primarily because of the variability in size of the two

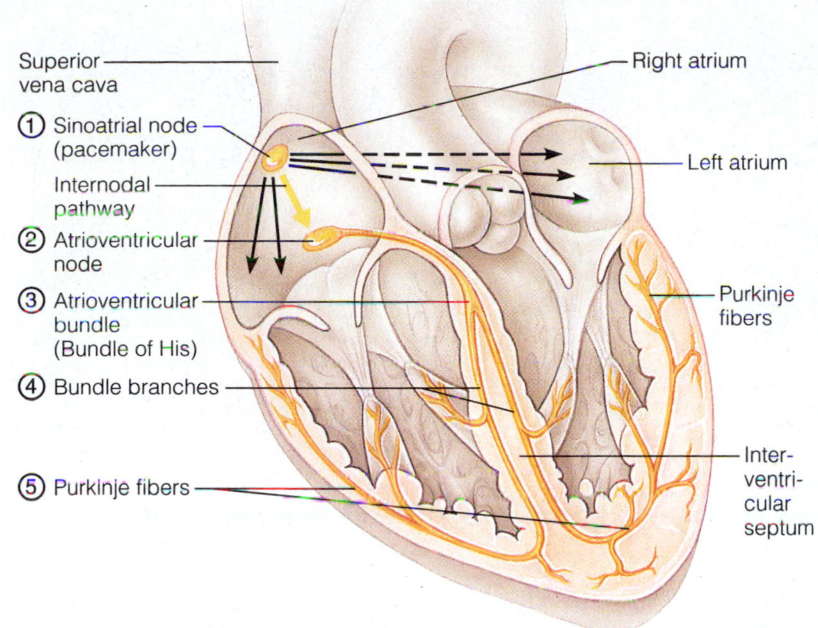

Figure 31.1 The intrinsic conduction system of the heart. Dashed-line arrows indicate transmission of the impulse from the SA node through the atria. Solid arrows indicate transmission of the impulse from the SA node to the AV node via the internodal pathway.

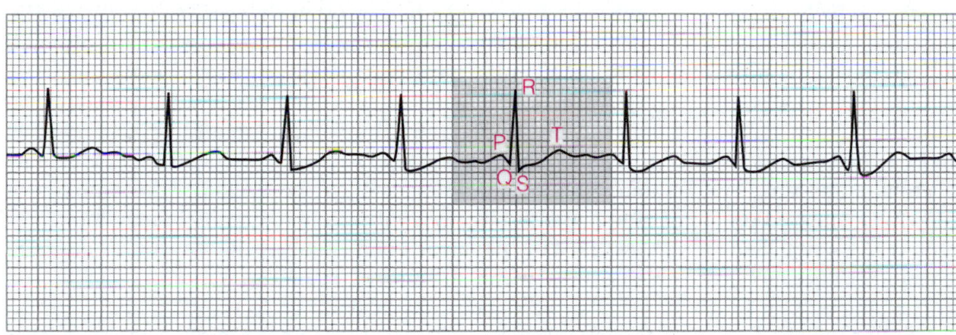

(a)

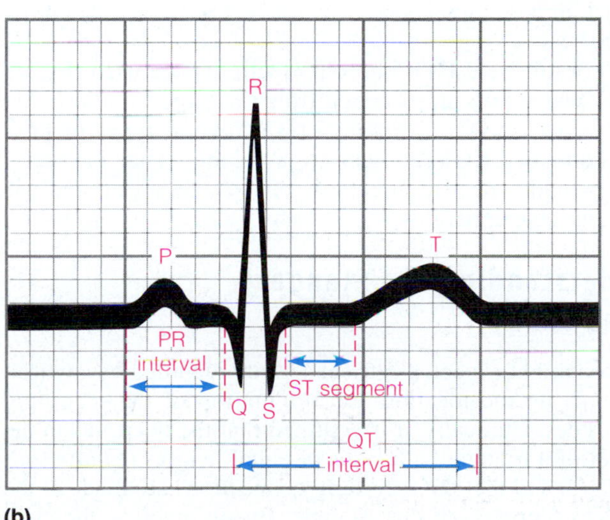

(b)

Time: small squares = 0.04 sec
1 large square = 0.20 sec
5 large squares = 1.00 sec

Figure 31.2 The normal electrocardiogram. **(a)** Regular sinus rhythm. **(b)** Waves, segments, and intervals of a normal ECG.

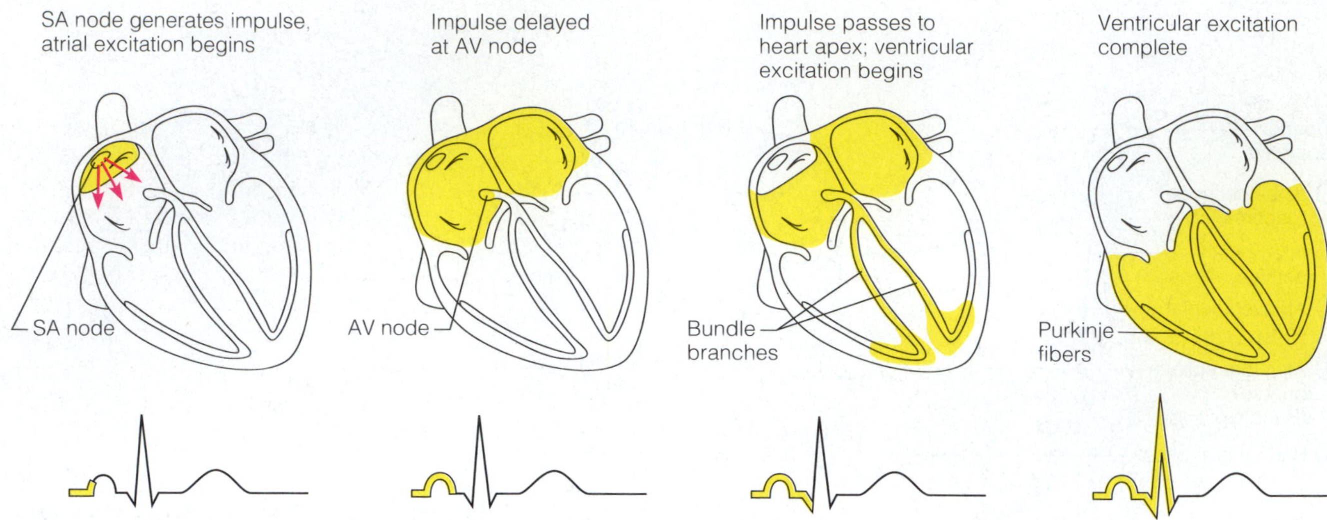
SA node generates impulse, atrial excitation begins
Impulse delayed at AV node
Impulse passes to heart apex; ventricular excitation begins
Ventricular excitation complete
SA node
AV node
Bundle branches
Purkinje fibers

Figure 31.3 The sequence of excitation of the heart related to the deflection waves of an ECG tracing.

ventricles and the time differences required for these chambers to depolarize). It precedes ventricular contraction. The **T wave** results from currents propagated during ventricular repolarization. The repolarization of the atria, which occurs during the QRS interval, is generally obscured by the large QRS complex. The relationship between the deflection waves of an ECG and sequential excitation of the heart is shown in Figure 31.3.

It is important to understand what an ECG does and does not show: First, an ECG is a record of voltage and time—nothing else. Although we can and do infer that muscle contraction follows its excitation, sometimes it does not. Secondly, an ECG records electrical events occurring in relatively large amounts of muscle tissue (i.e., the bulk of the heart muscle), *not* the activity of nodal tissue which, like muscle contraction, can only be inferred. Nonetheless, abnormalities of the deflection waves and changes in the time intervals of the ECG are useful in detecting myocardial infarcts or problems with the conduction system of the heart. The P-R (P-Q) interval represents the time between the beginning of atrial depolarization and ventricular depolarization. Thus, it typically includes the period during which the depolarization wave passes to the AV node, atrial systole, and the passage of the excitation wave to the balance of the conducting system. Generally, the P-R interval is about 0.16 to 0.18 sec. A longer interval may suggest a partial AV heart block caused by damage to the AV node. In total heart block, no impulses are transmitted through the AV node, and the atria and ventricles beat independently of one another—the atria at the SA node rate and the ventricles at their intrinsic rate, which is considerably slower.

If the QRS interval (normally 0.08 sec) is prolonged, it may indicate a right or left bundle branch block in which one ventricle is contracting later than the other. The Q-T interval is the period from the beginning of ventricular depolarization through repolarization and includes the time of ventricular contraction (the S-T segment). With a heart rate of 70 beats/min, this interval is normally 0.31 to 0.41 sec. As the rate increases, this interval becomes shorter; conversely, when the heart rate drops, the interval is longer.

A heart rate over 100 beats/min is referred to as **tachycardia;** a rate below 60 beats/min is **bradycardia.** Although neither condition is pathological, prolonged tachycardia may progress to **fibrillation,** a condition of rapid uncoordinated heart contractions which makes the heart useless as a pump. Bradycardia in athletes is a positive finding; that is, it indicates an increased efficiency of cardiac functioning. Because *stroke volume* (the amount of blood ejected by a ventricle with each contraction) increases with physical conditioning, the heart can contract more slowly and still meet circulatory demands.

Twelve standard leads are used to record an ECG for diagnostic purposes. Three of these are bipolar leads that measure the voltage difference between the arms, or an arm and a leg, and nine are unipolar leads. Together the 12 leads provide a fairly comprehensive picture of the electrical activity of the heart.

For this investigation, four electrodes are used (Figure 31.4), and results are obtained from the three *standard limb leads* (also shown in Figure 31.4). Several types of physiographs or ECG recorders are available. Your instructor will provide specific directions on how to set up and use the available apparatus.

Understanding the Standard Limb Leads

As you might expect, electrical activity recorded by any lead depends on the location and orientation of the recording electrodes. Clinically, it is assumed that the heart lies in the center of a triangle with sides of equal lengths (*Einthoven's triangle*) and that the recording connections are made at the vertices (corners) of that triangle. But in practice, the electrodes connected to each arm and to the left leg are considered to connect to the triangle vertices. The standard limb

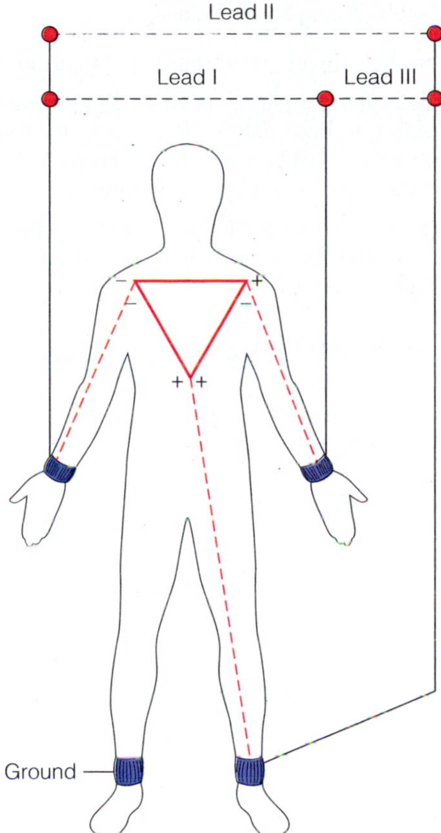

Figure 31.4 ECG recording positions for the standard limb leads.

leads record the voltages generated in the extracellular fluids surrounding the heart by the ion flows occurring simultaneously in many cells between any two of the connections. A recording using lead I (RA-LA), which connects the right arm (RA) and the left arm (LA), is most sensitive to electrical activity spreading horizontally across the heart. Lead II (RA-LL) and lead III (LA-LL) record activity along the vertical axis (from the base of the heart to its apex), but from different orientations. The significance of Einthoven's triangle is that the sum of the voltages of leads I and III equals that in lead II (Einthoven's law). Hence, if the voltages of two of the standard leads are recorded, that of the third lead can be determined mathematically.

Activity: Recording ECGs

Once the subject is prepared, the ECG will be recorded first under baseline (resting) conditions and then under conditions of fairly strenuous activity. Finally, recordings will be made while the subject holds his or her breath. The activity and breath-holding recordings will be compared to the baseline recordings, and you will be asked to determine the reasons for the observed differences in the recordings.

Using a Standard ECG Apparatus

Preparing the Subject

1. Place electrode paste on four electrode plates and position each electrode as follows after scrubbing the skin at the attachment site with an alcohol swab. Attach an electrode to the anterior surface of each forearm, about 5–8 cm or 2 to 3 in. above the wrist, and secure them with rubber straps. In the same manner, attach an electrode to each leg, approximately 2 to 3 in. above the medial malleolus (inner aspect of the ankle).

2. Attach the appropriate tips of the patient cable to the electrodes. The cable leads are marked RA (right arm), LA (left arm), LL (left leg), and RL (right leg, the ground).

Making a Baseline Recording

1. Position the subject comfortably in a supine position on a cot (if available), or sitting relaxed on a laboratory chair.

2. Turn on the power switch and adjust the sensitivity knob to 1. Set the paper speed to 25 mm/sec and the lead selector to the position corresponding to recording from lead I (RA-LA).

3. Set the control knob at the RUN position and record the subject's at-rest ECG from lead I for 2 to 3 minutes or until the recording stabilizes. (You will need a tracing long enough to provide each student with a representative segment.) The subject should try to relax and not move unnecessarily, because the skeletal muscle action potentials will also be picked up and recorded.

4. Stop the recording and mark it "lead I."

5. Repeat the recording procedure for leads II (RA-LL) and III (LA-LL).

6. Each student should take a representative segment of one of the lead recordings and label the record with the name of the subject and the lead used. Identify and label the P, QRS, and T waves. The calculations you perform for your recording should be based on the following information: Because the paper speed was 25 mm/sec, each millimeter of paper corresponds to a time interval of 0.04 sec. Thus, if an interval requires 4 mm of paper, its duration is 4 mm × 0.04 sec/mm = 0.16 sec.

7. Compute the heart rate. Measure the distance (mm) from the beginning of one QRS complex to the beginning of the next QRS complex, and plug this value into the following equation to find the time for one heartbeat.

_______ mm/beat × 0.04 sec/mm = _______ sec/beat

Now find the beats per minute, or heart rate, by using the figure just computed for seconds per beat in the following equation:

$$\text{Beats/min} = \frac{1}{\_\_\_\_\_\_\_\ \text{sec/beat}} \times 60\ \text{sec/min}$$

Beats/min = _______

Is the obtained value within normal limits? ____________

Measure the QRS interval and compute its duration.

____

Measure the Q-T interval and compute its duration.

____

Measure the P-R interval and compute its duration.

____

Are the computed values within normal limits?

____

8. At the bottom of this page, attach segments of the ECG recordings from leads I through III. Make sure you indicate the paper speed, lead, and subject's name on each tracing. To the recording on which you based your previous computations, add your calculations for the duration of the QRS, P-R intervals, and Q-T intervals above the respective area of tracing. Also record the heart rate on that tracing.

Recording the ECG while Running in Place

1. Make sure the electrodes are securely attached to prevent electrode movement while recording the ECG.
2. Set the paper speed to 25 mm/sec, and prepare to make the recording using lead I.
3. Record the ECG while the subject is running in place for 3 min. Then have the subject sit down, but continue to record the ECG for an additional 4 minutes. *Mark the recording* at the end of the 3 minutes of running and at 1 minute after cessation of activity.
4. Stop the recording. Compute the beats/min during the third minute of running, at 1 minute after exercise, and at 4 minutes after exercise. Record below:

_______ beats/min while running in place

_______ beats/min at 1 minute after exercise

_______ beats/min at 4 minutes after exercise

5. Compare this recording with the previous recording from lead I. Which intervals are shorter in the "running" recording?

____

Does the subject's heart rate return to resting level by 4 minutes after exercise?

____

Recording the ECG during Breath Holding

1. Position the subject comfortably in the sitting position.
2. Using lead I and a paper speed of 25 mm/sec, *begin* the recording. After approximately 10 seconds, instruct the subject to begin breath holding and mark the record to indicate the onset of the 1-min breath-holding interval.
3. Stop the recording after 1 minute and remind the subject to breathe. Compute the beats/minute during the 1-minute experimental (breath-holding) period.

Beats/min during breath holding ____________________

4. Compare this recording with the lead I recording obtained under resting conditions.

What differences are seen? ____________________

____

Attempt to *explain* the physiological reason for the differences you have seen. (Hint: a good place to start might be to check hypoventilation or the role of the *respiratory* system in acid-base balance of the blood.)

____

____

____

____ ■

Using PowerLab® to Record the ECG

Making a Baseline Recording

1. Make sure the computer is off, and connect the PowerLab® unit. Start the computer.
2. Plug the ECG cable into the Dual Bio port on the PowerLab® unit (Figure 31.5).
3. Ask a volunteer to remove all jewelry from the wrists and ankles and to sit quietly with hands in lap in front of the computer.
4. Attach the + electrode to the left wrist and the − electrode to the right wrist. Attach the ground electrode to one ankle (Figure 31.5).
5. Open the PowerLab® folder and select the file named Experiment 2—ECG. The Chart program will open. CHANNEL 2 is labeled *Event* and CHANNEL 3 is labeled *ECG* (Figure 31.6).
6. Click on the CHANNEL 3 (ECG) pop-up menu, select *Input Amplifier*, and examine the ECG (Figure 31.6).

If the tracing is upside down, click on *Invert*.

If the tracing is too small, adjust *Range* in the Input Amplifier.

Click on *OK* to return to the Chart window.

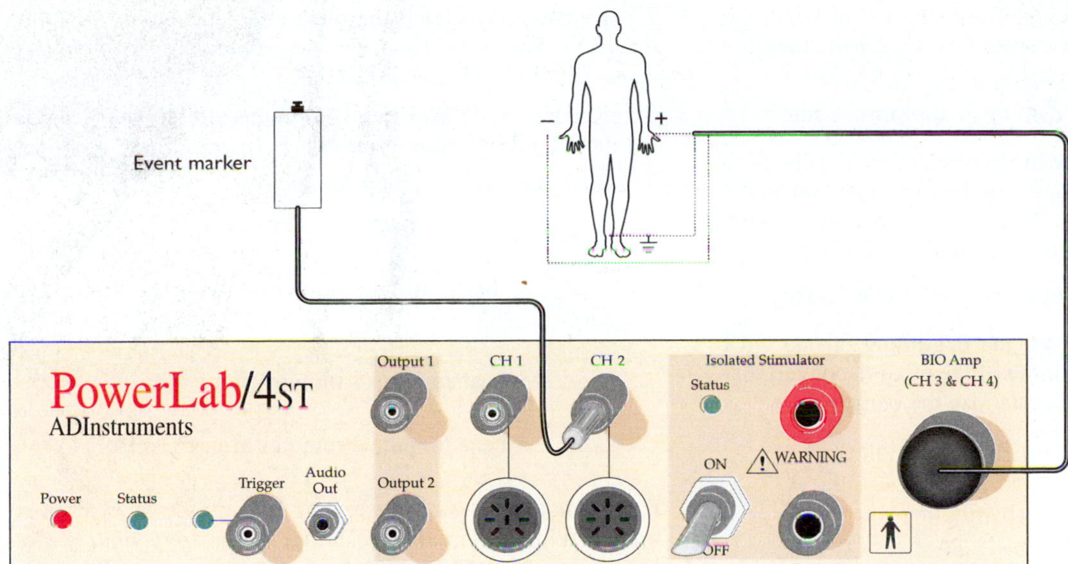

Figure 31.5 Equipment used to record the ECG from a volunteer.

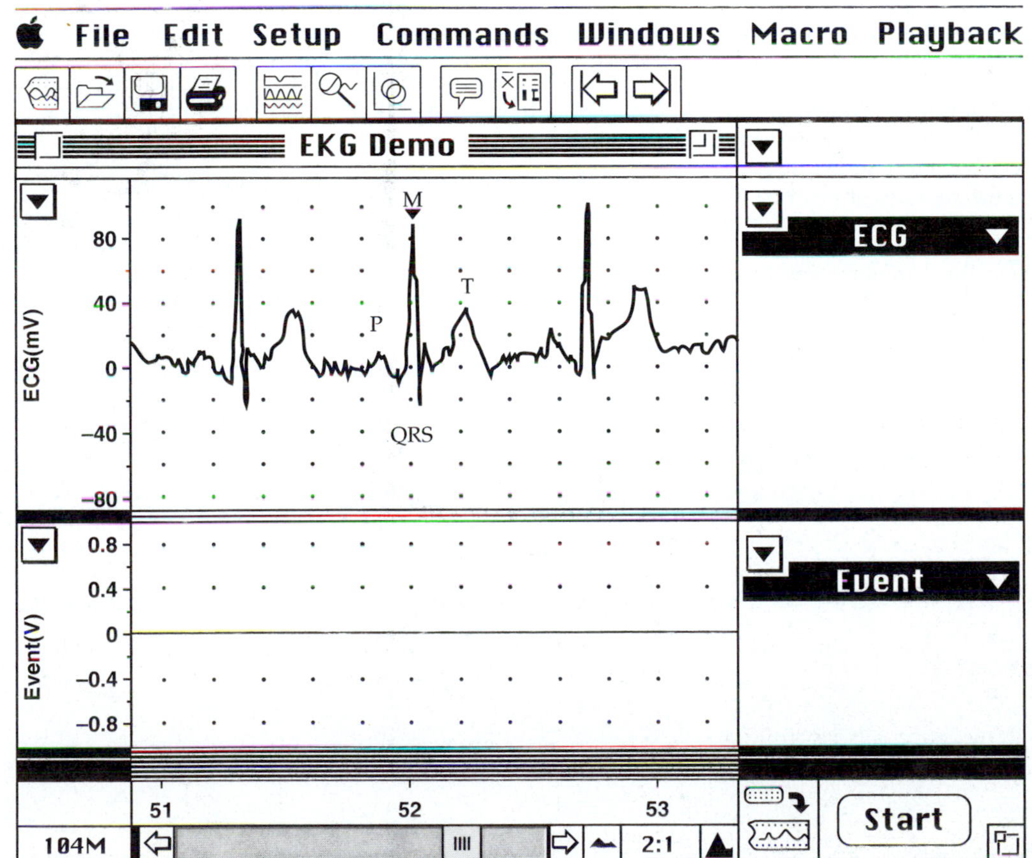

Figure 31.6 A Chart recording of an ECG trace. Shows the P, QRS, and T waves; the marker and cursor are positioned to measure the amplitude (in volts) of the QRS wave.

7. Click *Start* to begin tracing. Label the tracing with the subject's name and press the Return key. *Stop* halts the tracing.

8. Select *File* and save the data to an appropriate file.

Analyzing the Data Note: To use the marker, move the cursor to the marker box, click and drag the marker to a location on the ECG, and release the mouse. As the cursor is moved, measurements relative to the position of the marker are given.

1. Use the marker and cursor to measure the following:

   a. The time interval from the beginning of one QRS complex to the beginning of the next; to convert that measurement to heart rate, use the equation

   Heart rate (beats/minute) = 60/time intervals

   Heart rate: ______ beats/minute

   b. The QRS interval QRS ______

   c. The Q-T interval Q-T ______

   d. The P-R interval P-R ______

2. Evaluate the data. Are all the measurements within normal range or limits?

Recording the ECG after Running in Place Follow the directions on p. 314, Recording the ECG while Running in Place, using the PowerLab® setup. Secure the lead wires to the subject with the clips provided to increase the stability of the recording. If the recording is unstable:

1. Ask the subject to run in place for 3 minutes and then have the subject sit down.

2. Immediately attach the electrodes and begin recording the ECG. Label the first recording *Immediately after exercise*.

3. Record the ECG at 1 and 4 minutes after exercise and label the recordings appropriately.

4. Save the data to an appropriate file.

5. Calculate and compare the subject's heart rates.

   ______ beats/minute immediately after running in place

   ______ beats/minute 1 minute after exercise

   ______ beats/minute 4 minutes after exercise

6. Compare the subject's resting ECG to the ECG after exercise. Which intervals are shorter in the *Immediately after running* recording?

____

Does the subject's heart rate return to resting level by 4 minutes after exercise? ______

Recording the ECG for Breath Holding

1. Use the setup for recording a resting ECG, and follow the instructions on p. 314 for Recording the ECG during Breath Holding.

2. Label and save the data to an appropriate file.

Anatomy of Blood Vessels

exercise

32

Objectives

1. To describe the tunics of blood vessel walls and state the function of each layer.
2. To correlate differences in artery, vein, and capillary structure with the functions of these vessels.
3. To recognize a cross-sectional view of an artery and vein when provided with a microscopic view or appropriate diagram.
4. To list and/or identify the major arteries arising from the aorta, and to indicate the body region supplied by each.
5. To list and/or identify the major veins draining into the superior and inferior venae cavae, and to indicate the body regions drained.
6. To point out and/or discuss the unique features of special circulations (hepatic portal system, circle of Willis, pulmonary circulation, fetal circulation) in the body.

Materials

- ❑ Anatomical charts of human arteries and veins (or a three-dimensional model of the human circulatory system)
- ❑ Anatomical charts of the following specialized circulations: pulmonary circulation, hepatic portal circulation, arterial supply and circle of Willis of the brain (or a brain model showing this circulation), fetal circulation
- ❑ Compound microscope
- ❑ Prepared microscope slides showing cross sections of an artery and vein
- ❑ *Human Cardiovascular System: The Blood Vessels* videotape*

For instructions on animal dissections, see the dissection exercises starting on p. 709 in the cat and fetal pig editions of this manual.

AIA See Appendix C, Exercise 32 for links to A.D.A.M.® Interactive Anatomy.

*Available to qualified adopters from Benjamin Cummings.

The blood vessels constitute a closed transport system. As the heart contracts, blood is propelled into the large arteries leaving the heart. It moves into successively smaller arteries and then to the arterioles, which feed the capillary beds in the tissues. Capillary beds are drained by the venules, which in turn empty into veins that ultimately converge on the great veins entering the heart.

Arteries, carrying blood away from the heart, and veins, which drain the tissues and return blood to the heart, function simply as conducting vessels or conduits. Only the tiny capillaries that connect the arterioles and venules and ramify throughout the tissues directly serve the needs of the body's cells. It is through the capillary walls that exchanges are made between tissue cells and blood. Respiratory gases, nutrients, and wastes move along diffusion gradients. Thus, oxygen and nutrients diffuse from the blood to the tissue cells, and carbon dioxide and metabolic wastes move from the cells to the blood.

In this exercise you will examine the microscopic structure of blood vessels and identify the major arteries and veins of the systemic circulation and other special circulations.

Microscopic Structure of the Blood Vessels

Except for the microscopic capillaries, the walls of blood vessels are constructed of three coats, or *tunics* (Figure 32.1). The **tunica interna,** or **intima,** which lines the lumen of a vessel, is a single thin layer of *endothelium* (squamous cells underlain by a scant basal lamina) that is continuous with the endocardium of the heart. Its cells fit closely together, forming an extremely smooth blood vessel lining that helps decrease resistance to blood flow.

The **tunica media** is the more bulky middle coat and is composed primarily of smooth muscle and elastin. The smooth muscle, under the control of the sympathetic nervous system, plays an active role in regulating the diameter of blood vessels, which in turn alters peripheral resistance and blood pressure.

The **tunica externa,** or **adventitia,** the outermost tunic, is composed of areolar or fibrous connective tissue. Its function is basically supportive and protective.

In general, the walls of arteries are thicker than those of veins. The tunica media in particular tends to be much heavier and contains substantially more smooth muscle and elastic tissue. This anatomical difference reflects a functional difference in the two types of vessels. Arteries, which are closer to the pumping action of the heart, must be able to expand as an increased volume of blood is propelled into them and then

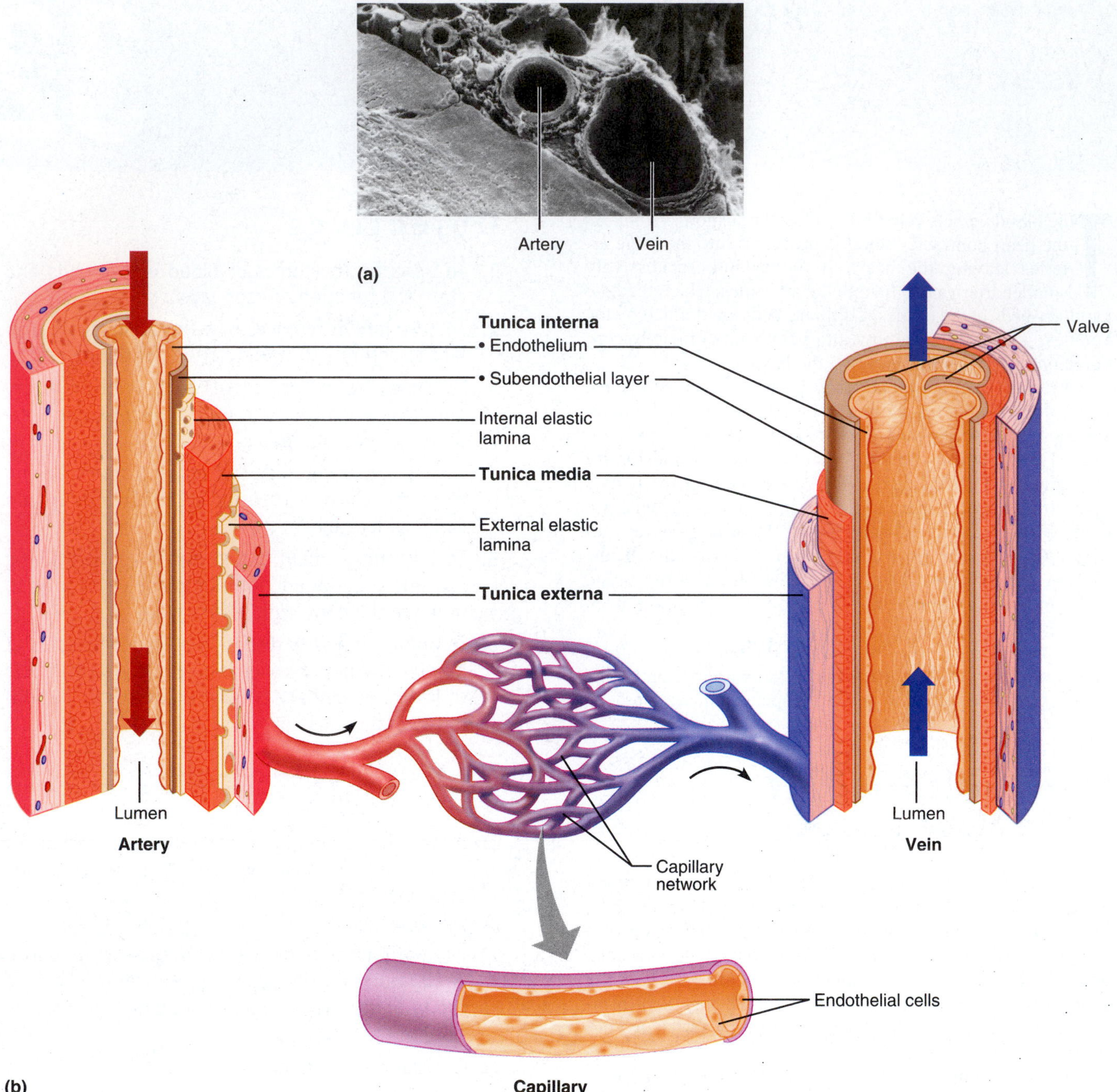

Figure 32.1 Generalized structure of arteries, veins, and capillaries. (a) Scanning electron micrograph of a muscular artery and the corresponding vein in cross section (120×). (From R. G. Kessel and R. H. Kardon, *Tissues and Organs.* © 1979 W. H. Freeman.) **(b)** The walls of the arteries and veins have three layers, the tunica interna, tunica media, and tunica externa. Capillaries have only endothelium and a sparse basal lamina.

recoil passively as the blood flows off into the circulation during diastole. Their walls must be sufficiently strong and resilient to withstand such pressure fluctuations. Since these larger arteries have such large amounts of elastic tissue in their media, they are often referred to as *elastic arteries.* Smaller arteries, further along in the circulatory pathway, are exposed to less extreme pressure fluctuations. They have less elastic tissue but still have substantial amounts of smooth muscle in their media. For this reason, they are called *muscular arteries.*

By contrast, veins, which are far removed from the heart in the circulatory pathway, are not subjected to such pressure fluctuations and are essentially low-pressure vessels. Thus, veins may be thinner-walled without jeopardy. However, the low-pressure condition itself and the fact that blood returning to the heart often flows against gravity require structural

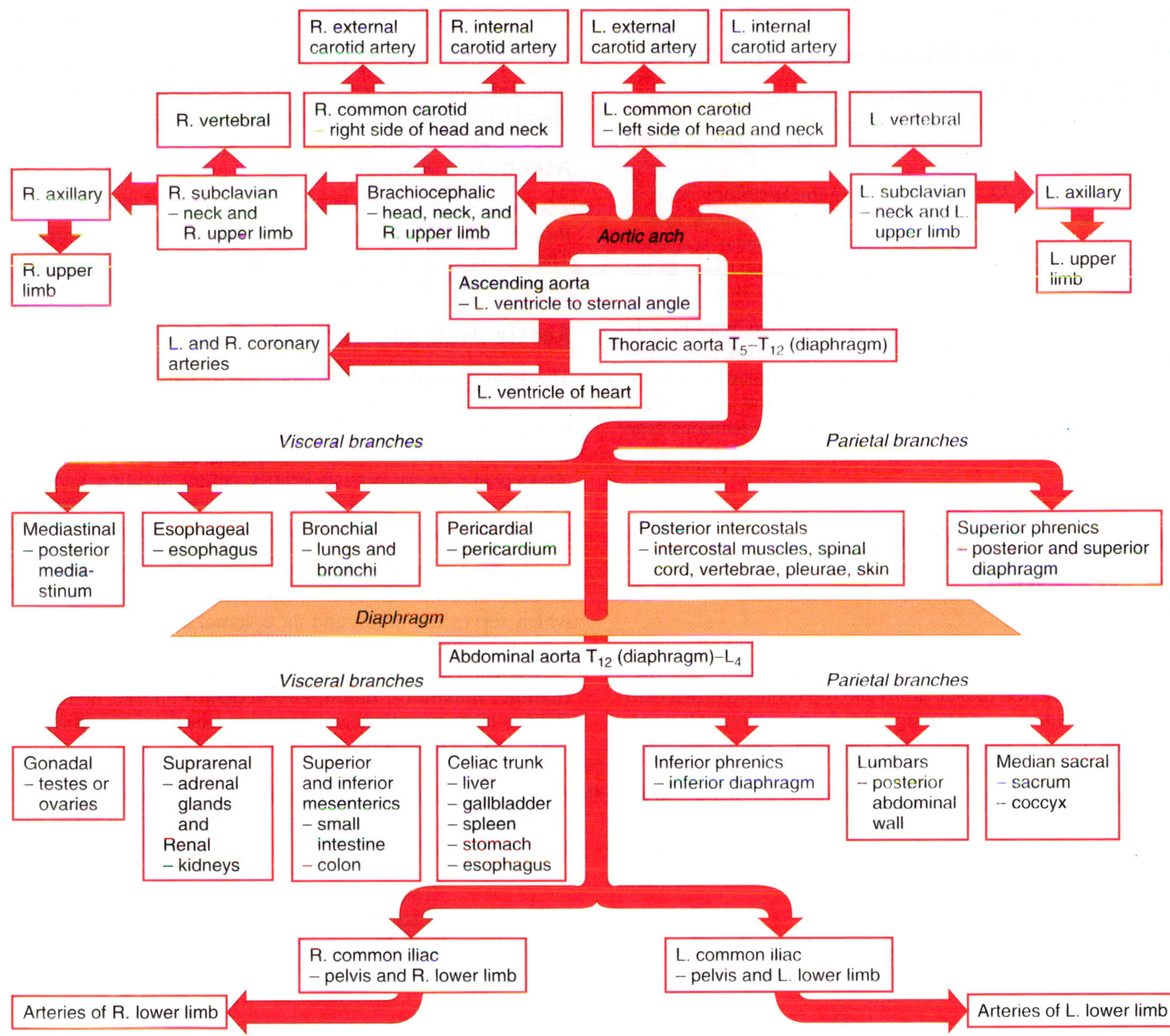

Figure 32.2 Schematic of the systemic arterial circulation.

modifications to ensure that venous return equals cardiac output. Thus, the lumens of veins tend to be substantially larger than those of corresponding arteries, and valves in larger veins act to prevent backflow of blood in much the same manner as the semilunar valves of the heart. The skeletal muscle "pump" also promotes venous return; as the skeletal muscles surrounding the veins contract and relax, the blood is milked through the veins toward the heart. (Anyone who has been standing relatively still for an extended time will be happy to show you their swollen ankles, caused by blood pooling in their feet during the period of muscle inactivity!) Pressure changes that occur in the thorax during breathing also aid the return of blood to the heart.

• To demonstrate how efficiently venous valves prevent backflow of blood, perform the following simple experiment. Allow one hand to hang by your side until the blood vessels on the dorsal aspect become distended. Place two fingertips against one of the distended veins and, pressing firmly, move the superior finger proximally along the vein and then release this finger. The vein will remain flattened and collapsed despite gravity. Then remove the distal fingertip and observe the rapid filling of the vein.

The transparent walls of the tiny capillaries are only one cell layer thick, consisting of just the endothelium underlain by a basal lamina, that is, the tunica intima. Because of this exceptional thinness, exchanges are easily made between the blood and tissue cells.

Fetal Circulation

In a developing fetus, the lungs and digestive system are not yet functional, and all nutrient, excretory, and gaseous exchanges occur through the placenta (see Figure 32.14a). Nutrients and oxygen move across placental barriers from the mother's blood into fetal blood, and carbon dioxide and other metabolic wastes move from the fetal blood supply to the mother's blood.

Fetal blood travels through the umbilical cord, which contains three blood vessels: two smaller umbilical arteries and one large umbilical vein. The **umbilical vein** carries blood rich in nutrients and oxygen to the fetus; the **umbilical arteries** carry carbon dioxide and waste-laden blood from the fetus to the placenta. The umbilical arteries, which transport blood away from the fetal heart, meet the umbilical vein at the *umbilicus* (navel, or belly button) and wrap around the vein within the cord en route to their placental attachments. Newly oxygenated blood flows in the umbilical vein superiorly toward the fetal heart. En route, some of this blood perfuses the liver, but the larger proportion is ducted through the relatively nonfunctional liver to the inferior vena cava via a shunt vessel called the **ductus venosus,** which carries the blood to the right atrium of the heart.

Because fetal lungs are nonfunctional and collapsed, two shunting mechanisms ensure that blood almost entirely bypasses the lungs. Much of the blood entering the right atrium is shunted into the left atrium through the **foramen ovale,** a flaplike opening in the interatrial septum. The left ventricle then pumps the blood out the aorta to the systemic circulation. Blood that does enter the right ventricle and is pumped out of the pulmonary trunk encounters a second shunt, the **ductus arteriosus,** a short vessel connecting the pulmonary trunk and the aorta. Because the collapsed lungs present an extremely high-resistance pathway, blood more readily enters the systemic circulation through the ductus arteriosus.

The aorta carries blood to the tissues of the body; this blood ultimately finds its way back to the placenta via the umbilical arteries. The only fetal vessel that carries highly oxygenated blood is the umbilical vein. All other vessels contain varying degrees of oxygenated and deoxygenated blood.

At birth, or shortly after, the foramen ovale closes and becomes the **fossa ovalis,** and the ductus arteriosus collapses and is converted to the fibrous **ligamentum arteriosum** (Figure 32.14b). Lack of blood flow through the umbilical vessels leads to their eventual obliteration, and the circulatory pattern becomes that of the adult. Remnants of the umbilical arteries persist as the **medial umbilical ligaments** on the inner surface of the anterior abdominal wall, of the umbilical vein as the **ligamentum teres** or **round ligament** of the liver, and of the ductus venosus as a fibrous band called the **ligamentum venosum** on the inferior surface of the liver.

Activity 6: Tracing the Pathway of Fetal Blood Flow

The pathway of fetal blood flow is indicated with arrows on Figure 32.14a. Appropriately *label* all specialized fetal circulatory structures provided with leader lines. ■

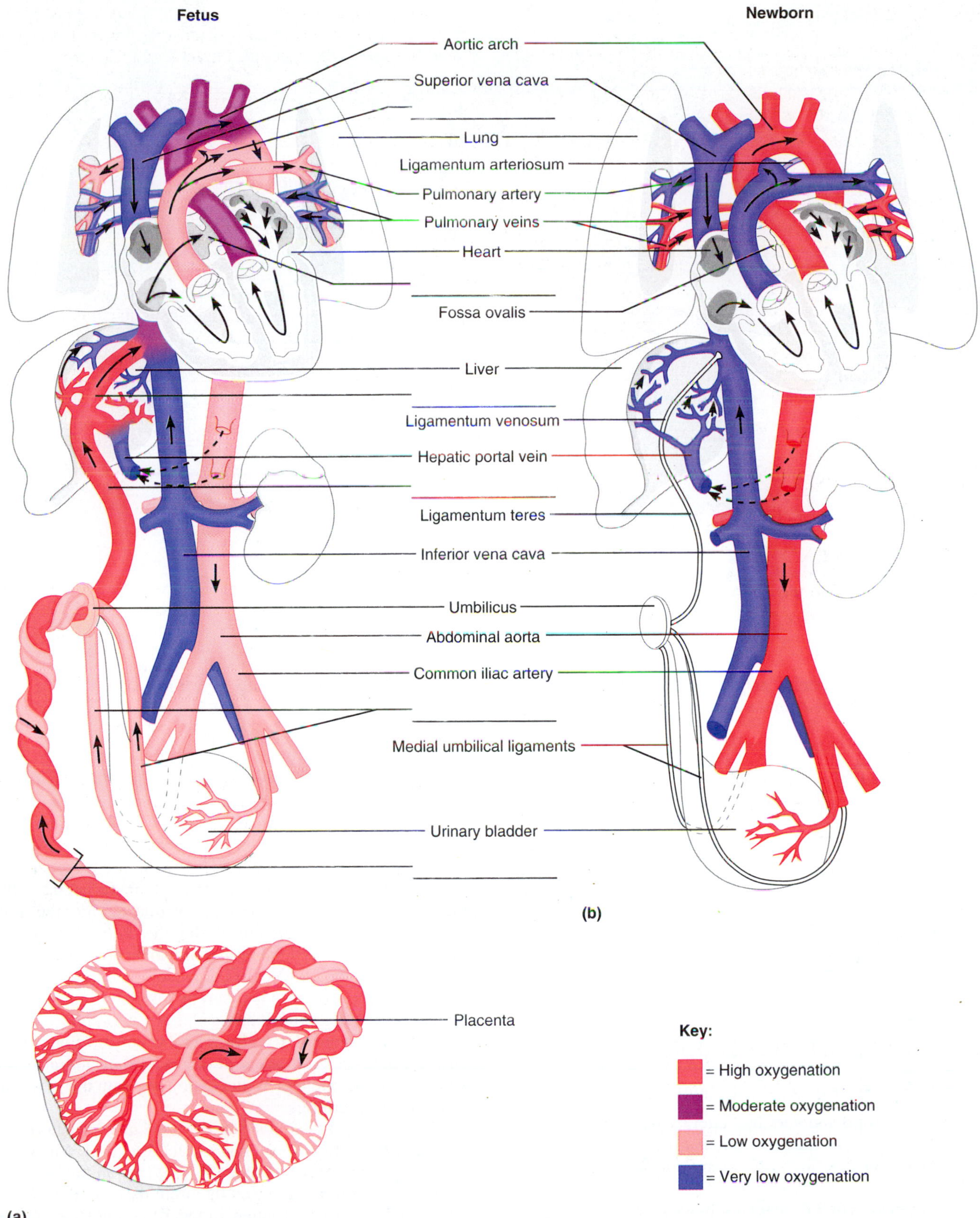

Figure 32.14 Circulation in fetus and newborn. Arrows indicate direction of blood flow. **(a)** Special adaptations for embryonic and fetal life. The umbilical vein carries oxygen- and nutrient-rich blood from the placenta to the fetus. The umbilical arteries carry waste-laden blood from the fetus to the placenta; the ductus arteriosus and foramen ovale bypass the nonfunctional lungs; and the ductus venosus allows blood to partially bypass the liver. **(b)** Changes in the cardiovascular system at birth. The umbilical vessels are occluded, as are the liver and lung bypasses (ductus venosus and arteriosus, and the foramen ovale).

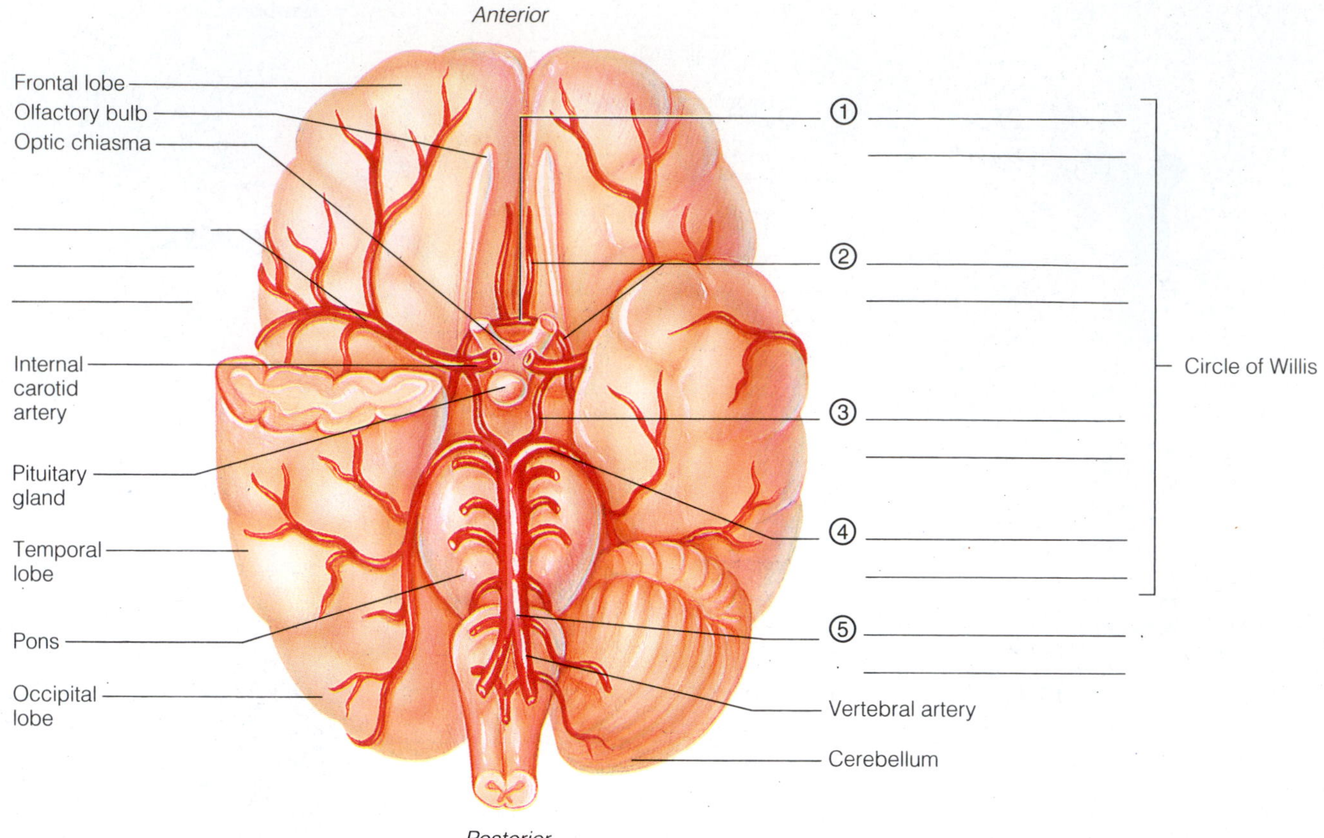

Figure 32.15 Arterial supply of the brain. Cerebellum is not shown on the left side of the figure. (See also Plate F in the Human Anatomy Atlas.)

Arterial Supply of the Brain and the Circle of Willis

A continuous blood supply to the brain is crucial because oxygen deprivation for even a few minutes causes irreparable damage to the delicate brain tissue. The brain is supplied by two pairs of arteries arising from the region of the aortic arch—the *internal carotid arteries* and the *vertebral arteries*. Figure 32.15 is a diagram of the brain's arterial supply.

Activity 7: Tracing the Arterial Supply of the Brain

The internal carotid and vertebral arteries are labeled in Figure 32.15. As you read the description of the blood supply below, *complete the labeling of this diagram.* ■

The **internal carotid arteries,** branches of the common carotid arteries, follow a deep course through the neck and along the pharynx, entering the skull through the carotid canals of the temporal bone. Within the cranium, each divides into **anterior** and **middle cerebral arteries,** which supply the bulk of the cerebrum. The internal carotid arteries also contribute to the formation of the **circle of Willis,** an arterial anastomosis at the base of the brain surrounding the pituitary gland and the optic chiasma, by contributing to a **posterior communicating artery** on each side. The circle is completed by the **anterior communicating artery,** a short shunt connecting the right and left anterior cerebral arteries.

The paired **vertebral arteries** diverge from the subclavian arteries and pass superiorly through the foramina of the transverse process of the cervical vertebrae to enter the skull through the foramen magnum. Within the skull, the vertebral arteries unite to form a single **basilar artery,** which continues superiorly along the ventral aspect of the brain stem, giving off branches to the pons, cerebellum, and inner ear. At the base of the cerebrum, the basilar artery divides to form the **posterior cerebral arteries.** These supply portions of the temporal and occipital lobes of the cerebrum and also become part of the circle of Willis by joining with the posterior communicating arteries.

The uniting of the blood supply of the internal carotid arteries and the vertebral arteries via the circle of Willis is a protective device that theoretically provides an alternate set of pathways for blood to reach the brain tissue in the case of arterial occlusion or impaired blood flow anywhere in the system. In actuality, the communicating arteries are tiny, and in many cases the communicating system is defective.

For instructions on animal dissections, see the dissection exercises starting on p. 709 in the cat and pig editions of this manual.

Human Cardiovascular Physiology—Blood Pressure and Pulse Determinations

exercise **33A**

Objectives

1. To define *systole, diastole*, and *cardiac cycle.*
2. To indicate the normal length of the cardiac cycle, the relative pressure changes occurring within the atria and ventricles during the cycle, and the timing of valve closure.
3. To use the stethoscope to auscultate heart sounds and to relate heart sounds to cardiac cycle events.
4. To describe the clinical significance of heart sounds and heart murmurs.
5. To demonstrate the thoracic locations where the first and second heart sounds are most accurately auscultated.
6. To define *pulse, pulse deficit, blood pressure*, and *sounds of Korotkoff.*
7. To accurately determine a subject's apical and radial pulse.
8. To accurately determine a subject's blood pressure with a sphygmomanometer, and to relate systolic and diastolic pressures to events of the cardiac cycle.
9. To investigate the effects of exercise on blood pressure, pulse, and cardiovascular fitness.
10. To note factors affecting and/or determining blood flow and skin color.

Materials

- ❑ Stethoscope
- ❑ Sphygmomanometer
- ❑ Watch (or clock) with a second hand
- ❑ Step stools (0.4 m or 16 in. and .05 m or 20 in. in height)
- ❑ Cot (if available)
- ❑ Alcohol swabs
- ❑ Millimeter ruler
- ❑ Ice
- ❑ Small basin suitable for the immersion of one hand
- ❑ Laboratory thermometer (°C)
- ❑ Record or audiotape: "Interpreting Heart Sounds" (available on free loan from the local chapters of the American Heart Association)
- ❑ Phonograph or tape deck
- ❑ Felt marker
- ❑ PowerLab® unit with cable, computer with Chart software installed, plethysmograph, plethysmograph cable or piezoelectric pulse transducer with cable

PhysioEx 4.0™ Computer Simulation on p. P-49

A1A See Appendix C, Exercise 33 for links to A.D.A.M.® Interactive Anatomy.

Any comprehensive study of human cardiovascular physiology takes much more time than a single laboratory period. However, it is possible to conduct investigations of a few phenomena such as pulse, heart sounds, and blood pressure, all of which reflect the heart in action and the function of blood vessels. (The electrocardiogram is studied separately in Exercise 31.) A discussion of the cardiac cycle will provide a basis for understanding and interpreting the various physiological measurements taken.

Cardiac Cycle

In a healthy heart, the two atria contract simultaneously. As they begin to relax, simultaneous contraction of the ventricles occurs. However, according to general usage, the terms **systole** and **diastole** refer to events of ventricular contraction and relaxation, respectively. The **cardiac cycle** is equivalent to one complete heartbeat—during which both atria and ventricles contract and then relax. It is marked by a succession of changes in blood volume and pressure within the heart.

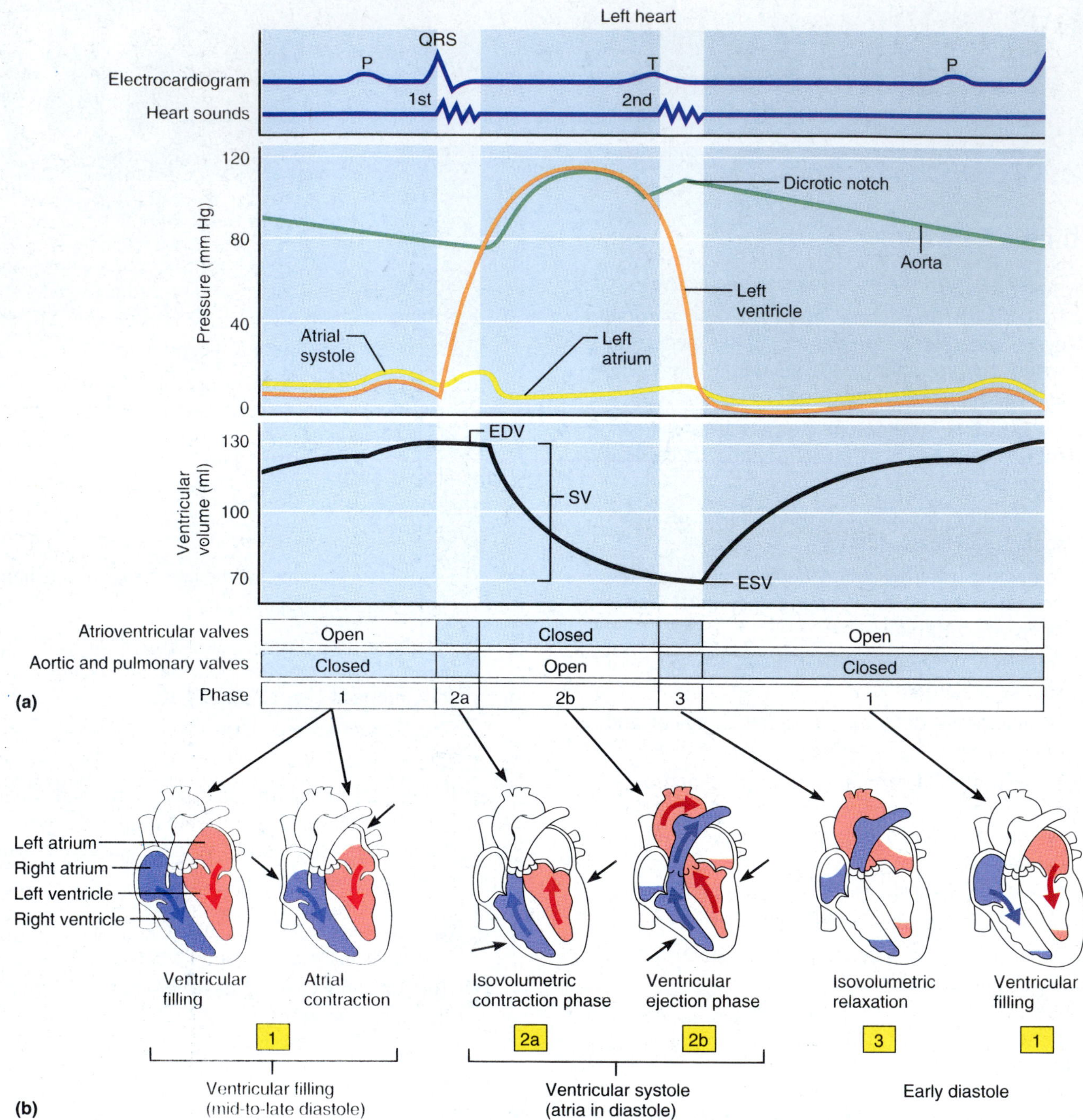

Figure 33A.1 Summary of events occurring in the heart during the cardiac cycle. **(a)** Events in the left side of the heart. An ECG tracing is superimposed on the graph (top) so that pressure and volume changes can be related to electrical events occurring at any point. Time occurrence of heart sounds is also indicated. **(b)** Events of phases 1 through 3 of the cardiac cycle are depicted in diagrammatic views of the heart.

Figure 33A.1 is a graphic representation of the events of the cardiac cycle for the left side of the heart. Although pressure changes in the right side are less dramatic than those in the left, the same relationships apply.

We will begin the discussion of the cardiac cycle with the heart in complete relaxation (diastole). At this point, pressure in the heart is very low, blood is flowing passively from the pulmonary and systemic circulations into the atria and on through to the ventricles; the semilunar valves are closed, and the AV valves are open. Shortly, atrial contraction occurs and atrial pressure increases, forcing residual blood into the ventricles. Then ventricular systole begins and intraventricular pressure increases rapidly, closing the AV valves. When ventricular pressure exceeds that of the large arteries leaving the heart, the semilunar valves are forced open; and the blood in the ventricular chambers is expelled through the valves. During this phase, the aortic pressure reaches approximately 120 mm Hg. During ventricular systole, the atria relax and their chambers fill with blood, which results in gradually increasing atrial pressure. At the end of ventricular systole, the ventricles relax; the semilunar valves snap shut, preventing backflow, and momentarily, the ventricles are closed chambers. When the aortic semilunar valve snaps shut, a momentary increase in the aortic pressure results from the elastic recoil of the aorta after valve closure. This event results in the pressure fluctuation called the *dicrotic notch* (see Figure 33A.1a). As the ventricles relax, the pressure within them begins to drop. When intraventricular pressure is again less than atrial pressure, the AV valves are forced open, and the ventricles again begin to fill with blood. Atrial and aortic pressures decrease, and the ventricles rapidly refill, completing the cycle.

The average heart beats approximately 75 beats per minute, and so the length of the cardiac cycle is about 0.8 second. Of this time period, atrial contraction occupies the first 0.1 second, which is followed by atrial relaxation and ventricular contraction for the next 0.3 second. The remaining 0.4 second is the quiescent, or ventricular relaxation, period. When the heart beats at a more rapid pace than normal, this last period decreases.

Notice that two different types of phenomena control the movement of blood through the heart: the alternate contraction and relaxation of the myocardium, and the opening and closing of valves (which is entirely dependent on the pressure changes within the heart chambers).

Study Figure 33A.1 carefully to make sure you understand what has been discussed before continuing with the next portion of the exercise.

Heart Sounds

Two distinct sounds can be heard during each cardiac cycle. These heart sounds are commonly described by the monosyllables "lub" and "dup"; and the sequence is designated lub-dup, pause, lub-dup, pause, and so on. The first heart sound (lub) is associated with closure of the AV valves at the beginning of systole. The second heart sound (dup) occurs as the semilunar valves close and corresponds with the end of systole. Figure 33A.1a indicates the timing of heart sounds in the cardiac cycle.

- Listen to the recording "Interpreting Heart Sounds" so that you may hear both normal and abnormal heart sounds.

Abnormal heart sounds are called **murmurs** and often indicate valvular problems. In valves that do not close tightly, closure is followed by a swishing sound due to the backflow of blood (regurgitation). Distinct sounds, often described as high-pitch "screeching," are associated with the tortuous flow of blood through constricted, or stenosed, valves. ■

Activity 1:
Auscultating Heart Sounds

In the following procedure, you will auscultate your partner's heart sounds with an ordinary stethoscope. A number of more sophisticated heart-sound amplification systems are on the market, and your instructor may prefer to use one such if it is available. If so, directions for the use of this apparatus will be provided by the instructor.

1. Obtain a stethoscope and some alcohol swabs. Heart sounds are best auscultated (listened to) if the subject's outer clothing is removed, so a male subject is preferable.

2. With an alcohol swab, clean the earpieces of the stethoscope. Allow the alcohol to dry. Notice that the earpieces are angled. For comfort and best auscultation, the earpieces should be angled in a *forward* direction when placed into the ears.

3. Don the stethoscope. Place the diaphragm of the stethoscope on your partner's thorax, just to the sternal side of the left nipple at the fifth intercostal space, and listen carefully for heart sounds. The first sound will be a longer, louder (more booming) sound than the second, which is short and sharp. After listening for a couple of minutes, try to time the pause between the second sound of one heartbeat and the first sound of the subsequent heartbeat.

How long is this interval? ______________ sec

How does it compare to the interval between the first and second sounds of a single heartbeat?

____

____

4. To differentiate individual valve sounds somewhat more precisely, auscultate the heart sounds over specific thoracic regions. Refer to Figure 33A.2 for the positioning of the stethoscope.

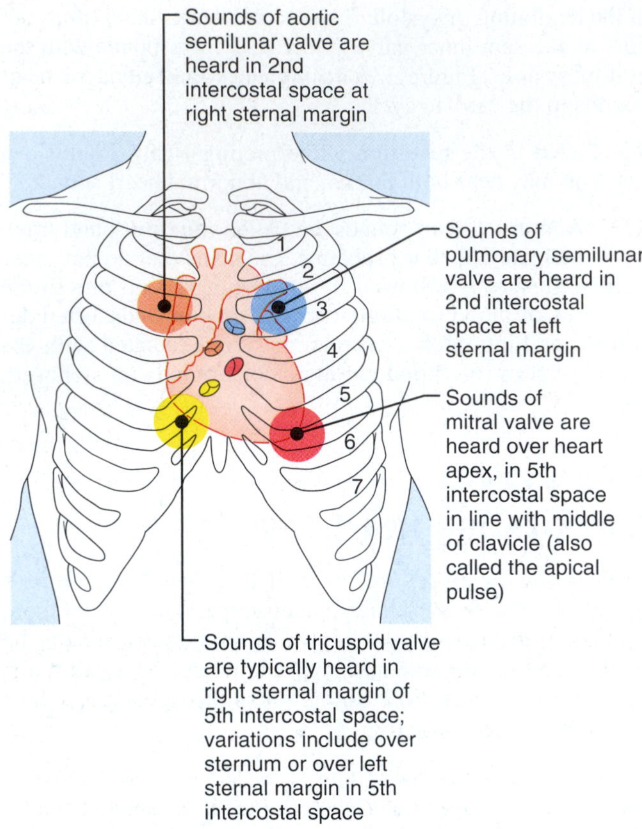
Sounds of aortic semilunar valve are heard in 2nd intercostal space at right sternal margin
Sounds of pulmonary semilunar valve are heard in 2nd intercostal space at left sternal margin
Sounds of mitral valve are heard over heart apex, in 5th intercostal space in line with middle of clavicle (also called the apical pulse)
Sounds of tricuspid valve are typically heard in right sternal margin of 5th intercostal space; variations include over sternum or over left sternal margin in 5th intercostal space
1
2
3
4
5
6
7

Figure 33A.2 Areas of the thorax where valvular sounds can best be detected.

Auscultation of AV Valves

As a rule, the mitral valve closes slightly before the tricuspid valve. You can hear the mitral valve more clearly if you place the stethoscope over the apex of the heart, which is at the fifth intercostal space, approximately in line with the middle region of the left clavicle. Listen to the heart sounds at this region; then move the stethoscope medially to the right margin of the sternum to auscultate the tricuspid valve. Can you detect the slight lag between the closure of the mitral and tricuspid valves?

____

There are normal variations in the site for "best" auscultation of the tricuspid valve. These range from the site depicted in Figure 33A.2 (right sternal margin over fifth intercostal space) to over the sternal body in the same plane, to the left sternal margin over the fifth intercostal space. If you have difficulty hearing closure of the tricuspid valve, try one of these other locations.

Auscultation of Semilunar Valves

Again there is a slight dissynchrony of valve closure; the aortic semilunar valve normally snaps shut just ahead of the pulmonary semilunar valve. If the subject inhales deeply but gently, filling of the right ventricle will be delayed slightly (due to the compression of the thoracic blood vessels by the increased intrapulmonary pressure); and the two sounds can be heard more distinctly. Position the stethoscope over the second intercostal space, just to the *right* of the sternum. The aortic valve is best heard at this position. As you listen, have your partner take a deep breath. Then move the stethoscope to the *left* side of the sternum in the same line; and auscultate the pulmonary valve. Listen carefully; try to hear the "split" between the closure of these two valves in the second heart sound.

Although at first it may seem a bit odd that the pulmonary valve issuing from the *right* heart is heard most clearly to the *left* of the sternum and the aortic valve of the left heart is best heard at the right sternal border, this is easily explained by reviewing heart anatomy. Because the heart is twisted, with the right ventricle forming most of the anterior ventricular surface, the pulmonary trunk actually crosses to the right as it issues from the right ventricle. Similarly, the aorta issues from the left ventricle at the left side of the pulmonary trunk before arching up and over that vessel. ■

The Pulse

The term **pulse** refers to the alternating surges of pressure (expansion and then recoil) in an artery that occur with each contraction and relaxation of the left ventricle. This difference between systolic and diastolic pressure is called the **pulse pressure.** (See p. 340.) Normally the pulse rate (pressure surges per minute) equals the heart rate (beats per minute), and the pulse averages 70 to 76 beats per minute in the resting state.

Parameters other than pulse rate are also useful clinically. You may also assess the regularity (or rhythmicity) of the pulse, and its amplitude and/or tension—does the blood vessel expand and recoil (sometimes visibly) with the pressure waves? Can you feel it strongly, or is it difficult to detect? Is it regular like the ticking of a clock, or does it seem to skip beats?

Activity 2:
Palpating Superficial Pulse Points

The pulse may be felt easily on any superficial artery when the artery is compressed over a bone or firm tissue. Palpate the following pulse or pressure points on your partner by placing the fingertips of the first two or three fingers of one hand over the artery. It helps to compress the artery firmly as you begin your palpation and then immediately ease up on the pressure slightly. In each case, notice the regularity of the pulse, and assess the degree of tension or amplitude. Figure 33A.3 illustrates the superficial pulse points to be palpated.

Common carotid artery: At the side of the neck

Temporal artery: Anterior to the ear, in the temple region

Facial artery: Clench the teeth, and palpate the pulse just anterior to the masseter muscle on the mandible (in line with the corner of the mouth)

Brachial artery: In the antecubital fossa, at the point where it bifurcates into the radial and ulnar arteries

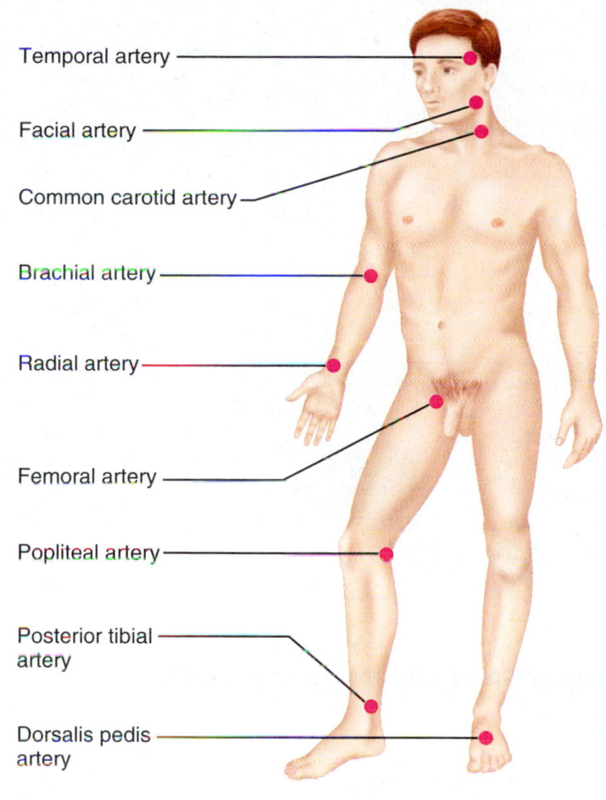

Figure 33A.3 Body sites where the pulse is most easily palpated.

Radial artery: At the lateral aspect of the wrist, above the thumb

Femoral artery: In the groin

Popliteal artery: At the back of the knee

Posterior tibial artery: Just above the medial malleolus

Dorsalis pedis artery: On the dorsum of the foot

Which pulse point had the greatest amplitude?

____

Which the least? ______________________________

Can you offer any explanation for this? ______________

____

____

____

Because of its easy accessibility, the pulse is most often taken on the radial artery. With your partner sitting quietly, practice counting the radial pulse for 1 minute. Make three counts and average the results.

count 1 ______________ count 2 ______________

count 3 ______________ average ______________ ■

Due to the elasticity of the arteries, blood pressure decreases and smooths out as blood moves farther away from the heart. A pulse, however, can still be felt in the fingers. A device called a plethysmograph or a piezoelectric pulse transducer can measure this pulse.

Activity 3: Measuring the Pulse Using a Plethysmograph and PowerLab®

1. Make sure the computer is off and connect the PowerLab® unit. Start the computer.
2. Connect the plethysmograph to the CHANNEL 2 BNC fitting on the PowerLab® unit. Rotate the plug to lock it into the socket (Figure 33A.4).
3. Open the PowerLab® folder and select the file named *Experiment 3—Pulse*. The Chart program will open and CHANNEL 1 will be labeled *Volume pulse*. CHANNEL 2 measures rate of blood flow, which is converted to volume pulse and displayed in CHANNEL 1.
4. Place the plethysmograph on the distal finger of the subject, and secure the plethysmograph with the Velcro tape. Both hands should be in the subject's lap for the test.
5. Select the CHANNEL 1 (labeled *Volume pulse*) pop-up menu and choose *Computed Input*. Adjust the range to produce the largest signal possible.
6. Click *Start* and run the experiment for about 10 seconds. Click *Stop* to halt the chart.
7. Click on the *File* menu and save the data to an appropriate file.
8. Examine the volume pulse in the Zoom window, and inspect the tracing for evidence of a dicrotic notch. (See Figure 33A.5.)
9. Use the marker and cursor to measure the time in seconds from peak to peak. Calculate your pulse rate and record.

____

10. If there is time, investigate the effects of hot and cold on pulse rate.

Place a bag of ice and water on the subject's forearm and record the pulse rate after 2 minutes.

Place a bag of hot water on the subject's forearm and record the pulse rate after 2 minutes.

Record your observations.

Effect of cold on pulse rate ______________________

Effect of heat on pulse rate ______________________ ■

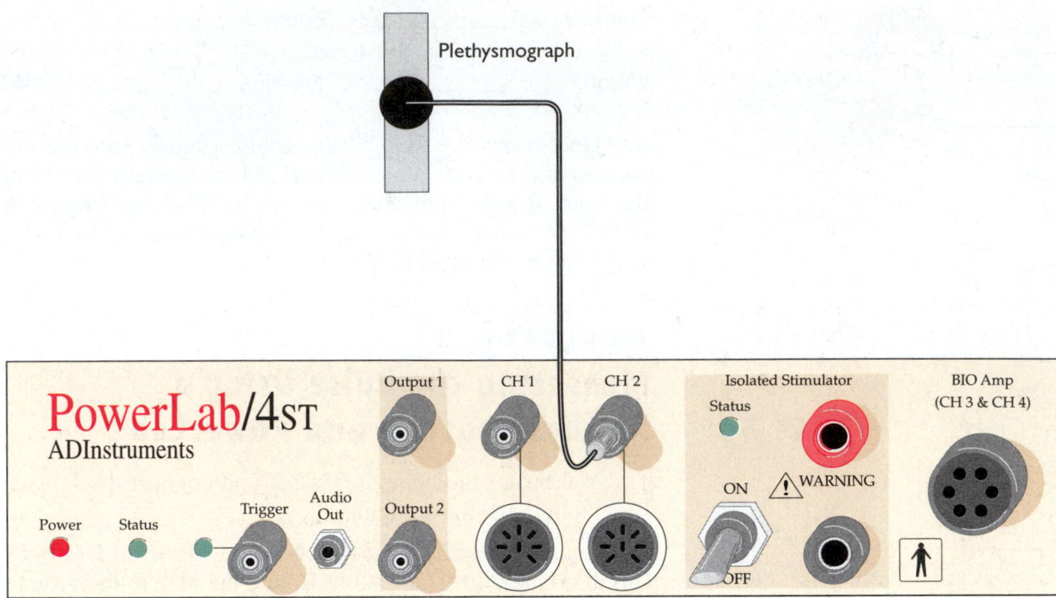

Figure 33A.4 The connection between the plethysmograph and the PowerLab® unit.

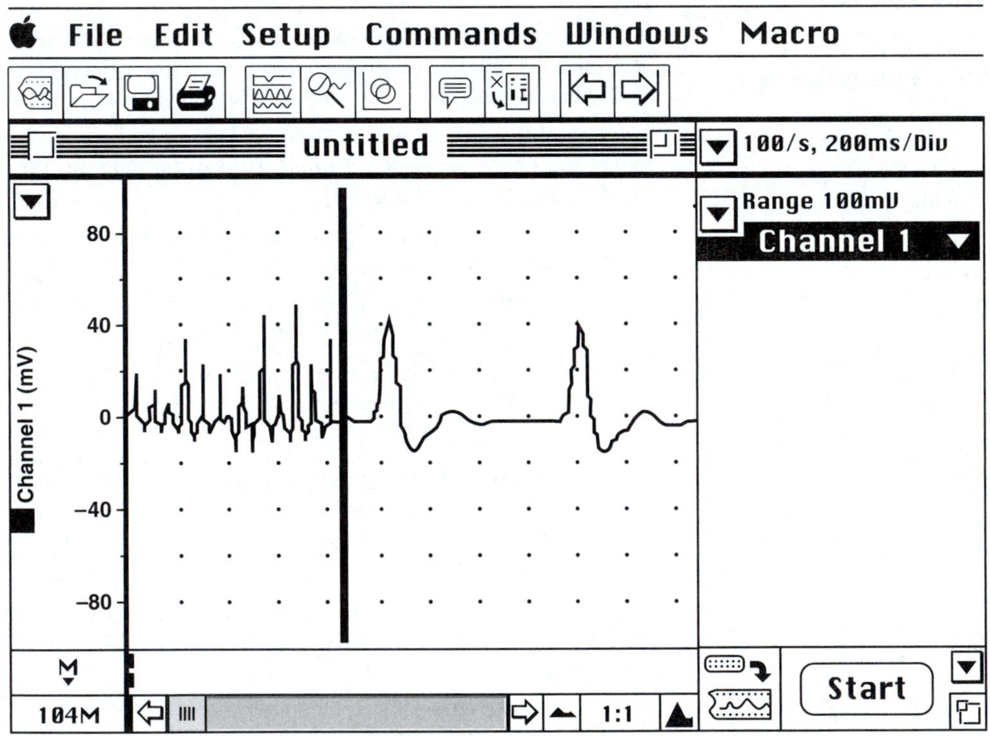

Figure 33A.5 The Chart window with a finger pulse trace.

Apical-Radial Pulse

The correlation between the apical and radial pulse rates can be determined by simultaneously counting them. The **apical pulse** (actually the counting of heartbeats) may be slightly faster than the radial because of a slight lag in time as the blood rushes from the heart into the large arteries where it can be palpated. However, any *large* difference between the values observed, referred to as a **pulse deficit,** may indicate cardiac impairment (a weakened heart that is unable to pump blood into the arterial tree to a normal extent—low cardiac output) or abnormal heart rhythms. In the case of atrial fibrillation or ectopic heartbeats, for instance, the second beat may follow the first so quickly that no second pulse is felt even though the apical pulse can still be heard (auscultated). Apical pulse counts are routinely ordered for those with cardiac decompensation.

Activity 4:
Taking an Apical Pulse

With the subject sitting quietly, one student, using a stethoscope, should determine the apical pulse rate while another simultaneously counts the radial pulse rate. The stethoscope should be positioned over the fifth left intercostal space. The person taking the radial pulse should determine the starting point for the count and give the stop-count signal exactly 1 minute later. Record your values below.

apical count ______________ beats/min

radial count ______________ pulses/min

pulse deficit ______________ /min ■

Blood Pressure Determinations

Blood pressure is defined as the pressure the blood exerts against any unit area of the blood vessel walls, and it is generally measured in the arteries. Because the heart alternately contracts and relaxes, the resulting rhythmic flow of blood into the arteries causes the blood pressure to rise and fall during each beat. Thus you must take two blood pressure readings: the **systolic pressure,** which is the pressure in the arteries at the peak of ventricular ejection, and the **diastolic pressure,** which reflects the pressure during ventricular relaxation. Blood pressures are reported in millimeters of mercury (mm Hg), with the systolic pressure appearing first; 120/80 translates to 120 over 80, or a systolic pressure of 120 mm Hg and a diastolic pressure of 80 mm Hg. Normal blood pressure varies considerably from one person to another.

In this procedure, you will measure arterial and venous pressures by indirect means and under various conditions. You will investigate and demonstrate factors affecting blood pressure, the rapidity of blood pressure changes, and the large differences between arterial and venous pressures.

Activity 5:
Using a Sphygmomanometer to Measure Arterial Blood Pressure Indirectly

The **sphygmomanometer,** commonly called a *blood pressure cuff,* is an instrument used to obtain blood pressure readings by the auscultatory method (Figure 33A.6). It consists of an inflatable cuff with an attached pressure gauge. The cuff is placed around the arm and inflated to a pressure higher than systolic pressure to occlude circulation to the forearm. As cuff pressure is gradually released, the examiner listens with a stethoscope for characteristic sounds called the **sounds of Korotkoff,** which indicate the resumption of blood flow into the forearm. The pressure at which the first soft tapping sounds can be detected is recorded as the systolic pressure. As the pressure is reduced further, blood flow becomes more turbulent, and the sounds become louder. As the pressure is reduced still further, below the diastolic pressure, the artery is no longer compressed; and blood flows freely and without turbulence. At this point, the sounds of Korotkoff can no longer be detected. The pressure at which the sounds disappear is recorded as the diastolic pressure.

1. Work in pairs to obtain radial artery blood pressure readings. Obtain a stethoscope, alcohol swabs, and a sphygmomanometer. Clean the earpieces of the stethoscope with the alcohol swabs, and check the cuff for the presence of trapped air by compressing it against the laboratory table. (A partially inflated cuff will produce erroneous measurements.)

2. The subject should sit in a comfortable position with one arm resting on the laboratory table (approximately at heart level if possible). Wrap the cuff around the subject's arm, just above the elbow, with the inflatable area on the medial arm surface. The cuff may be marked with an arrow; if so, the arrow should be positioned over the brachial artery (Figure 33A.6). Secure the cuff by tucking the distal end under the wrapped portion or by bringing the Velcro areas together.

3. Palpate the brachial pulse, and lightly mark its position with a felt pen. Don the stethoscope, and place its diaphragm over the pulse point.

⚠ *The cuff should not be kept inflated for more than 1 minute.* If you have any trouble obtaining a reading within this time, deflate the cuff, wait 1 or 2 minutes, and try again. (A prolonged interference with BP homeostasis can lead to fainting.)

Posture

	Trial 1		Trial 2	
	BP	**Pulse**	**BP**	**Pulse**
Sitting quietly	______	______	______	______
Reclining (after 2 to 3 min)	______	______	______	______
Immediately on standing from the reclining position ("at attention" stance)	______	______	______	______
After standing for 3 min	______	______	______	______

For each of the following tests, students should formulate hypotheses, collect data, and write lab reports. (See Getting Started: Writing a Lab Report, p. xii.) Conclusions should be shared with the class.

Posture

To monitor circulatory adjustments to changes in position, take blood pressure and pulse measurements under the conditions noted in the chart above. Record your results on the chart also.

Exercise

Blood pressure and pulse changes during and after exercise provide a good yardstick for measuring one's overall cardiovascular fitness. Although there are more sophisticated and more accurate tests that evaluate fitness according to a specific point system, the *Harvard step test* described here is a quick way to compare the relative fitness level of a group of people.

You will be working in groups of four, duties assigned as indicated above, except that student 4, in addition to recording the data, will act as the timer and call the cadence.

⚠ Any student with a known heart problem should refuse to participate as the subject.

All four students may participate as the subject in turn, if desired, but the bench stepping is to be performed *at least twice* in each group—once with a well-conditioned person acting as the subject, and once with a poorly conditioned subject.

Bench stepping is the following series of movements repeated sequentially:

1. Place one foot on the step.
2. Step up with the other foot so that both feet are on the platform. Straighten the legs and the back.
3. Step down with one foot.
4. Bring the other foot down.

The pace for the stepping will be set by the "timer" (student 4), who will repeat "Up-2-3-4, up-2-3-4" at such a pace that each "up-2-3-4" sequence takes 2 sec (i.e., 30 cycles/min).

1. Student 4 should obtain the step (0.5 m or 20-in. height for male subject, or 0.4 m or 16 in. for a female subject) while baseline measurements are being obtained on the subject.

2. Once the baseline pulse and blood pressure measurements have been recorded on the exercise chart on p. 343, the subject is to stand quietly at attention for 2 min to allow his or her blood pressure to stabilize before beginning to step.

3. The subject is to perform the bench stepping for as long as possible, up to a maximum of 5 min, according to the cadence called by the timer. The subject is to be watched for and warned against crouching (posture must remain erect). If he or she is unable to keep the pace for a span of 15 sec, the test is to be terminated.

4. When the subject is stopped by the pacer, stops voluntarily because he or she is unable to continue, or has completed 5 min of bench stepping, he or she is to sit down. The duration of exercise (in seconds) is to be recorded, and the blood pressure and pulse are to be measured immediately and thereafter at 1-min intervals for 3 min post-exercise.

Duration of exercise: _______ sec

5. The subject's *index of physical fitness* is to be calculated using the formula given below:

$$\text{Index} = \frac{\text{duration of exercise in seconds} \times 100}{2 \times \text{sum of the 3 pulse counts in recovery}}$$

Scores are interpreted according to the following scale:

below 55	poor physical condition
55 to 62	low average
63 to 71	average
72 to 79	high average
80 to 89	good
90 and over	excellent

6. Record the test values on the chart, and repeat the testing and recording procedure with the second subject.

Exercise

Harvard step test for 5 min at 30/min	Baseline		Interval Following Test: Immediately		1 min		2 min		3 min	
	BP	P	BP	P	BP	P	BP	P	BP	P
Well-conditioned individual	___	___	___	___	___	___	___	___	___	___
Poorly conditioned individual	___	___	___	___	___	___	___	___	___	___

Nicotine

Baseline		After smoking for 2 min		After smoking for 3 min		After smoking for 4 min	
BP	P	BP	P	BP	P	BP	P
___	___	___	___	___	___	___	___

When did you notice a greater elevation of blood pressure and pulse?

___

Explain: ___

___

Was there a sizable difference between the after-exercise values for well-conditioned and poorly conditioned individuals?

___ Explain: ___

___

Did the diastolic pressure also increase? ___

Explain: ___

___

Nicotine (Optional)

To investigate the effects of nicotine on blood pressure and pulse, take a baseline blood pressure and pulse of a smoker who is in a standing position. Ask the subject to light up and smoke in the usual manner. After 2 min, take blood pressure and pulse readings. Repeat the measurements at 1-min intervals for the next 2 min, and record your values on the nicotine chart above.

How do the effects of nicotine bring about the changes noted? (Hint: nicotine initially has vasoconstrictor effects.)

___

___

___

A Noxious Sensory Stimulus (Cold)

There is little question that blood pressure is affected by emotions and pain. This lability of blood pressure will be investigated through use of the **cold pressor test,** in which one hand will be immersed in unpleasantly (even painfully) cold water.

Measure the blood pressure and pulse of the subject as he or she sits quietly. Obtain a basin and thermometer, fill the basin with ice cubes, and add water. When the temperature of the ice bath has reached 5°C, immerse the subject's other hand (the noncuffed limb) in the ice water. With the hand still immersed, take blood pressure and pulse readings at 1-min intervals for a period of 3 min, and record the values on the chart on p. 344.

A Noxious Sensory Stimulus (Cold)

Baseline BP	P	1 min BP	P	2 min BP	P	3 min BP	P
______	______	______	______	______	______	______	______

How did the blood pressure change during cold exposure?

____

____

Was there any change in pulse? ______________

Subtract the respective baseline readings of systolic and diastolic blood pressure from the highest single reading of systolic and diastolic pressure obtained during cold immersion. (For example, if the highest experimental reading is 140/88 and the baseline reading is 120/70, then the differences in blood pressure would be systolic pressure, 20 mm Hg, and diastolic pressure, 18 mm Hg.) These differences are called the index of response. According to their index of response, subjects can be classified as follows:

Hyporeactors (stable blood pressure): Exhibit a rise of diastolic and/or systolic pressure ranging from 0 to 22 mm Hg; or a drop in pressures

Hyperreactors (labile blood pressure): Exhibit a rise of 23 mm Hg or more in the diastolic and/or systolic blood pressure

Is the subject tested a hypo- or hyperreactor?

______________________________ ■

Skin Color as an Indicator of Local Circulatory Dynamics

Skin color reveals with surprising accuracy the state of the local circulation, and allows inferences concerning the larger blood vessels and the circulation as a whole. The experiments on local circulation outlined below illustrate a number of factors that affect blood flow to the tissues.

Clinical expertise often depends upon good observation skills, accurate recording of data, and logical interpretation of the findings. A single example will be given to demonstrate this statement: A massive hemorrhage may be internal and hidden (thus, not obvious) but will still threaten the blood delivery to the brain and other vital organs. One of the earliest compensatory responses of the body to such a threat is constriction of cutaneous blood vessels, which reduces blood flow to the skin and diverts it into the circulatory mainstream to serve other, more vital tissues. As a result, the skin of the face and particularly of the extremities becomes pale, cold, and eventually moist with perspiration. Therefore, pale, cold, clammy skin should immediately lead the careful diagnostician to suspect that the circulation is dangerously inefficient. Other conditions, such as local arterial obstruction and venous congestion, as well as certain pathologies of the heart and lungs, also alter skin texture, color, and circulation in characteristic ways.

Activity 8: Examining the Effect of Local Chemical and Physical Factors on Skin Color

The local blood supply to the skin (indeed, to any tissue) is influenced by (1) local metabolites, (2) oxygen supply, (3) local temperature, (4) autonomic nervous system impulses, (5) local vascular reflexes, (6) certain hormones, and (7) substances released by injured tissues. A number of these factors are examined in the simple experiments that follow. Each experiment should be conducted by students in groups of three or four. One student will act as the subject; the others will conduct the tests and make and record observations.

Vasodilation and Flushing of the Skin Due to Local Metabolites

1. Obtain a blood pressure cuff (sphygmomanometer) and stethoscope. You will also need a watch with a second hand.

2. The subject should bare both arms by rolling up the sleeves as high as possible and then lay the forearms side by side on the bench top.

3. Observe the general color of the subject's forearm skin, and the normal contour and size of the veins. Notice whether skin color is bilaterally similar. Record your observations:

____

____

____

4. Apply the blood pressure cuff to one arm, and inflate it to 250 mm Hg. Keep it inflated for 1 min. During this period, repeat the observations made above and record the results:

___

___

5. Release the pressure in the cuff (leaving the deflated cuff in position), and again record the forearm skin color and the condition of the forearm veins. Make this observation immediately after deflation and then again 30 sec later.

Immediately after deflation ___

___

30 sec after deflation ___

___

The above observations constitute your baseline information. Now conduct the following tests.

6. Instruct the subject to raise the cuffed arm above his or her head and to clench the fist as tightly as possible. While the hand and forearm muscles are tightly contracted, rapidly inflate the cuff to 240 mm Hg or more. This maneuver partially empties the hand and forearm of blood, and stops most blood flow to the hand and forearm. Once the cuff has been inflated, the subject is to relax the fist and return the forearm to the bench top so that it can be compared to the other forearm.

7. Leave the cuff inflated for exactly 1 min. During this interval, compare the skin color in the "ischemic" (blood-deprived) hand to that of the "normal" (non-cuffed-limb) hand. Quickly release the pressure immediately after 1 min.

What are the subjective effects* of stopping blood flow to the arm and hand for 1 min?

___

___

What are the objective effects (color of skin and condition of veins)?

___

___

How long does it take for the subject's ischemic hand to regain its normal color?

___

*Subjective effects are sensations—such as pain, coldness, warmth, tingling, and weakness—experienced by the subject. They are "symptoms" of a change in function.

Effects of Venous Congestion

1. Again, but with a different subject, observe and record the appearance of the skin and veins on the forearms resting on the bench top. This time, pay particular attention to the color of the fingers, particularly the distal phalanges, and the nail beds. Record this information:

___

___

2. Wrap the blood pressure cuff around one of the subject's arms, and inflate it to 40 mm Hg. Maintain this pressure for 5 min. Make a record of the subjective and objective findings just before the 5 min are up, and then again immediately after release of the pressure at the end of 5 min.

Subjective (arm cuffed) ___

___

Objective (arm cuffed) ___

___

Subjective (pressure released) ___

___

Objective (pressure released) ___

___

3. With still another subject, conduct the following simple experiment: Raise one arm above the head, and let the other hang by the side for 1 min. After 1 min, quickly lay both arms on the bench top, and compare their color.

Color of raised arm ___

Color of dependent arm ___

From this and the two preceding observations, analyze the factors that determine tint of color (pink or blue) and intensity of skin color (deep pink or blue as opposed to light pink or blue). Record your conclusions.

___

___

___

___

___

Collateral Blood Flow

In some diseases, blood flow to an organ through one or more arteries may be completely and irreversibly obstructed. Fortunately, in most cases a given body area is supplied both by one main artery and by anastomosing channels connecting the main artery with one or more neighboring blood vessels. Consequently, an organ may remain viable even though its main arterial supply is occluded, as long as the **collateral vessels** are still functional.

The effectiveness of collateral blood flow in preventing ischemia can be easily demonstrated.

1. Check the subject's hands to be sure they are *warm* to the touch. If not, choose another, "warm-handed" subject, or warm the subject's hands in 35°C water for 10 min before beginning.

2. Palpate the subject's radial and ulnar arteries approximately 2.5 cm or 1 in. above the wrist flexure, and mark their locations with a felt marker.

3. Instruct the subject to supinate one forearm and to hold it in a partially flexed (about a 30° angle) position, with the elbow resting on the bench top.

4. Face the subject and grasp his or her forearm with both of your hands, the thumb and fingers of one hand compressing the marked radial artery and the thumb and fingers of the other hand compressing the ulnar artery. Maintain the pressure for 5 min, noticing the progression of the subject's hand to total ischemia.

5. At the end of 5 min, release the pressure abruptly. Record the subject's sensations, as well as the intensity and duration of the flush in the previously occluded hand. (Use the other hand as a baseline for comparison.)

___

___

___

6. Allow the subject to relax for 5 min; then repeat the maneuver, but this time *compress only the radial artery.* Record your observations.

___

___

How do the results of the first test differ from those of the second test with respect to color changes during compression, and the intensity and duration of reactive hyperemia (redness of the skin)?

___

___

___

___

7. Once again allow the subject to relax for 5 min. Then repeat the maneuver, *with only the ulnar artery compressed.* Record your observations:

___

___

What can you conclude about the relative sizes of, and hand areas served by, the radial and ulnar arteries?

___

___

Effect of Mechanical Stimulation of Blood Vessels of the Skin

With moderate pressure, draw the blunt end of your pen across the skin of a subject's forearm. Wait 3 min to observe the effects, and then repeat with firmer pressure.

What changes in skin color do you observe with light-to-moderate pressure?

___

With heavy pressure? ___

The redness, or *flare,* observed after mechanical stimulation of the skin results from a local inflammatory response promoted by chemical mediators released by injured tissues. These mediators stimulate increased blood flow into the area and leaking of fluid (from the capillaries) into the local tissues. (Note: People differ considerably in skin sensitivity. Those most sensitive will show **dermographism,** a condition in which the direct line of stimulation will swell quite obviously. This excessively swollen area is called a *wheal.*) ■

Frog Cardiovascular Physiology: Wet Lab

exercise **34A**

Objectives

1. To list the properties of cardiac muscle as automaticity and rhythmicity and define each.
2. To explain the statement "Cardiac muscle has an intrinsic ability to beat."
3. To compare the intrinsic rate of contraction of the "pacemaker" of the frog heart (sinus venosus) to that of the atria and ventricle.
4. To compare the relative length of the refractory period of cardiac muscle with that of skeletal muscle, and explain why it is not possible to tetanize cardiac muscle.
5. To define *extrasystole* and to explain at what point in the cardiac cycle (and on an ECG tracing) an extrasystole can be induced.
6. To describe the effect of the following on heart rate: cold, heat, vagal stimulation, pilocarpine, digitalis, atropine sulfate, epinephrine, and potassium, sodium, and calcium ions.
7. To define *vagal escape* and discuss its value.
8. To define *ectopic pacemaker*.
9. To define *partial* and *total heart block*.
10. To describe how heart block was induced in the laboratory and explain the results, and to explain how heart block might occur in the body.
11. To list the components of the microcirculatory system.
12. To identify an arteriole, venule, and capillaries in a frog's web, and to cite differences between relative size, rate of blood flow, and regulation of blood flow in these vessels.
13. To describe the effect of heat, cold, local irritation, and histamine on the rate of blood flow in the microcirculation, and to explain how these responses help maintain homeostasis.

Materials

- ❑ Frogs*
- ❑ Dissecting tray and instruments
- ❑ Disposable plastic gloves
- ❑ Petri dishes
- ❑ Medicine dropper
- ❑ Millimeter ruler
- ❑ Frog Ringer's solutions (at room temperature, 5°C, and 32°C)
- ❑ Thread, large rubber bands, fine common pins
- ❑ Frog board
- ❑ Cotton balls
- ❑ Freshly prepared solutions (using frog Ringer's solution as the solvent) of the following in dropper bottles:
  - 2.5% pilocarpine
  - 2% digitalis
  - 5% atropine sulfate
  - 1% epinephrine
  - 5% potassium chloride (KCl)
  - 2% calcium chloride ($CaCl_2$)
  - 0.7% sodium chloride (NaCl)
  - 0.01% histamine
  - 0.01 *N* HCl
- ❑ Compound microscope
- ❑ Apparatus A or B:
  - A: PowerLab® unit and cable, computer with Chart software installed, displacement transducer, transducer cable, transducer stand, stimulator, stimulator cable
  - B: Physiograph (polygraph), physiograph paper and ink, myograph (force) transducer, transducer cable, transducer stand, stimulator output extension cable, electrodes

PhysioEx 4.0™ Computer Simulation on p. P-60

*Instructor will double-pith frogs as required for student experimentation.

Investigations of human cardiovascular physiology are very interesting, but many areas obviously do not lend themselves to experimentation. It would be tantamount to murder to inject a human subject with various drugs to observe their effects on heart activity, or to expose the human heart in order to study the length of its refractory period. However, this type of investigation can be done on frogs or small laboratory animals and provides valuable data, because the physiological mechanisms in these animals are similar if not identical to those in humans.

In this exercise, you will conduct the cardiac investigations just mentioned and others. In addition, you will observe the microcirculation in a frog's web and subject it to various chemical and thermal agents to demonstrate their influence on local blood flow.

Special Electrical Properties of Cardiac Muscle: Automaticity and Rhythmicity

Cardiac muscle differs from skeletal muscle both functionally and in its fine structure. Skeletal muscle must be electrically stimulated to contract. In contrast, heart muscle can and does depolarize spontaneously in the absence of external stimulation. This property, called **automaticity,** is due to plasma membranes that have reduced permeability to potassium ions, but still allow sodium ions to slowly leak into the cells. This leakage causes the muscle cells to gradually depolarize until the action potential threshold is reached and *fast calcium channels* open, allowing Ca^{2+} entry from the extracellular fluid. Shortly thereafter, contraction occurs.

Also, the spontaneous depolarization-repolarization events occur in a regular and continuous manner in cardiac muscle, a property referred to as **rhythmicity.**

In the following experiment, you will observe these properties of cardiac muscle in an *in vitro* (outside, or removed from the body) situation. Work together in groups of three or four. (The instructor may choose to demonstrate this procedure if time or frogs are at a premium.)

Activity 1:
Investigating the Automaticity and Rhythmicity of Heart Muscle

1. Obtain a dissecting tray and instruments, disposable plastic gloves, two petri dishes, frog Ringer's solution, and a medicine dropper, and bring them to your laboratory bench.

2. Don the gloves, and then request and obtain a doubly pithed frog from your instructor. Quickly open the thoracic cavity and observe the heart rate *in situ* (at the site or within the body).

Record the heart rate: ______________ beats/min

3. Dissect out the heart and the gastrocnemius muscle of the calf and place the removed organs in separate petri dishes containing frog Ringer's solution. (Note: Just in case you don't remember, the procedure for removing the gastrocnemius muscle is provided on pp. 168–171 in Exercise 16A. The extreme care used in that procedure for the removal of the gastrocnemius muscle need not be exercised here.)

4. Observe the activity of the two organs for a few seconds.

Which is contracting? ______________________________

At what rate? ______________ beats/min

Is the contraction rhythmic? ______________

5. Sever the sinus venosus from the heart (see Figure 34A.1). The sinus venosus of the frog's heart corresponds to the SA node of the human heart.

Does the sinus venosus continue to beat? ______________

If not, lightly touch it with a probe to stimulate it. Record its rate of contraction.

Rate: ______________ beats/min

6. Sever the right atrium from the heart; then remove the left atrium. Does each atrium continue to beat?

______________ Rate: ______________ beats/min

Does the ventricle continue to beat? ______________

Rate: ______________ beats/min

7. Continue to fragment the ventricle to determine how small the ventricular fragments must be before the automaticity of ventricular muscle is abolished. Measure these fragments and record their approximate size.

_______ mm × _______ mm × _______ mm

Which portion of the heart exhibited the most marked automaticity?

____

Which the least? ______________________________

8. Properly dispose of the frog and heart fragments before continuing with the next procedure. ■

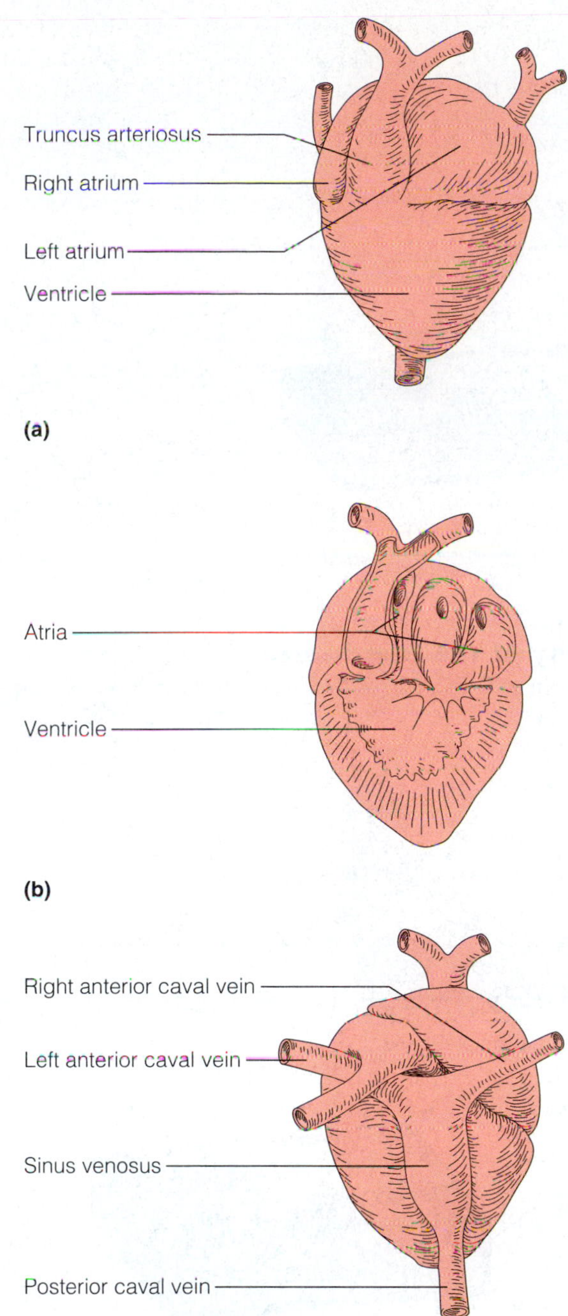

Figure 34A.1 Anatomy of the frog heart. **(a)** Ventral view showing the single truncus arteriosus leaving the undivided ventricle. **(b)** Longitudinal section showing the two atrial and single ventricular chambers. **(c)** Dorsal view showing the sinus venosus (pacemaker).

Baseline Frog Heart Activity

The heart's effectiveness as a pump is dependent both on intrinsic (within the heart) and extrinsic (external to the heart) controls. In this experiment, you will investigate some of these factors.

The nodal system, in which the "pacemaker" imposes its depolarization rate on the rest of the heart, is one intrinsic factor that influences the heart's pumping action. If its impulses fail to reach the ventricles (as in heart block), the ventricles continue to beat but at their own inherent rate, which is much slower than that usually imposed on them. Although heart contraction does not depend on nerve impulses, its rate can be modified by extrinsic impulses reaching it through the autonomic nerves. Additionally, cardiac activity is modified by various chemicals, hormones, ions, and metabolites. The effects of several of these chemical factors are examined in the next experimental series.

The frog heart has two atria and a single, incompletely divided ventricle (see Figure 34A.1). The pacemaker is located in the sinus venosus, an enlarged region between the venae cavae and the right atrium. The SA node of mammals may have evolved from the sinus venosus.

Activity 2: Recording Baseline Frog Heart Activity

To record baseline frog heart activity, work in groups of four—two students handling the equipment setup and two preparing the frog for experimentation. Two sets of instructions are provided for apparatus setup—one for PowerLab® (Figure 34A.2), the other for the physiograph (Figure 34A.3). Follow the procedure outlined for the apparatus you will be using.

PowerLab® Setup

1. Make sure the computer is off, and connect the PowerLab® unit. Start the computer.

2. Connect the transducer to the PowerLab® unit. Do not record while adjusting the transducer.

3. Connect stimulator probe leads to the positive and negative output terminals on the stimulator. (See Figure 34A.2.)

4. Open the *Setup Channel Settings* dialog box, and turn off unwanted channels. Change the title of CHANNEL 1 to *Contraction.* Hit *OK.*

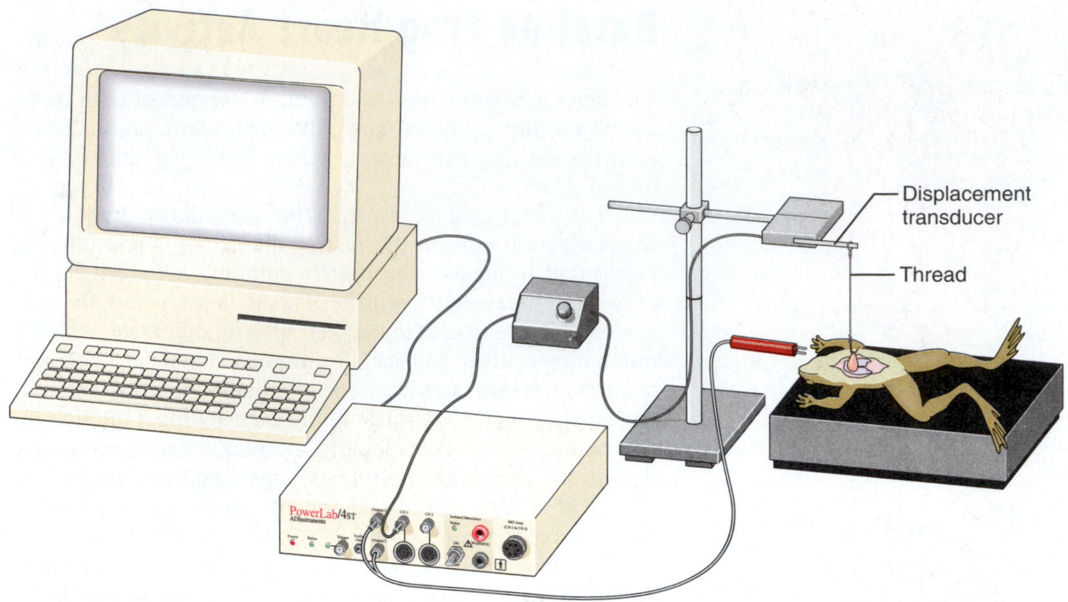

Figure 34A.2 PowerLab® setup for recording the activity of the frog heart.

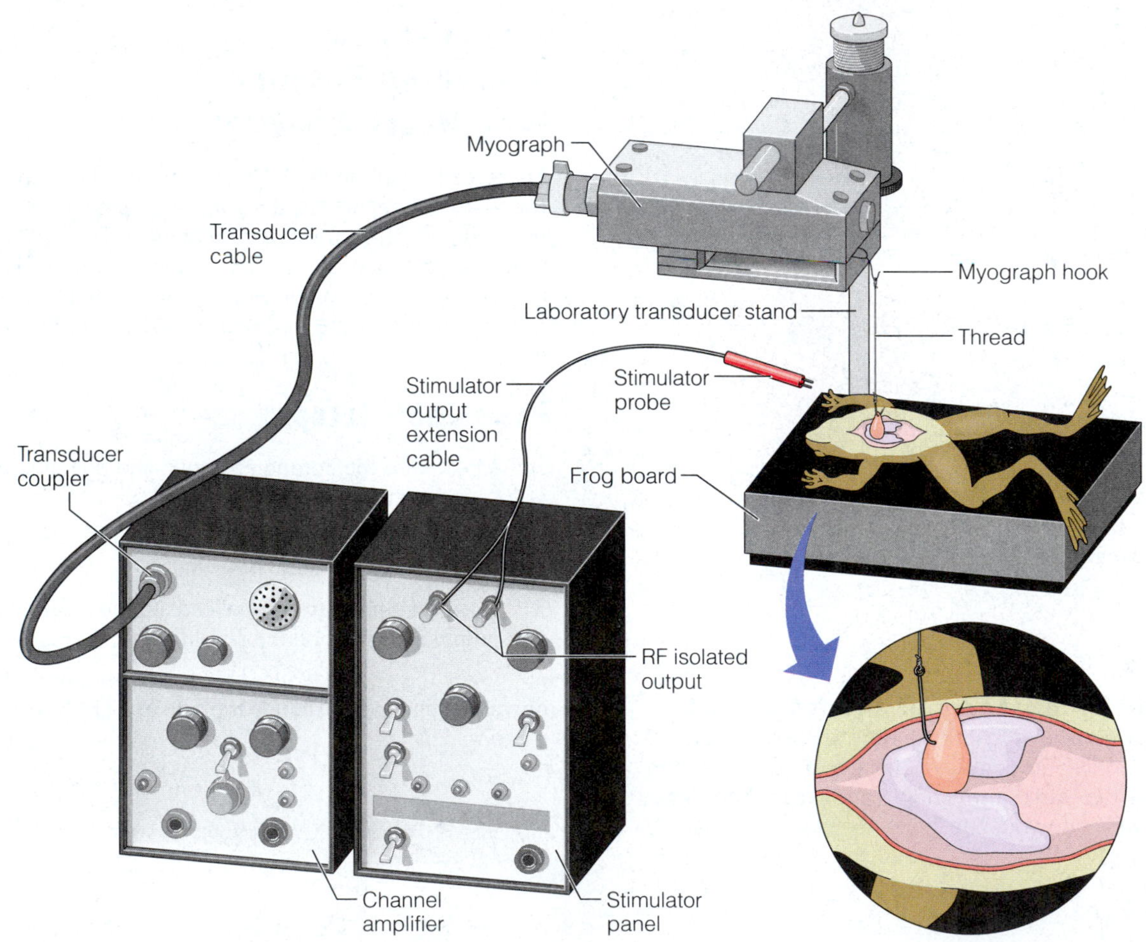

Figure 34A.3 Physiograph setup for recording the activity of the frog heart.

5. Select the CHANNEL 1 *(Contraction)* pop-up menu, and open *Input Amplifier*.

6. Adjust the scroll speed so that the peaks of ventricular contraction are about 2 mm apart. Begin with a scroll speed at 5 sec/Div. Adjust the range to a value that gives a recording about 2 cm high. Adjust the baseline so that it is close to the bottom of the screen.

7. Be sure to label the recordings, and save the data to an appropriate file after each experiment is conducted.

8. Once the setup is complete, begin with instruction number 3 under Making the Baseline Recording.

Physiograph Apparatus Setup

1. Obtain a myograph transducer, transducer cable, and stand, and bring them to the recording site.

2. Attach the myograph transducer to the transducer stand as shown in Figure 34A.3.

3. Then attach the transducer cable to the transducer coupler (input) on the amplifier of the physiograph and to the transducer.

4. Attach the stimulator output extension cable to output on the stimulator panel (red to red, black to black).

Preparation of the Frog

1. Obtain room-temperature frog Ringer's solution, a medicine dropper, dissecting instruments and pan, disposable gloves, fine common pins, and some thread, and bring them to your bench.

2. Don the gloves and obtain a doubly pithed frog from your instructor.

3. Make a longitudinal incision through the abdominal and thoracic walls with scissors, and then cut through the sternum to expose the heart.

4. Grasp the pericardial sac with forceps, and cut it open so that the beating heart can be observed.

Is the sequence an atrial-ventricular one? ______________

5. Locate the vagus nerve, which runs down the lateral aspect of the neck and parallels the trachea and carotid artery. (In good light, it appears to be striated.) Slip an 18-inch length of thread under the vagus nerve so that it can later be lifted away from the surrounding tissues by the thread. Then place a saline-soaked cotton ball over the nerve to keep it moistened until you are ready to stimulate it later in the procedure.

6. Using a medicine dropper, flush the heart with the saline (Ringer's) solution. *From this point on the heart must be kept continually moistened with room-temperature Ringer's solution unless other solutions are being used for the experimentation.*

7. Attach the frog to the frog board.

8. Bend a common pin to a 90° angle, and tie an 0.46 m-0.5 m or 18- to 20-inch length of thread to its head. Force the pin through the apex of the heart (but do not penetrate the ventricular chamber) until the apex is well secured in the angle of the pin.

9. Tie the thread to the hook on the myograph or displacement transducer. Do not pull the thread too tightly. It should be taut enough to lift the heart apex upward, away from the thorax, but should *not* stretch the heart. Adjust the myograph or displacement transducer as necessary.

Making the Baseline Recording

1. *If using the physiograph,* turn the amplifier on and balance the apparatus according to instructions provided by your instructor. Set the paper speed at 0.5 cm/sec. Press the record and paper advance buttons.

2. Set the signal magnet or time marker at 1/sec.

3. Record several normal heartbeats (12 to 15). Be sure you can distinguish atrial and ventricular contractions (see Figure 34A.4). Then adjust the paper or scroll speed so that the peaks of ventricular contractions are approximately 2 cm apart. (Peaks indicate systole; troughs indicate diastole.) Pay attention to the relative force of heart contractions while recording.

4. Count the number of ventricular contractions per minute and record:

______________ beats/min

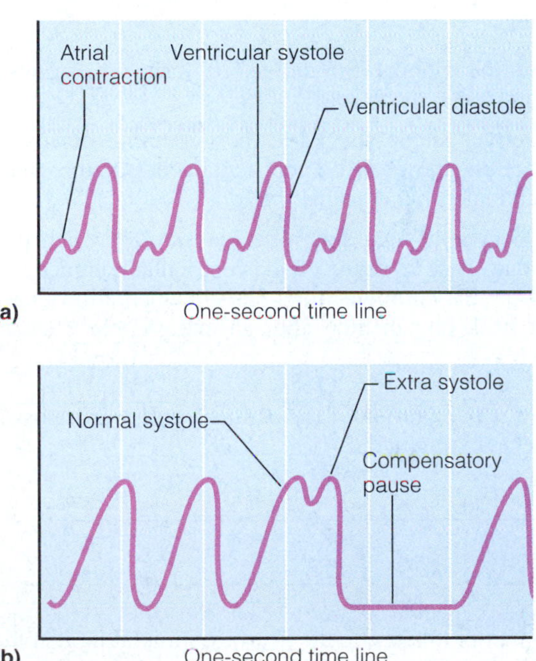

Figure 34A.4 Recording of contractile activity of a frog heart. **(a)** Normal heartbeat. **(b)** Induction of an extrasystole.

5. Compute the A-V interval (period from the beginning of atrial contraction to the beginning of ventricular contraction).

______________ sec

How do the two tracings compare in time?

____

____

____

6. Mark the atrial and ventricular systoles on the record. *Remember to keep the heart moistened with Ringer's solution.* ■

Activity 3:
Investigating the Refractory Period of Cardiac Muscle

In conducting Exercise 16A, you saw that repeated rapid stimuli could cause skeletal muscle to remain in a contracted state. In other words, the muscle could be tetanized. This was possible because of the relatively short refractory period of skeletal muscle. In this experiment, you will investigate the refractory period of cardiac muscle and its response to stimulation. During the procedure, one student should keep the stimulating electrodes in constant contact with the frog heart ventricle.

1. Set the stimulator to deliver 20-V shocks of 2-ms duration, and begin recording.

2. Deliver single shocks at the beginning of ventricular contraction, the peak of ventricular contraction, and then later and later in the cardiac cycle.

3. Observe the recording for **extrasystoles,** which are extra beats that show up riding on the ventricular contraction peak. Also note the **compensatory pause,** which allows the heart to get back on schedule after an extrasystole. (See Figure 34A.4b.)

During which portion of the cardiac cycle was it possible to induce an extrasystole?

____

____

4. Attempt to tetanize the heart by stimulating it at the rate of 20 to 30 impulses per second. What is the result?

____

____

Considering the function of the heart, why is it important that heart muscle cannot be tetanized?

____

____ ■

Activity 4:
Assessing Physical and Chemical Modifiers of Heart Rate

Now that you have observed normal frog heart activity, you will have an opportunity to investigate the effects of various factors that modify heart activity. In each case, record a few normal heartbeats before introducing the modifying factor. After removing the agent, allow the heart to return to its normal rate before continuing with the testing. On each record, indicate the point of introduction and removal of the modifying agent.

Temperature

1. Obtain 5°C and 32°C frog Ringer's solutions and medicine droppers.

2. Increase the scroll or paper speed so that heartbeats appear as spikes 4 to 5 mm apart.

3. Bathe the heart with 5°C saline, and continue to record until the recording indicates a change in cardiac activity.

4. Stop recording, pipette off the cold Ringer's solution (remove the fluid by sucking it into the barrel of a medicine dropper), and flood the heart with room-temperature saline.

5. Start recording again to determine the resumption of the normal heart rate. When this has been achieved, flood the heart with 32°C Ringer's solution and again record until a change is noted.

6. Stop the recording, pipette off the warm saline, and bathe the heart with room-temperature Ringer's solution once again.

What change occurred with the cold (5°C) Ringer's solution?

____

What change occurred with the warm (32°C) Ringer's solution?

____

7. Count the heart rate at the two temperatures and record below.

________ beats/min at 5°C; ________ beats/min at 32°C

Vagus Nerve Stimulation

The vagus nerve carries parasympathetic impulses to the heart, which modify heart activity.

1. Remove the cotton placed over the vagus nerve. Using the previously tied thread, lift the nerve away from the tissues and place the nerve on the stimulating electrodes.
2. Using a duration of 0.5 msec at a voltage of 1 mV, stimulate the nerve at a rate of 50/sec. Continue stimulation until the heart stops momentarily and then begins to beat again **(vagal escape).** If no effect is observed, increase stimulus intensity and try again. If no effect is observed after a substantial increase in stimulus voltage, reexamine your "vagus nerve" to make sure that it is not simply strands of connective tissue.
3. Discontinue stimulation after you observe vagal escape, and flush the heart with saline until the normal heart rate resumes. What is the effect of vagal stimulation on heart rate?

____

The phenomenon of vagal escape demonstrates that many factors are involved in heart regulation and that any deleterious factor (in this case, excessive vagal stimulation) will be overcome, if possible, by other physiological mechanisms such as activation of the sympathetic division of the autonomic nervous system (ANS).

Pilocarpine

Flood the heart with a 2.5% solution of pilocarpine. Record until a change in the pattern of the ECG is noticed. Pipette off the excess pilocarpine solution, and proceed immediately to the next test, which uses atropine as the testing solution. What happened when the heart was bathed in the pilocarpine solution?

____

Pilocarpine simulates the effect of parasympathetic nerve (hence, vagal) stimulation by enhancing acetylcholine release; such drugs are called parasympathomimetic drugs.

Atropine Sulfate

Apply a few drops of atropine sulfate to the frog's heart and observe the recording. If no changes are observed within 2 minutes, apply a few more drops. When you observe a response, pipette off the excess atropine sulfate and flood the heart with Ringer's solution. What happens when the atropine sulfate is added?

____

____

Atropine is a drug that blocks the effect of the neurotransmitter acetylcholine, which is liberated by the parasympathetic nerve endings. Do your results accurately reflect this effect of atropine?

____

Are pilocarpine and atropine agonists or antagonists in their effects on heart activity?

____

Epinephrine

Flood the frog heart with epinephrine solution, and continue to record until a change in heart activity is noted.

What are the results? ______________________________

____

Which division of the autonomic nervous system does its effect imitate?

____

Digitalis

Pipette off the excess epinephrine solution, and rinse the heart with room-temperature Ringer's solution. Continue recording, and when the heart rate returns to baseline values, bathe it in digitalis solution. What is the effect of digitalis on the heart?

____

____

____

Digitalis is a drug commonly prescribed for heart patients with congestive heart failure. It slows heart rate, providing more time for venous return and decreasing the load on the weakened heart. These effects are thought to be due to inhibition of the Na^+-K^+ pump and enhancement of Ca^{2+} entry into the myocardial fibers.

Various Ions

To test the effect of various ions on the heart, apply the designated solution until you observe a change in heart rate or in strength of contraction. Pipette off the solution, flush with Ringer's solution, and allow the heart to resume its normal rate before continuing. *Do not allow the heart to stop.* If the rate should decrease dramatically, flood the heart with room-temperature Ringer's solution.

Effect of Ca^{2+} (use 2% $CaCl_2$) ______________________

____

Effect of Na^+ (use 0.7% NaCl) ______________________

____

Effect of K^+ (use 5% KCl) ______________________

____

Potassium ion concentration is normally higher within cells than in the extracellular fluid. *Hyperkalemia* decreases the resting potential of plasma membranes, thus decreasing the force of heart contraction. In some cases, the conduction rate of the heart is so depressed that **ectopic pacemakers** (pacemakers appearing erratically and at abnormal sites in the heart muscle) appear in the ventricles, and fibrillation may occur. Was there any evidence of premature beats in the recording of potassium ion effects?

____

Was arrhythmia produced with any of the ions tested?

_______ If so, which? ______________________

Intrinsic Conduction System Disturbance (Heart Block)

1. Moisten a 25 cm or 10-inch length of thread and make a Stannius ligature (loop the thread around the heart at the junction of the atria and ventricle).

2. Decrease the scroll or paper speed to achieve intervals of approximately 2 cm between the ventricular contractions, and record a few normal heartbeats.

3. Tighten the ligature in a stepwise manner while observing the atrial and ventricular contraction curves. As heart block occurs, the atria and ventricle will no longer show a 1:1 contraction ratio. Record a few beats each time you observe a different degree of heart block—a 2:1 atria to ventricle, 3:1, 4:1, and so on. As long as you can continue to count a whole number ratio between the two chamber types, the heart is in **partial heart block.** When you can no longer count a whole number ratio, the heart is in **total,** or **complete, heart block.**

4. When total heart block occurs, release the ligature to see if the normal A-V rhythm is reestablished. What is the result?

____

____

5. Attach properly labeled recordings (or copies of the recordings) made during this procedure to the last page of this exercise for future reference.

6. Dispose of the frog remains and gloves in appropriate containers, and dismantle the experimental apparatus before continuing. ■

The Microcirculation and Local Blood Flow

The thin web of a frog's foot provides an excellent opportunity to observe the flow of blood to, from, and within the capillary beds, where the real business of the circulatory system occurs. The collection of vessels involved in the exchange mechanism is referred to as the **microcirculation;** it consists of arterioles, venules, capillaries, and vascular shunts called *metarteriole–thoroughfare channels* (Figure 34A.5).

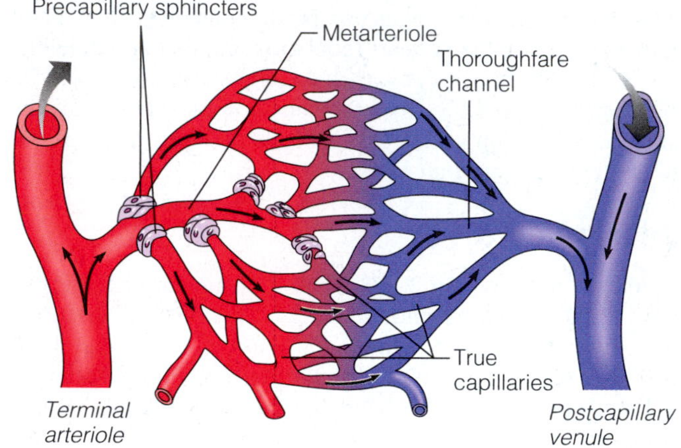

(a) Sphincters open

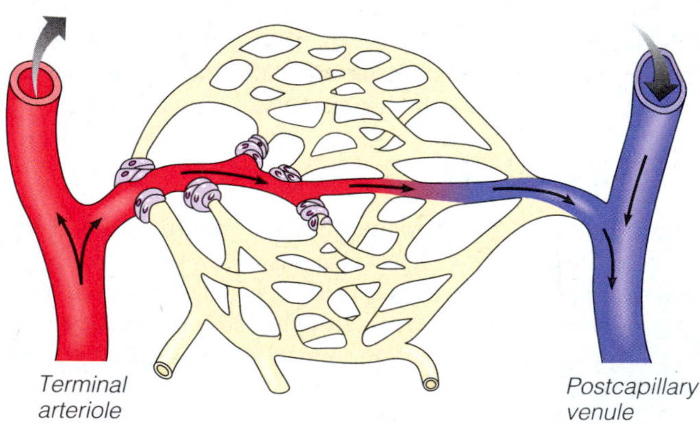

(b) Sphincters closed

Figure 34A.5 Anatomy of a capillary bed. The composite metarteriole–thoroughfare channels act as shunts to bypass the true capillaries when precapillary sphincters controlling blood entry into the true capillaries are constricted.

The total cross-sectional area of the capillaries in the body is much greater than that of the veins and arteries combined. Thus, the flow through the capillary beds is quite slow. Capillary flow is also intermittent, because if all capillary beds were filled with blood at the same time, there would be no blood at all in the large vessels. The flow of blood into the capillary beds is regulated by the activity of muscular *terminal arterioles,* which feed the beds, and by *precapillary sphincters* at entrances to the true capillaries. The amount of blood flowing into the true capillaries of the bed is regulated most importantly by local chemical controls (local concentrations of carbon dioxide, histamine, pH). Thus a capillary bed may be flooded with blood or almost entirely bypassed depending on what is happening within the body or in a particular body region at any one time. You will investigate some of the local controls in the next group of experiments.

Activity 5:
Investigating the Effect of Various Factors on the Microcirculation

1. Obtain a frog board (with a hole at one end), dissecting pins, disposable gloves, frog Ringer's solution, 0.01 *N* HCl, 0.01% histamine solution, 1% epinephrine solution, a large rubber band, and some paper towels.

2. Put on the gloves and obtain a frog (alive and hopping, *not* pithed). Moisten several paper towels with room-temperature Ringer's solution, and wrap the frog's body securely with them. One hind leg should be left unsecured and extending beyond the paper cocoon.

3. Attach the frog to the frog board (or other supporting structure) with a large rubber band and then carefully spread (but do not stretch) the web of the exposed hindfoot over the hole in the support. Have your partners hold the edges of the web firmly for viewing. Alternatively, secure the toes to the board with dissecting pins.

4. Obtain a compound microscope, and observe the web under low power to find a capillary bed. Focus on the vessels in high power. Keep the web moistened with Ringer's solution as you work. If the circulation seems to stop during your observations, massage the hind leg of the frog gently to restore blood flow.

5. Observe the red blood cells of the frog. Notice that, unlike human RBCs, they are nucleated. Watch their movement through the smallest vessels—the capillaries. Do they move in single file or do they flow through two or three cells abreast?

____

Are they flexible? ________ Explain. ____________________

Can you see any white blood cells in the capillaries?

________ If so, which types? ____________________________

6. Notice the relative speed of blood flow through the blood vessels. Differentiate between the arterioles, which feed the capillary bed, and the venules, which drain it. This may be tricky, because images are reversed in the microscope. Thus, the vessel that appears to feed into the capillary bed will actually be draining it. You can distinguish between the vessels, however, if you consider that the flow is more pulsating and turbulent in the arterioles and smoother and steadier in the venules. How does the *rate* of flow in the arterioles compare with that in the venules?

____

In the capillaries? ______________________________

What is the relative difference in the diameter of the arterioles and capillaries?

____

Temperature

To investigate the effect of temperature on blood flow, flood the web with cold saline two or three times to chill the entire area. Is a change in vessel diameter noticeable?

________ Which vessels are affected? ____________________

How? ____

Blot the web gently with a paper towel, and then bathe the web with warm saline. Record your observations.

____

Inflammation

Pipette 0.01 *N* HCl onto the frog's web. Hydrochloric acid will act as an irritant and cause a localized inflammatory response. Is there an increase or decrease in the blood flow into the capillary bed following the application of HCl?

____

What purpose do these local changes serve during a localized inflammatory response?

____

____

____

Flush the web with room-temperature Ringer's solution and blot.

Histamine

Histamine, which is released in large amounts during allergic responses, causes extensive vasodilation. Investigate this effect by adding a few drops of histamine solution to the frog web. What happens?

How does this response compare to that produced by HCl?

Blot the web and flood with warm saline as before. Now add a few drops of 1% epinephrine solution, and observe the web. What are epinephrine's effects on the blood vessels?

Epinephrine is used clinically to reverse the vasodilation seen in severe allergic attacks (such as asthma) which are mediated by histamine and other vasoactive molecules.

Return the dropper bottles to the supply area and the frog to the terrarium. Properly clean your work area before leaving the lab. ■

The Lymphatic System and Immune Response

exercise 35

Objectives

1. To name the components of the lymphatic system.
2. To relate the function of the lymphatic system to that of the blood vascular system.
3. To describe the formation and composition of lymph, and to describe how it is transported through the lymphatic vessels.
4. To relate immunological memory, specificity, and differentiation of self from nonself to immune function.
5. To differentiate between the roles of B cells and T cells in the immune response.
6. To describe the structure and function of lymph nodes, and to indicate the localization of T cells, B cells, and macrophages in a typical lymph node.
7. To describe (or draw) the structure of the immunoglobulin monomer, and to name the five immunoglobulin subclasses.
8. To differentiate between antigen and antibody.
9. To understand the use of antigen-antibody reaction using the Ouchterlony test.

Materials

- ❑ Large anatomical chart of the human lymphatic system
- ❑ Prepared microscope slides of lymph node, spleen, and tonsil
- ❑ Compound microscope
- ❑ Wax marking pencil
- ❑ Petri dish with simple saline agar
- ❑ Medicine dropper
- ❑ Dropper bottles of red and green food color
- ❑ Dropper bottles of goat antibody to horse serum albumin, goat antibody to bovine serum albumin, goat antibody to swine serum albumin, horse serum albumin diluted to 20% with physiologic saline, unknown albumin sample diluted to 20% (prepared from horse, swine, and/or bovine albumin)

For instructions on animal dissections, see the dissection exercises starting on p. 709 in the cat and fetal pig editions of this manual.

See Appendix C, Exercise 35 for links to A.D.A.M.® Interactive Anatomy.

The Lymphatic System

General Description

The **lymphatic system** consists of a network of lymphatic vessels (lymphatics), lymphatic tissue, lymph nodes, and a number of other lymphoid organs, such as the tonsils, thymus, and spleen. We will focus on the lymphatic vessels and lymph nodes in this section. The overall function of the lymphatic system is twofold. It transports tissue fluid (lymph) to the blood vessels. In addition, it protects the body by removing foreign material such as bacteria from the lymphatic stream and by serving as a site for lymphocyte "policing" of body fluids and lymphocyte multiplication. These white blood cells are the central actors in body immunity described later in this exercise.

Distribution and Function of Lymphatic Vessels and Lymph Nodes

As blood circulates through the body, the hydrostatic and osmotic pressures operating at the capillary beds result in fluid outflow at the arterial end of the bed and in its return at the venous end. However, not all of the lost fluid is returned to the bloodstream by this mechanism, and the fluid that lags behind in the tissue spaces must eventually return to the blood if the vascular system is to operate properly. (If it does not, fluid accumulates in the tissues, producing a condition called edema.) It is the microscopic, blind-ended **lymphatic capillaries,** which ramify through nearly all the tissues of the body, that pick up this leaked fluid (primarily water and a

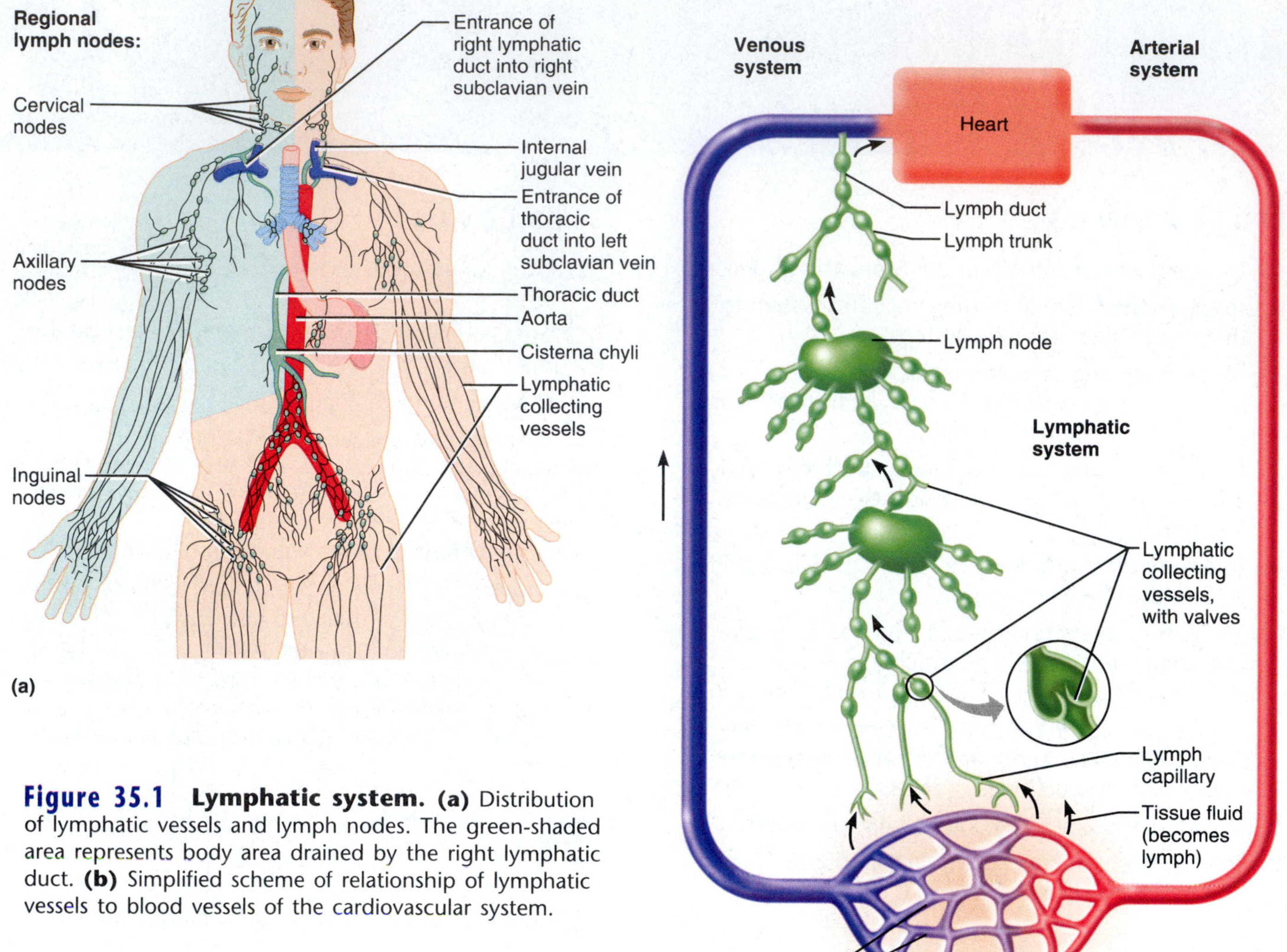

Figure 35.1 Lymphatic system. (a) Distribution of lymphatic vessels and lymph nodes. The green-shaded area represents body area drained by the right lymphatic duct. **(b)** Simplified scheme of relationship of lymphatic vessels to blood vessels of the cardiovascular system.

small amount of dissolved proteins) and carry it through successively larger vessels—**lymphatic collecting vessels** to **lymphatic trunks**—until the lymph finally returns to the blood vascular system through one of the two large ducts in the thoracic region (Figure 35.1). The **right lymphatic duct** drains lymph from the right upper extremity, head, and thorax; the large **thoracic duct** receives lymph from the rest of the body (see Figure 35.1a). In humans, both ducts empty the lymph into the venous circulation at the junction of the internal jugular vein and the subclavian vein, on their respective sides of the body. Notice that the lymphatic system, lacking both a contractile "heart" and arteries, is a one-way system; it carries lymph only toward the heart.

Like veins of the blood vascular system, the lymphatic collecting vessels have three tunics and are equipped with valves (see Plate 29 in the Histology Atlas). However, the lymphatics tend to be thinner-walled, to have *more* valves, and to anastomose (form branching networks) more than veins. Since the lymphatic system is a pumpless system, lymph transport depends largely on the milking action of the skeletal muscles and on pressure changes within the thorax that occur during breathing.

As lymph is transported, it filters through bean-shaped **lymph nodes,** which cluster along the lymphatic vessels of the body. There are thousands of lymph nodes; but because they are usually embedded in connective tissue, they are not ordinarily seen. Within the lymph nodes are **macrophages,** phagocytes that destroy bacteria, cancer cells, and other foreign matter in the lymphatic stream, thus rendering many harmful substances or cells harmless before the lymph enters the bloodstream. Particularly large collections of lymph nodes are found in the inguinal, axillary, and cervical regions of the body. Although we are not usually aware of the filtering and protective nature of the lymph nodes, most of us have experienced "swollen glands" during an active infection. This swelling is a manifestation of the trapping function of the nodes.

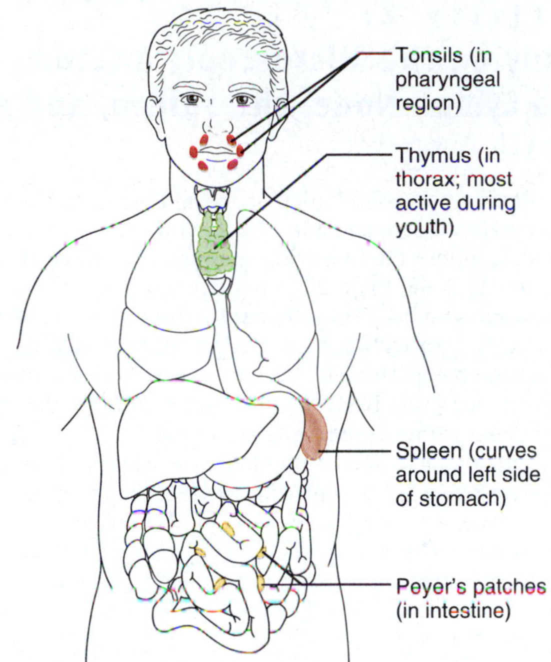

Figure 35.2 Body location of the tonsils, thymus, spleen, and Peyer's patches.

Other lymphoid organs—the tonsils, thymus, and spleen (Figure 35.2)—resemble the lymph nodes histologically, and house similar cell populations (lymphocytes and macrophages).

Activity 1: Identifying the Organs of the Lymphatic System

Study the large anatomical chart to observe the general plan of the lymphatic system. Notice the distribution of lymph nodes, various lymphatics, the lymphatic trunks, and the location of the right lymphatic duct and the thoracic duct. Also identify the **cisterna chyli,** the enlarged terminus of the thoracic duct that receives lymph from the digestive viscera. ■

For instructions on animal dissections, see the dissection exercises starting on p. 709 in the cat and fetal pig editions of this manual.

The Immune Response

The **adaptive immune system** is a functional system that recognizes something as foreign and acts to destroy or neutralize it. This response is known as the **immune response.** It is a systemic response and is not restricted to the initial infection site. When operating effectively, the immune response protects us from bacterial and viral infections, bacterial toxins, and cancer. When it fails or malfunctions, the body is quickly devastated by pathogens or its own assaults.

Major Characteristics of the Immune Response

The most important characteristics of the immune response are its (1) **memory,** (2) **specificity,** and (3) **ability to differentiate self from nonself.** Not only does the immune system have a "memory" for previously encountered foreign antigens (the chicken pox virus for example), but this memory is also remarkably accurate and highly specific.

An almost limitless variety of macromolecules is *antigenic*—that is, capable of provoking an immune response and reacting with its products. Nearly all foreign proteins, many polysaccharides, and many small molecules (haptens), when linked to our own body proteins, exhibit this capability. The cells that recognize antigens and initiate the immune response are lymphocytes. (As described in Exercise 29, lymphocytes are the second most numerous members of the leukocyte or white blood cell [WBC] population.) Each immunocompetent lymphocyte is virtually monospecific; that is, it has receptors on its surface allowing it to bind with only one or a few very similar antigens.

As a rule, our own proteins are tolerated, a fact that reflects the ability of the immune system to distinguish our own tissues (self) from foreign antigens (nonself). Nevertheless, an inability to recognize self can and does occasionally happen and our own tissues are attacked by the immune system. This phenomenon is called *autoimmunity.* Autoimmune diseases include multiple sclerosis (MS), myasthenia gravis, Graves' disease, systemic lupus erythematosus (SLE), glomerulonephritis, rheumatoid arthritis (RA), and insulin-dependent diabetes mellitus (IDDM), which is also called type I or juvenile diabetes.

Organs, Cells, and Cell Interactions of the Immune Response

The immune system utilizes as part of its arsenal the **lymphoid organs,** including the thymus, lymph nodes, spleen, tonsils, and bone marrow. Of these, the thymus and bone marrow are considered to be the *primary lymphoid organs.* The others are *secondary lymphoid areas.*

The stem cells that give rise to the immune system arise in the bone marrow. Their subsequent differentiation into one of the two populations of immunocompetent lymphocytes occurs in the primary lymphoid organs. The **B cells** differentiate in bone marrow, and the **T cells** differentiate in the thymus. While in their "programming organs" the lymphocytes become *immunocompetent,* an event indicated by the appearance of specific cell-surface proteins that enable the lymphocytes to respond (by binding) to a particular antigen.

After differentiation, the B and T cells leave the bone marrow and thymus, respectively; enter the bloodstream; and travel to peripheral (secondary) lymphoid organs, where clonal selection occurs. **Clonal selection** is triggered when an antigen binds to the specific cell-surface receptors of a T or B cell. This event causes the lymphocyte to proliferate rapidly, forming a clone of like cells, all bearing the same antigen-specific receptors. Then, in the presence of certain regulatory signals, the members of the clone specialize, or differentiate—some forming memory cells and others becoming effector or regulatory cells. Upon subsequent meetings with the same antigen, the immune response proceeds considerably faster because the troops are already mobilized and awaiting further orders, so to speak.

In the case of B cell clones, some become memory B cells; the others form antibody-producing **plasma cells.** Because the B cells act indirectly through the antibodies that their progeny release into the bloodstream (or other body fluids), they are said to provide **humoral immunity.** T cell clones are more diverse. Although all T cell clones also contain memory cells, some clones contain *cytotoxic* T cells (effector cells that directly attack virus-infected tissue cells). Others contain regulatory cells such as the *helper cells* (that help activate the B cells and cytotoxic T cells) and still others contain *suppressor cells* that can inhibit the immune response. Because certain T cells act directly to destroy cells infected with viruses, certain bacteria or parasites, and cancer cells, and to reject foreign grafts, T cells are said to mediate **cellular immunity.**

Absence or failure of thymic differentiation of T lymphocytes results in a marked depression of both antibody and cell-mediated immune functions. Additionally, the observation that the thymus naturally involutes with age has been correlated with the relatively immune-deficient status of elderly individuals. ■

All lymphoid tissues except the thymus and bone marrow contain both T and B cell–dependent regions.

Activity 2:

Studying the Microscopic Anatomy of a Lymph Node, the Spleen, and a Tonsil

1. Obtain a compound microscope and prepared slides of a lymph node, spleen, and a tonsil. As you examine the lymph node slide, notice the following anatomical features, depicted in Figure 35.3 and Plate 28 in the Histology Atlas. The node is enclosed within a fibrous **capsule,** from which connective tissue septa **(trabeculae)** extend inward to divide the node into several compartments. Very fine strands of reticular connective tissue issue from the trabeculae, forming the stroma of the gland within which cells are found.

In the outer region of the node, the **cortex,** some of the cells are arranged in globular masses, referred to as germinal centers. The **germinal centers** contain rapidly dividing B lymphocytes. The rest of the cortical cells are primarily T lymphocytes that circulate continuously, moving from the blood into the node and then exiting from the node in the lymphatic stream.

In the internal portion of the gland, the **medulla,** the cells are arranged in cordlike fashion. Most of the medullary cells are macrophages. Macrophages are important not only for their phagocytic function but also because they play an essential role in "presenting" the antigens to the T cells.

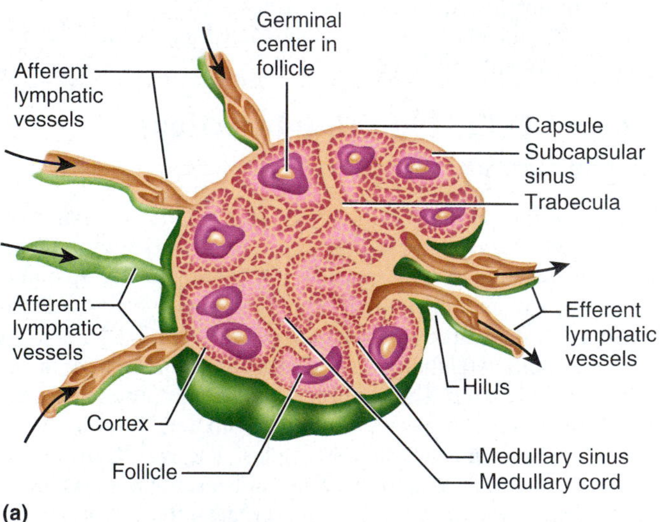

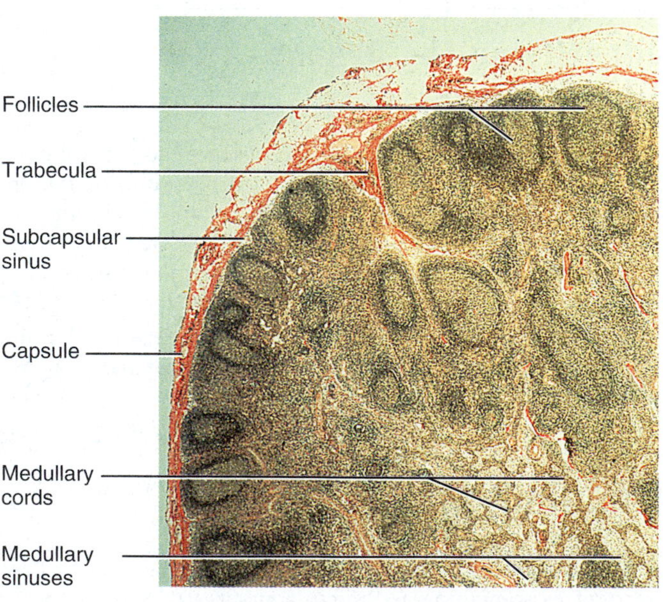

Figure 35.3 Structure of lymph node. **(a)** Cross section of a lymph node, diagrammatic view. Notice that the afferent vessels outnumber the efferent vessels, which slows the rate of lymph flow. The arrows indicate the direction of the lymph flow. **(b)** Photomicrograph of a part of a lymph node (28×). (See also Plate 28 in the Histology Atlas.)

Lymph enters the node through a number of afferent vessels, circulates through *lymph sinuses* within the node, and leaves the node through efferent vessels at the **hilus.** Since each node has fewer efferent than afferent vessels, the lymph flow stagnates somewhat within the node. This allows time for the generation of an immune response and for the macrophages to remove debris from the lymph before it reenters the blood vascular system.

In the space provided here, draw a pie-shaped section of a lymph node, showing the detail of cells in a germinal center, sinusoids, and afferent and efferent vessels. Label all elements.

2. As you observe the slide of the spleen, look for the areas of lymphocytes suspended in reticular fibers, the **white pulp,** clustered around central arteries. The remaining tissue in the spleen is the **red pulp,** which is composed of venous sinuses and areas of reticular tissue and macrophages called the **splenic cords.** The white pulp, composed primarily of lymphocytes, is responsible for the immune functions of the spleen. Macrophages remove worn-out red blood cells, debris, bacteria, viruses, and toxins from blood flowing through the sinuses of the red pulp. (See Plate 30 in the Histology Atlas.)

3. As you examine the tonsil slide, notice the **follicles** containing **germinal centers** surrounded by scattered lymphocytes. The characteristic **crypts** (invaginations of the mucosal epithelium) of the tonsils trap bacteria and other foreign material. Eventually the bacteria work their way into the lymphoid tissue and are destroyed. (See Plate 31 in the Histology Atlas.)

4. Compare and contrast the structure of the lymph node, spleen, and tonsils. ■

Antibodies and Tests for Their Presence

Antibodies, or **immunoglobulins (Igs),** produced by sensitized B cells and their plasma cell offspring in response to an antigen, are a heterogeneous group of proteins that comprise the general class of plasma proteins called **gamma globulins.** Antibodies are found not only in plasma but also (to greater or lesser extents) in all body secretions. Five major classes of immunoglobulins have been identified: IgM, IgG, IgD, IgA, and IgE. The immunoglobulin classes share a common basic structure, but differ functionally and in their localization in the body.

All Igs are composed of one or more monomers (structural units). A monomer consists of four protein chains bound together by disulfide bridges (Figure 35.4). Two of the chains are quite large and have a high molecular weight; these are the **heavy chains.** The other two chains are only half as long and have a low molecular weight. These are called **light chains.** The two heavy chains have a *constant (C) region,* in which the amino acid sequence is identical in both chains, and a *variable (V) region,* which differs in the Igs formed in

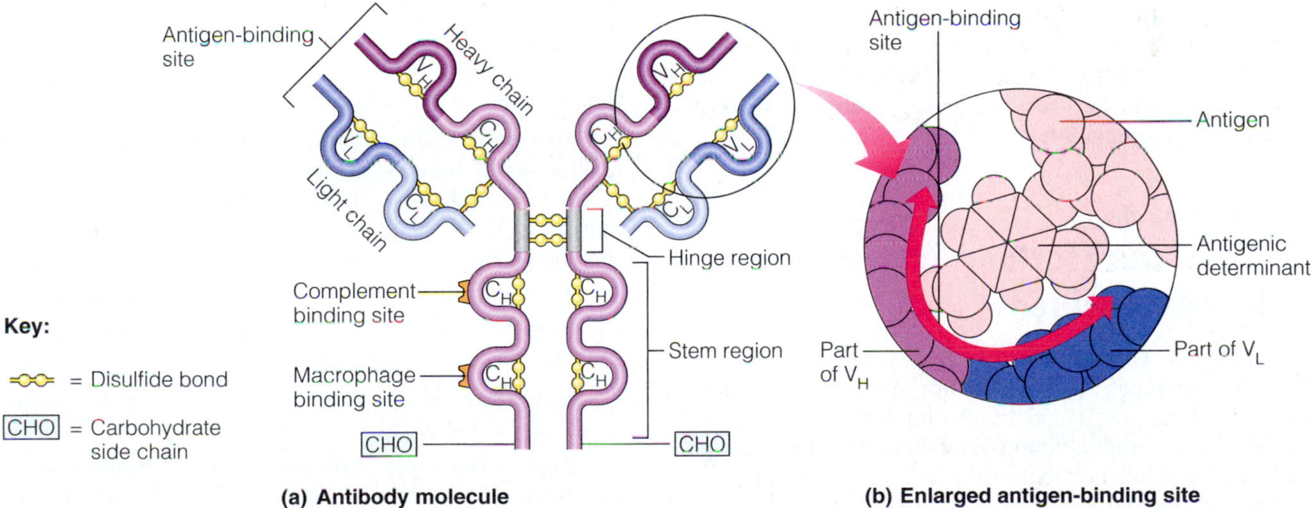

Figure 35.4 Structure of an immunoglobulin monomer. (a) Each monomer is composed of four protein chains (two heavy chains and two light chains) connected by disulfide bonds. Both the heavy and light chains have regions of constant amino acid sequence (C regions) and regions of variable amino acid sequence (V regions). The variable regions differ in each type of antibody and construct the antigen-binding sites. Each immunoglobulin monomer has two such antigen-specific sites. **(b)** Enlargement of an antigen-binding site of an immunoglobulin.

response to different antigens. The same is true of the two light chains; each has a constant and a variable region.

The intact Ig molecule has a three-dimensional shape that is generally Y-shaped. Together, the variable regions of the light and heavy chains in each "arm" construct one **antigen-binding site** uniquely shaped to "fit" a specific *antigenic determinant* (portion) of an antigen. Thus, each Ig monomer bears two identical sites that bind to a specific (and the same) antigen. Binding of the immunoglobulins to their complementary antigen(s) effectively immobilizes the antigens until they can be phagocytized or lysed by complement fixation.

Although the role of the immune system is to protect the body, symptoms of certain diseases involve excessively high antibody synthesis (as in multiple myeloma, a cancer of the bone marrow and adjacent bony structures) and/or the production of abnormal antibodies.

The antigen-antibody reaction is used diagnostically in a variety of ways. One of the most familiar is blood typing. (See Exercise 29 for instructions on ABO and Rh blood typing.) Pregnancy tests also use the antigen-antibody reaction to test for the presence of human chorionic gonadotropin (hCG), a hormone produced early in pregnancy. Another technique for detecting antigens is the enzyme-linked immunoabsorbent assay (ELISA). Originally designed to measure antibody titer, ELISA has been modified for use in HIV-1 blood screening. The antigen-antibody test that we will investigate in this laboratory session is the Ouchterlony technique, which is used mainly for rapid screening of suspected antigens.

Ouchterlony Double-Gel Diffusion, An Immunological Technique

The Ouchterlony double-gel diffusion technique was developed in 1948 to detect the presence of particular antigens in sera or extracts. Antigens and antibodies are placed in wells in a gel and allowed to diffuse toward each other. If an antigen reacts with an antibody, a thin white line called a *precipitin line* forms. In the following activity, the double-gel diffusion technique will be used to identify antigens. Work in groups of no more than three.

Activity 3: Using the Ouchterlony Technique to Identify Antigens

1. Obtain one each of the materials for conducting the Ouchterlony test: petri dish with saline agar; medicine dropper; wax marking pencil; and dropper bottles of red and green food dye, horse serum albumin, an unknown serum albumin sample, and antibodies to horse, bovine, and swine albumin. Put your initials and the number of the unknown albumin sample used on the bottom of the petri dish near the edge.

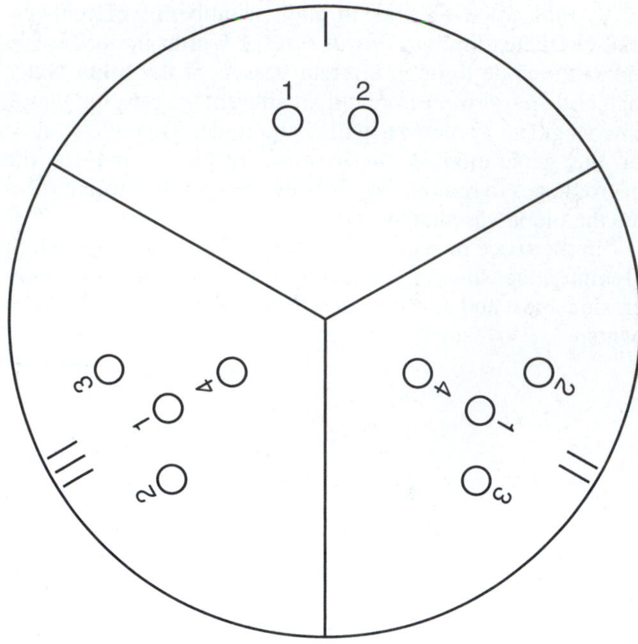

Figure 35.5 Template for well preparation for the Ouchterlony double-gel diffusion experiment.

2. Use the wax marking pencil and the template (Figure 35.5) to divide the dish into three sections, and mark them I, II, and III.

3. Prepare sample wells, again using the template in Figure 35.5. Squeeze the medicine dropper bulb, and gently touch the tip to the surface of the agar. While releasing the bulb, push the tip down through the agar to the bottom of the dish. Lift the dropper vertically; this should leave a straight-walled well in the agar.

4. Repeat step 3 so that section I has two wells and sections II and III have four wells each.

5. To observe diffusion through the gel, fill one well in section I with red dye and the other well with green dye. Be careful not to overfill the wells. Observe periodically for 30 to 45 minutes as the dyes diffuse through the agar. Draw your observations, and answer the appropriate questions.

6. To demonstrate positive and negative results, fill the wells in section II as listed in Table 35.1. A precipitin line should form only between wells 1 and 2.

7. To test the unknown sample, fill the wells in section III as listed in Table 35.2.

8. Replace the cover on the petri dish, and incubate at room temperature for at least 16 hours. Make arrangements to observe the agar for precipitin lines after 16 hours. Note that the lines may begin to fade after 48 hours. Draw the results, indicating the location of all precipitin lines that form.

TABLE 35.1

Well	Solution
1	Horse serum albumin
2	Goat anti-horse albumin
3	Goat anti-bovine albumin
4	Goat anti-swine albumin

TABLE 35.2

Well	Solution
1	Unknown # ______
2	Goat anti-horse albumin
3	Goat anti-bovine albumin
4	Goat anti-swine albumin

Results

1. Demonstration of diffusion using dye. Draw the appearance of section I of your dish after 30 to 45 minutes. Use colored pencils.

2. Draw section II as it appears after incubation, 16 to 48 hours. Be sure to number the wells.

3. Draw section III as it appears after incubation, 16 to 48 hours. Be sure to number the wells.

Unknown # ______

4. What evidence for diffusion did you observe in section I?

____

____

5. Is there any evidence of a precipitate in section I?

____

____

6. Which of the sera functioned as an antigen in section II?

____

____

7. a. Which antibody reacted with the antigen in section II?

____

b. How do you know? (Be specific about your observations.)

____

____

8. If swine albumin had been placed in well 1, what would you expect to happen? Explain.

____

____

9. If chicken albumin had been placed in well 1, what would you expect to happen? Explain.

____

____

10. a. What antigens were present in the unknown solution?

____

b. How do you know? (Be specific about your observations.)

____

____ ■

exercise 36

Anatomy of the Respiratory System

Objectives

1. To define the following terms: *respiratory system, pulmonary ventilation, external respiration,* and *internal respiration.*
2. To label the major respiratory system structures on a diagram (or identify them on a model) and to describe the function of each.
3. To recognize the histologic structure of the trachea (cross section) and lung tissue on prepared slides, and describe the functions the observed structural modifications serve.

Materials

- ❑ Resin cast of the respiratory tree (if available)
- ❑ Human torso model
- ❑ Respiratory organ system model and/or chart of the respiratory system
- ❑ Larynx model (if available)
- ❑ Preserved inflatable lung preparation (obtained from a biological supply house) or sheep pluck fresh from the slaughterhouse
- ❑ Source of compressed air*
- ❑ 0.6 m or 2-foot length of laboratory rubber tubing
- ❑ Dissection tray
- ❑ Disposable gloves
- ❑ Disposable autoclave bag
- ❑ Histologic slides of the following (if available): trachea (cross section), lung tissue, both normal and pathological specimens (e.g., sections taken from lung tissues exhibiting bronchitis, pneumonia, emphysema, or lung cancer)
- ❑ Compound and dissecting microscopes

For instructions on animal dissections, see the dissection exercises starting on p. 709 in the cat and fetal pig editions of this manual.

A1A See Appendix C, Exercise 36 for links to A.D.A.M.® Interactive Anatomy.

*If a compressed air source is not available, cardboard mouthpieces that fit the cut end of the rubber tubing should be available for student use. Disposable autoclave bags should also be provided for discarding the mouthpiece.

Body cells require an abundant and continuous supply of oxygen. As the cells use oxygen, they release carbon dioxide, a waste product that the body must get rid of. These oxygen-using cellular processes, collectively referred to as *cellular respiration,* are more appropriately described in conjunction with the topic of cellular metabolism. The major role of the **respiratory system,** our focus in this exercise, is to supply the body with oxygen and dispose of carbon dioxide. To fulfill this role, at least four distinct processes, collectively referred to as **respiration,** must occur:

Pulmonary ventilation: The tidelike movement of air into and out of the lungs so that the gases in the alveoli are continuously changed and refreshed. Also more simply called *ventilation,* or *breathing.*

External respiration: The gas exchange between the blood and the air-filled chambers of the lungs (oxygen loading/carbon dioxide unloading).

Transport of respiratory gases: The transport of respiratory gases between the lungs and tissue cells of the body accomplished by the cardiovascular system, using blood as the transport vehicle.

Internal respiration: Exchange of gases between systemic blood and tissue cells (oxygen unloading and carbon dioxide loading).

Only the first two processes are the exclusive province of the respiratory system, but all four must occur for the respiratory system to "do its job." Hence, the respiratory and circulatory systems are irreversibly linked. If either system fails, cells begin to die from oxygen starvation and accumulation of carbon dioxide. Uncorrected, this situation soon causes death of the entire organism.

Upper Respiratory System Structures

The upper respiratory system structures—the nose, pharynx, and larynx—are shown in Figure 36.1 and described below. As you read through the descriptions, identify each structure in the figure.

Air generally passes into the respiratory tract through the **external nares** (nostrils), and enters the **nasal cavity** (divided by the **nasal septum**). It then flows posteriorly over three pairs of lobelike structures, the **inferior, superior,** and **middle nasal conchae,** which increase the air turbulence. As the air passes through the nasal cavity, it is also warmed, moistened, and filtered by the nasal mucosa. The air that flows directly beneath the superior part of the nasal cavity may chemically stimulate the olfactory receptors located in

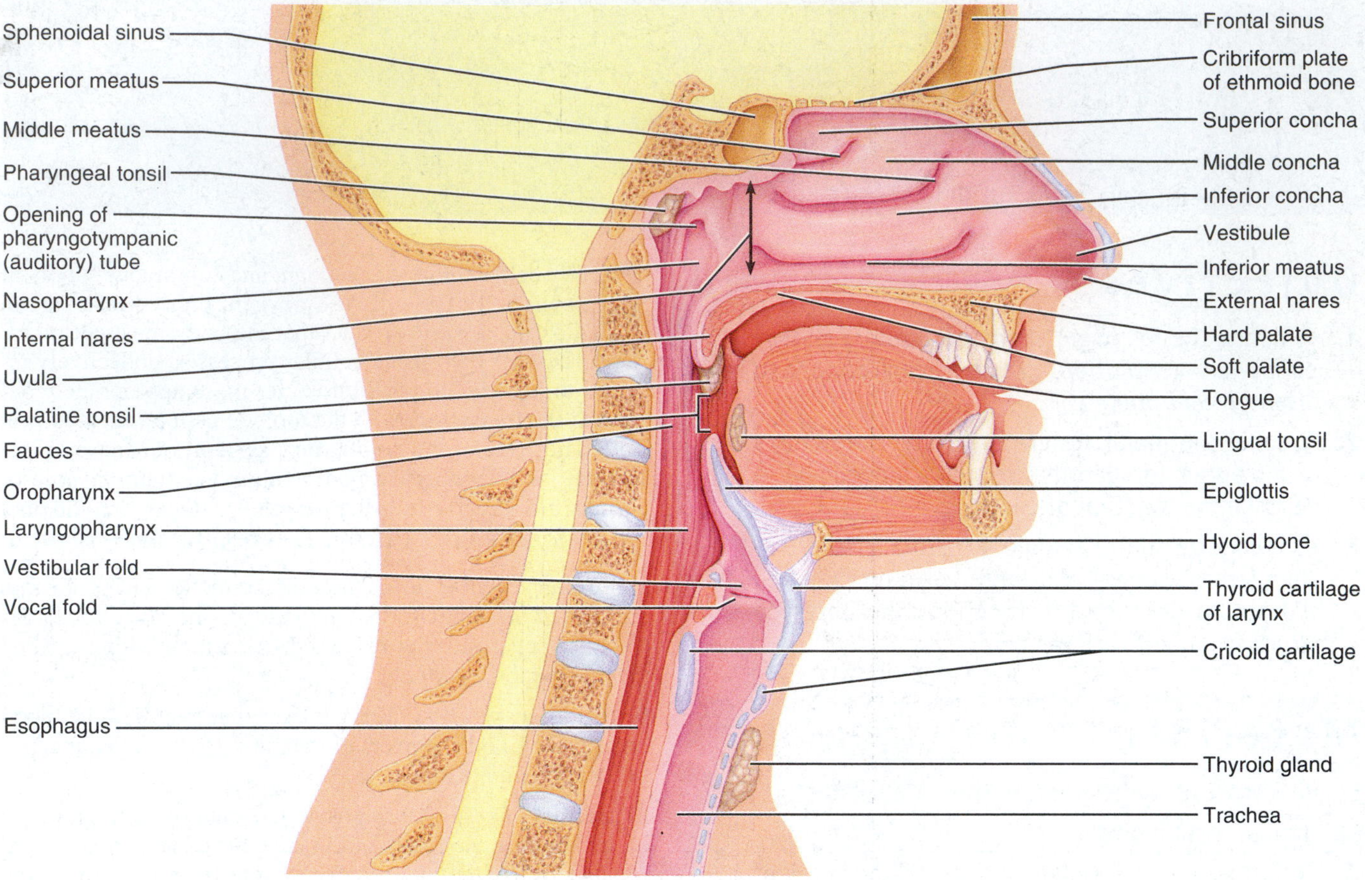

Sphenoidal sinus
Superior meatus
Middle meatus
Pharyngeal tonsil
Opening of pharyngotympanic (auditory) tube
Nasopharynx
Internal nares
Uvula
Palatine tonsil
Fauces
Oropharynx
Laryngopharynx
Vestibular fold
Vocal fold
Esophagus
Frontal sinus
Cribriform plate of ethmoid bone
Superior concha
Middle concha
Inferior concha
Vestibule
Inferior meatus
External nares
Hard palate
Soft palate
Tongue
Lingual tonsil
Epiglottis
Hyoid bone
Thyroid cartilage of larynx
Cricoid cartilage
Thyroid gland
Trachea

(a)

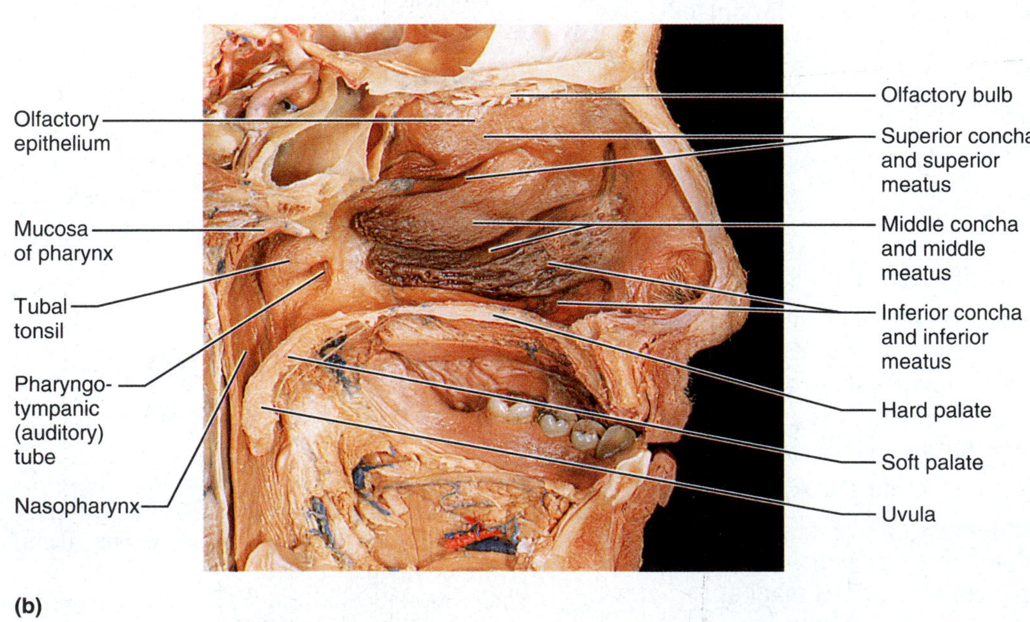

(b)

Figure 36.1 Structures of the upper respiratory tract (sagittal section). **(a)** Diagrammatic view. **(b)** Photograph.

the mucosa of that region. The nasal cavity is surrounded by the **paranasal sinuses** in the frontal, sphenoid, ethmoid, and maxillary bones. These sinuses act as resonance chambers in speech and their mucosae, like that of the nasal cavity, warm and moisten the incoming air.

The nasal passages are separated from the oral cavity below by a partition composed anteriorly of the **hard palate** and posteriorly by the **soft palate.**

The genetic defect called **cleft palate** (failure of the palatine bones and/or the palatine processes of the maxillary bones to fuse medially) causes difficulty in breathing and oral cavity functions such as sucking and, later, mastication and speech. ■

Of course, air may also enter the body via the mouth. From there it passes through the oral cavity to move into the pharynx posteriorly, where the oral and nasal cavities are joined temporarily.

Commonly called the *throat,* the funnel-shaped **pharynx** connects the nasal and oral cavities to the larynx and esophagus inferiorly. It has three named parts (Figure 36.1):

1. The **nasopharynx** lies posterior to the nasal cavity and is continuous with it via the **internal nares.** It lies above the soft palate; hence, it serves only as an air passage. High on its posterior wall are the *pharyngeal tonsils,* paired masses of lymphoid tissue that help to protect the respiratory passages from invading pathogens. The *pharyngotympanic (auditory) tubes,* which allow middle ear pressure to become equalized to atmospheric pressure, drain into the lateral aspects of the nasopharynx. The *tubule tonsils* surround the openings of these tubes into the nasopharynx.

Because of the continuity of the middle ear and nasopharyngeal mucosae, nasal infections may invade the middle ear cavity and cause *otitis media,* which is difficult to treat. ■

2. The **oropharynx** is continuous posteriorly with the oral cavity. Since it extends from the soft palate to the epiglottis of the larynx inferiorly, it serves as a common conduit for food and air. In its lateral walls are the *palatine tonsils.* The *lingual tonsil* covers the base of the tongue.

3. The **laryngopharynx,** like the oropharynx, accommodates both ingested food and air. It lies directly posterior to the upright epiglottis and extends to the larynx, where the common pathway divides into the respiratory and digestive channels. From the laryngopharynx, air enters the lower respiratory passageways by passing through the larynx (voice box) and into the trachea below.

The **larynx** (Figure 36.2) consists of nine cartilages. The two most prominent are the large shield-shaped **thyroid cartilage,** whose anterior medial laryngeal prominence is commonly referred to as *Adam's apple,* and the inferiorly located, ring-shaped **cricoid cartilage,** whose widest dimension faces posteriorly. All the laryngeal cartilages are composed of hyaline cartilage except the flaplike **epiglottis,** a flexible elastic cartilage located superior to the opening of the larynx. The epiglottis, sometimes referred to as the "guardian of the airways," forms a lid over the larynx when we swallow. This closes off the respiratory passageways to incoming food or drink, which is routed into the posterior esophagus, or food chute.

- Palpate your larynx by placing your hand on the anterior neck surface approximately halfway down its length. Swallow. Can you feel the cartilaginous larynx rising?

If anything other than air enters the larynx, a cough reflex attempts to expel the substance. Note that this reflex operates only when a person is conscious. Therefore, you should never try to feed or pour liquids down the throat of an unconscious person.

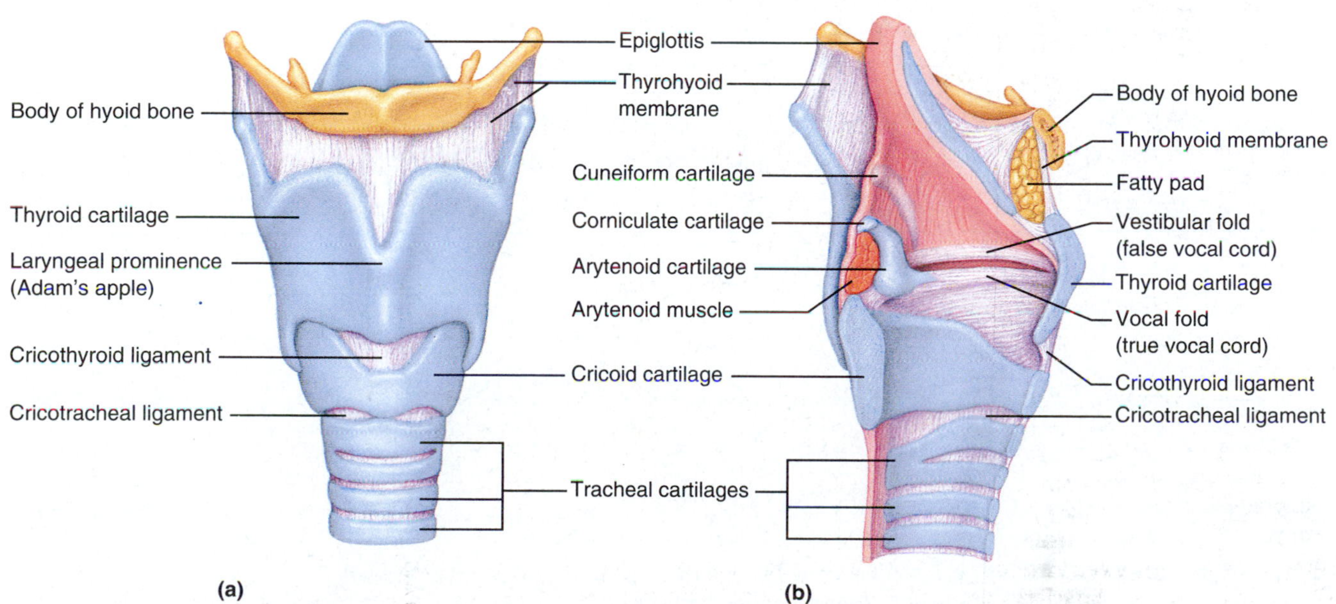

Figure 36.2 Structure of the larynx. **(a)** Anterior view. **(b)** Sagittal section.

The mucous membrane of the larynx is thrown into two pairs of folds—the upper **vestibular folds,** also called the **false vocal cords,** and the lower **vocal folds,** or **true vocal cords,** which vibrate with expelled air for speech. The vocal cords are attached posterolaterally to the small triangular **arytenoid cartilages** by the *vocal ligaments*. The slitlike passageway between the folds is called the **glottis.**

Lower Respiratory System Structures

Air entering the **trachea,** or windpipe, from the larynx travels down its length (about 11.0 cm) to the level of the *sternal angle* (or the disc between the fourth and fifth thoracic vertebrae). There the passageway divides into the right and left **primary bronchi** (Figure 36.3), which plunge into their respective lungs at an indented area called the **hilus** (see Figure 36.5c). The right primary bronchus is wider, shorter, and more vertical than the left. As a result, foreign objects that enter the respiratory passageways are more likely to become lodged in it.

The trachea is lined with a ciliated mucus-secreting, pseudostratified columnar epithelium, as are many of the other respiratory system passageways. The cilia propel mucus (produced by goblet cells) laden with dust particles, bacteria, and other debris away from the lungs and toward the throat, where it can be expectorated or swallowed. The walls of the trachea are reinforced with C-shaped cartilage rings, the incomplete portion located posteriorly. These C-shaped cartilages serve a double function: The incomplete parts allow the esophagus to expand anteriorly when a large food bolus is swallowed. The solid portions reinforce the trachea walls to maintain its open passageway regardless of the pressure changes that occur during breathing.

The primary bronchi further divide into smaller and smaller branches (the secondary, tertiary, on down), finally becoming the **bronchioles,** which have terminal branches called **respiratory bronchioles** (Figure 36.3b). All but the most minute branches have cartilaginous reinforcements in their walls, usually in the form of small plates of hyaline cartilage rather than cartilaginous rings. As the respiratory tubes get smaller and smaller, the relative amount of smooth muscle in their walls increases as the amount of cartilage declines and finally disappears. The complete layer of smooth muscle

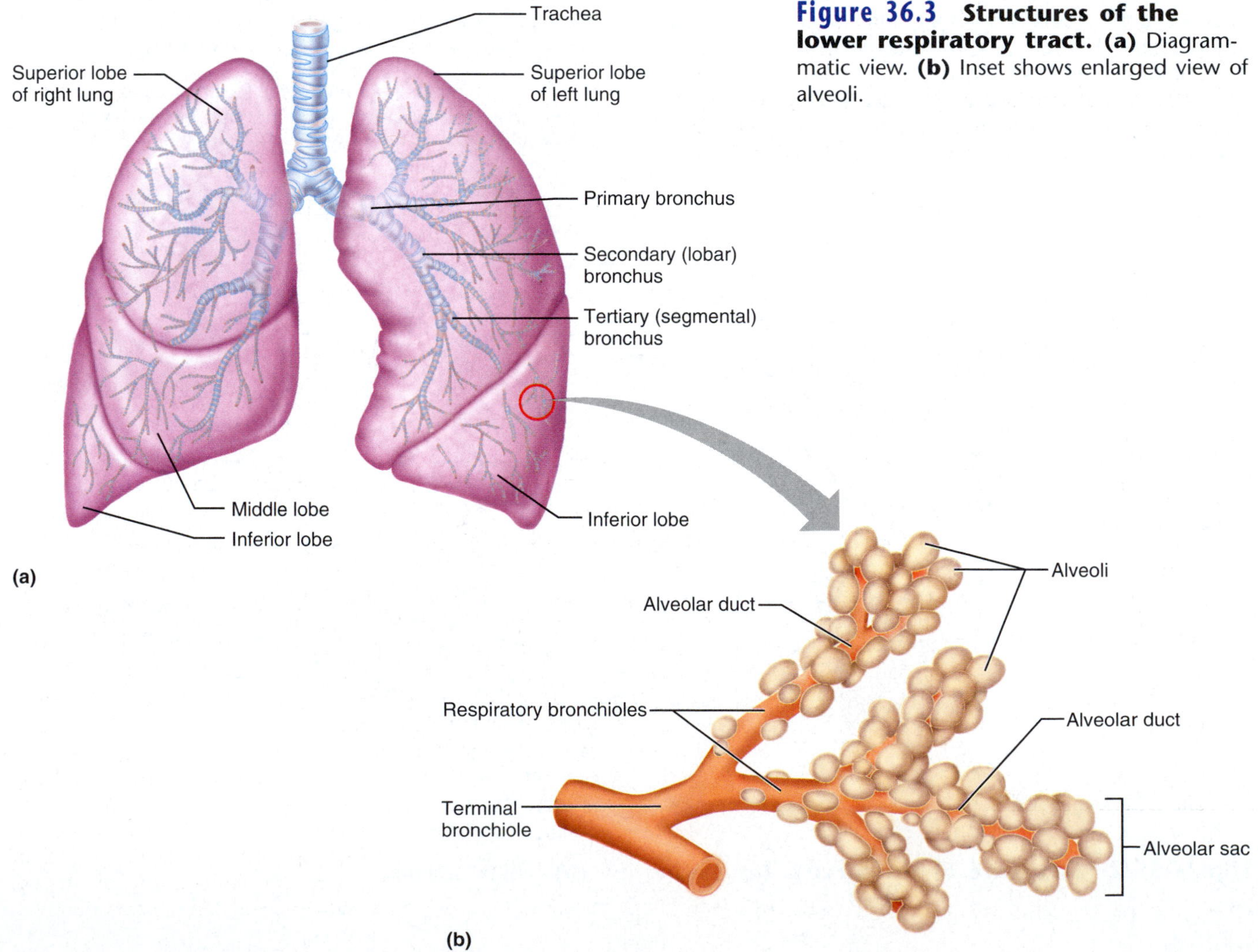

Figure 36.3 Structures of the lower respiratory tract. (a) Diagrammatic view. **(b)** Inset shows enlarged view of alveoli.

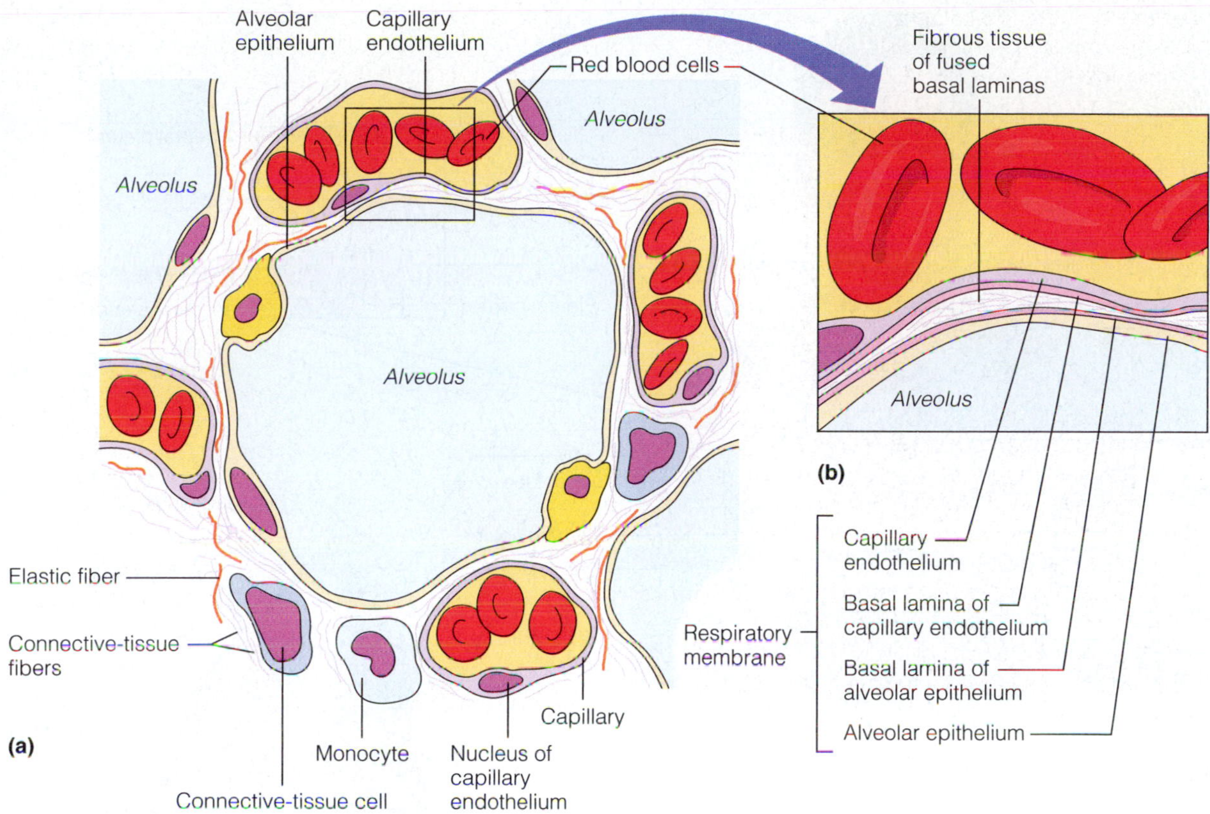

Figure 36.4 Diagrammatic view of the relationship between the alveoli and pulmonary capillaries involved in gas exchange. **(a)** One alveolus surrounded by capillaries. **(b)** Enlargement of the respiratory membrane.

present in the bronchioles enables them to provide considerable resistance to air flow under certain conditions (asthma, hay fever, etc.). The continuous branching of the respiratory passageways in the lungs is often referred to as the **respiratory tree.** The comparison becomes much more meaningful if you observe a resin cast of the respiratory passages. Do so, if one is available for observation in the laboratory.

The respiratory bronchioles in turn subdivide into several **alveolar ducts,** which terminate in alveolar sacs that rather resemble clusters of grapes. **Alveoli,** tiny balloonlike expansions along the alveolar sacs, and occasionally found protruding from the alveolar ducts and respiratory bronchioles, are composed of a single thin layer of squamous epithelium overlying a wispy basal lamina. The external surfaces of the alveoli are densely spiderwebbed with a network of pulmonary capillaries (Figure 36.4). Together, the alveolar and capillary walls and their fused basal laminas form the **respiratory membrane,** also called the **air-blood barrier.** Because gas exchanges occur by simple diffusion across the respiratory membrane—oxygen passing from the alveolar air to the capillary blood and carbon dioxide leaving the capillary blood to enter the alveolar air—the alveolar sacs, alveolar ducts, and respiratory bronchioles are referred to collectively as **respiratory zone structures.** All other respiratory passageways (from the nasal cavity to the terminal bronchioles) simply serve as access or exit routes to and from these gas exchange chambers, and are called **conducting zone structures.** Because the conducting zone structures have no exchange function, they are also referred to as *anatomical dead space.*

The Lungs and Their Pleural Coverings

The paired lungs are soft, spongy organs that occupy the entire thoracic cavity except for the *mediastinum,* which houses the heart, bronchi, esophagus, and other organs (Figure 36.5). Each lung is connected to the mediastinum by a *root* containing its vascular and bronchial attachments. The structures of the root enter (or leave) the lung via a medial indentation called the *hilus.* All structures distal to the primary bronchi are found within the lung substance. A lung's *apex,* the narrower superior aspect, lies just deep to the clavicle, and its *base,* the inferior concave surface, rests on the diaphragm. Anterior, lateral, and posterior lung surfaces are in close contact with the ribs and, hence, are collectively called the *costal surface.* The medial surface of the left lung exhibits a concavity called the *cardiac notch (impression),* which accommodates the heart where it extends left from the body midline. Fissures divide the lungs into a number of *lobes*—two in the left lung and three in the right. Other than the respiratory passageways and air spaces that make up the bulk of their volume, the lungs are mostly elastic connective tissue, which allows them to recoil passively during expiration.

Trachea
Apex of lung
Thymus
Right upper lobe
Horizontal fissure
Right middle lobe
Oblique fissure
Right lower lobe
Heart (in mediastinum)
Diaphragm
Base of lung
Lung
Pleural cavity
Parietal pleura
Rib
Intercostal muscle
Visceral pleura
Left upper lobe
Cardiac notch
Oblique fissure
Left lower lobe

(a)

Pulmonary apex
Pulmonary artery
Left primary bronchus
Pulmonary hilus
Pulmonary vein
Aortic impression
Impression of heart
Oblique fissure
Lobules

(b)

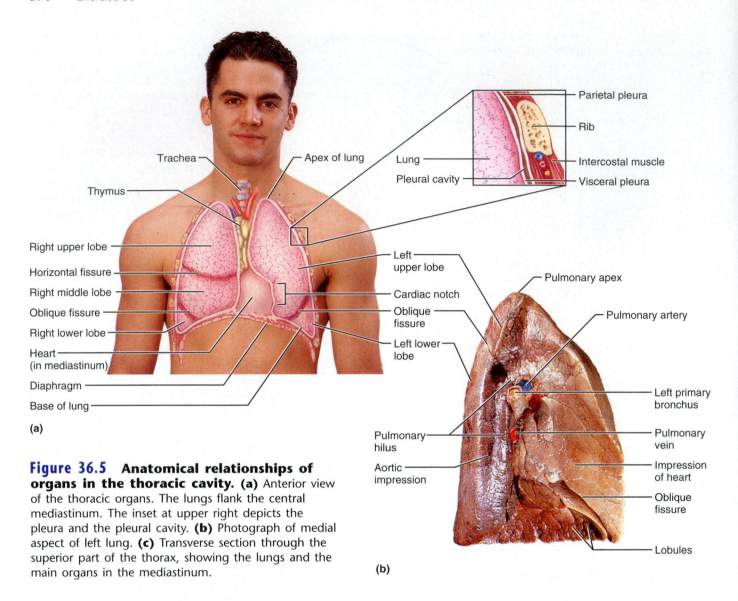

Figure 36.5 Anatomical relationships of organs in the thoracic cavity. (a) Anterior view of the thoracic organs. The lungs flank the central mediastinum. The inset at upper right depicts the pleura and the pleural cavity. **(b)** Photograph of medial aspect of left lung. **(c)** Transverse section through the superior part of the thorax, showing the lungs and the main organs in the mediastinum.

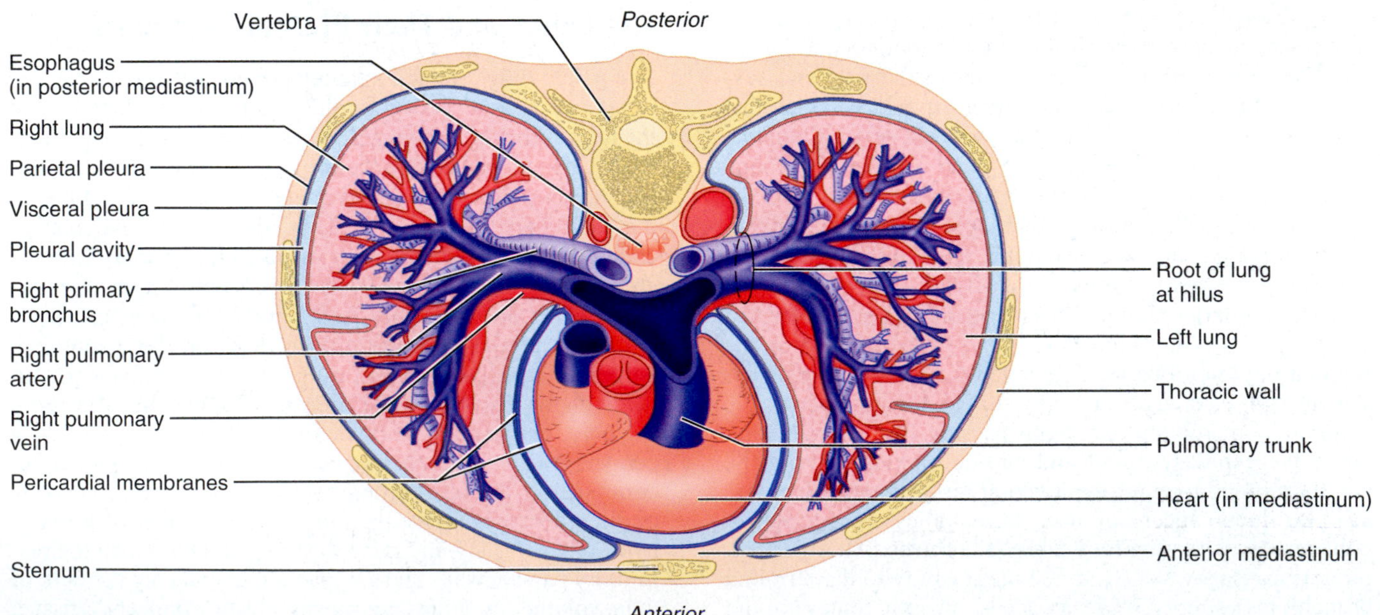

(c)

Each lung is enclosed in a double-layered sac of serous membrane called the **pleura.** The outer layer, the **parietal pleura,** is attached to the thoracic walls and the **diaphragm;** the inner layer, covering the lung tissue, is the **visceral pleura.** The two pleural layers are separated by the *pleural cavity,* which is more of a potential space than an actual one. The pleural layers produce lubricating serous fluid that causes them to adhere closely to one another, holding the lungs to the thoracic wall and allowing them to move easily against one another during the movements of breathing.

Activity 1: Identifying Respiratory System Organs

Before proceeding, be sure to locate on the torso model, thoracic cavity structures model, or an anatomical chart all the respiratory structures described—both upper and lower respiratory system organs. ■

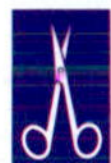
For instructions on animal dissections, see the dissection exercises starting on p. 709 in the cat and fetal pig editions of this manual.

Activity 2: Demonstrating Lung Inflation in a Sheep Pluck

A *sheep pluck* includes the larynx, trachea with attached lungs, the heart and pericardium, and portions of the major blood vessels found in the mediastinum (aorta, pulmonary artery and vein, venae cavae). If a sheep pluck is not available, a good substitute is a preserved inflatable pig lung.

Don plastic gloves, obtain a fresh sheep pluck (or a preserved pluck of another animal), and identify the lower respiratory system organs. Once you have completed your observations, insert a hose from an air compressor (vacuum pump) into the trachea and alternately allow air to flow in and out of the lungs. Notice how the lungs inflate. This observation is educational in a preserved pluck but it is a spectacular sight in a fresh one. Another advantage of using a fresh pluck is that the lung pluck changes color (becomes redder) as hemoglobin in trapped RBCs becomes loaded with oxygen.

If air compressors are not available, the same effect may be obtained by using a length of laboratory rubber tubing to blow into the trachea. Obtain a cardboard mouthpiece and fit it into the cut end of the laboratory tubing before attempting to inflate the lungs.

Dispose of the mouthpiece and gloves in the autoclave bag immediately after use. ■

Activity 3: Examining Prepared Slides of Trachea and Lung Tissue

1. Obtain a compound microscope and a slide of a cross section of the trachea wall. Identify the smooth muscle layer, the hyaline cartilage supporting rings, and the pseudostratified ciliated epithelium. Using Figure 36.6 as a guide, also try to identify a few goblet cells in the epithelium. In the space

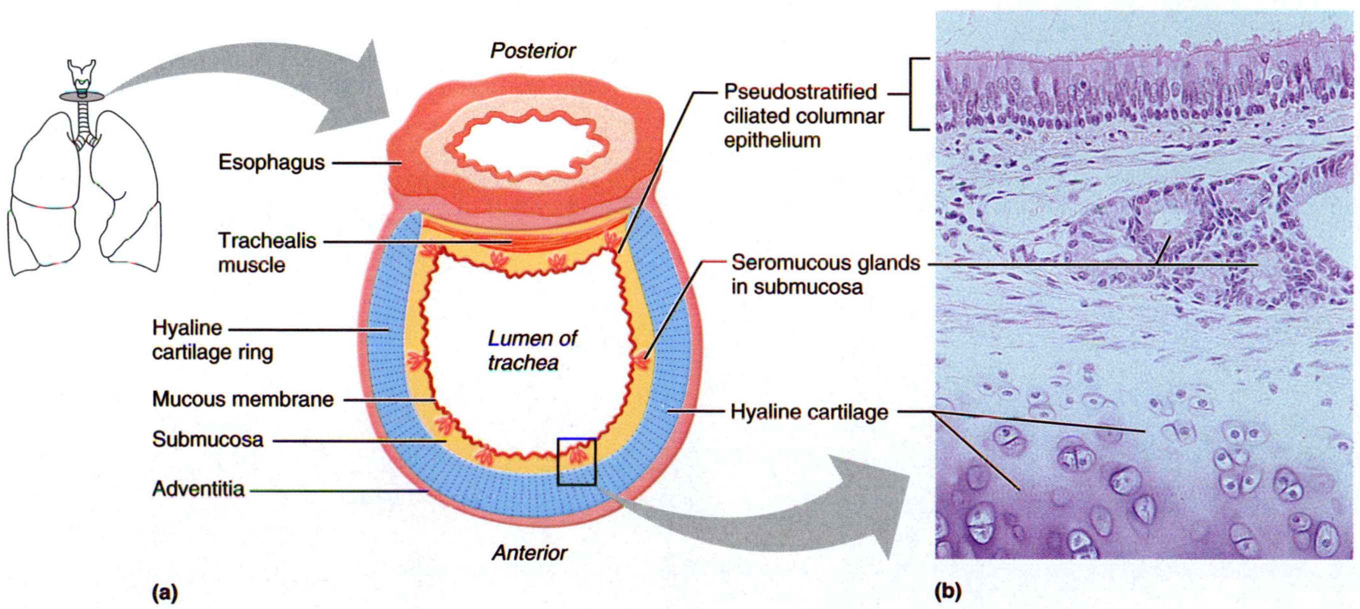

Figure 36.6 Microscopic structure of the trachea. **(a)** Cross-sectional view of the trachea. **(b)** Photomicrograph of a portion of the tracheal wall (212×). (See also Plate 33 in the Histology Atlas.)

below, draw a section of the trachea wall and label all tissue layers.

2. Obtain a slide of lung tissue for examination. The alveolus is the main structural and functional unit of the lung and is the actual site of gas exchange. Identify a bronchiole (Figure 36.7a) and the thin squamous epithelium of the alveolar walls (Figure 36.7b). Draw your observations of a small section of the alveolar tissue in the space below and label the alveoli.

3. Examine slides of pathological lung tissues, and compare them to the normal lung specimens. Record your observations in the Exercise 36 Review Sheets. ■

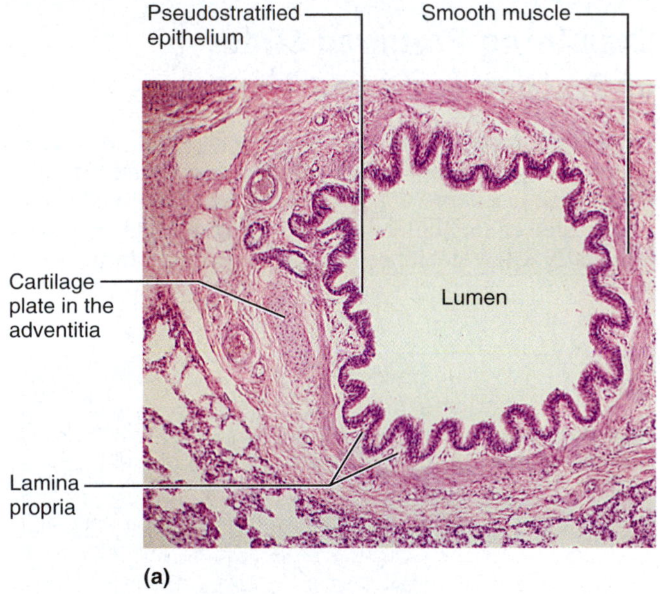

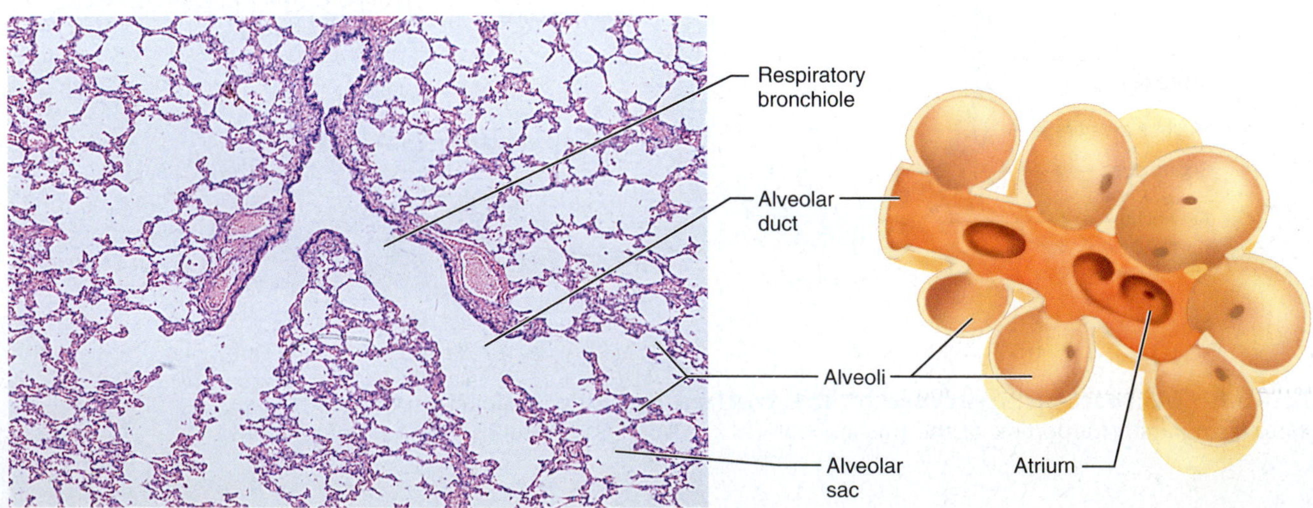

Figure 36.7 Microscopic structure of a bronchiole and alveoli. (a) Photomicrograph of a section of a bronchiole. (See also Plate 35 in the Histology Atlas.) **(b)** Photomicrograph and diagrammatic view of alveoli. (See also Plate 32 in the Histology Atlas.)

Respiratory System Physiology

exercise **37A**

Objectives

1. To define the following (and be prepared to provide volume figures if applicable):
   inspiration, expiration, tidal volume, vital capacity, expiratory reserve volume, inspiratory reserve volume, minute respiratory volume
2. To explain the role of muscles and volume changes in the mechanical process of breathing.
3. To demonstrate proper usage of the spirometer.
4. To explain the relative importance of various mechanical and chemical factors in producing respiratory variations.
5. To describe bronchial and vesicular breathing sounds.
6. To explain the importance of the carbonic acid–bicarbonate buffer system in maintaining blood pH.

Materials

- ❑ Model lung (bell jar demonstrator)
- ❑ Tape measure
- ❑ Spirometer
- ❑ Disposable mouthpieces
- ❑ Nose clips
- ❑ Alcohol swabs
- ❑ 70% ethanol solution in a battery jar
- ❑ Disposable autoclave bag
- ❑ Clear adhesive tape
- ❑ Paper bag
- ❑ Respiratory variations apparatus A or B:
  A: Physiograph, pneumograph, and recording attachments for physiograph
  B: PowerLab® unit, computer with Chart software installed, respiratory belt transducer and cable
- ❑ Stethoscope
- ❑ Table (on chalkboard) for recording class data
- ❑ pH meter (standardized with buffer of pH 7)
- ❑ Buffer solution (pH 7)
- ❑ Concentrated HCl and NaOH in dropper bottles
- ❑ 0.01 *M* HCl
- ❑ 250- and 50-ml beakers
- ❑ Graduated cylinder (100 ml)
- ❑ 0.05 *M* NaOH
- ❑ Phenol red in dropper bottle
- ❑ 100-ml beakers
- ❑ Straws
- ❑ Glass stirring rod
- ❑ Plastic wash bottles containing distilled water
- ❑ Animal plasma
- ❑ *Human Respiratory System* Videotape*

PhysioEx™ 4.0 Computer Simulation on p. P–67

*Available to qualified adopters from Benjamin Cummings.

Mechanics of Respiration

Pulmonary ventilation, or **breathing,** consists of two phases: **inspiration,** during which air is taken into the lungs, and **expiration,** during which air passes out of the lungs. As the inspiratory muscles (external intercostals and diaphragm) contract during inspiration, the size of the thoracic cavity increases. The diaphragm moves from its relaxed dome shape to a flattened position, increasing the superoinferior volume. The external intercostals lift the rib cage, increasing the anteroposterior and lateral dimensions (Figure 37A.1). Since the lungs adhere to the thoracic walls like flypaper because of the presence of serous fluid in the pleural cavity, the intrapulmonary volume (volume within the lungs) also increases, lowering the air (gas) pressure inside the lungs. The gases then expand to fill the available space, creating a partial vacuum that causes air to flow into the lungs—constituting the act of inspiration. During expiration, the inspiratory muscles relax, and the natural tendency of the elastic lung tissue to recoil acts to decrease the intrathoracic and intrapulmonary volumes. As the gas molecules within the lungs are forced closer together, the intrapulmonary pressure rises to a point higher than atmospheric pressure. This causes gases to flow from the lungs to equalize the pressure inside and outside the lungs—the act of expiration.

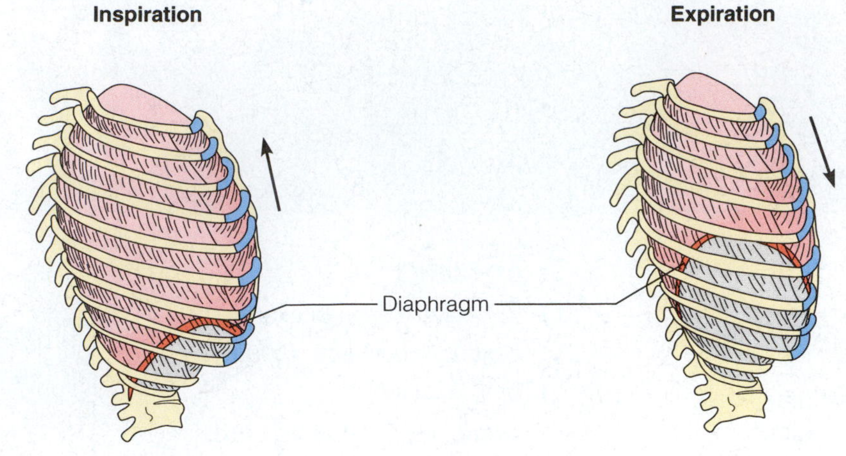

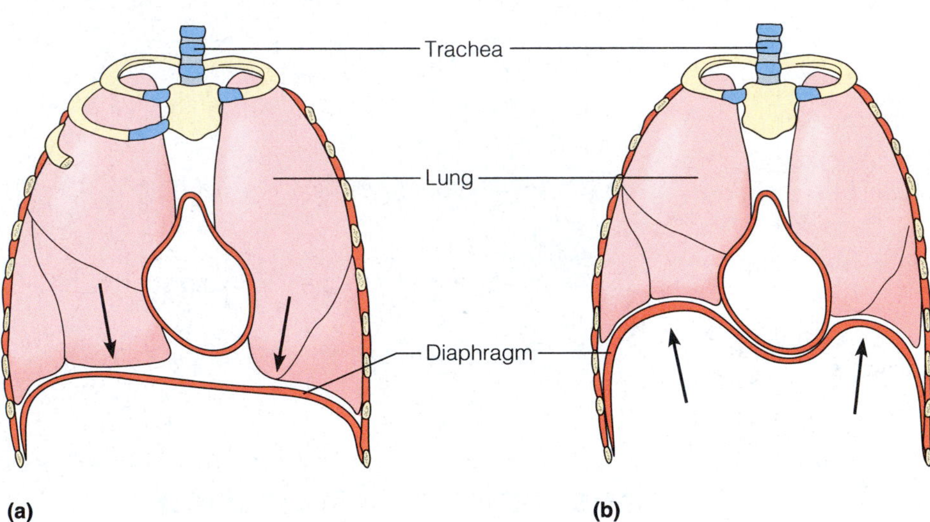

Figure 37A.1 Rib cage and diaphragm positions during breathing. **(a)** At the end of a normal inspiration; chest expanded, diaphragm depressed. **(b)** At the end of a normal expiration; chest depressed, diaphragm elevated.

Activity 1: Operating the Model Lung

Observe the model lung, which demonstrates the principles involved in gas flows into and out of the lungs. It is a simple apparatus with a bottle "thorax," a rubber membrane "diaphragm," and balloon "lungs."

1. Go to the demonstration area and work the model lung by moving the rubber diaphragm up and down. Notice the changes in balloon (lung) size as the volume of the thoracic cavity is alternately increased and decreased.

2. Check the appropriate columns in the chart concerning these observations in the Exercise 37A Review Sheet on p. 000 at the back of this book.

3. Simulate a pneumothorax. Inflate the balloon lungs by pulling down on the diaphragm. Ask your lab partner to let air into the bottle "thorax" by loosening the rubber stopper.

What happens to the balloon lungs?

____

____

4. After observing the operation of the model lung, conduct the following tests on your lab partner. Use the tape measure to determine his or her chest circumference by placing the tape around the chest as high up under the armpits as possible. Record the measurements in inches in the appropriate space for each of the conditions below.

Quiet breathing:

Inspiration ______________ Expiration ______________

Forced breathing:

Inspiration ______________ Expiration ______________

Do the results coincide with what you expected on the basis of what you have learned thus far? ______________ ■

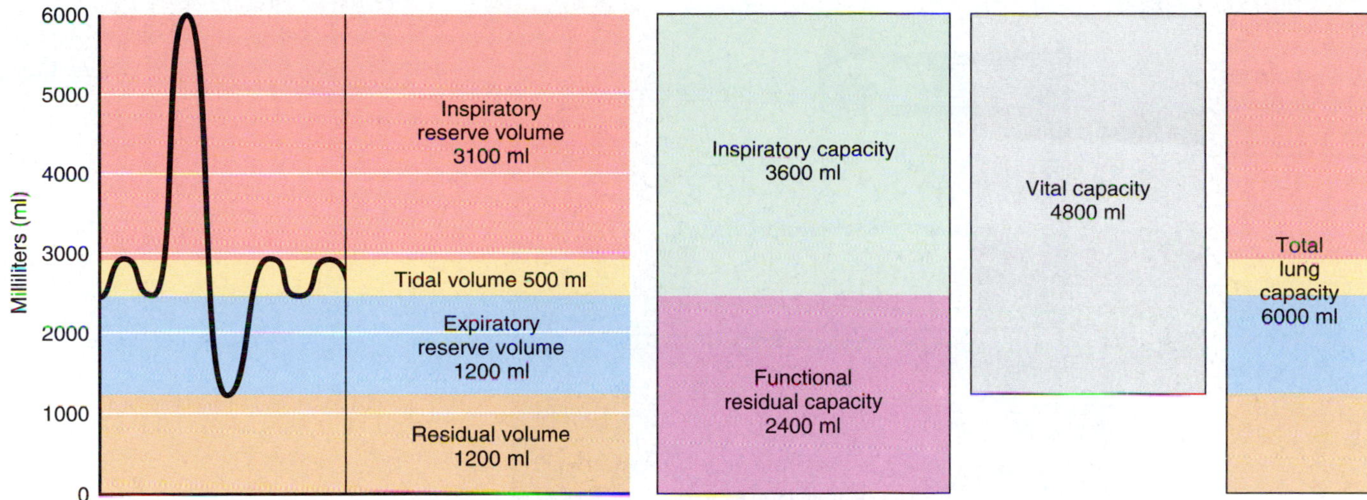

Figure 37A.2 Spirographic record for a male.

Respiratory Volumes and Capacities—Spirometry

A person's size, sex, age, and physical condition produce variations in respiratory volumes. Normal quiet breathing moves about 500 ml of air in and out of the lungs with each breath. As you have seen in the previous experiment, a person can usually forcibly inhale or exhale much more air than is exchanged in normal quiet breathing. The terms given to the measurable respiratory volumes are defined next. These terms and their normal values for an adult male should be memorized.

Tidal volume (TV): Amount of air inhaled or exhaled with each breath under resting conditions (500 ml)

Inspiratory reserve volume (IRV): Amount of air that can be forcefully inhaled after a normal tidal volume inhalation (3100 ml)

Expiratory reserve volume (ERV): Amount of air that can be forcefully exhaled after a normal tidal volume exhalation (1200 ml)

Vital capacity (VC): Maximum amount of air that can be exhaled after a maximal inspiration (4800 ml)

$$VC = TV + IRV + ERV$$

An idealized tracing of the various respiratory volumes and their relationships to each other are shown in Figure 37A.2.

Respiratory volumes will be measured with an apparatus called a **spirometer.** There are two major types of spirometers, which give comparable results—the handheld dry, or wheel, spirometers (such as the Wright spirometer illustrated in Figure 37A.3) and "wet" spirometers, such as the Phipps and Bird spirometer and the Collins spirometer (which is available in both recording and nonrecording varieties). The somewhat more sophisticated wet spirometer consists of a plastic or metal *bell* within a rectangular or cylindrical tank that air can be added to or removed from (Figure 37A.4). The outer tank contains water and has a tube running through it to carry air above the water level. The floating bottomless bell is inverted over the water-containing tank and connected to a volume indicator.

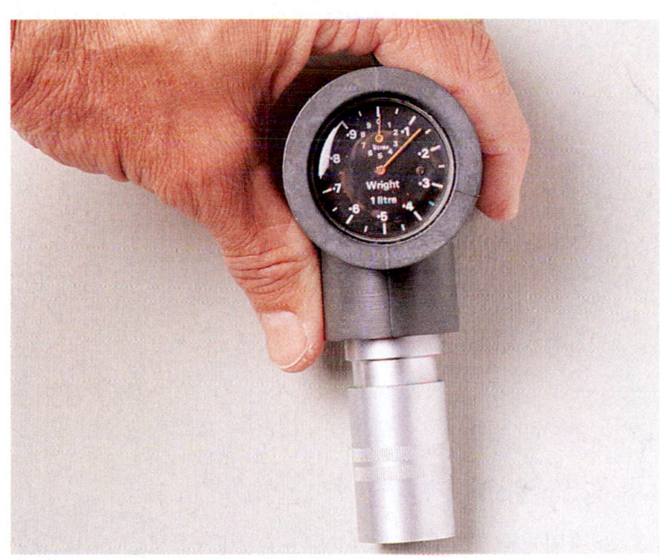

Figure 37A.3 The Wright handheld dry spirometer. Reset to zero prior to each test.

In nonrecording spirometers, an indicator moves as air is *exhaled,* and only expired air volumes can be measured directly. By contrast, recording spirometers allow both inspired and expired gas volumes to be measured. Directions for both types of apparatus are provided on pp. 377-378.

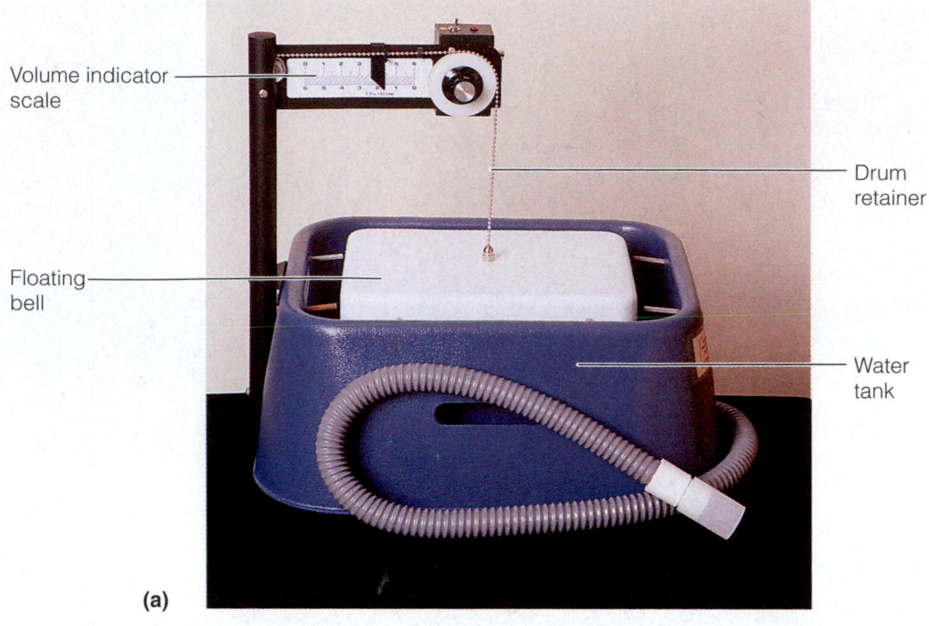

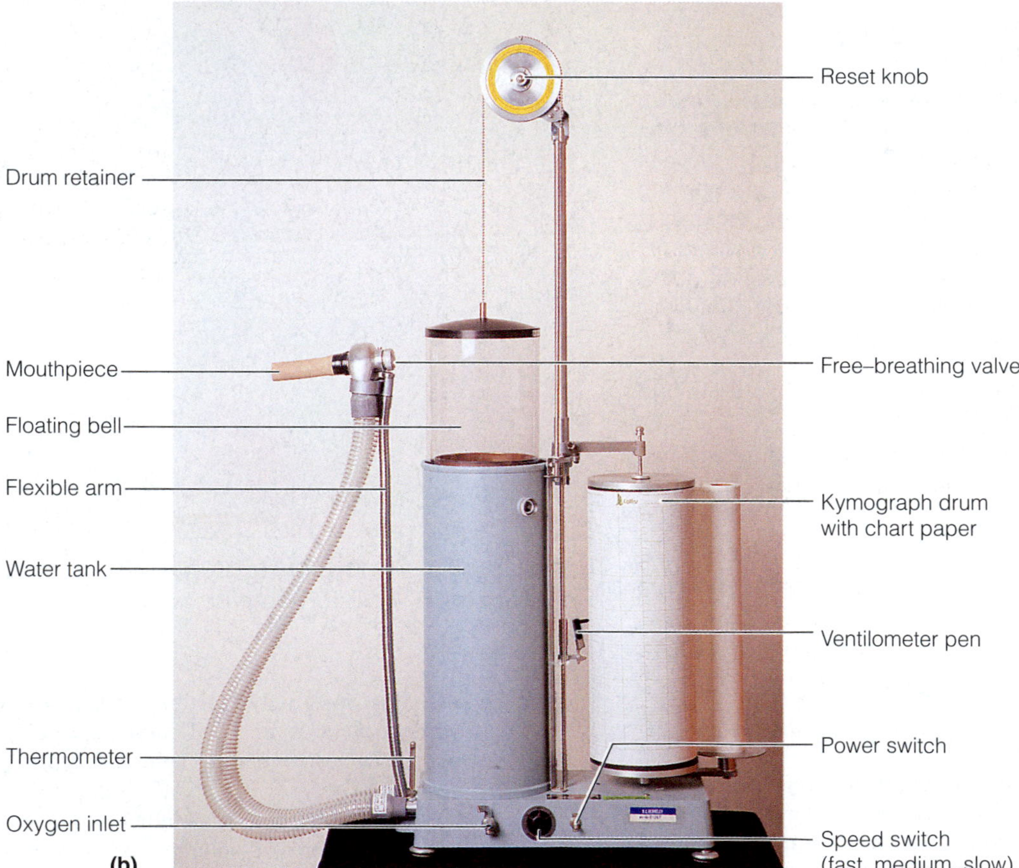

Figure 37A.4 Wet spirometers. (a) The Phipps and Bird "wet" spirometer. **(b)** The Collins-9L "wet" recording spirometer.

Activity 2: Measuring Respiratory Volumes

Follow the steps in Procedure A if using a nonrecording spirometer, and those in Procedure B if using a wet recording spirometer.

Procedure A: Using a Nonrecording Spirometer

1. Without using the spirometer, count and record the subject's normal respiratory rate.

Respirations per minute _______

2. Identify the parts of the spirometer you will be using by comparing it to the illustration in Figure 37A.3 or 37A.4. Examine the spirometer volume indicator *before beginning* to make sure you know how to read the scale. Work in pairs, with one person acting as the subject while the other records the data of the volume determinations. The subject should stand erect during testing. Reset the indicator to zero before beginning each trial. If you are using the handheld spirometer, make sure its dial faces upward so that the volumes can be easily read during the tests.

Obtain a disposable cardboard mouthpiece. Insert it in the open end of the valve assembly (attached to the flexible tube) of the wet spirometer or over the fixed stem of the handheld dry spirometer. Before beginning, the subject should practice exhaling through the mouthpiece without exhaling through the nose, or prepare to use the nose clips (clean them first with an alcohol swab).

3. Conduct the test three times for each required measurement. Record the data here, and then find the average volume figure for that respiratory measurement. After you have completed the trials and computed the averages, enter the average values on the table prepared on the chalkboard for tabulation of class data,* and copy all averaged data onto the Exercise 37A Review Sheet, p. 000.

4. Tidal volume (TV). The volume of air inhaled and exhaled with each normal respiration is approximately 500 ml. To conduct the test, inhale a normal breath, and then exhale a normal breath of air into the spirometer mouthpiece. (Do not force the expiration!) Record the volume and repeat the test twice.

trial 1 _____________ ml trial 2 _______________ ml

trial 3 _____________ ml average TV ___________ ml

* Note to the Instructor: The format of class data tabulation can be similar to that shown here. However, it would be interesting to divide the class into smokers and nonsmokers and then compare the mean average VC and ERV for each group. Such a comparison might help to determine if smokers are handicapped in any way. It also might be a good opportunity for an informal discussion of the early warning signs of bronchitis and emphysema, which are primarily smokers' diseases.

5. Compute the subject's **minute respiratory volume (MRV)** using the following formula:

$$\text{MRV} = \text{TV} \times \text{respirations/min}$$

MRV _______ ml/min

6. Expiratory reserve volume (ERV). The volume of air that can be forcibly exhaled after a normal expiration ranges between 1000 and 1200 ml.

Inhale and exhale normally two or three times, then insert the spirometer mouthpiece and exhale forcibly as much of the additional air as you can. Record your results, and repeat the test twice again.

trial 1 _____________ ml trial 2 _______________ ml

trial 3 _____________ ml average ERV ___________ ml

The ERV is dramatically reduced in conditions in which the elasticity of the lungs is decreased by a chronic obstructive pulmonary disease (COPD) such as **emphysema.** Since energy must be used to *deflate* the lungs in such conditions, expiration is physically exhausting to individuals suffering from COPD. ■

7. Vital capacity (VC). The total exchangeable air of the lungs (the sum of TV + IRV + ERV) is normally around 4500 ml with a range of 3600 ml to 4800 ml. Breathe in and out normally two or three times, and then bend over and exhale all the air possible. Then, as you raise yourself to the upright position, inhale as fully as possible. (It is very important to *strain* to inhale the maximum amount of air that you can.) Quickly insert the mouthpiece, and exhale as forcibly as you can. Record your results and repeat the test twice again.

trial 1 _____________ ml trial 2 ___________ ml

trial 3 _____________ ml average VC ___________ ml

8. Inspiratory reserve volume (IRV). The IRV, or volume of air that can be forcibly inhaled following a normal inspiration, can now be computed using the average values obtained for TV, ERV, and VC and plugging them into the equation:

$$\text{IRV} = \text{VC} - (\text{TV} + \text{ERV})$$

Record your average IRV: _______________________ ml

The normal IRV is substantial, ranging from 2100 to 3100 ml. How does your computed value compare?

____

Steps 9 and 10, which provide common directions for both nonrecording and recording spirometers, continue after the Procedure B directions on p. 380.

Procedure B: Using a Recording Spirometer

1. In preparation for recording, familiarize yourself with the spirometer by comparing it to the equipment illustrated in Figure 37A.4.

2. Examine the chart paper, noting that its horizontal lines represent milliliter units. To apply the chart paper to the recording drum, lift the drum retainer and then remove the kymograph drum. Wrap a sheet of chart paper around the drum, making sure that the right edge overlaps the left. Fasten it with tape, and then replace the kymograph drum and lower the drum retainer into its original position in the hole in the top of the drum.

3. Raise and lower the floating bell several times, noting as you do so that the *ventilometer pen* moves up and down on the drum. This pen, which writes in black ink, will be used for recording and should be positioned or adjusted so that it records in the approximate middle of the chart paper. This adjustment is made by repositioning the floating bell using the *reset knob* on the metal pulley at the top of the spirometer apparatus. The other pen, the respirometer pen, which records in red ink, will not be used for these tests and should be moved away from the drum's recording surface.

4. Clean the nose clips with an alcohol swab. While you wait for the alcohol to air dry, count and record your normal respiratory rate.

Respirations per minute _______

After the alcohol has air dried, apply the nose clips to your nose. This will enforce mouth breathing.

5. Open the *free-breathing valve.* Insert a disposable cardboard mouthpiece into the end (valve assembly) of the breathing tube, and then insert the mouthpiece into your mouth. Practice breathing for several breaths to get used to the apparatus. At this time, you are still breathing room air.

6. Set the spirometer switch to SLOW (32 mm/min). Close the free-breathing valve, and breathe in a normal manner for 2 minutes to record your tidal volume—the amount of air inspired or expired with each normal respiratory cycle. This recording should show a regular pattern of inspiration-expiration spikes and should gradually move upward on the chart paper. (A downward slope indicates that there is an air leak somewhere in the system—most likely at the mouthpiece.) Notice that on an apparatus using a counterweighted pen, such as the Collins Vitalometer shown in Fig. 37A.4b, inspirations are recorded by upstrokes and expirations are recorded by downstrokes.*

7. To record your vital capacity, take the deepest possible inspiration you can and then exhale to the greatest extent possible (really *push* the air out). The recording obtained should resemble that shown in Figure 37A.5. Repeat the vital capacity maneuver twice again. Then turn off the spirometer and remove the chart paper from the kymograph drum.

8. Determine and record your measured, averaged, and corrected respiratory volumes here. Because the pressure and temperature inside the spirometer are influenced by room temperature and differ from those in the body, all measured values are to be multiplied by a **BTPS** (body temperature, atmospheric pressure, and water saturation) **factor.** At room temperature, the BTPS factor is typically 1.1 or very close to that value. Hence, you will multiply your measured values by 1.1 to obtain your corrected respiratory volume values. Copy the averaged and corrected values onto the Exercise 37A Review Sheet.

- Tidal volume (TV). Select a typical resting tidal breath recording. Subtract the millimeter value of the trough (exhalation) from the millimeter value of the peak (inspiration). Record this value below as *measured TV 1*. Select two other TV tracings to determine the TV values for the TV 2 and TV 3 measurements. Then, determine your average TV and multiply it by 1.1 to obtain the BTPS-corrected average TV value.

measured TV 1 _________ ml   average TV _________ ml

measured TV 2 _________ ml   corrected average TV

measured TV 3 _________ ml   _________ ml

Also compute your **minute respiratory volume (MRV)** using the following formula:

$$MRV = TV \times \text{respirations/min}$$

MRV _______ ml/min

- Inspiratory capacity (IC). In the first vital capacity recording, find the expiratory trough immediately preceding the maximal inspiratory peak achieved during vital capacity determination. Subtract the milliliter value of that expiration from the value corresponding to the peak of the maximal inspiration that immediately follows. For example, according to Figure 37A.5, these values would be

$$6600 - 3650 = 2950 \text{ ml}$$

Record your computed value and the results of the two subsequent tests on the appropriate lines below. Then calculate the measured and corrected inspiratory capacity averages and record.

measured IC 1 _________ ml   average IC _________ ml

measured IC 2 _________ ml   corrected average IC

measured IC 3 _________ ml   _________ ml

- Inspiratory reserve volume (IRV). Subtract the corrected average tidal volume from the corrected average for the inspiratory capacity and record below.

$$IRV = \text{corrected average IC} - \text{corrected average TV}$$

corrected average IRV _________ ml

* If a Collins survey spirometer is used, the situation is exactly opposite: Upstrokes are expirations and downstrokes are inspirations.

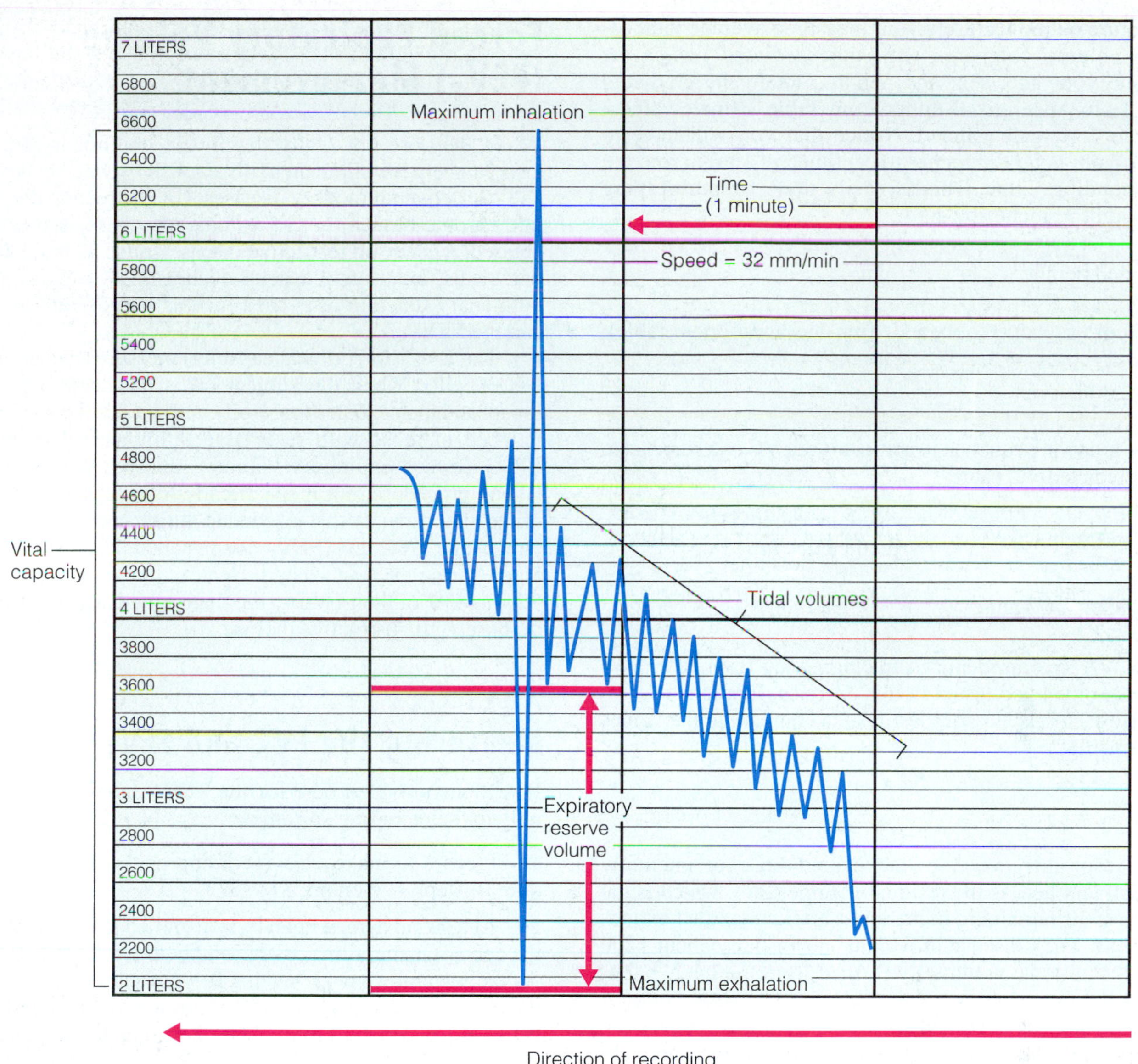

Figure 37A.5 A typical spirometry recording of tidal volume, inspiratory capacity, expiratory reserve volume, and vital capacity. At a drum speed of 32 mm/min, each vertical column of the chart represents a time interval of 1 minute.

• Expiratory reserve volume (ERV). Subtract the number of milliliters corresponding to the trough of the maximal expiration obtained during the vital capacity maneuver from milliliters corresponding to the last *normal* expiration before the VC maneuver is performed. For example, according to Figure 37A.5, these values would be

3650 ml − 2050 ml = 1600 ml

Record your measured and averaged values (three trials) below.

measured ERV 1_______ ml average ERV _________ ml

measured ERV 2_______ ml corrected average ERV

measured ERV 3_________ ml _________ ml

• Vital capacity (VC). Add your corrected values for ERV and IC to obtain the corrected average VC. Record below and on the Exercise 37A Review Sheet.

corrected average VC _________ ml

[*Now continue with step 9, whether you are following Procedure A or Procedure B.*]

9. Figure out how closely your measured average vital capacity volume compares with the *predicted values* for someone your age, sex, and height. Obtain the predicted figure from Appendix B either from Table 1 (male values) or Table 2 (female values). Notice that you will have to convert your height in inches to centimeters (cm) to find the corresponding value. This is easily done by multiplying your height in inches by 2.54.

Computed height: _________ cm

Predicted VC value (obtained from the appropriate table):

_______ ml

Use the following equation to compute your VC as a percentage of the predicted VC value:

$$\% \text{ of predicted VC} = \frac{\text{averaged measured VC}}{\text{predicted value}} \times 100$$

% predicted VC value: _________ %

Figure 37A.2 is an idealized tracing of the respiratory volumes described and tested in this exercise. Examine it carefully. How closely do your test results compare to the values in the tracing?

____

10. A respiratory volume that cannot be experimentally demonstrated here is the residual volume (RV), which is the amount of air remaining in the lungs after a maximal expiratory effort. The presence of residual air (usually about 1200 ml) that cannot be voluntarily flushed from the lungs is important because it allows gas exchange to go on continuously—even between breaths.

Although the residual volume cannot be measured directly, it can be approximated by using one of the following factors:

For ages 16–34 Factor = 1.250
For ages 35–49 Factor = 1.305
For ages 50–69 Factor = 1.445

Compute your predicted RV using the following equation:

$$RV = VC \times \text{factor}$$

11. Recording is finished for this subject. Before continuing with the next member of your group:

- Dispose of used cardboard mouthpieces in the autoclave bag.
- Swish the valve assembly (if removable) in the 70% ethanol solution, then rinse with tap water.
- Put a fresh mouthpiece into the valve assembly (or on the stem of the handheld spirometer). Using the procedures outlined above, measure and record the respiratory volumes for all members of your group. ■

Forced Expiratory Volume (FEV_T) Measurement

While they are not really diagnostic, pulmonary function tests can help the clinician to distinguish between obstructive and restrictive pulmonary diseases. (In obstructive disorders, like chronic bronchitis and asthma, airway resistance is increased, whereas in restrictive diseases, such as polio and tuberculosis, total lung capacity declines.) Two highly useful pulmonary function tests used for this purpose are the FVC and the FEV_T.

The **FVC** (forced vital capacity) measures the amount of gas expelled when the subject takes the deepest possible breath and then exhales forcefully and rapidly. This volume is reduced in those with restrictive pulmonary disease. The **FEV_T** (forced expiratory volume) involves the same basic testing procedure, but it specifically looks at the percentage of the vital capacity that is exhaled during specific time intervals of the FVC test. FEV_1, for instance, is the amount exhaled during the first second. Healthy individuals can expire 75% to 85% of their FVC in the first second. The FEV_1 is low in those with obstructive disease.

Activity 3: Measuring the FVC and FEV_1

Directions provided here for the FEV_T determination apply only to the recording spirometer.

1. Prepare to make your recording as described in Procedure B, steps 1–6 on pp. 378–379.

2. At a signal agreed upon by you and your lab partner, take the deepest inspiration possible and hold it for 1 to 2 seconds. As the inspiratory peak levels off, your partner is to change the drum speed to FAST (1920 mm/min) so that the distance between the vertical lines on the chart represents a time of 1 second.

3. Once the drum speed has been changed, exhale as much air as rapidly and forcibly as possible.

4. When the tracing plateaus (bottoms out), stop recording and determine your FVC. Subtract the milliliter reading in the expiration trough (the bottom plateau) from the preceding inhalation peak (the top plateau). Record this value.

FVC = _______ ml

5. Prepare to calculate the FEV_1. Draw a vertical line intersecting with the spirogram tracing at the precise point that exhalation began. Identify this line as *line 1*. From line 1, measure 32 mm horizontally to the left, and draw a second vertical line. Label this as *line 2*. The distance between the two lines represents 1 second, and the volume exhaled in the first second is read where line 2 intersects the spirogram tracing. Subtract that milliliter value from the milliliter value of the inhalation peak (at the intersection of line 1), to determine the volume of gas expired in the first second. According to the values given in Figure 37A.6, that figure would be 3400 ml (6800 ml − 3400 ml). Record your measured value on p. 382.

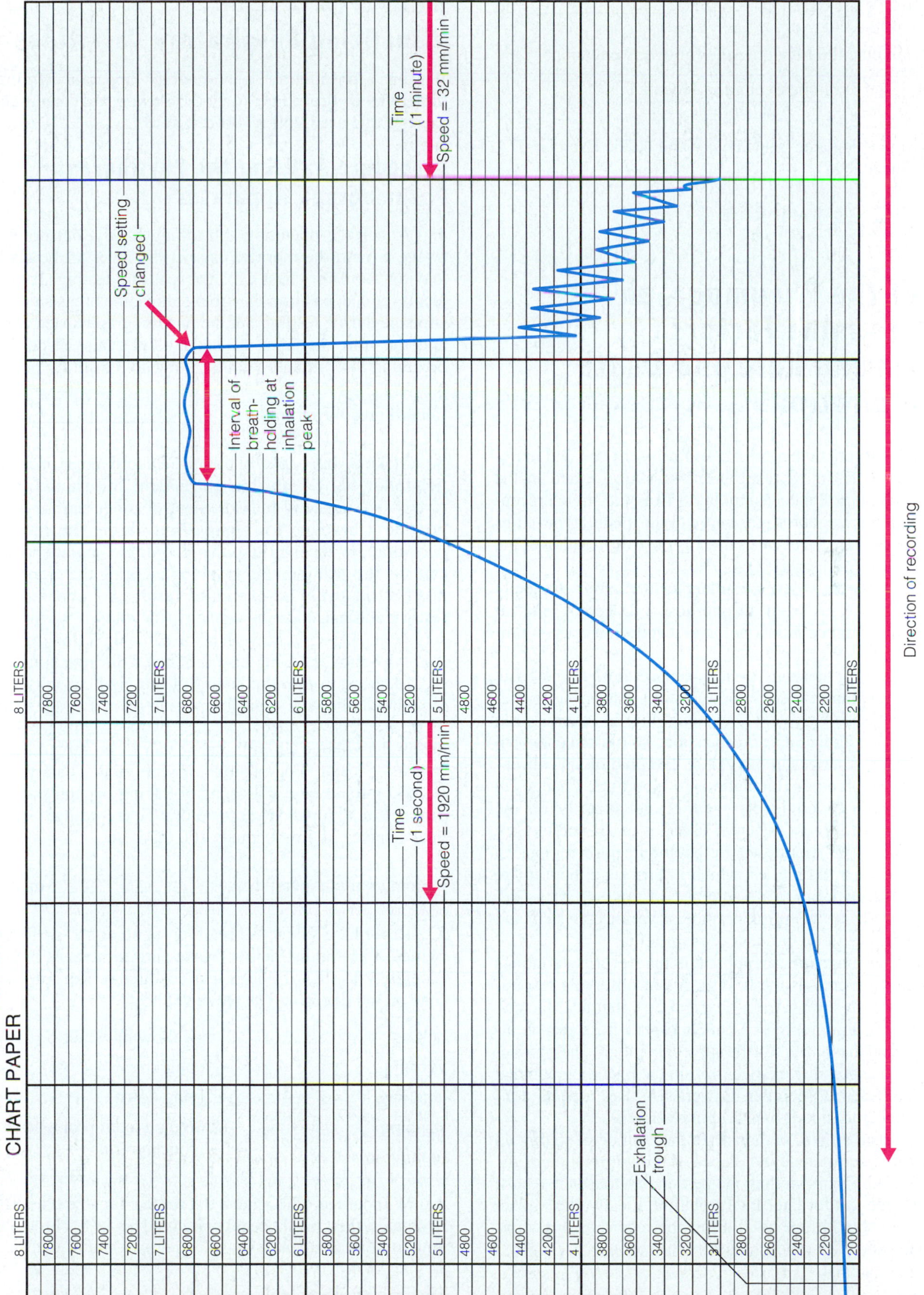

Figure 37A.6 A recording of the forced vital capacity (FVC) and forced expiratory volume (FEV) or timed vital capacity test.

Milliliters of gas expired in second 1: _______________ ml

6. To compute the FEV_1 use the following equation:

$$FEV_1 = \frac{\text{volume expired in second 1}}{\text{FVC volume}} \times 100\%$$

Record your calculated value below and on the Exercise 37A Review Sheet.

FEV_1 = _______ % of FVC ■

Use of the Pneumograph to Determine Factors Influencing Rate and Depth of Respiration*

The neural centers that control respiratory rhythm and maintain a rate of 12 to 18 respirations/min are located in the medulla and pons. On occasion, input from the stretch receptors in the lungs (via the vagus nerve to the medulla) modifies the respiratory rate, as in cases of extreme overinflation of the lungs (Hering-Breuer reflex).

Death occurs when medullary centers are completely suppressed, as from an overdose of sleeping pills or gross overindulgence in alcohol, and respiration ceases completely. ■

Although the nervous system centers initiate the basic rhythm of breathing, there is no question that physical phenomena such as talking, yawning, coughing, and exercise can modify the rate and depth of respiration. So too can chemical factors such as changes in oxygen or carbon dioxide concentrations in the blood or fluctuations in blood pH. Changes in carbon dioxide blood levels seem to act directly on the medulla control centers, whereas changes in pH and oxygen concentrations are monitored by chemoreceptor regions in the aortic and carotid bodies, which in turn send input to the medulla. The experimental sequence in this section is designed to test the relative importance of various physical and chemical factors in the process of respiration.

The **pneumograph,** an apparatus that records variations in breathing patterns, is the best means of observing respiratory variations resulting from physical and chemical factors. The chest pneumograph is a coiled rubber hose that is attached around the thorax. As the subject breathes, chest movements produce pressure changes within the pneumograph that are transmitted to a recorder. PowerLab® uses a respiratory belt transducer to record breathing movements.

The instructor will demonstrate the method of setting up the recording equipment and discuss the interpretation of the results. Work in pairs so that one person can mark the record to identify the test for later interpretation. Ideally, the student being tested should face away from the recording apparatus to prevent voluntary modification of the record.

* Note to the Instructor: This exercise may also be done without using the recording apparatus by simply having the students count the respiratory rate visually.

Activity 4: Visualizing Respiratory Variations

If using a physiograph-pneumograph apparatus, follow the directions in Procedure A. If the PowerLab® pneumograph apparatus is used, follow Procedure B.

Procedure A: Using the Physiograph-Pneumograph Apparatus

1. Attach the pneumograph tubing firmly, but not restrictively, around the thoracic cage at the level of the sixth rib, leaving room for chest expansion during testing. If the subject is female, position the tubing above the breasts to prevent slippage during testing. Set the pneumograph speed at 1 or 2, and the time signal at 10-second intervals. Record quiet breathing for 1 minute with the subject in a sitting position.

Record breaths per minute. _______________

2. Make a vital capacity tracing: Record a maximal inhalation followed by a maximal exhalation. This should correlate to the vital capacity measurement obtained earlier and will provide a baseline for comparison during the rest of the pneumograph testing. Stop the recording apparatus and indicate the following on the graph by marking the graph appropriately: tidal volume, expiratory reserve volume, inspiratory reserve volume, and vital capacity (the total of the three measurements). Also mark, with arrows, the direction the recording stylus moves during inspiration and during expiration.

Measure in mm the height of the vital capacity recording. Divide the vital capacity measurement average recorded on the Exercise 37A Review Sheet by the millimeter figure to obtain the volume (in milliliters of air) represented by 1 mm on the recording. For example, if your vital capacity reading is 4000 ml and the vital capacity tracing occupies a vertical distance of 40 mm on the pneumograph recording, then a vertical distance of 1 mm equals 100 ml of air.

Record your computed value. _______________ ml air/mm

3. Record the subject's breathing as he or she performs activities from the list below. Make sure the record is marked accurately to identify each test conducted. Record your results on the Exercise 37A Review Sheet.

talking	swallowing water
yawning	coughing
laughing	lying down
standing	doing a math problem (concentrating)
running in place	

4. Without recording, have the subject breathe normally for 2 minutes, then inhale deeply and hold his or her breath for as long as he or she can.

Time the breath-holding interval. _______________ sec

As the subject exhales, turn on the recording apparatus and record the recovery period (time to return to normal breathing—usually slightly over 1 minute):

Time of recovery period. ______________________ sec

Did the subject have the urge to inspire *or* expire during breath holding?

____

Without recording, repeat the above experiment, but this time exhale completely and forcefully *after* taking the deep breath. What was observed this time?

____

____

Explain the results. (Hint: the vagus nerve is the sensory nerve of the lungs and plays a role here.)

____

____

____

5. Have the subject hyperventilate (breathe deeply and forcefully at the rate of 1 breath/4 sec) for about 30 seconds.* Record both during and after hyperventilation. How does the pattern obtained during hyperventilation compare with that recorded during the vital capacity tracing?

____

Is the respiratory rate after hyperventilation faster *or* slower than during normal quiet breathing?

____

6. Repeat the above test, but do not record until after hyperventilating. After hyperventilation, the subject is to hold his or her breath as long as he or she can. Can the breath be held for a longer or shorter time after hyperventilating?

____

7. Without recording, have the subject breathe into a paper bag for 3 minutes, then record his or her breathing movements.

* A sensation of dizziness may develop. (As the carbon dioxide is washed out of the blood by overventilation, the blood pH increases, leading to a decrease in blood pressure and reduced cerebral circulation.) The subject may experience a lack of desire to breathe after forced breathing is stopped. If the period of breathing cessation—apnea—is extended, cyanosis of the lips may occur.

⚠ *Caution*: During the bag-breathing exercise the subject's partner should watch the subject carefully for any untoward reactions.

Is the breathing rate faster *or* slower than that recorded during normal quiet breathing?

____

After hyperventilating? ______________________

8. Run in place for 2 minutes, and then have your partner determine the length of time that you can hold your breath.

Length of breath-holding. ________ sec

9. To prove that respiration has a marked effect on circulation, conduct the following test. Have your lab partner record the rate and relative force of your radial pulse before you begin.

Rate __________ beats/min   Relative force __________

Inspire forcibly. Immediately close your mouth and nose to retain the inhaled air, and then make a forceful and prolonged expiration. Your lab partner should observe and record the condition of the blood vessels of your neck and face, and again immediately palpate the radial pulse.

Observations ______________________

____

Radial pulse ________ beats/min   Relative force ________

Explain the changes observed. ______________________

____

____

⚠ Dispose of the paper bag in the autoclave bag. Keep the pneumograph records to interpret results and hand them in if requested by the instructor. Observation of the test results should enable you to determine which chemical factor, carbon dioxide or oxygen, has the greatest effect on modifying the respiratory rate and depth. ■

Procedure B: Using the PowerLab® Respiratory Belt Transducer

1. Make sure the computer is off and connect the PowerLab® unit. Start the computer.

2. Open the PowerLab® folder and select the file named *Experiment 11—Breathing*. The Chart program will open and the Chart window will appear. CHANNEL 1 is the raw breathing signal from the respiratory belt, and CHANNEL 2 is the computed breathing rate from the raw signal in breaths per minute (BPM).

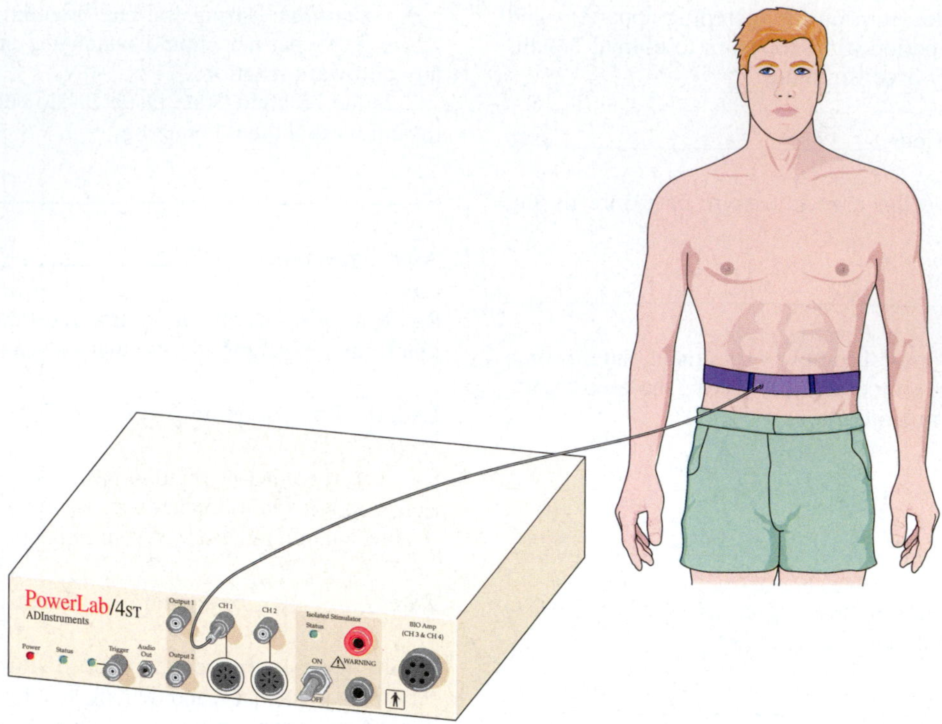

Figure 37A.7 PowerLab® setup for respiratory recording. Connecting the respiratory belt transducer to the subject and to PowerLab®.

3. Fasten the respiratory belt around the upper abdomen of a subject. (See Figure 37A.7). The transducer should be at the front of the body and level with the navel. The belt should be firm but not uncomfortably tight.

4. Attach the respiratory belt transducer to CHANNEL 1 on the PowerLab® unit.

5. Choose the *Input Amplifier* from CHANNEL 1 of the Channel function pop-up menu.

6. Ask the subject to take deep, strong breaths. Adjust the value in the *Range* pop-up menu of the Input Amplifier so that the breathing record occupies about a half to two-thirds of full scale. (See Figure 37A.8.) Click *OK* to close the dialog box.

7. Record the respiratory movements for exactly 30 seconds, and type a comment on the graph to indicate the 30-second timing period. Once this has been done, *do not adjust the chart speed*. This horizontal distance can then be used to measure time.

8. Be sure to type comments on the recordings, and save the data in an appropriate file. You will have to print the recordings to complete the exercise.

9. Begin by recording quiet breathing for 1 minute with the subject in a sitting position.

Record breaths per minute. ________

10. For instructions on the balance of the required tracings, go to step 2 of Procedure A and continue as indicated through step 9. Write comments as you record, and save data in an appropriate file. ■

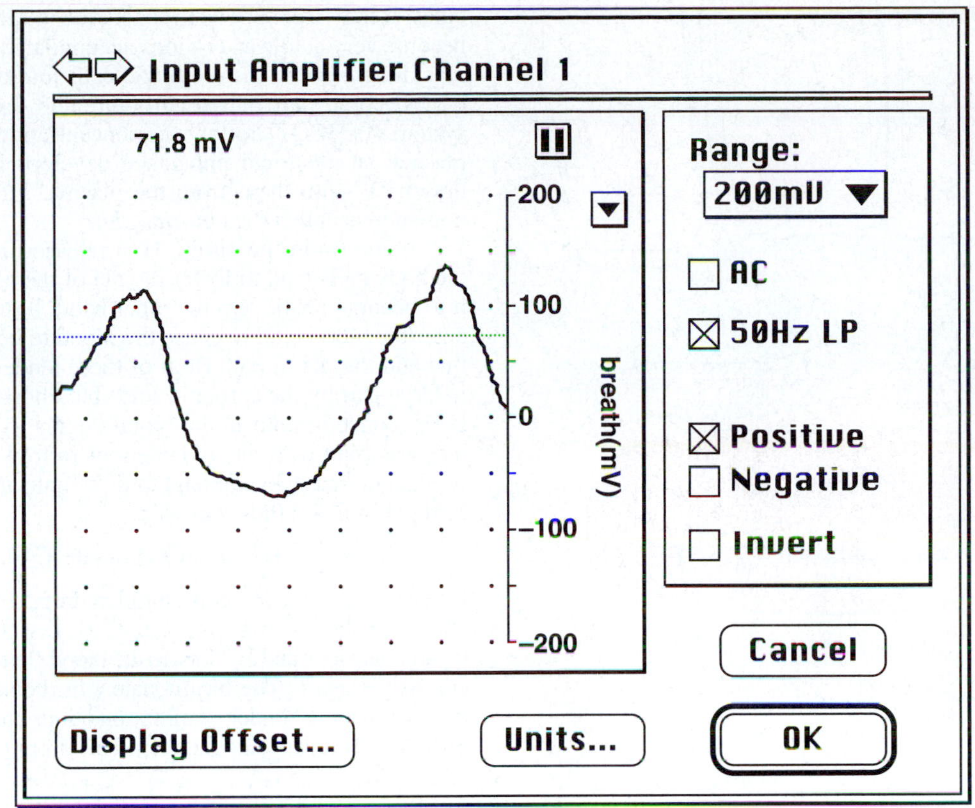

Figure 37A.8 PowerLab® range adjustment. The Input Amplifier dialog box showing the "breath" signal: the range has been adjusted so that the signal is the correct size.

Respiratory Sounds

As air flows in and out of the respiratory tree, it produces two characteristic sounds that can be picked up with a stethoscope (auscultated). The **bronchial sounds** are produced by air rushing through the large respiratory passageways (the trachea and the bronchi). The second sound type, **vesicular breathing sounds,** apparently results from air filling the alveolar sacs and resembles the sound of a rustling or muffled breeze.

Activity 5: Auscultating Respiratory Sounds

1. Obtain a stethoscope and clean the earpieces with an alcohol swab. Allow the alcohol to dry before donning the stethoscope.

2. Place the diaphragm of the stethoscope on the throat of the test subject just below the larynx. Listen for bronchial sounds on inspiration and expiration. Move the stethoscope down toward the bronchi until you can no longer hear sounds.

3. Place the stethoscope over the following chest areas and listen for vesicular sounds during respiration (heard primarily during inspiration).

- At various intercostal spaces
- At the *triangle of auscultation* (a small depressed area of the back where the muscles fail to cover the rib cage; located just medial to the inferior part of the scapula)
- Under the clavicle ■

Diseased respiratory tissue, mucus, or pus can produce abnormal chest sounds such as rales (a rasping sound) and wheezing (a whistling sound). ■

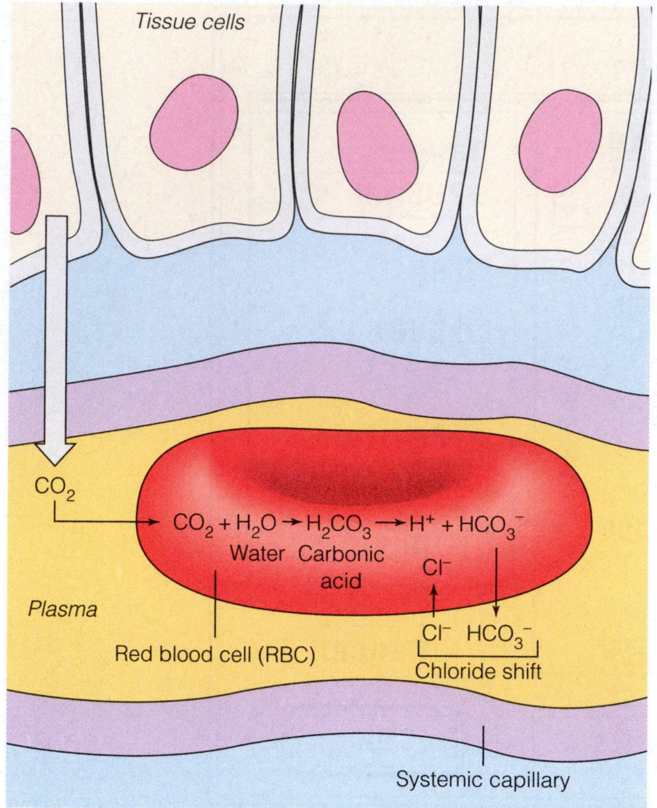

Figure 37A.9 Carbon dioxide loading in the tissues. As carbon dioxide is loaded into the blood, it enters red blood cells where it combines with water to yield carbonic acid, which immediately splits to HCO_3^- (bicarbonate ions) and H^+. HCO_3^- diffuses out of the RBC into the plasma, and chloride ions enter the RBC in exchange (the chloride shift).

Role of the Respiratory System in Acid-Base Balance of Blood

As you have already learned, pulmonary ventilation is necessary for continuous oxygenation of the blood and removal of carbon dioxide (a waste product of cellular respiration) from the blood. Blood pH must be relatively constant for the cells of the body to function optimally. The carbonic acid–bicarbonate buffer system of the blood is extremely important because it helps stabilize arterial blood pH at 7.4 ± 0.02.

When carbon dioxide diffuses into the blood from the tissue cells, much of it enters the red blood cells, where it combines with water to form carbonic acid (Figure 37A.9):

$$H_2O + CO_2 \xrightarrow[\text{enzyme present in RBC}]{\text{Carbonic anhydrase}} H_2CO_3$$

Some carbonic acid is also formed in the plasma, but its formation there is very slow because of the lack of the carbonic anhydrase enzyme. Shortly after it forms, carbonic acid dissociates to release bicarbonate (HCO_3^-) and hydrogen (H^+) ions. The hydrogen ions that remain in the cells are neutralized, or buffered, when they combine with hemoglobin molecules. If they were not neutralized, the intracellular pH would become very acidic as H^+ ions accumulated. The bicarbonate ions diffuse out of the red blood cells into the plasma, where they become part of the carbonic acid–bicarbonate buffer system. As HCO_3^- follows its concentration gradient into the plasma, an electrical imbalance develops in the RBCs that draws Cl^- into them from the plasma. This exchange phenomenon is called the *chloride shift*.

Acids (more precisely, H^+) released into the blood by the body cells tend to lower the pH of the blood and to cause it to become acidic. On the other hand, basic substances that enter the blood tend to cause the blood to become more alkaline and the pH to rise. Both of these tendencies are resisted in large part by the carbonic acid–bicarbonate buffer system. If H^+ concentration in the blood begins to increase, the H^+ ions combine with bicarbonate ions to form carbonic acid (a weak acid that does not tend to dissociate at physiological or acid pH) and are thus removed.

$$H^+ + HCO_3^- \rightarrow H_2CO_3$$

Likewise, as blood H^+ concentration drops below what is desirable and blood pH rises, H_2CO_3 dissociates to release bicarbonate ions and H^+ ions to the blood. The released H^+ lowers the pH again. The bicarbonate ions, being *weak* bases, are poorly functional under alkaline conditions and have little effect on blood pH unless and until blood pH drops toward acid levels.

$$H_2CO_3 \rightarrow H^+ + HCO_3^-$$

In the case of excessively slow or shallow breathing (hypoventilation) or fast deep breathing (hyperventilation), the amount of carbonic acid in the blood can be greatly modified—increasing dramatically during hypoventilation and decreasing substantially during hyperventilation. In either situation, if the buffering ability of the blood is inadequate, respiratory acidosis or alkalosis can result. Therefore, maintaining the normal rate and depth of breathing is important for proper control of blood pH.

Activity 6:
Demonstrating the Reaction Between Carbon Dioxide in Exhaled Air and Water

1. Fill a beaker with 100 ml of distilled water.
2. Add 5 ml of 0.05 *M* NaOH and five drops of phenol red. Phenol red is a pH indicator that turns yellow in acidic solutions.
3. Blow through a straw into the solution.

What do you observe?

___

___

What chemical reaction is taking place in the beaker?

___

___

4. Discard the straw in the autoclave bag. ■

Activity 7: Observing the Operation of Standard Buffers

1. To observe the ability of a buffer system to stabilize the pH of a solution, obtain five 250-ml beakers, and a wash bottle containing distilled water. Set up the following experimental samples:

Beaker 1
(150 ml distilled water) pH ______

Beaker 2
(150 ml distilled water and
1 drop concentrated HCl) pH ______

Beaker 3
(150 ml distilled water and
1 drop concentrated NaOH) pH ______

Beaker 4
(150 ml standard buffer solution
[pH 7] and 1 drop concentrated HCl) pH ______

Beaker 5
(150 ml standard buffer solution
[pH 7] and 1 drop concentrated NaOH) pH ______

2. Using a pH meter standardized with a buffer solution of pH 7, determine the pH of the contents of each beaker and record above. After *each and every* pH recording, the pH meter switch should be turned to *standby,* and the electrodes rinsed thoroughly with a stream of distilled water from the wash bottle.

3. Add 3 more drops of concentrated HCl to beaker 4,

stir, and record the pH: ______

4. Add 3 more drops of concentrated NaOH to beaker 5,

stir, and record the pH: ______

How successful was the buffer solution in resisting pH changes when a strong acid (HCl) or a strong base (NaOH) was added?

______________________________ ■

Activity 8: Exploring the Operation of the Carbonic Acid–Bicarbonate Buffer System

To observe the ability of the carbonic acid–bicarbonate buffer system of blood to resist pH changes, perform the following simple experiment.

1. Obtain two small beakers (50 ml), animal plasma, graduated cylinder, glass stirring rod, and a dropper bottle of 0.01 *M* HCl. Using the pH meter standardized with the buffer solution of pH 7.0, measure the pH of the animal plasma. Use only enough plasma to allow immersion of the electrodes and measure the volume used carefully.

pH of the animal plasma: ______

2. Add 2 drops of the 0.01 *M* HCl solution to the plasma; stir and measure the pH again.

pH of plasma plus 2 drops of HCl: ______

3. Turn the pH meter switch to *standby,* rinse the electrodes, and then immerse them in a quantity of distilled water (pH 7) exactly equal to the amount of animal plasma used. Measure the pH of the distilled water.

pH of distilled water: ______

4. Add 2 drops of 0.01 *M* HCl, swirl, and measure the pH again.

pH of distilled water plus the two drops of HCl: ______

Is the plasma a good buffer? ______

What component of the plasma carbonic acid–bicarbonate buffer system was acting to counteract a change in pH when HCl was added?

______________________________ ■

exercise

38 Anatomy of the Digestive System

Objectives

1. To state the overall function of the digestive system.
2. To identify on an appropriate diagram or torso model the organs comprising the alimentary canal; and to name their subdivisions if any.
3. To name and/or identify the accessory digestive organs.
4. To describe the general functions of the digestive system organs or structures.
5. To describe the general histologic structure of the wall of the alimentary canal and/or label a cross-sectional diagram of the wall with the following terms: mucosa, submucosa, muscularis externa, and serosa or adventitia.
6. To list and explain the specializations of anatomical structure of the stomach and small intestine that contribute to their functional roles.
7. To list the major enzymes or enzyme groups produced by each of the following organs: salivary glands, stomach, small intestine, pancreas.
8. To name human deciduous and permanent teeth and describe the anatomy of the generalized tooth.
9. To recognize by microscopic inspection, or by viewing an appropriate diagram or photomicrograph, the histologic structure of the following organs:

small intestine	pancreas	stomach
salivary glands	tooth	liver

Materials

- ❑ Dissectible torso model
- ❑ Anatomical chart of the human digestive system
- ❑ Model of a villus and liver lobules (if available)
- ❑ Jaw model or human skull
- ❑ Prepared microscope slides of the liver, pancreas, and mixed salivary glands; of longitudinal sections of the gastroesophageal junction and a tooth; and of cross sections of the stomach, duodenum, and ileum
- ❑ Compound microscope
- ❑ *Human Digestive System* videotape*

For instructions on animal dissections, see the dissection exercises starting on p. 709 in the cat and fetal pig editions of this manual.

See Appendix C, Exercise 38 for links to A.D.A.M.® Interactive Anatomy.

*Available to qualified adopters from Benjamin Cummings.

The **digestive system** provides the body with the nutrients, water, and electrolytes essential for health. The organs of this system ingest, digest, and absorb food and eliminate the undigested remains as feces.

The digestive system consists of a hollow tube extending from the mouth to the anus, into which various accessory organs or glands empty their secretions (Figure 38.1). Food material within this tube, the *alimentary canal,* is technically outside the body because it has contact only with the cells lining the tract. For ingested food to become available to the body cells, it must first be broken down *physically* (chewing, churning) and *chemically* (enzymatic hydrolysis) into its smaller diffusible molecules—a process called **digestion.** The digested end products can then pass through the epithelial cells lining the tract into the blood for distribution to the body cells—a process termed **absorption.** In one sense, the digestive tract can be viewed as a disassembly line, in which food is carried from one stage of its digestive processing to the next by muscular activity, and its nutrients are made available to the cells of the body en route.

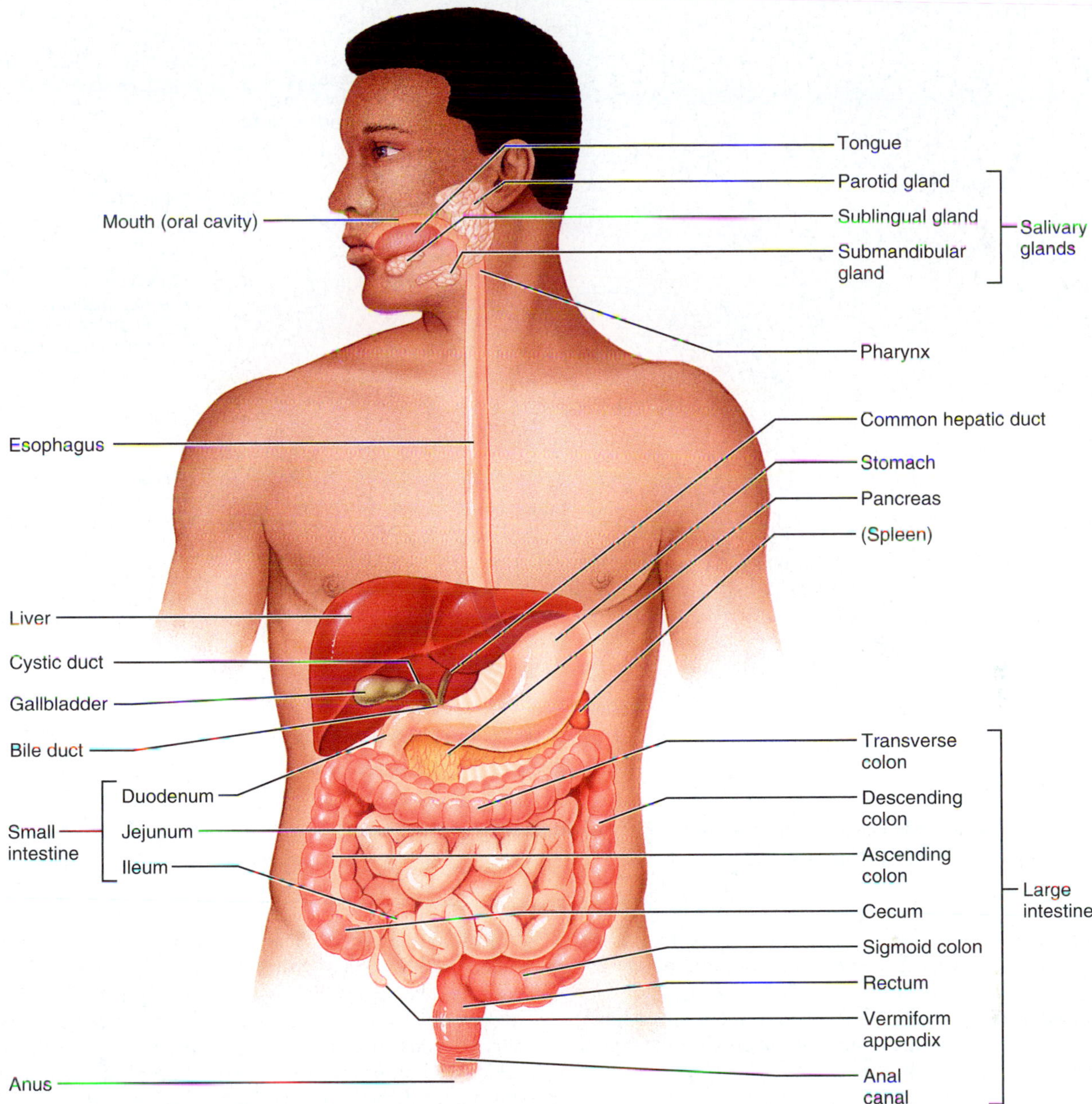

Figure 38.1 The human digestive system: alimentary tube and accessory organs. (Liver and gallbladder are reflected superiorly and to the right.)

The organs of the digestive system are traditionally separated into two major groups: the **alimentary canal,** or **gastrointestinal (GI) tract,** and the **accessory digestive organs.** The alimentary canal is approximately 9 meters long in a cadaver but is considerably shorter in a living person. It consists of the mouth, pharynx, esophagus, stomach, and small and large intestines. The accessory structures include the salivary glands, gallbladder, liver, and pancreas, which secrete their products into the alimentary canal. These individual organs are described shortly.

General Histological Plan of the Alimentary Canal

Because the alimentary canal has a shared basic structural plan (particularly from the esophagus to the anus), it makes sense to review that structure as we begin studying this group of organs. Once done, all that need be emphasized as the individual organs are described is their specializations for unique functions in the digestive process.

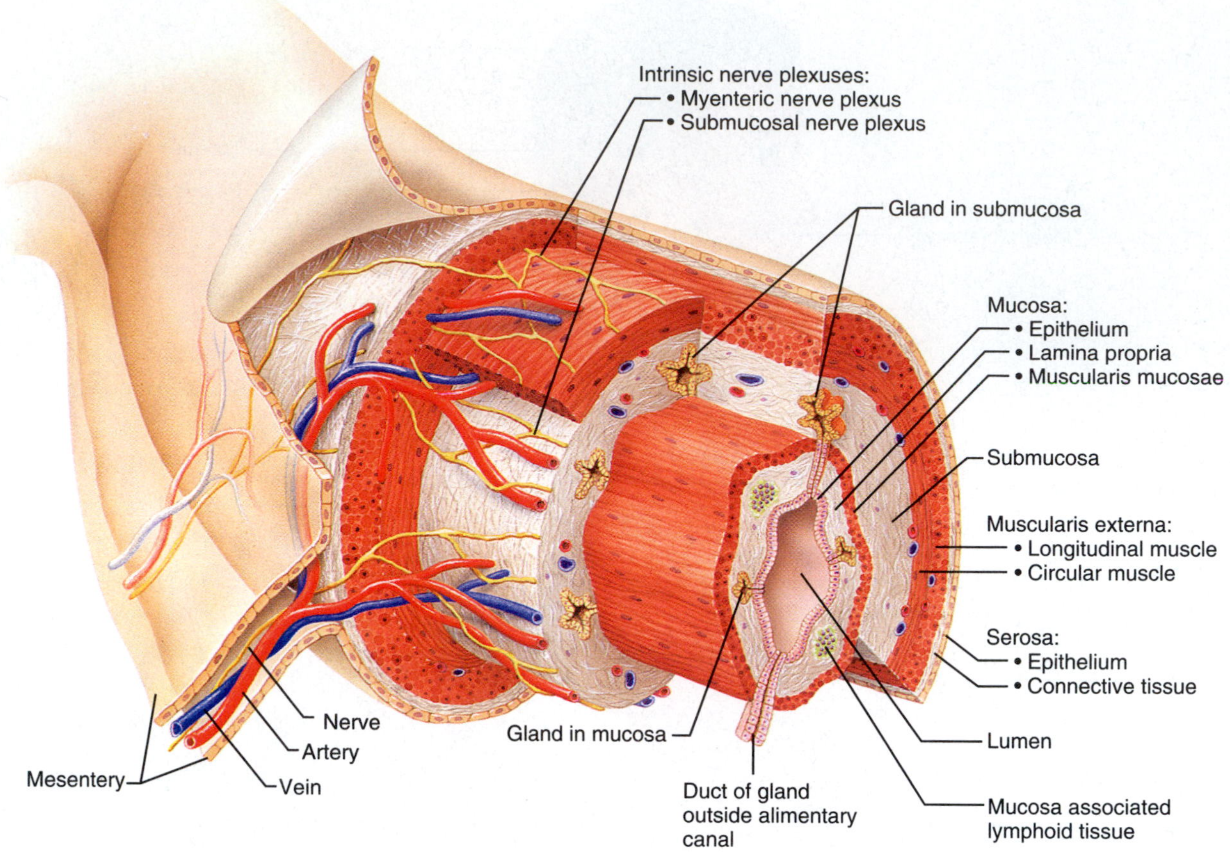

Figure 38.2 Basic structural pattern of the alimentary canal wall. (See also Plate 37 in the Histology Atlas.)

Essentially the alimentary canal walls have four basic **tunics** (layers). From the lumen outward, these are the *mucosa*, the *submucosa*, the *muscularis externa*, and the *serosa* or *adventitia* (Figure 38.2). Each of these tunics has a predominant tissue type and a specific function in the digestive process.

Mucosa (mucous membrane): The mucosa is the wet epithelial membrane abutting the alimentary canal lumen. It consists of a surface *epithelium* (in most cases, a simple columnar), a *lamina propria* (areolar connective tissue on which the epithelial layer rests), and a *muscularis mucosae* (a scant layer of smooth muscle fibers that enable local movements of the mucosa). The major functions of the mucosa are secretion (of enzymes, mucus, hormones, etc.), absorption of digested foodstuffs, and protection (against bacterial invasion). A particular mucosal region may be involved in one or all three functions.

Submucosa: Superficial to the mucosa, the submucosa is moderately dense connective tissue containing blood and lymphatic vessels, scattered lymph nodules, and nerve fibers. Its intrinsic nerve supply is called the *submucosal plexus.* Its major functions are nutrition and protection.

Muscularis externa: The muscularis externa, also simply called the *muscularis,* typically is a bilayer of smooth muscle, with the deeper layer running circularly and the superficial layer running longitudinally. Another important intrinsic nerve plexus, the *myenteric plexus,* is associated with this tunic. By controlling the smooth muscle of the muscularis, this plexus is the major regulator of GI motility.

Serosa: The outermost serosa is the *visceral peritoneum.* It consists of mesothelium associated with a thin layer of areolar connective tissue. In areas *outside* the abdominopelvic cavity, the serosa is replaced by an **adventitia,** a layer of coarse fibrous connective tissue that binds the organ to surrounding tissues. (This is the case with the esophagus.) The serosa reduces friction as the mobile GI tract organs work and slide across one another and the cavity walls. The adventitia anchors and protects the surrounded GI tract organ.

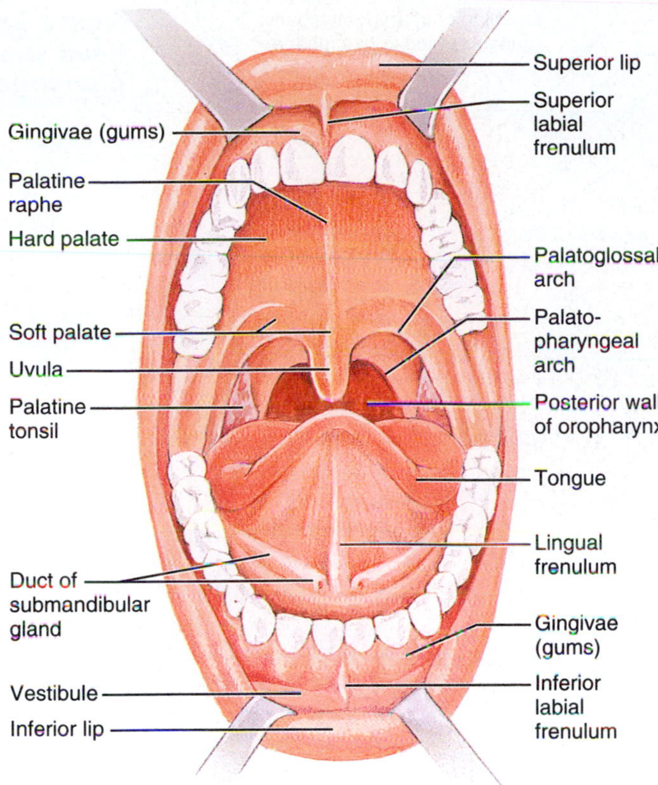

Figure 38.3 Anterior view of the oral cavity.

Organs of the Alimentary Canal

Activity 1: Identifying Alimentary Canal Organs

The sequential pathway and fate of food as it passes through the alimentary canal organs are described in the next sections. Identify each structure in Figure 38.1 and on the torso model as you work. ■

Oral Cavity or Mouth

Food enters the digestive tract through the **oral cavity,** or **mouth** (Figure 38.3). Within this mucous membrane–lined cavity are the gums, teeth, tongue, and openings of the ducts of the salivary glands. The **lips (labia)** protect the opening of the chamber anteriorly, the **cheeks** form its lateral walls, and the **palate,** its roof. The anterior portion of the palate is referred to as the **hard palate** because bone (the palatine processes of the maxillae and the palatine bones) underlies it. The posterior **soft palate** is a fibromuscular structure that is unsupported by bone. The **uvula,** a fingerlike projection of the soft palate, extends inferiorly from its posterior margin. The soft palate rises to close off the oral cavity from the nasal and pharyngeal passages during swallowing. The floor of the oral cavity is occupied by the muscular **tongue,** which is largely supported by the **mylohyoid muscle** (Figure 38.4) and attaches to the hyoid bone, mandible, styloid processes, and pharynx. A membrane called the **lingual frenulum** secures the inferior midline of the tongue to the floor of the mouth. The space between the lips and cheeks and the teeth is the **vestibule;** the area that lies within the teeth and gums is the **oral cavity** proper.

On each side of the mouth at its posterior end are masses of lymphoid tissue, the **palatine tonsils** (Figure 38.3). Each lies in a concave area bounded anteriorly and posteriorly by membranes, the **palatoglossal arch** (anterior membrane) and the **palatopharyngeal arch** (posterior membrane). Another mass of lymphoid tissue, the **lingual tonsil** (Figure 38.4), covers the base of the tongue, posterior to the oral cavity proper. The tonsils, in common with other lymphoid tissues, are part of the body's defense system.

Very often in young children, the palatine tonsils become inflamed and enlarge, partially blocking the entrance to the pharynx posteriorly and making swallowing difficult and painful. This condition is called **tonsilitis.** ■

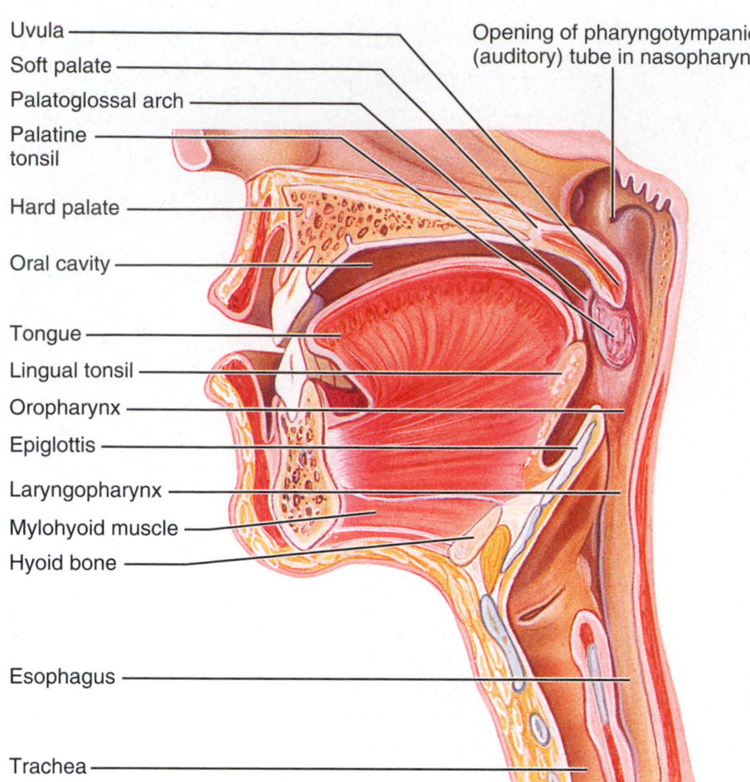

Figure 38.4 Sagittal view of the head showing oral, nasal, and pharyngeal cavities.

Three pairs of salivary glands duct their secretion, saliva, into the oral cavity. One component of saliva, salivary amylase, begins the digestion of starchy foods within the oral cavity. (The salivary glands are discussed in more detail on p. 401.)

As food enters the mouth, it is mixed with saliva and masticated (chewed). The cheeks and lips help hold the food between the teeth during mastication, and the highly mobile tongue manipulates the food during chewing and initiates swallowing. Thus the mechanical and chemical breakdown of food begins before the food has left the oral cavity. As noted in Exercise 26, the surface of the tongue is covered with papillae, many of which contain taste buds, receptors for taste sensation. So, in addition to its manipulative function, the tongue provides for the enjoyment and appreciation of the food ingested.

Pharynx

When the tongue initiates swallowing, the food passes posteriorly into the pharynx, a common passageway for food, fluid, and air (Figure 38.4). The pharynx is subdivided anatomically into three parts—the **nasopharynx** (behind the nasal cavity), the **oropharynx** (behind the oral cavity extending from the soft palate to the epiglottis overlying the larynx), and the **laryngopharynx** (extending from the epiglottis to the base of the larynx), which is continuous with the esophagus.

The walls of the pharynx consist largely of two layers of skeletal muscles: an inner layer of longitudinal muscle (the levator muscles) and an outer layer of circular constrictor muscles, which initiate wavelike contractions that propel the food inferiorly into the esophagus. Its mucosa, like that of the oral cavity, contains a friction-resistant stratified squamous epithelium.

Esophagus

The **esophagus,** or gullet, extends from the pharynx through the diaphragm to the gastroesophageal sphincter in the superior aspect of the stomach. It is approximately 25 cm long in humans and is essentially a food passageway that conducts food to the stomach in a wavelike peristaltic motion. The esophagus has no digestive or absorptive function. The walls at its superior end contain skeletal muscle, which is replaced by smooth muscle in the area nearing the stomach. The **gastroesophageal sphincter,** a thickening of the smooth muscle layer at the esophagus-stomach junction, controls food passage into the stomach (see Figure 38.5). Since the esophagus is located in the thoracic rather than the abdominal cavity, its outermost layer is an *adventitia,* rather than the serosa.

Stomach

The **stomach** (Figures 38.1 and 38.5) is on the left side of the abdominal cavity and is hidden by the liver and diaphragm. Different regions of the saclike stomach are the **cardiac region** (the area surrounding the cardiac orifice through which food enters the stomach from the esophagus), the **fundus** (the expanded portion of the stomach, superolateral to the cardiac region), the **body** (midportion of the stomach, inferior to the fundus), and the **pyloric region** (the terminal part of the stomach, which is continuous with the small intestine through the **pyloric sphincter**).

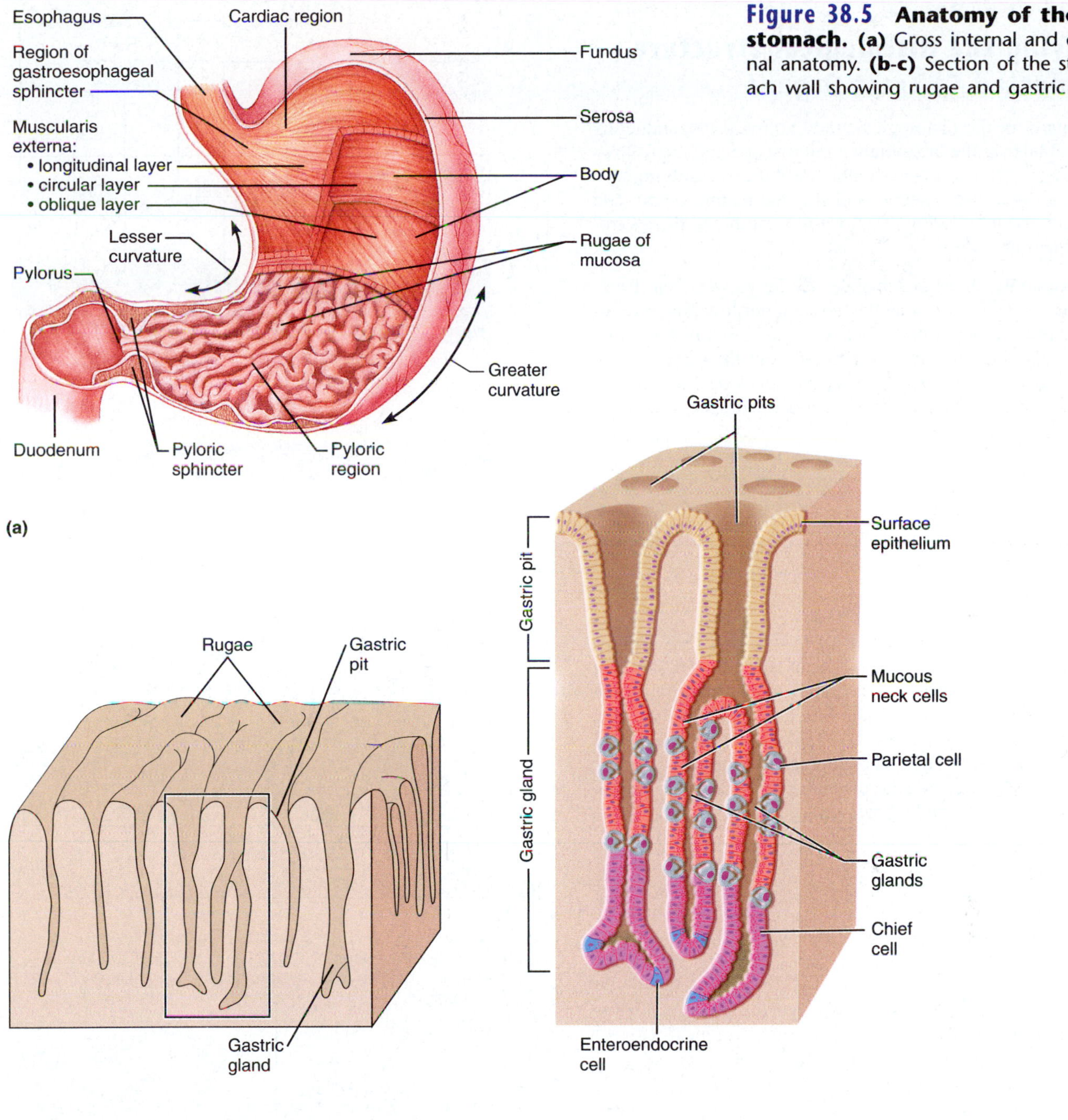

Figure 38.5 Anatomy of the stomach. **(a)** Gross internal and external anatomy. **(b-c)** Section of the stomach wall showing rugae and gastric pits.

The concave medial surface of the stomach is called the **lesser curvature;** its convex lateral surface is the **greater curvature.** Extending from these curvatures are two mesenteries, called *omenta.* The **lesser omentum** extends from the liver to attach to the lesser curvature of the stomach. The **greater omentum,** a saclike mesentery, extends from the greater curvature of the stomach, reflects downward over the abdominal contents to cover them in an apronlike fashion, and then blends with the **mesocolon** attaching the transverse colon to the posterior body wall. Figure 38.7 (p. 395) illustrates the omenta as well as the other peritoneal attachments of the abdominal organs.

The stomach is a temporary storage region for food as well as a site for mechanical and chemical breakdown of food. It contains a third *obliquely* oriented layer of smooth muscle in its muscularis externa that allows it to churn, mix, and pummel the food, physically reducing it to smaller fragments. Gastric glands of the mucosa secrete hydrochloric acid (HCl) and hydrolytic enzymes (primarily pepsinogen, the inactive form of *pepsin,* a protein-digesting enzyme), which begin the enzymatic, or chemical, breakdown of protein foods. The mucosal glands also secrete a viscous mucus that helps prevent the stomach itself from being digested by the proteolytic enzymes. Most digestive activity occurs in the pyloric region of the stomach. After the food is processed in the stomach, it resembles a creamy mass (chyme), which enters the small intestine through the pyloric sphincter.

Activity 2:
Studying the Histological Structure of Selected GI Tract Organs

To prepare for the histological study you will be conducting now and later in the lab, obtain a microscope and the following slides: salivary glands (submandibular or sublingual); pancreas; liver; cross sections of the duodenum, ileum, and stomach; and longitudinal sections of a tooth and the gastroesophageal junction.

1. **Stomach:** The stomach slide will be viewed first. Refer to Figure 38.6a as you scan the tissue under low power to locate the muscularis externa; then move to high power to more closely examine this layer. Try to pick out the three smooth muscle layers. How does the extra (oblique) layer of smooth muscle found in the stomach correlate with the stomach's churning movements?

___

___

___

Identify the gastric glands and the gastric pits (see Figures 38.5 and 38.6b). If the section is taken from the stomach fundus and is appropriately stained, you can identify, in the gastric glands, the blue-staining **chief,** or **zymogenic, cells,** which produce pepsinogen, and the red-staining **parietal cells,** which secrete HCl. The enteroendocrine cells that release hormones are indistinguishable. Draw a small section of the stomach wall and label it appropriately.

2. **Gastroesophageal junction:** Scan the slide under low power to locate the mucosal junction between the end of the esophagus and the beginning of the stomach, the gastroesophageal junction. Compare your observations to Figure 38.6c. What is the functional importance of the epithelial differences seen in the two organs?

___

___

___ ■

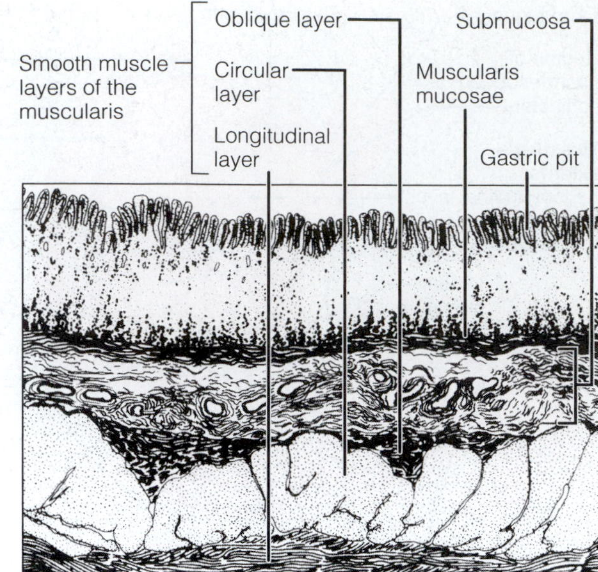

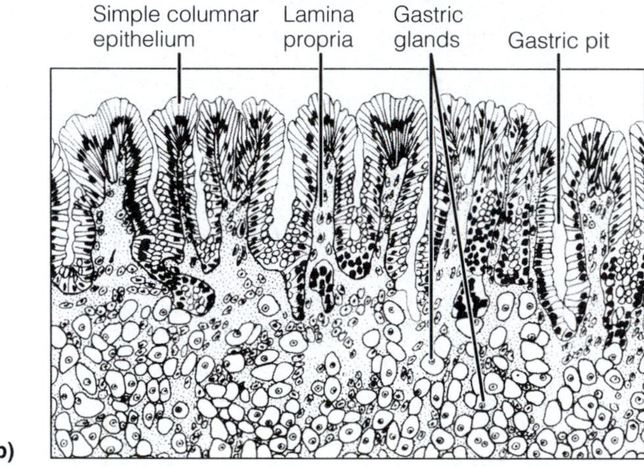

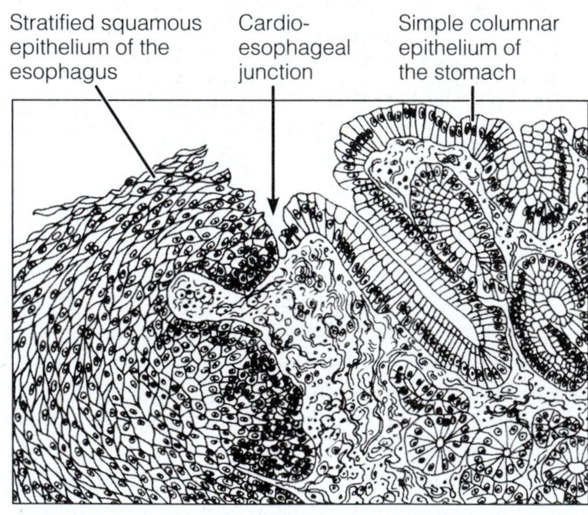

Figure 38.6 Histology of selected regions of the stomach and gastroesophageal junction. **(a)** Stomach wall. **(b)** Gastric pits and glands. **(c)** Gastroesophageal junction, longitudinal section. (See also Plates 37, 38, and 36 in the Histology Atlas.)

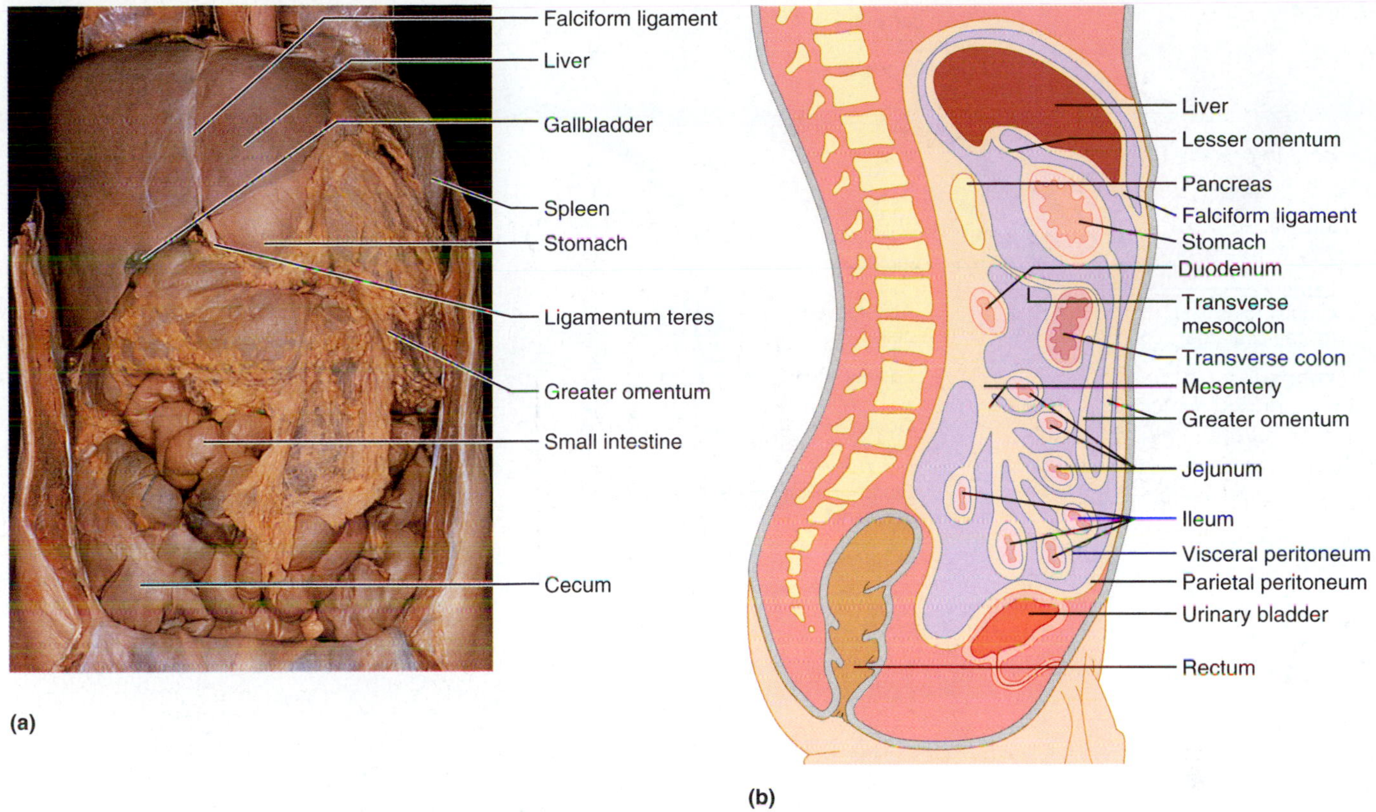

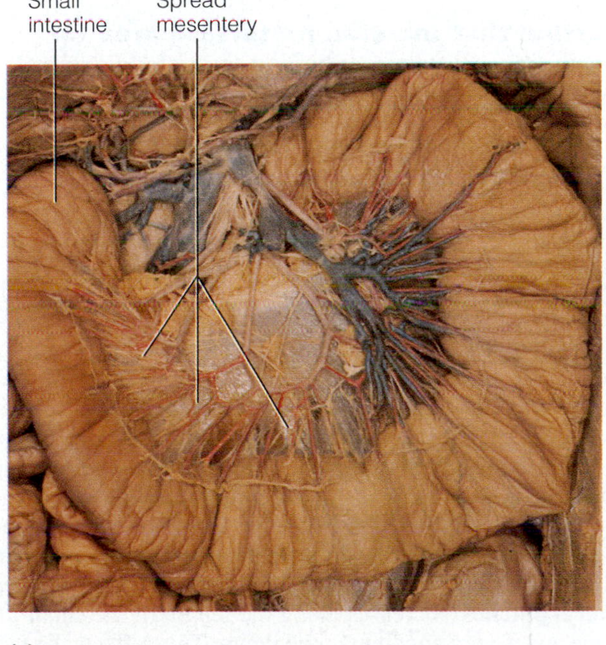

Figure 38.7 Peritoneal attachments of the abdominal organs. (a) Superficial anterior view of abdominal cavity with the greater omentum in place. **(b)** Sagittal view of a male torso. **(c)** Mesentery of the small intestine.

Small Intestine

The **small intestine** is a convoluted tube, 6 to 7 meters (about 20 feet) long in a cadaver but only about 2 m long during life because of its muscle tone. It extends from the pyloric sphincter to the ileocecal valve. The small intestine is suspended by a double layer of peritoneum, the fan-shaped **mesentery,** from the posterior abdominal wall (Figure 38.7), and it lies, framed laterally and superiorly by the large intestine, in the abdominal cavity. The small intestine has three subdivisions (see Figure 38.1): (1) the **duodenum** extends from the pyloric sphincter for about 25 cm (10 inches) and curves around the head of the pancreas; most of the duodenum lies in a retroperitoneal position. (2) The **jejunum,** continuous with the duodenum, extends for 2.5 m (about 8 feet). Most of the jejunum occupies the umbilical region of the abdominal cavity. (3) The **ileum,** the terminal portion of the small intestine, is about 3.6 m (12 feet) long and joins the large intestine at the **ileocecal valve.** It is located inferiorly and somewhat to the right in the abdominal cavity, but its major portion lies in the hypogastric region.

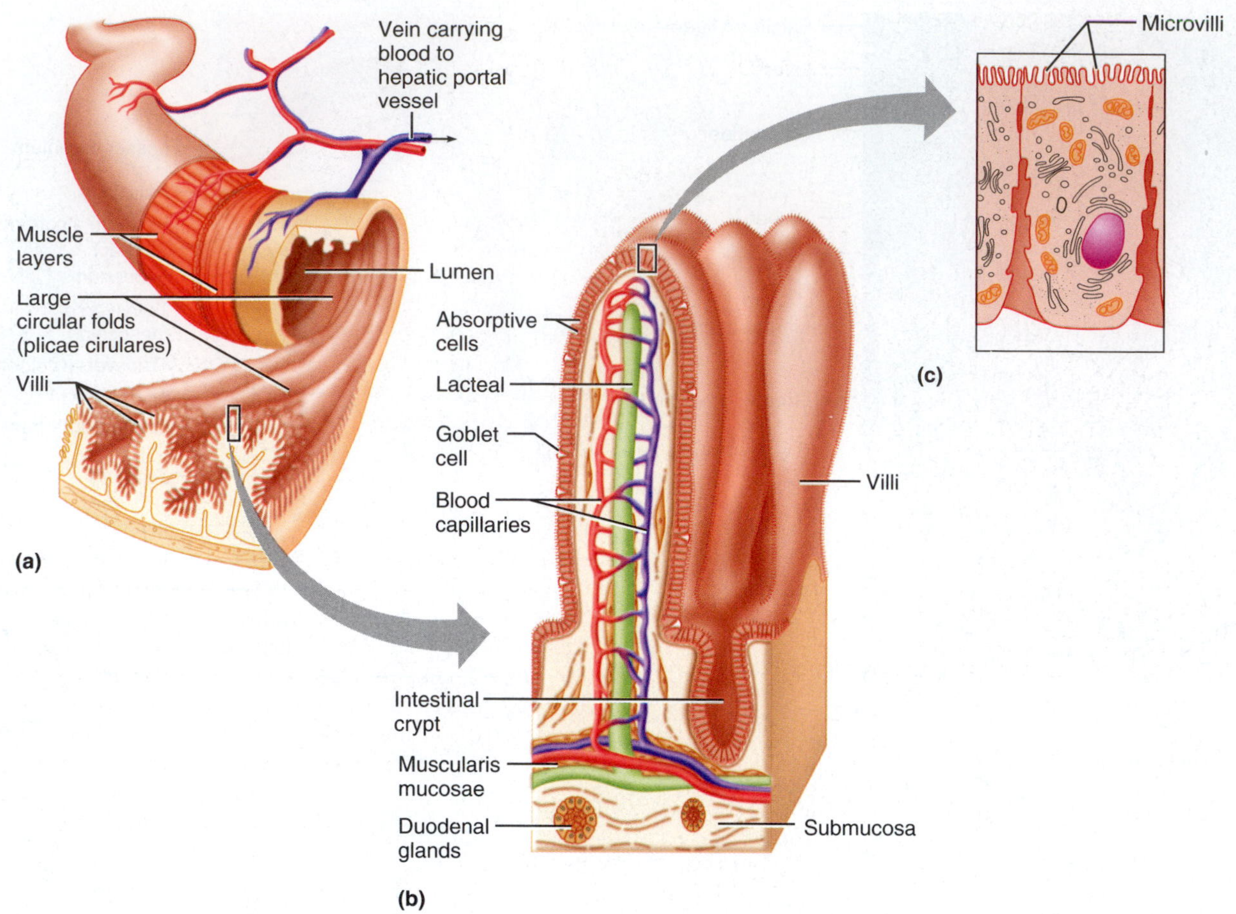

Figure 38.8 Structural modifications of the small intestine that increase its surface area for digestion and absorption. (a) Enlargement of a few circular folds (plicae circulares), showing associated fingerlike villi. **(b)** Diagrammatic view of the structure of a villus. **(c)** Two absorptive cells that exhibit microvilli on their free (luminal) surface.

Brush border enzymes, hydrolytic enzymes bound to the microvilli of the columnar epithelial cells, and, more importantly, enzymes produced by the pancreas and ducted into the duodenum via the **pancreatic duct** complete the enzymatic digestion process in the small intestine. Bile (formed in the liver) also enters the duodenum via the **bile duct** in the same area. At the duodenum, the ducts join to form the bulblike **hepatopancreatic ampulla** and empty their products into the duodenal lumen through the **major duodenal papilla,** an orifice controlled by a muscular valve called the **hepatopancreatic sphincter (sphincter of Oddi).**

Nearly all nutrient absorption occurs in the small intestine, where three structural modifications that increase the mucosa absorptive area appear—the microvilli, villi, and plicae circulares (Figure 38.8). **Microvilli** are minute projections of the surface plasma membrane of the columnar epithelial lining cells of the mucosa. **Villi** are the fingerlike projections of the mucosa tunic that give it a velvety appearance and texture. The **plicae circulares** are deep folds of the mucosa and submucosa layers that force chyme to spiral through the intestine, mixing it and slowing its progress. These structural modifications, which increase the surface area, decrease in frequency and elaboration toward the end of the small intestine. Any residue remaining undigested and unabsorbed at the terminus of the small intestine enters the large intestine through the ileocecal valve. In contrast, the amount of lymphoid tissue in the submucosa of the small intestine (especially the aggregated lymphoid nodules called **Peyer's patches**) increases along the length of the small intestine and is very apparent in the ileum. This reflects the fact that the remaining undigested food residue contains large numbers of bacteria that must be prevented from entering the bloodstream.

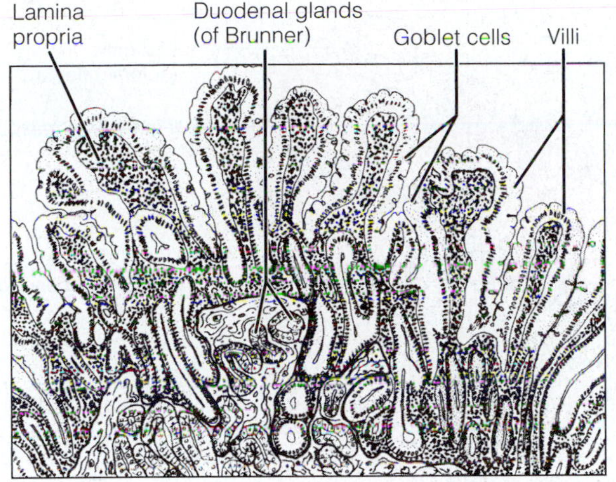

Submucosa
Muscularis
Peyer's patches (in the submucosa)
Villi
Lumen
(b)

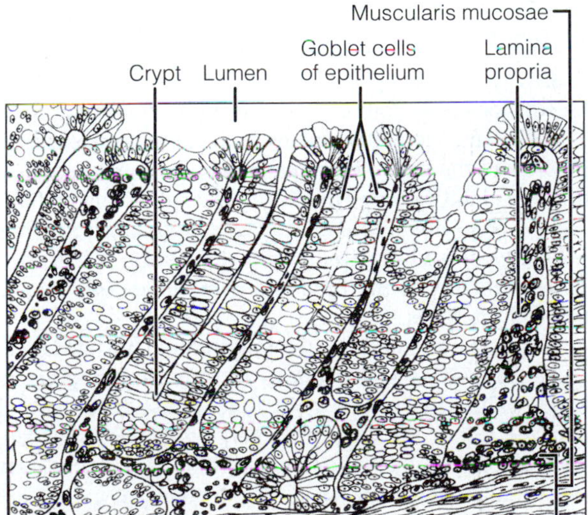

Figure 38.9 Histology of selected regions of the small and large intestines (cross-sectional views). **(a)** Duodenum of the small intestine. **(b)** Ileum of the small intestine. **(c)** Large intestine. (See also Plates 40, 41, and 42 in the Histology Atlas.)

Activity 3:

Observing the Histological Structure of the Small Intestine

1. **Duodenum:** Secure the slide of the duodenum (cross section) to the microscope stage. Observe the tissue under low power to identify the four basic tunics of the intestinal wall—that is, the **mucosa** lining (and its three sublayers), the **submucosa** (areolar connective tissue layer deep to the mucosa), the **muscularis externa** (composed of circular and longitudinal smooth muscle layers), and the **serosa** (the outermost layer, also called the *visceral peritoneum*). Consult Figure 38.9a to help you identify the scattered mucus-producing **duodenal glands** (Brunner's glands) in the submucosa.

What type of epithelium do you see here? ______________

____

Examine the large leaflike *villi*, which increase the surface area for absorption. Note also the *intestinal crypts* (crypts of Lieberkühn), invaginated areas of the mucosa between the villi containing the cells that produce intestinal juice, a watery mucus-containing mixture that serves as a carrier fluid for absorption of nutrients from the chyme. Sketch and label a small section of the duodenal wall, showing all layers and villi.

2. **Ileum:** The structure of the ileum resembles that of the duodenum, except that the villi are less elaborate (most of the absorption has occurred by the time the ileum is reached). Obtain and secure a slide of the ileum for viewing. Observe the villi, and identify the four layers of the wall and the large, generally spherical Peyer's patches (Figure 38.9b). What tissue composes Peyer's patches?

____

3. If a *villus model* is available, identify the following cells or regions before continuing: absorptive epithelium, goblet cells, lamina propria, slips of the muscularis mucosae, capillary bed, and lacteal. If possible, also identify the intestinal crypts that lie between the villi. ■

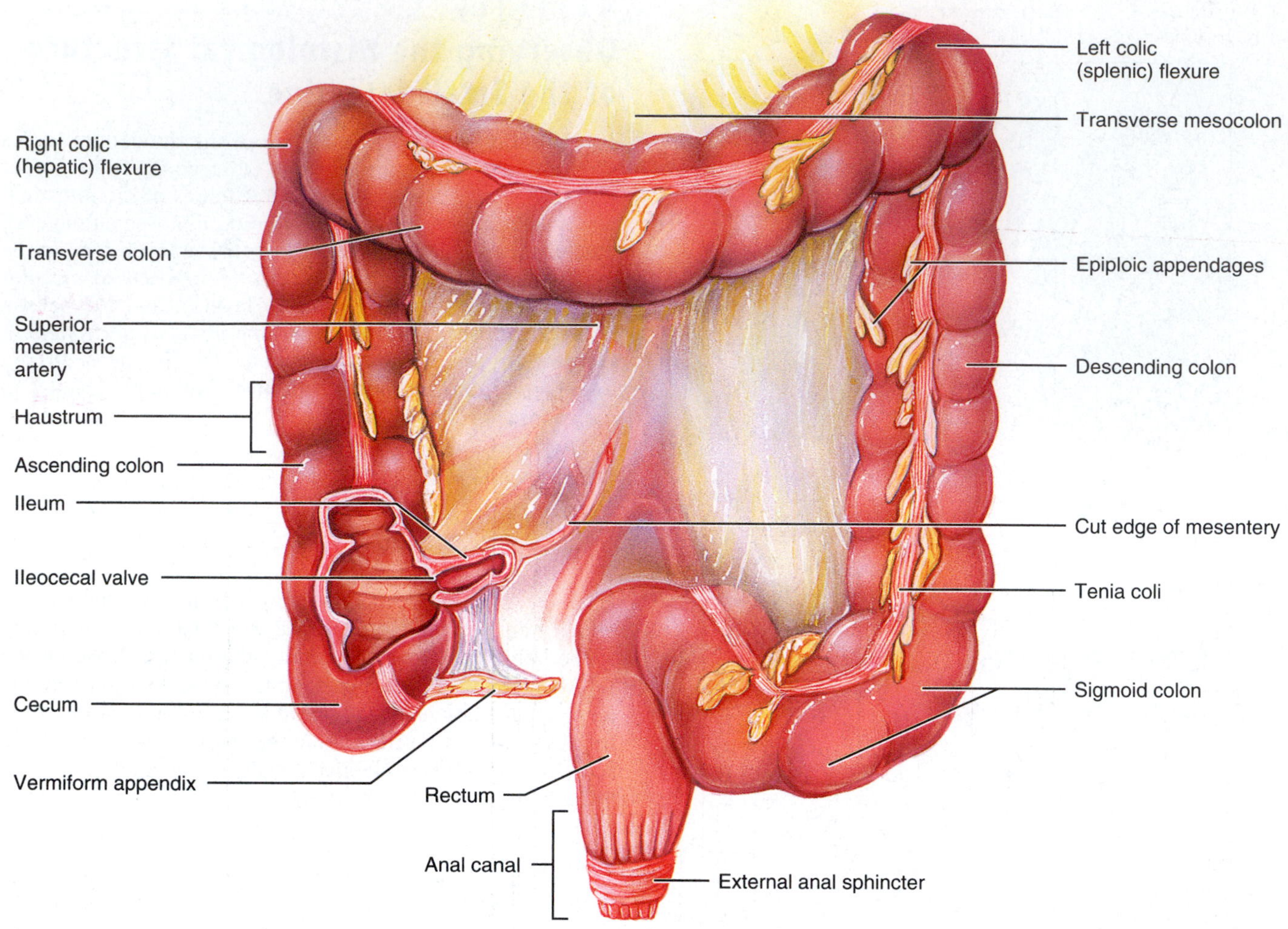

Figure 38.10 The large intestine. (Section of the cecum removed to show the ileocecal valve.)

Large Intestine

The **large intestine** (see Figure 38.10) is about 1.5 m (5 feet) long and extends from the ileocecal valve to the anus. It encircles the small intestine on three sides and consists of the following subdivisions: the **cecum, appendix, colon, rectum,** and **anal canal.**

The blind tubelike appendix, which hangs from the cecum, is a trouble spot in the large intestine. Since it is generally twisted, it provides an ideal location for bacteria to accumulate and multiply. Inflammation of the appendix, or appendicitis, is the result. ■

The colon is divided into several distinct regions. The **ascending colon** travels up the right side of the abdominal cavity and makes a right-angle turn at the **right colic (hepatic) flexure** to cross the abdominal cavity as the **transverse colon.** It then turns at the **left colic (splenic) flexure** and continues down the left side of the abdominal cavity as the **descending colon,** where it takes an S-shaped course as the **sigmoid colon.** The sigmoid colon, rectum, and the anal canal lie in the pelvis anterior to the sacrum and thus are not considered abdominal cavity structures. Except for the transverse and sigmoid colons, which are secured to the dorsal body wall by mesocolons (see Figure 38.7), the colon is retroperitoneal.

The anal canal terminates in the **anus,** the opening to the exterior of the body. The anus, which has an external sphincter of skeletal muscle (the voluntary sphincter) and an internal sphincter of smooth muscle (the involuntary sphincter), is normally closed except during defecation when the undigested remains of the food and bacteria are eliminated from the body as feces.

In the large intestine, the longitudinal muscle layer of the muscularis externa is reduced to three longitudinal muscle bands called the **teniae coli.** Since these bands are shorter than the rest of the wall of the large intestine, they cause the wall to pucker into small pocketlike sacs called **haustra.** Fat-filled pouches of visceral peritoneum, called *epiploic appendages,* hang from the colon's surface.

The major function of the large intestine is to consolidate and propel the unusable fecal matter toward the anus and eliminate it from the body. While it does that chore, it (1) provides a site for the manufacture, by intestinal bacteria, of some vitamins (B and K), which it then absorbs into the bloodstream; and (2) reclaims most of the remaining water (and some of the electrolytes) from undigested food, thus conserving body water.

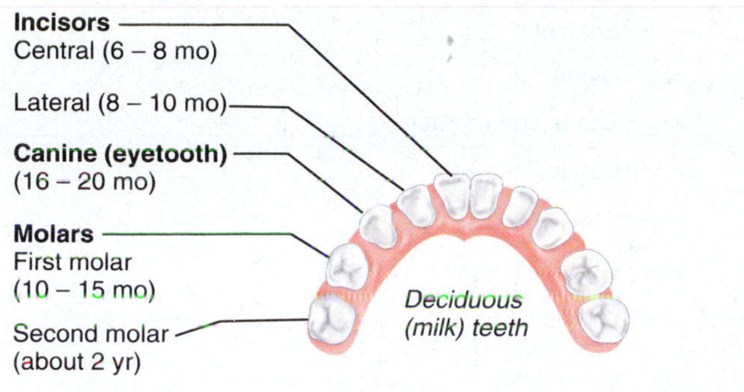

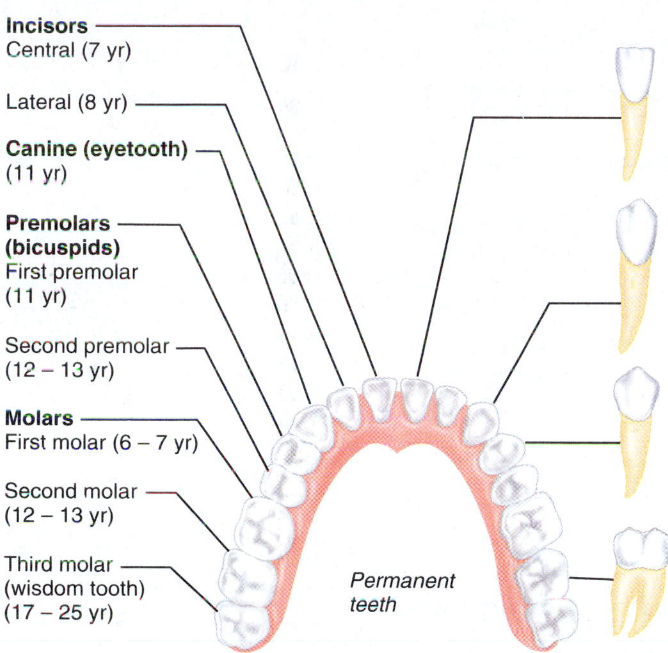

Figure 38.11 Human deciduous teeth and permanent teeth. (Approximate time of teeth eruption shown in parentheses.)

Watery stools, or diarrhea, result from any condition that rushes undigested food residue through the large intestine before it has had sufficient time to absorb the water (as in irritation of the colon by bacteria). Conversely, when food residue remains in the large intestine for extended periods (as with atonic colon or failure of the defecation reflex), excessive water is absorbed and the stool becomes hard and difficult to pass (constipation). ■

Activity 4: Examining the Histological Structure of the Large Intestine

Examine Figure 38.9c to compare the histology of the large intestine to that of the small intestine just studied. ■

Accessory Digestive Organs

Teeth

By the age of 21, two sets of teeth have developed (Figure 38.11). The initial set, called the **deciduous,** or **milk teeth,** normally appears between the ages of 6 months and 2½ years. The first of these to erupt are the lower central incisors. The child begins to shed the deciduous teeth around the age of 6, and a second set of teeth, the **permanent teeth,** gradually replaces them. As the deeper permanent teeth progressively enlarge and develop, the roots of the deciduous teeth are resorbed, leading to their final shedding. During years 6 to 12, the child has mixed dentition—both permanent and deciduous teeth. Generally, by the age of 12, all of the deciduous teeth have been shed, or exfoliated.

Teeth are classified as **incisors, canines** (eye teeth), **premolars** (bicuspids), and **molars.** Teeth names reflect differences in relative structure and function. The incisors are chisel-shaped and exert a shearing action used in biting. Canines are cone-shaped or fanglike, the latter description being much more applicable to the canines of animals whose teeth are used for the tearing of food. Incisors, canines, and premolars typically have single roots, though the first upper premolars may have two. The lower molars have two roots but the upper molars usually have three. The premolars have two cusps (grinding surfaces); the molars have broad crowns with rounded cusps specialized for the fine grinding of food.

Dentition is described by means of a **dental formula,** which designates the numbers, types, and position of the teeth in one side of the jaw. (Because tooth arrangement is bilaterally symmetrical, it is only necessary to designate one side of the jaw.) The complete dental formula for the deciduous teeth from the medial aspect of each jaw and proceeding posteriorly is as follows:

$$\frac{\text{Upper teeth: 2 incisors, 1 canine, 0 premolars, 2 molars}}{\text{Lower teeth: 2 incisors, 1 canine, 0 premolars, 2 molars}} \times 2$$

This formula is generally abbreviated to read as follows:

$$\frac{2,1,0,2}{2,1,0,2} \times 2 = 20 \text{ (number of deciduous teeth)}$$

The 32 permanent teeth are then described by the following dental formula:

$$\frac{2,1,2,3}{2,1,2,3} \times 2 = 32 \text{ (number of permanent teeth)}$$

Although 32 is designated as the normal number of permanent teeth, not everyone develops a full complement. In many people, the No. 3 molars, commonly called *wisdom teeth,* never erupt.

Activity 5: Identifying Types of Teeth

Identify the four types of teeth (incisors, canines, premolars, and molars) on the jaw model or human skull. ■

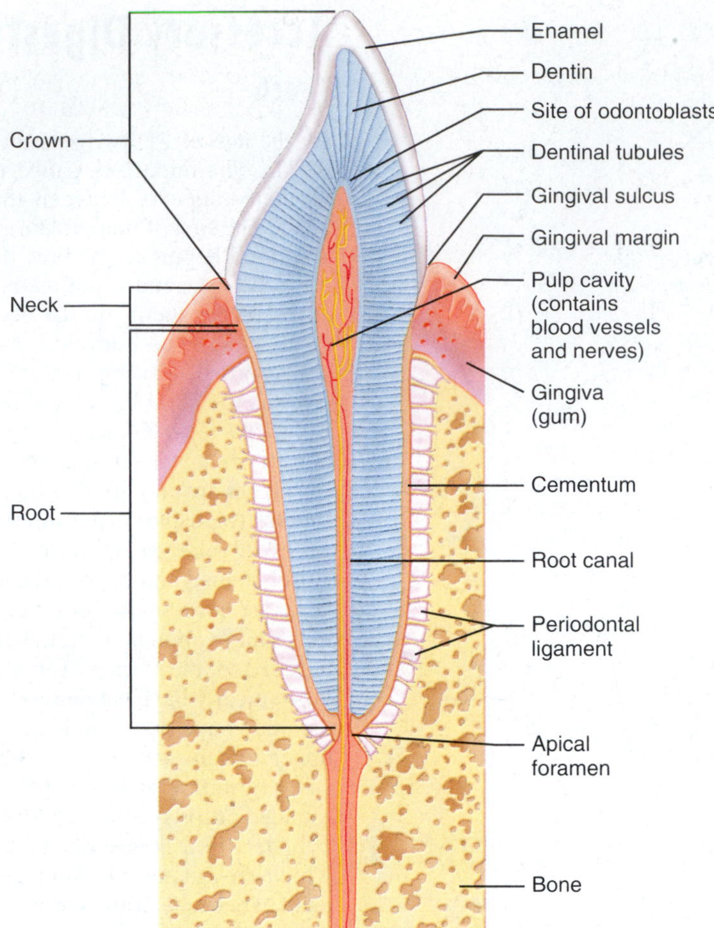

Figure 38.12 Longitudinal section of human canine tooth.

A tooth consists of two major regions, the **crown** and the **root.** A longitudinal section made through a tooth shows the following basic anatomical plan (Figure 38.12). The crown is the superior portion of the tooth. The portion of the crown visible above the **gum,** or **gingiva,** is referred to as the **clinical crown.** The entire area covered by **enamel** is called the **anatomical crown.** Enamel is the hardest substance in the body and is fairly brittle. It consists of 95% to 97% inorganic calcium salts (chiefly $CaPO_4$) and thus is heavily mineralized. The crevice between the end of the anatomical crown and the upper margin of the gingiva is referred to as the **gingival sulcus** and its apical border is the **gingival margin.**

That portion of the tooth embedded in the alveolar portion of the jaw is the root, and the root and crown are connected by a slight constriction, the **neck.** The outermost surface of the root is covered by **cementum,** which is similar to bone in composition and less brittle than enamel. The cementum attaches the tooth to the **periodontal ligament,** which holds the tooth in the alveolar socket and exerts a cushioning effect. **Dentin,** which composes the bulk of the tooth, is the bonelike material medial to the enamel and cementum.

The **pulp cavity** occupies the central portion of the tooth. **Pulp,** connective tissue liberally supplied with blood vessels, nerves, and lymphatics, occupies this cavity and provides for tooth sensation and supplies nutrients to the tooth tissues. **Odontoblasts,** specialized cells which reside in the outer margins of the pulp cavity, produce the dentin. The pulp cavity extends into distal portions of the root and becomes the **root canal.** An opening at the root apex, the **apical foramen,** provides a route of entry into the tooth for blood vessels, nerves, and other structures from the tissues beneath.

Activity 6: Studying Microscopic Tooth Anatomy

Observe a slide of a longitudinal section of a tooth, and compare your observations with the structures detailed in Figure 38.12. Identify as many of these structures as possible. ■

Salivary Glands

Three pairs of major **salivary glands** (see Figure 38.1) empty their secretions into the oral cavity.

Parotid glands: Large glands located anterior to the ear and ducting into the mouth over the second upper molar through the parotid duct

Submandibular glands: Located along the medial aspect of the mandibular body in the floor of the mouth, and ducting under the tongue to the base of the lingual frenulum

Sublingual glands: Small glands located most anteriorly in the floor of the mouth and emptying under the tongue via several small ducts

Food in the mouth and mechanical pressure (even chewing rubber bands or wax) stimulate the salivary glands to secrete saliva. Saliva consists primarily of mucin (a viscous glycoprotein), which moistens the food and helps to bind it together into a mass called a **bolus,** and a clear serous fluid containing the enzyme *salivary amylase*. Salivary amylase begins the digestion of starch (a large polysaccharide), breaking it down into disaccharides, or double sugars, and glucose. Parotid gland secretion is mainly serous, whereas the submandibular and sublingual glands are mixed glands that produce both mucin and serous components.

Activity 7: Examining Salivary Gland Tissue

Examine salivary gland tissue under low power and then high power to become familiar with the appearance of a glandular tissue. Notice the clustered arrangement of the cells around their ducts. The cells are basically triangular, with their pointed ends facing the duct orifice. If possible, differentiate between mucus-producing cells, which look hollow or have a clear cytoplasm, and serous cells, which produce the clear, enzyme-containing fluid and have granules in their cytoplasm. The serous cells often form *demilunes* ("caps") around the more central mucous cells. The photomicrograph below (Figure 38.13) and that in the Histology Atlas, Plate 43, may be helpful in this task. In the space below, draw your version of a small portion of the salivary gland tissue and label it appropriately. ■

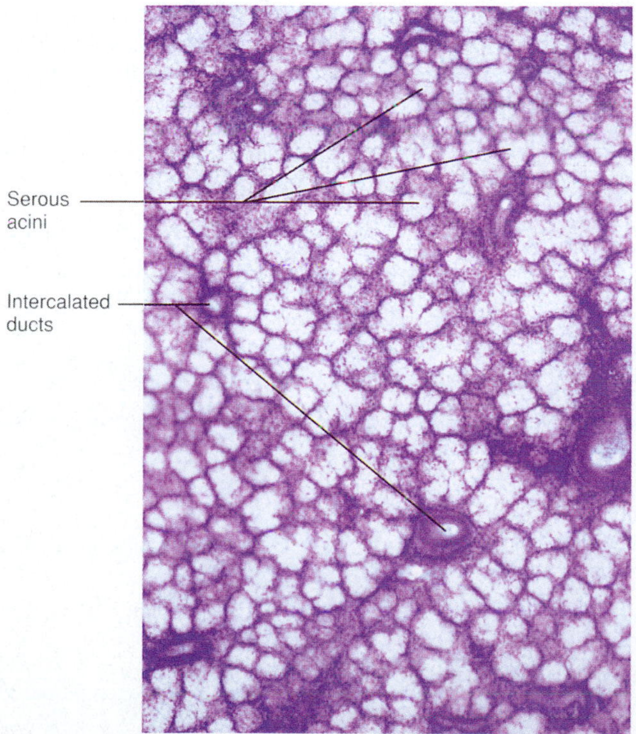

Figure 38.13 Photomicrograph of a cross section through a parotid salivary gland.

Liver and Gallbladder

The **liver** (see Figure 38.1), the largest gland in the body, is located inferior to the diaphragm, more to the right than the left side of the body. As noted earlier, it hides the stomach from view in a superficial observation of abdominal contents. The human liver has four lobes and is suspended from the diaphragm and anterior abdominal wall by the **falciform ligament** (see Figure 38.7a).

The liver is one of the body's most important organs, and it performs many metabolic roles. However, its digestive function is to produce bile, which leaves the liver through the

common hepatic duct and then enters the duodenum through the **bile duct.** Bile has no enzymatic action but emulsifies fats (breaks up large fat particles into smaller ones), thus creating a larger surface area for more efficient lipase activity. Without bile, very little fat digestion or absorption occurs.

When digestive activity is not occurring in the digestive tract, bile backs up the **cystic duct** and enters the **gallbladder,** a small, green sac on the inferior surface of the liver. It is stored there until needed for the digestive process. While in the gallbladder, bile is concentrated by the removal of water and some ions. When fat-rich food enters the duodenum, a hormonal stimulus causes the gallbladder to contract, releasing the stored bile and making it available to the duodenum.

If the common hepatic or bile duct is blocked (for example, by wedged gallstones), bile is prevented from entering the small intestine, accumulates, and eventually backs up into the liver. This exerts pressure on the liver cells, and bile begins to enter the bloodstream. As the bile circulates through the body, the tissues become yellow or jaundiced. ■

Blockage of the ducts is just one cause of jaundice. More often it results from actual liver problems such as **hepatitis** (an inflammation of the liver) or **cirrhosis,** a condition in which the liver is severely damaged and becomes hard and fibrous. Cirrhosis is almost guaranteed in those who drink excessive alcohol for many years. ■

As demonstrated by its highly organized anatomy, the liver (Figure 38.14) is very important in the initial processing of the nutrient-rich blood draining the digestive organs. Its structural and functional units are called **lobules.** Each lobule is a basically cylindrical structure consisting of cordlike arrays of parenchyma cells, which radiate outward from a central vein running upward in the longitudinal axis of the lobule. At each of the six corners of the lobule is a **portal triad,** (portal tract) so named because three basic structures are always present there: a branch of the *hepatic artery* (the functional blood supply of the liver), a branch of the *hepatic portal vein* (carrying nutrient-rich blood from the digestive viscera), and a *bile duct.* Between the liver parenchyma cells are blood-filled spaces, or **sinusoids,** through which blood from the hepatic portal vein and hepatic artery percolates past the parenchyma cells. Special phagocytic cells, **Kupffer cells,** line the sinusoids and remove debris such as bacteria from the blood as it flows past, while the parenchyma cells pick up oxygen and nutrients. Much of the glucose transported to the liver from the digestive system is stored as glycogen in the liver for later use, and amino acids are taken from the blood by the liver cells and utilized to make plasma proteins. The sinusoids empty into the central vein, and the blood ultimately drains from the liver via the *hepatic vein.*

Bile is continuously being made by the parenchyma cells. It flows through tiny canals, the **bile canaliculi,** which run between adjacent parenchyma cells toward the bile duct branches in the triad regions, where the bile eventually leaves the liver. Notice that the directions of blood and bile flow in the liver lobule are exactly opposite.

Activity 8:
Examining the Histology of the Liver

Examine a slide of liver tissue and identify as many of the structural features illustrated in Figure 38.14 and Histology Atlas Plates 45 and 46 as possible. Also examine a three-dimensional model of liver lobules if this is available. ■

Pancreas

The **pancreas** is a soft, triangular gland that extends horizontally across the posterior abdominal wall from the spleen to the duodenum (see Figure 38.1). Like the duodenum, it is a retroperitoneal organ (see Figure 38.7). As noted in Exercise 27, the pancreas has both an endocrine function (it produces the hormones insulin and glucagon) and an exocrine (enzyme-producing) function. It produces a whole spectrum of hydrolytic enzymes, which it secretes in an alkaline fluid into the duodenum through the pancreatic duct. Pancreatic juice is very alkaline. Its high concentration of bicarbonate ion (HCO_3^-) neutralizes the acidic chyme entering the duodenum from the stomach, enabling the pancreatic and intestinal enzymes to operate at their optimal pH. (Optimal pH for digestive activity to occur in the stomach is very acidic and results from the presence of HCl; that for the small intestine is slightly alkaline.)

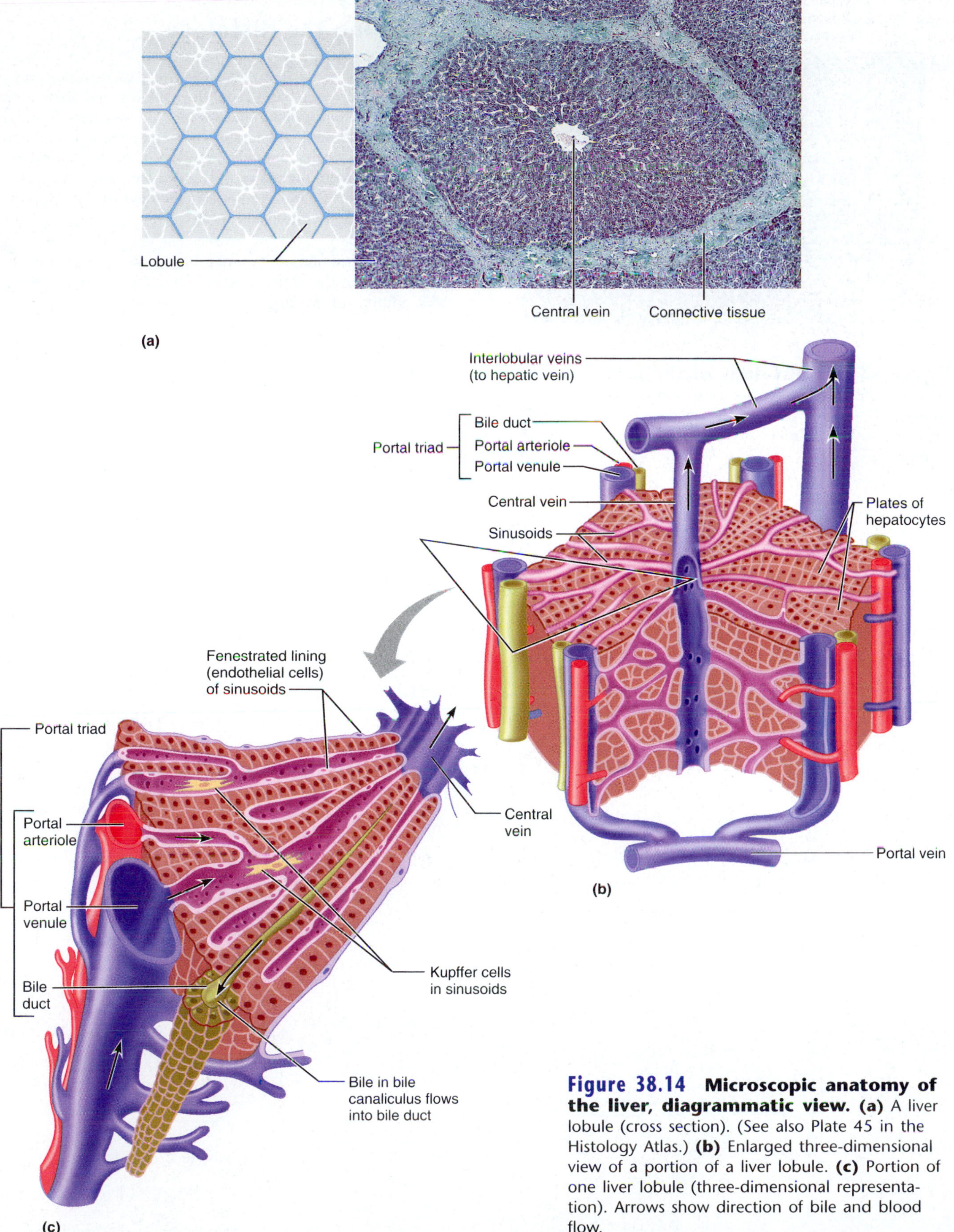

Figure 38.14 Microscopic anatomy of the liver, diagrammatic view. (a) A liver lobule (cross section). (See also Plate 45 in the Histology Atlas.) **(b)** Enlarged three-dimensional view of a portion of a liver lobule. **(c)** Portion of one liver lobule (three-dimensional representation). Arrows show direction of bile and blood flow.

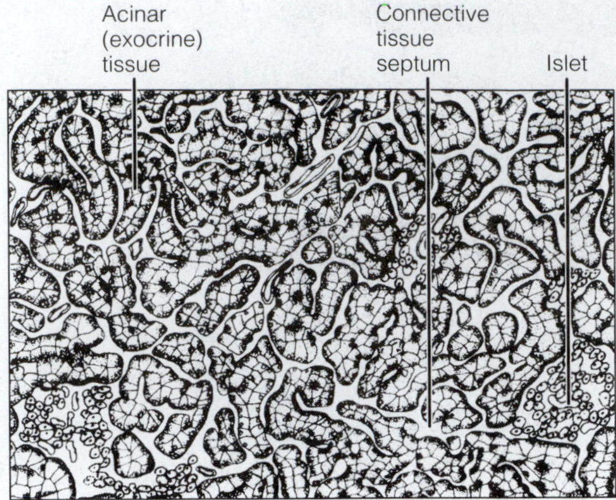

Figure 38.15 Histology of the pancreas. The pancreatic islet cells produce insulin and glucagon (hormones). The acinar cells synthesize digestive enzymes for "export" to the duodenum. (See also Plate 44 in the Histology Atlas.)

Activity 9:
Examining the Histology of the Pancreas

Observe pancreas tissue under low power and then high power to distinguish between the lighter-staining, endocrine-producing clusters of cells (pancreatic islets) and the deeper-staining acinar cells, which produce the hydrolytic enzymes and form the major portion of the pancreatic tissue (see Figure 38.15). Notice the arrangement of the exocrine cells around their central ducts. ■

For instructions on animal dissections, see the dissection exercises starting on p. 709 in the cat and fetal pig editions of this manual.

Chemical and Physical Processes of Digestion: Wet Lab

exercise

39A

Objectives

1. To list the digestive system enzymes involved in the digestion of proteins, fats, and carbohydrates; to state their site of origin; and to summarize the environmental conditions promoting their optimal functioning.
2. To recognize the variation between different types of enzyme assays.
3. To name the end products of protein, fat, and carbohydrate digestion.
4. To perform the appropriate chemical tests to determine if digestion of a particular foodstuff has occurred.
5. To cite the function(s) of bile in the digestive process.
6. To discuss the possible role of temperature and pH in the regulation of enzyme activity.
7. To define *enzyme, catalyst, control, substrate,* and *hydrolase.*
8. To explain why swallowing is both a voluntary and a reflex activity.
9. To discuss the role of the tongue, larynx, and gastroesophageal sphincter in swallowing.
10. To compare and contrast segmentation and peristalsis as mechanisms of propulsion.

Materials

Part I: Enzyme Action

General supply area:

- ❑ Test tubes and test tube rack
- ❑ Wax markers
- ❑ Hot plates
- ❑ 250-ml beakers
- ❑ Boiling chips
- ❑ Water bath set at 37°C (if not available, incubate at room temperature and double the time)
- ❑ Ice water bath
- ❑ Chart on chalkboard for recording class results

Supply area 1:

- ❑ Spot plates
- ❑ Dropper bottles of distilled water
- ❑ Dropper bottles of
  - 1% alpha-amylase solution*
  - 1% boiled starch solution, freshly prepared†
  - 1% maltose solution
  - Lugol's IKI (Lugol's iodine)
  - Benedict's solution

Supply area 2:

- ❑ Dropper bottles of
  - 1% trypsin
  - 0.01% BAPNA solution

Supply area 3:

- ❑ Dropper bottles of
  - 1% pancreatin solution
  - Litmus cream (fresh cream to which powdered litmus is added to achieve a blue color)
  - 0.1 *N* HCl
  - Vegetable oil
- ❑ Bile salts (sodium taurocholate)
- ❑ Parafilm (small squares to cover the test tubes)

Part II: Physical Processes

Supply area 4:

- ❑ Water pitcher
- ❑ Paper cups
- ❑ Stethoscope
- ❑ Alcohol swabs
- ❑ Disposable autoclave bag

Videotape viewing station:

- ❑ VHS cassette illustrating deglutition and peristalsis (*Passage of Food Through the Digestive Tract,* available from Ward's Biology)
- ❑ Television and VCR for independent viewing of videocassette by student.

PhysioEx™ 4.0 Computer Simulation on p. P-75

*The alpha-amylase must be a low-maltose preparation for good results.

†Prepare by adding 1 g starch to 100 ml distilled water; boil and cool; add a pinch of salt (NaCl). Prepare fresh daily.

Path of absorption | **Foodstuff** | **Enzyme(s) and source** | **Site of action**

Carbohydrate digestion

Absorption: Monosaccharides—glucose, galactose, and fructose—are absorbed into the capillary blood in the villi and transported to the liver via the hepatic portal vein.

Starch and disaccharides

- Salivary amylase — Mouth
- Pancreatic amylase — Small intestine

Oligosaccharides and disaccharides

Lactose Maltose Sucrose

Galactose Glucose Fructose

- Brush border enzymes in small intestine (dextrinase, lactase, maltase, and sucrase) — Small intestine

Protein digestion

Absorption: Amino acids are absorbed into the capillary blood in the villi and transported to the liver via the hepatic portal vein.

Proteins

- Pepsin (stomach glands) in the presence of HCl — Stomach

Large polypeptides

- Pancreatic enzymes (trypsin, chymotrypsin, carboxypeptidase) — Lumen of small intestine

Small polypeptides, small peptides

- Brush border enzymes (aminopeptidases, carboxypeptidase, and dipeptidases) — Brush border of small intestine

Amino acids (some dipeptides and tripeptides)

Fat digestion

Absorption: Absorbed primarily into the lacteals of the villi and transported to the systemic circulation via the lymph in the thoracic duct. (Glycerol and short-chain fatty acids are absorbed into the capillary blood in the villi and transported to the liver via the hepatic portal vein.)

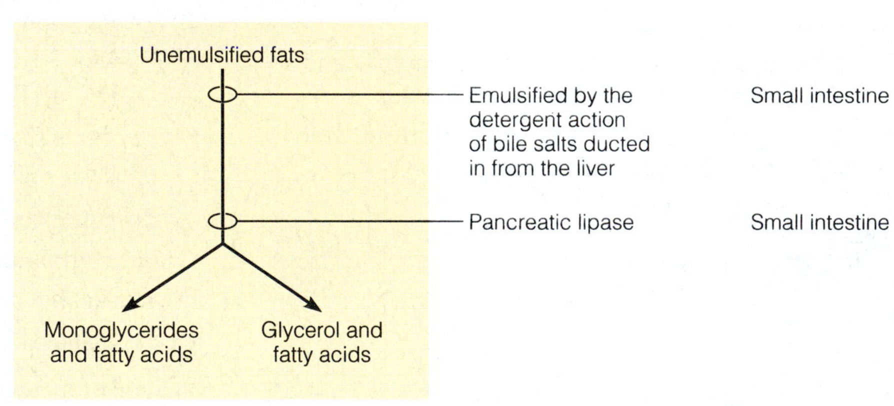

Figure 39A.1 Flowchart of digestion and absorption of foodstuffs.

Chemical Digestion of Foodstuffs: Enzymatic Action

Because nutrients can be absorbed only when broken down to their monomers, food digestion is a prerequisite to food absorption. You have already studied mechanisms of passive and active absorption in Exercise 5A and/or 5B. Before proceeding, review that material on pages 41–48 or P-4–P-14.

Enzymes are large protein molecules produced by body cells. They are biological **catalysts,** meaning they increase the rate of a chemical reaction without themselves becoming part of the product. The digestive enzymes are hydrolytic enzymes, or **hydrolases.** Their **substrates,** or the molecules on which they act, are organic food molecules which they break down by adding water to the molecular bonds, thus cleaving the bonds between the subunits or monomers.

The various hydrolytic enzymes are highly specific in their action. Each enzyme hydrolyzes only one or a small group of substrate molecules, and specific environmental conditions are necessary for it to function optimally. Since digestive enzymes actually function outside the body cells in the digestive tract, their hydrolytic activity can also be studied in a test tube. Such an *in vitro* study provides a convenient laboratory environment for investigating the effect of such variations on enzymatic activity.

Figure 39A.1 is a flowchart of the progressive digestion of proteins, fats, and carbohydrates. It indicates specific enzymes involved, their site of formation, and their site of action. Acquaint yourself with the flowchart before beginning this experiment, and refer to it as necessary during the laboratory session.

Activity 1: Developing Hypotheses for Experiments on Enzymatic Digestion

Work in groups of 4, with each group taking responsibility for setting up and conducting one of the following experiments. In each of the digestive procedures being studied (starch, protein, and fat digestion) and in the amylase assay, you are directed to boil the contents of one or more test tubes. To do this, obtain a 250-ml beaker, boiling chips, and a hot plate from the general supply area. Place a few boiling chips into the beaker and add about 125 ml of water, and bring to a boil. Place the specimen in the boiling water for the time (min.) specified in the directions.

Upon completion of the experiments, each group should communicate its results to the rest of the class by recording them in a chart on the chalkboard. All members of the class should observe the **controls** (the specimens or standards against which experimental samples are compared) as well as the positive and negative examples of all experimental results. Additionally, all members of the class should be able to explain the tests used and the results observed and anticipated for each experiment. Note that water baths and hot plates are at the general supply area.

1. Develop a hypothesis for each of the enzymatic digestions studied in Activities 2–4: starch, protein, and fat.

2. Write down your hypotheses, including your reasoning behind each hypothesis. ■

Activity 2: Assessing Starch Digestion by Salivary Amylase

1. From the general supply area, obtain a test tube rack, 10 test tubes, and a wax marking pencil. From supply area 1, obtain a dropper bottle of distilled water and dropper bottles of maltose, amylase, and starch solutions.

2. In this experiment you will investigate the hydrolysis of starch to maltose by **salivary amylase** (the enzyme produced by the salivary glands and secreted into the mouth), so you will need to be able to identify the presence of starch and maltose to determine to what extent the enzymatic activity has occurred. Thus controls must be prepared to provide a known standard against which comparisons can be made. Starch decreases and sugar increases as digestion occurs, according to the following formula:

$$\text{Starch} + \text{Water} \xrightarrow{\text{Amylase}} \text{Maltose}$$

Two students should prepare the controls (tubes 1A to 3A) while the other two prepare the experimental samples (tubes 4A to 6A).

- Mark each tube with a wax pencil and load the tubes as indicated in the chart on p. 408, using 3 drops (gtt) of each indicated substance.

- Place all tubes in a rack in the appropriate water bath for approximately 1 hour. Shake the rack gently from time to time to keep the contents evenly mixed.

- While these tubes are incubating, proceed to Physical Processes: Mechanisms of Food Propulsion and Mixing (p. 411).

Amylase Assay

1. From supply area 1 obtain a spot plate and dropper bottles of IKI and Benedict's solution. Set up your boiling water bath using a hot plate, boiling chips, and a 250-ml beaker obtained from the general supply area.

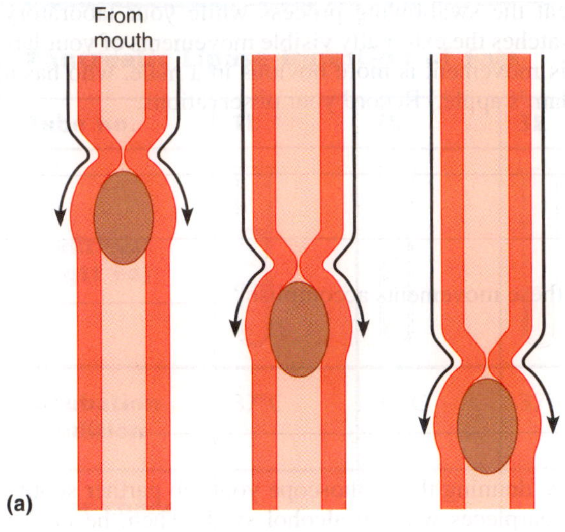

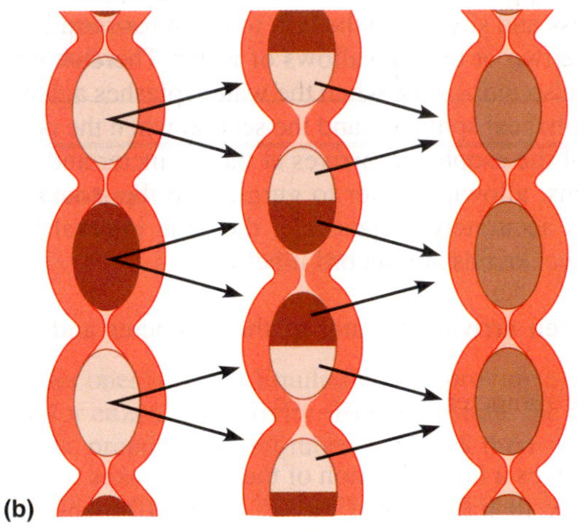

Figure 39A.2 Peristaltic and segmental movements of the digestive tract. (a) Peristalsis: neighboring segments of the intestine alternately contract and relax, moving food along the tract. **(b)** Segmentation: single segments of intestine alternately contract and relax. Because inactive segments exist between active segments, food mixing occurs to a greater degree than food movement. Peristalsis is superimposed on segmentation.

Segmentation and Peristalsis

Although several types of movements occur in the digestive tract organs, peristalsis and segmentation are most important as mixing and propulsive mechanisms (Figure 39A.2).

Peristaltic movements are the major means of propelling food through most of the digestive viscera. Essentially they are waves of contraction followed by waves of relaxation that squeeze foodstuffs through the alimentary canal, and they are superimposed on segmental movements.

Segmental movements are local constrictions of the organ wall that occur rhythmically. They serve mainly to mix the foodstuffs with digestive juices and to increase the rate of absorption by continually moving different portions of the chyme over adjacent regions of the intestinal wall. However, segmentation is also an important means of food propulsion in the small intestine, and slow segmenting movements called haustral contractions are frequently seen in the large intestine.

Activity 7: Viewing Segmental and Peristaltic Movements

If a videotape showing some of the propulsive movements is available, go to a viewing station to view it before leaving the laboratory. ■

exercise 40

Anatomy of the Urinary System

Metabolism of nutrients by the body produces wastes (carbon dioxide, nitrogenous wastes, ammonia, and so on) that must be eliminated from the body if normal function is to continue. Although excretory processes involve several organ systems (the lungs excrete carbon dioxide and skin glands excrete salts and water), it is the **urinary system** that is primarily concerned with the removal of nitrogenous wastes from the body. In addition to this purely excretory function, the kidney maintains the electrolyte, acid-base, and fluid balances of the blood and is thus a major, if not *the* major, homeostatic organ of the body.

To perform its functions, the kidney acts first as a blood "filter," and then as a blood "processor." It allows toxins, metabolic wastes, and excess ions to leave the body in the urine, while simultaneously retaining needed substances and returning them to the blood. Malfunction of the urinary system, particularly of the kidneys, leads to a failure in homeostasis which, unless corrected, is fatal.

Objectives

1. To describe the function of the urinary system.
2. To identify, on an appropriate diagram or torso model, the urinary system organs and to describe the general function of each.
3. To compare the course and length of the urethra in males and females.
4. To identify these regions of the dissected kidney (longitudinal section): hilus, cortex, medulla, medullary pyramids, major and minor calyces, pelvis, renal columns, and renal and adipose capsules.
5. To trace the blood supply of the kidney from the renal artery to the renal vein.
6. To define the nephron as the physiological unit of the kidney and to describe its anatomy.
7. To define *glomerular filtration, tubular reabsorption,* and *tubular secretion,* and to indicate the nephron areas involved in these processes.
8. To define *micturition,* and to explain pertinent differences in the control of the two bladder sphincters (internal and external).
9. To recognize microscopic or diagrammatic views of the histologic structure of the kidney and bladder.

Materials

- ❑ Human dissectible torso model and/or anatomical chart of the human urinary system
- ❑ 3-dimensional model of the cut kidney and of a nephron (if available)
- ❑ Dissection tray and instruments
- ❑ Pig or sheep kidney, doubly or triply injected
- ❑ Disposable gloves
- ❑ Prepared histologic slides of a longitudinal section of kidney and cross sections of the bladder
- ❑ Compound microscope

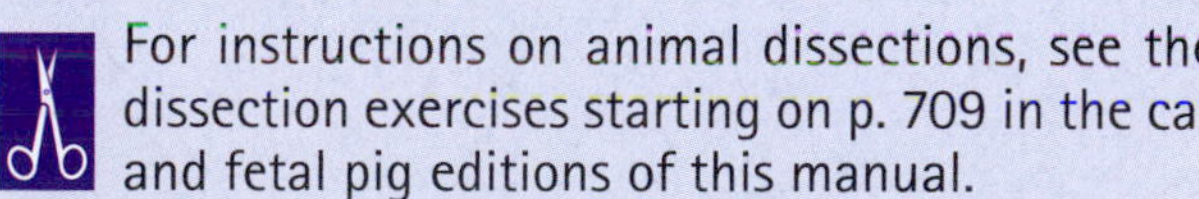

For instructions on animal dissections, see the dissection exercises starting on p. 709 in the cat and fetal pig editions of this manual.

A1A See Appendix C, Exercise 40 for links to A.D.A.M.® Interactive Anatomy.

Gross Anatomy of the Human Urinary System

The urinary system (Figure 40.1) consists of the paired kidneys and ureters and the single urinary bladder and urethra. The **kidneys** perform the functions described above and manufacture urine in the process. The remaining organs of the system provide temporary storage reservoirs or transportation channels for urine.

Activity 1: Identifying Urinary System Organs

Examine the human torso model, a large anatomical chart, or a three-dimensional model of the urinary system to locate and study the anatomy and relationships of the urinary organs.

1. Locate the paired kidneys on the dorsal body wall in the superior lumbar region. Note that they are not positioned at exactly the same level. Because it is "crowded" by the liver, the right kidney is slightly lower than the left kidney. In a living person, fat deposits (the *adipose capsules*) and the fibrous renal capsules hold the kidneys in place in a retroperitoneal position.

When the fatty material surrounding the kidneys is reduced or too meager in amount (in cases of rapid weight loss or in very thin individuals), the kidneys are less securely anchored to the body wall and may drop to a lower or more inferior position in the abdominal cavity. This phenomenon is called **ptosis.** ■

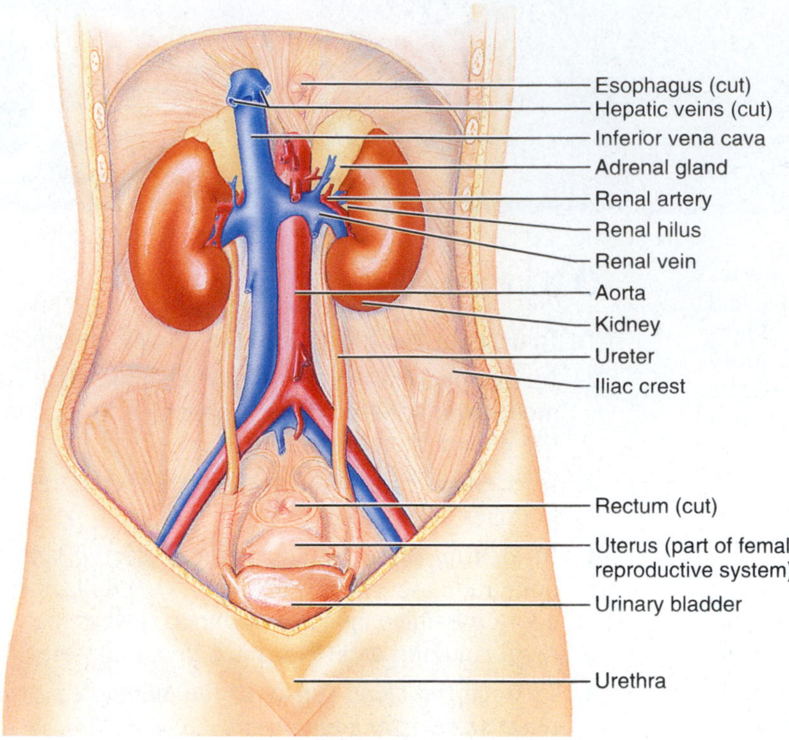

Figure 40.1 Organs of the urinary system. **(a)** Anterior view of the female urinary organs. Most unrelated abdominal organs have been removed. **(b)** Posterior in situ view showing the position of the kidneys relative to the twelfth rib pair. **(c)** Transverse section of the abdomen showing the kidneys in retroperitoneal position.

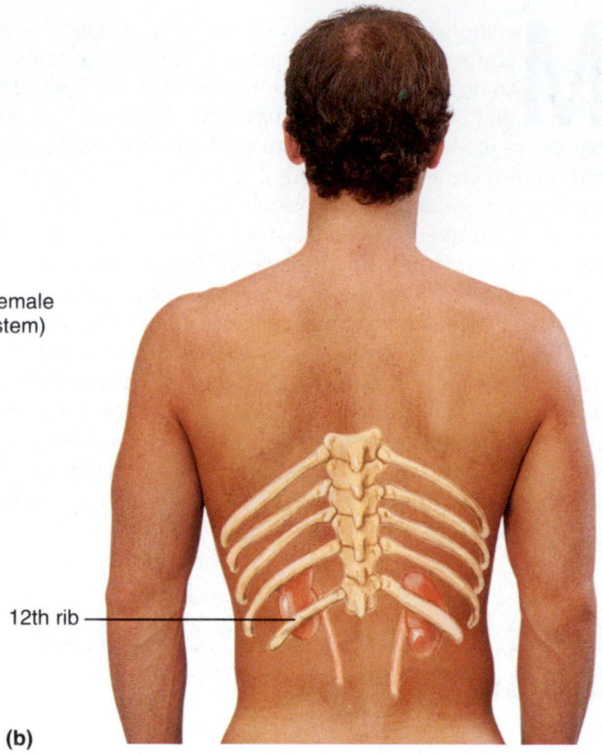

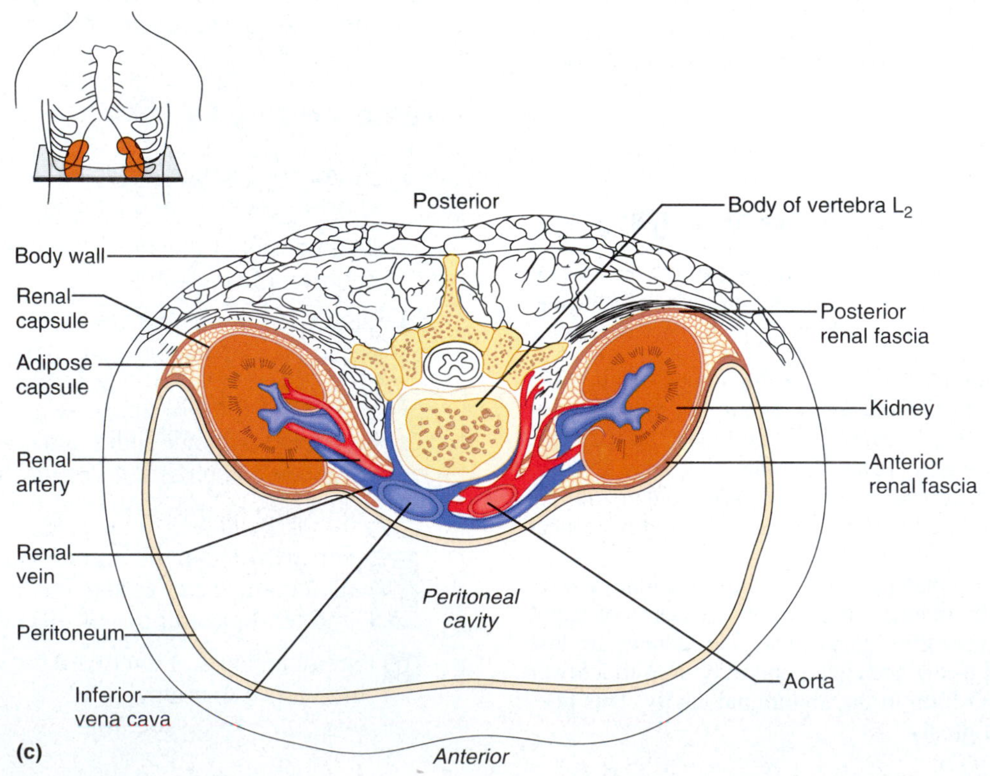

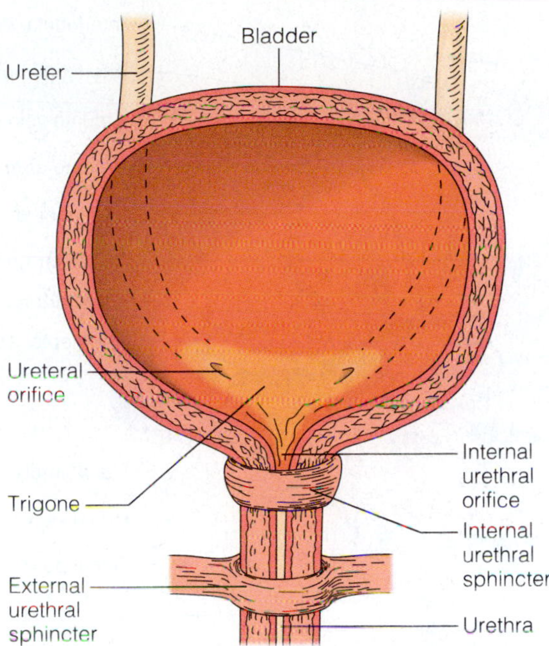

Figure 40.2 Detailed structure of the urinary bladder and urethral sphincters.

2. Observe the renal arteries as they diverge from the descending aorta and plunge into the indented medial region (**hilus**) of each kidney. Note also the **renal veins,** which drain the kidneys (circulatory drainage) and the two **ureters,** which drain urine from the kidneys and conduct it by peristalsis to the bladder for temporary storage.

3. Locate the urinary bladder, and observe the point of entry of the two ureters into this organ. Also locate the single **urethra,** which drains the bladder. The triangular region of the bladder, which is delineated by these three openings (two ureteral and one urethral orifice), is referred to as the **trigone** (Figure 40.2).

4. Follow the course of the urethra to the body exterior. In the male, it is approximately 20 cm (8 inches) long, travels the length of the **penis,** and opens at its tip. Its three named regions—the *prostatic, membranous,* and *spongy (penile) urethrae*—are described in more detail in Exercise 42 and illustrated in Figure 42.1 (pp. 429 and 430). The urethra of males has a dual function: it is a urine conduit to the body exterior, and it provides a passageway for the ejaculation of semen. Thus, in the male, the urethra is part of both the urinary and reproductive systems. In females, the urethra is very short, approximately 4 cm (1½ inches) long. There are no common urinary-reproductive pathways in the female, and the female's urethra serves only to transport urine to the body exterior. Its external opening, the **external urethral orifice,** lies anterior to the vaginal opening. ■

Dissection:

Gross Internal Anatomy of the Pig or Sheep Kidney

1. In preparation for dissection, don gloves. Obtain a preserved sheep or pig kidney, dissecting tray, and instruments. Observe the kidney to identify the **renal capsule,** a smooth transparent membrane that adheres tightly to the external aspect of the kidney.

2. Find the ureter, renal vein, and renal artery at the hilus (indented) region. The renal vein has the thinnest wall and will be collapsed. The ureter is the largest of these structures and has the thickest wall.

3. Make a cut through the longitudinal axis (frontal section) of the kidney and locate the anatomical areas described below and depicted in Figure 40.3.

Kidney cortex: The superficial kidney region, which is lighter in color. If the kidney is doubly injected with latex, you will see a predominance of red and blue latex specks in this region indicating its rich vascular supply.

Medullary region: Deep to the cortex; a darker, reddish-brown color. The medulla is segregated into triangular regions that have a striped, or striated, appearance—the **medullary (renal) pyramids.** The base of each pyramid faces toward the cortex. Its more pointed **apex,** or **papilla,** points to the innermost kidney region.

Renal columns: Areas of tissue, more like the cortex in appearance, which segregate and dip inward between the pyramids.

Renal pelvis: Medial to the hilus; a relatively flat, basinlike cavity that is continuous with the **ureter,** which exits from the hilus region. Fingerlike extensions of the pelvis should be visible. The larger, or primary, extensions are called the **major calyces;** subdivisions of the major calyces are the **minor calyces.** Notice that the minor calyces terminate in cuplike areas that enclose the apexes of the medullary pyramids and collect urine draining from the pyramidal tips into the pelvis.

4. If the preserved kidney is doubly or triply injected, follow the renal blood supply from the renal artery to the **glomeruli.** The glomeruli appear as little red and blue specks in the cortex region. (See Figures 40.3 and 40.4.)

Approximately a fourth of the total blood flow of the body is delivered to the kidneys each minute by the large **renal arteries.** As a renal artery approaches the kidney, it breaks up into five branches called **segmental arteries,** which enter the hilus. Each segmental artery, in turn, divides into several **lobar arteries.** The lobar arteries branch to form **interlobar arteries,** which ascend toward the cortex in the renal column areas. At the top of the medullary region, these arteries give off arching branches, the **arcuate arteries,** which curve over the bases of the medullary pyramids. Small

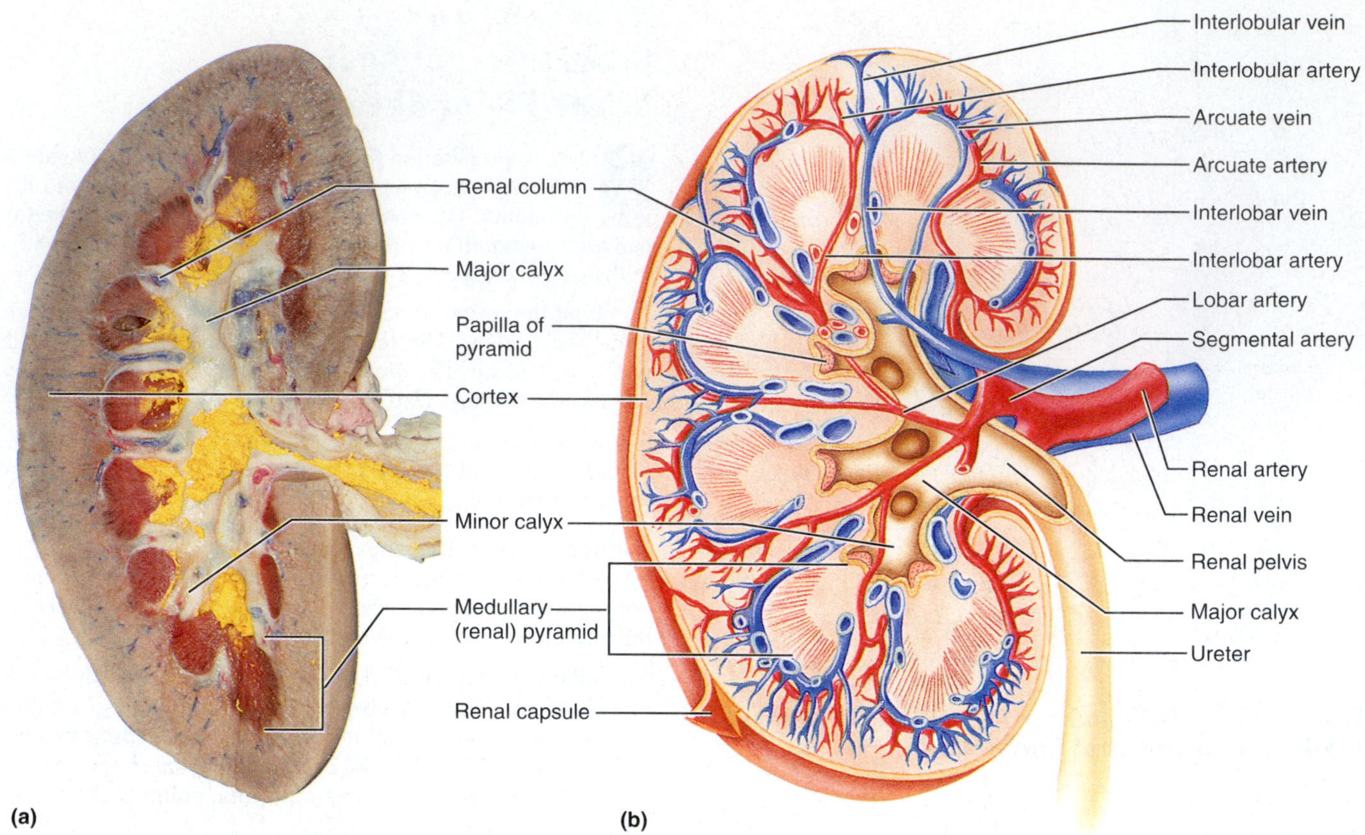

Figure 40.3 Frontal section of a kidney. (a) Photograph of a triple-injected pig kidney. Arteries and veins injected with red and blue latex respectively; pelvis and ureter injected with yellow latex. **(b)** Diagrammatic view showing the larger arteries supplying the kidney tissue.

interlobular arteries branch off the arcuate arteries and ascend into the cortex, giving off the individual **afferent arterioles,** which provide the capillary networks (glomeruli and peritubular capillary beds) that supply the nephrons, or functional units, of the kidney. Blood draining from the nephron capillary networks in the cortex enters the **interlobular veins** and then drains through the **arcuate veins** and the **interlobar veins** to finally enter the **renal vein** in the pelvis region. (There are no lobar or segmental veins.) ■

Functional Microscopic Anatomy of the Kidney and Bladder

Kidney

Each kidney contains over a million nephrons, which are the anatomical units responsible for forming urine. Figure 40.4 depicts the detailed structure and the relative positioning of the nephrons in the kidney.

Each nephron consists of two major structures: a **glomerulus** (a capillary knot) and a renal tubule. During embryologic development, each **renal tubule** begins as a blind-ended tubule that gradually encloses an adjacent capillary cluster, or glomerulus. The enlarged end of the tubule encasing the glomerulus is the **glomerular (Bowman's) capsule,** and its inner, or visceral, wall consists of highly specialized cells called **podocytes.** Podocytes have long, branching processes *(foot processes)* that interdigitate with those of other podocytes and cling to the endothelial wall of the glomerular capillaries, thus forming a very porous epithelial membrane surrounding the glomerulus. The glomerulus-capsule complex is sometimes called the **renal corpuscle.**

The rest of the tubule is approximately 3 cm (1.25 inches) long. As it emerges from the glomerular capsule, it becomes highly coiled and convoluted, drops down into a long hairpin loop, and then again coils and twists before entering a collecting duct. In order from the glomerular capsule, the anatomical areas of the renal tubule are: the **proximal convoluted tubule, loop of Henle** (descending and ascending limbs), and the **distal convoluted tubule.** The wall of the renal tubule is composed almost entirely of cuboidal epithelial cells, with the exception of part of the descending limb (and sometimes part of the ascending limb) of the loop of Henle, which is simple squamous epithelium. The lumen surfaces of the cuboidal cells in the proximal convoluted tubule have dense microvilli (a cellular modification that greatly increases the surface area exposed to the lumen contents, or filtrate). Microvilli also occur on cells of the distal convoluted tubule but in greatly reduced numbers, revealing its less significant role in reclaiming filtrate contents.

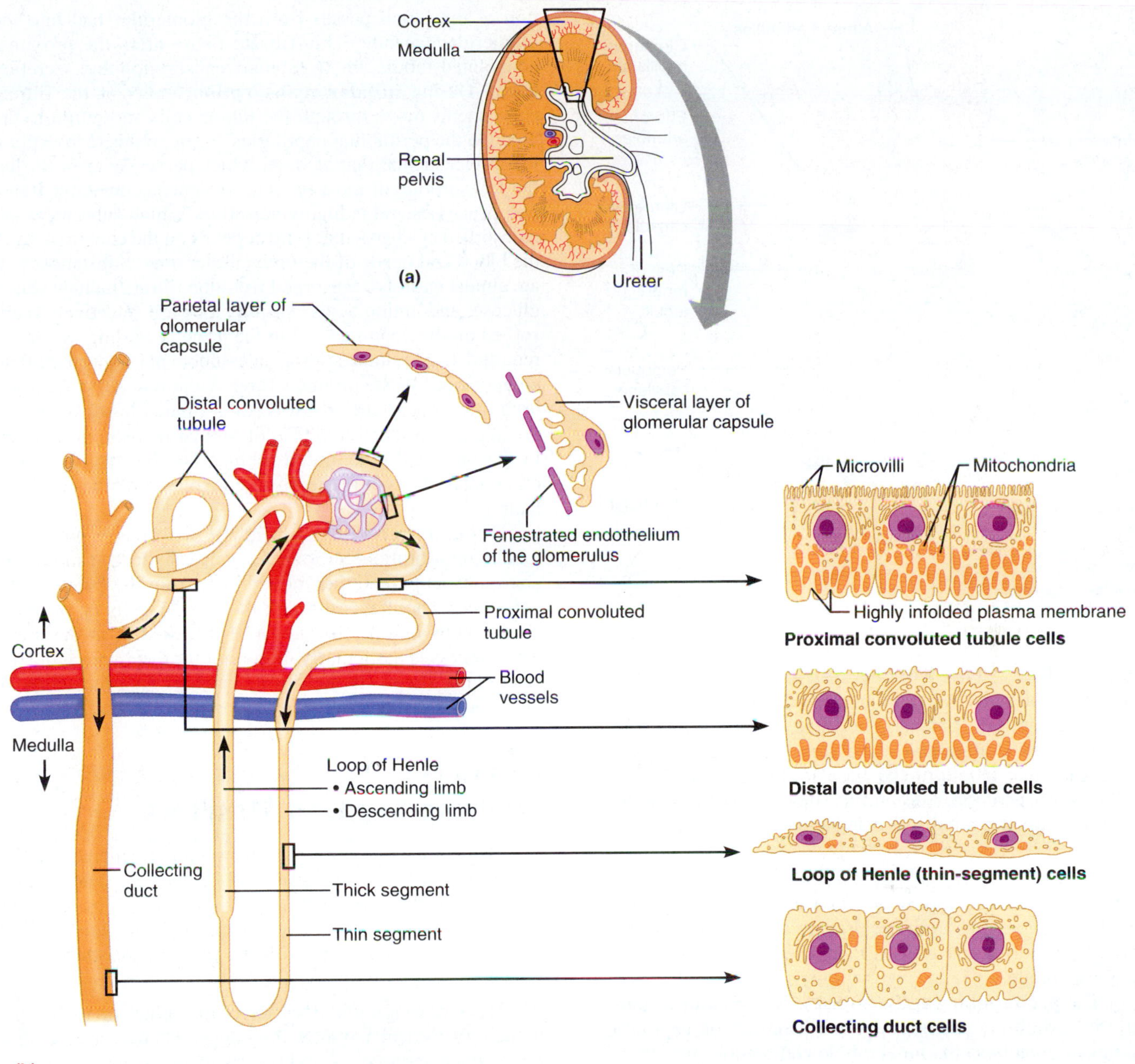

Figure 40.4 **Structure of a nephron.** **(a)** Wedge-shaped section of kidney tissue, indicating the position of the nephrons in the kidney. Note, however, that the section marked by the wedge contains hundreds of thousands of nephrons. **(b)** Detailed nephron anatomy and associated blood supply.

Most nephrons, called **cortical nephrons,** are located entirely within the cortex. However, parts of the loops of Henle of the **juxtamedullary nephrons** (located close to the cortex-medulla junction) penetrate well into the medulla. The **collecting ducts,** each of which receives urine from many nephrons, run downward through the medullary pyramids, giving them their striped appearance. As the collecting ducts approach the renal pelvis, they fuse to form larger *papillary ducts,* which empty the final urinary product into the calyces and pelvis of the kidney.

The function of the nephron depends on several unique features of the renal circulation. The capillary vascular supply consists of two distinct capillary beds, the *glomerulus* and the *peritubular capillary bed.* Vessels leading to and from the glomerulus, the first capillary bed, are both arterioles: the **afferent arteriole** feeds the bed while the **efferent arteriole** drains it. The glomerular capillary bed has no parallel elsewhere in the body. It is a high-pressure bed along its entire length. Its high pressure is a result of two major factors: (1) the bed is *fed and drained* by arterioles (arterioles are high-resistance vessels as opposed to venules, which are low-resistance vessels), and (2) the afferent feeder arteriole is larger in diameter than the efferent arteriole draining the bed. The high hydrostatic pressure created by these two

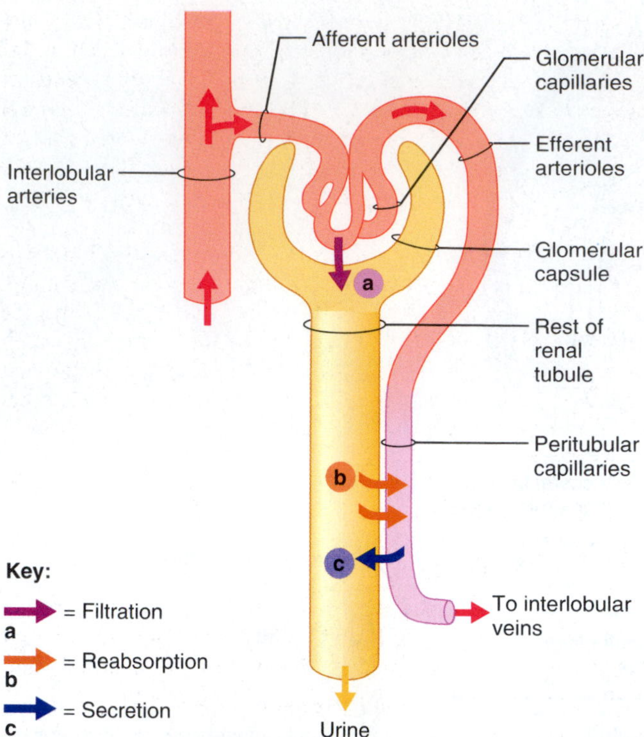

Figure 40.5 The kidney depicted as a single, large nephron. A kidney actually has millions of nephrons acting in parallel. The three major mechanisms by which the kidneys adjust the composition of plasma are **(a)** glomerular filtration, **(b)** tubular reabsorption, and **(c)** tubular secretion. Red arrows show the path of blood flow through the renal microcirculation.

anatomical features forces out fluid and blood components smaller than proteins from the glomerulus into the glomerular capsule. That is, it forms the filtrate which is processed by the nephron tubule.

The **peritubular capillary bed** arises from the efferent arteriole draining the glomerulus. This set of capillaries clings intimately to the renal tubule and empties into the interlobular veins that leave the cortex. The peritubular capillaries are *low-pressure* very porous capillaries adapted for absorption rather than filtration and readily take up the solutes and water reabsorbed from the filtrate by the tubule cells. The juxtamedullary nephrons have additional looping vessels, called the **vasa recta** ("straight vessels"), that parallel the long loops of Henle in the medulla. Hence, the two capillary beds of the nephron have very different, but complementary, roles: The glomerulus produces the filtrate and the peritubular capillaries reclaim most of that filtrate.

Additionally, each nephron has a region called a **juxtaglomerular apparatus** (JGA), which plays an important role in filtrate formation. The JGA consists of (1) *juxtaglomerular* (JG) cells, (2) pressure sensors in the walls of the arterioles near the glomerulus, and (3) a *macula densa,* a specialized group of columnar cells in the distal convoluted tubule abutting the JG cells (Figure 40.6).

Urine formation is a result of three processes: *filtration, reabsorption,* and *secretion* (Figure 40.5). **Filtration,** the role of the glomerulus, is largely a passive process in which a portion of the blood passes from the glomerular bed into the glomerular capsule. This filtrate then enters the proximal convoluted tubule where tubular reabsorption and secretion begin. During **tubular reabsorption,** many of the filtrate components move through the tubule cells and return to the blood in the peritubular capillaries. Some of this reabsorption is passive, such as that of water which passes by osmosis, but the reabsorption of most substances depends on active transport processes and is highly selective. Which substances are reabsorbed at a particular time depends on the composition of the blood and needs of the body at that time. Substances that are almost entirely reabsorbed from the filtrate include water, glucose, and amino acids. Various ions are selectively reabsorbed or allowed to go out in the urine according to what is required to maintain appropriate blood pH and electrolyte composition. Waste products (urea, creatinine, uric acid, and drug metabolites) are reabsorbed to a much lesser degree or not at all. Most (75% to 80%) of tubular reabsorption occurs in the proximal convoluted tubule. The balance occurs in other areas, especially the distal convoluted tubules and collecting ducts.

Tubular secretion is essentially the reverse process of tubular reabsorption. Substances such as hydrogen and potassium ions and creatinine move either from the blood of the peritubular capillaries through the tubular cells or from the tubular cells into the filtrate to be disposed of in the urine. This process is particularly important for the disposal of substances not already in the filtrate (such as drug metabolites), and as a device for controlling blood pH.

Activity 2: Studying Nephron Structure

1. Begin your study of nephron structure by identifying the glomerular capsule, proximal and distal convoluted tubule regions, and the loop of Henle on a model of the nephron. Then, obtain a compound microscope and a prepared slide of kidney tissue to continue with the microscope study of the kidney.

2. Hold the longitudinal section of the kidney up to the light to identify cortical and medullary areas. Then secure the slide on the microscope stage and scan the slide under low power.

3. Move the slide so that you can see the cortical area. Identify a glomerulus, which appears as a ball of tightly packed material containing many small nuclei (Figure 40.6). It is usually delineated by a vacant-appearing region (corresponding to the space between the visceral and parietal layers of the glomerular capsule) that surrounds it.

4. Notice that the renal tubules are cut at various angles. Also try to differentiate between the thin-walled loop of Henle portion of the tubules and the cuboidal epithelium of the proximal convoluted tubule, which has dense microvilli. ■

Bladder

Although urine production by the kidney is a continuous process, urine is usually removed from the body when voiding is convenient. In the meantime the **urinary bladder,** which receives urine via the ureters and discharges it via the urethra, stores it temporarily.

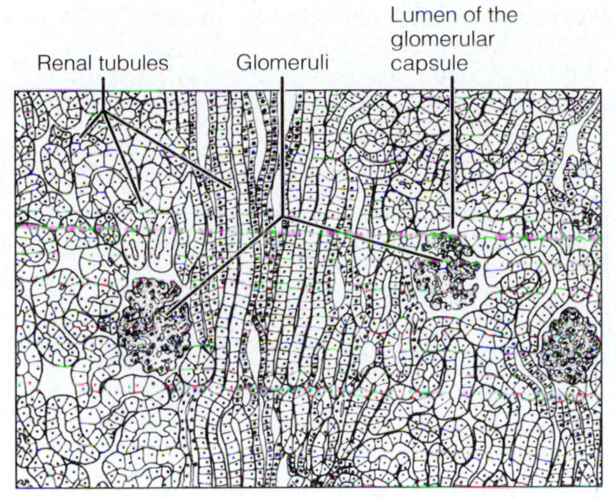

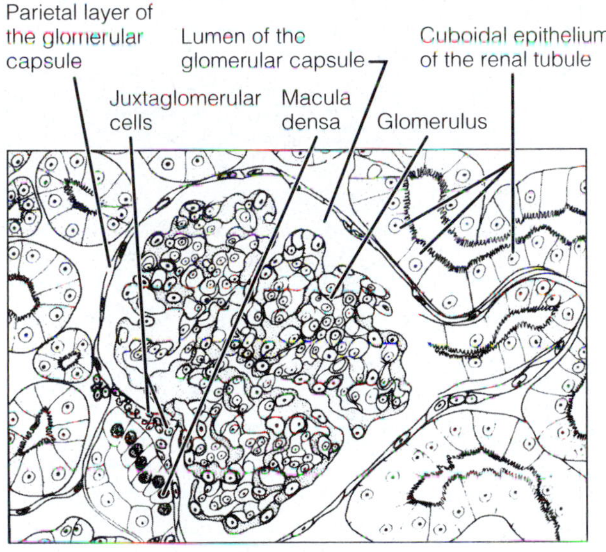

Figure 40.6 Microscopic structure of kidney tissue. (a) Low-power view of the renal cortex. **(b)** Detailed structure of the glomerulus. (See also Plates 47 and 48 in the Histology Atlas.)

Voiding, or **micturition,** is the process in which urine empties from the bladder. Two sphincter muscles or valves (Figure 40.2), the **internal urethral sphincter** (more superiorly located) and the **external urethral sphincter** (more inferiorly located) control the outflow of urine from the bladder. Ordinarily, the bladder continues to collect urine until about 200 ml have accumulated, at which time the stretching of the bladder wall activates stretch receptors. Impulses transmitted to the central nervous system subsequently produce reflex contractions of the bladder wall through parasympathetic nervous system pathways (i.e., via the pelvic splanchnic nerves). As the contractions increase in force and frequency, the stored urine is forced past the internal sphincter, which is a smooth muscle involuntary sphincter, into the superior part of the urethra. It is then that a person feels the urge to void. The inferior external sphincter consists of skeletal muscle and is voluntarily controlled. If it is not convenient to void, the opening of this sphincter can be inhibited. Conversely, if the time is convenient, the sphincter may be relaxed and the stored urine flushed from the body. If voiding is inhibited, the reflex contractions of the bladder cease temporarily and urine continues to accumulate in the bladder. After another 200 to 300 ml of urine have been collected, the *micturition reflex* will again be initiated.

Lack of voluntary control over the external sphincter is referred to as **incontinence.** Incontinence is normal in children 2 years old or younger, as they have not yet gained control over the voluntary sphincter. In adults and older children, incontinence is generally a result of spinal cord injury, emotional problems, bladder irritability, or some other pathology of the urinary tract. ■

Activity 3: Studying Bladder Structure

1. Return the kidney slide to the supply area and obtain a slide of bladder tissue. Scan the bladder tissue. Identify its three layers: mucosa, muscular layer, and fibrous adventitia.

2. Study the mucosa with its highly specialized transitional epithelium. The plump, transitional epithelial cells have the ability to slide over one another, thus decreasing the thickness of the mucosa layer as the bladder fills and stretches to accommodate the increased urine volume. Depending on the degree of stretching of the bladder, the mucosa may be three to eight cell layers thick. Compare the transitional epithelium of the mucosa to that shown in Figure 6A.3h (p. 55).

3. Examine the heavy muscular wall (detrusor muscle), which consists of three irregularly arranged muscular layers. The innermost and outermost muscle layers are arranged longitudinally; the middle layer is arranged circularly. Attempt to differentiate the three muscle layers.

4. Draw a small section of the bladder wall, and label all regions or tissue areas.

5. Compare your sketch of the bladder wall to the structure of the ureter wall shown in Plate 49 in the Histology Atlas. How are the two organs similar histologically?

____

____

____

What is/are the most obvious differences?

____

______________________________ ■

For instructions on animal dissections, see the dissection exercises starting on p. 709 in the cat and fetal pig editions of this manual.

Urinalysis

exercise **41A**

Objectives

1. To list the physical characteristics of urine, and to indicate the normal pH and specific gravity ranges.
2. To list substances that are normal urinary constituents.
3. To conduct various urinalysis tests and procedures and use them to determine the substances present in a urine specimen.
4. To define the following urinary conditions:

albuminuria	hematuria
calculi	hemoglobinuria
casts	ketonuria
glycosuria	pyuria

5. To explain the implications and possible causes of conditions listed in Objective 4.

Materials

- ❑ "Normal" artificial urine provided by the instructor*
- ❑ Numbered "pathological" urine specimens provided by the instructor*
- ❑ Urinometer
- ❑ Hot plate
- ❑ 500-ml beakers
- ❑ Wide-range pH paper
- ❑ Dip sticks: individual (Clinistix, Ketostix, Albustix, Hemastix, Bilistix) or combination (Chemstrip or Multistix)
- ❑ Ictotest reagent
- ❑ Test reagents for sulfates: 10% barium chloride solution, dilute HCl (hydrochloric acid)
- ❑ Test reagent for glucose (Clinitest tablets)
- ❑ Test reagent for phosphates (dilute nitric acid, dilute ammonium molybdate)
- ❑ Test reagent for urea (concentrated nitric acid in dropper bottle)
- ❑ Test reagent for Cl: 3.0% silver nitrate solution ($AgNO_3$), freshly prepared
- ❑ Test tubes, test tube rack, and test tube holders
- ❑ 10-cc graduated cylinders
- ❑ Glass stirring rods
- ❑ Medicine droppers
- ❑ Wax marking pencils
- ❑ Microscope slides
- ❑ Coverslips
- ❑ Compound microscope
- ❑ Centrifuge and centrifuge tubes
- ❑ Sedi-stain
- ❑ Flasks and laboratory buckets containing 10% bleach solution
- ❑ Disposable plastic gloves
- ❑ Disposable autoclave bags
- ❑ *Demonstration:* Instructor-prepared specimen of urine sediment set up for microscopic analysis

PhysioEx™ 4.0 Computer Simulation on p. P-86

*Directions for making artificial urine are provided in the Instructor's Guide for this manual.

Blood composition depends on three major factors: diet, cellular metabolism, and urinary output. In 24 hours, the kidneys' 2 million nephrons filter approximately 150 to 180 liters of blood plasma through their glomeruli into the tubules, where it is selectively processed by tubular reabsorption and secretion. In the same period, urinary output, which contains by-products of metabolism and excess ions, is 1.0 to 1.8 liters. In healthy individuals, the kidneys can maintain blood constancy despite wide variations in diet and metabolic activity. With certain pathological conditions, urine composition often changes dramatically.

Characteristics of Urine

Freshly voided urine is generally clear and pale yellow to amber in color. This normal yellow color is due to *urochrome*, a pigment metabolite arising from the body's destruction of hemoglobin (via bilirubin or bile pigments). As a rule, color variations from pale yellow to deeper amber indicate the relative concentration of solutes to water in the urine. The greater the solute concentration, the deeper the color. Abnormal urinary color may be due to certain foods, such as beets, various drugs, bile, or blood.

The odor of freshly voided urine is characteristic and slightly aromatic, but bacterial action gives it an ammonia-like odor when left standing. Some drugs, vegetables (such as asparagus), and various disease processes (such as diabetes mellitus) alter the characteristic odor of urine. For example, the urine of a person with uncontrolled diabetes mellitus (and elevated levels of ketones) smells fruity or acetonelike.

The pH of urine ranges from 4.5 to 8.0, but its average value, 6.0, is slightly acidic. Diet may markedly influence the pH of the urine. For example, a diet high in protein (meat, eggs, cheese) and whole wheat products increases the acidity of urine. Such foods are called *acid ash foods*. On the other hand, a vegetarian diet (*alkaline ash diet*) increases the alkalinity of the urine.* A bacterial infection of the urinary tract may also result in urine with a high pH.

Specific gravity is the relative weight of a specific volume of liquid compared with an equal volume of distilled water. The specific gravity of distilled water is 1.000, because 1 ml weighs 1 g. Since urine contains dissolved solutes, it weighs more than water, and its customary specific gravity ranges from 1.001 to 1.030. Urine with a specific gravity of 1.001 contains few solutes and is considered very dilute. Dilute urine commonly results when a person drinks excessive amounts of water, uses diuretics, or suffers from diabetes insipidus or chronic renal failure. Conditions that produce urine with a high specific gravity include limited fluid intake, fever, and kidney inflammation, called *pyelonephritis*. If urine becomes excessively concentrated, some of the substances normally held in solution begin to precipitate or crystallize, forming **kidney stones,** or **renal calculi.**

Normal constituents of urine (in order of decreasing concentration) include water; urea;† sodium,‡ potassium, phosphate, and sulfate ions; creatinine;† and uric acid.† Much smaller but highly variable amounts of calcium, magnesium, and bicarbonate ions are also found in the urine. Abnormally high concentrations of any of these urinary constituents may indicate a pathological condition.

* The naming of foods as acid or alkaline *ash* derives from the fact that if an acid ash food is burned to ash, the pH of the ash is acidic. The ash of alkaline ash foods is alkaline.

† Urea, uric acid, and creatinine are the most important nitrogenous wastes found in urine. Urea is an end product of protein breakdown; uric acid is a metabolite of purine breakdown; and creatinine is associated with muscle metabolism of creatine phosphate.

‡ Sodium ions appear in relatively high concentration in the urine because of reduced urine volume, not because large amounts are being secreted. Sodium is the major positive ion in the plasma; under normal circumstances, most of it is actively reabsorbed.

Abnormal Urinary Constituents

Abnormal urinary constituents are substances not normally present in the urine when the body is operating properly.

Glucose

The presence of glucose in the urine, a condition called **glycosuria,** indicates abnormally high blood sugar levels. Normally, blood sugar levels are maintained between 80 and 100 mg/100 ml of blood. At this level all glucose in the filtrate is reabsorbed by the tubular cells and returned to the blood. Glycosuria may result from carbohydrate intake so excessive that normal physiological and hormonal mechanisms cannot clear it from the blood quickly enough. In such cases, the active transport reabsorption mechanisms of the renal tubules for glucose are exceeded, but only temporarily.

Pathological glycosuria occurs in conditions such as uncontrolled diabetes mellitus, in which the body cells are unable to absorb glucose from the blood because the pancreatic islet cells produce inadequate amounts of the hormone insulin, or there is some abnormality of the insulin receptors. Under such circumstances, the body cells increase their metabolism of fats, and the excess and unusable glucose spills out in the urine.

Albumin

Albuminuria, or the presence of albumin in urine, is an abnormal finding. Albumin is the single most abundant blood protein and is very important in maintaining the osmotic pressure of the blood. Albumin, like other blood proteins, normally is too large to pass through the glomerular filtration membrane. Thus, albuminuria is generally indicative of an abnormally increased permeability of the glomerular membrane. Certain nonpathological conditions, such as excessive exertion, pregnancy, or overabundant protein intake, can temporarily increase the membrane permeability, leading to **physiological albuminuria.** Pathological conditions resulting in albuminuria include events that damage the glomerular membrane, such as kidney trauma due to blows, the ingestion of heavy metals, bacterial toxins, glomerulonephritis, and hypertension.

Ketone Bodies

Ketone bodies (acetoacetic acid, beta-hydroxybutyric acid, and acetone) normally appear in the urine in very small amounts. **Ketonuria,** the presence of these intermediate products of fat metabolism in excessive amounts, usually indicates that abnormal metabolic processes are occurring. The result may be *acidosis* and its complications. Ketonuria is an expected finding during starvation, or diets very low in carbohydrates, when inadequate food intake forces the body to use its fat stores. Ketonuria coupled with a finding of glycosuria is generally diagnostic for diabetes mellitus.

Red Blood Cells

Hematuria, the appearance of red blood cells, or erythrocytes, in the urine, almost always indicates pathology of the urinary tract, because erythrocytes are too large to pass through the glomerular pores. Possible causes include irritation of the urinary tract organs by calculi (kidney stones), which produces frank bleeding; infections; or physical trauma to the urinary organs. In healthy menstruating females, it may reflect accidental contamination of the urine sample with the menstrual flow.

Hemoglobin

Hemoglobinuria, the presence of hemoglobin in the urine, is a result of the fragmentation, or hemolysis, of red blood cells. As a result, hemoglobin is liberated into the plasma and subsequently appears in the kidney filtrate. Hemoglobinuria indicates various pathological conditions including hemolytic anemias, transfusion reactions, burns, or renal disease.

Nitrites

The presence of urinary nitrites might indicate a bacterial infection, particularly *E. coli* or other gram-negative rods. Nitrites are valuable for early detection of bladder infections.

Bile Pigments

Bilirubinuria, the appearance of bilirubin (bile pigments) in urine, is an abnormal finding and usually indicates liver pathology, such as hepatitis or cirrhosis. Bilirubinuria is signaled by a yellow foam that forms when the urine sample is shaken.

Urobilinogen is produced in the intestine from bilirubin and gives feces a brown color. Some urobilinogen is reabsorbed into the blood and either excreted back into the intestine by the liver or excreted by the kidneys in the urine. Complete absence of urobilinogen may indicate renal disease or obstruction of bile flow in the liver. Increased levels may indicate hepatitis A, cirrhosis, or biliary disease.

White Blood Cells

Pyuria is the presence of white blood cells or other pus constituents in the urine. It indicates inflammation of the urinary tract.

Casts

Any complete discussion of the varieties and implications of casts is beyond the scope of this exercise. However, because they always represent a pathological condition of the kidney or urinary tract, they should at least be mentioned. **Casts** are hardened cell fragments, usually cylindrical, which are formed in the distal convoluted tubules and collecting ducts and then flushed out of the urinary tract. Hyaline casts are formed from a mucoprotein secreted by tubule cells. These casts form when the filtrate flow rate is slow, the pH is low, or the salt concentration is high, all conditions which cause protein to denature. Red blood cell casts are typical in glomerulonephritis, as red blood cells leak through the filtration membrane and stick together in the tubules. White blood cell casts form when the kidney is inflamed, which is typically a result of pyelonephritis but sometimes occurs with glomerulonephritis. Degenerated renal tubule cells form granular or waxy casts. Broad waxy casts may indicate end-stage renal disease. Several cast types are shown in Figure 41A.2.

Activity 1:
Analyzing Urine Samples

In this part of the exercise, you will use prepared dip sticks and perform chemical tests to determine the characteristics of normal urine as well as to identify abnormal urinary components. You will investigate both "normal" urine (designated as the *standard urine specimen* in the chart below) *and* an unknown specimen of urine provided by your instructor. Make the following determinations on both samples and record your results by circling the appropriate item or description or

Urinalysis Results*

Observation or test	Normal values	Standard urine specimen	Unknown specimen (#)
Physical characteristics			
Color	Pale yellow	Yellow: pale medium dark other ________	Yellow: pale medium dark other ________
Transparency	Transparent	Clear Slightly cloudy Cloudy	Clear Slightly cloudy Cloudy
Odor	Characteristic	Describe ________	Describe ________
pH	4.5–8.0	________	________
Specific gravity	1.001–1.030	________	________

Chart continues on next page

Urinalysis Results* (continued)

Observation or test	Normal values	Standard urine specimen		Unknown specimen (#)	
Inorganic components					
Sulfates	Present	Present	Absent	Present	Absent
Phosphates	Present	Present	Absent	Present	Absent
Chlorides	Present	Present	Absent	Present	Absent
Nitrites	Absent	Present	Absent	Present	Absent
Organic components					
Urea	Present	Present	Absent	Present	Absent
Glucose					
Dip stick:	Negative	Record results: ________		Record results: ________	
Clinitest	Negative	Record results: ________		Record results: ________	
Albumin					
Dip stick:	Negative	Record results: ________		Record results: ________	
Ketone bodies					
Dip stick:	Negative	Record results: ________		Record results: ________	
RBCs/hemoglobin					
Dip stick:	Negative	Record results: ________		Record results: ________	
Bilirubin					
Dip stick:	Negative	Record results: ________		Record results: ________	
Ictotest	Negative (no color change)	Negative	Positive (purple)	Negative	Positive (purple)
Urobilinogen	Present	Present	Absent	Present	Absent
Leukocytes	Absent	Present	Absent	Present	Absent

* In recording urinalysis data, circle the appropriate description if provided; otherwise, record the results you observed. Identify dip sticks used.

by adding data to complete the chart. If you have more than one unknown sample, accurately identify each sample by number.

⚠ *Obtain and wear plastic disposable gloves throughout this laboratory session.* Although the instructor-provided urine samples are actually artificial urine (concocted in the laboratory to resemble real urine), the techniques of safe handling of body fluids should still be observed as part of your learning process. When you have completed the laboratory procedures: (1) dispose of the gloves and used pH paper strips and dip sticks in the autoclave bag; (2) put used glassware in the bleach-containing laboratory bucket; (3) wash the lab bench down with 10% bleach solution.

Determination of the Physical Characteristics of Urine

1. Determine the color, transparency, and odor of your "normal" sample and one of the numbered pathological samples, and circle the appropriate descriptions in the chart on pp. 423-424.

2. Obtain a roll of wide-range pH paper to determine the pH of each sample. Use a fresh piece of paper for each test, and dip the strip into the urine to be tested two or three times before comparing the color obtained with the chart on the dispenser. Record your results in the chart. (If you will be using one of the combination dip sticks—Chemstrip or Multistix—this pH determination can be done later.)

3. To determine specific gravity, obtain a urinometer cylinder and float. Mix the urine well, and fill the urinometer cylinder about two-thirds full with urine.

4. Examine the urinometer float to determine how to read its markings. In most cases, the scale has numbered lines separated by a series of unnumbered lines. The numbered lines give the reading for the first two decimal places. You must determine the third decimal place by reading the lower edge of the meniscus—the curved surface representing the urine-air junction—on the stem of the float.

5. Carefully lower the urinometer float into the urine. Make sure it is floating freely before attempting to take the reading. Record the specific gravity of both samples in the chart. *Do not dispose of this urine if the samples that you have are less than 200 ml in volume* because you will need to make several more determinations.

Determination of Inorganic Constituents in Urine

Sulfates Add 5 ml of urine to a test tube, and then add a few drops of dilute hydrochloric acid and 2 ml of 10% barium chloride solution. The appearance of a white precipitate (barium sulfate) indicates the presence of sulfates in the sample. Clean the test tubes well after use. Record your results.

Are sulfates a *normal* constituent of urine? ______________

Phosphates Obtain a hot plate and a 500 ml beaker. To prepare the hot water bath, half fill the beaker with tap water and heat it on the hot plate. Add 5 ml of urine to a test tube, and then add three or four drops of dilute nitric acid and 3 ml of ammonium molybdate. Mix well with a glass stirring rod, and then heat gently in a hot water bath. Formation of a yellow precipitate indicates the presence of phosphates in the sample. Record your results.

Chlorides Place 5 ml of urine in a test tube, and add several drops of silver nitrate ($AgNO_3$). The appearance of a white precipitate (silver chloride) is a positive test for chlorides. Record your results.

Nitrites Use a combination dip stick to test for nitrites. Record your results.

Determination of Organic Constituents in Urine

Individual dip sticks or combination dip sticks (Chemstrip or Multistix) may be used for many of the tests in this section. If combination dip sticks are used, be prepared to take the readings on several factors (pH, protein [albumin], glucose, ketones, blood/hemoglobin, leukocytes, urobilinogen, bilirubin, and nitrites) at the same time. Generally speaking, results for all of these tests may be read *during* the second minute after immersion, but readings taken after 2 minutes have passed should be considered invalid. Pay careful attention to the directions for method and time of immersion and disposal of excess urine from the strip, regardless of the dip stick used.

Urea Put two drops of urine on a clean microscope slide and *carefully* add one drop of concentrated nitric acid to the urine. Slowly warm the mixture on a hot plate until it begins to dry at the edges, but do not allow it to boil or to evaporate to dryness. When the slide has cooled, examine the edges of the preparation under low power to identify the rhombic or hexagonal crystals of urea nitrate, which form when urea and nitric acid react chemically. Keep the light low for best contrast. Record your results.

Glucose Use a combination dip stick or obtain a vial of Clinistix, and conduct the dip stick test according to the instructions on the vial. Record your results in the chart on pp. 423-424.

Because the Clinitest reagent is routinely used in clinical agencies for glucose determinations in pediatric patients (children), it is worthwhile to conduct this test as well. Obtain the Clinitest tablets and the associated color chart. Using a medicine dropper, put 5 drops of urine into a test tube; then rinse the dropper and add 10 drops of water to the tube. Add a Clinitest tablet. Wait 15 seconds and then compare the color obtained to the color chart. Record your results.

Albumin Use a combination dip stick or obtain the Albustix dip sticks, and conduct the determinations as indicated on the vial. Record your results.

Ketones Use a combination dip stick or obtain the Ketostix dip sticks. Conduct the determinations as indicated on the vial. Record your results.

Blood/Hemoglobin Test your urine samples for the presence of hemoglobin by using a Hemastix dip stick or a combination dip stick according to the directions on the vial. Usually a short drying period is required before making the reading, so read the directions carefully. Record your results.

Bilirubin Using a Bilistix dip stick or a combination dip stick, determine if there is any bilirubin in your urine samples. Record your results.

Also conduct the Ictotest for the presence of bilirubin. Using a medicine dropper, place one drop of urine in the center of one of the special test mats provided with the Ictotest reagent tablets. Place one of the reagent tablets over the drop of urine, and then add two drops of water directly to the tablet. If the mixture turns purple when you add water, bilirubin is present. Record your results.

Leukocytes Use a combination dip stick to test for leukocytes. Record your results.

Urobilinogen Use a combination dip stick to test for urobilinogen. Record your results. ■

Activity 2: Analyzing Urine Sediment Microscopically (Optional)

If desired by your instructor, a microscopic analysis of urine sediment (of "real" urine) can be done. The urine sample to be analyzed microscopically has been centrifuged to spin the more dense urine components to the bottom of a tube, and some of the sediment (mounted on a slide) has been stained with Sedi-stain to make the components more visible.

Go to the demonstration microscope to conduct this study. Using the lowest light source possible, examine the slide under low power to determine if any of the sediments illustrated in Figures 41A.1 and 41A.2 (and described here) can be seen.

Unorganized sediments (Figure 41A.1): Chemical substances that form crystals or precipitate from solution; for example, calcium oxalates, carbonates, and phosphates; uric acid; ammonium ureates; and cholesterol. Also, if one has been taking antibiotics or certain drugs such as sulfa drugs, these may be detectable in the urine in crystalline form. Normal urine contains very small amounts of crystals, but conditions such as urinary retention or urinary tract infection may cause the appearance of much larger amounts (and their possible consolidation into calculi). The high-power lens may be needed to view the various crystals, which tend to be much more minute than the organized (cellular) sediments.

Organized sediments (Figure 41A.2): Include epithelial cells (rarely of any pathological significance), pus cells (white blood cells), red blood cells, and casts. Urine is normally negative for organized sediments, and the presence of the last three categories mentioned, other than trace amounts, always indicates kidney pathology.

Draw a few of your observations below and attempt to identify them by comparing them with Figures 41A.1 and 41A.2. ■

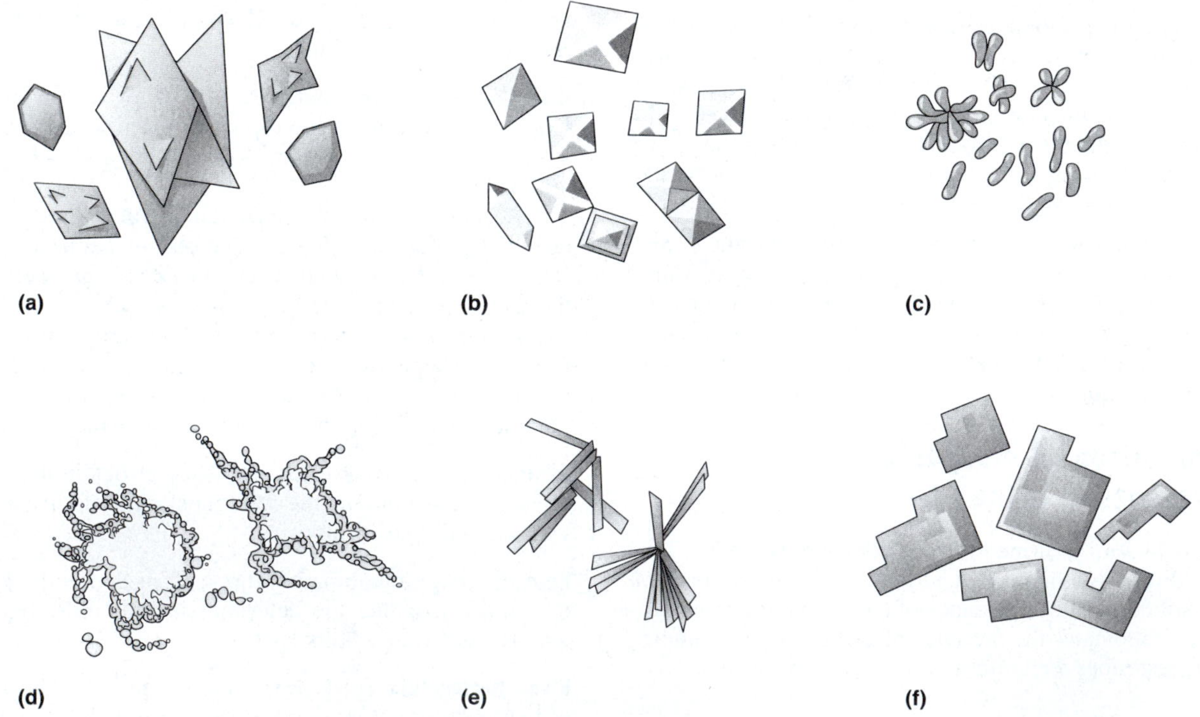

Figure 41A.1 Examples of unorganized sediments. **(a)** Uric acid crystals, **(b)** calcium oxalate crystals, **(c)** calcium carbonate crystals, **(d)** ammonium ureate crystals, **(e)** calcium phosphate crystals, **(f)** cholesterol crystals.

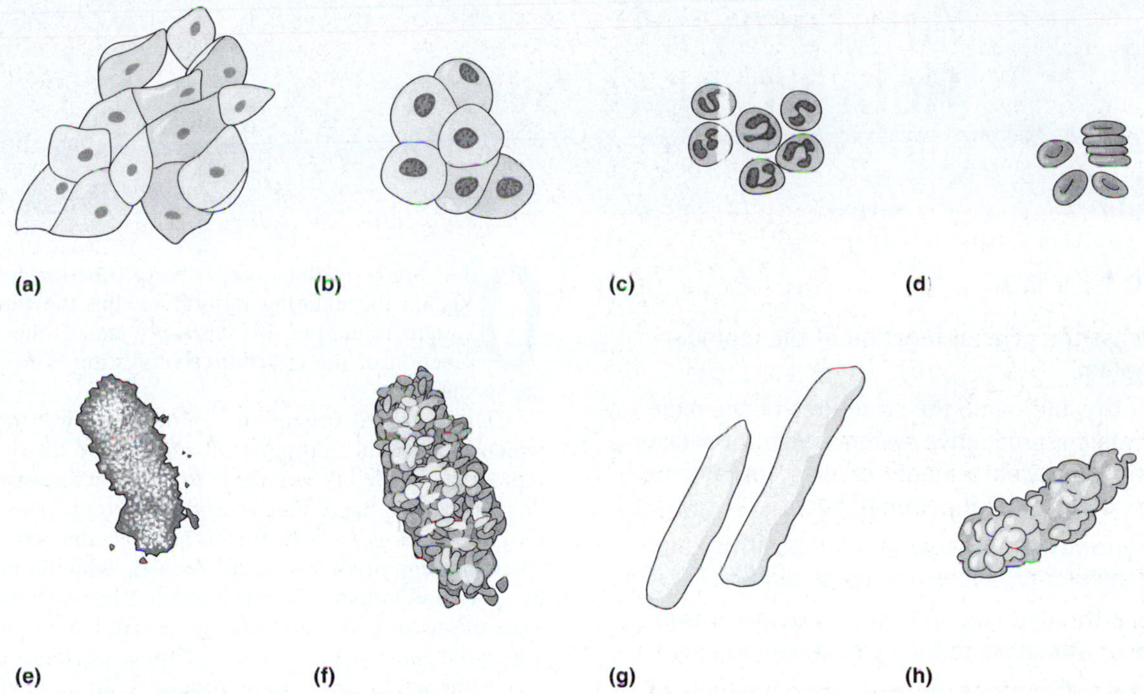

Figure 41A.2 Examples of organized sediments. **(a)** Squamous epithelial cells, **(b)** transitional epithelial cells, **(c)** white blood cells, largely neutrophils (pus), **(d)** red blood cells, **(e)** granular cast, **(f)** red blood cell cast, **(g)** hyaline cast, **(h)** waxy cast.

exercise

42 Anatomy of the Reproductive System

Objectives

1. To discuss the general function of the reproductive system.
2. To identify and name the structures of the male and female reproductive systems when provided with an appropriate model or diagram, and to discuss the general function of each.
3. To define *semen,* discuss its composition, and name the organs involved in its production.
4. To trace the pathway followed by a sperm from its site of formation to the external environment.
5. To name the exocrine and endocrine products of the testes and ovaries, indicating the cell types or structures responsible for the production of each.
6. To identify homologous structures of the male and female systems.
7. To discuss the microscopic structure of the penis, epididymis, uterine tube, and uterus, and to relate structure to function.
8. To define *ejaculation, erection,* and *gonad.*
9. To discuss the function of the fimbriae and ciliated epithelium of the uterine (fallopian) tubes.
10. To identify the fundus, body, and cervical regions of the uterus.
11. To define *endometrium, myometrium,* and *ovulation.*

Materials

- ❑ Models or large laboratory charts of the male and female reproductive tracts
- ❑ Prepared microscope slides of cross sections of the penis, epididymis, uterine tube, and uterus showing endometrium (proliferative phase)
- ❑ Compound microscope

For instructions on animal dissections, see the dissection exercises starting on p. 709 in the cat and fetal pig editions of this manual.

A1A See Appendix C, Exercise 42 for links to A.D.A.M.® Interactive Anatomy.

Other organ systems of the body function primarily to sustain the existing individual, but the reproductive system is unique. Most simply stated, the biological function of the **reproductive system** is to perpetuate the species.

The essential organs of reproduction are the **gonads,** which produce the germ cells, the testes and the ovaries. The reproductive role of the male is to manufacture sperm and to deliver them to the female reproductive tract. The female, in turn, produces eggs. If the time is suitable, the combination of sperm and egg produces a fertilized egg, which is the first cell of a new individual. Once fertilization has occurred, the female uterus provides a nurturing, protective environment in which the embryo, later called the fetus, develops until birth.

Although the drive to reproduce is strong in all animals, in humans this drive is also intricately related to nonbiological factors. Emotions and social considerations often enhance or thwart its expression.

Gross Anatomy of the Human Male Reproductive System

The primary reproductive organs of the male are the **testes,** the male *gonads,* which have both an exocrine (sperm production) and an endocrine (testosterone production) function. All other reproductive structures are conduits or sources of secretions, which aid in the safe delivery of the sperm to the body exterior or female reproductive tract.

Activity 1: Identifying Male Reproductive Organs

As the following organs and structures are described, locate them on Figure 42.1, and then identify them on a three-dimensional model of the male reproductive system or on a large laboratory chart. ■

The paired oval testes lie in the **scrotum** outside the abdominopelvic cavity. The temperature there (approximately 94°F, or 34°C) is slightly lower than body temperature, a requirement for producing viable sperm.

The accessory structures forming the *duct system* are the epididymis, the ductus deferens, the ejaculatory duct, and the urethra. The **epididymis** is an elongated structure running up the posterolateral aspect of the testis and capping its superior aspect. The epididymis forms the first portion of the duct system and provides a site for immature sperm entering it from the testis to complete their maturation process. The **ductus deferens** or **vas deferens** (sperm duct) arches superiorly

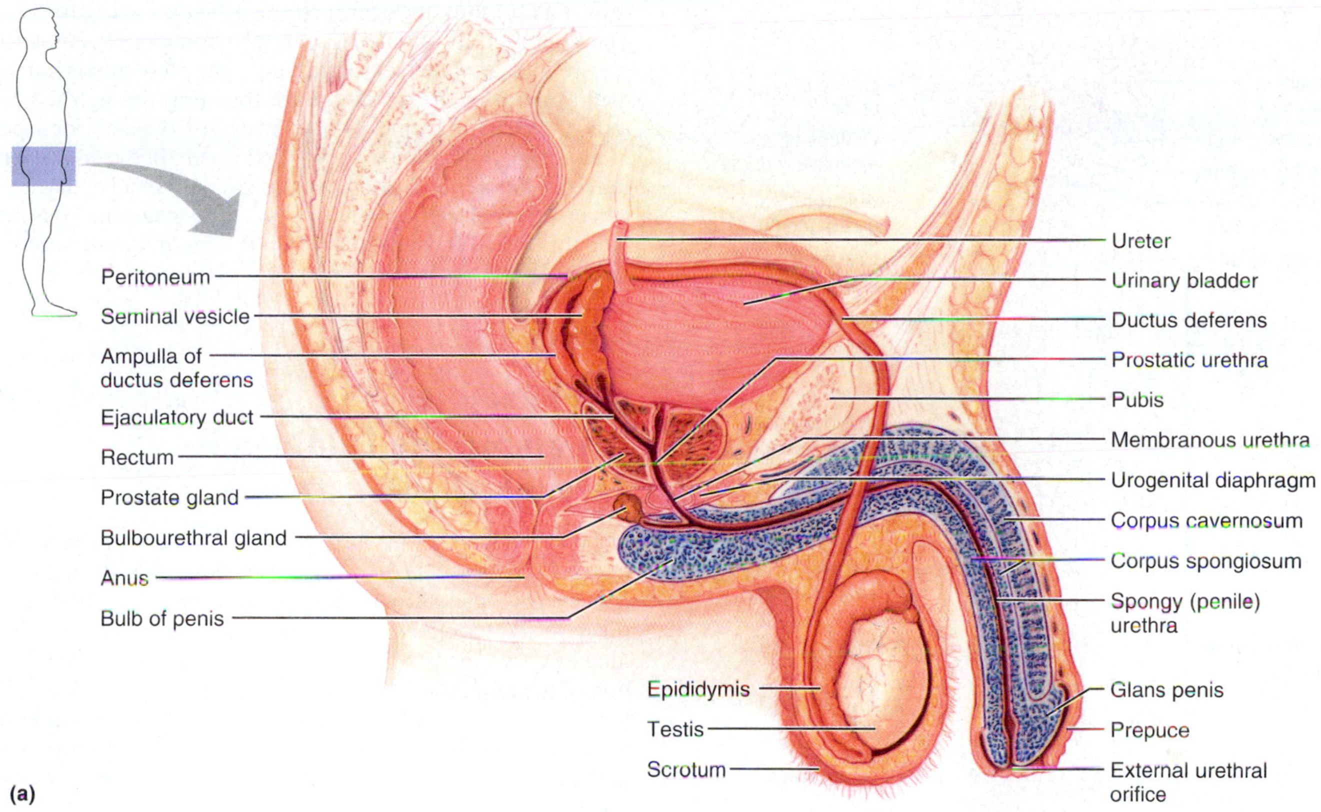

Figure 42.1 Reproductive system of the human male. **(a)** Midsagittal section.

from the epididymis, passes through the inguinal canal into the pelvic cavity, and courses over the superior aspect of the urinary bladder. In life, the ductus deferens is enclosed along with blood vessels and nerves in a connective tissue sheath called the **spermatic cord.** The terminus of the ductus deferens enlarges to form the region called the **ampulla,** which empties into the **ejaculatory duct.** During **ejaculation,** contraction of the ejaculatory duct propels the sperm through the prostate gland to the **prostatic urethra,** which in turn empties into the **membranous urethra** and then into the **spongy (penile) urethra,** which runs through the length of the penis to the body exterior.

The spermatic cord is easily palpated through the skin of the scrotum. When a *vasectomy* is performed, a small incision is made in each side of the scrotum, and each ductus deferens is cut through or cauterized. Although sperm are still produced, they can no longer reach the body exterior; thus a man is sterile after this procedure (and 12 to 15 ejaculations to clear the conducting tubules).

The *accessory glands* include the prostate gland, the paired seminal vesicles, and the bulbourethral glands. These glands produce **seminal fluid,** the liquid medium in which sperm leave the body. The **seminal vesicles,** which produce about 60% of seminal fluid, lie at the posterior wall of the urinary bladder close to the terminus of the ductus deferens. They produce a viscous alkaline secretion containing fructose (a simple sugar) and other substances that nourish the sperm passing through the tract or promote the fertilizing capability of sperm in some way. The duct of each seminal vesicle merges with a ductus deferens to form the ejaculatory duct (mentioned above); thus sperm and seminal fluid enter the urethra together.

The **prostate gland** encircles the urethra just inferior to the bladder. It secretes a milky fluid into the urethra, which plays a role in activating the sperm.

Hypertrophy of the prostate gland, a troublesome condition commonly seen in elderly men, constricts the urethra so that urination is difficult. ■

The **bulbourethral glands** are tiny, pea-shaped glands inferior to the prostate. They produce a thick, clear, alkaline mucus that drains into the membranous urethra. This secretion acts to wash residual urine out of the urethra when ejaculation of **semen** (sperm plus seminal fluid) occurs. The relative alkalinity of seminal fluid also buffers the sperm against the acidity of the female reproductive tract.

The **penis,** part of the external genitalia of the male along with the scrotal sac, is the copulatory organ of the male. Designed to deliver sperm into the female reproductive tract, it consists of a shaft, which terminates in an enlarged tip, the **glans penis** (see Figure 42.1a and b). The skin covering the penis is loosely applied, and it reflects downward to form a circular fold of skin, the **prepuce,** or **foreskin,** around the proximal end of the glans. (The foreskin is removed in the surgical procedure called *circumcision.*) Internally, the penis consists primarily of three elongated cylinders of erectile tissue, which engorge with blood during sexual excitement. This causes the penis to become rigid and enlarged so that it may more adequately serve as a penetrating device. This event is called **erection.** The paired dorsal cylinders are the **corpora cavernosa.** The single ventral **corpus spongiosum** surrounds the penile urethra (see Figure 42.1c).

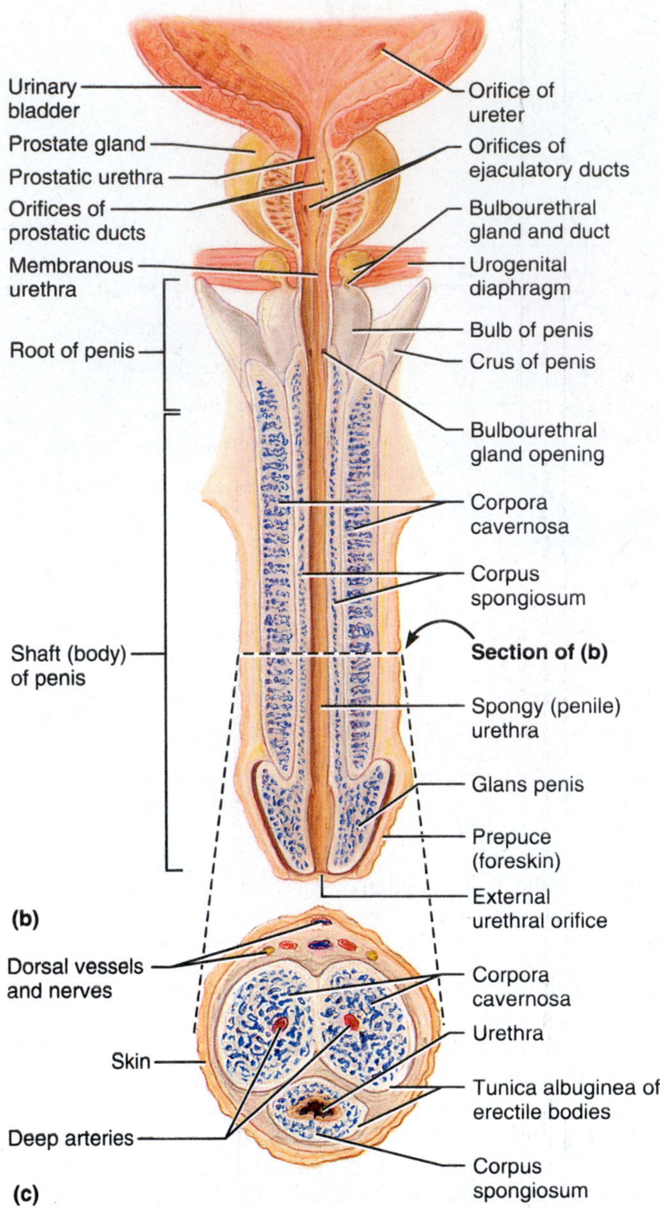

Figure 42.1 *(continued)* **Reproductive system of the human male. (b)** Longitudinal section of the penis. **(c)** Transverse section of the penis.

Microscopic Anatomy of Selected Male Reproductive Organs

Testis

Each testis is covered by a dense connective tissue capsule called the **tunica albuginea** (literally, "white tunic"). Extensions of this sheath enter the testis, dividing it into a number of lobes, each of which houses one to four highly coiled **seminiferous tubules,** the sperm-forming factories (Figure 42.2). The seminiferous tubules of each lobe converge to empty the sperm into another set of tubules, the **rete testis,** at the mediastinum of the testis. Sperm traveling through the rete testis then enter the epididymis, located on the exterior aspect of the testis, as previously described. Lying between the seminiferous tubules and softly padded with connective tissue are the **interstitial cells,** which produce testosterone, the hormonal product of the testis. You will conduct a microscopic study of testis tissue in Exercise 43.

Epididymis

Activity 2: Examining the Epididymis Through the Microscope

Obtain a microscope and a cross section of the epididymis. Notice the abundant tubule cross sections resulting from the fact that the coiling epididymis tubule has been cut through many times in the specimen. Look for sperm in the lumen of the tubule. Using Figure 42.3a as a guide, examine the composition of the tubule wall carefully. Identify the *stereocilia* of the pseudostratified columnar epithelial lining. These nonmotile microvilli absorb excess fluid and pass nutrients to the sperm in the lumen. Now identify the smooth muscle layer. What do you think the function of the smooth muscle is?

___

___

___ ■

Penis

Activity 3: Examining the Penis Microscopically

Obtain a cross section of the penis. Scan the tissue under low power to identify the urethra and the cavernous bodies. Compare your observations to Figure 42.3b. Observe the lumen of the urethra carefully. What type of epithelium do you see?

___

Explain the function of this type of epithelium.

___

___

___

___ ■

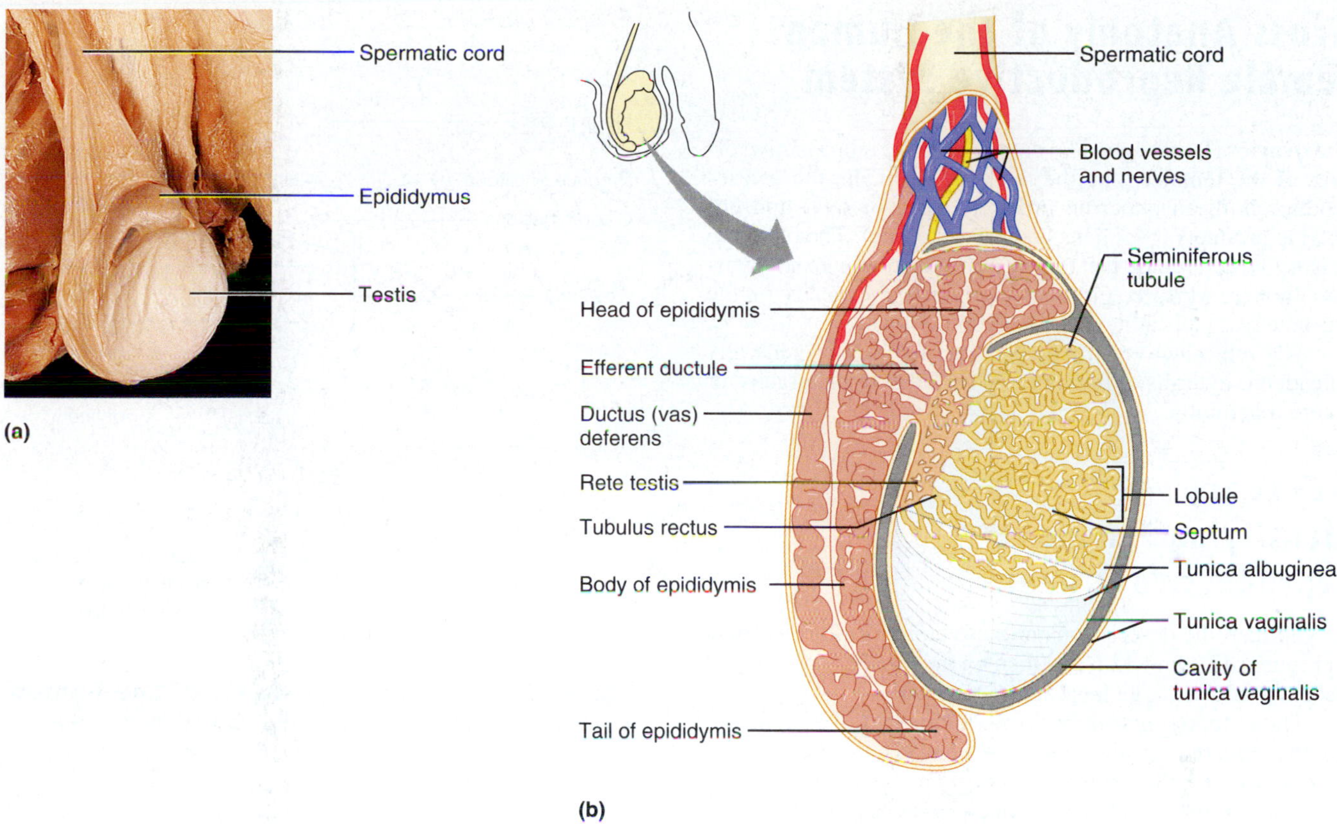

Figure 42.2 Structure of the testis. (a) External view of a testis from a cadaver. **(b)** Longitudinal section of the testis showing seminiferous tubules. Same orientation as (a). (Epididymis and part of the ductus deferens also shown.)

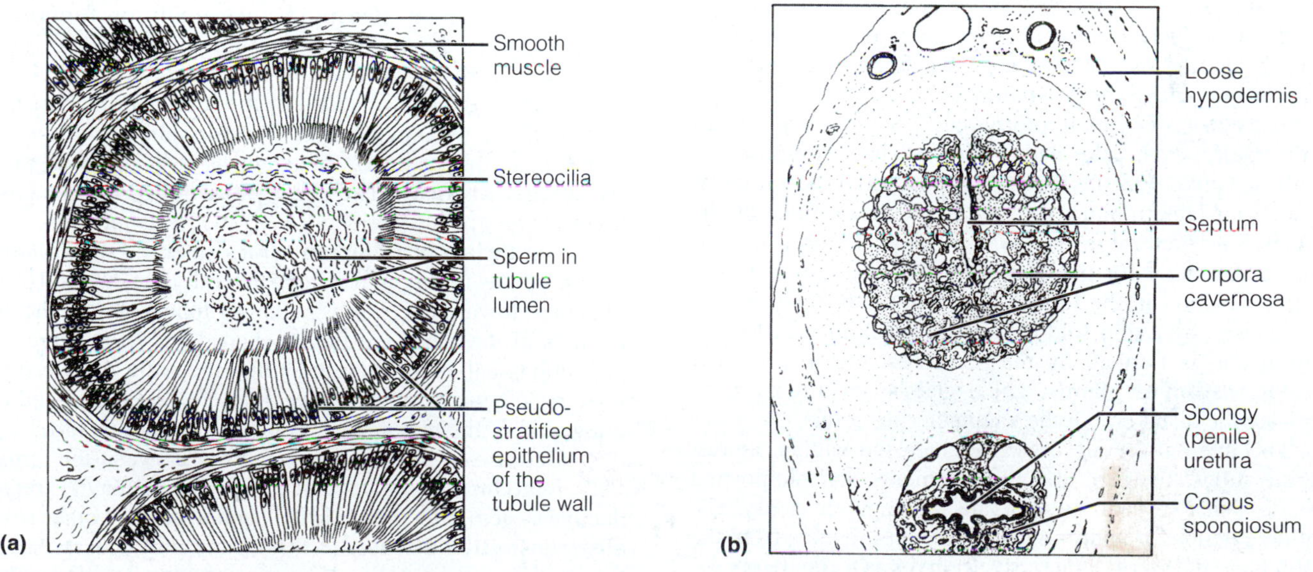

Figure 42.3 Microscopic anatomy of selected organs of the male reproductive duct system. (a) Epididymis. **(b)** Cross-sectional view of the penis. (See also Plates 50 and 51 in the Histology Atlas.)

Gross Anatomy of the Human Female Reproductive System

The **ovaries** (female gonads) are the primary reproductive organs of the female. Like the testes of the male, the ovaries produce both an exocrine product (eggs, or ova) and endocrine products (estrogens and progesterone). The other accessory structures of the female reproductive system transport, house, nurture, or otherwise serve the needs of the reproductive cells and/or the developing fetus.

The reproductive structures of the female are generally considered in terms of internal organs and external organs, or external genitalia.

Activity 4: Identifying Female Reproductive Organs

As you read the descriptions of these structures, locate them on Figures 42.4 and 42.5 and then on the female reproductive system model or large laboratory chart. ■

The **external genitalia (vulva)** consist of the mons pubis, the labia majora and minora, the clitoris, the urethral and vaginal orifices, the hymen, and the greater vestibular glands. The **mons pubis** is a rounded fatty eminence overlying the pubic symphysis. Running inferiorly and posteriorly from the mons pubis are two elongated, pigmented, hair-covered skin folds, the **labia majora,** which are homologous to the scrotum of the male. These enclose two smaller hair-free folds, the **labia minora.** (Terms indicating only one of the two folds in each case are *labium majus* and *minus,* respectively.) The labia minora, in turn, enclose a region called the **vestibule,** which contains many structures—the clitoris, most anteriorly, followed by the urethral orifice and the vaginal orifice. The diamond-shaped region between the anterior end of the labial folds, the ischial tuberosities laterally, and the anus posteriorly is called the **perineum.**

The **clitoris** is a small protruding structure, homologous to the male penis. Like its counterpart, it is composed of highly sensitive, erectile tissue. It is hooded by skin folds of the anterior labia minora, referred to as the **prepuce of the clitoris.** The urethral orifice, which lies posterior to the clitoris, is the outlet for the urinary system and has no reproductive function in the female. The vaginal opening is partially closed by a thin fold of mucous membrane called the **hymen** and is flanked by the pea-sized, mucus-secreting **greater vestibular glands.** These glands (see Figure 42.5) lubricate the distal end of the vagina during coitus.

The internal female organs include the vagina, uterus, uterine tubes, ovaries, and the ligaments and supporting structures that suspend these organs in the pelvic cavity. The **vagina** extends for approximately 10 cm (4 inches) from the vestibule to the uterus superiorly. It serves as a copulatory organ and birth canal, and permits passage of the menstrual flow. The pear-shaped **uterus,** situated between the bladder and the rectum, is a muscular organ with its narrow end, the **cervix,** directed inferiorly. The major portion of the uterus is referred to as the **body;** its superior rounded region above the entrance of the uterine tubes is called the **fundus.** A fertilized

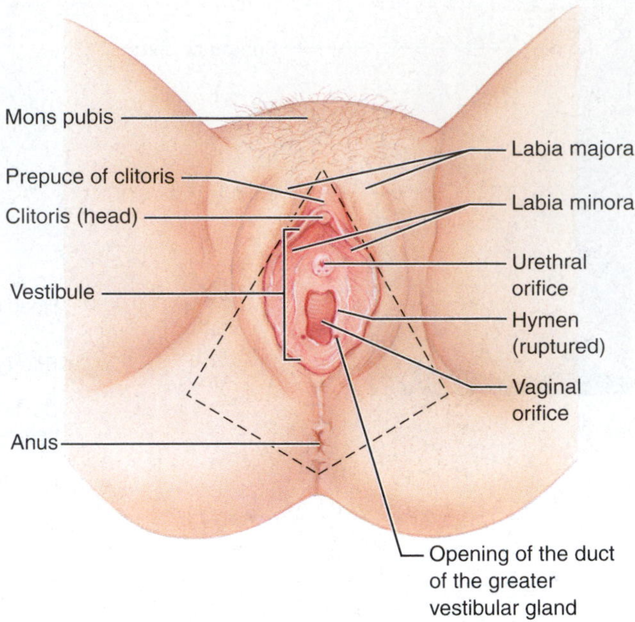

Figure 42.4 External genitalia of the human female. (The region enclosed by dashed lines is the perineum.)

egg is implanted in the uterus, which houses the embryo or fetus during its development.

In some cases, the fertilized egg may implant in a uterine tube or even on the abdominal viscera, creating an **ectopic pregnancy.** Such implantations are usually unsuccessful and may even endanger the mother's life because the uterine tubes cannot accommodate the increasing size of the fetus. ■

The superficial **stratum functionalis** or **functional layer** region of the **endometrium,** the thick mucosal lining of the uterus, sloughs off periodically (about every 28 days) in response to cyclic changes in the levels of ovarian hormones in the woman's blood. This sloughing-off process, which is accompanied by bleeding, is referred to as **menstruation,** or **menses.** The deeper **stratum basalis** or **basal layer** forms a new functionalis after menstruation ends.

The **uterine,** or **fallopian, tubes** enter the superolateral region of the uterus and extend laterally for about 10 cm (4 inches) toward the ovaries in the peritoneal cavity. The distal ends of the tubes are funnel-shaped and have fingerlike projections called **fimbriae.** Unlike the male duct system, there is no actual contact between the female gonad and the initial part of the female duct system—the uterine tube.

Because of this open passageway between the female reproductive organs and the peritoneal cavity, reproductive system infections, such as gonorrhea and other **sexually transmitted diseases (STDs),** can cause widespread inflammations of the pelvic viscera, a condition called **pelvic inflammatory disease** or **PID.** ■

The internal female organs are all retroperitoneal, except the ovaries. They are supported and suspended somewhat freely by ligamentous folds of peritoneum. The peritoneum takes an undulating course. From the pelvic cavity

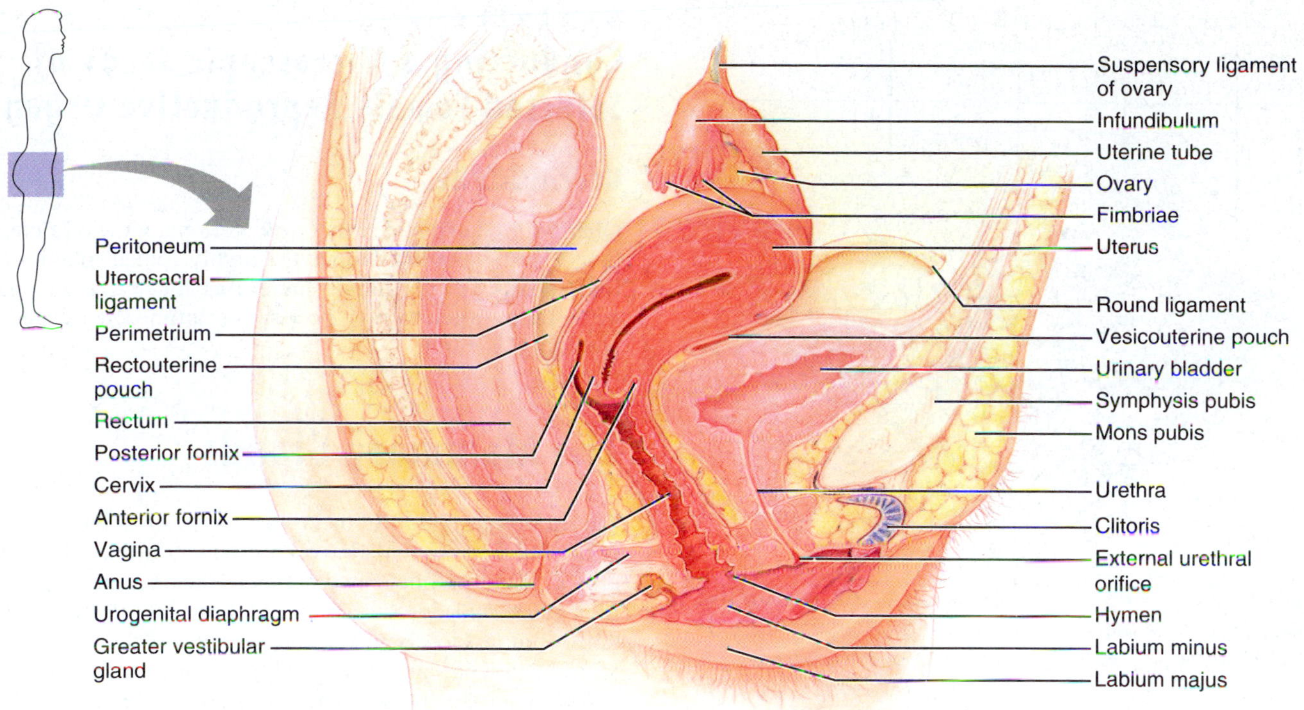

Figure 42.5 Midsagittal section of the human female reproductive system.

floor it moves superiorly over the top of the bladder, reflects over the anterior and posterior surfaces of the uterus, and then over the rectum, and up the posterior body wall. The fold that encloses the uterine tubes and uterus and secures them to the lateral body walls is the **broad ligament.** The part of the broad ligament specifically anchoring the uterus is called the **mesometrium** and that anchoring the uterine tubes, the **mesosalpinx.** The **round ligaments,** fibrous cords that run from the uterus to the labia majora, and the **uterosacral ligaments,** which course posteriorly to the sacrum, also help attach the uterus to the body wall. The ovaries are supported medially by the **ovarian ligament** (extending from the uterus to the ovary), laterally by the **suspensory ligaments,** and posteriorly by a fold of the broad ligament, the **mesovarium.**

Within the ovaries, the female gametes (eggs) begin their development in saclike structures called *follicles.* The growing follicles also produce *estrogens.* When a developing egg has reached the appropriate stage of maturity, it is ejected from the ovary in an event called **ovulation.** The ruptured follicle is then converted to a second type of endocrine gland, called a *corpus luteum,* which secretes progesterone (and some estrogens).

The flattened almond-shaped ovaries lie adjacent to the uterine tubes but are not connected to them; consequently, an ovulated egg* enters the pelvic cavity. The waving fimbriae of the uterine tubes create fluid currents that, if successful, draw the egg into the lumen of the uterine tube, where it begins its passage to the uterus, propelled by the cilia of the tubule walls. The usual and most desirable site of fertilization is the uterine tube, because the journey to the uterus takes about 3 to 4 days and an egg is viable for up to 24 hours after it is expelled from the ovary. Thus, sperm must swim upward through the vagina and uterus and into the uterine tubes to reach the egg. This must be an arduous journey, because they must swim against the downward current created by ciliary action—rather like swimming against the tide!

Wall of the Uterus

Obtain a cross-sectional view of the uterine wall. Identify the three layers of the uterine wall—the endometrium, myometrium, and serosa. Also identify the two strata of the endometrium: the functional layer (stratum functionalis) and

* To simplify this discussion, the ovulated cell is called an egg. What is actually expelled from the ovary is an earlier stage of development called a secondary oocyte. These matters are explained in more detail in Exercise 43.

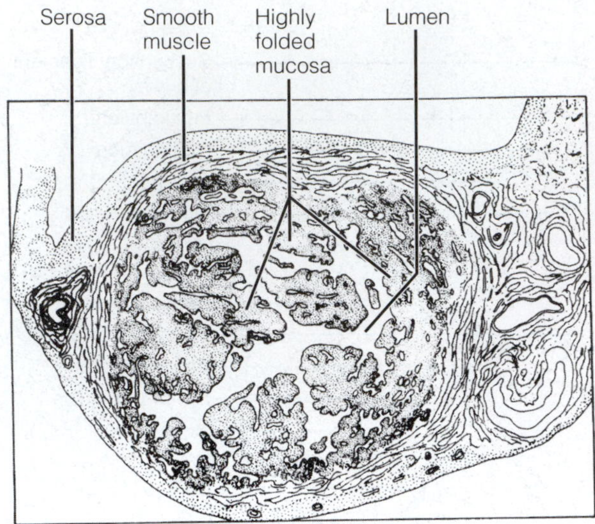

Figure 42.6 Cross-sectional view of a uterine tube. Notice its highly folded mucosa. (See also Plate 54 in the Histology Atlas.)

the basal layer (stratum basalis), which forms a new functional layer each month. Figure 43.6a (p. 443) of the proliferative endometrium may be of some help in this study.

As you study the slide, notice that the bundles of smooth muscle are oriented in several different directions. What is the function of the **myometrium** (smooth muscle layer) during the birth process?

___

___

Sketch a small portion of the uterine wall in the space below. Label the functional and basal layers. ■

Activity 5: Conducting a Microscopic Study of Selected Female Reproductive Organs[†]

Uterine Tube

Obtain a prepared slide of a cross-sectional view of a uterine tube for examination. Notice the highly folded mucosa (the folds nearly fill the tubule lumen) as illustrated in Figure 42.6. Then switch to high power to examine the ciliated secretory epithelium.

The Mammary Glands

The **mammary glands** or breasts exist, of course, in both sexes, but they normally have a reproduction-related function only in females. Since the function of the mammary glands is to produce milk to nourish the newborn infant, their importance is more closely associated with events that occur when reproduction has already been accomplished. Periodic stimulation by the female sex hormones, especially estrogens, increases the size of the female mammary glands at puberty. During this period, the duct system becomes more elaborate, and fat is deposited—fat deposition being the more important contributor to increased breast size.

The rounded, skin-covered mammary glands lie anterior to the pectoral muscles of the thorax, attached to them by connective tissue. Slightly below the center of each breast is a pigmented area, the **areola,** which surrounds a centrally protruding **nipple** (Figure 42.7).

Internally each mammary gland consists of 15 to 25 **lobes** which radiate around the nipple and are separated by fibrous connective tissue and adipose, or fatty, tissue. Within each lobe are smaller chambers called **lobules,** containing the glandular **alveoli** that produce milk during lactation. The alveoli of each lobule pass the milk into a number of **lactiferous ducts,** which join to form an expanded storage chamber, the **lactiferous sinus,** as they approach the nipple. The sinuses open to the outside at the nipple.

For instructions on animal dissections, see the dissection exercises starting on p. 709 in the cat and fetal pig editions of this manual.

[†] A microscopic study of the ovary is described in Exercise 43, p. 442. If that exercise is not to be conducted in its entirety, the instructor might want to include the ovary study in the scope of this exercise.

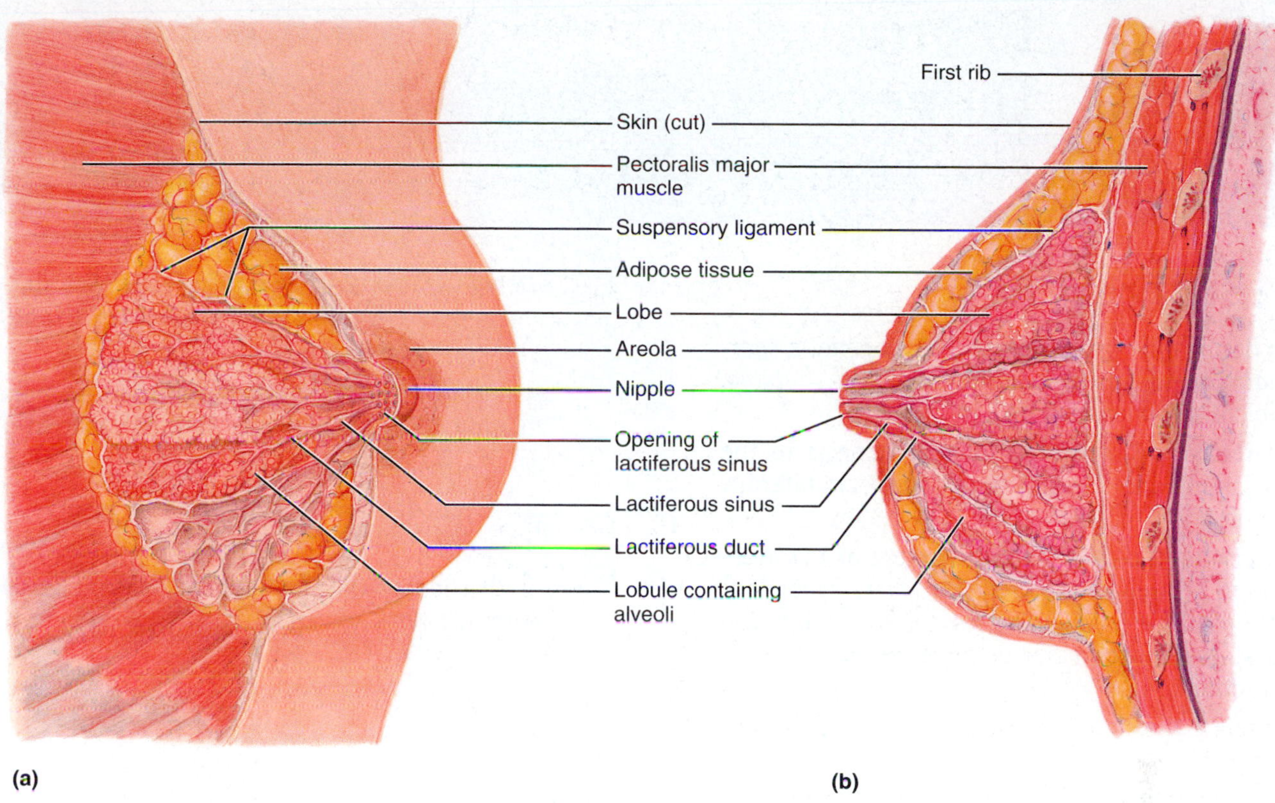

Figure 42.7 **Female mammary gland.** **(a)** Anterior view. **(b)** Sagittal section.

exercise 43 Physiology of Reproduction: Gametogenesis and the Female Cycles

Objectives

1. To define *meiosis, gametogenesis, oogenesis, spermatogenesis, spermiogenesis, synapsis, haploid, diploid,* and *menses.*
2. To relate the stages of spermatogenesis to the cross-sectional structure of the seminiferous tubule.
3. To discuss the microscopic structure of the ovary and to be prepared to identify primary, secondary, and vesicular follicles and the corpus luteum, and to state the hormonal products of the last two structures.
4. To relate the stages of oogenesis to follicle development in the ovary.
5. To cite similarities and differences between mitosis and meiosis, and between spermatogenesis and oogenesis.
6. To describe sperm anatomy and relate it to function.
7. To discuss the phases and control of the menstrual cycle.
8. To discuss the effect of FSH and LH on the ovary and to describe the feedback relationship between anterior pituitary gonadotropins and ovarian hormones.
9. To describe the effect of FSH and LH (ICSH) on testicular function.

Materials

- ❑ Prepared microscope slides of testis, ovary, human sperm, and uterine endometrium (showing menstrual, proliferative, and secretory stages)
- ❑ Three-dimensional models illustrating meiosis, spermatogenesis, and oogenesis
- ❑ Compound microscope
- ❑ Sets of "pop it" beads in two colors with magnetic centromeres (from Ward's Natural Science, Chromosome Simulation Activity Kit)
- ❑ *Demonstration Area:* microscopes set up to demonstrate the following stages of oogenesis in *Ascaris megalocephala:*

  1. Primary oocyte with fertilization membrane, sperm nucleus, and aligned tetrads apparent
  2. Formation of the first polar body
  3. Secondary oocyte with dyads aligned
  4. Formation of the ovum and second polar body
  5. Fusion of the male and female pronuclei to form the fertilized egg

Meiosis

Every human being, so far, has developed from the union of gametes. Gametes, produced only in the testis or ovary, are unique cells. They have only half the normal chromosome number (designated as ***n***, or the **haploid complement**) seen in all other body cells. In humans, gametes have 23 chromosomes instead of the 46 of other tissue cells. Theoretically, every gamete has a full set of genetic instructions, a conclusion borne out by the observation that some animals can develop from an egg that is artificially stimulated, as by a pinprick, rather than by sperm entry. (This is less true of the sperm, because some have an incomplete sex chromosome.)

Gametogenesis, the process of gamete formation, involves reduction of the chromosome number by half. This event is important to maintain the characteristic chromosomal number of the species generation after generation. Otherwise there would be a doubling of chromosome number with each succeeding generation and the cells would become so chock-full of genetic material there would be little room for anything else.

Egg and sperm chromosomes that carry genes for the same traits are called **homologous chromosomes.** When the sperm and egg fuse to form the **zygote,** or fertilized egg, it is said to contain 23 pairs of homologous chromosomes, or the **diploid (2*n*)** chromosome number of 46. The zygote, once formed, then divides to produce the cells needed to construct the multicellular human body. All cells of the developing human body have a chromosome content exactly identical in quality and quantity to that of the fertilized egg. This is assured by the nuclear division process called **mitosis.** (Mitosis was considered in depth in Exercise 4. You may want to review it at this time.)

To produce gametes with the reduced (haploid) chromosomal number, **meiosis,** a specialized type of nuclear division, occurs in the ovaries and testes during gametogenesis. Before meiosis begins, the chromosomes are replicated in the *mother cell* or stem cell just as they are before mitosis. As a result, the mother cell briefly has double the normal diploid genetic complement. The stem cell then undergoes two consecutive nuclear divisions, termed *meiosis I* and *II*, or the first and second maturation divisions, *without replicating the chromosomes before the second division.* Cytokinesis produces four haploid daughter cells, rather than the two diploid daughter cells produced after mitotic division.

The entire process of meiosis is complex and is dealt with here only to the extent necessary to reveal important differences between this type of nuclear division and mitosis. Essentially, each meiotic division involves the same phases and events seen in mitosis (prophase, metaphase, anaphase, and telophase), but during the first maturation division (meiosis I) an event not seen in mitosis occurs during prophase. The homologous chromosomes, each now a duplicated structure, begin to pair so that they become closely aligned along their entire length. This pairing is called **synapsis.** As a result, 23 **tetrads** (groupings of four chromatids) form, become attached to the spindle fibers, and begin to align themselves on the spindle equator. While in synapsis, the "arms" of adjacent homologous chromosomes coil around each other, forming many points of **crossover,** or **chiasmata.** (Perhaps this could be called the conjugal bed of the cell!) When anaphase of meiosis I begins, the homologues separate from one another, breaking and exchanging parts at points of crossover, and move apart toward opposite poles of the cell. The centromeres holding the "sister" *chromatids* (threads of chromatin) or **dyads** together do *not* break at this point (Figure 43.1).

During the second maturation division, events parallel those in mitosis, except that the daughter cells do *not* replicate their chromosomes before this division, and each daughter cell has only half of the homologous chromosomes rather than a complete set. The crossover events and the way in which the homologues align on the spindle equator during the first maturation division introduce an immense variability in the resulting gametes, which explains why we are all unique.

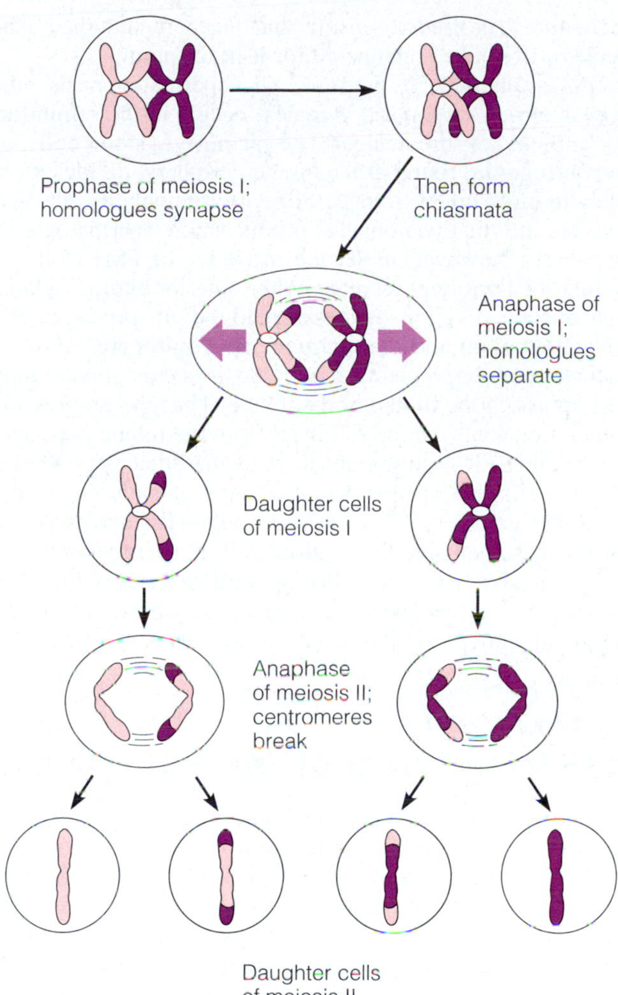

Figure 43.1 Events of meiosis involving one pair of homologous chromosomes. (Male homologue is purple; female homologue is pink.)

Activity 1: Identifying Meiotic Phases and Structures

1. Obtain a model depicting the events of meiosis, and follow the sequence of events during the first and second maturation divisions. Identify prophase, metaphase, anaphase, and telophase in each. Also identify tetrads and chiasmata during the first maturation division and **dyads** (groupings of two chromatids connected by centromeres) in the second maturation division. Note ways in which the daughter cells resulting from meiosis I differ from the mother cell and how the gametes differ from both cell populations. (Use the key on the model, your textbook, or an appropriate reference as necessary to aid you in these observations.)

2. Using strings of colored "pop it" beads with magnetic centromeres, demonstrate the phases of meiosis, including crossing over, for a cell with a diploid ($2n$) number of 4. Use one bead color for the male chromosomes and another color for the female chromosomes. ■

Spermatogenesis

Human sperm production or **spermatogenesis** begins at puberty and continues without interruption throughout life. The average male ejaculation contains about a quarter-billion sperm. Because only one sperm fertilizes an ovum, it seems that nature has tried to ensure that the perpetuation of the species will not be endangered for lack of sperm.

As explained in Exercise 42, spermatogenesis (the process of gametogenesis in males) occurs in the seminiferous tubules of the testes. The primitive stem cells or **spermatogonia,** found at the tubule periphery, divide extensively to build up the stem cell line. Before puberty, all divisions are mitotic divisions that produce more spermatogonia. At puberty, however, under the influence of FSH (follicle-stimulating hormone) secreted by the anterior pituitary gland, each mitotic division of a spermatogonium produces one spermatogonium and one **primary spermatocyte,** which is destined to undergo meiosis. As meiosis occurs, the dividing cells approach the lumen of the tubule. Thus the progression of meiotic events can be followed from the tubule periphery to the lumen. It is important to recognize that **spermatids,** haploid cells that are the actual product of meiosis, are not functional gametes. They are nonmotile cells and have too much excess baggage to function well in a reproductive capacity. Another process, called **spermiogenesis,** which follows meiosis, strips away the extraneous cytoplasm from the spermatid, converting it to a motile, streamlined sperm.

Activity 2:
Examining Events of Spermatogenesis

1. Obtain a slide of the testis and a microscope. Examine the slide under low power to identify the cross-sectional views of the cut seminiferous tubules. Then rotate the high power lens into position and observe the wall of one of the cut tubules. As you work, refer to Figure 43.2 and to Plate 52 in the Histology Atlas to make the following identifications.

2. Scrutinize the cells at the periphery of the tubule. The cells in this area are the spermatogonia. About half of these will form primary spermatocytes, which begin meiosis. These are recognizable by their pale-staining nuclei with centrally located nucleoli. The remaining daughter cells resulting from mitotic divisions of spermatogonia stay at the tubule periphery to maintain the germ cell line.

3. Observe the cells in the middle of the tubule wall. There you should see a large number of cells (spermatocytes) that are obviously undergoing a nuclear division process. Look for coarse clumps of chromatin or threadlike chromosomes (visible only during nuclear division) that have the appearance of coiled springs. Attempt to differentiate between the larger primary spermatocytes and the somewhat smaller secondary spermatocytes. Once formed, the secondary spermatocytes quickly undergo mitosis and so are more difficult to find.

Can you see tetrads? ______________________________

Evidence of crossover? ______________________________

In which location would you expect to see cells containing tetrads, closer to the spermatogonia or closer to the lumen?

____

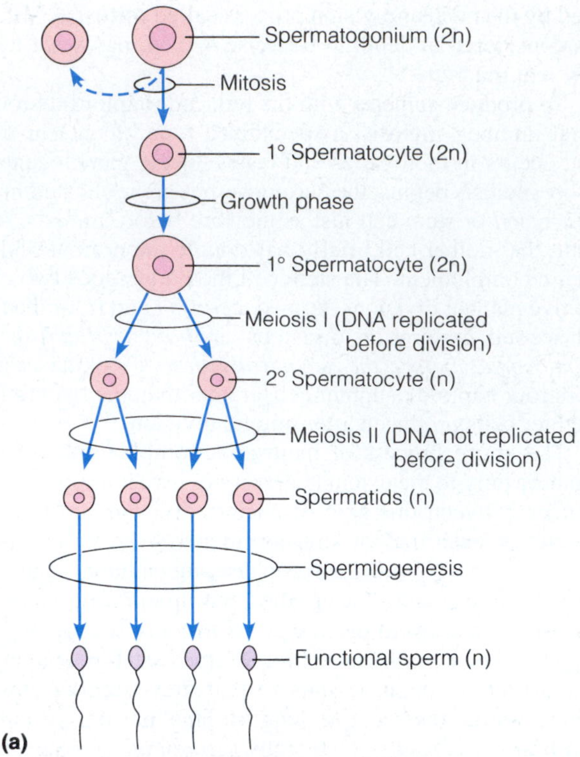

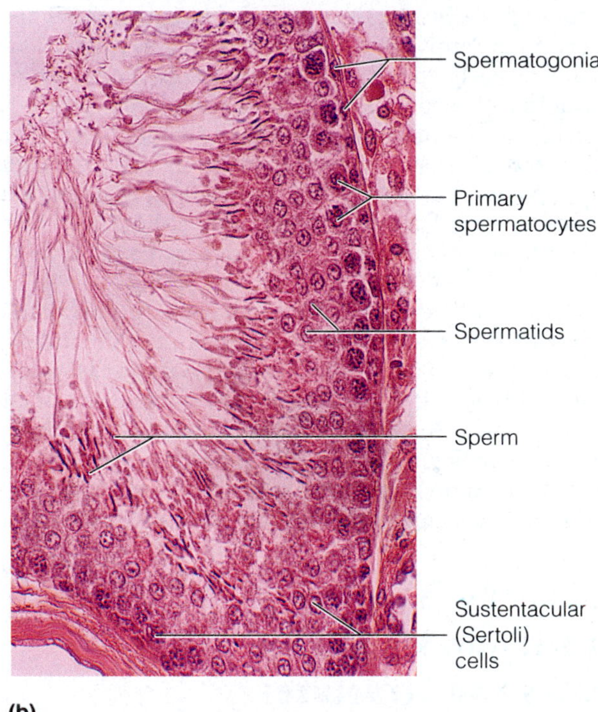

Figure 43.2 Spermatogenesis. **(a)** Flowchart of meiotic and spermiogenesis events. **(b)** Micrograph of an active seminiferous tubule.

Would these cells be primary or secondary spermatocytes?

____

____

4. Examine the cells at the tubule lumen. Identify the small round-nucleated spermatids, many of which may appear lopsided and look as though they are starting to lose their cytoplasm. See if you can find a spermatid embedded in an elongated cell type, a **sustentacular (Sertoli) cell,** which extends inward from the periphery of the tubule. The sustentacular cells nourish the spermatids as they begin their transformation into sperm. Also in the adluminal area, locate sperm, which can be identified by their tails. The sperm develop directly from the spermatids by the loss of extraneous cytoplasm and the development of a propulsive tail.

5. Identify the **interstitial cells** or **Leydig cells** lying external to and between the seminiferous tubules. LH (luteinizing hormone), also called *interstitial cell-stimulating hormone (ICSH)* in males, prompts these cells to produce testosterone, which acts synergistically with FSH to stimulate sperm production.

In the next stage of sperm development, spermiogenesis, all the superficial cytoplasm is sloughed off, and the remaining cell organelles are compacted into the three regions of the mature sperm. At the risk of oversimplifying, these anatomical regions are the *head,* the *midpiece,* and the *tail,* which correspond roughly to the activating and genetic region, the metabolic region, and the locomotor region respectively. The mature sperm is a streamlined cell equipped with an organ of locomotion and a high rate of metabolism that enable it to move long distances quickly to get to the egg. It is a prime example of the correlation of form and function.

The pointed sperm head contains the DNA, or genetic material, of the chromosomes. Essentially it is the nucleus of the spermatid. Anterior to the nucleus is the **acrosome,** which contains enzymes involved in sperm penetration of the egg.

In the midpiece of the sperm is a centriole which gives rise to the filaments that structure the sperm tail. Wrapped tightly around the centriole are mitochondria that provide the ATP needed for contractile activity of the tail.

The tail is a typical flagellum produced by a centriole. When powered by ATP, the tail propels the sperm.

6. Obtain a prepared slide of human sperm and view it with the oil immersion lens. Compare what you see to the photograph of sperm in Plate 53 in the Histology Atlas. Identify the head, acrosome, and tail regions. Draw and appropriately label two or three sperm in the space below.

Deformed sperm, for example sperm with multiple heads or tails, are sometimes present in such preparations. Did you observe any?

_______ If so, describe them. ______________________

____

____

7. Examine the model of spermatogenesis to identify the spermatogonia, the primary and secondary spermatocytes, the spermatids, and the functional sperm.

If your instructor wishes you to observe meiosis in *Ascaris* to provide cellular material for comparison, continue with the microscopic study described next. Otherwise, skip to the study of human oogenesis. ■

Demonstration of Oogenesis in *Ascaris* (Optional)

Generally speaking, oogenesis (the process of gametogenesis resulting in egg production) in mammals is difficult to demonstrate in a laboratory situation. However, the process of oogenesis and mechanics of meiosis* may be studied rather easily in the transparent eggs of *Ascaris megalocephala,* an invertebrate roundworm parasite found in the intestine of mammals. Since its diploid chromosome number is 4, the chromosomes are easily counted.

Activity 3: Examining Meiotic Events Microscopically

Go to the demonstration area where the slides are set up and make the following observations:

1. Scan the first demonstration slide to identify a *primary oocyte,* the cell type that begins the meiotic process. It will have what appears to be a relatively thick cell membrane; this is the *fertilization membrane* that the oocyte produces after sperm penetration. Find and study a primary oocyte that is undergoing the first maturation division. Look for a barrel-shaped spindle with two tetrads (two groups of four beadlike chromosomes) in it. Most often the spindle is located at the periphery of the cell. (The sperm nucleus may or may not be seen, depending on how the cell was cut.)

* In *Ascaris,* meiosis does not begin until the sperm has penetrated the primary oocyte, whereas in humans, meiosis I occurs before sperm penetration.

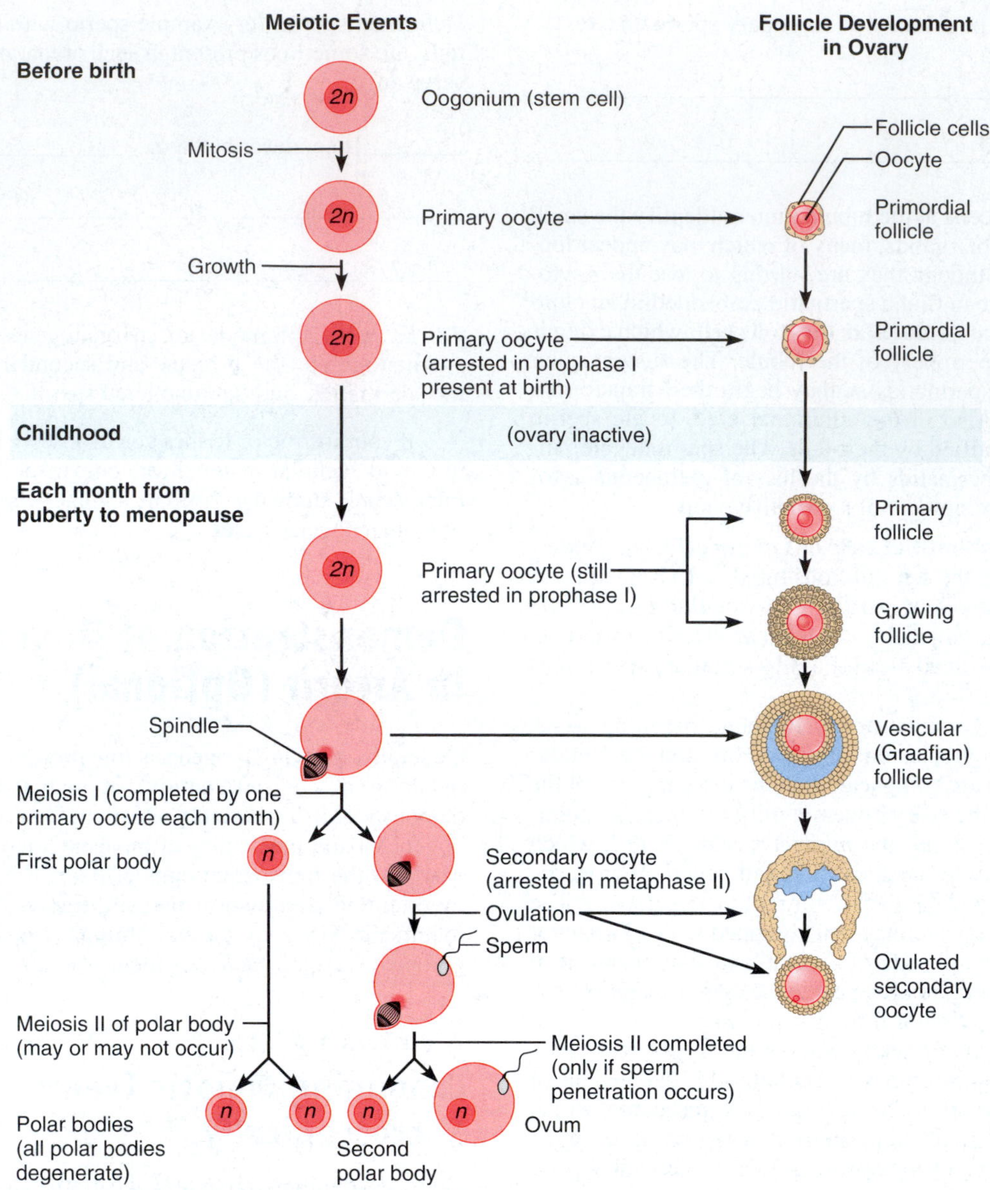

Figure 43.3 **Oogenesis.** Left, flowchart of meiotic events. Right, correlation with follicular development and ovulation in the ovary.

2. Observe slide 2. Locate a cell in which half of each tetrad (a dyad) is being extruded from the cell surface into a smaller cell called the *first polar body.*

3. On slide 3, attempt to locate a *secondary oocyte* (a daughter cell produced during meiosis I) undergoing the second maturation division. In this view, two dyads (two groups of two beadlike chromosomes) will be seen on the spindle.

4. On slide 4, locate a cell in which the *second polar body* is being formed. In this case, both it and the ovum will now contain two chromosomes, the haploid number for *Ascaris.*

5. On the fifth slide, identify a *fertilized egg* or a cell in which the sperm and ovum nuclei (actually *pronuclei*) are fusing to form a single nucleus containing four chromosomes. ■

Human Oogenesis and the Ovarian Cycle

Once the adult ovarian cycle is established, gonadotropic hormones produced by the anterior pituitary influence the development of ova in the ovaries and their cyclic production of female sex hormones. Within an ovary, each immature ovum develops within a saclike structure called a *follicle,* where it is encased by one or more layers of smaller cells called **follicle cells** (when one layer is present) or **granulosa cells** (when there is more than one layer).

The process of **oogenesis,** or female gamete formation, which occurs in the ovary, is similar to spermatogenesis occurring in the testis, but there are some important differences. The process, schematically outlined in Figure 43.3, begins

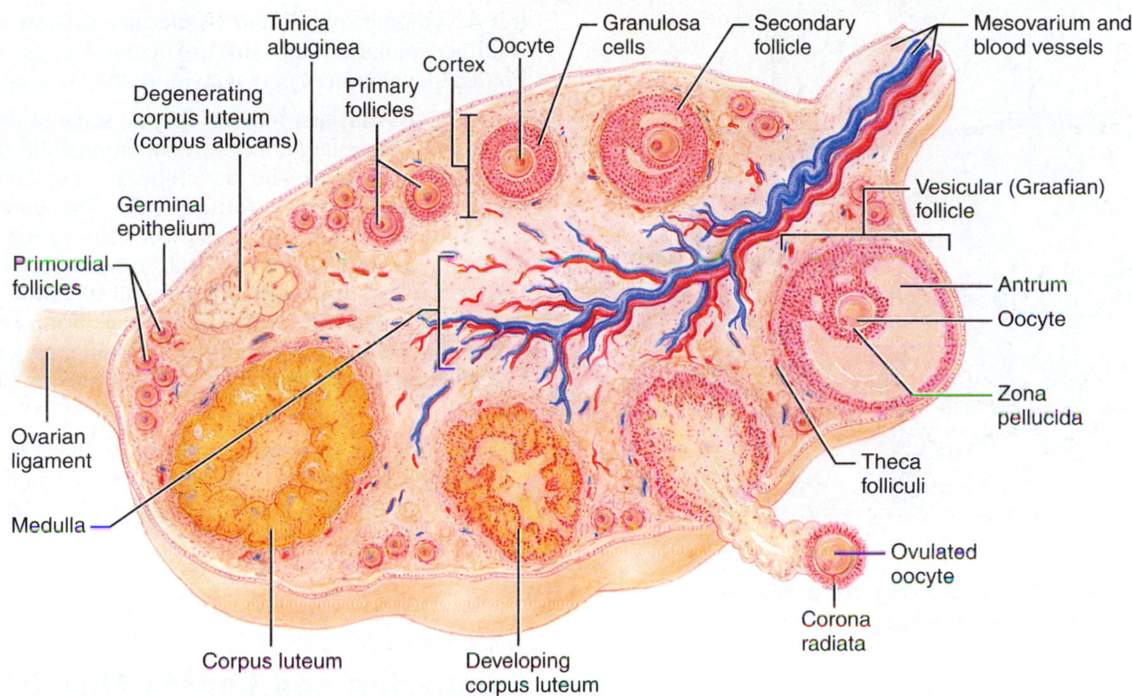

Figure 43.4 Anatomy of the human ovary.

with primitive stem cells called **oogonia,** located in the ovarian cortices of the developing female fetus. During fetal development, the oogonia undergo mitosis thousands of times until their number reaches 700,000 or more. They then become encapsulated by a single layer of squamouslike follicle cells and form the **primordial follicles** of the ovary. By the time the female child is born, most of her oogonia have increased in size and have become **primary oocytes,** which are in the prophase stage of meiosis I. Thus at birth, the total potential for producing germ cells in the female is already determined; the primitive stem-cell line no longer exists or will exist for only a brief period after birth.

From birth until puberty, the primary oocytes are quiescent. Then, under the influence of FSH, one or sometimes more of the follicles begin to undergo maturation approximately every 28 days.

As a follicle grows, its epithelium changes from squamous to cuboidal cells and it comes to be called a **primary follicle** (see also Figure 43.4). The primary follicle begins to produce estrogens, and the primary oocyte completes its first maturation division, producing two haploid daughter cells that are very disproportionate in size. One of these is the **secondary oocyte,** which contains nearly all of the cytoplasm in the primary oocyte. The other is the tiny **first polar body.** The first polar body then completes the second maturation division, producing two more polar bodies. These eventually disintegrate for lack of sustaining cytoplasm.

As the follicle containing the secondary oocyte continues to enlarge, blood levels of estrogens rise. Initially, estrogen exerts a negative feedback influence on the release of gonadotropins by the anterior pituitary. However, approximately in the middle of the 28-day cycle, as the follicle reaches the mature **vesicular,** or **Graafian, follicle** stage, rising estrogen levels become highly stimulatory and a sudden burstlike release of LH (and, to a lesser extent, FSH) by the anterior pituitary triggers ovulation. The secondary oocyte is extruded and begins its journey down the uterine tube to the uterus. If penetrated en route by a sperm, the secondary oocyte will undergo meiosis II, producing one large **ovum** and a tiny **second polar body.** When the second maturation division is complete, the chromosomes of the egg and sperm combine to form the diploid nucleus of the fertilized egg. If sperm penetration does not occur, the secondary oocyte simply disintegrates without ever producing the female gamete in human females.

Thus in the female, meiosis produces only one functional gamete, in contrast to the four produced in the male. Another major difference is in the relative size and structure of the functional gametes. Sperm are tiny and equipped with tails for locomotion. They have few organelles and virtually no nutrient-containing cytoplasm; hence the nutrients contained in semen are essential to their survival. In contrast, the egg is a relatively large nonmotile cell, well stocked with cytoplasmic reserves that nourish the developing embryo until implantation can be accomplished. Essentially all the zygote's organelles are "delivered" by the egg.

Once the secondary oocyte has been expelled from the ovary, LH transforms the ruptured follicle into the **corpus luteum,** which begins producing progesterone and estrogen. Rising blood levels of the two ovarian hormones inhibit FSH release by the anterior pituitary. As FSH declines, its stimulatory effect on follicular production of estrogens ends, and estrogen blood levels begin to decline. Since rising estrogen levels triggered LH release by the anterior pituitary, falling

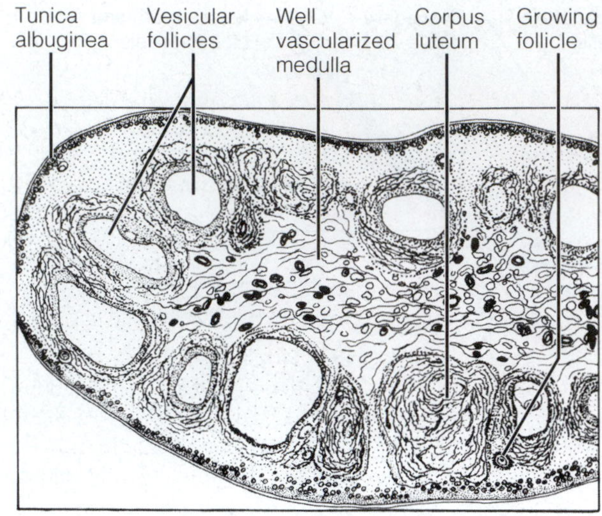

Figure 43.5 Line drawing of a photomicrograph of the human ovary. (See also Plate 55 in the Histology Atlas.)

estrogen levels result in declining levels of LH in the blood. Corpus luteum secretory function is maintained by high blood levels of LH. Thus as LH blood levels begin to drop toward the end of the 28-day cycle, progesterone production ends and the corpus luteum begins to degenerate and is replaced by scar tissue **(corpus albicans).** The graphs in Figure 43.7 depict the hormone relationships described here.

Activity 4: Examining Oogenesis in the Ovary

Because many different stages of ovarian development exist within the ovary at any one time, a single microscopic preparation will contain follicles at many different stages of development. Obtain a cross section of ovary tissue, and identify the following structures. Refer to Figures 43.4 and 43.5 as you work.

Germinal epithelium: Outermost layer of the ovary.

Primary follicle: One or a few layers of cuboidal follicle cells surrounding the larger central developing ovum.

Secondary (growing) follicles: Follicles consisting of several layers of follicle (granulosa) cells surrounding the central developing ovum, and beginning to show evidence of fluid accumulation and **antrum** (central cavity) formation. Follicle development may take more than one cycle.

Vesicular (Graafian) follicle: At this stage of development, the follicle has a large antrum containing fluid produced by the granulosa cells. The developing secondary oocyte is pushed to one side of the follicle and is surrounded by a capsule of several layers of granulosa cells called the **corona radiata** (radiating crown). When the secondary oocyte is released, it enters the uterine tubes with its corona radiata intact. The connective tissue stroma (background tissue) adjacent to the mature follicle forms a capsule, called the **theca,** that encloses the follicle. (See also Plate 25 in the Histology Atlas.)

Corpus luteum: A solid glandular structure or a structure containing a scalloped lumen that develops from the ruptured follicle. (See Plate 26 in the Histology Atlas.) ■

Activity 5: Comparing and Contrasting Oogenesis and Spermatogenesis

Examine the model of oogenesis and compare it with the spermatogenesis model. Note differences in the size and structure of the functional gametes.

How do the gametes differ in size and structure?

____

____

How are the gametogenic processes similar?

____

____

How are those processes different?

____

____ ■

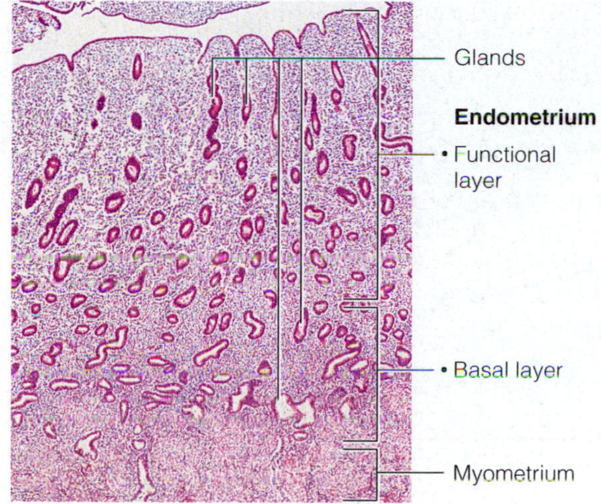

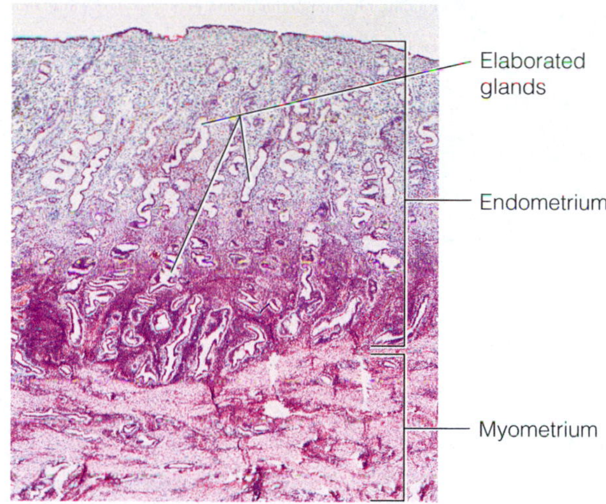

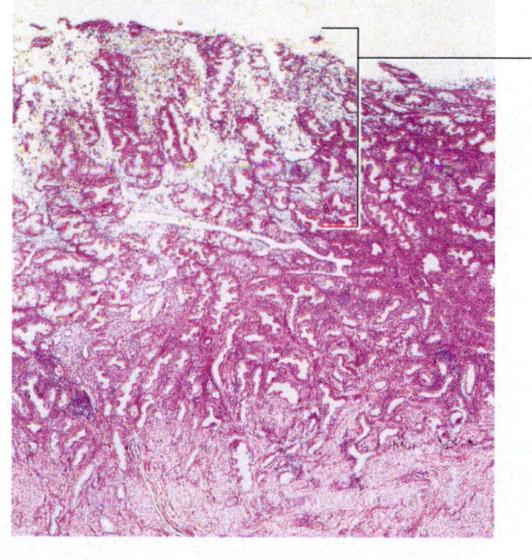

Figure 43.6 Endometrial changes during the menstrual cycle. **(a)** Early proliferative phase. **(b)** Early secretory phase. **(c)** Onset of menstruation.

The Menstrual Cycle

The **uterine cycle,** or **menstrual cycle,** is hormonally controlled by estrogens and progesterone secreted by the ovary. It is normally divided into three stages: menstrual, proliferative, and secretory. The endometrial changes are described next and shown in Figure 43.6. Figure 43.7 shows how they correlate with hormonal and ovarian changes.

Menstrual phase (menses): Approximately days 1 to 5. Sloughing off of the thick functional layer of the endometrial lining of the uterus, accompanied by bleeding.

Proliferative phase: Approximately days 6 to 14. Under the influence of estrogens produced by the growing follicle of the ovary, the endometrium is repaired, glands and blood vessels proliferate, and the endometrium thickens. Ovulation occurs at the end of this stage.

Secretory phase: Approximately days 15 to 28. Under the influence of progesterone produced by the corpus luteum, the vascular supply to the endometrium increases further. The glands increase in size and begin to secrete nutrient substances to sustain a developing embryo, if present, until implantation can occur. If fertilization has occurred, the embryo will produce a hormone much like LH, which will maintain the function of the corpus luteum. Otherwise, as the corpus luteum begins to deteriorate, lack of ovarian hormones in the blood causes blood vessels supplying the endometrium to kink and become spastic, setting the stage for menses to begin by the 28th day.

Although the foregoing explanation assumes a classic 28-day cycle, the length of the menstrual cycle is highly variable, sometimes as short as 21 days or as long as 38. Only one interval is relatively constant in all females: the time from ovulation to the onset of menstruation is almost always 14 days.

Activity 6: Observing Histological Changes in the Endometrium During the Menstrual Cycle

Obtain slides showing the menstrual, secretory, and proliferative phases of the uterine endometrium. Observe each carefully, comparing their relative thicknesses and vascularity. As you work, refer to the corresponding photomicrographs in Figure 43.6. ■

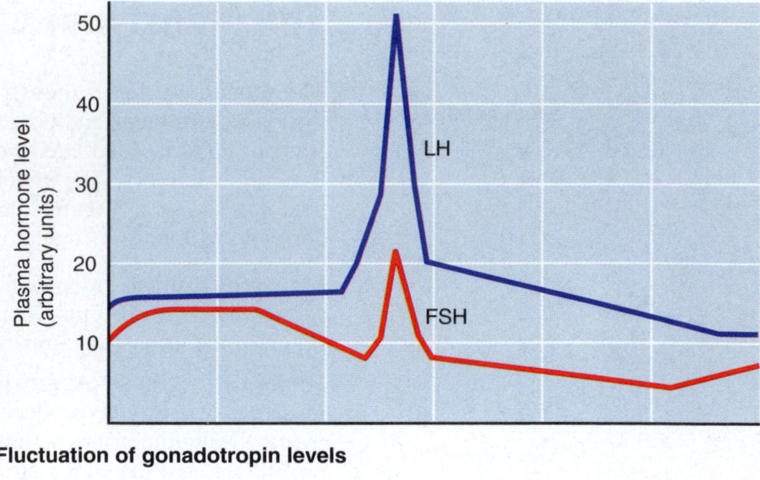

(a) Fluctuation of gonadotropin levels

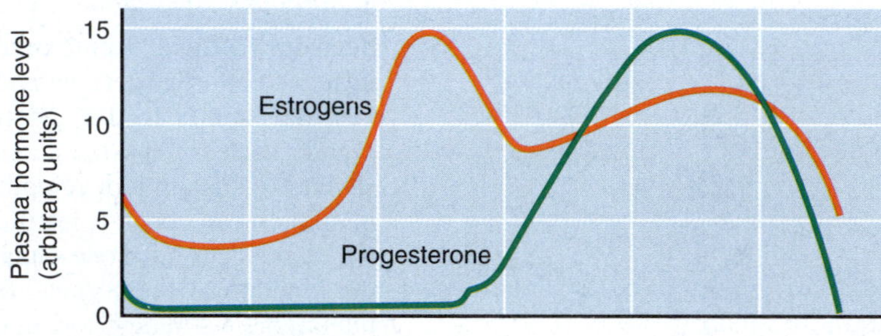

(b) Fluctuation of ovarian hormone levels

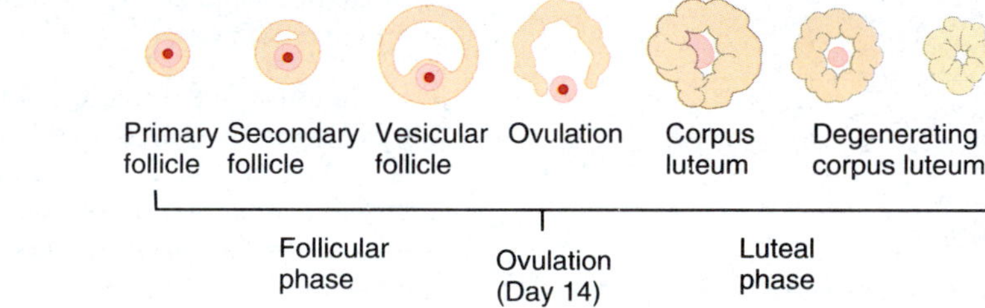

(c) Ovarian cycle

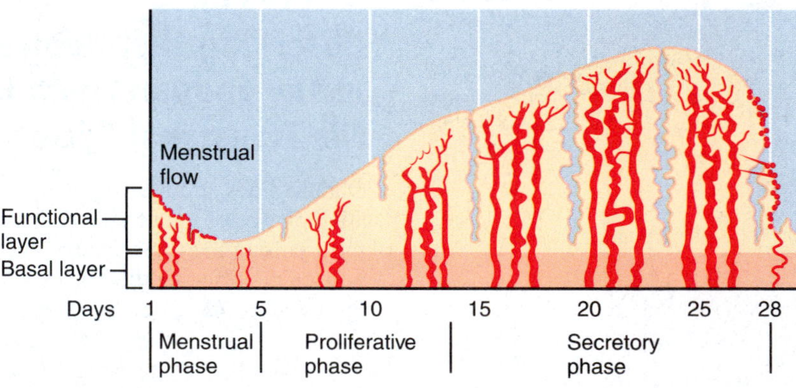

(d) Uterine cycle

Figure 43.7 Hormonal interactions of the female cycles. Relative level of anterior pituitary hormones correlated with follicular and hormonal changes in the ovary. The menstrual cycle is also depicted.

exercise 44

Survey of Embryonic Development

Objectives

1. To define *fertilization* and *zygote.*
2. To define and discuss the function of *cleavage* and *gastrulation.*
3. To name the three primary germ layers and discuss the importance of each.
4. To identify the following structures of a human chorionic vesicle when provided with an appropriate diagram, and to state the function of each.

inner cell mass	trophoblast
amnion	allantois
yolk sac	chorionic villi

5. To describe the process and timing of implantation in the human.
6. To define *decidua basalis* and *decidua capsularis.*
7. To state the germ-layer origin of several body organs and organ systems of the human.
8. To describe developmental direction.
9. To describe the gross anatomy and general function of the human placenta.

Materials

- ❑ Human development models or plaques (if available)
- ❑ Pregnant cat, rat, or pig uterus (one per laboratory session) with uterine wall dissected to allow student examination
- ❑ Dissecting instruments
- ❑ Disposable gloves
- ❑ Model of pregnant human torso
- ❑ Fresh or formalin-preserved placenta (obtained from a clinical agency)
- ❑ Microscope slide of placenta tissue
- ❑ Compound microscope
- ❑ Demonstration: Phases of human development in *A Colour Atlas of Life Before Birth: Normal Fetal Development*

Because reproduction is such a familiar event, we tend to lose sight of the wonder of the process. One part of that process, the development of the embryo, is the concern of embryologists who study the changes in structure that occur from the time of fertilization until the time of birth.

Early development in all animals involves three basic types of activities, which are integrated to ensure the formation of a viable offspring: (1) an increase in cell number and subsequent cell growth; (2) cellular specialization; and (3) morphogenesis, the formation of functioning organ systems. This exercise provides a rather broad overview of the changes in structure that take place during embryonic development in humans.

Developmental Stages of the Human

Activity 1: Examining the Stages of Human Development

Go to the demonstration area where the models of human development are on display. If these are not available, use Figure 44.1 for this study. Observe the models to identify the various stages of human development as they are described, and respond to the questions posed below.

1. Observe the fertilized egg, or **zygote,** which appears as a single cell immediately surrounded by a jellylike *zona pellucida* and then a crown of granulosa cells (the *corona radiata*). After a secondary oocyte is entered by a sperm, and meiosis is completed to yield the ovum (egg) nucleus, the egg and the sperm nuclei form female and male pronuclei and fuse to form a single nucleus. This process is called **fertilization.** Shortly after sperm entry, a granular barrier forms beneath the zona pellucida to prevent the entry of additional sperm.

2. Next, observe the cleavage stages. Once fertilization has occurred, the zygote begins to divide, forming a mass of successively smaller and smaller cells, called **blastomeres.** This series of mitotic divisions without intervening growth periods is referred to as **cleavage,** and it provides a large number of building blocks (cells) with which to build the forming body. If this is a little difficult to understand, consider trying

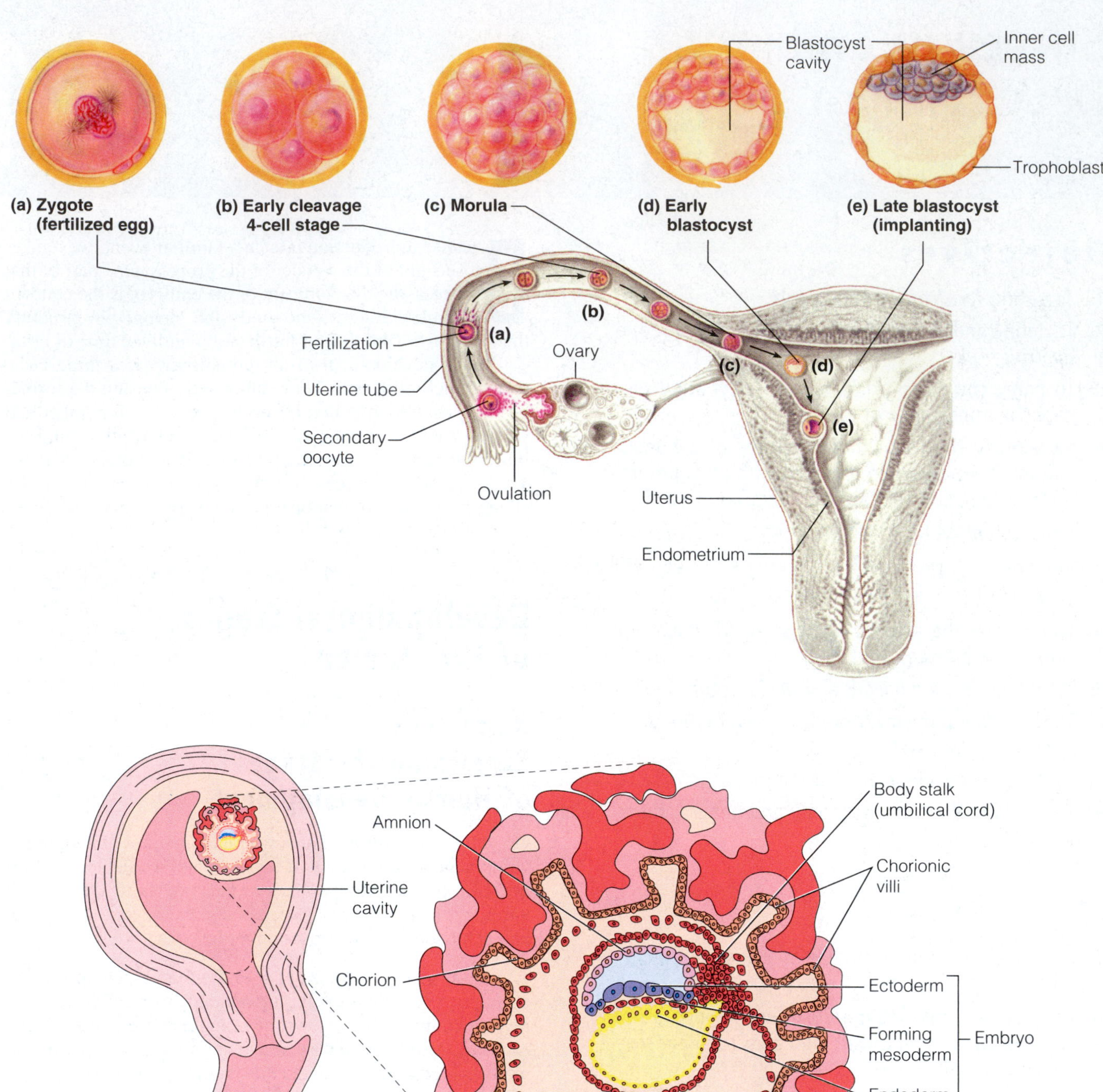

Figure 44.1 Early embryonic development of the human. Top **(a–e)**, from fertilization to blastocyst implantation in the uterus. Below, embryo of approximately 22 days. Embryonic membranes and germ layers present.

to erect a building with one huge block of granite rather than with small bricks. The product of early cleavage is a solid ball of cells. At the 16-cell stage, it is called the **morula,** and the embryo resembles a raspberry in form. Then the cell mass hollows out to become the embryonic form called the **blastula,** which is a ball of cells surrounding a central cavity. The blastula, more commonly called the **blastocyst** in humans, is the final product of cleavage.

Only a portion of the human blastocyst cells contribute to the formation of the body—those seen at one side of the blastocyst forming the so-called **inner cell mass (ICM).** The rest of the blastocyst—that portion enclosing the central cavity and overriding the ICM—is referred to as the **trophoblast.** The trophoblast becomes an extraembryonic membrane called the **chorion,** which forms the fetal portion of the *placenta.*

3. Observe the *implanting* blastocyst shown on the model or in the figure. By approximately the seventh day after ovulation, a developing human embryo (blastocyst) is floating free in the uterine cavity. About that time, it adheres to the uterine wall over the ICM area, and implantation begins. The trophoblast cells secrete enzymes that erode the uterine mucosa at the point of attachment to reach the vascular supply in the submucosa. By the fourteenth day after ovulation, implantation is completed and the uterine mucosa has grown over the burrowed-in embryo. The portion of the uterine wall beneath the ICM, destined to take part in placenta formation, is called the **decidua basalis** and that surrounding the rest of the blastocyst is called the **decidua capsularis.** Identify these regions.

By the time implantation is complete, the blastocyst has undergone **gastrulation.** As a result of gastrulation, a three-layered embryo called a **gastrula** forms. Each of the gastrula's three layers corresponds to a **primary germ layer** from which specific body tissues develop. Within the next 6 weeks, virtually all of the body organ systems will have been laid down at least in rudimentary form by the germ layers. The outermost layer, **ectoderm,** gives rise to the epidermis of the skin and the nervous system. The deepest layer, the **endoderm,** forms the mucosa of the digestive and respiratory tracts and associated glands. **Mesoderm,** the middle layer, forms virtually everything lying between the two (skeleton, walls of the digestive organs, urinary system, skeletal muscles, circulatory system, and others).

All the groundwork has been completed by the eighth week. By the ninth week of development, the embryo is referred to as a **fetus,** and from this point on, the major activities are growth and tissue and organ specialization.

4. Again observe the blastocyst to follow the formation of the embryonic membranes and the placenta (see Figure 44.1). Notice the villus extensions of the trophoblast. By the time implantation is complete, the trophoblast has differentiated into the chorion, and its large elaborate villi are lying in the blood-filled sinusoids in the uterine tissue. This composite of uterine tissue and **chorionic villi** is called the **placenta,** and all exchanges to and from the embryo occur through the chorionic membranes.

Three embryonic membranes (originating in the ICM) have also formed by this time—the amnion, the allantois, and the yolk sac. Attempt to identify each.

- The **amnion** encases the young embryonic body in a fluid-filled chamber that protects the embryo against mechanical trauma and temperature extremes.
- The **yolk sac** in humans has lost its original function, which was to pass nutrients to the embryo after digesting the yolk mass. The placenta has taken over that task; also, the human egg has very little yolk. However, the yolk sac is not totally useless. The embryo's first blood cells originate here, and the primordial germ cells migrate from it into the embryo's body to seed the gonadal tissue.
- The **allantois,** which protrudes from the posterior end of the yolk sac, is also largely redundant in humans because of the placenta. In birds and reptiles, it is a repository for embryonic wastes. In humans, it is the structural basis on which the mesoderm migrates to form the body stalk, or **umbilical cord,** which attaches the embryo to the placenta. (Refer ahead to p. 449, Figure 44.2 to refresh your memory of the structure of the umbilical cord if necessary.)

5. Go to the demonstration area to view the photographic series, *A Colour Atlas of Life Before Birth: Normal Fetal Development.* These photographs illustrate human development in a way you will long remember. After viewing them, respond to the following questions.

In your own words, what do the chorionic villi look like?

____

____

____

What organs or organ systems appear *very* early in embryonic development?

____

____

____

Does development occur in a rostral to caudal (head to toe) direction, or vice versa?

____

Does development occur in a distal to proximal direction, or vice versa?

____

Does spontaneous movement occur *in utero*? ____________

How does the mother recognize this? ____________

____

exercise

45 Principles of Heredity

Objectives

1. To define *allele, dominance, genotype, heterozygous, homozygous, incomplete dominance, phenotype,* and *recessiveness.*
2. To gain practice working out simple genetics problems, using a Punnett square.
3. To become familiar with basic laws of probability.
4. To observe selected human phenotypes and determine their genotype basis.
5. To separate variants of hemoglobin using agarose gel electrophoresis.

Materials

- ❑ Pennies (for coin tossing)
- ❑ PTC (phenylthiocarbamide) taste strips
- ❑ Sodium benzoate taste strips
- ❑ Chart drawn on chalkboard for tabulation of class results of human phenotype/genotype determinations

Blood typing supplies:

- ❑ Anti-A and Anti-B sera, slides, toothpicks, wax pencils, sterile lancets, alcohol swabs
- ❑ Beaker containing 10% bleach solution
- ❑ Disposable autoclave bag

Hemoglobin phenotyping supplies:

- ❑ Disposable gloves
- ❑ Plastic baggies
- ❑ HbA
- ❑ HbS
- ❑ HbA-HbS mixed solution
- ❑ Electrophoresis equipment and power supply
- ❑ Metric ruler
- ❑ 1.2% agarose gels
- ❑ TBE buffer pH 8.4
- ❑ TBE solubilization buffer with bromophenol blue
- ❑ Coomassie protein stain solution
- ❑ Staining tray
- ❑ Coomassie de-stain solution
- ❑ 100-ml gradulated cylinder
- ❑ Micropipet
- ❑ Distilled water

The field of genetics is bristling with excitement. Complex gene-splicing techniques have allowed researchers to precisely isolate genes coding for specific proteins and then to use those genes to harvest large amounts of particular proteins and even to cure some dreaded human diseases. At present, growth hormone, insulin, erythropoietin, and interferon produced by these genetic engineering techniques are available for clinical use, and the list is growing daily.

Comprehending genetics relative to such studies requires arduous training. However, a basic understanding and appreciation of how genes regulate our various traits (dimples and hair color, for example) can be gained by anyone. The thrust of this exercise is to provide a "genetics sampler" or relatively simple introduction to the principles of heredity.

Introduction to the Language of Genetics

In humans all cells, except eggs and sperm, contain 46 chromosomes, that is, the diploid number. This number is established when fertilization occurs and the egg and sperm fuse, combining the 23 chromosomes (or haploid complement) each is carrying. The diploid chromosomal number is maintained throughout life in nearly all cells of the body by the precise process of mitosis. As explained in Exercise 43, the diploid chromosomal number actually represents two complete (or nearly complete) sets of genetic instructions—one from the egg and the other from the sperm—or 23 pairs of *homologous chromosomes.*

Genes coding for the same traits on each pair of homologous chromosomes are called **alleles.** The alleles may be identical or different in their influence. For example, the members of the gene pair, or alleles, coding for hairline shape on your forehead may specify either straight across or widow's peak. When both alleles in a homologous chromosome pair have the same expression, the individual is **homozygous** for that trait. When the alleles differ in their expression, the individual is **heterozygous** for the given trait; and typically only one of the alleles, called the **dominant gene,** will exert its effects. The allele with less potency, the **recessive gene,** will be present but suppressed. Whereas dominant genes, or alleles, exert their effects in both homozygous and heterozygous conditions, as a rule recessive alleles *must* be present in double dose (homozygosity) to exert their influence. An individual's actual genetic makeup, that is, whether he is homozygous or heterozygous for the various alleles, is called **genotype.** The manner in which genotype is expressed (for example, the presence of a widow's peak (see Figure 45.2 on p. 457) or not, blue vs. brown eyes) is referred to as **phenotype.**

The complete story of heredity is much more complex than just outlined, and in actuality the expression of many traits (for example, eye color) is determined by the interaction of many allele pairs. However, our emphasis here will be to investigate only the less complex aspects of genetics.

Dominant-Recessive Inheritance

One of the best ways to master the terminology and learn the principles of heredity is to work out the solutions to some genetic crosses in much the same manner Gregor Mendel did in his classic experiments on pea plants. (Mendel, an Austrian monk of the mid-1800s, found evidence in these experiments that each gamete contributes just one allele to each pair in the zygote.)

To work out the various simple monohybrid (one pair of alleles) crosses in this exercise, you will be given the genotype of the parents. You will then determine the possible genotypes of their offspring by using a grid called the *Punnett square,* and you will record both genotype and phenotype percentages. To illustrate the procedure, an example of one of Mendel's pea plant crosses is outlined next.

Alleles: *T* (determines *tallness;* dominant)
t (determines *dwarfness;* recessive)

Genotypes of parents: *TT* (♂) × *tt* (♀)

Phenotypes of parents: Tall × dwarf

To use the Punnett, or checkerboard, square, write the alleles (actually gametes) of one parent across the top and the gametes of the other parent down the left side. Then combine the gametes across and down to achieve all possible combinations as shown below:

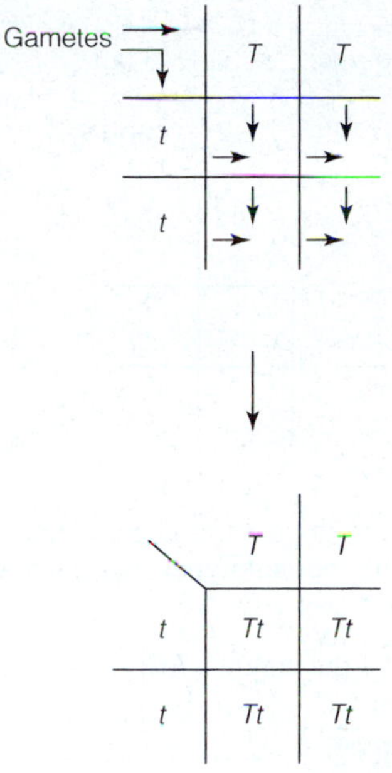

Results: Genotypes 100% *Tt* (all heterozygous)
Phenotypes 100% tall (because *T,* which determines tallness, is dominant and all contain the *T* allele)

Activity 1:
Working Out Crosses Involving Dominant and Recessive Genes

1. Using the technique outlined above, determine the genotypes and phenotypes of the offspring of the following crosses:

   a. Genotypes of parents: *Tt* (♂) × *tt* (♀)

   % of each genotype: ______________________

   % of each phenotype: ________ % tall

   ________ % dwarf

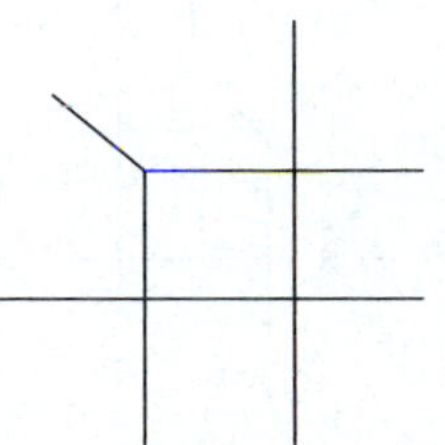

b. Genotypes of parents: *Tt* (♂) × *Tt* (♀)

% of each genotype: ______________________________

% of each phenotype: _______ % tall

_______ % dwarf

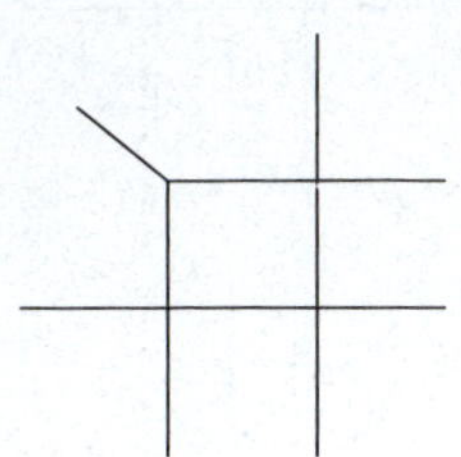

c. Genotypes of the parents: *RR* × *rr*

% of each genotype: ______________________________

% of each phenotype: _______ % rough

_______ % smooth

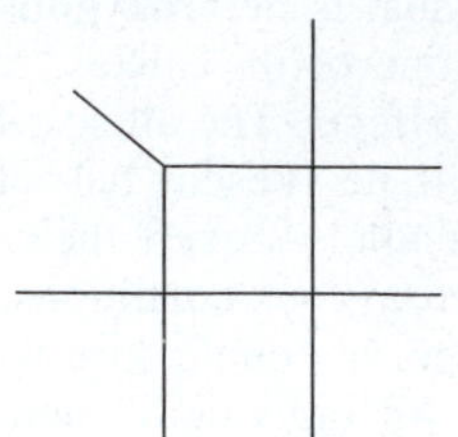

2. In guinea pigs, rough coat (*R*) is dominant over smooth coat (*r*). What will be the genotypes and phenotypes of the following monohybrid crosses?

a. Genotypes of the parents: *RR* × *rR*

% of each genotype: ______________________________

% of each phenotype: _______ % rough

_______ % smooth

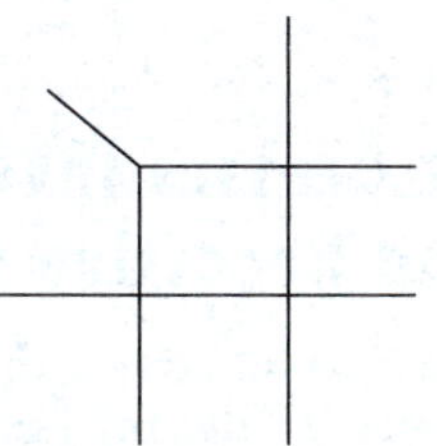

b. Genotypes of the parents: *Rr* × *rr*

% of each genotype: ______________________________

% of each phenotype: _______ % rough

_______ % smooth

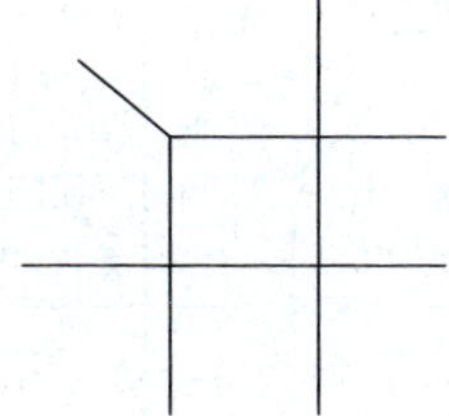

Incomplete Dominance

The concepts of dominance and recessiveness are somewhat arbitrary and artificial in some instances because so-called dominant genes may be expressed differently in homozygous and heterozygous individuals. This produces a condition called **incomplete dominance** or *intermediate inheritance*. In such cases, both alleles express themselves in the offspring. The crosses are worked out in the same manner as indicated previously, but heterozygous offspring exhibit a phenotype intermediate between that of the homozygous individuals. Some examples follow.

Activity 2: Working Out Crosses Involving Incomplete Dominance

1. The inheritance of flower color in snapdragons illustrates the principle of incomplete dominance. The genotype *RR* is expressed as a red flower, *Rr* yields pink flowers, and *rr* produces white flowers. Work out the following crosses to determine phenotypes seen and both genotype and phenotype percentages.

a. Genotypes of parents: *RR* × *rr*

Genotypes and %: ______________________________

Phenotypes and %: ______________________________

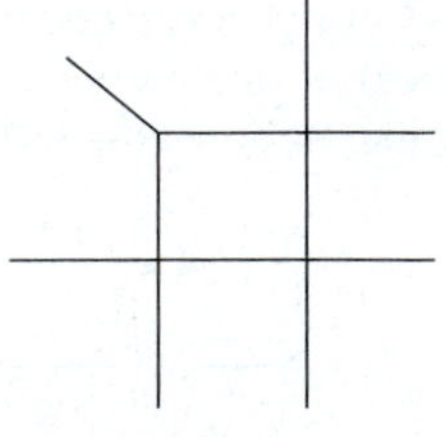

b. Genotypes of parents: *Rr* × *rr*

Genotypes and %: ______________________________

Phenotypes and %: ______________________________

____

c. Genotypes of parents *Rr* × *Rr*

Genotypes and %: ______________________________

Phenotypes and %: ______________________________

____

2. In humans, the inheritance of sickle-cell anemia/trait is determined by a single pair of alleles that exhibit incomplete dominance. Individuals homozygous for the sickling gene (*s*) have *sickle-cell anemia* (SCA). In double dose (*ss*) the sickling gene causes production of a very abnormal hemoglobin, which crystallizes and becomes sharp and spiky under conditions of oxygen deficit. This, in turn, leads to clumping and hemolysis of red blood cells in the circulation, which causes a great deal of pain and can be fatal. Heterozygous individuals (*Ss*) have the *sickle-cell trait* (SCT), which is much less severe. However, they are carriers for the abnormal gene and may pass it on to their offspring. Individuals with the genotype *SS* form normal hemoglobin. Work out the following crosses:

a. Parental genotypes: *SS* × *ss*

Genotypes and %: ______________________________

Phenotypes and %: ______________________________

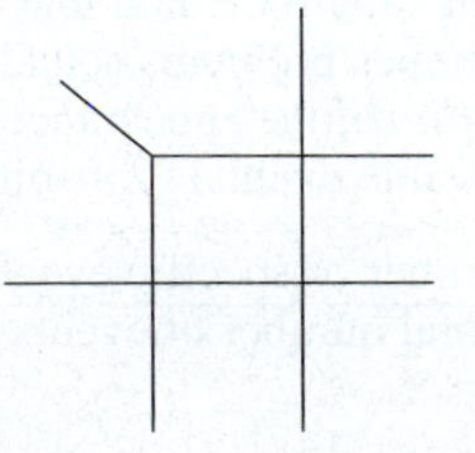

b. Parental genotypes: *Ss* × *Ss*

Genotypes and %: ______________________________

Phenotypes and %: ______________________________

____

c. Parental genotypes: *ss* × *Ss*

Genotypes and %: ______________________________

Phenotypes and %: ______________________________

____

Sex-Linked Inheritance

Of the 23 pairs of homologous chromosomes, 22 pairs are referred to as **autosomes.** Autosomes contain genes that determine most body (somatic) characteristics. The 23rd pair, the **sex chromosomes,** determine the sex of an individual, that is, whether an individual will be male or female. Normal females possess two sex chromosomes that look alike, the X chromosomes. Males possess two dissimilar sex chromosomes, referred to as X and Y. Possession of the Y chromosome determines maleness. A photomicrograph of part of a male's chromosome complement (male karyotype) is shown in Figure 45.1. The Y sex chromosome is only about a third the size of the X sex chromosome, and it lacks many of the genes (directing characteristics other than sex) that are found on the X.

Genes present *only* on the X sex chromosome are called *sex-linked* (or X-linked) genes. Some examples of X-linked genes include those that determine normal color vision (or, conversely, color blindness), and normal clotting ability (as opposed to hemophilia, or bleeder's disease). The alleles that determine color blindness and hemophilia are recessive alleles. In females, *both* X chromosomes must carry the recessive alleles for a woman to express either of these conditions, and thus they tend to be infrequently seen. However, should

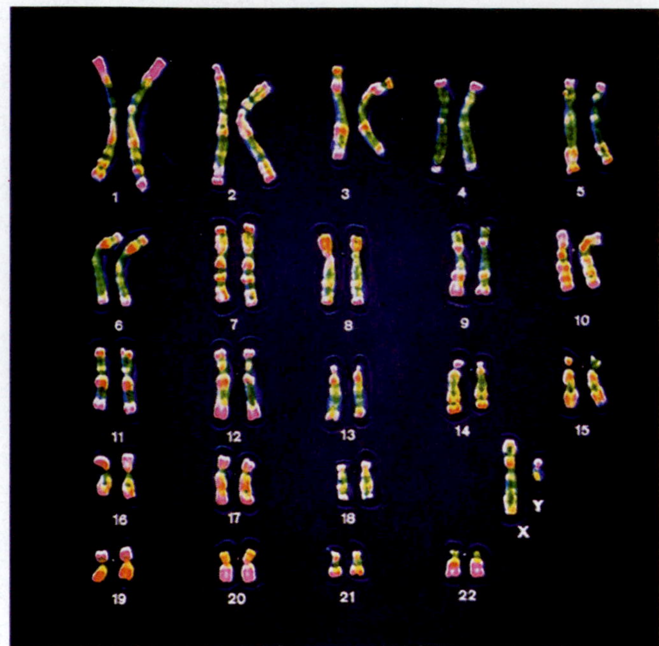

Figure 45.1 Karyotype (chromosomal complement) of human male. Each pair of homologous chromosomes is numbered except the sex chromosomes, which are identified by their letters, X and Y.

a male receive a sex-linked recessive allele for these conditions, he will exhibit the recessive phenotype because his Y chromosome does not contain alleles for that gene.

The critical understanding of X-linked inheritance is the *absence* of male to male (that is, father to son) transmission of X-linked genes. The X of the father *will* pass to each of his daughters but to none of his sons. Males always inherit sex-linked conditions from their mothers (via the X chromosome).

Activity 3: Working Out Crosses Involving Sex-Linked Inheritance

1. A heterozygous woman carrying the recessive gene for color blindness marries a man who is color-blind. Assume the dominant gene is X^C (allele for normal color vision) and the recessive gene is X^c (determines color blindness). The mother's genotype is X^CX^c and the father's X^cY. Do a Punnett square to determine the answers to the following questions.

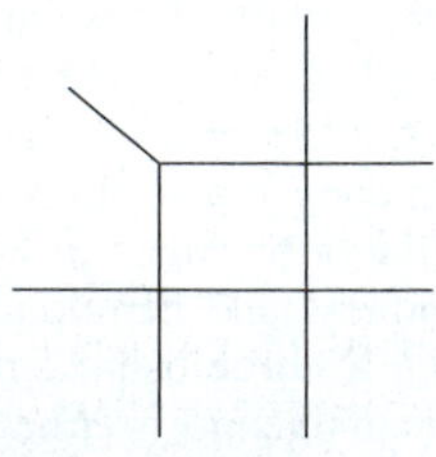

According to the laws of probability, what percent of all their children will be color-blind?

_______ %

What is the percent of color-blind individuals by sex?

_______ % males; _______ % females

What percentage will be carriers? _______________ %

What is the sex of the carriers? ____________________

2. A heterozygous woman carrying the recessive gene for hemophilia marries a man who is not a hemophiliac. Assume the dominant gene is X^H and the recessive gene is X^h. The woman's genotype is X^HX^h and her husband's genotype is X^HY. What is the potential percentage and sex of their offspring that will be hemophiliacs?

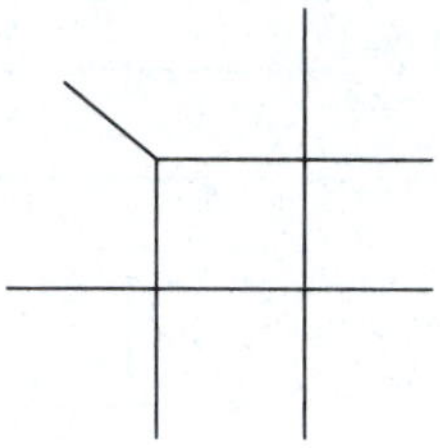

_______ % males; _______ % females

What percentage can be expected to neither exhibit nor carry the allele for hemophilia?

_______ % males; _______ % females

What is the anticipated sex and percentage of individuals that will be carriers for hemophilia?

_______ %; _______________ sex ■

Probability

Segregation (or parceling out) of chromosomes to daughter cells (gametes) during meiosis and the combination of egg and sperm are random or chance events. Hence, the possibility that certain genomes will arise and be expressed is based on the laws of probability. The randomness of gene recombination from each parent determines individual uniqueness and explains why siblings, however similar, never have totally corresponding traits (unless, of course, they are identical twins). The Punnett square method that you have been using to work out the genetics problems actually provides information on the *probability* of the appearance of certain genotypes considering all possible events. Probability (P) is defined as:

$$P = \frac{\text{number of specific events or cases}}{\text{total number of events or cases}}$$

If an event is certain to happen, its probability is 1. If it happens one out of every two times, its probability is ½; if one out of four times, its probability is ¼, and so on.

When figuring the probability of separate events occurring together (or consecutively), the probability of each event must be multiplied together to get the final probability figure. For example, the probability of a penny coming up "heads" in each toss is ½ (because it has two sides—heads and tails). But the probability of a tossed penny coming up heads four times in a row is $\frac{1}{2} \times \frac{1}{2} \times \frac{1}{2} \times \frac{1}{2} = \frac{1}{16}$.

Activity 4: Exploring Probability

1. Obtain two pennies and perform the following simple experiment to explore the laws of probability.

a. Toss one penny into the air 10 times, and record the number of heads/tails observed.

_______________ heads _______________ tails

Probability: _______ /10 tails; _______ /10 heads

b. Now simultaneously toss two pennies into the air for 24 tosses, and record the results of each toss below. In each case, report the probability in the lowest fractional terms.

#HH _____________ Probability _____________

#HT _____________ Probability _____________

#TT _____________ Probability _____________

Does the first toss have any influence on the second?

____

Does the third toss have any influence on the fourth?

____

c. Do a Punnett square using HT for one coin and HT for the "alleles" of the other.

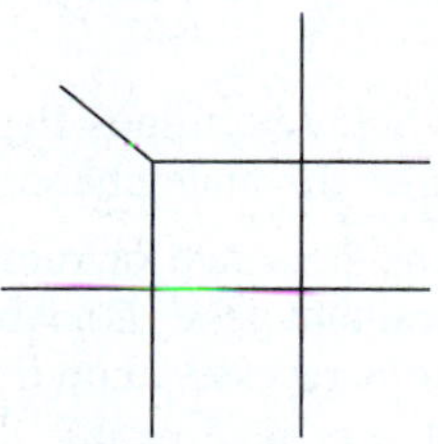

Probability of HH: _______

Probability of HT: _______

Probability of TT: _______

How closely do your coin-tossing results correlate with the percentages obtained from the Punnett square results?

____

____

2. Determine the probability of having a boy or girl offspring for each conception.

Parental genotypes: XY × XX

Probability of males: _______ %

Probability of females: _______ %

3. Dad wants a baseball team! What are the chances of his having nine sons in a row?

_______ (Sorry, Dad! Let the girls play.) ■

Genetic Determination of Selected Human Characteristics

Most human traits are determined by multiple alleles or the interaction of several gene pairs. However, a few visible human traits or phenotypes can be traced to a single gene pair. It is some of these that will be investigated here.

Activity 5: Using Phenotype to Determine Genotype

For each of the characteristics described here, determine (as best you can) both your own phenotype and genotype, and record this information on the chart on p. 456. Since it is impossible to know if you are homozygous or heterozygous for a nondetrimental trait when you exhibit its dominant expression, you are to record your genotype as *A*— (or *B*—, and so on, depending on the letter used to indicate the alleles) in such cases. As you will see, all the traits examined here are nonharmful characteristics. Generally speaking, in humans dominant gene disorders are presumed to be heterozygous (one dominant and one recessive allele), because having two such defective mutated alleles is usually not compatible with life. On the other hand, if you exhibit the recessive trait, you

Record of Human Genotypes/Phenotypes

Characteristic	Phenotype	Genotype
Interlocking fingers (*I,i*)		
PTC taste (*P,p*)		
Sodium benzoate taste (*S,s*)		
Sex (X,Y)		
Dimples (*D,d*)		
Tongue rolling (*T,t*)		
Attached earlobes (*E,e*)		
Widow's peak (*W,w*)		
Double-jointed thumb (*J,j*)		
Bent little finger (*L,l*)		
Middigital hair (*H,h*)		
Freckles (*F,f*)		
Blaze (*B,b*)		
ABO blood type (I^A,I^B,*i*)		

are homozygous for the recessive allele and should record it accordingly as *aa* (*bb, cc,* and so on). When you have completed your observations, also record your data on the chalkboard chart for tabulation of class results.

Interlocking fingers: Clasp your hands together by interlocking your fingers. Observe your clasped hands. Which thumb is uppermost? If the left thumb is uppermost, you possess a dominant allele (*I*) for this trait. If you clasped your right over your left thumb, you are illustrating the homozygous recessive (*ii*) phenotype.

PTC taste: Obtain a PTC taste strip. PTC or phenylthiocarbamide is a harmless chemical that some people can taste and others find tasteless. Chew the strip. If it tastes slightly bitter, you are a "taster" and possess the dominant gene (*P*) for this trait. If you cannot taste anything, you are a nontaster and are homozygous recessive (*pp*) for the trait. Approximately 70% of the people in the United States are tasters.

Sodium benzoate taste: Obtain a sodium benzoate taste strip and chew it. A different pair of alleles (from that determining PTC taste) determines the ability to taste sodium benzoate. If you can taste it, you have at least one of the dominant alleles (*S*). If not, you are homozygous recessive (*ss*) for the trait. Also record whether sodium benzoate tastes salty, bitter, or sweet to you (if a taster). Even though PTC and sodium benzoate taste are inherited independently, they interact to determine a person's taste sensations. Individuals who find PTC bitter and sodium benzoate salty tend to be devotees of sauerkraut, buttermilk, spinach, and other slightly bitter or salty foods.

Sex: The genotype XX determines the female phenotype, whereas XY determines the male phenotype.

Dimpled cheeks: The presence of dimples in one or both cheeks is due to a dominant gene (*D*). Absence of dimples indicates the homozygous recessive condition (*dd*).

Tongue rolling: Extend your tongue and attempt to roll it into a U shape longitudinally. People with this ability have the dominant allele for this trait. Use *T* for the dominant allele, and *t* for the recessive allele.

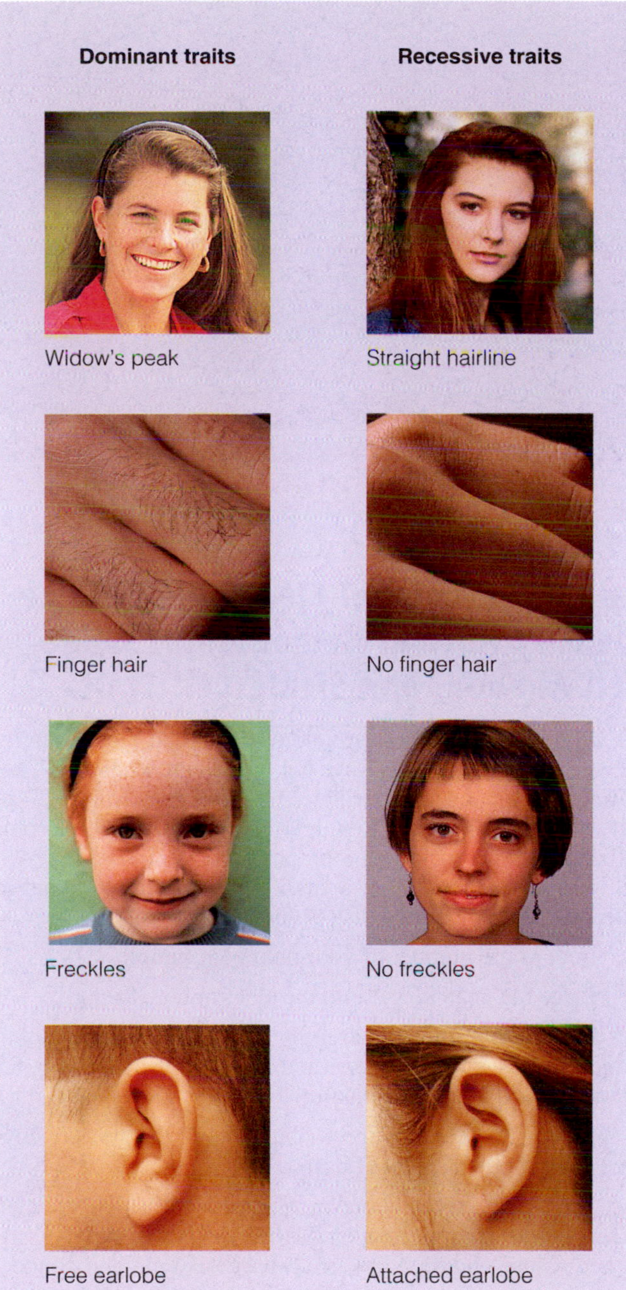

Figure 45.2 Selected examples of human phenotypes.

Attached earlobes: Have your lab partner examine your earlobes. If no portion of the lobe hangs free inferior to its point of attachment to the head, you are homozygous recessive (*ee*) for attached earlobes. If part of the lobe hangs free below the point of attachment, you possess at least one dominant gene (*E*) (see Figure 45.2).

Widow's peak: A distinct downward V-shaped hairline at the middle of the forehead is referred to as a widow's peak. It is determined by a dominant allele (*W*), whereas the straight or continuous forehead hairline is determined by the homozygous recessive condition (*ww*) (see Figure 45.2).

Table 45.1 Blood Groups

ABO blood group	Genotype
A	I^AI^A or I^Ai
B	I^BI^B or I^Bi
AB	I^AI^B
O	ii

Double-jointed thumb: A dominant gene determines a condition of loose ligaments that allows one to throw the thumb out of joint. The homozygous recessive condition determines tight joints. Use *J* for the dominant allele and *j* for the recessive allele.

Bent little finger: Examine your little finger on each hand. If its terminal phalanx angles toward the ring finger, you are dominant for this trait. If one or both terminal digits are essentially straight, you are homozygous recessive for the trait. Use *L* for the dominant allele and *l* for the recessive allele.

Middigital finger hair: Critically examine the dorsum of the middle segment (phalanx) of fingers 3 and 4. If no hair is obvious, you are recessive (*hh*) for this condition. If hair is seen, you have the dominant gene (*H*) for this trait (which, however, is determined by multigene inheritance) (see Figure 45.2).

Freckles: Freckles are the result of a dominant gene. Use *F* as the dominant allele and *f* as the recessive allele (see Figure 45.2).

Blaze: A lock of hair different in color from the rest of scalp hair is called a blaze; it is determined by a dominant gene. Use *B* for the dominant gene and *b* for the recessive gene.

Blood type: Inheritance of the ABO blood type is based on the existence of three alleles designated as I^A, I^B, and *i*. Both I^A and I^B are dominant over *i*, but neither is dominant over the other. Thus the possession of I^A and I^B will yield type AB blood, whereas the possession of the I^A and *i* alleles will yield type A blood, and so on as explained in Exercise 29. There are four ABO blood groups or phenotypes, A, B, AB, and O, and their correlation to genotype is indicated in Table 45.1 above.

Assuming you have previously typed your blood, record your phenotype and genotype in the chart on p. 456. If not, type your blood following the instructions on pp. 297–298, and then enter your results in the table.

Dispose of any blood-soiled supplies by placing the glassware in the bleach-containing beaker and all other items in the autoclave bag. ■

Once class data have been tabulated, scrutinize the results. Is there a single trait that is expressed in an identical manner by all members of the class?

____

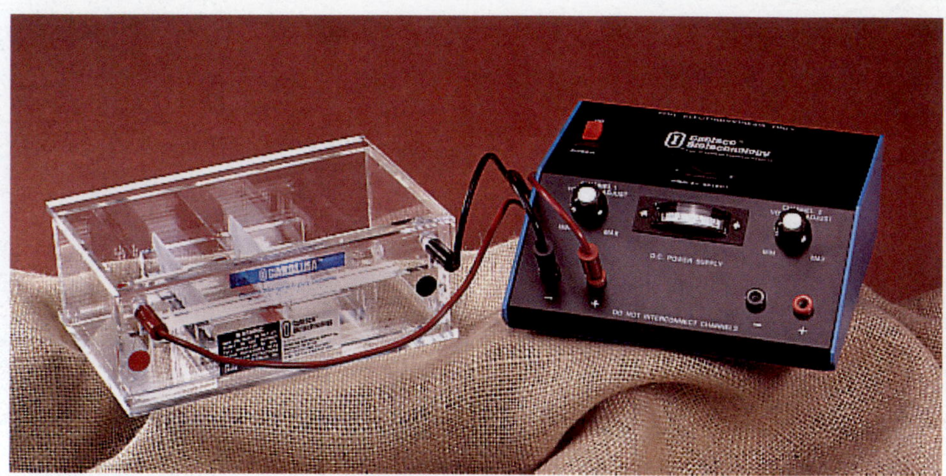

Figure 45.3 Agarose gel electrophoresis equipment and power supply.

Because all human beings have 23 pairs of homologues and each pair segregates independently at meiosis, the number of possible combinations at segregation is more than 8 million! On the basis of this information, what would you guess are the chances of any two individuals in the class having identical phenotypes for all 14 traits investigated?

____ ■

Hemoglobin Phenotype Identification using Agarose Gel Electrophoresis

Agarose gel electrophoresis separates molecules based on charge. In the appropriate buffer with an alkaline pH, hemoglobin molecules will move toward the anode of the apparatus at different speeds, based on the number of negative charges on the molecules. Agarose gel provides a medium for travel and slows the migration down a bit.

Sickle-cell anemia slides were observed in Exercise 29 and is discussed earlier in this exercise. The beta chains of hemoglobin S (HbS) contain a base substitution where a valine replaces glutamic acid. As a result of the substitution, HbS has fewer negative charges than HbA and can be separated from HbA using agarose gel electrophoresis.

Activity 6:

Using Agarose Gel Electrophoresis to Identify Normal Hemoglobin, Sickle-Cell Anemia, and Sickle-Cell Trait

1. You will need an electrophoresis unit and power supply (Figure 45.3), a 1.2% agarose gel with eight wells, 1X TBE buffer, micropipets or a variable automatic micropipet (2–20 μl) with tips, samples of hemoglobin dissolved in TBE solubilizing buffer containing bromophenol blue marked AA, AS, SS, and unknown #____, a marking pen, goggles, metric ruler, plastic baggie, and disposable gloves.

2. Record the number of your unknown sample. ________

3. Place the agarose gel into the electrophoresis unit.

4. Using a micropipet, carefully add 15 μl of hemoglobin sample AA to wells number 1 and 5, AS to wells 2 and 6, SS to wells 3 and 7, and the unknown samples to wells 4 and 8.

5. Slowly add electrophoresis buffer until the gels are covered with about 0.25 cm of buffer.

6. The electrophoresis unit runs with high voltage. Do not attempt to open it while the power supply is attached. Close and lock the electrophoresis unit, and connect the unit to the power source, red to red and black to black.

7. Run the unit about 50 minutes at 120 volts until the bromophenol blue is about 0.25 cm from the anode.

Agarose Gel with 8 Wells

Sample genotype	Well	Banding pattern
1. AA	1. ☐	
2. AS	2. ☐	
3. SS	3. ☐	
4. ______	4. ☐	
5. AA	5. ☐	
6. AS	6. ☐	
7. SS	7. ☐	
8. ______	8. ☐	

8. Turn the power supply *off*. Disconnect the cables.

9. Open the electrophoresis unit, carefully remove the gel, and slide the gel into a plastic baggie. You should be able to see the hemoglobin bands on the gel. Mark each of the bands on the gel with the marking pen.

Alternatively the gels may be stained with Coomassie blue. Obtain a flask of Coomassie blue stain, a flask of de-staining solution, a flask of distilled water, a staining dish, and a 100-ml graduated cylinder.

a. Carefully remove the gel from the plastic plate, and place the gel into a staining dish.

b. Add about 30 ml of stain (enough stain to cover the gel), and be sure the agarose is not stuck to the dish.

c. Allow the gel to remain in the stain for at least an hour (more time might be necessary) and then remove the stain. Rinse the gel and dish with distilled water.

d. Add about 100 ml of de-staining solution. Change the solution after a day. If the background stain has been reduced enough to see the bands, place the staining dish over a light source, and observe the bands. If the stain is still too dark, repeat the de-staining process until the bands can be observed.

e. To store the gels, refrigerate in a baggie with a small amount of de-stain solution or dry on a glass plate.

10. Draw the banding patterns for Samples 1 through 8 on the chart above. Based on the banding patterns of the known samples, what are the genotypes of your unknown samples? Record on the chart.

11. Rinse the electrophoresis unit in distilled or de-ionized water, and clean the glass plates with soap and water. ■

exercise

46 Surface Anatomy Roundup

Objectives

1. Define *surface anatomy,* and explain why it is an important field of study. Define *palpation.*
2. Describe and palpate the major surface features of the cranium, face, and neck.
3. Describe the easily palpated bony and muscular landmarks of the back. Locate the vertebral spines on the living body.
4. List the bony surface landmarks of the thoracic cage, and explain how they relate to the major soft organs of the thorax. Explain how to find the second to eleventh ribs.
5. Name and palpate the important surface features on the anterior abdominal wall, and explain how to palpate a full bladder.
6. Define and explain the following: *linea alba, umbilical hernia,* examination for an inguinal hernia, *linea semilunaris,* and *McBurney's point.*
7. Locate and palpate the main surface features of the upper limb.
8. Explain the significance of the antecubital fossa, pulse points in the distal forearm, and the anatomical snuff box.
9. Describe and palpate the surface landmarks of the lower limb.
10. Explain exactly where to administer an injection in the gluteal region and in the other major sites of intramuscular injection.

Materials

- ❑ Articulated skeletons
- ❑ Charts or models of the skeletal muscles of the body
- ❑ Washable markers
- ❑ Hand mirror
- ❑ Stethoscope
- ❑ Alcohol swabs

Surface anatomy is a valuable branch of anatomical and medical science. True to its name, **surface anatomy** does indeed study the *external surface* of the body, but more importantly, it also studies *internal* organs as they relate to external surface landmarks and as they are seen and felt through the skin. Feeling internal structures through the skin with the fingers is called **palpation** (literally, "touching").

Surface anatomy is living anatomy, better studied in live people than in cadavers. It can provide a great deal of information about the living skeleton (almost all bones can be palpated) and about the muscles and blood vessels that lie near the body surface. Furthermore, a skilled examiner can learn a good deal about the heart, lungs, and other deep organs by performing a surface assessment. Thus, surface anatomy serves as the basis of the standard physical examination. For those planning a career in the health sciences or physical education, a study of surface anatomy will show you where to take pulses, where to insert tubes and needles, where to locate broken bones and inflamed muscles, and where to listen for the sounds of the lungs, heart, and intestines.

We will take a regional approach to surface anatomy, exploring the head first and proceeding to the trunk and the limbs. You will be observing and palpating your own body as you work through the exercise, because your body is the best learning tool of all. To aid your exploration of living anatomy, skeletons and muscle models or charts are provided around the lab so that you can review the bones and muscles you will encounter. Where you are asked to mark with a washable marker areas of skin that are personally unreachable, it probably would be best to choose a male student as a subject.

Activity 1: Palpating Landmarks of the Head

The head (Figures 46.1 and 46.2) is divided into the cranium and the face.

Cranium

1. Run your fingers over the superior surface of your head. Notice that the underlying cranial bones lie very near the surface. Proceed to your forehead and palpate the *superciliary arches* ("brow ridges") directly superior to your orbits (Figure 46.1).

2. Move your hand to the posterior surface of your skull, where you can feel the knoblike *external occipital protuberance.* Run your finger directly laterally from this projection to feel the ridgelike *superior nuchal line* on the occipital bone. This line, which marks the superior extent of the muscles of the posterior neck, serves as the boundary between the head and the neck. Now feel the prominent *mastoid process* on each side of the cranium just posterior to your ear.

3. Place a hand on your temple, and clench your teeth together in a biting action. You should be able to feel the *temporalis muscle* bulge as it contracts. Next, raise your eyebrows, and feel your forehead wrinkle, an action produced

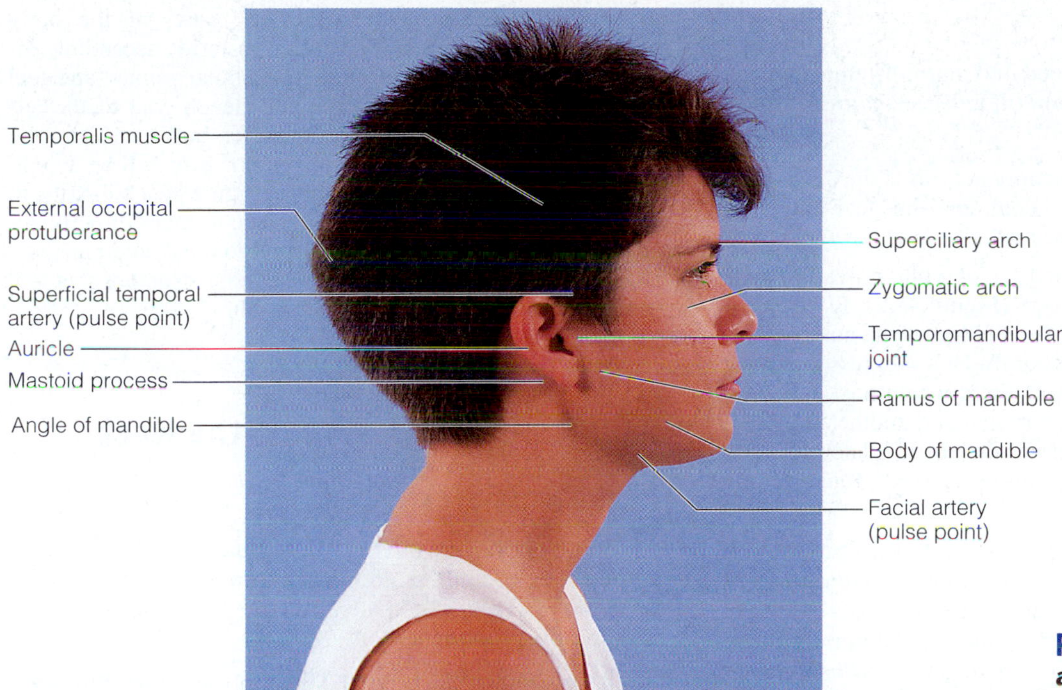

Figure 46.1 Surface anatomy of the lateral aspect of the head.

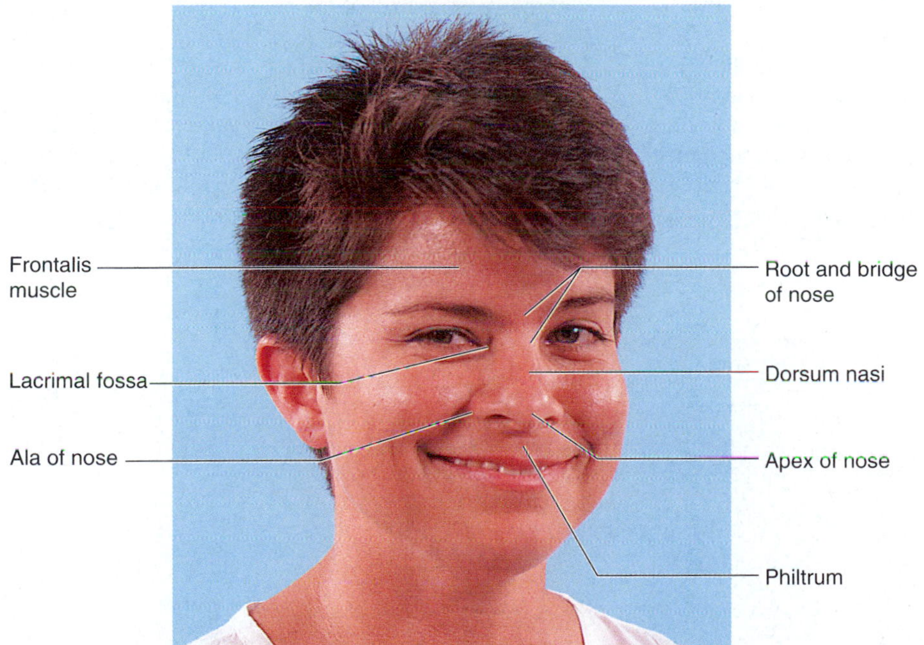

Figure 46.2 Surface structures of the face.

by the *frontalis muscle*. The frontalis inserts superiorly onto the broad aponeurosis called the *galea aponeurotica* (Table 15.1, pp. 140 and 142) that covers the superior surface of the cranium. This aponeurosis binds tightly to the overlying subcutaneous tissue and skin to form the true **scalp.** Push on your scalp, and confirm that it slides freely over the underlying cranial bones. Because the scalp is only loosely bound to the skull, people can easily be "scalped" (in industrial accidents, for example). The scalp is richly vascularized by a large number of arteries running through its subcutaneous tissue. Most arteries of the body constrict and close after they are cut or torn, but those in the scalp are unable to do so because they are held open by the dense connective tissue surrounding them.

What do these facts suggest about the amount of bleeding that accompanies scalp wounds?

____

____

____

Face

The surface of the face is divided into many different regions, including the *orbital, nasal, oral* (mouth), and *auricular* (ear) areas (Figure 46.2).

1. Trace a finger around the entire margin of the bony orbit. The *lacrimal fossa,* which contains the tear-gathering lacrimal sac, may be felt on the medial side of the eye socket.

2. Touch the most superior part of your nose, its **root,** which lies between the eyebrows (Figure 46.2). Just inferior to this, between your eyes, is the **bridge** of the nose formed by the nasal bones. Continue your finger's progress inferiorly along the nose's anterior margin, the **dorsum nasi,** to its tip, the **apex.** Place one finger in a nostril and another finger on the flared winglike **ala** that defines the nostril's lateral border. Then feel the **philtrum,** the shallow vertical groove on the upper lip below the nose.

3. Grasp your *auricle,* the shell-like part of the external ear that surrounds the opening of the *external auditory canal* (Figure 46.1). Now trace the ear's outer rim, or *helix,* to the *lobule* (earlobe) inferiorly. The lobule is easily pierced, and since it is not highly sensitive to pain, it provides a convenient place to hang an earring or obtain a drop of blood for clinical blood analysis. Next, place a finger on your temple just anterior to the auricle (Figure 46.1). There, you will be able to feel the pulsations of the *superficial temporal artery,* which ascends to supply the scalp.

4. Run your hand anteriorly from your ear toward the orbit, and feel the *zygomatic arch* just deep to the skin. This bony arch is easily broken by blows to the face. Next, place your fingers on the skin of your face, and feel it bunch and stretch as you contort your face into smiles, frowns, and grimaces. You are now monitoring the action of several of the subcutaneous *muscles of facial expression* (Table 15.1, pp. 140 and 142).

5. On your lower jaw, palpate the parts of the bony *mandible:* its anterior body and its posterior ascending *ramus.* Press on the skin over the mandibular ramus, and feel the *masseter muscle* bulge when you clench your teeth. Palpate the anterior border of the masseter, and trace it to the mandible's inferior margin. At this point, you will be able to detect the pulse of your *facial artery* (Figure 46.1). Finally, to feel the *temporomandibular joint,* place a finger directly anterior to the external auditory canal of your ear, and open and close your mouth several times. The bony structure you feel moving is the *head of the mandible.* ■

Activity 2:

Palpating Landmarks of the Neck

Bony Landmarks

1. Run your fingers inferiorly along the back of your neck, in the posterior midline, to feel the *spinous processes* of the cervical vertebrae. The spine of G_7, the *vertebra prominens,* is especially prominent.

2. Now, beginning at your chin, run a finger inferiorly along the anterior midline of your neck (Figure 46.3). The first hard structure you encounter will be the U-shaped *hyoid bone,* which lies in the angle between the floor of the mouth and the vertical part of the neck (Figure 46.4). Directly inferior to this, you will feel the *laryngeal prominence* (Adam's apple) of the thyroid cartilage. Just inferior to the laryngeal prominence, your finger will sink into a soft depression (formed by the *cricothyroid ligament*) before proceeding onto the rounded surface of the *cricoid cartilage.* Now swallow several times, and feel the whole larynx move up and down.

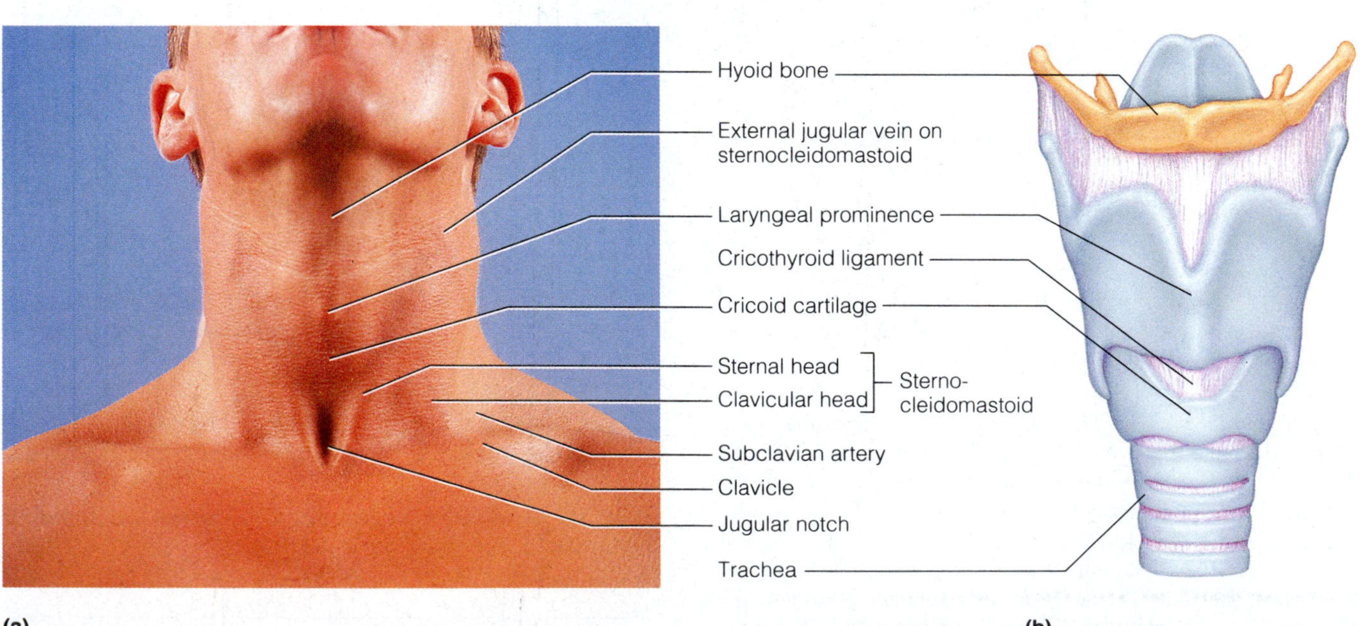

Figure 46.3 Anterior surface of the neck. (a) Photograph. **(b)** Diagram of the underlying skeleton of the larynx.

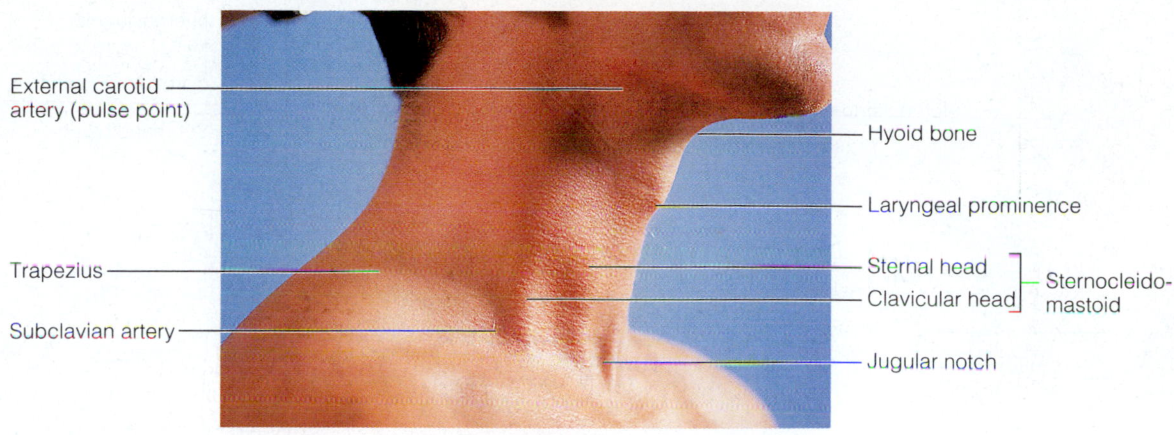

Figure 46.4 Lateral surface of the neck.

3. Continue inferiorly to the trachea. Attempt to palpate the *isthmus of the thyroid gland,* which feels like a spongy cushion over the second to fourth tracheal rings (see Figure 46.3b). Then, try to palpate the two soft lateral *lobes* of your thyroid gland along the sides of the trachea.

4. Move your finger all the way inferiorly to the root of the neck, and rest it in the *jugular notch,* the depression in the superior part of the sternum between the two clavicles. By pushing deeply at this point, you can feel the cartilage rings of the trachea.

Muscles

The *sternocleidomastoid* is the most prominent muscle in the neck and the neck's most important surface landmark. You can best see and feel it when you turn your head to the side.

1. Obtain a hand mirror, hold it in front of your face, and turn your head sharply from right to left several times. You will be able to see both heads of this muscle, the *sternal head* medially and the *clavicular head* laterally (Figures 46.3 and 46.4). Several important structures lie beside or beneath the sternocleidomastoid:

- The *cervical lymph nodes* lie both superficial and deep to this muscle. (Swollen cervical nodes provide evidence of infections or cancer of the head and neck.)
- The *common carotid artery* and *internal jugular vein* lie just deep to the sternocleidomastoid, a relatively superficial location that exposes these vessels to danger in slashing wounds to the neck.
- Just lateral to the inferior part of the sternocleidomastoid is the large *subclavian artery* on its way to supply the upper limb. By pushing on the subclavian artery at this point, one can stop the bleeding from a wound anywhere in the associated limb.
- Just anterior to the sternocleidomastoid, superior to the level of your larynx, you can feel a carotid pulse—the pulsations of the *external carotid artery* (Figure 46.4).
- The *external jugular vein* descends vertically, just superficial to the sternocleidomastoid and deep to the skin (Figure 46.3). To make this vein "appear" on your neck, stand before the mirror, and gently compress the skin superior to your clavicle with your fingers.

2. Another large muscle in the neck, on the posterior aspect, is the *trapezius* (Figure 46.4). You can feel this muscle contracting just deep to the skin as you shrug your shoulders.

Triangles of the Neck

The sternocleidomastoid muscles divide each side of the neck into the posterior and anterior triangles (Figure 46.5a).

1. The **posterior triangle** is defined by the sternocleidomastoid anteriorly, the trapezius posteriorly, and the clavicle inferiorly. Palpate the borders of the posterior triangle.

The **anterior triangle** is defined by the inferior margin of the mandible superiorly, the midline of the neck anteriorly, and the sternocleidomastoid posteriorly.

2. The contents of these two triangles are shown in Figure 46.5b. The posterior triangle contains many important nerves and blood vessels, including the *accessory nerve* (cranial nerve XI), most of the *cervical plexus,* and the *phrenic nerve.* In the inferior part of the triangle are the external jugular vein, the trunks of the *brachial plexus,* and the subclavian artery. These structures are relatively superficial and are easily cut or injured by wounds to the neck. In the neck's anterior triangle, important structures include the *submandibular gland,* the *suprahyoid* and *infrahyoid muscles,* and parts of the carotid arteries and jugular veins that lie superior to the sternocleidomastoid.

- Palpate your carotid pulse. ■

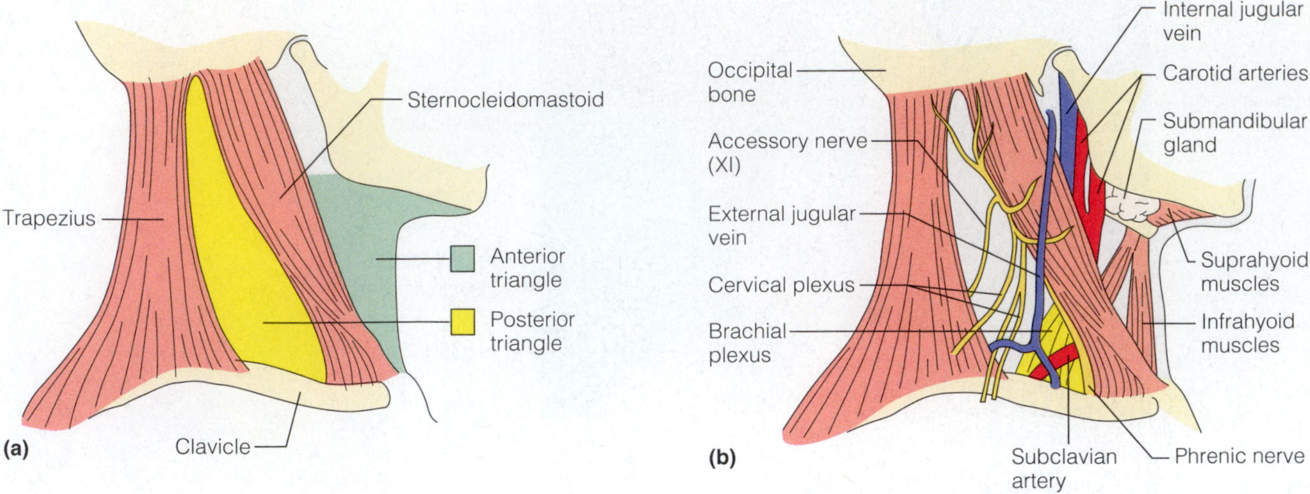

Figure 46.5 Anterior and posterior triangles of the neck. **(a)** Boundaries of the triangles. **(b)** Some contents of the triangles.

A wound to the posterior triangle of the neck can lead to long-term loss of sensation in the skin of the neck and shoulder, as well as partial paralysis of the sternocleidomastoid and trapezius muscles. Can you explain these effects?

____

____

____

Activity 3:
Palpating Landmarks of the Trunk

The trunk of the body consists of the thorax, abdomen, pelvis, and perineum. The *back* includes parts of all of these regions, but for convenience it is treated separately.

The Back

Bones 1. The vertical groove in the center of the back is called the **posterior median furrow** (Figure 46.6a). The *spinous processes* of the vertebrae are visible in the furrow when the spinal column is flexed. Palpate a few of these processes on your partner's back (C_7 and T_1 are the most prominent and the easiest to find). Also palpate the posterior parts of some ribs, as well as the prominent *spine of the scapula* and the scapula's long *medial border.* The scapula lies superficial to ribs 2 to 7; its *inferior angle* is at the level of the spinous process of vertebra T_7. The medial end of the scapular spine lies opposite the T_3 spinous process.

2. Now feel the *iliac crests* (superior margins of the iliac bones) in your own lower back. You can find these crests effortlessly by resting your hands on your hips. Locate the most superior point of each crest, a point that lies roughly halfway between the posterior median furrow and the lateral side of the body (Figure 46.6a). A horizontal line through these two superior points, the **supracristal line,** intersects L_4, providing a simple way to locate that vertebra. The ability to locate L_4 is essential for performing a *lumbar puncture,* a procedure in which the clinician inserts a needle into the vertebral canal of the spinal column directly superior or inferior to L_4 and withdraws cerebrospinal fluid.

3. The *sacrum* is easy to palpate just superior to the cleft in the buttocks. You can feel the *coccyx* in the extreme inferior part of that cleft, just posterior to the anus.

Muscles The largest superficial muscles of the back are the *trapezius* superiorly and *latissimus dorsi* inferiorly (Figure 46.6b). Furthermore, the deeper *erector spinae* muscles are very evident in the lower back, flanking the vertebral column like thick vertical cords.

1. Feel your partner's erector spinae muscles contract and bulge as he straightens his spine from a slightly bent-over position.

The superficial muscles of the back fail to cover a small area of the rib cage called the **triangle of auscultation** (Figure 46.6a). This triangle lies just medial to the inferior part of the scapula. Its three boundaries are formed by the trapezius medially, the latissimus dorsi inferiorly, and the scapula laterally. The physician places a stethoscope over the skin of this triangle to listen for lung sounds (*auscultation* = listening). To hear the lungs clearly, the doctor first asks the patient to fold the arms together in front of the chest and then flex the trunk.

What do you think is the precise reason for having the patient take this action?

____

____

____

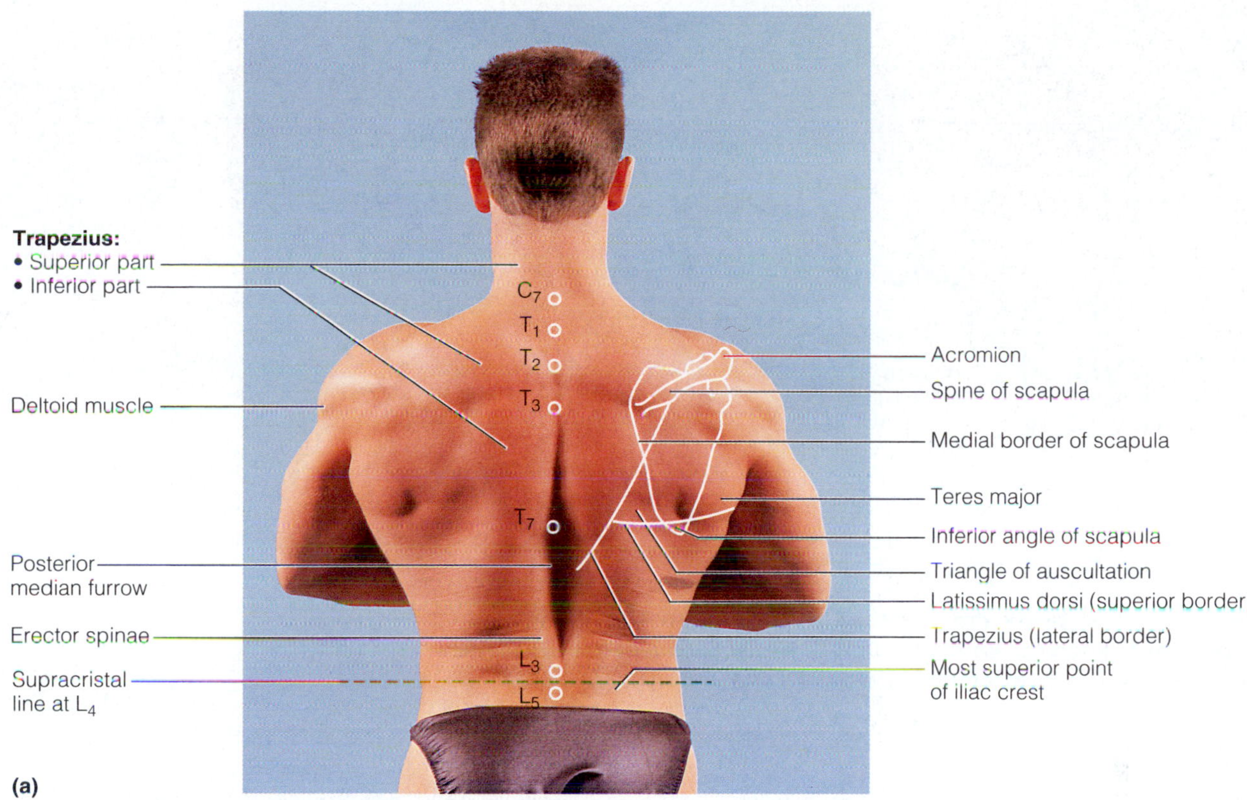

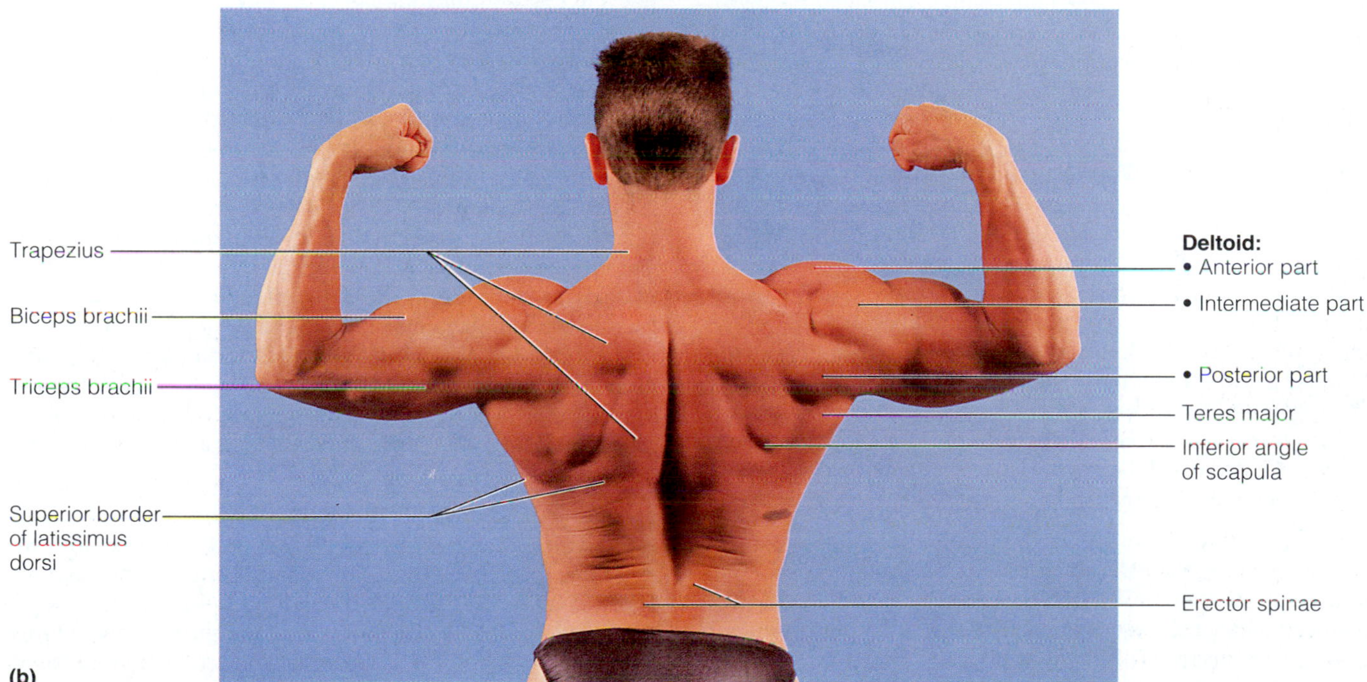

Figure 46.6 Surface anatomy of the back. Two different poses, **(a)** and **(b)**.

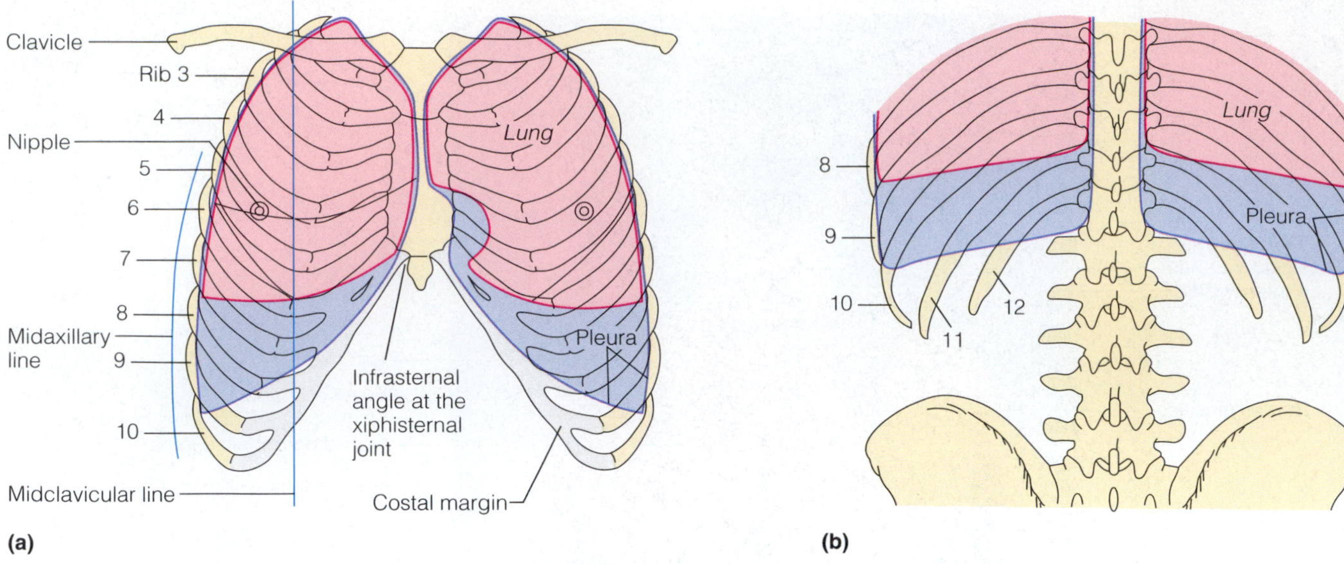

Figure 46.7 The bony rib cage as it relates to the underlying lungs and pleural cavities. Both the pleural cavities (blue) and the lungs (red) are outlined. **(a)** Anterior view. **(b)** Posterior view.

2. Have your partner assume the position just described. After cleaning the earpieces with an alcohol swab, use the stethoscope to auscultate the lung sounds. Compare the clarity of the lung sounds heard over the triangle of auscultation to that over other areas of the back.

The Thorax

Bones 1. Start exploring the anterior surface of your partner's bony *thoracic cage* (Figures 46.7 and 46.8) by defining the extent of the *sternum.* Use a finger to trace the sternum's triangular *manubrium* inferior to the jugular notch, its flat *body,* and the tongue-shaped *xiphoid process.* Now palpate the ridgelike *sternal angle,* where the manubrium meets the body of the sternum. Locating the sternal angle is important because it directs you to the second ribs (which attach to it). Once you find the second rib, you can count down to identify every other rib in the thorax (except the first and sometimes the twelfth rib, which lie too deep to be palpated). The sternal angle is a highly reliable landmark—it is easy to locate, even in overweight people.

2. By locating the individual ribs, you can mentally "draw" a series of horizontal lines of "latitude" that you can use to map and locate the underlying visceral organs of the thoracic cavity. Such mapping also requires lines of "longitude," so let us construct some vertical lines on the wall of your partner's trunk. As he lifts an arm straight up in the air, extend a line inferiorly from the center of the axilla onto his lateral thoracic wall. This is the **midaxillary line** (Figure 46.7a). Now estimate the midpoint of his *clavicle,* and run a vertical line inferiorly from that point toward the groin. This is the **midclavicular line,** and it will pass about 1 cm medial to the nipple.

3. Next, feel along the V-shaped inferior edge of the rib cage, the *costal margin.* At the **infrasternal angle,** the superior angle of the costal margin, lies the *xiphisternal joint.* Deep to the xiphisternal joint, the heart lies on the diaphragm.

4. The thoracic cage provides many valuable landmarks for locating the vital organs of the thoracic and abdominal cavities. On the anterior thoracic wall, ribs 2–6 define the superior-to-inferior extent of the female breast, and the fourth intercostal space indicates the location of the *nipple* in men, children, and small-breasted women. The right costal margin runs across the anterior surface of the liver and gallbladder. Surgeons must be aware of the inferior margin of the *pleural cavities* because if they accidentally cut into one of these cavities, a lung collapses. The inferior pleural margin lies adjacent to vertebra T_{12} near the posterior midline (Figure 46.7b) and runs horizontally across the back to reach rib 10 at the midaxillary line. From there, the pleural margin ascends to rib 8 in the midclavicular line (Figure 46.7a) and to the level of the xiphisternal joint near the anterior midline. The *lungs* do not fill the inferior region of the pleural cavity. Instead, their inferior borders run at a level that is two ribs superior to the pleural margin, until they meet that margin near the xiphisternal joint.

5. The relationship of the *heart* to the thoracic cage is considered in Exercise 30 (p. 300). We will review that information here. In essence, the superior right corner of the heart lies at the junction of the third rib and the sternum; the superior left corner lies at the second rib, near the sternum; the inferior left corner lies in the fifth intercostal space in the midclavicular line; and the inferior right corner lies at the sternal border of the sixth rib. You may wish to outline the heart on your chest or that of your lab partner by connecting the four corner points with a washable marker.

Muscles The main superficial muscles of the anterior thoracic wall are the *pectoralis major* and the anterior slips of the *serratus anterior.*

Using Figure 46.8 as a guide, try to palpate these two muscles on your chest. They both contract during push-ups, and you can confirm this by pushing yourself up from your

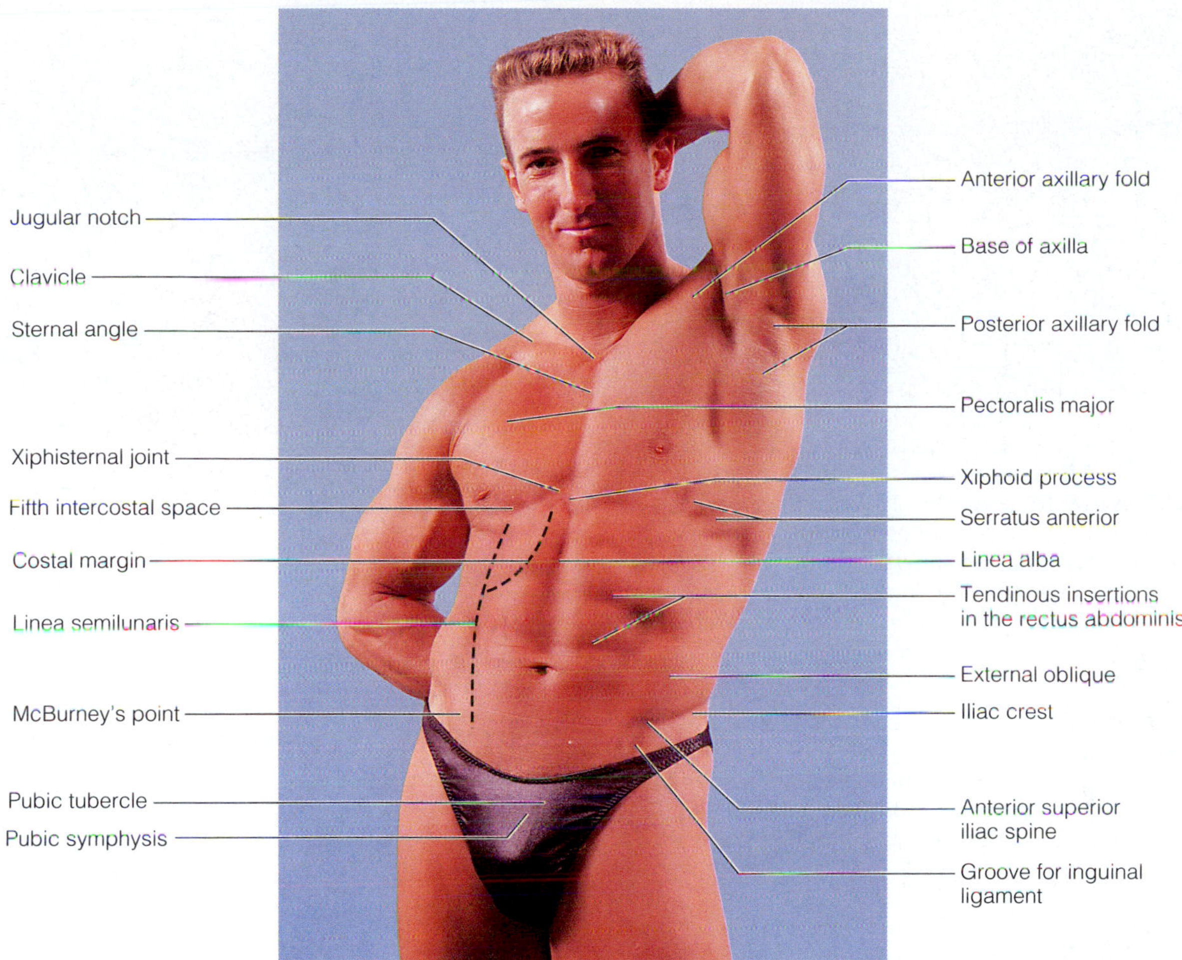

Figure 46.8 The anterior thorax and abdomen.

desk with one arm while palpating the muscles with your opposite hand. ■

Activity 4: Palpating Landmarks of the Abdomen

Bony Landmarks

The anterior abdominal wall (Figure 46.8) extends inferiorly from the costal margin to an inferior boundary that is defined by several landmarks. Palpate these landmarks as they are described below.

1. **Iliac crest.** Recall that the iliac crests are the superior margins of the iliac bones, and you can locate them by resting your hands on your hips.

2. **Anterior superior iliac spine.** Representing the most anterior point of the iliac crest, this spine is a prominent landmark. It can be palpated in everyone, even those who are overweight. Run your fingers anteriorly along the iliac crest to its end.

3. **Inguinal ligament.** The inguinal ligament, indicated by a groove on the skin of the groin, runs medially from the anterior superior iliac spine to the pubic tubercle of the pubic bone.

4. **Pubic crest.** You will have to press deeply to feel this crest on the pubic bone near the median *pubic symphysis*. The *pubic tubercle,* the most lateral point of the pubic crest, is easier to palpate, but you will still have to push deeply.

Inguinal hernias occur immediately superior to the inguinal ligament and may exit from a medial opening called the *superficial inguinal ring*. To locate this ring, one would palpate the pubic tubercle (Figure 46.9). The procedure used by the physician to test whether a male has an inguinal hernia is depicted in Figure 46.9. ■

Muscles and Other Surface Features

The central landmark of the anterior abdominal wall is the *umbilicus* (navel). Running superiorly and inferiorly from the umbilicus is the *linea alba* ("white line"), represented in the skin of lean people by a vertical groove (Figure 46.8). The linea alba is a tendinous seam that extends from the xiphoid process to the pubic symphysis, just medial to the rectus abdominis muscles (Table 15.3, pp. 145-146). The linea alba is a favored site for surgical entry into the abdominal cavity

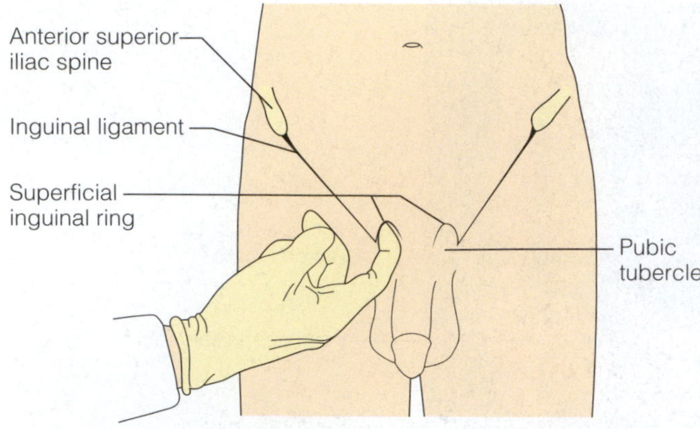

Figure 46.9 Clinical examination for an inguinal hernia in a male. The examiner palpates the patient's pubic tubercle, pushes superiorly to invaginate the scrotal skin into the superficial inguinal ring, and asks the patient to cough. If an inguinal hernia exists, it will push inferiorly and touch the examiner's fingertip.

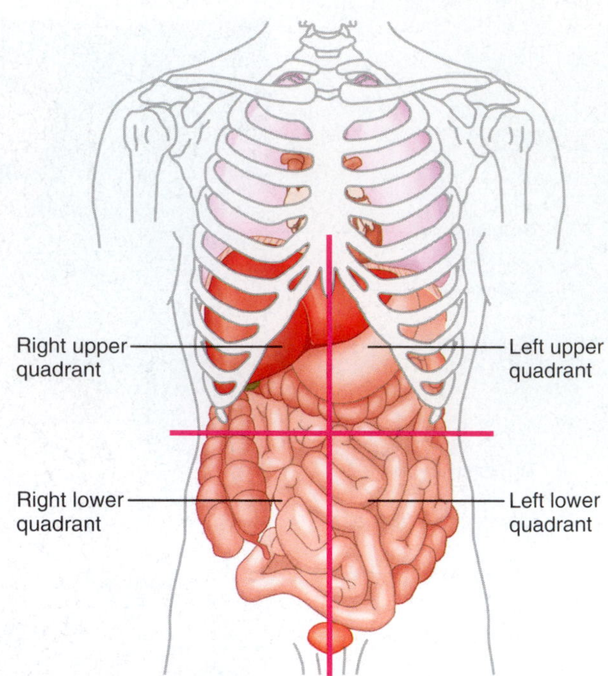

Figure 46.10 The four abdominopelvic quadrants.

because the surgeon can make a long cut through this line with no muscle damage and minimal bleeding.

Several kinds of hernias involve the umbilicus and the linea alba. In an **acquired umbilical hernia,** the linea alba weakens until intestinal coils push through it just superior to the navel. The herniated coils form a bulge just deep to the skin.

Another type of umbilical hernia is a **congenital umbilical hernia,** present in some infants: The umbilical hernia is seen as a cherry-sized bulge deep to the skin of the navel that enlarges whenever the baby cries. Congenital umbilical hernias are usually harmless, and most correct themselves automatically before the child's second birthday. ■

1. **McBurney's point** is the spot on the anterior abdominal skin that lies directly superficial to the base of the appendix (Figure 46.8). It is located one-third of the way along a line between the right anterior superior iliac spine and the umbilicus. Try to find it on your body.

McBurney's point is the most common site of incision in appendectomies, and it is often the place where the pain of appendicitis is experienced most acutely. Pain at McBurney's point after the pressure is removed (rebound tenderness) can indicate appendicitis. This is not a *precise* method of diagnosis, however.

2. Flanking the linea alba are the vertical straplike *rectus abdominis* muscles (Figure 46.8). Feel these muscles contract just deep to your skin as you do a bent-knee sit-up (or as you bend forward after leaning back in your chair). In the skin of lean people, the lateral margin of each rectus muscle makes a groove known as the **linea semilunaris** ("half-moon line"). On your right side, estimate where your linea semilunaris crosses the costal margin of the rib cage. The *gallbladder* lies just deep to this spot, so this is the standard point of incision for gallbladder surgery. In muscular people, three horizontal grooves can be seen in the skin covering the rectus abdominis. These grooves represent the *tendinous insertions* (or *intersections* or *inscriptions*), fibrous bands that subdivide the rectus muscle. Because of these subdivisions, each rectus abdominis muscle presents four distinct bulges. Try to identify these insertions on yourself or your partner.

3. The only other major muscles that can be seen or felt through the anterior abdominal wall are the lateral *external obliques*. Feel these muscles contract as you cough, strain, or raise your intra-abdominal pressure in some other way.

4. Recall that the anterior abdominal wall can be divided into four quadrants (Figure 46.10). A clinician listening to a patient's **bowel sounds** places the stethoscope over each of the four abdominal quadrants, one after another. Normal bowel sounds, which result as peristalsis moves air and fluid through the intestine, are high-pitched gurgles that occur every 5 to 15 seconds.

- Use the stethoscope to listen to your own or your partner's bowel sounds.

Abnormal bowel sounds can indicate intestinal disorders. Absence of bowel sounds indicates a halt in intestinal activity, which follows long-term obstruction of the intestine, surgical handling of the intestine, peritonitis, or other conditions. Loud tinkling or splashing sounds, by contrast, indicate an increase in intestinal activity. Such loud sounds may accompany gastroenteritis (inflammation and upset of the GI tract) or a partly obstructed intestine. ■

The Pelvis and Perineum

The bony surface features of the *pelvis* are considered with the bony landmarks of the abdomen (p. 467) and the gluteal region (p. 473). Most *internal* pelvic organs are not palpable through the skin of the body surface. A full *bladder,* however, becomes firm and can be felt through the abdominal wall just superior to the pubic symphysis. A bladder that can be palpated more than a few centimeters above this symphysis is

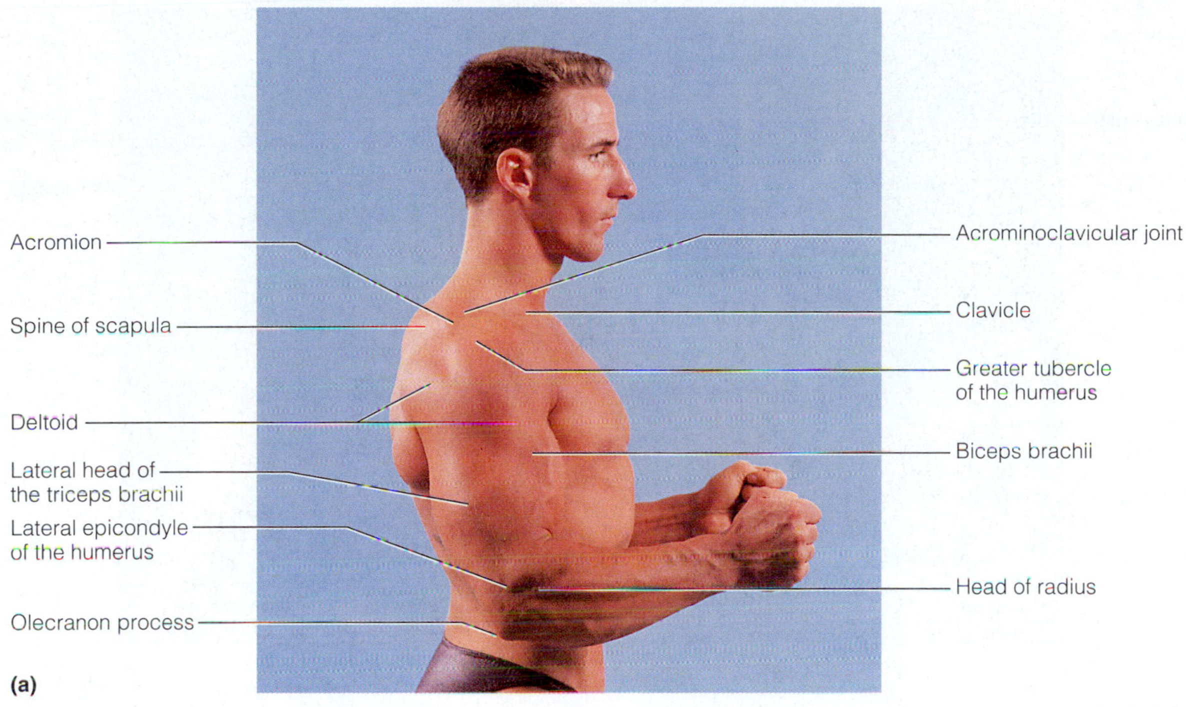

Figure 46.11 Shoulder and arm. (a) Lateral view.

retaining urine and dangerously full, and it should be drained by catheterization. ■

Activity 5: Palpating Landmarks of the Upper Limb

Axilla

The **base of the axilla** is the groove in which the underarm hair grows (Figure 46.8). Deep to this base lie the axillary *lymph nodes* (which swell and can be palpated in breast cancer), the large *axillary vessels* serving the upper limb, and much of the brachial plexus. The base of the axilla forms a "valley" between two thick, rounded ridges, the **axillary folds.** Just anterior to the base, clutch your **anterior axillary fold,** formed by the pectoralis major muscle. Then grasp your **posterior axillary fold.** This fold is formed by the latissimus dorsi and teres major muscles of the back as they course toward their insertions on the humerus.

Shoulder

1. Relocate the prominent spine of the scapula posteriorly (Figure 46.11). Follow the spine to its lateral end, the flattened *acromion* on the shoulder's summit. Then, palpate the *clavicle* anteriorly, tracing this bone from the sternum to the shoulder. Notice the clavicle's curved shape.

2. Now locate the junction between the clavicle and the acromion on the superolateral surface of your shoulder, at the *acromioclavicular joint.* To find this joint, thrust your arm anteriorly repeatedly until you can palpate the precise point of pivoting action.

3. Next, place your fingers on the *greater tubercle* of the humerus. This is the most lateral bony landmark on the superior surface of the shoulder. It is covered by the thick *deltoid muscle,* which forms the rounded superior part of the shoulder. Intramuscular injections are often given into the deltoid, about 5 cm (2 inches) inferior to the greater tubercle (refer to Figure 46.19a, p. 474).

Arm

Remember, according to anatomists, the arm runs only from the shoulder to the elbow, and not beyond.

1. In the arm, palpate the humerus along its entire length, especially along its medial and lateral sides.

2. Feel the *biceps brachii* muscle contract on your anterior arm when you flex your forearm against resistance. The medial boundary of the biceps is represented by the **medial bicipital furrow** (Figure 46.11b). This groove contains the large *brachial artery,* and by pressing on it with your fingertips you can feel your *brachial pulse.* Recall that the brachial artery is the artery routinely used in measuring blood pressure with a sphygmomanometer.

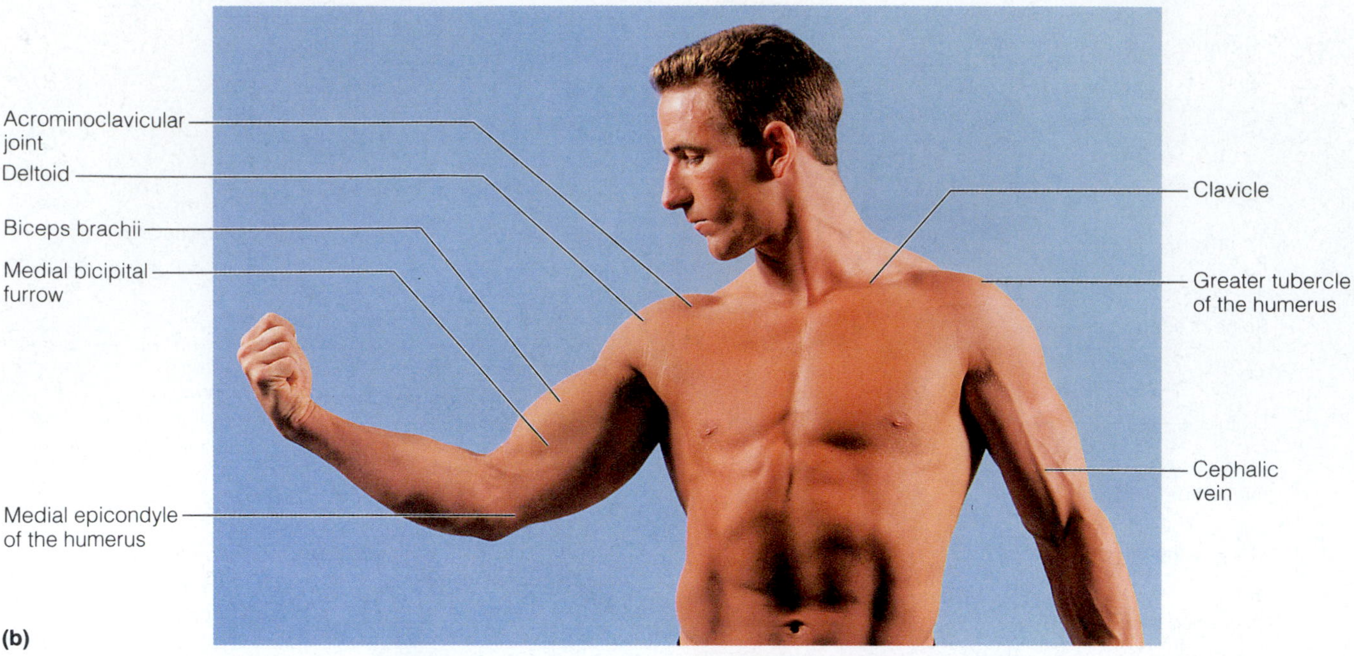

Figure 46.11 *(continued)* Shoulder and arm. (b) Anterior and medial view.

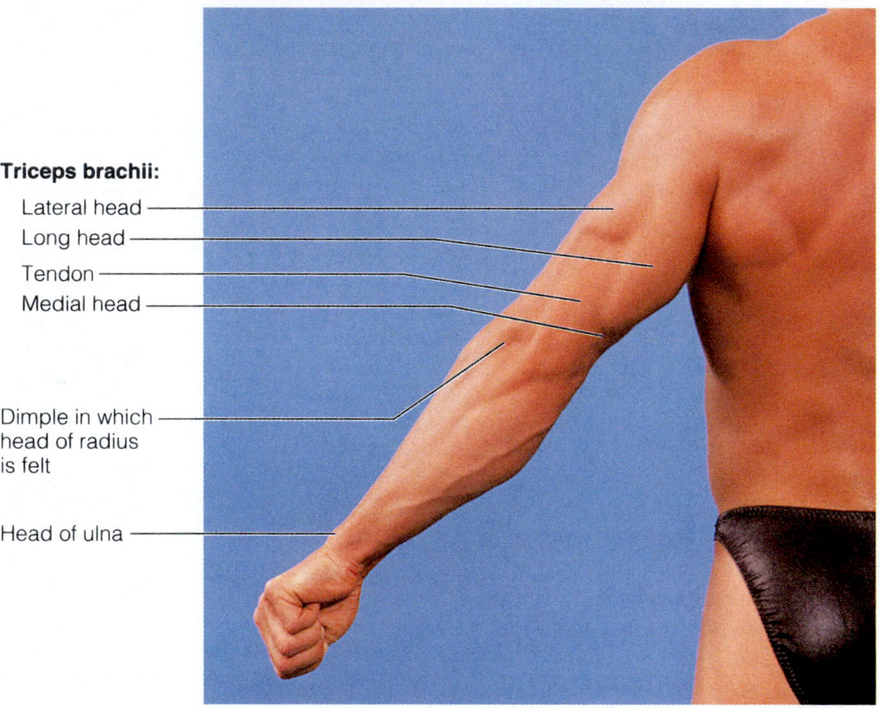

Figure 46.12 Surface anatomy of the upper limb, posterior view.

3. Extend your forearm against resistance, and feel your *triceps brachii* muscle bulge in the posterior arm. All three heads of the triceps (lateral, long, and medial) are visible through the skin of a muscular person (Figure 46.12).

Elbow Region

1. In the distal part of your arm, near the elbow, palpate the two projections of the humerus, the *lateral* and *medial epicondyles* (Figure 46.11). Midway between the epicondyles, on the posterior side, feel the *olecranon process* of the ulna, which forms the point of the elbow.

2. Confirm that the two epicondyles and the olecranon all lie in the same horizontal line when the elbow is extended. If these three bony processes do not line up, the elbow is dislocated.

3. Now feel along the posterior surface of the medial epicondyle. You are palpating your ulnar nerve.

4. On the anterior surface of the elbow is a triangular depression called the **antecubital fossa** or **cubital fossa** (Figure 46.13). The triangle's superior *base* is formed by a horizontal line between the humeral epicondyles; its two inferior sides are defined by the *brachioradialis* and *pronator teres* muscles (Figure 46.13). Try to define these boundaries on your

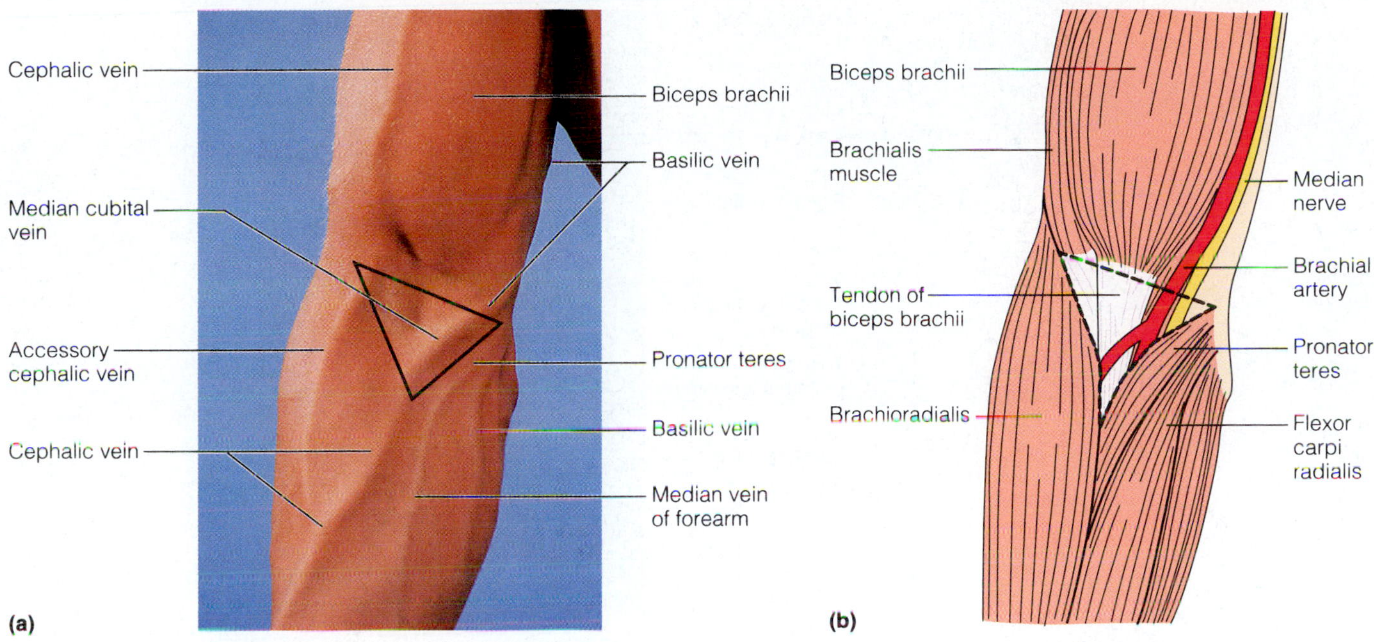

Figure 46.13 The antecubital (cubital) fossa on the anterior surface of the right elbow (outlined by the triangle). **(a)** Photograph. **(b)** Diagram of deeper structures in the fossa.

own limb. To find the brachioradialis muscle, flex your forearm against resistance, and watch this muscle bulge through the skin of your lateral forearm. To feel your pronator teres contract, palpate the antecubital fossa as you pronate your forearm against resistance. (Have your partner provide the resistance.)

Superficially, the antecubital fossa contains the *median cubital* vein (Figure 46.13a). Clinicians often draw blood from this superficial vein and insert intravenous (IV) catheters into it to administer medications, transfused blood, and nutrient fluids. The large *brachial artery* lies just deep to the median cubital vein (Figure 46.13b), so a needle must be inserted into the vein from a shallow angle (almost parallel to the skin) to avoid puncturing the artery. Other structures that lie deep in the fossa are also shown in Figure 46.13b.

5. The median cubital vein interconnects the larger cephalic and basilic veins of the upper limb. Recall from Exercise 32 that the *cephalic vein* ascends along the lateral side of the forearm and arm, whereas the *basilic vein* ascends through the limb's medial side. These veins are visible through the skin of lean people (Figure 46.13a). Examine your arm to see if your cephalic and basilic veins are visible.

Forearm and Hand

The two parallel bones of the forearm are the medial *ulna* and the lateral *radius*.

1. Feel the ulna along its entire length as a sharp ridge on the posterior forearm (confirm that this ridge runs inferiorly from the olecranon process). As for the radius, you can feel its distal half, but most of its proximal half is covered by muscle. You can, however, feel the rotating *head* of the radius. To do this, extend your forearm, and note that a dimple forms on the posterior lateral surface of the elbow region (Figure 46.12). Press three fingers into this dimple, and rotate your free hand as if you were turning a doorknob. You will feel the head of the radius rotate as you perform this action.

2. Both the radius and ulna have a knoblike *styloid process* at their distal ends. Figure 46.14 shows a way to locate these processes. Do not confuse the ulna's styloid process with the conspicuous *head of the ulna,* from which the styloid process stems. Confirm that the styloid process of the radius lies about 1 cm (0.4 inch) distal to that of the ulna.

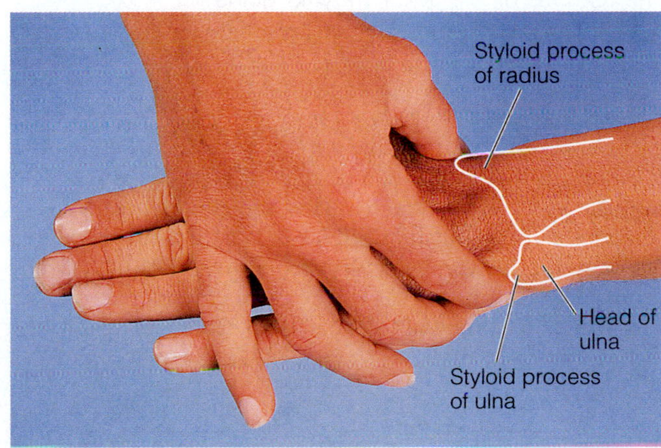

Figure 46.14 A way to locate the styloid processes of the ulna and radius. The right hand is palpating the left hand in this picture. Note that the head of the ulna is not the same as its styloid process. The styloid process of the radius lies about 1 cm distal to the styloid process of the ulna.

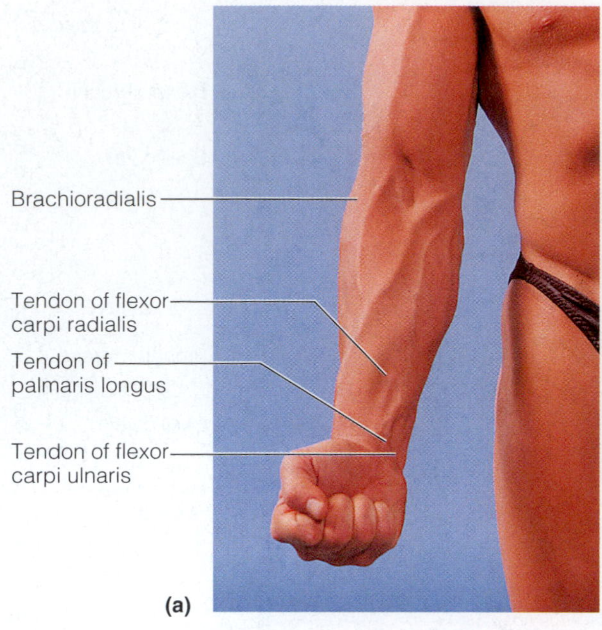

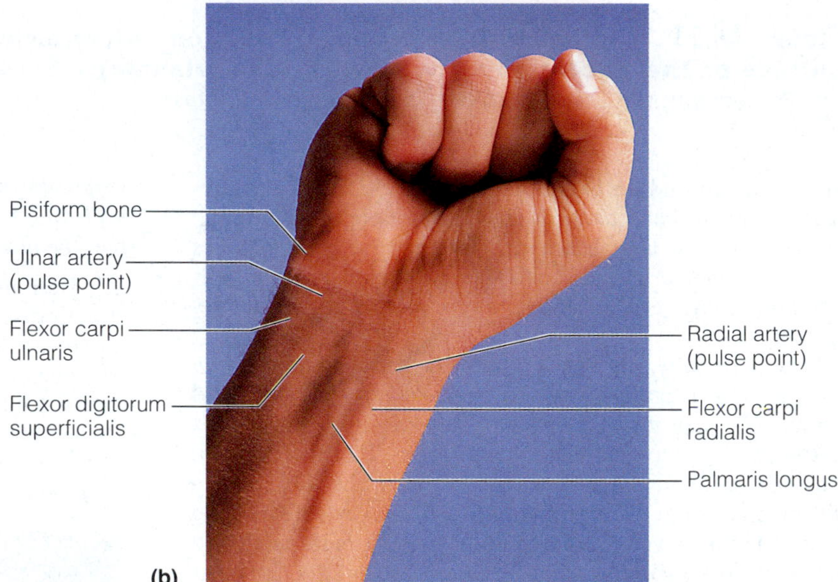

Figure 46.15 The anterior surface of the forearm and fist. **(a)** The entire forearm. **(b)** Enlarged view of the distal forearm and hand. The tendons of the flexor muscles guide the clinician to several sites for pulse taking.

Colles' fracture of the wrist is an impacted fracture in which the distal end of the radius is pushed proximally into the shaft of the radius. This sometimes occurs when someone falls on outstretched hands, and it most often happens to elderly women with osteoporosis. Colles' fracture bends the wrist into curves that resemble those on a fork. ■ Can you deduce how physicians use palpation to diagnose a Colles' fracture?

___

___

3. Next, feel the major groups of muscles within your forearm. Flex your hand and fingers against resistance, and feel the anterior *flexor muscles* contract. Then extend your hand at the wrist, and feel the tightening of the posterior *extensor muscles*.

4. Near the wrist, the anterior surface of the forearm reveals many significant features (Figure 46.15). Flex your fist against resistance; the tendons of the main wrist flexors will bulge the skin of the distal forearm. The tendons of the *flexor carpi radialis* and *palmaris longus* muscles are most obvious. (The palmaris longus, however, is absent from at least one arm in 30% of all people, so your forearm may exhibit just one prominent tendon instead of two.) The *radial artery* lies just lateral to (on the thumb side of) the flexor carpi radialis tendon, where the pulse is easily detected (Figure 46.15b). Feel your radial pulse here. The *median nerve* (which innervates the thumb) lies deep to the palmaris longus tendon. Finally, the *ulnar artery* lies on the medial side of the forearm, just lateral to the tendon of the *flexor carpi ulnaris*. Using Figure 46.15b as a guide, locate and feel your ulnar arterial pulse.

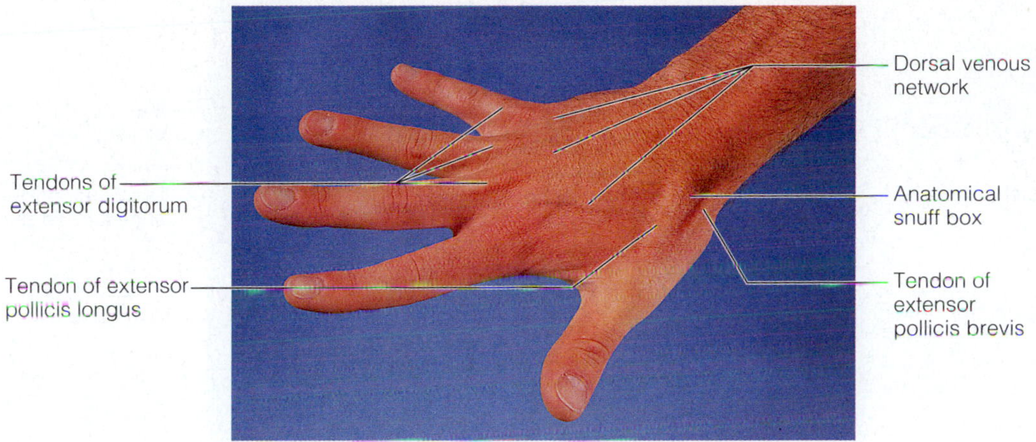

Figure 46.16 The dorsum of the hand. Note especially the anatomical snuff box and dorsal venous network.

5. Extend your thumb and point it posteriorly to form a triangular depression in the base of the thumb on the back of your hand. This is the **anatomical snuff box** (Figure 46.16). Its two elevated borders are defined by the tendons of the thumb extensor muscles, *extensor pollicis brevis* and *extensor pollicis longus.* The radial artery runs within the snuff box, so this is another site for taking a radial pulse. The main bone on the floor of the snuff box is the scaphoid bone of the wrist, but the styloid process of the radius is also present here. (If displaced by a bone fracture, the radial styloid process will be felt outside of the snuff box rather than within it.) The "snuff box" took its name from the fact that people once put snuff (tobacco for sniffing) in this hollow before lifting it up to the nose.

6. On the dorsum of your hand, observe the superficial veins just deep to the skin. This is the *dorsal venous network,* which drains superiorly into the cephalic vein. This venous network provides a site for drawing blood and inserting intravenous catheters and is preferred over the median cubital vein for these purposes. Next, extend your hand and fingers, and observe the tendons of the *extensor digitorum* muscle.

7. The anterior surface of the hand also contains some features of interest (Figure 46.17). These features include the *epidermal ridges* ("fingerprints") and many *flexion creases* in the skin. Grasp your *thenar eminence* (the bulge on the palm that contains the thumb muscles) and your *hypothenar eminence* (the bulge on the medial palm that contains muscles that move the little finger). ■

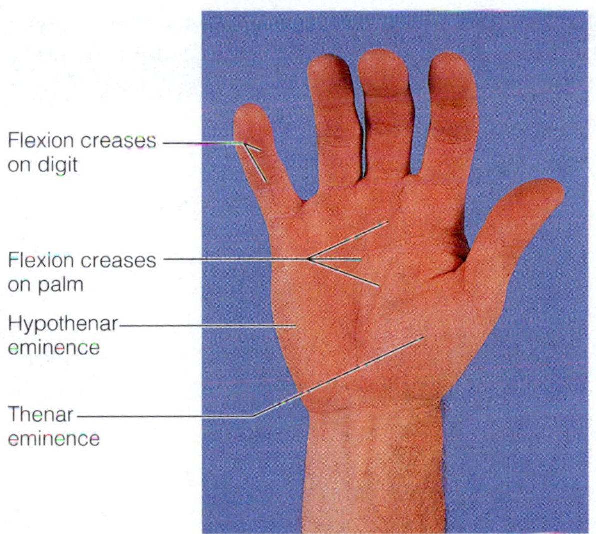

Figure 46.17 The palmar surface of the hand.

Activity 6: Palpating Landmarks of the Lower Limb

Gluteal Region

Dominating the gluteal region are the two *prominences* ("cheeks") of the buttocks. These are formed by subcutaneous fat and by the thick *gluteus maximus* muscles. The midline groove between the two prominences is called the **natal cleft** (*natal* = rump) or *gluteal cleft.* The inferior margin of each prominence is the horizontal **gluteal fold,** which roughly corresponds to the inferior margin of the gluteus maximus.

1. Try to palpate your *ischial tuberosity* just above the medial side of each gluteal fold (it will be easier to feel if you sit down or flex your thigh first). The ischial tuberosities are the robust inferior parts of the ischial bones, and they support the body's weight during sitting.

2. Next, palpate the *greater trochanter* of the femur on the lateral side of your hip (Figures 46.18 and 46.20). This trochanter lies just anterior to a hollow and about 10 cm (one hand's breadth) inferior to the iliac crest. To confirm that you have found the greater trochanter, alternately flex and extend your thigh. Because this trochanter is the most superior point on the lateral femur, it moves with the femur as you perform this movement.

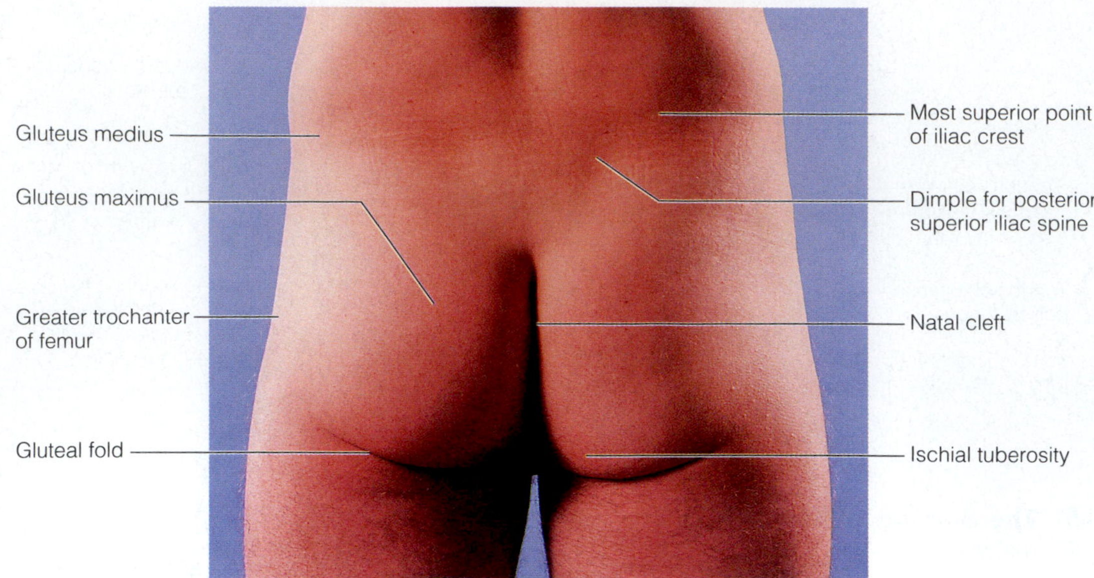

Figure 46.18 The gluteal region. The region extends from the iliac crests superiorly to the gluteal folds inferiorly. Therefore, it includes more than just the prominences of the buttock.

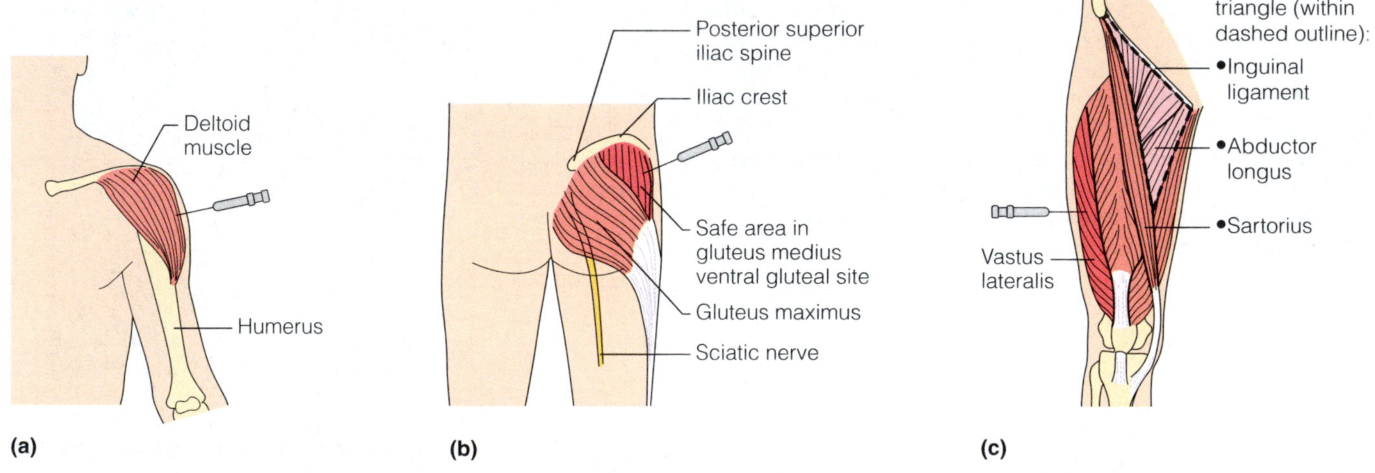

Figure 46.19 Three major sites of intramuscular injections. **(a)** Deltoid muscle of the arm. **(b)** Ventral gluteal site (gluteus medius). **(c)** Vastus lateralis in the lateral thigh. The femoral triangle is also shown.

3. To palpate the sharp *posterior superior iliac spine* (Figure 46.18), locate your iliac crests again, and trace each to its most posterior point. You may have difficulty feeling this spine, but it is indicated by a distinct dimple in the skin that is easy to find. This dimple lies two to three fingers' breadths lateral to the midline of the back. The dimple also indicates the position of the *sacroiliac joint*, where the hip bone attaches to the sacrum of the spinal column. (You can check *your* "dimples" out in the privacy of your home.)

The gluteal region is a major site for administering intramuscular injections. When giving such injections, extreme care must be taken to avoid piercing a major nerve that lies just deep to the gluteus maximus muscle. Can you guess what nerve this is?

It is the thick *sciatic nerve*, which innervates much of the lower limb. Furthermore, the needle must avoid the gluteal nerves and gluteal blood vessels, which also lie deep to the gluteus maximus.

To avoid harming these structures, the injections are most often applied to the gluteus *medius* (not maximus) muscle superior to the cheeks of the buttocks, in a safe area called the **ventral gluteal site** (Figure 46.19b). To locate this site, mentally draw a line laterally from the posterior superior iliac spine (dimple) to the greater trochanter; the injection would be given 5 cm (2 inches) superior to the midpoint of that line. Another safe way to locate the ventral gluteal site is to approach the lateral side of the patient's left hip with your extended right hand (or the right hip with your left hand). Then,

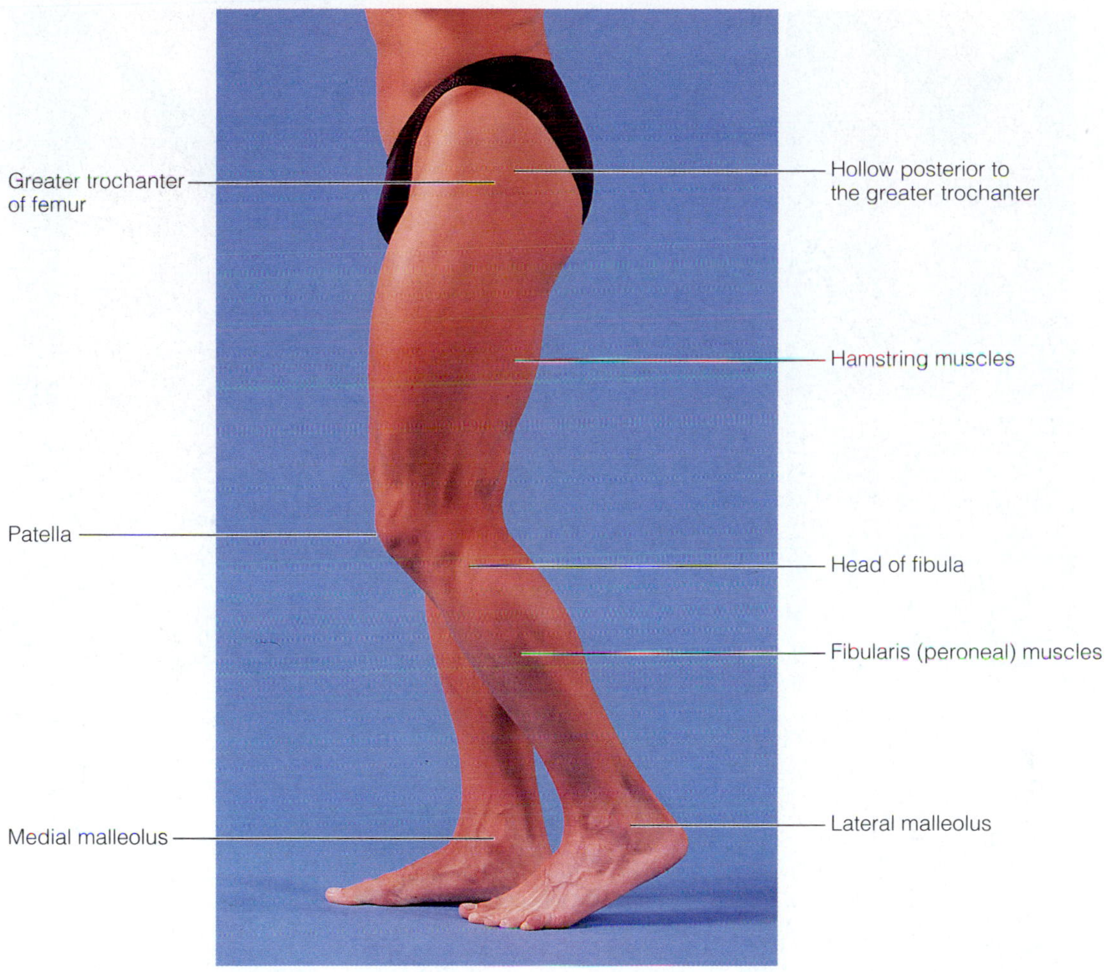

Figure 46.20 Lateral surface of the lower limb.

place your thumb on the anterior superior iliac spine and your index finger as far posteriorly on the iliac crest as it can reach. The heel of your hand comes to lie on the greater trochanter, and the needle is inserted in the angle of the V formed between your thumb and index finger about 4 cm (1.5 inches) inferior to the iliac crest.

Gluteal injections are not given to small children because their "safe area" is too small to locate with certainty and because the gluteal muscles are thin at this age. Instead, infants and toddlers receive intramuscular shots in the prominent vastus lateralis muscle of the thigh.

Thigh

The thigh is pictured in Figures 46.20, 46.21, and 46.22. Much of the femur is clothed by thick muscles, so the thigh has few palpable bony landmarks.

1. Distally, feel the *medial* and *lateral condyles of the femur* and the *patella* anterior to the condyles (see Figure 46.21c and a).

2. Next, palpate your three groups of thigh muscles (Figures 46.20, 46.21a, and 46.21b)—the *quadriceps femoris muscles* anteriorly, the *adductor muscles* medially, and the *hamstrings* posteriorly. The *vastus lateralis,* the lateral muscle of the quadriceps group, is a site for intramuscular injections. Such injections are administered about halfway down the length of this muscle (see Figure 46.19c).

3. The anterosuperior surface of the thigh exhibits a three-sided depression called the **femoral triangle** (Figure 46.21a). As shown in Figure 46.19c, the superior border of this triangle is formed by the inguinal ligament, and its two inferior borders are defined by the *sartorius* and *adductor longus* muscles. The large *femoral artery* and *vein* descend vertically through the center of the femoral triangle. To feel the pulse of your femoral artery, press inward just inferior to your midinguinal point (halfway between the anterior superior iliac spine and the pubic tubercle). Be sure to push hard, because the artery lies somewhat deep. By pressing very hard on this point, one can stop the bleeding from a hemorrhage in the lower limb. The femoral triangle also contains most of the *inguinal lymph nodes* (which are easily palpated if swollen).

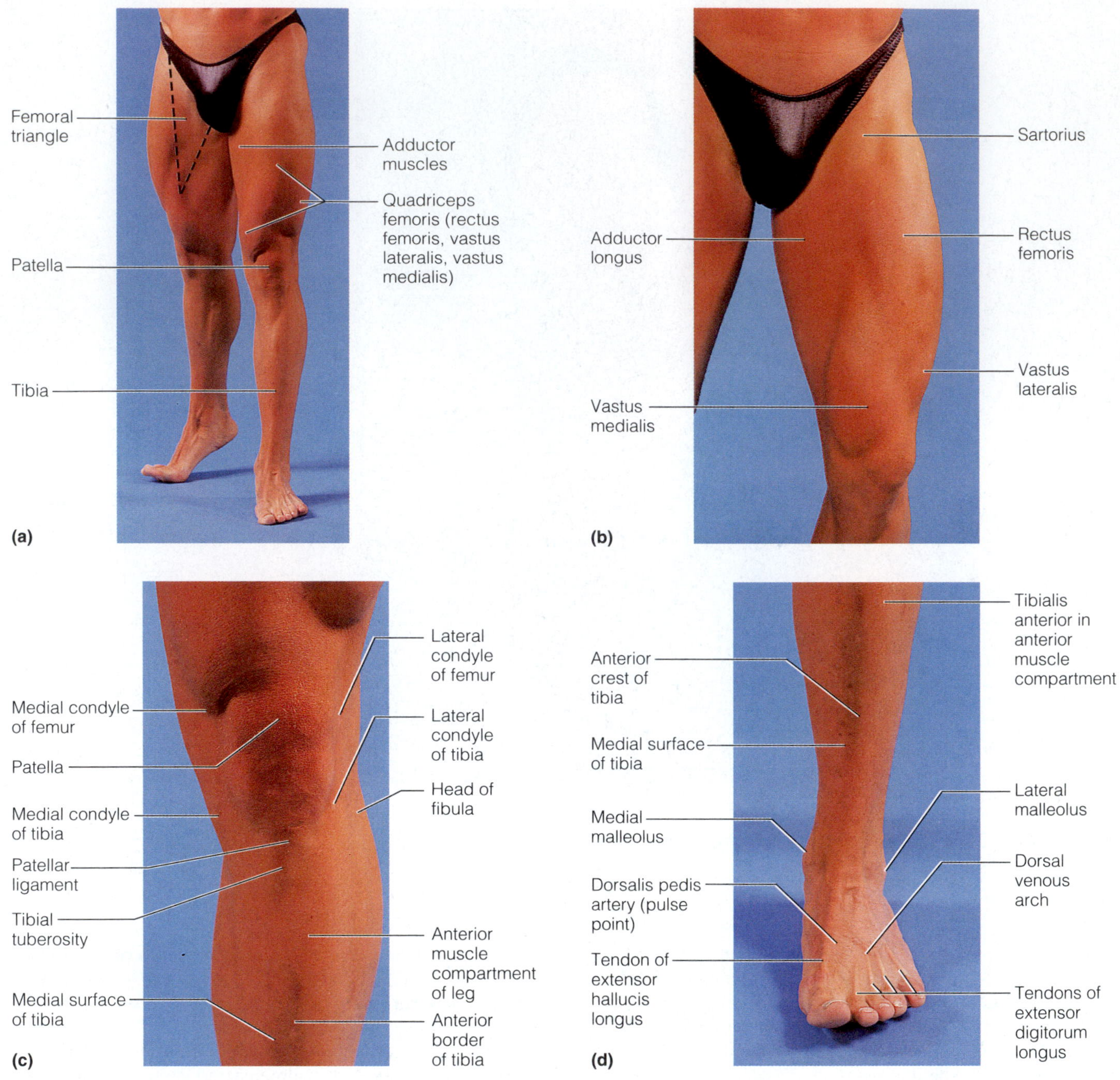

Figure 46.21 Anterior surface of the lower limb. (a) Both limbs, with the right limb revealing its medial aspect. The femoral triangle is outlined on the right limb. **(b)** Enlarged view of the left thigh. **(c)** The left knee region. **(d)** The dorsum of the left foot.

Leg and Foot

1. Locate your patella again, then follow the thick *patellar ligament* inferiorly from the patella to its insertion on the superior tibia (Figure 46.21c). Here you can feel a rough projection, the *tibial tuberosity.* Continue running your fingers inferiorly along the tibia's sharp *anterior crest* and its flat *medial surface*—bony landmarks that lie very near the surface throughout their length.

2. Now, return to the superior part of your leg, and palpate the expanded *lateral* and *medial condyles of the tibia* just inferior to the knee. (You can distinguish the tibial condyles from the femoral condyles because you can feel the tibial condyles move with the tibia during knee flexion.) Feel the bulbous *head of the fibula* in the superolateral region of the leg (Figures 46.20 and 46.21c). Try to feel the *common fibular nerve* (nerve to the anterior leg and foot) where it wraps around the fibula's *neck* just inferior to its head. This nerve is often bumped against the bone here and damaged.

3. In the most distal part of the leg, feel the *lateral malleolus* of the fibula as the lateral prominence of the ankle (Figure 46.21d). Notice that this lies slightly inferior to the *medial*

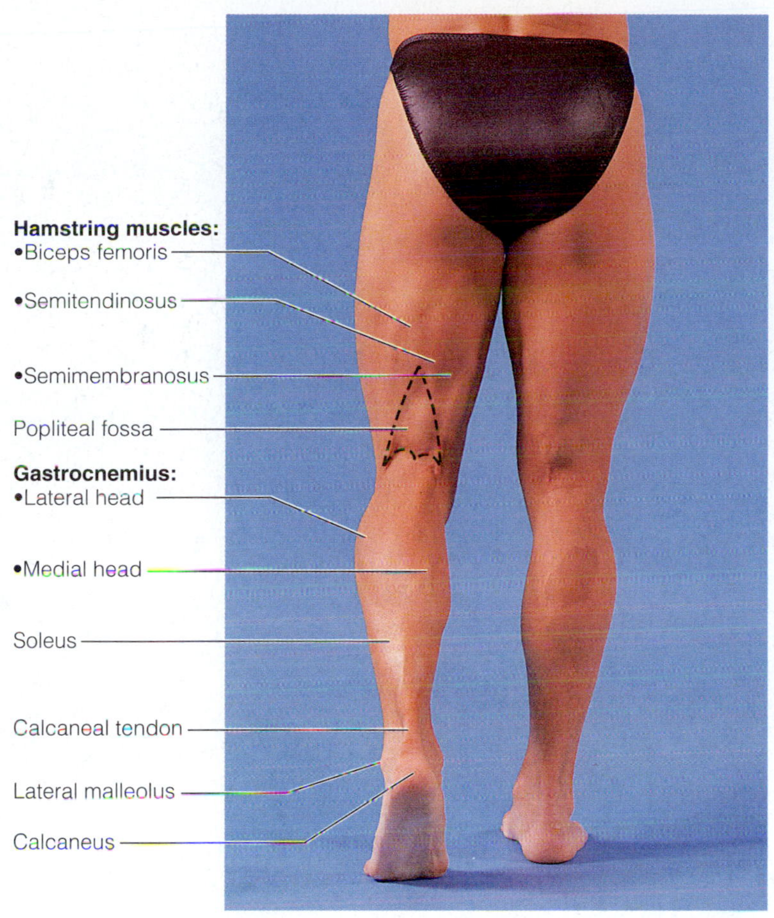

Figure 46.22 Posterior surface of the lower limb. Notice the diamond-shaped popliteal fossa posterior to the knee.

malleolus of the tibia, which forms the ankle's medial prominence. Place your finger just posterior to the medial malleolus to feel the pulse of your *posterior tibial artery.*

4. On the posterior aspect of the knee is a diamond-shaped hollow called the **popliteal fossa** (Figure 46.22). Palpate the large muscles that define the four borders of this fossa: The *biceps femoris* forming the superolateral border, the *semitendinosus* and *semimembranosus* defining the superomedial border, and the two heads of the *gastrocnemius* forming the inferior border. The *popliteal artery* and *vein* (main vessels to the leg) lie deep within this fossa. To feel a popliteal pulse, flex your leg at the knee and push your fingers firmly into the popliteal fossa. If a physician is unable to feel a patient's popliteal pulse, the femoral artery may be narrowed by atherosclerosis.

5. Next, palpate the main muscle groups of your leg, starting with the calf muscles posteriorly (Figure 46.22). Standing on tiptoes will help you feel the *lateral* and *medial heads of the gastrocnemius* and, inferior to these, the broad *soleus* muscle. Also feel the tension in your *calcaneal (Achilles) tendon* and at the point of insertion of this tendon onto the calcaneus bone of the foot.

6. Return to the anterior surface of the leg, and palpate the *anterior muscle compartment* (Figure 46.21c and d) while alternately dorsiflexing then plantar flexing your foot. You will feel the tibialis anterior and extensor digitorum muscles contracting then relaxing. Then, palpate the *fibularis (peroneal) muscles* that cover most of the fibula laterally (Figure 46.20). The tendons of these muscles pass posterior to the lateral malleolus and can be felt at a point posterior and slightly superior to that malleolus.

7. Observe the dorsum (superior surface) of your foot. You may see the superficial *dorsal venous arch* overlying the proximal part of the metatarsal bones (Figure 46.21d). This arch gives rise to both saphenous veins (the main superficial veins of the lower limb). Visible in lean people, the *great saphenous vein* ascends along the medial side of the entire limb (see Figure 32.8, p. 325). The *small saphenous vein* ascends through the center of the calf.

As you extend your toes, observe the tendons of the *extensor digitorum longus* and *extensor hallucis longus* muscles on the dorsum of the foot. Finally, place a finger on the extreme proximal part of the space between the first and second metatarsal bones. Here you should be able to feel the pulse of the *dorsalis pedis artery.* ■

NAME ______________________ LAB TIME/DATE ______________________

REVIEW SHEET
exercise
1

The Language of Anatomy

Surface Anatomy

1. Match each of the following descriptions with a key equivalent, and record the key letter or term in front of the description.

Key: a. buccal
b. calcaneal
c. cephalic
d. digital
e. patellar
f. scapular

__________ 1. cheek

__________ 2. pertaining to the fingers

__________ 3. shoulder blade region

__________ 4. anterior aspect of knee

__________ 5. heel of foot

__________ 6. pertaining to the head

2. Indicate the following body areas on the accompanying diagram by placing the correct key letter at the end of each line.

Key:

a. abdominal
b. antecubital
c. axillary
d. brachial
e. cervical
f. femoral
g. gluteal
h. inguinal
i. lumbar
j. occipital
k. oral
l. popliteal
m. pubic
n. sural
o. thoracic
p. umbilical

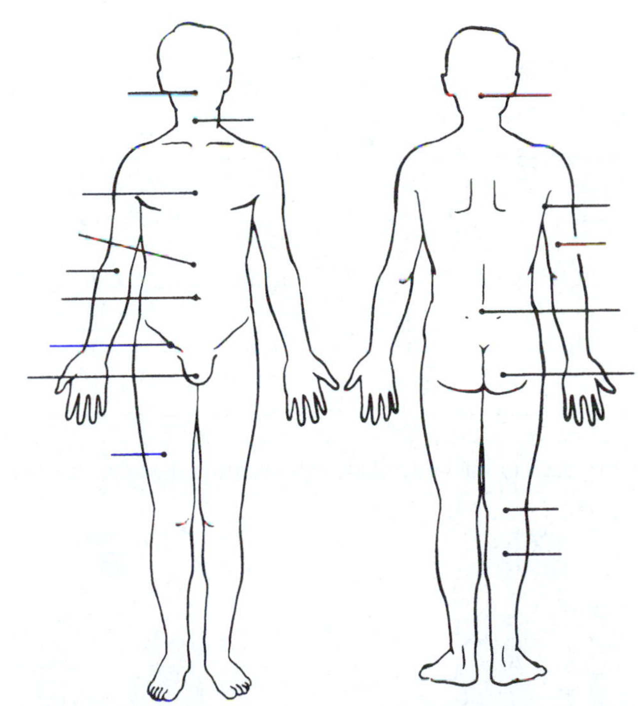

3. Classify each of the surface anatomy terms in the key of question 2 above, into one of the body regions indicated below. Insert the appropriate key letters on the answer blanks.

__________ 1. Appendicular

__________ 2. Axial

Body Orientation, Direction, Planes, and Sections

1. Describe completely the standard human anatomical position. ______________________________

____

2. Define *section:* ______________________________

3. Several incomplete statements are listed below. Correctly complete each statement by choosing the appropriate anatomical term from the key. Record the key letters and/or terms on the correspondingly numbered blanks below.

Key:		
a. anterior	e. lateral	i. sagittal
b. distal	f. medial	j. superior
c. frontal	g. posterior	k. transverse
d. inferior	h. proximal	

In the anatomical position, the face and palms are on the __1__ body surface; the buttocks and shoulder blades are on the __2__ body surface; and the top of the head is the most __3__ part of the body. The ears are __4__ and __5__ to the shoulders and __6__ to the nose. The heart is __7__ to the vertebral column (spine) and __8__ to the lungs. The elbow is __9__ to the fingers but __10__ to the shoulder. The abdominopelvic cavity is __11__ to the thoracic cavity and __12__ to the spinal cavity. In humans, the dorsal surface can also be called the __13__ surface; however, in quadruped animals, the dorsal surface is the __14__ surface.

If an incision cuts the heart into right and left parts, the section is a __15__ section; but if the heart is cut so that superior and inferior portions result, the section is a __16__ section. You are told to cut a dissection animal along two planes so that the kidneys are observable in both sections. The two sections that meet this requirement are the __17__ and __18__ sections.

1. ______________ 7. ______________ 13. ______________

2. ______________ 8. ______________ 14. ______________

3. ______________ 9. ______________ 15. ______________

4. ______________ 10. ______________ 16. ______________

5. ______________ 11. ______________ 17. ______________

6. ______________ 12. ______________ 18. ______________

4. Correctly identify each of the body planes by inserting the appropriate term for each on the answer line below the drawing.

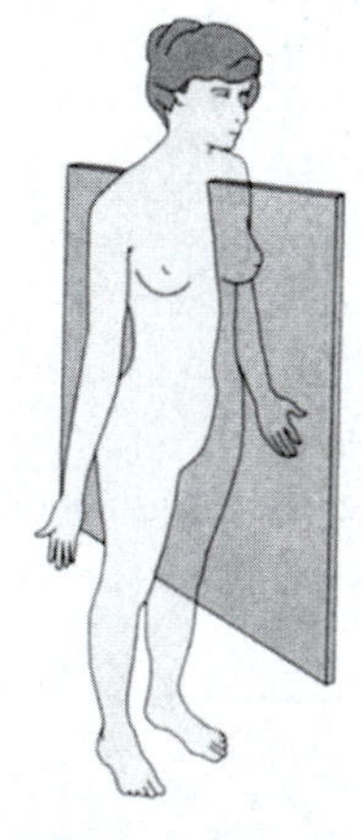

(a) ______________

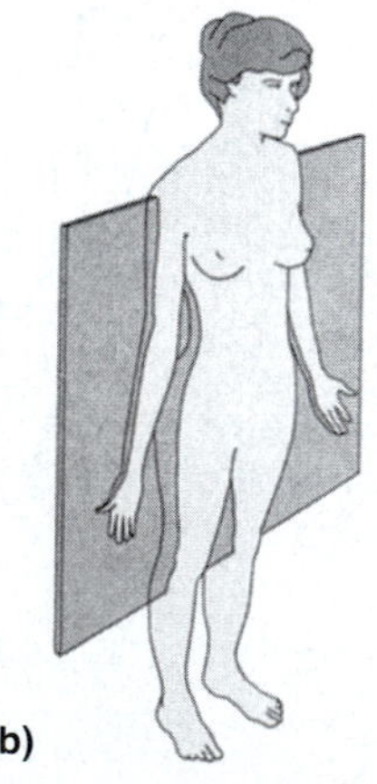

(b) ______________

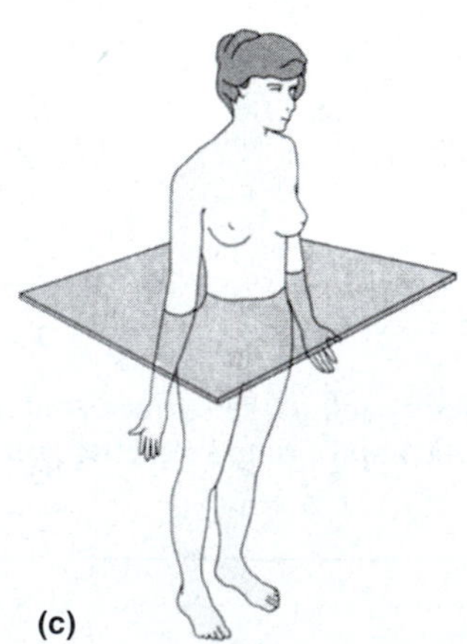

(c) ______________

5. Draw a kidney as it would appear sectioned in three different planes.

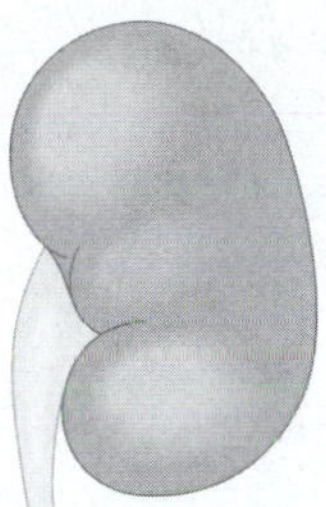

6. Correctly identify each of the nine areas of the abdominal surface by inserting the appropriate term for each of the letters indicated in the drawing.

a. ______________________

b. ______________________

c. ______________________

d. ______________________

e. ______________________

f. ______________________

g. ______________________

h. ______________________

i. ______________________

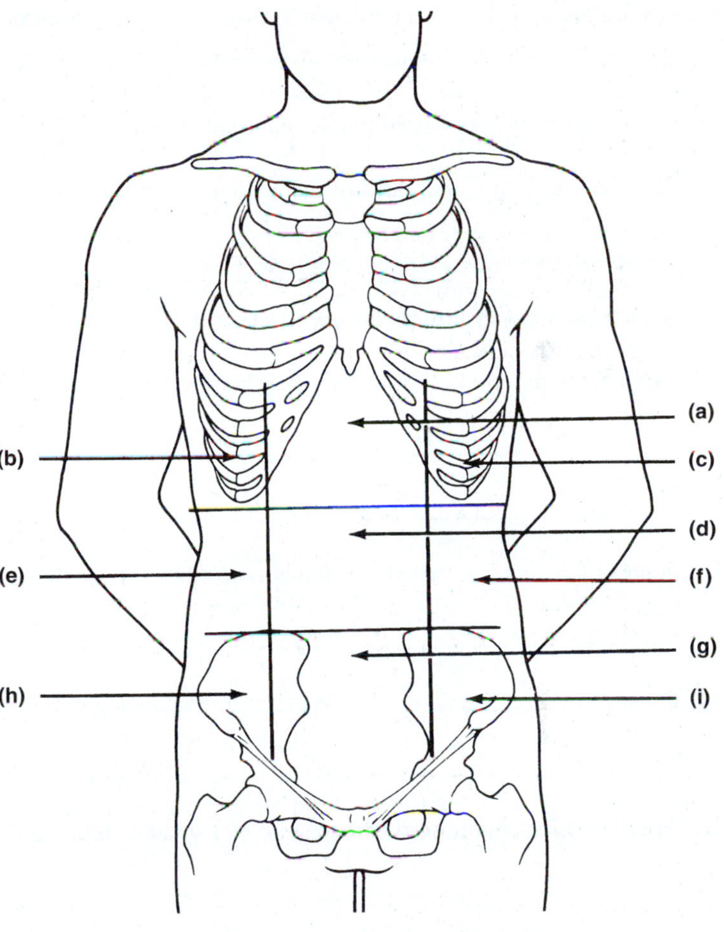

Body Cavities

1. Which body cavity would have to be opened for the following types of surgery? (Insert letter of key choice in same-numbered blank. More than one choice may apply.)

Key: a. abdominopelvic   c. dorsal   e. thoracic
b. cranial   d. spinal   f. ventral

1. surgery to remove a cancerous lung lobe
2. removal of the uterus or womb
3. removal of a brain tumor
4. appendectomy
5. stomach ulcer operation

The abdominopelvic and thoracic cavities are subdivisions of the __6__ body cavity, and the cranial and spinal cavities are subdivisions of the __7__ body cavity. The __8__ body cavity is totally surrounded by bone, and thus affords its contained structures very good protection.

1. __________
2. __________
3. __________
4. __________
5. __________
6. __________
7. __________
8. __________

2. a. The serous membranes covering the lungs are the ______________________ and ______________________.

b. The serous membranes covering the heart are the ______________________ and ______________________.

c. The serous membranes covering the organs of the abdominopelvic cavity are the ______________________ and ______________________.

3. Name the muscle that subdivides the ventral body cavity. ______________________

4. Which of the following organ systems are represented in all three subdivisions of the ventral body cavity? (Circle all appropriate responses.)

respiratory	circulatory	reproductive	lymphatic
nervous	excretory (urinary)	muscular	integumentary

5. Which organ system would not be represented in any of the body cavities? ______________________

6. What are the bony landmarks of the abdominopelvic cavity? ______________________

____

7. Which body cavity affords the least protection to its internal structures? ______________________

8. What is the function of the serous membranes of the body? ______________________

____

____

9. A nurse informs you that she is about to take blood from the antecubital region. What portion of your body should you present to her? ______________________

10. What do peritonitis, pleurisy, and pericarditis (pathological conditions) have in common? ______________________

____

11. Why are these conditions accompanied by a great deal of pain? ______________________

____

12. The mouth, or buccal cavity, and its extension, which stretches through the body to the anus, is not listed as an internal body cavity. Why is this so? ______________________

____

____

NAME ______________________ LAB TIME/DATE ______________________

REVIEW SHEET
exercise
2

Organ Systems Overview

1. Use the key below to indicate the body systems that perform the following functions for the body.

Key: a. cardiovascular
b. digestive
c. endocrine
d. integumentary
e. lymphatic/immunity
f. muscular
g. nervous
h. reproductive
i. respiratory
j. skeletal
k. urinary

______________ 1. rids the body of nitrogen-containing wastes

______________ 2. is affected by removal of the thyroid gland

______________ 3. provides support and levers on which the muscular system acts

______________ 4. includes the heart

______________ 5. causes the onset of the menstrual cycle

______________ 6. protects underlying organs from drying out and from mechanical damage

______________ 7. protects the body; destroys bacteria and tumor cells

______________ 8. breaks down ingested food into its building blocks

______________ 9. removes carbon dioxide from the blood

______________ 10. delivers oxygen and nutrients to the tissues

______________ 11. moves the limbs; facilitates facial expression

______________ 12. conserves body water or eliminates excesses

______________ and ______________ 13. facilitate conception and childbearing

______________ 14. controls the body by means of chemical molecules called hormones

______________ 15. is damaged when you cut your finger or get a severe sunburn

2. Using the above key, choose the *organ system* to which each of the following sets of organs or body structures belong:

______________ 1. thymus, spleen, lymphatic vessels

______________ 2. bones, cartilages, tendons

______________ 3. pancreas, pituitary, adrenals

______________ 4. trachea, bronchi, alveoli

______________ 5. kidneys, bladder, ureters

______________ 6. testis, vas deferens, urethra

______________ 7. esophagus, large intestine, rectum

______________ 8. arteries, veins, heart

3. Using the key below, place the following organs in their proper body cavity.

Key: a. abdominopelvic   b. cranial   c. spinal   d. thoracic

_______ 1. stomach   _______ 5. spinal cord   _______ 9. brain

_______ 2. esophagus   _______ 6. urinary bladder   _______ 10. rectum

_______ 3. large intestine   _______ 7. heart

_______ 4. liver   _______ 8. trachea

4. Using the organs listed in item 3 above, record, by number, which would be found in the abdominal regions listed below:

_____________ 1. hypogastric region   _____________ 4. epigastric region

_____________ 2. right lumbar region   _____________ 5. left iliac region

_____________ 3. umbilical region   _____________ 6. left hypochondriac region

5. The levels of organization of a living body are chemical, ____________________, ____________________,

____________________, ____________________, and organism.

6. Define *organ:* ____

____

____

7. Using the terms provided, correctly identify all of the body organs provided with leader lines in the drawings shown below. Then name the organ systems by entering the name of each on the answer blank below each drawing.

Blood vessels	Heart	Nerves	Spinal cord	Urethra
Brain	Kidney	Sensory receptor	Ureter	Urinary bladder

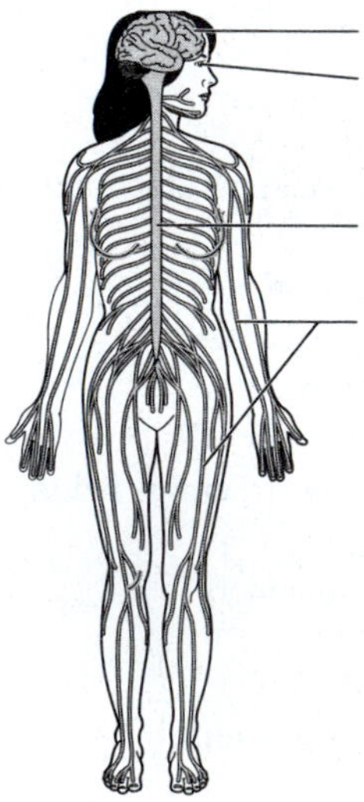

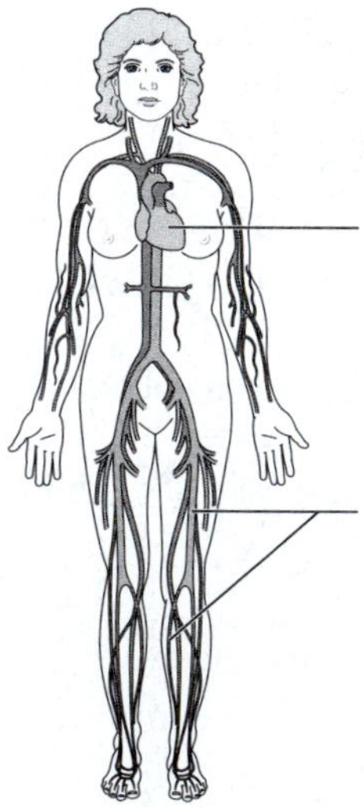

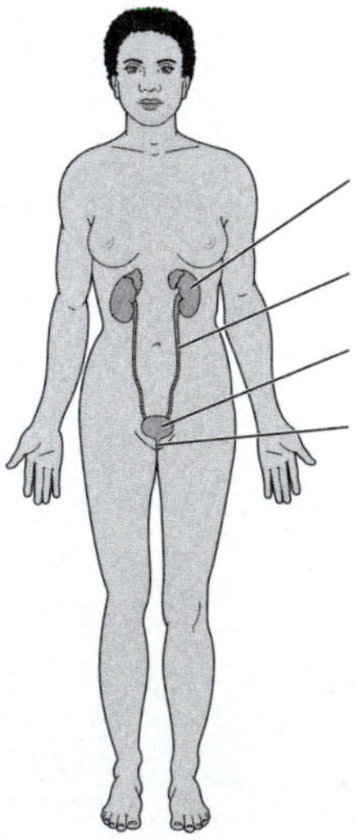

1. ____________________   2. ____________________   3. ____________________

8. Why is it helpful to study the external and internal structures of the rat? ____________________

____

NAME ______________________ LAB TIME/DATE ______________________

REVIEW SHEET
exercise
3

The Microscope

Care and Structure of the Compound Microscope

1. Label all indicated parts of the microscope.

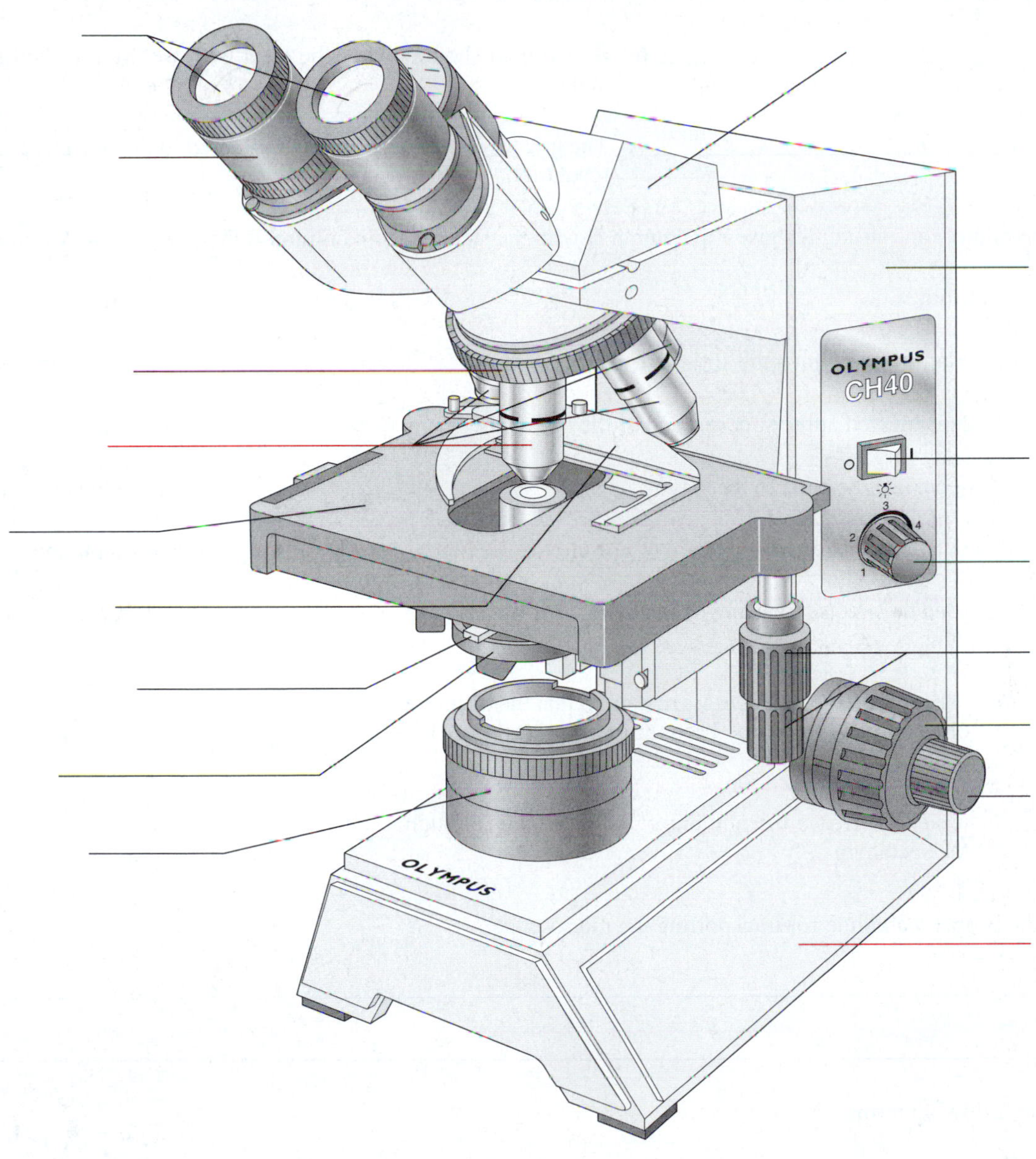

2. The following statements are true or false. If true, write *T* on the answer blank. If false, correct the statement by writing on the blank the proper word or phrase to replace the one that is underlined.

________ 1. The microscope lens may be cleaned with any soft tissue.

________ 2. The coarse adjustment knob may be used in focusing with all objective lenses.

________ 3. The microscope should be stored with the oil immersion lens in position over the stage.

________ 4. When beginning to focus, the lowest power lens should be used.

________ 5. When focusing, always focus toward the specimen.

________ 6. A coverslip should always be used with wet mounts and the high-power and oil lenses.

________ 7. The greater the amount of light delivered to the objective lens, the less the resolution.

3. Match the microscope structures given in column B with the statements in column A that identify or describe them.

Column A	Column B
____ 1. platform on which the slide rests for viewing	a. coarse adjustment knob
____ 2. lens located at the superior end of the body tube	b. condenser
____ 3. secure(s) the slide to the stage	c. fine adjustment knob
____ 4. delivers a concentrated beam of light to the specimen	d. iris diaphragm
____ 5. used for precise focusing once initial focusing has been done	e. mechanical stage or spring clips
____ 6. carries the objective lenses; rotates so that the different objective lenses can be brought into position over the specimen	f. movable nosepiece
____ 7. used to increase the amount of light passing through the specimen	g. objective lenses
	h. ocular
	i. stage

4. Explain the proper technique for transporting the microscope.

____

____

5. Define the following terms.

real image: ____

____

virtual image: ____

____

6. Define *total magnification:* ______________________________

____

7. Define *resolution:* ______________________________

Viewing Objects Through the Microscope

1. Complete, or respond to, the following statements:

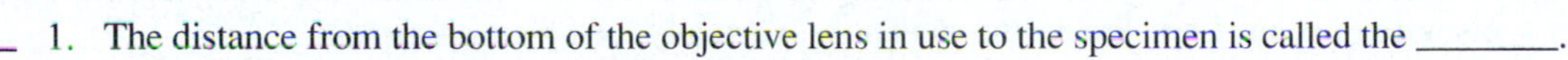

______________ 1. The distance from the bottom of the objective lens in use to the specimen is called the _______.

______________ 2. The resolution of the human eye is approximately _______ μm.

______________ 3. The area of the specimen seen when looking through the microscope is the _______.

______________ 4. If a microscope has a 10× ocular and the total magnification at a particular time is 950×, the objective lens in use at that time is _______ ×.

______________ 5. Why should the light be dimmed when looking at living (nearly transparent) cells?

______________ 6. If, after focusing in low power, only the fine adjustment need be used to focus the specimen at the higher powers, the microscope is said to be _______.

______________ 7. If, when using a 10× ocular and a 15× objective, the field size is 1.5 mm, the approximate field size with a 30× objective is _______ mm.

______________ 8. If the size of the high-power field is 1.2 mm, an object that occupies approximately a third of that field has an estimated diameter of _______ mm.

______________ 9. Assume there is an object on the left side of the field that you want to bring to the center (that is, toward the apparent right). In what direction would you move your slide?

______________ 10. If the object is in the top of the field and you want to move it downward to the center, you would move the slide _______.

2. You have been asked to prepare a slide with the letter *k* on it (as below). In the circle below, draw the *k* as seen in the low-power field.

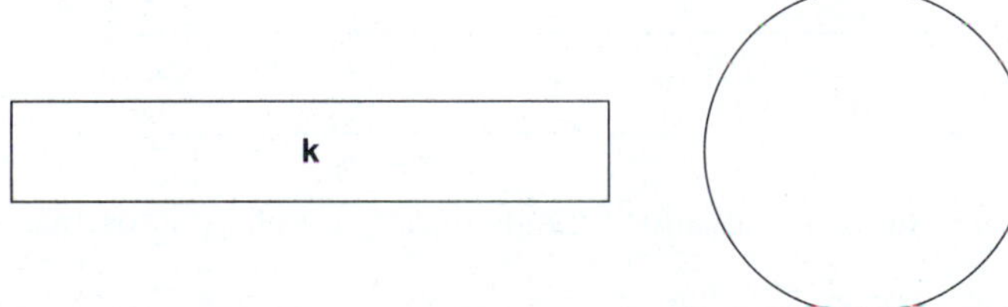

3. The numbers for the field sizes below are too large to represent the typical compound microscope lens system, but the relationships depicted are accurate. Figure out the magnification of fields 1 and 3, and the field size of 2. (*Hint:* Use your ruler.)

1.	2.	3.
5 mm	____ mm	0.5 mm
____ ×	100×	____ ×

4. Say you are observing an object in the low-power field. When you switch to high-power, it is no longer in your field of view.

Why might this occur? ______________________________

____

What should be done initially to prevent this from happening? ______________________

____

5. Do the following factors increase or decrease as one moves to higher magnifications with the microscope?

resolution ______________________ amount of light needed ______________________

working distance ______________________ depth of field ______________________

6. A student has the high-dry lens in position and appears to be intently observing the specimen. The instructor, noting a working distance of about 1 cm, knows the student isn't actually seeing the specimen.

How so? ______________________

7. Why is it important to be able to use your microscope to perceive depth when studying slides?

____

____

8. If you are observing a slide of tissue two cell layers thick, how can you determine which layer is superior?

____

____

____

9. Describe the proper procedure for preparing a wet mount.

____

____

____

10. Give two reasons why the light should be dimmed when viewing living or unstained material.

____

____

11. Indicate the probable cause of the following situations arising during use of a microscope.

a. Only half of the field is illuminated: ______________________

____

b. Field does not change as mechanical stage is moved: ______________________

____

12. Under what circumstances is the stereomicroscope used to study biological specimens? ______________________

____

NAME ____________________ LAB TIME/DATE ____________________

REVIEW SHEET

exercise

4

The Cell—Anatomy and Division

Anatomy of the Composite Cell

1. Define the following:

organelle: ____________________

____

cell: ____________________

2. Although cells have differences that reflect their specific functions in the body, what functions do they have in common?

____

3. Identify the following cell parts:

____________________ 1. external boundary of cell; regulates flow of materials into and out of the cell; site of cell signaling

____________________ 2. contains digestive enzymes of many varieties; "suicide sac" of the cell

____________________ 3. scattered throughout the cell; major site of ATP synthesis

____________________ 4. slender extensions of the plasma membrane that increase its surface area

____________________ 5. stored glycogen granules, crystals, pigments, and so on

____________________ 6. membranous system consisting of flattened sacs and vesicles; packages proteins for export

____________________ 7. control center of the cell; necessary for cell division and cell life

____________________ 8. two rod-shaped bodies near the nucleus; direct formation of the mitotic spindle

____________________ 9. dense, darkly staining nuclear body; packaging site for ribosomes

____________________ 10. contractile elements of the cytoskeleton

____________________ 11. membranous system; involved in intracellular transport of proteins and synthesis of membrane lipids

____________________ 12. attached to membrane systems or scattered in the cytoplasm; synthesize proteins

____________________ 13. threadlike structures in the nucleus; contain genetic material (DNA)

____________________ 14. site of free radical detoxification

4. In the following diagram, label all parts provided with a leader line.

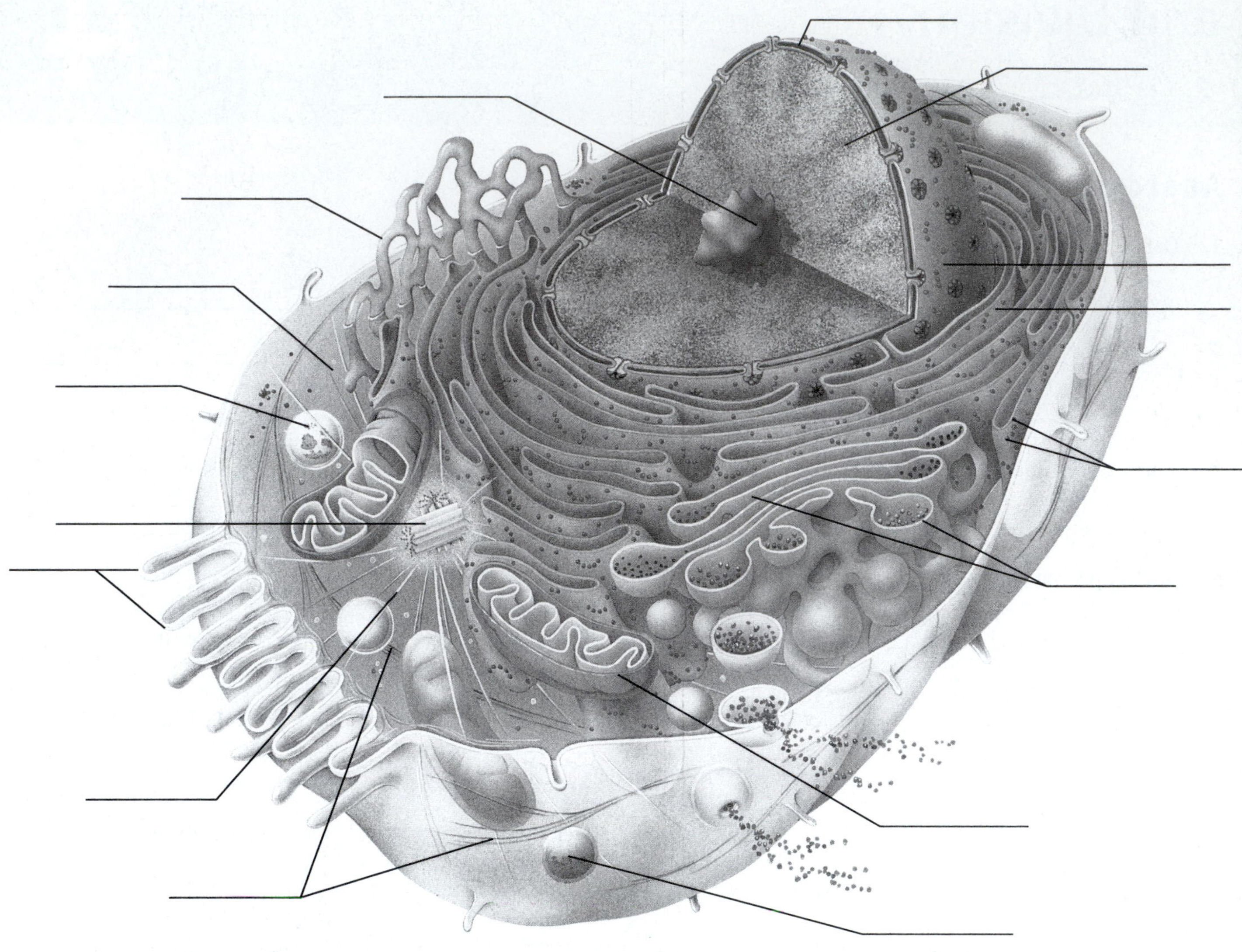

Differences and Similarities in Cell Structure

1. For each of the following cell types, list (a) *one* important structural characteristic observed in the laboratory, and (b) the function that the structure complements or ensures.

squamous epithelium a. ______________________________

b. ______________________________

sperm a. ______________________________

b. ______________________________

smooth muscle a. ______________________________

b. ______________________________

red blood cells a. ______________________________

b. ______________________________

2. Construct a concept map that separates these cells based on observable features. (Refer to Exercise 6A, p. 67, Constructing a Concept Map.)

3. What is the significance of the red blood cell being anucleate (without a nucleus)? ______________________________

____

Did it ever have a nucleus? ________ When? ______________________________

4. List the four cells observed from largest to smallest. ____________ ____________ ____________ ____________

Cell Division: Mitosis and Cytokinesis

1. Identify the three phases of mitosis in the following photomicrographs.

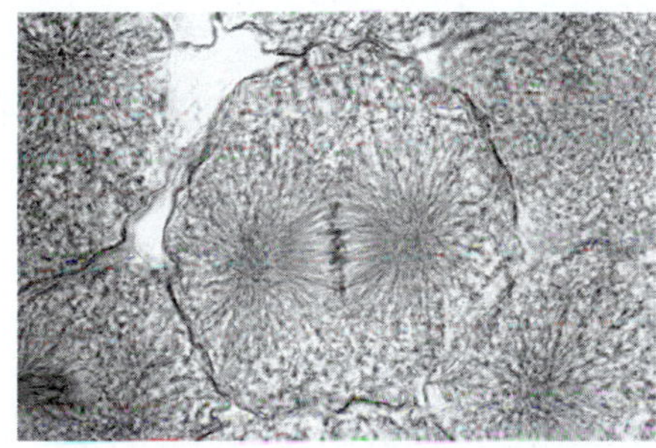

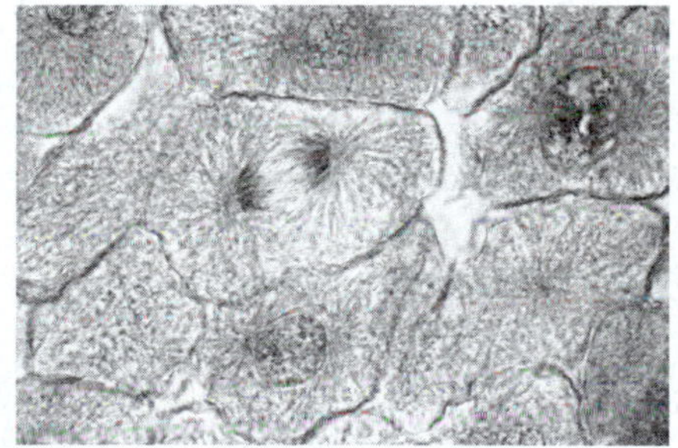

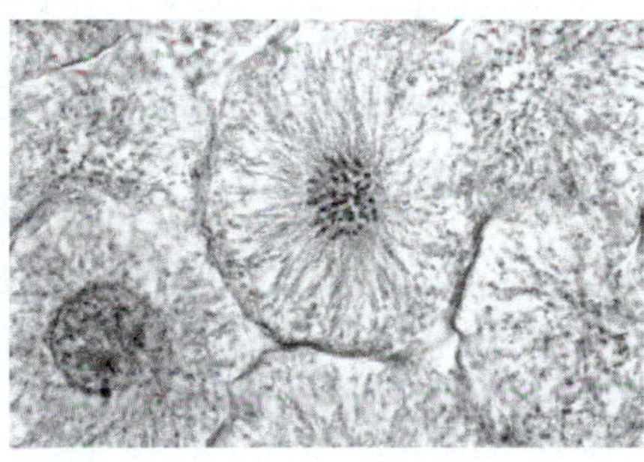

1. ____________________ 2. ____________________ 3. ____________________

2. What is the importance of mitotic cell division? ______________________________

____

3. Draw the phases of mitosis for a cell with a diploid (2N) number of 4.

4. Complete or respond to the following statements:

Division of the __1__ is referred to as mitosis. Cytokinesis is division of the __2__. The major structural difference between chromatin and chromosomes is that the latter is __3__. Chromosomes attach to the spindle fibers by undivided structures called __4__. If a cell undergoes mitosis but not cytokinesis, the product is __5__. The structure that acts as a scaffolding for chromosomal attachment and movement is called the __6__. __7__ is the period of cell life when the cell is not involved in division. Two cell populations in the body that do not undergo cell division are __8__ and __9__. The implication of an inability of a cell population to divide is that when some of its members die, they are replaced by __10__.

1. ____________
2. ____________
3. ____________
4. ____________
5. ____________
6. ____________
7. ____________
8. ____________
9. ____________
10. ____________

5. Using the key, categorize each of the events described below according to the phase in which it occurs.

Key: a. prophase b. anaphase c. telophase d. metaphase e. interphase

____________ 1. Chromatin coils and condenses, forming chromosomes.

____________ 2. The chromosomes are V-shaped.

____________ 3. The nuclear membrane re-forms.

____________ 4. Chromosomes stop moving toward the poles.

____________ 5. Chromosomes line up in the center of the cell.

____________ 6. The nuclear membrane fragments.

____________ 7. The mitotic spindle forms.

____________ 8. DNA synthesis occurs.

____________ 9. Centrioles replicate.

____________ 10. Chromosomes first appear to be duplex structures.

____________ 11. Chromosomal centromeres are attached to the kinetochore fibers.

____________ 12. Cleavage furrow forms.

____________, ____________ 13. The nuclear membrane(s) is absent.

6. What is the physical advantage of the chromatin coiling and condensing to form short chromosomes at the onset of mitosis?

____

____

NAME ______________________ LAB TIME/DATE ______________________

REVIEW SHEET
exercise
5A

The Cell—Transport Mechanisms and Permeability: Wet Lab

Choose all answers that apply to items 1 and 2, and place their letters on the response blanks to the right.

1. Molecular motion ______________________

a. reflects the kinetic energy of molecules
b. reflects the potential energy of molecules
c. is ordered and predictable
d. is random and erratic

2. Velocity of molecular movement ______________________

a. is higher in larger molecules
b. is lower in larger molecules
c. increases with increasing temperature
d. decreases with increasing temperature
e. reflects kinetic energy

3. The following refer to the laboratory experiment using dialysis sacs 1 through 4 to study diffusion through nonliving membranes:

Sac 1: 40% glucose suspended in distilled water

Did glucose pass out of the sac? ______________________

Test used to determine presence of glucose: ______________________

Did the sac weight change? ______________________

If so, explain the reason for its weight change: ______________________

____

____

Sac 2: 40% glucose suspended in 40% glucose

Was there net movement of glucose in either direction? ______________________

Explanation: ______________________

Did the sac weight change? ______________________ Explanation: ______________________

____

Sac 3: 10% NaCl in distilled water

Was there net movement of NaCl out of the sac? ______________________

Test used to determine the presence of NaCl: ______________________

____

Direction of net osmosis: ______________________

Sac 4: Boiled starch in distilled water

Was there net movement of starch out of the sac? ______________________________

Test used to determine presence of starch: ______________________________

Direction of net osmosis: ______________________________

4. What single characteristic of the differentially permeable membranes used in the laboratory determines the substances that can pass through them? ______________________________

In addition to this characteristic, what other factors influence the passage of substances through living membranes?

____

____

5. A semipermeable sac containing 4% NaCl, 9% glucose, and 10% albumin is suspended in a solution with the following composition: 10% NaCl, 10% glucose, and 40% albumin. Assume that the sac is permeable to all substances except albumin. State whether each of the following will (a) move into the sac, (b) move out of the sac, or (c) not move.

glucose ______________________________ albumin ______________________________

water ______________________________ NaCl ______________________________

6. The diagrams below represent three microscope fields containing red blood cells. Arrows show the direction of net osmosis.

Which field contains a hypertonic solution? _______ The cells in this field are said to be ______________________________.

Which field contains an isotonic bathing solution? _______ Which field contains a hypotonic solution? _______ What is happening to the cells in this field? ______________________________

____

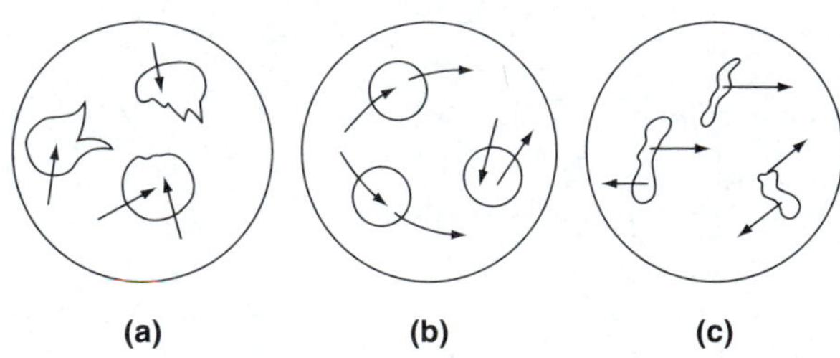

7. Assume you are conducting the experiment illustrated in the next figure. Both hydrochloric acid (HCl) with a molecular weight of about 36.5 and ammonium hydroxide (NH_4OH) with a molecular weight of 35 are volatile and easily enter the gaseous state. When they meet, the following reaction will occur:

$$HCl + NH_4OH \rightarrow H_2O + NH_4Cl$$

Ammonium chloride (NH_4Cl) will be deposited on the glass tubing as a smoky precipitate where the two gases meet. Predict which gas will diffuse more quickly and indicate to which end of the tube the smoky precipitate will be closer.

a. The faster diffusing gas is ______________________.

b. The precipitate forms closer to the ______________________ end.

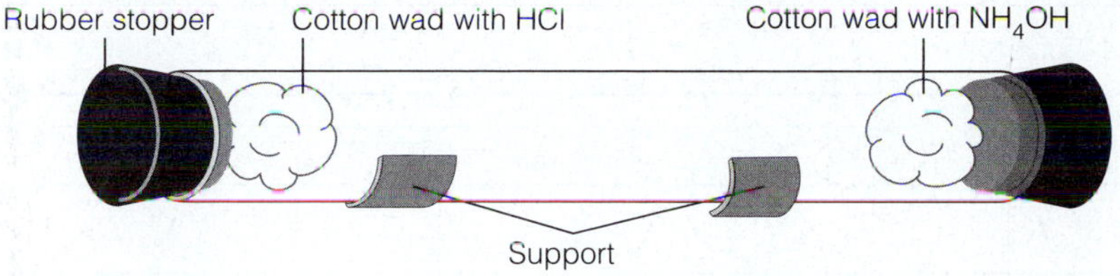

Rubber stopper Cotton wad with HCl Cotton wad with NH_4OH

Support

8. When food is pickled for human consumption, as much water as possible is removed from the food. What method is used to achieve this dehydrating effect? ____

____

9. What determines whether a transport process is active or passive? ____

____

10. Characterize membrane transport as fully as possible by choosing all the phrases that apply and inserting their letters on the answer blanks.

Passive processes: ______________ Active processes: ______________

a. account for the movement of fats and respiratory gases through the plasma membrane
b. explain solute pumping, phagocytosis, and pinocytosis
c. include osmosis, simple diffusion, and filtration
d. may occur against concentration and/or electrical gradients
e. use hydrostatic pressure or molecular energy as the driving force
f. move ions, amino acids, and some sugars across the plasma membrane

11. For the osmometer demonstration (#2), explain why the level of the water column rose during the laboratory session.

____

____

____

____

____

12. Define the following terms:

diffusion: ____

____

____

osmosis: ______________________________

____

____

simple diffusion: ______________________________

____

____

filtration: ______________________________

____

____

active transport: ______________________________

____

____

phagocytosis: ______________________________

____

____

bulk-phase endocytosis: ______________________________

____

____

NAME ______________________ LAB TIME/DATE ______________________

REVIEW SHEET
exercise
6A

Classification of Tissues

Tissue Structure and Function—General Review

1. Define *tissue:* ____

____

2. Use the key choices to identify the major tissue types described below.

Key: a. connective tissue b. epithelium c. muscle d. nervous tissue

______________ 1. lines body cavities and covers the body's external surface

______________ 2. pumps blood, flushes urine out of the body, allows one to swing a bat

______________ 3. transmits electrochemical impulses

______________ 4. anchors, packages, and supports body organs

______________ 5. cells may absorb, secrete, and filter

______________ 6. most involved in regulating and controlling body functions

______________ 7. major function is to contract

______________ 8. synthesizes hormones

______________ 9. the most durable tissue type

______________ 10. abundant nonliving extracellular matrix

______________ 11. most widespread tissue in the body

______________ 12. forms nerves and the brain

Epithelial Tissue

1. Describe five general characteristics of epithelial tissue. ____

____

____

____

2. On what bases are epithelial tissues classified? ____

3. List the six major functions of epithelium in the body, and give examples of each. ______

______

______

______

______

______

______

4. How is the function of epithelium reflected in its arrangement? ______

______

______

5. Where is ciliated epithelium found? ______

______

What role does it play? ______

______

______

6. Transitional epithelium is actually stratified squamous epithelium, but there is something special about it.

How does it differ structurally from other stratified squamous epithelia? ______

______

How does this reflect its function in the body? ______

______

______

7. How do the endocrine and exocrine glands differ in structure and function? ______

______

______

______

8. Respond to the following with the key choices.

Key: a. pseudostratified ciliated columnar
b. simple columnar
c. simple cuboidal
d. simple squamous
e. stratified squamous
f. transitional

______ 1. lining of the esophagus

_______________ 2. lining of the stomach

_______________ 3. alveolar sacs of lungs

_______________ 4. tubules of the kidney

_______________ 5. epidermis of the skin

_______________ 6. lining of bladder; peculiar cells that have the ability to slide over each other

_______________ 7. forms the thin serous membranes; a single layer of flattened cells

Connective Tissue

1. What are three general characteristics of connective tissues? _______________

____

____

____

____

2. What functions are performed by connective tissue? _______________

____

3. How are the functions of connective tissue reflected in its structure? _______________

____

____

4. Using the key, choose the best response to identify the connective tissues described below.

Key:
a. adipose connective tissue
b. areolar connective tissue
c. dense connective tissue
d. elastic cartilage
e. fibrocartilage
f. hematopoietic tissue
g. hyaline cartilage
h. osseous tissue

_______________ 1. attaches bones to bones and muscles to bones

_______________ 2. acts as a storage depot for fat

_______________ 3. the dermis of the skin

_______________ 4. makes up the intervertebral discs

_______________ 5. forms your hip bone

_______________ 6. composes basement membranes; a soft packaging tissue with a jellylike matrix

_______________ 7. forms the larynx, the costal cartilages of the ribs, and the embryonic skeleton

_______________ 8. provides a flexible framework for the external ear

_______________ 9. firm, structurally amorphous matrix heavily invaded with fibers; appears glassy and smooth

_______________ 10. matrix hard owing to calcium salts; provides levers for muscles to act on

_______________ 11. insulates against heat loss

5. Why do adipose cells remind people of a ring with a single jewel? ___

___

Muscle Tissue

The three types of muscle tissue exhibit similarities as well as differences. Check the appropriate space in the chart to indicate which muscle types exhibit each characteristic.

Characteristic	Skeletal	Cardiac	Smooth
Voluntarily controlled			
Involuntarily controlled			
Striated			
Has a single nucleus in each cell			
Has several nuclei per cell			
Found attached to bones			
Allows you to direct your eyeballs			
Found in the walls of the stomach, uterus, and arteries			
Contains spindle-shaped cells			
Contains branching cylindrical cells			
Contains long, nonbranching cylindrical cells			
Has intercalated discs			
Concerned with locomotion of the body as a whole			
Changes the internal volume of an organ as it contracts			
Tissue of the heart			

Nervous Tissue

1. What two physiological characteristics are highly developed in neurons (nerve cells)? ___

___

2. In what ways are neurons similar to other cells? ___

___

How are they different? ___

3. Sketch a neuron, recalling in your diagram the most important aspects of its structure. Below the diagram, describe how its particular structure relates to its function in the body.

____

____

For Review

Label the following tissue types here and on the next pages, and identify all major structures.

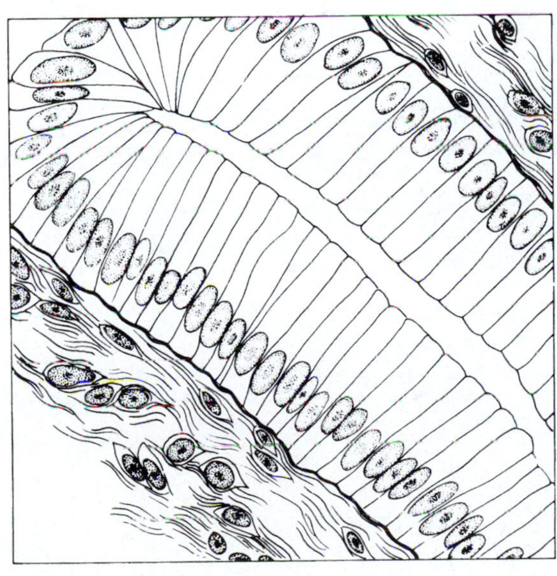

(a) ____

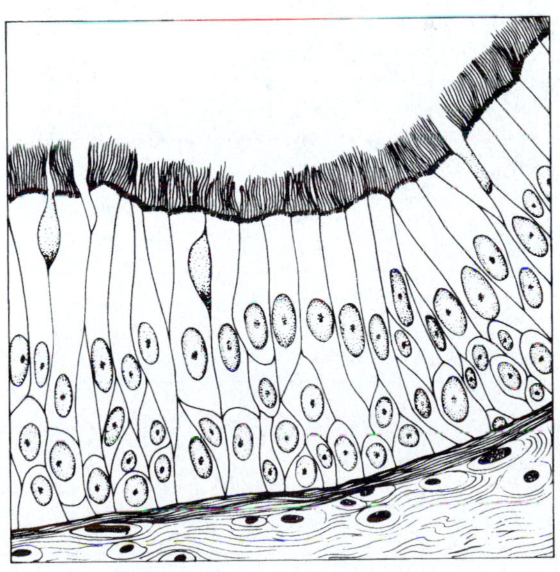

(b) ____

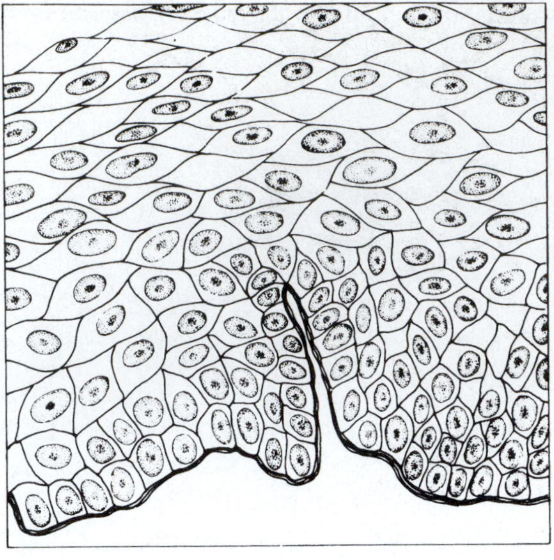

(c) ______________________________

(d) ______________________________

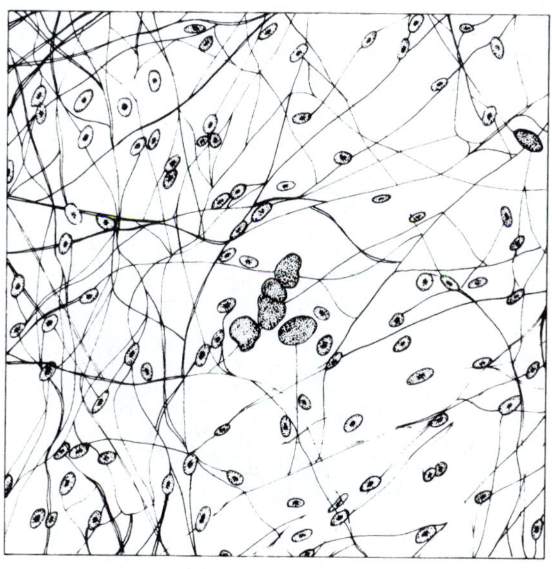

(e) ______________________________

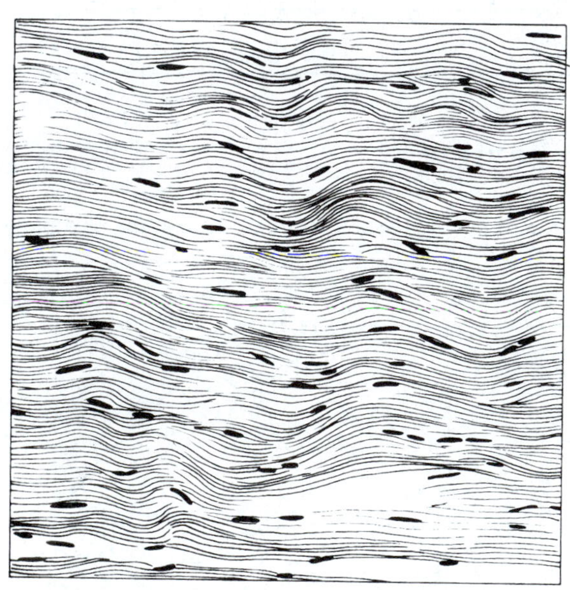

(f) ______________________________

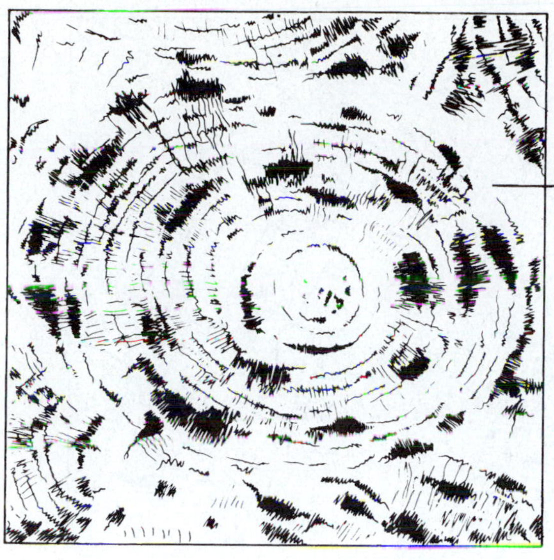

(g) ______________________________

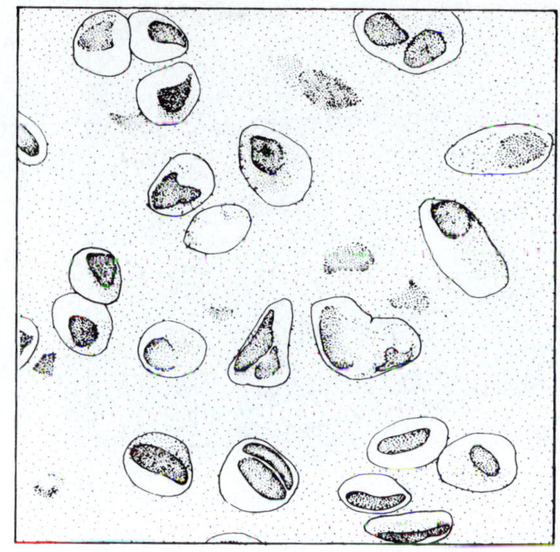

(h) ______________________________

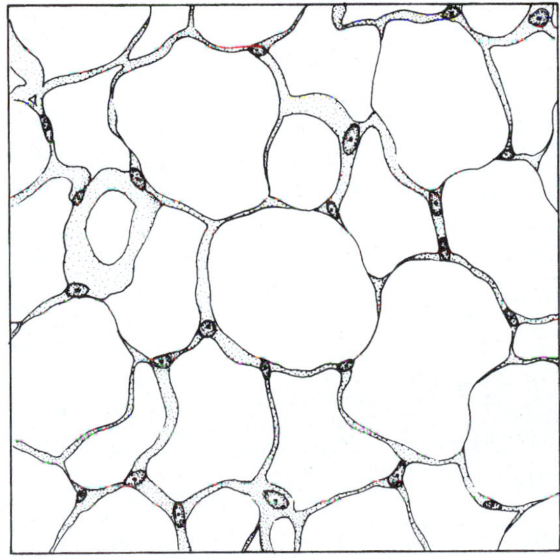

(i) ______________________________

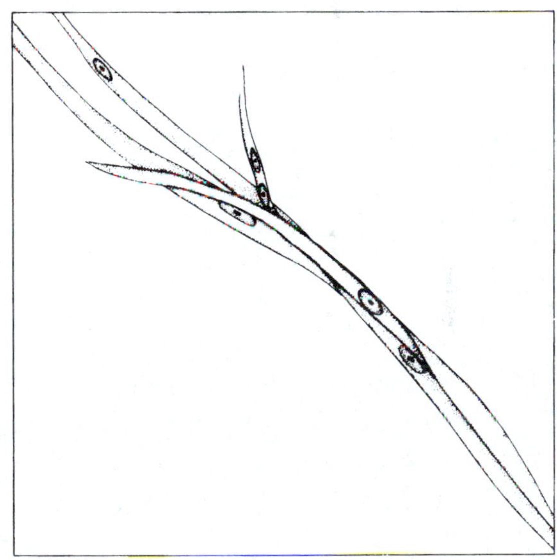

(j) ______________________________

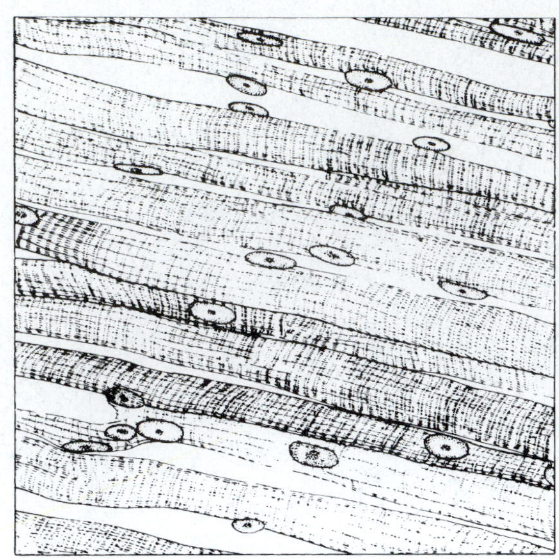

(k) ______________________________

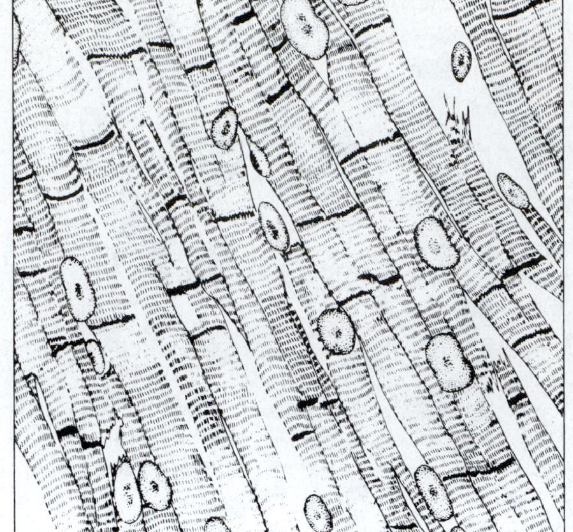

(l) ______________________________

NAME ______________________ LAB TIME/DATE ______________________

REVIEW SHEET
exercise
7

The Integumentary System

Basic Structure of the Skin

1. Complete the following statements by writing the appropriate word or phrase on the correspondingly numbered blank:

The two basic tissues of which the skin is composed are dense irregular connective tissue, which makes up the dermis, and __1__, which forms the epidermis. The tough water-repellent protein found in the epidermal cells is called __2__. The pigments melanin and __3__ contribute to skin color. A localized concentration of melanin is referred to as a __4__.

1. ______________________
2. ______________________
3. ______________________
4. ______________________

2. Four protective functions of the skin are ____

____.

3. Using the key choices, choose all responses that apply to the following descriptions.

Key:
a. stratum basale
b. stratum corneum
c. stratum granulosum
d. stratum lucidum
e. stratum spinosum
f. papillary layer
g. reticular layer
h. epidermis as a whole
i. dermis as a whole

______________________ 1. translucent cells in thick skin containing keratin fibrils

______________________ 2. dead cells

______________________ 3. dermis layer responsible for fingerprints

______________________ 4. vascular region

______________________ 5. major skin area that produces derivatives (nails and hair)

______________________ 6. epidermal region exhibiting the most rapid cell division

______________________ 7. scalelike dead cells, full of keratin, that constantly slough off

______________________ 8. mitotic cells filled with intermediate filaments

______________________ 9. has abundant elastic and collagenic fibers

______________________ 10. location of melanocytes and Merkel cells

______________________ 11. area where weblike prekeratin filaments first appear

4. Label the skin structures and areas indicated in the accompanying diagram of thin skin. Then, complete the statements that follow.

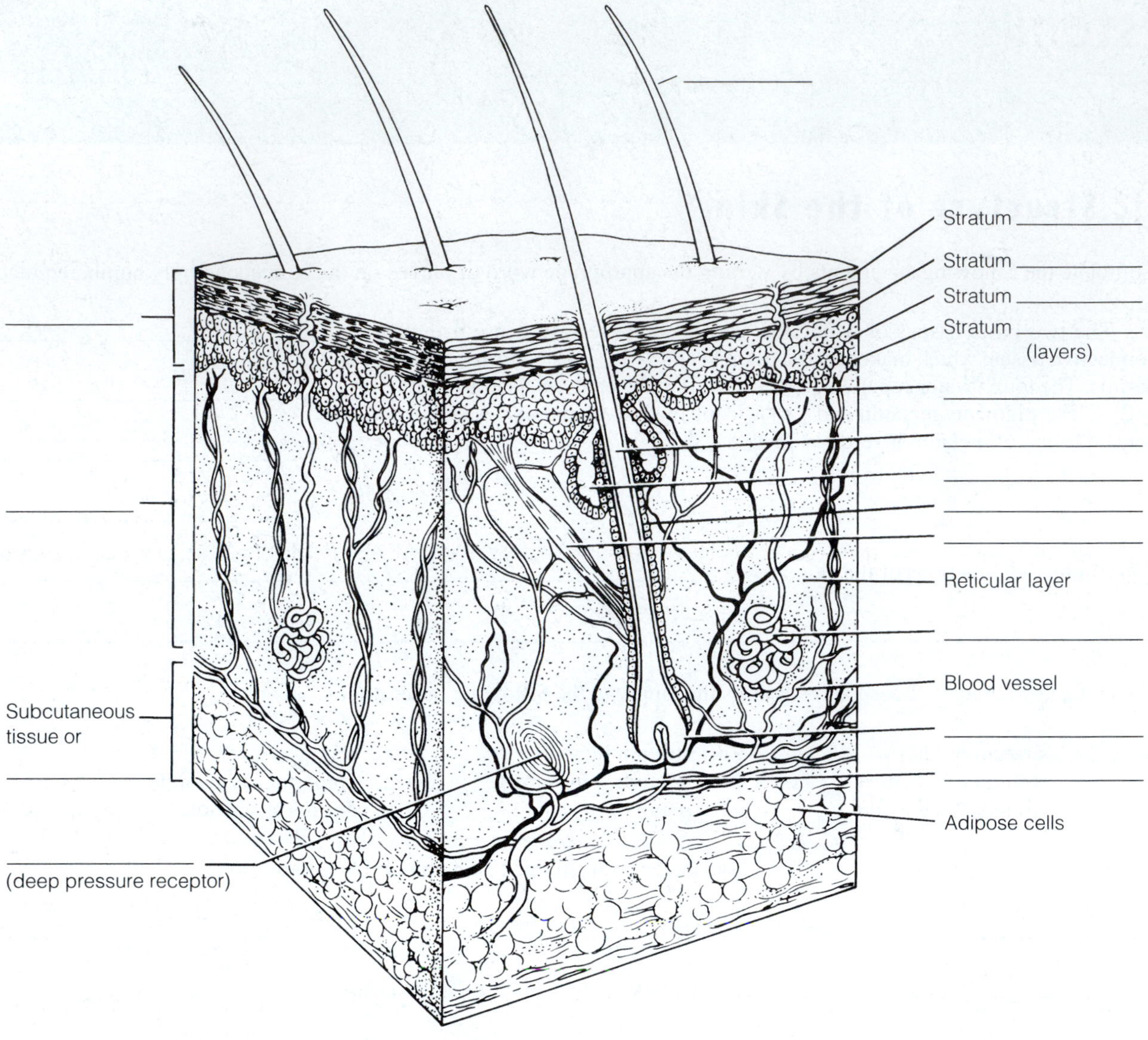

a. ______________________________ granules extruded from the keratinocytes prevent water loss by diffusion through the epidermis.

b. Fibers in the dermis are produced by ______________________________.

c. Glands that respond to rising androgen levels are the ______________________________ glands.

d. Phagocytic cells that occupy the epidermis are called ______________________________.

e. A unique touch receptor formed from a stratum basale cell and a nerve fiber is a ______________________________.

f. What layer is present in thick skin but not in thin skin? ______________________________

g. What cell-to-cell structures hold the cells of the stratum spinosum tightly together? ______________________________

5. What substance is manufactured in the skin (but is not a secretion) to play a role elsewhere in the body?

____

6. List the sensory receptors found in the dermis of the skin. ______________________________

____

____

7. A nurse tells a doctor that a patient is cyanotic. Define cyanosis. ______________________________

What does its presence imply? ______________________________

8. What is the mechanism of a suntan? ______________________________

____

9. What is a bedsore (decubitus ulcer) ? ______________________________

Why does it occur? ______________________________

____

10. Some injections hurt more than others. On the basis of what you have learned about skin structure, can you determine why this is so? ______________________________

____

Appendages of the Skin

1. Using key choices, respond to the following descriptions.

Key: a. arrector pili
b. cutaneous receptors
c. hair
d. hair follicle
e. nail
f. sebaceous glands
g. sweat gland—apocrine
h. sweat gland—eccrine

______________ 1. A blackhead is an accumulation of oily material that is produced by _______.

______________ 2. Tiny muscles, attached to hair follicles, that pull the hair upright during fright or cold.

______________ 3. Perspiration glands with a role in temperature control.

______________ 4. Sheath formed of both epithelial and connective tissues.

______________ 5. Less numerous type of perspiration-producing gland; found mainly in the pubic and axillary regions.

______________ 6. Found everywhere on body except palms of hands and soles of feet.

______________ 7. Primarily dead/keratinized cells.

______________ 8. Specialized nerve endings that respond to temperature, touch, etc.

______________ 9. Its secretion is a lubricant for hair and skin.

______________ 10. "Sports" a lunula and a cuticle.

2. Describe two integumentary system mechanisms that help in regulating body temperature. ______________________

____

____

____

____

3. Several structures or skin regions are listed below. Identify each by matching its letter with the appropriate area on the figure.

a. Adipose cells

b. Dermis

c. Epidermis

d. Hair follicle

e. Hair shaft

f. Sloughing stratum corneum cells

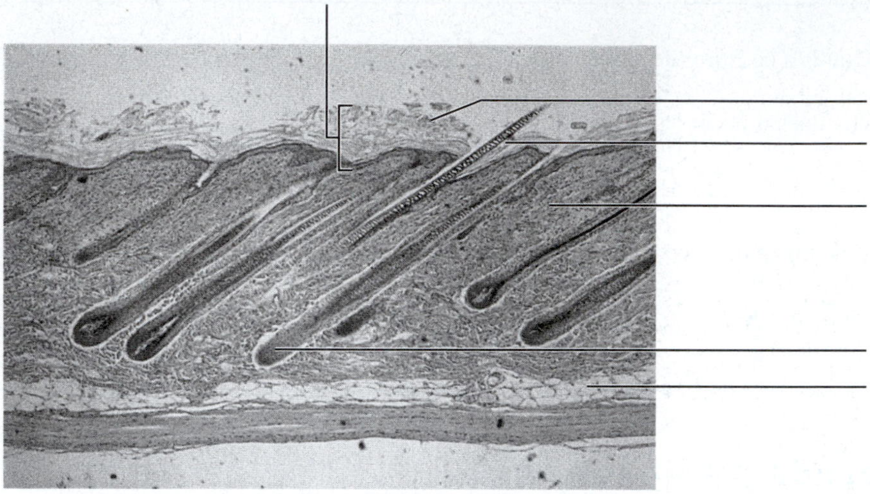

Plotting the Distribution of Sweat Glands

1. With what substance in the bond paper does the iodine painted on the skin react? ______________________

2. Based on class data, which skin area—the forearm or palm of hand—has more sweat glands? ______________________

Was this an expected result? _______ Explain. ______________________

____

____

Which other body areas would, if tested, prove to have a high density of sweat glands? ______________________

____

3. What organ system controls the activity of the eccrine sweat glands? ______________________

NAME ______________________ LAB TIME/DATE ______________________

REVIEW SHEET
exercise
8

Classification of Body Membranes

Classification of Body Membranes

1. Complete the following chart:

Membrane	Tissue types (epithelial/connective)	Common locations	Functions
cutaneous			
mucous			
serous			
synovial			

2. Respond to the following statements by choosing an answer from the key.

Key: a. cutaneous b. mucous c. serous d. synovial

______________ 1. membrane type in joints, bursae, and tendon sheaths

______________ 2. epithelium is always simple squamous epithelium

______________, ______________ 3. membrane types *not* found in the ventral body cavity

______________ 4. the only membrane type in which goblet cells are found

______________ 5. the dry membrane with keratinizing epithelium

______________ 6. "wet" membranes

______________ 7. adapted for absorption and secretion

______________ 8. has parietal and visceral layers

3. Using terms from the list above the figure, identify specifically the different types of body membranes (cutaneous, mucous, serous, and synovial) by writing in the terms at the end of the appropriate leader lines.

a. cutaneous membrane (skin)
b. esophageal mucosa
c. gastric mucosa
d. nasal mucosa
e. oral mucosa
f. mucosa of lung bronchi
g. parietal pericardium
h. parietal pleura
i. synovial membrane of joint
j. tracheal mucosa
k. visceral pericardium
l. visceral pleura

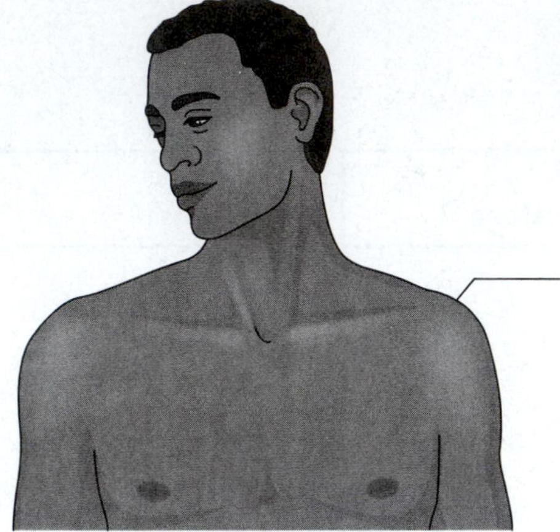

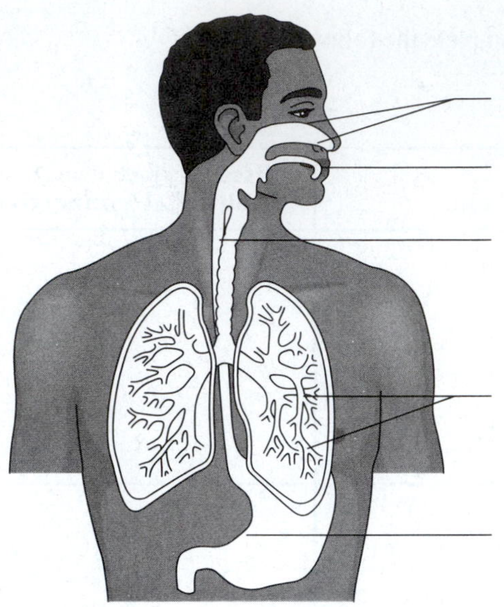

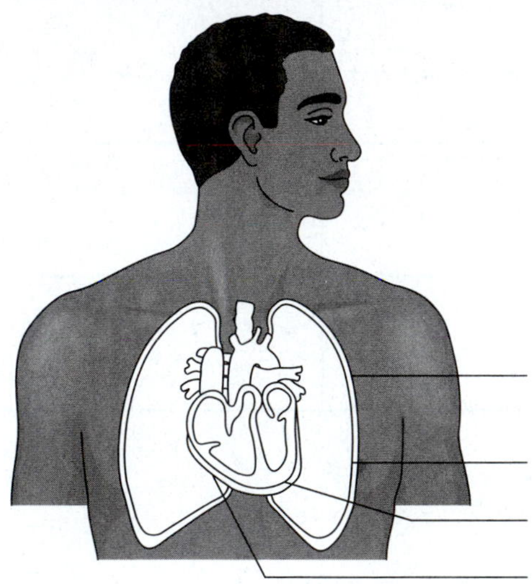

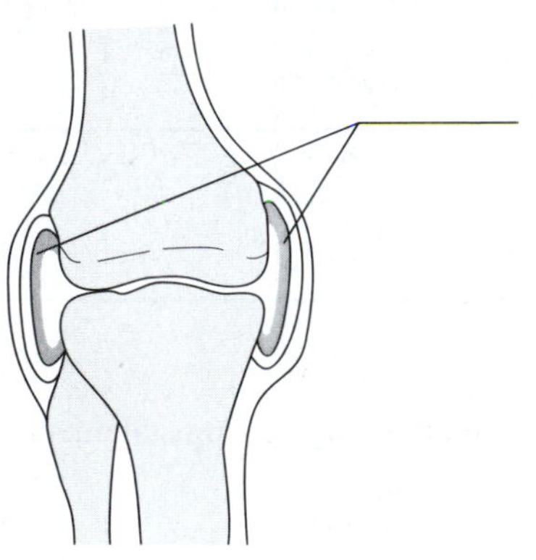

4. Use the terms in question 3 to create two lists:

a. The mucous membranes are ______________________________

____

b. The serous membranes are ______________________________

____

NAME ____________________ LAB TIME/DATE ____________________

REVIEW SHEET
exercise
9

Overview of the Skeleton: Classification and Structure of Bones and Cartilages

Bone Markings

Match the terms in column B with the appropriate description in column A:

	Column A	Column B
________	1. sharp, slender process*	a. condyle
________	2. small rounded projection*	b. crest
________	3. narrow ridge of bone*	c. epicondyle
________	4. large rounded projection*	d. fissure
________	5. structure supported on neck†	e. foramen
________	6. armlike projection†	f. fossa
________	7. rounded, convex projection†	g. head
________	8. narrow depression or opening‡	h. meatus
________	9. canal-like structure‡	i. process
________	10. opening through a bone‡	j. ramus
________	11. shallow depression†	k. sinus
________	12. air-filled cavity	l. spine
________	13. large, irregularly shaped projection*	m. trochanter
________	14. raised area of a condyle*	n. tubercle
________	15. projection or prominence	o. tuberosity

* A site of muscle attachment.
† Takes part in joint formation.
‡ A passageway for nerves or blood vessels.

Classification of Bones

1. The four major anatomical classifications of bones are long, short, flat, and irregular. Which category has the least amount of spongy bone relative to its total volume? ____________________

2. Classify each of the bones in the next chart into one of the four major categories by checking the appropriate column. Use appropriate references as necessary.

	Long	Short	Flat	Irregular
humerus				
phalanx				
parietal				
calcaneus				
rib				
vertebra				
ulna				

Gross Anatomy of the Typical Long Bone

1. Use the terms below to identify the structures marked by leader lines and braces in the diagrams (some terms are used more than once).

Key:
a. articular cartilage
b. compact bone
c. diaphysis
d. endosteum
e. epiphyseal line
f. epiphysis
g. medullary cavity
h. nutrient artery
i. periosteum
j. red marrow cavity
k. trabeculae of spongy bone
l. yellow marrow

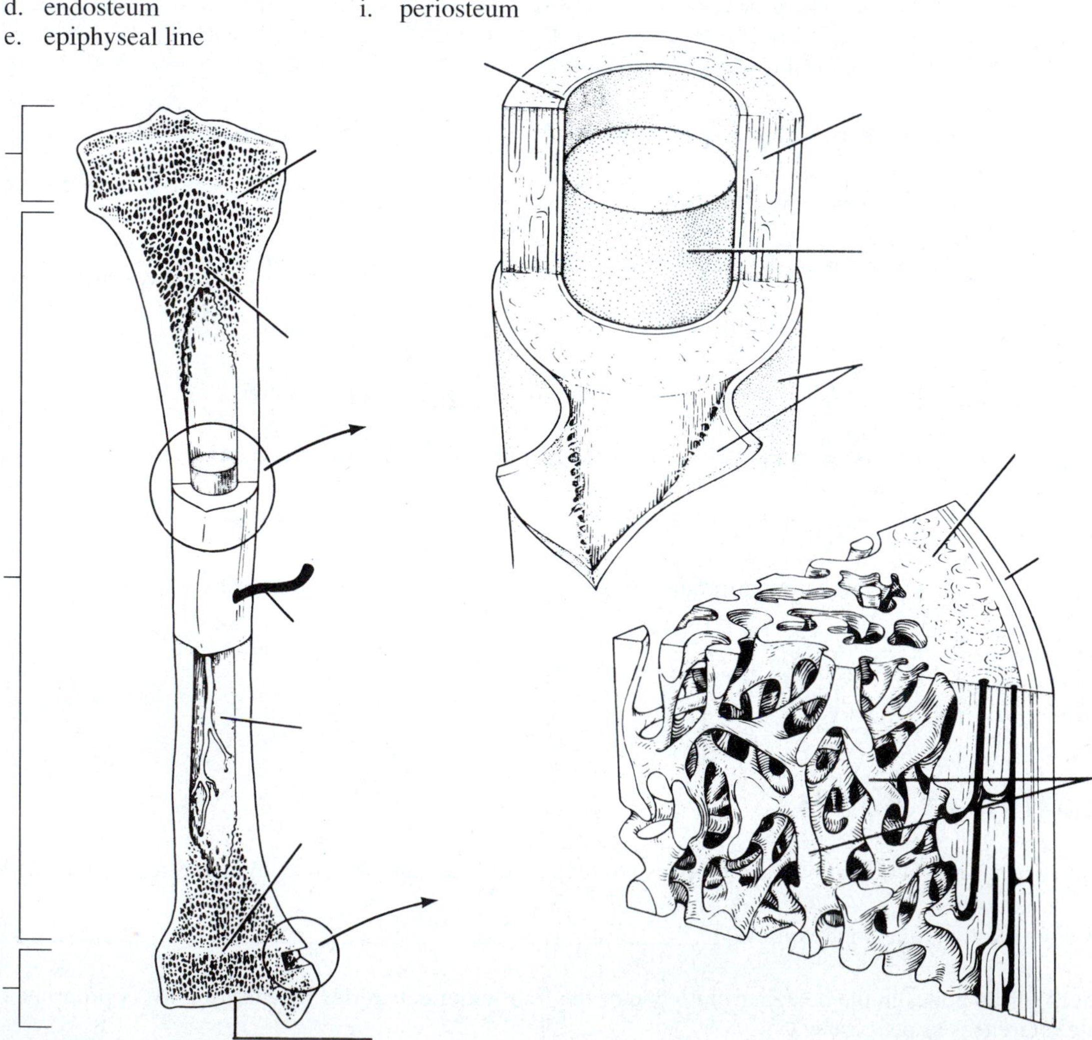

2. Match the terms in question 1 with the information below.

_______ 1. contains spongy bone in adults

_______ 2. made of compact bone

_______ 3. site of blood cell formation

_______, _______ 4. major submembranous site of osteoclasts

_______ 5. scientific term for bone shaft

_______ 6. contains fat in adult bones

_______ 7. growth plate remnant

_______ 8. major submembranous site of osteoblasts

3. What differences between compact and spongy bone can be seen with the naked eye? _______________

____

4. What is the function of the periosteum? _______________

____

Microscopic Structure of Compact Bone

1. Trace the route taken by nutrients through a bone, starting with the periosteum and ending with an osteocyte in a lacuna.

Periosteum → _______________ → _______________

_______________ → _______________ → osteocyte

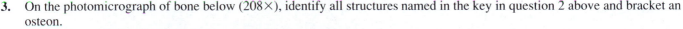

2. Several descriptions of bone structure are given below. Identify the structure involved by choosing the appropriate term from the key and placing its letter in the blank.

Key: a. canaliculi b. central canal c. concentric lamellae d. lacunae e. matrix

_______ 1. layers of bony matrix around a central canal

_______ 2. site of osteocytes

_______ 3. longitudinal canal carrying blood vessels, lymphatics, and nerves

_______ 4. minute canals connecting osteocytes of an osteon

_______ 5. inorganic salts deposited in organic ground substance

3. On the photomicrograph of bone below (208×), identify all structures named in the key in question 2 above and bracket an osteon.

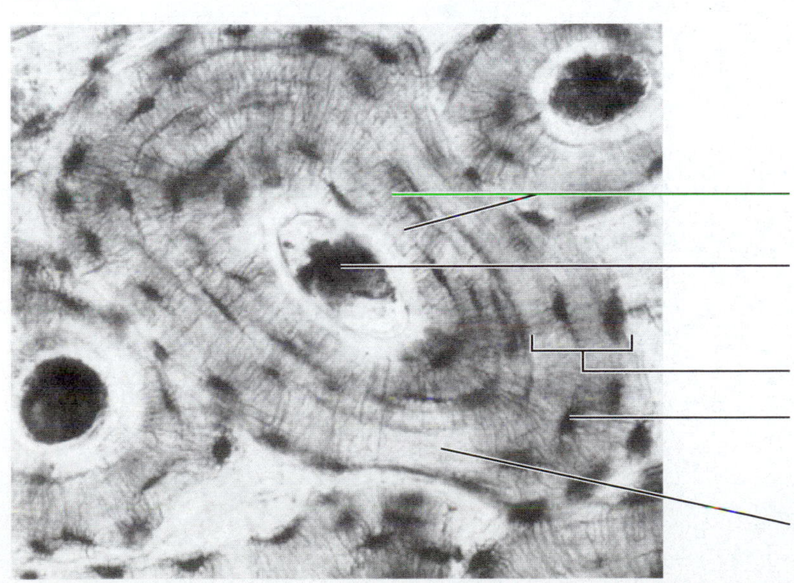

Chemical Composition of Bone

1. What is the function of the organic matrix in bone? ____________________

2. Name the important organic bone components. ____________________

3. Calcium salts form the bulk of the inorganic material in bone. What is the function of the calcium salts?

____

4. Baking removes ______________ from bone. Soaking bone in acid removes ______________.

5. Which is responsible for bone structure? (circle the appropriate response)

inorganic portion organic portion both contribute

Cartilages of the Skeleton

1. Using key choices, identify each type of cartilage described (in terms of its body location or function) below.

Key: a. elastic b. fibrocartilage c. hyaline

________________ 1. supports the external ear

________________ 2. between the vertebrae

________________ 3. forms the walls of the voice box (larynx)

________________ 4. the epiglottis

________________ 5. articular cartilages

________________ 6. meniscus in a knee joint

________________ 7. connects the ribs to the sternum

________________ 8. most effective at resisting compression

________________ 9. most springy and flexible

________________ 10. most abundant

2. Identify the two types of cartilage diagrammed below. On each, label the *chondrocytes in lacunae* and the *matrix*.

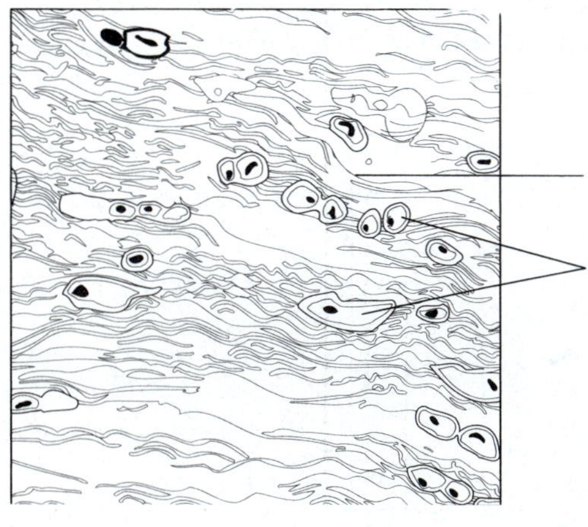

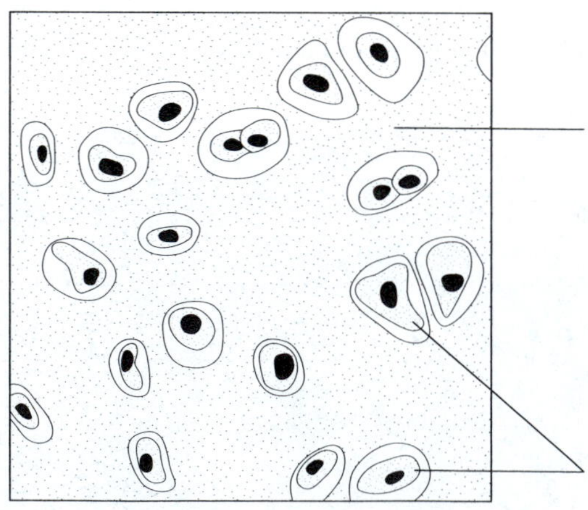

(a) ____________________ (b) ____________________

NAME ______________________ LAB TIME/DATE ______________________

REVIEW SHEET
exercise
10

The Axial Skeleton

The Skull

1. The skull is one of the major components of the axial skeleton. Name the other two:

   ______________________ and ______________________

   What structures do each of these areas protect? ______________________

2. Define *suture:* ______________________

3. With one exception, the skull bones are joined by sutures. Name the exception. ______________________

4. What are the four major sutures of the skull, and what bones do they connect?

   a. ______________________

   b. ______________________

   c. ______________________

   d. ______________________

5. Name the eight bones of the cranium.

   ______ ______ ______ ______

   ______ ______ ______ ______

6. Give two possible functions of the sinuses. ______________________

7. What is the orbit? ______________________

   What bones contribute to the formation of the orbit? ______________________

8. Why can the sphenoid bone be called the keystone of the cranial floor? ______________________

9. What is a cleft palate? ____

____

10. Match the bone names in column B with the descriptions in column A.

Column A

__________ 1. forehead bone

__________ 2. cheekbone

__________ 3. lower jaw

__________ 4. bridge of nose

__________ 5. posterior bones of the hard palate

__________ 6. much of the lateral and superior cranium

__________ 7. most posterior part of cranium

__________ 8. single, irregular, bat-shaped bone forming part of the cranial floor

__________ 9. tiny bones bearing tear ducts

__________ 10. anterior part of hard palate

__________ 11. superior and medial nasal conchae formed from its projections

__________ 12. site of mastoid process

__________ 13. site of sella turcica

__________ 14. site of cribriform plate

__________ 15. site of mental foramen

__________ 16. site of styloid processes

__________, __________, __________, and

__________ 17. four bones containing paranasal sinuses

__________ 18. condyles here articulate with the atlas

__________ 19. foramen magnum contained here

__________ 20. small U-shaped bone in neck, where many tongue muscles attach

__________ 21. middle ear found here

__________ 22. nasal septum

__________ 23. bears an upward protrusion, the "cock's comb," or crista galli

__________, __________ 24. contain alveoli bearing teeth

Column B

a. ethmoid

b. frontal

c. hyoid

d. lacrimal

e. mandible

f. maxilla

g. nasal

h. occipital

i. palatine

j. parietal

k. sphenoid

l. temporal

m. vomer

n. zygomatic

11. Using choices from the numbered key to the right, identify all bones and bone markings provided with leader lines in the two diagrams below.

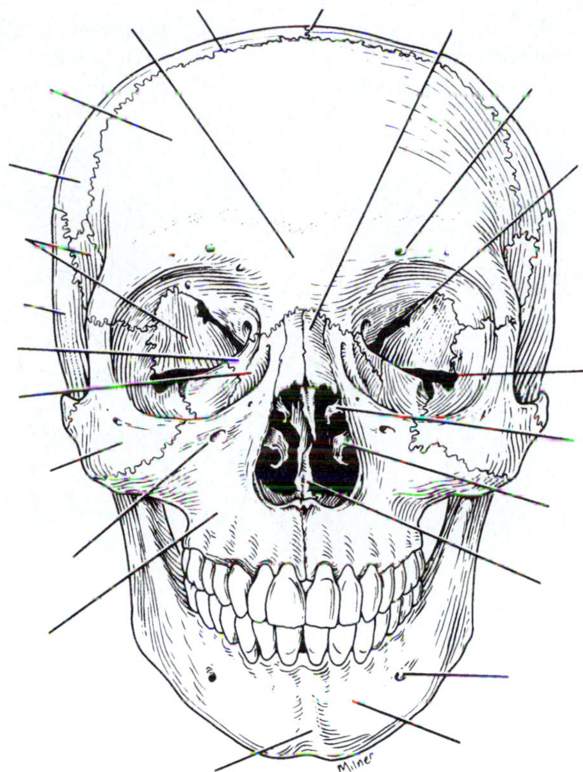

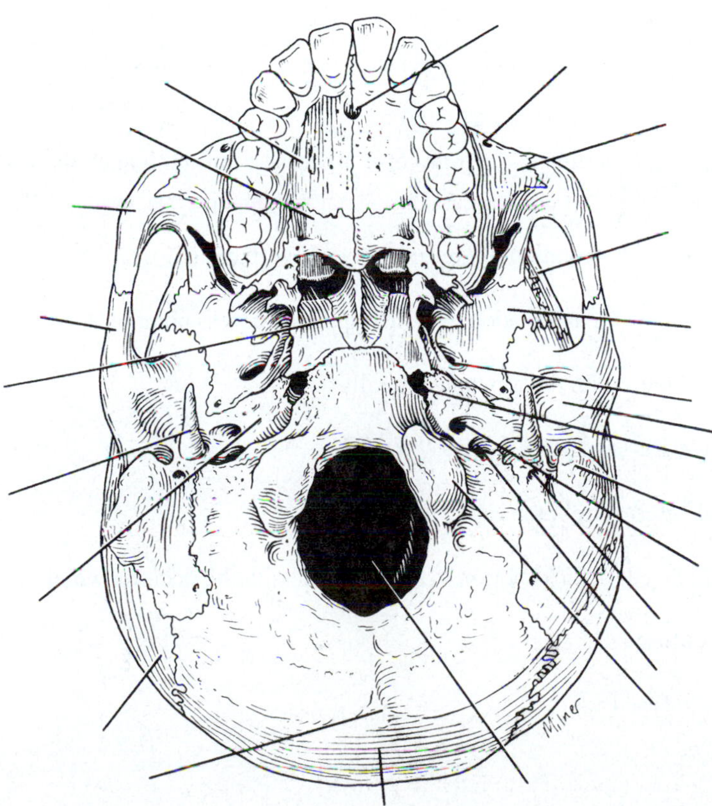

1. carotid canal
2. coronal suture
3. ethmoid bone
4. external occipital protuberance
5. foramen lacerum
6. foramen magnum
7. foramen ovale
8. frontal bone
9. glabella
10. incisive fossa
11. inferior nasal concha
12. inferior orbital fissure
13. infraorbital foramen
14. jugular foramen
15. lacrimal bone
16. mandible
17. mandibular fossa
18. mandibular symphysis
19. mastoid process
20. maxilla
21. mental foramen
22. middle nasal concha of ethmoid
23. nasal bone
24. occipital bone
25. occipital condyle
26. palatine bone
27. palatine process of maxilla
28. parietal bone
29. sagittal suture
30. sphenoid bone
31. styloid process
32. stylomastoid foramen
33. superior orbital fissure
34. supraorbital foramen
35. temporal bone
36. vomer
37. zygomatic bone
38. zygomatic process of temporal bone

The Vertebral Column

1. Using the key, correctly identify the vertebral parts/areas described below. (More than one choice may apply in some cases.) Also use the key letters to correctly identify the vertebral areas in the diagram.

Key: a. body
b. intervertebral foramina
c. lamina
d. pedicle
e. spinous process
f. superior articular process
g. transverse process
h. vertebral arch
i. vertebral foramen

_______ 1. cavity enclosing the nerve cord

_______ 2. weight-bearing portion of the vertebra

_______ 3. provide levers against which muscles pull

_______ 4. provide an articulation point for the ribs

_______ 5. openings providing for exit of spinal nerves

_______ 6. structures that form an enclosure for the spinal cord

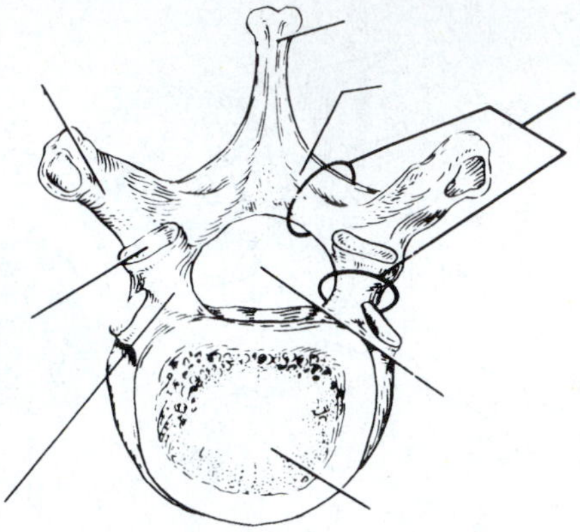

2. The distinguishing characteristics of the vertebrae composing the vertebral column are noted below. Correctly identify each described structure/region by choosing a response from the key.

Key: a. atlas
b. axis
c. cervical vertebra—typical
d. coccyx
e. lumbar vertebra
f. sacrum
g. thoracic vertebra

_________________ 1. vertebral type containing foramina in the transverse processes, through which the vertebral arteries ascend to reach the brain

_________________ 2. dens here provides a pivot for rotation of the first cervical vertebra (C_1)

_________________ 3. transverse processes faceted for articulation with ribs; spinous process pointing sharply downward

_________________ 4. composite bone; articulates with the hip bone laterally

_________________ 5. massive vertebrae; weight-sustaining

_________________ 6. "tail bone"; vestigial fused vertebrae

_________________ 7. supports the head; allows a rocking motion in conjunction with the occipital condyles

_________________ 8. seven components; unfused

_________________ 9. twelve components; unfused

3. Identify specifically each of the vertebra types shown in the diagrams below. Also identify and label the following markings on each: transverse processes, spinous process, body, superior articular processes.

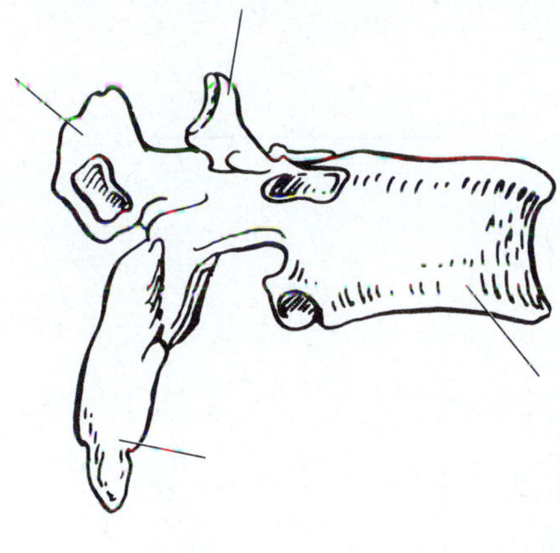

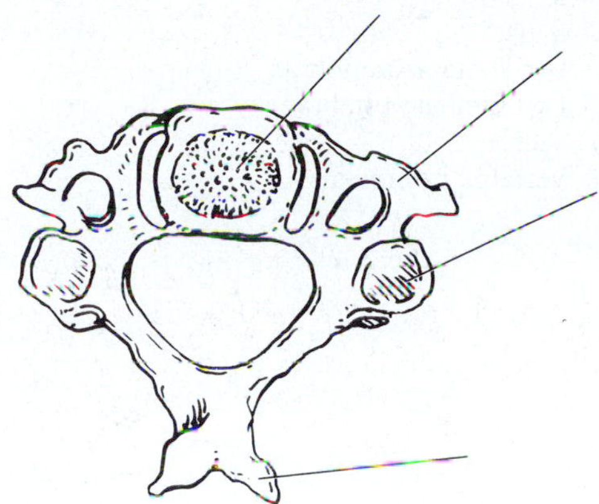

________________________ ________________________

4. Describe how a spinal nerve exits from the vertebral column. ________________________

____

5. Name two factors/structures that allow for flexibility of the vertebral column.

________________________ and ________________________

6. What kind of tissue composes the intervertebral discs? ________________________

7. What is a herniated disc? ________________________

What problems might it cause? ________________________

____

8. Which two spinal curvatures are obvious at birth? ________________________ and

____

Under what conditions do the secondary curvatures develop? ________________________

____

____

9. On this illustration of an articulated vertebral column, identify each curvature indicated and label it as a primary or a secondary curvature. Also identify the structures provided with leader lines, using the letters of the terms listed below.

a. atlas
b. axis
c. a disc
d. two thoracic vertebrae
e. two lumbar vertebrae
f. sacrum
g. vertebra prominens

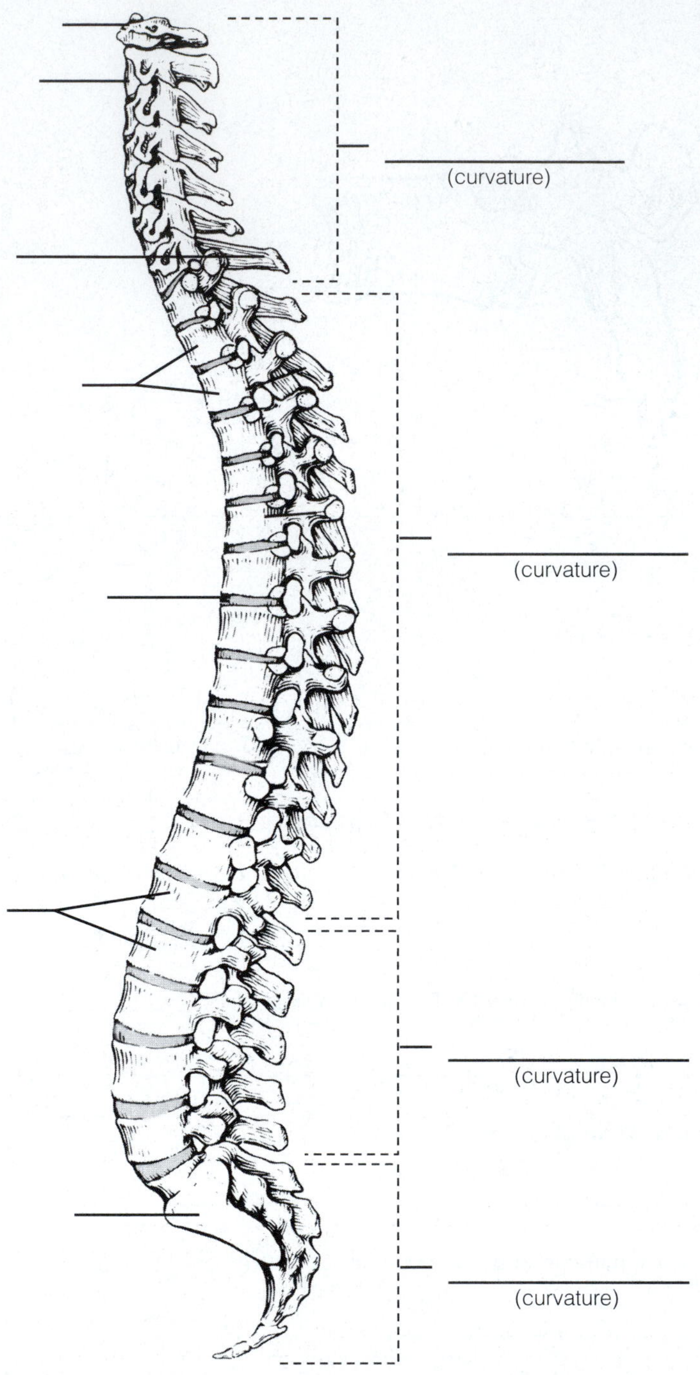

10. Diagram the abnormal spinal curvatures named below. (Use posterior or lateral views as necessary.)

Lordosis | Scoliosis | Kyphosis

The Bony Thorax

1. The major components of the thorax (excluding the vertebral column) are the ______________________

   and the ______________.

2. Differentiate between a true rib and a false rib. ______________________

   ____

   ____

   Is a floating rib a true or a false rib? ______________

3. What is the general shape of the thoracic cage? ______________________

   ____

   ____

4. Provide the more scientific name for the following rib types.

   a. True ribs ______________________

   b. False ribs (not including c) ______________________

   c. Floating ribs ______________________

5. Using the terms at the right, identify the regions and landmarks of the bony thorax.

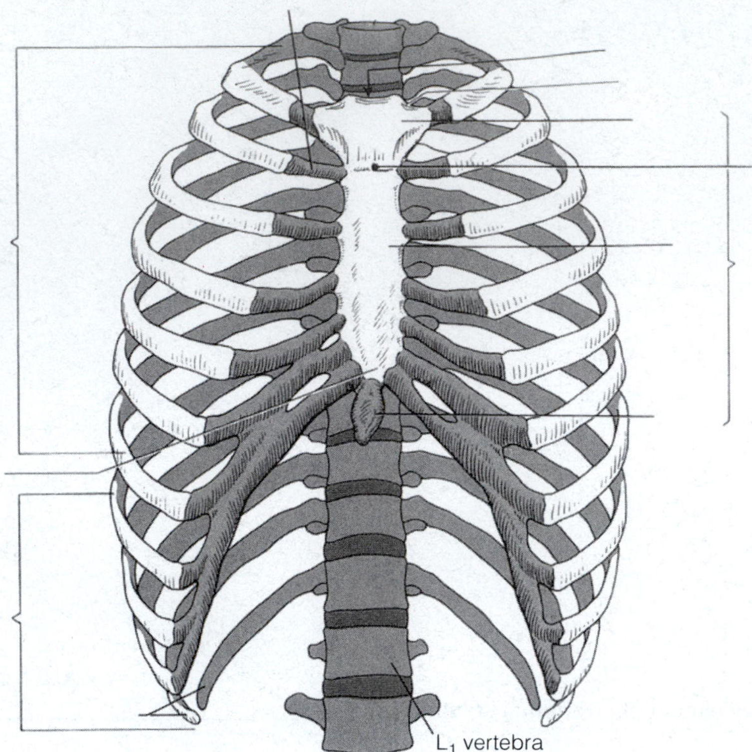

a. body

b. clavicular notch

c. costal cartilage

d. false ribs

e. floating ribs

f. jugular notch

g. manubrium

h. sternal angle

i. sternum

j. true ribs

k. xiphisternal joint

l. xiphoid process

NAME ______________________ LAB TIME/DATE ________ ________

REVIEW SHEET
exercise
11

The Appendicular Skeleton

Bones of the Pectoral Girdle and Upper Extremity

1. Match the bone names or markings in column B with the descriptions in column A.

Column A

__________ 1. raised area on lateral surface of humerus to which deltoid muscle attaches

__________ 2. arm bone

__________, __________ 3. bones of the shoulder girdle

__________, __________ 4. forearm bones

__________ 5. scapular region to which the clavicle connects

__________ 6. shoulder girdle bone that is unattached to the axial skeleton

__________ 7. shoulder girdle bone that transmits forces from the upper limb to the bony thorax

__________ 8. depression in the scapula that articulates with the humerus

__________ 9. process above the glenoid cavity that permits muscle attachment

__________ 10. the "collarbone"

__________ 11. distal condyle of the humerus that articulates with the ulna

__________ 12. medial bone of forearm in anatomical position

__________ 13. rounded knob on the humerus; adjoins the radius

__________ 14. anterior depression, superior to the trochlea, which receives part of the ulna when the forearm is flexed

__________ 15. forearm bone involved in formation of the elbow joint

__________ 16. wrist bones

__________ 17. finger bones

__________ 18. heads of these bones form the knuckles

__________, __________ 19. bones that articulate with the clavicle

Column B

a. acromion
b. capitulum
c. carpals
d. clavicle
e. coracoid process
f. coronoid fossa
g. deltoid tuberosity
h. glenoid cavity
i. humerus
j. metacarpals
k. olecranon fossa
l. olecranon process
m. phalanges
n. radial tuberosity
o. radius
p. scapula
q. sternum
r. styloid process
s. trochlea
t. ulna

2. Why is the clavicle at risk to fracture when a person falls on his or her shoulder? ______

______

3. Why is it generally no problem for the arm to clear the widest dimension of the thoracic cage?

______

______

4. What is the total number of phalanges in the hand? ______

5. What is the total number of carpals in the wrist? ______

In the proximal row, the carpals are (medial to lateral) ______

In the distal row, they are (medial to lateral) ______

6. Using items from the list at the right, identify the anatomical landmarks and regions of the scapula.

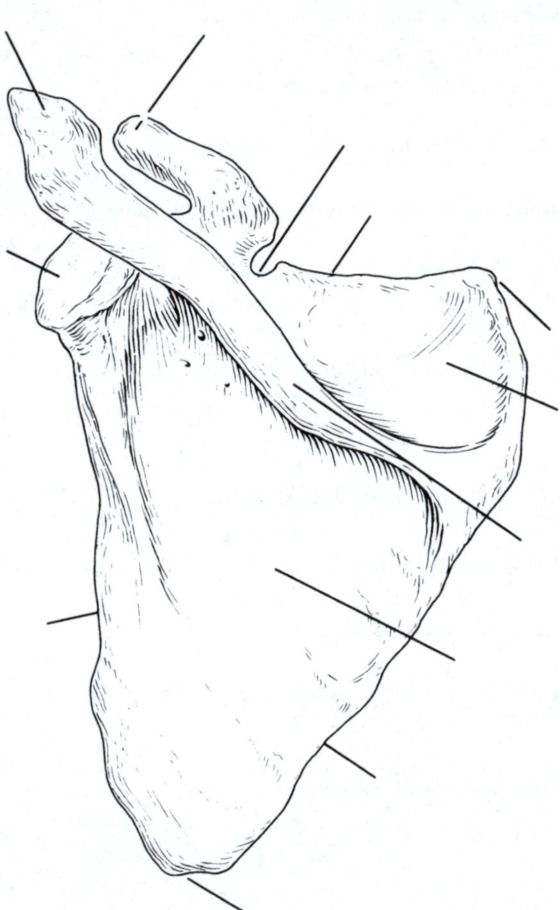

a. acromion
b. coracoid process
c. glenoid cavity
d. inferior angle
e. infraspinous fossa
f. lateral border
g. medial border
h. spine
i. superior angle
j. superior border
k. suprascapular notch
l. supraspinous fossa

7. Match the terms in the key with the appropriate leader lines on the drawings of the humerus and the radius and ulna. Also decide whether these bones are right or left bones.

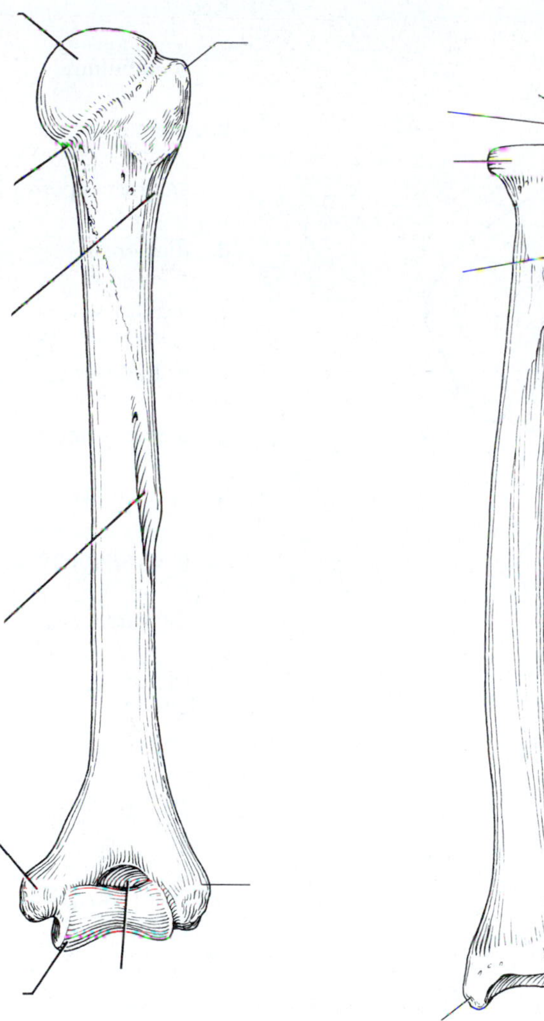

Key:

a. anatomical neck
b. coronoid process
c. distal radioulnar joint
d. greater tubercle
e. head of humerus
f. head of radius
g. head of ulna
h. lateral epicondyle
i. medial epicondyle
j. olecranon fossa
k. olecranon process
l. proximal radioulnar joint
m. radial groove
n. radial notch
o. radial tuberosity
p. styloid process of radius
q. styloid process of ulna
r. surgical neck
s. trochlea
t. trochlear notch

The humerus is a ____________________ bone; the radius and ulna are ____________________ bones.

Bones of the Pelvic Girdle and Lower Limb

1. Compare the pectoral and pelvic girdles by choosing appropriate descriptive terms from the key.

Key: a. flexibility most important
b. massive
c. lightweight
d. insecure axial and limb attachments
e. secure axial and limb attachments
f. weight-bearing most important

Pectoral: _______, _______, _______ Pelvic: _______, _______, _______

2. What organs are protected, at least in part, by the pelvic girdle? ____________________

____

3. Distinguish between the true pelvis and the false pelvis. ____________________

____

____

4. Use letters from the key to identify the bone markings on this illustration of an articulated pelvis. Make an educated guess as to whether the illustration shows a male or female pelvis and provide two reasons for your decision.

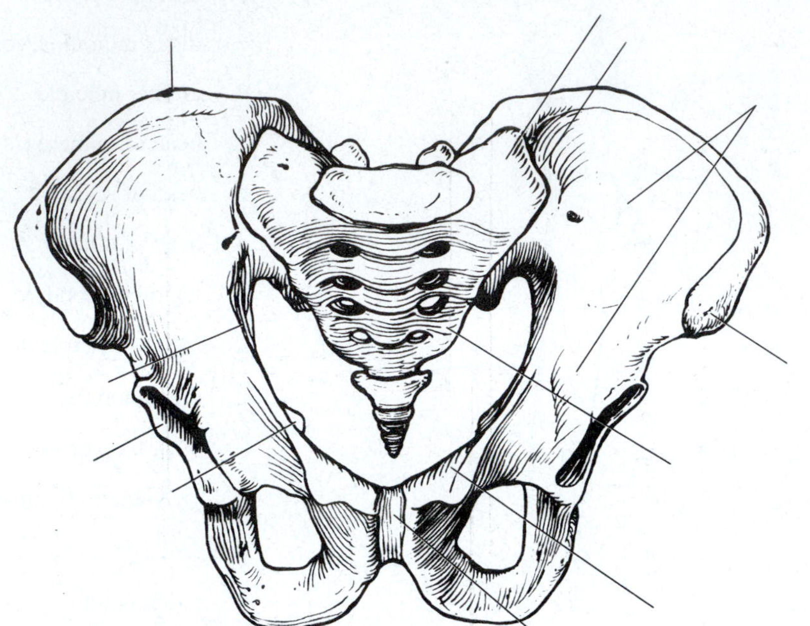

Key:

a. acetabulum

b. ala

c. anterior superior iliac spine

d. iliac crest

e. iliac fossa

f. ischial spine

g. pelvic brim

h. pubic crest

i. pubic symphysis

j. sacroiliac joint

k. sacrum

This is a ______________ (female/male) pelvis because:

____

____

5. Deduce why the pelvic bones of a four-legged animal such as the cat or pig are much less massive than those of the human.

____

____

6. A person instinctively curls over his abdominal area in times of danger. Why? ______________________________

____

7. For what anatomical reason do many women appear to be slightly knock-kneed? ______________________________

____

____

8. What does *fallen arches* mean? ____

____

9. Match the bone names and markings in column B with the descriptions in column A.

Column A	Column B
______________, ______________, and	a. acetabulum
______________ 1. fuse to form the coxal bone	b. calcaneus
______________ 2. inferoposterior "bone" of the coxal bone	c. femur
______________ 3. point where the coxal bones join anteriorly	d. fibula
______________ 4. superiormost margin of the coxal bone	e. gluteal tuberosity
______________ 5. deep socket in the coxal bone that receives the head of the thigh bone	f. greater sciatic notch
______________ 6. joint between axial skeleton and pelvic girdle	g. greater and lesser trochanters
______________ 7. longest, strongest bone in body	h. iliac crest
______________ 8. thin lateral leg bone	i. ilium
______________ 9. heavy medial leg bone	j. ischial tuberosity
______________, ______________ 10. bones forming knee joint	k. ischium
______________ 11. point where the patellar ligament attaches	l. lateral malleolus
______________ 12. kneecap	m. lesser sciatic notch
______________ 13. shin bone	n. linea aspera
______________ 14. medial ankle projection	o. medial malleolus
______________ 15. lateral ankle projection	p. metatarsals
______________ 16. largest tarsal bone	q. obturator foramen
______________ 17. ankle bones	r. patella
______________ 18. bones forming the instep of the foot	s. pubic symphysis
______________ 19. opening in hip bone formed by the pubic and ischial rami	t. pubis
______________ and ______________ 20. sites of muscle attachment on the proximal femur	u. sacroiliac joint
______________ 21. tarsal bone that "sits" on the calcaneus	v. talus
	w. tarsals
	x. tibia
	y. tibial tuberosity

10. Match the terms in the key with the appropriate leader lines on the drawings of the femur and the tibia and fibula. Also decide if these bones are right or left bones.

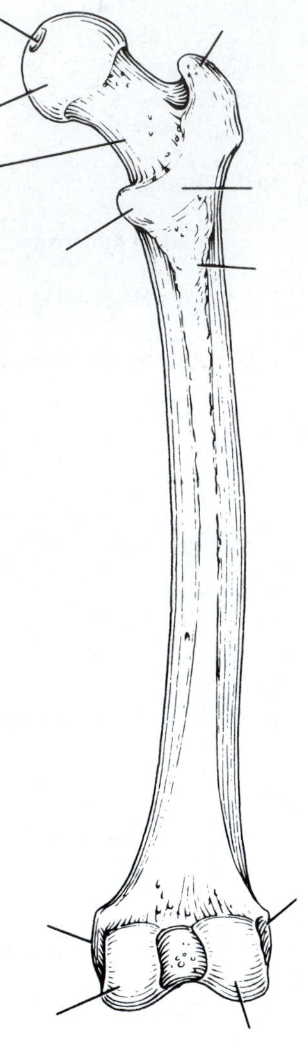

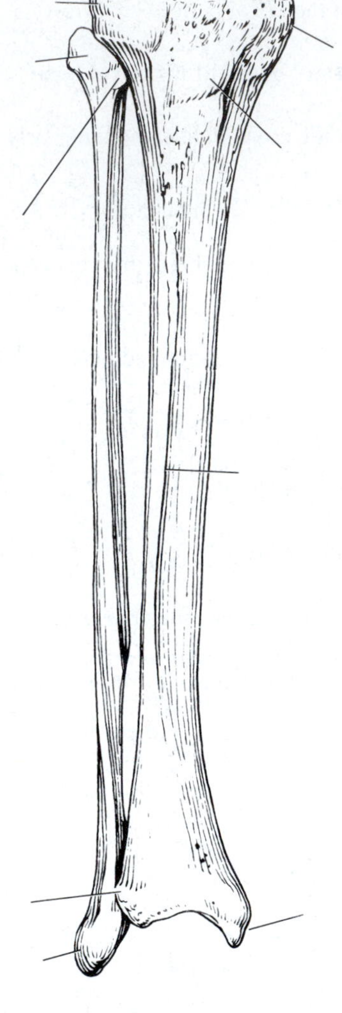

Key:

a. distal tibiofibular joint
b. fovea capitis
c. gluteal tuberosity
d. greater trochanter
e. head of femur
f. head of fibula
g. intercondylar eminence
h. intertrochanteric crest
i. lateral condyle
j. lateral epicondyle
k. lateral malleolus
l. lesser trochanter
m. medial condyle
n. medial epicondyle
o. medial malleolus
p. neck of femur
q. proximal tibiofibular joint
r. tibial anterior crest
s. tibial tuberosity

The femur (the diagram on the ____________________ side) is the ____________________ member of the two femurs.

The tibia and fibula (the diagram on the ____________________ side) are ____________________ bones.

Summary of Skeleton

1. Identify all indicated bones (or groups of bones) in the diagram of the articulated skeleton on page 529.

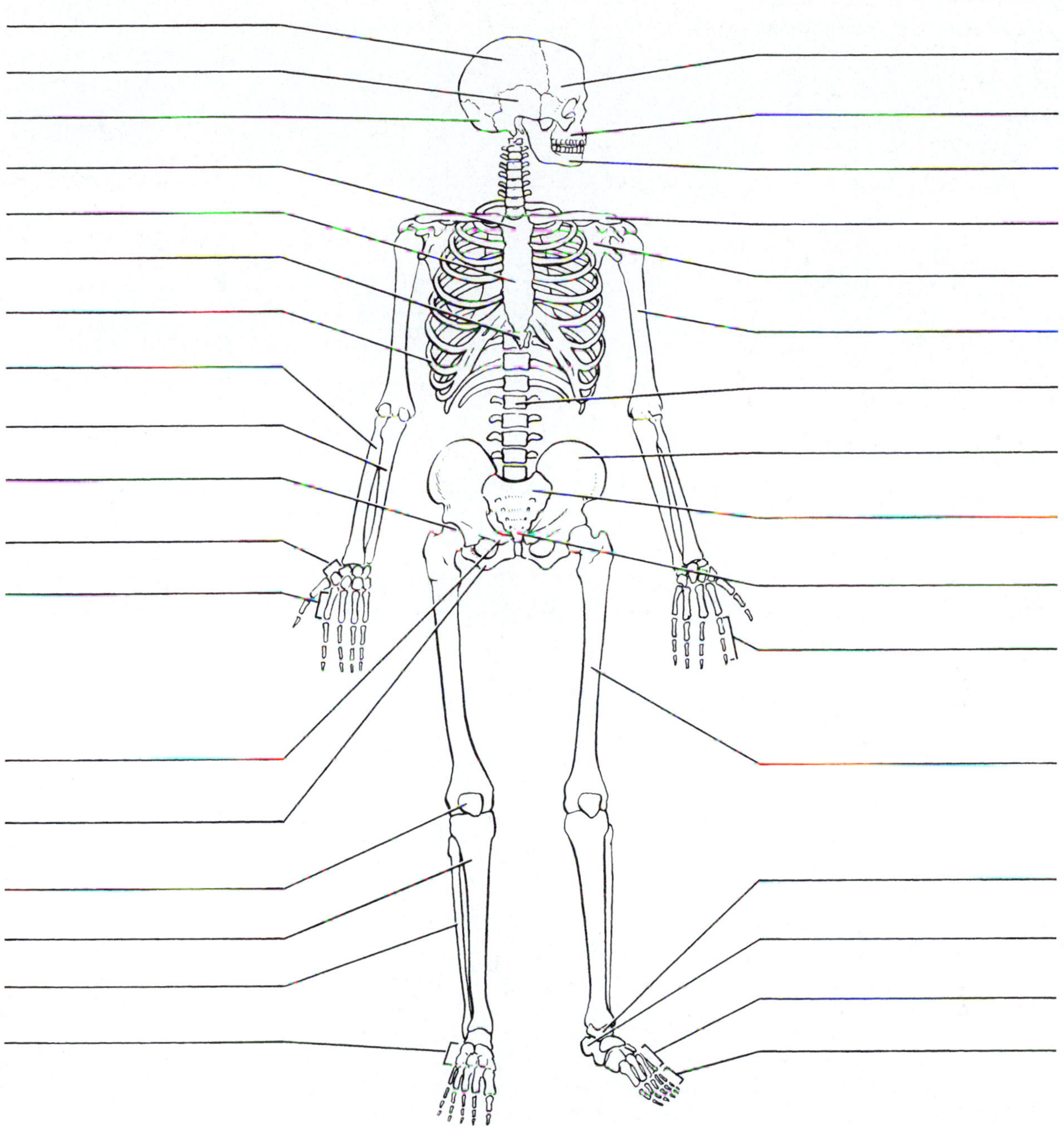

NAME ______________________ LAB TIME/DATE ______________________

REVIEW SHEET
exercise
12

The Fetal Skeleton

1. Are the same skull bones seen in the adult found in the fetal skull? ______________________

2. How does the size of the fetal face compare to its cranium? ______________________

____

How does this compare to the adult skull? ______________________

____

____

3. What are the outward conical projections in some of the fetal cranial bones? ______________________

____

4. What is a fontanel? ______________________

What is its fate? ______________________

What is the function of the fontanels in the fetal skull? ______________________

____

5. Describe how the fetal skeleton compares with the adult skeleton in the following areas:

vertebrae ______________________

____

os coxae ______________________

____

carpals and tarsals ______________________

sternum ______________________

frontal bone ______________________

patella ______________________

rib cage ______________________

____

6. How does the size of the fetus's head compare to the size of its body? ______________________________

____

7. Using the terms listed, identify each of the fontanels shown on the fetal skull below.

a. anterior fontanel   b. mastoid fontanel   c. posterior fontanel   d. sphenoidal fontanel

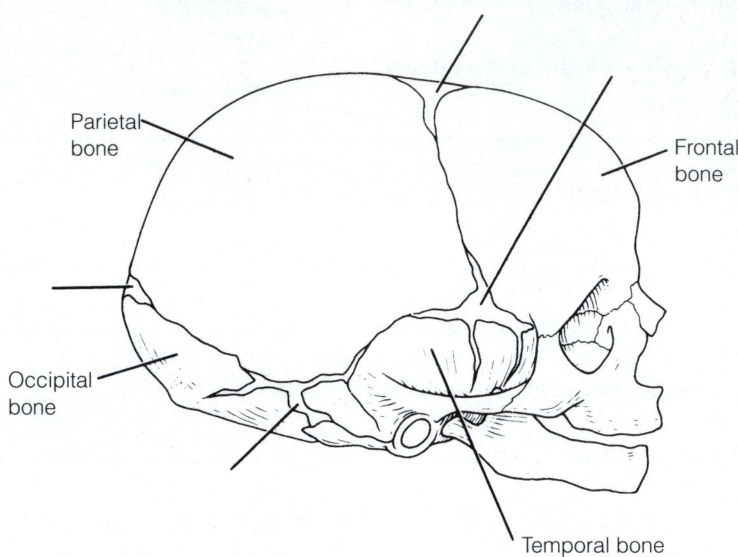

NAME ____________________ LAB TIME/DATE ____________________

REVIEW SHEET
exercise
13

Articulations and Body Movements

Types of Joints

1. Use key responses to identify the joint types described below.

Key: a. cartilaginous b. fibrous c. synovial

__________ 1. typically allows a slight degree of movement

__________ 2. includes joints between the vertebral bodies and the pubic symphysis

__________ 3. essentially immovable joints

__________ 4. sutures are the most remembered examples

__________ 5. characterized by cartilage connecting the bony portions

__________ 6. all characterized by a fibrous articular capsule lined with a synovial membrane surrounding a joint cavity

__________ 7. all are freely movable or diarthrotic

__________ 8. bone regions are united by fibrous connective tissue

__________ 9. include the hip, knee, and elbow joints

2. Describe the structure and function of the following structures or tissues in relation to a synovial joint and label the structures indicated by leader lines in the diagram.

ligament ____________________

____

tendon ____________________

____

articular cartilage ____________________

____

synovial membrane ____________________

____

____

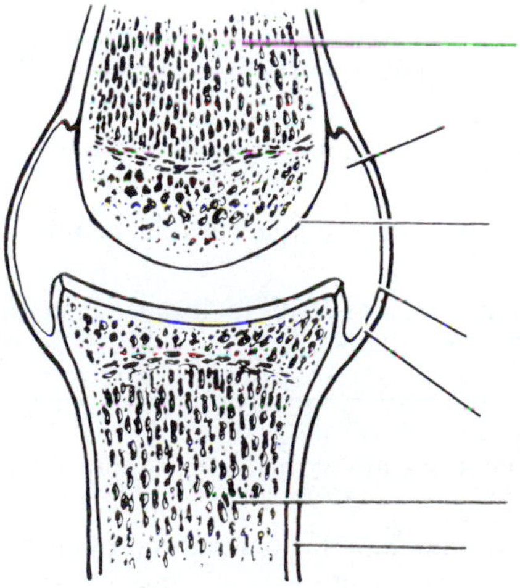

bursa ____________________

3. Match the joint subcategories in column B with their descriptions in column A, and place an asterisk (*) beside all choices that are examples of synovial joints.

	Column A	Column B
________________	1. joint between skull bones	a. ball and socket
________________	2. joint between the axis and atlas	b. condyloid
________________	3. hip joint	c. gliding
________________	4. intervertebral joints (between articular processes)	d. hinge
________________	5. joint between forearm bones and wrist	e. pivot
________________	6. elbow	f. saddle
________________	7. interphalangeal joints	g. suture
________________	8. intercarpal joints	h. symphysis
________________	9. joint between tarsus and tibia/fibula	i. synchondrosis
________________	10. joint between skull and vertebral column	j. syndesmosis
________________	11. joint between jaw and skull	
________________	12. joints between proximal phalanges and metacarpal bones	
________________	13. epiphyseal plate of a child's long bone	
________________	14. a multiaxial joint	
________________, ________________	15. biaxial joints	
________________, ________________	16. uniaxial joints	

4. When considering movement,

What do all uniaxial joints have in common? ________________________________

What do all biaxial joints have in common? ________________________________

What do all multiaxial joints have in common? ________________________________

5. What characteristics do all joints have in common? ________________________________

____

6. Which joint, the hip or the knee, is more stable? ________________________________

Name two important factors that contribute to the stability of the hip joint.

________________________________ and ________________________________

Name two important factors that contribute to the stability of the knee.

________________________________ and ________________________________

7. The diagram shows a frontal section of the hip joint. Identify its major structural elements by using the key letters.

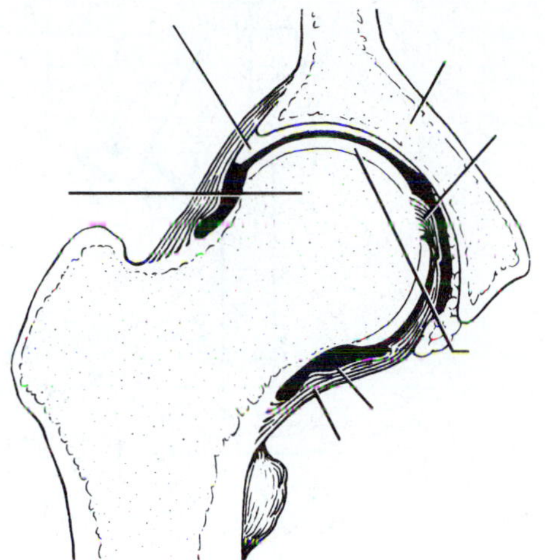

Key:

a. acetabular labrum

b. articular capsule

c. articular cartilage

d. coxal bone

e. head of femur

f. ligamentum teres

g. synovial cavity

Movements Allowed by Synovial Joints

1. Label the *origin* and *insertion* points on the diagram below and complete the following statement:

During muscle contraction, the ________________

moves toward the ___________________________.

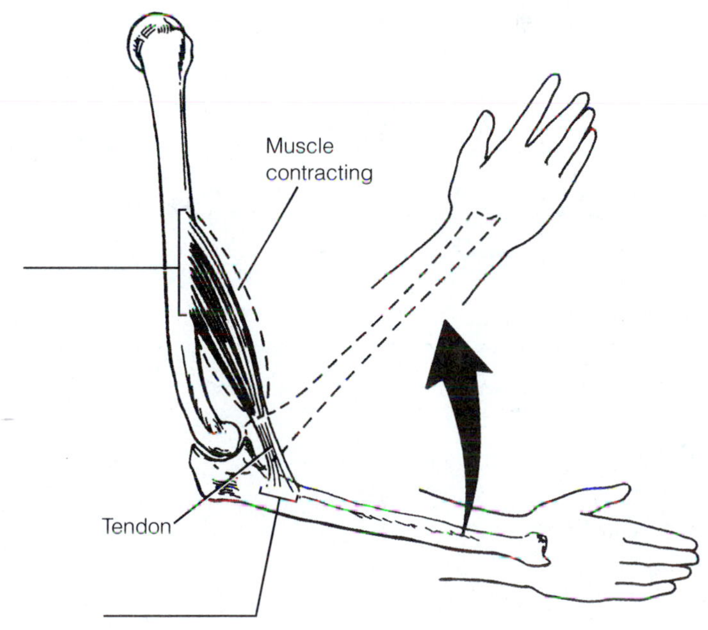

2. Complete the statements below the stick diagrams by inserting the missing words in the answer blanks.

1. ____________________________

2. ____________________________

(continues on next page)

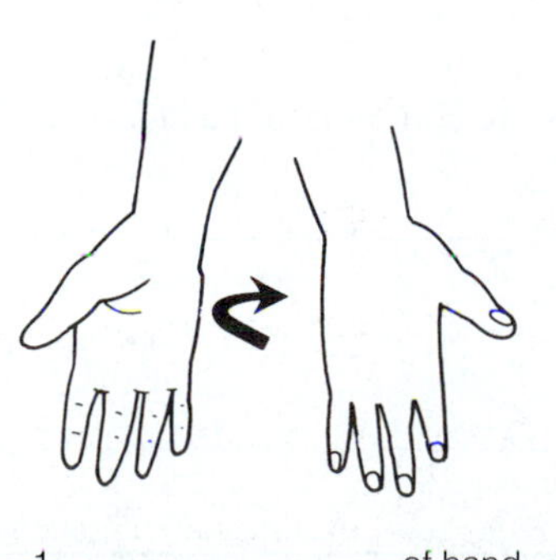

1. ______________ of hand

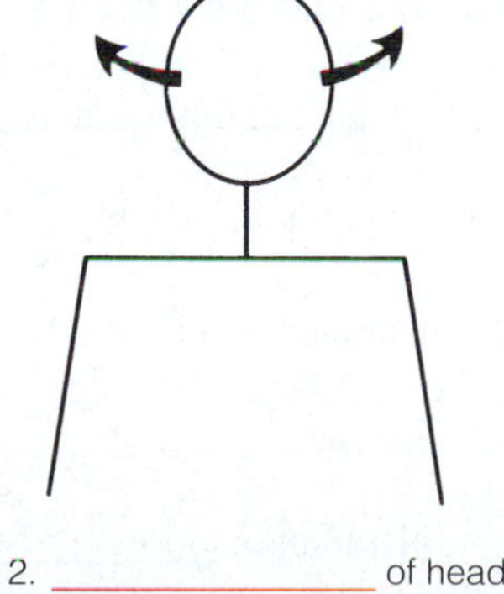

2. ______________ of head

3. ______________

4. ______________

5. ______________

6. ______________

7. ______________

8. ______________

9. ______________

10. ______________

11. ______________

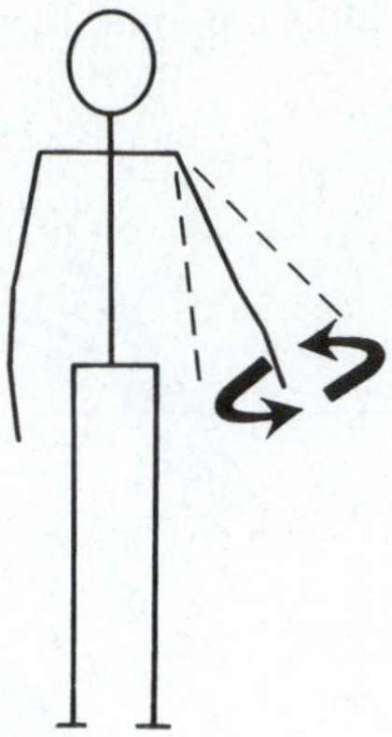

3. ______________ of the arm

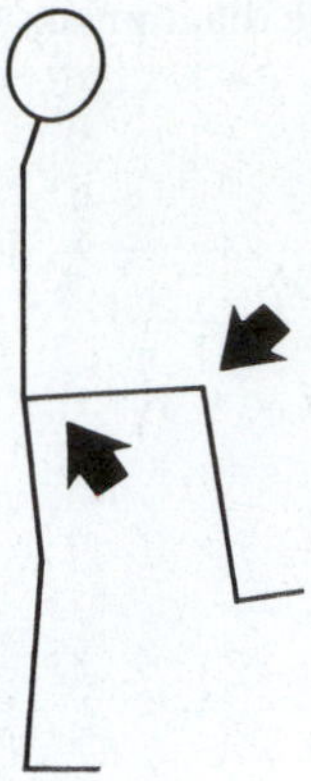

4. ______________ of hip
5. ______________ of knee

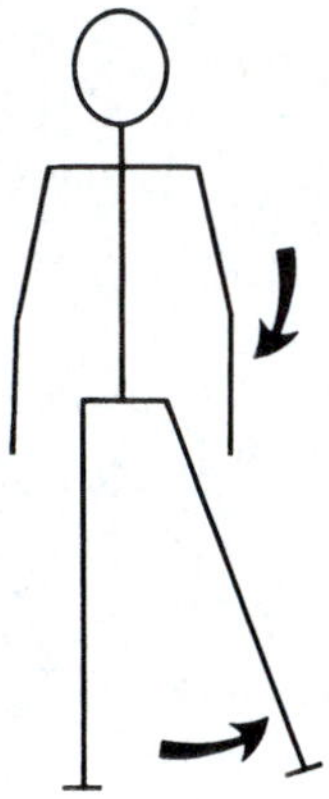

6. ______________ of thigh
7. ______________ of arm

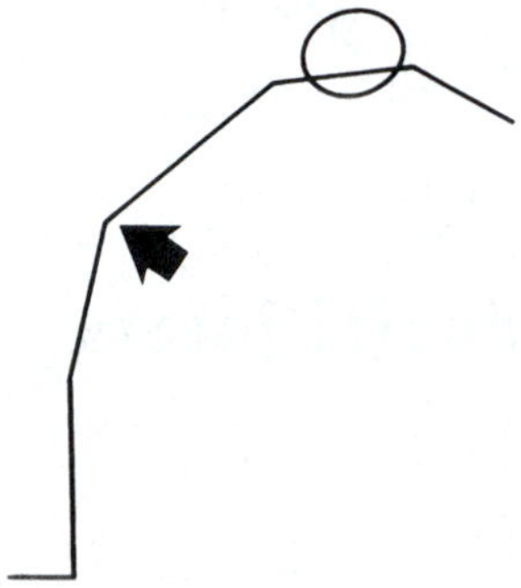

8. ______________ of hip

9. ________ of foot

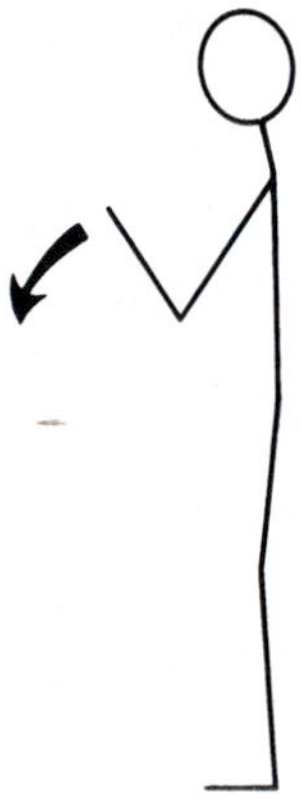

10. ______________ of elbow

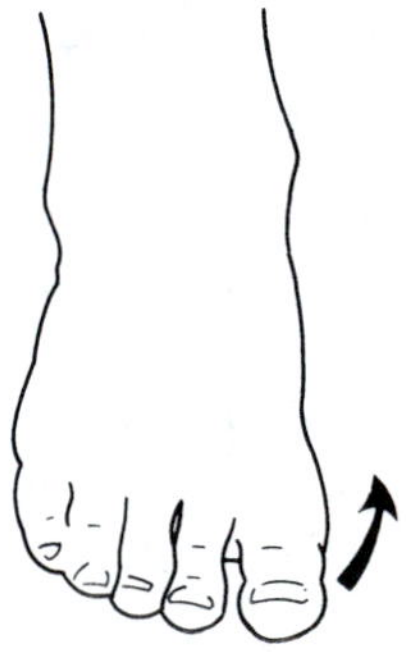

11. ______________ of foot

Joint Disorders

1. What structural joint changes are common to the elderly? ______________

____

2. Define:

sprain ______________

dislocation ______________

NAME ______________________ LAB TIME/DATE ______________________

REVIEW SHEET
exercise
14

Microscopic Anatomy, Organization, and Classification of Skeletal Muscle

Skeletal Muscle Cells and Their Packaging into Muscles

1. What capability is most highly expressed in muscle tissue? ______________________

2. Use the items on the right to correctly identify the structures described on the left.

__________	1. connective tissue ensheathing a bundle of muscle cells	a. endomysium
__________	2. bundle of muscle cells	b. epimysium
__________	3. contractile unit of muscle	c. fascicle
__________	4. a muscle cell	d. fiber
__________	5. thin reticular connective tissue investing each muscle cell	e. myofilament
__________	6. plasma membrane of the muscle fiber	f. myofibril
__________	7. a long filamentous organelle with a banded appearance found within muscle cells	g. perimysium
__________	8. actin- or myosin-containing structure	h. sarcolemma
__________	9. cord of collagen fibers that attaches a muscle to a bone	i. sarcomere
		j. sarcoplasm
		k. tendon

3. Why are the connective tissue wrappings of skeletal muscle important? (Give at least three reasons.)

____

____

____

____

4. Why are indirect—that is, tendinous—muscle attachments to bone seen more often than direct attachments?

____

____

____

5. How does an aponeurosis differ from a tendon? ______________________

____

____

6. The diagram illustrates a small portion of a muscle myofibril. Using letters from the key, correctly identify each structure indicated by a leader line or a bracket. Below the diagram make a sketch of how this segment of the myofibril would look if contracted.

Key:
a. actin filament
b. A band
c. I band
d. myosin filament
e. sarcomere
f. Z disc

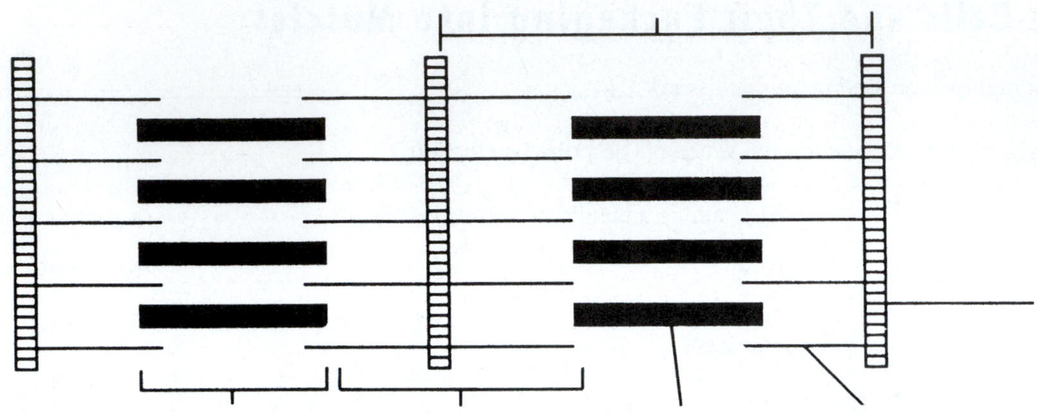

7. On the following figure, label blood vessel, endomysium, epimysium, fascicle, muscle cell, perimysium, and tendon.

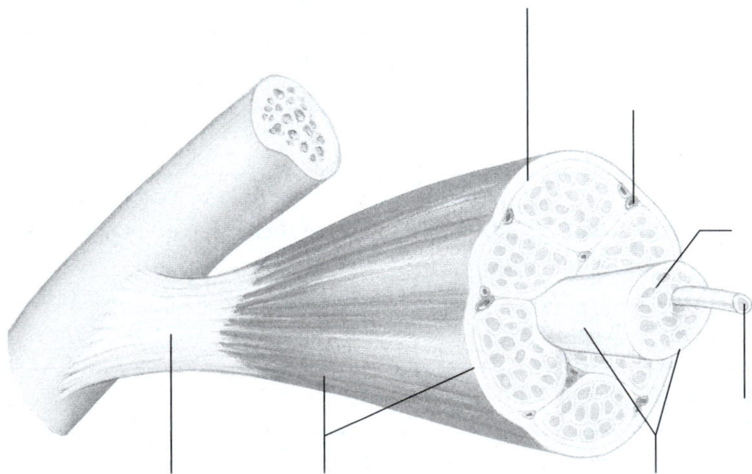

The Neuromuscular Junction

Complete the following statements:

The junction between a motor neuron's axon and the muscle cell membrane is called a neuromuscular junction or a __1__ junction. A motor neuron and all of the skeletal muscle cells it stimulates is called a __2__. The actual gap between the axonal terminal and the muscle cell is called a __3__. Within the axonal terminal are many small vesicles containing a neurotransmitter substance called __4__. When the __5__ reaches the ends of the axon, the neurotransmitter is released and diffuses to the muscle cell membrane to combine with receptors there. The combining of the neurotransmitter with the muscle membrane receptors causes the membrane to become permeable to sodium, which results in the influx of sodium ions and __6__ of the membrane. Then contraction of the muscle cell occurs. Before a muscle cell can be stimulated to contract again, __7__ must occur.

1. ______________________
2. ______________________
3. ______________________
4. ______________________
5. ______________________
6. ______________________
7. ______________________

Classification of Skeletal Muscles

1. Several criteria were given relative to the naming of muscles. Match the criteria (column B) to the muscle names (column A). Note that more than one criterion may apply in some cases.

Column A	Column B
________ 1. gluteus maximus	a. action of the muscle
________ 2. adductor magnus	b. shape of the muscle
________ 3. biceps femoris	c. location of the origin and/or insertion of the muscle
________ 4. transversus abdominis	d. number of origins
________ 5. extensor carpi ulnaris	e. location of the muscle relative to a bone or body region
________ 6. trapezius	f. direction in which the muscle fibers run relative to some imaginary line
________ 7. rectus femoris	g. relative size of the muscle
________ 8. external oblique	

2. When muscles are discussed relative to the manner in which they interact with other muscles, the terms shown in the key are often used. Match the key terms with the appropriate definitions.

Key: a. antagonist   b. fixator   c. prime mover   d. synergist

________________ 1. agonist

________________ 2. postural muscles, for the most part

________________ 3. reverses and/or opposes the action of a prime mover

________________ 4. stabilizes a joint so that the prime mover may act at more distal joints

________________ 5. performs the same movement as the prime mover

________________ 6. immobilizes the origin of a prime mover

NAME ______________________ LAB TIME/DATE ______________________

REVIEW SHEET
exercise
15

Gross Anatomy of the Muscular System

Muscles of the Head and Neck

1. Using choices from the list at the right, correctly identify muscles provided with leader lines on the diagram.

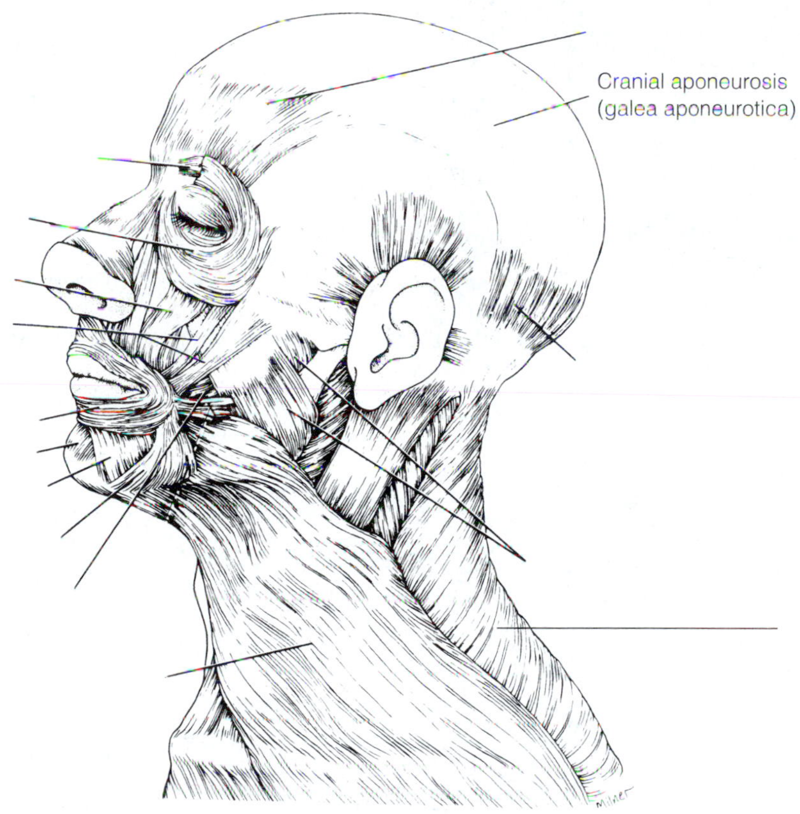

a. buccinator

b. corrugator supercilii

c. depressor anguli oris

d. depressor labii inferioris

e. frontalis

f. levator labii superioris

g. masseter

h. mentalis

i. platysma

j. occipitalis

k. orbicularis oculi

l. orbicularis oris

m. trapezius

n. zygomaticus major and minor

2. Using the terms provided above, identify the muscles described next.

_______ 1. used in smiling

_______ 2. used to suck in your cheeks

_______ 3. used in blinking and squinting

_______ 4. used to pout (pulls the corners of the mouth downward)

_______ 5. raises your eyebrows for a questioning expression

_______ 6. used to form the vertical frown crease on the forehead

_______ 7. your "kisser"

_______ 8. prime mover to raise the lower jawbone

_______ 9. tenses skin of the neck during shaving

Muscles of the Trunk

1. Correctly identify both intact and transected (cut) muscles depicted in the diagram, using the terms given at the right. (Not all terms will be used in this identification.)

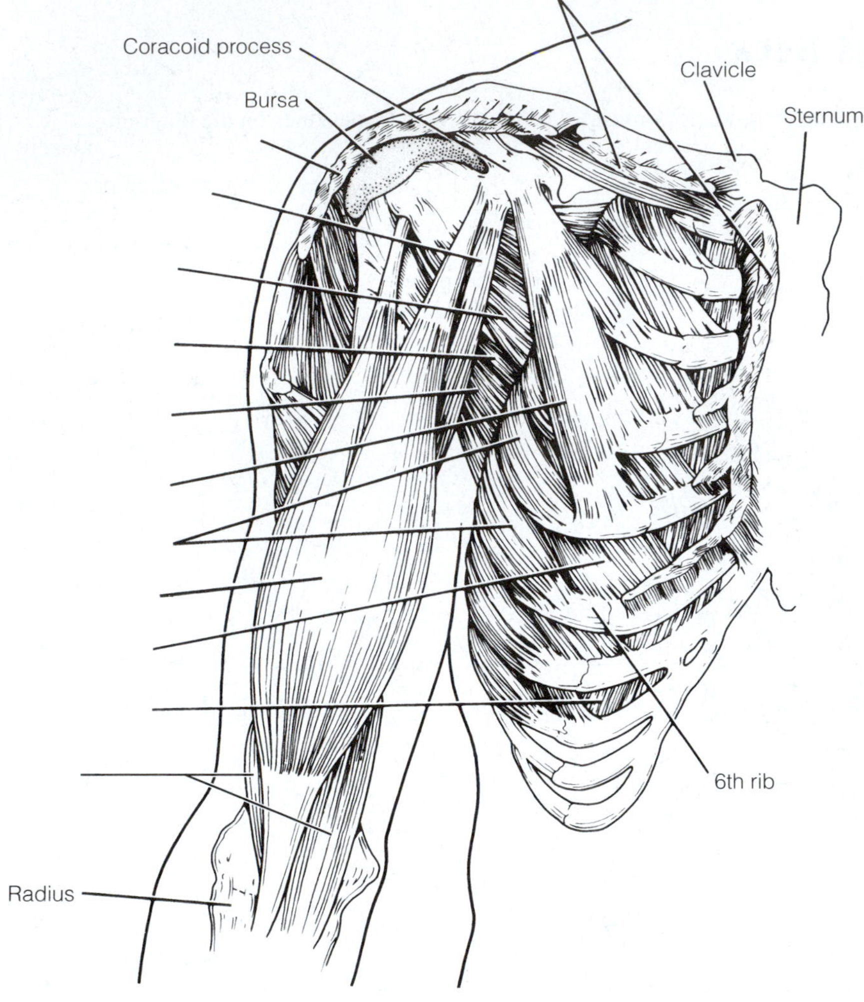

a. biceps brachii
b. brachialis
c. coracobrachialis
d. deltoid (cut)
e. external intercostals
f. external oblique
g. internal intercostals
h. internal oblique
i. latissimus dorsi
j. pectoralis major (cut)
k. pectoralis minor
l. rectus abdominis
m. rhomboids
n. serratus anterior
o. subscapularis
p. teres major
q. teres minor
r. transversus abdominis
s. trapezius

2. Using choices from the terms provided in question 1 above, identify the major muscles described next:

_______ 1. a major spine flexor

_______ 2. prime mover for pulling the arm posteriorly

_______ 3. prime mover for shoulder flexion

_______ 4. assume major responsibility for forming the abdominal girdle (three pairs of muscles)

_______ 5. pulls the shoulder backward and downward

_______ 6. prime mover of shoulder abduction

_______ 7. important in shoulder adduction; antagonists of the shoulder abductor (two muscles)

_______ 8. moves the scapula forward and downward

_______ 9. small, inspiratory muscles between the ribs; elevate the ribs

_______ 10. extends the head

_______ 11. pull the scapulae medially

Muscles of the Upper Limb

1. Using terms from the list on the right, correctly identify all muscles provided with leader lines in the diagram. Note that not all the listed terms will be used in this exercise.

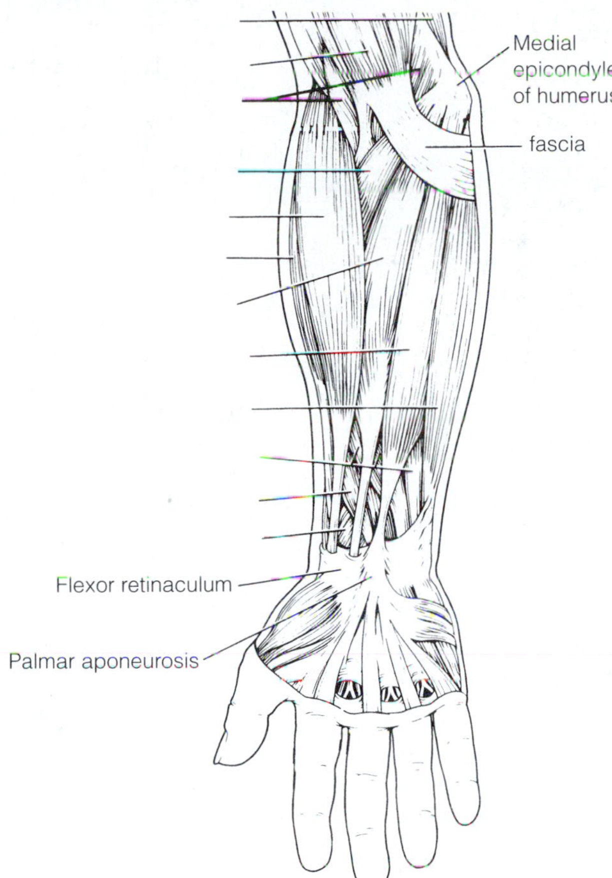

a. biceps brachii
b. brachialis
c. brachioradialis
d. extensor carpi radialis longus
e. extensor digitorum
f. flexor carpi radialis
g. flexor carpi ulnaris
h. flexor digitorum superficialis
i. flexor pollicis longus
j. palmaris longus
k. pronator quadratus
l. pronator teres
m. supinator
n. triceps brachii

2. Use the terms provided in question 1 to identify the muscles described next.

_______ 1. places the palm upward (two muscles)

_______ 2. flexes the forearm and supinates the hand

_______ 3. forearm flexors; no role in supination (two muscles)

_______ 4. elbow extensor

_______ 5. power wrist flexor and abductor

_______ 6. flexes wrist and middle phalanges

_______ 7. pronate the hand (two muscles)

_______ 8. flexes the thumb

_______ 9. extends and abducts the wrist

_______ 10. extends the wrist and digits

_______ 11. flat muscle that is a weak wrist flexor

Muscles of the Lower Limb

1. Using the terms listed to the right, correctly identify all muscles provided with leader lines in the diagram below. Not all listed terms will be used in this exercise.

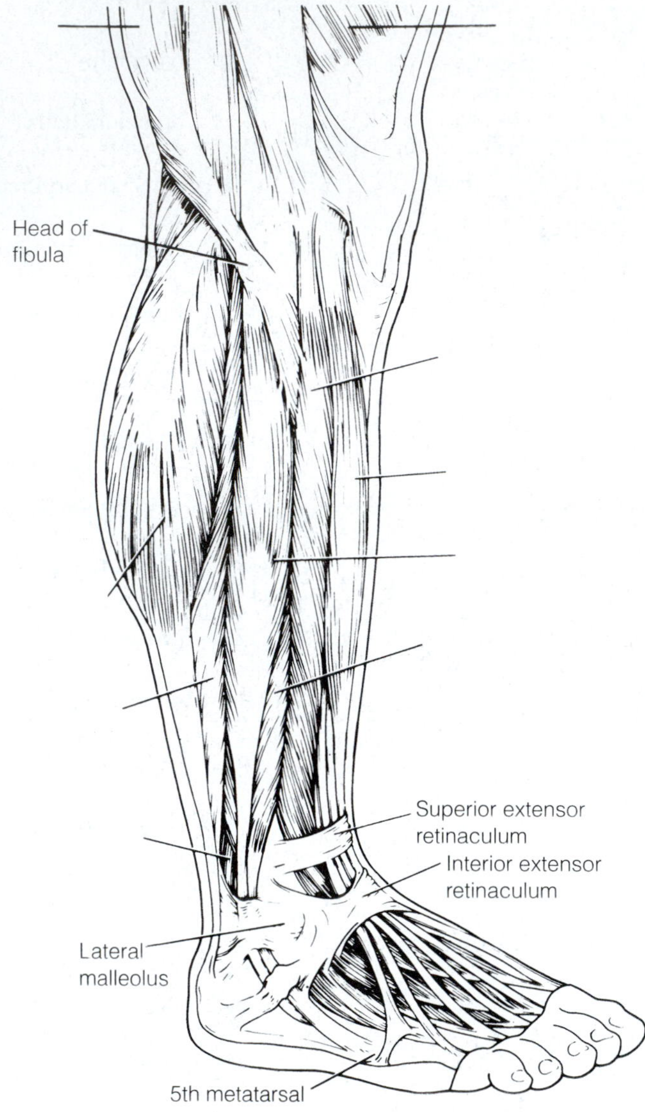

a. adductor group

b. biceps femoris

c. extensor digitorum longus

d. fibularis brevis

e. fibularis longus

f. flexor hallucis longus

g. gastrocnemius

h. gluteus maximus

i. gluteus medius

j. rectus femoris

k. semimembranosus

l. semitendinosus

m. soleus

n. tensor fasciae latae

o. tibialis anterior

p. tibialis posterior

q. vastus lateralis

2. Use the key terms in exercise 1 to respond to the descriptions below.

_______ 1. flexes the great toe and inverts the ankle

_______ 2. lateral compartment muscles that plantar flex and evert the ankle (two muscles)

_______ 3. move the thigh laterally to take the "at ease" stance (two muscles)

_______ 4. used to extend the hip when climbing stairs

_______ 5. prime movers of ankle plantar flexion (two muscles)

_______ 6. major foot inverter

_______ 7. prime mover of ankle dorsiflexion

_______ 8. allow you to draw your legs to the midline of your body, as when standing at attention

_______ 9. extends the toes

_______ 10. extend thigh and flex knee (three muscles)

_______ 11. extends knee and flexes thigh

General Review: Muscle Recognition

1. Identify the lettered muscles in the diagram of the human anterior superficial musculature by matching the letter with one of the following muscle names:

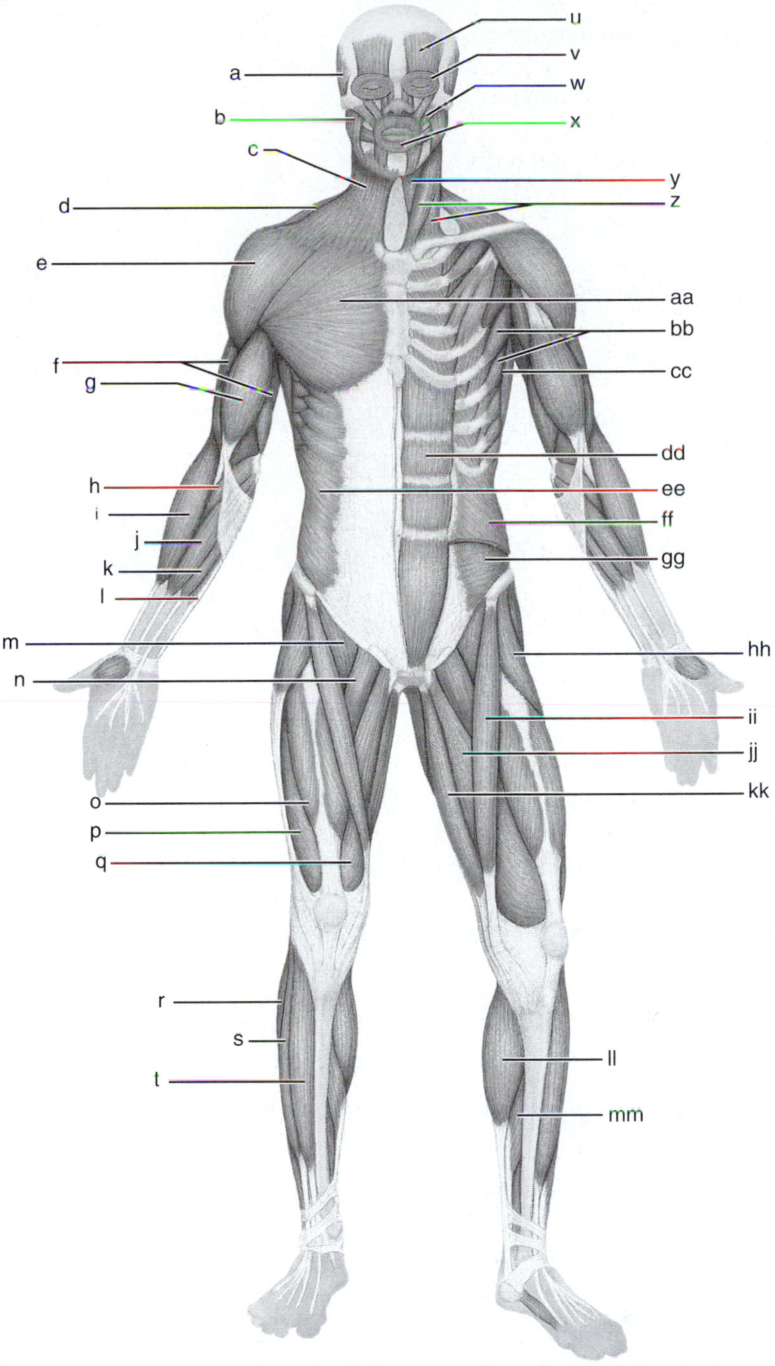

_____ 1. orbicularis oris
_____ 2. pectoralis major
_____ 3. external oblique
_____ 4. sternocleidomastoid
_____ 5. biceps brachii
_____ 6. deltoid
_____ 7. vastus lateralis
_____ 8. brachioradialis
_____ 9. frontalis
_____ 10. rectus femoris
_____ 11. pronator teres
_____ 12. rectus abdominis
_____ 13. sartorius
_____ 14. gracilis
_____ 15. flexor carpi ulnaris
_____ 16. adductor longus
_____ 17. palmaris longus
_____ 18. flexor carpi radialis
_____ 19. latissimus dorsi
_____ 20. orbicularis oculi
_____ 21. gastrocnemius
_____ 22. masseter
_____ 23. trapezius
_____ 24. tibialis anterior
_____ 25. extensor digitorum longus
_____ 26. tensor fasciae latae
_____ 27. pectineus
_____ 28. sternohyoid
_____ 29. serratus anterior
_____ 30. platysma
_____ 31. vastus medialis
_____ 32. transversus abdominis
_____ 33. fibularis longus
_____ 34. iliopsoas
_____ 35. temporalis
_____ 36. zygomaticus
_____ 37. soleus
_____ 38. triceps brachii
_____ 39. internal oblique

2. Identify each of the lettered muscles in this diagram of the human posterior superficial musculature by matching its letter to one of the following muscle names:

_______ 1. gluteus maximus

_______ 2. semimembranosus

_______ 3. gastrocnemius

_______ 4. latissimus dorsi

_______ 5. deltoid

_______ 6. iliotibial tract (tendon)

_______ 7. teres major

_______ 8. semitendinosus

_______ 9. trapezius

_______ 10. biceps femoris

_______ 11. triceps brachii

_______ 12. external oblique

_______ 13. gluteus medius

_______ 14. gracilis

_______ 15. flexor carpi ulnaris

_______ 16. extensor carpi ulnaris

_______ 17. extensor digitorum

_______ 18. extensor carpi radialis longus

_______ 19. occipitalis

_______ 20. sternocleidomastoid

_______ 21. adductor magnus

_______ 22. brachialis

_______ 23. infraspiratus

_______ 24. brachioradialis

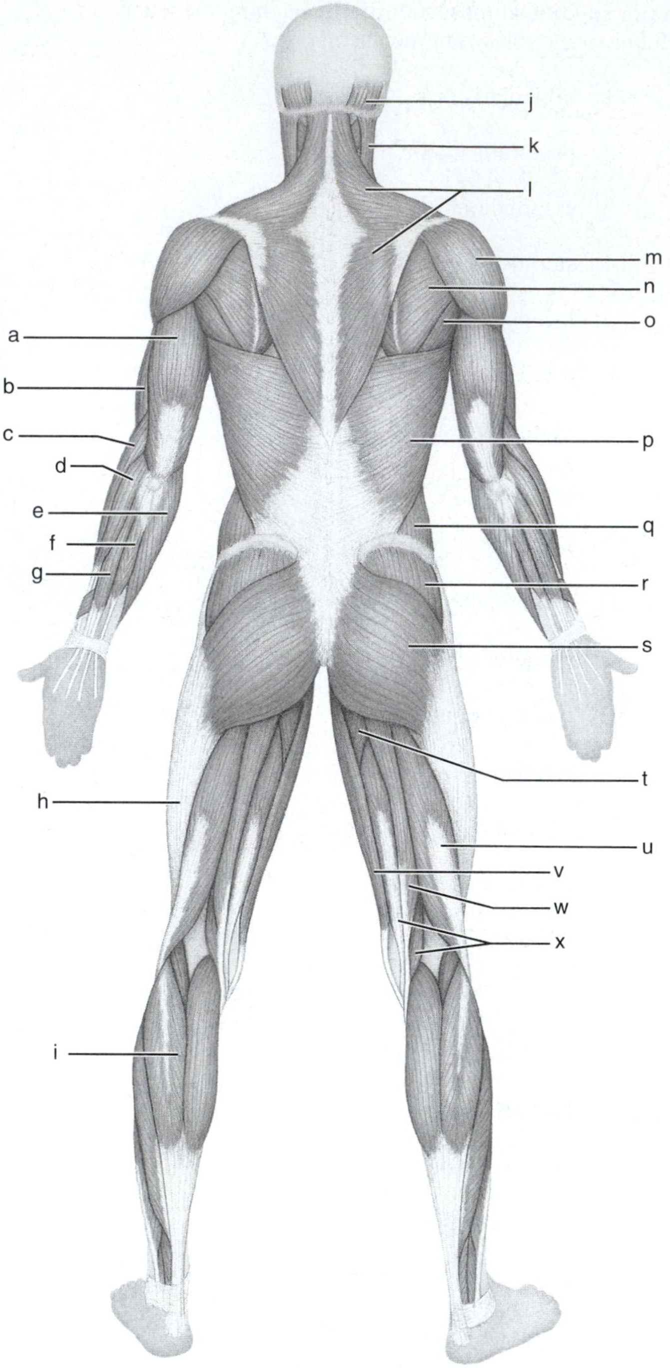

General Review: Muscle Descriptions

1. Identify the muscles described below by completing the statements:

   ________________________________, ________________________________, and ________________________________

   are commonly used for intramuscular injections (three muscles).

2. The insertion tendon of the ______________________________ group contains a large sesamoid bone, the patella.

3. The triceps surae insert in common into the ______________________________ tendon.

4. The bulk of the tissue of a muscle tends to lie ______________________________ to the part of the body it causes to move.

5. The extrinsic muscles of the hand originate on the ____.

6. Most flexor muscles are located on the ____ aspect of the body; most extensors are located ____. An exception to this generalization is the extensor-flexor musculature of the ____.

NAME ____________________ LAB TIME/DATE ____________________

REVIEW SHEET

exercise

16A

Skeletal Muscle Physiology—Frog Experimentation: Wet Lab

Muscle Activity

1. The following group of incomplete statements refers to a muscle cell in the resting or polarized state just before stimulation. Complete each statement by choosing the correct response from the key items below.

Key:
a. Na^+ diffuses out of the cell
b. K^+ diffuses out of the cell
c. Na^+ diffuses into the cell
d. K^+ diffuses into the cell
e. inside the cell
f. outside the cell
g. relative ionic concentrations on the two sides of the membrane
h. electrical conditions
i. activation of the sodium-potassium pump, which moves K^+ into the cell and Na^+ out of the cell
j. activation of the sodium-potassium pump, which moves Na^+ into the cell and K^+ out of the cell

There is a greater concentration of Na^+ ________; there is a greater concentration of K^+ ________. When the stimulus is delivered, the permeability of the membrane at that point is changed; and ________, initiating the depolarization of the membrane. Almost as soon as the depolarization wave has begun, a repolarization wave follows it across the membrane. This occurs as ________. Repolarization restores the ________ of the resting cell membrane. The ________ is (are) reestablished by ________.

2. Number the following statements in the proper sequence to describe the contraction mechanism in a skeletal muscle cell. Number 1 has already been designated.

___1___ Acetylcholine is released into the neuromuscular junction by the axonal terminal.

________ The action potential, carried deep into the cell by the T system, triggers the release of calcium ions from the sarcoplasmic reticulum.

________ The muscle cell relaxes and lengthens.

________ Acetylcholine diffuses across the neuromuscular junction and binds to receptors on the sarcolemma.

________ The calcium ion concentrations at the myofilaments increase; the myofilaments slide past one another, and the cell shortens.

________ Depolarization occurs, and the action potential is generated.

________ The concentration of the calcium ions at the myofilaments decreases as they are actively transported into the sarcoplasmic reticulum.

3. Muscle contraction is commonly explained by the sliding filament hypothesis. What are the essential points of this hypothesis? ______________________________

____

____

4. Relative to your observations of muscle fiber contraction (pp. 166–168):

a. What percentage of contraction was observed with the solution containing ATP, K^+, and Mg^{2+}? _______%

With *just* ATP? _______% With *just* Mg^{2+} and K^+? _______%

b. Did your data support your hypothesis? _______

c. *Explain* your observations fully. ______________________________

____

d. What zones or bands disappear when the muscle cell contracts? ______________________________

e. Draw a relaxed and a contracted sarcomere below.

Relaxed Contracted

Induction of Contraction in the Frog Gastrocnemius Muscle

1. Why is it important to destroy the brain and spinal cord of a frog before conducting physiological experiments on muscle contraction? ______________________________

____

2. What sources of stimuli, other than electrical shocks, cause a muscle to contract? ______________________________

____

3. What is the most common stimulus for muscle contraction in the body? ______________________________

4. Give the name and duration of each of the three phases of the muscle twitch, and describe what is happening during each phase:

(1) ______________, _______ sec, ______________________________

____

(2) ______________, _______ sec, ______________________________

(3) ______________, _______ sec, ______________________________

5. Use the terms given on the right to identify the conditions described on the left:

______________	1. sustained contraction without any evidence of relaxation	a. maximal stimulus
______________	2. stimulus that results in no perceptible contraction	b. multiple motor unit summation
______________	3. stimulus at which the muscle first contracts perceptibly	c. subthreshold stimulus
______________	4. increasingly stronger contractions in the absence of increased stimulus intensity	d. tetanus
______________	5. increasingly stronger contractions owing to stimulation at a rapid rate	e. threshold stimulus
______________	6. increasingly stronger contractions owing to increased stimulus strength	f. treppe
______________	7. weakest stimulus at which all muscle cells in the muscle are contracting	g. wave summation

6. With brackets and labels, identify the portions of the tracing below that best correspond to three of the phenomena listed in the preceding key.

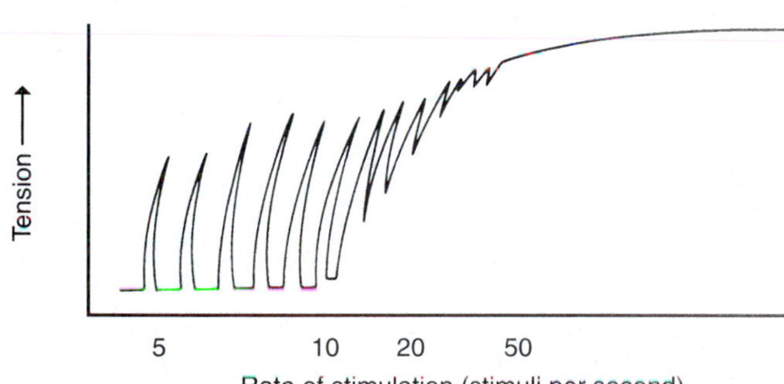

7. Complete the following statements by writing the appropriate words on the correspondingly numbered blanks at the right.

When a weak but smooth muscle contraction is desired, a few motor units are stimulated at a __1__ rate. Treppe is referred to as the "warming up" process. It is believed that muscles contract more strongly after the first few contractions because the __2__ become more efficient. If blue litmus paper is pressed to the cut surface of a fatigued muscle, the paper color changes to red, indicating low pH. This situation is caused by the accumulation of __3__ in the muscle. Within limits, as the load on a muscle is increased, the muscle contracts __4__ strongly. The relative refractory period is the time when the muscle cell will not respond to a stimulus because __5__ is occurring.

1. ______________
2. ______________
3. ______________
4. ______________
5. ______________

8. During the experiment on muscle fatigue, how did the muscle contraction pattern change as the muscle began to fatigue?

____

____

How long was stimulation continued before fatigue was apparent? ______________________

If the sciatic nerve that stimulates the living frog's gastrocnemius muscle had been left attached to the muscle and the stimulus had been applied to the nerve rather than the muscle, would fatigue have become apparent sooner or later?

____

Explain your answer. ____

____

9. Explain how the weak but smooth sustained muscle contractions of precision movements are produced.

____

10. What do you think happens to a muscle in the body when its nerve supply is destroyed or badly damaged?

____

____

11. Explain the relationship between the load on a muscle and its strength of contraction. ______________

____

____

12. The skeletal muscles are maintained in a slightly stretched condition for optimal contraction. How is this accomplished?

____

Why does overstretching a muscle drastically reduce its ability to contract? (Include an explanation of the events at the level of the myofilaments.) ____

____

____

13. If the length but not the tension of a muscle is changed, the contraction is called an isotonic contraction. In an isometric contraction the tension is increased but the muscle does not shorten. Which type of contraction did you observe most often during the laboratory experiments? ______________________________

What is the role of isometric contractions in normal body functioning? ______________________

____

____

NAME ______________________ LAB TIME/DATE ______________________

REVIEW SHEET
exercise
17

Histology of Nervous Tissue

1. The cellular unit of the nervous system is the neuron. What is the major function of this cell type?

____

2. Name four types of neuroglia and list at least four functions of these cells. (You will need to consult your textbook for this.)

Types	**Functions**
a. ______________	a. ______________
	______________
b. ______________	b. ______________
	______________
c. ______________	c. ______________
	______________
d. ______________	d. ______________
	______________

3. Match each statement with a response chosen from the key.

Key:

a. afferent neuron	e. ganglion	i. nuclei
b. association neuron	f. neuroglia	j. peripheral nervous system
c. central nervous system	g. neurotransmitters	k. synapse
d. efferent neuron	h. nerve	l. tract

_______ 1. the brain and spinal cord collectively

_______ 2. specialized supporting cells in the CNS

_______ 3. junction or point of close contact between neurons

_______ 4. a bundle of nerve processes inside the central nervous system

_______ 5. neuron serving as part of the conduction pathway between sensory and motor neurons

_______ 6. spinal and cranial nerves and ganglia

_______ 7. collection of nerve cell bodies found outside the CNS

_______ 8. neuron that conducts impulses away from the CNS to muscles and glands

_______ 9. neuron that conducts impulses toward the CNS from the body periphery

_______ 10. chemicals released by neurons that stimulate or inhibit other neurons or effectors

Neuron Anatomy

1. Match the following anatomical terms (column B) with the appropriate description or function (column A).

	Column A	Column B
________	1. region of the cell body from which the axon originates	a. axon
________	2. secretes neurotransmitters	b. axonal terminal
________	3. receptive region of a neuron	c. axon hillock
________	4. insulates the nerve fibers	d. dendrite
________	5. is site of the nucleus and the most important metabolic area	e. myelin sheath
________	6. may be involved in the transport of substances within the neuron	f. neuronal cell body
________	7. essentially rough endoplasmic reticulum, important metabolically	g. neurofibril
________	8. impulse generator and transmitter	h. Nissl bodies

2. Draw a "typical" neuron in the space below. Include and label the following structures on your diagram: cell body, nucleus, nucleolus, Nissl bodies, dendrites, axon, axon collateral branch, myelin sheath, nodes of Ranvier, axonal terminals, and neurofibrils.

3. How is one-way conduction at synapses ensured? ____

____

4. What anatomical characteristic determines whether a particular neuron is classified as unipolar, bipolar, or multipolar?

____

Make a simple line drawing of each type here.

Unipolar neuron   Bipolar neuron   Multipolar neuron

5. Correctly identify the sensory (afferent) neuron, association neuron (interneuron), and motor (efferent) neuron in the figure below.

Which of these neuron types is/are unipolar? ______________________________

Which is/are most likely multipolar? ______________________________

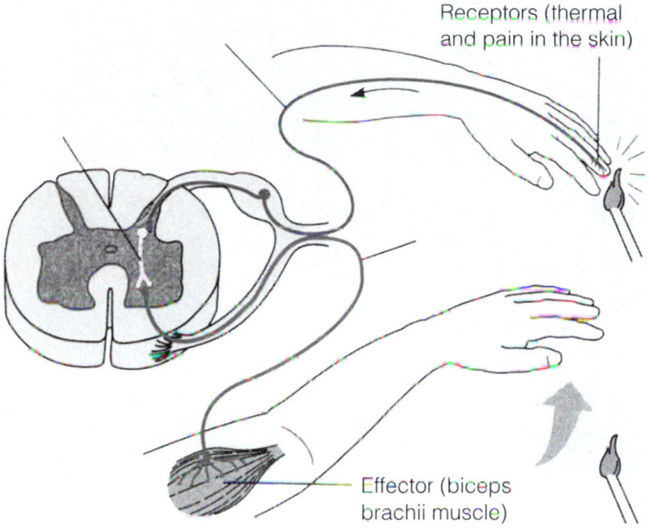

Receptors (thermal and pain in the skin)

Effector (biceps brachii muscle)

6. Describe how the Schwann cells form the myelin sheath and the neurilemma encasing the nerve processes. (You may want to diagram the process.)

____

____

____

Structure of a Nerve

1. What is a nerve? ______________________________

____

2. State the location of each of the following connective tissue coverings:

endoneurium ______________________________

perineurium ______________________________

epineurium ______________________________

3. What is the value of the connective tissue wrappings found in a nerve? ______________________________

____

4. Define *mixed nerve:* ____

____

5. Identify all indicated parts of the nerve section.

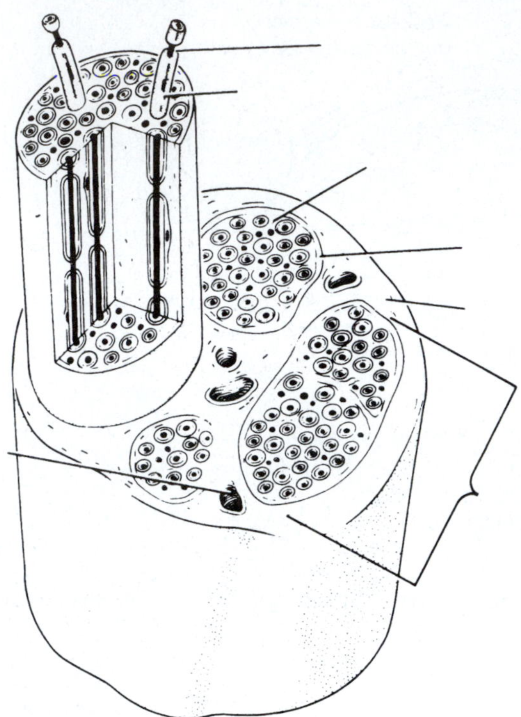

NAME ____________________ LAB TIME/DATE ____________________

REVIEW SHEET
exercise
18A

Neurophysiology of Nerve Impulses

The Nerve Impulse

1. Match each of the terms in column B to the appropriate definition in column A.

	Column A	Column B
______	1. period of depolarization of the neuron membrane during which it cannot respond to a second stimulus	a. action potential
______	2. reversal of the resting potential owing to an influx of sodium ions	b. absolute refractory period
______	3. period during which potassium ions diffuse out of the neuron owing to a change in membrane permeability	c. depolarization
______	4. self-propagated transmission of the depolarization wave along the neuronal membrane	d. relative refractory period
______	5. mechanism in which ATP is used to move sodium out of the cell and potassium into the cell; restores the resting membrane voltage and intracellular ionic concentrations	e. repolarization
		f. sodium-potassium pump

2. Respond appropriately to each statement below either by completing the statement or by answering the question raised. Insert your responses in the corresponding numbered blanks on the right.

1. The cellular unit of the nervous system is the neuron. What is the major function of this cell type? — 1. ____________________

2 and 3. What characteristics are highly developed to allow the neuron to perform this function? — 2. ____________________ 3. ____________________

4. Would a substance that decreases membrane permeability to sodium increase or decrease the probability of generating a nerve impulse? — 4. ____________________

3. Why don't the terms *depolarization* and *action potential* mean the same thing? (*Hint:* under which conditions will a local depolarization *not* lead to the action potential?) ____________________

____

____

4. Below is a drawing of a section of an axon. Complete the figure by illustrating the resting membrane potential, an area of depolarization and local current flow. Indicate the direction of the depolarization wave.

$[Na^+]$ $[K^+]$

$[K^+]$ $[Na^+]$

5. A nerve generally contains many thickly myelinated fibers that typically exhibit nodes of Ranvier. An action potential is generated along these fibers by "saltatory conduction." Use an appropriate reference to explain how saltatory conduction differs from conduction along unmyelinated fibers.

____

____

____

Physiology of Nerve Fibers: Eliciting and Inhibiting the Nerve Impulse

1. Respond appropriately to each question posed below. Insert your responses in the corresponding numbered blanks to the right.

1–3. Name three types of stimuli that resulted in action potential generation in the sciatic nerve of the frog during the laboratory experiments.

4. Which of the stimuli resulted in the most effective nerve stimulation?

5. Which of the stimuli employed in that experiment might represent types of stimuli to which nerves in the human body are subjected?

6. What is the usual mode of stimulus transfer in neuron-to-neuron interactions?

7. Since the action potentials themselves were not visualized with an oscilloscope during this initial set of experiments, how did you recognize that impulses were being transmitted?

1. ______________________________

2. ______________________________

3. ______________________________

4. ______________________________

5. ______________________________

6. ______________________________

7. ______________________________

2. Describe the observed effects of ether on nerve-muscle interaction. ______________________________

____

Does ether exert its blocking effects on nerve *or* muscle? ____________ What observations made during the experiment support your conclusion? ______________________________

____

____

3. At what site did the tubocurarine block the impulse transmission? ____________________ Provide evidence from the experiment to substantiate your conclusion. ____________________

____

____

Why was one of the frog's legs ligated in this experiment? ____________________

____

____

Visualizing the Action Potential with an Oscilloscope

1. What is a stimulus artifact? ____________________

____

2. Explain why the amplitude of the action potential recorded from the frog sciatic nerve increased when the voltage of the stimulus was increased above the threshold value. ____________________

____

3. What was the effect of cold temperature (flooding the nerve with iced Ringer's solution) on the functioning of the sciatic nerve tested? ____________________

____

4. When the nerve was reversed in position, was the impulse conducted in the opposite direction? ____________________

How can this result be reconciled with the concept of one-way conduction in neurons? ____________________

____

____

____

____

____

NAME ______________________ LAB TIME/DATE ______________________

REVIEW SHEET
exercise
19

Gross Anatomy of the Brain and Cranial Nerves

The Human Brain

1. Match the letters on the diagram of the human brain (right lateral view) to the appropriate terms listed at the left:

_______ 1. frontal lobe

_______ 2. parietal lobe

_______ 3. temporal lobe

_______ 4. precentral gyrus

_______ 5. parieto-occipital sulcus

_______ 6. postcentral gyrus

_______ 7. lateral sulcus

_______ 8. central sulcus

_______ 9. cerebellum

_______ 10. medulla

_______ 11. occipital lobe

_______ 12. pons

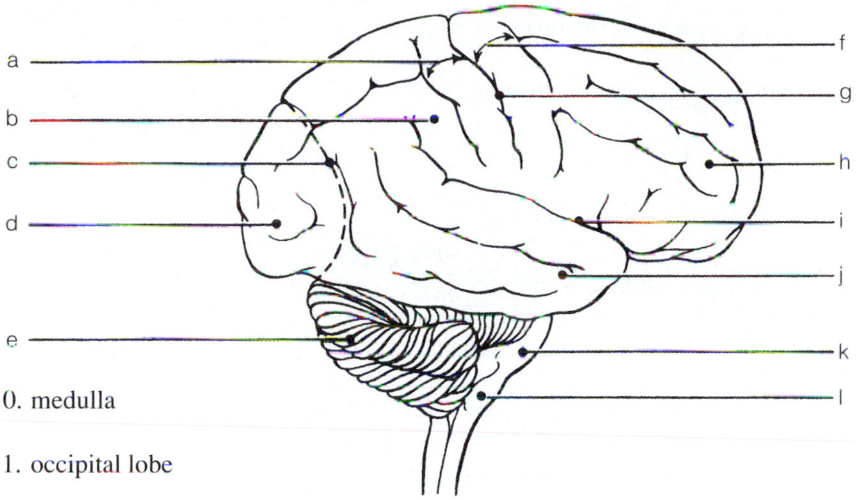

a
b
c
d
e
f
g
h
i
j
k
l

2. In which of the cerebral lobes would the following functional areas be found?

auditory area ______________________ olfactory area ______________________

primary motor area ______________________ visual area ______________________

primary sensory area ______________________ Broca's area ______________________

3. Which of the following structures are not part of the brain stem? (Circle the appropriate response or responses.)

cerebral hemispheres pons midbrain cerebellum medulla diencephalon

4. Complete the following statements by writing the proper word or phrase on the corresponding blanks at the right.

A(n) __1__ is an elevated ridge of cerebral tissue. The convolutions seen in the cerebrum are important because they increase the __2__. Gray matter is composed of __3__. White matter is composed of __4__. A fiber tract that provides for communication between different parts of the same cerebral hemisphere is called a(n) __5__, whereas one that carries impulses to the cerebrum from, and from the cerebrum to, lower CNS areas is called a(n) __6__ tract. The lentiform nucleus along with the amygdaloid and caudate nuclei are collectively called the __7__.

1. ______________________
2. ______________________
3. ______________________
4. ______________________
5. ______________________
6. ______________________
7. ______________________

5. Identify the structures on the following sagittal view of the human brain by matching the numbered areas to the proper terms in the list.

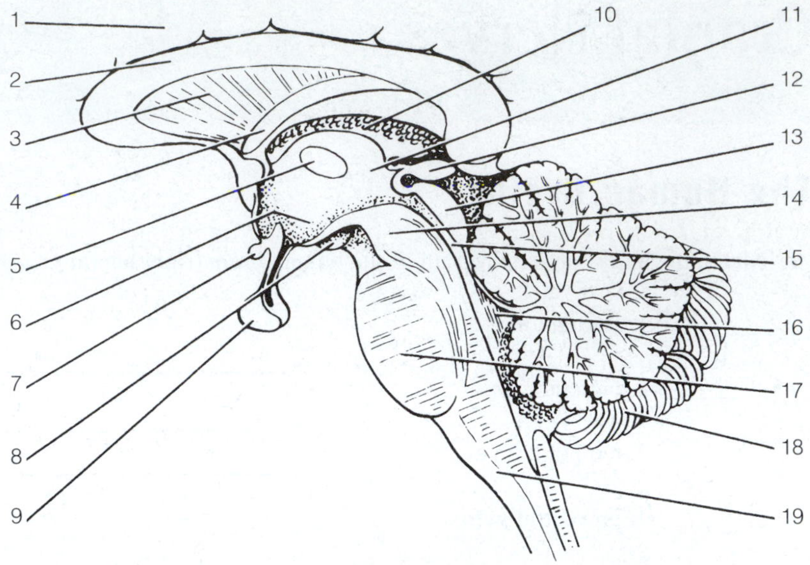

_______ a. cerebellum

_______ b. cerebral aqueduct

_______ c. cerebral hemisphere

_______ d. cerebral peduncle

_______ e. choroid plexus

_______ f. corpora quadrigemina

_______ g. corpus callosum

_______ h. fornix

_______ i. fourth ventricle

_______ j. hypothalamus

_______ k. mammillary bodies

_______ l. massa intermedia

_______ m. medulla oblongata

_______ n. optic chiasma

_______ o. pineal body

_______ p. pituitary gland

_______ q. pons

_______ r. septum pellucidum

_______ s. thalamus

1
2
3
4
5
6
7
8
9
10
11
12
13
14
15
16
17
18
19

6. Using the terms from item 5, match the appropriate structures with the descriptions given below:

_______ 1. site of regulation of body temperature and water balance; most important autonomic center

_______ 2. consciousness depends on the function of this part of the brain

_______ 3. located in the midbrain; contains reflex centers for vision and audition

_______ 4. responsible for regulation of posture and coordination of complex muscular movements

_______ 5. important synapse site for afferent fibers traveling to the sensory cortex

_______ 6. contains autonomic centers regulating blood pressure, heart rate, and respiratory rhythm, as well as coughing, sneezing, and swallowing centers

_______ 7. large commissure connecting the cerebral hemispheres

_______ 8. fiber tract involved with olfaction

_______ 9. connects the third and fourth ventricles

_______ 10. encloses the third ventricle

7. Embryologically, the brain arises from the rostral end of a tubelike structure that quickly becomes divided into three major regions. Groups of structures that develop from the embryonic brain are listed below. Designate the embryonic origin of each group as the hindbrain, midbrain, or forebrain.

_______________ 1. the diencephalon, including the thalamus, optic chiasma, and hypothalamus

_______________ 2. the medulla, pons, and cerebellum

_______________ 3. the cerebral hemispheres

8. What is the function of the basal nuclei? _______________

____

9. What is the corpus striatum, and how is it related to the fibers of the internal capsule? _______________

____

____

10. A brain hemorrhage within the region of the right internal capsule results in paralysis of the left side of the body.

Explain why the left side (rather than the right side) is affected. _______________

____

11. Explain why trauma to the base of the brain is often much more dangerous than trauma to the frontal lobes. (*Hint:* Think about the relative functioning of the cerebral hemispheres and the brain stem structures. Which contain centers more vital to life?)

____

____

____

12. In "split brain" experiments, the main commissure connecting the cerebral hemispheres is cut. First, name this commissure:

____

Then, describe what results (in terms of behavior) can be anticipated in such experiments. (Use an appropriate reference if you need help with this one!)

____

____

____

____

____

Meninges of the Brain

Identify the meningeal (or associated) structures described below:

______________________ 1. outermost meninx covering the brain; composed of tough fibrous connective tissue

______________________ 2. innermost meninx covering the brain; delicate and highly vascular

______________________ 3. structures instrumental in returning cerebrospinal fluid to the venous blood in the dural sinuses

______________________ 4. structure that forms the cerebrospinal fluid

______________________ 5. middle meninx; like a cobweb in structure

______________________ 6. its outer layer forms the periosteum of the skull

______________________ 7. a dural fold that attaches the cerebrum to the crista galli of the skull

______________________ 8. a dural fold separating the cerebrum from the cerebellum

Cerebrospinal Fluid

Fill in the following flowchart by delineating the circulation of cerebrospinal fluid from its formation site (assume that this is one of the lateral ventricles) to the site of its reabsorption into the venous blood:

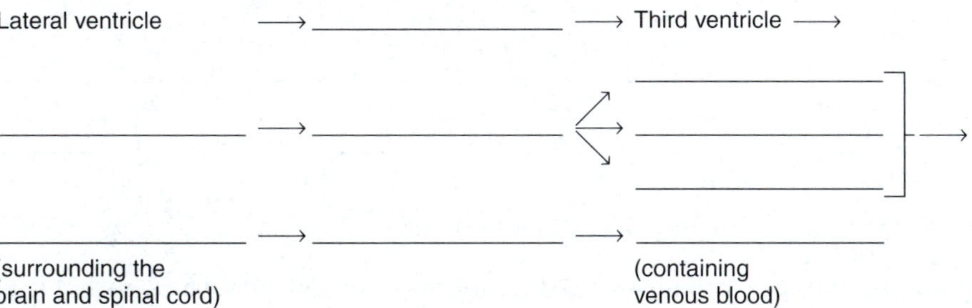

Now label appropriately the structures involved with circulation of CS fluid on the accompanying diagram. (These structures are identified by leader lines.)

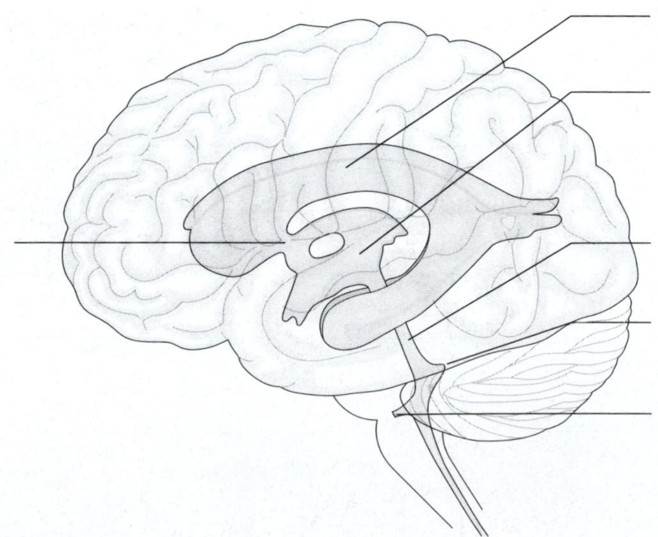

Cranial Nerves

1. Using the terms below, correctly identify all structures indicated by leader lines on the diagram below.

a. abducens nerve (VI)

b. accessory nerve (XI)

c. cerebellum

d. cerebral peduncle

e. decussation of the pyramids

f. facial nerve (VII)

g. frontal lobe of cerebral hemisphere

h. glossopharyngeal nerve (IX)

i. hypoglossal nerve (XII)

j. longitudinal fissure

k. mammillary body

l. medulla oblongata

m. oculomotor nerve (III)

n. olfactory bulb

o. olfactory tract

p. optic chiasma

q. optic nerve (II)

r. optic tract

s. pituitary gland

t. pons

u. spinal cord

v. temporal lobe of cerebral hemisphere

w. trigeminal nerve (V)

x. trochlear nerve (IV)

y. vagus nerve (X)

z. vestibulocochlear nerve (VIII)

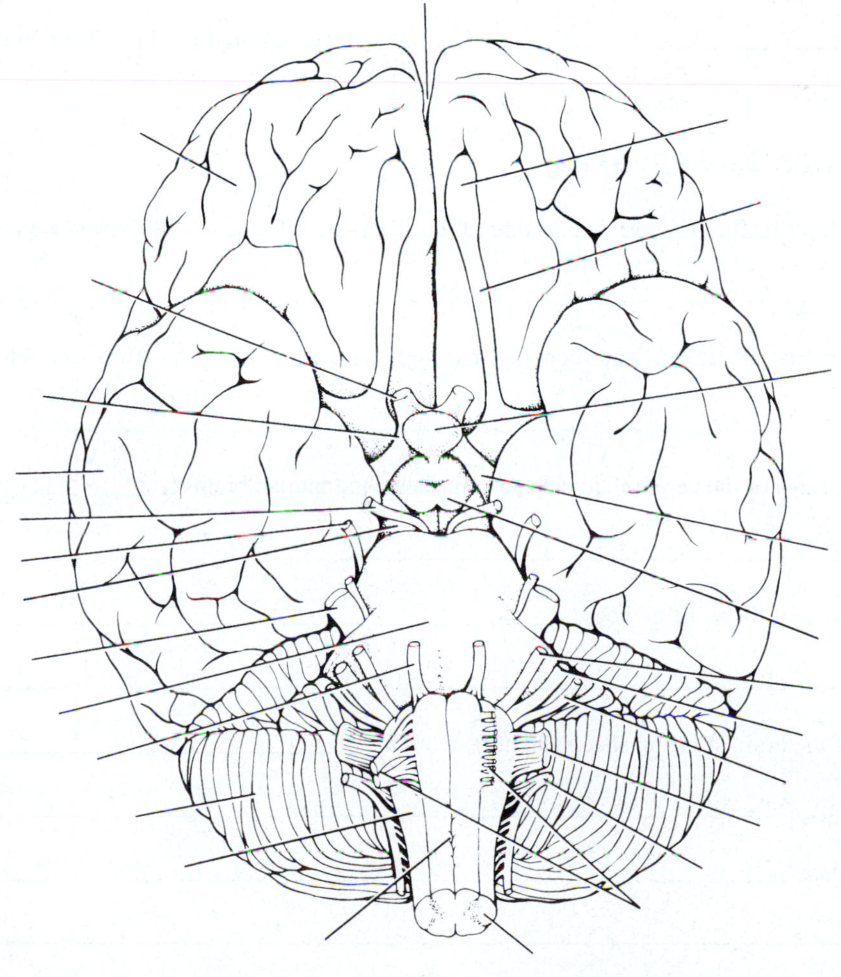

2. Provide the name and number of the cranial nerves involved in each of the following activities, sensations, or disorders:

________________ 1. shrugging the shoulders

________________ 2. smelling a flower

________________ 3. raising the eyelids; focusing the lens of the eye for accommodation; and pupillary constriction

________________ 4. slows the heart; increases the mobility of the digestive tract

________________ 5. involved in Bell's palsy (facial paralysis)

________________ 6. chewing food

________________ 7. listening to music; seasickness

________________ 8. secretion of saliva; tasting well-seasoned food

________________ 9. involved in "rolling" the eyes (three nerves—provide numbers only)

________________ 10. feeling a toothache

________________ 11. reading *Mad* magazine

________________ 12. purely sensory in function (three nerves—provide numbers only)

Dissection of the Sheep Brain

1. In your own words, describe the firmness and texture of the sheep brain tissue as observed when cutting into it.

____

Because formalin hardens all tissue, what conclusions might you draw about the firmness and texture of living brain tissue? ________________

2. Compare the relative sizes of the cerebral hemispheres in sheep and human brains? ________________

____

What is the significance of these differences? ________________

____

3. Compare the sizes of the brain stems in sheep and human brains. ________________

____

What is the significance? ________________

____

4. Why are the olfactory bulbs much larger in the sheep brain than in the human brain? ________________

____

NAME ______________________ LAB TIME/DATE ______________________

REVIEW SHEET
exercise
20

Electroencephalography

Brain Wave Patterns and the Electroencephalogram

1. Define *EEG*. ______________________

2. What are the four major types of brain wave patterns? ______________________

   Match each statement below to a type of brain wave pattern:

   __________ below 4 Hz; slow, large waves; normally seen during deep sleep

   __________ rhythm generally apparent when an individual is in a relaxed, nonattentive state with the eyes closed

   __________ correlated to the alert state; usually about 15 to 25 Hz

3. What is meant by the term *alpha block*? ______________________

   ____

   ____

4. List at least four types of brain lesions that may be determined by EEG studies. ______________________

   ____

5. What is the common result of hypoactivity or hyperactivity of the brain neurons? ______________________

Observing Brain Wave Patterns

1. How was alpha block demonstrated in the laboratory experiment? ______________________

   ____

2. What was the effect of mental concentration on the brain wave pattern? ______________________

   ____

   ____

3. What effect on the brain wave pattern did hyperventilation have? ______________________

   Why? ______________________

   ____

Brain Wave Patterns and the Electroencephalogram

Observing Brain Wave Patterns

NAME ______________________ LAB TIME/DATE ______________________

REVIEW SHEET
exercise
21

Spinal Cord and Spinal Nerves

Anatomy of the Spinal Cord

1. Match the descriptions given below to the proper anatomical term:

a. cauda equina b. conus medullaris c. filum terminale d. foramen magnum

_______ 1. most superior boundary of the spinal cord

_______ 2. meningeal extension beyond the spinal cord terminus

_______ 3. spinal cord terminus

_______ 4. collection of spinal nerves traveling in the vertebral canal below the terminus of the spinal cord

2. Match the key letters on the diagram with the following terms.

_______ 1. anterior (ventral) horn

_______ 2. arachnoid mater

_______ 3. central canal

_______ 4. dorsal ramus of spinal nerve

_______ 5. dorsal root ganglion

_______ 6. dorsal root of spinal nerve

_______ 7. dura mater

_______ 8. gray commissure

_______ 9. lateral horn

_______ 10. pia mater

_______ 11. posterior (dorsal) horn

_______ 12. spinal nerve

_______ 13. ventral ramus of spinal nerve

_______ 14. ventral root of spinal nerve

_______ 15. white matter

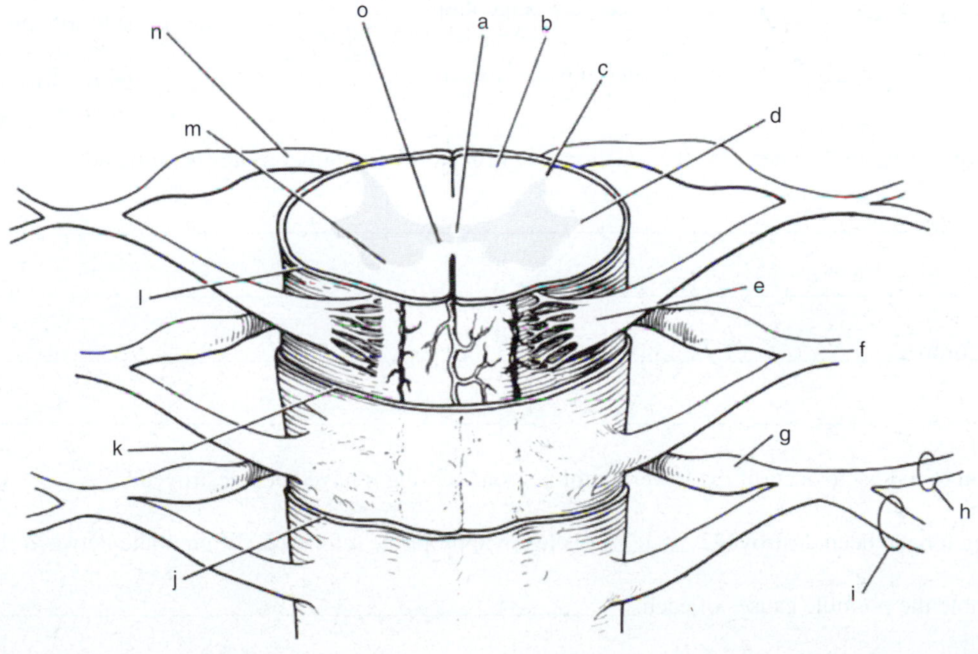

3. Choose the proper answer from the following key to respond to the descriptions relating to spinal cord anatomy.

Key: a. afferent b. efferent c. both afferent and efferent d. association

_______ 1. neuron type found in posterior horn

_______ 2. neuron type found in anterior horn

_______ 3. neuron type in dorsal root ganglion

_______ 4. fiber type in ventral root

_______ 5. fiber type in dorsal root

_______ 6. fiber type in spinal nerve

4. Where in the vertebral column is a lumbar puncture generally done? _______________

Why is this the site of choice? _______________

____

5. The spinal cord is enlarged in two regions, the _______________ and the _______________ regions.

What is the significance of these enlargements? _______________

____

6. How does the position of the gray and white matter differ in the spinal cord and the cerebral hemispheres?

____

____

7. Choose the name of the tract, from the following key, that might be damaged when the following conditions are observed. (More than one choice may apply.)

_______________ 1. uncoordinated movement

_______________ 2. lack of voluntary movement

_______________ 3. tremors, jerky movements

_______________ 4. diminished pain perception

_______________ 5. diminished sense of touch

Key:
a. fasciculus gracilis
b. fasciculus cuneatus
c. lateral corticospinal tract
d. anterior corticospinal tract
e. tectospinal tract
f. rubrospinal tract
g. lateral spinothalamic tract
h. anterior spinothalamic tract
i. posterior spinocerebellar tract
j. vestibulospinal tract
k. anterior spinocerebellar tract

8. Use an appropriate reference to describe the functional significance of an upper motor neuron and a lower motor neuron:

upper motor neuron _______________

____

lower motor neuron _______________

____

Will contraction of a muscle occur if the lower motor neurons serving it have been destroyed? _______ If the upper motor neurons serving it have been destroyed? _______ Using an appropriate reference, differentiate between flaccid and spastic paralysis and note the possible causes of each. _______________

____

____

____

____

Spinal Nerves and Nerve Plexuses

1. In the human, there are 31 pairs of spinal nerves, named according to the region of the vertebral column from which they issue. The spinal nerves are named below. Indicate how they are numbered.

cervical nerves ______________________ sacral nerves ______________________

lumbar nerves ______________________ thoracic nerves ______________________

2. The ventral rami of spinal nerves C_1 through T_1 and T_{12} through S_4 take part in forming ______________________, which serve the ______________________ of the body. The ventral rami of T_2 through T_{12} run between the ribs to serve the ______________________. The dorsal rami of the spinal nerves serve ______________________.

3. What would happen if the following structures were damaged or transected? (Use key choices for responses.)

Key: a. loss of motor function b. loss of sensory function c. loss of both motor and sensory function

_______ 1. dorsal root of a spinal nerve _______ 3. anterior ramus of a spinal nerve

_______ 2. ventral root of a spinal nerve

4. Define *plexus:* ____

____

5. Name the major nerves that serve the following body areas:

______________________ 1. head, neck, shoulders (name plexus only)

______________________ 2. diaphragm

______________________ 3. posterior thigh

______________________ 4. leg and foot (name two)

______________________ 5. anterior forearm muscles (name two)

______________________ 6. arm muscles (name two)

______________________ 7. abdominal wall (name plexus only)

______________________ 8. anterior thigh

______________________ 9. medial side of the hand

Dissection of the Spinal Cord

1. Compare and contrast the meninges of the spinal cord and the brain. ______________________________

____

2. How can you distinguish between the anterior and posterior horns? ______________________________

____

3. How does the position of gray and white matter differ from that in the cerebral hemispheres of the sheep brain? ________

____

NAME ______________________ LAB TIME/DATE ______________________

REVIEW SHEET
exercise
22

Human Reflex Physiology

The Reflex Arc

1. Define *reflex:* ______________________

____

2. Name five essential components of a reflex arc: ______________________, ______________________,

______________________, ______________________, and ______________________

3. In general, what is the importance of reflex testing in a routine physical examination? ______________________

____

____

Somatic and Autonomic Reflexes

1. Use the key terms to complete the statements given below.

Key:
a. abdominal reflex
b. Achilles jerk
c. ciliospinal reflex
d. corneal reflex
e. crossed extensor reflex
f. gag reflex
g. patellar reflex
h. plantar reflex
i. pupillary light reflex

Reflexes classified as somatic reflexes include a ________, ________, ________, ________, ________, ________, and ________.

Of these, the simple stretch reflexes are ________ and ________, and the superficial cord reflexes are ________ and ________.

Reflexes classified as autonomic reflexes include ________ and ________.

2. In what way do cord-mediated reflexes differ from those involving higher brain centers? ______________________

____

Name two cord-mediated reflexes: ______________________ and ______________________

____

Name two somatic reflexes in which the higher brain centers participate: ______________________

and ______________________

3. Can the stretch reflex be elicited in a pithed animal? ______________________

Explain your answer. ______________________

____

4. Trace the reflex arc, naming efferent and afferent nerves, receptors, effectors, and integration centers, for the following reflexes:

patellar reflex ____

____

____

Achilles reflex ____

____

____

5. Three factors that influence the rapidity and effectiveness of reflex arcs were investigated in conjunction with patellar reflex testing—mental distraction, effect of simultaneous muscle activity in another body area, and fatigue.

Which of these factors increases the excitatory level of the spinal cord? ____________________

Which factor decreases the excitatory level of the muscles? ____________________

When the subject was concentrating on an arithmetic problem, did the change noted in the patellar reflex indicate that brain activity is necessary for the patellar reflex or only that it may modify it? ____________________

____

6. Name the division of the autonomic nervous system responsible for each of the following reflexes:

ciliospinal reflex ____________________ salivary reflex ____________________

pupillary light reflex ____________________

7. The pupillary light reflex, the crossed extensor reflex, and the corneal reflex illustrate the purposeful nature of reflex activity. Describe the protective aspect of each:

pupillary light reflex ____

corneal reflex ____

crossed extensor reflex ____

____

____

____

____

8. Was the pupillary consensual response contralateral or ipsilateral? ____________________

Why would such a response be of significant value in this particular reflex? ____________________

____

9. Differentiate between the types of activities accomplished by somatic and autonomic reflexes. ______

______

______

______

10. Several types of reflex activity were not investigated in this exercise. The most important of these are autonomic reflexes, which are difficult to illustrate in a laboratory situation. To rectify this omission, complete the following chart, using references as necessary.

Reflex	**Organ involved**	**Receptors stimulated**	**Action**
Micturition (urination)			
Hering-Breuer			
Defecation			
Carotid sinus			

Reaction Time of Basic and Learned or Acquired Reflexes

1. How do basic and learned or acquired reflexes differ? ______________________________

____

2. Name at least three factors that may modify reaction time to a stimulus. ______________________________

____

3. In general, how did the response time for the unlearned activity performed in the laboratory compare to that for the simple patellar reflex? ______________________________

4. Did the response time without verbal stimuli decrease with practice? _______ Explain the reason for this.

____

5. Explain, in detail, why response time increased when the subject had to react to a word stimulus.

____

____

NAME ____________________ LAB TIME/DATE ____________________

REVIEW SHEET
exercise
23

General Sensation

Structure of General Sensory Receptors

1. Differentiate between interoceptors and exteroceptors relative to location and stimulus source:

   Interoceptor: ____________________

   Exteroceptor: ____________________

2. A number of activities and sensations are listed in the chart below. For each, check whether the receptors would be exteroceptors or interoceptors; and then name the specific receptor types. (Because visceral receptors were not described in detail in this exercise, you need only indicate that the receptor is a visceral receptor if it falls into that category.)

Activity or sensation	Exteroceptor	Interoceptor	Specific receptor type
Backing into a sun-heated iron railing			
Someone steps on your foot			
Reading a book			
Leaning on your elbows			
Doing sit-ups			
The "too full" sensation			
Seasickness			

Receptor Physiology

1. Explain how the sensory receptors act as transducers: ____________________

   ____

2. Define *stimulus:* ____________________

3. What was demonstrated by the two-point discrimination test? ____________________

   ____

   How well did your results correspond to your predictions? ____________________

   ____

   Correlate the accuracy of the subject's tactile localization with the results of the two-point discrimination test.

   ____

4. Define *punctate distribution:* ______________________________

____

5. Several questions regarding general sensation are posed below. Answer each by placing your response in the appropriately numbered blanks to the right.

1. Which cutaneous receptors are the most numerous?

2–3. Which two body areas tested were most sensitive to touch?

4–5. Which two body areas tested were least sensitive to touch?

6. Which appear to be more numerous—receptors that respond to cold or to heat?

7–9. Where would referred pain appear if the following organs were receiving painful stimuli—(7) gallbladder, (8) kidneys, and (9) appendix? (Use your textbook if necessary.)

10. Where was referred pain felt when the elbow was immersed in ice water during the laboratory experiment?

11. What region of the cerebrum interprets the kind and intensity of stimuli that cause cutaneous sensations?

1. ______________________________
2. ______________________________
3. ______________________________
4. ______________________________
5. ______________________________
6. ______________________________
7. ______________________________
8. ______________________________
9. ______________________________
10. ______________________________
11. ______________________________

6. Define *adaptation of sensory receptors:* ______________________________

____

7. Why is it advantageous to have pain receptors that are sensitive to all vigorous stimuli, whether heat, cold, or pressure?

____

Why is the nonadaptability of pain receptors important? ______________________________

____

8. Imagine yourself without any cutaneous sense organs. Why might this be very dangerous? ______________________________

____

____

9. Define *referred pain:* ______________________________

____

What is the probable explanation for referred pain? (Consult your textbook or an appropriate reference if necessary.)

____

____

____

NAME ____________________ LAB TIME/DATE ____________________

REVIEW SHEET
exercise
24

Special Senses: Vision

Anatomy of the Eye

1. Name five accessory eye structures that contribute to the formation of tears and/or aid in lubrication of the eyeball, and then name the major secretory product of each. Indicate which has antibacterial properties by circling the correct secretory product.

Accessory structures	Product

2. The eyeball is wrapped in adipose tissue within the orbit. What is the function of the adipose tissue?

____

What seven bones form the bony orbit? (Think! If you can't remember, check a skull or your text.)

____________ ____________ ____________

____________ ____________

____________ ____________

3. Why does one often have to blow one's nose after crying? ____________________

____

4. Identify the extrinsic eye muscle predominantly responsible for the actions described below.

____________________ 1. turns the eye laterally

____________________ 2. turns the eye medially

____________________ 3. turns the eye up and laterally

____________________ 4. turns the eye inferiorly

____________________ 5. turns the eye superiorly

____________________ 6. turns the eye down and laterally

5. What is a sty? ____

____

Conjunctivitis? ____

6. Using the terms listed on the right, correctly identify all structures provided with leader lines in the diagram.

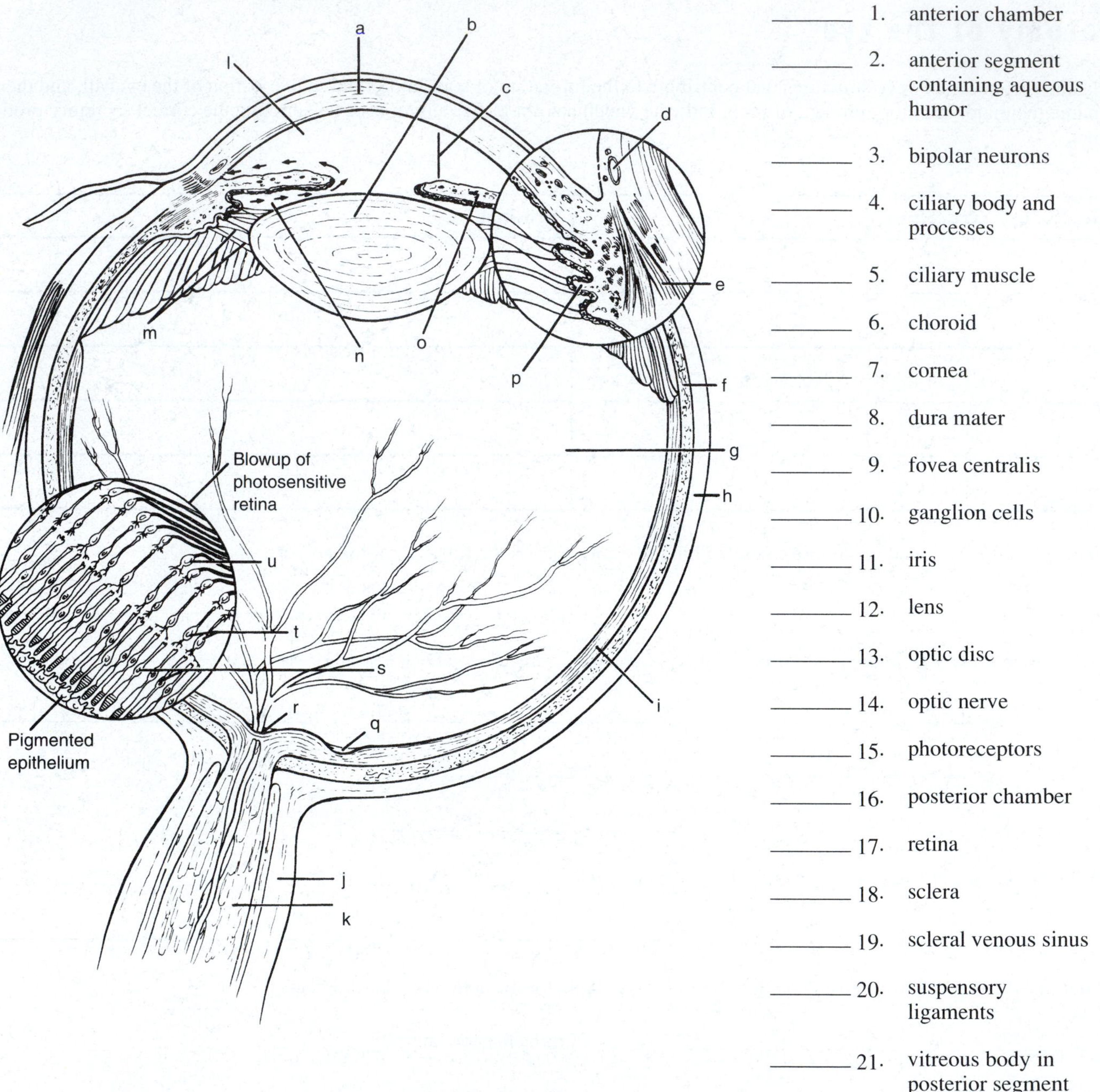

_______ 1. anterior chamber

_______ 2. anterior segment containing aqueous humor

_______ 3. bipolar neurons

_______ 4. ciliary body and processes

_______ 5. ciliary muscle

_______ 6. choroid

_______ 7. cornea

_______ 8. dura mater

_______ 9. fovea centralis

_______ 10. ganglion cells

_______ 11. iris

_______ 12. lens

_______ 13. optic disc

_______ 14. optic nerve

_______ 15. photoreceptors

_______ 16. posterior chamber

_______ 17. retina

_______ 18. sclera

_______ 19. scleral venous sinus

_______ 20. suspensory ligaments

_______ 21. vitreous body in posterior segment

Notice the arrows drawn close to the left side of the iris in the diagram above. What do they indicate?

____

____

____

7. Match the key responses with the descriptive statements that follow.

Key:
a. aqueous humor
b. choroid
c. ciliary body
d. ciliary processes of the ciliary body
e. cornea
f. fovea centralis
g. iris
h. lens
i. optic disc
j. retina
k. sclera
l. scleral venous sinus
m. suspensory ligament
n. vitreous humor

________________ 1. attaches the lens to the ciliary body

________________ 2. fluid filling the anterior segment of the eye

________________ 3. the "white" of the eye

________________ 4. part of the retina that lacks photoreceptors

________________ 5. modification of the choroid that controls the shape of the crystalline lens

________________ 6. contains the ciliary muscle

________________ 7. drains the aqueous humor from the eye

________________ 8. tunic containing the rods and cones

________________ 9. substance occupying the posterior segment of the eyeball

________________ 10. forms the bulk of the heavily pigmented vascular tunic

________________, ________________ 11. smooth muscle structures

________________ 12. area of critical focusing and discriminatory vision

________________ 13. form (by filtration) the aqueous humor

________________, ________________, ________________,

________________ 14. light-bending media of the eye

________________ 15. anterior continuation of the sclera—your "window on the world"

________________ 16. composed of tough, white, opaque, fibrous connective tissue

8. The iris is composed primarily of two smooth muscle layers, one arranged radially and the other circularly.

Which of these dilates the pupil? ________________

9. You would expect the pupil to be dilated in which of the following circumstances? Circle the correct response(s).

a. in brightly lit surroundings

b. in dimly lit surroundings

c. during focusing for near vision

d. in observing distant objects

10. The intrinsic eye muscles are under the control of which of the following? (Circle the correct response.)

autonomic nervous system   somatic nervous system

Microscopic Anatomy of the Retina

1. The two major layers of the retina are the epithelial and nervous layers. In the nervous layer, the neuron populations are arranged as follows from the epithelial layer to the vitreous humor. (Circle all proper responses.)

   bipolar cells, ganglion cells, photoreceptors | photoreceptors, ganglion cells, bipolar cells

   ganglion cells, bipolar cells, photoreceptors | photoreceptors, bipolar cells, ganglion cells

2. The axons of the ______________________________ cells form the optic nerve, which exits from the eyeball.

3. Complete the following statements by writing either *rods* or *cones* on each blank:

   The dim light receptors are the ______________________________. Only ______________________________ are found in the fovea centralis, whereas mostly ______________________________ are found in the periphery of the retina. ______________________________ are the photoreceptors that operate best in bright light and allow for color vision.

Visual Pathways to the Brain

1. The visual pathway to the occipital lobe of the brain consists most simply of a chain of five neurons. Beginning with the photoreceptor cell of the retina, name them and note their location in the pathway.

   (1) ______________________________ (4) ______________________________

   (2) ______________________________ (5) ______________________________

   (3) ______________________________ ______________________________

2. Visual field tests are done to reveal destruction along the visual pathway from the retina to the optic region of the brain. Note where the lesion is likely to be in the following cases:

   Normal vision in left eye visual field; absence of vision in right eye visual field: ______________________________

   Normal vision in both eyes for right half of the visual field; absence of vision in both eyes for left half of the visual field: ______________________________

3. How is the right optic *tract* anatomically different from the right optic *nerve*? ______________________________

Dissection of the Cow (Sheep) Eye

1. What modification of the choroid that is not present in humans is found in the cow eye? ______________________________

   What is its function? ______________________________

2. What does the retina look like? ____

____

At what point is it attached to the posterior aspect of the eyeball? ____________________________

Visual Tests and Experiments

1. Match the terms in column B with the descriptions in column A:

	Column A		Column B
______________	1. light bending	a.	accommodation
______________	2. ability to focus for close (under 20 ft) vision	b.	astigmatism
______________	3. normal vision	c.	convergence
______________	4. inability to focus well on close objects (farsightedness)	d.	emmetropia
______________	5. nearsightedness	e.	hyperopia
______________	6. blurred vision due to unequal curvatures of the lens or cornea	f.	myopia
______________	7. medial movement of the eyes during focusing on close objects	g.	refraction

2. Complete the following statements:

In farsightedness, the light is focused __1__ the retina. The lens required to treat myopia is a __2__ lens. The "near point" increases with age because the __3__ of the lens decreases as we get older. A convex lens, like that of the eye, produces an image that is upside down and reversed from left to right. Such an image is called a __4__ image.

1. ______________________

2. ______________________

3. ______________________

4. ______________________

3. Use terms from the key to complete the statements concerning near and distance vision.

Key: a. contracted b. decreased c. increased d. relaxed e. taut

During distance vision: The ciliary muscle is _______, the suspensory ligament is _______, the convexity of the lens is _______, and light refraction is _______. During close vision: The ciliary muscle is _______, the suspensory ligament is _______, lens convexity is _______, and light refraction is _______.

4. Explain why vision is lost when light hits the blind spot. ______________________________________

5. What is meant by the term *negative afterimage* and what does this phenomenon indicate? ____________________

____

____

____

6. Record your Snellen eye test results below:

Left eye (without glasses) ______________________ (with glasses) ______________________

Right eye (without glasses) ______________________ (with glasses) ______________________

Is your visual acuity normal, less than normal, or better than normal? ______________________

Explain. ______________________

____

Explain why each eye is tested separately when using the Snellen eye chart. ______________________

____

Explain 20/40 vision. ______________________

____

Explain 20/10 vision. ______________________

____

7. Define *astigmatism:* ______________________

____

How can it be corrected? ______________________

8. Record the distance of your near point of accommodation as tested in the laboratory:

right eye ______________________ left eye ______________________

Is your near point within the normal range for your age? ______________________

9. Define *presbyopia:* ______________________

____

What causes it? ______________________

10. To which wavelengths of light do the three cone types of the retina respond maximally?

______________________, ______________________, and ______________________

11. How can you explain the fact that we see a great range of colors even though only three cone types exist?

____

____

12. What is the usual cause of color blindness? ______________________

____

13. Record the results of the demonstration of the relative positioning of rods and cones in the circle below (use appropriately colored pencils).

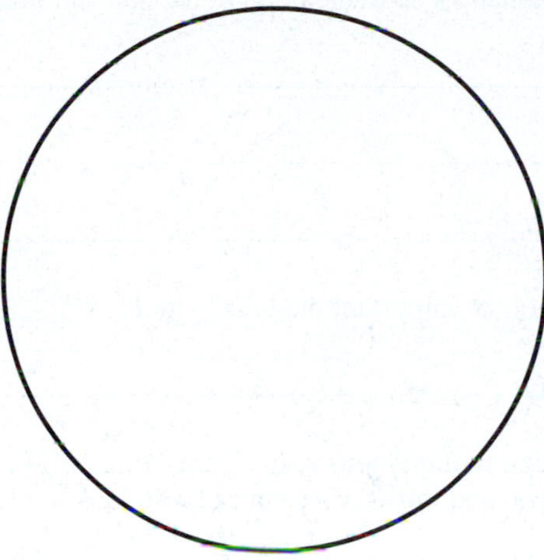

14. Explain the difference between binocular and panoramic vision. ______________________________

____

____

____

____

What is the advantage of binocular vision? ______________________________

What factor(s) are responsible for binocular vision? ______________________________

____

____

15. In the experiment on the convergence reflex, what happened to the position of the eyeballs as the object was moved closer to the subject's eyes? ______________________________

What extrinsic eye muscles control the movement of the eyes during this reflex? ______________________________

What is the value of this reflex? ______________________________

____

What would be the visual result of an inability of these muscles to function? ______________________________

____

16. In the experiment on the photopupillary reflex, what happened to the pupil of the eye exposed to light?

_______________ What happened to the pupil of the nonilluminated eye? _______________

Explanation? _______________

____

____

17. Why is the ophthalmoscopic examination an important diagnostic tool? _______________

____

18. Many college students struggling through mountainous reading assignments are told that they need glasses for "eyestrain." Why is it more of a strain on the extrinsic and intrinsic eye muscles to look at close objects than at far objects?

____

____

NAME ______________________ LAB TIME/DATE ______________________

REVIEW SHEET
exercise
25

Special Senses: Hearing and Equilibrium

Anatomy of the Ear

1. Select the terms from column B that apply to the column A descriptions. Some terms are used more than once.

Column A

1. structures composing the outer or external ear ______________, ______________, ______________
2. structures composing the inner ear ______________, ______________, ______________
3. collectively called the ossicles ______________, ______________, ______________
4. ear structures not involved with audition ______________, ______________
5. involved in equalizing the pressure in the middle ear with atmospheric pressure ______________
6. vibrates at the same frequency as sound waves hitting it; transmits the vibrations to the ossicles ______________
7. contain receptors for the sense of balance ______________, ______________
8. transmits the vibratory motion of the stirrup to the fluid in the scala vestibuli of the inner ear ______________
9. acts as a pressure relief valve for the increased fluid pressure in the scala tympani; bulges into the tympanic cavity ______________
10. passage between the throat and the tympanic cavity ______________
11. fluid contained within the membranous labyrinth ______________
12. fluid contained within the osseous labyrinth and bathing the membranous labyrinth ______________

Column B

a. auditory (pharyngotympanic) tube
b. cochlea
c. endolymph
d. external auditory canal
e. incus (anvil)
f. malleus (hammer)
g. oval window
h. perilymph
i. pinna
j. round window
k. semicircular canals
l. stapes (stirrup)
m. tympanic membrane
n. vestibule

2. Identify all indicated structures and ear regions in the following diagram.

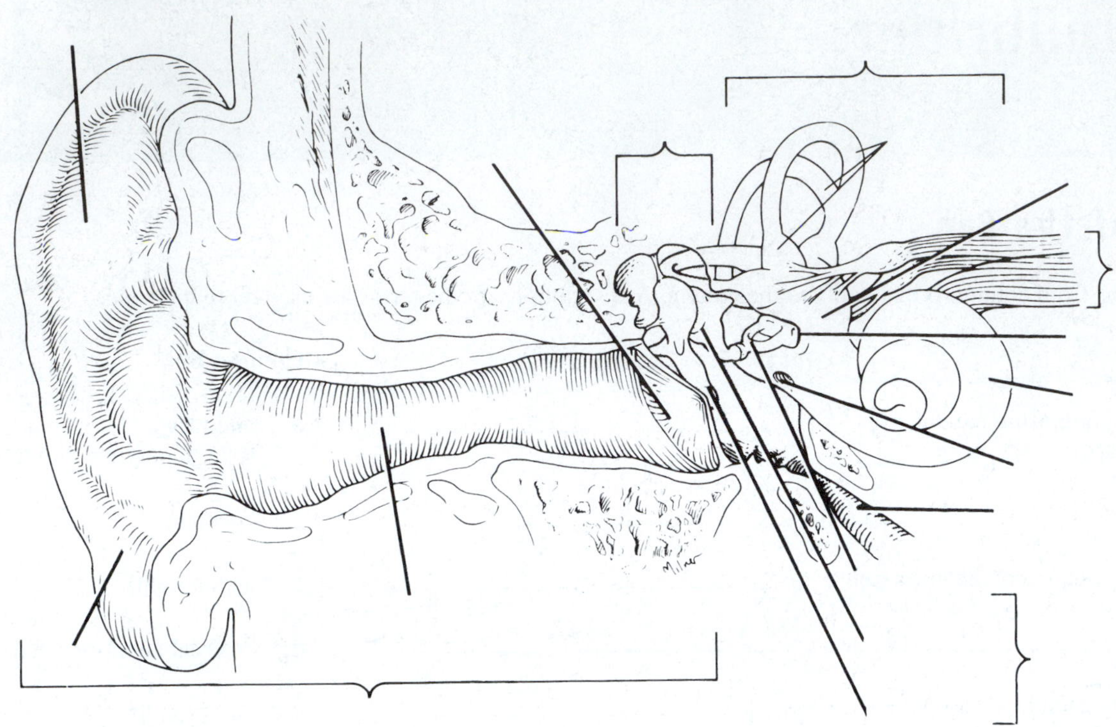

3. Match the membranous labyrinth structures listed in column B with the descriptive statements in column A:

Column A	Column B
_______, _______ 1. sacs found within the vestibule	a. ampulla
_______ 2. contains the organ of Corti	b. basilar membrane
_______, _______ 3. sites of the maculae	c. cochlear duct
_______ 4. positioned in all spatial planes	d. cochlear nerve
_______ 5. hair cells of organ of Corti rest on this membrane	e. cupula
_______ 6. gelatinous membrane overlying the hair cells of the organ of Corti	f. otoliths
_______ 7. contains the crista ampullaris	g. saccule
_______, _______, _______, _______ 8. function in static equilibrium	h. semicircular ducts
_______, _______, _______, _______ 9. function in dynamic equilibrium	i. tectorial membrane
_______ 10. carries auditory information to the brain	j. utricle
_______ 11. gelatinous cap overlying hair cells of the crista ampullaris	k. vestibular nerve
_______ 12. grains of calcium carbonate in the maculae	

4. Sound waves hitting the eardrum initiate its vibratory motion. Trace the pathway through which vibrations and fluid currents are transmitted to finally stimulate the hair cells in the organ of Corti. (Name the appropriate ear structures in their correct sequence.) Eardrum → ______________________________

____

____

5. Describe how sounds of different frequency (pitch) are differentiated in the cochlea. ______________________________

____

____

____

6. Explain the role of the endolymph of the semicircular canals in activating the receptors during angular motion.

____

____

____

7. Explain the role of the otoliths in perception of static equilibrium (head position). ______________________________

____

____

____

Laboratory Tests

1. Was the auditory acuity measurement made during the experiment on page 259 the same or different for both ears?

______________________________ What factors might account for a difference in the acuity of the two ears?

____

____

2. During the sound localization experiment on pages 259 and 260, in which position(s) was the sound least easily located?

____

How can this phenomenon be explained? ______________________________

____

3. In the frequency experiment on pages 260 and 261, which tuning fork was the most difficult to hear? ______________ Hz

What conclusion can you draw? ______________________________

____

4. When the tuning fork handle was pressed to your forehead during the Weber test, where did the sound seem to originate?

____

Where did it seem to originate when one ear was plugged with cotton? ______________________

How do sound waves reach the cochlea when conduction deafness is present? ______________________

____

5. Indicate whether the following conditions relate to conduction deafness (C) or sensorineural deafness (S):

_______ 1. can result from the fusion of the ossicles

_______ 2. can result from a lesion on the cochlear nerve

_______ 3. sound heard in one ear but not in the other during bone and air conduction

_______ 4. can result from otitis media

_______ 5. can result from impacted cerumen or a perforated eardrum

_______ 6. can result from a blood clot in the auditory cortex

6. The Rinne test evaluates an individual's ability to hear sounds conducted by air or bone. Which is more indicative of normal hearing? ______________________

7. Define *nystagmus*: ______________________

____

Define *vertigo*: ______________________

____

8. The Barany test investigated the effect that rotatory acceleration had on the semicircular canals. Explain *why* the subject still had the sensation of rotation immediately after being stopped. ______________________

____

9. What is the usual reason for conducting the Romberg test? ______________________

____

Was the degree of sway greater with the eyes open or closed? Why? ______________________

____

10. Normal balance, or equilibrium, depends on input from a number of sensory receptors. Name them.

____

11. What effect does alcohol consumption have on balance and equilibrium? Explain. ______________________

____

NAME ______________________ LAB TIME/DATE ______________________

REVIEW SHEET
exercise
26

Special Senses: Olfaction and Taste

Localization and Anatomy of Taste Buds

1. Name five sites where receptors for taste are found, and circle the predominant site:

______________________, ______________________, ______________________,

______________________, and ______________________

2. Describe the cellular makeup and arrangement of a taste bud. (Use a diagram, if helpful.) ______________________

____

Localization and Anatomy of the Olfactory Receptors

1. Describe the cellular composition and the location of the olfactory epithelium. ______________________

____

____

2. How and why does sniffing improve your sense of smell? ______________________

____

Laboratory Experiments

1. Taste and smell receptors are both classified as ______________________, because they both respond to ______________________

2. Why is it impossible to taste substances with a dry tongue? ______________________

____

3. State the most important sites of your taste-specific receptors, as determined during the plotting exercise in the laboratory:

salt ______________________ sour ______________________

bitter ______________________ sweet ______________________

4. The basic taste sensations are mediated by specific chemical substances or groups. Name them:

salt ______________________ sour ______________________

bitter ______________________ sweet ______________________

5. Name three factors that influence our appreciation of foods. Substantiate each choice with an example from the laboratory experience.

____________________ Substantiation ____________________

____________________ Substantiation ____________________

____________________ Substantiation ____________________

Which of the factors chosen is most important? ____________________

Substantiate your choice with an example from everyday life. ____________________

Expand on your explanation and choices by explaining why a cold, greasy hamburger is unappetizing to most people.

6. Babies tend to favor bland foods, whereas adults tend to like highly seasoned foods. What is the basis for this phenomenon?

7. How palatable is food when you have a cold? ____________________

Explain. ____________________

8. What is the mechanism of olfactory adaptation? ____________________

In your opinion, is olfactory adaptation desirable? ____________________ Explain your answer.

NAME ______________________ LAB TIME/DATE ______________________

REVIEW SHEET
exercise
27

Anatomy and Basic Function of the Endocrine Glands

Gross Anatomy and Basic Function of the Endocrine Glands

1. Both the endocrine and nervous systems are major regulating systems of the body; however, the nervous system has been compared to an airmail delivery system and the endocrine system to the pony express. Briefly explain this comparison.

____

____

____

2. Define *hormone:* ____

____

3. Chemically, hormones belong chiefly to two molecular groups, the ______________________

and the ____.

4. What do all hormones have in common? ____

____

5. Define *target organ:* ____

____

6. If hormones travel in the bloodstream, why don't all tissues respond to all hormones? ______________________

____

7. Identify the endocrine organ described by the following statements:

______________ 1. located in the throat; bilobed gland connected by an isthmus

______________ 2. found close to the kidney

______________ 3. a mixed gland, located close to the stomach and small intestine

______________ 4. paired glands suspended in the scrotum

______________ 5. ride "horseback" on the thyroid gland

______________ 6. found in the pelvic cavity of the female, concerned with ova and female hormone production

______________ 7. found in the upper thorax overlying the heart; large during youth

______________ 8. found in the roof of the third ventricle

8. For each statement describing hormonal effects, identify the hormone(s) involved by choosing a number from key A, and note the hormone's site of production with a letter from key B. More than one hormone may be involved in some cases.

Key A:

1. ACTH
2. ADH
3. aldosterone
4. cortisone
5. epinephrine
6. estrogens
7. FSH
8. glucagon
9. GH
10. insulin
11. LH
12. melatonin
13. MSH
14. oxytocin
15. progesterone
16. prolactin
17. PTH
18. serotonin
19. testosterone
20. thymosin
21. thyrocalcitonin/calcitonin
22. T_4 / T_3
23. TSH

Key B:

a. adrenal cortex
b. adrenal medulla
c. anterior pituitary
d. hypothalamus
e. ovaries
f. pancreas
g. parathyroid glands
h. pineal gland
i. posterior pituitary
j. testes
k. thymus gland
l. thyroid gland

_______, _______ 1. basal metabolism hormone

_______, _______ 2. programming of T lymphocytes

_______, _______ and _______, _______ 3. regulate blood calcium levels

_______, _______ and _______, _______ 4. released in response to stressors

_______, _______ and _______, _______ 5. drive development of secondary sexual characteristics

_______, _______; _______, _______; _______, _______; and

_______, _______ 6. regulate the function of another endocrine gland

_______, _______ 7. mimics the sympathetic nervous system

_______, _______ and _______, _______ 8. regulate blood glucose levels; produced by the same "mixed" gland

_______, _______ and _______, _______ 9. directly responsible for regulation of the menstrual cycle

_______, _______ and _______, _______ 10. regulate the ovarian cycle

_______, _______ and _______, _______ 11. maintenance of salt and water balance in the ECF

_______, _______ and _______, _______ 12. directly involved in milk production and ejection

_______, _______ 13. questionable function; may stimulate the melanocytes of the skin

9. Although the pituitary gland is often referred to as the master gland of the body, the hypothalamus exerts some control over the pituitary gland. How does the hypothalamus control both anterior and posterior pituitary functioning?

____

____

____

____

____

10. Indicate whether the release of the hormones listed below is stimulated by (A) another hormone; (B) the nervous system (neurotransmitters, or releasing factors); or (C) humoral factors (the concentration of specific nonhormonal substances in the blood or extracellular fluid):

_______ 1. T_4 / T_3

_______ 2. insulin

_______ 3. estrogens

_______ 4. parathyroid hormone

_______ 5. testosterone

_______ 6. norepinephrine

_______ 7. ADH

_______ 8. TSH, FSH

_______ 9. aldosterone

11. Name the hormone(s) produced in *inadequate* amounts that directly result in the following conditions. (Use your textbook as necessary.)

_______________________ 1. sexual immaturity

_______________________ 2. tetany

_______________________ 3. excessive diuresis without high blood glucose levels

_______________________ 4. polyurea, polyphagia, and polydipsia

_______________________ 5. abnormally small stature, normal proportions

_______________________ 6. miscarriage

_______________________ 7. lethargy, hair loss, low BMR, obesity

12. Name the hormone(s) produced in *excessive* amounts that directly result in the following conditions. (Use your textbook as necessary.)

_______________________ 1. lantern jaw and large hands and feet in the adult

_______________________ 2. bulging eyeballs, nervousness, increased pulse rate

_______________________ 3. demineralization of bones, spontaneous fractures

Microscopic Anatomy of Selected Endocrine Glands (Optional)

1. Choose a response from the key below to name the hormone(s) produced by the cell types listed:

Key:
a. insulin
b. GH, prolactin
c. T_4 / T_3
d. calcitonin
e. TSH, ACTH, FSH, LH
f. mineralocorticoids
g. glucagon
h. PTH
i. glucocorticoids

_______ 1. parafollicular cells of the thyroid

_______ 2. follicular epithelial cells of the thyroid

_______ 3. beta cells of the islets of Langerhans

_______ 4. alpha cells of the islets of Langerhans

_______ 5. basophil cells of the anterior pituitary

_______ 6. zona fasciculata cells

_______ 7. zona glomerulosa cells

_______ 8. chief cells

_______ 9. acidophil cells of the anterior pituitary

2. Five diagrams of the microscopic structures of the endocrine glands are presented here. Identify each and name all indicated structures.

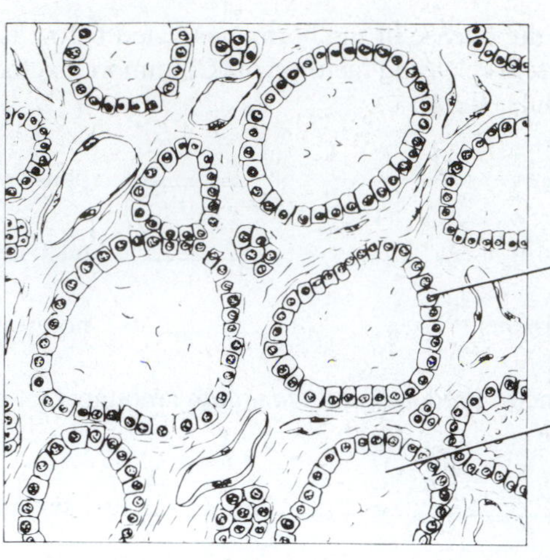

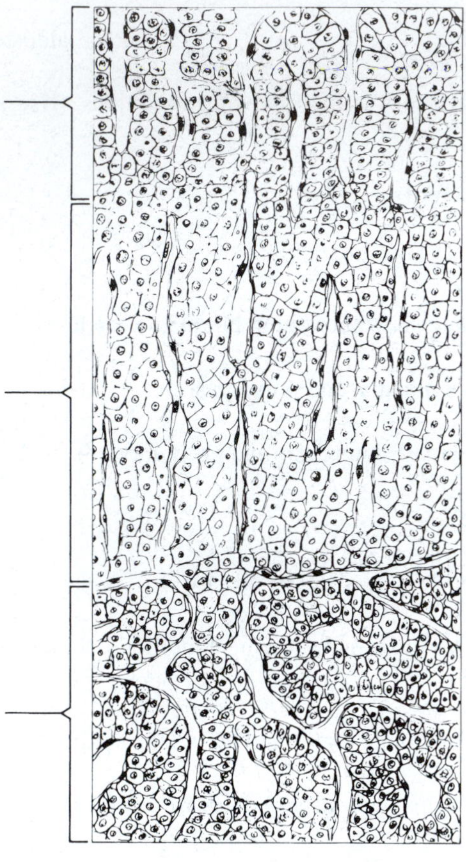

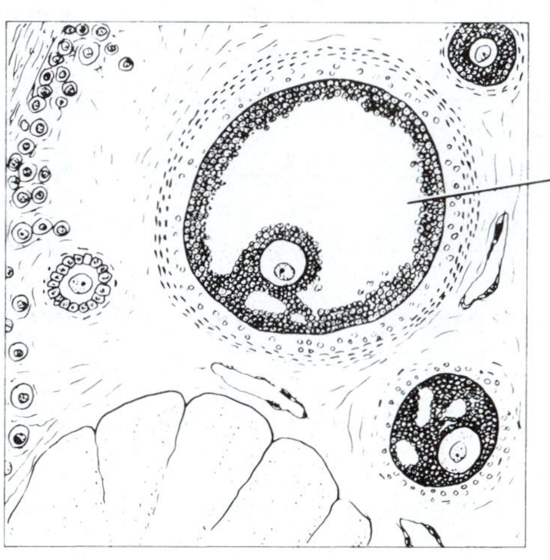

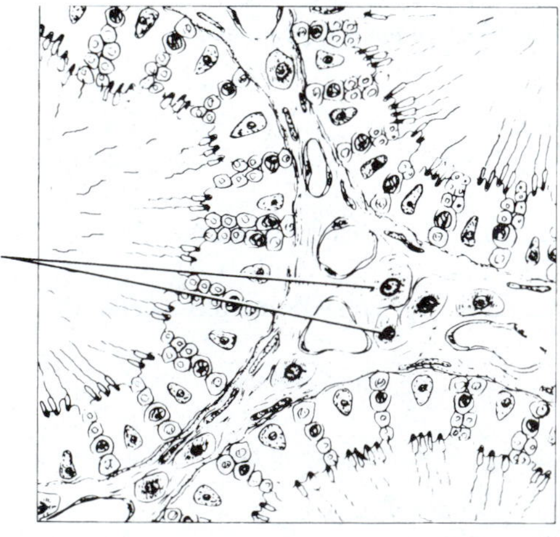

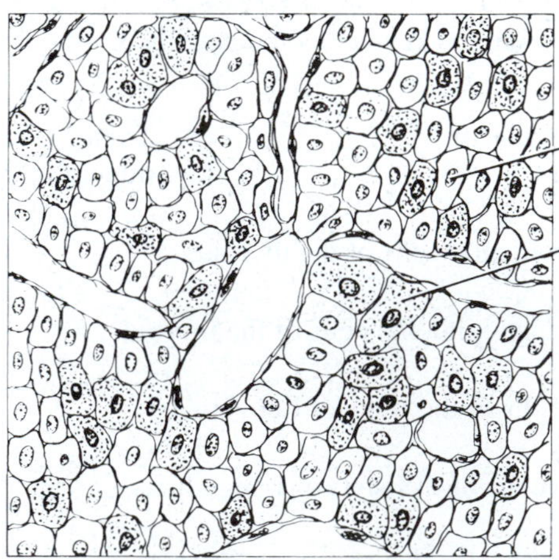

NAME ______________________ LAB TIME/DATE ______________________

REVIEW SHEET
exercise

28A

Experiments on Hormonal Action: Wet Lab

Determining the Effect of Thyroid Hormone on Metabolic Rate

1. Relative to the measurement of oxygen consumption in rats, which group had the highest metabolic rate?

______________________ Which group had the lowest metabolic rate? ______________________

Correlate these observations with the pretreatment these animals received. ______________________

____

____

Which group of rats was hyperthyroid? ______________________

Which euthyroid? ______________________ Which hypothyroid? ______________________

2. Since oxygen used = carbon dioxide evolved, how were you able to measure the oxygen consumption in the experiments?

____

____

3. What did changes in the fluid levels in the manometer arms indicate? ______________________

____

4. The techniques used in this set of laboratory experiments probably allowed for several inaccuracies. One was the inability to control the activity of the rats. How would changes in their activity levels affect the results observed?

____

____

Another possible source of error was the lack of control over the amount of food consumed by the rats in the 14-day period preceding the laboratory session. If each of the rats had been force-fed equivalent amounts of food in that 14-day period, which group (do you think) would have gained the most weight?

______________________ Which the least? ______________________ Explain your answers.

____

____

____

5. TSH, produced by the anterior pituitary, prods the thyroid gland to release thyroid hormone to the blood. Which group of rats can be assumed to have the *highest* blood levels of TSH? ______________________________

Which the lowest? ____________________ Explain your reasoning. ____________________

____

____

6. Use an appropriate reference to determine how each of the following factors modifies metabolic rate. Indicate increase by ↑ and decrease by ↓.

increased exercise ________ aging ________ infection/fever ________

small/slight stature ________ obesity ________ sex (♂ or ♀) ________

Determining the Effect of Pituitary Hormones on the Ovary

1. In the experiment on the effects of pituitary hormones, two anterior pituitary hormones caused ovulation to occur in the experimental animal. Which of these actually triggered ovulation or egg expulsion?

____________________ The normal function of the second hormone involved, ____________________,

is to ______________________________

2. Why was a second frog injected with saline? ______________________________

Observing the Effects of Hyperinsulinism

1. Briefly explain what was happening within the fish's system when the fish was immersed in the insulin solution.

____

____

2. What is the mechanism of the recovery process observed? ______________________________

____

3. What would you do to help a friend who had inadvertently taken an overdose of insulin? ____________________

____________________ Why? ____________________

____

4. What is a glucose tolerance test? (Use an appropriate reference, as necessary, to answer this question.)

____

____

____

5. How does diabetic coma differ from insulin shock? ______________________________

____

____

____

____

Testing the Effect of Epinephrine on the Heart

1. Based on your observations, what is the effect of epinephrine on the force and rate of the heartbeat?

____

2. What is the role of this effect in the fight-or-flight response?

____

NAME ____________________ LAB TIME/DATE ____________________

REVIEW SHEET
exercise
29

Blood

Composition of Blood

1. What is the blood volume of an average-size adult male? __________ liters An average adult female? __________ liters

2. What determines whether blood is bright red or a dull brick-red? ____________________

____

3. Use the key to identify the cell type(s) or blood elements that fit the following descriptive statements.

Key: a. red blood cell
b. megakaryocyte
c. eosinophil
d. basophil
e. monocyte
f. neutrophil
g. lymphocyte
h. formed elements
i. plasma

__________ 1. most numerous leukocyte

__________, and __________

__________ 2. granulocytes

__________ 3. also called an erythrocyte; anucleate formed element

__________, __________ 4. actively phagocytic leukocytes

__________, __________ 5. agranulocytes

__________ 6. ancestral cell of platelets

__________ 7. (a) through (g) are all examples of these

__________ 8. number rises during parasite infections

__________ 9. releases histamine; promotes inflammation

__________ 10. many formed in lymphoid tissue

__________ 11. transports oxygen

__________ 12. primarily water, noncellular; the fluid matrix of blood

__________ 13. increases in number during prolonged infections

__________, __________, __________,

__________, __________ 14. also called white blood cells

4. List four classes of nutrients normally found in plasma: ______________________________,

______________________________, ______________________________, and ______________________________

Name two gases. ______________________________ and ______________________________

Name three ions. ______________________________, ______________________________, and ______________________________

5. Describe the consistency and color of the plasma you observed in the laboratory. ______________________________

____

6. What is the average life span of a red blood cell? How does its anucleate condition affect this life span?

____

____

____

7. From memory, describe the structural characteristics of each of the following blood cell types as accurately as possible, and note the percentage of each in the total white blood cell population.

eosinophils ______________________________

____

neutrophils ______________________________

____

lymphocytes ______________________________

____

basophils ______________________________

____

monocytes ______________________________

____

8. Correctly identify the blood pathologies described in column A by matching them with selections from column B:

	Column A	Column B
______________	1. abnormal increase in the number of WBCs	a. anemia
______________	2. abnormal increase in the number of RBCs	b. leukocytosis
______________	3. condition of too few RBCs or of RBCs with hemoglobin deficiencies	c. leukopenia
		d. polycythemia
______________	4. abnormal decrease in the number of WBCs	

Hematologic Tests

1. Broadly speaking, why are hematologic studies of blood so important in the diagnosis of disease?

___

___

2. In the chart below, record information from the blood tests you read about or conducted. Complete the chart by recording values for healthy male adults and indicating the significance of high or low values for each test.

Test	Student test results	Normal values (healthy male adults)	Significance High values	Significance Low values
Total WBC count	No data			
Total RBC count	No data			
Hematocrit				
Hemoglobin determination				
Bleeding time	No data			
Sedimentation rate				
Coagulation time				

3. Why is a differential WBC count more valuable than a total WBC count when trying to pin down the specific source of pathology? ___

___

___

4. What name is given to the process of RBC production? ______________________________

What hormone acts as a stimulus for this process? ______________________________

What organ provides this stimulus and under what conditions? ______________________________

____

____

5. Discuss the effect of each of the following factors on RBC count. Consult an appropriate reference as necessary, and explain your reasoning.

long-term effect of athletic training (for example, running 4 to 5 miles per day over a period of 6 to 9 months)

____

____

a permanent move from sea level to a high-altitude area ______________________________

____

____

6. Define *hematocrit:* ______________________________

7. If you had a high hematocrit, would you expect your hemoglobin determination to be high or low? ______________

Why? ______________________________

____

8. What is an anticoagulant? ______________________________

Name two anticoagulants used in conducting the hematologic tests. ______________________________

and ______________________________

What is the body's natural anticoagulant? ______________________________

9. If your blood clumped with both anti-A and anti-B sera, your ABO blood type would be ______________

To what ABO blood groups could you give blood? ______________________________

From which ABO donor types could you receive blood? ______________________________

Which ABO blood type is most common? ______________ Least common? ______________

10. What blood type is theoretically considered the universal donor? __________ Why? ______________

____

11. Explain why an Rh-negative person does not have a transfusion reaction on the first exposure to Rh-positive blood but *does* have a reaction on the second exposure. ____

____

What happens when an ABO blood type is mismatched for the first time? ____

____

12. Record your observations of the five demonstration slides viewed.

a. Macrocytic hypochromic anemia: ____

b. Microcytic hypochromic anemia: ____

____

c. Sickle-cell anemia: ____

d. Lymphocytic leukemia (chronic): ____

____

e. Eosinophilia: ____

Which of slides a through e above corresponds with the following conditions?

_______ 1. iron-deficient diet

_______ 2. a type of bone marrow cancer

_______ 3. genetic defect that causes hemoglobin to become sharp/spiky

_______ 4. lack of vitamin B_{12}

_______ 5. a tapeworm infestation in the body

_______ 6. a bleeding ulcer

13. Provide the normal, or at least "desirable," range for plasma cholesterol concentration:

______________ mg/100 ml

14. Describe the relationship between high blood cholesterol levels and cardiovascular diseases such as hypertension, heart attacks, and strokes.

____

____

____

____

____

NAME ______________________ LAB TIME/DATE ______________________

REVIEW SHEET
exercise
30

Anatomy of the Heart

Gross Anatomy of the Human Heart

1. An anterior view of the heart is shown here. Match each structure listed on the left with the correct key letter:

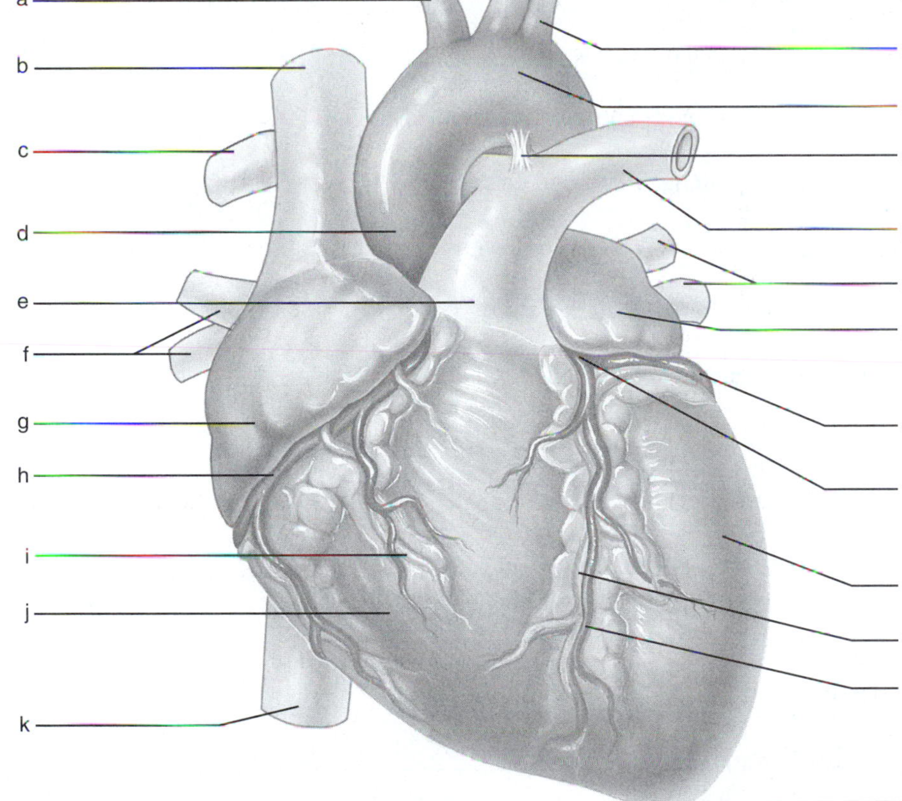

_______ 1. right atrium

_______ 2. right ventricle

_______ 3. left atrium

_______ 4. left ventricle

_______ 5. superior vena cava

_______ 6. inferior vena cava

_______ 7. ascending aorta

_______ 8. aortic arch

_______ 9. brachiocephalic artery

_______ 10. left common carotid artery

_______ 11. left subclavian artery

_______ 12. pulmonary trunk

_______ 13. right pulmonary artery

_______ 14. left pulmonary artery

_______ 15. ligamentum arteriosum

_______ 16. right pulmonary veins

_______ 17. left pulmonary veins

_______ 18. right coronary artery

_______ 19. anterior cardiac vein

_______ 20. left coronary artery

_______ 21. circumflex artery

_______ 22. anterior interventricular artery

_______ 23. apex of heart

_______ 24. great cardiac vein

2. What is the function of the fluid that fills the pericardial sac? ______________________________

____

3. Match the terms in the key to the descriptions provided below.

Key:

a. atria
b. coronary arteries
c. coronary sinus
d. endocardium
e. epicardium
f. mediastinum
g. myocardium
h. ventricles

_______ 1. location of the heart in the thorax

_______ 2. superior heart chambers

_______ 3. inferior heart chambers

_______ 4. visceral pericardium

_______ 5. "anterooms" of the heart

_______ 6. equals cardiac muscle

_______ 7. provide nutrient blood to the heart muscle

_______ 8. lining of the heart chambers

_______ 9. actual "pumps" of the heart

_______ 10. drains blood into the right atrium

4. What is the function of the valves found in the heart? ______________________________

____

5. Can the heart function with leaky valves? (Think! Can a water pump function with leaky valves?) __________

6. What is the role of the chordae tendineae? ______________________________

____

7. Define:

angina pectoris ______________________________

____

pericarditis ______________________________

Pulmonary, Systemic, and Cardiac Circulations

1. A simple schematic of a so-called general circulation is shown on the opposite page. What part of the circulation is missing from this diagram? ______________________________

Add to the diagram as best you can to make it depict a complete systemic/pulmonary circulation and reidentify "general circulation" as the correct subcirculation.

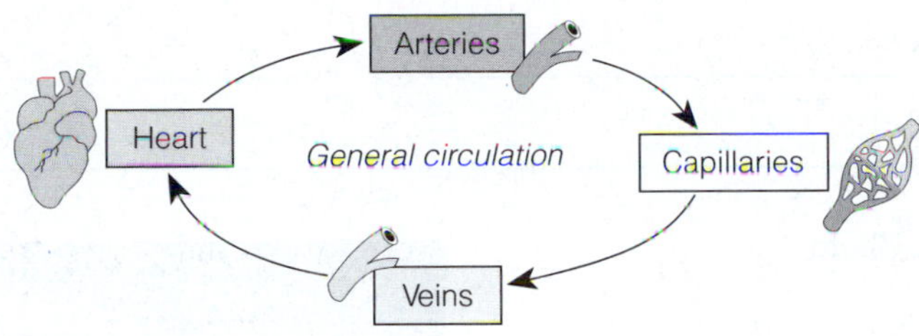

2. Differentiate clearly between the roles of the pulmonary and systemic circulations. ______________________________

____

____

3. Complete the following scheme of circulation of a red blood cell in the human body:

Right atrium through the tricuspid valve to the ______________________ through the ______________________ valve to the pulmonary trunk to the ______________________ to the capillary beds of the lungs to the ______________________ to the ______________________ of the heart through the ______________________ valve to the ______________________ through the ______________________ valve to the ______________________ to the systemic arteries to the ______________________ of the tissues to the systemic veins to the ______________________, ______________________, and ______________________ entering the right atrium of the heart.

4. If the mitral valve does not close properly, which circulation is affected? ______________________

5. Why might a thrombus (blood clot) in the anterior descending branch of the left coronary artery cause sudden death?

____

____

Microscopic Anatomy of Cardiac Muscle

1. How would you distinguish the structure of cardiac muscle from the structure of skeletal muscle? ______________________

____

____

____

____

2. Add the following terms to the photo of cardiac muscle at the right:

   a. intercalated disc

   b. nucleus of cardiac fiber

   c. striations

   d. cardiac muscle fiber

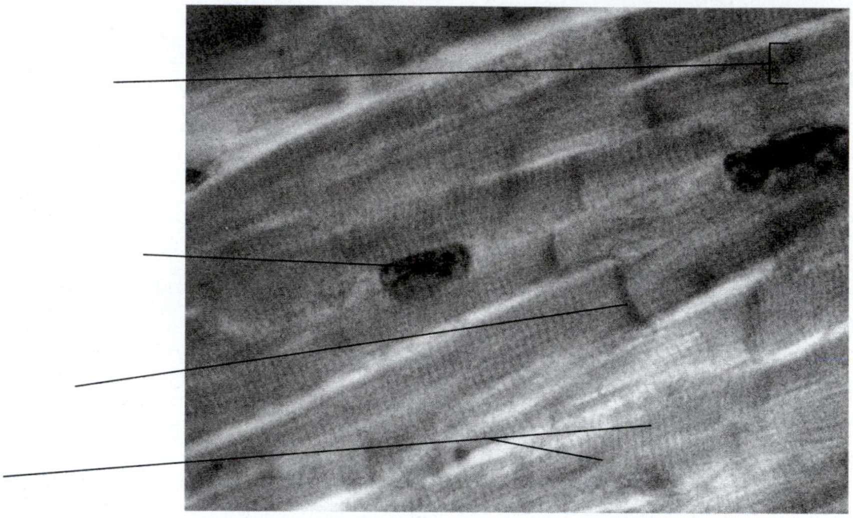

3. What role does the unique structure of cardiac muscle play in its function? (Note: before attempting a response, *describe* the unique anatomy.) ____

____

____

____

Dissection of the Sheep Heart

1. During the sheep heart dissection, you were asked initially to identify the right and left ventricles without cutting into the heart. During this procedure, what differences did you observe between the two chambers?

____

____

Knowing that structure and function are related, how would you say this structural difference reflects the relative functions of these two heart chambers? ______________________________

____

____

____

2. Semilunar valves prevent backflow into the ______________________; AV valves prevent backflow into the ______________________. Using your own observations, explain how the operation of the semilunar valves differs from that of the AV valves. ______________________________

____

____

____

____

3. Differentiate clearly between the location and appearance of pectinate muscle and trabeculae carneae. ______________

____

____

4. Compare and contrast the structure of the right and left atrioventricular valves. ______________________

____

____

____

____

5. Two remnants of fetal structures are observable in the heart—the ligamentum arteriosum and the fossa ovalis. What were they called in the fetal heart, where was each located, and what common purpose did they serve as functioning fetal structures?

____

____

____

____

NAME ______________________ LAB TIME/DATE ______________________

REVIEW SHEET
exercise
31

Conduction System of the Heart and Electrocardiography

The Intrinsic Conduction System

1. List the elements of the intrinsic conduction system in order starting from the SA node.

   SA node ⟶ ______________________ ⟶ ______________________ ⟶

   ______________________ ⟶ ______________________

   Which of those structures is replaced when an artificial pacemaker is installed? ______________________

   At what structure in the transmission sequence is the impulse temporarily delayed? ______________________

   Why? ______________________

   ____

2. Even though cardiac muscle has an inherent ability to beat, the nodal system plays a critical role in heart physiology. What is that role? ______________________

   ____

   ____

3. How does the "all-or-none" law apply to normal heart operation? ______________________

   ____

   ____

Electrocardiography

1. Define *ECG:* ______________________

   ____

2. Draw an ECG wave form representing one heartbeat. Label the P, QRS, and T waves; the P-R interval; the S-T segment, and the Q-T interval.

3. What is the normal length of the P-R interval? ______________ QRS interval? ______________

When the heart rate increases, which interval becomes shorter? ______________

4. What changes from *baseline* were noted in the ECG recorded during running? ______________

Explain. ______________

In that recorded during breath holding? ______________

Explain. ______________

5. Describe what happens in the cardiac cycle in the following situations:

   1. during the P wave ______________
   2. immediately before the P wave ______________
   3. immediately after the P wave ______________
   4. during the QRS wave ______________
   5. immediately after the QRS wave (S-T interval) ______________
   6. during the T wave ______________

6. Define the following terms:

   1. *tachycardia* ______________
   2. *bradycardia* ______________
   3. *flutter* ______________
   4. *fibrillation* ______________
   5. *myocardial infarction* ______________

7. Which would be more serious, atrial or ventricular fibrillation? ____________________

Why? ____________________

____

8. Abnormalities of heart valves can be detected more accurately by auscultation than by electrocardiography. Why is this so?

____

____

____

NAME ____________________ LAB TIME/DATE ____________________

REVIEW SHEET
exercise
32

Anatomy of Blood Vessels

Microscopic Structure of the Blood Vessels

1. Use key choices to identify the blood vessel tunic described.

Key: a. tunica interna b. tunica media c. tunica externa

__________ 1. innermost tunic

__________ 2. bulky middle tunic contains smooth muscle and elastin

__________ 3. its smooth surface decreases resistance to blood flow

__________ 4. tunic(s) of capillaries

__________, __________, __________ 5. tunic(s) of arteries and veins

__________ 6. is especially thick in elastic arteries

__________ 7. most superficial tunic

2. Servicing the capillaries is the essential function of the organs of the circulatory system. Explain this statement.

____

____

____

3. Cross-sectional views of an artery and of a vein are shown here. Identify each; and on the lines beneath, note the structural details that enabled you to make these identifications:

__________ (vessel type)

__________ (a)

__________ (b)

__________ (vessel type)

__________ (a)

__________ (b)

4. Why are valves present in veins but not in arteries? ______________________________

____

____

5. Name two events *occurring within the body* that aid in venous return:

______________________________ and ______________________________

6. Why are the walls of arteries proportionately thicker than those of the corresponding veins? ____________________

____

____

Major Systemic Arteries and Veins of the Body

1. Use the key on the right to identify the arteries or veins described on the left.

_______ 1. the arterial system has one of these; the venous system has two

_______ 2. these arteries supply the myocardium

_______, _______ 3. two paired arteries serving the brain

_______ 4. longest vein in the lower limb

_______ 5. artery on the dorsum of the foot checked after leg surgery

_______ 6. serves the posterior thigh

_______ 7. supplies the diaphragm

_______ 8. formed by the union of the radial and ulnar veins

_______, _______ 9. two superficial veins of the arm

_______ 10. artery serving the kidney

_______ 11. veins draining the liver

_______ 12. artery that supplies the distal half of the large intestine

_______ 13. drains the pelvic organs

_______ 14. what the external iliac artery becomes on entry into the thigh

_______ 15. major artery serving the arm

_______ 16. supplies most of the small intestine

_______ 17. join to form the inferior vena cava

_______ 18. an arterial trunk that has three major branches, which run to the liver, spleen, and stomach

_______ 19. major artery serving the tissues external to the skull

_______, _______, _______ 20. three veins serving the leg

_______ 21. artery generally used to take the pulse at the wrist

Key:

a. anterior tibial
b. basilic
c. brachial
d. brachiocephalic
e. celiac trunk
f. cephalic
g. common carotid
h. common iliac
i. coronary
j. deep femoral
k. dorsalis pedis
l. external carotid
m. femoral
n. fibular
o. greater saphenous
p. hepatic
q. inferior mesenteric
r. internal carotid
s. internal iliac
t. phrenic
u. posterior tibial
v. radial
w. renal
x. subclavian
y. superior mesenteric
z. vertebral

2. The human arterial and venous systems are diagrammed on the next two pages. Identify all indicated blood vessels.

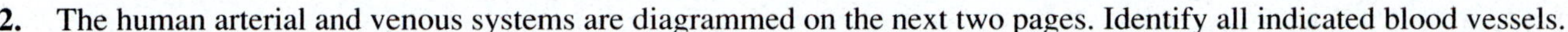

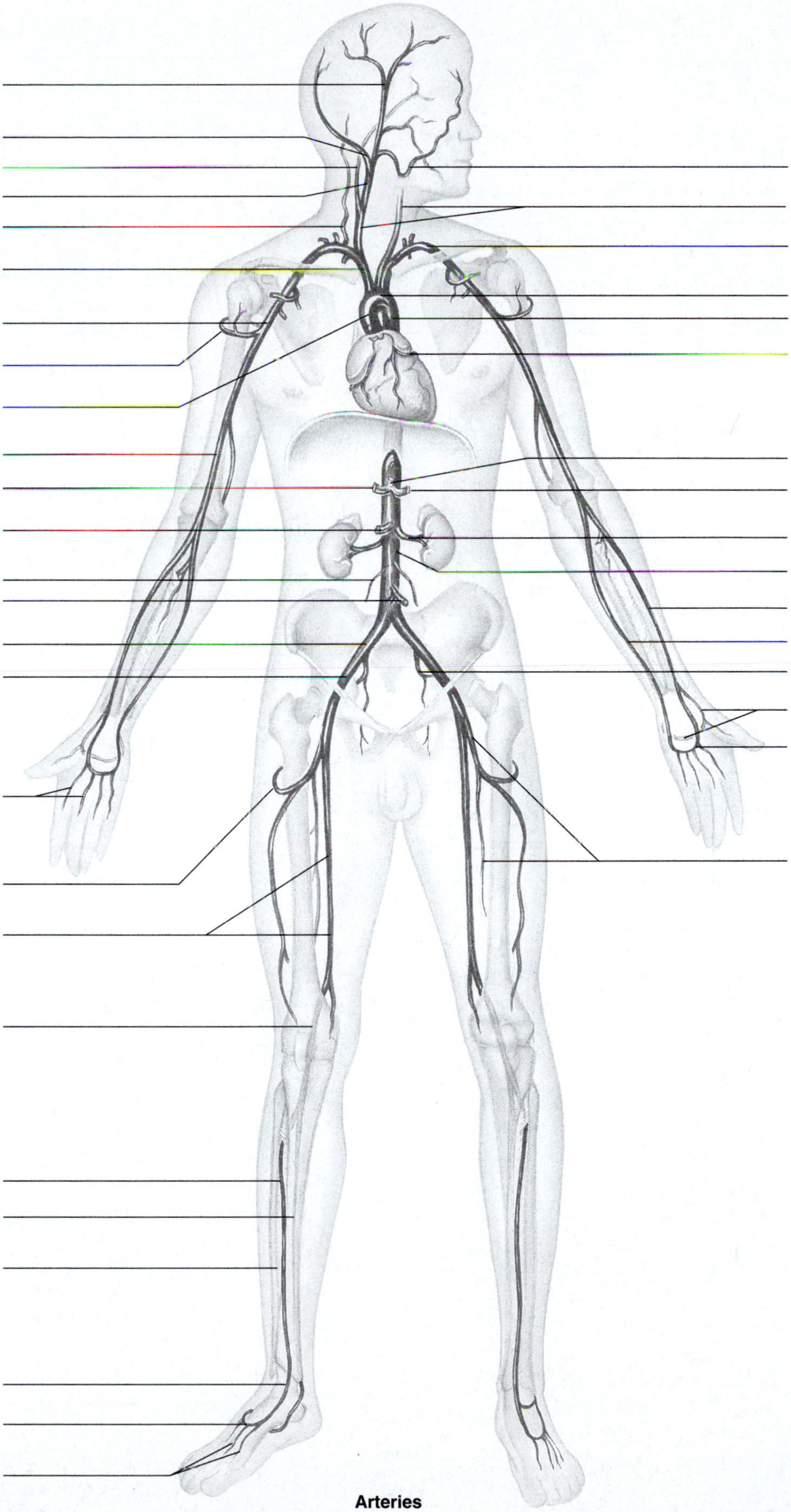

Arteries

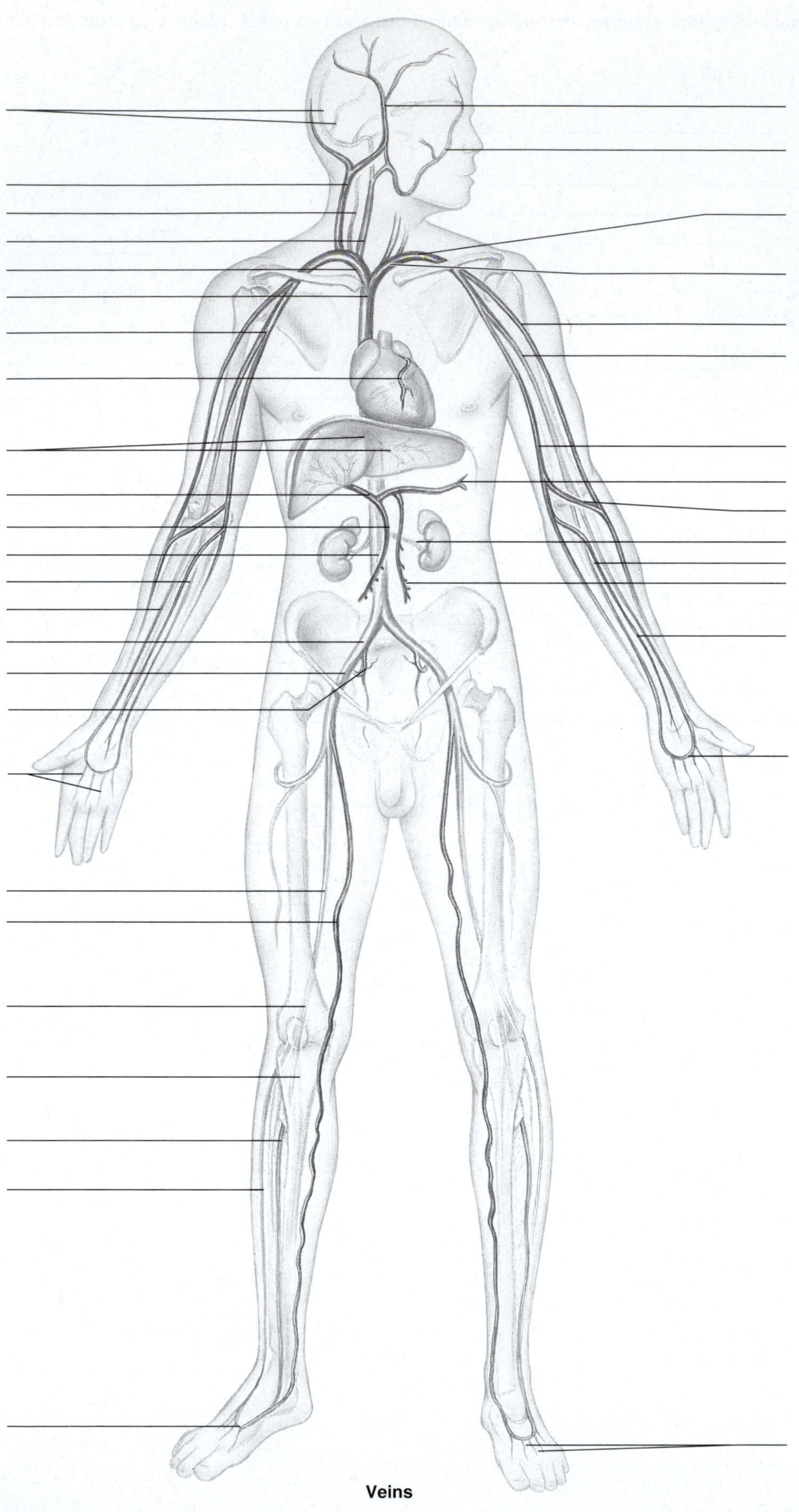

Veins

3. Trace the blood flow for the following situations:

a. From the capillary beds of the left thumb to the capillary beds of the right thumb ______

______

______

______

______

______

______

b. From the bicuspid valve to the tricuspid valve by way of the great toe ______

______

______

______

______

______

c. From the pulmonary vein to the pulmonary artery by way of the right side of the brain ______

______

______

______

______

______

______

______

Special Circulations

Pulmonary circulation:

1. Trace the pathway of a carbon dioxide gas molecule in the blood from the inferior vena cava until it leaves the bloodstream. Name all structures (vessels, heart chambers, and others) passed through en route.

______

______

______

2. Trace the pathway of oxygen gas molecules from an alveolus of the lung to the right atrium of the heart. Name all structures through which it passes. Circle the areas of gas exchange. ____________________

____

____

3. Most arteries of the adult body carry oxygen-rich blood, and the veins carry oxygen-depleted, carbon dioxide–rich blood. How does this differ in the pulmonary arteries and veins? ____________________

____

____

4. How do the arteries of the pulmonary circulation differ structurally from the systemic arteries? What condition is indicated by this anatomical difference? ____________________

____

____

Hepatic portal circulation:

1. What is the source of blood in the hepatic portal system? ____________________

____

2. Why is this blood carried to the liver before it enters the systemic circulation? ____________________

____

____

3. The hepatic portal vein is formed by the union of the ____________________, which drains the ____________________, ____________________, ____________________, and the ____________________, which drains the ____________________ and ____________________. The ____________________ vein, which drains the lesser curvature of the stomach, empties directly into the hepatic portal vein.

4. Trace the flow of a drop of blood from the small intestine to the right atrium of the heart, noting all structures encountered or passed through on the way. ____________________

____

____

Arterial supply of the brain and the circle of Willis:

1. What two paired arteries enter the skull to supply the brain?

_______________ and _______________

2. Branches of the paired arteries just named cooperate to form a ring of blood vessels encircling the pituitary gland, at the base of the brain. What name is given to this communication network? _______________

What is its function? _______________

____

3. What portion of the brain is served by the anterior and middle cerebral arteries? _______________

____

Both the anterior and middle cerebral arteries arise from the _______________ arteries.

4. Trace the pathway of a drop of blood from the aorta to the left occipital lobe of the brain, noting all structures through which it flows. _______________

____

Fetal circulation:

1. The failure of two of the fetal bypass structures to become obliterated after birth can cause congenital heart disease, in which the youngster would have improperly oxygenated blood. Which two structures are these?

_______________ and _______________

2. For each of the following structures, first indicate its function in the fetus; and then note what happens to it or what it is converted to after birth. Circle the blood vessel that carries the most oxygen-rich blood.

Structure	**Function in fetus**	**Fate**
Umbilical artery		
Umbilical vein		
Ductus venosus		
Ductus arteriosus		
Foramen ovale		

3. What organ serves as a respiratory/digestive/excretory organ for the fetus? _______________

NAME ______________________ LAB TIME/DATE ______________________

REVIEW SHEET
exercise
33A

Human Cardiovascular Physiology—Blood Pressure and Pulse Determinations

Cardiac Cycle

1. Correctly identify valve closings and openings, chamber pressures, and volume lines, and the ECG and heart sound scan lines on the diagram below by matching the diagram labels with the terms to the right of the diagram.

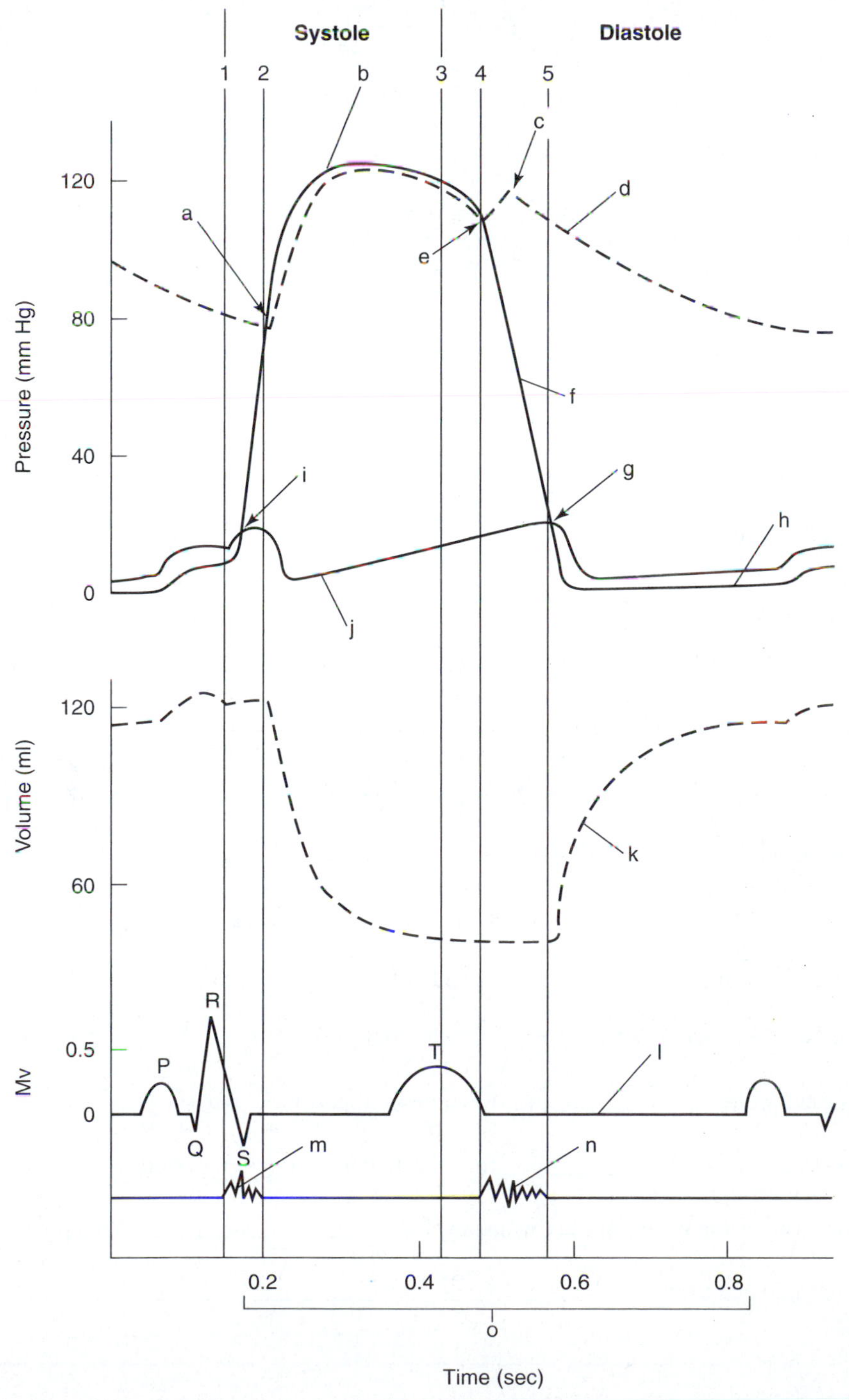

______ 1. aortic and semilunar valves closed

______ 2. aortic pressure

______ 3. aortic valve closes

______ 4. aortic valve opens

______ 5. atrial pressure

______ 6. AV valve closes

______ 7. AV valve opens

______ 8. cardiac cycle

______ 9. dicrotic notch

______ 10. ECG

______ 11. first heart sound

______ 12. second heart sound

______ 13. ventricular diastole

______ 14. peak of ventricular systole

______ 15. ventricular volume

2. Define the following terms:

systole ___

diastole ___

cardiac cycle ___

3. Answer the following questions, which concern events of the cardiac cycle:

When are the AV valves closed? ___

Open? ___

What event within the heart causes the AV valves to open? ___

What causes them to close? ___

When are the semilunar valves closed? ___

___

Open? ___

What event causes the semilunar valves to open? ___

___

To close? ___

Are both sets of valves closed during any part of the cycle? ___

If so, when? ___

Are both sets of valves open during any part of the cycle? ___

At what point in the cardiac cycle is the pressure in the heart highest? ___

Lowest? ___

What event results in the pressure deflection called the dicrotic notch? ___

___

4. Using the key below, indicate the time interval occupied by the following events of the cardiac cycle.

Key: a. 0.4 sec b. 0.3 sec c. 0.1 sec d. 0.8 sec

___ 1. the length of the normal cardiac cycle

___ 2. the time interval of atrial systole

___ 3. the quiescent period, or pause

___ 4. the ventricular contraction period

5. If an individual's heart rate is 80 beats/min, what is the length of the cardiac cycle? ___ What portion of the cardiac cycle is shortened by this more rapid heart rate? ___

___

6. What two factors promote the movement of blood through the heart? ______________________

______________________ and ______________________

____

Heart Sounds

1. Complete the following statements:

The monosyllables describing the heart sounds are __1__. The first heart sound is a result of closure of the __2__ valves, whereas the second is a result of closure of the __3__ valves. The heart chambers that have just been filled when you hear the first heart sound are the __4__, and the chambers that have just emptied are the __5__. Immediately after the second heart sound, the __6__ are filling with blood, and the __7__ are empty.

1. ______________________
2. ______________________
3. ______________________
4. ______________________
5. ______________________
6. ______________________
7. ______________________

2. As you listened to the heart sounds during the laboratory session, what differences in pitch, length, and amplitude (loudness) of the two sounds did you observe? ______________________

____

____

3. Indicate where you would place your stethoscope to auscultate most accurately the following:

closure of the tricuspid valve ______________________

closure of the aortic semilunar valve ______________________

apical heartbeat ______________________

Which valve is heard most clearly when the apical heartbeat is auscultated? ______________________

4. No one expects you to be a full-fledged physician on such short notice; but on the basis of what you have learned about heart sounds, how might abnormal sounds be used to diagnose heart problems?

____

____

____

The Pulse

1. Define *pulse:* ______________________

____

2. Describe the procedure used to take the pulse. ______________________________

____

____

3. Identify the artery palpated at each of the following pressure points:

at the wrist ______________________ on the dorsum of the foot ______________________

in front of the ear ______________________ at the side of the neck ______________________

4. When you were palpating the various pulse or pressure points, which appeared to have the greatest amplitude or tension?

______________________ Why do you think this was so? ______________________

____

5. Assume someone has been injured in an auto accident and is hemorrhaging badly. What pressure point would you compress to help stop bleeding from each of the following areas?

the thigh ______________________ the calf ______________________

the forearm ______________________ the thumb ______________________

6. How could you tell by simple observation whether bleeding is arterial or venous? ______________________

____

7. You may sometimes observe a slight difference between the value obtained from an apical pulse (beats/min) and that from an arterial pulse taken elsewhere on the body. What is this difference called?

____

Blood Pressure Determinations

1. Define *blood pressure:* ______________________________

2. Identify the phase of the cardiac cycle to which each of the following apply:

systolic pressure ______________________ diastolic pressure ______________________

3. What is the name of the instrument used to compress the artery and record pressures in the auscultatory method of determining blood pressure? ______________________________

4. What are the sounds of Korotkoff? ______________________________

____

What causes the systolic sound? ______________________________

____

The disappearance of sound? ______________________________

5. Interpret 145/85/82. ______________________________

____

____

6. Assume the following BP measurement was recorded for an elderly patient with severe arteriosclerosis: 170/110/–. Explain the inability to obtain the third reading.

____

____

7. Define *pulse pressure:* ______________________________

Why is this measurement important? ______________________________

____

8. How do venous pressures compare to arterial pressures? ______________________________

Why? ______________________________

9. What maneuver to increase the thoracic pressure illustrates the effect of external factors on venous pressure?

______________________________ How was it performed? ______________________________

____

____

10. What might an abnormal increase in venous pressure indicate? (Think!) ______________________________

____

Observing the Effect of Various Factors on Blood Pressure and Heart Rate

1. What effect do the following have on blood pressure? (Indicate increase by I and decrease by D.)

_______ 1. increased diameter of the arterioles

_______ 2. increased blood viscosity

_______ 3. increased cardiac output

_______ 4. hemorrhage

_______ 5. arteriosclerosis

_______ 6. increased pulse rate

2. In which position (sitting, reclining, or standing) is the blood pressure normally the highest?

______________________________ The lowest? ______________________________

What immediate changes in blood pressure did you observe when the subject stood up after being in the sitting or reclining position? ______________________________

____

What changes in the blood vessels might account for the change? ______________________________

____

____

After the subject stood for 3 minutes, what changes in blood pressure were observed? ______________________________

____

How do you account for this change? ______________________________

3. What was the effect of exercise on blood pressure? ______________________________

On pulse? ______________________________ Do you think these effects reflect changes in cardiac output *or* in peripheral resistance? ______________________________

Why are there normally no significant increases in diastolic pressure after exercise? ______________________________

____

4. What effects of the following did you observe on blood pressure in the laboratory?

nicotine ______________________________ cold temperature ______________________________

What do you think the effect of heat would be? ______________________________

Why? ______________________________

5. Differentiate between a hypo- and a hyperreactor relative to the cold pressor test. ______________________________

____

Skin Color as an Indicator of Local Circulatory Dynamics

1. Describe normal skin color and the appearance of the veins in the subject's forearm before any testing was conducted.

____

2. What changes occurred when the subject emptied his forearm of blood (by raising his arm and making a fist) and the flow was occluded with the cuff? ______________________________

What changes occurred during venous congestion? ______________________________

____

3. What is the importance of collateral blood supplies? ______________________________

____

4. Explain the mechanism by which mechanical stimulation of the skin produced a flare. ______________________________

____

NAME ______________________ LAB TIME/DATE ______________________

REVIEW SHEET
exercise
34A

Frog Cardiovascular Physiology: Wet Lab

Special Electrical Properties of Cardiac Muscle: Automaticity and Rhythmicity

1. Define the following terms:

   automaticity ______________________

   rhythmicity ______________________

2. Discuss the anatomical differences you observed between frog and human hearts. ______________________

3. Which region of the dissected frog heart had the highest intrinsic rate of contraction? ______________________

   The greatest automaticity? ______________________

   The greatest regularity or rhythmicity? ______________________ How do these properties correlate with the duties of a pacemaker? ______________________

   Is this region the pacemaker of the frog heart? ______________________

   Which region had the lowest intrinsic rate of contraction? ______________________

Investigating the Refractory Period of Cardiac Muscle

1. Define *extrasystole:* ______________________

2. In responding to the following questions, refer to the recordings you made during this exercise:

   What was the effect of stimulation of the heart during ventricular contraction? ______________________

   During ventricular relaxation (first portion)? ______________________

   During the pause interval? ______________________

   What does this indicate about the refractory period of cardiac muscle? ______________________

   Can cardiac muscle be tetanized? ________ Why or why not? ______________________

Why is this important to the normal function of the heart? ______________________________

____

Assessing Physical and Chemical Modifiers of Heart Rate

1. Describe the effect of thermal factors on the frog heart:

   cold ______________________________ heat ______________________________

2. What was the effect of vagal stimulation on heart rate? ______________________________

   Which of the following factors cause the same (or very similar) heart rate–reducing effects? Epinephrine, acetylcholine, atropine sulfate, pilocarpine, sympathetic nervous system activity, digitalis, potassium ions?

   ____

   Which of the factors listed above would reverse or antagonize vagal effects? ______________________________

   ____

3. What is vagal escape? ______________________________

   ____

   Why is vagal escape valuable in maintaining homeostasis? ______________________________

   ____

4. Once again refer to your recordings. Did the administration of the following produce any changes in force of contraction (shown by peaks of increasing or decreasing height)? If so, explain the mechanism.

   epinephrine ______________________________

   ____

   acetylcholine ______________________________

   ____

   ____

   calcium ions ______________________________

   ____

   ____

5. Excessive amounts of each of the following ions would most likely interfere with normal heart activity. Note the type of changes caused in each case.

   K^+ ______________________________

   Ca^{2+} ______________________________

   Na^+ ______________________________

6. How does the Stannius ligature used in the laboratory produce heart block? ______________________________

____

____

7. Define *partial heart block*, and describe how it was recognized in the laboratory. ______________________

____

____

8. Define *total heart block*, and describe how it was recognized in the laboratory. ________________________

____

____

____

9. What do your heart block experiment results indicate about the spread of impulses from the atria to the ventricles?

____

____

Observing the Microcirculation Under Various Conditions

1. In what way are the red blood cells of the frog different from those of the human? ______________________

____

On the basis of this one factor, would you expect their life spans to be longer or shorter? __________________

2. The following statements refer to your observation of one or more of the vessel types observed in the microcirculation in the frog's web. Characterize each statement by choosing one or more responses from the key.

Key: a. arteriole b. venule c. capillary

________ 1. smallest vessels observed

________ 2. vessel within which the blood flow is rapid, pulsating

________ 3. vessel in which blood flow is least rapid

________ 4. red blood cells pass through these vessels in single file

________ 5. blood flow smooth and steady

________ 6. most numerous vessels

________ 7. vessels that deliver blood to the capillary bed

________ 8. vessels that serve the needs of the tissues via exchanges

________ 9. vessels that drain the capillary beds

3. Which of the vessel diameters changed most? ______________________________

What division of the nervous system controls the vessels? ______________________________

4. Discuss the effects of the following on blood vessel diameter (state specifically the blood vessels involved) and rate of blood flow. Then explain the importance of the reaction observed to the general well-being of the body.

local application of cold ______________________________

____

____

local application of heat ______________________________

____

____

inflammation (or application of HCl) ______________________________

____

____

histamine ______________________________

____

____

NAME ________________________ LAB TIME/DATE ____________________

REVIEW SHEET
exercise
35

The Lymphatic System and Immune Response

The Lymphatic System

1. Match the terms below with the correct letters on the diagram.

_______ 1. axillary lymph nodes

_______ 2. bone marrow

_______ 3. cervical lymph nodes

_______ 4. cisterna chyli

_______ 5. inguinal lymph nodes

_______ 6. lymphatic vessels

_______ 7. Peyer's patches (in intestine)

_______ 8. right lymphatic duct

_______ 9. spleen

_______ 10. thoracic duct

_______ 11. thymus gland

_______ 12. tonsils

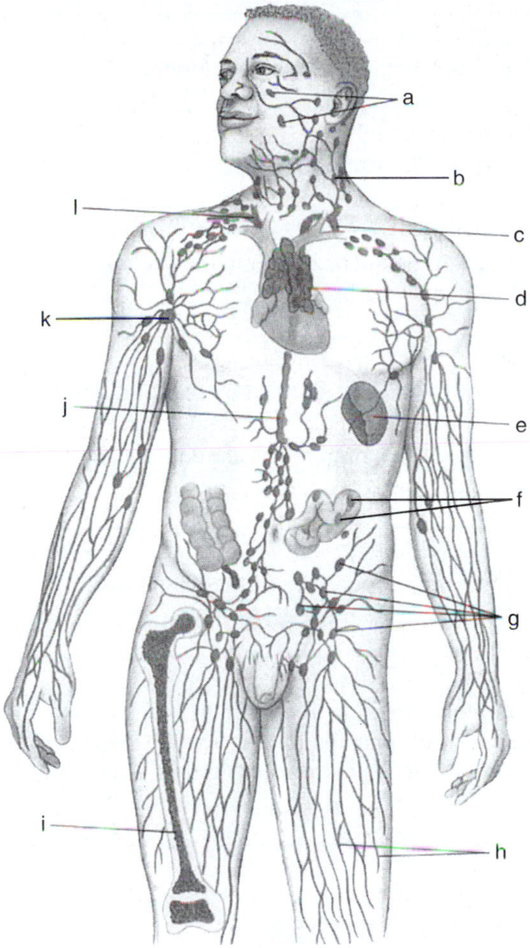

2. Explain why the lymphatic system is a one-way system, whereas the blood vascular system is a two-way system.

____

____

3. How do lymphatic vessels resemble veins? ____

How do lymphatic capillaries differ from blood capillaries? ___

___

4. What is the function of the lymphatic vessels? ___

___

5. What is lymph? ___

6. What factors are involved in the flow of lymphatic fluid? ___

___

7. What name is given to the terminal duct draining most of the body? ___

8. What is the cisterna chyli? ___

___

How does the composition of lymph in the cisterna chyli differ from that in the general lymphatic stream?

___

9. Which portion of the body is drained by the right lymphatic duct? ___

10. Note three areas where lymph nodes are densely clustered: ___,

___, and ___

11. What are the two major functions of the lymph nodes? ___

___

12. The radical mastectomy is an operation in which a cancerous breast, surrounding tissues, and the underlying muscles of the anterior thoracic wall, plus the axillary lymph nodes, are removed. After such an operation, the arm usually swells, or becomes edematous, and is very uncomfortable—sometimes for months. Why?

___

The Immune Response

1. What is the function of B cells in the immune response? ___

___

2. What is the role of T cells? ___

___

___

3. Define the following terms related to the operation of the immune system:

Immunological memory ____

____

Specificity ____

____

Recognition of self from nonself ____

____

Autoimmune disease ____

____

Studying the Microscopic Anatomy of a Lymph Node, the Spleen, and a Tonsil

1. In the space below, make a rough drawing of the structure of a lymph node. Identify the cortex area, germinal centers, and medulla. For each identified area, note the cell type (T cell, B cell, or macrophage) most likely to be found there.

2. What structural characteristic ensures a *slow* flow of lymph through a lymph node? ______________________________

____

Why is this desirable? ____

____

3. What similarities in structure and function are found in the lymph nodes, spleen, and tonsils? ______________________

____

____

Antibodies and Tests for Their Presence

1. Distinguish between antigen and antibody. ______

2. Describe the structure of the immunoglobulin monomer. ______

3. Are the genes coding for one antibody entirely different from those coding for a different antibody? ______

   Explain. ______

4. In the Ouchterlony test, what happened when the antibody to horse serum albumin mixed with horse serum albumin?

   ______

5. If the unknown antigen contained bovine and swine serum albumin, what would you expect to happen in the Ouchterlony test, and why? ______

NAME ____________________ LAB TIME/DATE ____________________

REVIEW SHEET
exercise
36

Anatomy of the Respiratory System

Upper and Lower Respiratory System Structures

1. Complete the labeling of the diagram of the upper respiratory structures (sagittal section).

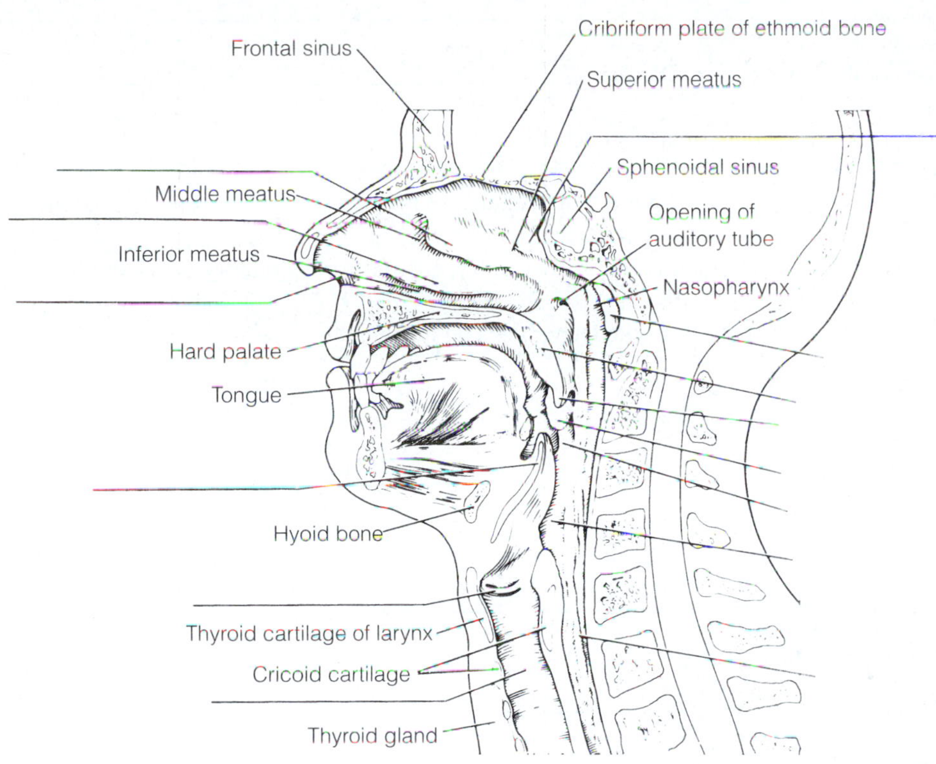

2. Two pairs of vocal folds are found in the larynx. Which pair are the true vocal cords (superior or inferior)?

____

3. What is the significance of the fact that the human trachea is reinforced with cartilage rings?

____

Of the fact that the rings are incomplete posteriorly? ____________________

____

4. Name the specific cartilages in the larynx that correspond to the following descriptions:

1. forms the Adam's apple ____________________
2. a "lid" for the larynx ____________________
3. shaped like a signet ring ____________________
4. vocal cord attachment ____________________

5. Trace a molecule of oxygen from the external nares to the pulmonary capillaries of the lungs: External nares →

____

____

____

6. What is the function of the pleural membranes? ____

____

____

7. Name two functions of the nasal cavity mucosa: ____

8. The following questions refer to the primary bronchi:

Which is longer? ______________ Larger in diameter? ______________ More horizontal? ______________

The more common site for lodging of a foreign object that has entered the respiratory passageways? ______________

9. Appropriately label all structures provided with leader lines on the diagrams below.

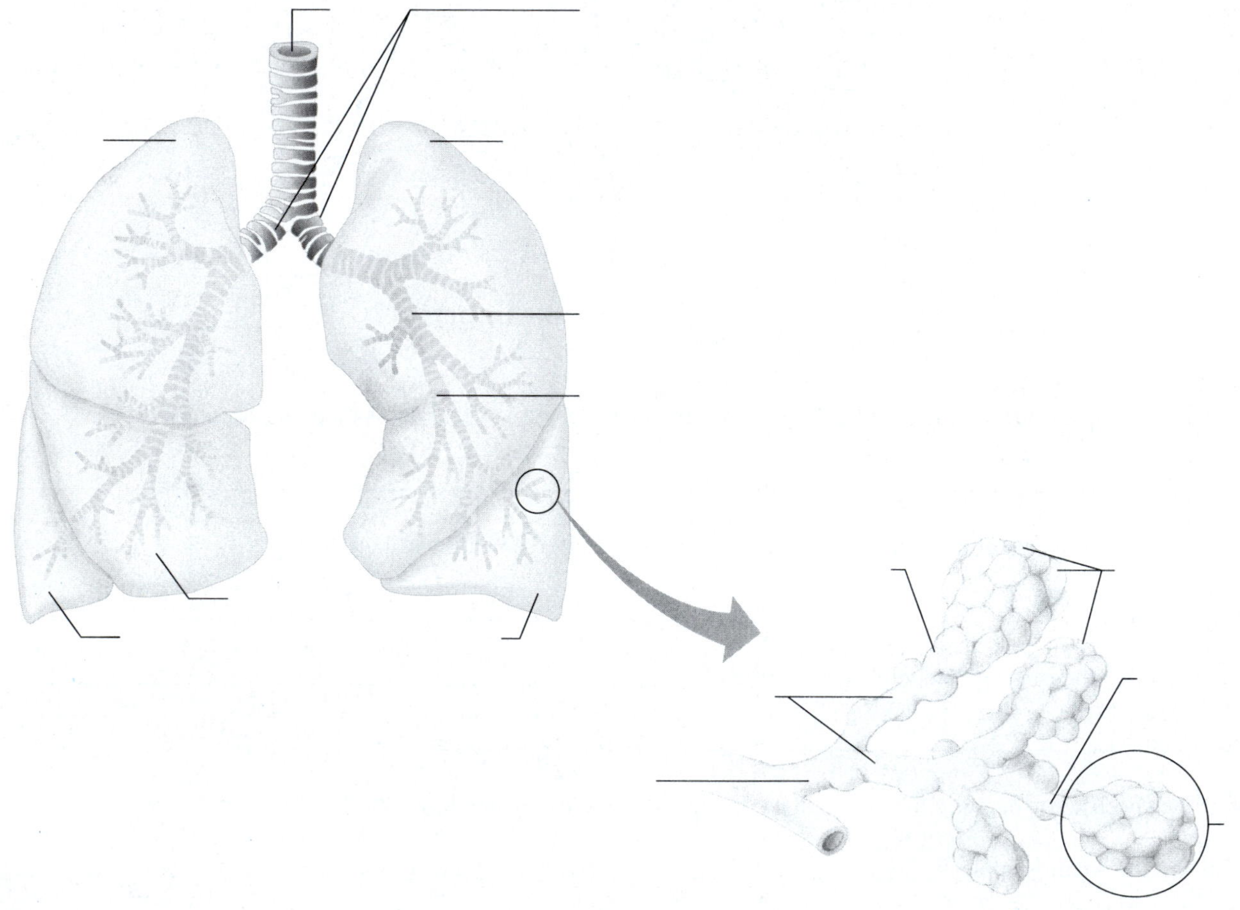

10. Match the terms in column B to the descriptions in column A.

Column A	Column B
_______ 1. connects the larynx to the primary bronchi	a. alveolus
_______ 2. site of tonsils	b. bronchiole
_______ 3. food passageway posterior to the trachea	c. concha
_______ 4. flaps over the glottis during swallowing of food	d. epiglottis
_______ 5. contains the vocal cords	e. esophagus
_______ 6. nerve that activates the diaphragm during inspiration	f. glottis
_______ 7. pleural layer lining the walls of the thorax	g. larynx
_______ 8. site from which oxygen enters the pulmonary blood	h. opening of auditory tube
_______ 9. connects the middle ear to the nasopharynx	i. palate
_______ 10. opening between the vocal folds	j. parietal pleura
_______ 11. increases air turbulence in the nasal cavity	k. pharynx
_______ 12. separates the oral cavity from the nasal cavity	l. phrenic nerve
	m. primary bronchi
	n. trachea
	o. vagus nerve
	p. visceral pleura

11. What portions of the respiratory system are referred to as anatomical dead space? ____________________

____

____

Why? ____

12. Define *external respiration:* ____

____

internal respiration: ____

____

cellular respiration: ____

____

13. On the diagram below identify alveolar epithelium, capillary endothelium, alveoli, and red blood cells. Bracket the respiratory membrane.

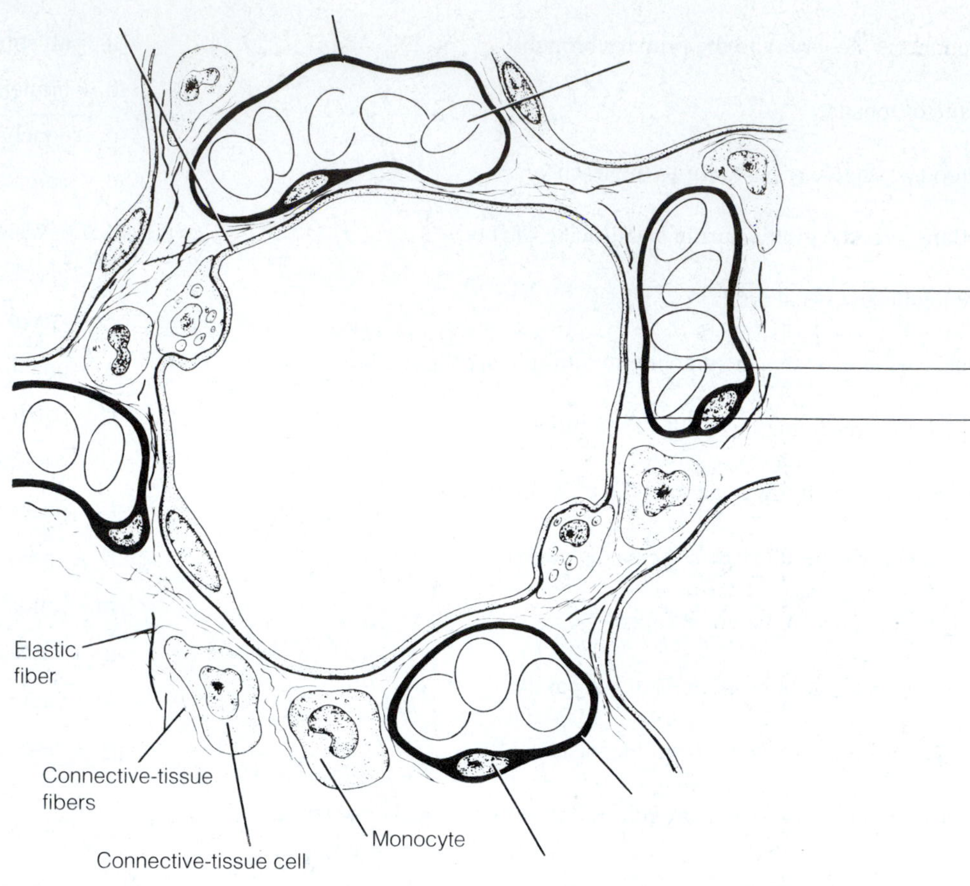

Demonstrating Lung Inflation in a Sheep Pluck

Does the lung inflate part by part or as a whole, like a balloon? ______________________________

What happened when the pressure was released? ______________________________

What type of tissue ensures this phenomenon? ______________________________

Examining Prepared Slides of Lung and Trachea Tissue

1. The tracheal epithelium is ciliated and has goblet cells. What is the function of each of these modifications?

   Cilia? ______________________________

   Goblet cells? ______________________________

2. The tracheal epithelium is said to be pseudostratified. Why? ______________________________

   ____

   ____

3. What structural characteristics of the alveoli make them an ideal site for the diffusion of gases?

____

Why does oxygen move from the alveoli into the pulmonary capillary blood? ______________________

____

____

4. If you observed pathological lung sections, what were the responsible conditions and how did the tissue differ from normal lung tissue?

Slide type	Observations

NAME ______________________ LAB TIME/DATE ______________________

REVIEW SHEET
exercise
37A

Respiratory System Physiology

Mechanics of Respiration

1. For each of the following cases, check the column appropriate to your observations on the operation of the model lung.

	Diaphragm pushed up		Diaphragm pulled down	
Change	**Increased**	**Decreased**	**Increased**	**Decreased**
In internal volume of the bell jar (thoracic cage)				
In internal pressure				
In the size of the balloons (lungs)				
In direction of air flow	Into lungs	Out of lungs	Into lungs	Out of lungs

2. Base your answers to the following on your observations in question 1.

Under what internal conditions does air tend to flow into the lungs? ______________________

____

Under what internal conditions does air tend to flow out of the lungs? Explain. ______________________

____

3. Activation of the diaphragm and the external intercostal muscles begins the inspiratory process. What effect does contraction of these muscles have on thoracic volume, and how is this accomplished? ______________________

____

____

____

4. What was the approximate increase in diameter of chest circumference during a quiet inspiration?

______________ inches During forced inspiration? ______________ inches

What temporary physiological advantage does the substantial increase in chest circumference during forced inspiration create? ______________________________

____

____

5. The presence of a partial vacuum between the pleural membranes is integral to normal breathing movements. What would happen if an opening were made into the chest cavity, as with a puncture wound?

____

How is this condition treated medically? ______________________________

Respiratory Volumes and Capacities—Spirometry

1. Write the respiratory volume term and the normal value that is described by the following statements:

Volume of air present in the lungs after a forceful expiration ______________________________

Volume of air that can be expired forcibly after a normal expiration ______________________________

Volume of air that is breathed in and out during a normal respiration ______________________________

Volume of air that can be inspired forcibly after a normal inspiration ______________________________

Volume of air corresponding to TV + IRV + ERV ______________________________

2. Record experimental respiratory volumes as determined in the laboratory. (Corrected values are for Procedure B only.)

Average tidal volume ______________ ml

Corrected value for TV ______________ ml

Average IRV ______________ ml

Corrected value for IRV ______________ ml

Minute respiratory volume ______________ ml/min

Average ERV ______________ ml

Corrected value for ERV ______________ ml

Average VC ______________ ml

Corrected value for VC ______________ ml

% predicted VC ______________ %

FEV_1 ______________ % FVC

3. Would your vital-capacity measurement differ if you performed the test while standing? _______ While lying down? _______ Explain. ______________________________

____

4. Which respiratory ailments can respiratory volume tests be used to detect?

____

____

5. Using an appropriate reference, complete the chart below:

		O_2	CO_2	N_2
% of composition of air	Inspired			
	Expired			

Use of the Pneumograph to Determine Factors Influencing Rate and Depth of Respiration

1. Where are the neural control centers of respiratory rhythm? ______________________ and ______________________

2. Based on pneumograph reading of respiratory variation, what was the rate of quiet breathing?

Initial testing ______________________ breaths/min

Record observations of how the initial pneumograph or respiratory belt transducer recording was modified during the various testing procedures described below. Indicate the respiratory rate, and include comments on the relative depth of the respiratory peaks observed.

Test performed	Observations
Talking	
Yawning	
Laughing	
Standing	
Concentrating	
Swallowing water	
Coughing	
Lying down	
Running in place	

3. Student data:

Breath-holding interval after a deep inhalation _______ sec   length of recovery period _______ sec

Breath-holding interval after a forceful expiration _______ sec   length of recovery period _______ sec

After breathing quietly and taking a deep breath (which you held), was your urge to inspire or expire? _______________

After exhaling and then holding one's breath, was the desire for inspiration or expiration? _______________

Explain these results. (*Hint:* what reflex is involved here?) _______________

____

____

4. Observations after hyperventilation: _______________

____

5. Length of breath holding after hyperventilation: _______ sec

Why does hyperventilation produce apnea or a reduced respiratory rate? _______________

____

____

6. Observations for rebreathing breathed air: _______________

____

Why does rebreathing breathed air produce an increased respiratory rate? _______________

____

7. What was the effect of running in place (exercise) on the duration of breath holding? _______________

Explain: _______________

____

8. Relative to the test illustrating the effect of respiration on circulation: *(student data)*

Radial pulse before beginning test _______ /min   Radial pulse after testing _______ /min

Relative pulse force before beginning test _______   Relative force of radial pulse after testing _______

Condition of neck and facial veins after testing _______________

Explain: _______________

____

9. Do the following factors generally increase (indicate with I) or decrease (indicate with D) the respiratory rate and depth?

1. increase in blood CO_2 _______
2. decrease in blood O_2 _______
3. increase in blood pH _______
4. decrease in blood pH _______

Did it appear that CO_2 or O_2 had a more marked effect on modifying the respiratory rate? _______________

10. Where are sensory receptors sensitive to changes in blood pressure located? _______________

____

11. Where are sensory receptors sensitive to changes in O_2 levels in the blood located? _______________

____

12. What is the primary factor that initiates breathing in a newborn infant? _______________

____

13. Blood CO_2 levels and blood pH are related. When blood CO_2 levels increase, does the pH increase or decrease?

_______________ Explain why. _______________

____

____

Respiratory Sounds

1. Which of the respiratory sounds is heard during both inspiration and expiration? _______________

   Which is heard primarily during inspiration? _______________

2. Where did you best hear the vesicular respiratory sounds? _______________

Role of the Respiratory System in Acid-Base Balance of Blood

1. Define *buffer:* _______________

2. How successful was the laboratory buffer (pH 7) in resisting changes in pH when the acid was added? _______________

   ____

   When the base was added? _______________

   How successful was the buffer in resisting changes in pH when the additional aliquots (3 more drops) of the acid and base were added to the original samples? _______________

3. What buffer system operates in blood plasma? _______________

   Which of its species resists a *drop* in pH? _______________

   Which resists a *rise* in pH? _______________

4. Explain how the carbonic acid–bicarbonate buffer system of the blood operates. ______________________________

____

____

____

____

____

5. What happened when the carbon dioxide in exhaled air mixed with water? ______________________________

____

____

What role does exhalation of carbon dioxide play in maintaining relatively constant blood pH? ______________________________

____

____

NAME ______________________ LAB TIME/DATE ______________________

REVIEW SHEET
exercise
38

Anatomy of the Digestive System

General Histological Plan of the Alimentary Canal

The general anatomical features of the digestive tube are listed below. Fill in the table to complete the information.

Wall layer	Subdivisions of the layer if applicable	Major functions
mucosa		
submucosa		
muscularis externa		
serosa or adventitia		

Organs of the Alimentary Canal

1. The tubelike digestive system canal that extends from the mouth to the anus is the ______________________ canal.

2. How is the muscularis externa of the stomach modified? ______________________

   How does this modification relate to the function of the stomach? ______________________

3. What transition in epithelium type exists at the gastroesophageal junction? ______________________

   How do the epithelia of these two organs relate to their specific functions? ______________________

4. Differentiate between the colon and the large intestine. ______________________

5. Match the items in column B with the descriptive statements in column A.

Column A

_______ 1. structure that suspends the small intestine from the posterior body wall

_______ 2. fingerlike extensions of the intestinal mucosa that increase the surface area for absorption

_______ 3. large collections of lymphoid tissue found in the submucosa of the small intestine

_______ 4. deep folds of the mucosa and submucosa that extend completely or partially around the circumference of the small intestine

_______, _______ 5. regions that break down foodstuffs mechanically

_______ 6. mobile organ that manipulates food in the mouth and initiates swallowing

_______ 7. conduit for both air and food

_______, _______, _______ 8. three structures continuous with and representing modifications of the peritoneum

_______ 9. the "gullet"; no digestive/absorptive function

_______ 10. folds of the gastric mucosa

_______ 11. sacculations of the large intestine

_______ 12. projections of the plasma membrane of a mucosal epithelial cell

_______ 13. valve at the junction of the small and large intestines

_______ 14. primary region of food and water absorption

_______ 15. membrane securing the tongue to the floor of the mouth

_______ 16. absorbs water and forms feces

_______ 17. area between the teeth and lips/cheeks

_______ 18. wormlike sac that outpockets from the cecum

_______ 19. initiates protein digestion

_______ 20. structure attached to the lesser curvature of the stomach

_______ 21. organ distal to the stomach

_______ 22. valve controlling food movement from the stomach into the duodenum

_______ 23. posterosuperior boundary of the oral cavity

_______ 24. location of the hepatopancreatic sphincter through which pancreatic secretions and bile pass

_______ 25. serous lining of the abdominal cavity wall

_______ 26. principal site for the synthesis of vitamin K by microorganisms

_______ 27. region containing two sphincters through which feces are expelled from the body

_______ 28. bone-supported anterosuperior boundary of the oral cavity

Column B

a. anus
b. appendix
c. circular folds
d. esophagus
e. frenulum
f. greater omentum
g. hard palate
h. haustra
i. ileocecal valve
j. large intestine
k. lesser omentum
l. mesentery
m. microvilli
n. oral cavity
o. parietal peritoneum
p. Peyer's patches
q. pharynx
r. pyloric valve
s. rugae
t. small intestine
u. soft palate
v. stomach
w. tongue
x. vestibule
y. villi
z. visceral peritoneum

6. Correctly identify all organs depicted in the diagram below.

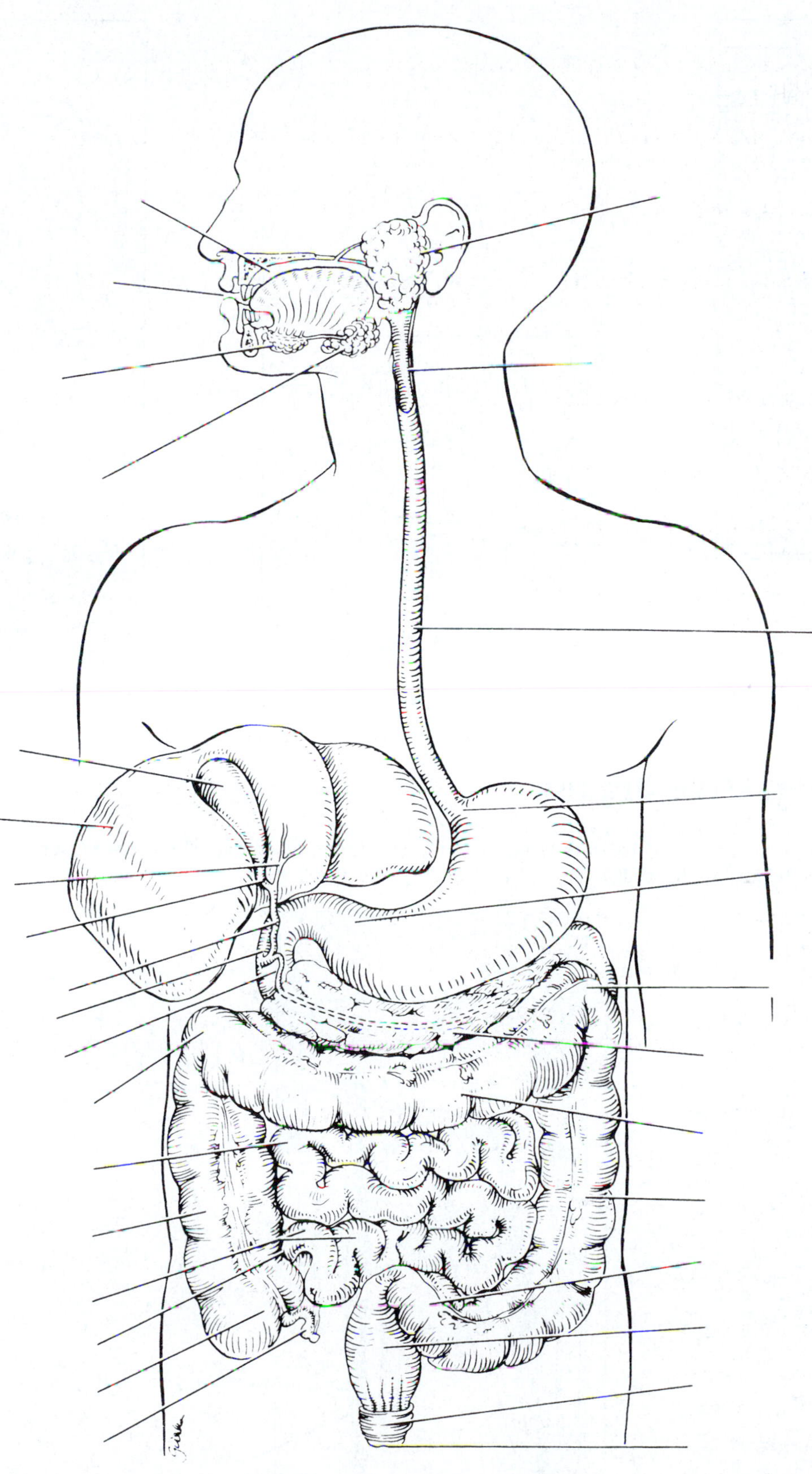

7. You have studied the histological structure of a number of organs in this laboratory. Three of these are diagrammed below. Identify and correctly label each.

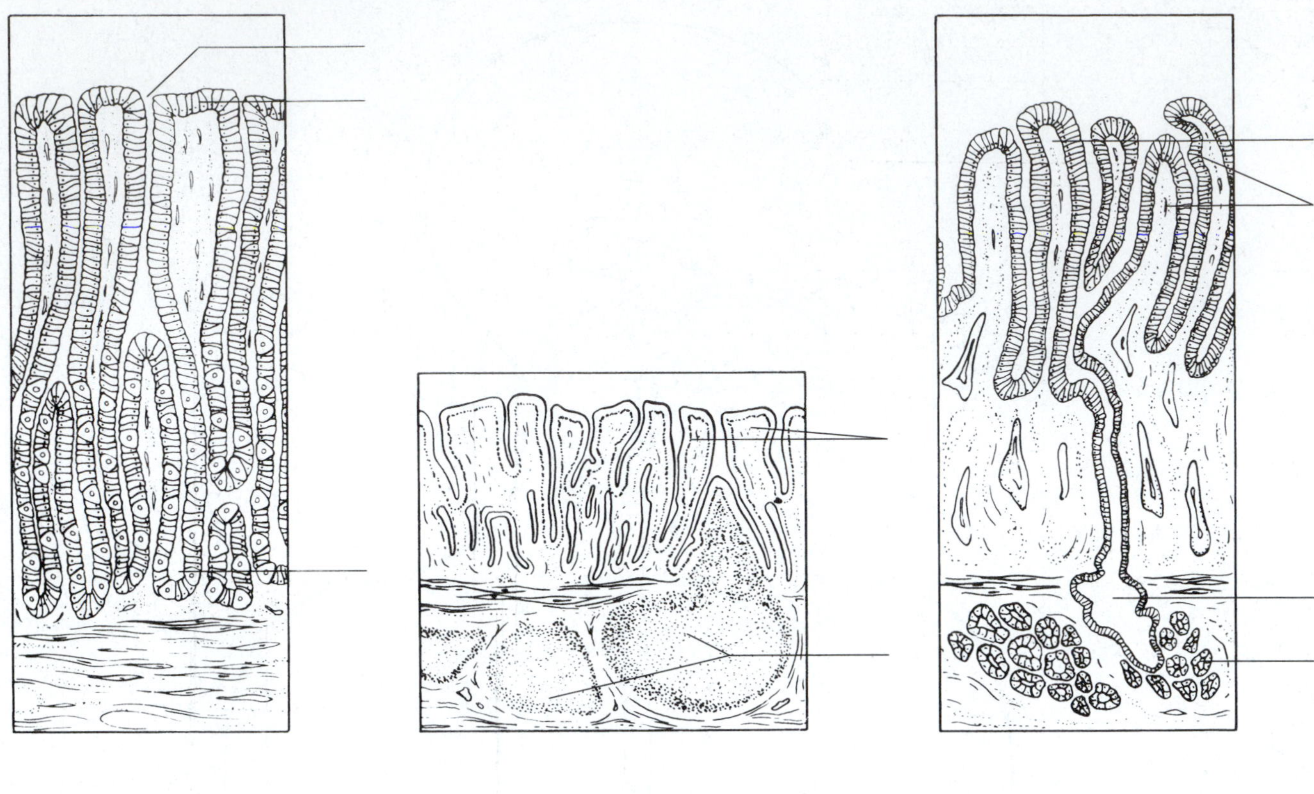

Accessory Digestive Organs

1. Correctly label all structures provided with leader lines in the diagram of a molar below. (Note: some of the terms in the key for item 2 may be helpful in this task.)

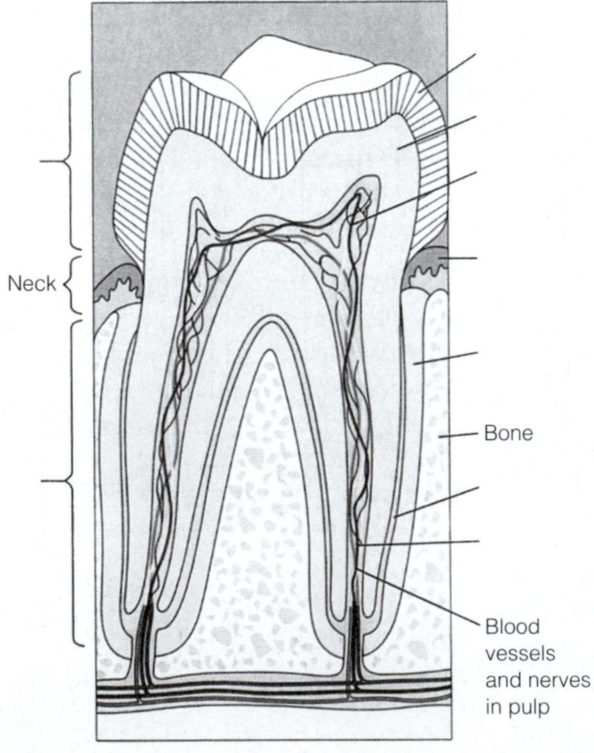

2. Use the key to identify each tooth area described below.

Key:
a. anatomical crown
b. cementum
c. clinical crown
d. dentin
e. enamel
f. gingiva
g. odontoblast
h. periodontal ligament
i. pulp
j. root

_______ 1. visible portion of the tooth *in situ*

_______ 2. material covering the tooth root

_______ 3. hardest substance in the body

_______ 4. attaches the tooth to bone and surrounding alveolar structures

_______ 5. portion of the tooth embedded in bone

_______ 6. forms the major portion of tooth structure; similar to bone

_______ 7. produces the dentin

_______ 8. site of blood vessels, nerves, and lymphatics

_______ 9. entire portion of the tooth covered with enamel

3. In the human, the number of deciduous teeth is _______; the number of permanent teeth is _______________.

4. The dental formula for permanent teeth is $\frac{2, 1, 2, 3}{2, 1, 2, 3} \times 2 = 32$

Explain what this means: ____

____

What is the dental formula for the deciduous teeth? _______ × _______ = _______

5. What teeth are the "wisdom teeth"? ____

____

____

6. Various types of glands form a part of the alimentary tube wall or duct their secretions into it. Match the glands listed in column B with the function/locations described in column A.

Column A

_______ 1. produce(s) mucus; found in the submucosa of the small intestine

_______ 2. produce(s) a product containing amylase that begins starch breakdown in the mouth

_______ 3. produce(s) a whole spectrum of enzymes and an alkaline fluid that is secreted into the duodenum

_______ 4. produce(s) bile that it secretes into the duodenum via the bile duct

_______ 5. produce(s) HCl and pepsinogen

_______ 6. found in the mucosa of the small intestine; produce(s) intestinal juice

Column B

a. duodenal glands
b. gastric glands
c. intestinal crypts
d. liver
e. pancreas
f. salivary glands

7. Which of the salivary glands produces a secretion that is mainly serous? _______________________

8. What is the role of the gallbladder? ____

9. Name three structures always found in the portal triad regions of the liver. ____________________,

_________________________________ and ____.

10. Where would you expect to find the Kupffer cells of the liver? ______________________________

What is their function? ______________________________

11. Why is the liver so dark red in the living animal? ______________________________

____

12. The pancreas has two major populations of secretory cells—those in the islets and the acinar cells. Which population serves the digestive process? ______________________________

NAME ______________________ LAB TIME/DATE ______________________

Chemical and Physical Processes of Digestion: Wet Lab

Chemical Digestion of Foodstuffs: Enzymatic Action

1. Match the following definitions with the proper choices from the key.

Key: a. catalyst b. control c. enzyme d. substrate

__________ 1. increases the rate of a chemical reaction without becoming part of the product

__________ 2. provides a standard of comparison for test results

__________ 3. biologic catalyst: protein in nature

__________ 4. substance on which a catalyst works

2. List the three characteristics of enzymes. ______________________

____

____

3. The enzymes of the digestive system are classified as hydrolases. What does this mean?

____

____

4. Fill in the following chart about the various digestive system enzymes encountered in this exercise.

Enzyme	Organ producing it	Site of action	Substrate(s)	Optimal pH
Salivary amylase				
Trypsin				
Lipase (pancreatic)				

5. Name the end products of digestion for the following types of foods:

proteins: ______________ carbohydrates: ______________

fats: ______________ and ______________

6. You used several indicators or tests in the laboratory to determine the presence or absence of certain substances. Choose the correct test or indicator from the key to correspond to the condition described below.

Key: a. IKI (Lugol's iodine) b. Benedict's solution c. litmus d. BAPNA

_______ 1. used to test for protein hydrolysis, which was indicated by a yellow color

_______ 2. used to test for the presence of starch, which was indicated by blue-black color

_______ 3. used to test for the presence of fatty acids, which was evidenced by a color change from blue to pink

_______ 4. used to test for the presence of reducing sugars (maltose, sucrose, glucose) as indicated by a blue to green color change

7. What conclusions can you draw when an experimental sample gives both a positive starch test and a positive maltose test after incubation? _______________

Why was 37°C the optimal incubation temperature? _______________

Why did very little, if any, starch digestion occur in test tube 4A? _______________

____

Why did very little, if any, starch digestion occur in test tube 6A? _______________

____

Assume you have made the statement to a group of your peers that amylase is capable of starch hydrolysis to maltose. If you had not done control tube 1A, what objection to your statement could be raised? _______________

____

What if you had not done tube 2A? _______________

8. In the exercise concerning trypsin function, why was an enzyme assay like Benedict's or Lugol's IKI (which test for the presence of a reaction product) not necessary? _______________

Why was tube 1T necessary? _______________

Why was tube 2T necessary? _______________

Trypsin is a protease similar to pepsin, the protein-digesting enzyme in the stomach. Would trypsin work well in the stomach? _______ Why? _______________

9. In the procedure concerning pancreatic lipase digestion of fats and the action of bile salts, how did the appearance of tubes 1E and 2E differ? _______________

____

Can you explain the difference? _______________

____

Why did the litmus indicator change from blue to pink during fat hydrolysis? ______

______

Why is bile not considered an enzyme? ______

How did the tubes containing bile compare with those not containing bile? ______

What role does bile play in fat digestion? ______

______

10. The three-dimensional structure of a functional protein is altered by intense heat or nonphysiological pH even though peptide bonds may not break. Such inactivation is called denaturation, and denatured enzymes are nonfunctional. Explain why.

______

______

What specific experimental conditions resulted in denatured enzymes? ______

______

11. Pancreatic and intestinal enzymes operate optimally at a pH that is slightly alkaline, yet the chyme entering the duodenum from the stomach is very acid. How is the proper pH for the functioning of the pancreatic-intestinal enzymes ensured?

______

______

12. Assume you have been chewing a piece of bread for 5 or 6 minutes. How would you expect its taste to change during this interval? ______

Why? ______

13. Note the mechanism of absorption (passive or active transport) of the following food breakdown products, and indicate by a check mark (✓) whether the absorption would result in their movement into the blood capillaries or the lymph capillaries (lacteals).

Substance	Mechanism of absorption	Blood	Lymph
Monosaccharides			
Fatty acids and glycerol			
Amino acids			
Water			
Na^{+}, Cl^{-}, Ca^{2+}			

14. People on a strict diet to lose weight begin to metabolize stored fats at an accelerated rate. How does this condition affect blood pH? ______

15. Using a flow chart, trace the pathway of a ham sandwich (ham = protein and fat; bread = starch) from the mouth to the site of absorption of its breakdown products, noting where digestion occurs and what specific enzymes are involved.

16. Some of the digestive organs have groups of secretory cells that liberate hormones (parahormones) into the blood. These exert an effect on the digestive process by acting on other cells or structures and causing them to release digestive enzymes, expel bile, or increase the mobility of the digestive tract. For each hormone below, note the organ producing the hormone and its effects on the digestive process. Include the target organs affected.

Hormone	Produced by	Target organ(s) and effects
Secretin		
Gastrin		
Cholecystokinin		

Physical Processes: Mechanisms of Food Propulsion and Mixing

Complete the following statements.

Swallowing, or__1__, occurs in two phases—the__2__and __3__. One of these phases, the__4__phase, is voluntary. During the voluntary phase, the__5__is used to push the food into the back of the throat. During swallowing, the__6__rises to ensure that its passageway is covered by the epiglottis so that the ingested substances don't enter the respiratory passageways. It is possible to swallow water while standing on your head because the water is carried along the esophagus involuntarily by the process of__7__. The pressure exerted by the foodstuffs on the__8__sphincter causes it to open, allowing the foodstuffs to enter the stomach.

The two major types of propulsive movements that occur in the small intestine are__9__and__10__. One of these movements, the __11__, acts to continually mix the foods and to increase the absorption rate by moving different parts of the chyme mass over the intestinal mucosa, but has less of a role in moving foods along the digestive tract.

1. ____________________
2. ____________________
3. ____________________
4. ____________________
5. ____________________
6. ____________________
7. ____________________
8. ____________________
9. ____________________
10. ____________________
11. ____________________

NAME ______________________________ LAB TIME/DATE ______________________

REVIEW SHEET
exercise
40

Anatomy of the Urinary System

Gross Anatomy of the Human Urinary System

1. Complete the following statements:

The kidney is referred to as an excretory organ because it excretes __1__ wastes. It is also a major homeostatic organ because it maintains the electrolyte, __2__ , and __3__ balance of the blood.

Urine is continuously formed by the __4__ and is routed down the __5__ by the mechanism of __6__ to a storage organ called the __7__ . Eventually, the urine is conducted to the body __8__ by the urethra. In the male, the urethra is __9__ centimeters long and transports both urine and __10__ . The female urethra is __11__ centimeters long and transports only urine.

Voiding or emptying the bladder is called __12__ . Voiding has both voluntary and involuntary components. The voluntary sphincter is the __13__ sphincter. An inability to control this sphincter is referred to as __14__ .

1. ______________________
2. ______________________
3. ______________________
4. ______________________
5. ______________________
6. ______________________
7. ______________________
8. ______________________
9. ______________________
10. ______________________
11. ______________________
12. ______________________
13. ______________________
14. ______________________

2. What is the function of the fat cushion that surrounds the kidneys in life? ______________________

____

3. Define *ptosis:* ____

____

4. Why is incontinence a normal phenomenon in the child under 1½ to 2 years old? ______________________

____

What events may lead to its occurrence in the adult? ______________________

____

____

5. Complete the labeling of the diagram to correctly identify the urinary system organs.

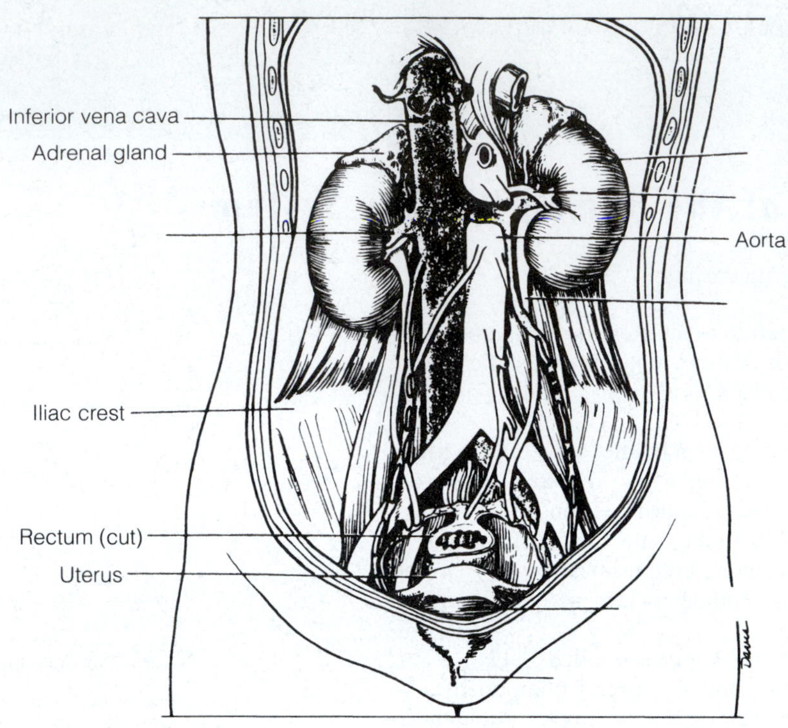

Gross Internal Anatomy of the Pig or Sheep Kidney

Match the appropriate structure in column B to its description in column A.

	Column A	Column B
________	1. smooth membrane, tightly adherent to the kidney surface	a. cortex
________	2. portion of the kidney containing mostly collecting ducts	b. medulla
________	3. portion of the kidney containing the bulk of the nephron structures	c. minor calyx
________	4. superficial region of kidney tissue	d. renal capsule
________	5. basinlike area of the kidney, continuous with the ureter	e. renal column
________	6. a cup-shaped extension of the pelvis that encircles the apex of a pyramid	f. renal pelvis
________	7. area of cortical tissue running between the medullary pyramids	

Functional Microscopic Anatomy of the Kidney and Bladder

1. Use the key letters to identify the diagram of the nephron (and associated renal blood supply) on the left.

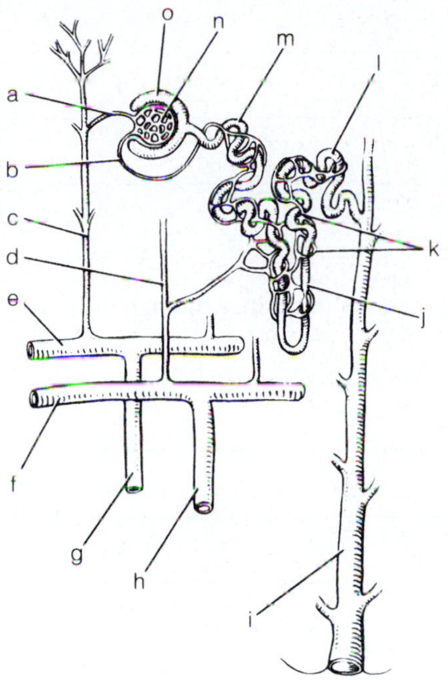

_______ 1. arcuate artery

_______ 2. arcuate vein

_______ 3. afferent arteriole

_______ 4. collecting duct

_______ 5. distal convoluted tubule

_______ 6. efferent arteriole

_______ 7. glomerular capsule

_______ 8. glomerulus

_______ 9. interlobar artery

_______ 10. interlobar vein

_______ 11. interlobular artery

_______ 12. interlobular vein

_______ 13. loop of Henle

_______ 14. peritubular capillaries

_______ 15. proximal convoluted tubule

2. Using the terms provided in item 1, identify the following:

_______________ 1. site of filtrate formation

_______________ 2. primary site of tubular reabsorption

_______________ 3. secondarily important site of tubular reabsorption

_______________ 4. structure that conveys the processed filtrate (urine) to the renal pelvis

_______________ 5. blood supply that directly receives substances from the tubular cells

_______________ 6. its inner (visceral) membrane forms part of the filtration membrane

3. Explain *why* the glomerulus is such a high-pressure capillary bed. _______________

____

How does its high-pressure condition aid its function of filtrate formation? _______________

____

4. What structural modification of certain tubule cells enhances their ability to reabsorb substances from the filtrate?

____

5. Explain the mechanism of tubular secretion and explain its importance in the urine formation process. ______________

____

____

6. Compare and contrast the composition of blood plasma and glomerular filtrate. ______________

____

7. Trace a drop of blood from the time it enters the kidney in the renal artery until it leaves the kidney through the renal vein.

Renal artery → ____

____

____ → renal vein

8. Define *juxtaglomerular apparatus:* ____

____

9. Label the figure using the key letters of the correct terms.

Key: a. juxtaglomerular cells
b. cuboidal epithelium
c. macula densa
d. glomerular capsule (parietal layer)
e. distal convoluted tubule

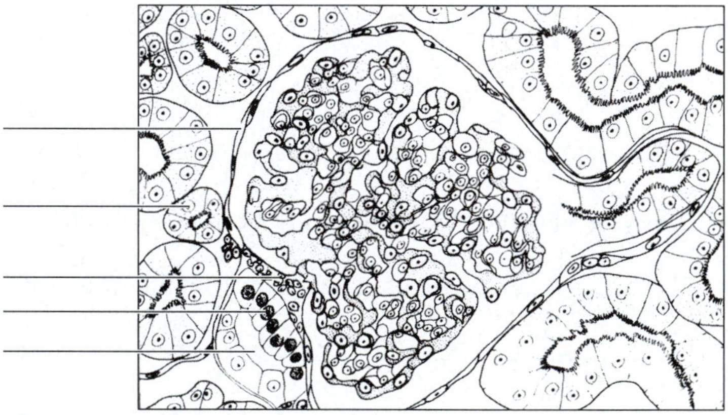

10. Trace the anatomical pathway of a molecule of creatinine (metabolic waste) from the glomerular capsule to the urethra. Note each microscopic and/or gross structure it passes through in its travels. Name the subdivisions of the renal tubule.

Glomerular capsule → ____

____

____ → urethra

11. What is important functionally about the specialized epithelium (transitional epithelium) in the bladder?

____

NAME ______________________ LAB TIME/DATE ______________________

REVIEW SHEET
exercise
41A

Urinalysis

Characteristics of Urine

1. What is the normal volume of urine excreted in a 24-hour period? __________

   a. 0.1–0.5 liters b. 0.5–1.2 liters c. 1.0–1.8 liters

2. Assuming normal conditions, note whether each of the following substances would be (a) in greater relative concentration in the urine than in the glomerular filtrate, (b) in lesser concentration in the urine than in the glomerular filtrate, or (c) absent in both the urine and the glomerular filtrate.

______	1. water	______	6. amino acids	______	11. uric acid
______	2. phosphate ions	______	7. glucose	______	12. creatinine
______	3. sulfate ions	______	8. albumin	______	13. pus (WBC)
______	4. potassium ions	______	9. red blood cells	______	14. nitrites
______	5. sodium ions	______	10. urea		

3. Explain why urinalysis is a routine part of any good physical examination. ______________________

   ____

4. What substance is responsible for the normal yellow color of urine? ______________________

5. Which has a greater specific gravity: 1 ml of urine or 1 ml of distilled water? ______________________

   Explain. ______________________

   ____

6. Explain the relationship between the color, specific gravity, and volume of urine. ______________________

   ____

Abnormal Urinary Constituents

1. A microscopic examination of urine may reveal the presence of certain abnormal urinary constituents.

   Name three constituents that might be present if a urinary tract infection exists. ______________________,

   ______________________, and ______________________

2. How does a urinary tract infection influence urine pH? ______________________

   How does starvation influence urine pH? ______________________

3. Several specific terms have been used to indicate the presence of abnormal urine constituents. Identify each of the abnormalities described below by inserting a term from the list at the right that names the condition.

______ 1.	presence of erythrocytes in the urine	a. albuminuria
______ 2.	presence of hemoglobin in the urine	b. glycosuria
______ 3.	presence of glucose in the urine	c. hematuria
______ 4.	presence of albumin in the urine	d. hemoglobinuria
______ 5.	presence of ketone bodies (acetone and others) in the urine	e. ketonuria
		f. pyuria
______ 6.	presence of pus (white blood cells) in the urine	

4. What are renal calculi and what conditions favor their formation? ______________________________

____

5. All urine specimens become alkaline and cloudy on standing at room temperature. Explain. ______________________________

____

6. Glucose and albumin are both normally absent in the urine, but the reason for their exclusion differs. Explain the reason for

the absence of glucose. ______________________________

Explain the reason for the absence of albumin. ______________________________

7. Several conditions (both pathological and nonpathological) are named below. Using the key provided, characterize the probable abnormal constituents or conditions of the urinary product of each. More than one choice is necessary to fully characterize the condition in most cases.

______ 1.	glomerulonephritis	______ 7.	starvation	Key:
				a. albumin
______ 2.	diabetes mellitus	______ 8.	diabetes insipidus	b. hemoglobin
				c. blood cells
______ 3.	pregnancy, exertion	______ 9.	kidney stones	d. glucose
______ 4.	hepatitis, cirrhosis of the liver	______ 10.	eating a 5-lb box of candy at one sitting	e. ketone bodies
				f. bilirubin
				g. pus
______ 5.	pyelonephritis	______ 11.	hemolytic anemias	h. high specific gravity
				i. low specific gravity
______ 6.	gonorrhea	______ 12.	cystitis (inflammation of the bladder)	j. casts

8. Name the three major nitrogenous wastes found in the urine. ______________________________,

______________________________, and ______________________________

9. Explain the difference between organized and unorganized sediments. ______________________________

____

____

NAME ______________________ LAB TIME/DATE ______________________

REVIEW SHEET
exercise
42

Anatomy of the Reproductive System

Gross Anatomy of the Human Male Reproductive System

1. List the two principal functions of the testis: ______________________

2. Identify all indicated structures or portions of structures on the diagrammatic view of the male reproductive system below.

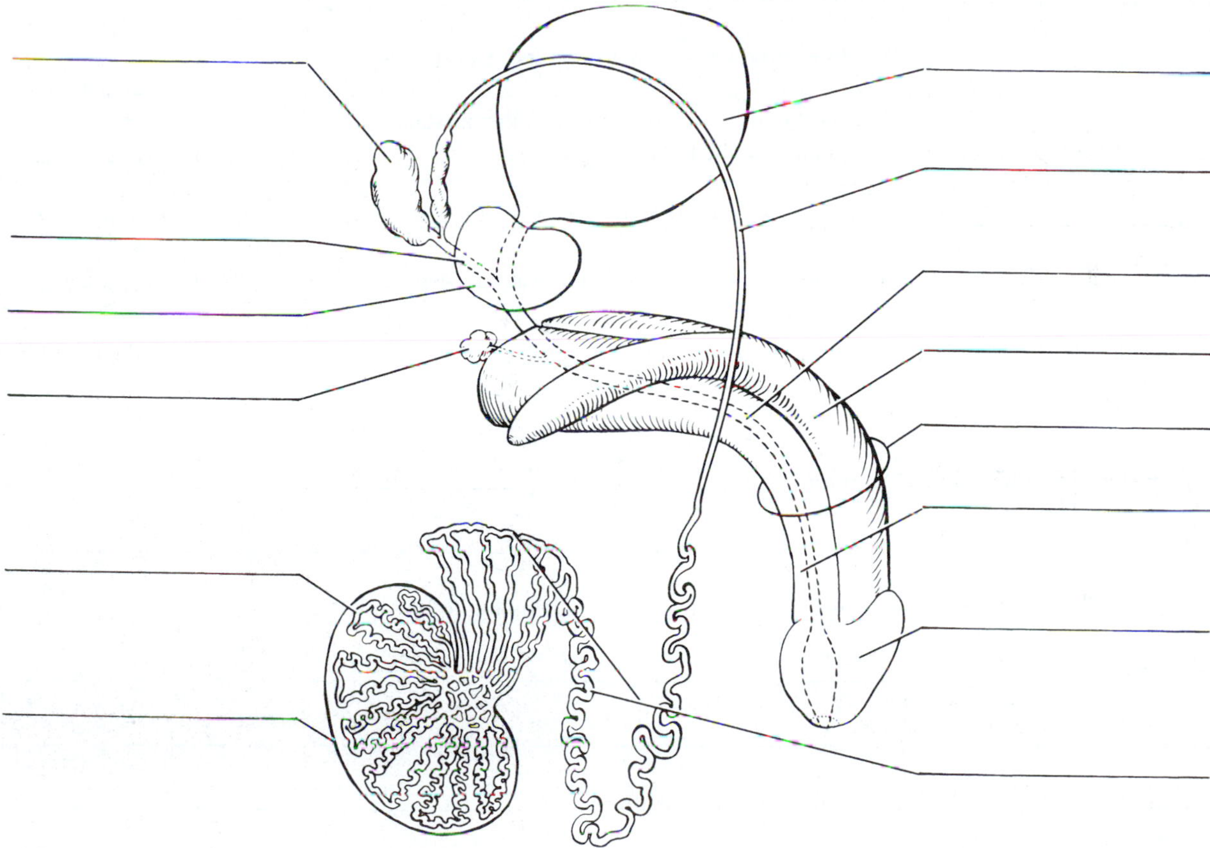

3. A common part of any physical examination of the male is palpation of the prostate gland. How is this accomplished? (Think!) ______________________

4. How might enlargement of the prostate gland interfere with urination or the reproductive ability of the male?

____

____

5. Match the terms in column B to the descriptive statements in column A.

	Column A	Column B
____________	1. copulatory organ/penetrating device	a. bulbourethral glands
____________	2. site of sperm/androgen production	b. ductus deferens
____________	3. muscular passageway conveying sperm to the ejaculatory duct; in the spermatic cord	c. epididymis
		d. glans penis
____________	4. transports both sperm and urine	e. membranous urethra
____________	5. sperm maturation site	f. penis
____________	6. location of the testis in adult males	g. prepuce
____________	7. loose fold of skin encircling the glans penis	h. prostate gland
____________	8. portion of the urethra between the prostate gland and the penis	i. prostatic urethra
____________	9. empties a secretion into the prostatic urethra	j. seminal vesicles
____________	10. empties a secretion into the membranous urethra	k. scrotum
		l. spongy urethra
		m. testes

6. Why are the testes located in the scrotum? ____

____

7. Describe the composition of semen and name all structures contributing to its formation. ____________________

____

____

8. Of what importance is the fact that seminal fluid is alkaline? ____

____

9. What structures compose the spermatic cord? ____

____

Where is it located? ____

____

10. Using the following terms, trace the pathway of sperm from the testes to the urethra: rete testis, epididymis, seminiferous tubule, ductus deferens.

____________ → ____________ → ____________ → ____________

11. Using an appropriate reference, define *cryptorchidism* and discuss its significance.

____

____

Gross Anatomy of the Human Female Reproductive System

1. On the diagram below of a frontal section of a portion of the female reproductive system, identify all indicated structures.

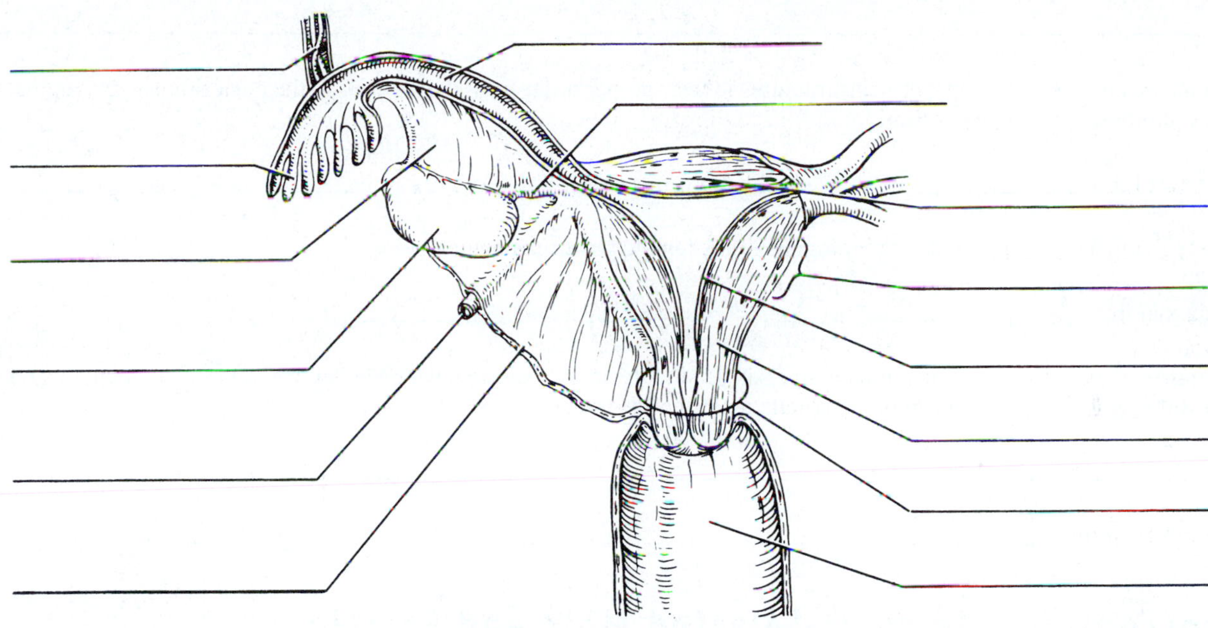

2. Identify the female reproductive system structures described below:

________________ 1. site of fetal development

________________ 2. copulatory canal

________________ 3. "fertilized egg" typically formed here

________________ 4. becomes erectile during sexual excitement

________________ 5. duct extending superolaterally from the uterus

________________ 6. partially closes the vaginal canal; a membrane

________________ 7. produces oocytes, estrogens, and progesterone

________________ 8. fingerlike ends of the fallopian tube

3. Do any sperm enter the pelvic cavity of the female? Why or why not? ________________

____

4. What is an ectopic pregnancy, and how can it happen? ______________________________

____

____

____

5. Name the structures composing the external genitalia, or vulva, of the female. ______________________________

____

____

6. Put the following vestibular-perineal structures in their proper order from the anterior to the posterior aspect: vaginal orifice, anus, urethral opening, and clitoris.

Anterior limit: ______________ → ______________ → ______________ → ______________

7. Name the male structure that is homologous to the female structures named below.

labia majora ______________ clitoris ______________

8. Assume a couple has just consummated the sex act and the male's sperm have been deposited in the woman's vagina. Trace the pathway of the sperm through the female reproductive tract.

____

9. Define *ovulation:* ______________________________

Microscopic Anatomy of Selected Male and Female Reproductive Organs

1. The testis is divided into a number of lobes by connective tissue. Each of these lobes contains one to four ______________

______________, which converge on a tubular region at the testis hilus called the ______________

______________.

2. What is the function of the cavernous bodies seen in the male penis? ______________________________

____

____

3. Name the three layers of the uterine wall from the inside out.

______________, ______________, ______________

Which of these is sloughed during menses? ______________________________

Which contracts during childbirth? ______________________________

4. What is the function of the stereocilia exhibited by the epithelial cells of the mucosa of the epididymis? ______________

____

5. On the diagram showing the sagittal section of the human testis, correctly identify all structures provided with leader lines.

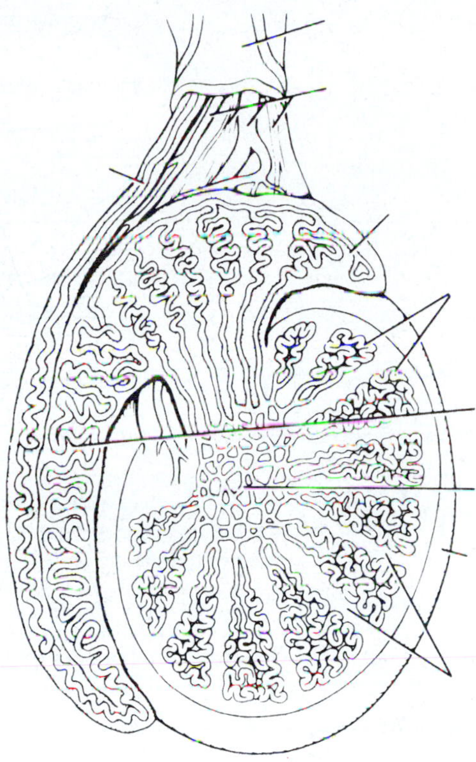

The Mammary Glands

1. Match the term with the correct description.

_______ glands that produce milk during lactation	a. alveoli
_______ subdivisions of mammary lobes that contain alveoli	b. areola
_______ enlarged storage chambers for milk	c. lactiferous duct
_______ ducts connecting alveoli to the lactiferous sinus	d. lactiferous sinus
_______ pigmented area surrounding the nipple	e. lobule
_______ releases milk to the outside	f. nipple

2. Using the key terms, correctly identify breast structures.

Key:
a. adipose tissue
b. lobule containing alveoli
c. areola
d. lactiferous duct
e. lactiferous sinus
f. nipple

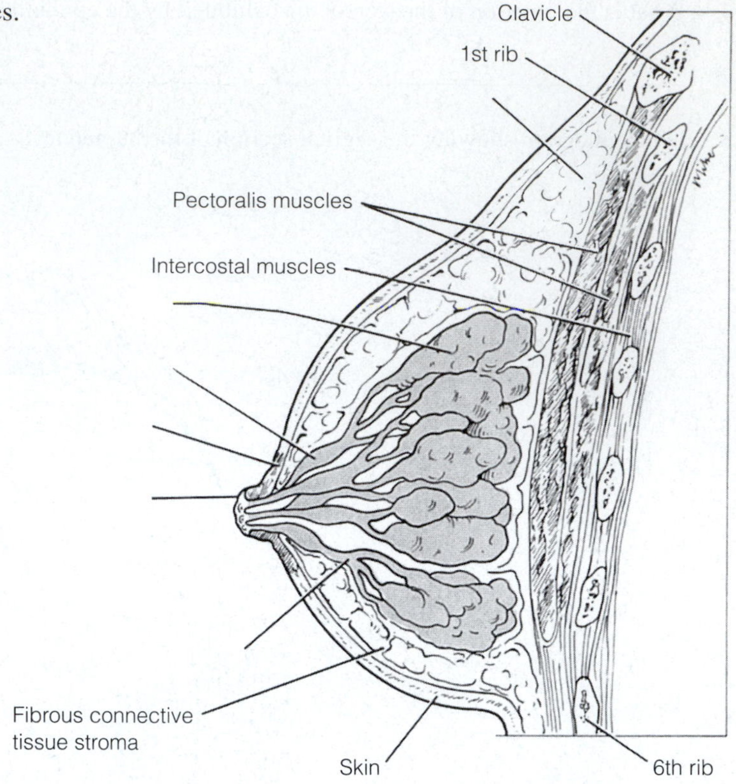

Clavicle
1st rib
Pectoralis muscles
Intercostal muscles
Fibrous connective tissue stroma
Skin
6th rib

3. Describe the procedure for self-examination of the breasts. (Men are not exempt from breast cancer, you know!)

____

____

____

____

____

____

NAME ______________________________ LAB TIME/DATE ____________________

REVIEW SHEET
exercise
43

Physiology of Reproduction: Gametogenesis and the Female Cycles

Meiosis

1. The following statements refer to events occurring during mitosis and/or meiosis. For each statement, decide if the event occurs in (a) mitosis only, (b) meiosis only, or (c) both mitosis and meiosis.

_______ 1. dyads are visible

_______ 2. tetrads are visible

_______ 3. product is two diploid daughter cells

_______ 4. product is four haploid daughter cells

_______ 5. involves the phases prophase, metaphase, anaphase, and telophase

_______ 6. occurs throughout the body

_______ 7. occurs only in the ovaries and testes

_______ 8. provides cells for growth and repair

_______ 9. homologues synapse and chiasmata are seen

_______ 10. daughter cells are quantitatively and qualitatively different from the mother cell

_______ 11. daughter cells are genetically identical to the mother cell

_______ 12. chromosomes are replicated before the division process begins

_______ 13. provides cells for replication of the species

_______ 14. consists of two consecutive nuclear divisions, without chromosomal replication occurring before the second division

2. Describe the process of synapsis. ____

____

3. How does crossover introduce variability in the daughter cells? ______________________

____

____

4. Define *homologous chromosomes:* ____

____

Spermatogenesis

1. The cell types seen in the seminiferous tubules are listed in the key. Match the correct cell type(s) with the descriptions given below.

Key: a. primary spermatocyte
b. secondary spermatocyte
c. spermatogonium
d. sustentacular cell
e. spermatid
f. sperm

__________ 1. primitive stem cell

__________ 2. haploid

__________ 3. provides nutrients to developing sperm

__________ 4. products of meiosis II

__________ 5. product of spermiogenesis

__________ 6. product of meiosis I

2. Why are spermatids not considered functional gametes? __________

____

3. Differentiate between *spermatogenesis* and *spermiogenesis*. __________

____

____

4. Draw a sperm below and identify the *acrosome, head, midpiece,* and *tail.* Then beside each label, note the composition and function of each of these sperm structures.

5. The life span of a sperm is very short. What anatomical characteristics might lead you to suspect this even if you didn't know its life span? __________

Oogenesis, the Ovarian Cycle, and the Menstrual Cycle

1. The sequence of events leading to germ cell formation in the female begins during fetal development. By the time the child is born, all viable oogonia have been converted to __________.

In view of this fact, how does the total germ cell potential of the female compare to that of the male?

____

2. The female gametes develop in structures called *follicles.* What is a follicle? __________

____

How are primary and vesicular follicles anatomically different? ____

____

____

____

What is a corpus luteum? ____

____

3. What is the major hormone produced by the vesicular follicle? ____________________________

By the corpus luteum? ____

4. Use the key to identify the cell type you would expect to find in the following structures.

Key: a. oogonium b. primary oocyte c. secondary oocyte d. ovum

________ 1. forming part of the primary follicle in the ovary

________ 2. in the uterine tube before fertilization

________ 3. in the mature vesicular follicle of the ovary

________ 4. in the uterine tube shortly after sperm penetration

5. The cellular product of spermatogenesis is four ______________________; the final product of oogenesis is one

______________________ and three ______________________. What is the function of this unequal cytoplasmic

division seen during oogenesis in the female? ____

____

____

What is the fate of the three tiny cells produced during oogenesis? ______________________________

Why? ____

6. The following statements deal with anterior pituitary, ovarian hormones, and hormonal interrelationships. Name the hormone(s) described in each statement.

______________________ 1. produced by primary follicles in the ovary

______________________ 2. ovulation occurs after its burstlike release

______________________ and ______________________ 3. exert negative feedback on the anterior pituitary relative to FSH secretion

______________________ 4. stimulates LH release by the anterior pituitary

______________________ 5. stimulates the corpus luteum to produce progesterone and estrogen

______________________ 6. maintains the hormonal production of the corpus luteum in a nonpregnant woman

7. Why does the corpus luteum deteriorate toward the end of the ovarian cycle? ______________________

____

8. For each statement below dealing with hormonal blood levels during the female ovarian and menstrual cycles, decide whether the condition in column A is usually (a) greater than, (b) less than, or (c) essentially equal to the condition in column B.

	Column A		Column B
________	1. amount of estrogen in the blood during menses	. . . . . .	amount of estrogen in the blood at ovulation
________	2. amount of progesterone in the blood on the fourteenth day	. . . .	amount of progesterone in the blood on the twenty-third day
________	3. amount of LH in the blood during menses	. . . .	amount of LH in the blood at ovulation
________	4. amount of FSH in the blood on day 6 of the cycle	. . . .	amount of FSH in the blood on day 20 of the cycle
________	5. amount of estrogen in the blood on the tenth day	. . . . . .	amount of progesterone in the blood on the tenth day

9. Ovulation and menstruation usually cease by the age of ____________.

10. What uterine tissue undergoes dramatic changes during the menstrual cycle? ____________________________

11. When during the female menstrual cycle would fertilization be unlikely? Explain why. ____________________________

____

12. Assume that a woman could be an "on demand" ovulator like the rabbit, in which copulation stimulates the hypothalamic-anterior pituitary axis and causes LH release, and an oocyte was ovulated and fertilized on day 26 of her 28-day cycle. Why would a successful pregnancy be unlikely at this time? ____________________________

____

____

13. The menstrual cycle depends on events within the female ovary. The stages of the menstrual cycle are listed below. For each, note its approximate time span and the related events in the uterus; and then to the right, record the ovarian events occurring simultaneously. Pay particular attention to hormonal events.

Menstrual cycle stage	Uterine events	Ovarian events
Menstruation		
Proliferative		
Secretory		

NAME ______________________ LAB TIME/DATE ______________________

REVIEW SHEET
exercise
44

Survey of Embryonic Development

Developmental Stages of the Human

1. Use the key choices to identify the embryonic stage or process described below.

Key: a. cleavage c. zygote e. blastula
b. morula d. fertilization f. gastrulation

________ 1. fusion of male and female pronuclei

________ 2. solid ball of embryonic cells

________ 3. process of rapid mitotic cell division without intervening growth periods

________ 4. combination of egg and sperm

________ 5. process involving cell rearrangements to form the three primary germ layers

________ 6. embryonic stage in which the embryo consists of a hollow ball of cells

2. What is the importance of cleavage in embryonic development? ______________________

____

How is cleavage different from mitotic cell division, which occurs later in life? ______________________

____

____

3. The cells of the human blastula (blastocyst) have various fates. Which blastocyst structures have the following fates?

________________ 1. produces the embryonic body

________________ 2. becomes the chorion and cooperates with uterine tissues to form the placenta

________________ 3. produces the amnion, yolk sac, and allantois

________________ 4. produces the primordial germ cells (an embryonic membrane)

________________ 5. an embryonic membrane that provides the structural basis for the body stalk or umbilical cord

4. Using the letters on the diagram, correctly identify each of the following maternal or embryonic structures.

_______ amnion

_______ chorion

_______ decidua basalis

_______ endoderm

_______ body stalk

_______ chorionic villi

_______ decidua capsularis

_______ mesoderm

_______ ectoderm

_______ uterine cavity

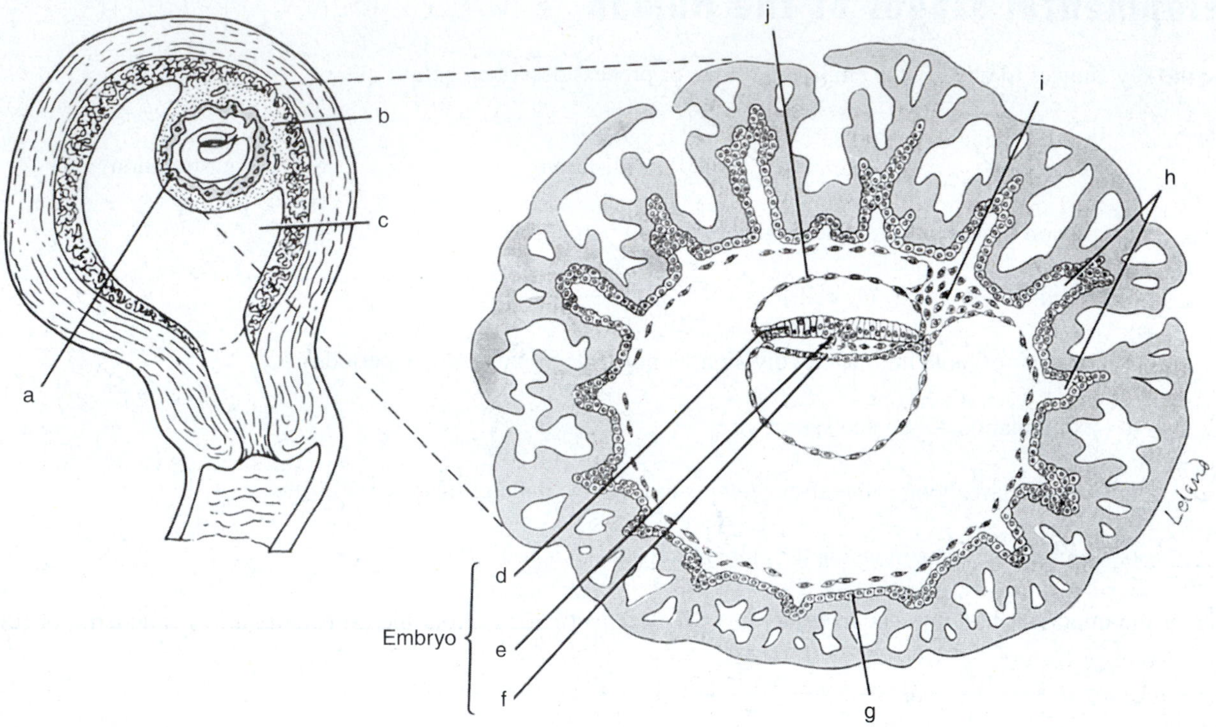

5. Explain the importance of gastrulation. ____

____

____

6. What is the function of the amnion and the amniotic fluid? ____

____

____

7. Describe the process of implantation, noting the role of the trophoblast cells. ____

____

____

8. How many days after fertilization is implantation generally completed? _______ What event in the female menstrual cycle ordinarily occurs just about this time if implantation does not occur? ____

____

9. What name is given to the part of the uterine wall directly under the implanting embryo? ____________________

That surrounding the rest of the embryonic structure? ____________________

10. Using an appropriate reference, find out what *decidua* means and state the definition. ____________________

____

How is this terminology applicable to the deciduas of pregnancy? ____________________

____

11. Referring to the illustrations and text of *A Colour Atlas of Life Before Birth: Normal Fetal Development*, answer the following:

Which two organ systems are extensively developed in the *very young* embryo?

____________________ and ____________________

Describe the direction of development by circling the correct descriptions below:

proximal-distal   distal-proximal   caudal-rostral   rostral-caudal

Does bodily control during infancy develop in the same directions? Think! Can an infant pick up a common pin (pincer grasp) or wave his arms earlier? Is arm-hand or leg-foot control achieved earlier?

____

12. Note whether each of the following organs or organ systems develops from the (a) ectoderm, (b) endoderm, or (c) mesoderm. Use an appropriate reference as necessary.

_______ 1. skeletal muscle   _______ 4. respiratory mucosa   _______ 7. nervous system

_______ 2. skeleton   _______ 5. circulatory system   _______ 8. serosa membrane

_______ 3. lining of gut   _______ 6. epidermis of skin   _______ 9. liver, pancreas

In Utero Development

1. Make the following comparisons between a human and the pregnant dissected animal structures.

Comparison object	**Human**	**Dissected animal**
Shape of the placenta		
Shape of the uterus		

2. Where in the human uterus do implantation and placentation ordinarily occur? ____________________

____

3. Describe the function(s) of the placenta. ____________________

____

What embryonic membranes has the placenta more or less "put out of business"? ______

______

4. When does the human embryo come to be called a fetus? ______

5. What is the usual and most desirable fetal position in utero? ______

Why is this the most desirable position? ______

______

______

______

Gross and Microscopic Anatomy of the Placenta

1. Describe fully the gross structure of the human placenta as observed in the laboratory. ______

______

______

2. What is the tissue origin of the placenta: fetal, maternal, or both? ______

3. What are the placental barriers that must be crossed to exchange materials? ______

______

NAME ______________________ LAB TIME/DATE ______________________

REVIEW SHEET

exercise

45

Principles of Heredity

Introduction to the Language of Genetics

1. Match the key choices with the definitions given below.

Key:

a. alleles	d. genotype	g. phenotype
b. autosomes	e. heterozygous	h. recessive
c. dominant	f. homozygous	i. sex chromosomes

______________ 1. actual genetic makeup

______________ 2. chromosomes determining maleness/femaleness

______________ 3. situation in which an individual has identical alleles for a particular trait

______________ 4. genes not expressed unless they are present in homozygous condition

______________ 5. expression of a genetic trait

______________ 6. situation in which an individual has different alleles making up his genotype for a particular trait

______________ 7. genes for the same trait that may have different expressions

______________ 8. chromosomes regulating most body characteristics

______________ 9. the more-potent gene allele; masks the expression of the less-potent allele

Dominant-Recessive Inheritance

1. In humans, farsightedness is inherited by possession of a dominant gene. If a man who is homozygous for normal vision (*aa*) marries a woman who is heterozygous for farsightedness, what proportion of their children would be expected to be farsighted? ________%

2. A metabolic disorder called PKU is due to an abnormal recessive gene (*p*). Only homozygous recessive individuals exhibit this disorder. What percentage of the offspring will be anticipated to have PKU if the parents are *Pp* and *pp*? ________%

3. A man obtained 32 spotted and 10 solid-color rabbits from a mating of two spotted rabbits.

Which trait is dominant? ______________________ Recessive? ______________________

What is the probable genotype of the rabbit parents? ________ × ________

4. Assume that the allele controlling brown eyes (*B*) is dominant over that controlling blue eyes (*b*) in human beings. (In actuality, eye color in humans is an example of multigene inheritance, which is much more complex than this.) A blue-eyed man marries a brown-eyed woman; and they have six children, all brown-eyed. What is the most likely genotype of the father?

_______ Of the mother? _______ If the seventh child had *blue* eyes, what could you conclude about the parents' genotypes?

___

___

Incomplete Dominance

1. Tail length on a bobcat is controlled by incomplete dominance. The alleles are *T* for normal tail length and *t* for tail-less.

   What name could/would you give to the tails of heterozygous (*Tt*) cats? ___

   How would their tail length compare with that of *TT* or *tt* bobcats? ___

2. If curly-haired individuals are genotypically *CC*, straight-haired individuals are *cc*, and wavy-haired individuals are heterozygotes (*Cc*), what percentage of the various phenotypes would be anticipated from a cross between a *CC* woman and a *cc* man?

   _________% curly   _________% wavy   _________% straight

Sex-Linked Inheritance

1. What does it mean when someone says a particular characteristic is sex-linked? ___

   ___

2. You are a male, and you have been told that hemophilia "runs in your genes." Whose ancestors, your mother's or your father's, should you investigate? _________ Why? ___

   ___

   ___

3. An X^CX^c female marries an X^CY man. Do a Punnett square for this match.

   What is the probability of producing a color-blind son? _________

   A color-blind daughter? _________

   A daughter that is a carrier for the color-blind gene? _________

4. Why are consanguineous marriages (marriages between blood relatives) prohibited in most cultures?

   ___

   ___

Probability

1. What is the probability of having three daughters in a row? ______________________

2. A man and a woman, each of seemingly normal intellect, marry. Although neither is aware of the fact, each is a heterozygote for the allele for mental retardation. Is the allele for mental retardation dominant or recessive? ______________________

   What are the chances of their having one mentally retarded child? ______________________

   What are the chances that all of their children (they plan a family of four) will be mentally retarded? ______________________

Genetic Determination of Selected Human Characteristics

1. Look back at your data to complete this section. For each of the situations described here, determine if an offspring with the characteristics noted is possible with the parental genotypes listed. Check (✓) the appropriate column.

		Possibility	
Parental genotypes	**Phenotype of child**	**Yes**	**No**
Jj × *jj*	Double-jointed thumbs		
FF × *Ff*	Straight little finger		
EE × *ee*	Detached ear lobes		
HH × *Hh*	Middigital hair		
$I^Ai \times I^Bi$	Type O blood		
$I^AI^B \times ii$	Type B blood		

2. You have dimples, and you would like to know if you are homozygous or heterozygous for this trait. You have six brothers and sisters. By observing your siblings, how could you tell, with some degree of certainty, that you are a heterozygote?

____

____

____

Identifying Hemoglobin Phenotypes Using Agar Gel Electrophoresis

1. Draw the banding patterns you obtained on the figure below. Indicate the genotype of each band.

Sample genotype	Well	Banding pattern
1. ______	1.	
2. ______	2.	
3. ______	3.	
4. ______	4.	
5. ______	5.	
6. ______	6.	
7. ______	7.	
8. ______	8.	

2. What is the genotype of sickle cell anemia? _______ Sickle cell trait? _______

3. Why does sickle-cell hemoglobin behave differently from normal hemoglobin during agarose gel electrophoresis?

____

____

NAME ______________________ LAB TIME/DATE ______________________

REVIEW SHEET
exercise
46

Surface Anatomy Roundup

__________ **1.** A blow to the cheek is most likely to break what superficial bone or bone part? (a) superciliary arches, (b) the philtrum, (c) zygomatic arch, (d) the tragus.

__________ **2.** Rebound tenderness (a) occurs in appendicitis, (b) is whiplash of the neck, (c) is a sore foot from playing basketball, (d) occurs when the larynx falls back into place after swallowing.

__________ **3.** The anatomical snuff box (a) is in the nose, (b) contains the styloid process of the radius, (c) is defined by tendons of the flexor carpi radialis and palmaris longus, (d) cannot really hold snuff.

__________ **4.** Some landmarks on the body surface can be seen or felt, but others are abstractions that you must construct by drawing imaginary lines. Which of the following pairs of structures is abstract and invisible? (a) umbilicus and costal margin, (b) anterior superior iliac spine and natal cleft, (c) linea alba and linea semilunaris, (d) McBurney's point and midaxillary line, (e) philtrum and sternocleidomastoid.

__________ **5.** Many pelvic organs can be palpated by placing a finger in the rectum or the vagina, but only one pelvic organ is readily palpated through the skin. This is the (a) nonpregnant uterus, (b) prostate gland, (c) full bladder, (d) ovaries, (e) rectum.

__________ **6.** A muscle that contributes to the posterior axillary fold is the (a) pectoralis major, (b) latissimus dorsi, (c) trapezius, (d) infraspinatus, (e) pectoralis minor, (f) a and e.

__________ **7.** Which of the following is not a pulse point? (a) anatomical snuff box, (b) inferior margin of mandible anterior to masseter muscle, (c) center of distal forearm at palmaris longus tendon, (d) medial bicipital furrow on arm, (e) dorsum of foot between the first two metatarsals.

__________ **8.** Which pair of ribs inserts on the sternum at the sternal angle? (a) first, (b) second, (c) third, (d) fourth, (e) fifth.

__________ **9.** The inferior angle of the scapula is at the same level as the spinous process of which vertebra? (a) C_5, (b) C_7, (c) T_3, (d) T_7, (e) L_4.

__________ **10.** An important bony landmark that can be recognized by a distinct dimple in the skin is the (a) posterior superior iliac spine, (b) styloid process of the ulna, (c) shaft of the radius, (d) acromion.

__________ **11.** A nurse missed a patient's median cubital vein while trying to withdraw blood and then inserted the needle far too deeply into the cubital fossa. This error could cause any of the following problems, except this one: (a) paralysis of the ulnar nerve, (b) paralysis of the median nerve, (c) bruising the insertion tendon of the biceps brachii muscle, (d) blood spurting from the brachial artery.

__________ **12.** Which of these organs is almost impossible to study with surface anatomy techniques? (a) heart, (b) lungs, (c) brain, (d) nose.

__________ **13.** A preferred site for inserting an intravenous medication line into a blood vessel is the (a) medial bicipital furrow on arm, (b) external carotid artery, (c) dorsal venous arch of hand, (d) popliteal fossa.

__________ **14.** One listens for bowel sounds with a stethoscope placed (a) on the four quadrants of the abdominal wall; (b) in the triangle of auscultation; (c) in the right and left midaxillary line, just superior to the iliac crests; (d) inside the patient's bowels (intestines), on the tip of an endoscope.

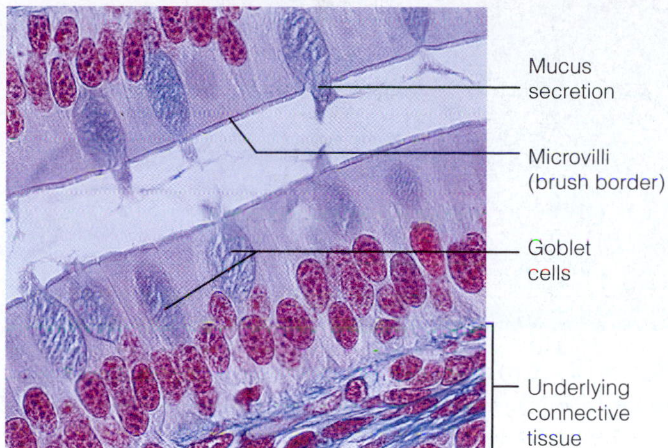

Plate 1 Simple columnar epithelium containing goblet cells, which are secreting mucus (1092×). (Exercise 6, p. 51)

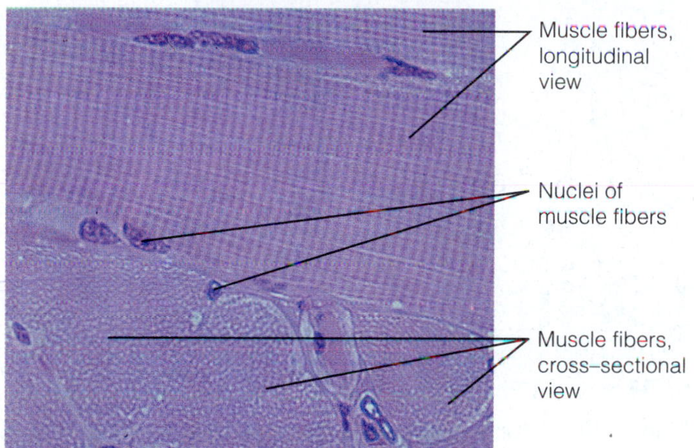

Plate 2 Skeletal muscle, transverse and longitudinal views shown (576×). (Exercise 6, p. 64; Exercise 14, p. 134)

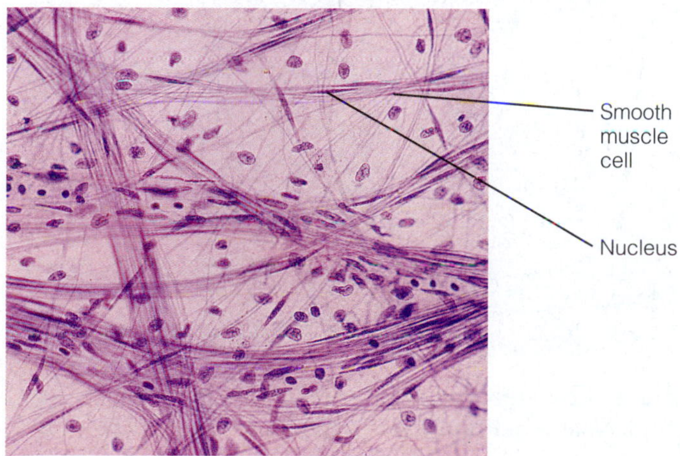

Plate 3 Teased smooth muscle. (Exercise 6, p. 65)

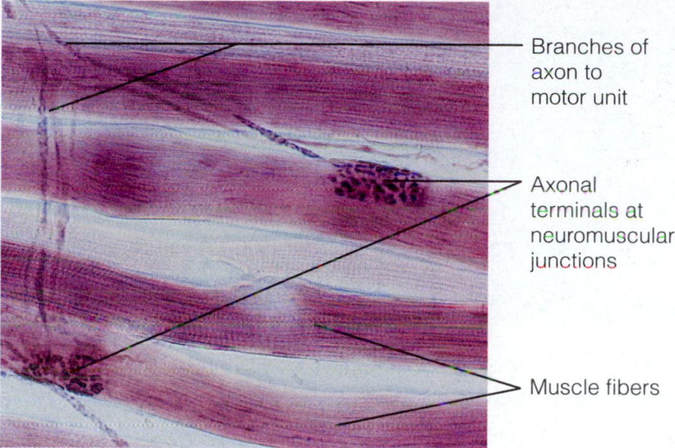

Plate 4 Part of a motor unit (273×). (Exercise 14, p. 136)

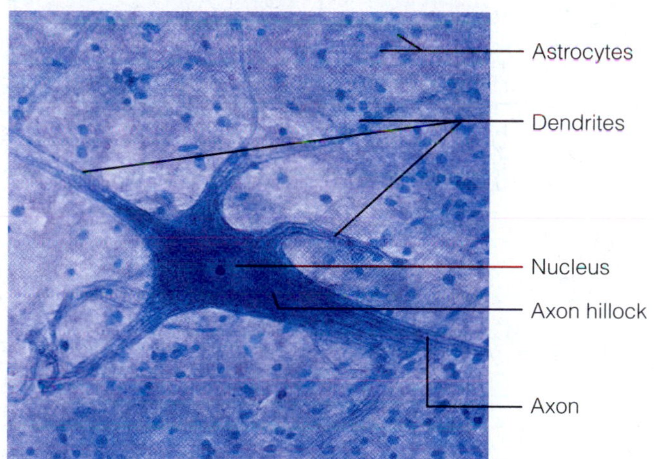

Plate 5 Light micrograph of a multipolar neuron (3000×). (Exercise 17, p. 179)

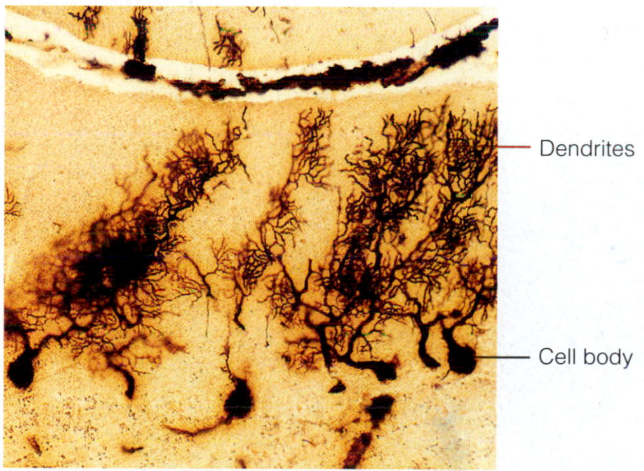

Plate 6 Silver-stained Purkinje cells of the cerebellum (2730×). (Exercise 17, p. 182)

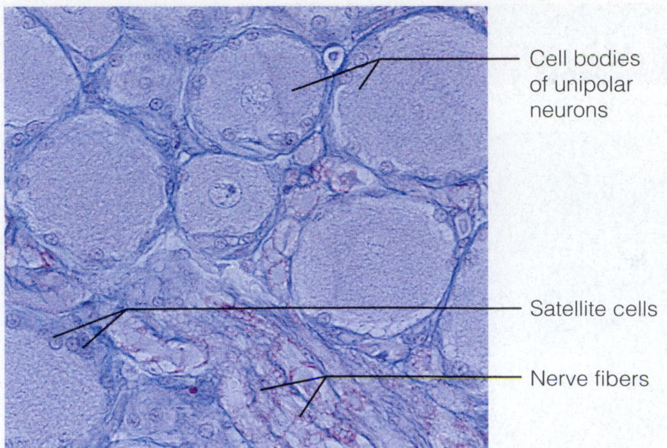

Plate 7 Dorsal root ganglion displaying neuron cell bodies and satellite cells (1092×). (Exercise 17, p. 182)

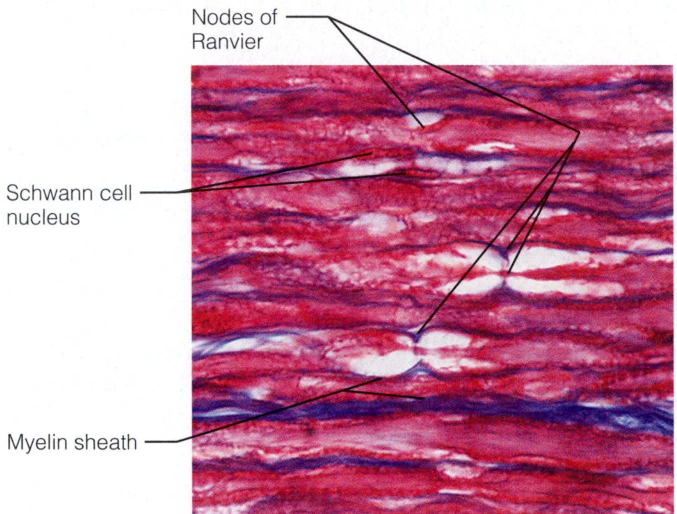

Plate 8 Longitudinal view of myelinated axons (1000×). Myelin sheaths appear "bubbly" because some fatty myelin dissolved during slide preparation. (Exercise 17, p. 180)

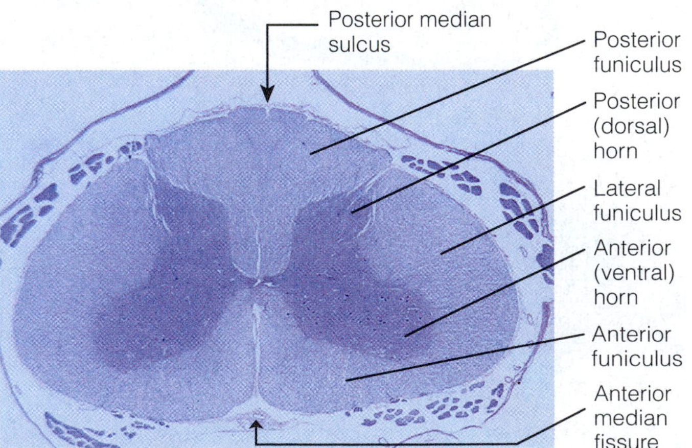

Plate 9 Adult spinal cord, cross-sectional view (453×). (Exercise 21, p. 218)

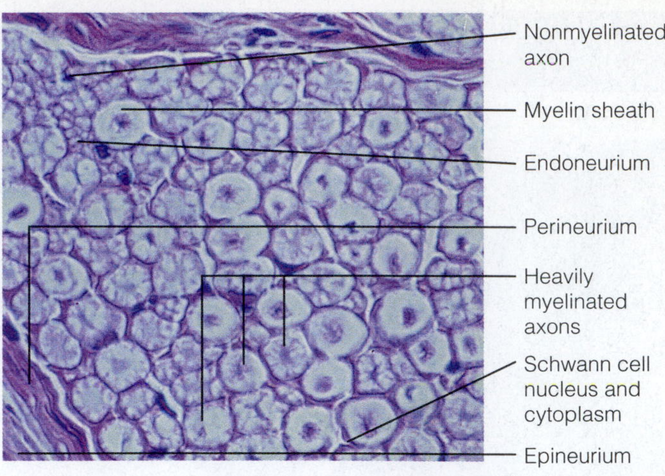

Plate 10 Cross section of a portion of a peripheral nerve (560×). Heavily myelinated fibers are identified by a centrally located axon, surrounded by an unstained ring of myelin. (Exercise 17, p. 184)

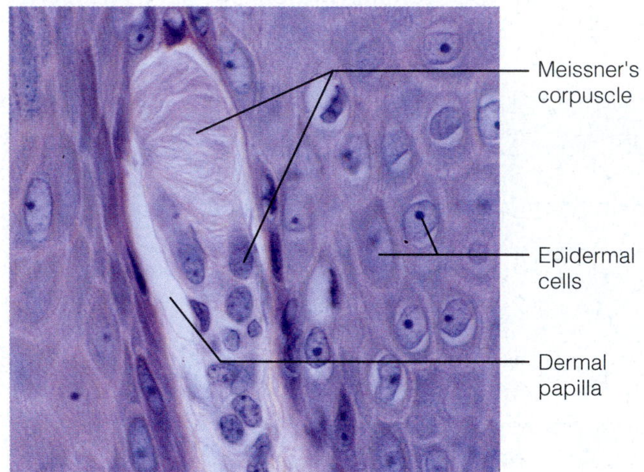

Plate 11 Meissner's corpuscle in a dermal papilla (2730×). (Exercise 23, p. 236)

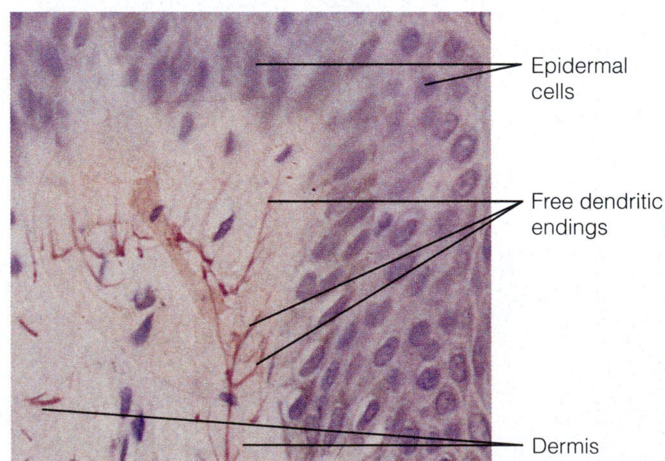

Plate 12 Free dendritic endings at dermal-epidermal junction (251×). (Exercise 23, p. 236)

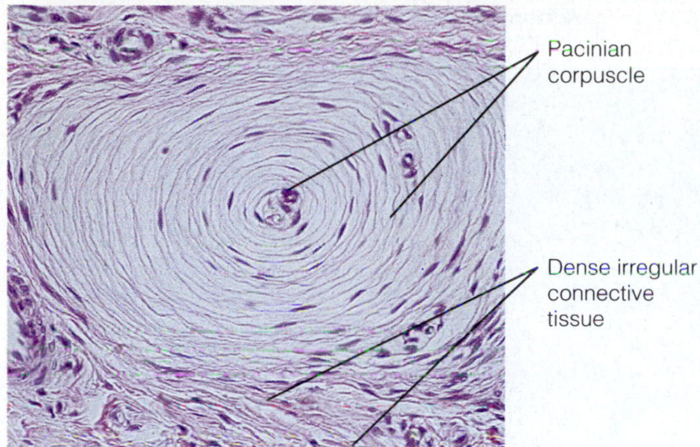

Plate 13 Pacinian corpuscle in the hypodermis (450×). (Exercise 23, p. 236)

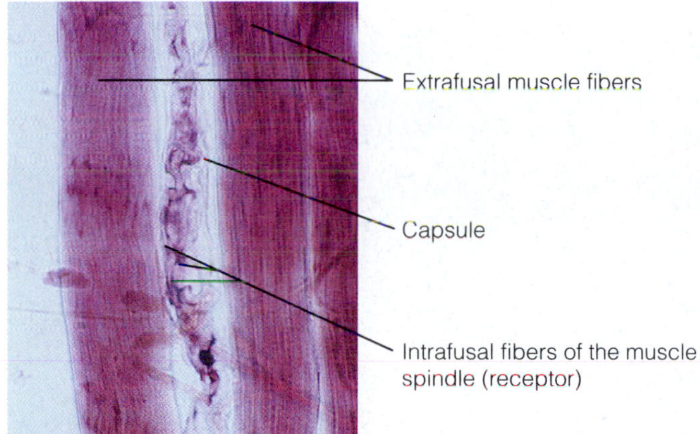

Plate 14 Longitudinal section of a muscle spindle (382×). (Exercise 23, p. 237)

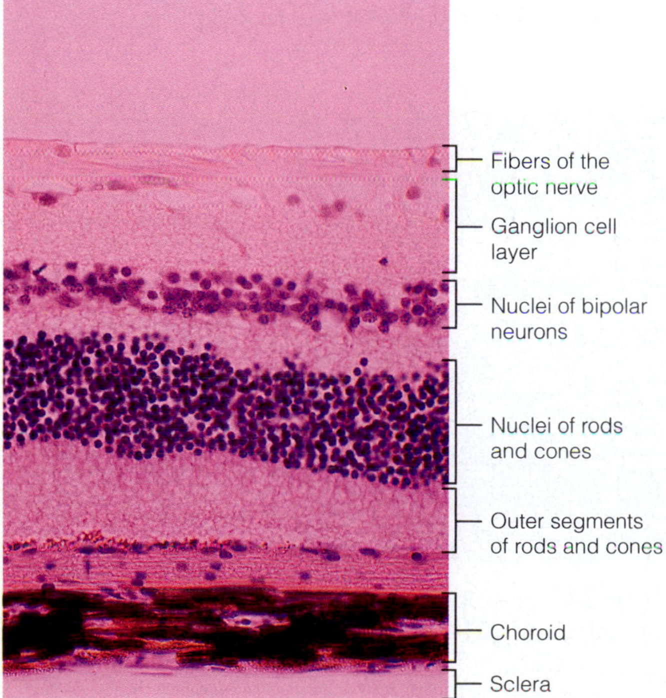

Plate 15 Structure of retina of the eye (1092×). (Exercise 24, p. 246)

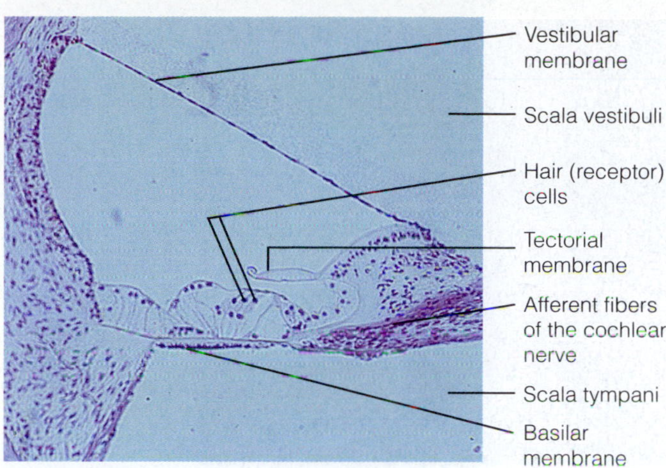

Plate 16 The organ of Corti (109×). (Exercise 25, p. 258)

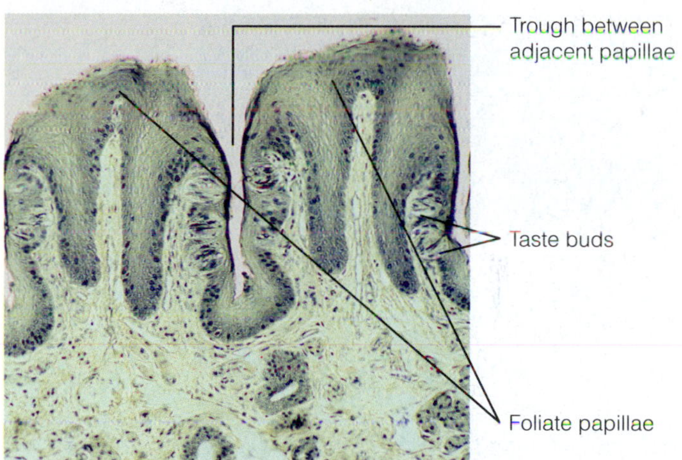

Plate 17 Location of taste buds on lateral aspects of foliate papillae of tongue (130×). (Exercise 26, p. 268)

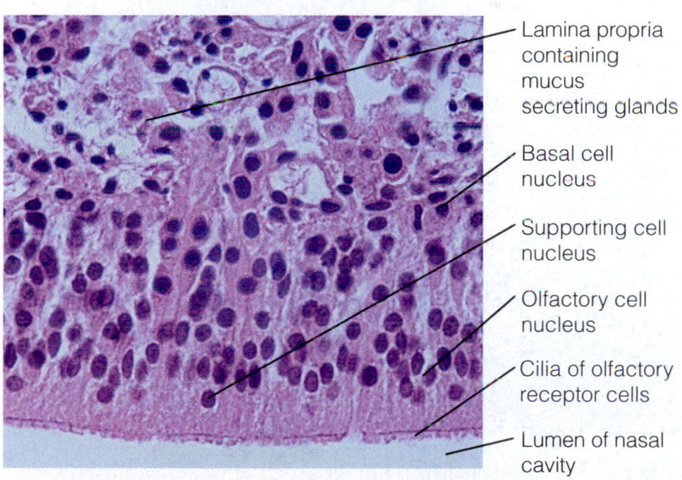

Plate 18 Olfactory epithelium. From lamina propria to nasal cavity, the general arrangement of cells in this pseudostratified epithelium: basal cells, olfactory receptor cells, and supporting cells (560×). (Exercise 26, p. 267)

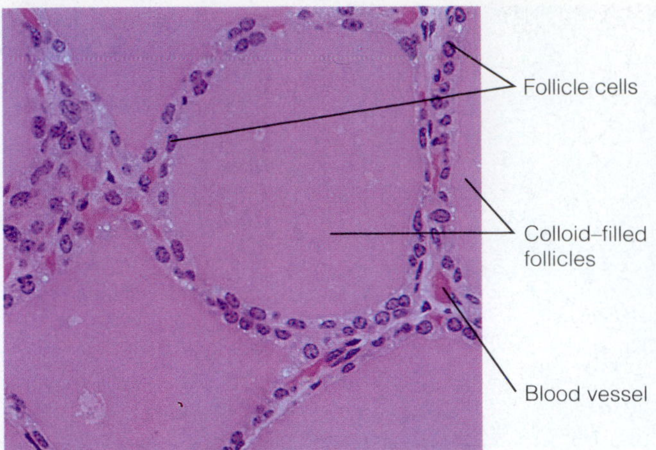

Plate 19 Thyroid gland (355×). (Exercise 27, p. 276)

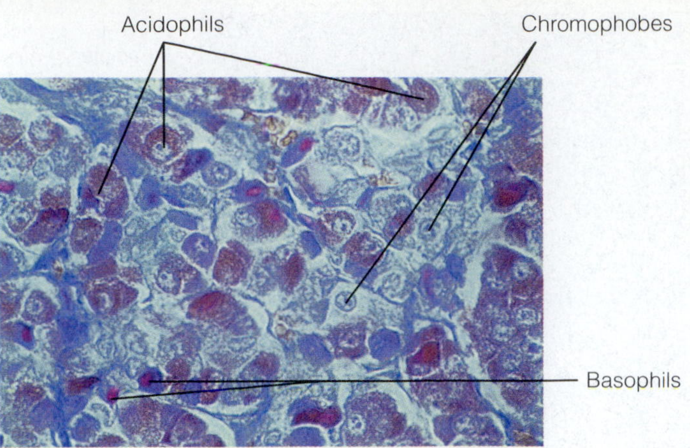

Plate 22 Anterior pituitary gland. Differential staining distinguishes acidophils, basophils, and chromophobes (560×). (Exercise 27, p. 276)

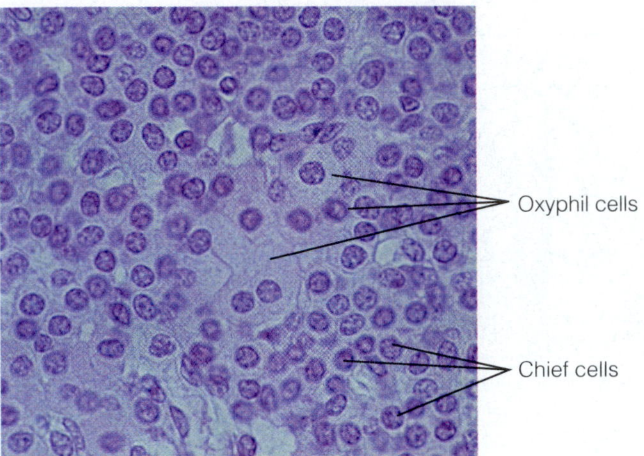

Plate 20 Parathyroid gland tissue (560×). (Exercise 27, p. 276)

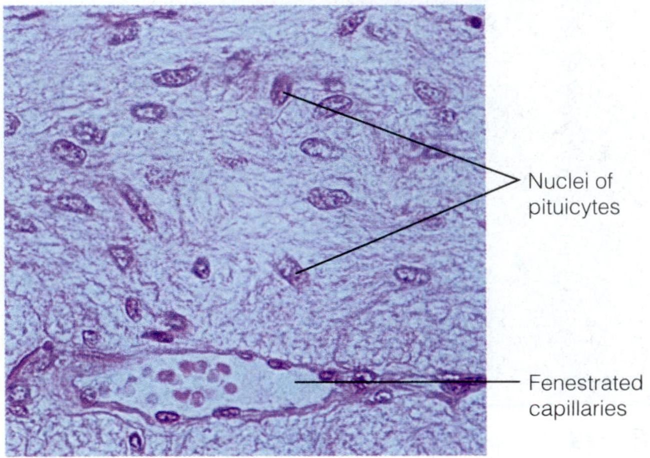

Plate 23 Posterior pituitary. Axons of neurosecretory cells are indistinguishable from cytoplasm of pituicytes (449×). (Exercise 27, p. 276)

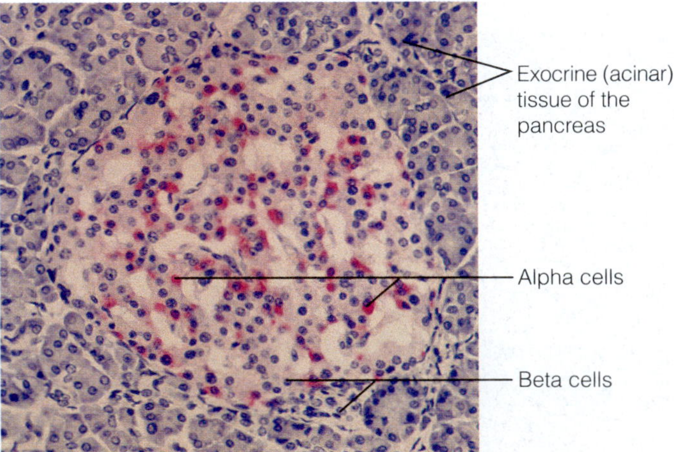

Plate 21 Pancreatic islet stained differentially to allow identification of the glucagon-secreting alpha cells and the insulin-secreting beta cells (206×). (Exercise 27, p. 276)

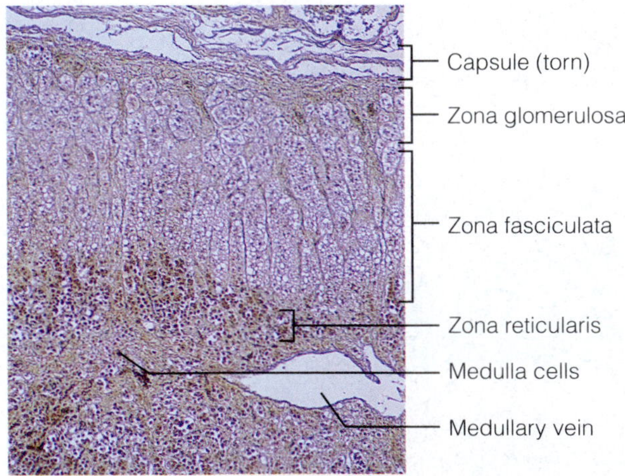

Plate 24 Histologically distinct regions of adrenal gland (214×). (Exercise 27, p. 276)

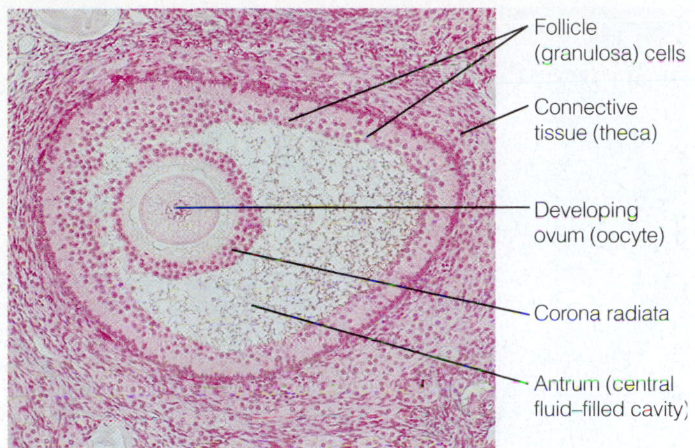

Plate 25 A vesicular follicle of ovary (108×). (Exercise 27, p. 278; Exercise 43, p. 442)

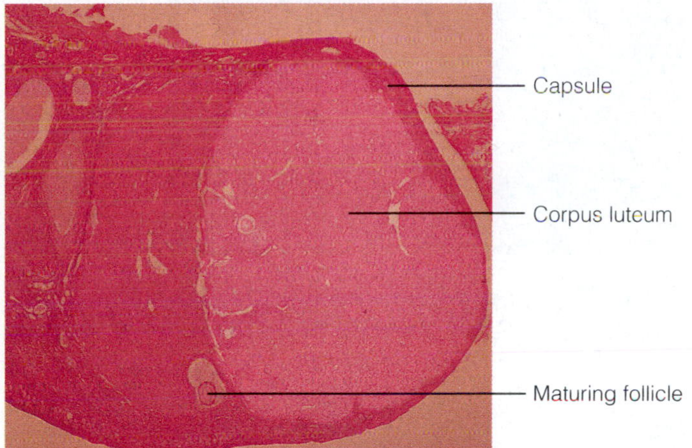

Plate 26 Glandular corpus luteum of an ovary (55×). (Exercise 27, p. 278; Exercise 43, p. 442)

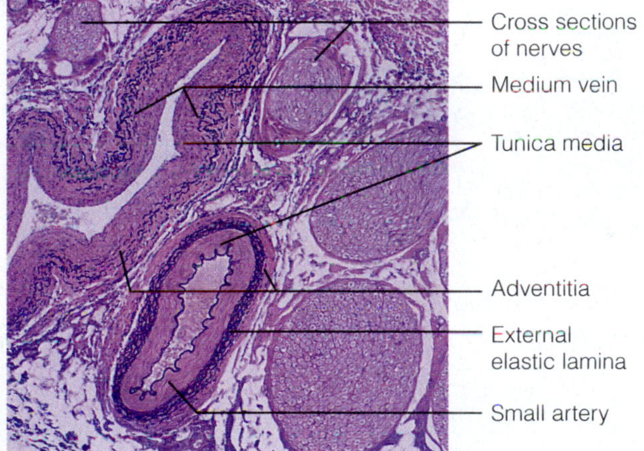

Plate 27 Cross-sectional view of an artery, a vein, and nerves (273×). (Exercise 32, p. 320)

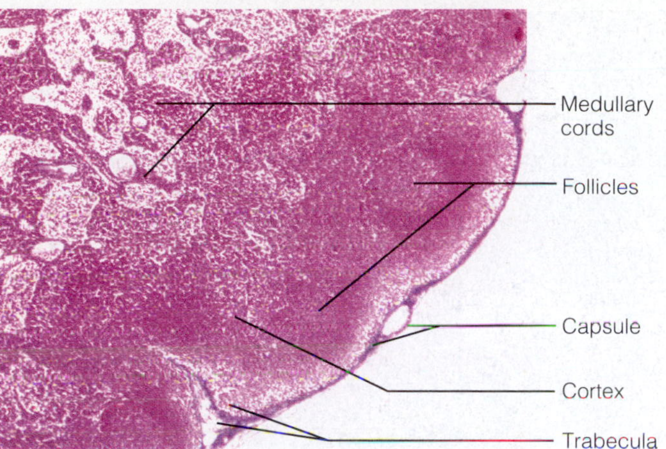

Plate 28 Main structural features of a lymph node (68×). (Exercise 35, p. 360)

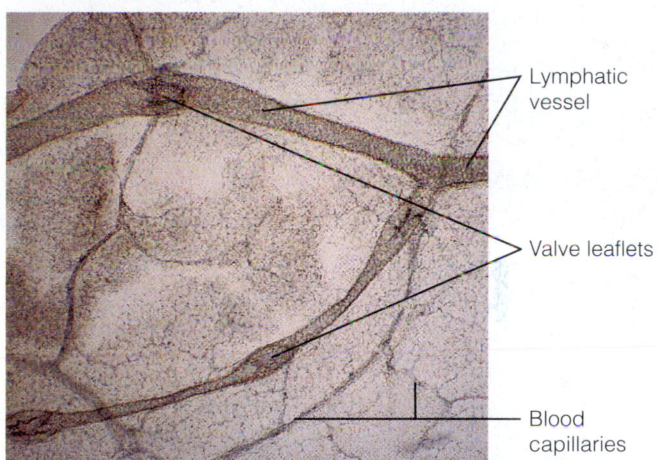

Plate 29 Lymphatic vessels and blood capillaries. Valves in lymphatic vessels are clearly visible (109×). (Exercise 35, p. 358)

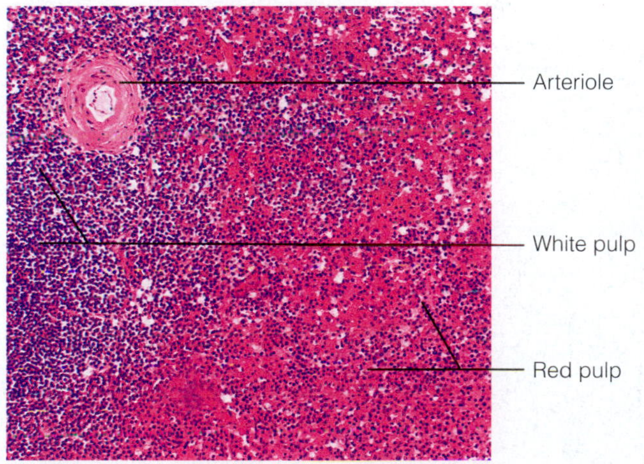

Plate 30 Microscopic portion of spleen showing red and white pulp regions (273×). (Exercise 35, p. 361)

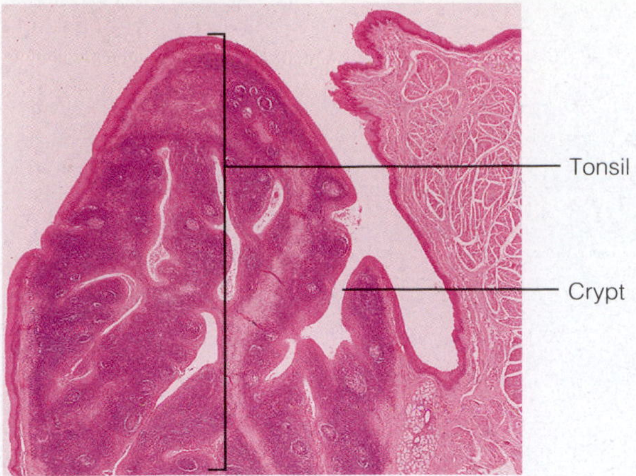

Plate 31 Histology of a palatine tonsil. The luminal surface is covered with epithelium that invaginates deeply to form crypts (50×). (Exercise 35, p. 361)

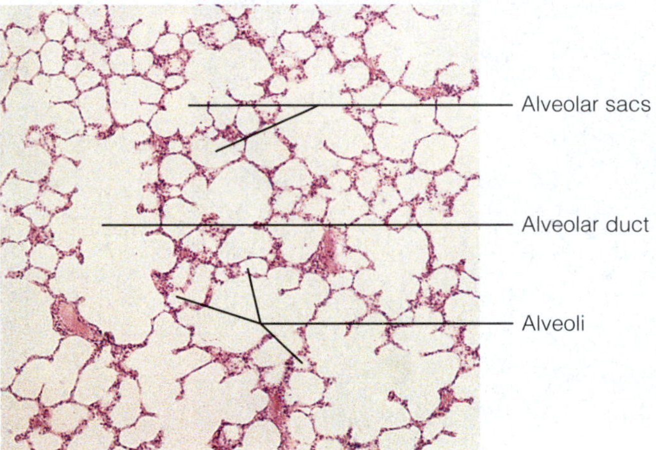

Plate 32 Photomicrograph of part of the lung showing alveoli and alveolar ducts and sacs (36×). (Exercise 36, p. 372)

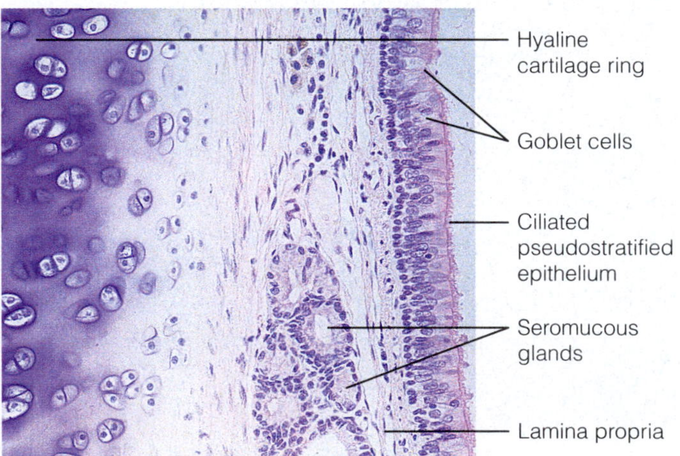

Plate 33 Cross section through the trachea showing the pseudostratified ciliated epithelium, glands, and part of the supporting ring of hyaline cartilage (155×). (Exercise 36, p. 371)

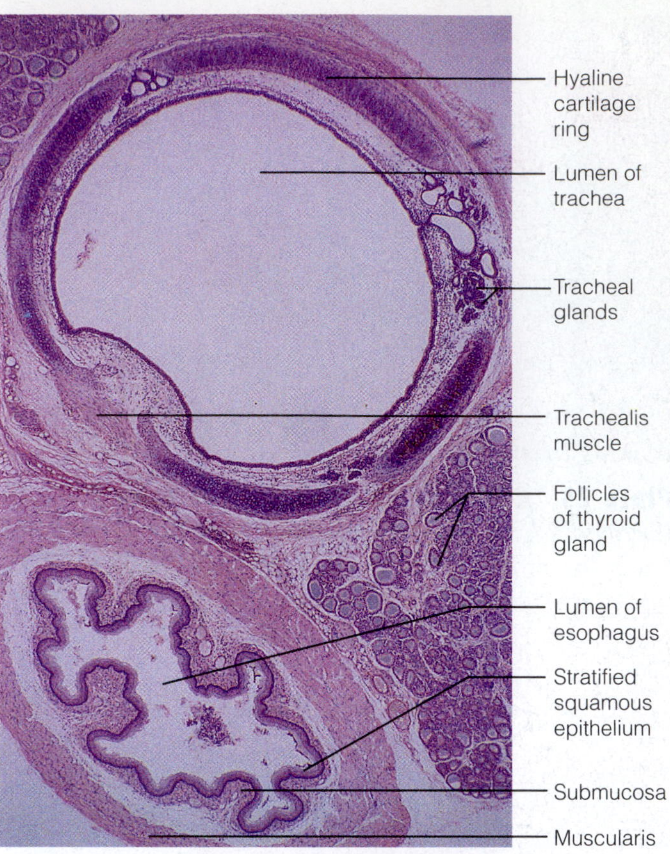

Plate 34 Cross section through trachea and esophagus (124×). (Exercise 36, p. 371)

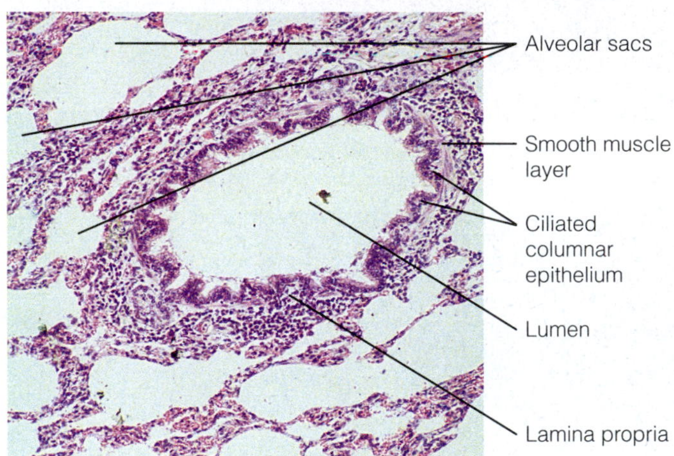

Plate 35 Bronchiole, cross-sectional view (109×). (Exercise 36, p. 372)

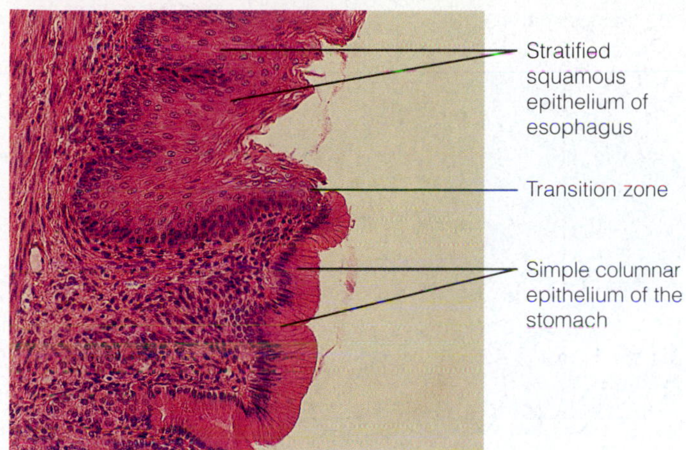

Plate 36 Gastroesophageal junction showing simple columnar epithelium of stomach meeting stratified squamous epithelium of esophagus (633×). (Exercise 38, p. 394)

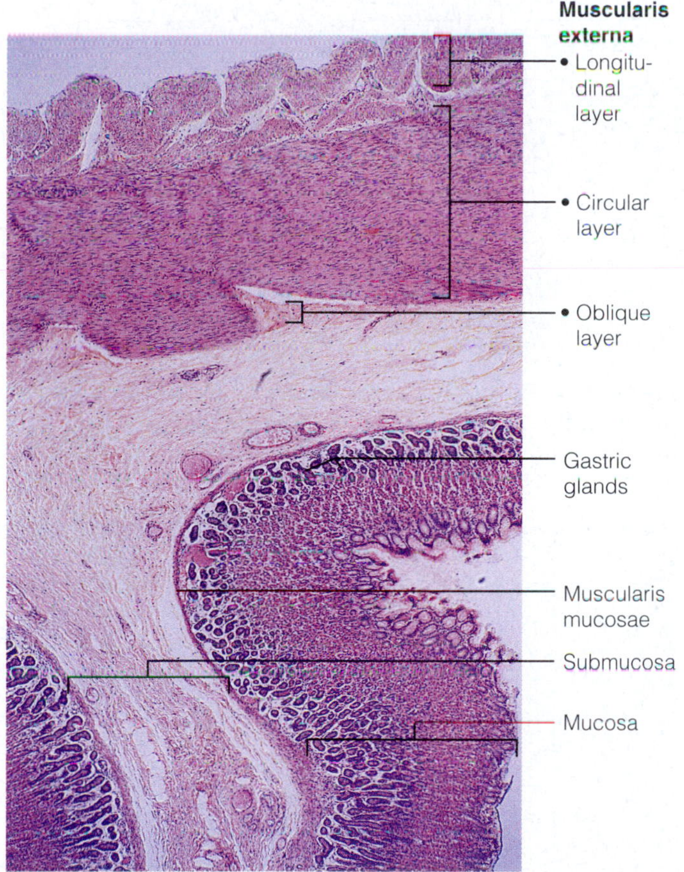

Plate 37 Stomach. Longitudinal view through wall showing four tunics (124×). (Exercise 38, p. 394)

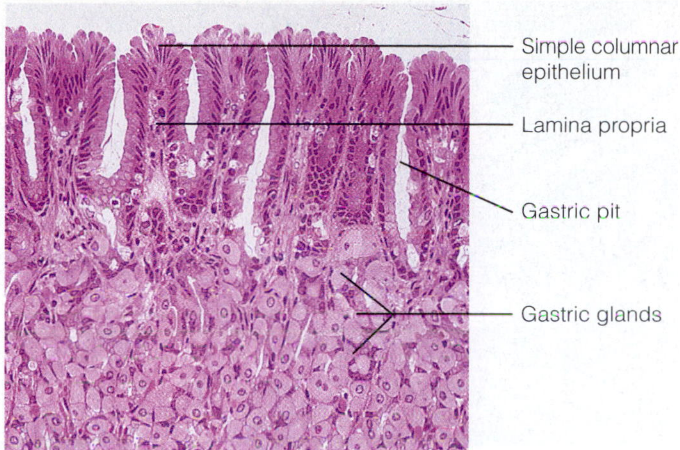

Plate 38 Detailed structure of gastric glands and pits (218×). (Exercise 38, p. 394)

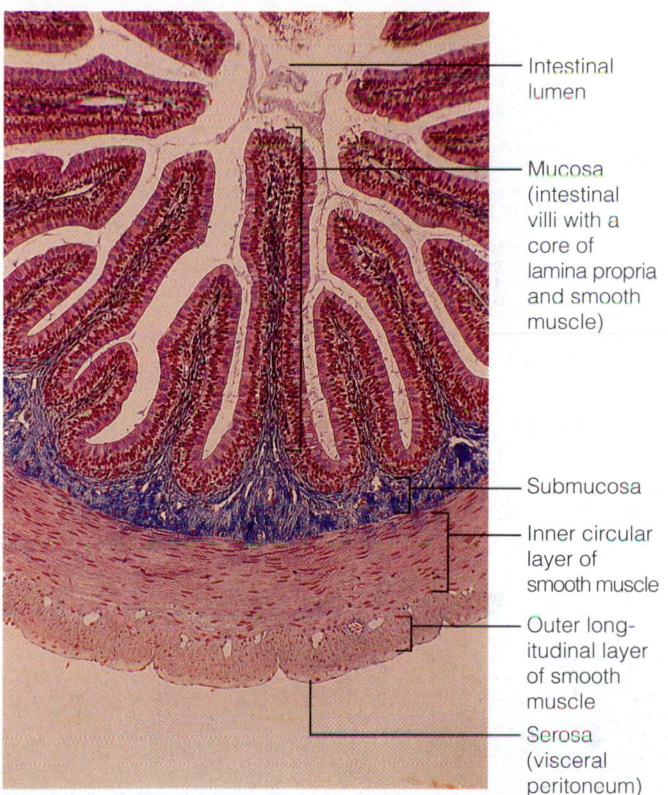

Plate 39 Cross section through wall of small intestine showing arrangement of layers or tunics (129×). Villi of mucosa are large and obvious. (Exercise 38, p. 398)

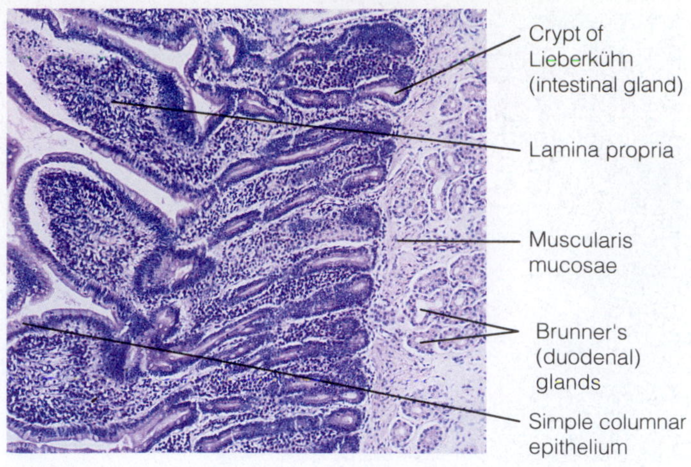

Plate 40 Cross-sectional view of duodenum showing villi and duodenal glands (273×). (Exercise 38, p. 397)

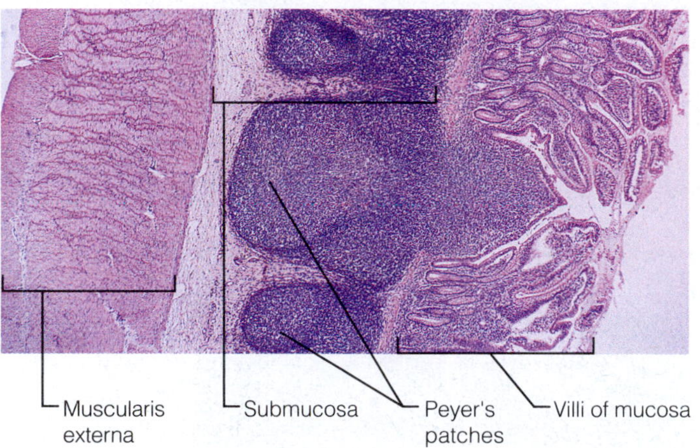

Plate 41 Cross section through ileum, showing Peyer's patches (109×). (Exercise 38, p. 397)

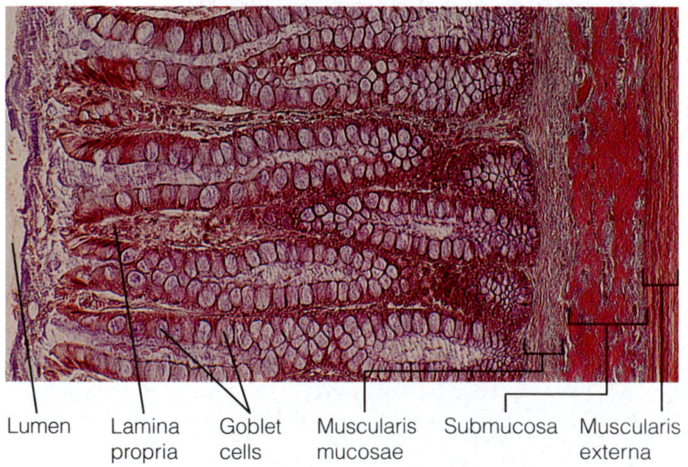

Plate 42 Large intestine. Cross-sectional view showing the abundant goblet cells of the mucosa (546×). (Exercise 38, p. 397)

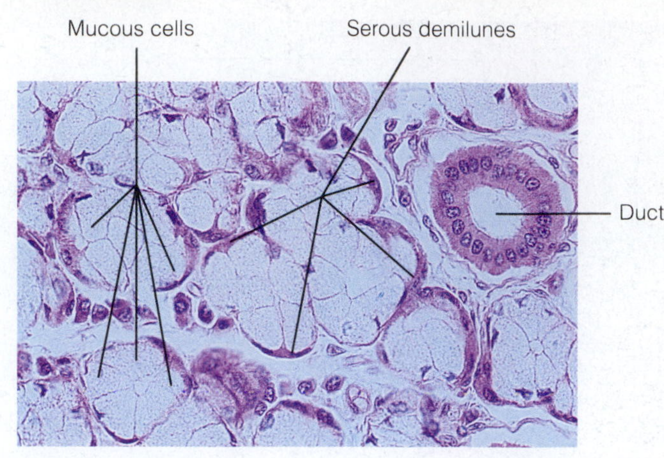

Plate 43 Sublingual salivary glands (361×). (Exercise 38, p. 401)

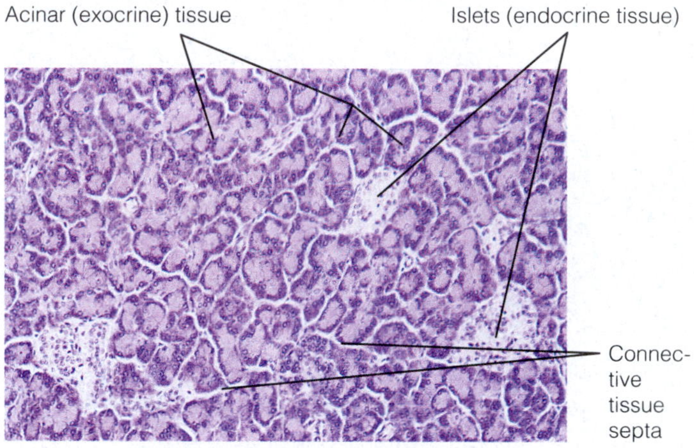

Plate 44 Pancreas tissue. Exocrine and endocrine (islets) areas clearly visible (110×). (Exercise 27, p. 277, and Exercise 38, p. 404)

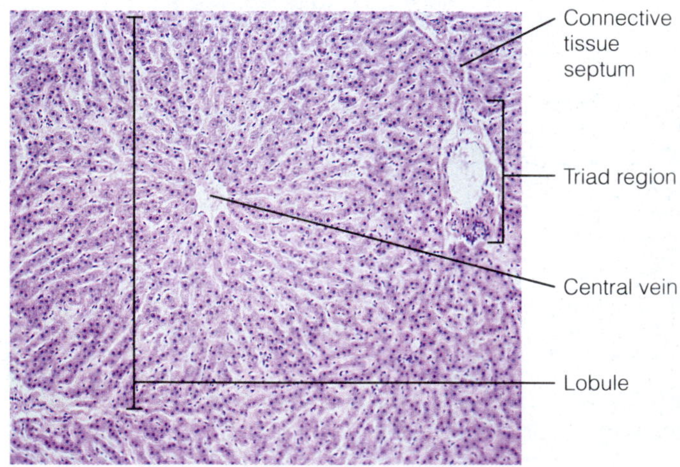

Plate 45 Pig liver. Structure of liver lobules (64×). (Exercise 38, p. 403)

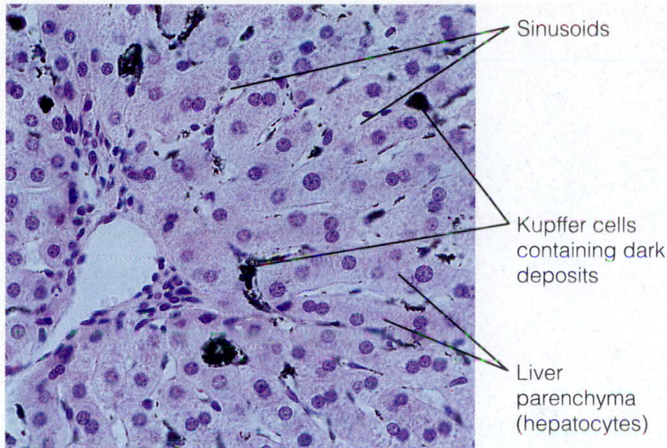

Plate 46 Liver stained to show location of phagocytic cells (Kupffer cells) lining sinusoids (255×). (Exercise 38, p. 402)

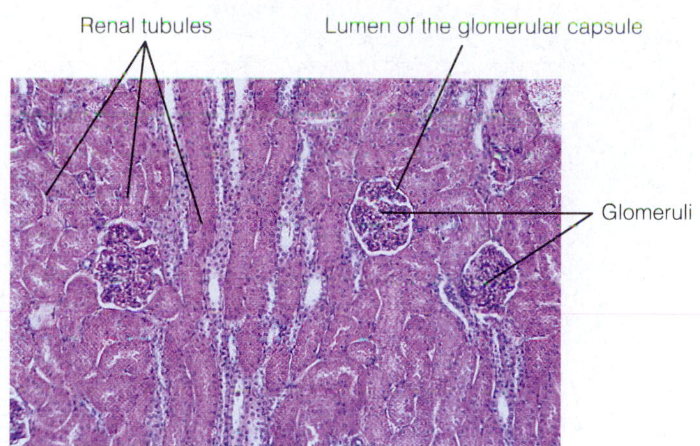

Plate 47 Renal cortex of kidney (86×). (Exercise 40, p. 419)

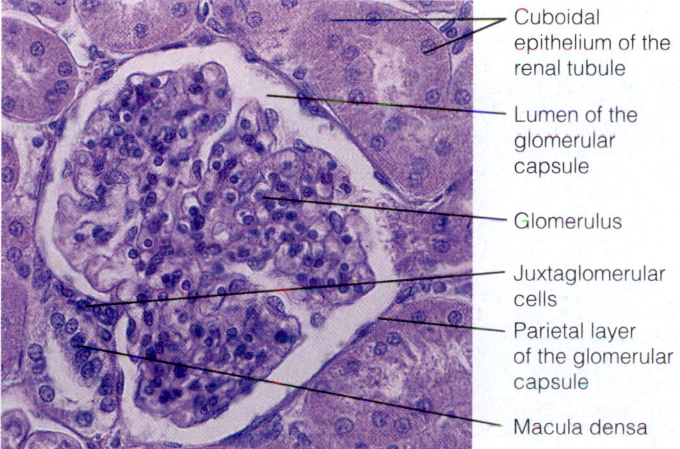

Plate 48 Detailed structure of a glomerulus (337×). (Exercise 40, p. 419)

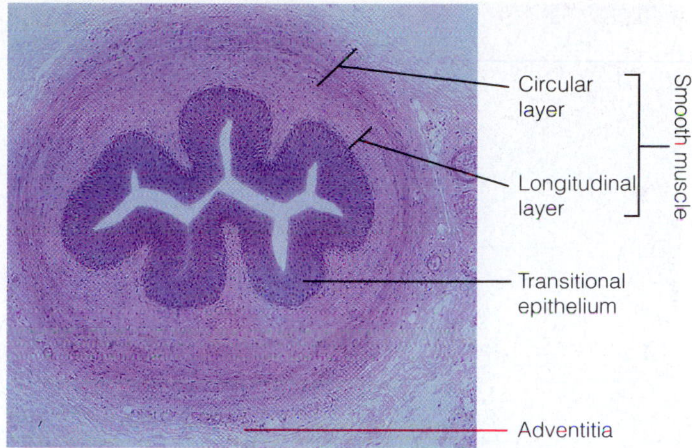

Plate 49 Cross section of ureter (43×). (Exercise 40, p. 420)

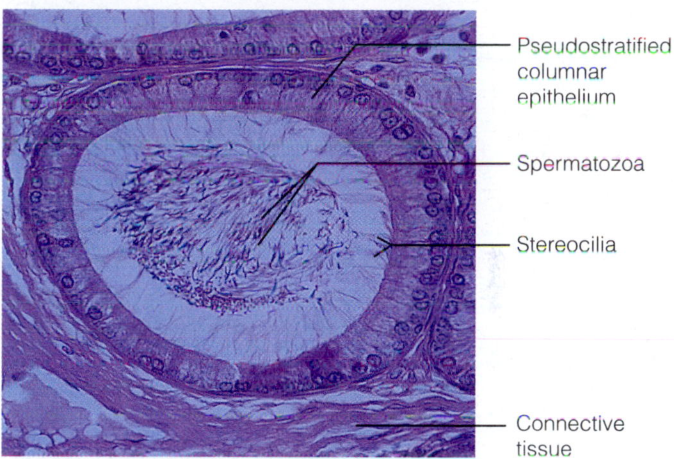

Plate 50 Cross section of epididymis (1092×). Stereocilia of the epithelial lining are obvious. (Exercise 42, p. 431)

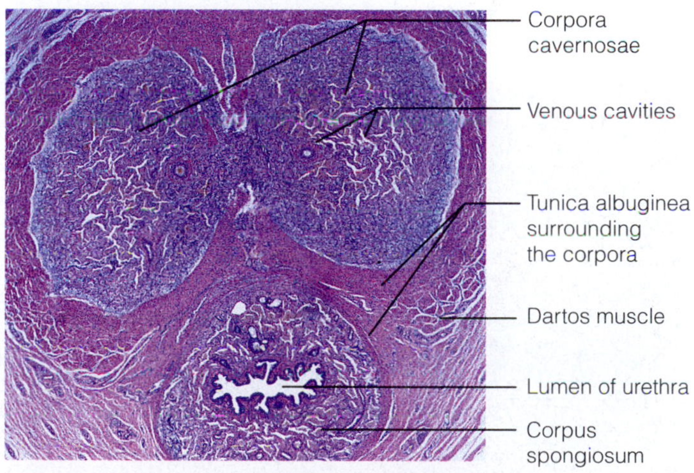

Plate 51 Penis, transverse section (51×). (Exercise 42, p. 431)

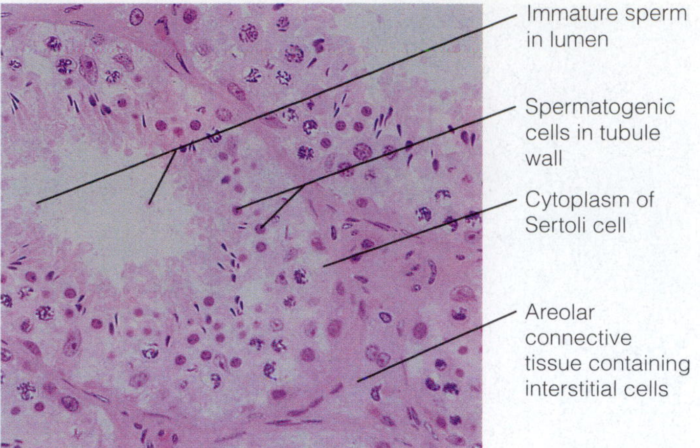

Plate 52 Cross section of a portion of a seminiferous tubule (36×). (Exercise 27, p. 278, and Exercise 43, p. 438)

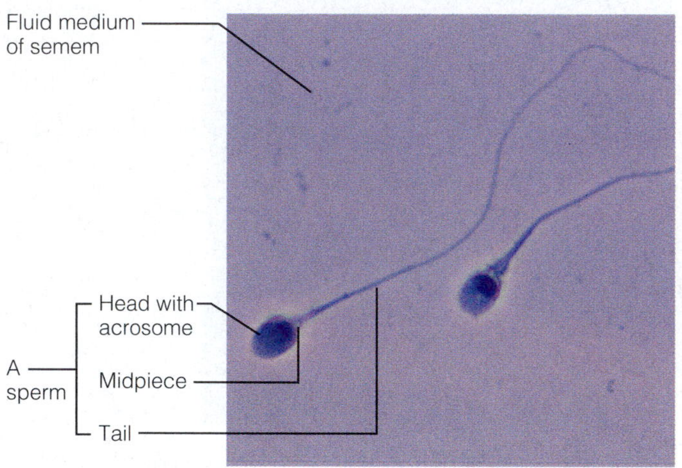

Plate 53 Semen, the product of ejaculation, consisting of sperm and fluids secreted by the accessory glands (particularly the prostate and seminal vesicles) (250×). (Exercise 43, p. 439)

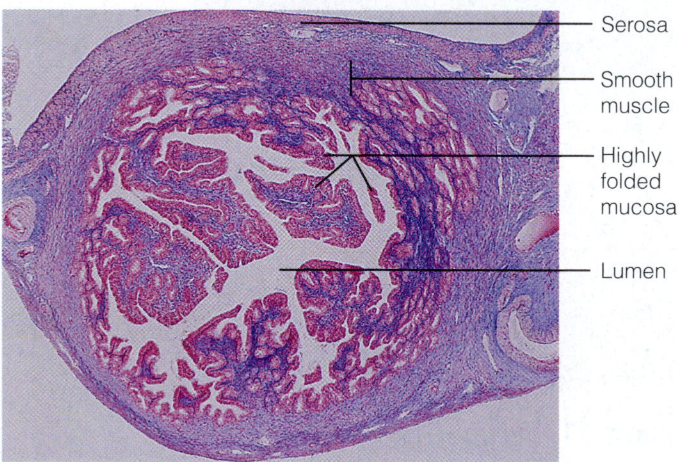

Plate 54 Cross-sectional view of the uterine tube (523×). (Exercise 42, p. 434)

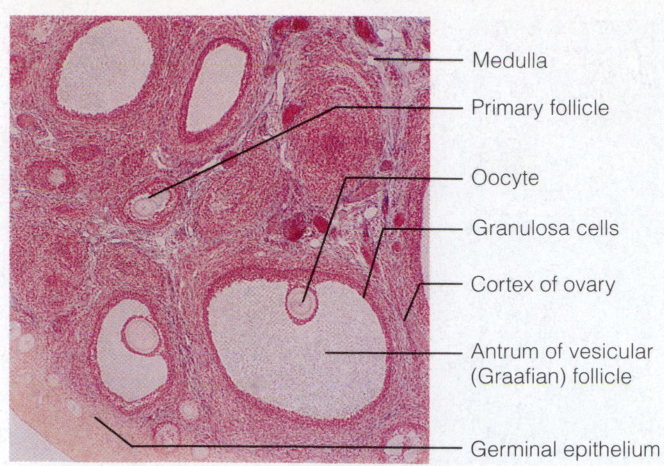

Plate 55 The ovary showing its follicles in various stages of development (104×). (Exercise 43, p. 442)

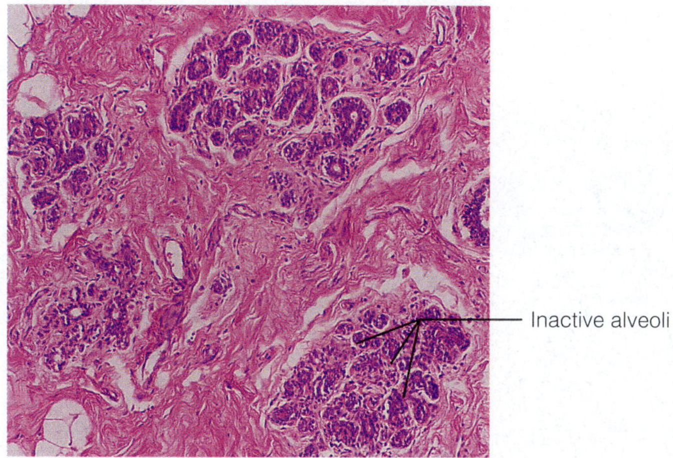

Plate 56 Breast alveoli of nonpregnant woman (273×). (Exercise 42, p. 434)

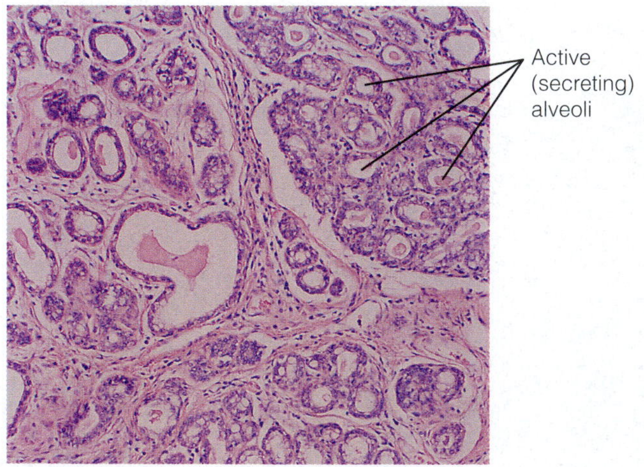

Plate 57 Breast alveoli (lactating) of pregnant woman (273×). (Exercise 42, p. 434)

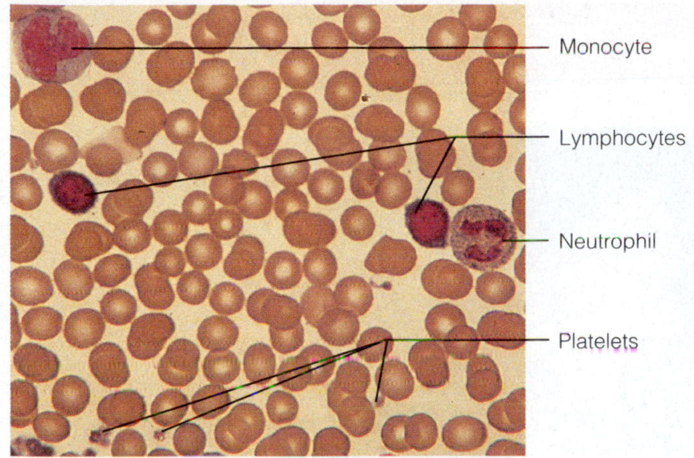

Plate 58 Human blood smear (656×). (Exercise 29, p. 289)

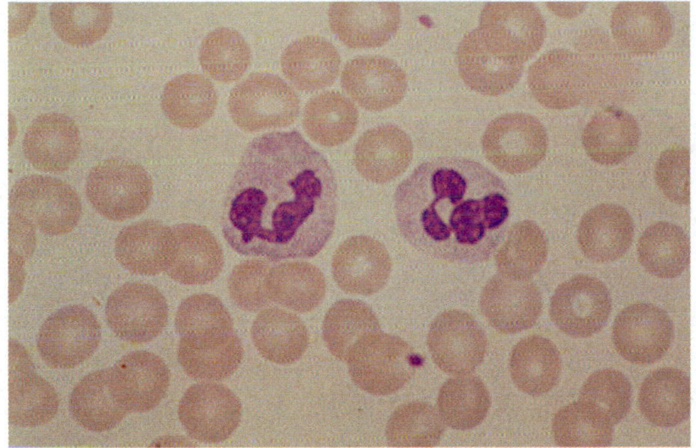

Plate 59 Two neutrophils surrounded by erythrocytes (842×). (Exercise 29, p. 290)

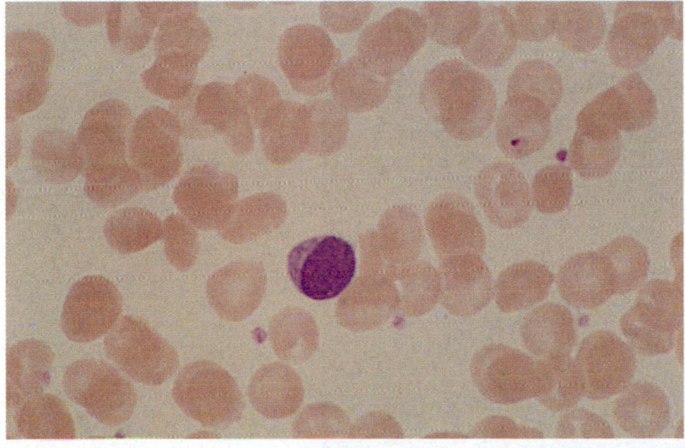

Plate 60 Lymphocyte surrounded by erythrocytes (842×). (Exercise 29, p. 290)

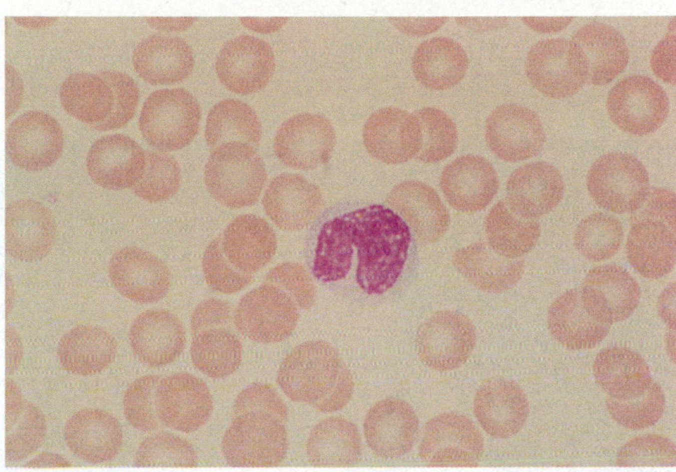

Plate 61 Monocyte surrounded by erythrocytes (842×). (Exercise 29, p. 290)

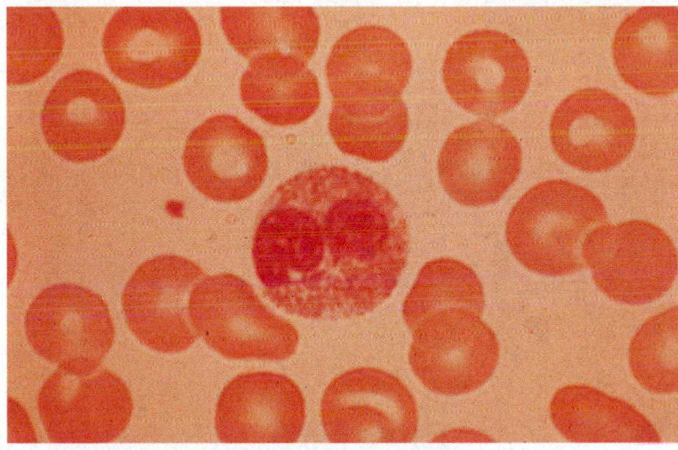

Plate 62 An eosinophil surrounded by erythrocytes (640×). (Exercise 29, p. 290)

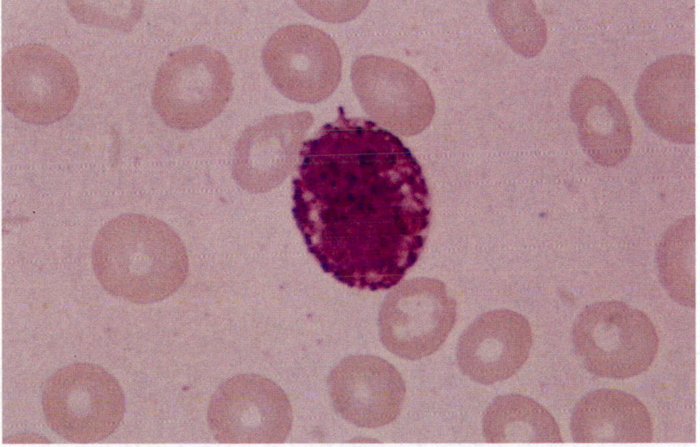

Plate 63 A basophil surrounded by erythrocytes (640×). (Exercise 29, p. 290)

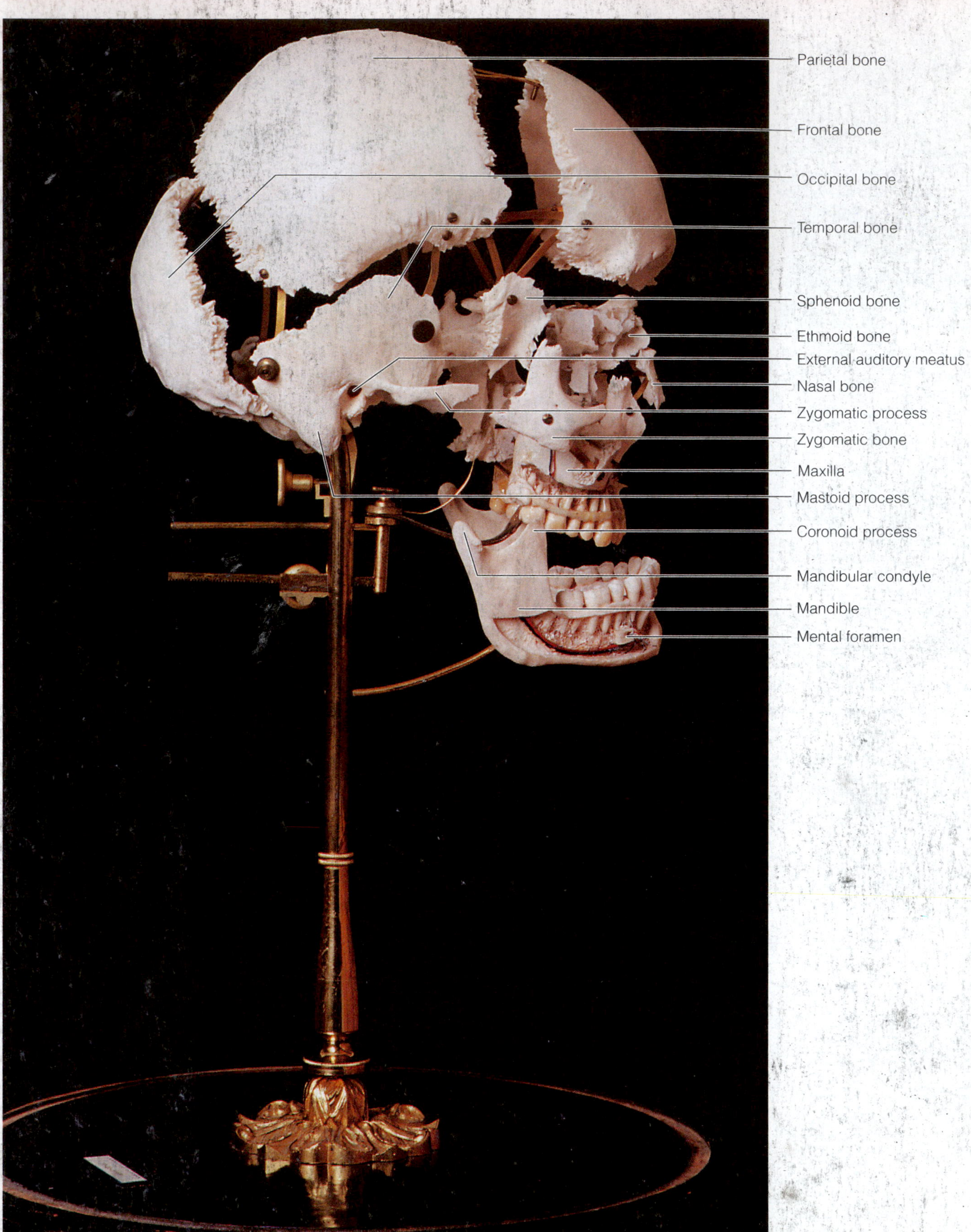

Plate A Beauchene skull, lateral view. (See Exercise 10, p. 90.)

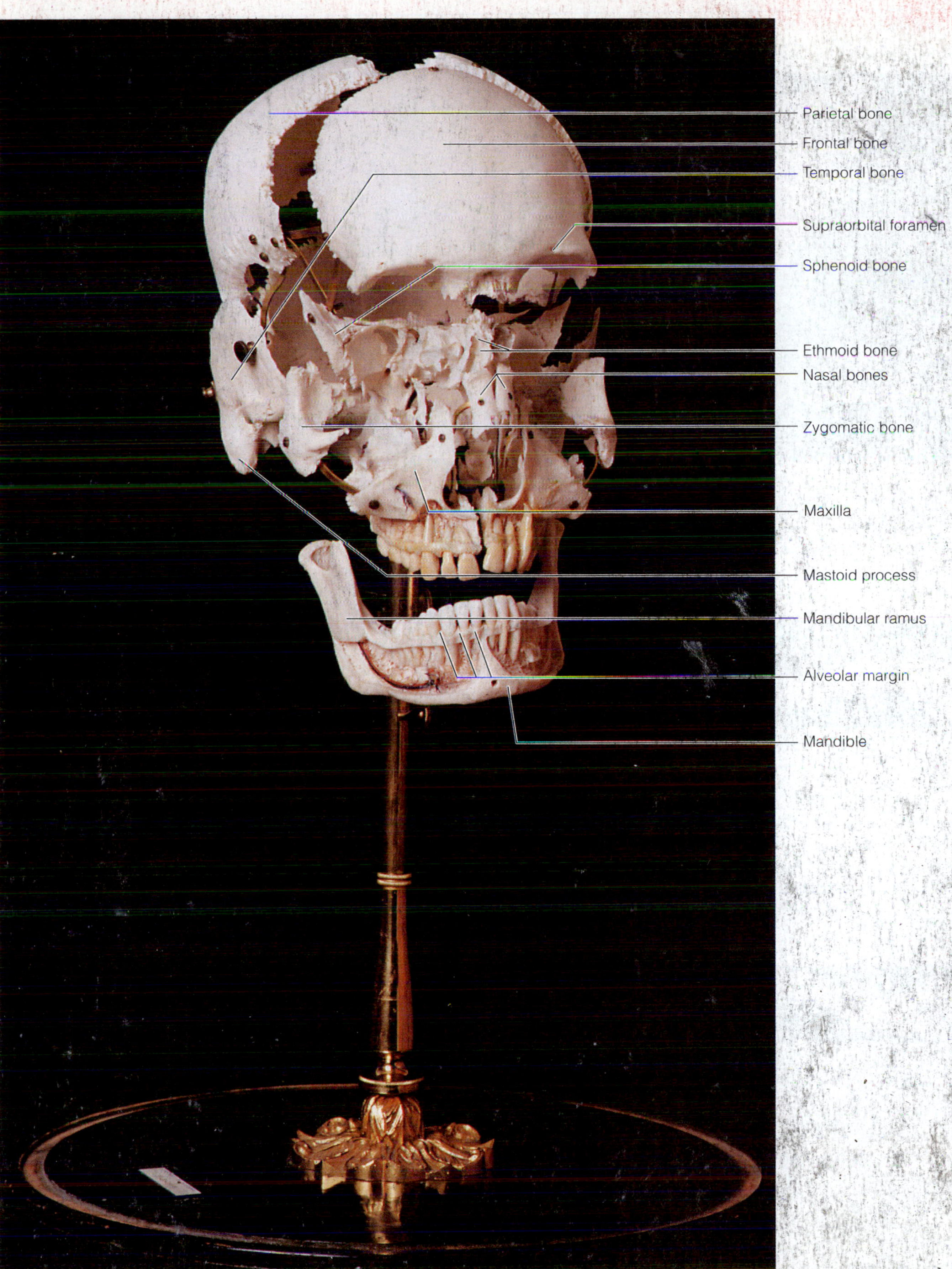

Plate B Beauchene skull, frontal view. (See Exercise 10, p. 90.)

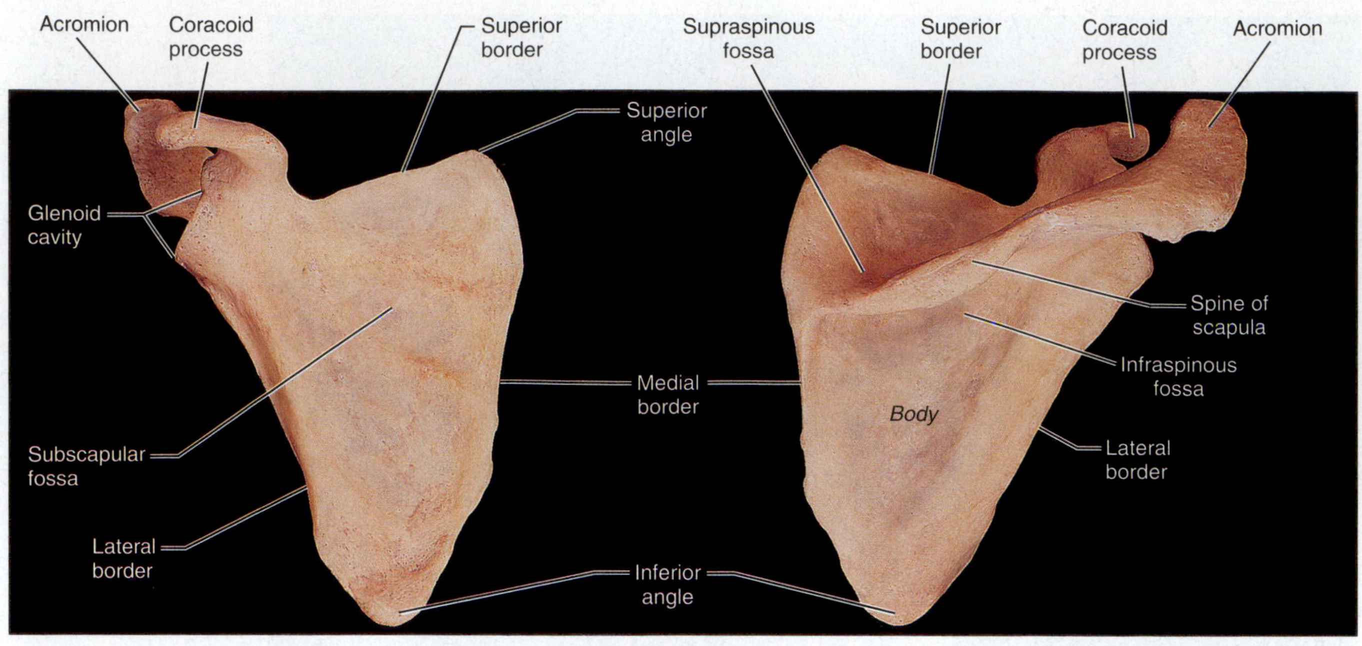

Anterior view **Posterior view**

Scapula

Posterior

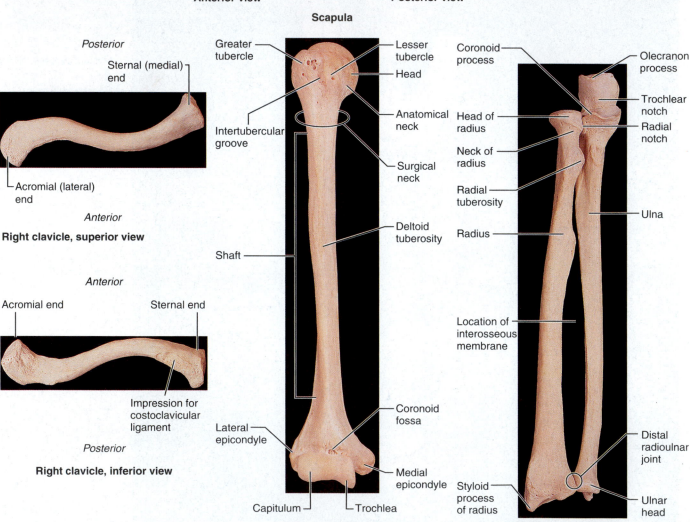

Sternal (medial) end

Acromial (lateral) end

Anterior

Right clavicle, superior view

Anterior

Acromial end

Sternal end

Impression for costoclavicular ligament

Posterior

Right clavicle, inferior view

Greater tubercle

Lesser tubercle

Head

Intertubercular groove

Anatomical neck

Surgical neck

Deltoid tuberosity

Shaft

Coronoid fossa

Lateral epicondyle

Medial epicondyle

Capitulum

Trochlea

Humerus, Anterior view

Coronoid process

Olecranon process

Trochlear notch

Head of radius

Radial notch

Neck of radius

Radial tuberosity

Ulna

Radius

Location of interosseous membrane

Distal radioulnar joint

Styloid process of radius

Ulnar head

Radius (left) and Ulna (right), anterior view

Plate C Photographs of selected bones of the pectoral girdle and right upper limb. (See Exercise 11, pp. 106–111.)

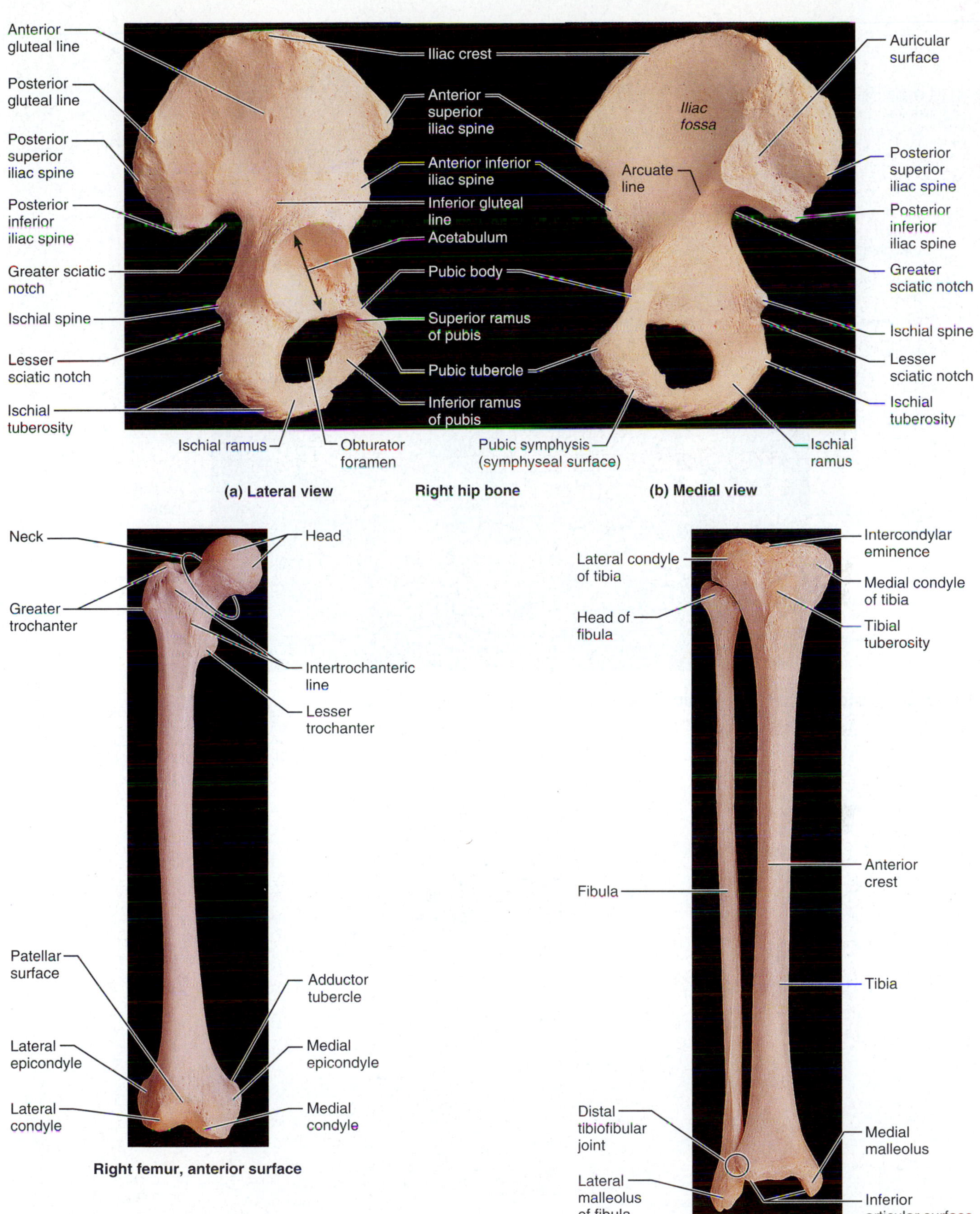
Anterior gluteal line
Posterior gluteal line
Posterior superior iliac spine
Posterior inferior iliac spine
Greater sciatic notch
Ischial spine
Lesser sciatic notch
Ischial tuberosity
Iliac crest
Anterior superior iliac spine
Anterior inferior iliac spine
Inferior gluteal line
Acetabulum
Pubic body
Superior ramus of pubis
Pubic tubercle
Inferior ramus of pubis
Ischial ramus
Obturator foramen
Pubic symphysis (symphyseal surface)
Iliac fossa
Arcuate line
Auricular surface
Posterior superior iliac spine
Posterior inferior iliac spine
Greater sciatic notch
Ischial spine
Lesser sciatic notch
Ischial tuberosity
Ischial ramus
(a) Lateral view
Right hip bone
(b) Medial view
Neck
Head
Greater trochanter
Intertrochanteric line
Lesser trochanter
Patellar surface
Lateral epicondyle
Lateral condyle
Adductor tubercle
Medial epicondyle
Medial condyle
Right femur, anterior surface
Lateral condyle of tibia
Head of fibula
Intercondylar eminence
Medial condyle of tibia
Tibial tuberosity
Anterior crest
Fibula
Tibia
Distal tibiofibular joint
Lateral malleolus of fibula
Medial malleolus
Inferior articular surface
Right tibia and fibula, anterior view

Plate D Photographs of selected bones of the pelvic girdle and right lower limb. (See Exercise 11, pp. 111–115.)

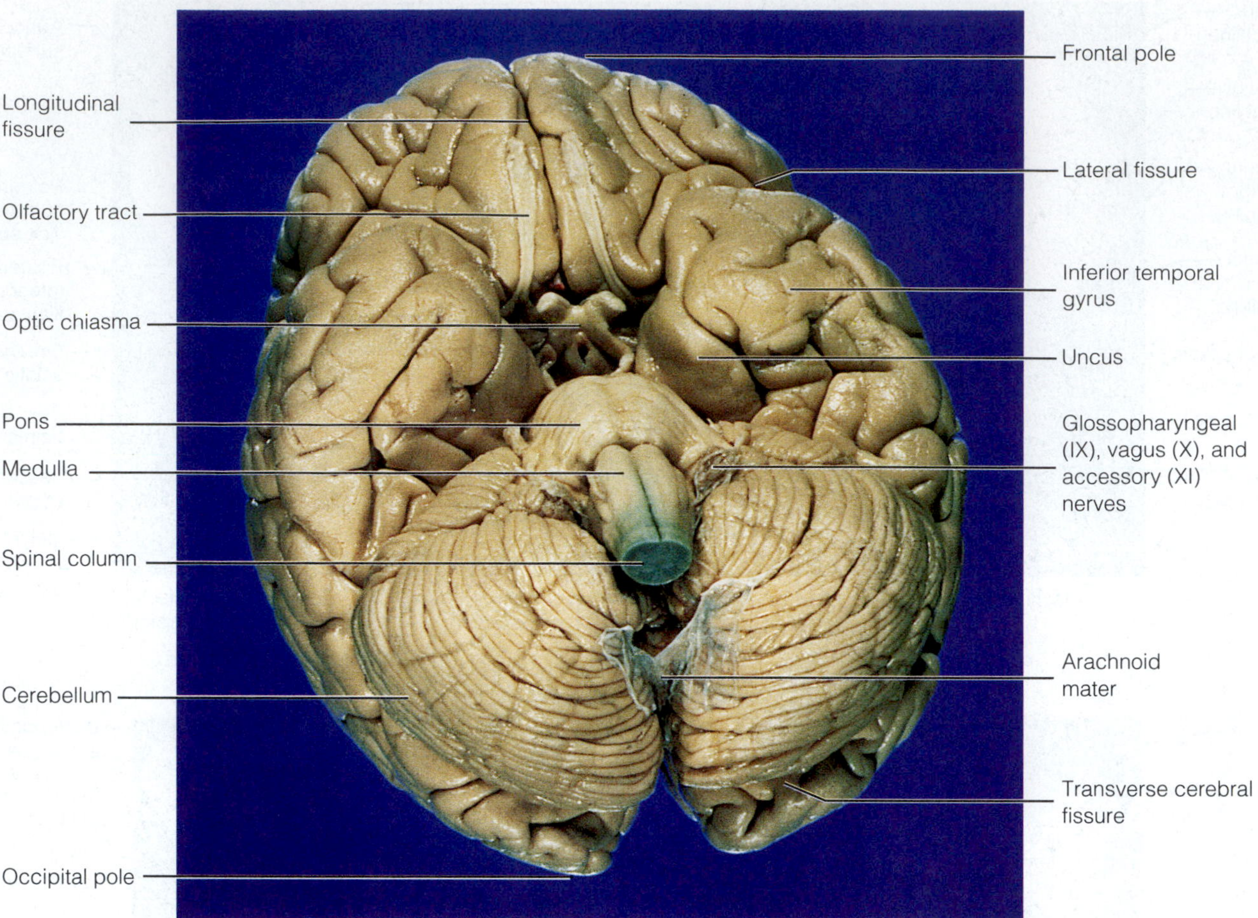

Plate E Ventral view of the brain. (See Exercise 19, p. 195.)

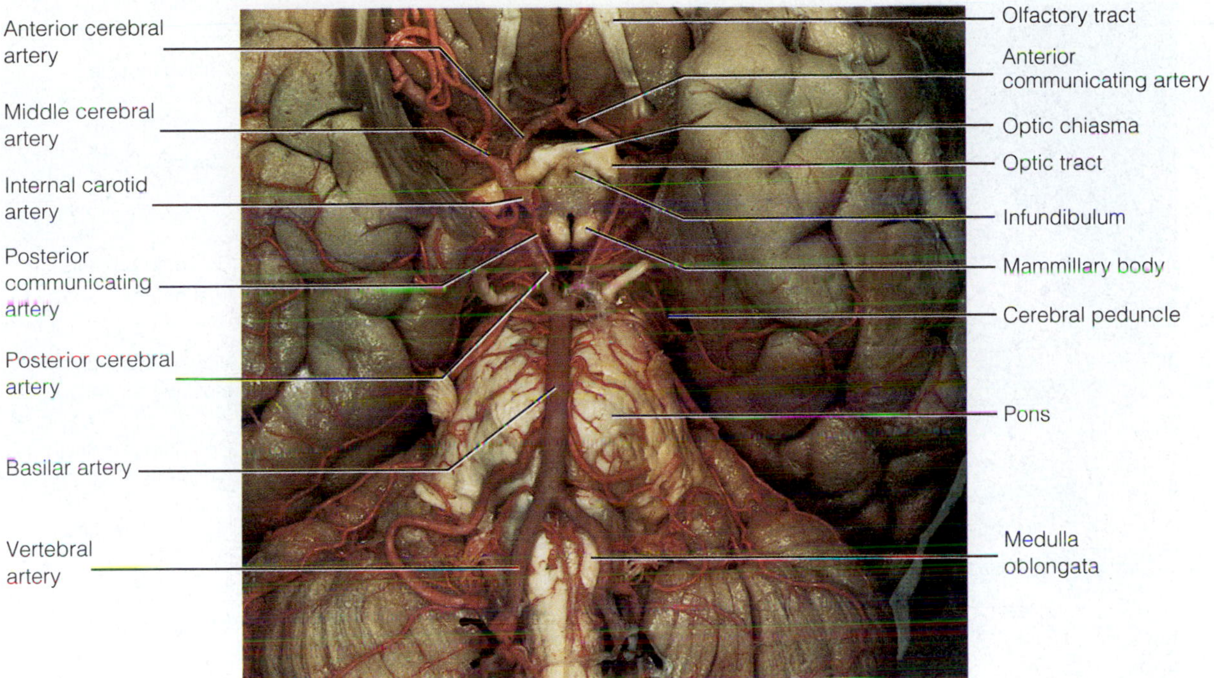

Plate F Circle of Willis. (See Exercise 32, p. 332.)

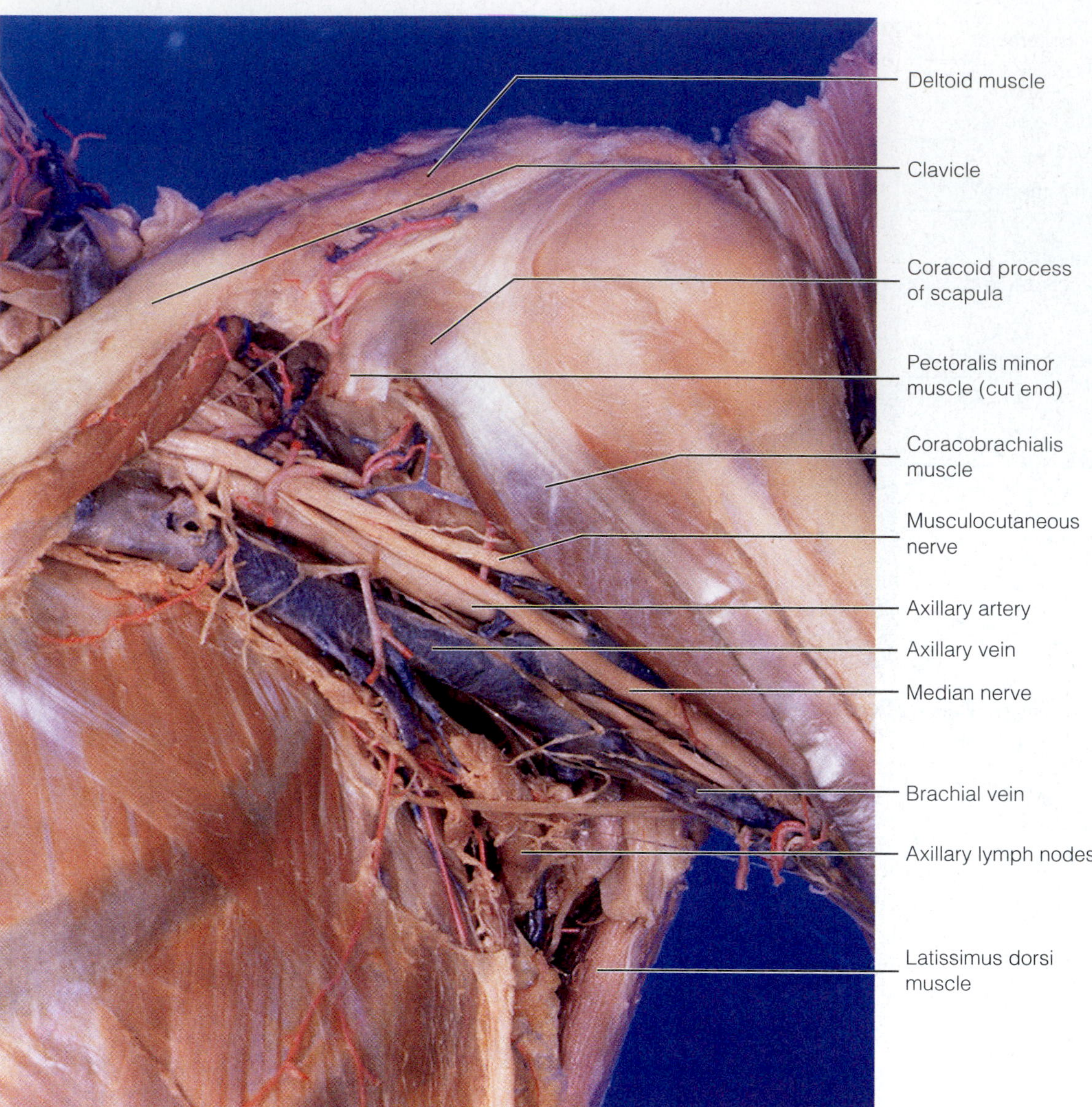

Plate G Brachial plexus and axilla. (See Exercise 21, pp. 221–222.)

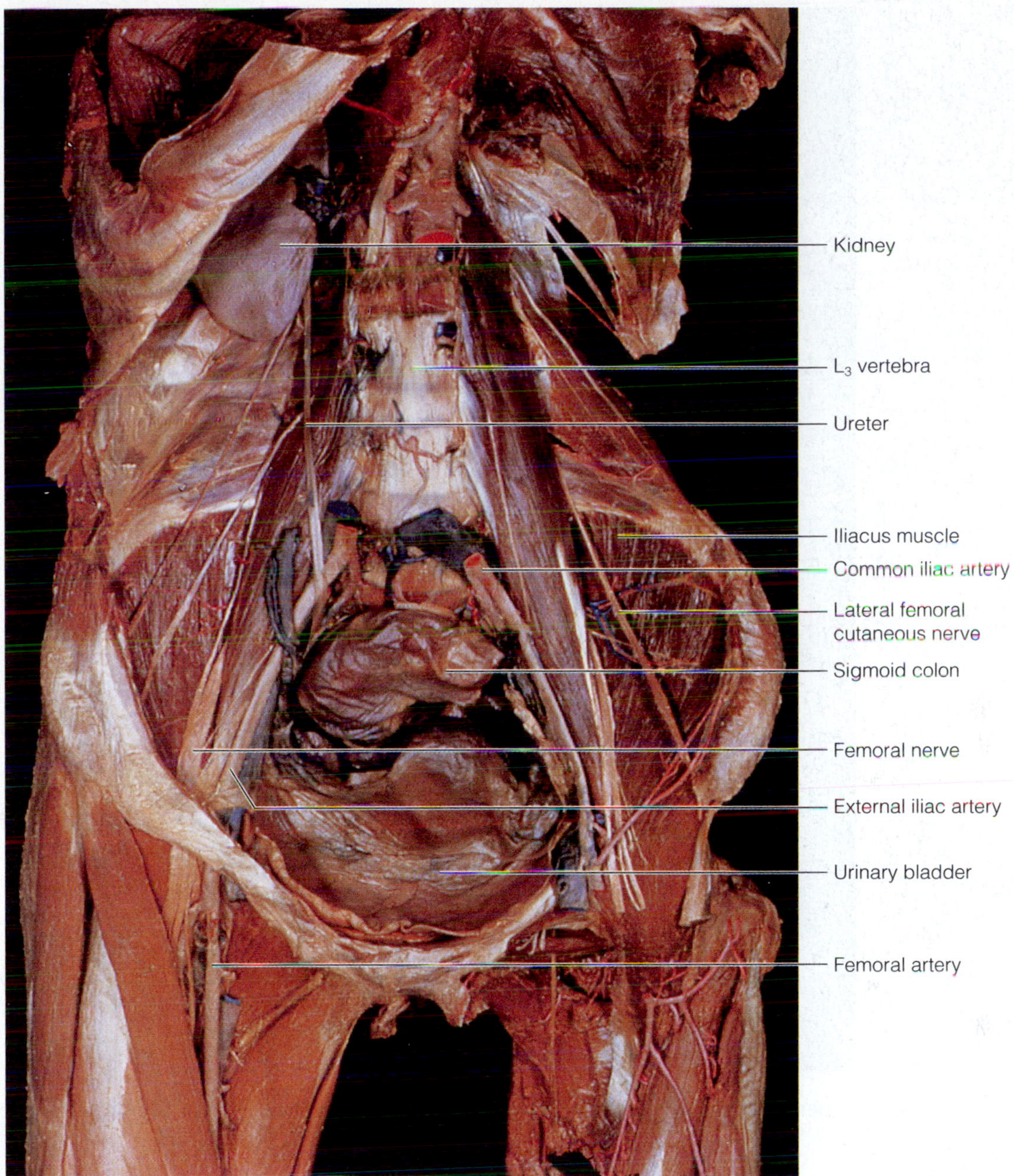

Plate H Lumbar plexus. (See Exercise 21, pp. 223–224.)

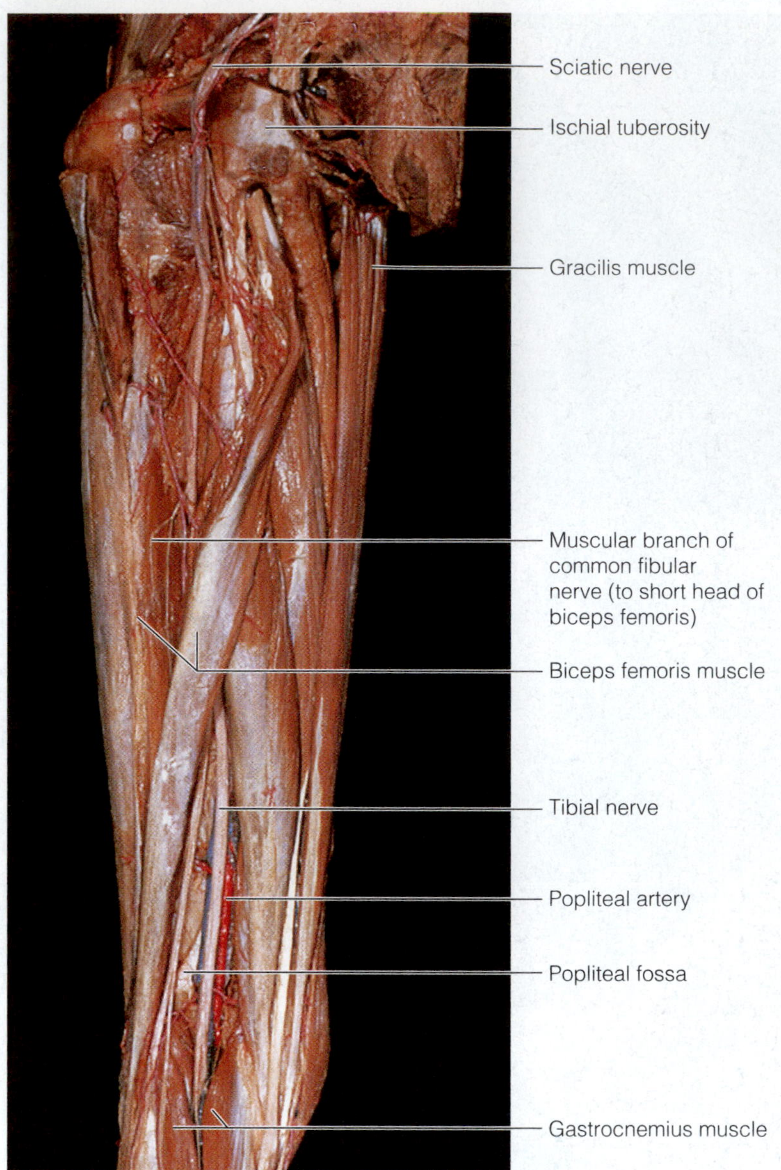

Plate I Course of sciatic nerve along posterior thigh and knee. (See Exercise 21, pp. 224–225.)

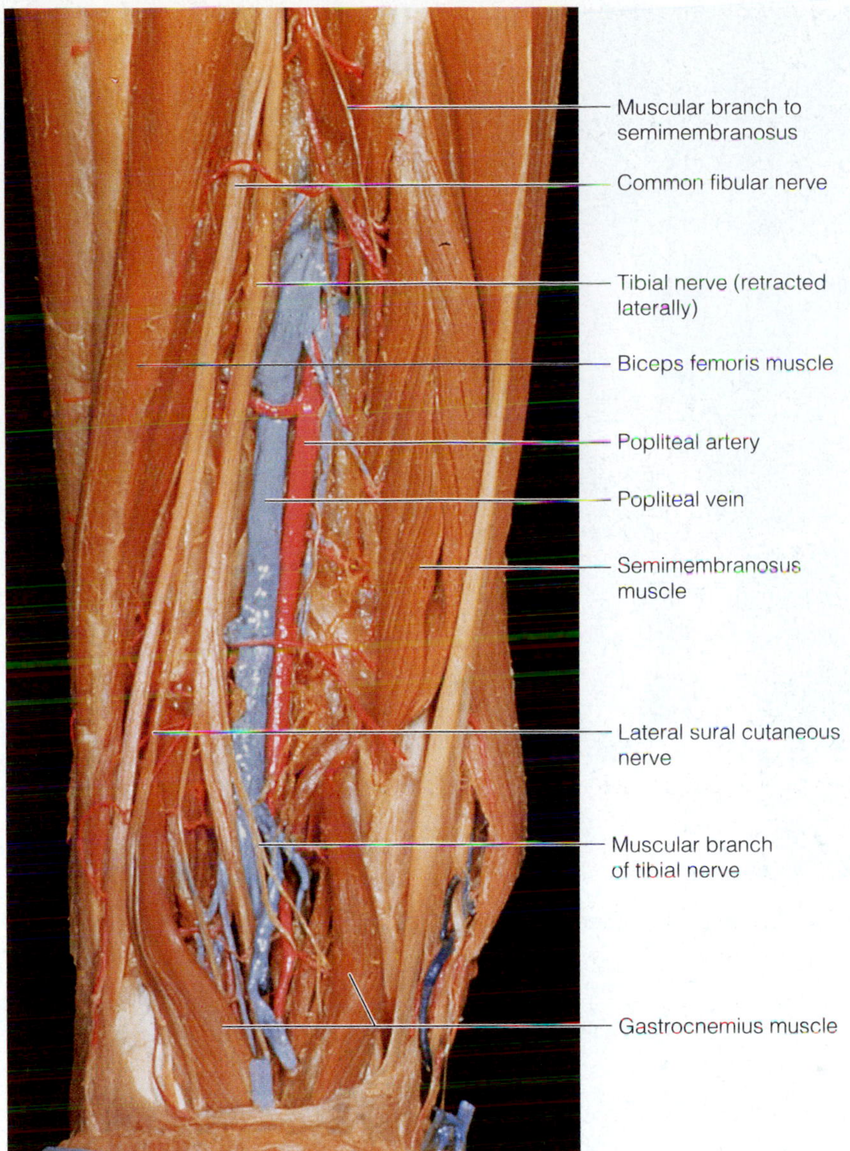

Plate J Deep relations of sciatic nerve in popliteal fossa. (See Exercise 21, p. 225.)

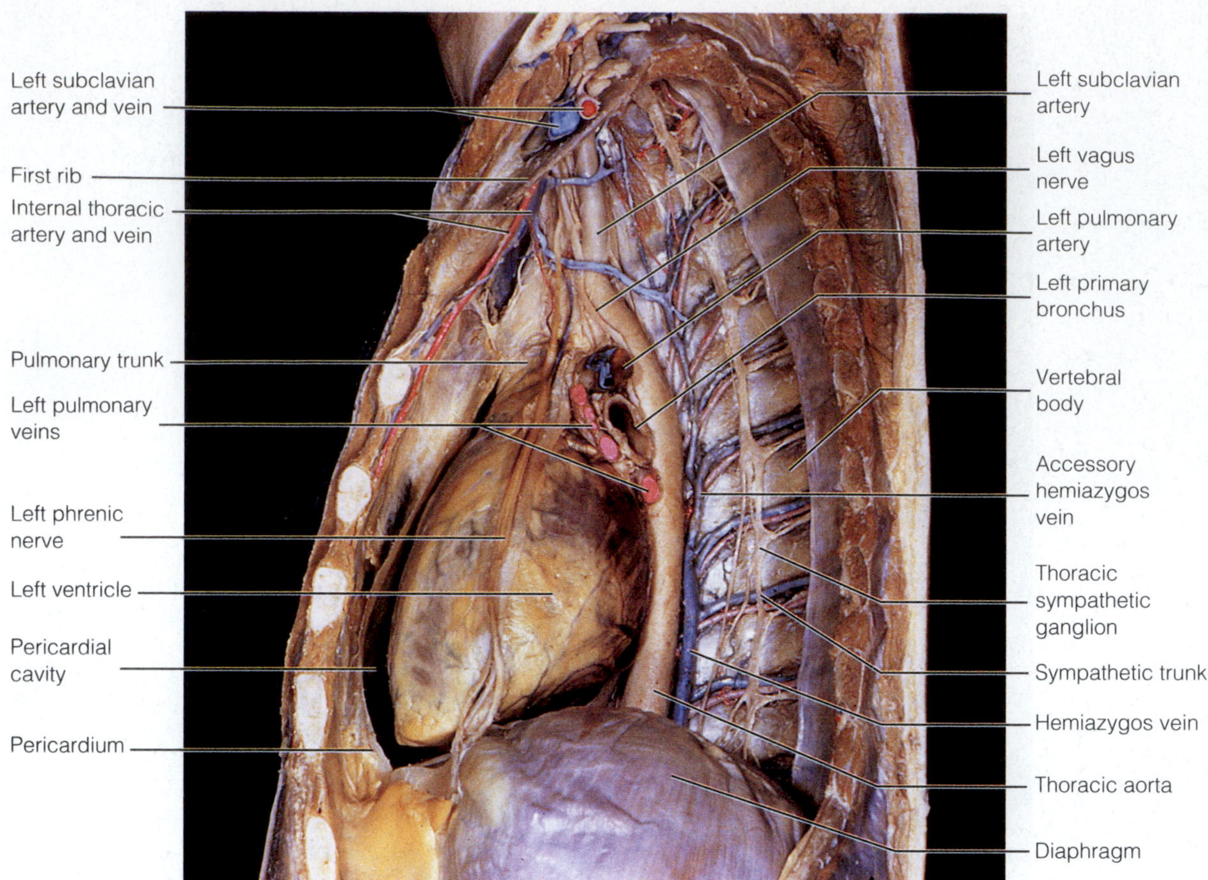

Plate K Lateral view of mediastinum. (See Exercise 30, p. 301.)

PHYSIOEX 4.0
by

Peter Zao, North Idaho College
Timothy Stabler, Indiana University Northwest
Marcia C. Gibson, University of Wisconsin–Madison (Histology Review Supplement)

CONTENTS

PhysioEx Version 4.0 consists of ten physiology lab simulations that may be used to supplement or replace wet labs. This easy-to-use software allows you to repeat labs as often as you like, perform experiments without harming live animals, and conduct experiments that may be difficult to perform in a wet lab environment due to time, cost, or safety concerns. You also have the flexibility to change the parameters of an experiment and observe how outcomes are affected. In addition, PhysioEx includes an extensive histology tutorial that allows you to study histology images at various magnifications. This manual will walk you through each lab step-by-step. You will also find Review Sheets in the back of your manual to test your understanding of the key concepts in each lab.

New to Version 4.0

Note to instructors: If you have used previous versions of PhysioEx, here is a summary of what you will find new to 4.0:

- A new lab on acid-base balance has been added.
- A new Histology Review Supplement has been added, consisting of forty histology slides that relate to topics covered in the PhysioEx lab simulations. In addition, written review worksheets are provided to accompany the slides.
- PhysioEx is now available in two formats: CD-ROM or online at www.physioex.com. Passcodes to access PhysioEx online are included with every new copy of *Human Anatomy and Physiology Laboratory Manual* by Elaine Marieb and Linda Kollett.

Topics in This Edition

Exercise 5B The Cell—Transport Mechanisms and Cell Permeability: Computer Simulation. Explores how substances cross the cell's membrane. Simple and facilitated diffusion, osmosis, filtration, and active transport are covered.

Exercise 6B Histology Review: Computer Simulation. Includes over 200 histology images, viewable at various magnifications, with accompanying descriptions and labels.

Exercise 16B Skeletal Muscle Physiology: Computer Simulation. Provides insights into the complex physiology of skeletal muscle. Electrical stimulation, isometric contractions, and isotonic contractions are investigated.

Exercise 18B Neurophysiology of Nerve Impulses: Computer Simulation. Investigates stimuli that elicit action potentials, stimuli that inhibit action potentials, and factors affecting nerve conduction velocity.

Exercise 28B Endocrine System Physiology: Computer Simulation. Investigates the relationship between hormones and metabolism; the effect of estrogen replacement therapy; and the effects of insulin on diabetes.

Exercise 33B Cardiovascular Dynamics: Computer Simulation. Topics of inquiry include vessel resistance and pump (heart) mechanics.

Exercise 34B Frog Cardiovascular Physiology: Computer Simulation. Variables influencing heart activity are examined. Topics include setting up and recording baseline heart activity, the refractory period cardiac muscle, and an investigation of physical and chemical factors that affect enzyme activity.

Exercise 37B Respiratory System Mechanics: Computer Simulation. Investigates physical and chemical aspects of pulmonary function. Students collect data simulating normal lung volumes. Other activities examine factors such as airway resistance and the effect of surfactant on lung function.

Exercise 39B Chemical and Physical Processes of Digestion: Computer Simulation. Turns the student's computer into a virtual chemistry lab where enzymes, reagents, and incubation conditions can be manipulated (in compressed time) to examine factors that affect enzyme activity.

Exercise 41B Renal Physiology—The Function of the Nephron: Computer Simulation. Simulates the function of a single nephron. Topics include factors influencing glomerular filtration, the effect of hormones on urine function, and glucose transport maximum.

Exercise 47 Acid-Base Balance: Computer Simulation. Topics include respiratory and metabolic acidosis/alkalosis, as well as renal and respiratory compensation.

Getting Started

To use PhysioEx version 4.0, your computer should meet the following minimum requirements (regardless of whether you are using the CD or accessing PhysioEx via the web):

- **IBM/PC:** Windows 95, 98, NT, 2000, Millennium Edition or higher; Pentium I/266 MHz or faster.
- **Macintosh:** Macintosh 8.6 and above; 604/300 MHz or G3/233 MHz.
- 64 MB RAM (128 MB recommended)
- 800 × 600 screen resolution, millions of colors
- Internet Explorer 5.0 (or higher) *or* Netscape 4.6 (or higher)*
- Flash 6[†] plug-in
- 4× CD-ROM drive (if using CD-ROM)
- Printer

*Although you do not need a live Internet connection to run the CD, you do need to have a browser installed on your computer. If you do not have a browser, the CD includes a free copy of Netscape which you may install. See instructions on the CD-liner notes.

[†] If you do not have Flash 6 installed on your computer, the CD includes a free Flash installer. See instructions on the CD-liner notes.

Instructions for Getting Started—Mac Users (CD version)

1. Put the PhysioEx CD in the CD-ROM drive. The program should launch automatically. If autorun is disabled on your computer, double click on the PhysioEx icon that appears on your desktop.

2. Although you do not need a live Internet connection to run PhysioEx, you do need to have a browser (such as Netscape or Internet Explorer) installed on your computer. If you already have a browser installed, proceed to step 3. If you do not have a browser installed, follow the instructions for installing Netscape found on the liner notes that are packaged with your CD.

3. If you can see a clock onscreen with the clock hands moving, click Proceed. If you cannot see the clock, or if you can see the clock but the hands are not moving, follow the instructions for installing Flash 6 found on the liner notes that are packaged with your CD.

4. On the License Agreement screen, click "Agree" to proceed.

5. On the screen with the PhysioEx icon at the top, click the License Agreement link to read the full agreement. Then close the License Agreement window and click the Main Menu link.

6. From the Main Menu, click on the lab you wish to enter.

Instructions for Getting Started—IBM/PC Users (CD version)

1. Put the PhysioEx CD in the CD-ROM drive. The program should launch automatically. If autorun is disabled on your machine, double click the My Computer icon on your Windows desktop, and then double click the PhysioEx icon.

2. Although you do not need a live Internet connection to run PhysioEx, you do need to have a browser (such as Netscape or Internet Explorer) installed on your computer. If you already have a browser installed, proceed to step 3. If you do not have a browser installed, follow the instructions for installing Netscape found on the liner notes that are packaged with your CD.

3. If you can see a clock onscreen with the clock hands moving, click Proceed. If you cannot see the clock, or if you can see the clock but the hands are not moving, follow the instructions for installing Flash 6 found on the liner notes that are packaged with your CD.

4. On the License Agreement screen, click "Agree" to proceed.

5. On the screen with the PhysioEx icon at the top, click the License Agreement link to read the full agreement. Then close the License Agreement window and click the Main Menu link.

6. From the Main Menu, click on the lab you wish to enter.

Instructions for Getting Started—Web Users

Follow the instructions for accessing www.physioex.com that appear at the very front of your lab manual.

Technical Support

Phone: 800.677.6337
Email: media.support@pearsoned.com
Hours: 8 A.M. to 5 P.M. CST, Monday–Friday

exercise

5B The Cell—Transport Mechanisms and Permeability: Computer Simulation

Objectives

1. To define *differential permeability; diffusion (simple diffusion, facilitated diffusion,* and *osmosis); isotonic, hypotonic,* and *hypertonic solutions; passive* and *active processes* of transport; *bulk-phase endocytosis; phagocytosis;* and *solute pump.*
2. To describe the processes that account for the movement of substances across the plasma membrane and to indicate the driving force for each.
3. To determine which way substances will move passively through a differentially permeable membrane (given the appropriate information on concentration differences).

The molecular composition of the plasma membrane allows it to be selective about what passes through it. It allows nutrients to enter the cell but keeps out undesirable substances. By the same token, valuable cell proteins and other substances are kept within the cell, and excreta or wastes pass to the exterior. This property is known as **differential,** or **selective, permeability.** Transport through the plasma membrane occurs in two basic ways. In **active processes,** the cell provides energy (ATP) to power the transport process. In the other, **passive processes,** the transport process is driven by concentration or pressure differences between the interior and exterior of the cell.

Passive Processes

The two key passive processes of membrane transport are diffusion and filtration. Diffusion is an important transport process for every cell in the body. By contrast, filtration usually occurs only across capillary walls. Each of these will be considered in turn.

Diffusion

Recall that all molecules possess *kinetic energy* and are in constant motion. As molecules move about randomly at high speeds, they collide and ricochet off one another, changing direction with each collision. For a given temperature, all matter has about the same average kinetic energy. Because kinetic energy is directly related to both mass and velocity ($KE = \frac{1}{2} mv^2$), smaller molecules tend to move faster.

When a **concentration gradient** (difference in concentration) exists, the net effect of this random molecular movement is that the molecules eventually become evenly distributed throughout the environment, i.e., the process called diffusion occurs. Hence, **diffusion** is the movement of molecules from a region of their higher concentration to a region of their lower concentration. Diffusion's driving force is the kinetic energy of the molecules themselves.

The diffusion of particles into and out of cells is modified by the plasma membrane, which constitutes a physical barrier. In general, molecules diffuse passively through the plasma membrane if they are small enough to pass through its pores (and are aided by an electrical gradient), or if they can dissolve in the lipid portion of the membrane as in the case of CO_2 and O_2. The diffusion of solute particles dissolved in water through a differentially permeable membrane is called **simple diffusion.** The diffusion of water through a differentially permeable membrane is called *osmosis*. Both simple diffusion and osmosis involve movement of a substance from an area of its higher concentration to one of its lower concentration, i.e., down its concentration gradient.

Solute Transport Through Nonliving Membranes

This computerized simulation provides information on the passage of water and solutes through semipermeable membranes, which may be applied to the study of transport mechanisms in living membrane-bound cells.

Activity 1: Simulating Dialysis (Simple Diffusion)

Choose **Cell Transport Mechanisms and Permeability** from the main menu. The opening screen will appear in a few seconds (Figure 5B.1). The primary features on the screen when the program starts are a pair of glass beakers perched atop a solutions dispenser, a dialysis membranes cabinet at the right side of the screen, and a data collection unit at the bottom of the display.

The beakers are joined by a membrane holder, which can be equipped with any of the dialysis membranes from the cabinet. Each membrane is represented by a thin colored line suspended in a gray supporting frame. The solute concentration of dispensed solutions is displayed at the side of each beaker. As you work through the experiments, keep in mind that membranes are three-dimensional; thus what appears as a slender line is actually the edge of a membrane sheet.

The solutions you can dispense are listed beneath each beaker. You can choose more than one solution, and the amount to be dispensed is controlled by clicking (+) to increase concentration, or (−) to decrease concentration. The chosen solutions are then delivered to their beaker by

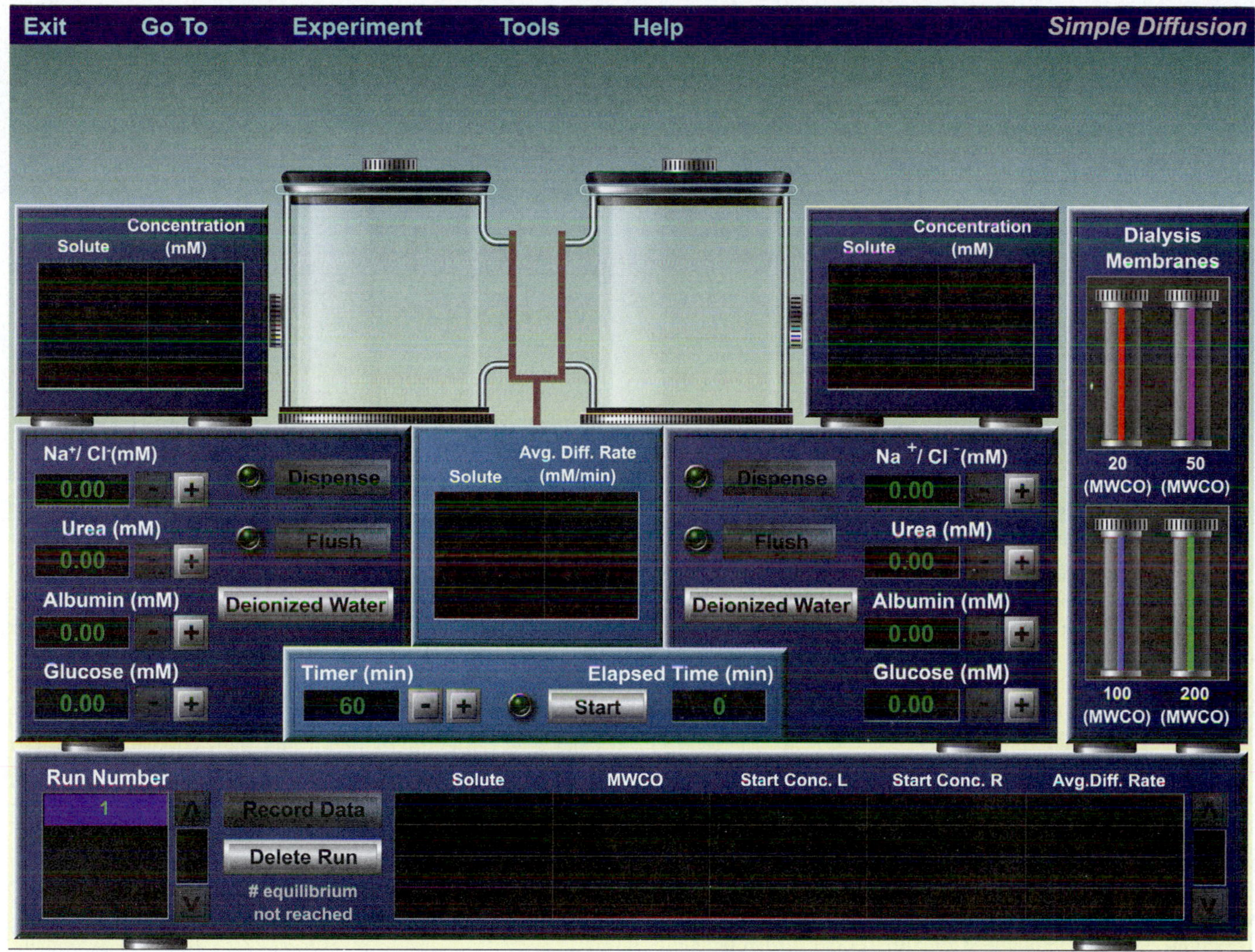

Figure 5B.1 Opening screen of the Simple Diffusion experiment.

clicking the **Dispense** button on the same side. Clicking the **Start** button opens the membrane holder and begins the experiment. The **Start** button will become a **Pause** button after it is clicked. To clean the beakers and prepare them for the next run, click **Flush.** Clicking **Pause** and then **Flush** during a run stops the experiment and prepares the beakers for another run. You can adjust the timer for any interval between 5 and 300; the elapsed time is shown in the small window to the right of the timer.

To move dialysis membranes from the cabinet to the membrane holder, click and hold the mouse on the selected membrane, drag it into position between the beakers, and then release the mouse button to drop it into place. Each membrane possesses a different molecular weight cutoff (MWCO), indicated by the number below it. You can think of MWCO in terms of pore size; the larger the MWCO number, the larger the pores in the membrane.

The Run Number window in the data collection unit at the bottom of the screen displays each experimental trial (run). When you click the **Record Data** button, your data is recorded in the computer's memory and is displayed in the data grid at the bottom of the screen. Data displayed in the data grid include the solute (Solute) and membrane (MWCO) used in a run, the starting concentrations in the left and right beakers (Start Conc. L. and Start Conc. R.), and the average diffusion rate (Avg. Dif. Rate). If you are not satisfied with a run, you can click **Delete Run.** Note: Remember NaCl does not move as a molecule. It dissociates to Na^+ and Cl^- ions in water.

1. Click and hold the mouse on the 20 MWCO membrane and drag it to the membrane holder between the beakers. Release the mouse button to lock the membrane into place.

2. Now increase the NaCl concentration to be dispensed by clicking the (+) button under the left beaker until the display window reads 9.00 m*M*. Click **Dispense** to fill the left beaker with 9.00 m*M* NaCl solution.

3. Click the **Deionized Water** button under the right beaker and then click **Dispense** to fill the right beaker with deionized water.

4. Adjust the timer to 60 min (compressed time), then click the **Start** button. When Start is clicked, the barrier between the beakers descends, allowing the solutions in each beaker to have access to the dialysis membrane separating them. Notice that the Start button becomes a Pause button that allows

Chart 1 Dialysis Results

	Membrane (MWCO)			
Solute	20	50	100	200
NaCl				
Urea				
Albumin				
Glucose				

you to momentarily halt the progress of the experiment so you can see instantaneous diffusion or transport rates.

5. Watch the concentration windows at the side of each beaker for any activity. A level above zero in NaCl concentration in the right beaker indicates that Na^+ and Cl^- ions are diffusing from the left into the right beaker through the semipermeable dialysis membrane. Record your results (+ for diffusion, − for no diffusion) in Chart 1. Click the **Record Data** button to keep your data in the computer's memory.

6. Click the 20 MWCO membrane (in the membrane holder) again to automatically return it to the membranes cabinet and then click **Flush** beneath each beaker to prepare for the next run.

7. Drag the next membrane (50 MWCO) to the holder and repeat steps 2 through 6. Continue the runs until you have tested all four membranes. (Remember: click **Flush** beneath each beaker between runs.)

8. Now perform the same experiment for urea, albumin, and glucose by repeating steps 1 through 7 three times. In step 2 you will be dispensing first urea, then albumin, and finally glucose, instead of NaCl.

9. Click **Tools** → **Print Data** to print your data.

Which solute(s) were able to diffuse into the right beaker from the left?

____

Which solute(s) did not diffuse?

____

If the solution in the left beaker contained both urea and albumin, which membrane(s) could you choose to selectively remove the urea from the solution in the left beaker? How would you carry out this experiment?

____

____

____

Assume that the solution in the left beaker contained NaCl in addition to the urea and albumin. How could you set up an experiment so that you removed the urea, but left the NaCl concentration unchanged?

____

____

____ ■

Facilitated Diffusion

Some molecules are lipid insoluble or too large to pass through plasma membrane pores; instead, they pass through the membrane by a passive transport process called **facilitated diffusion.** In this form of transport, solutes combine with carrier protein molecules in the membrane and are then transported *along* or *down* their concentration gradient. Because facilitated diffusion relies on carrier proteins, solute transport varies with the number of available membrane transport proteins.

Activity 2:
Simulating Facilitated Diffusion

Click the **Experiment** menu and then choose **Facilitated Diffusion.** The opening screen will appear in a few seconds (Figure 5B.2). The basic screen layout is similar to that of the previous experiment with only a few modifications to the equipment. You will notice that only NaCl and glucose solutes are available in this experiment, and you will see a Membrane Builder on the right side of the screen.

The (+) and (−) buttons underneath each beaker adjust solute concentration in the solutions to be delivered into each beaker. Similarly, the buttons in the Membrane Builder allow you to control the number of carrier proteins implanted in the membrane when you click the **Build Membrane** button.

In this experiment, you will investigate how glucose transport is affected by the number of available carrier molecules.

1. The Glucose Carriers window in the Membrane Builder should read 500. If not, adjust to 500 by using the (+) or (−) button.

2. Now click **Build Membrane** to insert 500 glucose carrier proteins into the membrane. You should see the membrane appear as a slender line encased in a support structure within the Membrane Builder. Remember that we are looking at the edge of a three-dimensional membrane.

3. Click on the membrane and hold the mouse button down as you drag the membrane to the membrane holder between the beakers. Release the mouse to lock the membrane into place.

4. Adjust the glucose concentration to be delivered to the left beaker to 2.00 m*M* by clicking the (+) button next to the glucose window until it reads 2.00.

5. To fill the left beaker with the glucose solution, click the **Dispense** button just below the left beaker.

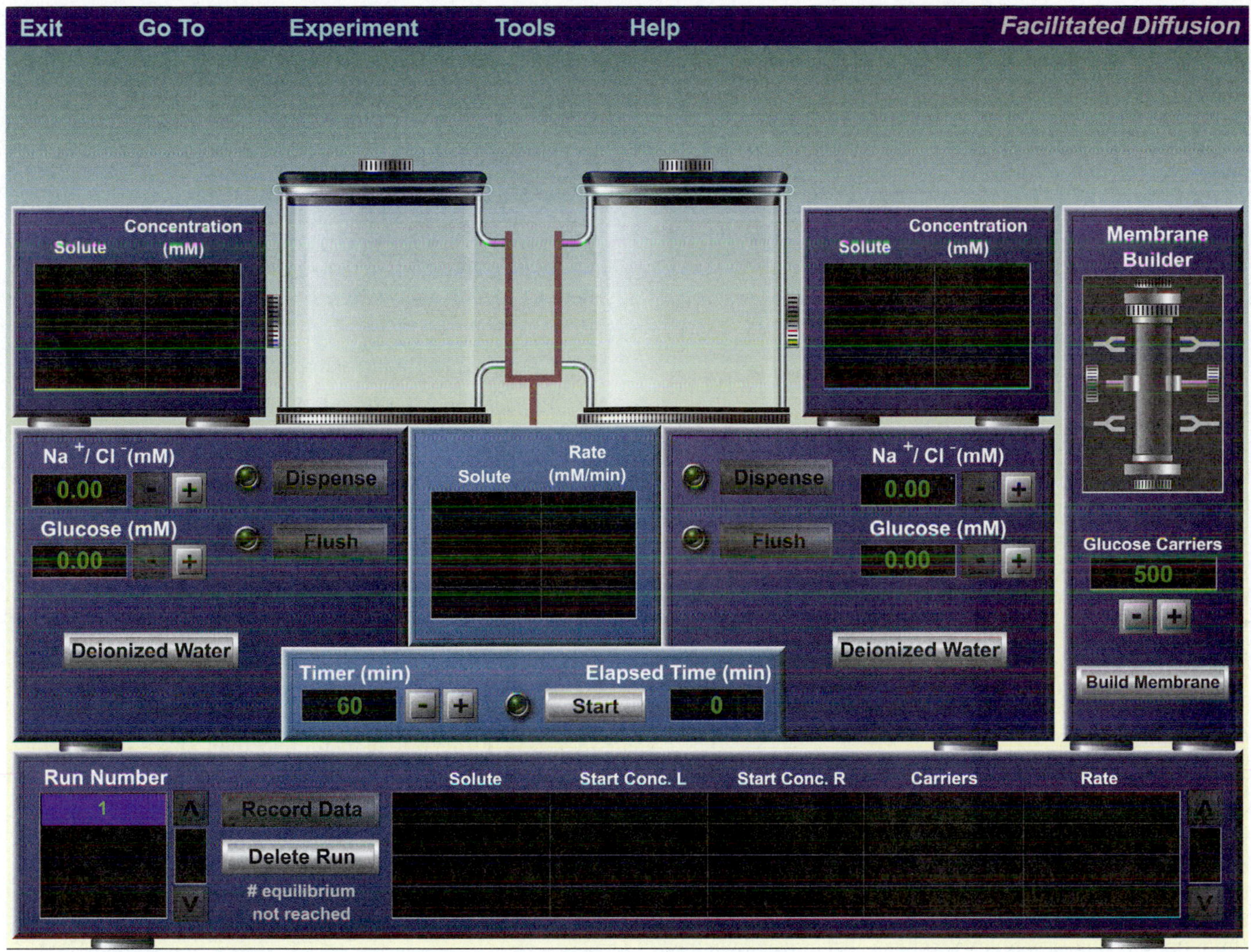

Figure 5B.2 Opening screen of the Facilitated Diffusion experiment.

6. Click the **Deionized Water** button below the right beaker, and then click the **Dispense** button. The right beaker will fill with deionized water.

7. Set the timer to 60 min and click **Start.** Watch the concentration windows next to the beakers. When the 60 minutes have elapsed, click the **Record Data** button to display glucose transport rate information in the grid at the lower edge of the screen. Record the glucose transport rate in Chart 2.

Chart 2 Facilitated Diffusion Results

Glucose concentration (mM)	No. of glucose carrier proteins		
	500	700	900
2.00			
8.00			

8. Click the **Flush** button beneath each beaker to remove any residual solution.

9. Click the membrane support to return it to the Membrane Builder. Increase the glucose carriers and repeat steps 2 through 8 using membranes with 700 and then 900 glucose carrier proteins. Record your results in Chart 2 each time.

10. Repeat steps 1 through 9 at 8.00 m*M* glucose concentration. Record your results in Chart 2.

11. Click **Tools** → **Print Data** to print your data.

What happened to the rate of facilitated diffusion as the number of protein carriers increased? Explain your answer.

What do you think would happen to the transport rate if you put the same concentration of glucose into both beakers instead of deionized water in the right beaker?

___

___

Should NaCl have an effect on glucose diffusion? Explain your answer. Use the simulation to see if it does.

___

___

___ ■

Osmosis

A special form of diffusion, the diffusion of water through a semipermeable membrane, is called **osmosis.** Because water can pass through the pores of most membranes, it can move from one side of a membrane to another relatively unimpeded. Osmosis occurs whenever there is a difference in water concentration on the two sides of a membrane.

If we place distilled water on both sides of a membrane, *net* movement of water will not occur; however, water molecules would still move between the two sides of the membrane. In such a situation, we would say that there is no *net* osmosis. The concentration of water in a solution depends on the number of solutes present. Therefore, increasing the solute concentration coincides with a decrease in water concentration. Because water moves down its concentration gradient, it will always move toward the solution with the highest concentration of solutes. Similarly, solutes also move down their concentration gradient. If we position a *fully* permeable membrane (permeable to solutes and water) between two solutions of differing concentrations, then all substances—solutes and water—will diffuse freely, and an equilibrium will be reached between the two sides of the membrane. However, if we use a semipermeable membrane that is impermeable to the solutes, then we have established a condition where water will move but solutes will not. Consequently, water will move toward the more concentrated solution, resulting in a volume increase. By applying this concept to a closed system where volumes cannot change, we can predict that the pressure in the more concentrated solution would rise.

Activity 3: Simulating Osmotic Pressure

Click the **Experiment** menu and then select **Osmosis.** The opening screen will appear in a few seconds (Figure 5B.3). The most notable difference in this experiment screen concerns meters atop the beakers that measure pressure changes in the beaker they serve. As before, (+) and (−) buttons control solute concentrations in the dispensed solutions.

1. Drag the 20 MWCO membrane to the holder between the two beakers.

2. Adjust the NaCl concentration to 8.00 m*M* in the left beaker, and then click the **Dispense** button.

3. Click **Deionized Water** under the right beaker and then click **Dispense.**

4. Set the timer to 60 min and then click **Start** to run the experiment. Pay attention to the pressure displays. Now click the **Record Data** button to retain your data in the computer's memory and also record the osmotic pressure in Chart 3 below.

5. Click the membrane to return it to the membrane cabinet.

6. Repeat steps 1 through 5 with the 50, 100, and 200 MWCO membranes.

Do you see any evidence of pressure changes in either beaker, using any of the four membranes? If so, which ones?

___

Does NaCl appear in the right beaker? If so, which membrane(s) allowed it to pass?

___

___

7. Now perform the same experiment for albumin and glucose by repeating steps 1 through 6. In step 2 you will be dispensing 9.00 m*M* albumin first, and then 10.00 m*M* glucose, instead of NaCl.

8. Click **Tools** → **Print Data** to print your data.

Answer the following questions using the results you recorded in Chart 3. Use the simulation if you need help formulating a response.

Chart 3 Osmosis Results (pressure in mm Hg)

	Membrane (MWCO)			
Solute	20	50	100	200
Na^+Cl^-				
Albumin				
Glucose				

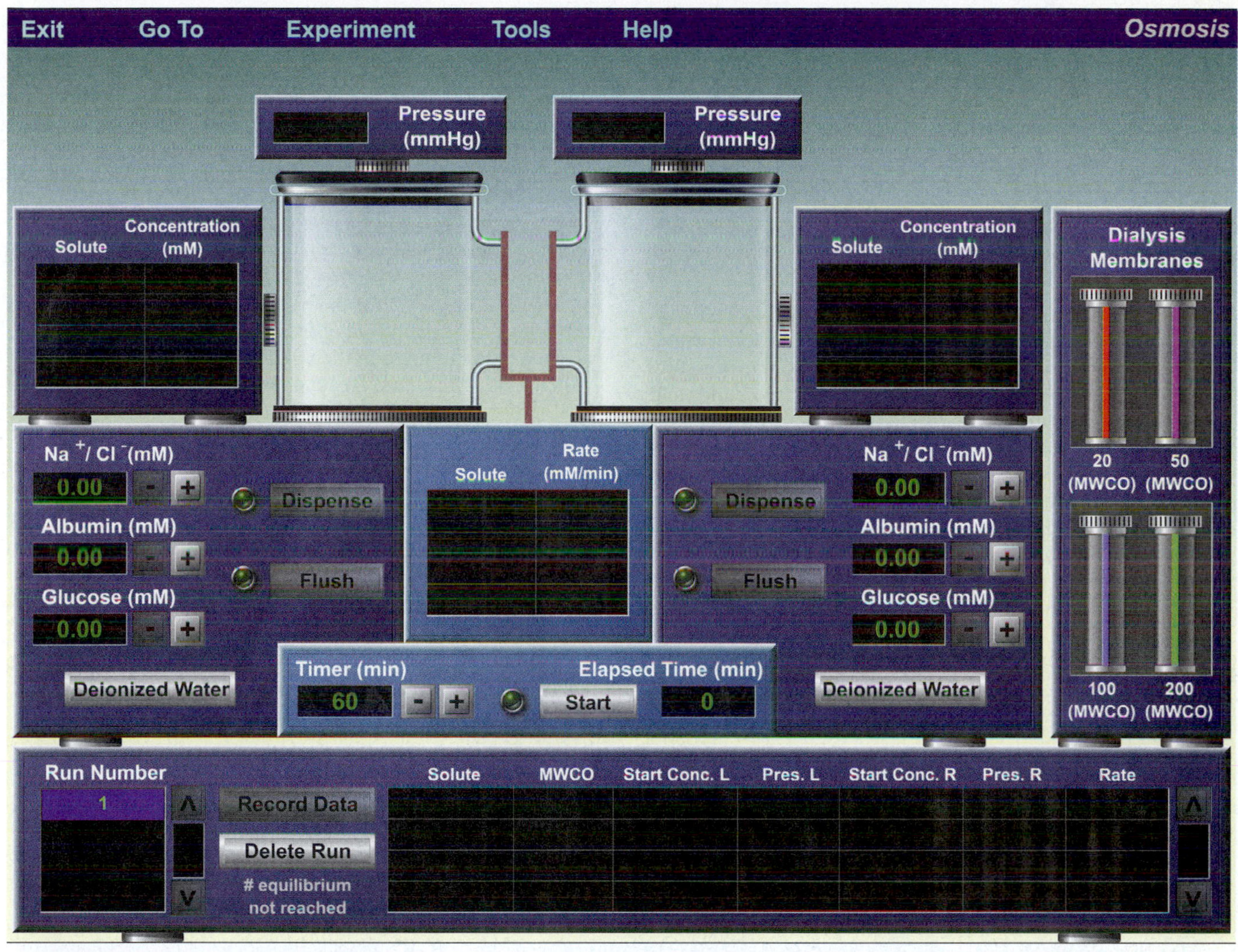

Figure 5B.3 Opening screen of the Osmosis experiment.

Explain the relationship between solute concentration and osmotic pressure.

____

____

Will osmotic pressure be generated if solutes are able to diffuse? Explain your answer.

____

____

Because the albumin molecule is much too large to pass through a 100 MWCO membrane, you should have noticed the development of osmotic pressure in the left beaker in the albumin run using the 100 MWCO membrane. What do you think would happen to the osmotic pressure if you replaced the deionized water in the right beaker with 9.00 m*M* albumin in that run? (Both beakers would contain 9.00 m*M* albumin.)

____

What would happen if you doubled the albumin concentration in the left beaker using any membrane?

____

In the albumin run using the 200 MWCO membrane, what would happen to the osmotic pressure if you put 10 m*M* glucose in the right beaker instead of deionized water? Explain your answer.

____

____

____

What if you used the 100 MWCO membrane in the albumin/glucose run described in the previous question?

____

____ ■

Activity 4: Simulating Filtration

Filtration is the process by which water and solutes pass through a membrane from an area of higher hydrostatic (fluid) pressure into an area of lower hydrostatic pressure. Like diffusion, it is a passive process. For example, fluids and solutes filter out of the capillaries in the kidneys into the kidney tubules because blood pressure in the capillaries is greater than the fluid pressure in the tubules. Filtration is not a selective process. The amount of filtrate—fluids and solutes—formed depends almost entirely on the pressure gradient (the difference in pressure on the two sides of the membrane) and on the size of the membrane pores.

Click the **Experiment** menu and then choose **Filtration.** The opening screen will appear in a few seconds (Figure 5B.4). The basic screen elements resemble the other simulations. The top beaker can be pressurized to force fluid through the filtration membrane into the bottom beaker. Any of the filtration membranes can be positioned in the holder between the beakers by drag-and-drop as in the previous experiments. The solutions you can dispense are listed to the right of the top beaker, and are adjusted by clicking the (+) and (−) buttons. The selected solutions are then delivered to the top beaker by clicking **Dispense.** The top beaker is cleaned and prepared for the next run by clicking **Flush.** You can adjust the timer for any interval between 5 and 300; the elapsed time is shown in the window to the right of the timer. When you click the **Record Data** button, your data is recorded in the computer's memory and is displayed in the data grid at the bottom of the screen.

Solute concentrations in the filtrate are automatically monitored by the *Filtrate Analysis Unit* to the right of the bottom beaker. After a run you can detect the presence of any solute remaining on a membrane by using the *Membrane Residue Analysis* unit located above the membrane cabinet.

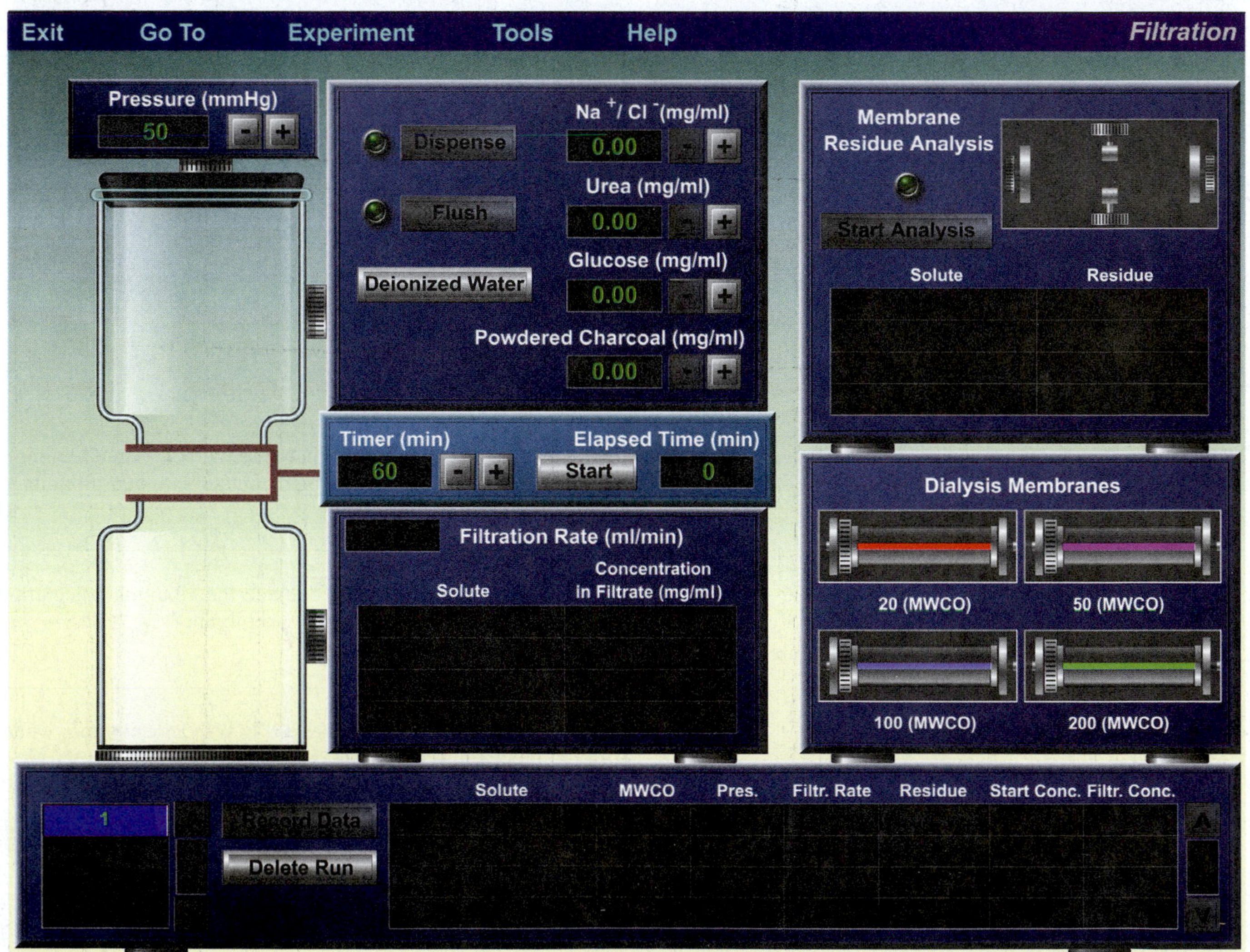

Figure 5B.4 Opening screen of the Filtration experiment.

Chart 4 Filtration Results (Filtration Rate, Solute Presence or Absence)

		Membrane (MWCO)			
Solute		20	50	100	200
	Rate				
NaCl	Filtrate				
	Residue				
Urea	Filtrate				
	Residue				
Glucose	Filtrate				
	Residue				
Powdered charcoal	Filtrate				
	Residue				

1. Click and hold the mouse on the 20 MWCO membrane and drag it to the holder below the top beaker. Release the mouse button to lock the membrane into place.
2. Now adjust the NaCl, urea, glucose, and powdered charcoal windows to 5.00 mg/ml each, and then click **Dispense.**
3. If necessary, adjust the pressure unit atop the beaker until its window reads 50 mm Hg.
4. Set the timer to 60 min and then click **Start.** When the Start button is clicked, the membrane holder below the top beaker retracts, and the solution will flow through the membrane into the beaker below.
5. Watch the Filtrate Analysis Unit for any activity. A rise in detected solute concentration indicates that the solute particles are moving through the filtration membrane. At the end of the run, record the amount of solute present in the *filtrate* (mg/ml) and the filtration *rate* in Chart 4.
6. Now drag the 20 MWCO membrane to the holder in the Membrane Residue Analysis unit. Click **Start Analysis** to begin analysis (and cleaning) of the membrane. Record your results for solute *residue* presence on the membrane (+ for present, − for not present) in Chart 4 and click the **Record Data** button to keep your data in the computer's memory.
7. Click the 20 MWCO membrane again to automatically return it to the membranes cabinet and then click **Flush** to prepare for the next run.
8. Repeat steps 1 through 7 using 50, 100, and 200 MWCO membranes.
9. Click **Tools** → **Print Data** to print your data.

Did the membrane's MWCO affect the filtration rate?

____

Which solute did not appear in the filtrate using any of the membranes?

____

What would happen if you increased the driving pressure? Use the simulation to arrive at an answer.

____

Explain how you can increase the filtration rate through living membranes.

____

____

By examining the filtration results, we can predict that the molecular weight of glucose must be:

greater than _______, but less than _______. ■

Active Transport

Whenever a cell expends cellular energy (ATP) to move substances across its membrane, the process is referred to as an *active transport process*. Substances moved across cell membranes by active means are generally unable to pass by diffusion. There are several possible reasons why substances may not be able to pass through a membrane by diffusion: they may be too large to pass through the membrane channels, they may not be lipid soluble, or they may have to move against rather than with a concentration gradient.

In one type of active transport, substances move across the membrane by combining with a protein carrier molecule; the process resembles an enzyme-substrate interaction. ATP provides the driving force, and in many cases the substances move against concentration or electrochemical gradients or both. Some of the substances that are moved into the cells by such carriers, commonly called **solute pumps,** are amino acids and some sugars. Both solutes are lipid-insoluble and too large to pass through the membrane channels, but are necessary for cell life. On the other hand, sodium ions (Na^+) are ejected from the cells by active transport. There is more Na^+ outside the cell than inside, so the Na^+ tends to remain in the cell unless actively transported out. In the body, the most common type of solute pump is the coupled Na^+-K^+ pump that moves Na^+ and K^+ in opposite directions across cellular membranes. 3 Na^+ are ejected for every 2 K^+ entering the cell.

Engulfment processes such as bulk-phase endocytosis and phagocytosis also require ATP. In **bulk-phase endocytosis,** the cell membrane sinks beneath the material to form a small vesicle, which then pinches off into the cell interior. Bulk-phase endocytosis is most common for taking in liquids containing protein or fat.

In **phagocytosis** (cell eating), parts of the plasma membrane and cytoplasm expand and flow around a relatively large or solid material such as bacteria or cell debris and engulf it, forming a membranous sac called a phagosome. The phagosome is then fused with a lysosome and its contents are digested. In the human body, phagocytic cells are mainly found among the white blood cells and macrophages that act as scavengers and help protect the body from disease-causing microorganisms and cancer cells.

You will examine various factors influencing the function of solute pumps in the following experiment.

Activity 5: Simulating Active Transport

Click the **Experiment** menu and then choose **Active Transport.** The opening screen will appear in a few seconds (Figure 5B.5). This experiment screen resembles the osmosis experiment screen, except that an ATP dispenser is substituted for the pressure meters atop the beakers. The (+) and (−) buttons control NaCl, KCl, and glucose concentrations in the dispensed solutions. You will use the Membrane Builder to build membranes containing glucose (facilitated diffusion) carrier proteins and active transport Na^+-K^+ pumps.

In this experiment, we will assume that the left beaker represents the cell's interior and the right beaker represents the extracellular space. The Membrane Builder will insert the Na^+-K^+ pumps into the membrane so Na^+ will be pumped toward the right (out of the cell) while K^+ is simultaneously moved to the left (into the cell).

1. In the Membrane Builder, adjust the number of glucose carriers and the number of Na^+-K^+ pumps to 500.
2. Click **Build Membrane,** and then drag the membrane to its position in the membrane holder between the beakers.
3. Adjust the NaCl concentration to be delivered to the left beaker to 9.00 m*M*, then click the **Dispense** button.
4. Adjust the KCl concentration to be delivered to the right beaker to 6.00 m*M*, then click **Dispense.**
5. Adjust the ATP dispenser to 1.00 m*M*, then click **Dispense ATP.** This action delivers the chosen ATP concentration to both sides of the membrane.
6. Adjust the timer to 60 min, and then click **Start.** Click **Record Data** after each run.
7. Click **Tools** → **Print Data** to print your data.

Watch the solute concentration windows at the side of each beaker for any changes in Na^+ and K^+ concentrations. The Na^+ transport rate slows and then stops before transport has completed. Why do you think that this happens?

____

What would happen if you did not dispense any ATP?

____

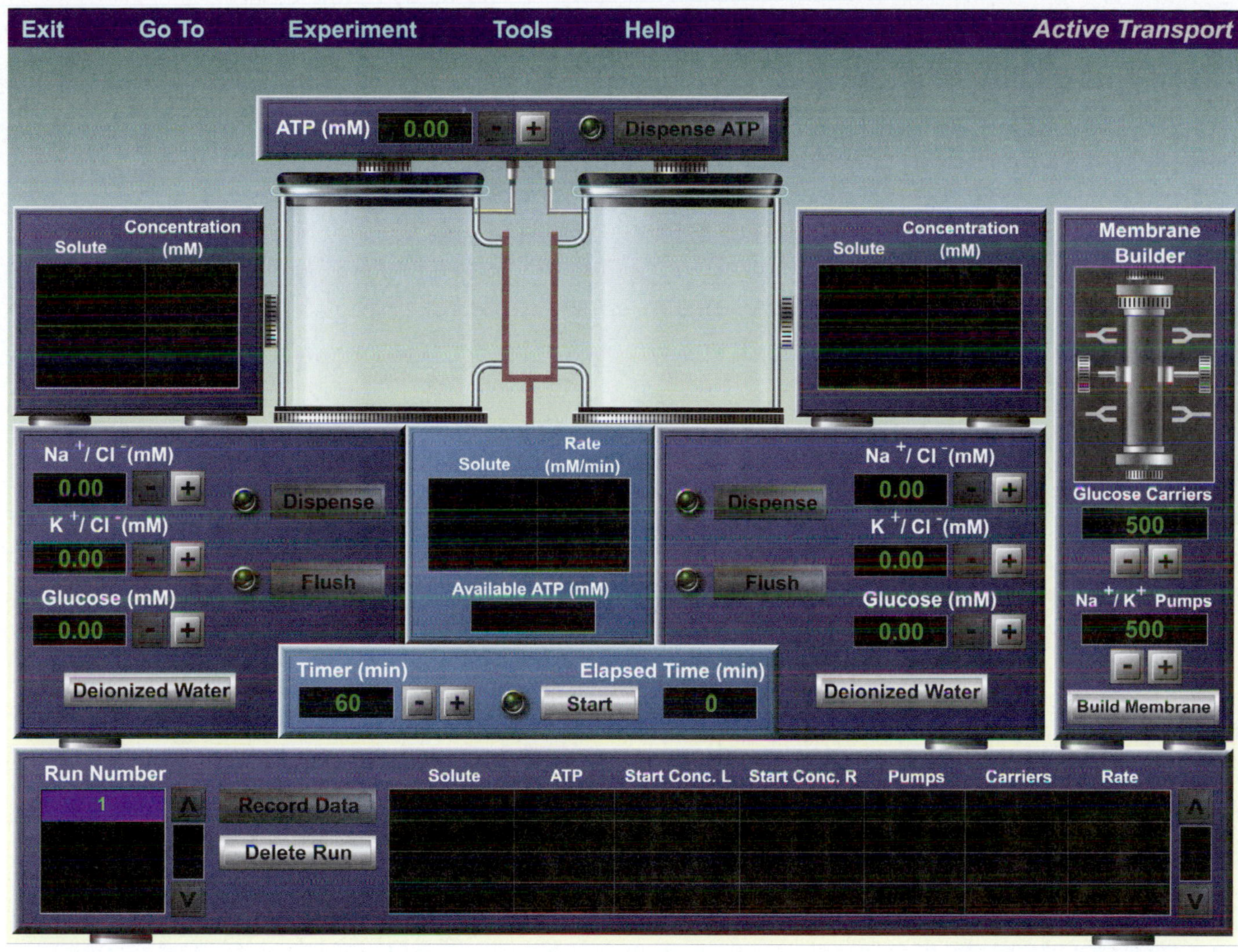

Figure 5B.5 Opening screen of the Active Transport experiment.

7. Click either Flush button to clean both beakers. Repeat steps 3 through 6, adjusting the ATP concentration to 3.00 m*M* in step 5.

Has the amount of Na^+ transported changed?

____

Do these results support your ideas in step 6 above? _______

What would happen if you decreased the number of Na^+-K^+ pumps?

____

Explain how you could show that this phenomenon is not just simple diffusion. (Hint: adjust the Na^+ concentration in the right beaker.)

____

____

8. Now repeat steps 1 through 6 dispensing 9.00 m*M* NaCl into the left beaker and 10.00 m*M* NaCl into the right beaker (instead of 6.00 m*M* KCl). Is Na^+ transport affected by this change? Explain your answer.

____

____

What would happen to the rate of ion transport if we increased the number of Na^+-K^+ pump proteins?

____

Would Na^+ and K^+ transport change if we added glucose solution?

____

9. Click **Tools** → **Print Data** to print your recorded data. ■

Try adjusting various membrane and solute conditions and attempt to predict the outcome of experimental trials. For example, you could dispense 10 m*M* glucose into the right beaker instead of deionized water.

Histology Tutorial

exercise

6B

Examining a specimen using a microscope accomplishes two goals: first, it gives you an understanding of the cellular organization of tissues, and second, perhaps even more importantly, it hones your observational skills. Because developing these skills is crucial in the understanding and eventual mastery of the way of thinking in science, using this histology module is not intended as a substitute for using the microscope. Instead, use the histology module to gain an overall appreciation of the specimens and then make your own observations using your microscope. The histology module is also an excellent review tool.

For a review of histology slides specific to topics covered in the PhysioEx lab simulations, turn to the Histology Review Supplement on p. P-141.

Figure 6B.1 Opening screen of the Histology Tutorial module.

Exploring Digital Histology

Choose **Histology Tutorial** from the main menu. The opening screen for the Histology module will appear in a few seconds (Figure 6B.1). The main features on the screen when the program starts are a pair of empty boxes at the left side of the screen, a large image-viewing window on the right side, and a set of magnification buttons below the image window.

Click **Select an Image.** You will see an A-to-Z index appear. Roll your mouse over each letter to see the list of images available. Highlight the image you want to view, and click on it. The image will appear in the image-viewing window.

The text description of the specimen is displayed in the large text box at the upper left and the image title underneath the image-viewing window. Occasionally the words in the text are highlighted in color; clicking those words displays a second image that you can compare with the main image. Click the **Close** button to exit the second image before selecting a new image.

Click and hold the mouse button (left button on a PC) on an image and then drag the mouse to move other areas of the specimen into the image-viewing window, much like moving a slide on a microscope stage.

Choose the magnification of the image in the viewing window by clicking one of the magnification buttons below the image window. Please note that a red magnification button indicates that the slide is not available for viewing at that magnification.

To see parts of the slide labeled, click **Labels On**. Click the button again to remove the labels.

Skeletal Muscle Physiology: Computer Simulation

exercise

16B

Objectives

1. To define these terms used in describing muscle physiology: *multiple motor unit summation, maximal stimulus, treppe, wave summation, tetanus.*
2. To identify two ways that the mode of stimulation can affect muscle force production.
3. To plot a graph relating stimulus strength and twitch force to illustrate graded muscle response.
4. To explain how slow, smooth, sustained contraction is possible in a skeletal muscle.
5. To graphically understand the relationships between passive, active, and total forces.
6. To identify the conditions under which muscle contraction is isometric or isotonic.
7. To describe in terms of length and force the transitions between isometric and isotonic conditions during a single muscle twitch.
8. To describe the effects of resistance and starting length on the initial velocity of shortening.
9. To explain why muscle force remains constant during isotonic shortening.
10. To explain experimental results in terms of muscle structure.

Skeletal muscles are composed of hundreds to thousands of individual cells, each doing their share of work in the production of force. As their name suggests, skeletal muscles move the skeleton. Skeletal muscles are remarkable machines; while allowing us the manual dexterity to create magnificent works of art, they are also capable of generating the brute force needed to lift a 100-lb sack of concrete. When a skeletal muscle from an experimental animal is electrically stimulated, it behaves in the same way as a stimulated muscle in the intact body, that is, *in vivo*. Hence, such an experiment gives us valuable insight into muscle behavior.

This set of computer simulations demonstrates many important physiological concepts of skeletal muscle contraction. The program graphically provides all the equipment and materials necessary for you, the investigator, to set up experimental conditions and observe the results. In student-conducted laboratory investigations there are many ways to approach a problem, and the same is true of these simulations. The instructions will guide you in your investigation, but you should also try out alternate approaches to gain insight into the logical methods used in scientific experimentation.

Try this approach: As you work through the simulations for the first time, follow the instructions closely and answer the questions posed as you go. Then try asking "What if . . . ?" questions to test the validity of your theories. The major advantages of these computer simulations are that the muscle cannot be accidentally damaged, lab equipment will not break down at the worst possible time, and you will have ample time to think critically about the processes being investigated.

Because you will be working with a simulated muscle and oscilloscope display, you need to watch both carefully during the experiments. Think about what is happening in each situation. You need to understand how you are experimentally manipulating the muscle in order to understand your results.

Electrical Stimulation

A contracting skeletal muscle will produce force and/or shortening when nervous or electrical stimulation is applied. The graded contractile response of a whole muscle reflects the number of motor units firing at a given time. Strong muscle contraction implies that many motor units are activated and each unit has maximally contracted. Weak contraction means that few motor units are active; however, the activated units are maximally contracted. By increasing the number of motor units firing, we can produce a steady increase in muscle force, a process called recruitment or motor unit summation.

Regardless of the number of motor units activated, a single contraction of skeletal muscle is called a muscle twitch. A tracing of a muscle twitch is divided into three phases: latent, contraction, and relaxation. The latent phase is a short period between the time of stimulation and the beginning of contraction. Although no force is generated during this interval, chemical changes occur intracellularly in preparation for contraction. During contraction, the myofilaments are sliding past each other and the muscle shortens. Relaxation takes place when contraction has ended and the muscle returns to its normal resting state and length.

The first activity you will conduct simulates an isometric, or fixed length, contraction of an isolated skeletal muscle. This activity allows you to investigate how the strength and frequency of an electrical stimulus affect whole muscle function. Note that these simulations involve indirect stimulation by an electrode placed on the surface of the muscle. This differs from the situation *in vivo* where each fiber in the muscle receives direct stimulation via a nerve ending. In other words, increasing the intensity of the electrical stimulation mimics how the nervous system increases the number of motor units activated.

Single Stimulus

Choose **Skeletal Muscle Physiology** from the main menu. The opening screen will appear in a few seconds (Figure 16B.1). The oscilloscope display, the grid at the top of the screen, is the most important part of the screen because it

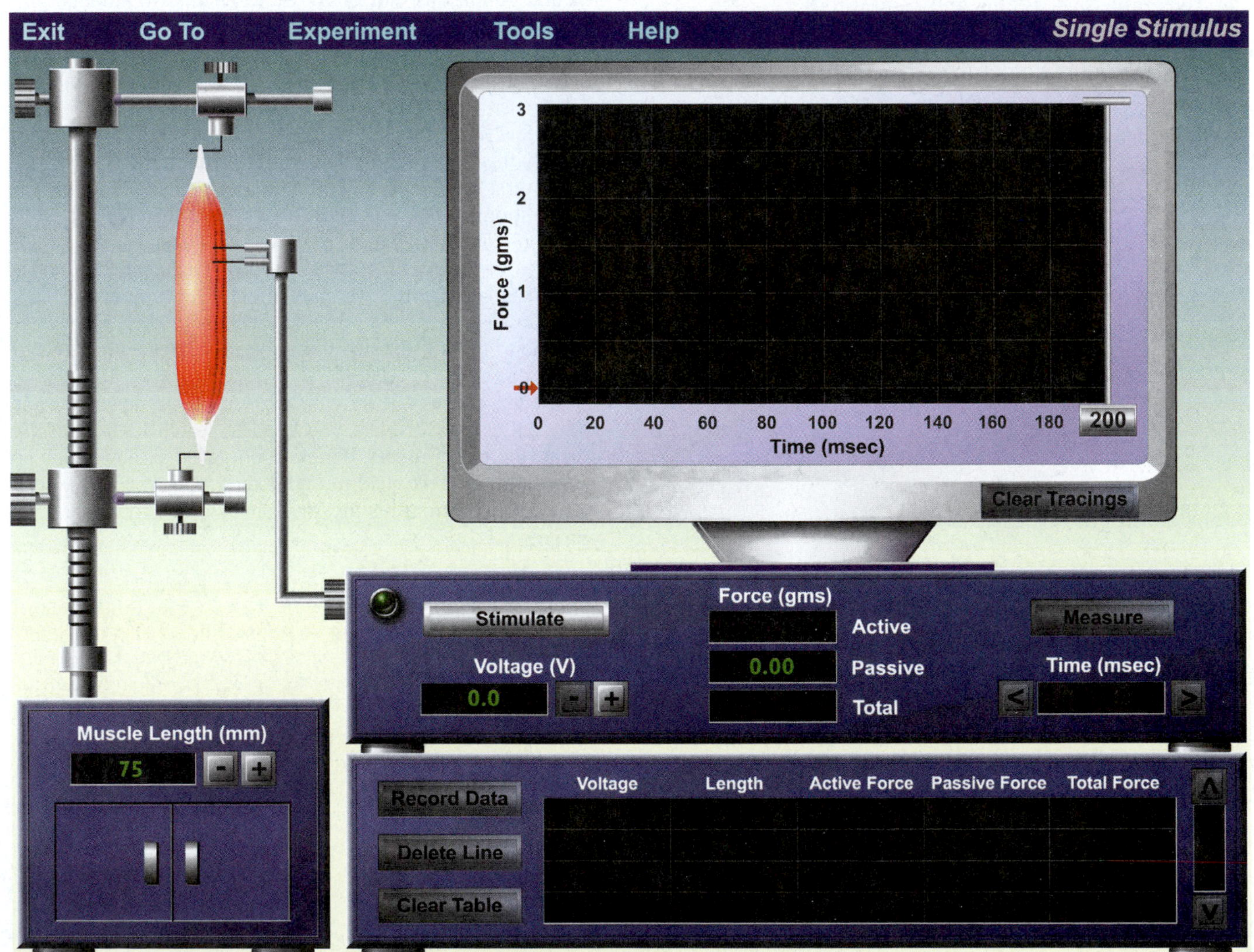

Figure 16B.1 Opening screen of the Single Stimulus experiment.

graphically displays the contraction data for analysis. Time is displayed on the horizontal axis. A full sweep is initially set at 200 msec. However, you can adjust the sweep time from 200 msec to 1000 msec by clicking and dragging the **200** msec button at the lower right corner of the oscilloscope display to the left to a new position on the time axis. The force (in grams) produced by muscle contraction is displayed on the vertical axis. Clicking the **Clear Tracings** button erases all muscle twitch tracings from the oscilloscope display.

The *electrical stimulator* is the equipment seen just beneath the oscilloscope display. Clicking **Stimulate** delivers the electrical shock to the muscle through the electrodes lying on the surface of the muscle. Stimulus voltage is set by clicking the (+) or (−) buttons next to the voltage window. Three small windows to the right of the Stimulate button display the force measurements. *Active force* is produced during muscle contraction, while *passive force* results from the muscle being stretched (much like a rubber band). The *total force* is the sum of active and passive forces. The red arrow at the left of the oscilloscope display is an indicator of passive force. After the muscle is stimulated, the **Measure** button at the right edge of the electrical stimulator becomes active. When the Measure button is clicked, a vertical orange line will be displayed at the left edge of the oscilloscope display. Clicking the arrow buttons below the Measure button moves the orange line horizontally across the screen. The Time window displays the difference in time between the zero point on the X-axis and the intersection between the orange measure line and the muscle twitch tracing.

The muscle is suspended in the support stand to the left of the oscilloscope display. The hook through the upper tendon of the muscle is part of the force transducer, which measures the force produced by the muscle. The hook through the lower tendon secures the muscle in place. The weight cabinet just below the muscle support stand is not active in this experiment; it contains weights you will use in the isotonic contraction part of the simulation. You can adjust the starting length of the muscle by clicking the (+) or (−) buttons located next to the Muscle Length display window.

When you click the **Record Data** button in the data collection unit below the electrical stimulator, your data is recorded in the computer's memory and is displayed in the data grid at the bottom of the screen. Data displayed in the data grid include the voltage, muscle length, and active, passive, and total force measurements. If you are not satisfied with a single run, you can click **Delete Line** to erase a single line of data. Clicking the **Clear Table** button will remove all accumulated data in the experiment and allow you to start over.

Activity 1:
Practicing Generating a Tracing

1. Click the **Stimulate** button once. Because the beginning voltage is set to zero, no muscle activity should result. You will see a blue line moving across the bottom of the oscilloscope display. This blue line will indicate muscle force in the experiments. If the tracings move too slowly across the screen, click and hold the **200** button at the lower right corner of the oscilloscope and drag it to the left to the 40 msec mark and release it. This action resets the total sweep time to 1000 msec to speed up the display time.

2. Click and hold the (+) button beneath the Stimulate button until the voltage window reads 3.0 volts. Click **Stimulate** once. You will see the muscle react, and a contraction tracing will appear on the screen. Notice that the muscle tracing color alternates between blue and yellow each time the Stimulate button is clicked to enhance the visual difference between twitch tracings. You can click the **Clear Tracings** button as needed to clean up the oscilloscope display. To retain your data, click the **Record Data** button at the end of each stimulus.

3. Change the voltage to 5.0 volts and click **Stimulate** again. Notice how the force of contraction also changes. Identify the latent, contraction, and relaxation phases in the tracings.

4. You may print your data by clicking **Tools → Print Data.** You may also print out hard copies of the graphs you generate by clicking on **Tools → Print Graph.**

Feel free to experiment with anything that comes to mind to get a sense of how whole muscle responds to an electrical stimulus. ■

Activity 2:
Determining the Latent Period

1. Click **Clear Tracings** to erase the oscilloscope display. The voltage should be set to 5.0 volts.

2. Drag the **200** msec button to the right edge of the oscilloscope.

3. Click the **Stimulate** button once and allow the tracing to complete.

4. When you measure the length of the latent period from a printed graph, you measure the time between the application of the stimulus and the beginning of the first observable response (increase in force). The computer can't "look ahead," anticipating the change in active force. To measure the length of the latent period using the computer, click the **Measure** button. Then click the right arrow button next to the **Time** window repeatedly until you notice the first increase in the Active Force window. This takes you beyond the actual length of the latent period. Now click the left arrow button next to the **Time** window until the Active Force window again reads zero. At this point the computer is measuring the time between the application of the stimulus and the last point where the active force is zero (just prior to contraction).

How long is the latent period? ______________ msec

What occurs in the muscle during this apparent lack of activity?

____

____ ■

The Graded Muscle Response to Increased Stimulus Intensity

As the stimulus to a muscle is increased, the amount of force produced by the muscle also increases. As more voltage is delivered to the whole muscle, more muscle fibers are activated and the total force produced by the muscle is increased. Maximal contraction occurs when all the muscle cells have been activated. Any stimulation beyond this voltage will not increase the force of contraction. This experiment mimics muscle activity *in vivo* where the recruitment of additional motor units increases the total force produced. This phenomenon is called *multiple motor unit summation*.

Activity 3:
Investigating Graded Muscle Response to Increased Stimulus Intensity

1. Click **Clear Tracings** if there are tracings on your screen.
2. Set the voltage to 0.0 and click **Stimulate.**
3. Click **Record Data.** If you decide to redo a single stimulus, choose the data line in the grid and click **Delete Line** to erase that single line of data. If you want to repeat the entire experiment, click the **Clear Table** button to erase all data recorded to that point.
4. Repeat steps 2 and 3, increasing the voltage by 0.5 each time until you reach the maximum voltage of 10.0. Be sure to select **Record Data** each time.
5. Observe the twitch tracings. Click on the **Tools** menu and then choose **Plot Data.**
6. Use the slider bars to display Active Force on the Y-axis and Voltage on the X-axis.
7. Use your graph to answer the following questions:

   What is the minimal or threshold stimulus? _______ V

   What is the maximal stimulus? _______ V

   How can you explain the increase in force that you observe?________________________________

   ____

   What type of summation is this? ________________

   ____

8. Click **Print Plot** at the top left corner of the Plot Data window to print a hard copy of the graph. When finished, click the X at the top right of the plot window.
9. Click **Tools** → **Print Data** to print your data.

Multiple Stimulus

Choose **Multiple Stimulus** from the **Experiment** menu. The opening screen will appear in a few seconds (Figure 16B.2).

The only significant change to the on-screen equipment is found in the electrical stimulator. The measuring equipment has been removed and other controls have been added: The **Multiple Stimulus** button is a toggle that allows you to alternately start and stop the electrical stimulator. When Multiple Stimulus is first clicked, its name changes to Stop Stimulus and electrical stimuli are delivered to the muscle at the rate specified in the Stimuli/sec window until the muscle completely fatigues or the stimulator is turned off. The stimulator is turned off by clicking the **Stop Stimulus** button. The stimulus rate is adjusted by clicking the (+) or (−) buttons next to the stimuli/sec window.

Activity 4:
Investigating Treppe

When a muscle first contracts, the force it is able to produce is less than the force it is able to produce in subsequent contractions within a relatively narrow time span. A myogram, a recording of a muscle twitch, reveals this phenomenon as the **treppe,** or staircase, effect. For the first few twitches, each successive stimulation produces slightly more force than the previous contraction as long as the muscle is allowed to fully relax between stimuli, and the stimuli are delivered relatively close together. Treppe is thought to be caused by increased efficiency of the enzyme systems within the cell and increased availability of intracellular calcium.

1. The voltage should be set to 8.2 volts and the muscle length should be 75 mm.
2. Drag the **200** msec button to the center of the X-axis time range.
3. Be sure that you fully understand the following three steps before you proceed.

- Click **Single Stimulus.** Watch the twitch tracing carefully.
- After the tracing shows that the muscle has completely relaxed, immediately click **Single Stimulus** again.
- When the second twitch completes, click **Single Stimulus** once more and allow the tracing to complete.

4. Click **Tools** → **Print Graph.**

What happens to force production with each subsequent stimulus?

____

________________________________ ■

Activity 5:
Investigating Wave Summation

As demonstrated in Activity 3 with single stimuli, multiple motor unit summation is one way to increase the amount of

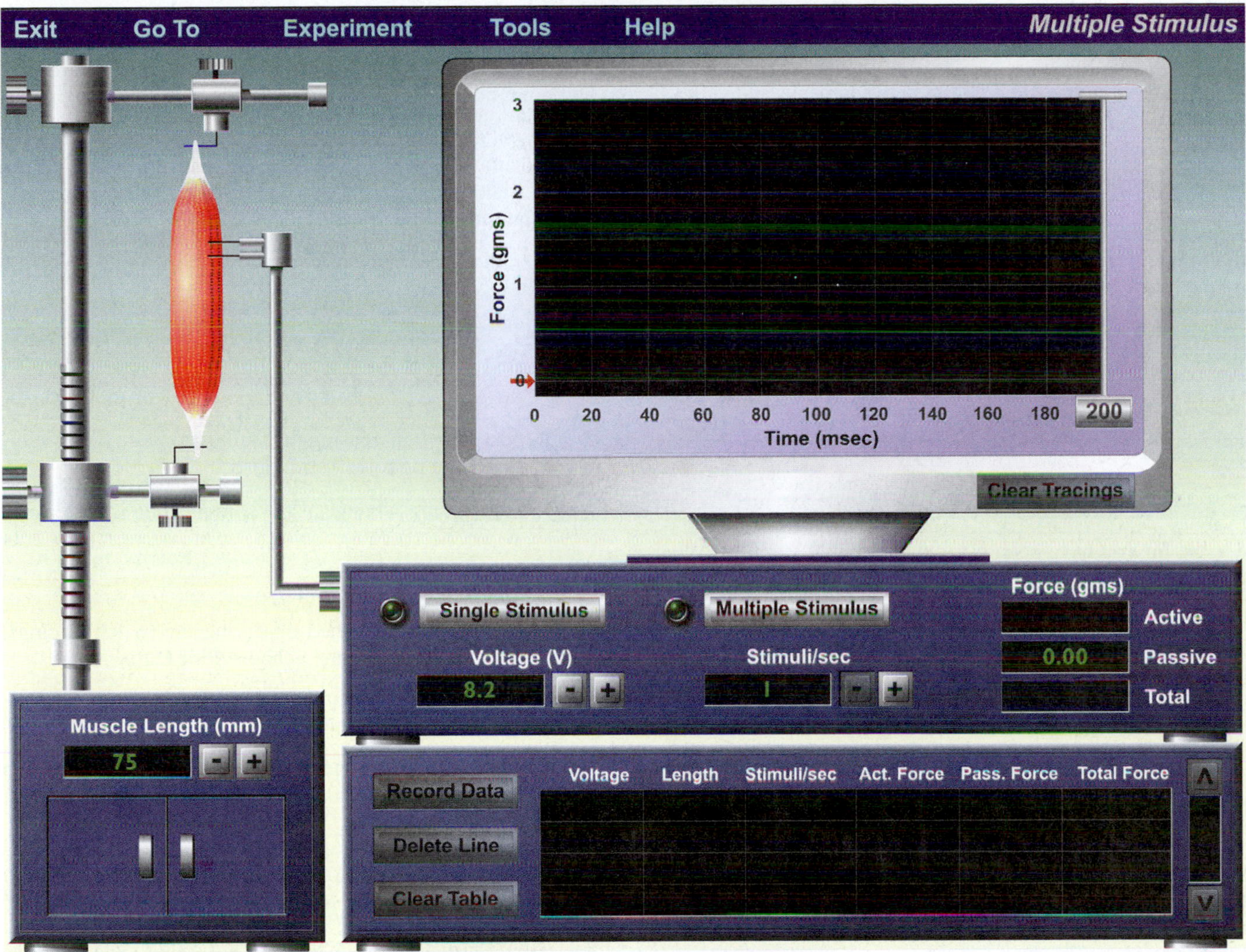

Figure 16B.2 Opening screen of the Multiple Stimulus experiment.

force produced by muscle. Multiple motor unit summation relied on increased stimulus *intensity* in that simulation. Another way to increase force is by wave, or temporal, summation. **Wave summation** is achieved by increasing the stimulus *frequency,* or rate of stimulus delivery to the muscle. Wave summation occurs because the muscle is already in a partially contracted state when subsequent stimuli are delivered.

Tetanus can be considered an extreme form of wave summation that results in a steady, sustained contraction. In effect, the muscle does not have any chance to relax because it is being stimulated at such a high frequency. This "fuses" the force peaks so that we observe a smooth tracing.

1. Click **Clear Tracings** to erase the oscilloscope display.
2. Set and keep the voltage at the maximal stimulus (8.2 volts) and the muscle length at 75 mm.
3. Drag the **200** msec button to the right edge of the oscilloscope display unless you are using a slower computer.
4. Click **Single Stimulus,** and then click **Single Stimulus** again when the muscle has relaxed about halfway.

You may click **Tools → Print Graph** any time you want to print graphs generated during this activity.

Is the peak force produced in the second contraction greater

than that produced by the first stimulus? _______

5. Try stimulating again at greater frequencies by clicking the Single Stimulus button several times in rapid succession.

Is the total force production even greater? _______

6. To see if you can produce smooth, sustained contraction at Active Force = 2 gms, try rapidly clicking **Single Stimulus** several times.

Is it possible to produce a smooth contraction (fused force peaks), or does the force rise and fall periodically?

____

7. Using the same method as in step 6, try to produce a smooth contraction at 3 gms.

Is the trace smoother (less height between peaks and valleys) this time?

____

Because there is a limit to how fast you can manually click the Single Stimulus button, what do you think would happen to the smoothness of the tracing if you could click even faster?

____

____

____

8. So far you have been using the maximal stimulus. What do you think would happen if you used a lower voltage?

____

____

____

____

9. Use the concepts of motor unit summation and stimulus frequency to explain how human skeletal muscles work to achieve smooth, steady contractions at all desired levels of force.

____

____

____

____ ■

Activity 6:
Investigating Fusion Frequency

1. Click **Clear Tracings** to erase the oscilloscope display.

2. The voltage should be set to 8.2 volts and the muscle length should be 75 mm.

3. Adjust the stimulus rate to 30 stimuli/sec.

4. The following steps constitute a single "run." Become familiar with the procedure for completing a run before continuing.

- Click **Multiple Stimulus.**
- When the tracing is close to the right side of the screen, click **Stop Stimulus** to turn off the stimulator.
- Click **Record Data** to retain your data in the grid in the bottom of the screen and the computer's memory. Click **Tools → Print Graph** to print a hard copy of your graph at any point during this activity.

If you decide to redo a single run, choose the data line in the grid and click **Delete Line** to erase that single line of data. If you want to repeat the entire experiment, click the **Clear Table** button to erase all data recorded thus far.

Describe the appearance of the tracing.

____

____

5. Repeat step 4, increasing the stimulation rate by 10 stimuli/sec each time up to 150 stimuli/sec.

How do the tracings change as the stimulus rate is increased?

____

6. When you have finished observing the twitch tracings, click the **Tools** menu and then choose **Plot Data.**

7. Set the Y-axis slider to display Active Force and the X-axis slider to display Stimuli/sec.

From your graph, estimate the stimulus rate above which there appears to be no significant increase in force.

______________ stimuli/sec

This rate is the fusion frequency, also called tetanus.

8. Click **Print Plot** at the top left of the Plot Data window. When finished, click the X at the top right of the plot window.

9. Reset the stimulus rate to the fusion frequency.

10. Try to produce a smooth contraction at Force = 2 gms and Force = 3 gms by adjusting only the stimulus intensity, or voltage, using the following procedure.

- Decrease the voltage to a starting point of 1.0 volt and click **Multiple Stimulus.**
- Click **Stop Stimulus** to turn off the stimulator when the tracing is near the right side of the oscilloscope display.
- If the force produced is not smooth and continuous at the desired level of force, increase the voltage in 0.1-volt increments and stimulate as above until you achieve a smooth force at 2 gms and again at 3 gms.

What stimulus intensity produced smooth force at Force = 2 gms?

______________ V

Which intensity produced smooth contraction at Force = 3 gms?

______________ V

Explain what must be happening in the muscle to achieve smooth contraction at different force levels.

____

____

____

____

_______________ ■

Activity 7: Investigating Muscle Fatigue

A prolonged period of sustained contraction will result in muscle fatigue, a condition in which the tissue has lost its ability to contract. Fatigue results when a muscle cell's ATP consumption is faster than its production. Consequently, increasingly fewer ATP molecules are available for the contractile parts within the muscle cell.

1. Click **Clear Tracings** to erase the oscilloscope display.
2. The voltage should be set to 8.2 volts and the muscle length should be 75 mm.
3. Adjust the stimulus rate to 120 stimuli/sec.
4. Click **Multiple Stimulus** and allow the tracing to sweep through three screens and then click **Stop Stimulus** to stop the stimulator.

Click **Tools** → **Print Graph** at any time to print graphs during this activity. Click **Tools** → **Print Data** to print your data.

Why does the force begin to fall with time? Note that a fall in force indicates muscle fatigue.

____

____

5. Click **Clear Tracings** to erase the oscilloscope display. Keep the same settings as before.
6. You will be clicking **Multiple Stimulus** on and off three times to demonstrate fatigue with recovery. Read the steps below before proceeding.

- Click **Multiple Stimulus.**
- When the tracing reaches the middle of the screen, briefly turn off the stimulator by clicking **Stop Stimulus,** then immediately click **Multiple Stimulus** again.
- You will see a dip in the force tracing where you turned the stimulator off and then on again. The force tracing will continue to drop as the muscle fatigues.
- Before the muscle fatigues completely, repeat the on/off cycle twice more without clearing the screen.

Turning the stimulator off allows a small measure of recovery. The muscle will produce force for a longer period if the stimulator is briefly turned off than if the stimulations were allowed to continue without interruption. Explain why.

____

____

____

7. To see the difference between continuous multiple stimulation and multiple stimulation with recovery, click **Multiple Stimulus** and let the tracing fall without interruption to zero force. This tracing will follow the original myogram exactly until the first "dip" is encountered, after which you will notice a difference in the amount of force produced between the two runs.

Describe the difference between the current tracing and the myogram generated in step 6.

____

____

____

____

____

____

_______________ ■

Isometric Contraction

Isometric contraction is the condition in which muscle length does not change regardless of the amount of force generated by the muscle (*iso* = same, *metric* = length). This is accomplished experimentally by keeping both ends of the muscle in a fixed position while stimulating it electrically. Resting length (length of the muscle before contraction) is an important factor in determining the amount of force that a muscle can develop. Passive force is generated by stretching the muscle, and is due to the elastic properties of the tissue itself. Active force is generated by the physiological contraction of the muscle. Think of the muscle as having two force properties: it exerts passive force when it is stretched (like a rubber band exerts passive force), and active force when it contracts. Total force is the sum of passive and active forces, and it is what we experimentally measure.

This simulation allows you to set the resting length of the experimental muscle and stimulate it with individual maximal stimulus shocks. A graph relating the forces generated to the length of the muscle will be automatically plotted as you stimulate the muscle. The results of this simulation can then be applied to human muscles in order to understand how

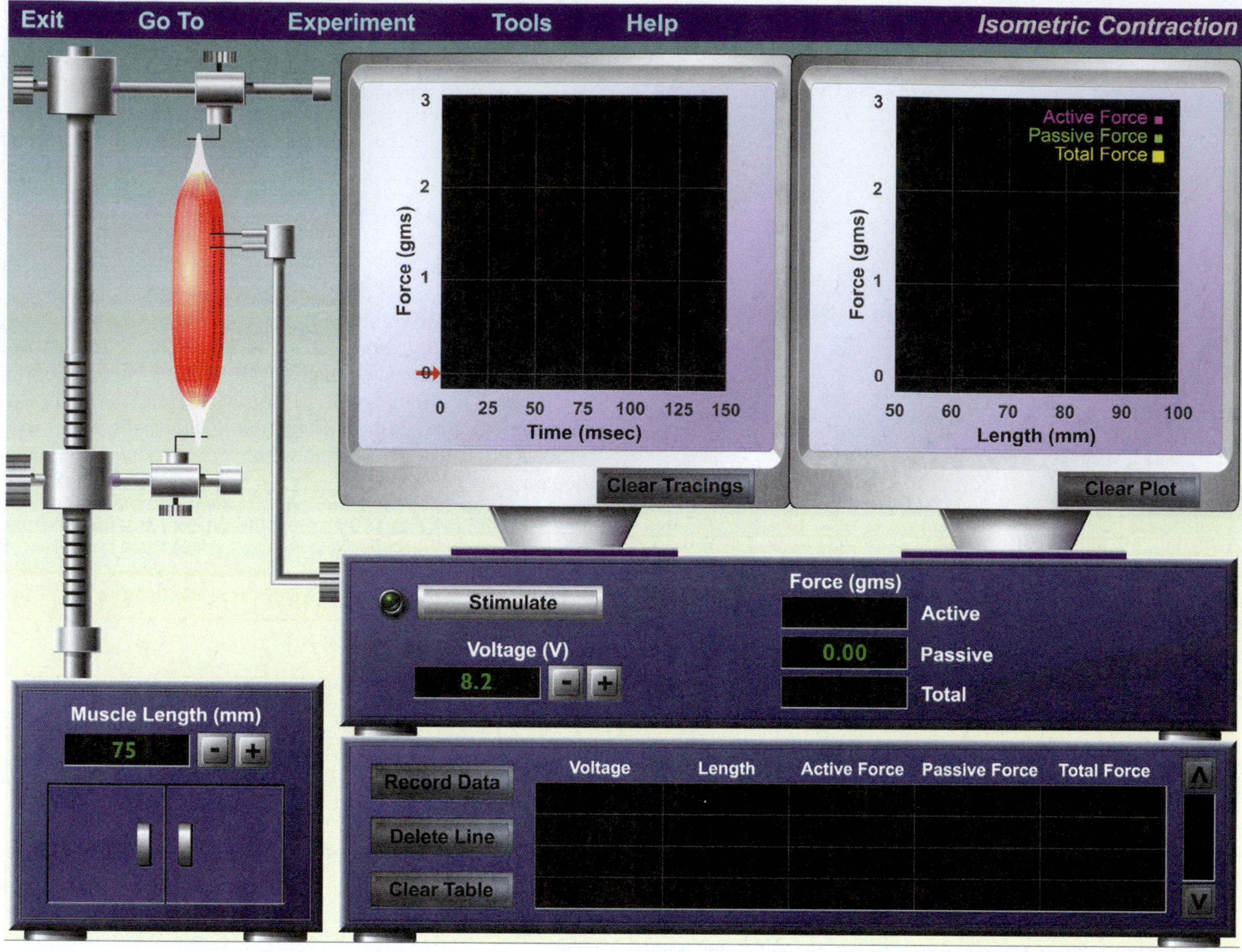

Figure 16B.3 Opening screen of the Isometric Contraction experiment.

optimum resting length will result in maximum force production. In order to understand why muscle tissue behaves as it does, it is necessary to comprehend contraction at the cellular level. Hint: If you have difficulty understanding the results of this exercise, review the sliding filament model of muscle contraction. Then think in terms of sarcomeres that are too short, too long, or just the right length.

Choose **Isometric Contraction** from the **Experiment** menu. The opening screen will appear in a few seconds (Figure 16B.3). Notice that the oscilloscope is now divided into two parts. The left side of the scope displays the muscle twitch tracing. The active, passive, and total force data points are plotted on the right side of the screen.

Activity 8:
Investigating Isometric Contraction

1. The voltage should be set to the maximal stimulus (8.2 volts) and the muscle length should be 75 mm.

2. To see how the equipment works, stimulate once by clicking **Stimulate.** You should see a single muscle twitch tracing on the left oscilloscope display, and three data points representing active, passive, and total force on the right display. The yellow box represents the total force and the red dot it contains symbolizes the superimposed active force. The green square represents the passive force data point.

3. Try adjusting the muscle length by clicking the (+) or (−) buttons located next to the Muscle Length window and watch the effect on the muscle.

4. When you feel comfortable with the equipment, click **Clear Tracings** and **Clear Plot.**

5. Now stimulate at different muscle lengths using the following procedure.

- Shorten the muscle to a length of 50 mm by clicking the (−) button next to the Muscle Length window.
- Click **Stimulate** and when the tracing is complete, click **Record Data.**
- Repeat the **Stimulate** and **Record Data** sequence, increasing the muscle length by 2 mm each time until you reach the maximum muscle length of 100 mm.

6. Carefully examine the active, passive, and total force plots in the right oscilloscope display.
7. Click **Tools** → **Print Data** to print your data.

What happens to the passive and active forces as the muscle length is increased from 50 mm to 100 mm?

Passive force:

____

____

Active force:

____

____

Total force:

____

____

Explain the dip in the total force curve. (Hint: keep in mind you are measuring the sum of active and passive forces.)

____

____

____

____

______________________________________ ■

Isotonic Contraction

During isotonic contraction, muscle length changes, but the force produced stays the same (*iso* = same, *tonic* = force). Unlike the isometric exercise in which both ends of the muscle are held in a fixed position, one end of the muscle remains free in the isotonic contraction exercise. Different weights can then be attached to the free end while the other end is fixed in position on the force transducer. If the weight is not too great, the muscle will be able to lift it with a certain velocity. You can think of lifting an object from the floor as an example: if the object is light it can be lifted quickly (high velocity), whereas a heavier weight will be lifted with a slower velocity. Try to transfer the idea of what is happening in the simulation to the muscles of your arm when you lift a weight. The two important variables in this exercise are starting length of the muscle and resistance (weight) applied. You have already examined the effect of starting length on muscle force production in the previous exercise. Now you will change both muscle length and resistance to investigate how such changes affect the speed of skeletal muscle shortening. Both variables can be independently altered and the results are graphically presented on the screen.

Choose **Isotonic Contraction** from the **Experiment** menu. The opening screen will appear in a few seconds (Figure 16B.4). The general operation of the equipment is the same as in the previous experiments. In this simulation, the weight cabinet doors are open. You will attach weights to the lower tendon of the muscle by clicking and holding the mouse on any weight in the cabinet and then dragging-and-dropping the weight's hook onto the lower tendon. The Muscle Length window displays the length achieved when the muscle is stretched by hanging a weight from its lower tendon. You can click the (+) and (−) buttons next to the Platform Height window to change the position of the platform on which the weight rests. Click on the weight again to automatically return it to the weight cabinet. The electrical stimulator displays the initial velocity of muscle shortening in the velocity window to the right of the Voltage control.

Activity 9:
Investigating the Effect of Load on Skeletal Muscle

1. Set the voltage to the maximal stimulus (8.2 volts).
2. Drag-and-drop the .5-g weight onto the muscle's lower tendon.
3. Platform height should be 75 mm.
4. Click **Stimulate** and simultaneously watch the muscle action and the oscilloscope tracing.
5. Click the **Record Data** button to retain and display the data in the grid.

What do you see happening to the muscle during the flat part of the tracing? Click **Stimulate** to repeat if you wish to see the muscle action again.

____

____

Does the force the muscle produces change during the flat part of the tracing (increase, decrease, or stay the same)?

____

Describe the muscle activity during the flat part of the tracing in terms of isotonic contraction and relaxation.

____

____

____

____

____

Circle the correct terms in parentheses in the following sequence:

The force rises during the first part of the muscle tracing due to (isometric, isotonic) contraction. The fall in force on the right side of the tracing corresponds to (isometric, isotonic) relaxation.

6. Return the .5-g weight to the cabinet. Drag the 1.5-g weight to the muscle. Click **Stimulate** and then click **Record Data.**

Which of the two weights used so far results in the highest initial velocity of shortening?

___

7. Repeat step 6 for the remaining two weights.
8. Choose **Plot Data** from the **Tools** menu.
9. Set Weight as the X-axis and Total Force as the Y-axis by dragging the slider bars. Click **Print Plot** if you wish to print the graph.

What does the plot reveal about the resistance and the initial velocity of shortening?

___

___

___

10. Close the plot window. If you wish, click **Tools** → **Print Data** to print your data.
11. Click **Clear Table** in the data control unit at the bottom of the screen. Click **Yes** when you are asked if you want to erase all data in the table.
12. Return the current weight to the weight cabinet.

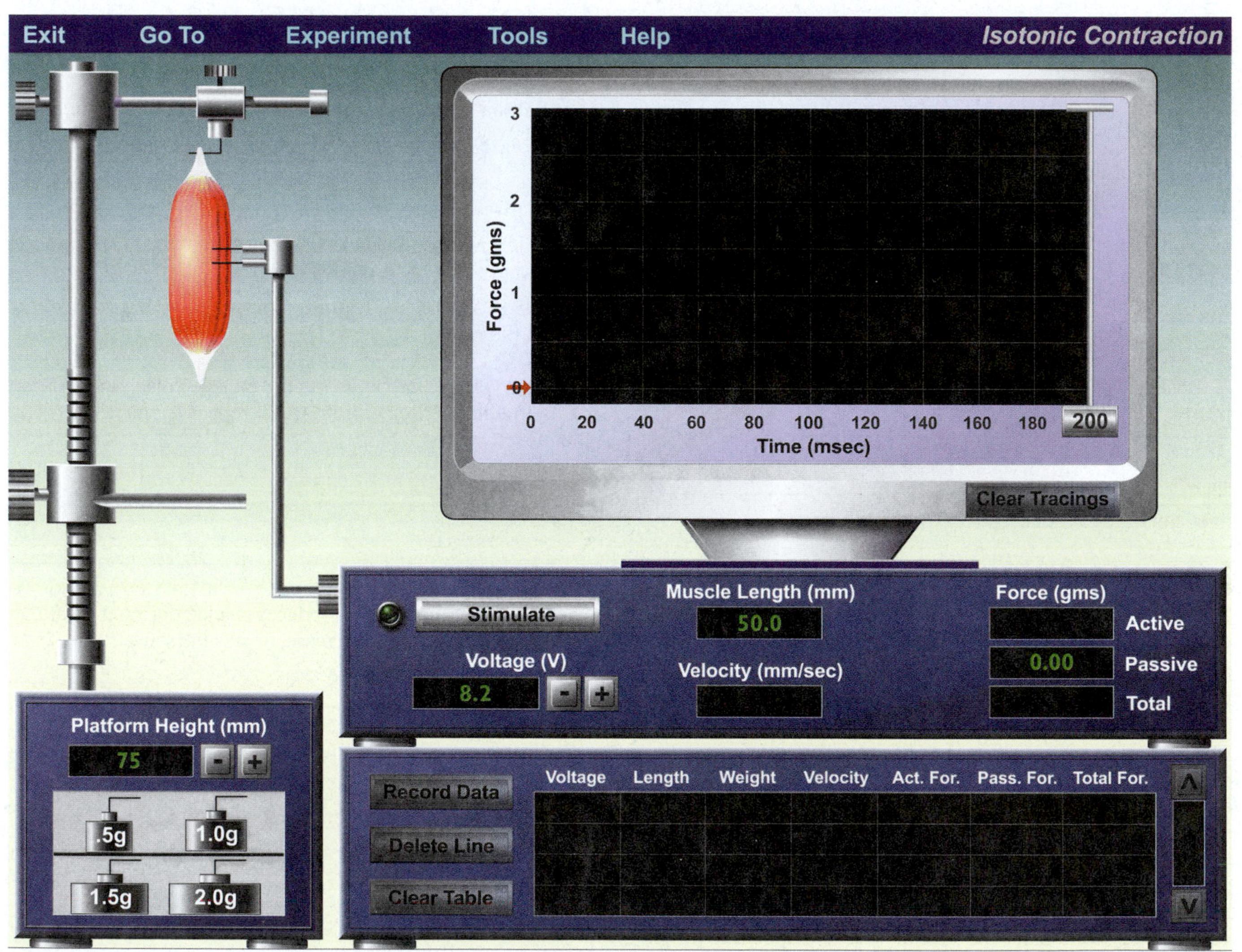

Figure 16B.4 Opening screen of the Isotonic Contraction experiment.

13. Attach the 1.5-g weight to the muscle and run through the range of starting lengths from 60–90 mm in 5-mm increments. Be sure to click **Record Data** after each stimulus.

14. After all runs have been completed, choose **Plot Data** from the **Tools** menu.

15. Set Length as the X-axis and Velocity as the Y-axis by dragging the slider bars. Click **Print Plot** if you wish to print the graph.

Describe the relationship between starting length and initial velocity of shortening.

____

____

____

____

Do these results support your ideas in the isometric contraction part of this exercise?

____

16. Close the plot window. If you wish, click **Tools → Print Data.**

Can you set up a contraction that is completely isometric?

_______ One that is completely isotonic? _______

Explain your answers.

____

____

____

____ ■

Histology Review Supplement

Turn to p. P-143 for a review of skeletal muscle tissue.

exercise

18B Neurophysiology of Nerve Impulses: Computer Simulation

Objectives

1. To define the following:
   irritability, conductivity, resting membrane potential, polarized, sodium-potassium pump, threshold stimulus, depolarization, action potential, repolarization, hyperpolarization, absolute refractory period, relative refractory period, nerve impulse, compound nerve action potential, and *conduction velocity*
2. To list at least four different stimuli capable of generating an action potential.
3. To list at least two agents capable of inhibiting an action potential.
4. To describe the relationship between nerve size and conduction velocity.
5. To describe the relationship between nerve myelination and conduction velocity.

The Nerve Impulse

Neurons have two major physiological properties: **irritability,** or the ability to respond to stimuli and convert them into nerve impulses, and **conductivity,** the ability to transmit an impulse (in this case, to take the neural impulse and pass it along the cell membrane). In the resting neuron (i.e., a neuron that does not have any neural impulses), the exterior of the cell membrane is positively charged and the interior of the neuron is negatively charged. This difference in electrical charge across the plasma membrane is referred to as the **resting membrane potential** and the membrane is said to be **polarized.** The **sodium-potassium pump** in the membrane maintains the difference in electrical charge established by diffusion of ions. This active transport mechanism moves 3 sodium ions out of the cell while moving in 2 potassium ions. Therefore, the major cation (positively charged ion) outside the cell in the extracellular fluid is sodium, while the major cation inside the cell is potassium. The inner surface of the cell membrane is more negative than the outer surface, mainly due to intracellular proteins, which, at body pH, tend to be negatively charged.

The resting membrane potential can be measured with a voltmeter by putting a recording electrode just inside the cell membrane with a reference, or ground, electrode outside the membrane (see Figure 18B.1). In the giant squid axon (where most early neural research was conducted), or in the frog axon that will be used in this exercise, the resting membrane potential is measured at −70 millivolts (mV). (In humans, the resting membrane potential typically measures between −40 mV and −90mV.)

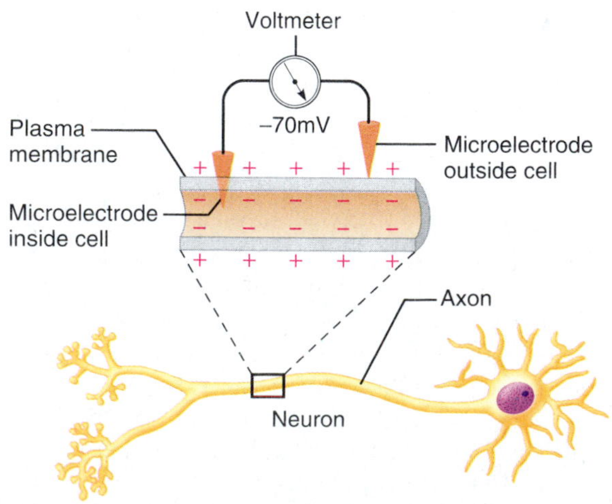

Figure 18B.1

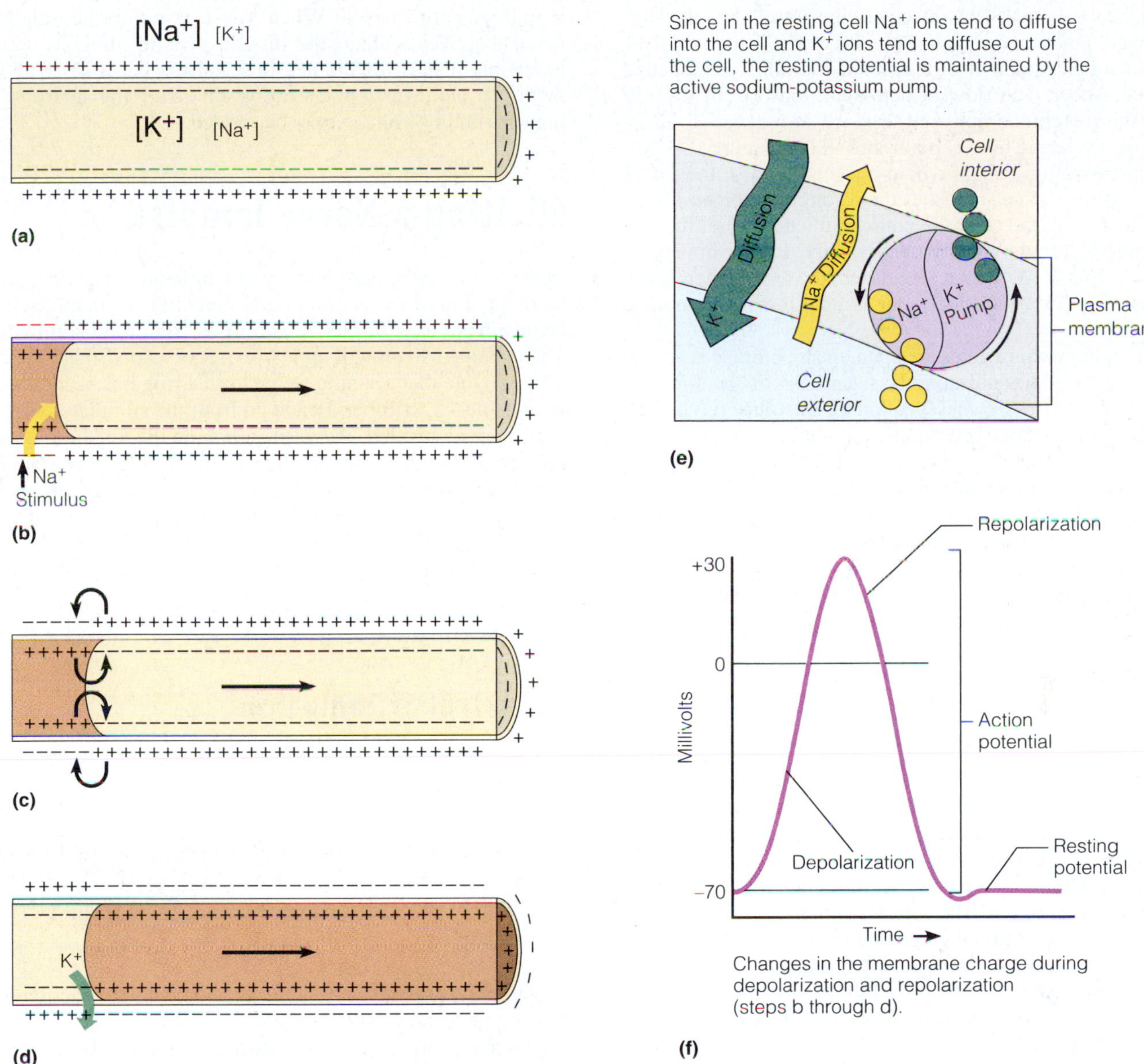

Figure 18B.2 The nerve impulse. (a) Resting membrane potential (~85 mV). There is an excess of positive ions outside the cell, with Na^+ the predominant extracellular fluid ion and K^+ the predominant intracellular ion. The plasma membrane has a low permeability to Na^+. **(b)** Depolarization—reversal of the resting potential. Application of a stimulus changes the membrane permeability, and Na^+ ions are allowed to diffuse rapidly into the cell. **(c)** Generation of the action potential or nerve impulse. If the stimulus is of adequate intensity, the depolarization wave spreads rapidly along the entire length of the membrane. **(d)** Repolarization—reestablishment of the resting potential. The negative charge on the internal plasma membrane surface and the positive charge on its external surface are reestablished by diffusion of K^+ ions out of the cell, proceeding in the same direction as in depolarization. **(e)** The original ionic concentrations of the resting state are restored by the sodium-potassium pump. **(f)** A tracing of an action potential.

When a neuron is activated by a stimulus of adequate intensity, known as a **threshold stimulus,** the membrane at its *trigger zone,* typically the axon hillock, briefly becomes more permeable to sodium ions (sodium gates in the cell membrane open). Sodium ions rush into the cell, increasing the number of positive ions inside the cell and changing the membrane polarity. The interior surface of the membrane becomes less negative and the exterior surface becomes less positive, a phenomenon called **depolarization** (see Figure 18B.2a). When depolarization reaches a certain point called **threshold,** an **action potential** is initiated (see Figure 18B.2c) and the polarity of the membrane reverses.

When the membrane depolarizes, the resting membrane potential of −70 mV becomes less negative. When the membrane potential reaches 0 mV, indicating there is no charge difference across the membrane, the sodium ion channels start to close and potassium ion channels open. By the time the sodium ions channels finally close, the membrane potential has reached +35 mV. The opening of the potassium ion channels allows potassium ions to flow out of the cell down their electrochemical gradient—remember, like ions are repelled from each other. The flow of potassium ions out of the cell causes the membrane potential to move in a negative direction. This is referred to as **repolarization** (see Figure 18B.2d). This repolarization occurs within a millisecond of the initial sodium influx and reestablishes the resting membrane potential. Actually, by the time the potassium gates close, the cell membrane has undergone a **hyperpolarization,** slipping to perhaps −75 mV. With the gates closed, the resting membrane potential is quickly returned to the normal resting membrane potential.

When the sodium gates are open, the membrane is totally insensitive to additional stimuli, regardless of the force of stimulus. The cell is in what is called the **absolute refractory period.** During repolarization, the membrane may be stimulated if a very strong stimulus is used. This period is called the **relative refractory period.**

The action potential, once started, is a self-propagating phenomenon, spreading rapidly along the neuron membrane. The action potential follows the *all-or-none* law, in which the neuron membrane either depolarizes 100% or not at all. In neurons, the action potential is also called a **nerve impulse.** When it reaches the axon terminal, it triggers the release of neurotransmitters into the synaptic cleft. Depending on the situation, the neurotransmitter will either excite or inhibit the postsynaptic neuron.

In order to study nerve physiology, we will use a frog nerve and several electronic instruments. The first instrument is the *electronic stimulator.* Nerves can be stimulated by chemicals, touch, or electric shock. The electronic stimulator administers an electric shock that is pure direct current (DC), and allows duration, frequency, and voltage of the shock to be precisely controlled. The stimulator has two output terminals; the positive terminal is red and the negative terminal is black. Voltage leaves the stimulator via the red terminal, passes through the item to be stimulated (in this case, the nerve), and returns to the stimulator at the black terminal to complete the circuit.

The second instrument is the **oscilloscope,** an instrument that measures voltage changes over a period of time. The face of the oscilloscope is similar to a black-and-white TV screen. The screen of the oscilloscope is the front of a tube with a filament at the other end. The filament is heated and gives off a beam of electrons. The beam passes to the front of the tube. Electronic circuitry allows for the electron beam to be brought across the screen in preset time intervals. When the electrons hit the phosphorescent material on the inside of the screen, a spot on the screen will glow. When we apply a stimulus to a nerve, the oscilloscope screen will display one of the following three results: no response, a flat line, or a graph with a peak. A graph with a peak indicates that an action potential has been generated.

While performing the following experiments, keep in mind that you are working with a nerve, which consists of many neurons—you are not working with just a single neuron. The action potential you will see on the oscilloscope screen reflects the cumulative action potentials of all the neurons in the nerve, called a **compound nerve action potential.** Although an action potential follows the all-or-none law within a single neuron, it does not necessarily follow this law within an entire nerve. When you electrically stimulate a nerve at a given voltage, the stimulus may result in the depolarization of most of the neurons, but not necessarily all of them. To achieve depolarization of *all* of the neurons, a higher stimulus voltage may be needed.

Eliciting a Nerve Impulse

In the following experiments, you will be investigating what kinds of stimuli trigger an action potential. To begin, select **Neurophysiology of Nerve Impulses** from the main menu. The opening screen will appear in a few seconds (see Figure 18B.3). Note that a sciatic nerve from a frog has been placed into the nerve chamber. Leads go from the stimulator output to the nerve chamber, the vertical box on the left side. Leads also go from the nerve chamber to the oscilloscope. Notice that these leads are red and black. The stimulus travels along the red lead to the nerve. When the nerve depolarizes, it will generate an electrical impulse that will travel along the red wire to the oscilloscope and back to the nerve along the black wire.

Activity 1: Electrical Stimulation

1. Set the voltage at 1.0 V by clicking the **(+)** button next to the **Voltage** display.
2. Click **Single Stimulus.**

Do you see any kind of response on the oscilloscope screen?

____

If you saw no response, or a flat line indicating no action potential, click the **Clear** button on the oscilloscope, increase the voltage, and click **Single Stimulus** again until you see a trace (deflection of the line) that indicates an action potential.

What was the *threshold voltage,* that is, the voltage at which you first saw an action potential?

________ V

Click **Record Data** on the data collection box to record your results.

3. If you wish to print your graph, click **Tools** and then **Print Graph.** You may do this each time you generate a graph on the oscilloscope screen.
4. Increase the voltage by 0.5 V and click **Single Stimulus.**

How does this tracing compare to the one trace that was generated at the threshold voltage? (*Hint:* look very carefully at the tracings.)

____

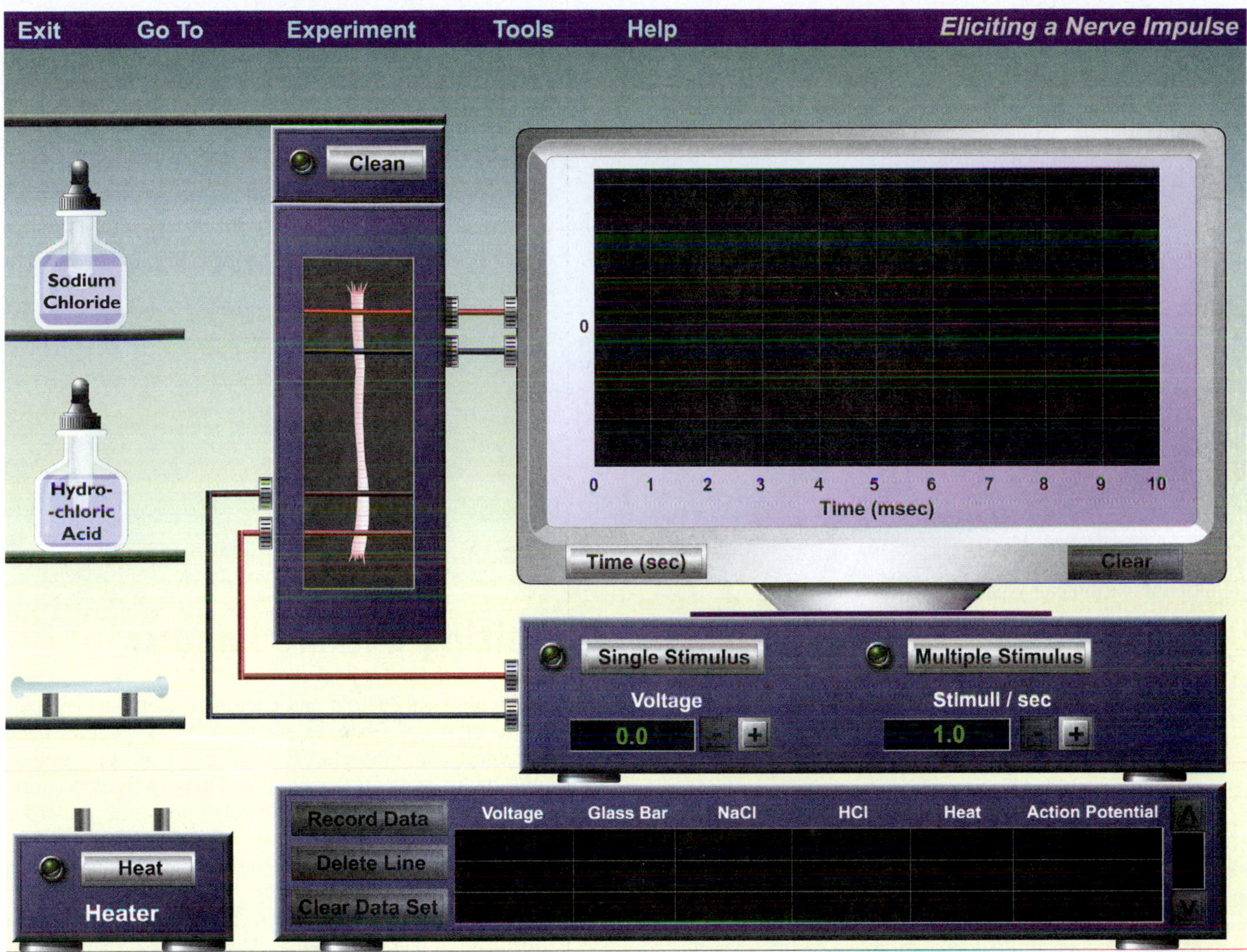

Figure 18B.3 **Opening screen of the Eliciting a Nerve Impulse experiment.**

What reason can you give for your answer?

___

___

___

Click **Record Data** on the data collection box to record your results.

5. Continue to increase the voltage by 0.5 V and to click **Single Stimulus** until you find the point beyond which no further increase occurs in the peak of the action potential trace.

Record this maximal voltage here: _______ V

Click **Record Data** to record your results. ■

Now that we have seen that an electrical impulse can cause an action potential, let's try some other methods of stimulating a nerve.

Activity 2: Mechanical Stimulation

1. Click the Clear button on the oscilloscope.

2. Using the mouse, click the glass rod located on the bottom shelf an the left side of the screen, and drag it over to the nerve. When the glass rod is over the nerve, release the mouse button to indicate that the rod is now touching the nerve. What do you see on the oscilloscope screen?

___

How does this tracing compare with the other tracings that you have generated?

___

___

Click **Record Data** to record your results. Leave the graph on the screen so that you can compare it to the graph you will generate in the next activity. ■

Activity 3:
Thermal Stimulation

1. Click on the glass rod and drag it to the heater, releasing the mouse button. Click on the **Heat** button. When the rod turns red, indicating that it has been heated, click and drag the rod over the nerve and release the mouse button. What happens?

____

____

How does this trace compare to the trace that was generated with the unheated glass rod?

____

____

What explanation can you provide for this?

____

____

Click **Record Data** to record your results. Then click **Clear** to clear the oscilloscope screen for the next activity. ■

Activity 4:
Chemical Stimulation

1. Click and drag the dropper from the bottle of sodium chloride (salt solution) over to the nerve in the chamber and then release the mouse button to dispense drops.

Does this generate an action potential? ______________

2. Using your threshold setting, stimulate the nerve.

Does this tracing differ from the original threshold stimulus tracing?

____

Click **Record Data** to record your results.

3. Click the **Clean** button on top of the nerve chamber. This will return the nerve to its original (nonsalted) state. Click **Clear** to clear the oscilloscope screen.

4. Click and drag the dropper from the bottle of hydrochloric acid over to the nerve, and release the mouse button to dispense drops.

Does this generate an action potential? ______________

5. Does this tracing differ from the one generated by the original threshold stimulus?

____

Click **Record Data** to record your results.

6. Click on the **Clean** button on the nerve chamber to clean the chamber and return the nerve to its untouched state.

7. Click **Tools** → **Print Data** to print the data you have recorded for this experiment.

To summarize your experimental results, what kinds of stimuli can elicit an action potential?

____

____

____ ■

Inhibiting a Nerve Impulse

Numerous physical factors and chemical agents can impair the ability of nerve fibers to function. For example, deep pressure and cold temperature both block nerve impulse transmission by preventing local blood supply from reaching the nerve fibers. Local anesthetics, alcohol, and numerous other chemicals are also effective in blocking nerve transmission. In this experiment, we will study the effects of various agents on nerve transmission.

To begin, click the **Experiment** pull-down menu and select **Inhibiting a Nerve Impulse.** The display screen for this activity is similar to the screen in the first activity (see Figure 18B.4). To the left are bottles of several agents that we will test on the nerve. Keep the tracings you printed out from the first activity close at hand for comparison.

Activity 5:
Testing the Effects of Ether

1. Using the mouse, click and drag the dropper from the bottle marked *ether* over to the nerve, in between the stimulating electrodes and recording electrodes. Release the mouse button to dispense drops.

2. Click **Stimulate,** using the voltage setting from the threshold stimulus you used in the earlier activities. What sort of trace do you see?

____

What has happened to the nerve? ______________

____

Click **Record Data** to record your results.

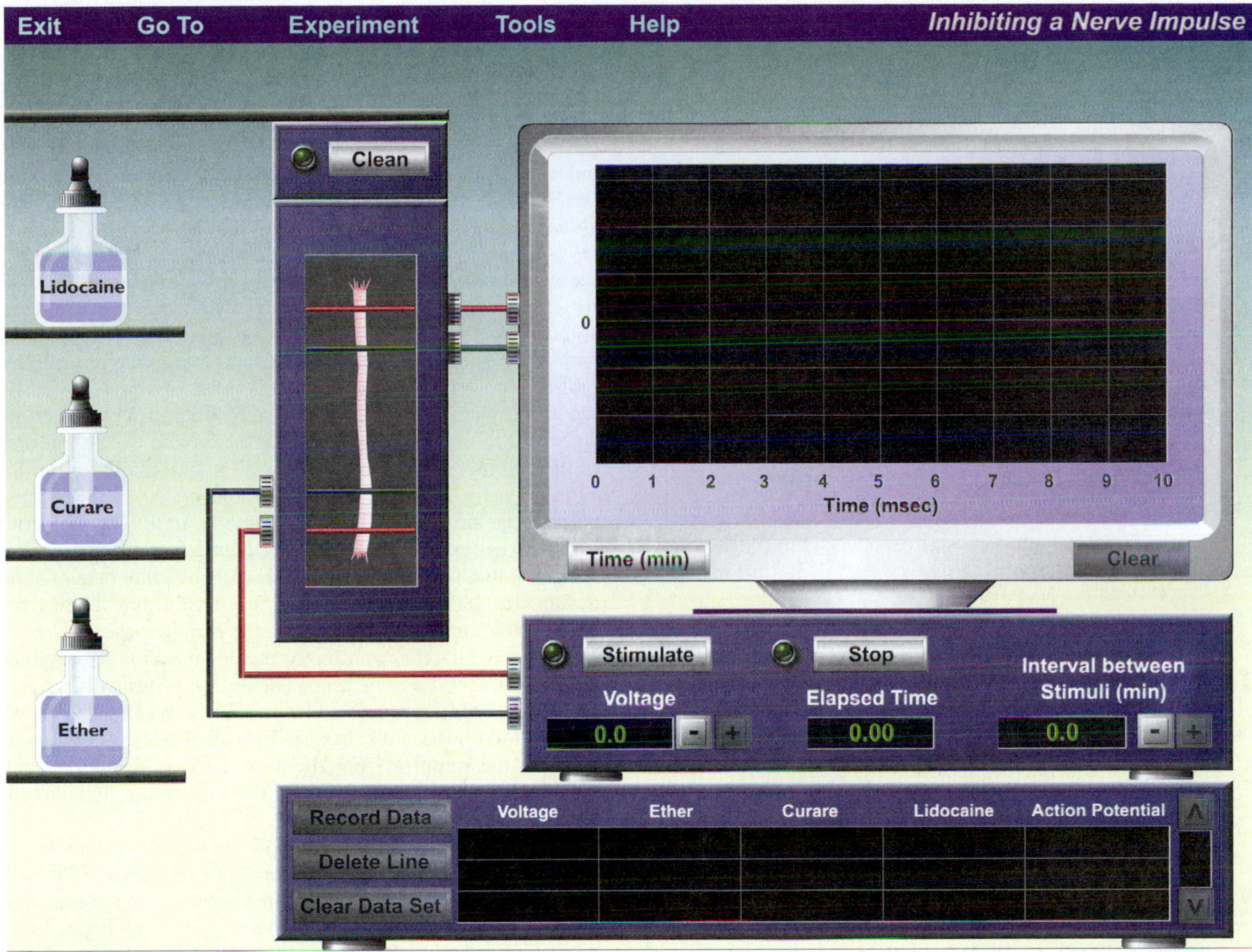

Figure 18B.4 Opening screen of the Inhibiting a Nerve Impulse experiment.

3. Click on the **Time (min.)** button on the oscilloscope. The screen will now display activity over the course of 10 minutes (the space between each vertical line representing 1 minute). Because of the change in time scale, an action potential will look like a sharp vertical spike on the screen.

4. Click the (+) button under **Interval between Stimuli** on the stimulator until the timer is set for 2.0 minutes. This will set the stimulus to stimulate the nerve every two minutes. Click on **Stimulate** to start the stimulations. Watch the **Elapsed Time** display.

How long does it take for the nerve to return to normal?

____

5. Click on the **Stop** button to stop this action and to return the Elapsed Time to 0.0.

6. Click the **Time (msec)** button on the oscilloscope to return it to its normal millisecond display.

7. Click **Clear** to clear the oscilloscope for the next activity.

8. Click the (−) button under **Interval between Stimuli** until it is reset to 0.00. ■

Activity 6:
Testing the Effects of Curare

Curare is a well-known plant extract that South American Indians used to paralyze their prey. It is an alpha-toxin that binds to acetylcholine binding sites on the postsynaptic cell membrane, which will prevent the acetylcholine from acting. Curare blocks synaptic transmission by preventing the flow of neural impulses from neuron to neuron.

1. Click and drag the dropper from the bottle marked *curare* and position the dropper on the nerve, in between the stimulating and recording electrodes. Release the mouse button to dispense drops.

2. Set the stimulator at the threshold voltage and stimulate the nerve. What effect on the action potential is noted?

____

What explains this effect? ____________________________

What do you think would be the overall effect of curare on the organism?

____

____

Click **Record Data** to record your results.

3. Click on the **Clean** button on the nerve chamber to remove the curare and return the nerve to its original untouched state.

4. Click **Clear** to clear the oscilloscope screen for the next activity. ■

Activity 7: Testing the Effects of Lidocaine

Note: lidocaine is a sodium-channel antagonist.

1. Click and drag the dropper from the bottle marked *lidocaine* and position it over the nerve, between the stimulating and recording electrodes. Release the mouse button to dispense drops. Does this generate a trace?

____

2. Stimulate the nerve at the threshold voltage. What sort of tracing is seen?

____

Why does lidocaine have this effect on nerve fiber transmission?

____

____

Click **Record Data** to record your results. Click **Tools → Print Data** if you wish to print your data.

3. Click on the **Clean** button on the nerve chamber to remove the lidocaine and return the nerve to its original untouched state. ■

Nerve Conduction Velocity

As has been pointed out, one of the major physiological properties of neurons is conductivity: the ability to transmit the nerve impulse to other neurons, muscles, or glands. The nerve impulse, or propagated action potential, occurs when sodium ions flood into the neuron, causing the membrane to depolarize. Although this event is spoken of in electrical terms, and is measured using instruments that measure electrical events, the velocity of the action potential along a neural membrane does not occur at the speed of light. Rather, this event is much slower. In certain nerves in the human, the velocity of an action potential may be as fast as 120 meters per second. In other nerves, conduction speed is much slower, occurring at a speed of less than 3 meters per second.

To see the setup for this experiment, click the **Experiment** pull-down menu and select **Nerve Conduction Velocity** (Figure 18B.5). In this exercise, the oscilloscope and stimulator will be used along with a third instrument, the bio-amplifier. The **bio-amplifier** is used to amplify any membrane depolarization so that the oscilloscope can easily record the event. Normally, when a membrane depolarization sufficient to initiate action potential is looked at, the interior of the cell membrane goes from −70 mV to about +40 mV. This is easily registered and viewable on an oscilloscope, without the aid of an amplifier. However, in this experiment, it is the change in the membrane potential on the *outside* of the nerve that is being observed. The change that occurs here during depolarization will be so minuscule that it must be amplified in order to be visible on the oscilloscope.

A nerve chamber (similar to the one used in the previous two experiments) will be used. The design is basically a plastic box with platinum electrodes running across it. The nerve will be laid on these electrodes. Two electrodes will be used to bring the impulse from the stimulator to the nerve and three will be used for recording the membrane depolarization.

In this experiment, we will determine and compare the conduction velocities of different types of nerves. We will examine four nerves: an earthworm nerve, a frog nerve, and two rat nerves. The earthworm nerve is the smallest of the four. The frog nerve is a medium-sized myelinated nerve (myelination is discussed in Exercise 17). Rat nerve 1 is a medium-sized unmyelinated nerve. Rat nerve 2 is a large, myelinated nerve—the largest nerve in this group. We will observe the effects of size and myelination on nerve conductivity.

The basic layout of the materials is shown in Figure 18B.5. The two wires (red and black) from the stimulator connect with the top right side of the nerve chamber. Three wires (red, black, and a bare wire cable) are attached to connectors on the other end of the nerve chamber and go to the bio-amplifier. The bare cable serves as a "ground reference" for the electrical circuit and provides the reference for comparison of any change in membrane potential. The bio-amplifier is connected to the oscilloscope so that any amplified membrane changes can be observed. The stimulator output, called the *pulse*, has been connected to the oscilloscope so that when the nerve is stimulated, the tracing will start across the oscilloscope screen. Thus, the time from the start of the trace on the left-hand side of the screen (when the nerve was stimulated) to the actual nerve deflection (from the recording electrodes) can be accurately measured. This amount of time, usually in milliseconds, is critical for determining conduction velocity.

Look closely at the screen. The wiring of the circuit may seem complicated but actually is not. First, look at the stimulator, found on top of the oscilloscope. On the left side, red

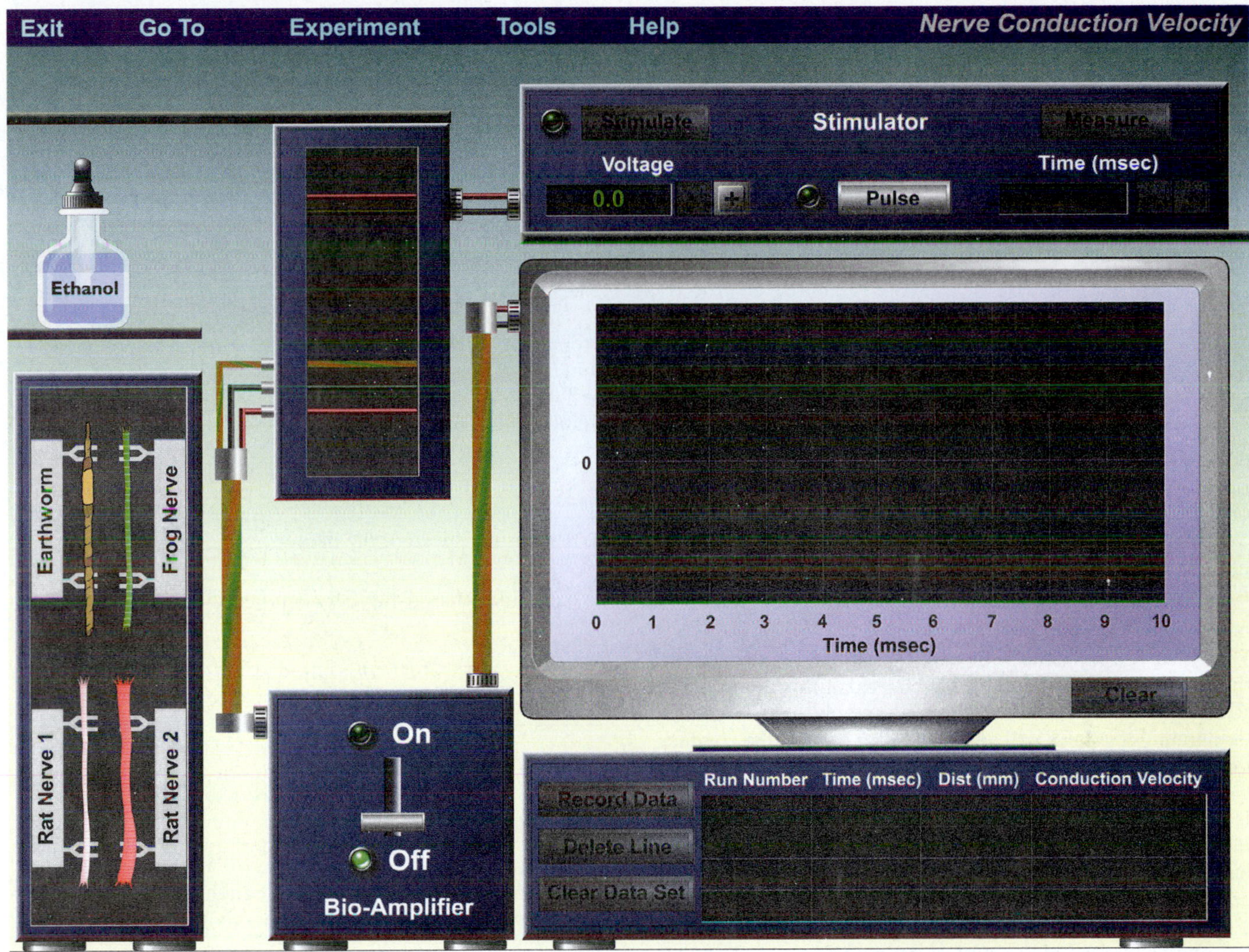

Figure 18B.5 Opening screen of the Nerve Conduction Velocity experiment.

and black wires leave the stimulator to go to the nerve chamber. Remember, the red wire is the *hot* wire that carries the impulse from the stimulator and the black wire is the return to the stimulator that completes the circuit. When the nerve is stimulated, the red recording wire (leaving the left side of the nerve chamber) will pick up the membrane impulse and bring it to the bio-amplifier. The black wire, as before, completes the circuit, and the bare cable wire simply acts as a reference electrode. The membrane potential, picked up by the red wire, is then amplified by the bio-amplifier and the output is carried to the oscilloscope. The oscilloscope then shows the trace of the nerve action potential.

Activity 8:
Measuring Nerve Conduction Velocity

1. On the stimulator, click the **Pulse** button.

2. Turn the bio-amplifier on by clicking the horizontal bar on the bio-amplifier and dragging it to the **On** setting.

On the left side of the screen are the four nerves that will be studied. The nerves included are the earthworm, a frog nerve, and two rat nerves of different sizes. The earthworm as a whole is used because it has a nerve running down its ventral surface. A frog nerve is used as the frog has long been the animal of choice in many physiology laboratories. The rat nerves are used so that you may compare (a) the conduction velocity of different sized nerves and (b) the conduction velocity of a myelinated versus unmyelinated nerve. Remember that the frog nerve is myelinated and that rat nerve 1 is the same size as the frog nerve but unmyelinated. Rat nerve 2, the largest nerve of the bunch, is myelinated.

3. Using the mouse, click and drag the dropper from the bottle labeled *ethanol* over the earthworm and release the mouse button to dispense drops of ethanol. This will narcotize the worm so it does not move around during the experiment but will not affect nerve conduction velocity. The alcohol is at a low enough percentage that the worm will be fine and back to normal within 15 minutes.

4. Click and drag the earthworm into the nerve chamber. Be sure the worm is over both of the stimulating electrodes and all three of the recording electrodes.

5. Using the (+) button next to the **Voltage** display, set the voltage to 1.0 V. Then click **Stimulate** to stimulate the nerve. Do you see an action potential? If not, increase the voltage by increments of 1.0 V until a trace is obtained.

At what threshold voltage do you first see an action potential generated?

_______ V

6. Next, click on the **Measure** button located on the stimulator. You will see a vertical yellow line appear on the far left edge of the oscilloscope screen. Now click the (+) button under the Measure button. This will move the yellow line to the right. This line lets you measure how much time has elapsed on the graph at the point that the line is crossing the graph. You will see the elapsed time appear on the **Time (msec)** display on the stimulator. Keep clicking (+) until the yellow line is right at the point in the graph where the graph ceases being a flat line and first starts to rise.

7. Once you have the yellow line positioned at the start of the graph's ascent, note the time elapsed at this point. Click **Record Data** to record the elapsed time on the data collection graph. PhysioEx will automatically compute the conduction velocity based on this data. Note that the data collection box includes a **Distance (mm)** column and that the distance is always 43 mm. This is the distance from the red stimulating wire to the red recording wire. In a wet lab, you would have to measure the distance yourself before you could proceed with calculating the conduction velocity.

It is important that you have the yellow vertical measuring line positioned at the start of the graph's rise before you click **Record Data**—otherwise, the conduction velocity calculated for the nerve will be inaccurate.

8. Fill in the data under the Earthworm column on the chart below:

9. Click and drag the earthworm to its original place. Click **Clear** to clear the oscilloscope screen.

10. Repeat steps 4 through 9 for the remaining nerves. Remember to click **Record Data** after each experimental run and to fill in the chart for question 8.

11. Click **Tools** → **Print Data** to print your data.

Which nerve in the group has the slowest conduction velocity?

____

What was the speed of the nerve? ______________

Which nerve in the group of four has the fastest conduction velocity?

____

What was the speed of the nerve? ______________

What is the relationship between nerve size and conduction velocity? ______________________________

____

____

____

Based on the results, what is your conclusion regarding conduction velocity and whether the nerve is myelinated or not?

____

____

____

____

Nerve	Earthworm (small nerve)	Frog (medium nerve, myelinated)	Rat nerve 1 (medium nerve, unmyelinated)	Rat nerve 2 (large nerve, myelinated)
Threshold voltage				
Elapsed time from stimulation to action potential				
Conduction velocity				

What is the major reason for the differences seen in conduction velocity between the myelinated nerves and the unmyelinated nerves? ______________________________

____

____

______________________________ ■

Histology Review Supplement

Turn to p. P-145 for a review of nervous tissue.

exercise

28B Endocrine System Physiology: Computer Simulation

Objectives

1. To define the following:

   metabolism, hormone replacement therapy, diabetes type I, diabetes type II, and *glucose standard curve.*
2. To explain the role of thyroxine in maintaining an animal's metabolic rate.
3. To explain the effects of thyroid-stimulating hormone on an animal's metabolic rate.
4. To understand how estrogen affects uterine tissue growth.
5. To explain how hormone replacement therapy works.
6. To explain why insulin is important, and how it can be used to treat diabetes.

The endocrine system exerts many complex and interrelated effects on the body as a whole, as well as on specific tissues and organs. Studying the effects of hormones on the body is difficult to do in a wet lab because experiments often take days, weeks, or even months to complete and are expensive. In addition, live animals may need to be sacrificed, and technically difficult surgical procedures are sometimes necessary. This computer simulation allows you to study the effects of given hormones on the body by using "virtual" animals rather than live ones. You can carry out delicate surgical techniques with the click of a button, and complete experiments in a fraction of the time that it would take in an actual wet lab environment.

Hormones and Metabolism

Metabolism is the broad term used for all biochemical reactions occurring in the body. Metabolism involves *catabolism,* a process by which complex materials are broken down into simpler substances, usually with the aid of enzymes found in the body cells. Metabolism also involves *anabolism,* in which the smaller materials are built up by enzymes to build larger, more complex molecules. When larger molecules are made, energy is stored in the various bonds formed. When bonds are broken in catabolism, energy that was stored in the bonds is released for use by the cell. Some of the energy liberated may go into the formation of ATP, the energy-rich material used by the body to run itself. However, not all of the energy liberated goes into this pathway; some is given off as body heat. Humans are *homeothermic* animals, meaning they have a fixed body temperature. Maintaining this temperature is important to maintaining the metabolic pathways found in the body.

The most important hormone in maintaining metabolism and body heat is *thyroxine,* the hormone of the thyroid gland, found in the neck. The thyroid gland secretes thyroxine, but the production of thyroxine is really controlled by the pituitary gland, which secretes *thyroid-stimulating hormone (TSH).* TSH is carried by the blood to the thyroid gland (its *target tissue*) and causes the thyroid to produce more thyroxine. So in an indirect way, an animal's metabolic rate is the result of pituitary hormones.

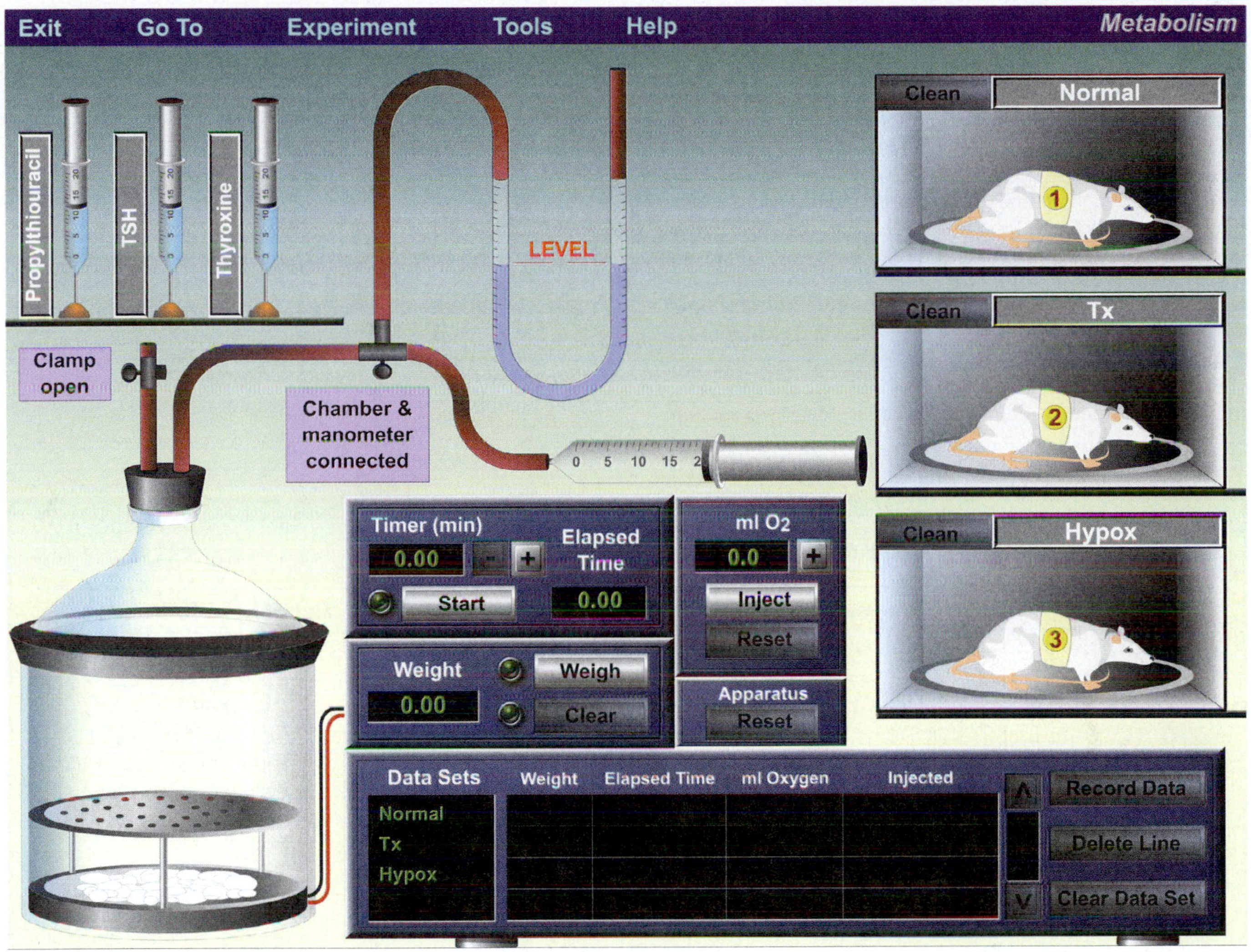

Figure 28B.1 Opening screen of the Metabolism experiment.

In the following experiments, you will investigate the effects of thyroxine and TSH on an animal's metabolic rate. To begin, select **Endocrine System Physiology** from the main menu. The opening screen will appear in a few seconds (see Figure 28B.1). Select **Balloons On** from the Help menu for help identifying the equipment on-screen (you will see labels appear as you roll over each piece of equipment). Select **Balloons Off** to turn this feature off before you begin the experiments.

Study the screen. You will see a jar-shaped chamber to the left, connected to a *respirometer-manometer apparatus* (consisting of a U-shaped tube, a syringe, and associated tubing.) You will be placing animals—in this case, rats—in the chamber in order to gather information about how thyroxine and TSH affect their metabolic rates. Note that the chamber also includes a weight scale, and next to the chamber is a timer for setting and timing the length of a given experiment. Under the timer is a weight display.

Two tubes are connected to the top of the chamber. The left tube has a clamp on it that can be opened or closed. Leaving the clamp open allows outside air into the chamber; closing the clamp creates a closed, airtight system. The other tube leads to a *T-connector*. One branch of the T leads to a fluid-containing U-shaped tube, called a *manometer*. As an animal uses up the air in the closed system, this fluid will rise in the left side of the U-shaped tube and fall in the right.

The other branch of the T-connector leads to a syringe filled with air. Using the syringe to inject air into the tube, you will measure the amount of air that is needed to return the fluid columns to their original levels. This measurement will be equal to the amount of oxygen used by the animal during the elapsed time of the experiment. Soda lime, found at the bottom of the chamber, absorbs the carbon dioxide given off by the animal so that the amount of oxygen used can easily be measured. The amount of oxygen used by the animal, along with its weight, will be used to calculate the animal's metabolic rate.

Also on the screen are three white rats in their individual cages. These are the specimens you will use in the following experiments. One rat is *normal;* the second is *thyroidectomized* (abbreviated on the screen as *Tx*)— meaning its thyroid has been removed; and the third is *hypophysectomized* (abbreviated on the screen as *Hypox*), meaning its pituitary gland has been removed. The pituitary gland is also known as the *hypophysis,* and removal of this organ is called a *hypophysectomy.*

To the top left of the screen are three syringes with various chemicals inside: propylthiouracil, thyroid-stimulating hormone (TSH), and thyroxine. TSH and thyroxine have been previously mentioned; propylthiouracil is a drug that inhibits the production of thyroxine. You will perform four experiments on each animal to: (1) determine its baseline metabolic rate, (2) determine its metabolic rate after it has been injected with thyroxine, (3) determine its metabolic rate after it has been injected with TSH, and (4) determine its metabolic rate after it has been injected with propylthiouracil.

You will be recording all of your data on the chart below. You may also record your data on-screen by using the equipment in the lower part of the screen, called the *data collection unit.* This equipment records and displays data you accumulate during the experiments. Check that the data set for **Normal** is highlighted in the **Data Sets** window, since you will be experimenting with the normal rat first. The **Record Data** button lets you record data after an experimental trial. Clicking the **Delete Line** or **Clear Data Set** buttons erases any data you want to delete.

Activity 1: Determining Baseline Metabolic Rates

First, you will determine the baseline metabolic rate for each rat.

1. Using the mouse, click and drag the **normal** rat into the chamber and place it on top of the scale. When the animal is in the chamber, release the mouse button.

Effects of Hormones on Metabolic Rate

	Normal rat	Thyroidectomized rat	Hypophysectomized rat
Baseline			
Weight	_______ grams	_______ grams	_______ grams
ml O_2 used in 1 minute	_______ ml	_______ ml	_______ ml
ml O_2 used per hour	_______ ml	_______ ml	_______ ml
Metabolic rate	_______ ml O_2/kg/hr	_______ ml O_2/kg/hr	_______ ml O_2/kg/hr
With thyroxine			
Weight	_______ grams	_______ grams	_______ grams
ml O_2 used in 1 minute	_______ ml	_______ ml	_______ ml
ml O_2 used per hour	_______ ml	_______ ml	_______ ml
Metabolic rate	_______ ml O_2/kg/hr	_______ ml O_2/kg/hr	_______ ml O_2/kg/hr
With TSH			
Weight	_______ grams	_______ grams	_______ grams
ml O_2 used in 1 minute	_______ ml	_______ ml	_______ ml
ml O_2 used per hour	_______ ml	_______ ml	_______ ml
Metabolic rate	_______ ml O_2/kg/hr	_______ ml O_2/kg/hr	_______ ml O_2/kg/hr
With propylthiouracil			
Weight	_______ grams	_______ grams	_______ grams
ml O_2 used in 1 minute	_______ ml	_______ ml	_______ ml
ml O_2 used per hour	_______ ml	_______ ml	_______ ml
Metabolic rate	_______ ml O_2/kg/hr	_______ ml O_2/kg/hr	_______ ml O_2/kg/hr

2. Be sure the clamp on the left tube (on top of the chamber) is open, allowing air to enter the chamber. If the clamp is closed, click on it to open it.

3. Be sure the indicator next to the T-connector reads "Chamber and manometer connected." If not, click on the **T-connector knob.**

4. Click on the **Weigh** button in the box to the right of the chamber to weigh the rat. Record this weight in the Baseline section of the chart on p. P-40 in the row labeled "Weight."

5. Click the (+) button on the **Timer** so that the Timer display reads 1 minute.

6. Click on the clamp to close it. This will prevent any outside air from entering the chamber and ensure that the only oxygen the rat is breathing is the oxygen inside the closed system.

7. Click **Start** on the Timer display. You will see the elapsed time appear in the "Elapsed Time" display. Watch what happens to the water levels in the U-shaped tube.

8. At the end of the 1-minute period, the timer will automatically stop. When it stops, click on the **T-connector knob** so that the indicator reads "Manometer and syringe connected."

9. Click on the clamp to open it so that the rat can once again breathe outside air.

10. Click the (+) button under ml O_2 so that the display reads "1.0". Then click **Inject,** and watch what happens to the fluid levels. Continue clicking the (+) button and injecting air until the fluid in the two arms of the U-tube are level again. How many milliliters of air need to be added to equalize the fluid in the two arms? (This is equivalent to the amount of oxygen that the rat used up during 1 minute in the closed chamber.) Record this measurement in the Baseline section of the chart on p. P-40 in the row labeled "ml O_2 used in 1 minute."

11. Determine the oxygen consumption per hour for the rat. Use the following formula:

$$\frac{\text{ml } O_2 \text{ consumed}}{1 \text{ minute}} \times \frac{60 \text{ minutes}}{1 \text{ hr}} = \text{ml } O_2/\text{hr}$$

Record this data in the Baseline section of the chart in the row labeled "ml O_2 used per hour."

12. Now that you have the amount of oxygen used per hour, determine the metabolic rate per kilogram of body weight by using the following formula. (Note that you will need to convert the weight data from g to kg before you can use the formula.)

$$\text{Metabolic Rate} = \frac{\text{ml } O_2/\text{hr}}{\text{wt. in kg}} = \underline{\hspace{3cm}} \text{ ml } O_2/\text{kg/hr}$$

Record this data in the Baseline section of the chart in the row labeled "Metabolic rate."

13. Click **Record Data.**

14. Click and drag the rat from the chamber back to its cage.

15. Click the **Reset** button in the box labeled *Apparatus.*

16. Now repeat steps 1–15 for the thyroidectomized (Tx) and hypophysectomized (Hypox) rats. Record your data in the Baseline section of the chart under the corresponding column for each rat. Be sure to highlight **Tx** under **Data Sets** (on the data collection box) before beginning the experiment on the thyroidectomized rat; likewise, highlight **Hypox** under **Data Sets** before beginning the experiment on the hypophysectomized rat.

How did the metabolic rates of the three rats differ?

___

___

___

Why did the metabolic rates differ?

___

___

___ ■

Activity 2: Determining the Effect of Thyroxine on Metabolic Rate

Next, you will investigate the effects of thyroxine injections on the metabolic rates of all three rats.

Note that in a wet lab environment, you would normally need to inject thyroxine (or any other hormone) into a rat *daily* for a minimum of 1–2 weeks in order for any response to be seen. However, in the following simulations, you will inject the rat only once and be able to witness the same results as if you had administered multiple injections over the course of several weeks. In addition, by clicking the **Clean** button while a rat is inside its cage, you can magically remove all residue of any previously injected hormone from the rat and perform a new experiment on the same rat. In a real wet lab environment, you would need to either wait weeks for hormonal residue to leave the rat's system or use a different rat.

1. Select a rat to test. You will eventually test all three, and it doesn't matter what order you test them in. Under **Data Sets,** highlight **Normal, Tx,** or **Hypox** depending on which rat you select.

2. Click the **Reset** button in the box labeled *Apparatus.*

3. Click on the syringe labeled **thyroxine** and drag it over to the rat. Release the mouse button. This will cause thyroxine to be injected into the rat.

4. Click and drag the rat into the chamber. Perform steps 1–12 of Activity 1 again, except this time record your data in the With Thyroxine section of the chart.

5. Click **Record Data.**

6. Click and drag the rat from the chamber back to its cage and click **Clean** to cleanse it of all traces of thyroxine.

7. Now repeat steps 1–6 for the remaining rats. Record your data in the With Thyroxine section of the chart under the corresponding column for each rat.

What was the effect of thyroxine on the normal rat's metabolic rate? How does it compare to the normal rat's baseline metabolic rate?

____

____

Why was this effect seen? ______________________

____

What was the effect of thyroxine on the thyroidectomized rat's metabolic rate? How does it compare to the thyroidectomized rat's baseline metabolic rate?

____

____

____

Why was this effect seen? ______________________

____

What was the effect of thyroxine on the hypophysectomized rat's metabolic rate? How does it compare to the hypophysectomized rat's baseline metabolic rate?

____

____

____

Why was this effect seen? ______________________

____ ■

Activity 3: Determining the Effect of TSH on Metabolic Rate

Next, you will investigate the effects of TSH injections on the metabolic rates of the three rats. Select a rat to experiment on first, and then proceed.

1. Under **Data Sets,** highlight **Normal, Tx,** or **Hypox** depending on which rat you are using.
2. Click the **Reset** button in the box labeled *Apparatus*.
3. Click and drag the syringe labeled **TSH** over to the rat and release the mouse button, injecting the rat.
4. Click and drag the rat into the chamber. Perform steps 1–12 of Activity 1 again. Record your data in the With TSH section of the chart.
5. Click **Record Data.**
6. Click and drag the rat from the chamber back to its cage and click **Clean** to cleanse it of all traces of TSH.
7. Now repeat this activity for the remaining rats. Record your data in the With TSH section of the chart under the corresponding column for each rat.

What was the effect of TSH on the normal rat's metabolic rate? How does it compare to the normal rat's baseline metabolic rate?

____

____

____

Why was this effect seen? ______________________

____

What was the effect of TSH on the thyroidectomized rat's metabolic rate? How does it compare to the thyroidectomized rat's baseline metabolic rate?

____

____

____

Why was this effect seen? ______________________

____

What was the effect of TSH on the hypophysectomized rat's metabolic rate? How does it compare to the hypophysectomized rat's baseline metabolic rate?

____

____

____

Why was this effect seen? ______________________

____ ■

Activity 4: Determining the Effect of Propylthiouracil on Metabolic Rate

Next, you will investigate the effects of propylthiouracil injections on the metabolic rates of the three rats. Keep in mind that propylthiouracil is a drug that inhibits the production of thyroxine. Select a rat to experiment on first, and then proceed.

1. Under **Data Sets,** highlight **Normal, Tx,** or **Hypox** depending on which rat you are using (**Normal** for the normal rat, **Tx** for thyroidectomized, **Hypox** for hypophysectomized.)

2. Click the **Reset** button in the box labeled *Apparatus*.

3. Click and drag the syringe labeled **propylthiouracil** over to the rat and release the mouse button, injecting the rat.

4. Click and drag the rat into the chamber. Perform steps 1–12 of Activity 1 again, except this time record your data in the With Propylthiouracil section of the chart.

5. Click **Record Data.**

6. Click and drag the rat from the chamber back to its cage and click **Clean** to cleanse the rat of all traces of propylthiouracil.

7. Now repeat this activity for the remaining rats. Record your data in the With Propylthiouracil section of the chart under the corresponding column for each rat.

What was the effect of propylthiouracil on the normal rat's metabolic rate? How does it compare to the normal rat's baseline metabolic rate?

____

____

____

Why was this effect seen? __________________________

____

What was the effect of propylthiouracil on the thyroidectomized rat's metabolic rate? How does it compare to the thyroidectomized rat's baseline metabolic rate?

____

____

____

Why was this effect seen? __________________________

____

What was the effect of propylthiouracil on the hypophysectomized rat's metabolic rate? How does it compare to the hypophysectomized rat's baseline metabolic rate?

____

____

____

Why was this effect seen? __________________________

____

8. If you wish, click **Tools → Print Data** to print all of your recorded data for this experiment. ■

Hormone Replacement Therapy

Ovaries are stimulated by *follicle-stimulating hormone (FSH)* to get ovarian follicles to develop so that they may be ovulated and fertilized. While the follicles are developing, they produce the hormone *estrogen*. The target tissue for estrogen is the uterus, and the action of estrogen is to enable the uterus to grow and develop so that it may receive fertilized eggs for implantation. *Ovariectomy,* the removal of ovaries, will remove the source of estrogen and cause the uterus to slowly atrophy.

In this activity, you will re-create a classic endocrine experiment and examine how estrogen affects uterine tissue growth. You will be working with two female rats, both of which have been ovariectomized and are thus no longer producing estrogen. You will administer **hormone replacement therapy** to one rat by giving it daily injections of estrogen. The other rat will serve as your control and receive daily injections of saline. You will then remove the uterine tissues from both rats, weigh the tissues, and compare them to determine the effects of hormone replacement therapy.

Start by selecting **Hormone Replacement Therapy** from the **Experiment** menu. A new screen will appear (Figure 28B.2), with the two ovariectomized rats in cages. (Note that if this were a wet lab, the ovariectomies would need to have been performed on the rats a month prior to the rest of the experiment in order to ensure that no residual hormones remained in the rats' systems.) Also on screen are a bottle of saline, a bottle of estrogen, a syringe, a box of weighing paper, and a weighing scale.

Proceed carefully with this experiment. Each rat will disappear from the screen once you remove its uterus, and cannot be brought back unless you restart the experiment. This replicates the situation you would encounter if working with live animals: once the uterus is removed, the animal would have to be sacrificed.

Activity 5: Hormone Replacement Therapy

1. Click on the syringe, drag it to the bottle of **saline,** and release the mouse button. The syringe will automatically fill with 1 ml of saline.

2. Drag the syringe to the **control** rat and place the tip of the needle in the rat's lower abdominal area. Injections into this area are considered *interperitoneal* and will quickly be picked up by the abdominal blood vessels. Release the mouse button—the syringe will empty into the rat and automatically return to its holder. Click **Clean** on the syringe holder to clean the syringe of all residue.

3. Click on the syringe again, this time dragging it to the bottle of **estrogen,** and release the mouse button. The syringe will automatically fill with 1 ml of estrogen.

4. Drag the syringe to the **experimental** rat and place the tip of the needle in the rat's lower abdominal area. Release the mouse button—the syringe will empty into the rat and automatically return to its holder. Click **Clean** on the syringe holder to clean the syringe of all residue.

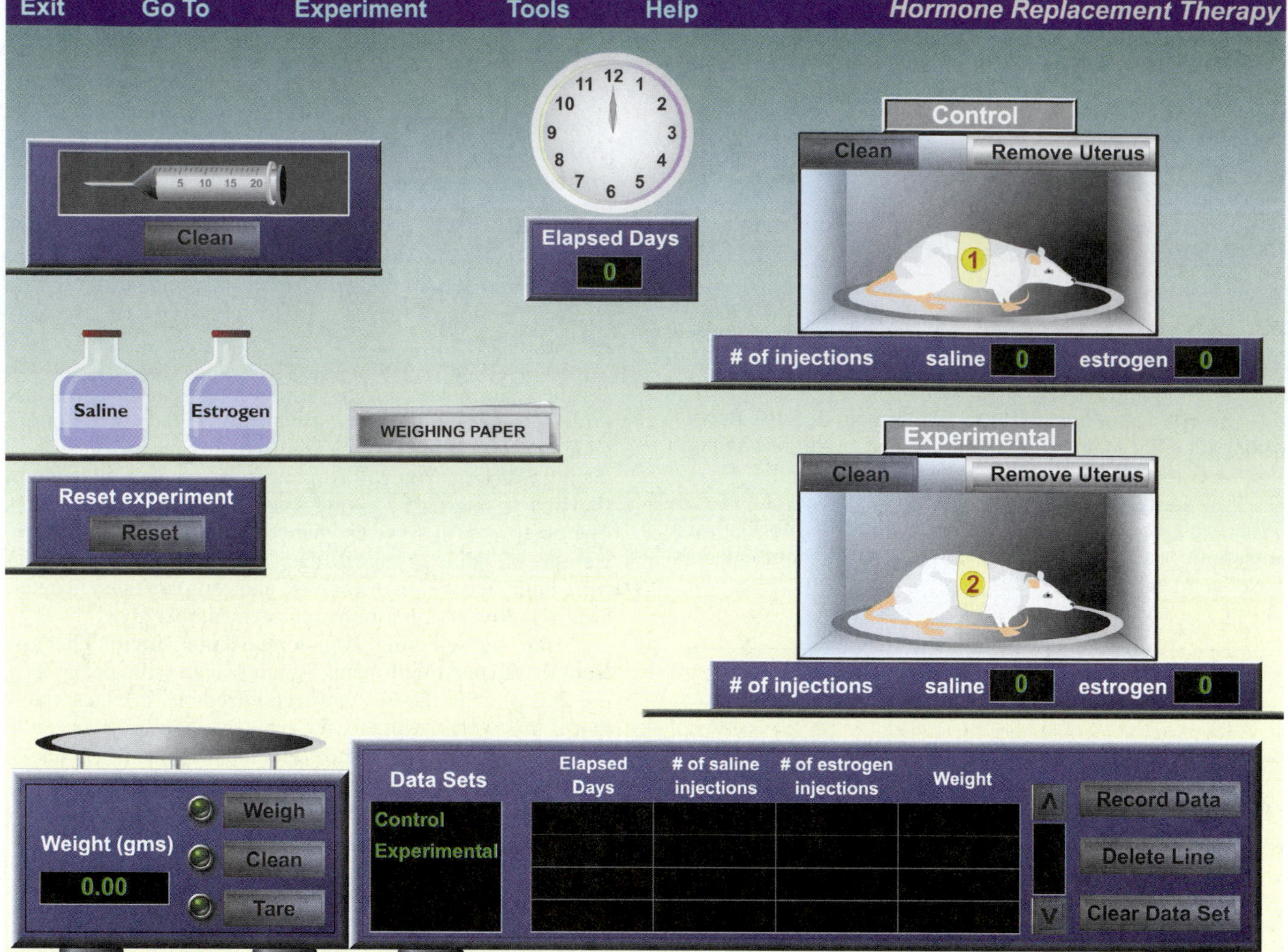

Figure 28B.2 Opening screen of the Hormone Replacement Therapy experiment.

5. Click on the **Clock.** You will notice the hands sweep the clock face twice, indicating that 24 hours have passed.

6. Repeat steps 1–5 until each rat has received a total of 7 injections over the course of 7 days (1 injection per day). Note that the **# of injections** displayed below each rat cage records how many injections the rat has received. The control rat should receive 7 injections of saline, whereas the experimental rat should receive 7 injections of estrogen.

7. Next, click on the box of weighing paper. You will see a small piece of paper appear. Click and drag this paper over to the top of the scale and release the mouse button.

8. Notice the scale will give you a weight for the paper. With the mouse arrow, click on the **Tare** button to tare the scale to zero (0.0000 g), adjusting for the weight of the paper.

9. You are now ready to remove the uteruses. In a wet lab, this would require surgery. Here, you will simply click on the **Remove Uterus** button found in each rat cage. The rats will disappear, and a uterus (consisting of a uterine body and two uterine horns) will appear in each cage.

10. Click and drag the uterus from the **control** rat over to the scale and release it on the weighing paper. Click on the **Weigh** button to obtain the weight. Record the weight here:

Uterus weight (control): _______________ g

11. Click **Record Data.**

12. Click **Clean** on the weight scale to dispense of the weighing paper and uterus.

13. Repeat steps 7 and 8. Then click and drag the uterus from the **experimental** rat over to the scale and release it on the weighing paper. Click **Weigh** to obtain the weight. Record the weight here:

Uterus weight (experimental): _______________ g

14. Click **Record Data.**

15. Click **Clean** on the weight scale to dispense of the weighing paper and uterus.

16. Click **Tools** → **Print Data** to print your recorded data for this experiment.

How does the control uterus weight compare to the experimental uterus weight?

____

What can you conclude about the administration of estrogen injections on the experimental animal?

____

____

What might be the effect if testosterone had been administered instead of estrogen? Explain your answer.

____

____

____ ■

Insulin and Diabetes

Insulin is produced by the beta cells of the endocrine portion of the pancreas. It is vital to the regulation of blood glucose levels because it enables the body's cells to absorb glucose from the bloodstream. When insulin is not produced by the pancreas, **diabetes mellitus Type I** results. When insulin is produced by the pancreas but the body fails to respond to it, **diabetes mellitus Type II** results. In either case, glucose remains in the bloodstream, unable to be taken up by the body's cells to serve as the primary fuel for metabolism.

In the following experiment, you will study the effects of insulin treatment for type I diabetes. The experiment is divided into two parts. In Part I, you will obtain a *glucose standard curve,* which will be explained shortly. In Part II, you will compare the glucose levels of a normal rat versus that of a diabetic rat, and then compare them again after each rat has been injected with insulin.

Part I

Activity 6:
Obtaining a Glucose Standard Curve

To begin, select **Insulin and Diabetes-Part 1** from the **Experiments** menu (see Figure 28B.3).

Select **Balloons On** from the Help menu for help identifying the equipment on-screen. (You will see labels appear as you roll over each piece of equipment.) Select **Balloons Off** to turn this feature off before you begin the experiments.

On the right side of the opening screen is a special spectrophotometer. The **spectrophotometer** is one of the most widely used research instruments in biology. It is used to measure the amounts of light of different wavelengths absorbed and transmitted by a pigmented solution. Inside the spectrophotometer is a source for white light, which is separated into various wavelengths (or colors) by a prism. The user selects a wavelength (color), and light of this color is passed through a tube, or *cuvette,* containing the sample being tested. (For this experiment, the spectrophotometer light source will be preset for a wavelength of 450 nm.) The light transmitted by the sample then passes onto a photoelectric tube, which converts the light energy into an electrical current. The current is then measured by a meter. Alternatively, the light may be measured before the sample is put into the light path, and the amount of light absorbed—called **optical density**—is measured. Using either method, the change in light transmittance or light absorbed can be used to measure the amount of a given substance in the sample being tested.

In Part II, you will use the spectrophotometer to determine how much glucose is present in blood samples taken from two rats. Before using the spectrophotometer, you must obtain a **glucose standard curve** so that you have a point of reference for converting optical density readings into glucose readings, which will be measured in mg/deciliter(mg/dl). To do this, you will prepare five test tubes that contain known amounts of glucose: 30 mg/dl, 60 mg/dl, 90 mg/dl, 120 mg/dl, and 150 mg/dl, respectively. You will then use the spectrophotometer to determine the corresponding optical density readings for each of these known amounts of glucose. Information obtained in Part I will be used to perform Part II.

Also on the screen are three dropper bottles, a test tube washer, a test tube dispenser (on top of the washer), and a test tube incubation unit that you will need to prepare the samples for analysis.

1. Click and drag the test tube (on top of the test tube washer) into slot 1 of the incubation unit. You will see another test tube pop up from the dispenser. Click and drag this second test tube into slot 2 of the incubation unit. Repeat until you have dragged a total of five test tubes into the five slots in the incubation unit.

2. Click and hold the mouse button on the dropper cap of the **glucose standard** bottle. Drag the dropper cap over to tube 1. Release the mouse button to dispense the glucose. You will see that one drop of glucose solution is dropped into the tube and that the dropper cap automatically returns to the bottle of glucose standard.

3. Repeat step 2 with the remaining four tubes. Notice that each subsequent tube will automatically receive one additional drop of glucose standard into the tube (i.e., tube 2 will receive two drops, tube 3 will receive three drops, tube 4 will receive four drops, and tube 5 will receive 5 drops).

4. Click and hold the mouse button on the dropper cap of the **deionized water** bottle. Drag the dropper cap over to tube 1. Release the mouse button to dispense the water. Notice that four drops of water are automatically added to the first tube.

5. Repeat step 4 with tubes 2, 3, and 4. Notice that each subsequent tube will receive one *less* drop of water than the previous tube (i.e., tube 2 will receive three drops, tube 3 will receive two drops, and tube 4 will receive one drop. Tube 5 will receive *no* drops of water.)

6. Click on the **Mix** button of the incubator to mix the contents of the tubes.

7. Click on the **Centrifuge** button. The tubes will descend into the incubator and be centrifuged.

8. When the tubes resurface, click on the **Remove Pellet** button. Any pellets from the centrifuging process will be removed from the test tubes.

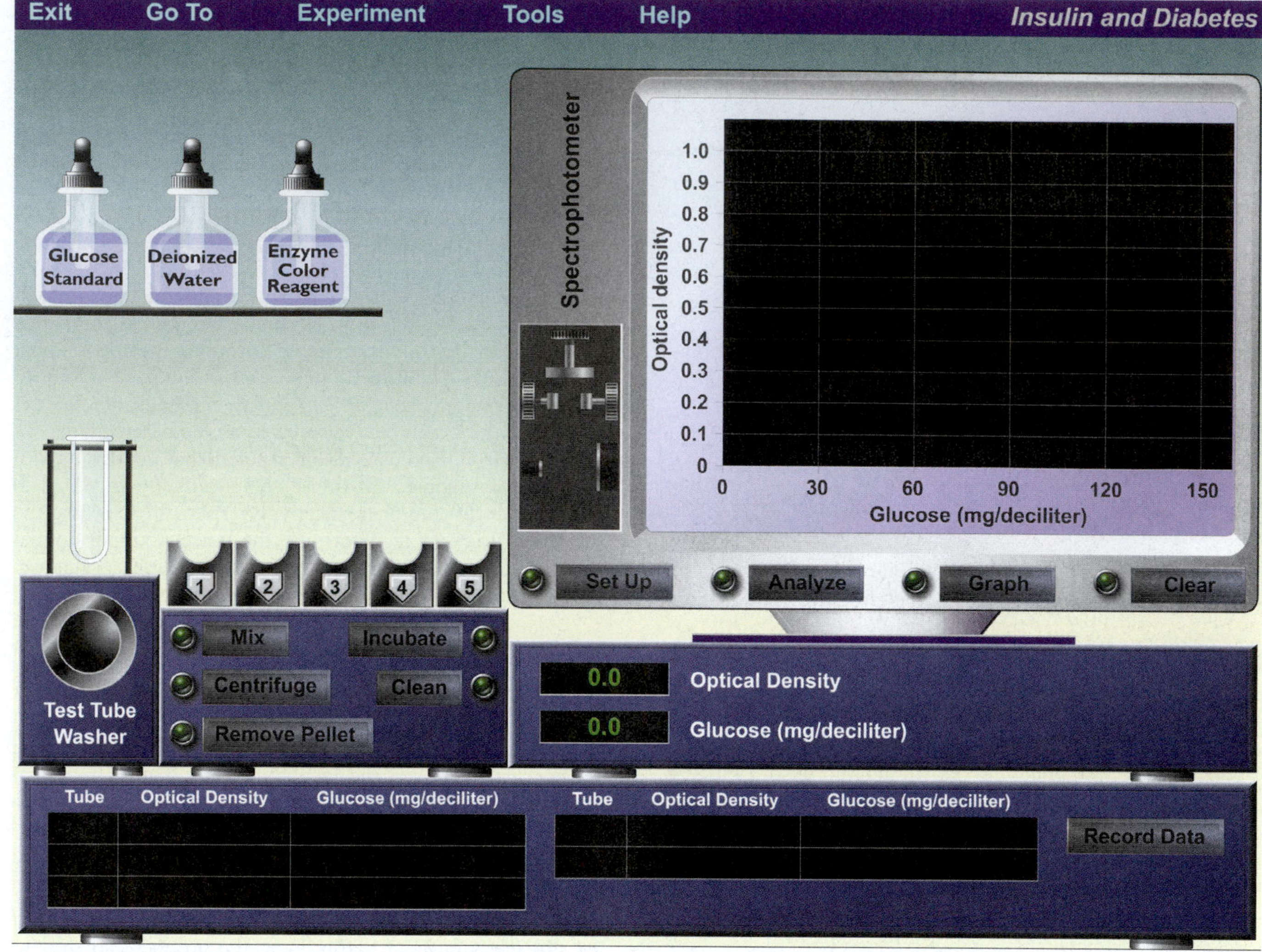

Figure 28B.3 Opening screen of the Insulin and Diabetes experiment, Part I.

9. Click and hold the mouse button on the dropper cap of the **enzyme-color reagent** bottle. Still holding the mouse button down, drag the dropper cap over to tube 1. When you release the mouse, you will note that 5 drops of reagent are added to the tube and that the stopper is returned to its bottle.

10. Repeat step 9 for the remaining tubes.

11. Now click **Incubate.** The tubes will descend into the incubator, incubate, and then resurface.

12. Using the mouse, click on **Set Up** on the spectrophotometer. This will warm up the instrument and get it ready for your readings.

13. Click and drag tube 1 into the spectrophotometer (right above the **Set Up** button) and release the mouse button. The tube will lock into place.

14. Click **Analyze.** You will see a spot appear on the screen and values appear in the **Optical Density** and **Glucose** displays.

15. Click **Record Data** on the data collection unit.

16. Click and drag the tube into the test tube washer.

17. Repeat steps 13–16 for the remaining test tubes.

18. When all five tubes have been analyzed, click on the **Graph** button. This is the glucose standard graph, which you will use in Part II of the experiment. Click **Tools** → **Print Data** to print your recorded data. ■

Part II

Activity 7:

Comparing Glucose Levels Before and After Insulin Injection

Select **Insulin and Diabetes-Part 2** from the **Experiments** menu.

The opening screen will look similar to the screen from Part I (see Figure 28B.4). Notice the two rats in their cages. One will be your control animal, the other your experimental animal. Also note the three syringes, containing insulin, saline, and alloxan, respectively. **Alloxan** is a rather nasty drug: when administered to an animal, it will selectively kill all the cells that produce insulin and render the animal instantly diabetic. It does this by destroying the beta cells of the pancreas, which are responsible for insulin production.

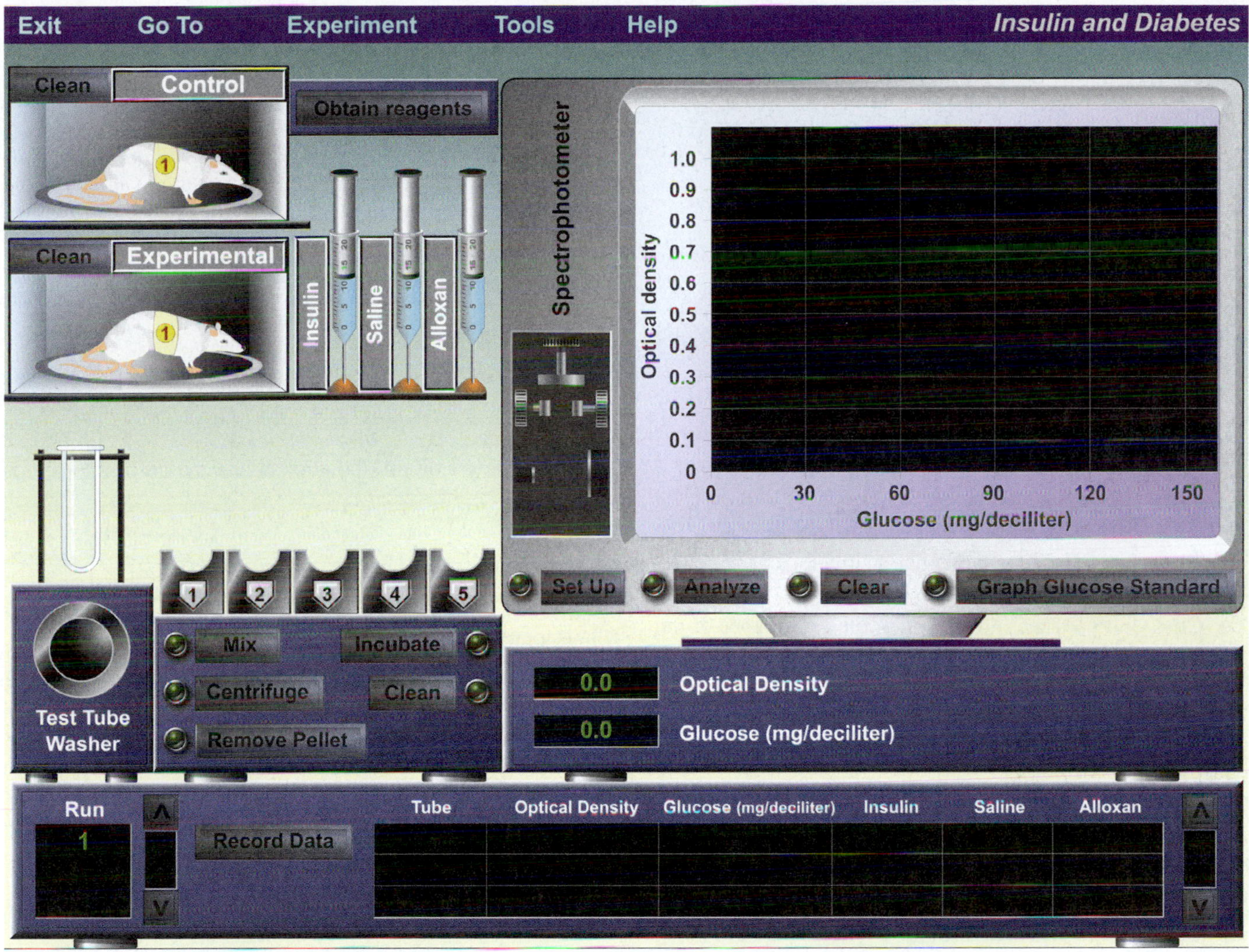

Figure 28B.4 Opening screen of the Insulin and Diabetes experiment, Part II.

In this experiment, you will inject the control rat with saline and the experimental rat with alloxan. (Normally, injections are given *daily* for a week. In this simulation, we will administer the injections only once but will be able to see results as though the injections had been given over a longer period of time.)

After administering the saline and alloxan injections, you will obtain blood samples from the two rats. You will then inject both rats with insulin, and obtain blood samples again. Finally, you will analyze all the blood samples in the spectrophotometer (described in Part I) comparing the amounts of glucose present.

1. Click and drag the **saline** syringe to the **control** rat and release the mouse button to inject the animal.

2. Click and drag the **alloxan** syringe to the **experimental** rat and release the mouse button to inject the animal.

3. Click and drag a new test tube (from the test tube dispenser) over to the tail of the **control** rat and release the mouse button. You will note three drops of blood being drawn from the tail into the tube. Next, click and drag the tube into test tube holder **1** in the incubator. (*Note:* The tail is a popular place to get blood from a rat. The end of the tail can easily be clipped and blood collected without really disturbing the rat. The tail heals quickly with no harm to the animal.)

4. Click and drag another new test tube (from the test tube dispenser) over to the tail of the **experimental** rat and release the mouse button. Again, you will note that three drops of blood are drawn from the tail into the tube. Click and drag the tube into test tube holder **2** in the incubator.

5. Click and drag the **insulin** syringe to the **control** rat and release the mouse button to inject the animal.

6. Repeat step 5 with the **experimental** rat.

7. Repeat steps 3 and 4 again, drawing blood samples from each rat and placing the samples into test tube holders **3** and **4.**

8. Click the **Obtain Reagents** button on the cabinet that currently displays the syringes. The syringes will disappear, and you will see four dropper bottles in their place.

9. Click and hold the mouse button on the dropper of the **deionized water** bottle. Drag the dropper cap over to tube 1. Release the mouse button to dispense. You will note that five drops of water are added to the tube.

10. Repeat step 9 for the remaining test tubes.

11. Click and hold the mouse button on the dropper of the **barium hydroxide.** Drag the dropper cap over to tube 1. Release the mouse button to dispense. Note that five drops of solution are added to the tube. (Barium hydroxide is used for clearing proteins and cells so that clear glucose readings may be obtained.)

12. Repeat step 11 for the remaining test tubes.

13. Click and hold the mouse button on the dropper of the **heparin** bottle. Still holding the mouse button down, drag the dropper cap over to tube 1. Release the mouse button to dispense. (Heparin is an anti-coagulant that prevents blood clotting.)

14. Repeat step 13 for the remaining test tubes.

15. Click on the **Mix** button of the incubator to mix the contents of the tubes.

16. Click on the **Centrifuge** button. The tubes will descend into the incubator, be centrifuged, and then resurface.

17. Click on the **Remove Pellet** button to remove any pellets from the centrifuging process.

18. Click and hold the mouse button on the dropper of the **enzyme-color reagent** bottle. Drag the dropper cap to tube 1. Release the mouse to dispense.

19. Repeat step 18 with the remaining test tubes. In an actual wet lab, you would also shake the test tubes after adding the enzyme-color reagent.

20. Click **Incubate** one more time. The tubes will descend into the incubator, incubate, and then resurface.

21. Click on **Set Up** on the spectrophotometer to warm up the instrument and get it ready for your readings.

22. Click **Graph Glucose Standard.** The graph from Part I of the experiment will appear on the monitor.

23. Click and drag tube 1 to the spectrophotometer and release the mouse button. The tube will lock into place.

24. Click **Analyze.** You will see a horizontal line appear on the screen and a value appear in the **Optical Density** display.

25. Drag the **movable rule** (the vertical line on the far right of the spectrophotometer monitor) over to where the horizontal line (from step 24) crosses the glucose standard line. Watch what happens to the Glucose display as you move the movable rule to the left.

What is the glucose reading where the horizontal line crosses the glucose standard line? Test tube 1: __________ mg/dl glucose

This is your glucose reading for the sample being tested.

26. Click **Record Data** on the data collection unit.

27. Click and drag the test tube from the spectrophotometer into the test tube washer, then click **Clear** underneath the oscilloscope display.

28. Repeat steps 22–27 for the remaining test tubes. Record your glucose readings for each test tube here:

Test tube 2: __________ mg/dl glucose

Test tube 3: __________ mg/dl glucose

Test tube 4: __________ mg/dl glucose

How does the glucose level in test tube 1 compare to the level in test tube 2? Recall that tube 1 contains a sample from your control rat (which received injections of saline) and that tube 2 contains a sample from your experimental rat (which received injections of alloxan).

____

Explain this result: ______________________________

____

What is the condition that alloxan has caused in the experimental rat?

____

How does the glucose level in test tube 3 compare to the level in test tube 1?

____

Explain this result: ______________________________

____

How does the glucose level in test tube 4 compare to the level in test tube 2?

____

Explain this result: ______________________________

____

What was the effect of administering insulin to the control animal?

____

What was the effect of administering insulin to the experimental animal?

____ ■

29. Click **Tools → Print Data** to print your recorded data.

Histology Review Supplement

Turn to p. P-147 for a review of endocrine tissue.

Cardiovascular Dynamics: Computer Simulation

exercise **33B**

Objectives

1. To define the following: *blood flow; peripheral resistance; viscosity; systole; diastole; end diastolic volume; end systolic volume; stroke volume; cardiac output.*
2. To explore cardiovascular dynamics using an experimental setup to simulate a human body function.
3. To understand that heart and blood vessel functions are highly coordinated.
4. To comprehend that pressure differences provide the driving force that moves blood through the blood vessels.
5. To recognize that body tissues may differ in their blood demands at a given time.
6. To identify the most important factors in control of blood flow.
7. To comprehend that changing blood vessel diameter can alter the pumping ability of the heart.
8. To examine the effect of stroke volume on blood flow.

The physiology of human blood circulation can be divided into two distinct but remarkably harmonized processes: (1) the pumping of blood by the heart, and (2) the transport of blood to all body tissues via the vasculature, or blood vessels. Blood supplies all body tissues with the substances needed for survival, so it is vital that blood delivery is ample for tissue demands.

The Mechanics of Circulation

To understand how blood is transported throughout the body, let's examine three important factors influencing how blood circulates through the cardiovascular system: blood flow, blood pressure, and peripheral resistance.

Blood flow is the amount of blood moving through a body area or the entire cardiovascular system in a given amount of time. While total blood flow is determined by cardiac output (the amount of blood the heart is able to pump per minute), blood flow to specific body areas can vary dramatically in a given time period. Organs differ in their requirements from moment to moment, and blood vessels constrict or dilate to regulate local blood flow to various areas in response to the tissue's immediate needs. Consequently, blood flow can increase to some regions and decrease to other areas at the same time.

Blood pressure is the force blood exerts against the wall of a blood vessel. Owing to cardiac activity, pressure is highest at the heart end of any artery. Because of the effect of peripheral resistance, which will be discussed shortly, pressure within the arteries (or any blood vessel) drops as the distance (vessel length) from the heart increases. This pressure gradient causes blood to move from and then back to the heart, always moving from high- to low-pressure areas.

Peripheral resistance is the opposition to blood flow resulting from the friction developed as blood streams through blood vessels. Three factors affect vessel resistance: blood viscosity, vessel radius, and vessel length.

Blood viscosity is a measure of the "thickness" of the blood, and is caused by the presence of proteins and formed elements in the plasma (the fluid part of the blood). As the viscosity of a fluid increases, its flow rate through a tube decreases. Blood viscosity in healthy people normally does not change, but certain conditions such as too many or too few blood cells may modify it.

Controlling *blood vessel radius* (one-half of the diameter) is the principal method of blood flow control. This is accomplished by contracting or relaxing the smooth muscle within the blood vessel walls. To see why radius has such a pronounced effect on blood flow, consider the physical relationship between blood and the vessel wall. Blood in direct contact with the vessel wall flows relatively slowly because of the friction, or drag, between the blood and the lining of the vessel. In contrast, fluid in the center of the vessel flows more freely because it is not rubbing against the vessel wall. Now picture a large- and a small-radius vessel: proportionately more blood is in contact with the wall of the small vessel; hence blood flow is notably impeded as vessel radius decreases.

Although *vessel length* does not ordinarily change in a healthy person, any increase in vessel length causes a corresponding flow decrease. This effect is principally caused by friction between blood and the vessel wall. Consequently, given two blood vessels of the same diameter, the longer vessel will have more resistance, and thus a reduced blood flow.

The Effect of Blood Pressure and Vessel Resistance on Blood Flow

Poiseuille's equation describes the relationship between pressure, vessel radius, viscosity, and vessel length on blood flow:

$$\text{Blood flow } (\Delta Q) = \frac{\pi \Delta P\, r^4}{8 \eta l}$$

In the equation, ΔP is the pressure difference between the two ends of the vessel and represents the driving force behind blood flow. Viscosity (η) and blood vessel length (l) are not commonly altered in a healthy adult. We can also see from the equation that blood flow is directly proportional to the fourth power of vessel radius (r^4), which means that small variations in vessel radius translate into large changes in blood flow. In the human body, changing blood vessel radius provides an extremely effective and sensitive method of blood flow control. Peripheral resistance is the most important factor in blood flow control, because circulation to individual organs can be independently regulated even though systemic pressure may be changing.

Vessel Resistance

Imagine for a moment that you are one of the first cardiovascular researchers interested in the physics of blood flow. Your first task as the principal investigator for this project is to plan an effective experimental design simulating a simple fluid pumping system that can be related to the mechanics of the cardiovascular system. The initial phenomenon you study is how fluids, including blood, flow through tubes or blood vessels. Questions you might ask include:

1. What role does pressure play in the flow of fluid?
2. How does peripheral resistance affect fluid flow?

The equipment required to solve these and other questions has already been designed for you in the form of a computerized simulation, which frees you to focus on the logic of the experiment. The first part of the computer simulation indirectly investigates the effects of pressure, vessel radius, viscosity, and vessel length on fluid flow. The second part of the experiment will explore the effects of several variables on the output of a single-chamber pump. Follow the specific guidelines in the exercise for collecting data. As you do so, also try to imagine alternate methods of achieving the same experimental goal.

Choose **Cardiovascular Dynamics** from the main menu. The opening screen will appear in a few seconds (Figure 33B.1).

The primary features on the screen when the program starts are a pair of glass beakers perched atop a simulated electronic device called the *equipment control unit*, which is used to set experiment parameters and to operate the equipment. When the **Start** button (beneath the left beaker) is clicked, the simulated blood flows from the left beaker (source) to the right beaker (destination) through the connecting tube. To relate this to the human body, think of the left beaker as the left side of your heart, the tube as your aorta, and the right beaker as any organ to which blood is flowing.

Clicking the **Refill** button refills the source beaker after an experimental trial. Experimental parameters can be adjusted by clicking the plus (+) or minus (−) buttons to the right of each display window.

The equipment in the lower part of the screen is called the *data collection unit*. This equipment records and displays data you accumulate during the experiments. The data set for the first experiment (Radius) is highlighted in the **Data Sets** window. You can add or delete a data set by clicking the appropriate button to the right of the **Data Sets** window. The **Record Data** button in the lower right part of the screen activates automatically after an experimental trial. Clicking the **Delete Line** or **Clear Data Set** buttons erases any data you want to delete.

You will record the data you accumulate in the experimental values grid in the lower middle part of the screen.

Activity 1: Studying the Effect of Flow Tube Radius on Fluid Flow

Our initial study will examine the effect of flow tube radius on fluid flow.*

1. Open the Vessel Resistance window if it isn't already open.

The Radius line in the data collection unit should be highlighted in bright blue. If it is not, choose it by clicking the **Radius** line. The data collection unit will now record flow variations due to changing flow tube radius.

* If you need help identifying any piece of equipment, choose Balloons On from the Help menu and move the mouse pointer onto any piece of equipment visible on the computer's screen. As the pointer touches the object, a pop-up window appears to identify the equipment. To close the pop-up window, move the mouse pointer away from the equipment. Repeat the process for all equipment on the screen until you feel confident with it. When finished, choose Balloons On again to turn off this help feature.

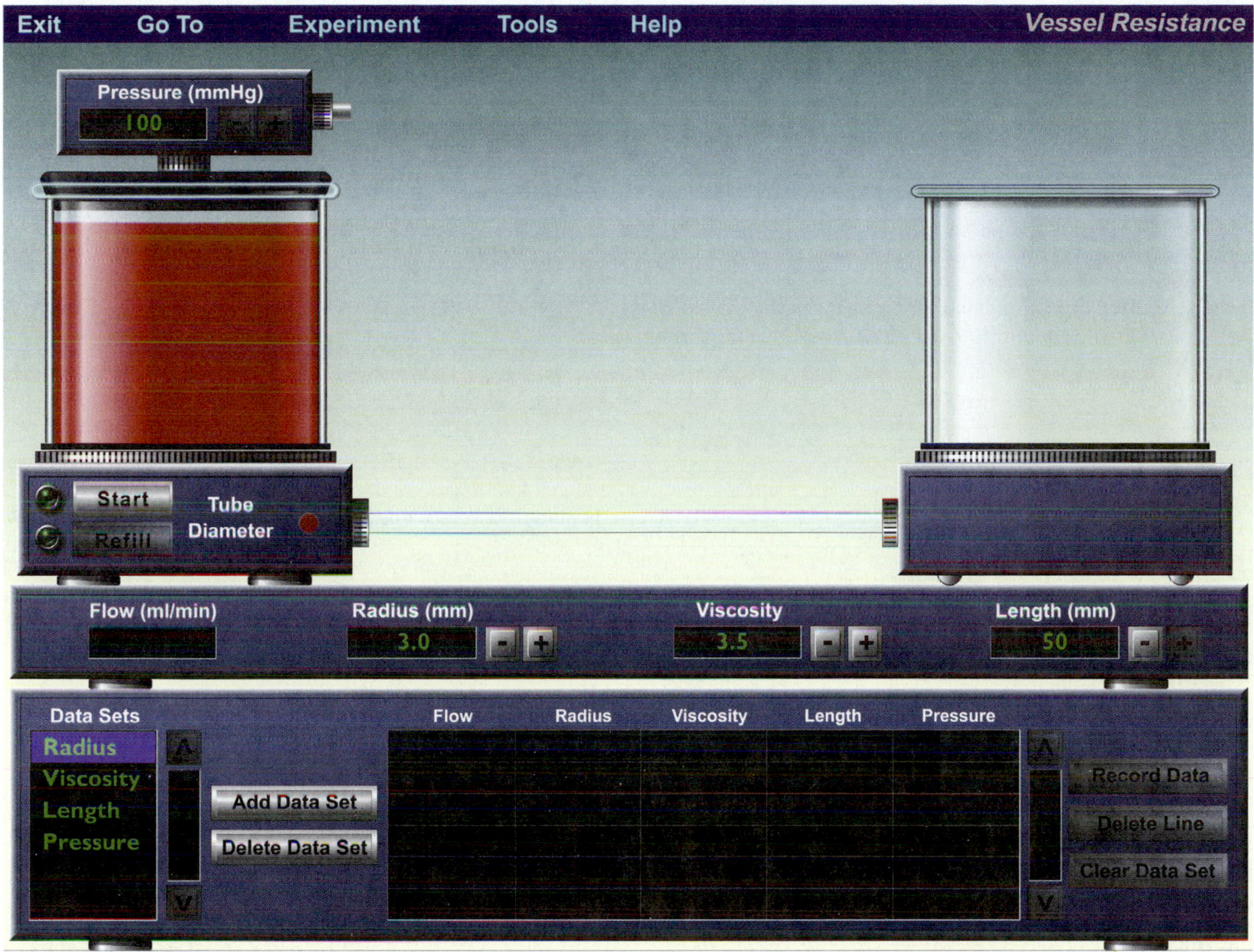

Figure 33B.1 Opening screen of the Vessel Resistance experiment.

If the data grid is not empty, click **Clear Data Set** to discard all previous values.

If the left beaker is not full, click **Refill.**

2. Adjust the flow tube radius to 1.5 mm and the viscosity to 1.0 by clicking the appropriate (+) or (−) button. During the course of this part of the experiment, maintain the other experiment conditions at:

100 mm Hg driving pressure (top left)

50-mm flow tube length (middle right)

3. Click **Start** and watch the fluid move into the right beaker. (Fluid moves slowly under some conditions—be patient!) Pressure (currently set to 100 mm Hg) propels fluid from the left beaker to the right beaker through the flow tube. The flow rate is displayed in the Flow window when the left beaker has finished draining. Now click **Record Data** to display the flow rate and experiment parameters in the experimental values grid (and retain the data in the computer's memory for printing and saving). Click **Refill** to replenish the left beaker.

4. Increase the radius in 0.5-mm increments and repeat step 3 until the maximum radius (6.0 mm) is achieved. Be sure to click **Record Data** after each fluid transfer. If you make an error and want to delete a data value, click the data line in the experimental values grid and then click **Delete Line.**

5. View your data graphically by choosing **Plot Data** from the **Tools** menu. Choose **Radius** as the data set to be graphed, and then use the slider bars to select the radius data to be plotted on the X-axis and the flow data to be plotted on the Y-axis. You can highlight individual data points by clicking a line in the data grid. When you are finished, click the close box at the top of the plot window.

What happened to fluid flow as the radius of the flow tube was increased?

____

____

____

Because fluid flow is proportional to the fourth power of the radius, ____________________ changes in tube radius cause ____________________ changes in fluid flow.

Is the relationship between fluid flow and flow tube radius linear or exponential? ______________________________

In this experiment, a simulated motor changes the diameter of the flow tube. Explain how our blood vessels alter blood flow.

____

____

____

____

After a heavy meal when we are relatively inactive, we might expect blood vessels in the skeletal muscles to be somewhat ________________, whereas blood vessels in the digestive organs are probably ____________________. ■

Activity 2: Studying the Effect of Viscosity on Fluid Flow

With a viscosity of 3 to 4, blood is much more viscous than water (1.0 viscosity). Although viscosity is altered by factors such as dehydration and altered blood cell numbers, a body in homeostatic balance has a relatively stable blood consistency. Nonetheless it is useful to examine the effects of viscosity changes on fluid flow, because we can then predict what might transpire in the human cardiovascular system under certain homeostatic imbalances.

1. Set the starting conditions as follows:

- 100 mm Hg driving pressure
- 5.0-mm flow tube radius
- 1.0 viscosity
- 50-mm flow tube length

2. Click the **Viscosity** data set in the data collection unit. (This action prepares the experimental values grid to record the viscosity data.)

3. **Refill** the left beaker if you have not already done so.

4. Click **Start** to begin the experiment. After all the fluid has drained into the right beaker, click **Record Data** to record this data point, and then click **Refill** to replenish the left beaker.

5. In 1.0-unit increments, increase the fluid viscosity and repeat step 4 until the maximum viscosity (10.0) is reached.

6. View your data graphically by choosing **Plot Data** from the **Tools** menu. Choose **Viscosity** as the data set to be graphed, and then use the slider bars to select the viscosity data to be plotted on the X-axis and the flow data to be plotted on the Y-axis. You can highlight individual data points by clicking a line in the data grid. When finished, click the close box at the top of the plot window.

How does fluid flow change as viscosity is modified?

____

____

____

Is fluid flow versus viscosity an inverse or direct relationship?

____

How does the effect of viscosity compare with the effect of radius on fluid flow?

____

____

Predict the effect of anemia (e.g., fewer red blood cells than normal) on blood flow.

____

____

What might happen to blood flow if we increased the number of blood cells?

____

____

____

Explain why changing blood viscosity would or would not be a reasonable method for the body to control blood flow.

____

____

____

____

Blood viscosity would ________________ in conditions of dehydration, resulting in ________________ blood flow. ■

Activity 3: Studying the Effect of Flow Tube Length on Fluid Flow

With the exception of the normal growth that occurs until the body reaches full maturity, blood vessel length does not significantly change. In this activity you will investigate the physical relationship between vessel length and blood movement; specifically, how blood flow changes in flow tubes (vessels) of constant radius but of different lengths.

1. Set the starting conditions as follows:

- 100 mm Hg driving pressure
- 5.0-mm flow tube radius
- 3.5 viscosity
- 10-mm flow tube length

2. Click the **Length** data set in the data collection unit. (This action prepares the experimental values grid to display the length data.)

3. **Refill** the left beaker if you have not already done so.

4. Click **Start** to begin the experiment. After all the fluid has drained into the right beaker, click **Record Data** to record this data point, and then click **Refill** to refill the left beaker.

5. In 5-mm increments, increase the flow tube length by clicking (+) next to the Length window and repeat step 4 until the maximum length (50 mm) has been reached.

6. View your data graphically by choosing **Plot Data** from the **Tools** menu. Choose **Length** as the data set to be graphed, and then use the slider bars to select the length data to be plotted on the X-axis and the flow data to be plotted on the Y-axis. You can highlight individual data points by clicking a line in the grid. When finished, click the close box at the top of the plot window.

How does flow tube length affect fluid flow?

____

____

____

Explain why altering blood vessel length would or would not be a good method of controlling blood flow in the body.

____

____

____

____ ■

Activity 4: Studying the Effect of Pressure on Fluid Flow

The pressure difference between the two ends of a blood vessel is the driving force behind blood flow. In comparison, our experimental setup pressurizes the left beaker, thereby providing the driving force that propels fluid through the flow tube to the right beaker. You will examine the effect of pressure on fluid flow in this part of the experiment.

1. Set the starting conditions as follows:

- 25 mm Hg driving pressure
- 5.0-mm flow tube radius
- 3.5 viscosity
- 50-mm flow tube length

2. Click the **Pressure** data set in the data collection unit. (This action prepares the experimental values grid for the pressure data.)

3. **Refill** the left beaker if you have not already done so.

4. Click **Start** to begin the experiment. After all the fluid has moved into the right beaker, click **Record Data** to record this data point. Click **Refill** to refill the left beaker.

5. In increments of 25 mm Hg, increase the driving pressure by clicking (+) next to the Pressure window and repeat step 4 until the maximum pressure (225 mm Hg) has been reached.

6. View your data graphically by choosing **Plot Data** from the **Tools** menu. Choose **Pressure** as the data set to be graphed, and then use the slider bars to select the pressure data to be plotted on the X-axis and the flow data to be plotted on the Y-axis. You can highlight individual data points by clicking a line in the data grid. When finished, click the close box at the top of the plot window.

How does driving pressure affect fluid flow?

____

____

____

How does this plot differ from the others?

____

____

____

Although changing pressure could be used as a means of blood flow control, explain why this approach would not be as effective as altering blood vessel radius.

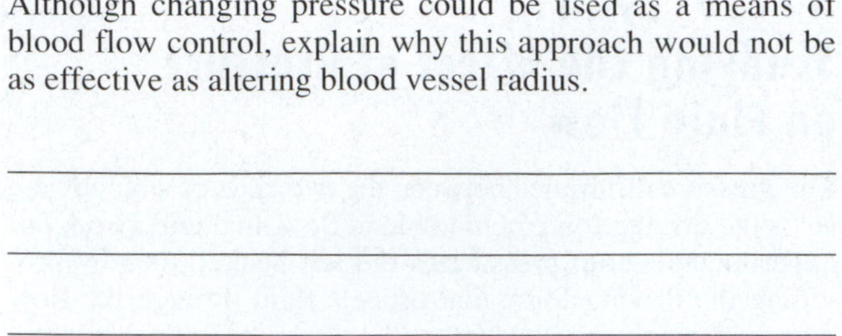

7. Click **Tools → Print Data** to print your recorded data for this experiment. ■

Pump Mechanics

In the human body, the heart beats approximately 70 strokes each minute. Each heartbeat consists of a filling interval during which blood moves into the chambers of the heart, and an ejection period when blood is actively pumped into the great arteries. The pumping activity of the heart can be described in terms of the phases of the cardiac cycle. Heart chambers fill during **diastole** (relaxation of the heart) and pump blood out during **systole** (contraction of the heart). As you can imagine, the length of time the heart is relaxed is one factor that determines the amount of blood within the heart at the end of the filling interval. Up to a point, increasing ventricular filling time results in a corresponding increase in ventricular volume. The volume in the ventricles at the end of diastole, just before cardiac contraction, is called the **end diastolic volume,** or **EDV.**

Blood moves from the heart into the arterial system when systolic pressure increases above the residual pressure (from the previous systole) in the great arteries leaving the heart. Although ventricular contraction causes blood ejection, the heart does not empty completely; a small quantity of blood—the **end systolic volume,** or **ESV**—remains in the ventricles at the end of systole.

Because the oxygen requirements of body tissue change depending on activity levels, we would expect the **cardiac output** (amount of blood pumped by each ventricle per minute) to vary correspondingly. We can calculate the **stroke volume** (**SV,** the amount of blood pumped per contraction of each ventricle) by subtracting the end systolic volume from the end diastolic volume (SV = EDV – ESV). We then compute cardiac output by multiplying the stroke volume by heart rate.

The human heart is a complex four-chambered organ, consisting of two individual pumps (the right and left sides) connected together in series. The right heart pumps blood through the lungs into the left heart, which in turn delivers blood to the systems of the body. Blood then returns to the right heart to complete the circuit.

Using the Pump Mechanics part of the program you will explore the operation of a simple one-chambered pump, and apply the physical concepts in the simulation to the operation of either of the two pumps composing the human heart.

In this experiment you can vary the starting and ending volumes of the pump (analogous to EDV and ESV, respectively), driving and resistance pressures, and the diameters of the flow tubes leading to and from the pump chamber. As you proceed through the exercise, try to apply the ideas of ESV, EDV, cardiac output, stroke volume, and blood flow to the on-screen simulated pump system. For example, imagine that the flow tube leading to the pump from the left represents the pulmonary veins, while the flow tube exiting the pump to the right represents the aorta. The pump would then represent the left side of the heart.

Select **Pump Mechanics** from the **Experiment** menu. The equipment for the Pump Mechanics part of the experiment appears (Figure 33B.2).

As in the previous experiment, there are two simulated electronic control units on the computer's screen. The upper apparatus is the *equipment control unit,* which is used to adjust experiment parameters and to operate the equipment. The lower apparatus is the *data collection* unit, in which you will record the data you accumulate.

This equipment differs slightly from that used in the Vessel Resistance experiment. There are still two beakers: the *source* beaker on the left and the *destination* beaker on the right. Now the pressure in each beaker is individually controlled by the small pressure units on top of the beakers. Between the two beakers is a simple pump, which can be thought of as one side of the heart, or even as a single ventricle (e.g., left ventricle). The left beaker and flow tube are analogous to the venous side of human blood flow, while arterial circulation is simulated by the right flow tube and beaker. One-way valves in the flow tubes supplying the pump ensure fluid movement in one direction—from the left beaker into the pump, and then into the right beaker. If you imagine that the pump represents the left ventricle, then think of the valve to the left of the pump as the bicuspid valve and the valve to the right of the pump as the aortic semilunar valve. The pump is driven by a pressure unit mounted on its cap. An important distinction between the pump's pressure unit and the pressure units atop the beakers is that the pump delivers pressure only during its downward stroke. Upward pump strokes are driven by pressure from the left beaker (the pump does not exert any resistance to flow from the left beaker during pump filling). In contrast, pressure in the right beaker works against the pump pressure, which means that the net pressure driving the fluid into the right beaker is calculated (automatically) by subtracting the right beaker pressure from the pump pressure. The resulting pressure difference between the pump and the right beaker is displayed in the experimental values grid in the data collection unit as **Pres.Dif.R.**

Clicking the **Auto Pump** button will cycle the pump through the number of strokes indicated in the Max. strokes window. Clicking the **Single** button cycles the pump through one stroke. During the experiment, the pump and flow rates are automatically displayed when the number of pump strokes is 5 or greater. The radius of each flow tube is individually controlled by clicking the appropriate button. Click (+) to increase flow tube radius or (–) to decrease flow tube radius.

The pump's stroke volume (the amount of fluid ejected in one stroke) is automatically computed by subtracting its ending volume from the starting volume. You can adjust starting and ending volumes, and thereby stroke volume, by clicking the appropriate (+) or (–) button in the equipment control unit.

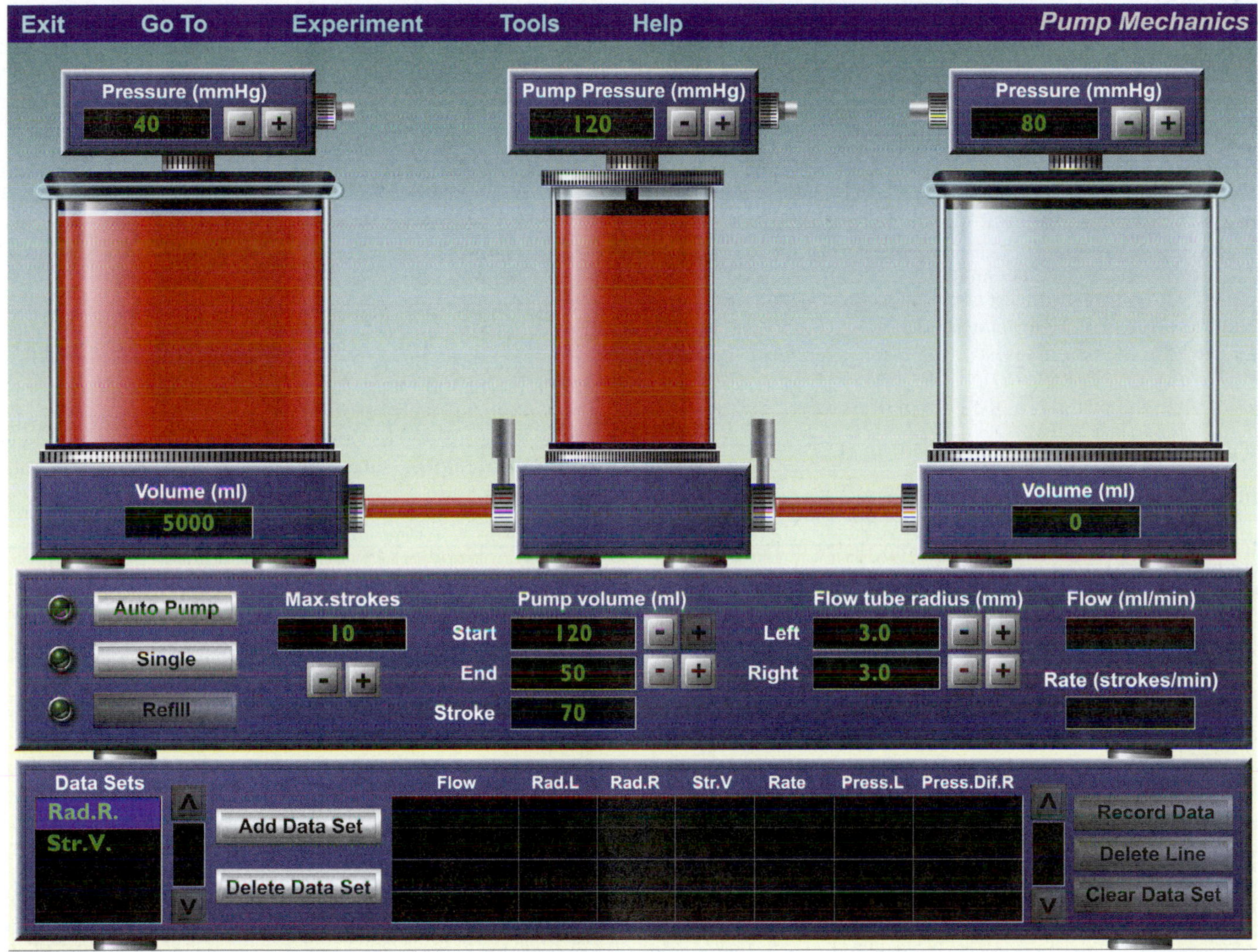

Figure 33B.2 Opening screen of the Pump Mechanics experiment.

The data collection unit records and displays data you accumulate during the experiments. The data set for the first experiment (**Rad.R.,** which represents right flow tube radius) is highlighted in the **Data Sets** window. You can add or delete a data set by clicking the appropriate button to the right of the **Data Sets** window. Clicking **Delete Data Set** will erase the data set itself, including all the data it contained. The **Record Data** button at the right edge of the screen activates automatically after an experimental trial. When clicked, the **Record Data** button displays the flow rate data in the experimental values grid and saves it in the computer's memory. Clicking the **Delete Line** or **Clear Data Set** button erases any data you want to delete.

Activity 5:

Studying the Effect of Radius on Pump Activity

Although you will be manipulating the radius of only the right flow tube in this part of the exercise, try to predict the consequence of altering the left flow tube radius as you collect the experimental data. (Remember that the left flow tube simulates the pulmonary veins and the right flow tube simulates the aorta.)

1. Open the pump mechanics window if it isn't already open.

Click the **Rad.R.** data set to activate it. The data collection unit is now ready to record flow variations due to changing flow tube radius.

If the experimental values grid is not empty, click **Clear Data Set** to discard all previous values.

If the left beaker is not full, click **Refill.**

2. Adjust the **right** flow tube radius to 2.5 mm, and the **left** flow tube radius to 3.0 mm by clicking and holding the appropriate button. During the entire radius part of the experiment, maintain the other experiment conditions at:

- 40 mm Hg for left beaker pressure (this pressure drives fluid into the pump, which offers no resistance to filling)
- 120 mm Hg for pump pressure (pump pressure is the driving force that propels fluid into the right beaker)

- 80 mm Hg for right beaker pressure (this pressure is the resistance to the pump's pressure)
- 120-ml Start volume in pump (analogous to EDV)
- 50-ml End volume in pump (analogous to ESV)
- 10 strokes in the Max. strokes window

Notice that the displayed 70-ml stroke volume is automatically calculated. Before starting, click the **Single** button one or two times and watch the pump action.

To be sure you understand how this simple mechanical pump can be thought of as a simulation of the human heart, complete the following statements by circling the correct term within the parentheses:

a. When the piston is at the bottom of its travel, the volume remaining in the pump is analogous to (EDV, ESV) of the heart.

b. The amount of fluid ejected into the right beaker by a single pump cycle is analogous to (stroke volume, cardiac output) of the heart.

c. The volume of blood in the heart just before systole is called (EDV, ESV), and is analogous to the volume of fluid present in the simulated pump when it is at the (top, bottom) of its stroke.

3. Click the **Auto Pump** button in the equipment control unit to start the pump. After the 10 stroke volumes have been delivered, the Flow and Rate windows will automatically display the experiment results. Now click **Record Data** to display the figures you just collected in the experimental values grid. Click **Refill** to replenish the left beaker.

4. Increase the *right* flow tube radius in 0.5-mm increments and repeat step 3 above until the maximum radius (6.0 mm) is achieved. Be sure to click **Record Data** after each trial.

5. When you have completed the experiment, view your data graphically by choosing **Plot Data** from the **Tools** menu. Choose **Rad.R.** as the data set to be graphed, and then use the slider bars to select the Rad.R. data to be plotted on the X-axis and the flow data to be plotted on the Y-axis. You can highlight individual data points by clicking a line in the data grid.

The total flow rates you just determined depend on the flow rate into the pump from the left and on the flow rate out of the pump toward the right. Consequently, the shape of the plot is different from what you might predict after viewing the vessel resistance radius graph.

Try to explain why this graph differs from the radius plot in the Vessel Resistance experiment. Remember that the flow rate into the pump did not change, whereas the flow rate out of the pump varied according to your radius manipulations. When you have finished, click the close box at the top of the plot window.

____

____

____

____

Complete the following statements by circling the correct term within the parentheses.

a. As the right flow tube radius is increased, fluid flow rate (increases, decreases). This is analogous to (dilation, constriction) of blood vessels in the human body.

b. Even though the pump pressure remains constant, the pump rate (increases, decreases) as the radius of the right flow tube is increased. This happens because the resistance to fluid flow is (increased, decreased).

Apply your observations of the simulated mechanical pump to complete the following statements about human heart function. If you are not sure how to formulate your response, use the simulation to arrive at an answer. Circle the correct term within the parentheses.

c. The heart must contract (more, less) forcefully to maintain cardiac output if the resistance to blood flow in the vessels exiting the heart is increased.

d. Increasing the resistance (that is constricting) the blood vessels entering the heart would (increase, decrease) the time needed to fill the heart chambers.

What do you think would happen to the flow rate and the pump rate if the left flow tube radius is changed (either increased or decreased)?

____

____

____ ■

Activity 6:
Studying the Effect of Stroke Volume on Pump Activity

Whereas the heart of a person at rest pumps about 60% of the blood in its chambers, 40% of the total amount of blood remains in the chambers after systole. The 60% of blood ejected by the heart is called the stroke volume, and is the difference between EDV and ESV. Even though our simple pump in this experiment does not work exactly like the human heart, you can apply the concepts to basic cardiac function. In this experiment, you will examine how the activity of the simple pump is affected by changing the pump's starting and ending volumes.

1. Click the **Str.V.** (stroke volume) data set to activate it. If the experimental values grid is not empty, click **Clear Data Set** to discard all previous values. If the left beaker is not full, click **Refill.**

2. Adjust the stroke volume to 10 ml by setting the Start volume (EDV) to 120 ml and the End volume (ESV) to 110 ml (stroke volume = start volume – end volume). During the entire stroke volume part of the experiment, keep the other experimental conditions at:

- 40 mm Hg for left beaker pressure
- 120 mm Hg for pump pressure
- 80 mm Hg for right beaker pressure
- 3.0 mm for left and right flow tube radius
- 10 strokes in the Max. strokes window

3. Click the **Auto Pump** button to start the experiment. After 10 stroke volumes have been delivered, the Flow and Rate windows will display the experiment results given the current parameters. Click **Record Data** to display the figures you just collected in the experimental values grid. Click **Refill** to replenish the left beaker.

4. Increase the stroke volume in 10-ml increments (by decreasing the End volume) and repeat step 3 until the maximum stroke volume (120 ml) is achieved. Be sure to click the **Record Data** button after each trial. Watch the pump action during each stroke to see how you can apply the concepts of starting and ending pump volumes to EDV and ESV of the heart.

5. View your data graphically by choosing **Plot Data** from the **Tools** menu. Choose **Str.V.** as the data set to be graphed, and then use the slider bars to select the Str.V. data to be plotted on the X-axis and the pump rate data to be plotted on the Y-axis. Answer the following questions. When you have finished, click the close box at the top of the plot window.

What happened to the pump's rate as its stroke volume was increased?

____

____

____

Using your simulation results as a basis for your answer, explain why an athlete's resting heart rate might be lower than that of the average person.

____

____

____

Applying the simulation outcomes to the human heart, predict the effect of increasing the stroke volume on cardiac output (at any given rate).

____

____

____

When heart rate is increased, the time of ventricular filling is (circle one: increased, decreased), which in turn (increases, decreases) the stroke volume.

What do you think might happen to the pressure in the pump during filling if the valve in the right flow tube became leaky? (Remember that the pump offers no resistance to filling.)

____

____

Applying this concept to the human heart, what might occur in the left heart and pulmonary blood vessels if the aortic valve became leaky?

____

____

What might occur if the aortic valve became slightly constricted?

____

____

In your simulation, increasing the pressure in the right beaker is analogous to the aortic valve becoming (circle one: leaky, constricted). ■

Activity 7:
Studying Combined Effects

In this section, you will set up your own experimental conditions to answer the following questions. Carefully examine each question and decide how to set experiment parameters to arrive at an answer. You can examine your previously collected data if you need additional information. Record several data points for each question as evidence for your answer (unless the question calls for a single pump stroke).

Click the **Add Data Set** button in the data collection unit. Next, create a new data set called Combined. Your newly created data set will be displayed beneath Str.V. Now click the **Combined** line to activate the data set. As you collect the supporting data for the following questions, be sure to click **Record Data** each time you have a data point you wish to keep for your records.

How is the flow rate affected when the right flow tube radius is kept constant (at 3.0 mm) and the left flow tube radius is modified (either up or down)?

____

____

____

____

How does decreasing left flow tube radius affect pump chamber filling time? Does it affect pump chamber emptying?

____

____

____

____

You have already examined the effect of changing the pump's end volume as a way of manipulating stroke volume. What happens to flow and pump rate when you keep the end volume constant and alter the start volume to manipulate stroke volume?

____

____

____

____

Try manipulating the pressure delivered to the left beaker. How does changing the left beaker pressure affect flow rate? (This change would be similar to changing pulmonary vein pressure.)

____

____

If the left beaker pressure is decreased to 10 mm Hg, how is pump-filling time affected?

____

____

____

What happens to the pump rate if the filling time is shortened?

____

____

What happens to fluid flow when the right beaker pressure equals the pump pressure?

____

____

____ ■

Activity 8:
Studying Compensation

In this activity you will explore the concept of cardiovascular compensation. Click the **Add Data Set** button in the data collection unit. Next, create a new data set called **Comp.** Your newly created data set will be displayed beneath Combined. Now click the **Comp. Line** to activate the data set. As you collect the supporting data for the following questions, be sure to click **Record Data** each time you have a data point you wish to keep for your records.

Adjust the experimental conditions to the following:

- 40 mm Hg for left beaker pressure
- 120 mm Hg for pump pressure
- 80 mm Hg for right beaker pressure
- 3.0 mm for left and right flow tube radius
- 10 strokes in the Max. strokes window
- 120-ml Start volume in pump
- 50-ml End volume in pump

Click **Auto Pump** and then record your flow rate data. Let's declare the value you just obtained to be the "normal" flow rate for the purpose of this exercise.

Now decrease the right flow tube radius to 2.5 mm, and run another trial. How does this flow rate compare with "normal"?

____

____

Leave the right flow tube radius at 2.5 mm radius and try to adjust one or more other conditions to return flow to "normal." Think logically about what condition(s) might compensate for a decrease in flow tube radius. How were you able to accomplish this? (Hint: There are several ways.)

____

____

____

Decreasing the right flow tube radius is similar to a partial (circle one: leakage, blockage) of the aortic valve or (increased, decreased) resistance in the arterial system.

Explain how the human heart might compensate for such a condition.

____

____

____

____

To increase (or decrease) blood flow to a particular body system (e.g., digestive), would it be better to adjust heart rate or blood vessel diameter? Explain.

____

____

____

____

Complete the following statements by circling the correct response. (If necessary, use the pump simulation to help you with your answers.)

a. If we decreased overall peripheral resistance in the human body (as in an athlete), the heart would need to generate (more, less) pressure to deliver an adequate amount of blood flow, and arterial pressure would be (higher, lower).

b. If the diameter of the arteries of the body were partly filled with fatty deposits, the heart would need to generate (more, less) force to maintain blood flow, and pressure in the arterial system would be (higher, lower) than normal.

Click **Tools** → **Print Data** to print your recorded data for this experiment. ■

Histology Review Supplement

Turn to p. P-149 for a review of cardiovascular tissue.

exercise

34B Frog Cardiovascular Physiology: Computer Simulation

Objectives

1. To list the properties of cardiac muscle as automaticity and rhythmicity and define each.
2. To explain the statement "Cardiac muscle has an intrinsic ability to beat."
3. To compare the relative length of the refractory period of cardiac muscle with that of skeletal muscle, and explain why it is not possible to tetanize cardiac muscle.
4. To define *extrasystole* and to explain at what point in the cardiac cycle (and on an ECG tracing) an extrasystole can be induced.
5. To describe the effect of the following on heart rate: cold, heat, vagal stimulation, pilocarpine, digitalis, atropine, epinephrine, and potassium, sodium, and calcium ions.
6. To define *vagal escape* and discuss its value.
7. To define *ectopic pacemaker.*

Investigation of human cardiovascular physiology is very interesting, but many areas obviously do not lend themselves to experimentation. It would be tantamount to murder to inject a human subject with various drugs to observe their effects on heart activity, or to expose the human heart in order to study the length of its refractory period. However, this type of investigation can be done on frogs or computer simulations, and provides valuable data because the physiological mechanisms in these animals, or programmed into the computer simulation, are similar if not identical to those in humans.

In this exercise, you will conduct the cardiac investigations just mentioned.

Special Electrical Properties of Cardiac Muscle: Automaticity and Rhythmicity

Cardiac muscle differs from skeletal muscle both functionally and in its fine structure. Skeletal muscle must be electrically stimulated to contract. In contrast, heart muscle can and does depolarize spontaneously in the absence of external stimulation. This property, called **automaticity,** is due to plasma membranes that have reduced permeability to potassium ions, but still allow sodium ions to slowly leak into the cells. This leakage causes the muscle cells to slowly depolarize until the action potential threshold is reached and *fast calcium channels* open, allowing Ca^{2+} entry from the extracellular fluid. Shortly thereafter, contraction occurs.

The spontaneous depolarization-repolarization events occur in a regular and continuous manner in cardiac muscle, a property referred to as **rhythmicity.**

In the following experiment, you will observe these properties of cardiac muscle in a computer simulation. Additionally, your instructor may demonstrate this procedure using a real frog.

Nervous Stimulation of the Heart

Both the parasympathetic and sympathetic nervous systems innervate the heart. Stimulation of the sympathetic nervous system increases the rate and force of contraction of the heart. Stimulation of the parasympathetic nervous system (vagal nerves) decreases the depolarization rhythm of the sinoatrial node and slows transmission of excitation through the atrioventricular node. If vagal stimulation is excessive, the heart will stop beating. After a short time, the ventricles will begin to beat again. This is referred to as **vagal escape** and may be the result of sympathetic reflexes or initiation of a rhythm by the Purkinje fibers.

Baseline Frog Heart Activity

The heart's effectiveness as a pump is dependent both on intrinsic (within the heart) and extrinsic (external to the heart) controls. In this experiment, you will investigate some of these factors.

The nodal system, in which the "pacemaker" imposes its depolarization rate on the rest of the heart, is one intrinsic factor that influences the heart's pumping action. If its impulses fail to reach the ventricles (as in heart block), the ventricles continue to beat but at their own inherent rate, which is much slower than that usually imposed on them. Although heart contraction does not depend on nerve impulses, its rate can be modified by extrinsic impulses reaching it through the autonomic nerves. Additionally, cardiac activity is modified by various chemicals, hormones, ions, and metabolites. The effects of several of these chemical factors are examined in the next experimental series, Activities 4–9.

The frog heart has two atria and a single, incompletely divided ventricle. The pacemaker is located in the sinus venosus, an enlarged region between the venae cavae and the right atrium. The SA node of mammals may have evolved from the sinus venosus.

Choose **Frog Cardiovascular Physiology** from the main menu. The opening screen will appear in a few seconds (Figure 34B.1). When the program starts, you will see a tracing of the frog's heartbeat on the *oscilloscope display* in the upper right part of the screen. Because the simulation automatically adjusts itself to your computer's speed, you may not see the heart tracing appear in real-time. If you want to increase the speed of the tracing (at the expense of tracing quality), click the **Tools** menu, choose **Modify Display,** and then select **Increase Speed.**

The oscilloscope display shows the ventricular contraction rate in the Heart Rate window. The *heart activity window*

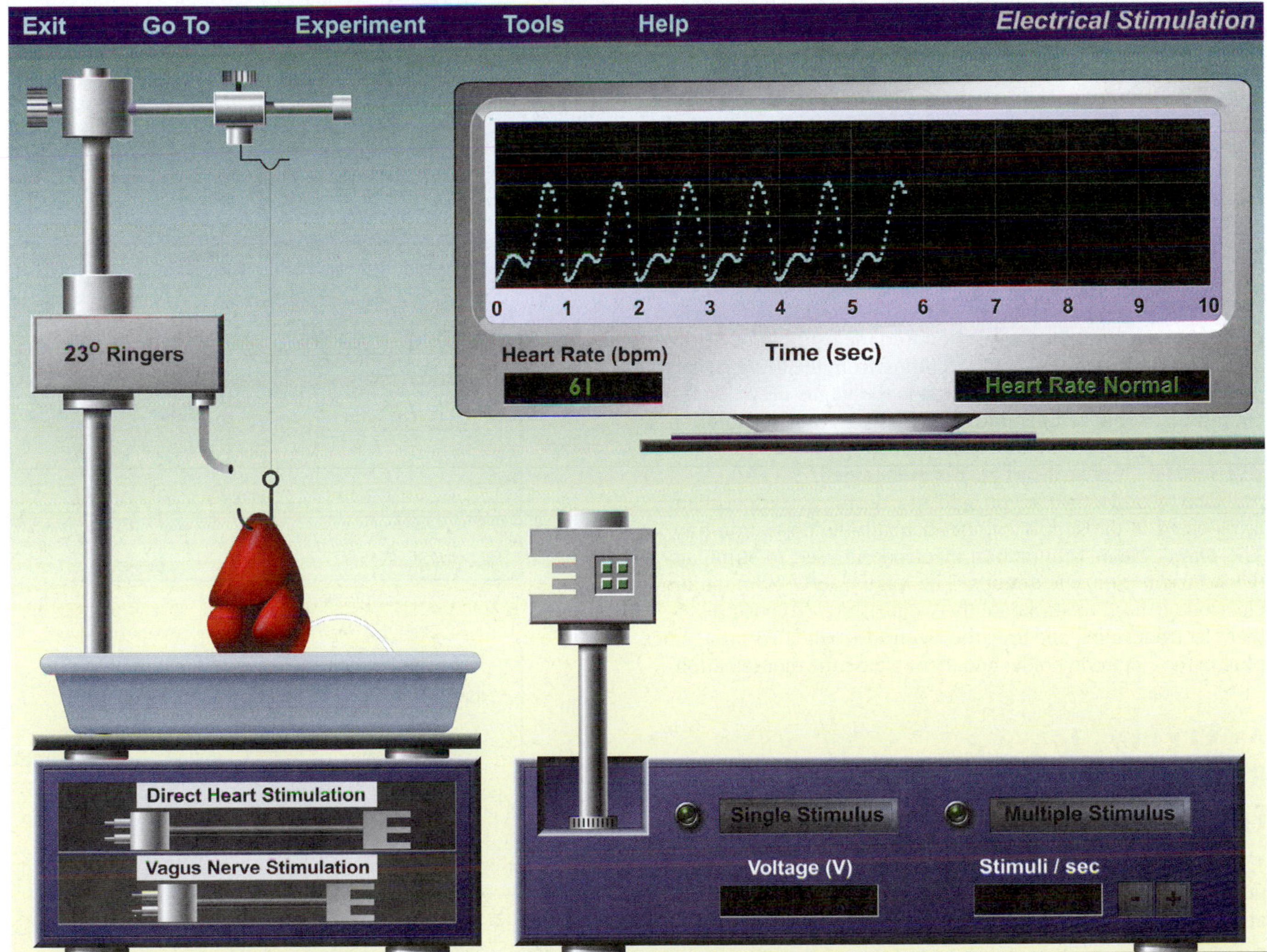

Figure 34B.1 Opening screen of the Electrical Stimulation experiment.

to the right of the Heart Rate display provides the following messages:

- Heart Rate Normal—displayed when the heart is beating under resting conditions.
- Heart Rate Changing—displayed when the heart rate is increasing or decreasing.
- Heart Rate Stable—displayed when the heart rate is steady, but higher or lower than normal. For example, if you applied a chemical that increased heart rate to a stable but higher-than-normal rate, you would see this message.

The *electrical stimulator* is below the oscilloscope display. In the experiment, clicking **Single Stimulus** delivers a single electrical shock to the frog heart. Clicking **Multiple Stimulus** delivers repeated electrical shocks at the rate indicated in the Stimuli/sec window just below the Multiple Stimulus button. When the **Multiple Stimulus** button is clicked, it changes to a **Stop Stimulus** button that allows you to stop electrical stimulation as desired. Clicking the (+) or (−) buttons next to the Stimuli/sec window adjusts the stimulus rate. The voltage delivered when Single Stimulus or Multiple Stimulus is clicked is displayed in the Voltage window. The simulation automatically adjusts the voltage for the experiment. The postlike apparatus extending upward from the electrical stimulator is the *electrode holder* into which you will drag-and-drop electrodes from the supply cabinet in the bottom left corner of the screen.

The left side of the screen contains the apparatus that sustains the frog heart. The heart has been lifted away from the body of the frog by a hook passed through the apex of the heart. Although the frog cannot be seen because it is in the dissection tray, its heart has not been removed from its circulatory system. A thin string connects the hook in the heart to the force transducer at the top of the support bracket. As the heart contracts, the string exerts tension on the force transducer that converts the contraction into the oscilloscope tracing. The slender white strand extending from the heart toward the right side of the dissection tray is the vagus nerve. In the simulation, room-temperature (23°C) frog Ringer's solution continuously drips onto the heart to keep it moist and responsive so that a regular heart beat is maintained.

The two electrodes you will use during the experiment are located in the supply cabinet beneath the dissection tray. The Direct Heart Stimulation electrode is used to stimulate the ventricular muscle directly. The Vagus Nerve Stimulation electrode is used to stimulate the vagus nerve. To position either electrode, click and drag the electrode to the two-pronged plug in the electrode holder and then release the mouse button.

Activity 1: Recording Baseline Frog Heart Activity

1. Before beginning to stimulate the frog heart experimentally, watch several heartbeats. Be sure you can distinguish atrial and ventricular contraction (Figure 34B.2a).

2. Record the number of ventricular contractions per minute displayed in the Heart Rate window under the oscilloscope.

_____________ bpm (beats per minute) ■

Activity 2: Investigating the Refractory Period of Cardiac Muscle

In Exercise 16B you saw that repeated rapid stimuli could cause skeletal muscle to remain in a contracted state. In other words, the muscle could be tetanized. This was possible because of the relatively short refractory period of skeletal muscle. In this experiment you will investigate the refractory period of cardiac muscle and its response to stimulation.

1. Click and hold the mouse button on the Direct Heart Stimulation electrode and drag it to the electrode holder.

2. Release the mouse button to lock the electrode in place. The electrode will touch the ventricular muscle tissue.

3. Deliver single shocks by clicking **Single Stimulus** at each of the following times. You may need to practice to acquire the correct technique.

- Near the beginning of ventricular contraction
- At the peak of ventricular contraction
- During the relaxation part of the cycle

Watch for **extrasystoles,** which are extra beats that show up riding on the ventricular contraction peak. Also note the compensatory pause, which allows the heart to get back on schedule after an extrasystole (Figure 34B.2b).

During which portion of the cardiac cycle was it possible to induce an extrasystole?

____

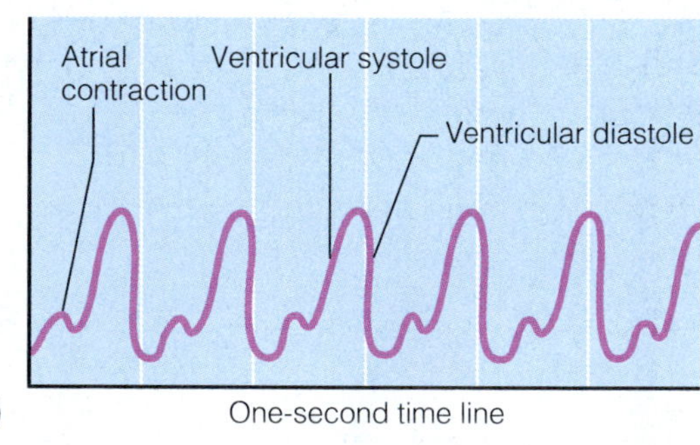

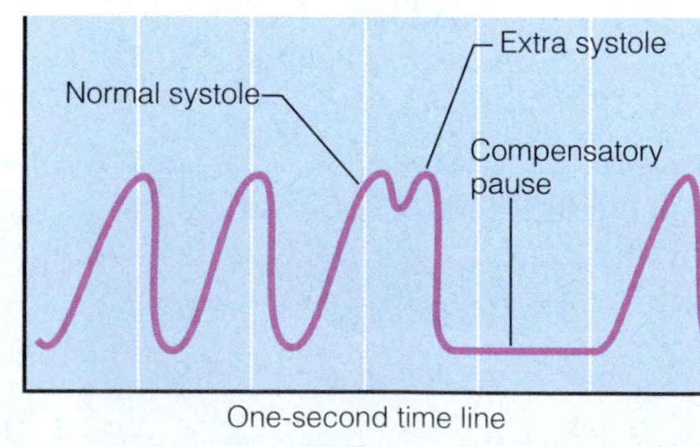

Figure 34B.2 Recording of contractile activity of a frog heart. (a) Normal heartbeat. **(b)** Induction of an extrasystole.

4. Attempt to tetanize the heart by clicking **Multiple Stimulus.** Electrical shocks will be delivered to the muscle at a rate of 20 stimuli/sec. What is the result?

Considering the function of the heart, why is it important that the heart muscle cannot be tetanized?

5. Click **Stop Stimulus** to stop the electrical stimulation. ■

Activity 3:

Examining the Effect of Vagus Nerve Stimulation

The vagus nerve carries parasympathetic impulses to the heart, which modify heart activity.

1. Click the Direct Heart Stimulation electrode to return it to the supply cabinet.
2. Click and drag the Vagus Nerve Stimulation electrode to the electrode holder.
3. Release the mouse button to lock the electrode in place. The vagus nerve will automatically be draped over the electrode contacts.
4. Adjust the stimulator to 50 stimuli/sec by clicking the (+) or (−) buttons.
5. Click **Multiple Stimulus.** Allow the vagal stimulation to continue until the heart stops momentarily and then begins to beat again (vagal escape), and then click **Stop Stimulus.**

What is the effect of vagal stimulation on heart rate?

The phenomenon of vagal escape demonstrates that many factors are involved in heart regulation and that any deleterious factor (in this case, excessive vagal stimulation) will be overcome, if possible, by other physiological mechanisms such as activation of the sympathetic division of the autonomic nervous system (ANS). ■

Assessing Physical and Chemical Modifiers of Heart Rate

Now that you have observed normal frog heart activity, you will have an opportunity to investigate the effects of various modifying factors on heart activity. After removing the agent in each activity, allow the heart to return to its normal rate before continuing with the testing.

Choose **Modifiers of Heart Rate** from the **Experiment** menu. The opening screen will appear in a few seconds (Figure 34B.3). The appearance and functionality of the *oscilloscope display* is the same as it was in the Electrical Stimulation experiment. The *solutions shelf* above the oscilloscope display contains the chemicals you'll use to modify heart rate in the experiment. You can choose the temperature of the Ringer's solution dispensed by clicking the appropriate button in the Ringer's dispenser at the left part of the screen. The doors to the supply cabinet are closed during this experiment because the electrical stimulator is not used.

When you click **Record Data** in the *data control unit* below the oscilloscope, your data is stored in the computer's memory and is displayed in the data grid at the bottom of the screen; data displayed include the solution used and the resulting heart rate. If you are not satisfied with a trial, you can click **Delete Line.** Click **Clear Table** if you wish to repeat the entire experiment.

Activity 4:

Assessing the Effect of Temperature

1. Click the **5°C Ringer's** button to bathe the frog heart in cold Ringer's solution. Watch the recording for a change in cardiac activity.
2. When the heart activity window displays the message Heart Rate Stable, click **Record Data** to retain your data in the data grid.

What change occurred with the cold (5°C) Ringer's solution?

3. Now click the **23°C Ringer's** button to flood the heart with fresh room-temperature Ringer's solution.
4. After you see the message Heart Rate Normal in the heart activity window, click the **32°C Ringer's** button.
5. When the heart activity window displays the message Heart Rate Stable, click **Record Data** to retain your data.

What change occurred with the warm (32°C) Ringer's solution?

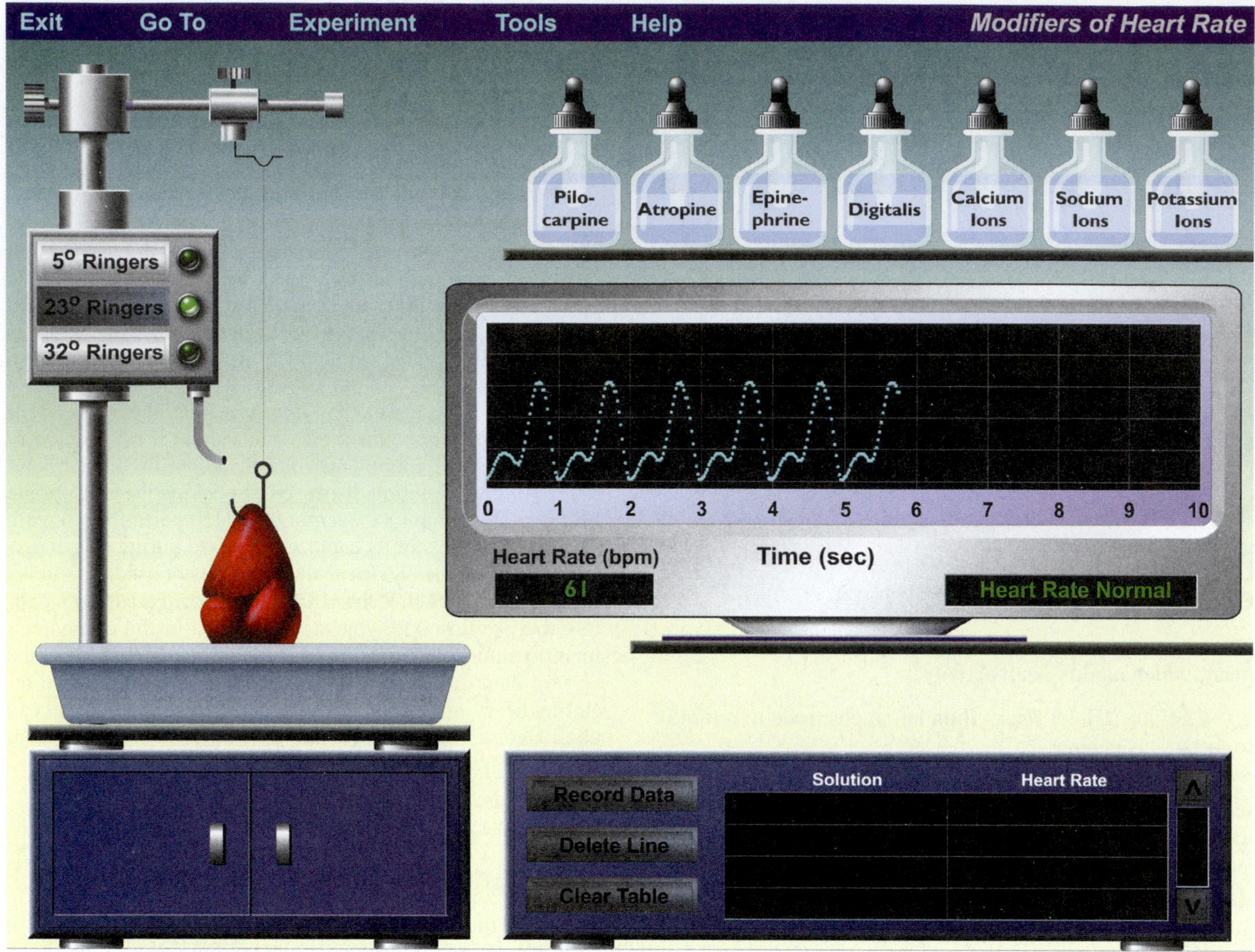

Figure 34B.3 Opening screen of the Modifiers of Heart Rate experiment.

Record the heart rate at the two temperatures below.

______________ bpm at 5°C; ______________ bpm at 32°C

What can you say about temperature and heart rate?

____

____

6. Click the **23°C Ringer's** button to flush the heart with fresh Ringer's solution. Watch the heart activity window for the message Heart Rate Normal before beginning the next test. ■

Activity 5:
Assessing the Effect of Pilocarpine

1. Click and hold the mouse on the Pilocarpine dropper cap.

2. Drag the dropper cap to a point about an inch above the heart and release the mouse.

3. Pilocarpine solution will be dispensed onto the heart and the dropper cap will automatically return to the Pilocarpine bottle.

4. Watch the heart activity window for the message Heart Rate Stable, indicating that the heart rate has stabilized under the effects of pilocarpine.

5. After the heart rate stabilizes, record the heart rate in the space provided below, and click **Record Data** to retain your data in the grid.

______________ bpm

What happened when the heart was bathed in the pilocarpine solution?

____

6. Click the **23°C Ringer's** button to flush the heart with fresh Ringer's solution. Watch the heart activity window for the message Heart Rate Normal, an indication that the heart is ready for the next test. ■

Pilocarpine simulates the effect of parasympathetic nerve (hence, vagal) stimulation by enhancing acetylcholine release; such drugs are called parasympathomimetic drugs.

Activity 6: Assessing the Effect of Atropine

1. Drag-and-drop the Atropine dropper cap to a point about an inch above the heart.
2. Atropine solution will automatically drip onto the heart and the dropper cap will return to its position in the Atropine bottle.
3. Watch the heart activity window for the message Heart Rate Stable.
4. After the heart rate stabilizes, record the heart rate in the space below, and click **Record Data** to retain your data in the grid.

______________ bpm

What is the effect of atropine on the heart?

____

Atropine is a drug that blocks the effect of the neurotransmitter acetylcholine, liberated by the parasympathetic nerve endings. Do your results accurately reflect this effect of atropine?

____

Are pilocarpine and atropine agonists or antagonists in their effects on heart activity?

____

5. Click the **23°C Ringer's** button to flush the heart with fresh Ringer's solution. Watch the heart activity window for the message Heart Rate Normal before beginning the next test. ■

Activity 7: Assessing the Effect of Epinephrine

1. Drag-and-drop the Epinephrine dropper cap to a point about an inch above the heart.
2. Epinephrine solution will be dispensed onto the heart and the dropper cap will return to the Epinephrine bottle.
3. Watch the heart activity window for the message Heart Rate Stable.
4. After the heart rate stabilizes, record the heart rate in the space provided below, and click **Record Data** to retain your data in the grid.

______________ bpm

What happened when the heart was bathed in the epinephrine solution?

____

Which division of the autonomic nervous system does its effect imitate?

____

5. Click the **23°C Ringer's** button to flush the heart with fresh Ringer's solution. Watch the heart activity window for the message Heart Rate Normal, meaning that the heart is ready for the next test. ■

Activity 8: Assessing the Effect of Digitalis

1. Drag-and-drop the Digitalis dropper cap to a point about an inch above the heart.
2. Digitalis solution will automatically drip onto the heart and then the dropper will return to the Digitalis bottle.
3. Watch the heart activity window to the right of the Heart Rate window for the message Heart Rate Stable.
4. After the heart rate stabilizes, record the heart rate in the space provided below, and click **Record Data** to retain your data in the grid.

______________ bpm

What is the effect of digitalis on the heart?

____

5. Click the **23°C Ringer's** button to flush the heart with fresh Ringer's solution. Watch the heart activity window for the message Heart Rate Normal, then proceed to the next test. ■

Digitalis is a drug commonly prescribed for heart patients with congestive heart failure. It slows heart rate, providing more time for venous return and decreasing the workload on the weakened heart. These effects are thought to be due to inhibition of the Na^+-K^+ pump and enhancement of Ca^{2+} entry into myocardial fibers.

Activity 9:
Assessing the Effect of Various Ions

To test the effect of various ions on the heart, apply the desired solution using the following method.

1. Drag-and-drop the Calcium Ions dropper cap to a point about an inch above the heart.

2. Calcium ions will automatically be dripped onto the heart and the dropper cap will return to the Calcium Ions bottle.

3. Watch the heart activity window for the message Heart Rate Stable.

4. After the heart rate stabilizes, record the heart rate in the space provided below, and click **Record Data** to retain your data in the grid.

5. Click the **23°C Ringer's** button to flush the heart with fresh Ringer's solution. Watch the heart activity window for the message Heart Rate Normal, which means that the heart is ready for the next test.

6. Repeat steps 1 through 5 for Sodium Ions and then Potassium Ions.

Effect of Ca^{2+}:

Does the heart rate stabilize and remain stable?

____

Describe your observations of force and rhythm of the heartbeat.

____

____

Effect of Na^+:

Does the heart rate stabilize and remain stable?

____

Describe your observations of force and rhythm of the heartbeat.

____

____

Effect of K^+:

Does the heart rate stabilize and remain stable?

____

Describe your observations of force and rhythm of the heartbeat.

____

____

Potassium ion concentration is normally higher within cells than in the extracellular fluid. *Hyperkalemia* decreases the resting potential of plasma membranes, thus decreasing the force of heart contraction. In some cases, the conduction rate of the heart is so depressed that **ectopic pacemakers** (pacemakers appearing erratically and at abnormal sites in the heart muscle) appear in the ventricle, and fibrillation may occur.

Was there any evidence of premature beats in the recording of potassium ion effects?

____

Was arrhythmia produced with any of the ions tested?

__________ If so, which?

____

7. Click **Tools → Print Data** to print your recorded data for this experiment. ■

Histology Review Supplement

Turn to p. P-149 for a review of cardiovascular tissue.

Respiratory System Mechanics: Computer Simulation

exercise **37B**

Objectives

1. To define:

   ventilation, inspiration, expiration, forced expiration, tidal volume, vital capacity, expiratory reserve volume, inspiratory reserve volume, residual volume, surfactant, minute respiratory volume, forced expiratory volume, and *pneumothorax*
2. To describe the role of muscles and volume changes in the mechanics of breathing.
3. To understand that the lungs do not contain muscle and that respirations are therefore caused by external forces.
4. To explore the effect of changing airway resistance on breathing.
5. To study the effect of surfactant on lung function.
6. To examine the factors that cause lung collapse.
7. To understand the effects of hyperventilation, rebreathing, and breath holding on the CO_2 level in the blood.

The two phases of **pulmonary ventilation** or **breathing** are **inspiration,** during which air is taken into the lungs, and **expiration,** during which air is expelled from the lungs. Inspiration occurs as the external intercostal muscles and the diaphragm contract. The diaphragm, normally a dome-shaped muscle, flattens as it moves inferiorly while the external intercostal muscles between the ribs lift the rib cage. These cooperative actions increase the thoracic volume. Because the increase in thoracic volume causes a partial vacuum, air rushes into the lungs. During normal expiration, the inspiratory muscles relax, causing the diaphragm to rise and the chest wall to move inward. The thorax returns to its normal shape due to the elastic properties of the lung and thoracic wall. Like a deflating balloon, the pressure in the lungs rises, which forces air out of the lungs and airways. Although expiration is normally a passive process, abdominal wall muscles and the internal intercostal muscles can contract to force air from the lungs. Blowing up a balloon is an example where such **forced expiration** would occur.

Simulating Spirometry: Measuring Respiratory Volumes and Capacities

This computerized simulation allows you to investigate the basic mechanical function of the respiratory system as you determine lung volumes and capacities. The concepts you will learn by studying this simulated mechanical lung can then be applied to help you understand the operation of the human respiratory system.

Normal quiet breathing moves about 500 ml (0.5 liter) of air (the tidal volume) in and out of the lungs with each breath, but this amount can vary due to a person's size, sex, age, physical condition, and immediate respiratory needs. The terms used for the normal respiratory volumes are defined next. The values are for the normal adult male and are approximate.

Normal Respiratory Volumes

Tidal volume (TV): Amount of air inhaled or exhaled with each breath under resting conditions (500 ml)

Expiratory reserve volume (ERV): Amount of air that can be forcefully exhaled after a normal tidal volume exhalation (1200 ml)

Inspiratory reserve volume (IRV): Amount of air that can be forcefully inhaled after a normal tidal volume inhalation (3100 ml)

Residual volume (RV): Amount of air remaining in the lungs after complete exhalation (1200 ml)

Vital capacity (VC): Maximum amount of air that can be exhaled after a normal maximal inspiration (4800 ml)

$$VC = TV + IRV + ERV$$

Total lung capacity (TLC): Sum of vital capacity and residual volume

An idealized tracing of the various respiratory volumes and their relationships to each other are shown in Figure 37A.2, p. 375.

Pulmonary Function Tests

Forced vital capacity (FVC): Amount of air that can be expelled when the subject takes the deepest possible breath and exhales as completely and rapidly as possible

Forced expiratory volume (FEV_1): Measures the percentage of the vital capacity that is exhaled during 1 second of the FVC test (normally 75% to 85% of the vital capacity)

Choose **Respiratory System Mechanics** from the main menu. The opening screen for the Respiratory Volumes experiment will appear in a few seconds (Figure 37B.1). The main features on the screen when the program starts are a pair of *simulated lungs within a bell jar* at the left side of the screen, an *oscilloscope* at the upper right part of the screen, a *data display* area beneath the oscilloscope, and a *data control unit* at the bottom of the screen.

The black rubber "diaphragm" sealing the bottom of the glass bell jar is attached to a rod in the pump just below the jar. The rod moves the rubber diaphragm up and down to change the pressure within the bell jar (comparable to the intrapleural pressure in the body). As the diaphragm moves inferiorly, the resulting volume increase creates a partial vacuum in the bell jar because of lowered pressure. This partial vacuum causes air to be sucked into the tube at the top of the bell jar and then into the simulated lungs.

Conversely, as the diaphragm moves up, the rising pressure within the bell jar forces air out of the lungs. The partition between the two lungs compartmentalizes the bell jar

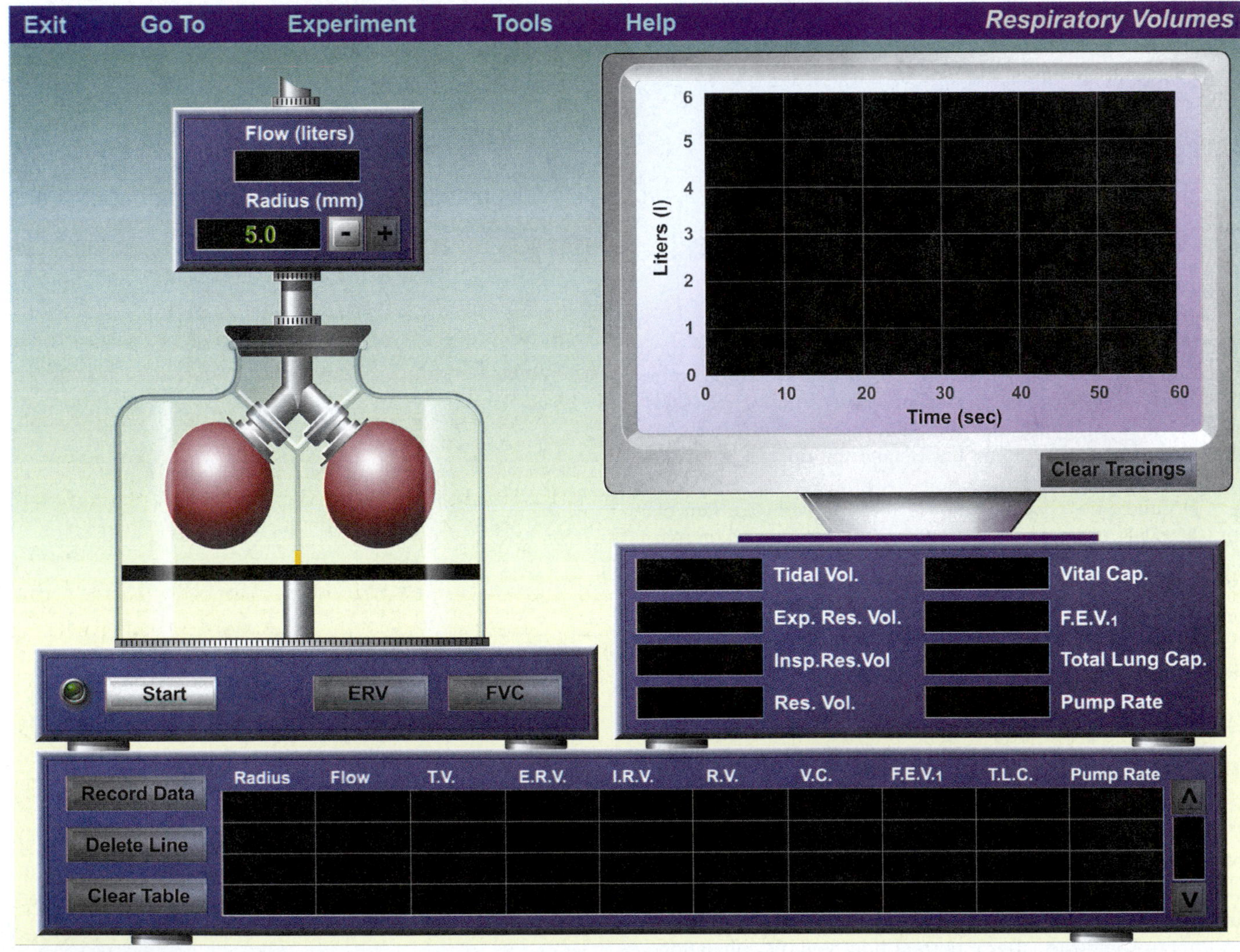

Figure 37B.1 Opening screen of the Respiratory Volumes experiment.

into right and left sides. The lungs are connected to an air-flow tube in which the diameter is adjustable by clicking the (+) and (−) buttons next to the Radius window in the equipment atop the bell jar. The volume of each breath that passes through the single air-flow tube above the bell jar is displayed in the Flow window. Clicking **Start** below the bell jar begins a trial run in which the simulated lungs will "breathe" in normal tidal volumes and the oscilloscope will display the tidal tracing. When **ERV** is clicked, the lungs will exhale maximally at the bottom of a tidal stroke and the expiratory reserve volume will be displayed in the Exp. Res. Vol. window below the oscilloscope. When **FVC** is clicked, the lungs will first inhale maximally and then exhale fully to demonstrate forced vital capacity. After ERV and FVC have been measured, the remaining lung values will be calculated and displayed in the small windows below the oscilloscope.

The data control equipment in the lower part of the screen records and displays data accumulated during the experiments. When you click **Record Data,** your data is recorded in the computer's memory and is displayed in the data grid. Data displayed in the data grid include the Radius, Flow, TV (tidal volume), ERV (expiratory reserve volume), IRV (inspiratory reserve volume), RV (residual volume), VC (vital capacity), FEV_1 (forced expiratory volume—1 second), TLC (total lung capacity), and Pump Rate. Clicking **Delete Line** allows you to discard the data for a single run; clicking **Clear Table** erases the entire experiment to allow you to start over.

If you need help identifying any piece of equipment, choose **Balloons On** from the Help menu and move the mouse pointer onto any piece of equipment visible on the computer's screen. As the pointer touches the object, a pop-up window appears to identify the equipment. To close the pop-up window, move the mouse pointer away from the equipment. Choose **Balloons On** again to turn off this help feature.

Activity 1: Measuring Respiratory Volumes

Your first experiment will establish the baseline respiratory values.

1. If the grid in the data control unit is not empty, click **Clear Table** to discard all previous data.

2. Adjust the radius of the airways to 5.00 mm by clicking the appropriate button next to the Radius window.

3. Click **Start** and allow the tracing to complete. Watch the simulated lungs begin to breathe as a result of the "contraction and relaxation" of the diaphragm. Simultaneously, the oscilloscope will display a tracing of the tidal volume for each breath. The Flow window atop the bell jar indicates the tidal volume for each breath, and the Tidal Vol. window below the oscilloscope shows the average tidal volume. The Pump Rate window displays the number of breaths per minute.

4. Click **Clear Tracings.**

5. Now click **Start** again. After a second or two, click **ERV,** wait 2 seconds and then click **FVC** to complete the measurement of respiratory volumes. The expiratory reserve volume, inspiratory reserve volume, and residual volume will be automatically calculated and displayed from the tests you have performed so far. Also, the equipment calculates and displays the total lung capacity.

6. Compute the **minute respiratory volume (MRV)** using the following formula (you can use the Calculator in the Tools menu):

$$\text{MRV} = \text{TV} \times \text{BPM (breaths per minute)}$$

MRV ____________ ml/min

7. Does expiratory reserve volume include tidal volume?

____

Explain your answer.

____

____

8. Now click **Record Data** to record the current experimental data in the data grid. Then click **Clear Tracings.**

9. If you want to print a tracing at any time, click **Tools** and then **Print Graph.** ■

Activity 2: Examining the Effect of Changing Airway Resistance on Respiratory Volumes

Lung diseases are often classified as obstructive or restrictive. With an obstructive problem, expiratory flow is affected, whereas a restrictive problem might indicate reduced inspiratory volume. Although they are not diagnostic, pulmonary function tests such as FEV_1 can help a clinician determine the difference between obstructive and restrictive problems. FEV_1 is the forced volume exhaled in one second. In obstructive disorders like chronic bronchitis and asthma, airway resistance is increased and FEV_1 will be low. Here you will explore the effect of changing the diameter of the airway on pulmonary function.

1. Do *not* clear the data table from the previous experiment.

2. Adjust the radius of the airways to 4.50 mm by clicking the appropriate button next to the Radius window.

3. Click **Start** to begin respirations.

4. Click **FVC.** As you saw in the previous test, the simulated lungs will inhale maximally and then exhale as forcefully as possible. FEV_1 will be displayed in the FEV_1 window below the oscilloscope.

5. When the lungs stop respiring, click **Record Data** to record the current data in the data grid.

6. Decrease the radius of the airways in 0.50-mm decrements and repeat steps 4 and 5 until the minimum radius (3.00 mm) is achieved. Be sure to click **Record Data** after each trial. Click **Clear Tracings** between trials. If you make an error and want to delete a single value, click the data line in the data grid and then click **Delete Line.**

7. A useful way to express FEV_1 is as a percentage of the forced vital capacity. Copy the FEV_1 and vital capacity values from the computer screen to the chart below and then calculate the FEV_1 (%) by dividing the FEV_1 volume by the vital capacity volume and multiply by 100. Record the FEV_1 (%) in the chart. You can use the Calculator under the Tools menu.

What happened to the FEV_1 (%) as the radius of the airways was decreased?

____

Explain your answer.

____

____

8. Click **Tools** → **Print Data** to print your data. ■

Simulating Factors Affecting Respirations

This part of the computer simulation allows you to explore the action of surfactant on pulmonary function and the effect of changing the intrapleural pressure.

Choose **Factors Affecting Respirations** from the **Experiment** menu. The opening screen will appear in a few seconds (Figure 37B.2). The basic features on the screen when the program starts are the same as in the Respiratory Volumes experiment screen. Additional equipment includes a surfactant dispenser atop the bell jar, and valves on each side of the bell jar. Each time **Surfactant** is clicked, a measured amount of surfactant is sprayed into the lungs. Clicking **Flush** washes surfactant from the lungs to prepare for another run.

Clicking the valve button (which currently reads **valve closed**) allows the pressure within that side of the bell jar to equalize with the atmospheric pressure. When **Reset** is clicked, the lungs are prepared for another run.

Data accumulated during a run are displayed in the windows below the oscilloscope. When you click **Record Data,** that data is recorded in the computer's memory and is displayed in the data grid. Data displayed in the data grid include the Radius, Pump Rate, the amount of Surfactant, Pressure Left (pressure in the left lung), Pressure Right (pressure in the right lung), Flow Left (air flow in the left lung), Flow Right (air flow in the right lung), and Total Flow. Clicking **Delete Line** allows you to discard data values for a single run, and clicking **Clear Table** erases the entire experiment to allow you to start over.

Activity 3: Examining the Effect of Surfactant

At any gas-liquid boundary, the molecules of the liquid are attracted more strongly to each other than they are to the air molecules. This unequal attraction produces tension at the liquid surface called *surface tension*. Because surface tension resists any force that tends to increase surface area, it acts to decrease the size of hollow spaces, such as the alveoli or microscopic air spaces within the lungs. If the film lining the air spaces in the lung were pure water, it would be very difficult, if not impossible, to inflate the lungs. However, the aqueous film covering the alveolar surfaces contains **surfactant,** a detergent-like lipoprotein that decreases surface tension by reducing the attraction of water molecules for each other. You will explore the action of surfactant in this experiment.

1. If the data grid is not empty, click **Clear Table** to discard all previous data values.
2. Adjust the airway radius to 5.00 mm by clicking the appropriate button next to the Radius window.
3. If necessary, click **Flush** to clear the simulated lungs of existing surfactant.
4. Click **Start** and allow a baseline run without added surfactant to complete.
5. When the run completes, click **Record Data.**
6. Now click **Surfactant** twice.
7. Click **Start** to begin the surfactant run.
8. When the lungs stop respiring, click **Record Data** to display the data in the grid.

FEV_1 as % of VC

Radius	FEV_1	Vital Capacity	FEV_1 (%)
5.00			
4.50			
4.00			
3.50			
3.00			

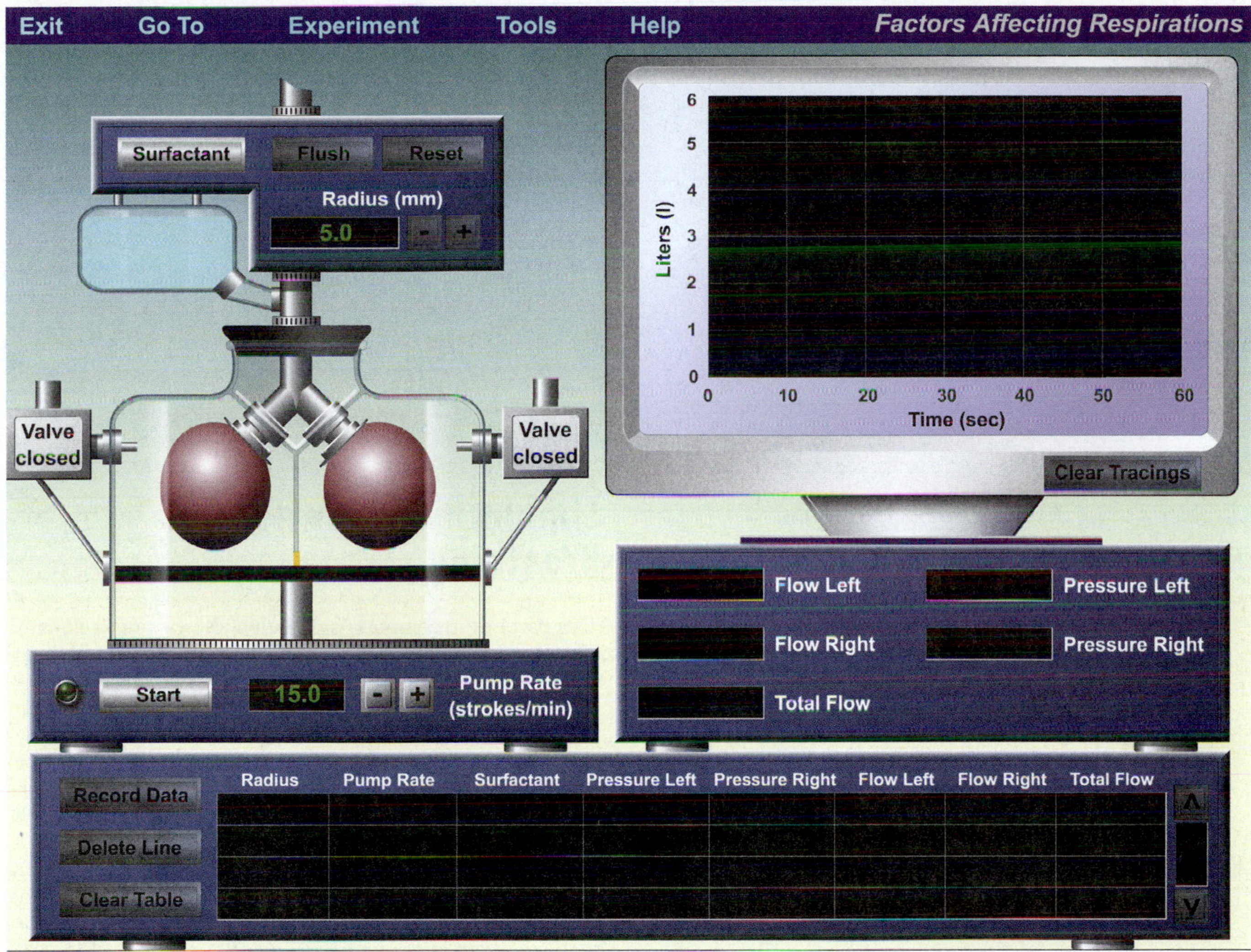

Figure 37B.2 Opening screen of the Factors Affecting Respirations experiment.

How has the air flow changed compared to the baseline run?

____

____

Premature infants often have difficulty breathing. Explain why this might be so. (Use your text as needed.)

____

____ ■

Activity 4:
Investigating Intrapleural Pressure

The pressure within the pleural cavity, **intrapleural pressure,** is less than the pressure within the alveoli. This negative pressure condition is caused by two forces, the tendency of the lung to recoil due to its elastic properties and the surface tension of the alveolar fluid. These two forces act to pull the lungs away from the thoracic wall, creating a partial vacuum in the pleural cavity. Because the pressure in the intrapleural space is lower than atmospheric pressure, any opening created in the thoracic wall equalizes the intrapleural pressure with the atmospheric pressure, allowing air to enter the pleural cavity, a condition called **pneumothorax.** Pneumothorax allows lung collapse, a condition called **atelectasis** (at″ĕ-lik′tah-sis).

In the simulated respiratory system on the computer's screen, the intrapleural space is the space between the wall of the bell jar and the outer wall of the lung it contains. The pressure difference between inspiration and expiration for the left lung and for the right lung is individually displayed in the Pressure Left and Pressure Right windows.

1. Do *not* discard your previous data.

2. Click **Clear Tracings** to clean up the screen and then click **Flush** to clear the lungs of surfactant from the previous run.

3. Adjust the radius of the airways to 5.00 mm by clicking the appropriate button next to the Radius window.

4. Click **Start** and allow one screen of respirations to complete. Notice the negative pressure condition displayed below the oscilloscope when the lungs inhale.

5. When the lungs stop respiring, click **Record Data** to display the data in the grid.

6. Now click the valve button (which currently reads valve closed) on the left side of the bell jar above the Start button to open the valve.

7. Click **Start** to begin the run.

8. When the run completes, click **Record Data** again.

What happened to the lung in the left side of the bell jar?

____

____

How did the pressure in the left lung differ from that in the right lung?

____

Explain your reasoning.

____

____

How did the total air flow in this trial compare with that in the previous trial in which the pleural cavities were intact?

____

____

What do you think would happen if the two lungs were in a single large cavity instead of separate cavities?

____

____

9. Now close the valve you opened earlier by clicking it and then click **Start** to begin a new trial.

10. When the run completes, click **Record Data** to display the data in the grid.

Did the deflated lung reinflate? _______________

Explain your answer.

____

____

11. Click the **Reset** button atop the bell jar. This action draws the air out of the intrapleural space and returns it to normal resting condition.

12. Click **Start** and allow the run to complete.

13. When the run completes, click **Record Data** to display the data in the grid.

Why did lung function in the deflated (left) lung return to normal after you clicked Reset?

____

____ ■

14. Click **Tools** → **Print Data** to print data.

Simulating Variations in Breathing

This part of the computer simulation allows you to examine the effects of hyperventilation, rebreathing, and breath holding on CO_2 level in the blood.

Choose **Variations in Breathing** from the **Experiment** menu. The opening screen will appear in a few seconds (Figure 37B.3). The basic features on the screen when the program starts are the same as in the lung volumes screen. The buttons beneath the oscilloscope control the various possible breathing patterns. Clicking **Rapid Breathing** causes the lungs to breathe faster than normal. A small bag automatically covers the airway tube when **Rebreathing** is clicked. Clicking **Breath Holding** causes the lungs to stop respiring. Click **Normal Breathing** at any time to resume normal tidal cycles. The window next to the Start button displays the breathing pattern being performed by the simulated lungs.

The windows below the oscilloscope display the PCO_2 (partial pressure of CO_2) of the air in the lungs, Maximum PCO_2, Minimum PCO_2, and Pump rate.

Data accumulated during a run are displayed in the windows below the oscilloscope. When you click **Record Data,** that data is recorded in the computer's memory and is displayed in the data grid. Data displayed in the data grid include the Condition, PCO_2, Max. PCO_2, Min. PCO_2, Pump Rate, Radius, and Total Flow. Clicking **Delete Line** allows you to discard data values for a single run, and clicking **Clear Table** erases the entire experiment to allow you to start over.

Activity 5:
Exploring Various Breathing Patterns

You will establish the baseline respiratory values in this first experiment.

1. If the grid in the data control unit is not empty, click **Clear Table** to discard all previous data.

2. Adjust the radius of the airways to 5.00 mm by clicking the appropriate button next to the Radius window. Now, read through steps 3–5 before attempting to execute them.

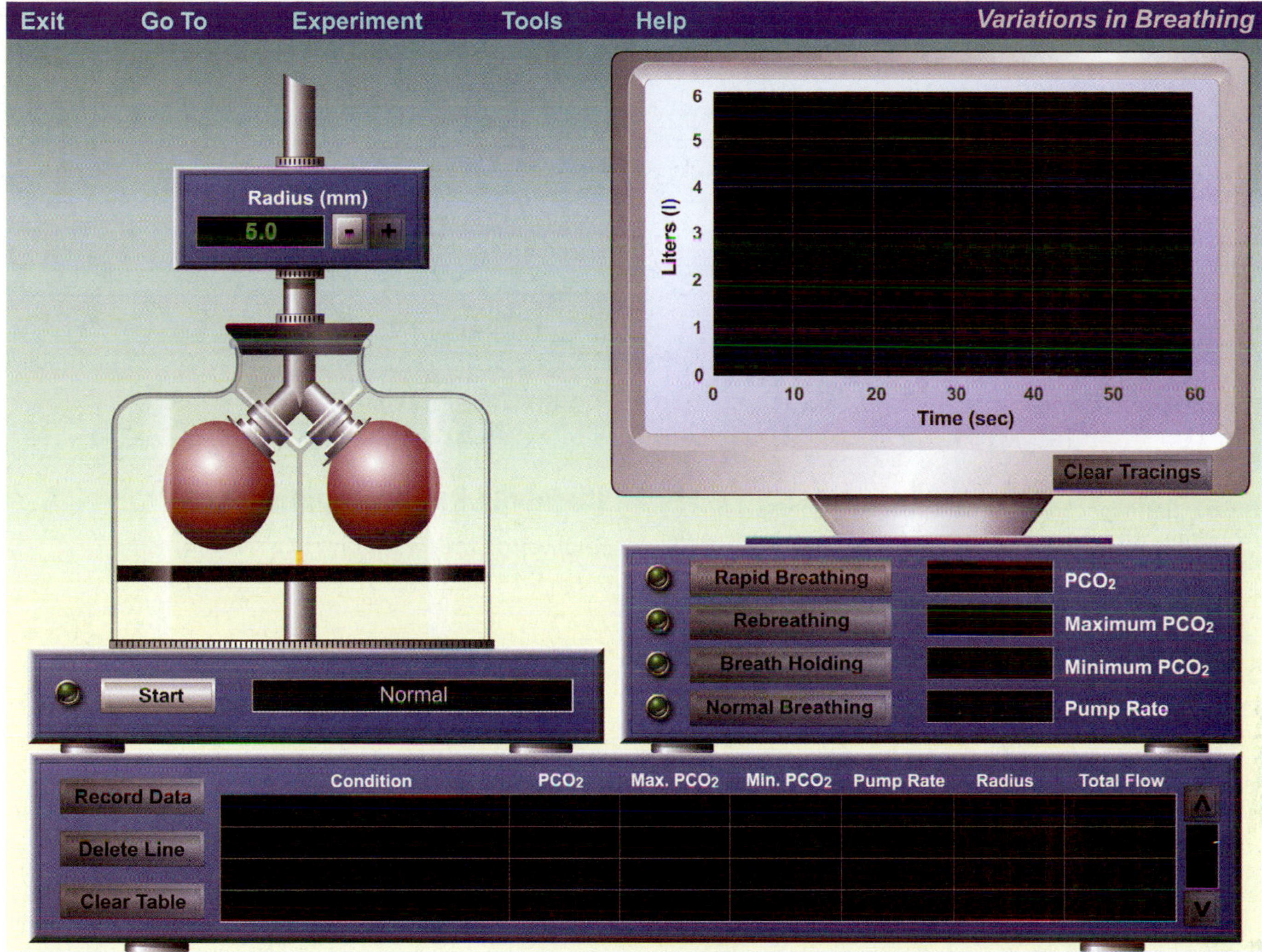

Figure 37B.3 Opening screen of the Variations in Breathing experiment.

3. Click **Start** and notice that it changes to **Stop** to allow you to stop the respiration. Watch the simulated lungs begin to breathe as a result of the external mechanical forces supplied by the pump below the bell jar. Simultaneously, the oscilloscope will display a tracing of the tidal volume for each breath.

4. After 2 seconds, click the **Rapid Breathing** button and watch the PCO_2 displays. The breathing pattern will change to short, rapid breaths. The PCO_2 of the air in the lungs will be displayed in the small window to the right of the Rapid Breathing button.

5. Watch the oscilloscope display and the PCO_2 window, and click **Stop** before the tracing reaches the end of the screen.

What happens to PCO_2 during rapid breathing? Explain your answer.

____

____

6. Click **Record Data.**

7. Now click **Clear Tracings** to prepare for the next run.

Rebreathing

When **Rebreathing** is clicked, a small bag will appear over the end of the air tube to allow the air within the lungs to be repeatedly inspired and expired.

1. Click **Start,** wait 2 seconds, and then click **Rebreathing.**

2. Watch the breathing pattern on the oscilloscope and notice the PCO_2 during the course of the run. Click **Stop** when the tracing reaches the right edge of the oscilloscope.

What happens to PCO_2 during the entire time of the rebreathing activity?

____

____

Did the depth of the breathing pattern change during rebreathing? (Carefully examine the tracing for rate and depth changes; the changes can be subtle.) Explain.

____

____

____

____

3. Click **Record Data** and then click **Clear Tracings** to prepare for the next run.

Breath Holding

Breath holding can be considered an extreme form of rebreathing in which there is no gas exchange between the outside atmosphere and the air within the lungs.

1. Click **Start,** wait a second or two, and then click **Breath Holding.**

2. Let the breath-holding activity continue for about 5 seconds and then click **Normal Breathing.**

3. Click **Stop** when the tracing reaches the right edge of the oscilloscope.

What happened to the PCO_2 during breath holding?

____

What happened to the breathing pattern when normal respirations resume?

____

____

4. Click **Record Data.**

5. Click **Tools** → **Print Data** to print your data. ■

Histology Review Supplement

Turn to p. P-151 for a review of respiratory tissue.

Chemical and Physical Processes of Digestion: Computer Simulation

exercise **39B**

Objectives

1. To list the digestive system enzymes involved in the digestion of proteins, fats, and carbohydrates; to state their site of origin; and to summarize the environmental conditions promoting their optimal functioning.
2. To recognize the variation between different types of enzyme assays.
3. To name the end products of digestion of proteins, fats, and carbohydrates.
4. To perform the appropriate chemical tests to determine if digestion of a particular food has occurred.
5. To cite the function(s) of bile in the digestive process.
6. To discuss the possible role of temperature and pH in the regulation of enzyme activity.
7. To define *enzyme, catalyst, control, substrate,* and *hydrolase.*
8. To explain why swallowing is both a voluntary and a reflex activity.
9. To discuss the role of the tongue, larynx, and gastroesophageal sphincter in swallowing.
10. To compare and contrast segmentation and peristalsis as mechanisms of propulsion.

The digestive system is a physiological marvel, composed of finely orchestrated chemical and physical activities. The food we ingest must be broken down to its molecular form for us to get the nutrients we need, and digestion involves a complex sequence of mechanical and chemical processes designed to achieve this goal as efficiently as possible. As food passes through the gastrointestinal tract, it is progressively broken down by the mechanical action of smooth muscle and the chemical action of enzymes until most nutrients have been extracted and absorbed into the blood.

Chemical Digestion of Foodstuffs: Enzymatic Action

Nutrients can only be absorbed when broken down into their monomer form, so food digestion is a prerequisite to food absorption. You have already studied mechanisms of passive and active absorption in Exercise 5. Before proceeding, review Exercise 5A p. 41.

Enzymes are large protein molecules produced by body cells. They are biological **catalysts** that increase the rate of a chemical reaction without becoming part of the product. The digestive enzymes are hydrolytic enzymes, or **hydrolases,** which break down organic food molecules or **substrates** by adding water to the molecular bonds, thus cleaving the bonds between the subunits or monomers.

A hydrolytic enzyme is highly specific in its action. Each enzyme hydrolyzes one or, at most, a small group of substrate molecules, and specific environmental conditions are necessary for an enzyme to function optimally. For example, temperature and pH have a large effect on the degree of enzymatic hydrolysis, and each enzyme has its preferred environment.

Because digestive enzymes actually function outside the body cells in the digestive tract lumen, their hydrolytic activity can also be studied in a test tube. Such *in vitro* studies provide a convenient laboratory environment for investigating the effect of various factors on enzymatic activity.

Figure 39A.1, is a flowchart of the progressive digestion of proteins, fats, and carbohydrates. It indicates the specific enzymes involved, their site of formation, and their site of action. Acquaint yourself with this flowchart before beginning this experiment, and refer to it as necessary during the laboratory session.

Starch Digestion by Salivary Amylase

In this experiment you will investigate the hydrolysis of starch to maltose by salivary amylase, the enzyme produced by the salivary glands and secreted into the mouth. For you to be able to detect whether or not enzymatic action has occurred, you need to be able to identify the presence of these substances to determine to what extent hydrolysis has occurred. Thus, **controls** must be prepared to provide a known standard against which comparisons can be made. The controls will vary for each experiment, and will be discussed in each enzyme section in this exercise.

Starch decreases and sugar increases as digestion proceeds according to the following equation:

$$\text{Starch} + \text{water} \xrightarrow{\text{amylase}} \text{X maltose}$$

Because the chemical changes that occur as starch is digested to maltose cannot be seen by the naked eye, you need to conduct an *enzyme assay,* the chemical method of detecting the presence of digested substances. You will perform two enzyme assays on each sample. The IKI assay detects the presence of starch and the Benedict's assay tests for the presence of maltose, which is the digestion product of starch. Normally a caramel-colored solution, IKI turns blue-black in the presence of starch. Benedict's reagent is a bright blue solution that changes to green to orange to reddish-brown with increasing amounts of maltose. It is important to understand that enzyme assays only indicate the presence or absence of substances. It is up to you to analyze the results of the experiments to decide if enzymatic hydrolysis has occurred.

Choose **Chemical and Physical Processes of Digestion** from the main menu. The opening screen will appear in a few seconds (Figure 39B.1). The *solutions shelf* in the upper right part of the screen contains the substances to be used in the experiment. The *incubation unit* beneath the solutions shelf contains a rack of test tube holders and the apparatus needed to run the experiments. Test tubes from the *test tube washer* on the left part of the screen are loaded into the rack in the incubation unit by clicking and holding the mouse button on the first tube, and then releasing (dragging-and-dropping) it into any position in the rack. The substances in the dropper bottles on the solutions shelf are dispensed by dragging-and-dropping the dropper cap to a position over any test tube in

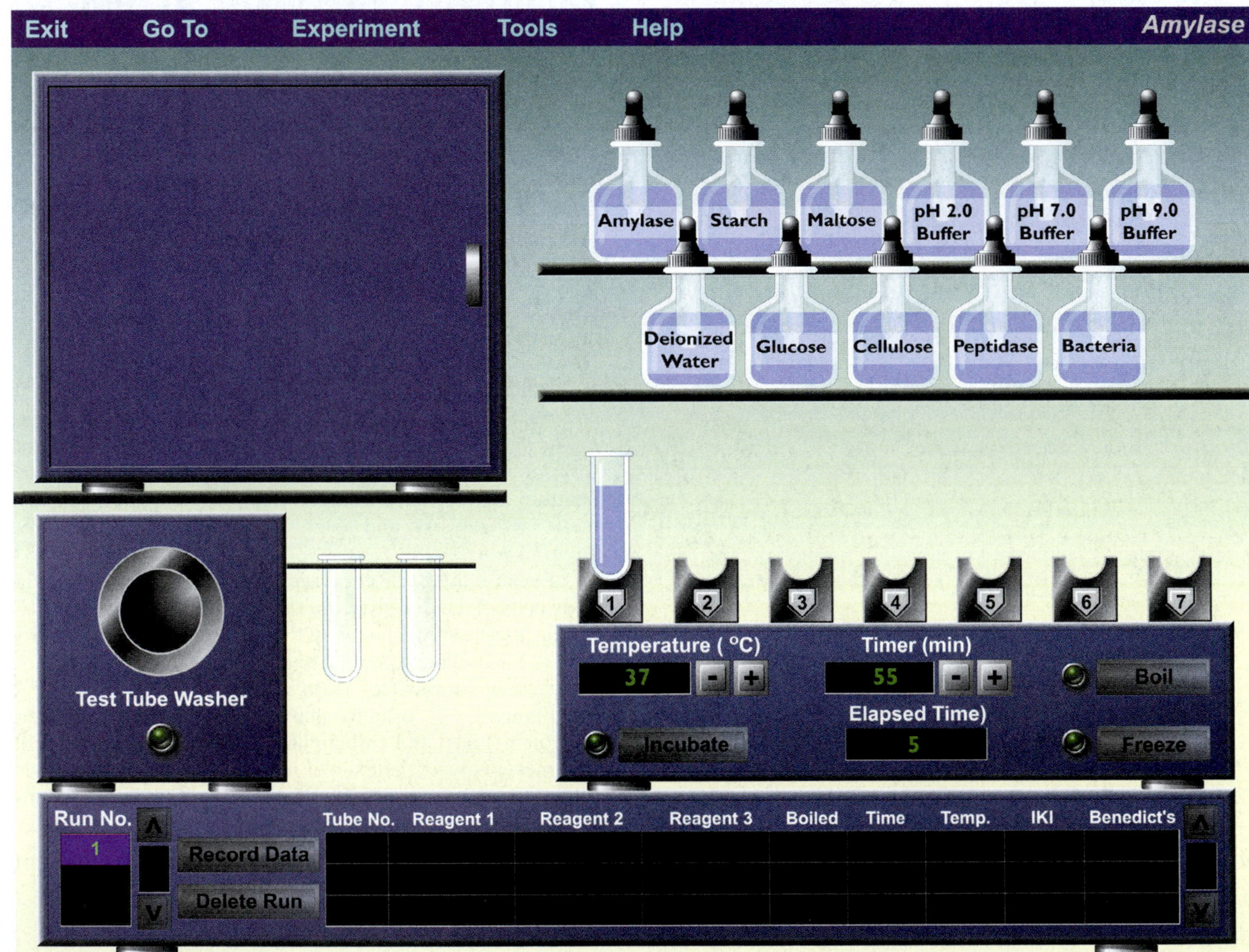

Figure 39B.1 Opening screen of the Amylase experiment.

the rack and then releasing it. During each dispensing event, five drops of solution drip into the test tube; then the dropper cap automatically returns to its position in the bottle.

Each test tube holder in the incubation unit not only supports but also allows you to boil the contents of a single test tube. Clicking the numbered button at the base of a test tube holder causes that single tube to descend into the incubation unit. To boil the contents of all tubes inside the incubation unit, click **Boil.** After they have been boiled, the tubes automatically rise. You can adjust the incubation temperature for the experiment by clicking the (+) or (−) buttons next to the Temperature window. Set the incubation time by clicking the (+) or (−) buttons next to the Timer window. Clicking the **Incubate** button starts the timer and causes the entire rack of tube holders to descend into the incubation unit where the tubes will be incubated at the temperature and the time indicated. While incubating, the tubes are automatically agitated to ensure that their contents are well mixed. During the experiment, elapsed time is displayed in the Elapsed Time window.

The cabinet doors in the *assay cabinet* above the test tube washer are closed at the beginning of the experiment, but they automatically open when the set time for incubation has elapsed. The assay cabinet contains the reagents and glassware needed to assay your experimental samples.

When you click the **Record Data** button in the *data control unit* at the bottom of the screen, your data is recorded in the computer's memory and displayed in the data grid at the bottom of the screen. Data displayed in the grid include the tube number, the three substances dispensed into each tube, the time and incubation temperature, and (+) or (−) marks indicating enzyme assay results and whether or not a sample was boiled. If you are not satisfied with a single run, you can click **Delete Run** to erase an experiment.

Once an experimental run is completed and you have recorded your data, discard the test tubes to prepare for a new run by dragging the used tubes to the large opening in the test tube washer. The test tubes will automatically be prepared for the next experiment.

Activity 1: Assessing Animal Starch Digestion by Salivary Amylase

Incubation

1. Individually drag seven test tubes to the test tube holders in the incubation unit.

2. Prepare tubes 1 through 7 with the substances indicated in the chart below using the following approach.

- Click and hold the mouse button on the dropper cap of the desired substance on the solutions shelf.
- While still holding the mouse button down, drag the dropper cap to the top of the desired test tube.
- Release the mouse button to dispense the substance. The dropper cap automatically returns to its bottle.

Note that the starch we are using here is animal starch.

3. When all tubes are prepared, click the number (**1**) under the first test tube. The tube will descend into the incubation unit. All other tubes should remain in the raised position.

4. Click **Boil** to boil the number 1 tube. After boiling for a few moments, the tube will automatically rise.

5. Now adjust the incubation temperature to 37°C and the timer to 60 min (compressed time) by clicking the (+) or (−) buttons.

6. Click **Incubate** to start the run. The incubation unit will gently agitate the test tube rack, evenly mixing the contents of all test tubes throughout the incubation. Notice that the computer compresses the 60-minute time period into 60 seconds of real time, so what would be a 60-minute incubation in real life will take only 60 seconds in the simulation. When the incubation time elapses, the test tube rack will automatically rise, and the doors to the assay cabinet will open.

Salivary Amylase Digestion of Animal Starch

Tube no.	1	2	3	4	5	6	7
Additives	Amylase Starch pH 7.0 buffer	Amylase Starch pH 7.0 buffer	Amylase Deionized water pH 7.0 buffer	Deionized water Starch pH 7.0 buffer	Deionized water Maltose pH 7.0 buffer	Amylase Starch pH 2.0 buffer	Amylase Starch pH 9.0 buffer
Incubation condition	Boil first, then incubate at 37°C	37°C	37°C	37°C	37°C	37°C	37°C
Benedict's test							
IKI test							

Assays

After the assay cabinet doors open, notice the two reagents in the assay cabinet. IKI tests for the presence of starch and Benedict's detects the presence of maltose, the digestion product of starch. Below the reagents are seven small assay tubes into which you will dispense a small amount of test solution from the incubated samples in the incubation unit, plus a drop of IKI.

1. Click and hold the mouse on the first tube in the incubation unit. Notice that the mouse pointer is now a miniature test tube tilted to the left.
2. While still holding the mouse button down, move the mouse pointer to the first small assay tube on the left side of the assay cabinet. Release the mouse button. Watch the first test tube automatically decant approximately half of its contents into the first assay tube on the left.
3. Repeat steps 1 and 2 for the remaining tubes in the incubation unit, moving to a fresh assay tube each time.
4. Next, click and hold the mouse on the IKI dropper cap and drag it to the first assay tube. Release the mouse button to dispense a drop of IKI into the first assay tube on the left. You will see IKI drip into the tube, which may cause a color change in the solution. A blue-black color indicates a **positive starch test.** If starch is not present, the mixture will look like diluted IKI, a **negative starch test.** Intermediate starch amounts result in a pale gray color.
5. Now dispense IKI into the remaining assay tubes. Record your results (+ for positive, − for negative) in the chart.
6. Dispense Benedict's reagent into the remaining mixture in each tube in the incubation unit by dragging-and-dropping the Benedict's dropper cap to the top of each test tube.
7. After Benedict's reagent has been delivered to each tube in the incubation unit, click **Boil.** The entire tube rack will descend into the incubation unit and automatically boil the tube contents for a few moments.
8. When the rack of tubes rises, inspect the tubes for color change. A green-to-reddish color indicates that maltose is present; this is a **positive sugar test.** An orange-colored sample contains more maltose than a green sample. A reddish-brown color indicates even more maltose. A negative sugar test is indicated by no color change from the original bright blue. Record your results in the chart.
9. Click **Record Data** to display your results in the grid and retain your data in the computer's memory for later analysis. To repeat the experiment, drag all test tubes to the test tube washer and start again.
10. Answer the following questions, referring to the chart (or the data grid in the simulation) as necessary. Hint: closely examine the IKI and Benedict's results for each tube.

What do tubes 2, 6, and 7 reveal about pH and amylase activity?

____

____

Which pH buffer allowed the highest amylase activity?

____

Which tube indicates that the amylase did not contain maltose? ______________________________

Which tubes indicate that the deionized water did not contain starch or maltose? ______________________

If we left out control tubes 3, 4, or 5, what objections could be raised to the statement: "Amylase digests starch to maltose"? (Hint: think about the purity of the chemical solutions.)

____

____

Would the amylase present in saliva be active in the stomach? Explain your answer.

____

____

What effect does boiling have on enzyme activity?

____

______________________________________ ■

Activity 2:

Assessing Plant Starch Digestion

If any test tubes are still in the incubator, click and drag them to the test tube washer before beginning this activity.

In the previous activity, we learned that salivary amylase can digest *animal* starch. In this activity, we will test to see whether amylase digests **cellulose,** a type of *plant* starch. We will also investigate whether bacteria (such as that found in the large intestine) will digest cellulose.

Incubation

1. Individually drag seven test tubes to the test tube holders in the incubation unit.
2. Prepare tubes 1 through 7 with the substances indicated in the chart on the next page using the following approach:

- Click and hold the mouse button on the dropper cap of the desired substance on the solutions shelf.
- While holding the mouse button down, drag the dropper cap to the top of the desired test tube.
- Release the mouse button to dispense the substance. The dropper cap automatically returns to its bottle.

Note that the Starch bottle contains animal starch whereas the Cellulose bottle contains plant starch.

3. When all tubes are prepared, click the number **(1)** under the first test tube. The tube will descend into the incubation unit.

4. Click **Freeze.** The tube's contents will be subjected to a temperature of −25°C. The tube will then automatically rise, with the contents of the tube frozen.

5. Adjust the incubation temperature to 37°C and the timer to 60 minutes by clicking the (+) and (−) buttons.

6. Click **Incubate** to start the run. The incubation unit will gently agitate the tubes as they incubate to thoroughly mix the tubes' contents. At the end of the incubation period, the tubes will ascend to their original positions on top of the incubator, and the doors to the assay cabinet will open.

Assays

When the assay cabinet opens, notice the two reagents in the cabinet. They are the same ones as in the previous activity: IKI will test for the presence of starch while Benedict's solution will test for the presence of maltose. On the floor of the cabinet are seven small tubes that you will use to test the results of your experiment. The procedure will be identical to the one from the previous activity:

1. Click and hold the mouse on the first tube in the incubation unit. Notice that the mouse pointer is now a miniature test tube tilted to the left.

2. While still holding the mouse button down, move the mouse pointer to the first small assay tube on the left side of the assay cabinet. Release the mouse button.

3. Repeat steps 1 and 2 for the remaining tubes in the incubation unit, moving to a fresh assay tube each time.

4. Next, click and hold the mouse on the IKI dropper cap and drag it to the first assay tube. Release the mouse button to dispense a drop of IKI into the first assay tube on the left. You will see IKI drip into the tube, which may cause a color change in the solution. A blue-black color indicates the presence of starch. If there is only a small amount of starch, you may see a pale gray color. If starch is not present, the mixture will look like diluted IKI.

5. Place IKI into the remaining assay tubes and note the color of each tube. Record your results in the chart.

6. Dispense Benedict's reagent into the remaining mixture in each tube in the incubation unit by dragging-and-dropping the Benedict's dropper cap to the top of each test tube.

7. After Benedict's reagent has been delivered to each tube in the incubation unit, click **Boil.** The entire tube rack will descend into the incubation unit and automatically boil the tube contents for a few moments.

8. When the rack of tubes rises, inspect the tubes for color change. A green to reddish color indicates maltose is present for a positive sugar test. An orange sample indicates more maltose than the green color, while a reddish-brown indicates the highest amounts of maltose. If there has been no color change from the original blue, no maltose is present in the tube. Record your results in the chart.

9. Click **Record Data** to display your results in the grid and retain your data in the computer's memory for later analysis. To repeat the experiment, drag all test tubes to the test tube washer and start again.

10. Answer the following questions, referring to the chart.

Which tubes showed a positive test for the IKI reagent?

____

Which tubes showed a positive test for the Benedict's reagent? ________________________________

What was the effect of freezing tube 1?

____

Enzyme Digestion of Animal Starch and Plant Starch

Tube no.	**1**	**2**	**3**	**4**	**5**	**6**	**7**
Additives	Amylase Starch pH 7.0 buffer	Amylase Starch pH 7.0 buffer	Amylase Glucose pH 7.0 buffer	Amylase Cellulose pH 7.0 buffer	Amylase Cellulose Deionized water	Peptidase Starch pH 7.0 buffer	Bacteria Cellulose pH 7.0 buffer
Incubation condition	Freeze first, then incubate at 37°C	37°C	37°C	37°C	37°C	37°C	37°C
Benedict's test							
IKI test							

How does the effect of freezing differ from the effect of boiling? ___

___

What was the effect of amylase on glucose in tube 3? Can you offer an explanation for this effect? ___

___

___

What was the effect of amylase on cellulose in tube 4?

___

___

Popcorn and celery are nearly pure plant starch or cellulose. What can you conclude about the digestion of cellulose, judging from the results of test tubes 4, 5, and 7?

___

___

What was the effect of the different enzyme, peptidase, used in tube 6? Explain your answer, based on what you know about peptidase.

___

___

11. Click **Tools** → **Print Data** to print your recorded data. ■

Protein Digestion by Pepsin

The chief cells of the stomach glands produce pepsin, a protein-digesting enzyme. Pepsin hydrolyzes proteins to small fragments (proteoses, peptones, and peptides). In this experi-

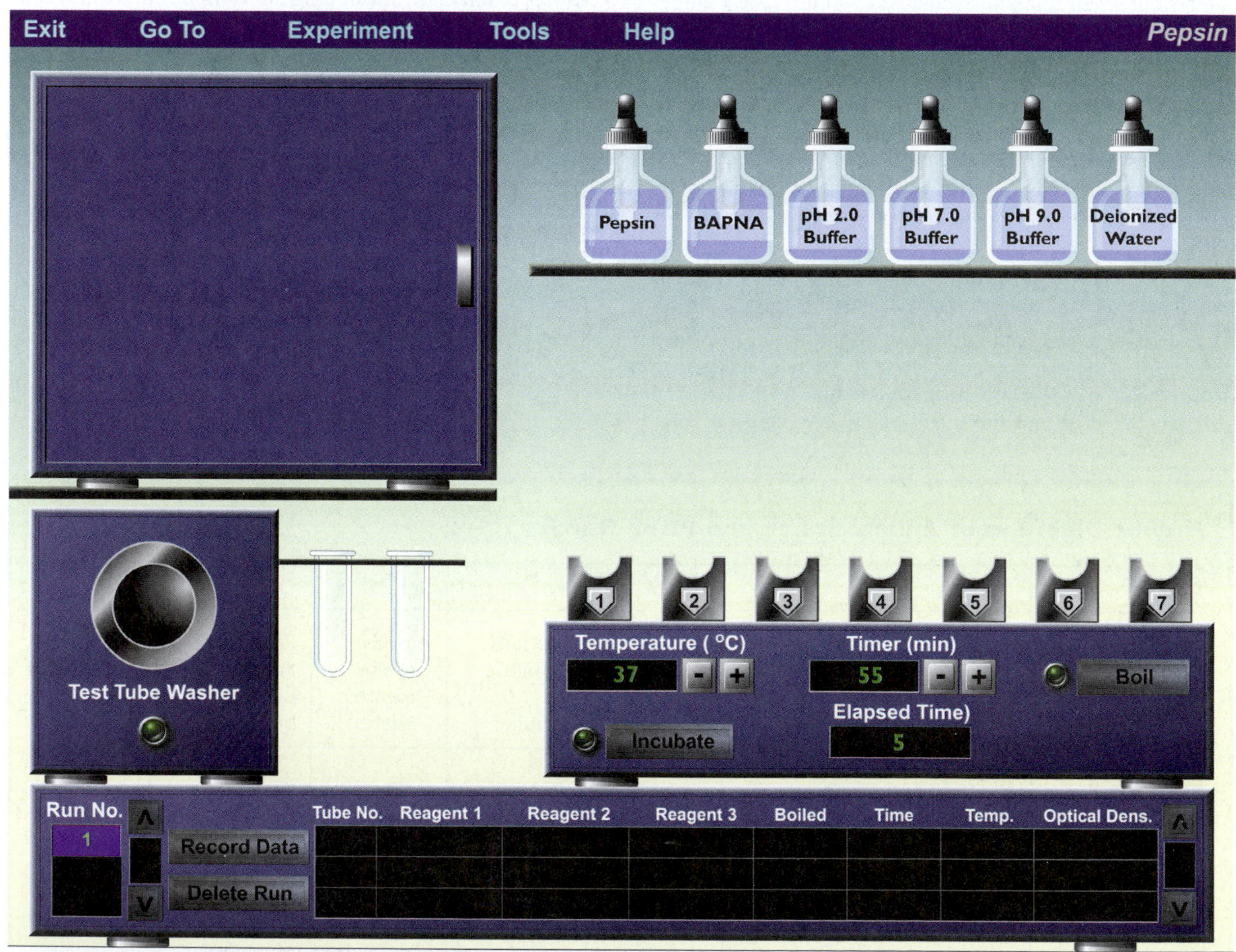

Figure 39B.2 Opening screen of the Pepsin experiment.

ment, you will use BAPNA, a synthetic "protein" that is transparent and colorless when in solution. However, if an active, protein-digesting enzyme such as pepsin is present, the solution will become yellow. You can use this characteristic to detect pepsin activity: the solution turns yellow if the enzyme digests the BAPNA substrate; it remains colorless if pepsin is not active or not present. One advantage of using a synthetic substrate is that you do not need any additional indicator reagents to see enzyme activity.

Choose **Pepsin** from the **Experiment** menu. The opening screen will appear in a few seconds (Figure 39B.2). The solutions shelf, test tube washer, and incubation equipment are the same as in the amylase experiment; only the solutions have changed.

Data displayed in the grid include the tube number, the three substances dispensed into each tube, a (+) or (−) mark indicating whether or not a sample was boiled, the time and temperature of the incubation, and the optical density measurement indicating enzyme assay results.

Activity 3:
Assessing Protein Digestion by Pepsin

Pepsin Incubation

1. Individually drag six test tubes to the test tube holders in the incubation unit.

2. Prepare the tubes with the substances indicated in the chart below using the following method.

- Click and hold the mouse button on the dropper cap of the desired substance and drag the dropper cap to the top of the desired test tube.
- Release the mouse button to dispense the substance.

3. Once all tubes are prepared, click the number **(1)** under the first test tube. The tube will descend into the incubation unit. All other tubes should remain in the raised position.

4. Click **Boil** to boil tube 1. After boiling for a few moments, the tube will automatically rise.

5. Adjust the incubation temperature to 37°C and the timer to 60 min (compressed time) by clicking the (+) or (−) buttons.

6. Click **Incubate** to start the run. The incubation unit will gently agitate the test tube rack, evenly mixing the contents of all test tubes throughout the incubation. The computer is compressing the 60-minute time period into 60 seconds of real time. When the incubation time elapses, the test tube rack will automatically rise, and the doors to the assay cabinet will open.

Pepsin Assay

After the assay cabinet doors open, you will see an instrument called a spectrophotometer, which you will use to measure how much yellow dye was liberated by pepsin digestion of BAPNA. When a test tube is dragged to the holder in the spectrophotometer and the **Analyze** button is clicked, the instrument will shine a light through a specimen to measure the amount of light absorbed by the sample within the tube. The measure of the amount of light absorbed by the solution is known as its *optical density.* A colorless solution does not absorb light, whereas a colored solution has a relatively high light absorbance. For example, a colorless solution has an optical density of 0.0. A colored solution, however, absorbs some of the light emitted by the spectrophotometer, resulting in an optical density reading greater than zero.

In this experiment a yellow-colored solution is a direct indication of the amount of BAPNA digested by pepsin.

Pepsin Digestion of Protein

Tube no.	1	2	3	4	5	6
Additives	Pepsin BAPNA pH 2.0 buffer	Pepsin BAPNA pH 2.0 buffer	Pepsin Deionized water pH 2.0 buffer	Deionized water BAPNA pH 2.0 buffer	Pepsin BAPNA pH 7.0 buffer	Pepsin BAPNA pH 9.0 buffer
Incubation condition	Boil first, then incubate at 37°C	37°C	37°C	37°C	37°C	37°C
Optical density						

Although you can visually estimate the yellow color produced by pepsin digestion of BAPNA, the spectrophotometer precisely measures how much BAPNA digestion occurred in the experiment.

1. Click and hold the mouse on the first tube in the incubation unit and drag it to the holder in the spectrophotometer.
2. Release the mouse button to drop the tube into the holder.
3. Click **Analyze.** You will see light shining through the solution in the test tube as the spectrophotometer measures its optical density. The optical density of the sample will be displayed in the optical density window below the Analyze button.
4. Record the optical density in the chart.
5. Drag the tube to its original position in the incubation unit and release the mouse button.
6. Repeat steps 1 through 5 for the remaining test tubes in the incubation unit.
7. Click **Record Data** to display your results in the grid and retain your data in the computer's memory for later analysis. To repeat the experiment, you must drag all test tubes to the test tube washer and start again.
8. Answer the following questions, referring to the chart (or the data grid in the simulation) as necessary.

Which pH provided the highest pepsin activity? _______

Would pepsin be active in the mouth? Explain your answer.

____

____

How did the results of tube 1 compare with those of tube 2?

____

Tubes 1 and 2 contained the same substances. Explain why their optical density measurements were different.

____

____

If you had not run the tube 2 and 3 samples, what argument could be made against the statement "Pepsin digests BAPNA"?

____

____

What do you think would happen if you reduced the incubation time to 30 minutes? Use the simulation to help you answer this question if you are not sure.

____

____

What do you think would happen if you decreased the temperature of incubation to 10°C? Use the simulation to help you answer this question if you are not sure.

____

____

9. Click **Tools** → **Print Data** to print your recorded data. ■

Fat Digestion by Pancreatic Lipase and the Action of Bile

The treatment that fats and oils undergo during digestion in the small intestine is a bit more complicated than that of carbohydrates or proteins. Fats and oils require pretreatment with bile to physically emulsify the fats. As a result, two sets of reactions must occur.

First:

$$\text{Fats/oils} \xrightarrow[\text{(emulsification)}]{\text{bile}} \text{minute fat/oil droplets}$$

Then:

$$\text{Fat/oil droplets} \xrightarrow{\text{lipase}} \text{monoglycerides and fatty acids}$$

Lipase hydrolyzes fats and oils to their component monoglycerides and two fatty acids. Occasionally lipase hydrolyzes fats and oils to glycerol and three fatty acids.

The fact that some of the end products of fat digestion (fatty acids) are organic acids that decrease the pH provides an easy way to recognize that digestion is ongoing or completed. You will be using a pH meter in the assay cabinet to record the drop in pH as the test tube contents become acid.

Choose **Lipase** from the **Experiment** menu. The opening screen will appear in a few seconds (Figure 39B.3). The solutions shelf, test tube washer, and incubation equipment are the same as in the previous two experiments; only the solutions have changed.

Data displayed in the grid include the tube number, the four reagents dispensed into each tube, a (+) or (−) mark indicating whether or not a sample was boiled, the time and temperature of the incubation, and the pH measurement indicating enzyme assay results.

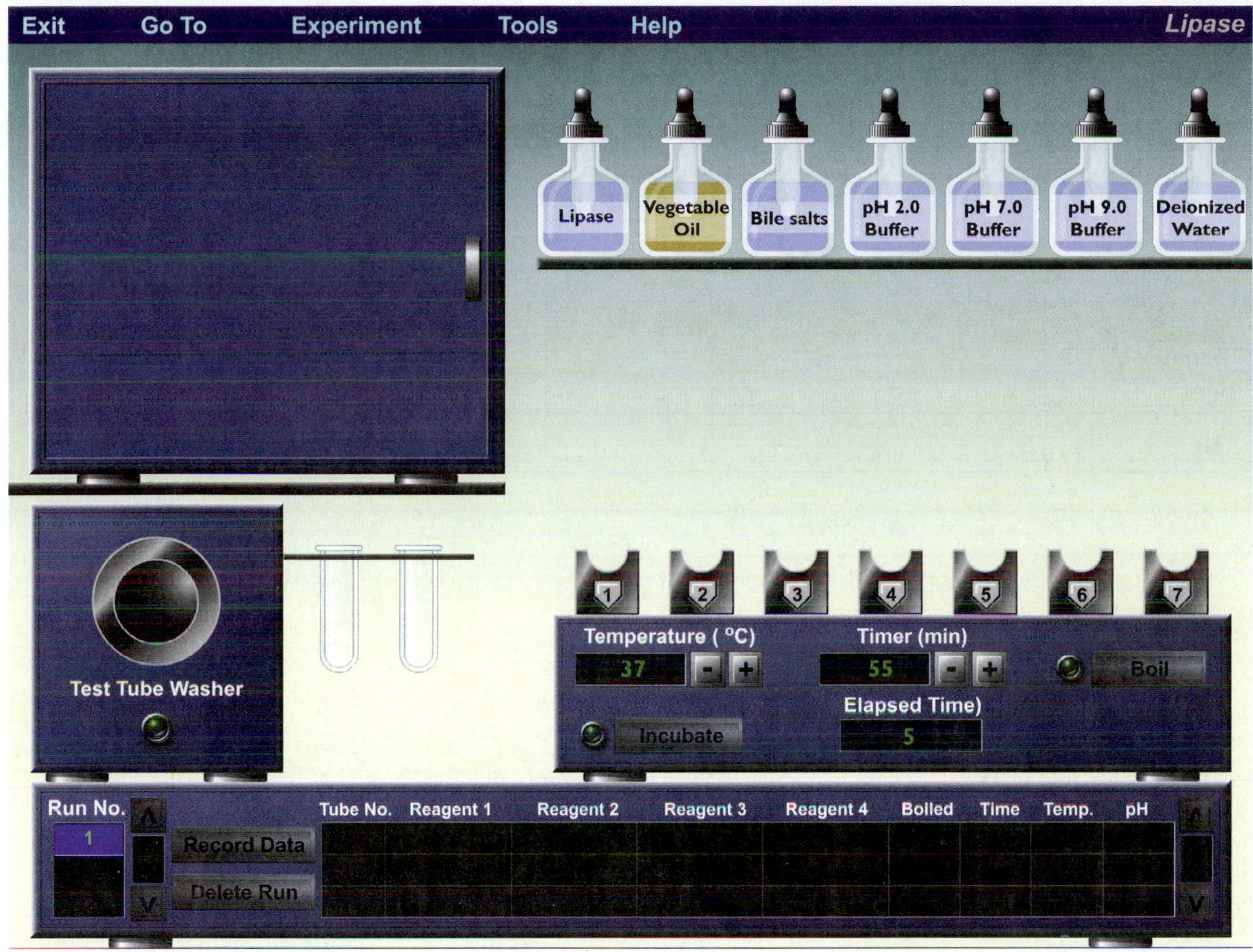

Figure 39B.3 Opening screen of the Lipase experiment.

Activity 4:

Assessing Fat Digestion by Pancreatic Lipase and the Action of Bile

Lipase Incubation

1. Individually drag 6 test tubes to the test tube holders in the incubation unit.

2. Prepare the tubes with the solutions indicated in the chart on p. P-84 by using the following method.

- Click and hold the mouse button on the dropper cap of the desired substance.
- While holding the mouse button down, drag the dropper cap to the top of the desired test tube.
- Release the mouse button to dispense the substance.

3. Adjust the incubation temperature to 37°C and the timer to 60 min (compressed time) by clicking the (+) or (−) buttons.

4. Click **Incubate** to start the run. The incubation unit will gently agitate the test tube rack, evenly mixing the contents of all test tubes throughout the incubation. The computer is compressing the 60-minute time period into 60 seconds of real time. When the incubation time elapses, the test tube rack automatically rises, and the doors to the assay cabinet open.

Lipase Assay

After the assay cabinet doors open, you will see a pH meter that you will use to measure the relative acidity of your test solutions. When a test tube is dragged to the holder in the pH meter and the Measure pH button is clicked, a probe will descend into the sample, take a pH reading, and then retract. The pH of the sample will be displayed in the pH window below the Measure pH button. A solution containing fatty acids liberated from fat by the action of lipase will exhibit a lower pH than one without fatty acids.

Pancreatic Lipase Digestion of Fats and the Action of Bile

Tube no.	1	2	3	4	5	6
Additives	Lipase Vegetable oil Bile salts pH 7.0 buffer	Lipase Vegetable oil Deionized water pH 7.0 buffer	Lipase Deionized water Bile salts pH 9.0 buffer	Deionized water Vegetable oil Bile salts pH 7.0 buffer	Lipase Vegetable oil Bile salts pH 2.0 buffer	Lipase Vegetable oil Bile salts pH 9.0 buffer
Incubation condition	37°C	37°C	37°C	37°C	37°C	37°C
pH						

1. Click and hold the mouse on the first tube in the incubation unit and drag it to the holder in the pH meter. Release the mouse button to drop the tube into the holder.
2. Click **Measure pH.**
3. In the chart above, record the pH displayed in the pH window.
4. Drag the test tube in the pH meter to its original position in the incubation unit and release the mouse button.
5. Repeat steps 1 through 4 for the remaining test tubes in the incubation unit.
6. Click **Record Data** to display your results in the grid and retain your data in the computer's memory for later analysis. To repeat the experiment, you must drag all test tubes to the test tube washer and begin again.
7. Answer the following questions, referring to the chart (or the data grid in the simulation) as necessary.

Explain the difference in activity between tubes 1 and 2.

___

___

Can we determine if fat hydrolysis has occurred in tube 6?

___ Explain your answer. ___

___

Which pH resulted in maximum lipase activity? ___

Is this method of assay sufficient to determine if the optimum activity of lipase is at pH 2.0? ___

In theory, would lipase be active in the mouth? ___

Would it be active in the stomach? ___

Explain your answers. ___

___

Based on the enzyme pH optima you determined, where in the body would we expect to find the enzymes in these experiments?

Amylase ___

Pepsin ___

Lipase ___

8. Click **Tools → Print Data** to print your recorded data. ■

Physical Processes: Mechanisms of Food Propulsion and Mixing

Although enzyme activity is an essential part of the overall digestion process, food must also be processed physically by churning and chewing, and moved by mechanical means along the tract if digestion and absorption are to be completed. Just about any time organs exhibit mobility, muscles are involved, and movements of and in the gastrointestinal tract are no exception. Although we tend to think only of smooth muscles for visceral activities, both skeletal and smooth muscles are necessary in digestion. This fact is demonstrated by the simple trials in the next activity. Obtain the following materials:

- Water pitcher
- Paper cups
- Stethoscope
- Alcohol swabs
- Disposable autoclave bag

Activity 5: Studying Mechanisms of Food Propulsion and Mixing

Deglutition (Swallowing)

Swallowing, or *deglutition,* which is largely the result of skeletal muscle activity, occurs in two phases: *buccal* (mouth) and *pharyngeal-esophageal.* The initial phase—the buccal—is voluntarily controlled and initiated by the tongue. Once begun, the process continues involuntarily in the pharynx and esophagus, through peristalsis, resulting in the delivery of the swallowed contents to the stomach.

1. While swallowing a mouthful of water, consciously note the movement of your tongue during the process. Record your observations.

____

____

____

2. Repeat the swallowing process while your laboratory partner watches the externally visible movements of your larynx. This movement is more obvious in a male, who has a larger Adam's apple. Record your observations.

____

____

What do these movements accomplish? ______________

____

____

3. Your lab partner should clean the ear pieces of a stethoscope with an alcohol swab and don the stethoscope. Then your lab partner should place the diaphragm of the stethoscope on your abdominal wall, approximately 1 inch below the xiphoid process and slightly to the left, to listen for sounds as you again take two or three swallows of water. There should be two audible sounds. The first sound occurs when the water splashes against the gastroesophageal sphincter. The second occurs when the peristaltic wave of the esophagus arrives at the sphincter and the sphincter opens, allowing water to gurgle into the stomach. Determine, as accurately as possible, the time interval between these two sounds and record it below.

Interval between arrival of water at the sphincter and the opening of the sphincter:

_______ sec.

This interval gives a fair indication of the time it takes for the peristaltic wave to travel down the 10-inch-long esophagus. Actually the time interval is slightly less than it seems because pressure causes the sphincter to relax before the peristaltic wave reaches it.

Dispose of the used paper cup in the autoclave bag. ■

Segmentation and Peristalsis

Although several types of movement occur in the digestive tract organs, segmentation and peristalsis are most important as mixing and propulsive mechanisms.

Segmental movements are local constrictions of the organ wall that occur rhythmically. They serve mainly to mix the foodstuffs with digestive juices and to increase the rate of absorption by continually moving different portions of the chyme over adjacent regions of the intestinal wall. However, segmentation is an important means of food propulsion in the small intestine, and slow segmenting movements called haustral contractions are common in the large intestine.

Peristaltic movements are the major means of propelling food through most of the digestive viscera. Essentially they are waves of contraction followed by waves of relaxation that squeeze foodstuffs through the alimentary canal, and they are superimposed on segmental movements.

Histology Review Supplement

Turn to p. P-153 for a review of digestive tissue.

exercise

41B Renal Physiology—The Function of the Nephron: Computer Simulation

Objectives

1. To define:

   glomerulus, glomerular capsule, renal corpuscle, renal tubule, nephron, proximal convoluted tubule, loop of Henle, and *distal convoluted tubule*
2. To describe the blood supply to each nephron.
3. To identify the regions of the nephron involved in glomerular filtration and tubular reabsorption.
4. To study the factors affecting glomerular filtration.
5. To explore the concept of carrier transport maximum.
6. To understand how the hormones aldosterone and ADH affect the function of the kidney.
7. To describe how the kidneys can produce urine that is four times more concentrated than the blood.

Metabolism produces wastes that must be eliminated from the body. This excretory function is the job of the renal system, most importantly the paired kidneys. Each kidney consists of about one million nephrons that carry out two crucial services, blood filtration and fluid processing.

Microscopic Structure and Function of the Kidney

Each of the million or so **nephrons** in each kidney is a microscopic tubule consisting of two major parts: a glomerulus and a renal tubule. The **glomerulus** is a tangled capillary knot that filters fluid from the blood into the lumen of the renal tubule. The function of the **renal tubule** is to process that fluid, also called the **filtrate.** The beginning of the renal tubule is an enlarged end called the **glomerular capsule,** which surrounds the glomerulus and serves to funnel the filtrate into the rest of the renal tubule. Collectively, the glomerulus and the glomerular capsule are called the **renal corpuscle.**

As the rest of the renal tubule extends from the glomerular capsule, it becomes twisted and convoluted, then dips sharply down to form a hairpin loop, and then coils again before entering a collecting duct. Starting at the glomerular capsule, the anatomical parts of the renal tubule are as follows: the **proximal convoluted tubule,** the **loop of Henle** (nephron loop), and the **distal convoluted tubule.**

Two arterioles supply each glomerulus: an afferent arteriole feeds the glomerular capillary bed and an efferent arteriole drains it. These arterioles are responsible for blood flow through the glomerulus. Constricting the afferent arteriole lowers the downstream pressure in the glomerulus, whereas constricting the efferent arteriole will increase the pressure in the glomerulus. In addition, the diameter of the efferent arteriole is smaller than the diameter of the afferent arteriole, restricting blood flow out of the glomerulus. Consequently, the pressure in the glomerulus forces fluid through the endothelium of the glomerulus into the lumen of the surrounding glomerular capsule. In essence, everything in the blood except the cells and proteins are filtered through the glomerular wall. From the capsule, the filtrate moves into the rest of the renal tubule for processing. The job of the tubule is to reabsorb all the beneficial substances from its lumen while allowing the wastes to travel down the tubule for elimination from the body.

The nephron performs three important functions to process the filtrate into urine: glomerular filtration, tubular reabsorption, and tubular secretion. **Glomerular filtration** is a passive process in which fluid passes from the lumen of the glomerular capillary into the glomerular capsule of the renal tubule. **Tubular reabsorption** moves most of the filtrate back into the blood, leaving principally salt water plus the wastes in the lumen of the tubule. Some of the desirable or needed solutes are actively reabsorbed, and others move passively from the lumen of the tubule into the interstitial spaces. **Tubular secretion** is essentially the reverse of tubular reabsorption and is a process by which the kidneys can rid the blood of additional unwanted substances such as creatinine and ammonia.

The reabsorbed solutes and water that move into the interstitial space between the nephrons need to be returned to the blood, or the kidneys will rapidly swell like balloons. The peritubular capillaries surrounding the renal tubule reclaim the reabsorbed substances and return them to general circulation. Peritubular capillaries arise from the efferent arteriole exiting the glomerulus and empty into the veins leaving the kidney.

Simulating Glomerular Filtration

This computerized simulation allows you to explore one function of a single simulated nephron, glomerular filtration. The concepts you will learn by studying a single nephron can then be applied to understand the function of the kidney as a whole.

Choose **Renal System Physiology** from the main menu. The opening screen for the Simulating Glomerular Filtration experiment will appear in a few seconds (Figure 41B.1). The main features on the screen when the program starts are a simulated blood supply at the left side of the screen, a simulated nephron within a supporting tank on the right side, and a data control unit at the bottom of the display.

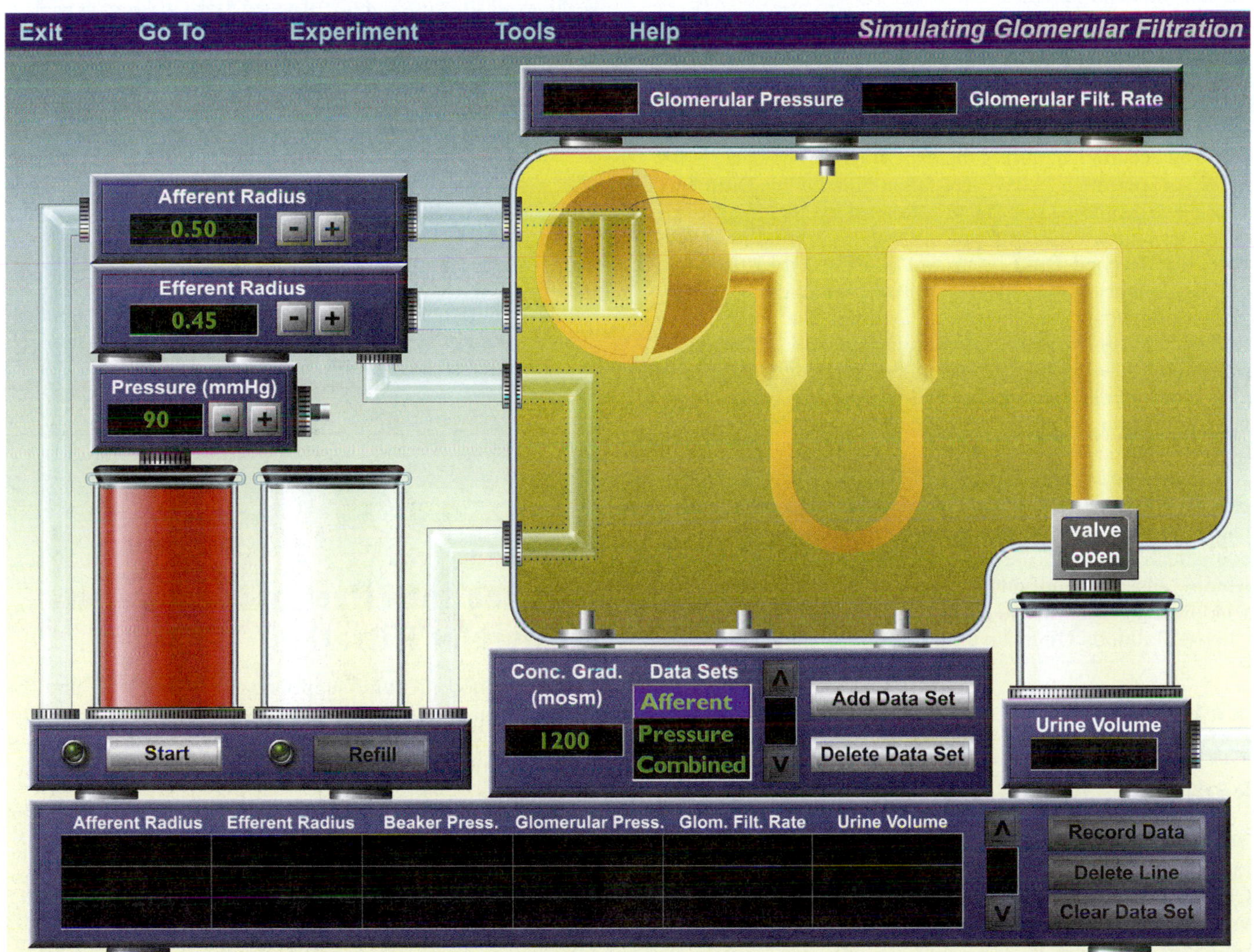

Figure 41B.1 **Opening screen of the Simulating Glomerular Filtration experiment.**

The left beaker is the "blood" source representing the general circulation supplying the nephron. The "blood pressure" in the beaker is adjustable by clicking the (+) and (–) buttons on top of the beaker. A tube with an adjustable radius called the *afferent flow* tube connects the left beaker to the simulated glomerulus. Another adjustable tube called the *efferent flow* tube drains the glomerulus. The afferent flow tube represents the afferent arteriole feeding the glomerulus of each nephron, and the efferent flow tube represents the efferent arteriole draining the glomerulus. The outflow of the nephron empties into a collecting duct, which in turn drains into another small beaker at the bottom right part of the screen. Clicking the valve at the end of the collecting duct stops the flow of fluid through the nephron and collecting duct.

The Glomerular Pressure window on top of the nephron tank displays the pressure within the glomerulus. The Glomerular Filt. Rate window indicates the flow rate of the fluid moving from the lumen of the glomerulus into the lumen of the renal tubule.

The concentration gradient bathing the nephron is fixed at 1200 mosm. Clicking **Start** begins the experiment. Clicking **Refill** resets the equipment to begin another run.

The equipment in the lower part of the screen is called the *data control unit*. This equipment records and displays data you accumulate during the experiments. The data set for the first experiment (Afferent) is highlighted in the **Data Sets** window. You can add or delete a data set by clicking the appropriate button to the right of the Data Sets window. When you click **Record Data,** your data is recorded in the computer's memory and is displayed in the data grid. Data displayed in the data grid include the Afferent Radius, Efferent Radius, Beaker Pressure, Glomerular Pressure, Glomerular Filtration Rate, and the Urine Volume. Clicking **Delete Line** allows you to discard data values for a single run, and clicking **Clear Data Set** erases the entire experiment to allow you to start over.

If you need help identifying any piece of equipment, choose **Balloons On** from the Help menu and move the mouse pointer onto any piece of equipment visible on the computer's screen. As the pointer touches the object, a pop-up window appears identifying the equipment. To close the pop-up window, move the mouse pointer away from the equipment. Choose **Balloons On** again to turn off this help feature.

Activity 1:
Investigating the Effect of Flow Tube Radius on Glomerular Filtration

Your first experiment will examine the effects of flow tube radii and pressures on the rate of glomerular filtration. Click **Start** to see the on-screen action. Continue when you understand how the simulation operates. Click **Refill** to reset the experiment.

1. The **Afferent** line in the Data Sets window of the data control unit should be highlighted in bright blue. If it is not, choose it by clicking the **Afferent** line. The data control unit will now record filtration rate variations due to changing afferent flow tube radius.

2. If the data grid is not empty, click **Clear Data Set** to discard all previous data.

3. Adjust the afferent radius to 0.35 mm, and the efferent radius to 0.40 mm by clicking the appropriate buttons.

4. If the left beaker is not full, click **Refill.**

5. Keep the beaker pressure at 90 mm Hg during this part of the experiment.

6. Click **Start** and watch the blood flow. Simultaneously, filtered fluid will be moving through the nephron and into the collecting duct. The Glomerular Filtration Rate window will display the fluid flow rate into the renal tubule when the left beaker has finished draining.

7. Now click **Record Data** to record the current experiment data in the data grid. Click **Refill** to replenish the left beaker and prepare the nephron for the next run.

8. Increase the afferent radius in 0.05-mm increments and repeat steps 6 through 8 until the maximum radius (0.60 mm) is achieved. Be sure to click **Record Data** after each trial. If you make an error and want to delete a single value, click the data line in the data grid and then click **Delete Line.**

What happens to the glomerular filtration rate as the afferent radius is increased?

____

____

Predict the effect of increasing or decreasing the efferent radius on glomerular filtration rate. Use the simulation to reach an answer if you are not sure.

____

____ ■

Activity 2:
Studying the Effect of Pressure on Glomerular Filtration

Both the blood pressure supplying the glomerulus and the pressure in the renal tubule have a significant impact on the glomerular filtration rate. In this activity, the data control unit will record filtration rate variations due to changing pressure.

1. Click the **Pressure** line in the Data Sets window of the data control unit.

2. If the data grid is not empty, click **Clear Data Set** to discard all previous data.

3. If the left beaker is not full, click **Refill.**

4. Adjust the pressure in the left beaker to 70 mm Hg by clicking the appropriate button.

5. During this part of the experiment, maintain the afferent flow tube radius at 0.55 mm and the efferent flow tube radius at 0.45 mm.

6. Click **Start** and watch the blood flow. Filtrate will move through the nephron into the collecting duct. At the end of the run, the Glomerular Filtration Rate window will display the filtrate flow rate into the renal tubule.

7. Now click **Record Data** to record the current experiment data in the data grid. Click **Refill** to replenish the left beaker.

8. Increase the pressure in the upper beaker in increments of 10 mm Hg and repeat steps 6 through 8 until the maximum pressure (100 mm Hg) is achieved. Be sure to click **Record Data** after each trial. If you make an error and want to delete a single value, click the data line in the data grid and then click **Delete Line.**

What happened to the glomerular filtration rate as the beaker pressure was increased?

____

Explain your answer.

____

____

____

____ ■

Activity 3:

Assessing Combined Effects on Glomerular Filtration

So far, you have examined the effects of flow tube radius and pressure on glomerular filtration rate. In this experiment you will be altering both variables to explore the combined effects on glomerular filtration rate and to see how one can compensate for the other to maintain an adequate glomerular filtration rate.

1. Click **Combined** in the Data Sets window of the data control unit.

2. If the data grid is not empty, click **Clear Data Set** to discard all previous data.

3. If the left beaker is not full, click **Refill.**

4. Set the starting conditions at

- 100 mm Hg beaker pressure
- 0.55 mm afferent radius
- 0.45 mm efferent radius

5. Click **Start.**

6. Now click **Record Data** to record the current baseline data in the data grid.

7. Click **Refill.**

You will use this baseline data to compare a run in which the valve at the end of the collecting duct is in the open position with a run in which the valve is in the closed position.

8. Use the simulation and your knowledge of basic renal anatomy to arrive at answers to the following questions. Be sure to click **Record Data** after each trial. If you make an error and want to delete a single value, click the data line in the data grid and then click **Delete Line.**

Click **valve open** on the end of the collecting duct. Note that it now reads **valve closed.** Click **Start** and allow the run to complete. How does this run compare to the runs in which the valve was open?

____

Expanding on this concept, what might happen to total glomerular filtration and therefore urine production in a human kidney if all of its collecting ducts were totally blocked?

____

____

Would kidney function as a whole be affected if a single nephron was blocked? Explain.

____

____

Would the kidney be functioning if glomerular filtration was zero? Explain.

____

____

Explain how the body could increase glomerular filtration rate in a human kidney.

____

____

____

____

If you increased the pressure in the beaker, what other condition(s) could you adjust to keep the glomerular filtration rate constant?

____

____

9. Click **Tools** → **Print Data** to print your recorded data. ■

Simulating Urine Formation

This part of the computer simulation allows you to explore some aspects of urine formation by manipulating the interstitial solute concentration. Other activities include investigating the effects of aldosterone and ADH (antidiuretic hormone), and the role that glucose carrier proteins play in renal function.

Choose **Simulating Urine Formation** from the **Experiment** menu. The opening screen will appear in a few seconds (Figure 41B.2). The basic features on the screen when the program starts are similar to the glomerular filtration screen. Most of the vascular controls have been moved off-screen to the left because they will not be needed in this set of experiments. Additional equipment includes a *supplies shelf* at the right side of the screen, a *glucose carrier control* located at the top of the nephron tank, and a concentration probe at the bottom left part of the screen.

The maximum concentration of the "interstitial gradient" to be dispensed into the tank surrounding the nephron is adjusted by clicking the (+) and (–) buttons next to the Conc. Grad. window. Click **Dispense** to fill the tank through the jets at the bottom of the tank with the chosen solute gradient. Click **Start** to begin a run. After a run completes, the concentration probe can be clicked and dragged over the nephron to display the solute concentration within.

Hormone is dispensed by dragging a hormone bottle cap to the gray cap button in the nephron tank at the top of the collecting duct and then letting go of the mouse button.

The (+) and (–) buttons in the glucose carrier control are used to adjust the number of glucose carriers that will be inserted into the simulated proximal convoluted tubule when the Add Carriers button is clicked.

Data displayed in the data grid will depend on which experiment is being conducted. Clicking **Delete Line** allows you to discard data values for a single run, and clicking **Clear Data Set** erases the entire experiment to allow you to start over.

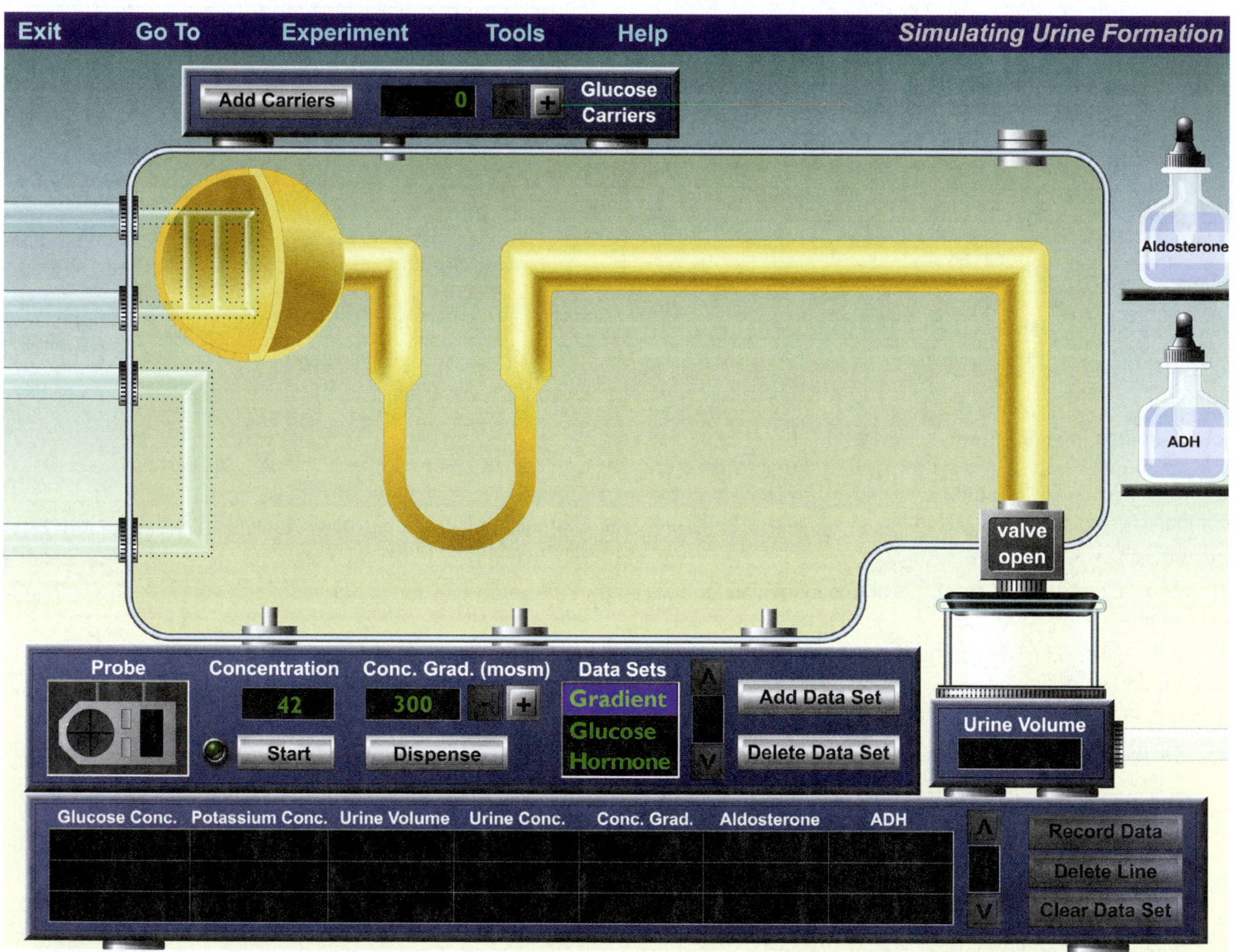

Figure 41B.2 **Opening screen of the Simulating Urine Formation experiment.**

Activity 4:
Exploring the Role of the Solute Gradient on Maximum Urine Concentration Achievable

In the process of urine formation, solutes and water move from the lumen of the nephron into the interstitial spaces. The passive movement of solutes and water from the lumen of the renal tubule into the interstitial spaces relies in part on the total solute gradient surrounding the nephron. When the nephron is permeable to solutes or water, an equilibrium will be reached between the interstitial fluid and the contents of the nephron. Antidiuretic hormone (ADH) increases the water permeability of the distal convoluted tubule and the collecting duct, allowing water to flow to areas of higher solute concentration, usually from the lumen of the nephron into the surrounding interstitial area. You will explore the process of passive reabsorption in this experiment. While doing this part of the simulation, assume that when ADH is present the conditions favor the formation of the most concentrated urine possible.

1. **Gradient** in the Data Sets window of the data control unit should be highlighted in bright blue. If it is not then click **Gradient.**

2. If the data grid is not empty, click **Clear Data Set** to discard all previous data.

3. Click and hold the mouse button on the ADH bottle cap and drag it to the gray cap at the top right side of the nephron tank. Release the mouse button to dispense ADH onto the collecting duct.

4. Adjust the maximum total solute concentration of the gradient **(Conc. Grad.)** to 300 mosm by clicking the appropriate button. Because the blood solute concentration is also 300 mosm, there is no osmotic difference between the lumen of the nephron and the surrounding interstitial fluid.

5. Click **Dispense.**

6. Click **Start** to begin the experiment. Filtrate will move through the nephron and then drain into the beaker below the collecting duct.

7. While the experiment is running, watch the Probe. When it turns red, click and hold the mouse on it and drag it to the urine beaker. Observe the total solute concentration in the Concentration window.

8. Now click **Record Data** to record the current experiment data in the data grid.

9. Increase the maximum concentration of the gradient in 300-mosm increments and repeat steps 3 through 8 until 1200 mosm is achieved. Be sure to click **Record Data** after each trial. If you make an error and want to delete a single value, click the data line in the data grid and then click **Delete Line.**

What happened to the urine concentration as the gradient concentration was increased?

____

____

What factor limits the maximum possible urine concentration?

____

____

The solute concentration of the blood is about 300 mosm, and the highest interstitial solute concentration in a human kidney is about 1200 mosm. This means that the maximum urine solute concentration is about four times that of the blood. What would be the maximum possible urine concentration if the maximum interstitial solute concentration were 3000 mosm instead of 1200 mosm? Explain. (Use the simulation to arrive at an answer if you are not sure.)

____

____

____

____ ■

Activity 5:
Studying the Effect of Glucose Carrier Proteins on Glucose Reabsorption

Because carrier proteins are needed to move glucose from the lumen of the nephron into the interstitial spaces, there is a limit to the amount of glucose that can be reabsorbed. When all glucose carriers are bound with the glucose they are transporting, excess glucose is eliminated in urine. In this experiment, you will examine the effect of varying the number of glucose transport proteins in the proximal convoluted tubule.

1. Click **Glucose** in the Data Sets window of the data control unit.

2. If the data grid is not empty, click **Clear Data Set** to discard all previous data.

3. Set the concentration gradient **(Conc. Grad.)** to 1200 mosm.

4. Click **Dispense.**

5. Adjust the number of glucose carriers to 100 (an arbitrary figure) by clicking the appropriate button.

6. Click **Add Carriers.** This action inserts the specified number of glucose carrier proteins per unit area into the membrane of the proximal convoluted tubule.

7. Click **Start** to begin the run after the carriers have been added.

8. Click **Record Data** to record the current experiment data in the data grid. Glucose presence in the urine will be displayed in the data grid.

9. Now increase the number of glucose carrier proteins in the proximal convoluted tubule in increments of 100 glucose carriers and repeat steps 6 through 8 until the maximum

number of glucose carrier proteins (500) is achieved. Be sure to click **Record Data** after each trial. If you make an error and want to delete a single value, click the data line in the data grid and then click **Delete Line.**

What happened to the amount of glucose present in the urine as the number of glucose carriers was increased?

____

The amount of glucose present in normal urine is minimal because there are normally enough glucose carriers present to handle the "traffic." Predict the consequence in the urine if there was more glucose than could be transported by the available number of glucose carrier proteins.

____

____

Explain why we would expect to find glucose in the urine of a diabetic person.

____

______________________________ ■

Activity 6:
Testing the Effect of Hormones on Urine Formation

The concentration of the urine excreted by our kidneys changes depending on our immediate needs. For example, if a person consumes a large quantity of water, the excess water will be eliminated, producing dilute urine. On the other hand, under conditions of dehydration, there is a clear benefit in being able to produce urine as concentrated as possible, thereby retaining precious water. Although the medullary gradient makes it possible to excrete concentrated urine, urine dilution or concentration is ultimately under hormonal control. In this experiment, you will investigate the effects of two different hormones on renal function, aldosterone produced by the adrenal gland and ADH manufactured by the hypothalamus and stored in the posterior pituitary gland. Aldosterone works to reabsorb Na^+ (and thereby water) at the expense of losing K^+. Its site of action is the distal convoluted tubule. ADH makes the distal tubule and collecting duct more permeable to water, thereby allowing the body to reabsorb more water from the filtrate when it is present.

1. Click **Hormone** in the Data Sets window of the data control unit.

2. If the data grid is not empty, click **Clear Data Set** to discard all previous data.

3. During this part of the experiment, keep the concentration gradient at 1200 mosm.

4. Click **Dispense** to add the gradient and then click **Start** to begin the experiment.

5. Now click **Record Data** to record the current experiment data in the data grid.

You will use this baseline data to compare with the conditions of the filtrate under the control of the two hormones.

6. Keeping all experiment conditions the same as before, do the following:

- Drag the aldosterone bottle cap to the gray cap on the top right side of the nephron tank and release the mouse to automatically dispense aldosterone into the tank surrounding the distal convoluted tubule and collecting duct.
- Click **Start** and allow the run to complete.
- Click **Record Data.**

In this run, how does the volume of urine differ from the previously measured baseline volume?

____

____

Explain the difference in the total amount of potassium in the urine between this run and the baseline run.

____

____

7. Drag the ADH bottle cap to the gray cap on the top right side of the nephron tank and release it to dispense ADH.

- Click **Start** and allow the run to complete
- Click **Record Data.**

In this run, how does the volume of urine differ from the baseline measurement?

____

____

Is there a difference in the total amount of potassium in this run and the total amount of potassium in the baseline run? Explain. (Hint: the urine volume with ADH present is about one-tenth the urine volume when it is not present.)

____

____

Are the effects of aldosterone and ADH similar or antagonistic? ______________________________

Consider this situation: we want to reabsorb sodium ions but do not want to increase the volume of the blood by reabsorbing water from the filtrate. Assuming that aldosterone and ADH are both present, how would you adjust the hormones to accomplish the task?

____

____

____

____

If the interstitial gradient ranged from 300 mosm to 3000 mosm and ADH was not present, what would be the maximum possible urine concentration? ____________________

8. Click **Tools** → **Print Data** to print your recorded data. ■

Histology Review Supplement

Turn to p. P-155 for a review of renal tissue.

exercise

47 Acid-Base Balance: Computer Simulation

Objectives

1. To define *pH* and identify the normal range of human blood pH levels.
2. To define *acid* and *base*, and explain what characterizes each of the following: *strong acid, weak acid, strong base, weak base.*
3. To explain how chemical and physiological buffering systems help regulate the body's pH levels.
4. To define the conditions of *acidosis* and *alkalosis.*
5. To explain the difference between *respiratory acidosis and alkalosis* and *metabolic acidosis and alkalosis.*
6. To understand the causes of respiratory acidosis and alkalosis.
7. To explain how the renal system compensates for respiratory acidosis and alkalosis.
8. To understand the causes of metabolic acidosis and alkalosis.
9. To explain how the respiratory system compensates for metabolic acidosis and alkalosis.

The term **pH** is used to denote the hydrogen ion concentration $[H^+]$ in body fluids. pH values are the reciprocal of $[H^+]$ and follow the formula

$$pH = \log(1/[H^+])$$

At a pH of 7.4, $[H^+]$ is about 40 nanomolars (n*M*) per liter. Because the relationship is reciprocal, $[H^+]$ is higher at *lower* pH values (indicating higher acid levels) and lower at *higher* pH values (indicating lower acid levels).

The pH of a body's fluids is also referred to as its **acid-base balance.** An **acid** is a substance that releases H^+ in solution (such as in body fluids). A **base,** often a hydroxyl ion (OH^-) or bicarbonate ion (HCO_3^-), is a substance that binds to H^+. A *strong acid* is one that completely dissociates in solution, releasing all of its hydrogen ions and thus lowering the solution's pH level. A *weak acid* dissociates incompletely and does not release all of its hydrogen ions in solution. A *strong base* has a strong tendency to bind to H^+, which has the effect of raising the pH value of the solution. A *weak base* binds less of the H^+, having a lesser effect on solution pH.

The body's pH levels are very tightly regulated. Blood and tissue fluids normally have pH values between 7.35 and 7.45. Under pathological conditions, blood pH values as low as 6.9 or as high as 7.8 have been recorded; however, values higher or lower than these cannot sustain human life. The narrow range of 7.35–7.45 is remarkable when one considers the vast number of biochemical reactions that take place in the body. The human body normally produces a large amount of H^+ as the result of metabolic processes, ingested acids, and the products of fat, sugar, and amino acid metabolism. The regulation of a relatively constant internal pH environment is one of the major physiological functions of the body's organ systems.

To maintain pH homeostasis, the body utilizes both *chemical* and *physiological* buffering sytems. Chemical buffers are composed of a mixture of weak acids and weak bases. They help regulate body pH levels by binding H^+ and removing it from solution as its concentration begins to rise, or releasing H^+ into solution as its concentration begins to fall. The body's three major chemical buffering systems are the *bicarbonate, phosphate,* and *protein buffer systems.* We will not focus on chemical buffering systems in this lab, but keep in mind that chemical buffers are the fastest form of compensation and can return pH to normal levels within a fraction of a second.

The body's two major physiological buffering systems are the renal and respiratory systems. The renal system is the slower of the two, taking hours to days to do its work. The respiratory system usually works within minutes, but cannot handle the amount of pH change that the renal system can. These physiological buffer systems help regulate body pH by controlling the output of acids, bases, or CO_2 from the body. For example, if there is too much acid in the body, the renal system may respond by excreting more H^+ from the body in urine. Similarly, if there is too much carbon dioxide in the blood, the respiratory system may respond by breathing faster to expel the excess carbon dioxide. Carbon dioxide levels have a direct effect on pH levels because the addition of carbon dioxide to the blood results in the generation of more H^+. The following reaction shows what happens in the respiratory system when carbon dioxide combines with water in the blood:

$$H_2O + CO_2 \rightleftarrows \underset{\text{carbonic acid}}{H_2CO_3} \rightleftarrows H^+ + \underset{\text{bicarbonate ion}}{HCO_3^-}$$

This is a reversible reaction and is useful for remembering the relationships between CO_2 and H^+. Note that as more CO_2 accumulates in the blood (which frequently is caused by reduced gas exchange in the lungs), the reaction moves to the right and more H^+ is produced, lowering the pH:

$$H_2O + \mathbf{CO_2} \rightarrow \underset{\text{carbonic acid}}{H_2CO_3} \rightarrow \mathbf{H^+} + \underset{\text{bicarbonate ion}}{HCO_3^-}$$

Conversely, as $[H^+]$ increases, more carbon dioxide will be present in the blood:

$$H_2O + \mathbf{CO_2} \leftarrow \underset{\text{carbonic acid}}{H_2CO_3} \leftarrow \mathbf{H^+} + \underset{\text{bicarbonate ion}}{HCO_3^-}$$

Disruptions of acid-base balance occur when the body's pH levels fall below or above the normal pH range of 7.35–7.45. When pH levels fall below 7.35, the body is said to be in a state of **acidosis.** When pH levels rise above 7.45, the body is said to be in a state of **alkalosis. Respiratory acidosis** and **respiratory alkalosis** are the result of the respiratory system accumulating too much or too little carbon dioxide in the blood. **Metabolic acidosis** and **metabolic alkalosis** refer to all other conditions of acidosis and alkalosis (i.e., those not caused by the respiratory system). The experiments in this lab will focus on these disruptions of acid-base balance, and on the physiological buffer systems (renal and respiratory) that compensate for such imbalances.

Respiratory Acidosis and Alkalosis

Respiratory acidosis is the result of impaired respiration, or *hypoventilation,* which leads to the accumulation of too much carbon dioxide in the blood. The causes of impaired respiration include airway obstruction, depression of the respiratory center in the brain stem, lung disease, and drug overdose. Recall that carbon dioxide acts as an acid by forming carbonic acid when it combines with water in the body's blood. The carbonic acid then forms hydrogen ions plus bicarbonate ions:

$$H_2O + \mathbf{CO_2} \rightarrow \underset{\text{carbonic acid}}{H_2CO_3} \rightarrow \mathbf{H^+} + \underset{\text{bicarbonate ion}}{HCO_3^-}$$

Because hypoventilation results in elevated carbon dioxide levels in the blood, the H^+ levels increase, and the pH value of the blood decreases.

Respiratory alkalosis is the condition of too little carbon dioxide in the blood. It is commonly the result of traveling to a high altitude (where the air contains less oxygen) or hyperventilation, which may be brought on by fever or anxiety. Hyperventilation removes more carbon dioxide from the blood, reducing the amount of H^+ in the blood and thus increasing the blood's pH level.

In this first set of activities, we focus on the causes of respiratory acidosis and alkalosis. Follow the instructions in the Getting Started section on pp. P-2 and P-3 in the PhysioEx Introduction to start PhysioEx. From the Main Menu, select **Acid-Base Balance.** You will see the opening screen for "Respiratory Acidosis/Alkalosis" (Figure 47.1). If you have already completed PhysioEx Exercise 37B on respiratory system mechanics, this screen should look familiar. At the left is a pair of simulated lungs, which look like balloons, connected by a tube that looks like an upside-down Y. Air flows in and out of this tube, which simulates the trachea and other air passageways into the lungs. Beneath the "lungs" is a black platform simulating the diaphragm. The long, U-shaped tube containing red fluid represents blood flowing through the lungs. At the top left of the U-shaped tube is a pH meter that will measure the pH level of the blood once the experiment is begun (experiments are begun by clicking the **Start** button at the left of the screen). To the right is an oscilloscope monitor, which will graphically display respiratory volumes. Note that respiratory volumes are measured in liters (l) along the Y-axis, and time in seconds is measured along the X-axis. Below the monitor are three buttons: **Normal Breathing, Hyperventilation,** and **Rebreathing.** Clicking any one of these buttons will induce the given pattern of breathing. Next to these buttons are three data displays for P_{CO_2} (partial pressure of carbon dioxide)—these will give us the levels of carbon dioxide in the blood over the course of an experimental run. At the very bottom of the screen is the data collection grid, where you may record and view your data after each activity.

Activity 1:
Normal Breathing

To get familiarized with the equipment, as well as to obtain baseline data for this experiment, we will first observe what happens during normal breathing.

1. Click **Start.** Notice that the **Normal Breathing** button dims, indicating that the simulated lungs are "breathing" normally. Also notice the reading in the pH meter at the top left, the readings in the P_{CO_2} displays, and the shape of the trace that starts running across the oscilloscope screen. As the trace runs, record the readings for pH at each of the following times:

At 20 seconds, pH = ______________

At 40 seconds, pH = ______________

At 60 seconds, pH = ______________

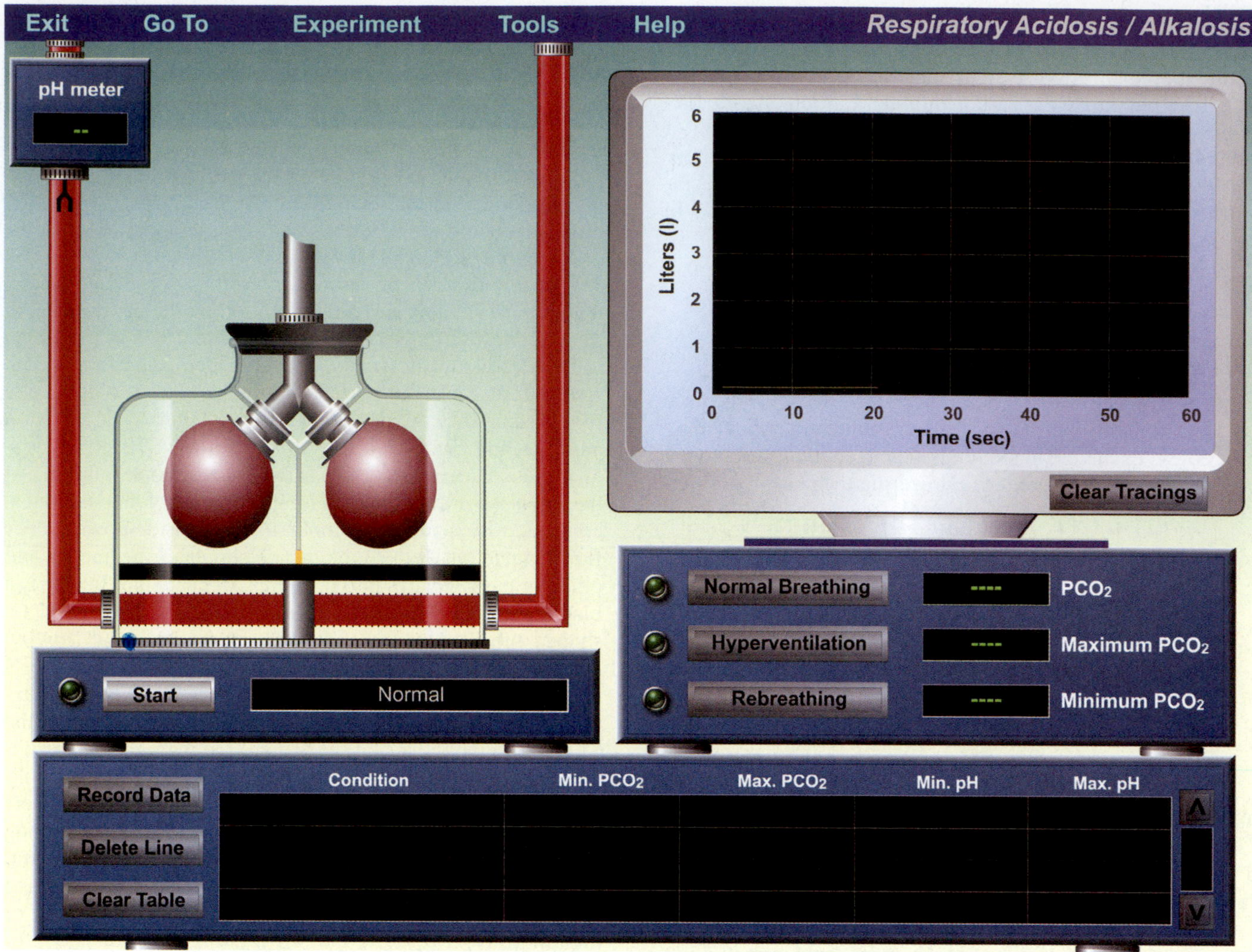

Figure 47.1 Opening screen of the Respiratory Acidosis/Alkalosis experiment.

2. Allow the trace to run all the way to the right side of the oscilloscope screen. At this point, the run will automatically end.

3. Click **Record Data** at the bottom left to record your results.

4. If you have printer access, click **Tools** at the top of the screen and select **Print Graph.** Otherwise, manually sketch what you see on the oscilloscope screen.

5. Click **Clear Tracings** to clear the oscilloscope screen.

Did the pH level of the blood change at all during normal breathing? If so, how?

____

Was the pH level always within the "normal" range for the human body?

____

Did the P_{CO_2} level change during the course of normal breathing? If so, how?

____ ■

Activity 2a: Hyperventilation—Run 1

Next, we will observe what happens to pH and carbon dioxide levels in the blood during hyperventilation.

1. Click **Start.** Allow the normal breathing trace to run for 10 seconds; then at the 10-second mark, click **Hyperventilation.** Watch the pH meter display, as well as the readings in the P_{CO_2} displays and the shape of the trace. As the trace runs, record the readings for pH at each of the following times:

At 20 seconds, pH = _______________

At 40 seconds, pH = _______________

At 60 seconds, pH = _______________

2. Allow the trace to run all the way across the oscilloscope screen and end.

3. Click **Record Data.**

4. If you have printer access, click **Tools** at the top of the screen and select **Print Graph.** Otherwise, manually sketch what you see on the oscilloscope screen on a separate sheet of paper.

5. Click **Clear Tracings** to clear the oscilloscope screen.

Did the pH level of the blood change at all during this run? If so, how?

____

Was the pH level always within the "normal" range for the

human body? ________________________________

If not, when was the pH value outside of the normal range, and what acid-base imbalance did this pH value indicate?

____

Did the P_{CO_2} level change during the course of this run? If so, how?

____

If you observed an acid-base imbalance during this run, how would you expect the renal system to compensate for this condition?

____

How did the hyperventilation trace differ from the trace for normal breathing? Did the tidal volumes change?

____

____

What might cause a person to hyperventilate?

____ ■

Activity 2b: Hyperventilation—Run 2

This activity is a variation on Activity 2a.

1. Click **Start.** Allow the normal breathing trace to run for 10 seconds, then click **Hyperventilation** at the 10-second mark. Allow the hyperventilation trace to run for 10 seconds, then click **Normal Breathing** at the 20-second mark. Allow the trace to finish its run across the oscilloscope screen. Observe the changes in the pH meter and the P_{CO_2} displays.

2. Click **Record Data.**

3. If you have printer access, click **Tools** at the top of the screen and select **Print Graph.** Otherwise, manually sketch what you see on the oscilloscope screen.

4. Click **Clear Tracings** to clear the oscilloscope screen.

What happened to the trace after the 20-second mark when you stopped the hyperventilation? Did the breathing return to normal immediately? Explain your observation.

____

____

____

____ ■

Activity 3: Rebreathing

Rebreathing is the action of breathing in air that was just expelled from the lungs. Breathing into a paper bag is an example of rebreathing. In this activity, we will observe what happens to pH and carbon dioxide levels in the blood during rebreathing.

1. Click **Start.** Allow the normal breathing trace to run for 10 seconds; then at the 10 second mark, click **Rebreathing.** Watch the pH meter display, as well as the readings in the P_{CO_2} displays and the shape of the trace. As the trace runs, record the readings for pH at each of the following times:

At 20 seconds, pH = ____________

At 40 seconds, pH = ____________

At 60 seconds, pH = ____________

2. Allow the trace to run all the way across the oscilloscope screen and end.

3. Click **Record Data.**

4. If you have printer access, click **Tools** at the top of the screen and select **Print Graph.** Otherwise, manually sketch what you see on the oscilloscope screen.

5. Click **Clear Tracings** to clear the oscilloscope screen.

Did the pH level of the blood change at all during this run? If so, how?

____

Was the pH level always within the "normal" range for the

human body? ________________________________

If not, when was the pH value outside of the normal range, and what acid-base imbalance did this pH value indicate?

____

Did the P_{CO_2} level change during the course of this run? If so, how?

____

If you observed an acid-base imbalance during this run, how would you expect the renal system to compensate for this condition?

____

How did the rebreathing trace differ from the trace for normal breathing? Did the tidal volumes change?

____

____

Give examples of respiratory problems that would result in pH and P_{CO_2} patterns similar to what you observed during rebreathing.

____

____

6. To print out all of the recorded data from this activity, click **Tools** and then **Print Data.** ■

In the next set of activities, we will focus on the body's primary mechanism of compensating for respiratory acidosis or alkalosis: renal compensation.

Renal System Compensation

The kidneys play a major role in maintaining fluid and electrolyte balance in the body's internal environment. By regulating the amount of water lost in the urine, the kidneys defend the body against excessive hydration or dehydration. By regulating the excretion of individual ions, the kidneys maintain normal electrolyte patterns of body fluids. By regulating the acidity of urine and the rate of electrolyte excretion, the kidneys maintain plasma pH levels within normal limits. Renal compensation is the body's primary method of compensating for conditions of respiratory acidosis or respiratory alkalosis. (Although the renal system also compensates for metabolic acidosis or metabolic alkalosis, a more immediate mechanism for compensating for metabolic acid-base imbalances is the respiratory system, as we will see in a later experiment.)

The activities in this section examine how the renal system compensates for respiratory acidosis or alkalosis. The primary variable we will be working with is P_{CO_2} (the partial pressure of carbon dioxide in the blood). We will observe how increases and decreases in P_{CO_2} affect the levels of $[H^+]$ and $[HCO_3^-]$ (bicarbonate) that the kidneys excrete in urine.

Click on **Experiment** at the top of the screen and select **Renal System Compensation.** You will see the screen shown in Figure 47.2. If you completed Exercise 41B on renal physiology, this screen should look familiar. There are two beakers on the left side of the screen, one of which is filled with blood, simulating the body's blood supply to the kidneys. Notice that the P_{CO_2} level is currently set to 40, and that the corresponding pH value is 7.4—both "normal" values. By clicking **Start,** you will initiate the process of delivering blood to the simulated nephron at the right side of the screen. As blood flows through the glomerulus of the nephron, you will see the filtration from the plasma of everything except proteins and cells (note that the moving red dots in the animation do *not* include red blood cells). Blood will then drain from the glomerulus to the beaker at the right of the original beaker. At the end of the nephron tube, you will see the collection of urine in a small beaker. Keep in mind that although only one nephron is depicted here, there are actually over a million nephrons in each human kidney. Below the urine beaker are displays for H^+ and HCO_3^-, which will tell us the relative levels of these ions present in the urine.

Activity 4:
Renal Response to Normal Acid-Base Balance

1. Set the P_{CO_2} value to 40, if it is not already. (To increase or decrease P_{CO_2}, click the (−) or (+) buttons. Notice that as P_{CO_2} changes, so does the blood pH level.)
2. Click **Start** and allow the run to finish.
3. At the end of the run, click **Record Data.**

At normal P_{CO_2} and pH levels, what level of H^+ was present in the urine? ______________________

What level of $[HCO_3^-]$ was present in the urine? __________

Why does the blood pH value change as P_{CO_2} changes?

____

____

4. Click **Refill** to prepare for the next activity. ■

Activity 5:
Renal Response to Respiratory Alkalosis

In this activity, we will simulate respiratory alkalosis by setting the P_{CO_2} to values lower than normal (thus, blood pH will be *higher* than normal). We will then observe the renal system's response to these conditions.

1. Set P_{CO_2} to 35 by clicking the (−) button. Notice that the corresponding blood pH value is 7.5.
2. Click **Start.**
3. At the end of the run, click **Record Data.**
4. Click **Refill.**

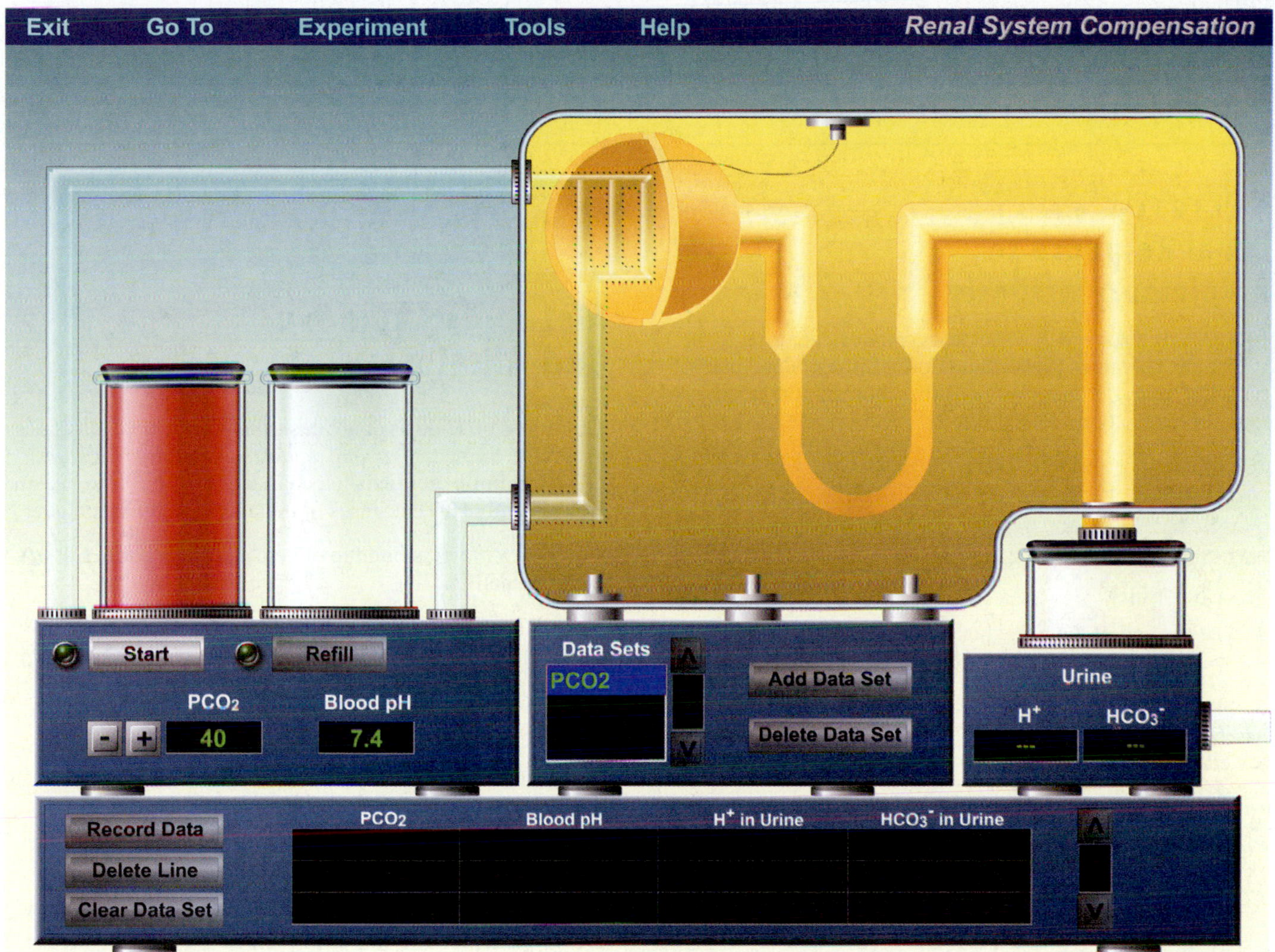

Figure 47.2 Opening screen of the Renal Compensation experiment.

5. Repeat steps 1–4, setting P_{CO_2} to increasingly lower values (i.e., set P_{CO_2} to 30 and then 20, the lowest value allowed).

What level of $[H^+]$ was present in the urine at each of these P_{CO_2}/pH levels?

____

What level of $[HCO_3^-]$ was present in the urine at each of these P_{CO_2}/pH levels?

____

Recall that it may take hours or even days for the renal system to respond to disruptions in acid-base balance. Assuming that enough time has passed for the renal system to fully compensate for respiratory alkalosis, would you expect P_{CO_2} levels to increase or decrease? Would you expect blood pH levels to increase or decrease?

____

____

Recall your activities in the first experiment on respiratory acidosis and alkalosis. Which type of breathing resulted in P_{CO_2} levels closest to the ones we experimented with in this activity—normal breathing, hyperventilation, or rebreathing?

____

Explain why this type of breathing resulted in alkalosis.

____

____ ■

Explain why this type of breathing resulted in acidosis.

____

____

7. Before going on to the next activity, select **Tools** and then **Print Data** in order to save a hard copy of your data results. ■

Activity 6:
Renal Response to Respiratory Acidosis

In this activity, we will simulate respiratory acidosis by setting the P_{CO_2} values higher than normal (thus, blood pH will be *lower* than normal). We will then observe the renal system's response to these conditions.

1. Make sure the left beaker is filled with blood. If not, click **Refill.**

2. Set P_{CO_2} to 60 by clicking the (+) button. Notice that the corresponding blood pH value is 7.3.

3. Click **Start.**

4. At the end of the run, click **Record Data.**

5. Click **Refill.**

6. Repeat steps 1–5, setting P_{CO_2} to increasingly higher values (i.e., set P_{CO_2} to 75 and then 90, the highest value allowed).

What level of $[H^+]$ was present in the urine at each of these P_{CO_2}/pH levels?

____

What level of $[HCO_3^-]$ was present in the urine at each of these P_{CO_2}/pH levels?

____

Recall that it may take hours or even days for the renal system to respond to disruptions in acid-base balance. Assuming that enough time has passed for the renal system to fully compensate for respiratory acidosis, would you expect P_{CO_2} levels to increase or decrease? Would you expect blood pH levels to increase or decrease?

____

____

Recall your activities in the first experiment on respiratory acidosis and alkalosis. Which type of breathing resulted in P_{CO_2} levels closest to the ones we experimented with in this activity—normal breathing, hyperventilation, or rebreathing?

____

Metabolic Acidosis and Alkalosis

Conditions of acidosis or alkalosis that do not have respiratory causes are termed *metabolic acidosis* or *metabolic alkalosis.*

Metabolic acidosis is characterized by low plasma HCO_3^- and pH. The causes of metabolic acidosis include:

- *Ketoacidosis,* a buildup of keto acids that can result from diabetes mellitus
- *Salicylate poisoning,* a toxic condition resulting from ingestion of too much aspirin or oil of wintergreen (a substance often found in laboratories)
- The ingestion of too much alcohol, which metabolizes to acetic acid
- Diarrhea, which results in the loss of bicarbonate with the elimination of intestinal contents
- Strenuous exercise, which may cause a buildup of lactic acid from anaerobic muscle metabolism

Metabolic alkalosis is characterized by elevated plasma HCO_3^- and pH. The causes of metabolic alkalosis include:

- Alkali ingestion, such as antacids or bicarbonate
- Vomiting, which may result in the loss of too much H^+
- Constipation, which may result in reabsorption of elevated levels of HCO_3^-

Increases or decreases in the body's normal metabolic rate may also result in metabolic acidosis or alkalosis. Recall that carbon dioxide—a waste product of metabolism—mixes with water in plasma to form carbonic acid, which in turn forms H^+:

$$H_2O + \mathbf{CO_2} \rightarrow \underset{\text{carbonic acid}}{H_2CO_3} \rightarrow \mathbf{H^+} + \underset{\text{bicarbonate ion}}{HCO_3^-}$$

Therefore, an increase in the normal rate of metabolism would result in more carbon dioxide being formed as a metabolic waste product, resulting in the formation of more H^+—lowering plasma pH and potentially causing acidosis. Other acids that are also normal metabolic waste products, such as ketone bodies and phosphoric, uric, and lactic acids, would likewise accumulate with an increase in metabolic rate. Conversely, a decrease in the normal rate of metabolism would result in less carbon dioxide being formed as a metabolic waste product, resulting in the formation of less H^+—raising

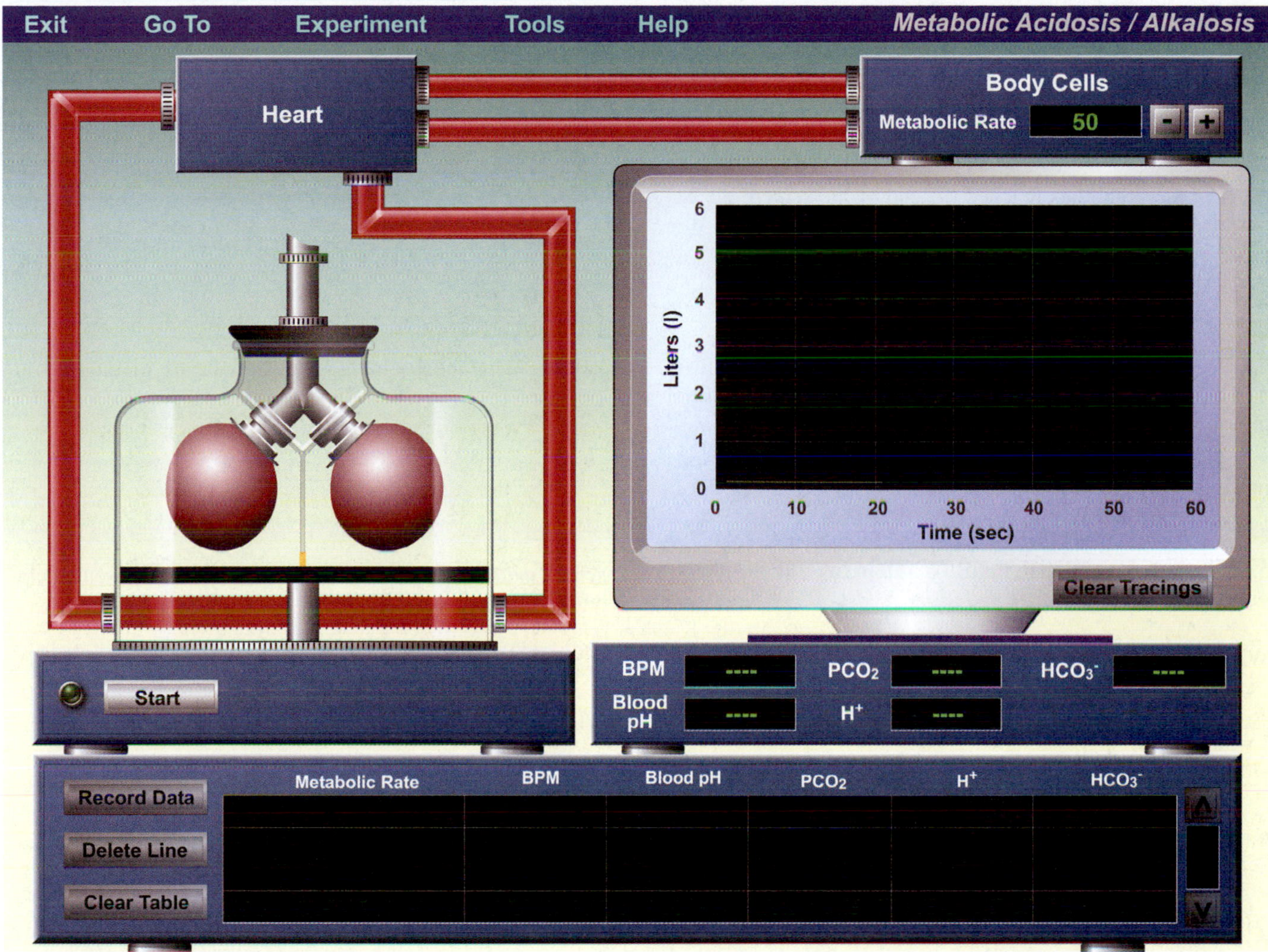

Figure 47.3 Opening screen of the Metabolic Acidosis/Alkalosis experiment.

plasma pH and potentially causing alkalosis. Many factors can affect the rate of cell metabolism. For example, fever, stress, or the ingestion of food all cause the rate of cell metabolism to increase. Conversely, a fall in body temperature or a decrease in food intake causes the rate of cell metabolism to decrease.

The respiratory system compensates for metabolic acidosis or alkalosis by expelling or retaining carbon dioxide in the blood. During metabolic acidosis, respiration increases to expel carbon dioxide from the blood and decrease $[H^+]$ in order to raise the pH level. During metabolic alkalosis, respiration decreases to promote the accumulation of carbon dioxide in the blood, thus increasing $[H^+]$ and decreasing the pH level.

The renal system also compensates for metabolic acidosis and alkalosis by conserving or excreting bicarbonate ions. However, in this set of activities we will focus on respiratory compensation of metabolic acidosis and alkalosis.

To begin, click **Experiment** at the top of the screen and select **Metabolic Acidosis/Alkalosis.** The screen shown in Figure 47.3 will appear. This screen is similar to the screen from the first experiment; the main differences are the addition of a box representing the heart; tubes showing the double circulation of the heart; and a box representing the body's cells. The default "normal" metabolic rate has been set to 50 kcal/h—an arbitrary value, given that "normal" metabolic rates vary widely from individual to individual. The (+) and (−) buttons in the Body Cells box allow you to increase or decrease the body's metabolic rate. In the following activities, we will observe the respiratory response to acidosis or alkalosis brought on by increases or decreases in the body's metabolic rate.

Activity 7: Respiratory Response to Normal Metabolism

We will begin by observing respiratory activity at normal metabolic conditions. This data will serve as a baseline against which we will compare our data in Activities 8 and 9.

1. Make sure the Metabolic Rate is set to 50, which for the purposes of this experiment we will consider the "normal" value.

2. Click **Start** to begin the experiment. Notice the arrows showing the direction of blood flow. A graph displaying respiratory activity will appear on the oscilloscope screen.

3. After the graph has reached the end of the screen, the experiment will automatically stop. Note the data in the displays below the oscilloscope screen:

- The **BPM** display gives you the *breaths-per-minute*—the rate at which respiration occurred.
- Blood pH tells you the pH value of the blood.
- PCO_2 (shown as P_{CO_2} in the text) tells you the partial pressure of carbon dioxide in the blood.
- $\mathbf{H^+}$ and $\mathbf{HCO_3^-}$ tell you the levels of each of these ions present.

4. Click **Record Data.**

5. Click **Tools** and then **Print Graph** in order to print your graph.

What is the respiratory rate? ______________________

Are the blood pH and P_{CO_2} values within normal ranges?

____

6. Click **Clear Tracings** before proceeding to the next activity. ■

Activity 8:
Respiratory Response to Increased Metabolism

1. Increase the metabolic rate to 60.
2. Click **Start** to begin the experiment.
3. Allow the graph to reach the end of the oscilloscope screen. Note the data in the displays below the oscilloscope screen.
4. Click **Record Data.**
5. Click **Tools** and then **Print Graph** in order to print your graph.
6. Repeat steps 1–5 with the metabolic rate set at 70, and then 80.

As the body's metabolic rate increased:

How did respiration change?

____

How did blood pH change?

____

How did P_{CO_2} change?

____

How did $[H^+]$ change?

____

How did $[HCO_3^-]$ change?

____

Explain why these changes took place as metabolic rate increased.

____

____

Which metabolic rates caused pH levels to decrease to a condition of metabolic acidosis?

____

What were the pH values at each of these rates?

____

By the time the respiratory system fully compensated for acidosis, how would you expect the pH values to change?

____

7. Click **Clear Tracings** before proceeding to the next activity. ■

Activity 9:
Respiratory Response to Decreased Metabolism

1. Decrease the metabolic rate to 40.
2. Click **Start** to begin the experiment.
3. Allow the graph to reach the end of the oscilloscope screen. Note the data in the displays below the oscilloscope screen.
4. Click **Record Data.**
5. Click **Tools** and then **Print Graph** in order to print your graph.
6. Repeat steps 1–5 with the metabolic rate set at 30, and then 20.

As the body's metabolic rate decreased:

How did respiration change?

____

How did blood pH change?

How did P_{CO_2} change?

How did $[H^+]$ change?

How did $[HCO_3^-]$ change?

Explain why these changes took place as the metabolic rate decreased.

Which metabolic rates caused pH levels to increase to a condition of metabolic alkalosis?

What were the pH values at each of these rates?

By the time the respiratory system fully compensated for acidosis, how would you expect the pH values to change?

---

7. Click **Tools** → **Print Data** to print your recorded data. ■

NAME ______________________ LAB TIME/DATE ______________________

REVIEW SHEET

exercise

5B

Cell Transport Mechanisms and Permeability: Computer Simulation

Choose all answers that apply to items 1 and 2, and place their letters on the response blanks to the right.

1. Differential permeability: ______________

a. is also called selective permeability

b. refers to the ability of the plasma membrane to select what passes through it

c. implies that all substances pass through membranes without hindrance

d. keeps wastes inside the cell and nutrients outside the cell

2. Passive transport includes: ______________

a. osmosis b. simple diffusion c. bulk-phase endocytosis d. pinocytosis e. facilitated diffusion

3. The following refer to the dialysis simulation.

Did the 20 MWCO membrane exclude any solute(s)? ______________________________

Which solute(s) passed through the 100 MWCO membrane? ______________________________

Which solute exhibited the highest diffusion rate through the 100 MWCO membrane? ______________________________

Explain why this is so: ______________________________

4. The following refer to the facilitated diffusion simulation.

Are substances able to travel against their concentration gradient? ______________________________

Name two ways to increase the rate of glucose transport. ______________________________

Did NaCl affect glucose transport? ______________________________

Does NaCl require a transport protein for diffusion? ______________________________

5. The following refer to the osmosis simulation.

Does osmosis require energy? ______________________________

Is water excluded by any of the dialysis membranes? ______________________________

Is osmotic pressure generated if solutes freely diffuse? ______________________________

Explain how solute concentration affects osmotic pressure. ______________________________

____

6. The following refer to the filtration simulation.

What does the simulated filtration membrane represent in a living organism? ______________________

____

What characteristic of a solute determines whether or not it passes through a filtration membrane?

____

Would filtration occur if we equalized the pressure on both sides of a filtration membrane? ______________

7. The following questions refer to the active transport simulation.

Does the presence of glucose carrier proteins affect Na^+ transport? ____________

Can Na^+ be transported against its concentration gradient? ____________

Are Na^+ and K^+ transported in the same direction? ____________

The ratio of Na^+ to K^+ transport is ______ Na^+ transported out of the cell for every ______ K^+ transported into the cell.

8. What single characteristic of the semipermeable membranes used in the simple diffusion and filtration experiments determines which substances pass through them? ______________________

In addition to this characteristic, what other factors influence the passage of substances through living membranes?

____

9. Assume the left beaker contains 4 m*M* NaCl, 9 m*M* glucose, and 10 m*M* albumin. The right beaker contains 10 m*M* NaCl, 10 m*M* glucose, and 40 m*M* albumin. Furthermore, the dialysis membrane is permeable to all substances except albumin. State whether the substance will move (a) to the right beaker, (b) to the left beaker, or (c) not move.

Glucose ______ Albumin ______

Water ______ NaCl ______

10. Assume you are conducting the experiment illustrated below. Both hydrochloric acid (HCl) with a molecular weight of about 36.5 and ammonium hydroxide (NH_4OH) with a molecular weight of 35 are volatile and easily enter the gaseous state. When they meet, the following reaction will occur:

$$HCl + NH_4OH \rightarrow H_2O + NH_4Cl$$

Ammonium chloride (NH_4Cl) will be deposited on the glass tubing as a smoky precipitate where the two gases meet. Predict which gas will diffuse more quickly and indicate to which end of the tube the smoky precipitate will be closer.

a. The faster diffusing gas is ______________________.

b. The precipitate forms closer to the ______________________ end.

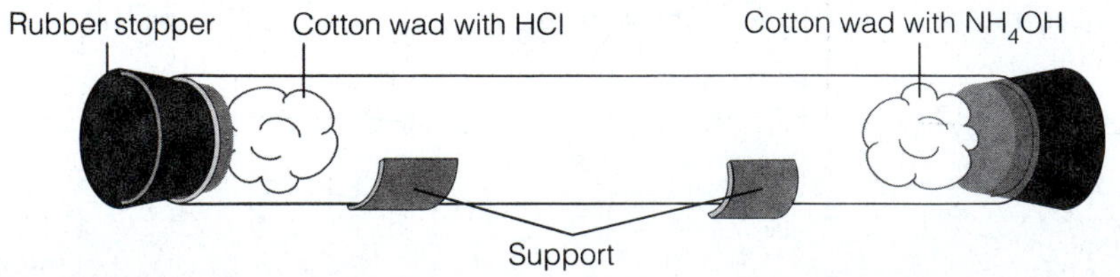

11. When food is pickled for human consumption, as much water as possible is removed from the food. What method is used to achieve this dehydrating effect? ______________________________

____

12. What determines whether a transport process is active or passive? ______________________________

____

13. Characterize passive and active transport as fully as possible by choosing all the phrases that apply and inserting their letters on the answer blanks.

Passive transport ______________ Active transport ______________

a. accounts for the movement of fats and respiratory gases through the plasma membrane
b. explains solute pumping, bulk-phase endocytosis, and pinocytosis
c. includes osmosis, simple diffusion, and filtration
d. may occur against concentration and/or electrical gradients
e. uses hydrostatic pressure or molecular energy as the driving force
f. moves ions, amino acids, and some sugars across the plasma membrane

14. Define the following:

diffusion: ______________________________

____

osmosis: ______________________________

____

simple diffusion: ______________________________

____

filtration: ______________________________

____

active transport: ______________________________

____

bulk-phase endocytosis: ______________________________

____

pinocytosis: ______________________________

____

facilitated diffusion: ______________________________

____

NAME ______________________ LAB TIME/DATE ______________________

REVIEW SHEET

exercise

16B

Skeletal Muscle Physiology: Computer Simulation

Electrical Stimulation

1. Complete the following statements by filling in your answer on the lines provided below.

A motor unit consists of a __a__ and all the __b__ it innervates. Whole muscle contraction is a(n) __c__ response. In order for muscles to work in a practical sense, __d__ is the method used to produce a slow, steady increase in muscle force.

When we see the slightest evidence of force production on a tracing, the stimulus applied must have reached __e__.

The weakest stimulation that will elicit the strongest contraction that a muscle is capable of is called the __f__. That level of contraction is called the __g__.

When the __h__ of stimulation is so high that the muscle tracing shows fused peaks, __i__ has been achieved.

a. ______________________ f. ______________________

b. ______________________ g. ______________________

c. ______________________ h. ______________________

d. ______________________ i. ______________________

e. ______________________

2. Name each phase of a typical muscle twitch and describe what is happening in each phase.

a. ______________________

____

b. ______________________

____

c. ______________________

____

3. Explain how the PhysioEx experimental muscle stimulation differs from the *in vivo* stimulation via the nervous system. (Note that the graded muscle response following both stimulation methods is similar.)

____

____

____

4. What are the two *experimental* ways in which mode of stimulation can affect the muscle force?

_______________________________ and _______________________________

Explain your answer.

___

___

___

Isometric Contraction

1. Identify the following conditions by choosing one of the key terms listed on the right.

Key:

_______ is generated by muscle tissue when it is being stretched — a. Total force

_______ requires the input of energy — b. Resting force

_______ is measured by recording instrumentation during contraction — c. Active force

2. Circle the correct response in the parentheses for each statement.

An increase in resting length results in a(n) (increase/decrease) in passive force.

The active force initially (increased/decreased) and then (increased/decreased) as the resting length was increased from minimum to maximum.

As the total force increased, the active force (increased/decreased).

3. Explain what happens to muscle force production at extremes of length (too short or too long). Hint: think about sarcomere structure.

Muscle too short: _______________________________

___

Muscle too long: _______________________________

___

Isotonic Contraction

1. Assuming a fixed starting length, describe the effect resistance has on the initial velocity of shortening, and explain why it has this effect.

___

___

___

2. A muscle has just been stimulated under conditions that will allow both isometric and isotonic contractions. Describe what is happening in terms of length and force.

Isometric: ____

____

Isotonic: ____

____

Terms

Select the condition from column B that most correctly identifies the term in column A.

	Column A	Column B
_______	1. muscle twitch	a. many cells responding to one neuron
_______	2. wave summation	b. affects the force a muscle can generate
_______	3. multiple motor unit summation	c. a single contraction of intact muscle
_______	4. resting length	d. recruitment
_______	5. resistance	e. increasing force produced by increasing stimulus frequency
_______	6. initial velocity of shortening	f. muscle length changing due to relaxation
_______	7. isotonic shortening	g. caused by application of maximal stimulus
_______	8. isotonic lengthening	h. weight
_______	9. motor unit	i. exhibits graded response
_______	10. whole muscle	j. high values with low resistance values
_______	11. tetanus	k. changing muscle length due to active forces
_______	12. maximal response	l. recording shows no evidence of muscle relaxation

NAME ______________________ LAB TIME/DATE ______________________

REVIEW SHEET
exercise
18B

Neurophysiology of Nerve Impulses: Computer Simulation

The Nerve Impulse

1. Match each of the terms in column B with the appropriate definition in column A.

Column A	Column B
_______ 1. term used to denote that a membrane potential is sitting at about −70 mV	a. threshold
_______ 2. reversal of membrane potential due to influx of sodium ions	b. sodium
_______ 3. major cation found outside the cell	c. potassium
_______ 4. minimal stimulus needed to elicit an action potential	d. resting membrane potential
_______ 5. period when cell membrane is totally insensitive to additional stimuli, regardless of the stimulus force used	e. absolute refractory period
_______ 6. major cation found inside the cell	f. depolarization

2. Fill in the blanks with the correct words or terms.

Neurons, as with other excitable cells of the body, have two major physiological properties: ______________ and ______________. A neuron has a positive charge on the outer surface of the cell membrane due in part to the action of an active transport system called the ______________________. This system moves ______________ out of the cell and ______________ into the cell. The inside of the cell membrane will be negative, not only due to the active transport system but also because of ______________, which remain negative due to intracellular pH and keep the inside of the cell membrane negative.

3. Why don't the terms *depolarization* and *action potential* mean the same thing?

____

____

____

4. What is the difference between membrane irritability and membrane conductivity?

____

____

____

Eliciting a Nerve Impulse

1. Why does the nerve action potential increase slightly when you add 1.0 V to the threshold voltage and stimulate the nerve?

2. If you were to spend a lot of time studying nerve physiology in the laboratory, what type of stimulus would you use and why?

3. Why does the addition of sodium chloride elicit an action potential?

Inhibiting a Nerve Impulse

1. What was the effect of ether on eliciting an action potential?

2. Does the addition of ether to the nerve cause any permanent alteration in neural response?

3. What was the effect of curare on eliciting an action potential?

4. Explain the reason for your answer to question 3 above.

5. What was the effect of lidocaine on eliciting an action potential?

Nerve Conduction Velocity

1. What is the relationship between size of the nerve and conduction velocity? ______________________________

____

____

2. Keeping your answer to question 1 in mind, how might you draw an analogy between the nerves in the human body and electrical wires?

____

3. Hypothesize what types of animals would have the fastest conduction velocities.

____

4. How does myelination affect nerve conduction velocity? Explain. ______________________________

____

____

____

5. If any of the nerves used were reversed in their placement on the stimulating and recording electrodes, would any differences be seen in conduction velocity? Explain.

____

____

____

NAME ____________________ LAB TIME/DATE ____________________

Endocrine System Physiology: Computer Simulation

1. In the following columns, match the hormone on the left with its source on the right.

_______ thyroxine	a. ovary
_______ estrogen	b. thyroid gland
_______ thyroid-stimulating hormone (TSH)	c. pancreas
_______ insulin	d. pituitary gland

2. Each hormone is known to have a specific target tissue. For each hormone listed, what is its target tissue and what is its specific action?

thyroxine ____________________

estrogen ____________________

thyroid-stimulating hormone (TSH) ____________________

insulin ____________________

follicle-stimulating hormone (FSH) ____________________

____

Metabolism

1. In the metabolism experiment, what was the effect of thyroxine on the overall metabolic rate of the animals?

____

____

2. Using the respirometer-manometer, you observed the amount of oxygen being used by animals in a closed chamber. What happened to the carbon dioxide the animals produced while in the chamber?

____

____

3. If the experimental animals in the chamber were engaged in physical activity (such as running in a wheel), how do you think this would change the results of the metabolism experiment?

____

What changes would you expect to see in fluid levels of the manometer? ____

____

4. Why didn't the administration of thyroid-stimulating hormone (TSH) have any effect on the metabolic rate of the thyroidectomized rat? ____

____

5. Why didn't the administration of propylthiouracil have any effect on the metabolic rate of either the thyroidectomized rat or the hypophysectomized rat? ____

____

Hormone Replacement Therapy

1. In the experiment with hormone replacement therapy, what was the effect of removing the ovaries from the animals?

____

____

2. Specifically, what hormone did the ovariectomies effectively remove from the animals, and what purpose does this hormone serve? ____

____

3. If a hormone such as testosterone were used in place of estrogen, would any effect be seen? Explain your answer.

____

____

4. In the experiment, you administered 7 injections of estrogen to the experimental rat over the course of 7 days. What do you think would happen if you administered one injection of estrogen per day for an additional week?

____

____

5. What do you think would happen if you administered 7 injections of estrogen to the experimental rat all in one day?

____

6. In a wet lab, why would you need to wait several weeks after the animals received their ovariectomies before you could perform this experiment on them? ____

____

Insulin and Diabetes

1. In the insulin and diabetes experiment, what was the effect of administering alloxan to the experimental animal?

____

2. When insulin travels to the cells of the body, the concentration of what compound will elevate within the cells?

____

What is the specific action of this compound, within the cells? ____

____

3. Name the diseases described below:

   a. the condition when insulin is not produced by the pancreas: ____

   b. the condition when insulin is produced by the pancreas, but the body fails to respond to the insulin:

____

4. What was the effect of administering insulin to the diabetic rat? ____

____

5. What is a *glucose standard curve*, and why did you need to obtain one for this experiment? ____

____

6. Would altering the light wavelength of the spectrophotometer have any bearing on the results obtained? Explain your answer. ____

____

7. What would you do to help a friend who had inadvertently taken an overdose of insulin? Why? ____

____

NAME ______________________ LAB TIME/DATE ______________________

Cardiovascular Dynamics: Computer Simulation

For numbers 1 and 2 below, choose all answers that apply and place their letters on the response blanks to the right of the statement.

1. The circulation of blood through the vascular system is influenced by ______________.

a. blood viscosity
b. the length of blood vessels
c. the driving pressure behind the blood
d. the radius or diameter of blood vessels

2. Peripheral resistance depends on ______________.

a. blood viscosity
b. blood pressure
c. vessel length
d. vessel radius

3. Complete the following statements.

The volume of blood remaining in the heart after ventricular contraction is called the ______________________.

Cardiac output is defined as ____.

The amount of blood pumped by the heart in a single beat is called the ______________________ volume.

The human heart is actually two individual pumps working in ______________________.

Stroke volume is calculated by ____.

If stroke volume decreased, heart rate would ______________________ in order to maintain blood flow.

4. How could the heart compensate to maintain proper blood flow for the following conditions?

High peripheral resistance: ____

A leaky atrioventricular valve: ____

A constricted semilunar valve: ____

5. How does the size of the heart change under conditions of chronic high peripheral resistance?

____

6. The following questions refer to the Vessel Resistance experiment.

How was the flow rate affected when the radius of the flow tube was increased? ______________________

Which of the adjustable parameters had the greatest effect on fluid flow? ______________________

How does vessel length affect fluid flow? ____

If you increased fluid viscosity, what parameter(s) could you adjust to keep fluid flow constant?

____

Explain your answer. ____

____

If the driving pressure in the left beaker was 100 mm Hg, how could you adjust the conditions of the experiment to completely stop fluid flow? ____

7. The following questions refer to the Pump Mechanics experiment.

What would happen if the right side of the heart pumped faster than the left side of the heart?

____

____

When you change the radius of the right flow tube in Pump Mechanics, the resulting plot looks different than the radius plot in the Vessel Resistance experiment. How would the plot look if you changed the radius of both flow tubes in Pump Mechanics instead of just the right flow tube?

____

____

Why are valves needed in the Pump Mechanics equipment? ____________________________________

What happens to blood flow if peripheral resistance equals pump pressure? ______________________

Theoretically, what would happen to the pumping ability of the heart if the end systolic volume was equal to the end diastolic volume? ____

8. Match the part in the simulation equipment to the analogous cardiac structure or physiological term listed in the key below.

Simulation equipment:		Key:
________	1. valve leading to the right beaker	a. pulmonary veins
________	2. valve leading to the pump	b. bicuspid valve
________	3. left flow tube	c. ventricular filling pressure
________	4. right flow tube	d. peripheral resistance
________	5. pump end volume	e. aortic valve
________	6. pump starting volume	f. end diastolic volume
________	7. pressure in the right beaker	g. aorta
________	8. pressure in the left beaker	h. end systolic volume
________	9. pump pressure	i. systolic pressure

9. Define the following terms:

blood flow: ____

____

peripheral resistance: ____

____

viscosity: ____

____

radius: ____

____

end diastolic volume: ____

____

systole: ____

____

diastole: ____

____

NAME ______________________ LAB TIME/DATE ______________________

REVIEW SHEET
exercise
34B

Frog Cardiovascular Physiology: Computer Simulation

Special Electrical Properties of Cardiac Muscle: Automaticity and Rhythmicity

1. Define the following terms:

   Automaticity ______________________

   Rhythmicity ______________________

2. Explain the anatomical differences between frog and human hearts.

   ____

Baseline Frog Heart Activity

1. Define the following terms:

   Intrinsic heart control ______________________

   Extrinsic heart control ______________________

2. Why is it necessary to keep the frog heart moistened with Ringer's solution? ______________________

Refractory Period of Cardiac Muscle

1. Define *extrasystole* ______________________

2. Refer to the exercise to answer the following questions.

   What was the effect of stimulating the heart during ventricular contraction? ______________________

   During ventricular relaxation ______________________

   During the pause interval ______________________

   What does this information indicate about the refractory period of cardiac muscle?

   ____

   Can cardiac muscle be tetanized? ____________ Why or why not? ______________________

The Effect of Vagus Nerve Stimulation

1. What was the effect of vagal stimulation on heart rate? ______________________.

2. What is vagal escape? ______________________

3. Why is the vagal escape valuable in maintaining homeostasis? ______________________

____

Physical and Chemical Modifiers of Heart Rate

1. Describe the effect of thermal factors on the frog heart.

   Cold ______________________ Heat ______________________

2. Which of the following factors caused the same, or very similar, heart rate-reducing effects: epinephrine, atropine, pilocarpine, digitalis, potassium ions.

   ____

   ____

   Which of the factors listed above would reverse or antagonize vagal effects? ______________________

   ____

3. Did administering any of the following produce any changes in force of contraction (shown by peaks of increasing or decreasing height)? If so, explain the mechanism.

   Epinephrine ______________________

   ____

   Calcium ions ______________________

   ____

4. Excessive amounts of each of the following ions would most likely interfere with normal heart activity. Explain the type of changes caused in each case.

   K^+ ______________________

   Ca^{2+} ______________________

   Na^+ ______________________

5. Define the following:

   Parasympathomimetic ______________________

   Ectopic pacemaker ______________________

6. Explain how digitalis works. ______________________

   ____

NAME ______________________ LAB TIME/DATE ______________________

REVIEW SHEET

exercise

37B

Respiratory System Mechanics: Computer Simulation

Define the following terms:

1. Ventilation ______________________

2. Inspiration ______________________

3. Expiration ______________________

Measuring Respiratory Volumes

1. Write the respiratory volume term and the normal value that is described by the following statements:

   Volume of air present in the lungs after a forceful expiration ______________________

   Volume of air that can be expired forcefully after a normal expiration ______________________

   Volume of air that is breathed in and out during a normal respiration ______________________

   Volume of air that can be inspired forcefully after a normal inspiration ______________________

   Volume of air corresponding to TV + IRV + ERV ______________________

2. Fill in the formula for minute respiratory volume:

   ____

Examining the Effect of Changing Airway Resistance on Respiratory Volumes

1. Even though pulmonary function tests are not diagnostic, they can help determine the difference between

   ______________ and ______________ disorders.

2. Chronic bronchitis and asthma are examples of ______________ disorders.

3. Describe FEV_1: ______________________

   ____

4. Explain the difference between FVC and FEV_1: ______________________

   ____

5. What effect would increasing airway resistance have on FEV_1? ______________________

Examining the Effect of Surfactant

1. Explain the term *surface tension.*____

____

2. Surfactant is a detergent-like __________________________.

3. How does surfactant work? ____

4. What might happen to ventilation if the watery film lining the alveoli did not contain surfactant?

____

Investigating Intrapleural Pressure

1. Complete the following statements.

The pressure within the pleural cavity, __1__, is __2__ than the pressure within the alveoli. This __3__ pressure condition is caused by two forces, the tendency of the lung to recoil due to its __4__ properties and the __5__ of the alveolar fluid. These two forces act to pull the lungs away from the thoracic wall, creating a partial __6__ in the pleural cavity. Because the pressure in the __7__ space is lower than __8__, any opening created in the thoracic wall equalizes the intrapleural pressure with the atmospheric pressure, allowing air to enter the pleural cavity, a condition called __9__. Pneumothorax allows __10__, a condition called __11__.

1. ________________________
2. ________________________
3. ________________________
4. ________________________
5. ________________________
6. ________________________
7. ________________________
8. ________________________
9. ________________________
10. ________________________
11. ________________________

2. Why is the intrapleural pressure negative rather than positive?________________

____

____

3. Would intrapleural pressure be positive or negative when blowing up a balloon? Explain your answer.

____

____

Exploring Various Breathing Patterns

1. Match the term listed in column B with the descriptive phrase in column A. (There may be more than one correct answer.)

Column A	Column B
_______ 1. causes a drop in carbon dioxide concentration in the blood	a. rebreathing
_______ 2. results in lower blood pH	b. rapid breathing
_______ 3. stimulates an increased respiratory rate	c. breath holding
_______ 4. results in a lower respiratory rate	
_______ 5. can be considered an extreme form of rebreathing	
_______ 6. causes a rise in blood carbon dioxide	

2. Because carbon dioxide is the main stimulus for respirations, what would happen to respiratory drive if you held your breath? ____

____

NAME ______________________ LAB TIME/DATE ______________________

Chemical and Physical Processes of Digestion: Compter Simulation

Chemical Digestion of Foodstuffs: Enzymatic Action

1. Match the following definitions with the proper choices from the key.

Key: a. catalyst b. control c. enzyme d. substrate

_______ 1. increases the rate of a chemical reaction without becoming part of the product

_______ 2. provides a standard of comparison for test results

_______ 3. biological catalyst: protein in nature

_______ 4. substance on which an enzyme works

2. Name three characteristics of enzymes. ____

____

3. Explain the following statement: The enzymes of the digestive system are classified as hydrolases.

____

____

4. Fill in the chart below with what you have learned about the various digestive system enzymes encountered in this exercise.

Enzyme	Organ producing it	Site of action	Substrate(s)	Optimal pH
Salivary amylase				
Pepsin				
Lipase (pancreatic)				

5. Name the end products of digestion for the following types of foods.

Proteins: ______________________ Carbohydrates: ______________________

Fats: ______________________ and ______________________

6. You used several different indicators or tests in the laboratory to determine the presence or absence of certain substances. Choose the correct test or indicator from the key to correspond to the condition described below.

Key: a. IKI b. Benedict's solution c. pH meter d. BAPNA

_______ 1. used to test for protein hydrolysis, which was indicated by a yellow color

_______ 2. used to test for the presence of starch, which was indicated by a blue-black color

_______ 3. used to test for the presence of fatty acids

_______ 4. used to test for the presence of maltose, which was indicated by a blue to green (or to rust) color change

7. The three-dimensional structure of a functional protein is altered by intense heat or nonphysiological pH even though peptide bonds may not break. Such a change in protein structure is called denaturation, and denatured enzymes are not functional. Explain why.

____

____

8. What experimental conditions in the simulation resulted in denatured enzymes? _______________________

____

9. Complete the mechanism of absorption section in the chart below for each of the substances listed. Use a check mark to indicate whether the absorption would result in the movement of a substance into the blood capillaries or the lymph capillaries (lacteals).

Substance	Mechanism of absorption	Blood	Lymph
Monosaccharides			
Fatty acids and glycerol			
Amino acids			

10. Imagine that you have been chewing a piece of bread for 5 to 6 minutes. How would you expect its taste to change during this time? ____

11. People on a strict diet to lose weight begin to metabolize stored fats at an accelerated rate. How could this condition affect blood pH? ____

Starch Digestion by Salivary Amylase

1. What conclusions can you draw when an experimental sample gives both a positive starch test and a positive maltose test?

____

2. Why was 37°C the optimal incubation temperature? ____

____

3. Why did very little, if any, starch digestion occur in tube 1? ____

4. Why did very little starch digestion occur in tubes 6 and 7? ____

5. Imagine that you have told a group of your peers that amylase is capable of digesting starch to maltose. If you had not run the experiment in control tubes 3, 4, and 5, what objections to your statement could be raised?

____

____

Assessing Plant Starch Digestion

1. What is the effect of freezing on enzyme activity?

____

2. What can you conclude about the ability of salivary amylase to digest plant starch?

____

3. What can you conclude about the ability of bacteria to digest plant starch?

____

Protein Digestion by Pepsin

1. Why is an indicator reagent such as IKI or Benedict's solution not necessary when using a substrate like BAPNA?

____

2. Trypsin is a pancreatic hydrolase present in the small intestine during digestion. Would trypsin work well in the stomach? Explain your answer.

____

3. How does the optical density of a solution containing BAPNA relate to enzyme activity? ____

____

4. What happens to pepsin activity as it reaches the small intestine? ____

Fat Digestion by Pancreatic Lipase and the Action of Bile

1. Why does the pH of a fatty solution decrease as enzymatic hydrolysis increases? ______________________

____

2. How does bile affect fat digestion? ______________________

3. Why is it not possible to determine the activity of lipase in the pH 2.0 buffer using the pH meter assay method?

____

4. Why is bile not considered an enzyme? ______________________

Physical Processes: Mechanisms of Food Propulsion and Mixing

Complete the following statements. Write your answers in the numbered spaces below.

Swallowing, or __1__, occurs in two phases—the __2__ and __3__. One of these phases, the __4__ phase, is voluntary. During the voluntary phase, the __5__ is used to push the food into the back of the throat. During swallowing, the __6__ rises to ensure that its passageway is covered by the epiglottis so that the ingested substances do not enter the respiratory passageways. It is possible to swallow water while standing on your head because the water is carried along the esophagus involuntarily by the process of __7__. The pressure exerted by the foodstuffs on the __8__ sphincter causes it to open, allowing the food to enter the stomach.

The two major types of propulsive movements that occur in the small intestine are __9__ and __10__. One of these movements, __11__, acts to continually mix the foods and to increase the absorption rate by moving different parts of the chyme mass over the intestinal mucosa, but it has less of a role in moving foods along the digestive tract.

1. ______________________
2. ______________________
3. ______________________
4. ______________________
5. ______________________
6. ______________________
7. ______________________
8. ______________________
9. ______________________
10. ______________________
11. ______________________

NAME ______________________ LAB TIME/DATE ______________________

Renal Physiology—The Function of the Nephron: Computer Simulation

Define the following terms:

1. Glomerulus ______________________

2. Renal tubule ______________________

3. Glomerular capsule ______________________

4. Renal corpuscle ______________________

5. Afferent arteriole ______________________

6. Efferent arteriole ______________________

Investigating the Effect of Flow Tube Radius on Glomerular Filtration

1. In terms of the blood supply to and from the glomerulus, explain why the glomerular capillary bed is unusual.

____

____

2. How would pressure in the glomerulus be affected by constricting the afferent arteriole? Explain your answer.

____

____

3. How would pressure in the glomerulus be affected by constricting the efferent arteriole? Explain your answer.

____

____

Assessing Combined Effects on Glomerular Filtration

1. If systemic blood pressure started to rise, what could the arterioles of the glomerulus do to keep glomerular filtration rate constant? ______________________

____

2. One of the experiments you performed in the simulation was to close the valve at the end of the collecting duct. Is closing that valve more like constricting an afferent arteriole or more like a kidney stone? Explain your answer.

____

____

3. Constricting the efferent arteriole would have the same effect on glomerular filtration as (constricting/dilating) the afferent arteriole.

____

____

Exploring the Role of the Solute Gradient on Maximum Urine Concentration Achievable

Complete the following statements.

In the process of urine formation, solutes and water move from the __1__ of the nephron into the __2__ spaces. The passive movement of solutes and water from the lumen of the renal tubule into the interstitial spaces relies in part on the __3__ surrounding the nephron. When the nephron is permeable to solutes or water, and __4__ will be reached between the interstitial fluid and the contents of the nephron. __5__ is a hormone that increases the water permeability of the __6__ and the collecting duct, allowing water to flow to areas of higher solute concentration, usually from the lumen of the nephron into the surrounding interstitial area. __7__ is hormone that causes __8__ reabsorption at the expense of __9__ loss into the lumen of the tubule.

1. ______________________
2. ______________________
3. ______________________
4. ______________________
5. ______________________
6. ______________________
7. ______________________
8. ______________________
9. ______________________

10. Would the passive movement of substances occur if the interstitial solute concentration was the same as the filtrate solute concentration? Explain your answer.

____

____

Studying the Effect of Glucose Carrier Proteins on Glucose Reabsorption

1. In terms of the function of the nephron, explain why one might find glucose in the urine exiting the collecting duct.

____

____

2. Imagine this scenario: a person has the normal number of glucose carriers in the nephrons yet has glucose in the urine. What could be the cause of this condition? (Hint: think about filtration rate.)

____

____

Testing the Effect of Hormones on Urine Formation

1. Complete the following statements.

The concentration of the ___1___ excreted by our kidneys changes depending on our immediate needs. For example, if a person consumes a large quantity of water, the excess water will be eliminated, producing ___2___ urine. On the other hand, under conditions of dehydration, there is a clear benefit to being able to produce urine as ___3___ as possible, thereby retaining precious water. Although the medullary gradient makes it possible to excrete concentrated urine, urine dilution or concentration is ultimately under ___4___ control. In this experiment, you will investigate the effects of two different hormones on renal function, aldosterone produced by the ___5___ and ADH manufactured by the ___6___ and stored in the ___7___. Aldosterone works to reabsorb ___8___ (and thereby water) at the expense of losing ___9___. Its site of action is the ___10___. ADH makes the distal tubule and collecting duct more permeable to ___11___, thereby allowing the body to reabsorb more water from the filtrate when it is present.

1. ______________________
2. ______________________
3. ______________________
4. ______________________
5. ______________________
6. ______________________
7. ______________________
8. ______________________
9. ______________________
10. ______________________
11. ______________________

2. Match the term listed in column B with the descriptive phrase in column A. (There may be more than one correct answer.)

Column A	Column B
_______1. causes production of dilute urine	a. increased ADH
_______2. results in increased sodium loss	b. increased aldosterone
_______3. causes the body to retain more potassium	c. decreased ADH
_______4. will cause water retention due to sodium movement	d. decreased aldosterone
_______5. causes water reabsorption due to increased membrane permeability	
_______6. increases sodium reabsorption	

NAME ______________________ LAB TIME/DATE ______________________

REVIEW SHEET
exercise
47

Acid-Base Balance

1. Match each of the terms in column A with the appropriate description in column B.

Column A		Column B	
________	1. pH	a.	condition in which the human body's pH levels fall below 7.35
________	2. acid	b.	condition in which the human body's pH levels rise above 7.45
________	3. base	c.	mixes with water in the blood to form carbonic acid
________	4. acidosis	d.	substance that binds to H^+ in solution
________	5. alkalosis	e.	substance that releases H^+ in solution
________	6. carbon dioxide	f.	term used to denote hydrogen ion concentration in body fluids

2. What is the normal range of pH levels of blood and tissue fluids in the human body?

____

3. What is the difference between a *strong acid* and a *weak acid*?

____

4. What is the difference between a *strong base* and a *weak base*?

____

5. What is the difference between respiratory acidosis/alkalosis and metabolic acidosis/alkalosis?

____

____

6. What are the body's two major physiological buffer systems for compensating for acid-base imbalances?

____

Respiratory Acidosis and Alkalosis

1. What are some of the causes of respiratory acidosis?

____

2. What are some of the causes of respiratory alkalosis?

____

3. What happens to blood pH levels during hyperventilation? Why?

____

4. What happens to blood pH levels during rebreathing? Why?

____

____

5. Circle the correct bolfaced terms:

As respiration increases, P_{CO_2} levels **increase / decrease** and pH levels **rise / fall.**

As respiration decreases, P_{CO_2} levels **increase / decrease** and pH levels **rise / fall.**

Renal Compensation

1. How does the renal system compensate for conditions of respiratory acidosis?

____

2. How does the renal system compensate for conditions of respiratory alkalosis?

____

Metabolic Acidosis and Alkalosis

1. What are some of the causes of metabolic acidosis?

____

2. What are some of the causes of metabolic alkalosis?

____

3. Explain how the respiratory system compensates for metabolic acidosis and alkalosis.

____

____

____

4. Explain how the renal system compensates for metabolic acidosis and alkalosis.

____

____

____

5. Circle the correct bolfaced terms:

As metabolic rate increases, respiration **increases / decreases,** P_{CO_2} levels **increase / decrease,** and pH levels **rise / fall.**

As metabolic rate decreases, respiration **increases / decreases,** P_{CO_2} levels **increase / decrease,** and pH levels **rise / fall.**

Histology Review Supplement

The slides in this section are designed to provide a basic histology review related to topics introduced in the PhysioEx lab simulations and in your anatomy and physiology textbook.

From the PhysioEx main menu, select **Histology Review Supplement.** When the screen comes up, click **Select an Image Group.** You will note that the slides in the histology module are grouped in the following categories:

Skeletal muscle slides

Nervous tissue slides

Endocrine tissue slides

Cardiovascular tissue slides

Respiratory tissue slides

Digestive tissue slides

Renal tissue slides

Select the group of slides you wish to view, and then refer to the relevant worksheet in this section for a step-by-step tutorial. For example, if you would like to review the skeletal muscle slides, click on **Skeletal muscle slides** and then turn to the next page of this lab manual for the worksheet entitled Skeletal Muscle Tissue Review to begin your review.

Since the slides in this module have been selected for their relevance to topics covered in the PhysioEx lab simulation, it is recommended that you complete the worksheets along with a related PhysioEx lab. For example, you might complete the Skeletal Muscle Tissue worksheet right before or after your instructor assigns you Exercise 16B, the PhysioEx lab simulation on Skeletal Muscle Physiology.

For additional histology review, turn to page P-15.

Skeletal Muscle Tissue Review

From the PhysioEx main menu, select **Histology Review Supplement.** When the screen comes up, click **Select an Image Group**. From Group Listing, click **Skeletal Muscle Slides.** To view slides without labels, click the **Labels Off** button at the bottom right of the monitor.

Click slide 1.
Skeletal muscle is composed of extremely large, cylindrical multinucleated cells called **myofibers.** The nuclei of the skeletal muscle cell (**myonuclei**) are located peripherally just subjacent to the muscle cell plasmalemma (sarcolemma). The interior of the cell is literally filled with an assembly of contractile proteins (myofilaments) arranged in a specific overlapping pattern oriented parallel to the long axis of the cell.

Click slides 2, 3.
Sarcomeres are the functional units of skeletal muscle. The organization of contractile proteins into a regular end-to-end repeating pattern of sarcomeres along the length of each cell accounts for the striated or striped appearance of skeletal muscle in longitudinal section.

Click slide 4.
The smooth endoplasmic reticulum (sarcoplasmic reticulum), modified into an extensive network of membranous channels that store, release, and take up the calcium necessary for contraction, also functions to further organize the myofilaments inside the cell into cylindrical bundles called myofibrils. The **stippled appearance of the cytoplasm** in cells cut in cross section represents the internal organization of myofilaments bundled into myofibrils by the membranous sarcoplasmic reticulum.

What is the functional unit of contraction in skeletal muscle?

____

What are the two principal contractile proteins that compose the functional unit of contraction?

____

What is the specific relationship of the functional unit of contraction to the striated appearance of a skeletal muscle fiber?

____

____

Click slide 5.
The neural stimulus for contraction arises from the **axon** of a motor neuron whose axon terminal comes into close apposition to the muscle cell sarcolemma.

Would you characterize skeletal muscle as voluntary or involuntary?

____

Name the site of close juxtaposition of an axon terminal with the muscle cell plasmalemma.

____

Skeletal muscle also has an extensive connective tissue component that, in addition to conducting blood vessels and nerves, becomes continuous with the connective tissue of its tendon. The tendon in turn is directly continuous with the connective tissue covering (the periosteum) of the adjacent bone. This connective tissue continuity from muscle to tendon to bone is the basis for movement of the musculoskeletal system.

What is the name of the loose areolar connective tissue covering of an individual muscle fiber?

____

The perimysium is a collagenous connective tissue layer that groups several muscle fibers together into bundles called

________________________.

Which connective tissue layer surrounds the entire muscle and merges with the connective tissue of tendons and aponeuroses?

____

Nervous Tissue Review

From the PhysioEx main menu, select **Histology Review Supplement.** When the screen comes up, click **Select an Image Group.** From Group Listing, click **Nervous Tissue Slides.** To view slides without labels, click the **Labels Off** button at the bottom right of the monitor.

Nervous tissue is composed of nerve cells (neurons) and a variety of support cells.

Click slide 1.
Each nerve cell consists of a **cell body** (perikaryon) and one or more **cellular processes** (axon and dendrites) extending from it. The cell body contains the **nucleus,** which is typically pale-staining and round or spherical in shape, and the usual assortment of cytoplasmic organelles. Characteristically, the nucleus features a prominent **nucleolus** often described as resembling the pupil of a bird's eye (**"bird's eye"** or "owl's eye" nucleolus).

Click slide 2.
The cytoplasm of the cell body is most often granular in appearance due to the presence of darkly stained clumps of ribosomes and rough endoplasmic reticulum **(Nissl bodies/ Nissl substance).** Generally, a single axon arises from the **cell body** at a pale-staining region (axon hillock), devoid of Nissl bodies. The location and number of dendrites arising from the cell body varies greatly.

Axons and dendrites are grouped together in the peripheral nervous system (PNS) to form peripheral nerves.

What is the primary unit of function in nervous tissue?

____

Name the pale-staining region of the cell body from which the axon arises.

____

The support cells of the nervous system perform extremely important functions including support, protection, insulation, and maintenance and regulation of the microenvironment that surrounds the nerve cells.

Click slides 3, 4.
In the PNS, support cells surround both cell bodies (satellite cells) and individual **axons** and **dendrites** (Schwann cells). Schwann cells, in particular, are responsible for wrapping their cell membrane jelly-roll style around axons and dendrites to form an insulating sleeve called the **myelin sheath.**

Click slide 5.
Because Schwann cells are aligned in series and myelinate only a small segment of a single axon, small gaps occur between the myelin sheaths of adjacent contiguous Schwann cells. The gaps, called **nodes of Ranvier,** together with the insulating properties of **myelin,** enhance the speed of conduction of electrical impulses along the length of the axon. Different support cells and myelinating cells are present in the CNS.

What is the general name for all support cells within the CNS?

____

Name the specific myelinating cell of the CNS.

____

In the PNS, connective tissue also plays a role in providing support and organization. In fact, the composition and organization of the connective tissue investments of peripheral nerves are similar to those of skeletal muscle.

Click slide 3.
Each individual axon or dendrite is surrounded by a thin and delicate layer of loose connective tissue called the endoneurium (not shown.) The **perineurium,** a slightly thicker layer of loose connective tissue, groups many axons and dendrites together into bundles (fascicles). The outermost **epineurium** surrounds the entire nerve with a thick layer of dense irregular connective tissue, often infiltrated with adipose tissue, that conveys blood and lymphatic vessels to the nerve. There is no connective tissue component within the nervous tissue of the CNS.

What is the relationship of the endoneurium to the myelin sheath?

____

____

Endocrine Tissue Review

From the PhysioEx main menu, select **Histology Review Supplement.** When the screen comes up, click **Select an Image Group.** From Group Listing, click **Endocrine Tissue Slides**. To view slides without labels, click the **Labels Off** button at the bottom right of the monitor.

Thyroid Gland

The thyroid gland regulates metabolism by regulating the secretion of the hormones T_3 and T_4 (thyroxine) into the blood.

Click slide 1.
The gland is composed of fluid-filled **(colloid)** spheres, called **follicles,** formed by a simple epithelium that can be squamous to columnar depending upon the gland's activity. The colloid stored in the follicles is primarily composed of a glycoprotein (thyroglobulin) that is synthesized and secreted by the follicular cells. Under the influence of the pituitary gland, the follicular cells take up the colloid, convert it into T_3 and T_4, and secrete the T_3 and T_4 into an extensive capillary network. A second population of cells, parafollicular (C) cells (not shown), may be found scattered through the follicular epithelium, but often are present in the connective tissue between follicles. The pale-staining parafollicular cells secrete the protein hormone calcitonin.

Why is the thyroid gland considered to be an endocrine organ?

____

What hormone secreted by the pituitary gland controls the synthesis and secretion of T_3 and T_4 (thyroxine)?

____

What is the function of calcitonin?

____

Ovary

The ovary is an organ that serves both an exocrine function in producing eggs (ova) and an endocrine function in secreting the hormones estrogen and progesterone.

Click slide 2.
Grossly, the ovary is divided into a peripherally located **cortex** in which the oocytes (precursors to the ovulated egg) develop, and a central **medulla** in which connective tissue surrounds blood vessels, lymphatic vessels and nerves. The oocytes, together with supporting cells (granulosa cells), form the **ovarian follicles** seen in the cortex at various stages of development.

Click slide 3.
As an individual oocyte grows, **granulosa cells** proliferate from a single layer of cuboidal cells that surround the oocyte to a multicellular layer that defines a fluid-filled spherical follicle. In a mature follicle (Graafian follicle) the **granulosa cells** are displaced to the periphery of the fluid-filled **antrum,** except for a thin rim of granulosa cells **(corona radiata)** that encircles the **oocyte,** and a pedestal of granulosa cells (cumulus oophorus) that attaches the oocyte to the inner wall of the antrum.

Which cells of the ovarian follicle secrete estrogen?

____

Uterus

Click slides 4, 5, 6.
The uterus is a hollow muscular organ with three major layers: the **endometrium, myometrium,** and either an adventitia or a serosa.

The middle, myometrial layer of the uterine wall is composed of several layers of smooth muscle oriented in different planes.

Click slide 6.
The innermost (nearest the lumen) endometrial layer is further divided functionally into a superficial functional layer **(stratum functionalis)** and a deep basal layer **(stratum basalis).**

Click slide 4.
A simple columnar **epithelium** with both ciliated and nonciliated cells lines the surface of the **endometrium.** The endometrial connective tissue features an abundance of tubular endometrial **glands** that extend from the base to the surface of the layer. During the proliferative phase of the menstrual cycle, shown here, the endometrium becomes thicker as the glands and blood vessels proliferate.

Click slide 5.
In the secretory phase, the **endometrium** and its **glands** and blood vessels are fully expanded.

Click slide 6.
In the menstrual phase, the glands and blood vessels degenerate as the functional layer of the endometrium sloughs away. The deep basal layer **(stratum basalis)** is not sloughed and will regenerate the endometrium during the next proliferative phase.

Which layer of the endometrium is shed during the menstrual phase of the menstrual cycle?

____

What is the function of the deep basal layer (stratum basalis) of the endometrium?

____

What composes a serosa?

____

How does the serosa of the uterus, where present, differ from visceral peritoneum?

____

Pancreas

The pancreas is both an endocrine and an exocrine gland.

Click slide 7.

The exocrine portion is characterized by glandular **secretory units** (acini) formed by a simple epithelium of triangular or pyramidal cells that encircle a small central lumen. The central lumen is the direct connection to the duct system that conveys the exocrine secretions out of the gland. Scattered among the exocrine secretory units are the pale-staining clusters of cells that compose the endocrine portion of the gland. The cells that form these clusters, called **islets of Langerhans** cells (pancreatic islets), secrete a number of hormones, including insulin and glucagon.

Do the islets of Langerhans cells secrete their hormones into the same duct system used by the exocrine secretory cells?

____

Cardiovascular Tissue Review

From the PhysioEx main menu, select **Histology Review Supplement.** When the screen comes up, click **Select an Image Group.** From Group Listing, click **Cardiovascular Tissue Slides.** To view slides without labels, click the **Labels Off** button at the bottom right of the monitor.

Heart

The heart is a four-chambered muscular pump. Although its wall can be divided into three distinct histological layers (endocardium, myocardium, and epicardium), the cardiac muscle of the myocardium composes the bulk of the heart wall.

Click slide 1.
Contractile **cardiac muscle cells** (myocytes, myofibers) have the same striated appearance as skeletal muscle, but are branched rather than cylindrical in shape and have one (occasionally two) **nucleus** (myonucleus) rather than many. The cytoplasmic **striations** represent the same organization of myofilaments (sarcomeres) and alignment of sarcomeres as in skeletal muscle, and the mechanism of contraction is the same. The **intercalated disk,** however, is a feature unique to cardiac muscle. The densely stained structure is a complex of intercellular junctions (desmosome, gap junction, fascia adherens) that structurally and functionally link cardiac muscle cells end to end.

A second population of cells in the myocardium composes the noncontractile intrinsic conduction system (nodal system). Although cardiac muscle is autorhythmic, meaning it has the ability to contract involuntarily in the absence of extrinsic innervation provided by the nervous system, it is the intrinsic conduction system that prescribes the rate and orderly sequence of contraction. Extrinsic innervation only modulates the inherent activity.

Click slide 2.
Of the various components of the noncontractile intrinsic conduction system, **Purkinje fibers** are the most readily observed histologically. They are particularly abundant in the ventricular myocardium and are recognized by their very pale-staining cytoplasm and larger diameter.

The connective tissue component of cardiac muscle is relatively sparse and lacks the organization present in skeletal muscle.

Which component of the intercalated disk is a strong intercellular junction that functions to keep cells from being pulled apart during contraction?

____

What is a functional syncytium?

____

____

Which component of the intercalated disk is a junction that provides the intercellular communication required for the myocardium to perform as a functional syncytium?

____

Blood Vessels

Blood vessels form a system of conduits through which life-sustaining blood is conveyed from the heart to all parts of the body and back to the heart again.

Click slide 3.
Generally, the wall of every vessel is described as being composed of three layers or *tunics.* The **tunica intima,** or *tunica interna,* a simple squamous endothelium and a small amount of subjacent loose connective tissue, is the innermost layer adjacent to the vessel lumen. Smooth muscle and elastin are the predominant constituents of the middle **tunica media,** and the outermost **tunica adventitia,** or *tunica externa,* is a connective tissue layer of variable thickness that provides support and transmits smaller blood and lymphatic vessels and nerves. The thickness of each tunic varies widely with location and function of the vessel. **Arteries,** subjected to considerable pressure fluctuations, have thicker walls overall, with the tunica media being thicker than the tunica adventitia. **Veins,** in contrast, are subjected to much lower pressures and have thinner walls overall, with the tunica adventitia often outsizing the tunica media. Because thin-walled veins conduct blood back to the heart against gravity, valves (not present in arteries) also are present at intervals to prevent backflow. In capillaries, where exchange occurs between the blood and tissues, the tunica intima alone composes the vessel wall.

The tunica media of the aorta would have a much greater proportion of what type of tissue than a small artery?

____

In general, which vessel would have a larger lumen, an artery or its corresponding vein?

____

Why would the tunica media and tunica adventitia not be present in a capillary?

____

Respiratory Tissue Review

From the PhysioEx main menu, select **Histology Review Supplement.** When the screen comes up, click **Select an Image Group.** From Group Listing, click **Respiratory Tissue Slides.** To view slides without labels, click the **Labels Off** button at the bottom right of the monitor.

The respiratory system serves both to conduct oxygenated air deep into the lungs and to exchange oxygen and carbon dioxide between the air and the blood. The trachea, bronchi, and bronchioles are the part of the system of airways that conduct air into the lungs.

Click slide 2.
The trachea and bronchi are similar in morphology. Their lumens are lined by **pseudostratified columnar ciliated epithelium** with **goblet cells** (respiratory epithelium), underlain by a connective tissue **lamina propria** and a deeper connective tissue submucosa with coiled sero-mucous glands that open onto the surface lining of the airway lumen.

Click slide 1.
Deep to the submucosa are the **hyaline cartilage rings** that add structure to the wall of the airway and prevent its collapse. Peripheral to the cartilage is a connective tissue adventitia. The **sero-mucous glands** are also visible in this slide.

Click slide 3.
The bronchioles, in contrast, are much smaller in diameter with a continuous layer of **smooth muscle** in place of the cartilaginous reinforcements. A gradual decrease in the height of the epithelium to **simple columnar** also occurs as the bronchioles decrease in diameter. Distally the bronchioles give way to the respiratory bronchioles, alveolar ducts, alveolar sacs, and alveoli in which gas exchange occurs. In the respiratory bronchiole the epithelium becomes simple cuboidal and the continuous smooth muscle layer is interrupted at intervals by the presence of alveoli inserted into the bronchiolar wall.

Click slide 4.
Although some exchange occurs in the respiratory bronchiole, it is within the **alveoli** of the alveolar ducts and **sacs** that the preponderance of gas exchange transpires. Here the walls of the alveoli, devoid of smooth muscle, are reduced in thickness to the thinnest possible juxtaposition of simple squamous alveolar cell to simple squamous capillary endothelial cell.

What are the primary functions of the respiratory epithelium?

____

____

Why doesn't gas exchange occur in bronchi?

____

What is the primary functional unit of the lung?

____

The alveolar wall is very delicate and subject to collapse. Why is there no smooth muscle present in its wall for support?

____

____

What are the three basic components of the air-blood barrier?

____

____

Digestive Tissue Review

From the PhysioEx main menu, select **Histology Review Supplement.** When the screen comes up, click **Select an Image Group.** From Group Listing, click **Digestive Tissue Slides.** To view slides without labels, click the **Labels Off** button at the bottom right of the monitor.

Salivary Gland

The digestive process begins in the mouth with the physical breakdown of food by mastication. At the same time salivary gland secretions moisten the food and begin to hydrolyze carbohydrates. The saliva that enters the mouth is a mix of serous secretions and mucus (mucin) produced by the three major pairs of salivary glands.

Click slide 1.
The **secretory units** of the salivary tissue shown here are composed predominantly of clusters of pale-staining mucus-secreting cells. More darkly stained serous cells cluster to form a **demilune** (half moon) adjacent to the **lumen** and contribute a clear fluid to the salivary secretion. Salivary secretions flow to the mouth from the respective glands through a well-developed **duct** system.

Are salivary glands endocrine or exocrine glands?

____

Which salivary secretion, mucous or serous, is more thin and watery in consistency?

____

Esophagus

Through contractions of its muscular wall (peristalsis), the esophagus propels food from the mouth to the stomach. Four major layers are apparent when the wall of the esophagus is cut in transverse section:

Click slide 2.
1. The **mucosa** layer adjacent to the lumen consists of a nonkeratinized **stratified squamous epithelium,** its immediately subjacent connective tissue (lamina propria) containing blood vessels, nerves, lymphatic vessels, and cells of the immune system, and a thin smooth muscle layer (muscularis mucosa) that forms the boundary between the mucosa and the submucosa. Because this slide is a low magnification view, it is not possible to discern all parts of the mucosa, nor the boundary between it and the submucosa.

2. The **submucosa** is a layer of connective tissue of variable density, traversed by larger caliber vessels and nerves, that houses the mucus-secreting esophageal **glands** whose secretions protect the epithelium and further lubricate the passing food bolus.

3. Much of the substance of the esophageal wall consists of both circumferentially and longitudinally oriented layers of muscle called the **muscularis externa.** The muscularis externa is composed of skeletal muscle nearest the mouth, smooth muscle nearest the stomach, and a mix of both skeletal and smooth muscle in between.

4. The outermost layer of the esophagus is an adventitia for the portion of the esophagus in the thorax, and a serosa after the esophagus penetrates the diaphragm and enters the abdominal cavity.

Click slide 3.
Here we can see the abrupt change in epithelium at the gastroesophageal junction, where the **esophagus** becomes continuous with the **stomach.**

Briefly explain the difference between an adventitia and a serosa.

____

____

Stomach

The wall of the stomach has the same basic four-layered organization as that of the esophagus.

Click slide 4.
The **mucosa** of the stomach consists of a simple columnar epithelium, a thin connective tissue lamina propria, and a thin **muscularis mucosa.** The most significant feature of the stomach mucosa is that the epithelium invaginates deeply into the lamina propria to form superficial **gastric pits** and deeper **gastric glands.** Although the epithelium of the stomach is composed of a variety of cell types, each with a unique and important function, only three are mentioned here.

Click slide 5.
The **surface mucous cells** are simple columnar cells that line the **gastric pits** and secrete mucus continuously onto the surface of the epithelium. The large round pink- to red-stained **parietal cells** that secrete HCl line the upper half of the gastric glands, and more abundant in the lower half of the gastric glands are the chief cells (not shown), usually stained blue, that secrete pepsinogen (a precursor to pepsin).

Click slide 4 again.
The submucosa is similar to that of the esophagus, but without glands. The muscularis externa has the two typical circumferential and longitudinal layers of smooth muscle, plus an extra layer of smooth muscle oriented obliquely. The stomach's outermost layer is a serosa.

What is the function of the mucus secreted by surface mucous cells?

____

Small Intestine

The key to understanding the histology of the small intestine lies in knowing that its major function is absorption. To that end, its absorptive surface area has been amplified greatly in the following ways:

1. The mucosa and submucosa are thrown into permanent folds (plicae circulares).

2. Fingerlike extensions of the lamina propria form **villi** (singular: **villus**) that protrude into the intestinal lumen *(click slide 7).*

3. The individual **simple columnar epithelial** cells (enterocytes) that cover the villi have **microvilli** (a **brush border**), tiny projections of apical plasma membrane to increase their absorptive surface area (*click slide 6).*

Click slide 7.
Although all three segments of the small intestine (duodenum, jejunum, and ileum) possess villi and tubular **crypts** of Lieberkühn that project deep into the mucosa between villi, some unique features are present in particular segments. For example, large mucous glands **(Brunner's glands)** are present in the submucosa of the duodenum. In addition, permanent aggregates of lymphatic tissue **(Peyer's patches)** are a unique characteristic of the ileum (*click slide 8).*

Aside from these specific features and the fact that the height of the villi vary from quite tall in the duodenum to fairly short in the terminal ileum, the overall morphology of mucosa, submucosa, muscularis externa, and serosa is quite similar in all three segments.

Why is it important for the duodenum to add large quantities of mucus (from Brunner's glands) to the partially digested food entering it from the stomach?

____

Colon

Click slide 9.
The four-layered organization is maintained in the wall of the colon, but the colon has no villi, only **crypts** of Lieberkühn. Simple columnar epithelial cells (enterocytes with microvilli) are present to absorb water from the digested food mass, and the numbers of mucous **goblet cells** are increased substantially, especially toward the distal end of the colon.

Why is it important to have an abundance of mucous goblet cells in the colon?

____

Liver

The functional tissue of the liver is organized into hexagonally shaped cylindrical lobules, each delineated by connective tissue.

Click slide 11.
Within the lobule, large rounded **hepatocytes** form linear cords that radiate peripherally from the center of the lobule at the **central vein** to the surrounding connective tissue. Blood **sinusoids** lined by simple squamous endothelial cells and darkly stained **phagocytic Kupffer cells** are interposed between cords of hepatocytes in the same radiating pattern.

Click slide 10.
Located in the surrounding connective tissue, roughly at the points of the hexagon where three lobules meet, is the **portal triad** (portal canal).

Click slide 12.
The three constituents of the portal triad include a branch of the **hepatic artery,** a branch of the hepatic **portal vein,** and a **bile duct.** Both the hepatic artery and portal vein empty their oxygen-rich blood and nutrient-rich blood, respectively, into the sinusoids. This blood mixes in the sinusoids and flows centrally in between and around the hepatocytes toward the central vein. Bile, produced by hepatocytes, is secreted into very small channels (bile canaliculi) and flows peripherally (away from the central vein) to the bile duct. Thus, the flow of blood is peripheral to central in a hepatic lobule, while the bile flow is central to peripheral.

What general type of cell is the phagocytic Kupffer cell?

____

Blood in the portal vein flows directly from what organs?

____

What is the function of bile in the digestive process?

____

Pancreas

Click slide 13.
The exocrine portion of the pancreas synthesizes and secretes pancreatic enzymes. The individual exocrine **secretory unit,** or acinus, is formed by a group of pyramidal-shaped pancreatic acinar cells clustered around a central lumen into which they secrete their products. A system of pancreatic ducts then transports the enzymes to the duodenum where they are added to the lumen contents to further aid digestion. The groups of pale-staining cells are the endocrine **islets of Langerhans** (pancreatic islet) cells.

Renal Tissue Review

From the PhysioEx main menu, select **Histology Review Supplement.** When the screen comes up, click **Select an Image Group.** From Group Listing, click **Renal Tissue Slides.** To view slides without labels, click the **Labels Off** button at the bottom right of the monitor.

The many functions of the kidney include filtration, absorption, and secretion. The kidney filters the blood of metabolic wastes, water, and electrolytes, and reabsorbs most of the water and sodium ions filtered to regulate and maintain the body's fluid volume and electrolyte balance. The kidney also plays an endocrine role in secreting compounds that increase blood pressure and stimulate red blood cell production.

The uriniferous tubule is the functional unit of the kidney. It consists of two components: the nephron to filter and the collecting tubules and ducts to carry away the filtrate.

Click slide 1.
The nephron itself consists of the **renal corpuscle,** an intimate association of the glomerular capillaries **(glomerulus)** with the cup-shaped Bowman's capsule, and a single elongated renal tubule consisting of segments regionally and sequentially named the **proximal convoluted tubule (PCT),** the descending and ascending segments of the loop of Henle, and the distal convoluted tubule (DCT).

Click slide 2.
A closer look at the renal corpuscle shows both the simple squamous epithelium of the outer layer (parietal layer) of the **Bowman's capsule** (glomerular capsule), and the specialized inner layer (visceral layer) of **podocytes** that extend footlike processes to completely envelop the capillaries of the renal glomerulus. Processes of adjacent podocytes interdigitate with one another, leaving only small slits (filtration slits) between the processes through which fluid from the blood is filtered. The filtrate then flows into the **urinary space** that is directly continuous with the first segment of the renal tubule, the **PCT.** The PCT is lined by robust cuboidal cells equipped with microvilli to greatly increase the surface area of the side of the cell facing the lumen.

Click slide 3.
In the **loop of Henle,** lining cells are simple squamous to simple cuboidal. The DCT cells are also simple cuboidal but are usually much smaller than those of the PCT. The sparse distribution of microvilli, if present at all, on the cells of the DCT relates to their lesser role in absorption. The DCT is continuous directly with the **collecting tubules** and collecting ducts that drain the filtrate out of the kidney.

The large renal artery and its many subdivisions provide an abundant blood supply to the kidney. The smallest distal branches of the renal artery become the afferent arterioles that directly supply the capillaries of the glomerulus. In a unique situation, blood from the glomerular capillaries passes into the efferent arteriole rather than into a venule. The efferent arteriole then perfuses two more capillary beds, the peritubular capillary bed and vasa recta that provide nutrient blood to the kidney tissue itself, before ultimately draining into the renal venous system.

In which segment of the renal tubule does roughly 75–80% of reabsorption occur?

____

How are proximal convoluted tubule (PCT) cells similar to enterocytes of the small intestine?

____

What are the three layers through which the filtrate must pass starting from inside the glomerular capillary through to the urinary space?

____

Under normal circumstances in a healthy individual, would red blood cells or any other cells be present in the renal filtrate?

____

In addition to providing nutrients to the kidney tubules, what is one other function of the capillaries in the peritubular capillary bed?

____

The Metric System

Appendix A

Measurement	Unit and abbreviation	Metric equivalent	Metric to English conversion factor	English to metric conversion factor
Length	1 kilometer (km)	= 1000 (10^3) meters	1 km = 0.62 mile	1 mile = 1.61 km
	1 meter (m)	= 100 (10^2) centimeters = 1000 millimeters	1 m = 1.09 yards 1 m = 3.28 feet 1 m = 39.37 inches	1 yard = 0.914 m 1 foot = 0.305 m
	1 centimeter (cm)	= 0.01 (10^{-2}) meter	1 cm = 0.394 inch	1 foot = 30.5 cm 1 inch = 2.54 cm
	1 millimeter (mm)	= 0.001 (10^{-3}) meter	1 mm = 0.039 inch	
	1 micrometer (μm) [formerly micron (μ)]	= 0.000001 (10^{-6}) meter		
	1 nanometer (nm) [formerly millimicron (mμ)]	= 0.000000001 (10^{-9}) meter		
	1 angstrom (Å)	= 0.0000000001 (10^{-10}) meter		
Area	1 square meter (m^2)	= 10,000 square centimeters	1 m^2 = 1.1960 square yards 1 m^2 = 10.764 square feet	1 square yard = 0.8361 m^2 1 square foot = 0.0929 m^2
	1 square centimeter (cm^2)	= 100 square millimeters	1 cm^2 = 0.155 square inch	1 square inch = 6.4516 cm^2
Mass	1 metric ton (t)	= 1000 kilograms	1 t = 1.103 ton	1 ton = 0.907 t
	1 kilogram (kg)	= 1000 grams	1 kg = 2.205 pounds	1 pound = 0.4536 kg
	1 gram (g)	= 1000 milligrams	1 g = 0.0353 ounce 1 g = 15.432 grains	1 ounce = 28.35 g
	1 milligram (mg)	= 0.001 gram	1 mg = approx. 0.015 grain	
	1 microgram (μg)	= 0.000001 gram		
Volume (solids)	1 cubic meter (m^3)	= 1,000,000 cubic centimeters	1 m^3 = 1.3080 cubic yards 1 m^3 = 35.315 cubic feet	1 cubic yard = 0.7646 m^3 1 cubic foot = 0.0283 m^3
	1 cubic centimeter (cm^3 or cc)	= 0.000001 cubic meter = 1 milliliter	1 cm^3 = 0.0610 cubic inch	1 cubic inch = 16.387 cm^3
	1 cubic millimeter (mm^3)	= 0.000000001 cubic meter		
Volume (liquids and gases)	1 kiloliter (kl or kL)	= 1000 liters	1 kL = 264.17 gallons	1 gallon = 3.785 L 1 quart = 0.946 L
	1 liter (l or L)	= 1000 milliliters	1 L = 0.264 gallons 1 L = 1.057 quarts	
	1 milliliter (ml or mL)	= 0.001 liter = 1 cubic centimeter	1 ml = 0.034 fluid ounce 1 ml = approx. $\frac{1}{4}$ teaspoon 1 ml = approx. 15–16 drops (gtt.)	1 quart = 946 ml 1 pint = 473 ml 1 fluid ounce = 29.57 ml 1 teaspoon = approx. 5 ml
	1 microliter (μl or μL)	= 0.000001 liter		
Time	1 second (s)	= $\frac{1}{60}$ minute		
	1 millisecond (ms)	= 0.001 second		
Temperature	Degrees Celsius (°C)		$°F = \frac{9}{5}°C + 32$	$°C = \frac{5}{9}(°F - 32)$

Appendix B Predicted Vital Capacities for Males

Age	146	148	150	152	154	156	158	160	162	164	166	168	170	172	174	176	178	180	182	184	186	188	190	192	194
	Height in centimeters																								
16	3765	3820	3870	3920	3975	4025	4075	4130	4180	4230	4285	4335	4385	4440	4490	4540	4590	4645	4695	4745	4800	4850	4900	4955	5005
18	3740	3790	3840	3890	3940	3995	4045	4095	4145	4200	4250	4300	4350	4405	4455	4505	4555	4610	4660	4710	4760	4815	4865	4915	4965
20	3710	3760	3810	3860	3910	3960	4015	4065	4115	4165	4215	4265	4320	4370	4420	4470	4520	4570	4625	4675	4725	4775	4825	4875	4930
22	3680	3730	3780	3830	3880	3930	3980	4030	4080	4135	4185	4235	4285	4335	4385	4435	4485	4535	4585	4635	4685	4735	4790	4840	4890
24	3635	3685	3735	3785	3835	3885	3935	3985	4035	4085	4135	4185	4235	4285	4330	4380	4430	4480	4530	4580	4630	4680	4730	4780	4830
26	3605	3655	3705	3755	3805	3855	3905	3955	4000	4050	4100	4150	4200	4250	4300	4350	4395	4445	4495	4545	4595	4645	4695	4740	4790
28	3575	3625	3675	3725	3775	3820	3870	3920	3970	4020	4070	4115	4165	4215	4265	4310	4360	4410	4460	4510	4555	4605	4655	4705	4755
30	3550	3595	3645	3695	3740	3790	3840	3890	3935	3985	4035	4080	4130	4180	4230	4275	4325	4375	4425	4470	4520	4570	4615	4665	4715
32	3520	3565	3615	3665	3710	3760	3810	3855	3905	3950	4000	4050	4095	4145	4195	4240	4290	4340	4385	4435	4485	4530	4580	4625	4675
34	3475	3525	3570	3620	3665	3715	3760	3810	3855	3905	3950	4000	4045	4095	4140	4190	4225	4285	4330	4380	4425	4475	4520	4570	4615
36	3445	3495	3540	3585	3635	3680	3730	3775	3825	3870	3920	3965	4010	4060	4105	4155	4200	4250	4295	4340	4390	4435	4485	4530	4580
38	3415	3465	3510	3555	3605	3650	3695	3745	3790	3840	3885	3930	3980	4025	4070	4120	4165	4210	4260	4305	4350	4400	4445	4495	4540
40	3385	3435	3480	3525	3575	3620	3665	3710	3760	3805	3850	3900	3945	3990	4035	4085	4130	4175	4220	4270	4315	4360	4410	4455	4500
42	3360	3405	3450	3495	3540	3590	3635	3680	3725	3770	3820	3865	3910	3955	4000	4050	4095	4140	4185	4230	4280	4325	4370	4415	4460
44	3315	3360	3405	3450	3495	3540	3585	3630	3675	3725	3770	3815	3860	3905	3950	3995	4040	4085	4130	4175	4220	4270	4315	4360	4405
46	3285	3330	3375	3420	3465	3510	3555	3600	3645	3690	3735	3780	3825	3870	3915	3960	4005	4050	4095	4140	4185	4230	4275	4320	4365
48	3255	3300	3345	3390	3435	3480	3525	3570	3615	3655	3700	3745	3790	3835	3880	3925	3970	4015	4060	4105	4150	4190	4235	4280	4325
50	3210	3255	3300	3345	3390	3430	3475	3520	3565	3610	3650	3695	3740	3785	3830	3870	3915	3960	4005	4050	4090	4135	4180	4225	4270
52	3185	3225	3270	3315	3355	3400	3445	3490	3530	3575	3620	3660	3705	3750	3795	3835	3880	3925	3970	4010	4055	4100	4140	4185	4230
54	3155	3195	3240	3285	3325	3370	3415	3455	3500	3540	3585	3630	3670	3715	3760	3800	3845	3890	3930	3975	4020	4060	4105	4145	4190
56	3125	3165	3210	3255	3295	3340	3380	3425	3465	3510	3550	3595	3640	3680	3725	3765	3810	3850	3895	3940	3980	4025	4065	4110	4150
58	3080	3125	3165	3210	3250	3290	3335	3375	3420	3460	3500	3545	3585	3630	3670	3715	3755	3800	3840	3880	3925	3965	4010	4050	4095
60	3050	3095	3135	3175	3220	3260	3300	3345	3385	3430	3470	3500	3555	3595	3635	3680	3720	3760	3805	3845	3885	3930	3970	4015	4055
62	3020	3060	3110	3150	3190	3230	3270	3310	3350	3390	3440	3480	3520	3560	3600	3640	3680	3730	3770	3810	3850	3890	3930	3970	4020
64	2990	3030	3080	3120	3160	3200	3240	3280	3320	3360	3400	3440	3490	3530	3570	3610	3650	3690	3730	3770	3810	3850	3900	3940	3980
66	2950	2990	3030	3070	3110	3150	3190	3230	3270	3310	3350	3390	3430	3470	3510	3550	3600	3640	3680	3720	3760	3800	3840	3880	3920
68	2920	2960	3000	3040	3080	3120	3160	3200	3240	3280	3320	3360	3400	3440	3480	3520	3560	3600	3640	3680	3720	3760	3800	3840	3880
70	2890	2930	2970	3010	3050	3090	3130	3170	3210	3250	3290	3330	3370	3410	3450	3480	3520	3560	3600	3640	3680	3720	3760	3800	3840
72	2860	2900	2940	2980	3020	3060	3100	3140	3180	3210	3250	3290	3330	3370	3410	3450	3490	3530	3570	3610	3650	3680	3720	3760	3800
74	2820	2860	2900	2930	2970	3010	3050	3090	3130	3170	3200	3240	3280	3320	3360	3400	3440	3470	3510	3550	3590	3630	3670	3710	3740

Courtesy of Warren E. Collins, Inc., Braintree, Mass.

Appendix B Predicted Vital Capacities for Females

Age	Height in centimeters																								
	146	148	150	152	154	156	158	160	162	164	166	168	170	172	174	176	178	180	182	184	186	188	190	192	194
16	2950	2990	3030	3070	3110	3150	3190	3230	3270	3310	3350	3390	3430	3470	3510	3550	3590	3630	3670	3715	3755	3800	3840	3880	3920
17	2935	2975	3015	3055	3095	3135	3175	3215	3255	3295	3335	3375	3415	3455	3495	3535	3575	3615	3655	3695	3740	3780	3820	3860	3900
18	2920	2960	3000	3040	3080	3120	3160	3200	3240	3280	3320	3360	3400	3440	3480	3520	3560	3600	3640	3680	3720	3760	3800	3840	3880
20	2890	2930	2970	3010	3050	3090	3130	3170	3210	3250	3290	3330	3370	3410	3450	3490	3525	3565	3605	3645	3695	3720	3760	3800	3840
22	2860	2900	2940	2980	3020	3060	3095	3135	3175	3215	3255	3290	3330	3370	3410	3450	3490	3530	3570	3610	3650	3685	3725	3765	3800
24	2830	2870	2910	2950	2985	3025	3065	3100	3140	3180	3220	3260	3300	3335	3375	3415	3455	3490	3530	3570	3610	3650	3685	3725	3765
26	2800	2840	2880	2920	2960	3000	3035	3070	3110	3150	3190	3230	3265	3300	3340	3380	3420	3455	3495	3530	3570	3610	3650	3685	3725
28	2775	2810	2850	2890	2930	2965	3000	3040	3070	3115	3155	3190	3230	3270	3305	3345	3380	3420	3460	3495	3535	3570	3610	3650	3685
30	2745	2780	2820	2860	2895	2935	2970	3010	3045	3085	3120	3160	3195	3235	3270	3310	3345	3385	3420	3460	3495	3535	3570	3610	3645
32	2715	2750	2790	2825	2865	2900	2940	2975	3015	3050	3090	3125	3160	3200	3235	3275	3310	3350	3385	3425	3460	3495	3535	3570	3610
34	2685	2725	2760	2795	2835	2870	2910	2945	2980	3020	3055	3090	3130	3165	3200	3240	3275	3310	3350	3385	3425	3460	3495	3535	3570
36	2655	2695	2730	2765	2805	2840	2875	2910	2950	2985	3020	3060	3095	3130	3165	3205	3240	3275	3310	3350	3385	3420	3460	3495	3530
38	2630	2665	2700	2735	2770	2810	2845	2880	2915	2950	2990	3025	3060	3095	3130	3170	3205	3240	3275	3310	3350	3385	3420	3455	3490
40	2600	2635	2670	2705	2740	2775	2810	2850	2885	2920	2955	2990	3025	3060	3095	3135	3170	3205	3240	3275	3310	3345	3380	3420	3455
42	2570	2605	2640	2675	2710	2745	2780	2815	2850	2885	2920	2955	2990	3025	3060	3100	3135	3170	3205	3240	3275	3310	3345	3380	3415
44	2540	2575	2610	2645	2680	2715	2750	2785	2820	2855	2890	2925	2960	2995	3030	3060	3095	3130	3165	3200	3235	3270	3305	3340	3375
46	2510	2545	2580	2615	2650	2685	2715	2750	2785	2820	2855	2890	2925	2960	2995	3030	3060	3095	3130	3165	3200	3235	3270	3305	3340
48	2480	2515	2550	2585	2620	2650	2685	2715	2750	2785	2820	2855	2890	2925	2960	2995	3030	3060	3095	3130	3160	3195	3230	3265	3300
50	2455	2485	2520	2555	2590	2625	2655	2690	2720	2755	2785	2820	2855	2890	2925	2955	2990	3025	3060	3090	3125	3155	3190	3225	3260
52	2425	2455	2490	2525	2555	2590	2625	2655	2690	2720	2755	2790	2820	2855	2890	2925	2955	2990	3020	3055	3090	3125	3155	3190	3220
54	2395	2425	2460	2495	2530	2560	2590	2625	2655	2690	2720	2755	2790	2820	2855	2885	2920	2950	2985	3020	3050	3085	3115	3150	3180
56	2365	2400	2430	2460	2495	2525	2560	2590	2625	2655	2690	2720	2755	2790	2820	2855	2885	2920	2950	2980	3015	3045	3080	3110	3145
58	2335	2370	2400	2430	2460	2495	2525	2560	2590	2625	2655	2690	2720	2750	2785	2815	2850	2880	2920	2945	2975	3010	3040	3075	3105
60	2305	2340	2370	2400	2430	2460	2495	2525	2560	2590	2625	2655	2685	2720	2750	2780	2810	2845	2875	2915	2940	2970	3000	3035	3065
62	2280	2310	2340	2370	2405	2435	2465	2495	2525	2560	2590	2620	2655	2685	2715	2745	2775	2810	2840	2870	2900	2935	2965	2995	3025
64	2250	2280	2310	2340	2370	2400	2430	2465	2495	2525	2555	2585	2620	2650	2680	2710	2740	2770	2805	2835	2865	2895	2925	2955	2990
66	2220	2250	2280	2310	2340	2370	2400	2430	2460	2495	2525	2555	2585	2615	2645	2675	2705	2735	2765	2800	2825	2860	2890	2920	2950
68	2190	2220	2250	2280	2310	2340	2370	2400	2430	2460	2490	2520	2550	2580	2610	2640	2670	2700	2730	2760	2795	2820	2850	2880	2910
70	2160	2190	2220	2250	2280	2310	2340	2370	2400	2425	2455	2485	2515	2545	2575	2605	2635	2665	2695	2725	2755	2780	2810	2840	2870
72	2130	2160	2190	2220	2250	2280	2310	2335	2365	2395	2425	2455	2480	2510	2540	2570	2600	2630	2660	2685	2715	2745	2775	2805	2830
74	2100	2130	2160	2190	2220	2245	2275	2305	2335	2360	2390	2420	2450	2475	2505	2535	2565	2590	2620	2650	2680	2710	2740	2765	2795

Courtesy of Warren E. Collins, Inc., Braintree, Mass.

Appendix C

A.D.A.M.® Interactive Anatomy Correlations

Appendix B lists correlations of Dissectible Anatomy images in the A.D.A.M.® Interactive Anatomy (AIA) program with exercises in this manual. To start the AIA program, insert the AIA CD in your CD-ROM drive. Double-click the Interactive Anatomy 3.0 application icon located in the Interactive Anatomy folder on your hard drive. From the Introduction screen, click Dissectible Anatomy. From the open dialog, click Open. To change anatomical views, click the View button drop-down menu in the tool bar, and select Anterior, Posterior, Lateral, or Medial. Click on the Structure List to launch the List Manager (Mac only) or choose from the list of available structures (Win only).

Note that AIA also offers Atlas Anatomy and 3-D Anatomy. Take some time to explore these additional features of AIA.

After you click Open from the open dialog, the following image appears:

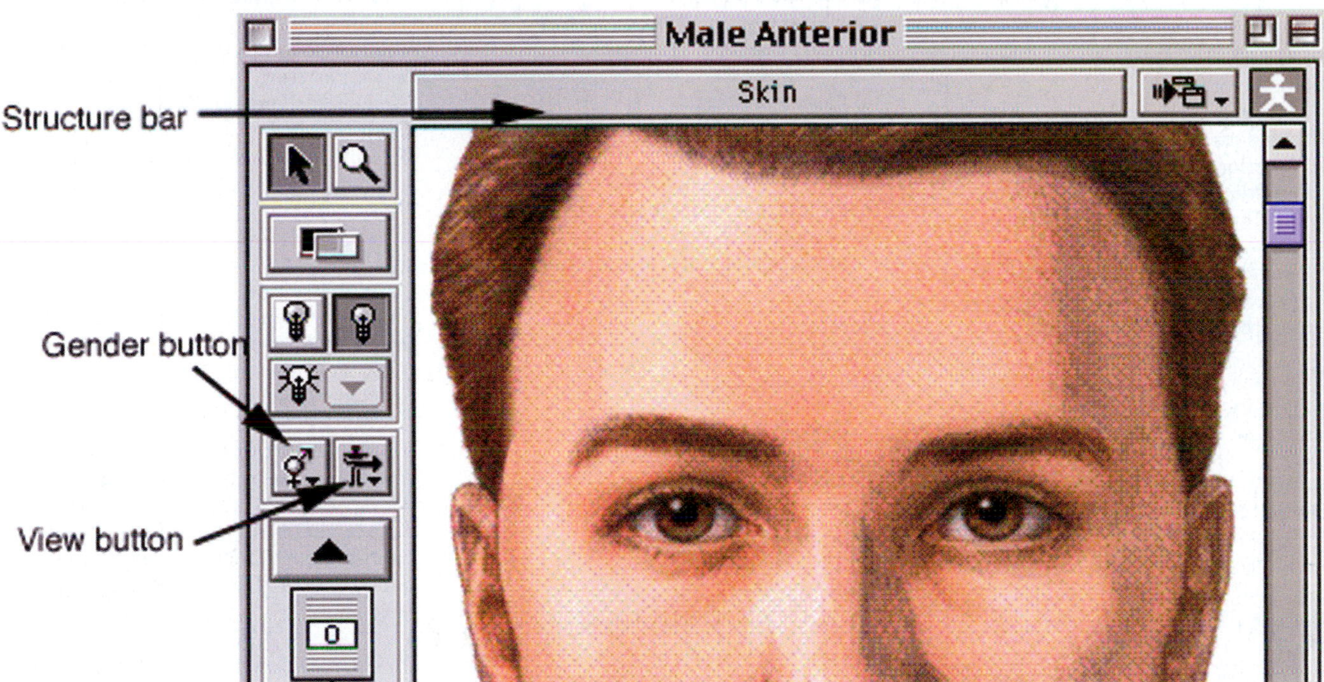

Exercise 1
The Language of Anatomy

To illustrate a coronal plane
View: Anterior
Structure: Skull-coronal section

To illustrate a midsagittal plane
View: Medial
Structure: Skin

To illustrate a ventral body cavity
View: Anterior
Structure: Diaphragm

To illustrate serous membranes of the ventral body cavity
View: Anterior
Structure: Pericardial sac, Parietal peritoneum, Peritoneum

To illustrate a parietal pleura
View: Medial
Structure: Costal parietal pleura diaphragmatic

Exercise 8
Classification of Body Membranes

To illustrate parietal pleural membranes
View: Anterior
Structure: Parietal pleura

Exercise 10
The Axial Skeleton

To illustrate bones of the cranium and face
View: Anterior
Structure: Skull

To illustrate bones of the cranium
View: Posterior
Structure: Bones—coronal section

To illustrate bones of the cranium and face
View: Lateral
Structure: Skull

To illustrate hyoid bone
View: Lateral
Structure: Hyoid bone

To illustrate paranasal sinuses
View: Anterior
Structure: Mucosa of maxillary, Frontal, Sphenoidal sinuses, Ethmoidal sinus

To illustrate a vertebral column
View: Anterior
Structure: Intervertebral disc

To illustrate a posterior vertebral column
View: Posterior
Structure: Intervertebral disc

To illustrate a vertebral column and intervertebral discs
View: Medial
Structure: Sacrum

To illustrate a sacrum and lumbar vertebrae
View: Posterior
Structure: Sacrum

To illustrate a bony thorax
View: Anterior
Structure: Ribs

Exercise 11
The Appendicular Skeleton

To illustrate a pectoral girdle
View: Anterior
Structure: Clavicle

To illustrate a pectoral girdle
View: Posterior
Structure: Scapula

To illustrate a humerus
View: Anterior
Structure: Humerus

To illustrate an ulna
View: Anterior
Structure: Ulna

To illustrate bones of the wrist and hand
View: Anterior
Structure: Bones of hand

To illustrate a female pelvis click on the Gender button and choose Female
View: Anterior
Structure: Bones—coronal section

To illustrate a male pelvis click on the Gender button and choose Male
View: Anterior
Structure: Bones—coronal section

To illustrate a femur
View: Anterior
Structure: Femur

To illustrate a tibia
View: Anterior
Structure: Tibia

To illustrate bones of the foot
View: Anterior
Structure: Bones of the foot

Exercise 13
Articulations and Body Movements

To illustrate a synovial joint of the knee
View: Anterior
Structure: Synovial capsule of knee joint

To illustrate a synovial joint capsule of the hip
View: Anterior
Structure: Synovial joint capsule of the hip

Exercise 15
Gross Anatomy of the Muscular System

To illustrate muscles of the face
View: Anterior
Structure: Orbicularis oculi muscle

To illustrate muscles of mastication
View: Anterior
Structure: Masseter muscle

To illustrate superficial muscles of the neck
View: Lateral
Structure: Platysma muscle, Sternocleidomastoid muscle

To illustrate deep muscles of the neck
View: Lateral
Structure: Strap muscles

To illustrate thorax and shoulder muscles
View: Lateral
Structure: Pectoralis major muscle, Serratus anterior muscle

To illustrate thorax muscles
View: Anterior
Structure: External intercostal muscles, Internal intercostal muscle

To illustrate abdominal wall
View: Anterior
Structure: Rectus abdominis muscle, External abdominal oblique muscle, Internal abdominal oblique muscle

To illustrate thorax muscles
View: Lateral
Structure: Latissimus dorsi muscle

To illustrate posterior muscles of the trunk
View: Posterior
Structure: Rhomboideus muscle

To illustrate muscles associated with the vertebral column
View: Posterior
Structure: Semispinalis muscle, Splenius muscle

To illustrate muscles of the humerus that act on the forearm
View: Posterior
Structure: Brachialis muscle

To illustrate anterior muscles of the forearm that act on the hand and fingers
View: Anterior
Structure: Flexor carpi radialis muscle, Palmaris longus muscle, Flexor carpi ulnaris muscle, Flexor digitorum superficialis muscle

To illustrate deep muscles of the forearm that act on the hand and fingers
View: Posterior
Structure: Abductor pollicis longus muscle

To illustrate muscles acting on the thigh
View: Anterior
Structure: Sartorius muscle

To illustrate quadriceps
View: Anterior
Structure: Rectus femoris muscle, Vastus lateralis muscle, Vastus medialis muscle, Vastus intermedius muscle, Tensor fasciae latae muscle

To illustrate muscles acting on the thigh and originating on the pelvis
View: Posterior
Structure: Gluteus maximus muscle, Gluteus medius muscle, Gluteus minimus muscle

To illustrate hamstrings
View: Posterior
Structure: Long head of the biceps femoris muscle, Semitendinosus muscle, Semimembranosus muscle

To illustrate superficial muscles acting on the foot and ankle
View: Posterior
Structure: Gastrocnemius muscle, Soleus muscle, Popliteus muscle, Tibialis posterior muscle

To illustrate muscles acting on the foot and ankle
View: Anterior
Structure: Tibialis anterior muscle, Extensor digitorum longus muscle, Extensor hallucis longus muscle

Exercise 19
Gross Anatomy of the Brain and Cranial Nerves

To illustrate a cerebrum
View: Anterior
Structure: Skull—coronal section

To illustrate a cerebrum
View: Lateral
Structure: Brain

To illustrate cranial nerves
View: Lateral
Structure: Cranial nerves

To illustrate a vagus nerve
View: Lateral
Structure: Vagus nerve {CN X}

To illustrate phrenic and vagus nerves
View: Anterior
Structure: Phrenic and vagus nerves {CN X}

Exercise 21
Spinal Cord and Spinal Nerves

To illustrate a spinal cord
View: Posterior
Structure: Spinal cord

To illustrate deep nerve plexuses and intercostal nerves
View: Anterior
Structure: Deep nerve plexuses and intercostal nerves

To illustrate a brachial plexus and branches
View: Anterior
Structure: Brachial plexus

To illustrate a central nervous system and sacral plexus
View: Medial
Structure: Central nervous system and sacral plexus

To illustrate a sciatic nerve and branches
View: Medial
Structure: Sciatic nerve and branches

To illustrate a tibial nerve
View: Posterior
Structure: Tibial nerve and branches

To illustrate autonomic nerve plexuses
View: Medial
Structure: Autonomic nerve plexuses

To illustrate a sympathetic trunk
View: Anterior
Structure: Sympathetic trunk

Exercise 24
Special Senses: Vision

To illustrate muscles and external anatomy of the eye
View: Anterior
Structure: Muscles of the eye

To illustrate eye muscles
View: Lateral
Structure: Eye muscles—medial

Exercise 27
Anatomy and Basic Function of the Endocrine Glands

To illustrate a pituitary gland
View: Lateral
Structure: Pituitary gland

To illustrate a thyroid gland
View: Anterior
Structure: Thyroid gland

To illustrate a suprarenal gland
View: Anterior
Structure: Suprarenal {Adrenal} gland

To illustrate a pancreas
View: Anterior
Structure: Pancreas

To illustrate ovaries click on the Gender button and choose Female
View: Medial
Structure: Ovary

To illustrate testes click on the Gender button and choose Male
View: Anterior
Structure: Testis

To illustrate a thymus gland
View: Anterior
Structure: Thymus gland

Exercise 30
Anatomy of the Heart

To illustrate a heart
View: Anterior
Structure: Heart

To illustrate fibrous pericardium
View: Anterior
Structure: Pericardiacophrenic vein

To illustrate visceral pericardium
View: Anterior
Structure: Epicardium

To illustrate heart chambers and heart valves
View: Anterior
Structure: Heart—cut section

To illustrate coronary arteries
View: Anterior
Structure: Coronary arteries

Exercise 32
Anatomy of Blood Vessels

To illustrate an aortic arch and branches
View: Anterior
Structure: Aortic arch and branches

To illustrate common carotid arteries
View: Anterior
Structure: Common carotid arteries

To illustrate major branches of the descending aorta
View: Anterior
Structure: Celiac trunk and branches

To illustrate a descending thoracic aorta
View: Anterior
Structure: Descending thoracic aorta

To illustrate an abdominal aorta and branches
View: Anterior
Structure: Abdominal aorta and branches

To illustrate an aortic arch and branches
View: Medial
Structure: Aortic

To illustrate veins draining into the vena cavae
View: Lateral
Structure: Venae Cavae and tributaries

To illustrate pulmonary circulation
View: Anterior
Structure: Pulmonary arteries

To illustrate arterial supply of the brain
View: Lateral
Structure: Aortic arch and branches

To illustrate hepatic portal circulation
View: Anterior
Structure: Portal vein and tributaries

Exercise 33 Human Cardiovascular Physiology—Blood Pressure and Pulse Determination

To illustrate the position of the heart and valves in the thoracic cavity
View: Anterior
Structure: Heart—cut section

To illustrate the pulse points
View: Anterior
Structure: Radial artery, Deep femoral artery

Exercise 35 The Lymphatic System and Immune Response

To illustrate lymphatic vessels and lymphoid organs
View: Anterior
Structure: Lymph vessels

To illustrate a thoracic duct
View: Anterior
Structure: Thoracic duct

To illustrate axillary and cervical lymph nodes
View: Anterior
Structure: Axillary and cervical lymph nodes

To illustrate a thymus gland
View: Anterior
Structure: Thymus gland

To illustrate a spleen
View: Anterior
Structure: Spleen

To illustrate palatine glands and tonsils
View: Anterior
Structure: Palatine glands and tonsils

Exercise 36 Anatomy of the Respiratory System

To illustrate an upper respiratory tract
View: Medial
Structure: Nasal conchae

To illustrate an upper respiratory tract
View: Lateral
Structure: Mediastinal parietal pleura of right pleural cavity

To illustrate a larynx
View: Anterior
Structure: Thyroid gland

To illustrate an epiglottis
View: Anterior
Structure: Epiglottis

To illustrate lower respiratory structures
View: Medial
Structure: Right lung

To illustrate lower respiratory structures
View: Anterior
Structure: Trachea

To illustrate lungs
View: Anterior
Structure: Lungs

To illustrate lungs—coronal section
View: Anterior
Structure: Lungs—coronal section

To illustrate a right lung
View: Lateral
Structure: Right lung

To illustrate a parietal pleura
View: Anterior
Structure: Parietal pleura

To illustrate a diaphragm
View: Anterior
Structure: Diaphragm

To illustrate a trachea
View: Lateral
Structure: Trachea

Exercise 38 Anatomy of the Digestive System

To illustrate an oral cavity
View: Lateral
Structure: Esophagus

To illustrate tonsils
View: Anterior
Structure: Palatine tonsil

To illustrate salivary glands
View: Lateral
Structure: Parotid gland, Parotid duct, Deep salivary glands

To illustrate an esophagus
View: Lateral
Structure: Esophagus

To illustrate a stomach
View: Anterior
Structure: Stomach, Stomach—coronal section

To illustrate a stomach
View: Lateral
Structure: Stomach

To illustrate an ileum
View: Anterior
Structure: Ileum

To illustrate a duodenum
View: Anterior
Structure: Duodenum

To illustrate a transverse colon
View: Anterior
Structure: Transverse colon

To illustrate an ascending and descending colon
View: Anterior
Structure: Ascending colon, Descending colon

To illustrate a colon
View: Medial
Structure: Colon

To illustrate teeth
View: Anterior
Structure: Skull

To illustrate a bile duct, liver, and gallbladder
View: Anterior
Structure: Bile duct {Common bile duct}, Liver—coronal section, Gallbladder

To illustrate a pancreas
View: Anterior
Structure: Pancreas

Exercise 40 Anatomy of the Urinary System

To illustrate kidneys
View: Anterior
Structure: Kidney, Kidney—longitudinal section

To illustrate a ureter
View: Anterior
Structure: Ureter

To illustrate a urinary bladder
View: Anterior
Structure: Urinary bladder

To illustrate a female urethra click on the Gender button and choose Female
View: Medial
Structure: Urethra

To illustrate a male urethra click on the Gender button and choose Male
View: Medial
Structure: Urinary bladder and prostate gland, Urethra

Exercise 42 Anatomy of the Reproductive System

To illustrate testes click on the Gender button and choose Male
View: Anterior
Structure: Testis

To illustrate a penis click on the Gender button and choose Male
View: Anterior
Structure: Penis

To illustrate a uterus click on the Gender button and choose Female
View: Anterior
Structure: Uterus

To illustrate mammary glands click on the Gender button and choose Female
View: Anterior
Structure: Breast

Credits

ILLUSTRATIONS

Exercise 1

1.2, 1.4, 1.6–1.8: Imagineering. 1.3, 1.5: Precision Graphics.

Exercise 2

2.3–2.6: Imagineering.

Exercise 3

3.1, 3.2–3.4, 3.6: Precision Graphics.

Exercise 4

4.1: Imagineering. 4.2: Carla Simmons/ Kristin Mount. 4.3: Tomo Narashima. 4.4: Adapted from Campbell, Mitchell, Reece Biology 5E, f 12.5 p 210. (San Francisco, Benjamin Cummings 1991) © Addison Wesley Longman, Inc., Table: Precision Graphics.

Exercise 5

5A.1: Precision Graphics. 5A.4: Nadine Sokol, Imagineering. 5B.1–5B.5: Steve McEntee.

Exercise 6

6.1, 6.3: Imagineering. 6.2, 6.6: Precision Graphics. 6.4: Kristin Otwell. 6.5–6.6: Karl Miyajima. 6.8: Precision Graphics.

Exercise 7

7.1: Tomo Narashima. 7.2, 7.4–7.5, 7.7: Imagineering.

Exercise 8

8.1–8.2: Imagineering.

Exercise 9

Table 9.1: Kristin Mount. 9.1: Laurie O'Keefe. 9.2 and 9.3: Barbara Cousins. 9.4: Precision Graphics.

Exercise 10

Table 10.1: Kristin Otwell. 10.1–10.10, 10.12: Nadine Sokol. 10.15: Precision Graphics. 10.13, 10.14, and 10.16: Laurie O'Keefe. 10.11: Vincent Perez/Kristin Mount.

Exercise 11

Table 11.1: Kristin Otwell/Kristin Mount. 11.1–11.8: Laurie O'Keefe/Kristin Mount.

Exercise 12

12.1: Nadine Sokol.

Exercise 13

13.1–13.2, 13.5–13.6: Precision Graphics. 13.7 and 13.8: Barbara Cousins/Kristin Mount. 13.4: Imagineering.

Exercise 14

14.1–14.2, 14.6: Imagineering. 14.3: Raychel Ciemma. 14.4: Raychel Ciemma/ Kristin Mount. 14.6: Imagineering.

Exercise 15

15.1–15.11: Raychel Ciemma. 15.12: Wendy Hiller-Gee. 15.13 and 15.14: Raychel Ciemma.

Exercise 16

16A.1–16A.5: Precision Graphics. 16B.1–16B.4: Steve McEntee.

Exercise 17

17.1–17.2, 17.5, 17.06, and 17.7b: Precision Graphics. 17.3: Imagineering. 17.7a: Charles W. Hoffman.

Exercise 18

18.1–18.4: Precision Graphics.

Exercise 19

19.2, 19.7–19.9, 19.11, and 19.13: Precision Graphics. 19.1, 19.4, 19.5: Imagineering.

Exercise 20

20.1: Imagineering.

Exercise 21

21.1–21.3, 21.6–21.9: Imagineering.

Exercise 22

22.1: Imagineering. 22.2, 22.3, and 22.7: Precision Graphics.

Exercise 23

23.2: Imagineering.

Exercise 24

24.1: Charles W. Hoffman/Kristin Mount. 24.2–24.4, 24.8–24.9: Imagineering. 24.7: Shirley Bortoli.

Exercise 25

25.1: Charles W. Hoffman/Kristin Mount. 25.2, 25.4, 25.5, 25.8: Imagineering.

Exercise 26

26.1 and 26.2: Imagineering.

Exercise 27

27.1, 27.2: Imagineering. 27.3: Precision Graphics.

Exercise 28

28.1: Precision Graphics.

Exercise 29

29.1–29.3: Precision Graphics. 29.6, Table: Imagineering.

Exercise 30

30.1: Wendy Hiller-Gee. 30.2, 30.3, and 30.7: Barbara Cousins. 30.8: Precision Graphics. 30.4: Imagineering. 30.5: Kristin Mount.

Exercise 31

31.1: Barbara Cousins. 31.2–31.4: Precision Graphics.

Exercise 32

32. 1: Adapted from Tortora and Grabowski, Principles of Anatomy and Physiology 9E, f 21.1 p 671 (New York: John Wiley & Sons, Inc. 2000) © 2000 Biological Sciences Textbooks, Inc. and Sandra Reynolds Grabowski. 32.3–32.6, 32.8–32.13: Barbara Cousins. 32.15: Kristin Mount. 32.2, 32.7: Imagineering.

Exercise 33

33.A1, 33.A3: Imagineering. 33.A4–33.A7: Precision Graphics.

Exercise 34

34.1–34.5: Precision Graphics.

Exercise 35

35.1–35.3: Imagineering. 35.4–35.5: Precision Graphics.

Exercise 36

36.1–36.7: Imagineering.

Exercise 37

37.A1–37.A9: Precision Graphics.

Exercise 38

38.1: Kristin Mount. 38.2: Adapted from Seeley, Stephens, Tate, Anatomy and Physiology 4E, f 24.2 p 778 (New York: WCB/McGraw-Hill, 1998) © 1998 The McGraw-Hill Companies, Inc. 38.3–38.4: Cyndie Wooley/Karl Miyajima. 38.5, 38.7–38.8, 38.11–38.12: Imagineering. 38.6, 38.9, 38.14: Precision Graphics. 38.10, 38.13: Karl Miyajima.

Exercise 39

39.A1–39.A3: Precision Graphics.

Exercise 40

40.1a: Linda McVay/Karl Miyajima. 40.2, 40.6: Precision Graphics. 40.3–40.5: Imagineering.

Exercise 41

41.A1–41.A2: Precision Graphics.

Exercise 42

42.1, 42.5, 42.7: Martha Blake/Kristin Mount. 42.3, 42.6: Precision Graphics. 42.4: Carla Simmons.

Exercise 43

43.1, 43.8: Precision Graphics. 43.4: Martha Blake/Karl Miyajima. 43.3, 43.8: Imagineering.

Exercise 44

44.1: Kristin Mount. 44.2: Martha Blake/Karl Miyajima.

Exercise 45

45.4: Precision Graphics.

Exercise 46

46.3: Raychel Ciemma. 46.5, 46.7, 46.9, 46.13: Precision Graphics. 46.10: Imagineering.

Fetal Pig Dissection Exercises

D1.3–D1.8, D2.2, D3.1, D3.2, D4.1–D4.5, D6.2, D7.1, D7.2, D8.1, D8.2, D9.1, and D9.2: Kristin Mount.

Cat Dissection Exercises

D1.1, D3.1, D3.3, D4.1, and D7.1: Precision Graphics. D1.4, D2.1, D2.3, D3.2, D4.3, D4.4, D4.6, D6.1, D7.2, D7.3, D8.1, D8.2, D9.1, and D9.2: Kristin Mount.

PHOTOGRAPHS

Special thanks to Christine Eckel, Elena Dorfman, Kirk Wuest, Dennis Strete, Ph.D., Merritt College, and University of California, Berkeley.

Exercise 1

1.1: © Benjamin/Cummings Publishing. Photo by BioMed Arts Associates, Inc. 1.4: © Jenny Thomas/Addison Wesley Longman.

Exercise 2

2.1a–d, 2.2, 2.3a, 2.4a, 2.5b–c: Elena Dorfman/Addison Wesley Longman. 2.3b, 2.4b, 2.4c, 2.5a, 2.6a–c © Dr. Robert A. Chase. 2.7: © Carolina Biological Supply Company/Phototake.

Exercise 3

3.5: © Benjamin/Cummings Publishing. Photo by Victor Eroschencko.

Exercise 4

4.1b: © Dr. Don Fawcett, Science Source/Photo Researchers. 4.4a–f: © Ed Reschke.

Exercise 5

5A.2: © Elena Dorfman/Addison Wesley Longman. 5A.3a: © Keith R. Porter, Science Source/Photo Researchers. 5A.3b: Courtesy of Dr. Mohnadas Narla, Lawrence Berkeley Laboratory. 5A.3a–c: © David M. Phillips/Visuals Unlimited.

Exercise 6

6.3a: Dennis Strete. 6.3b, c, e–h, 6.5c, d, f, h–j, 6.6c: © Benjamin/Cummings Publishing Company. Photo by Allen Bell, University of New England. 6.3d: © Calentine/Visuals Unlimited. 6.5a, b, g, and k, 6.6b, and 6.7: © Ed Reschke. 6.6a: © Eric Graves, Science Source/Photo Researchers. 6.6c: © Benjamin/Cummings Publishing Company. 6.7:Cabisco/Visuals Unlimited.

Exercise 7

7.3: © Addison Wesley Longman. 7.6: Courtesy of Marian Rice. 7.4: © Carolina Biological Supply Company/Phototake. 7.7a: Cabisco/Visuals Unlimited. 7.7b: John D. Cunningham/Visuals Unlimited. 7.8: From Gray's Anatomy, Henry Gray. Churchill Livingstone, UK.

Exercise 9

9.3c: © Benjamin/Cummings Publishing Company. Photo by Allen Bell, University of New England.

Exercise 10

10.4: From *A Stereoscopic Atlas of Human Anatomy,* by David L. Bassett. 10.5: © R. T. Hutchings. 10.9c: © Dr. Robert A. Chase.

Exercise 11

Table 11.1, 11.7a: From *A Stereoscopic Atlas of Human Anatomy,* by David L. Bassett. 11.5b: Department of Anatomy and Histology, University of California, San Francisco.

Exercise 12

12.2a–12.2b: Jack Scanlan, Holyoke Community College, MA.

Exercise 13

13.2: © Dr. Robert A. Chase. 13.7c, 13.8b: *From A Stereoscopic Atlas of Human Anatomy,* by David L. Bassett.

Exercise 14

14.4b: Courtesy of Marian Rice. 14.6: © Eric Graves/Photo Researchers, Inc.

Exercise 15

15.4a and15.8a: From *A Stereoscopic Atlas of Human Anatomy,* by David L. Bassett. 15.10f: © Benjamin/Cummings Publishing Company. Photo by Stephen Spector, courtesy of Dr. Charles Thomas, Kansas University Medical Center.

Exercise 17

17.2c: Triarch/Visuals Unlimited. 17.3: Don Fawcett/Photo Researchers, Inc. 17.4: © Dennis Strete, Ph.D.

Exercise 19

19.2c, and 19.4a: © Dr. Robert A. Chase. 19.3: © Benjamin/Cummings Publishing Company. Photo by Mark Nielsen, University of Utah. 19.5b: Pat Lynch, Photo Researchers, Inc. 19.6a–b, 19.7c, 19.10: From *A Stereoscopic Atlas of Human Anatomy,* by David L. Bassett. 19.11: © Elena Dorfman. 19.11b and 19.12: © Benjamin/Cummings Publishing Company. Photo by Dr. Sharon Cummings, University of California, Davis. 19.13b and 19.14: © Elena Dorfman/Addison Wesley Longman.

Exercise 20

20.1a: © Alexander Tsiaras, Science Source/Photo Researchers.

Exercise 21

21.1b–d: From *A Stereoscopic Atlas of Human Anatomy,* by David L. Bassett.

Exercise 22

22.4, 22.5, and 22.6: © Benjamin/Cummings Publishing Company. Photo by Richard Tauber.

Exercise 24

24.3: From *A Stereoscopic Atlas of Human Anatomy,* by David L. Bassett. 24.4: © Ed Reschke/Peter Arnold, Inc. 24.5a and b: © Benjamin/Cummings Publishing Company. Photo by Stephen Spector, courtesy of Dr. Charles Thomas, Kansas University Medical Center. 24.6: © Elena Dorfman/Addison Wesley Longman. 24.12a and b: © Benjamin/Cummings Publishing Company. Photo by Richard Tauber. 24.13: © Don Wong, Science Source/Photo Researchers.

Exercise 25

25.6a–c: © Benjamin/Cummings Publishing Company. Photo by Richard Tauber.

Exercise 29

29.6a–c, and 29.5a–d: © Elena Dorfman/Addison Wesley Longman. 29.6: © 1992 Dee Breger/Photo Researchers, Inc. 29.8: Carolina Biological Supply. 29.7: © Benjamin/Cummings Publishing Company. Photo by Jack Scanlan, Holyoke Community College.

Exercise 30

30.3b, d: From *A Stereoscopic Atlas of Human Anatomy,* by David L. Bassett. 30.3c: © Lennart Nilsson. From *The Body Victorious.* 30.8a, b, and 30.9: © Benjamin/Cummings Publishing Company. Photo by Wally Cash, Kansas State University. 30.7: Ed Reschke.

Exercise 32

32.1a: From R. G. Kessel and R. H. Kardon, Tissues and Organs. 1979 W. H. Freeman. 35.3b: © Biophoto Associates/Photo Researchers, Inc.

Exercise 36

36.1b, 36.5a: From *A Stereoscopic Atlas of Human Anatomy,* by David L. Bassett. 36.5b: © Benjamin/Cummings Publishing Company. Photo by Richard Tauber. 36.6b: Courtesy of the University of San Francisco. 36.7a: Ed Reschke/Peter Arnold, Inc. 36.7b: © Carolina Biological Supply/Phototake.

Exercise 37

37.A3–A4: © Elena Dorfman/Addison Wesley Longman.

Exercise 38

38.7a, c, 38.13a: From *A Stereoscopic Atlas of Human Anatomy,* by David L. Bassett.

Exercise 40

40.3: From *A Stereoscopic Atlas of Human Anatomy,* by David L. Bassett.

Exercise 42

42.2a: © Ed Reschke.

Exercise 43

43.2b: © Dennis Strete, Ph.D. 43.6a–c: © Benjamin/Cummings Publishing. Photo by Victor Eroschencko.

Exercise 45

45.1c: © CNRI/SPL/Photo Researchers, Inc. 45.1b: Courtesy Jane Cangleton and Reproductive Genetics Center. 45.2.1: © Llewellyn/Uniphoto Picture Agency. 45.2.2: © Marc Martinelli/The Image Bank. 45.2.3 and 45.2.4: © Ogust/Image Works. 45.2.5: © Boisvieux/Explorer/Photo Researchers. 45.2.6, 45.2.7, 45.2.8: © Benjamin/Cummings. Photo by Anthony Loveday. 45.5: Carolina Biological Supply Co.

Exercise 46

46.1, 46.2, 46.3a, 46.4, 46.6a, 46.6b, 46.8, 46.11a, 46.11b, 46.12, 46.13a, 46.14, 46.15a, 46.15b, 46.16, 46.17, 46.18, 46.20, 46.21a, 46.21b, 46.21c, 46.21d, and 46.22: © Paul Waring/BioMed Arts Associates, Inc.

Histology Atlas

PLATES 1, 5, 6, 7, 8, 11, 15, 24, 26, 29, 30, 34, 36, 37, 39, 40, 41, 42, 50, 51: © Dennis Strete, Ph.D.

PLATES 2, 4, 10, 12, 13, 14, 16, 17, 18, 19, 20, 22, 23, 25, 28, 33, 35, 38, 43, 44, 45, 46, 47, 48, 49, 52, 53, 54, 55, 56, 57, 58, 59, 60, 61, 62, 63: © Benjamin/Cummings Publishing. Photo by Victor Eroschencko.

PLATE 3: Manfred Kage/Peter Arnold, Inc. PLATE 8: Ed Reschke/Peter Arnold, Inc. PLATE 21: Carolina Biological Supply/Phototake. PLATE 28: Courtesy of LUMEN Histology, Loyola University Medical Education Network, http://www.lumen/MedEd/histo/frames/histo©frames.html. PLATE 31: © John Cunningham/Visuals Unlimited. PLATE 32: Courtesy of Marian Rice. PLATE 53: M. Abbey/Visuals Unlimited.

Human Anatomy Atlas

PLATES A and B: © Elena Dorfman/Addison Wesley Longman. PLATES C and D: © R. T. Hutchings. PLATES E–K: © Dr. Robert A. Chase.

Fetal Pig Dissection Exercises

D1.1, D1.2: © Benjamin/Cummings Publishing Company. Photo by Jack Scanlan, Holyoke Community College. D1.3b, D1.4b, D1.5b, D1.6b, D1.7b, D1.8b, D2.01, D4.01b, D4.02b, D4.03b, D4.04b, D6.01, D6.02b, D7.01b, D7.02b, D8.01b, D8.02b, D9.01b, D9.02b: © Elena Dorfman/Addison Wesley Longman.

TRADEMARK ACKNOWLEDGEMENTS

Adrenalin is a registered trademark of Warner-Lambert Company.

Albustix, Clinitest, Hemastix, Icotest, Ketostix, and Multistix are registered trademarks of Miles Laboratories, Inc.

Betadine is a registered trademark of The Purdue Frederick Company.

Chemstrip is a registered trademark of Boehringer Mannheim Corporation.

Clinistix is a registered trademark of Ames Company, Inc.

Harleco is a registered trademark of Hartman-Leddon Company, Inc.

Hefty is a registered trademark of Mobil Oil Company.

Landau is a registered trademark of Landau Uniform.

Macintosh and Power Macintosh are registered trademarks of Apple Computer, Inc.

Novocain is a registered trademark of Sterling Drug, Inc.

Parafilm is a registered trademark of American National Can Company.

Seal-Ease, Sedi-stain, and Unopette are registered trademarks of Becton, Dickinson, and Company.

Velcro is a registered trademark of Velcro, U.S.A., Inc.

Wampole is a registered trademark of Henry K. Wampole and Company.

Windows is a registered trademark of Microsoft Corporation.

Index

NOTE: Boldface page numbers indicate definitions. An *f* following a page number indicates a figure and a *t* indicates tabular material. The letter P preceding a page number indicates PhysioEx physiology lab computer simulations.